Reference Values for *Frequently* Assayed Clinical Chemistry Analytes* (continued)

Analyte	Specimen†	Reference Interval‡		Page
		Conventional Units	*Recommended SI Units*	
Gamma immunoglobulins	S	IgG 800–1200 mg/dL	IgG 8–12 g/L	176
		IgA 70–312 mg/dL	IgA 0.7–3.12 g/L	
		IgM 50–280 mg/dL	IgM 0.5–2.8 g/L	
		IgD 0.5–2.8 mg/dL	IgD 0.005–0.2 g/L	
		IgE 0.01–0.06 mg/dL	IgE 0.1–0.6 mg/L	
Glucose (fasting)	P	Fasting 70–110 mg/dL	3.9–6.0 mmol/L	302
		Non-fasting 70–150 mg/dL	3.9–8.3 mmol/L	
	CSF	40–70 mg/dL	2.2–3.9 mmol/L	
High-density lipoprotein (HDL) cholesterol	S	Male 29–60 mg/dL	0.75–1.6 mmol/L	325
		Female 38–75 mg/dL	1.0–1.94 mmol/L	
Iron	S	Male 65–170 µg/dL		284
		Female 50–170 µg/dL		
Lactate dehydrogenase (LD)	S	$(L \rightarrow P)$ 100–225 U/L		220
		$(P \rightarrow L)$ 80–280 U/L		
Lactate dehydrogenase isoenzymes (as percentage of total)	S	LD-1 14–26%		219
		LD-2 29–39%		
		LD-3 20–26%		
		LD-4 8–16%		
		LD-5 6–16%		
Lipase	S	0–1.0 U/mL		228
Magnesium (Mg^{2+})	S	1.6–2.4 mEq/L	0.65–1.0 mmol/L	267
	U	6.0–10.0 mEq/24 hr	3.00–5.00 mmol/24 hr	
Osmolality	S	275–295 mOsmol/kg		258
	U (24 hr)	300–900 mOsmol/kg		
Urine-serum ratio		1.0–3.0		
Phosphate	S	2.7–4.5 mg/dL	0.87–1.45 mmol/L	272
Potassium (K^+)	S	3.4–5.0 mEq/L	3.4–5.0 mmol/L	264
		Neonate 3.7–5.9 mEq/L	3.7–5.9 mmol/L	
	U (24 hr)	25–125 mEq/day	25–125 mmol/day	
Protein (total)	S	6.5–8.3 g/dL	65–83 g/L	185
	CSF	0.5% of plasma		
Resin T_3 uptake (RT_3U)	S	25%–30%		434
Sodium (Na^+)	S	135–145 mEq/L	135–145 mmol/L	261
	U (24 hr)	40–220 mEq/L	40–220 mmol/L	
	CSF	138–150 mEq/L	138–150 mmol/L	
Thyroid-stimulating hormone (TSH)	S	0.5–5.0 µU/mL		434
Thyroxine (T_4)	S	4.5–13 µg/dL	58–167 nmol/L	434
Triglycerides	S	10–190 mg/dL	0.11–2.15 mmol/L	323
Uric acid	S	Male 3.5–7.2 mg/dL	208–428 µmol/L	350
		Female 2.6–6.0 mg/dL	155–357 µmol/L	

*The list represents those analytes most frequently assayed in the clinical chemistry laboratory. For drug therapeutic ranges, refer to Appendix T, Summary Table of Pharmacokinetic Parameters.

†S = serum; P = plasma; U = urine; CSF = cerebrospinal fluid; WB = whole blood.

‡Values may vary according to method and population; values for enzymes are at 37°C unless otherwise noted.

CLINICAL CHEMISTRY

Lecture Test # 4:

Chapter - 12

GHi   GHi   TUV
ABC   MNO   GHi
PRS   MNO
              PRS
              DEF

Michael L. Bishop

MS, MT(ASCP), CLS(NCA)

Technical Services
Organon Teknika Corporation
Durham, North Carolina

Janet L. Duben-Engelkirk

EdD, MT(ASCP), CLS(NCA)

Director, Allied Health Education
and Program in Clinical Laboratory Science
Scott and White Memorial Hospital
Scott, Sherwood and Brindley Foundation
Temple, Texas

Edward P. Fody, MD

Medical Director
Department of Pathology
Bethesda Hospital
Cincinnati, Ohio

42 Contributors

THIRD EDITION

CLINICAL CHEMISTRY

Principles, Procedures, Correlations

last Lecture test : (written)

Ch. 12 - (255) electrolytes
" 11 - (237) pH, Blood Gs & Buffers

tests from test #5:

Ch. 27 - (539) therapeutic
Drug
monitoring

Ch. 28 (561) toxicology

Ch. 19 (349) Endocrinology

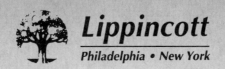

Lippincott
Philadelphia • New York

Acquisitions Editor: Kathleen P. Lyons
Assistant Editor: Stephanie Harold
Production Editor: Molly E. Dickmeyer
Production: Textbook Writers Associates
Cover Design: Larry Didona
Printer/Binder: Courier Westford
Cover Printer: Lehigh Press

Third Edition

Library of Congress Cataloging-in-Publications Data

Clinical chemistry: principles, procedures, correlations / edited by
  Michael L. Bishop, Janet L. Duben-Engelkirk, Edward P. Fody.—3rd ed.
     p.   cm.
  Includes bibliographical references and index.
  ISBN 0-397-55167-3
  1. Clinical chemistry.   I. Bishop, Michael L.   II. Duben-Engelkirk, Janet L.
III. Fody, Edward P.
  RB40.C576   1996
  616.07′56—dc20   95-41888
                                                        CIP

The material contained in this volume was submitted as previously unpublished material, except in the instances in which credit has been given to the source from which some of the illustrative material was derived.

Any procedure or practice described in this book should be applied by the health-care practitioner under appropriate supervision in accordance with professional standards of care used with regard to the unique circumstances that apply in each practice situation. Care has been taken to confirm the accuracy of information presented and to describe generally accepted practices. However, the authors, editors, and publisher cannot accept any responsibility for errors or omissions or for any consequences from application of the information in this book and make no warranty, express or implied, with respect to the contents of the book.

The authors and publisher have exerted every effort to ensure that drug selection and dosage set forth in this text are in accordance with current recommendations and practice at the time of publication. However, in view of ongoing research, changes in government regulations, and the constant flow of information relating to drug therapy and drug reactions, the reader is urged to check the package insert for each drug for any change in indications and dosage and for added warnings and precautions. This is particularly important when the recommended agent is a new or infrequently employed drug.

Materials appearing in this book prepared by individuals as part of their official duties as U.S. Government employees are not covered by the above-mentioned copyright.

9 8 7 6 5 4 3 2 1

Contributors

Mary Ruth Beckham, MT(ASCP)
Education Coordinator
Program in Clinical Laboratory Science
Research and Education
Scott & White Memorial Hospital, Scott, Sherwood and
    Brindley Foundation
Temple, TX

Laboratory Safety

Betsy D. Bennett, MD, PhD
Pathology
University of South Alabama
Mobile, AL

Endocrinology

Michael L. Bishop, MS, MT(ASCP), CLS(NCA)
Technical Service
Organon Teknika Corporation
Durham, NC

Appendices

John A. Bittikofer, PhD
Pathology and Laboratory Medicine Service
VA Medical Center
Durham, NC

Therapeutic Drug Monitoring
Toxicology

Anthony W. Butch, PhD, DABCC
Director, Clinical Chemistry
Assistant Professor, Department of Pathology
University of Arkansas for Medical Sciences
Little Rock, AR

Tumor Markers

Donald J. Cannon, PhD
Department of Pathology
Marshall University School of Medicine
Huntington, WV

Immunochemistry

Eileen Carreiro-Lewandowski, MS
Medical Laboratory Science
University of Massachusetts—Dartmouth
North Dartmouth, MA

Basic Principles and Practices of Clinical Chemistry
Pancreatic Function

George S. Cembrowski, MD, PhD
Director, Clinical Laboratory
Park Nicollet Clinic, Health System Minnesota
Minneapolis, MN

Clinical Assistant Professor
Department of Laboratory Medicine and Pathology
University of Minnesota School of Medicine
Minneapolis, MN

Quality Control and Statistics

Kenneth R. Copeland, PhD
Assistant Professor, Department of Pathology
Duke University Medical Center
Durham, NC

Trace Elements in Clinical Chemistry

James G. Donnelly, PhD
Division of Biochemistry, Department of
    Laboratory Medicine
Ottawa Civic Hospital

Assistant Professor, Department of Pathology and
    Laboratory Medicine
University of Ottawa
Ottawa, Ontario, Canada

Carbohydrates and Alterations in Glucose Metabolism

Susan A. Dovenbarger, BS, Phleb(NCA)
Former Supervisor, Pediatric Phlebotomy
Duke University Medical Center
Durham, NC

Specimen Collection and Processing

Janet L. Duben-Engelkirk, EdD, MT(ASCP), CLS(NCA)
Director, Allied Health Education
Director, Program in Clinical Laboratory Science
Scott & White Memorial Hospital, Scott, Sherwood and
   Brindley Foundation
Temple, TX

Geriatric Clinical Chemistry
Appendices

Sharon S. Ehrmeyer, PhD
Pathology and Laboratory Medicine
University of Wisconsin
Madison, WI

Blood Gases, pH, and Buffer Systems

Suzanne Elie-Cartledge, PhD
Head of Department
Biochemistry, Pathology Laboratories
Port of Spain General Hospital
Port of Spain, Trinidad, West Indies

Renal Function

Edward P. Fody, MD
Medical Director
Department of Pathology
Bethesda Hospital
Cincinnati, OH

Liver Function
Gastrointestinal Function
Appendices

Robin Gaynor Krefetz, MEd, CLS(NCA), MT(ASCP)
MLT Program
Community College of Philadelphia
Philadelphia, PA

Enzymes

H. Jesse Guiles, EdD
Department of Clinical Laboratory Sciences
University of Medicine & Dentistry of New Jersey
School of Health Related Professions
Newark, NJ

Thyroid Function

Carolyn Houghton, MS, MT(ASCP), DLM
Regulations Compliance Manager
Department of Pathology
Scott & White Memorial Hospital, Scott, Sherwood and
   Brindley Foundation
Temple, TX

Laboratory Safety

Patricia Hudson
Technical Services
Organon Teknika Corporation
Durham, NC

Assessment of Hemostasis

Bethany L. Hurtuk, PhD
Former Clinical Assistant Professor
Chemistry
Cleveland State University
Cleveland, OH

Enzymes

Dennis W. Jay, PhD, DABCC
Department of Pathology and Laboratory Medicine
Texas A&M University College of Medicine
Pathology and Laboratory Medicine Service
Olin E. Teague Veterans Center
Temple, TX

Porphyrins, Hemoglobin, and Myoglobin

Ioannis Laios, PhD
Lifescan, Inc.
Milpitas, CA

Analytical Techniques and Instrumentation

Louann W. Lawrence, DrPH, CLSpH(NCA), MT(ASCP)SH
Associate Professor and Department Head
Medical Technology
School of Allied Health Professions
Louisiana State University
New Orleans, LA

Porphyrins, Hemoglobin, and Myoglobin

Barbara J. Lindsey, MS, CLSp(C), C(ASCP)
Department of Clinical Laboratory Sciences
Medical College of Virginia Campus
Virginia Commonwealth University
Richmond, VA

Amino Acids and Proteins

Judith R. McNamara, MT
Lipid Research Laboratories
Jean Mayer USDA Human Nutrition Research
Center on Aging at Tufts University and New England
   Medical Center
Boston, MA

Lipids and Lipoproteins

Barbara Smith Michael, MS, JD, MT(ASCP)
Former Assistant Professor
Department of Clinical Lab Science
School of Allied Health Sciences
University of Texas—Houston
Health Science Center
Houston, TX

Laboratory Safety

Sharon Monahan Miller, PhC, CLS(NCA), MT(ASCP)
Associate Dean
College of Health and Human Sciences
Northern Illinois University
DeKalb, IL 60115

Vitamins

Beverly S. Oxford, BA, MT(ASCP)SC
Department of Pathology
Duke University Medical Center
Durham, NC

Specimen Collection and Processing

Alex A. Pappas, MD
Director, Clinical Laboratory
Professor, Department of Pathology
University of Arkansas for Medical Sciences
Little Rock, AR

Tumor Markers

Larry Schoeff, MS, MT(ASCP)
University of Utah
School of Medicine
Department of Pathology
Salt Lake City, UT

ARUP, Inc.
Salt Lake City, UT

Automated Techniques

Frank A. Sedor, PhD
Department of Pathology
Director, Clinical Chemistry Services
Duke University Medical Center
Durham, NC

Body Fluid Analysis

Georgia A. Sehon, MSHP
Information Resource Management Field Office
Department of Veterans Affairs
Grand Prairie, TX

Computer Interfacing in the Clinical Chemistry Laboratory

John E. Sherwin, PhD
Consultant
West Lake Village, CA

Pediatric Clinical Chemistry

Joan B. Shrout, MA
Johnson & Johnson Clinical Diagnostics
Rochester, NY

Blood Gases, pH, and Buffer Systems

Susan T. Smith, PhD
Clinical Laboratory Science
School of Allied Health Sciences
East Carolina University
Greenville, NC

Nonprotein Nitrogen

Juan B. Sobenes, MD
Associate Clinical Professor
Laboratory Medicine and Pathology
Valley Medical Center, Fresno
University of California, San Francisco
School of Medicine
Fresno, CA

Pediatric Clinical Chemistry

A. Michael Spiekerman
Department of Clinical Pathology
Scott & White Clinic and Memorial Hospital
Temple, TX

Nutritional Assessment

Anne M. Sullivan
Department of Biomedical Technologies
University of Vermont
Burlington, VT

Quality Control and Statistics

John Toffaletti, PhD
Associate Professor
Department of Pathology
Duke University Medical Center
Durham, NC

Electrolytes

G. Russell Warnick, MS, MBA
Pacific Biometrics
Seattle, WA

Lipids and Lipoproteins

David J. Wells, PhD
Pathology
University of South Alabama
Mobile, AL

Endocrinology

Alan H. B. Wu, PhD
Pathology and Laboratory Medicine
Hartford Hospital
Hartford, CT

Analytical Techniques and Instrumentation
Renal Function

Lily L. Wu, PhD
Department of Internal Medicine and Pathology
University of Utah
Salt Lake City, UT

Lipids and Lipoproteins

Foreword

Many new tests, methods, and measurement systems continue to be introduced in a healthcare market that is becoming more and more competitive. Laboratory tests seem to be getting easier to perform, but the testing processes often require more complex technology that is more difficult to understand. New tests are also evolving because of new research findings and new measurement processes that, again, often depend on complex technology. The one constant seems to be the increasing knowledge that is required to understand clinical laboratory science today.

This rapid change and evolution makes it increasingly difficult to capture the knowledge that defines the current state-of-the-practice for clinical laboratorians. The task becomes more daunting with each passing year and also becomes more critical for laboratory professionals. Fortunately, the editors and contributors of this text have been willing to tackle this seemingly impossible task in order to support and advance the profession. I want to both thank and congratulate them!

The third edition of *Clinical Chemistry: Principles, Procedures, Correlations* continues its mission of addressing the formal educational needs of our students in clinical laboratory science, as well as the ongoing needs of the professionals in the field. It facilitates the educational process by identifying the learning objectives, focusing on key concepts and ideas, and applying the theory through case studies. It covers the basics of laboratory testing, as well as the many special areas of clinical chemistry testing. And it is still possible to carry this text with you to class, to the laboratory, to the office, or home to study!

Having personally worked with some of the editors and contributors, I know they have high standards both in the laboratory and in the classroom. Their interests and backgrounds provide an excellent balance between the academic and the practical, ensuring that both we and our students are exposed to a well-developed base of knowledge that has been carefully refined by experience. They provide the "sifting and winnowing" that is necessary to separate the wheat from the chaff and to provide the real sustenance for our students and ourselves.

For the many students for whom this book is intended, let me offer some advice from my close friend and mentor, Hagar the Horrible. It seems his young Viking son was embarking on a voyage to the real world of work. Hagar's son asked him "How do I get to the top?" Hagar's advice was "You have to start at the bottom and work your way up!" After pondering this for a moment, his son asked "How do I get to the bottom?" Hagar replied, "You have to know somebody!" As students in clinical laboratory science, you should study and heed the advice offered in this book, but you should also search for mentors. The authors of this book, as well as the instructors in your courses and your teachers in the laboratory, are key to getting started in your careers. You need to seek them out to profit from their learning and experiences! They are the professionals who know the state of the art, possess the current knowledge of the field, and are dedicated to help you.

James O. Westgard, PhD
Professor, Pathology and Laboratory Medicine
University of Wisconsin
Madison, WI

Preface

Clinical chemistry continues to be one of the most rapidly evolving areas of the laboratory. Since the publication of the first edition of this textbook in 1985, many changes have taken place. New technologies and analytical techniques have been introduced with a dramatic impact on the practice of clinical chemistry. In addition, in the health care system, increased emphasis is being placed on improved quality of patient care or total quality management. Point-of-care (POC) testing is also at the forefront of health care practice, and has brought forth both challenges and opportunities to clinical laboratorians. Now, more than ever, clinical laboratorians need to be concerned with disease correlations, interpretations, problem-solving, quality assurance, and cost effectiveness; they need to know not only the *how* of tests, but also the *why* and *when*. The editors of *Clinical Chemistry: Principles, Procedures, Correlations* designed this third edition to be an even more valuable resource for both students and practitioners in the rapidly evolving area of clinical chemistry.

Like the previous two editions, the third edition of *Clinical Chemistry: Principles, Procedures, Correlations* is a comprehensive, up-to-date, and easy to understand textbook for students at all levels. It is also intended to serve as a practically organized resource for both instructors and practitioners. The editors have tried to maintain the book's readability and improve its content, while not adding significantly to its length. Because the interpretative and analytical skills of the clinical laboratorian are being increasingly relied upon, an effort has been made to maintain an appropriate balance between analytical principles and techniques and the correlation of results with disease states. This edition expands the division of chapters into four sections:

Section I   Basic Principles of Clinical Chemistry
Section II   Clinical Correlations and Analytical Procedures
Section III Clinical Correlations of Organ Systems
Section IV   Specialty Areas of Clinical Chemistry

In this third edition, the editors have made several significance changes in response to requests by our readers—students, instructors, and practitioners. In addition to the chapter outlines, we have added objectives, key words, and a summary to each chapter. Every chapter now includes case studies and/or exercises. Answers to the exercises and the case study discussions have been moved to a separate section of the book. A glossary of the key words has also been added to serve as a ready reference for the reader. To provide a thorough, up-to-date study of clinical chemistry, we have added chapters on Immunochemistry, Trace Elements in Clinical Chemistry, and Nutritional Assessment. All chapter material has been updated, improved, and rearranged for better continuity. The chapter on Quality Control and Statistics has been expanded and now includes discussions of total quality management and point-of-care testing.

The analytical procedures discussed in the chapters reflect the most recent or commonly performed techniques in the clinical chemistry laboratory. Generally, detailed procedures have been included only for reference methodologies or for those analytes most often measured by non-automated (manual) techniques. Step-by-step instructions for kit procedures have not been included because of differences in volumes, temperatures, and incubation times specified by various manufacturers. Details of kit procedures change frequently; therefore, the package insert is the reference of choice. The methodology sections include discussions of the analytical principle, special considerations, reagents, calculations, reference values, and sources of errors. Historical methodologies are mentioned only briefly to illustrate the advantages and development of the latest methodologies.

Acknowledgments

A project as large as this requires the assistance and support of many individuals. The editors wish to express their appreciation to the contributors of this third edition of *Clinical Chemistry: Principles, Procedures, Correlations*—the dedicated laboratory professionals and educators whom the editors have had the privilege of knowing and exchanging ideas with over the years. These individuals were selected because of their expertise in particular areas and their commitment to the education of clinical laboratorians. Many have spent their professional careers in the clinical laboratory, at the bench, teaching students, or consulting with clinicians. In these front-line positions, they have developed their perspective of what is important for the next generation of clinical laboratorians.

We extend appreciation to our colleagues, teachers, and mentors in the profession who have helped to shape our ideas about clinical chemistry practice and education. Also, we want to thank the many companies and professional organizations that provided product information and photographs or granted permission to reproduce diagrams and tables from their publications. Many National Committee for Clinical Laboratory Standards (NCCLS) documents have been important sources of information as well. These documents are referenced directly in the appropriate chapters.

The editors would like to acknowledge the contribution and effort of the following individuals to previous editions. Their efforts provided the framework for many of the current chapters. Contributors to the previous editions were: Shauna C. Anderson, Kathleen Becan-McBride, Nancy Blosser, Donald C. Cannon, Kent D. Cavender, Pauline D. J. Chantly, Susan Cockayne, Diane Coiner, Donna M. Corriveau, Carol L. Creech, Paul G. Engelkirk, Lee Ann F. Gillen, Linda S. Gorman, Lynda L. Hunter, Dianne F. Johnson, Frank J. Liu, Cheryl A. Lotspeich, Taofig A. Onigbinde, Sharon K. Palmer, Doris L. Ross, Larry Suddendorf, Edward O. Uthman, and Linda L. Woodward.

Finally, we gratefully acknowledge the cooperation and assistance of the staff at Lippincott–Raven Publishers, particularly, Dr. Andrew Allen, Executive Editor, Allied Health, Kathleen P. Lyons, Acquisitions Editor, Stephanie Harold, Assistant Editor, and Ms. Laura Dover, Editorial Assistant.

The editors are continually striving to improve future editions of this book. We again request and welcome our readers' comments, criticisms, and ideas for improvement.

Michael L. Bishop

Janet L. Duben-Engelkirk

Edward P. Fody

Contents

Part I

Basic Principles of Clinical Chemistry

Final exam Clin. Chem — ch's

9
10
11
12

13
14
15
16

19
25
27
28

Basic Principles and Practices of Clinical Chemistry

Eileen Carreiro-Lewandowski

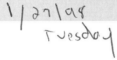

Objectives

Upon completion of this chapter, the clinical laboratorian should be able to:

- *Convert results from one unit format to another using the SI system.*

- *List and describe the types of thermometers used in the clinical laboratory.*

- *Identify the varying chemical grades used in reagent preparation and indicate their correct use.*

- *Define the following terms: primary standard, SRM, secondary standard.*

- *Describe each of the following terms that are associated with solutions and, when appropriate, provide the respective units: percent, molarity, normality, molality, saturation, colligative properties, redox potential, conductivity, density.*

- *Define a buffer and give the formula for the pH and pK calculations.*

- *When given either the pK and pH or the pK and concentration of the weak acid and its conjugate base, use the Henderson-Hasselbalch equation to determine the missing variable.*

- *Describe the specifications for each type of laboratory water.*

- *When given an actual pipet or its description, classify the type of pipet.*

- *Describe two ways to calibrate a pipetting device.*

- *Define a desiccant and discuss its use in the clinical laboratory.*

- *Correctly perform the laboratory mathematical calculations provided in this chapter.*

KEY WORDS

Analyte	Electrolyte	Pipet
Anhydrous	Equivalent Weight	Primary Standard
Beer's Law	Filtrate	Redox Potential
Buffer	Filtration	Secondary Standard
Buret	Henderson-	Serial Dilution
Centrifugation	Hasselbalch	Significant Figures
Character	Hydrate	Solute
Colligative Property	Hygroscopic	Solution
Conductivity	Ionic Strength	Solvent
Deionized Water	Mantissa	Specific Gravity
Deliquescent	Molality	Standard Reference
Substances	Molarity	Materials
Delta Absorbance	Normality	Standard
Density	One Point	Systeme
Desiccant	Calibration	International
Desiccator	Osmotic Pressure	(SI) d'Unités
Dialysis	Oxidizing Agent	Thermistor
Dilution	Percent Solution	Valence
Distilled Water	pH	

Michael L. Bishop, Janet L. Duben-Engelkirk, and Edward P. Fody.
CLINICAL CHEMISTRY. © 1996 Lippincott–Raven Publishers.

The primary goal of a clinical chemistry laboratory is to correctly perform analytical procedures that yield accurate and precise information to aid in patient diagnosis. In order to achieve reliable results, the clinical laboratorian must have the ability to use basic supplies and equipment correctly and have an understanding of fundamental concepts critical to any chemical test. This chapter includes such topics as units of measure, properties of a solution, classification of chemicals, reagents, glassware and plasticware, and laboratory mathematics.

UNITS OF MEASURE[1]

Any meaningful result requires both a number and a unit. The number describes the numeric value, while the unit defines the physical quantity or dimension, such as mass, length, time, or volume. While there are several systems of units that traditionally have been used by various scientific divisions, the Systeme International (SI) d'Unités was adopted internationally in 1960 and is the only system used in many countries. The units of the system are referred to as *SI units*. There are three types of SI units: basic, supplemental, and derived. Table 1–1 lists the basic units, which describe only one unit for fundamental, unrelated physical quantities. This avoids several different terms being used to describe the same physical quantity. Prefixes can be added to indicate decimal fractions or multiples of the basic SI unit. These prefixes are listed in Table 1–2. For example, 0.5 meters also could be expressed using the prefix *milli* (which is equivalent to $1/1000$, 0.001, or 10^{-3})—as 500 millimeters.

The derived unit is related mathematically to the basic or supplemental unit. Supplemental units are units that have not been classified as either basic or derived and currently include the radian (plane angle) and the steradian (solid angle). The derived units are listed in Table 1–3. It should be noted that some long-standing units, such as the hour, minute, day, plane angles expressed as degrees, gram, and liter, are not listed as basic SI units. With the exception of the gram, these units have been retained for use along with SI units. In

TABLE 1-1. Basic SI Units

Measure	Name	Symbol
Length	meter	m
Mass	kilogram	kg
Quantity of substance	mole	mol
Time	second	s
Electric current	ampere	A
Thermodynamic temperature	kelvin	K
Luminous intensity	candela	cd

Note: SI (Système International d'Unités) are units having a definition recognized by international agreement. Note that some SI units have capitalized symbols. This is to avoid confusion with SI prefixes using the same letter symbol.

TABLE 1-2. Prefixes to be Used with SI Units

Factor	Prefix	Symbol
10^{-18}	atto	a
10^{-15}	femto	f
10^{-12}	pico	p
10^{-9}	nano	n
10^{-6}	micro	μ
10^{-3}	milli	m
10^{-2}	centi	c
10^{-1}	deci	d
10^{1}	deka	da
10^{2}	hecto	h
10^{3}	kilo	k
10^{4}	mega	M
10^{9}	giga	G
10^{12}	tera	T
10^{15}	peta	P
10^{18}	exa	E

Note: Prefixes are used to indicate a subunit or multiple of a basic SI unit.

North America, it is acceptable to use the liter as the preferred unit of volume. In order to distinguish it from a lowercase letter "l," it is written as the capital letter L.

Reporting of laboratory results is often expressed in terms of substance concentration (*e.g.,* moles) or the mass of a substance (*e.g.,* mg/dL, g/dL, g/L, mEq/L, etc.). These familiar and traditional units can cause confusion during interpretation. It has been recommended that analytes be reported using moles of solute per volume of solution (substance concentration) and that the liter be used as the reference volume. In tables listing laboratory reference values, traditional values as well as SI recommended values are provided. Appendix E, *Conversion of Traditional Units to SI Units for Common Clinical Chemistry Analytes*, lists both units along with the conversion factor from traditional to SI units for common analytes.

TEMPERATURE[2]

The predominant practice is to make laboratory temperature readings using the Celsius (C) scale. However, Fahrenheit (F) and Kelvin (K) scales are also used. The SI designation for temperature is the Kelvin scale. Appendix D, *Basic Clinical Laboratory Conversions*, lists the various conversion formulas between each scale.

All analytical reactions take place at an optimal temperature. Some laboratory procedures, such as enzyme determinations, require precise temperature control, whereas others work well over a wide range of temperatures. Reactions that are very temperature dependent use some type of heating cell, heating block, or water/ice bath to provide the correct temperature environment. Temperatures of laboratory refrigerators often are critical and need periodic verification. Thermometers are either an integral part of an instrument or need

TABLE 1-3. Clinically Important SI-Derived Units

Quantity	Name	SI Symbol
Volume	cubic meter	m^3
Mass density	kilogram per cubic meter	kg/m^3
Concentration of amount of substance	mole per cubic meter	mol/m^3
Frequency	hertz	Hz
Force	newton	N
Pressure	pascal	Pa
Energy, work, quantity of heat	joule	J
Power	watt	W
Electric potential, potential difference, electromotive force	volt	V

to be placed in the device for temperature maintenance. The two major types of thermometers are the liquid (mercury)-in-glass and the electronic thermometer or thermistor probe.

Mercury-in-glass thermometers should be calibrated against a National Institute of Standards and Technology (NIST), U.S. Department of Commerce–certified thermometer. The NIST has a standard reference material (SRM) thermometer with various calibration points (0, 25, 30, and 37°C) for use with mercury-in-glass thermometers. Gallium, another SRM, has a known melting point and is also frequently used for thermometer verification. In these thermometers, the mercury column should be inspected to ensure that it is continuous and free from bubbles. If either condition is present, cooling the thermometer in ice may force the mercury toward the bulb. A circular "shake" of the thermometer also can help. Once the mercury column is intact, an ice point and desired bath/block temperature may be made in comparison with a certified SRM thermometer. It is important that the thermometer be immersed to its appropriate depth and calibrated in that fashion.

As automation advances and miniaturizes, the need for an accurate, fast-reading electronic thermometer (thermistor) has increased. The advantages of a thermistor over the more traditional mercury-in-glass thermometers are its size and millisecond response time. A disadvantage can be the initial cost, although the use of a thermistor probe attached to an already owned volt-ohm meter (VOM) can be very cost efficient. Like the mercury-in-glass thermometers, the thermistor can be calibrated against an SRM thermometer or the gallium melting-point cell.[3] Once the thermistor is calibrated against the gallium cell, it can then be used as a reference for any type of thermometer.

A special application of thermistor thermometers has been in making temperature verifications for flow-through cuvets.[4] Flow-through cuvets have become commonplace in the laboratory, but determining the temperature of a cuvet may be difficult unless there is a built-in temperature-detection system. From time to time it is necessary to recheck this mechanism. A procedure noting the change in absorbance using a temperature-sensitive solution has been recommended for checking flow-through cuvets.

REAGENT PREPARATION

In today's highly automated laboratory, there seems to be little need for reagent preparation by the laboratorian. Most instruments are sold by manufacturers who also make the reagents, usually in "kit" form (*i.e.,* all necessary reagents and respective storage containers are prepackaged as a unit). Because of heightened awareness of the hazards of certain chemicals and the existence of numerous regulatory agencies, clinical chemistry laboratories have readily eliminated their massive stocks of chemicals and have opted instead for the ease of using prepared reagents. From time to time, especially in hospital laboratories involved in research and development or specialized analyses, the laboratorian may still be faced with preparing various reagents. As a result of reagent deterioration, supply and demand, or the institution of cost-containment programs, the decision to prepare reagents in-house may be made. Therefore, a thorough knowledge of chemicals, standards, solutions, buffers, and water requirements is necessary.

Chemicals[5]

Chemicals exist in varying grades of purity: analytical reagent grade (AR), ultrapure, chemically pure (CP) or pure grade, United States Pharmacopeia (USP), National Formulary (NF), and technical or commercial grade.

The American Chemical Society (ACS) has established specifications for analytical reagent-grade chemicals. Labels on these reagents either state the actual impurities for each chemical lot or list the maximum allowable impurities. The label should reveal the percent impurities present and have the initials AR and/or ACS clearly printed. Chemicals of this category are suitable for use in most analytical laboratory procedures. Ultrapure chemicals have been put through additional purification steps for use in specific procedures such as chromatography, atomic absorption, fluorometry, standardization, or other techniques that require extremely pure chemicals.

Since USP and NF grade chemicals are used to manufacture drugs, the limitations established for this group of chemicals are based only on the criterion of not being injurious to the individual. Chemicals in this group may be pure enough for use in most chemical procedures, but it should be kept in mind that their purity standards are not based on the needs of the laboratory, and therefore, they may or may not meet all assay requirements.

Reagent designations of CP or pure grade indicate that the impurity limitations are not stated and that preparation of these chemicals is not uniform. Melting-point analysis is used quite often to ascertain the acceptable purity range. It is not recommended that clinical laboratories use these chemicals for reagent preparation unless further purification or a reagent blank is included.

Technical- or commercial-grade reagents are used primarily in manufacturing and should never be used in the clinical laboratory.

Reference Materials[1,6,7,8]

Unlike other areas of chemistry, clinical chemistry is involved in the analysis of substances found in serum that is difficult to purify and attest to its exact composition. For this reason, traditionally defined standards do not necessarily exist for use in clinical chemistry.

Recall that a *primary* standard is a highly purified chemical that can be measured directly to produce a substance of *exact* known concentration. The ACS purity tolerances for primary standards are $100 \pm 0.02\%$. Since most biologic constituents are not available within these limitations, NIST-certified *standard reference materials* (SRMs) are used in lieu of ACS primary standards.

The NIST developed certified SRMs for use in clinical chemistry laboratories. These substances may not have the properties and purity equivalent of a primary standard, but each one has been characterized for certain chemical or physical properties and can be used in place of an ACS primary standard in clinical work. A human serum SRM (no. 909a) is available and is certified for calcium, chloride, cholesterol, creatinine, glucose, lithium, magnesium, potassium, sodium, urea, uric acid, and the trace metals cadmium, chromium, copper, iron, lead, and vanadium.[9]

A *secondary standard* is defined as a substance of lower purity whose concentration is determined by comparison with a primary standard. The secondary standard depends not only on its composition, which cannot be determined directly, but also on the analytical reference method. Once again, because physiologic primary standards are generally unavailable, clinical chemists do not by definition have "true" secondary standards.

Solutions

In clinical chemistry, substances found in biologic fluids (*i.e.,* serum, urine, spinal fluid, etc.) are measured. A substance that is dissolved in a liquid is called a *solute,* and in laboratory science, these biologic solutes are also known as *analytes.* The liquid in which the solute is dissolved, in this instance a biologic fluid, is the *solvent.* Together they represent a *solution.* Any chemical or biologic solution has basic properties that describe it. These properties include concentration, saturation, colligative properties, redox potential, conductivity, density, pH, and ionic strength.

Concentration

Concentration of an analyte in solution can be expressed in many ways. Routinely, concentration is expressed as percent solution, molarity, molality, or normality.

Percent solutions are equal to parts per hundred or the amount of solute per 100 total units of solution. Three expressions of percent solutions are weight per weight (w/w), volume per volume (v/v), and most commonly, weight per volume (w/v). It is recommended that for v/v solutions, the units milliliters per liter (mL/L) be used instead of 70% (v/v).

Molarity is expressed as the number of moles per liter of solution. One mole of a substance equals its gram molecular weight. The SI representation for the traditional molar concentration is moles of solute per volume of solution, with the volume of the solution given in terms of liters. The SI expression for concentration should be represented as moles per liter (mol/L), millimoles per liter (mmol/L), micromoles per liter (μmol/L), and nanomoles per liter (nmol/L). The familiar concentration term of "molarity" has not been adopted by the SI as an expression of concentration.

Molality represents the amount of solute per one kilogram of solvent. Molality is sometimes confused with molarity, but it can be distinguished easily from molarity because molality is always expressed in terms of moles per kilogram or weight per weight and describes moles per 1000 g (1 kg) of solvent. The preferred expression for molality is moles per kilogram (mol/kg).

Normality is defined as the number of gram equivalent weights per liter of solution. An *equivalent weight* is equal to the molecular weight of a substance divided by its valence. The *valence* is the number of units that can combine with or replace one mole of hydrogen ions. Normality is no longer endorsed as an SI expression of concentration, and the expression milliequivalents per liter (mEq/L) for reporting sodium, potassium, and chloride concentration should be reported as millimoles per liter (mmol/L).

Saturation gives very little specific information about the concentration of a solution. Temperature, as well as the presence of other ions, can influence the solubility constant for a given solution and thus affect the saturation. Terms routinely used in the clinical laboratory to describe the extent of saturation are *dilute, concentrated, saturated,* and *supersaturated.* A *dilute solution* is one in which there is relatively little solute. In contrast, a *concentrated solution* has a very large quantity of solute in solution. A solution in which there is an excess of undissolved solute particles is called a *saturated solution.* As the name implies, a *supersaturated solution* has an even greater concentration of undissolved solute particles than does a saturated solution of the same substance. Because of the greater concentration of solute particles present, a supersaturated solution is thermodynamically unstable. Addition of a crystal of solute disturbs the supersaturated solution. The result is that any material in excess of what is required to saturate the solution will crystallize out. An example of this is seen when measuring serum osmolality by freezing point depression.

Colligative Properties

The behavior of particles in solution demonstrates four repeatable properties based only on the relative number of each kind of molecule present. The properties of osmotic pressure, freezing point, boiling point, and vapor pressure are called *colligative properties*. *Vapor pressure* is the pressure at which the liquid solvent is in equilibrium with the water vapor. *Freezing point* is the temperature at which the vapor pressures of the solid and liquid phases are the same. *Boiling point* is defined as the temperature at which the vapor pressure of the solvent reaches one atmosphere.

Osmotic pressure is the pressure that allows solvent flow through a semipermeable membrane to establish an equilibrium between compartments of different osmolality. The osmotic pressure of a dilute solution is proportional to the concentration of the molecules in solution. When a solute is dissolved in a solvent, these colligative properties change: the freezing point is lowered, the boiling point is raised, the vapor pressure is lowered, and the osmotic pressure is increased. In the clinical setting, freezing point and vapor pressure depression are measured as a function of osmolality.

Redox Potential

Redox potential, or *oxidation-reduction potential*, is a measure of a solution's ability to accept or donate electrons. Substances that donate electrons are called *reducing agents*, and those that accept electrons are considered *oxidizing agents*.

Conductivity

Conductivity is a measure of how well electricity passes through a solution. A solution's conductivity quality depends principally on the number of respective charges of the ions present. *Resistivity*, the reciprocal of conductivity, is a measure of a substance's resistance to the passage of electrical current. The primary application of resistivity in the clinical laboratory is for assessing the purity of water.

Buffers

Buffers are weak acids or bases and their related salts, which, because of their dissociation characteristics, minimize any changes in hydrogen ion concentration. Hydrogen ion concentration is often expressed as pH. A lowercase "p" in front of certain letters or abbreviations means the "negative logarithm of" or "inverse log of" that substance. In keeping with this convention, the term *pH* represents the negative or inverse log of the hydrogen ion concentration. Mathematically, pH is expressed as

$$pH = \log(1/[H^+])$$

$$pH = -\log[H^+] \qquad \textbf{\textit{Eq. 1-1}}$$

where $[H^+]$ is equal to the concentration of hydrogen ions in moles per liter.

The pH scale ranges from 0 to 14 and is a convenient way to express hydrogen ion concentration.

A buffer's capacity to minimize changes in pH is related to the dissociation characteristics of the weak acid or base in the presence of its respective salt. Unlike a strong acid or base, which dissociates almost completely, the dissociation constant for a weak acid or base solution tends to be very small, meaning little dissociation occurs.

The ionization of acetic acid (CH_3COOH), a weak acid, can be illustrated as follows:

$$[HA] \leftrightarrow [A^-] + [H^+]$$

$$[CH_3COOH] \leftrightarrow [CH_3COO^-] + [H^+] \qquad \textbf{\textit{Eq. 1-2}}$$

where: HA = weak acid, A^- = conjugate base, and H^+ = hydrogen ions.

Note that the dissociation constant, K_a, for a weak acid may be calculated using the following equation:

$$K_a = \frac{[A^-][H^+]}{[HA]} \qquad \textbf{\textit{Eq. 1-3}}$$

Rearrangement of this equation reveals

$$[H^+] = K_a \times \frac{[HA]}{[A^+]} \qquad \textbf{\textit{Eq. 1-4}}$$

Taking the log of each quantity and then multiplying by minus 1 (-1), the equation can be rewritten as

$$-\log[H^+] = -\log K_a \times -\log \frac{[HA]}{[A^+]} \qquad \textbf{\textit{Eq. 1-5}}$$

By definition, lower case "p" means "minus log of"; therefore, $-\log[H^+]$ may be written as pH, and $-\log K_a$ may be written as pK_a. The equation now becomes

$$pH = pK_a - \log \frac{[HA]}{[A^+]} \qquad \textbf{\textit{Eq. 1-6}}$$

Eliminating the minus sign in front of the log of the quantity $[HA]/[A^+]$ results in an equation known as the *Henderson-Hasselbalch* equation, which mathematically describes the dissociation characteristics of weak acids and bases and the effect on pH:

$$pH = pK_a - \log \frac{[A^+]}{[HA]} \qquad \textbf{\textit{Eq. 1-7}}$$

When the ratio of $[A^+]$ to $[HA]$ is 1, the pH equals the pK and the buffer has its greatest buffering capacity. The dissociation constant K_a, and therefore the pK_a, remains the same for a given substance. Any changes in pH are then due only to the ratio of base/salt $[A^+]$ concentration to weak acid $[HA]$ concentration.

Another important aspect of buffers is their ionic strength, particularly in separation techniques. *Ionic strength* is the concentration or activity of ions in a solution or buffer. It is defined[6] as follows:

$$\mu = I = \frac{1}{2} \sum C_i Z_i^2 \text{ or}$$

$$\frac{\sum \{(C_i) \times (Z_i)^2\}}{2} \qquad \textit{Eq. 1–8}$$

where C_i is the concentration of the ion, Z_i is the charge of the ion, and Σ is the sum of the quantity $(C_i) \times (Z_i)^2$ for each electrolyte present. It should be noted that in mixtures of substances, the degree of dissociation must be considered. It is known that increasing ionic strength can promote compounds to dissociate into ions and increase the solubility of some salts, which can affect electrophoretic migration.

Density

Density of a substance is expressed in terms of mass per unit volume, but this should not be confused with the previously defined concentration units. Another term used to express density is *specific gravity,* often associated with units such as grams per milliliter. Density or specific gravity is most often used with concentrated acids, such as hydrochloric or sulfuric acid.

Water Specifications

The most frequently used reagent in the laboratory is water. Since tap water is not suitable for laboratory applications, most procedures, including reagent and standard preparation, use water that has been substantially purified. Water solely purified by distillation results in *distilled water,* whereas water purified by ion exchange produces *deionized water.* Reverse osmosis, which pumps water across a semipermeable membrane, produces *RO water.* Laboratory requirements generally call for reagent grade water which, according to the National Committee for Clinical Laboratory Standards (NCCLS), belongs to one of three Types (I, II, and III). It is recommended that water be classified in terms of Type rather than the method of preparation. The NCCLS criteria for reagent grade water is discussed in NCCLS Document C3-A2, *Preparation and Testing of Reagent Water in the Clinical Laboratory.*[10]

Prefiltration can be used to remove particulate matter for municipal water supplies prior to any additional treatments. Filtration cartridges are composed of glass, cotton, activated charcoal which removes organic materials and chlorine, and submicron filters (~0.2 μm), which remove any substances larger than the filter's pores including bacteria. The manner in which these filters are used depends on the quality of the municipal water and the other purification methods employed. For example, hard water (contains calcium, iron, and other dissolved elements) may require prefiltration with a glass or cotton filter rather than activated charcoal or submicron filters, which would quickly become clogged and expensive to operate. The submicron filter might be better employed following distillation, deionization, or reverse osmosis treatment.

Distilled water has been purified to remove almost all organic materials using the technique of distillation, like that found in organic chemistry laboratory distillation experiments. Water is boiled and vaporized. The vapor rises and enters into the coil of a condenser, which is a glass tube that contains within it a glass coil. Cool water surrounds this condensing coil, resulting in a lowering of the temperature of the water vapor, which again forms a liquid and is then collected. Many impurities do not rise in the water vapor but remain in the boiling apparatus. The water collected after condensation has less contamination. Since laboratories use thousands of liters of water per day, stills are used instead of small condensing apparatuses, but the principles are basically the same. Water may be distilled more than once, and each distillation cycle will remove impurities.

Deionized water has some or all ions removed, although organic material may still be present, so it is neither pure nor sterile. In general, deionized water is purified from previously treated water, such as prefiltered or distilled water. Deionized water is produced using either an anion or a cation exchange resin, followed by replacement of the removed particles with hydroxyl or hydrogen ions. The anticipated ions to be removed from the water will dictate the type of ion-exchange column to be used. One column cannot service all ions present in water. A combination of several columns will produce different grades of deionized water. A two-bed system employs an anion followed by a cation column, while a mixed-bed system incorporates both in the same container.

Depending upon the quality of the feed water, Type I water can be obtained by initially filtering it to remove particulate matter, followed by reverse osmosis, deionization, and a 0.2 μm filter.

Type III water is acceptable for glassware washing but not for analysis or reagent preparation. Type II water is acceptable for most analytical requirements, including reagent, quality-control, and standard preparation. Type I water is used for test methods requiring minimum interference, such as trace metal, iron, and enzyme analyses. Use with high pressure liquid chromatography may require less than a 0.2 μm final filtration step. Storage of Type I water is discouraged, and Type II water should be stored in a manner that will reduce any chemical or bacterial contamination and for short periods of time.

Testing procedures to determine the quality of reagent-grade water include measurements of resistance, pH, colony counts on selective and nonselective media for the detection of coliforms, chlorine, ammonia, nitrate or nitrite, iron, hardness, phosphate, sodium, silica, carbon dioxide, chemical oxygen demand (COD), and metal detection. The College of American Pathologists (CAP) recommends that a laboratory document culture growth, pH, and specific resistance on water used in reagent preparation. Resistance is measured because pure water, devoid of ions, conducts electricity poorly. The relationship of water purity to resistance is a linear one: as the purity increases, so does the resistance. This one measurement does not suffice for determination of true water purity, because a nonionic contaminant may be present that has little effect on resistance.

CLINICAL LABORATORY SUPPLIES

Many different supplies are required in today's medical laboratory, but there are several items that are common to all. These items include pipets, flasks, beakers, burets, desiccators, and filtering material. A brief discussion of the composition and general use of these supplies follows.

Glass and Plasticware

Until recently, laboratory supplies consisted of some type of glass and could be correctly termed *glassware.* As plastic material was refined and made available to manufacturers, more and more plastic has been used to make laboratory utensils. Before the discussion of general laboratory supplies can begin, a brief summary of the types and uses of glass and plastic commonly seen in today's laboratories will be given.

Glassware used in the clinical laboratory may fall into one of the following categories: high thermal, high silica, high alkali resistant, low actinic, or soda lime glass. Appendix P, *Characteristics of Types of Glass,* describes some of the characteristics of several types of glass.[11,12] Whenever possible, routinely used clinical chemistry glassware should consist of high thermal borosilicate or aluminosilicate glass and meet class A tolerances prescribed by the NIST.[13] Glassware that does not meet Type A specifications may have twice the tolerance range despite the fact that it may appear to be identical to a piece of Type A glassware. The best source of information about specific uses, limitations, and accuracy specifications for glassware is the manufacturer.

Plasticware is beginning to replace glassware in the laboratory setting. Its unique high resistance to corrosion and breakage, as well as its varying flexibility, has made it most appealing. The major types of resins frequently used in the clinical chemistry laboratory are polystyrene, polyethylene, polypropylene, Tygon, Teflon, polycarbonate, and polyvinylchloride. Appendices Q, *Characteristics of Types of Plastic,* and R, *Chemical Resistance of Types of Plastic,* describe some important characteristics for each of these plastics. Once again, the individual manufacturer is the best source of information concerning the limitations of any plastic materials, and these appendices should be used only as guides.

In most laboratories, the glass or plastic directly associated with testing is almost always disposable. However, should the need arise, cleaning of glass or plastic may require special techniques. Immediately rinsing glass or plastic supplies after

using, followed by washing with a powder or liquid detergent designed for cleaning laboratory supplies and several distilled water rinses, may be sufficient. Presoaking glassware in soapy water is highly recommended whenever immediate cleaning is impractical. Many laboratories use automatic dishwashers and dryers for cleaning. Detergents and temperature levels should be compatible with the material and the manufacturer's recommendations. To ensure that all detergent has been removed from the labware, check the pH of the rinse water and compare it with the initial pH of the prerinse water. Detergent-contaminated water will have an alkaline pH. Visual inspection should reveal spotless vessel walls. Any biologically contaminated labware should be disposed of according to the precautions followed by that laboratory.

Some determinations, such as enzymes, iron, and other heavy metals, require scrupulously clean glassware. Cleaning solutions that have been used successfully are acid dichromate and nitric acid. It is suggested that disposable glass and plastic be used wherever possible. Appendix S, *Cleaning Labware,* lists a variety of cleaning techniques which may assist the laboratorian in obtaining clean labware.

Dirty pipets should be placed immediately in a container of soapy water with the pipet tips up. The container should be long enough to allow the pipet tips to be covered with solution. Use of a specially designed pipet soaking jar and washing/drying apparatus is recommended. For final water rinses, fresh Type I or II water should be provided for each rinse. If possible, set aside a pipet container for final rinses only. Cleaning brushes are available to fit almost any size glassware and are recommended for any articles that are washed routinely.

Plastic material is often easier to clean because of its nonwettable surface. A brush or harsh abrasive cleaner should not be used on plasticware. Acid rinses or washes are not required. The initial cleaning procedure described in Appendix S, *Cleaning Labware*, can be adapted for plasticware as well.

Pipets

Pipets are utensils made of glass or plastic that are used to transfer liquids, and they may be reusable or disposable. In many institutions, automatic pipetting devices have replaced the manual glass or plastic pipets. To minimize confusion, Table 1–4 lists the classification scheme further described here. Examples of pipets are found in Figure 1-1.

Pipets are designed to contain (TC) or to deliver (TD) a particular volume of liquid. Near the top of the pipet, most manufacturers stamp the initials *TC* or *TD* to alert the user as to which type of pipet it is. A TC pipet holds a particular volume but does not dispense that exact volume, whereas a TD pipet will dispense the volume indicated. When using either pipet, the tip of the pipe must be immersed in the liquid to be transferred to a level that will allow it to remain in solution after the volume of liquid has entered the pipet and

TABLE 1-4. Pipet Classification

I. Design
   A. To contain (TC)
   B. To deliver (TD)
II. Drainage characteristics
   A. Blow-out
   B. Self-draining
III. Type
   A. Measuring or graduated
      1. Serologic
      2. Mohr
      3. Bacteriologic
      4. Ball, Kolmer, or Kahn
      5. Micropipet
   B. Transfer
      1. Volumetric
      2. Ostwald-Folin
      3. Pasteur pipets
      4. Automatic macro- or micropipets

without touching the vessel walls. The pipet is held upright, not at an angle (Fig. 1–2). A slight suction using a pipet bulb or similar device is applied to the opposite end until the liquid enters the pipet and the meniscus is brought above the desired graduation line (Fig. 1–3A); suction is then stopped. While the meniscus level is held in place, the pipet tip is raised slightly out of the solution and wiped of any adhering liquid with a laboratory tissue, and the liquid is allowed to drain until the bottom of the meniscus touches the desired calibration mark (Fig. 1–3B). With the pipet held in a vertical position and the tip against the side of the receiving vessel, the pipet contents are allowed to drain into the vessel (*i.e.,* test tube, cuvet, flask, etc.). If the pipet has a continuous etched ring or two small continuous rings very close together located near the top of the pipet, it is called a *blow-out pipet.* This means that the last drop of liquid should be expelled into the receiving vessel. When these markings are absent from a pipet, it is *self-draining,* and the user allows the contents of the pipet to drain by gravity. The tip of the pipet should not be in contact with the accumulating fluid in the receiving vessel during drainage. With the exception of the Mohr pipet, the tip should remain in contact with the side of the vessel for several seconds after the liquid has drained.

▲
Figure 1-1. Types of pipets. From left to right: volumetric pipet, Ostwald-Folin pipet, serologic pipet—10 mL, serologic pipet—1 mL, Mohr pipet.

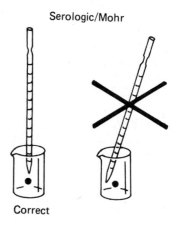

Serologic/Mohr

Correct

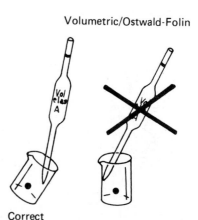

Volumetric/Ostwald-Folin

Correct

▲
Figure 1-2. Correct and incorrect pipet positions.

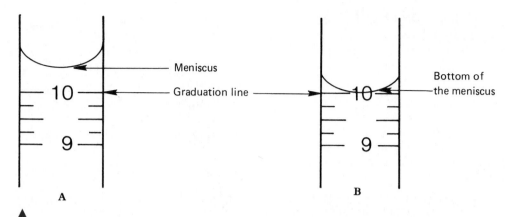

▲
Figure 1-3. Pipetting technique. (*A*) Meniscus is brought above the desired graduation line. (*B*) Liquid is allowed to drain until the bottom of the meniscus touches the desired calibration mark.

The pipet is then removed. Various examples of pipet bulbs are illustrated in Figure 1-4. Mouth pipetting is strictly *forbidden* and should never be done.

Measuring or graduated pipets are capable of dispensing several different volumes. Because the graduation lines located on the pipet may vary, they should be indicated on the top of each pipet. For example, a 5-mL pipet can be used to measure 5, 4, 3, 2, or 1 mL of liquid, with further graduations between each milliliter. The pipet is designated as a 5 in $\frac{1}{10}$ increments (Fig. 1-5), and could deliver any volume in tenths of a milliliter, up to 5 mL. Another pipet, such as a 1-mL pipet, may be designed to dispense 1 mL and have subdivisions of hundredths of a milliliter. The markings at the top of a measuring or graduated pipet indicate the volume(s) it is designed to dispense.

The subgroups of measuring or graduated pipets are Mohr, serologic, bacteriologic, and micropipets. A *Mohr pipet* does not have graduations to the tip. It is a self-draining pipet, but the tip should not be allowed to touch the vessel while the pipet is draining. A *serologic pipet* has calibration marks to the tip and is generally a blow-out pipet. A *micropipet* is a pipet with a total holding volume of less than 1 mL: it may be a Mohr or serologic type pipet.

The next major category of pipes consists of the *transfer pipets*. These pipets are designed to dispense one volume without further subdivisions. The *Ostwald-Folin* and *volumetric* subgroups are easily distinguished by the bulb-like enlargement in the pipet stem. Ostwald-Folin pipets are used

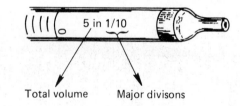

Total volume   Major divisons

▲
Figure 1-5. Volume indication of a pipet.

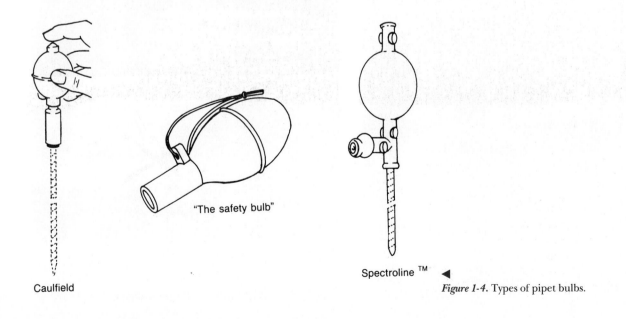

Caulfield

"The safety bulb"

Spectroline ™ ◀

Figure 1-4. Types of pipet bulbs.

with biologic fluids having a viscosity that is greater than that of water. They are blow-out pipets, indicated by two etched continuous rings at the top. The volumetric pipet is designed to dispense or transfer aqueous solutions and is always self-draining. This type of pipet usually has the greatest degree of accuracy and precision and should be used when diluting standards, calibrators, or quality-control material. *Pasteur pipets* do not have any calibration marks and are used to transfer solutions or biologic fluids without consideration of a specific volume. These pipets should not be used in any quantitative analytical techniques.

The *automatic pipet* is by far the most routinely used pipet in today's clinical chemistry laboratory. The major advantages of automatic and semi-automatic pipets are time savings, safety, ease of use, stability, increase in precision, and lack of required cleaning, since the contaminated portions of the pipet, such as the tips, are often disposable. Figure 1-6 illustrates many common automatic pipets. Many models have different volumes that may be selected and are termed *variable*, but only one volume at a time may be used. The range of available volumes is from 1 μL to 1 L. The widest volume range usually seen in a single pipet is 0 to 1 mL. Any pipet that has a pipetting capability of less than 1 mL is considered

a *micropipet*, and a pipet that dispenses greater than 1 mL is called an *automatic macropipet*.

The term *automatic* implies that the mechanism that draws up the liquid and dispenses it is an integral part of the pipet itself. There are three general types of automatic pipets: air-displacement, positive-displacement, and dispenser pipets. An *air-displacement pipet* relies on a piston for suction creation in order to draw the sample into a disposable tip. The piston does not come in contact with the liquid. A *positive-displacement pipet* operates by moving the piston in the pipet tip or barrel, much like a hypodermic syringe. It does not require a different tip for each use. Because of carryover concerns, rinsing and blotting between samples may be required. *Dispensers* and *dilutor/dispensers* are automatic pipets that obtain the liquid from a common reservoir and dispense it repeatedly. The dispensing pipets may be bottle-top, motorized, or hand-held, or attached to a dilutor. The dilutor often combines sampling and dispensing functions. Figure 1-7 provides examples of different types of automatic pipetting devices. These pipets should be used according to the individual manufacturer's directions. Many automated pipettors use a wash in between samples to eliminate carryover problems. However, in order to min-

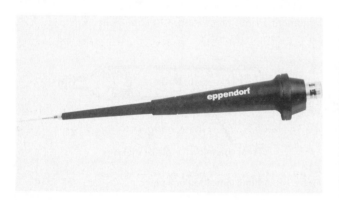

A

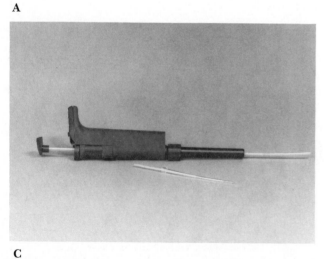

C

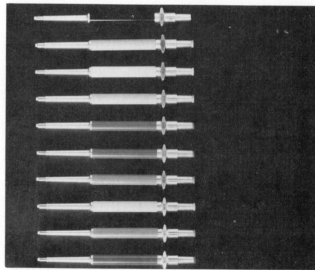

B

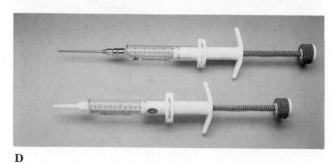

D

▲
Figure 1-6. (*A*) Fixed volume ultra micro-digital air displacement pipetter with tip ejector. (*B*) Fixed-volume air displacement pipet. (*C*) Digital electronic positive displacement pipetter. (*D*) Syringe pipetters. (Photos courtesy of Scientific Products Division, Baxter Healthcare Corporation)

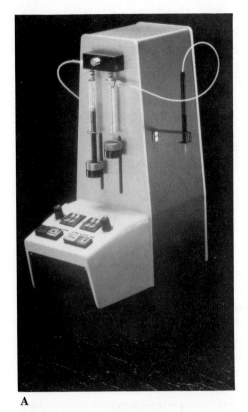
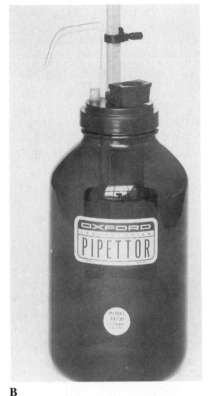
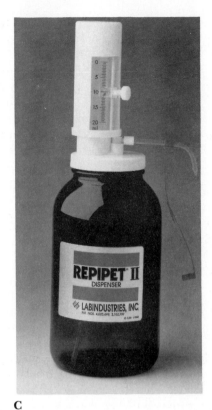

A   B   C

Figure 1-7. (*A*) Digital dilutor/dispenser. (*B*) Dispenser. (*C*) Dispenser. (Photos courtesy of Scientific Products Division, Baxter Healthcare Corporation)

imize carryover contamination with manual or semi-automated pipets, careful wiping of the tip may be necessary to remove any liquid that may have adhered to the outside of the tip before dispensing any liquid. Care should be taken to ensure that the orifice of the pipet tip is not blotted, thereby drawing sample from the tip. Another precaution in using manually operated semi-automatic pipets is to move the plunger in a continuous, slow manner.

Pipet tips are designed to be used with positive- and air-displacement pipets. The laboratorian needs to ensure that the pipet tip is seated snugly on the end of the pipet and is free from any deformity. Plastic tips used on air-displacement pipets are particularly likely to vary. Different brands can be used for one particular pipet but do not necessarily perform in an identical manner. Plastic burrs, which sometimes cannot be detected by the naked eye, can be present. A method using a 0.1% solution of phenol red in distilled water has been used to compare the reproducibility of different brands of pipet tips.[14] When using this method, the pipet and the operator should remain the same so that variation is due solely to changes in the pipet tips.

Class A pipets, like all other class A glassware, do not need to be recalibrated by the laboratory. Automatic pipetting devices, as well as non-Class A materials, do need recalibration. A gravimetric method can be used to accomplish this task. It is done by delivering and weighing a solution of

known specific gravity, such as water. It is suggested that a currently calibrated analytical balance and at least Class S weights be used.

Gravimetric Pipet Calibration[15]

Materials

Pipet.
10–20 pipet tips.
Balance capable of accuracy and resolution to ±0.1% of dispensed volumetric weight.
Weighing vessel large enough to hold volume of liquid.
Distilled water.
Thermometer.

Procedure

1. Record the weight of the vessel. Record the temperature of the water.
2. Take up the distilled water with the pipet. Carefully wipe the outside of tip. Care should be taken not to touch the end of the tip. This will cause liquid to be wicked out of the tip introducing an inaccuracy due to technique.
3. Dispense the water into the weighed vessel. Touch the tip to the side.
4. Record the weight of the vessel plus the water.

5. Subtract the weight obtained in step 1 above from that obtained in step 4. Record the result. This result is the weight of the first delivered water sample.
6. If plastic tips are used, change the tip between each dispensing. Repeat steps 1 through 5 at least 9 to 19 more times.
7. Obtain the average of the weights of the water.
8. Determine the accuracy or the ability of the pipet to dispense the selected or stated volume according to the following formula:

% deviation from expected volume

$$= \frac{\text{delivered volume} - \text{expected volume}}{\text{expected volume}} \times 100$$

Eq. 1–9

The delivered volume is calculated by dividing the mean ($\overline{x}$) or average weight of water by the specific gravity of water at the measured water temperature. Table 1–5 lists the specific gravity of water at probable laboratory temperatures.[6] Acceptable limitations for a particular pipet are usually given by the manufacturer, but they should not be used if the value differs by more than 1.5% from the expected value. Precision can be indicated as the percent coefficient of variation (%CV) or standard deviation (SD) for a series of repetitive pipetting steps. A discussion of %CV and SD can be found in Chapter 4. The equations to calculate the SD and %CV are as follows:

$$SD = \sqrt{\frac{\Sigma(x - \overline{x})^2}{n - 1}}$$

$$\%CV = \frac{SD}{\overline{x}} \times 100 \qquad Eq.\ 1–10$$

Required imprecision is usually ±1 SD. The %CV will vary with the expected volume of the pipet, but the smaller the %CV value, the greater is the precision. When *n* is large, the data are more statistically valid.

Pipet calibration also may be accomplished by using photometric methods. These procedures involve use of a spec-

trophotometer, a flame photometer, or a radioactive material read in a scintillation counter. When a spectrophotometer is used, the molar extinction coefficient of the compound used should be known. After an aliquot of diluent is pipetted, the change in concentration will reflect the volume of the pipet. Dilutions of various dyes added to volumetric glassware also can be used with a spectrophotometer for micropipets. A flame photometer can be used in the same manner, except that a known primary electrolyte standard is measured. The pipet to be calibrated is used to obtain an aliquot of diluent, and the expected electrolyte concentration can be compared with the actual value obtained. The radioactive method involves pipetting and counting a radioactive solution and comparing that value with the radioactivity obtained from the same material dispensed using a class A volumetric pipet.

These calibration techniques are time-consuming and therefore not practical for use in daily checks. It is recommended that pipets be checked initially and subsequently three to four times per year. A quick daily check for many larger-volume automatic pipetting devices uses volumetric flasks. For example, a bottle-top dispenser that routinely delivers 2.5 mL of a reagent may be checked by dispensing four aliquots of the reagent into a 10-mL class A volumetric flask. The bottom of the meniscus should meet with the calibration line on the volumetric flask.

Burets

A *buret* looks much like a wide, long, graduated pipet with a stopcock at one end. A buret usually holds a total volume ranging from 25 mL to 100 mL of solution and is used to dispense a particular volume of liquid during a titration (Fig. 1-8).

Laboratory Vessels

Flasks, beakers, and graduated cylinders are used to hold solutions. Volumetric and Erlenmeyer flasks are two types of containers in general use in the clinical laboratory.

A *volumetric flask*, like any volumetric utensil, is calibrated to hold one exact volume of liquid. The flask has a round, flat lower portion and a long, thin neck with the calibration line etched into the neck. Volumetric flasks are used to bring a given reagent to its final volume with the prescribed diluent and should be of Class A quality. When bringing the bottom of the meniscus up to the calibration mark, a pipet should be used when adding the final drops of diluent so that maximum control is maintained and the calibration line is not missed.

Erlenmeyer flasks and *Griffin beakers* are designed to hold different volumes rather than one exact amount. Because these pieces are often used in reagent preparation, the flask size, chemical inertness, and thermal stability should be considered. The Erlenmeyer flask has a wide bottom that gradually evolves into a smaller, short neck. The Griffin beaker

TABLE 1-5. Water Density	
°C	**Density**
20	0.9982
21	0.9980
22	0.9978
23	0.9976
24	0.9973
25	0.9971
26	0.9968
27	0.9965
28	0.9963
29	0.9960
30	0.9957

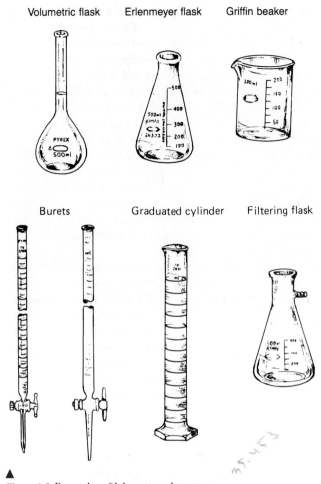

Volumetric flask   Erlenmeyer flask   Griffin beaker

Burets   Graduated cylinder   Filtering flask

Figure 1-8. Examples of laboratory glassware.

TABLE 1-6. Class A Tolerances

Size (mL)	Tolerances (mL)
BURETS	
5	±0.01
10	±0.02
25	±0.03
50	±0.05
100	±0.10
PIPETS (TRANSFER)	
0.5–2	±0.006
3–5	±0.01
10	±0.02
15–25	±0.03
50	±0.05
VOLUMETRIC FLASKS	
1–10	±0.02
25	±0.03
50	±0.05
100	±0.08
200	±0.10
250	±0.12
500	±0.20
1000	±0.30
2000	±0.50

has a flat bottom, straight sides, and an opening as wide as the flat base with a small spout in the lip.

Graduated cylinders are long, cylindrical tubes usually held upright by an octagonal or circular base. The cylinder has calibration marks along its length and is used to measure volumes of liquids. Graduated cylinders do not have the accuracy of volumetric glassware. The sizes routinely used are 10, 25, 50, 100, 500, 1000, and 2000 mL.

All laboratory utensils should be Class A whenever possible to maximize accuracy and precision and thus decrease calibration time. Figure 1-8 illustrates representative laboratory glassware. Table 1-6 lists the class A tolerances for some commonly used volumes.

Desiccators and Desiccants

Many compounds combine with water molecules to form loose chemical crystals. The compound and its associated water is called a *hydrate*. When the water of crystallization is removed from the compound, it is said to be *anhydrous*. Some substances take up water on exposure to atmospheric conditions and are called *hygroscopic*. Materials that are very hygroscopic can remove moisture from the air as well as from other materials. These make excellent drying substances and are sometimes used as *desiccants* (drying agents) to keep other chemicals from becoming hydrated. Commonly used desiccants[6] are listed in Table 1-7, beginning with the most hygroscopic and ending with the least effective drying agent. If these compounds absorb enough water from the atmosphere to cause dissolution, they are called *deliquescent substances*. Desiccants are most effective when placed in a closed chamber called a *desiccator* (Fig. 1-9). The desiccant is placed below the perforated platform inside the desiccator. A glass or plastic desiccator is sealed by placing a fine film of grease on the rim of the cover, and it is correctly opened or sealed by slowly sliding the lid sideways. The greased seal prevents the desiccator from being opened with a direct pull upward. Open a desiccator slowly and with caution because the air pressure inside the desiccator could be below atmospheric

TABLE 1-7. Common Desiccants

Agent	Formula
Magnesium perchlorate (Dehydrite)	$Mg(ClO_4)$
Barium oxide	BaO
Alumina	Al_2O_3
Phosphorous pentoxide	P_4O_{10}
Lithium perchlorate	$LiClO_4$
Calcium chloride	$CaCl_2$
Calcium sulfate (Drierite)	$CaSO_4$
Silica gel	SiO_2
Ascarite	NaOH on asbestos

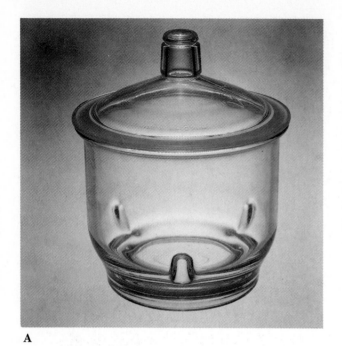

A   **B**

▲

Figure 1-9. (*A*) Glass desiccator. (*B*) Glass desiccator with dry seal ring (no grease is necessary to seal cover to body). (Photo courtesy of Scientific Products Division, Baxter Healthcare Corporation)

pressure. Desiccants containing indicators that signify desiccant exhaustion and that can be regenerated using heat are particularly helpful. Desiccants that produce dust should also be avoided. In the laboratory, desiccants are primarily used to prevent moisture absorption by chemicals, gases, and instrument components.

Balances

A properly operating balance is essential in producing high-quality reagents and standards Balances may be classified according to their design, number of pans (single or double), whether they are mechanical or electronic, or their operating ranges, as determined by precision balances, analytical balances, or microbalances (Fig. 1-10).

Analytical and electronic balances are the most popular balances used in today's clinical laboratories. Analytical balances are required for the preparation of any primary standards. The mechanical analytical balance is also known as a *substitution balance*. It has a single pan enclosed by sliding transparent doors, which minimize environmental influences on pan movement. The pan is attached to a series of calibrated weights, which are counterbalanced by a single weight at the opposite end of a knife-edge fulcrum. The operator adjusts the balance to the desired mass and places the material, contained within a tared weighing vessel, on the sample pan. An optical scale allows the operator to visualize the mass of the substance. The weight range for an analytical balance is from 0.01 mg up to 160 g on some models.

Electronic balances are single-pan balances that use an electromagnetic force to counterbalance the weighed sample's mass. Their measurements equal the accuracy and precision of any available mechanical balance, with the advantage of a very fast response time (less than 10 seconds).

It is recommended that NIST Class S weights be used for calibrating balances on a monthly basis. Balances should be kept scrupulously clean and located in an area away from heavy traffic, large pieces of electrical equipment, and open windows. A slab of marble separated from its supporting surface by a flexible material is sometimes placed under balances to minimize any vibration interference that may occur. The level checkpoint should always be corrected prior to any weighings.

BASIC SEPARATION TECHNIQUES

Centrifugation

Centrifugation is a process whereby centrifugal force is used to separate solid matter from a liquid suspension. The *centrifuge* is an apparatus that carries out this action. A centrifuge consists of a head or rotor, carriers, and/or shields (Fig. 1-11) that are attached to the vertical shaft of a motor and enclosed in a metal covering. The centrifuge always has a lid and an on/off switch, but many models also include a brake and/or a built-in tachometer, which indicates speed, and can be refrigerated. Centrifugal force is dependent on three variables: mass, speed, and radius. The speed is expressed in revolutions per minute (rpm), and the centrifugal force generated is expressed in terms of relative centrifugal force (RCF) or grav-

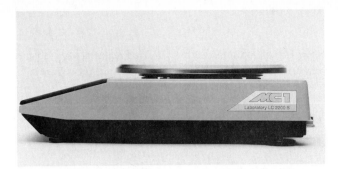

A

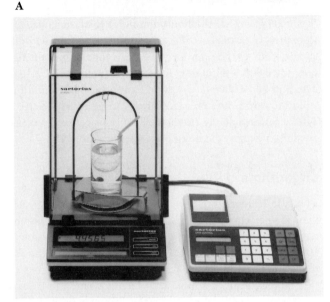

B

▲

Figure 1-10. (*A*) Electronic toploading balance. (*B*) Electronic analytical balance with printer. (Photos courtesy of Scientific Products Division, Baxter Healthcare Corporation)

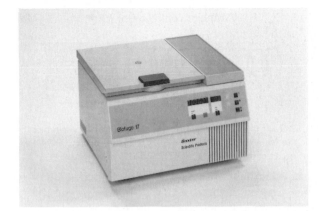

A

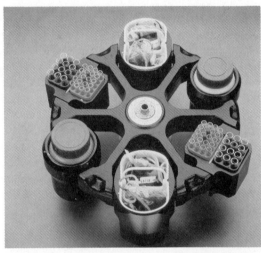

B

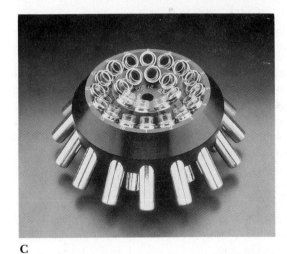

C

▲

Figure 1-11. (*A*) Benchtop centrifuge. (*B*) Swinging bucket rotor. (*C*) Fixed head rotor. (Photos courtesy of Scientific Products Division, Baxter Healthcare Corporation)

ities (g). The speed of the centrifuge is related to the RCF by the following equation:

$$RCF = 1.118 \times 10^{-5} \times r \times (rpm)^2 \qquad \textit{Eq. 1–11}$$

where 1.118×10^{-5} is a constant, determined from the angular velocity, and r is the radius in centimeters, measured from the center of the centrifuge axis to the bottom of the test-tube shield. The RCF value also may be obtained from a nomogram like that found in Appendix K, *Relative Centrifugal Force Nomograph*. Classification of centrifuges may be based on several criteria, including table-top or floor model, refrigeration, rotor head (*e.g.,* fixed, hematocrit, or angled; see Fig. 1-11), or maximum speed attainable (*i.e.,* ultracentrifuge).

Centrifuges are generally used either to separate serum from a blood clot as the blood is being processed, a supernatant from a precipitate during an analytical reaction, two immiscible liquids, or to expel air.

Care of a centrifuge includes daily cleaning of any spills or debris, such as blood or glass, and checking to see that the centrifuge is properly balanced and free from any excessive vibrations. The centrifuge cover should remain closed until the centrifuge has come to a complete stop to avoid any aerosol contamination. It is recommended that the timer, brushes (if present), and speed be checked periodically. The

brushes, which are graphite bars attached to a retainer spring (Fig. 1-12), create an electrical contact in the motor. The specific manufacturer's service manual should be consulted for details on how to change brushes and lubrication requirements. The speed of a centrifuge is easily checked using a tachometer or strobe light. The hole located in the lid of

Graphite centrifuge brush

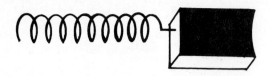

▲

Figure 1-12. The centrifuge brush creates an electrical contact in the motor.

many centrifuges is designed for speed verification using these devices.

Filtration

Filtration was used in place of centrifugation for the separation of solids from liquids. In today's laboratory, paper filtration is no longer routinely used, and thus only the basics will be discussed here. Filter material is made of paper, cellulose and its derivatives, polyester fibers, glass, and a variety of resin column materials. Traditionally, filter paper was folded in a manner that allowed it to fit into a funnel. In method A, round filter paper is folded like a fan (Fig. 1-13*A*), and in method B, the paper is folded into fourths (Fig. 1-13*B*).

Filter paper differs in terms of pore size and should be selected according to separation needs. Filter paper should not be employed when using strong acids or bases. Once the filter paper is placed inside the funnel, the solution slowly drains through the filter paper within the funnel and into a receiving vessel. The liquid that passes through the filter paper is called the *filtrate*.

Dialysis

Dialysis is another method for separating macromolecules from a solvent. It was made popular when used in conjunction with the Technicon Autoanalyzer™ system. Basically, a solution is put into a bag or is contained on one side of a semipermeable membrane. Larger molecules are retained within the sack or on one side of the membrane, while smaller molecules and solvents diffuse out. This process is very

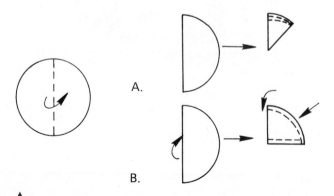

▲

Figure 1-13. Methods of folding a filter paper. (*A*) In a fan. (*B*) In fourths.

slow. Use of columns that contain a gel material have replaced manual dialysis separation in most analytical procedures.

LABORATORY MATHEMATICS AND CALCULATIONS

Significant Figures

Significant figures are the minimum number of digits needed to express a particular value in scientific notation without loss of accuracy. The number 814.2 has four significant figures, since in scientific notation it is written as 8.142×10^2. The number 0.000641 has three significant figures, and the scientific notation expression for this value is 6.41×10^{-4}. The zeros are merely holding decimal places and are not needed to express the number in scientific notation.

Logarithms

The base 10 logarithm (log) of a positive number N greater than zero is equal to the exponent to which 10 must be raised to produce N. Therefore, it can be stated that N equals 10^x, and the log of N is equal to x. The number N is the antilogarithm (antilog) of x.

The logarithm of a number, which is written in a decimal format, consists of two parts: the character, or characteristic, and the mantissa. The *character* is the number to the left of the decimal point in the log and is derived from the exponent, and the *mantissa* is that portion of the logarithm to the right of the decimal point and is derived from the number itself. While several approaches can be taken to determine the log, one approach is to write the number in scientific notation. The number 1424 expressed in scientific notation is 1.424×10^3, making the character a 3. The character also can be determined by adding the number of significant figures and subtracting 1 from the sum. The mantissa is derived from a log table or calculator having a log function for the remainder of the number. For 1.424, a calculator would give a mantissa of 0.1535, and a log table would reveal what is shown in Table 1–8. Some calculators with a log function do not require conversion to scientific notation.

When using a log table, the initial step is to determine N. N is obtained from the first two digits in the number. In this example, N is 14. Look under the N column in the log table (Table 1–8) until the number 14 is located. The next digit in our example, 1.424, is 2. Continue along the row and read the numbers under 142, which, in this case, are 1523. The last part of the mantissa is obtained from the proportional parts section of the log table and is added to 1523. In this instance (1.424), the remaining digit is 4. The value under 4 in the table is 12. The mantissa becomes 1523 + 12 = 1535. The logarithm of 1.425×10^3 is equal to 3.1535. The character is 3, and the mantissa is 1535. Since there are only four significant figures in the original number, only four significant

TABLE 1-8. Portion of a Logarithm Table

| | | | | | | | | | | | Proportional Parts | | | | | | | | |
N	0	1	2	3	4	5	6	7	8	9	1	2	3	4	5	6	7	8	9
14	1461	1492	1523	1553	1584	1614	1644	1673	1703	1732	3	6	9	12	15	18	21	24	27

figures should be written in the log. The log now becomes 3.154. For numbers less than 1, the character usually has a bar over the top. The number 2.12×10^{-2} has a log of $\overline{2}.3263$, or $\overline{2}.33$ to satisfy the number of significant figures.

To determine the original number from a log value, the process is basically done in reverse. This process is termed the *antilogarithm*. Using the previous example, determine the antilogarithm of $\overline{2}.33$. The character $\overline{2}$, which has a bar over the top of it, indicates 10^{-2}. The mantissa, 0.3263, or 0.33, may lead to the number in two ways. The mantissa may be located in a logarithm table, and N may be determined from it. A second approach is to use an antilogarithm table. Table 1-9 illustrates both methods. It should be noted that the rounding of numbers does affect the antilogarithm. Most calculators have logarithm and antilogarithm functions. Consult the specific manufacturer's directions to become acquainted with the proper use of these functions.

Negative or Inverse Logarithms

The character of a log may be positive or negative, but the mantissa is always positive. In some circumstances, the laboratorian must deal with inverse or negative logs. Such is the case with pH. As defined earlier, the pH of a solution is defined as minus the log of the hydrogen ion concentration.

The following is a convenient formula used to determine the negative logarithm when dealing with pH:

$$pH = x - \log N \qquad \text{Eq. 1–12}$$

For example, if the hydrogen ion concentration of a solution is 5.4×10^{-6}, then $x = 6$. The logarithm of N (5.4) is equal to 0.7324, or 0.73. The pH becomes

$$pH = 6 - 0.73 = 5.27 \qquad \text{Eq. 1–13}$$

The same formula can be applied to obtain the hydrogen ion concentration of a solution when only the pH is given. Using a pH of 5.27, the equation becomes

$$5.27 = x - \log N \qquad \text{Eq. 1–14}$$

The x term is always the next largest whole number. For this example, the next largest whole number is 6. Substituting for x the equation becomes

$$5.27 = 6 - \log N \qquad \text{Eq. 1–15}$$

Multiply all the variables by -1 and solve the equation for the unknown quantity, that is, log N. The result is -0.73, which is the antilogarithm value of N, which is 5.37, or 5.4. The hydrogen ion concentration for a solution with a pH of 5.27 is 5.4×10^{-6}.

TABLE 1-9. Antilogarithm Determination

A. Log Table

N	0	1	2	3
21	3222	3243	3263	3284

The number, as determined from the mantissa 3263, is 212.

B. Antilog Table

N	0	1	2	3	4	5	6	7	8	9	Proportional parts 1 2 3 4 5 6 7 8 9
.32	2089	2094	2099	2104	2109	2113	2118	2123	2128	2133	0 1 1 2 2 3 3 4 4
.33	2138	2143									

.3263 = 2118 plus 1 = 2119 = 2.12
.33 = 2138 = 2.14

Concentration

A detailed description of each concentration term (*i.e.,* molarity, normality, etc.) may be found at the beginning of this chapter. The following discussion will focus on basic mathematical expressions needed to prepare reagents of a stated concentration.

Percent Solution

A percent solution is determined in the same manner regardless of whether weight/weight, volume/volume, or weight/volume units are used. *Percent* means "parts per 100," which is represented as percent (%) and is independent of the molecular weight of a substance.

Example 1–1: Weight/Weight (w/w)

To make up 100 g of a 5% aqueous solution of hydrochloric acid (using 12 M HCl), multiply the total amount by the percent expressed as a decimal. The calculation becomes

$$5\% = \frac{5}{100} = 0.050 \qquad \textbf{\textit{Eq. 1–16}}$$

Therefore,

$$0.050 \times 100 = (5 \text{ g of HCl}) \qquad \textbf{\textit{Eq. 1–17}}$$

Another way of arriving at the answer is to set up a ratio so that

$$\frac{5}{100} = \frac{x}{100}$$

$$x = 5 \qquad \textbf{\textit{Eq. 1–18}}$$

Example 1–2: Weight/Volume (w/v)

The most frequently used term for a percent solution is weight per volume, which is often expressed as grams per 100 mL of diluent. To make up 1000 mL of a 10% (w/v) solution of NaOH, use the approach as indicated above. The calculations become

$$\underset{\text{(\% expressed as a decimal)}}{0.10} \times \underset{\text{(total amount)}}{1000} = 100\text{g}$$

or

$$\frac{10}{100} = \frac{x}{1000}$$

$$x = 100 \qquad \textbf{\textit{Eq. 1–19}}$$

Therefore, add 100 g of NaOH to a 1000-mL volumetric Class A flask and dilute to the calibration mark with Type II water.

Example 1–3: Volume/Volume (v/v)

Make up 50 mL of a 2% (v/v) concentrated hydrochloric acid solution.

$$0.02 \times 50 = 1 \text{ mL}$$

or

$$\frac{2}{100} = \frac{x}{50}$$

$$x = 1 \qquad \textbf{\textit{Eq. 1–20}}$$

Therefore, add 40 mL of water to a 50-mL Class A volumetric flask, add 1 mL of HCl, mix, and dilute up to the calibration mark with Type II water. Remember, always add acid to water!

Molarity

Molarity (M) is routinely expressed in units of moles per liter (mol/L) or sometimes millimoles per milliliter (mmol/mL). Remember that 1 mol of a substance is equal to the gram molecular weight of that substance. When trying to determine the amount of substance needed to yield a particular concentration, initially decide what final concentration *units* are needed. In the case of molarity, the final units will be moles per liter (mol/L) or millimoles per milliliter (mmol/mL). The second step is to consider the existing units and the relationship they have to the final desired units. Essentially, try to put as many units as possible into "like" terms so that the only units that are not identical are those wanted in the final answer.

Example 1–4

How many *grams* are needed to make 1 L of a 2 M solution of HCl?

Step 1: Which *units* are needed in the final answer? *Answer:* Grams per liter (g/L).
Step 2: Assess other mass/volume terms used in the problem. In this case, molarity is the concentration needed, so moles are needed for the calculation: How many grams are equal to 1 mol? The gram molecular weight of HCl, which can be determined from the periodic table, will be equal to 1 mol. For HCl, the gram molecular weight (gmw) is 36.5, so the equation may be written as:

$$\frac{36.5 \text{ g HCL}}{\cancel{\text{mol}}} \times \frac{2 \cancel{\text{mol}}}{\text{\textcircled{L}}} = \frac{73 \text{ g HCl}}{\text{L}} \qquad \textbf{\textit{Eq. 1–21}}$$

Cancel out "like" units, and the final units should be grams per liter. In this example, 73 g HCl per liter is needed to make up a 2 M solution of HCl.

Example 1–5

A solution of NaOH is contained within a Class A 1-L volumetric flask filled to the calibration mark. The content label reads 24 g of NaOH. Determine the molarity.

Step 1: What *units* are ultimately needed? *Answer:* Moles per liter (mol/L).
Step 2: The units that exist are grams and 1 L. NaOH may be expressed as moles and grams. The gmw of NaOH is calculated to equal 40 g/mol.
Step 3: The equation becomes

$$\frac{24 \cancel{\text{g}} \text{ NaOH}}{\text{\textcircled{L}}} \times \frac{1 \cancel{\text{mol}}}{40 \cancel{\text{g}} \text{ NaOH}} = 0.6 \frac{\text{mol}}{\text{L}} \qquad \textbf{\textit{Eq. 1–22}}$$

By canceling out "like" units and performing the appropriate calculations, the final answer of 0.6 M or 0.6 mol/L is derived.

Example 1–6

Make up 250 mL of a 4.8 *M* solution of HCl.

Step 1: *Units* needed? *Answer:* Grams (g).
Step 2: Determine the gmw of HCl (36.5 g), which is needed to calculate the molarity.
Step 3: Set up the equation, cancel out "like" units, and perform the appropriate calculations:

$$\frac{36.5 \text{ g HCL}}{\text{mol}} \times \frac{4.8 \text{ mol HCL}}{L} \times \frac{250 \text{ mL} \times 1 L}{1000 \text{ mL}} = 43.8 \text{ g HCl}$$

Eq. 1–23

In a 250-mL volumetric flask, add 200 mL of Type II water. Add 43.8 g of HCl and mix. Dilute up to the calibration mark with Type II water.

Although there are various methods to calculate laboratory mathematics problems, this technique of canceling "like" units can be used in most clinical chemistry situations, regardless of whether the problem requests molarity, normality, or exchanging one concentration term for another.

Normality

Normality (*N*) is expressed as the number of equivalent weights per liter (Eq/L) or milliequivalents per milliliter (mEq/mL). Equivalent weight is equal to the gram molecular weight divided by the valence *V*. Normality has often been used in acid-base calculations because an equivalent weight of a substance is also equal to its combining weight. Another advantage in using equivalent weight is that an equivalent weight of one substance is equal to the equivalent weight of any other chemical.

Example 1–7

A. Give the equivalent weight, in grams, for each substance listed below.

    1. NaCl (gmw = 58 g, valence = 1)

$$58/1 = 58 \text{ g per equivalent weight} \qquad \textit{Eq. 1–24}$$

    2. HCl (gmw = 36, valence = 1)

$$36/1 = 36 \text{ g per equivalent weight} \qquad \textit{Eq. 1–25}$$

    3. H_2SO_4 (gmw = 98, valence = 2)

$$98/2 = 49 \text{ g per equivalent weight} \qquad \textit{Eq. 1–26}$$

B. What is the normality of a 500-mL solution that contains 7 g of H_2SO_4? The approach used to calculate molarity can be used to solve this problem as well.

    Step 1: Units needed? *Answer:* Equivalents per liter (Eq/L).
    Step 2: Units you have? *Answer:* Milliliters and grams.
    Step 3: Change grams into equivalent weight terms and set up the final equation. This equation is

$$\frac{7 \text{ g } H_2SO_4}{500 \text{ mL}} \times \frac{1 \text{ Eq}}{49 \text{ g } H_2SO_4} \times \frac{1000 \text{ mL}}{1 \text{ L}}$$

$$= 0.285 \text{ Eq/L} = 0.285 \text{ } N \qquad \textit{Eq. 1–27}$$

Since 500 mL is equal to 0.5 L, the final equation could be written by substituting 0.5 L for 500 mL and thereby eliminating the need for including the 1000 mL/L conversion factor in the equation.

C. What is the normality of a 0.5 *M* solution of H_2SO_4? Continuing with the previous approach, the final equation is

$$\frac{0.5 \text{ mol } H_2SO_4}{L} \times \frac{98 \text{ g } H_2SO_4}{\text{mol } H_2SO_4} \times \frac{1 \text{ Eq } H_2SO_4}{49 \text{ g } H_2SO_4}$$

$$= 1 \text{ Eq/L} = 1 \text{ } N \qquad \textit{Eq. 1–28}$$

When changing molarity into normality or vice versa, the following conversion formula may be applied:

$$M \times V = N \qquad \textit{Eq. 1–29}$$

where *V* is the valence of the compound. Using this formula, Example 1–7C becomes

$$0.5 \text{ } M \times 2 = 1 \text{ } N \qquad \textit{Eq. 1–30}$$

Example 1–8

What is the molarity of a 2.5 *N* solution of HCl? This problem may be solved in several ways. One way is to use the stepwise approach in which existing units are exchanged for units needed. The equation is

$$\frac{2.5 \text{ Eq HCl}}{L} \times \frac{36 \text{ g HCl}}{1 \text{ Eq}} \times \frac{1 \text{ mol HCl}}{36 \text{ g HCl}}$$

$$= 2.5 \text{ mol/L HCl} \qquad \textit{Eq. 1–31}$$

The second approach is to use the normality to molarity conversion formula. The equation now becomes

$$M \times V = 2.5 \text{ } N$$
$$V = 1$$
$$M = \frac{2.5 \text{ } N}{1} = 2.5 \text{ } N \qquad \textit{Eq. 1–32}$$

When the valence of a substance is 1, the molarity will be equal to the normality.

Specific Gravity

The specific gravity is the ratio of the density of a material when compared to the density of water at a given density. The units for specific gravity are grams per milliliter. Specific gravity is often used with very concentrated materials, such as commercial acids.

The density of a concentrated acid also can be expressed in terms of an assay or percent purity. The actual concentration is equal to the *specific gravity* multiplied by the *assay or percent purity value* (expressed as a decimal) stated on the label of the container. (See Appendix G, *Concentrations Of Commonly Used Acids And Bases With Related Formulas*, for the specific gravity.)

Example 1–9

A. What is the actual weight of a supply of concentrated HCl whose label reads specific gravity 1.19 with an assay value of 37%?

$$1.19 \text{ g/mL} \times 0.37 = 0.44 \text{ g/mL of HCl} \qquad \textit{Eq. 1–33}$$

B. What is the molarity of this stock solution? The final units desired are moles per liter (mol/L). The molarity of the solution is

$$\frac{0.44 \, \cancel{g} \, HCl}{\cancel{mL}} \times \frac{1 \, \cancel{(mol)} \, HCl}{3.5 \, \cancel{g} \, HCl} \times \frac{1000 \, \cancel{mL}}{\cancel{(L)}}$$

$$= 12.05 \text{ mol/L or } 12 \, M \qquad \textbf{\textit{Eq. 1–34}}$$

Conversions

In order to convert one unit into another, the same approach of crossing out "like" units can be applied. In some instances, a chemistry laboratory may report a given analyte using two different concentration units. An example of this is calcium. The recommended SI unit for calcium is millimoles per liter. The better known and more traditional units are milligrams per deciliter (mg/dL).

Example 1–10

Convert 8.2 mg/dL calcium to millimoles per liter (mmol/L). The gmw of calcium is 40 g. So, if there are 40 g per mol, then it follows that there are 40 mg per mmol. The units wanted are mmol/L. The equation becomes

$$\frac{8.2 \, \cancel{mg}}{\cancel{dL}} \times \frac{1 \, \cancel{dL}}{100 \, \cancel{mL}} \times \frac{1000 \, \cancel{mL}}{\cancel{(L)}} \times \frac{1 \, \cancel{(mmol)}}{40 \, \cancel{mg}} = \frac{2.05 \text{ mmol}}{L}$$

$$\textbf{\textit{Eq. 1–35}}$$

Once again, the systematic stepwise approach of deleting similar units can be used for this conversion problem.

A frequently encountered conversion is when you need a weaker concentration or different volume than the substance available, but the concentration *terms* are the same. While the cross-out method can be applied, the following formula is often much faster:

$$V_1 \times C_1 = V_2 \times C_2 \qquad \textbf{\textit{Eq. 1–36}}$$

This formula is useful only if the concentration and volume units between the substances are the same and if three out of the four variables are known.

Example 1–11

What volume is needed to make 500 mL of a 0.1 M solution of tris buffer from a solution of 2 M tris buffer?

$$V_1 \times 2 \, M = 0.1 \, M \times 500 \text{ mL}$$

$$(V_1)(2 \, M) = (0.1)(500) = (V_1)(2) = 50 = V_1 = 50/2 = 25$$

$$\textbf{\textit{Eq. 1–37}}$$

It requires 25 mL of the 2 M solution to make up 500 mL of a 0.1 M solution. This problem differs from the other conversions in that it is actually a dilution of a stock solution. A more involved discussion of dilutions follows.

Dilutions

A *dilution* represents the ratio of concentrated or stock material to the total final volume of a solution and consists of the volume or weight of the concentrate plus the volume of the diluent, with the concentration units remaining the same. This ratio of concentrated or stock solution to the total solution volume equals the *dilution factor.* Because a dilution is made by adding concentrated stock to a diluent, it always represents a less concentrated substance. The relationship of the dilution factor to concentration is an inverse one; thus, the dilution factor increases as the concentration decreases. To determine the dilution factor required, simply take the amount needed and divide by the stock concentration, leaving it in a reduced-fraction form.

Example 1–12

What is the dilution factor needed to make a 100 mEq/L sodium solution from a 3000 mEq/L stock solution? The dilution factor becomes

$$\frac{\cancel{100}}{\cancel{3000}} = \frac{1}{30} \qquad \textbf{\textit{Eq. 1–38}}$$

The dilution factor indicates that the ratio of stock material is 1 part stock made to a *total volume* of 30. To actually make this dilution, 1 mL of stock is added to 29 mL of diluent. Note that the dilution factor indicates the parts per total amount, but in making the dilution, the sum of the amount of the stock material plus the amount of the diluent must equal the total volume or dilution fraction denominator. The dilution factor may be written as a fraction or can be expressed as 1:30. Either format is correct. Any total volume can be used as long as the fraction reduces to give the dilution factor.

Example 1–13

If in the preceding example 150 mL of the 100 mEq/L sodium solution was required, the dilution ratio of stock to total volume must be maintained. Set up a ratio between the desired total volume and the dilution factor to determine the amount of stock needed. The equation becomes

$$\frac{1}{30} = \frac{x}{150}$$

$$x = 5 \qquad \textbf{\textit{Eq. 1–39}}$$

Note that $5/150$ reduces to the dilution factor of $1/30$. To make up this solution, 5 mL of stock is added to 145 mL of the appropriate diluent, making the ratio of *stock volume* to *diluent volume* equal to $5/145$. Recall that the dilution factor includes the total volume of *both* stock plus diluent.

Simple Dilutions

When making a *simple dilution*, the laboratorian must decide on the total volume desired and the amount of stock to be used.

Example 1–14

A 1:10 ($1/10$) dilution of serum can be achieved by using any of the approaches listed below.

A. 100 microliters (µL) of serum and 900 µL of saline.
B. 20 µL of serum and 180 µL of saline.
C. 1 mL of serum and 9 mL of saline.
D. 2 mL of serum and 18 mL of saline.

Note that the ratio of serum to diluent needed to make up each dilution satisfies the dilution factor of stock material to total volume.

The dilution factor is also used to determine the concentration of a dilution or stock material by multiplying the original concentration by the dilution factor.

Example 1–15

Determine the concentration of a 200 N mol/mL human chorionic gonadotropin (hCG) standard that was diluted $\frac{1}{50}$. This value is obtained by multiplying the original concentration, 200 N mol/mL hCG, by the dilution factor, $\frac{1}{50}$. The result is 4 μ mol/mL hCG. Quite often, the concentration of the original material is needed.

When determining the original stockw or undiluted concentration, multiply the concentration of the dilution by the dilution factor denominator.

Example 1–16

A 1:2 dilution of serum with saline had a creatinine result of 8.6 mg/dL. Calculate the actual serum creatinine concentration.

Dilution factor: $\frac{1}{2}$
Dilution result = 8.6 mg/dL

Since this result represents $\frac{1}{2}$ of the concentration, the actual serum creatinine value is

$$2 \times 8.6 = 17.2 \text{ mg/dL} \qquad \textbf{\textit{Eq. 1–40}}$$

Serial Dilutions

A *serial dilution* may be defined as multiple progressive dilutions ranging from more concentrated solutions to less concentrated solutions. Serial dilutions are extremely useful when the volume of concentrate and/or diluent is in short supply and needs to be minimized. The volume of patient serum available to the laboratory may be very small (*e.g.,* pediatric samples), and a serial dilution may be needed to ensure that sufficient sample is available. The serial dilution is initially made in the same manner as a simple dilution. Subsequent dilutions will then be made from each preceding dilution. When a serial dilution is made, certain criteria may need to be satisfied. The criteria vary from one situation to the next but usually include such considerations as the total volume desired, the amount of diluent or concentrate available, the dilution factor, and the support materials needed.

Example 1–17

A serum sample is to be diluted 1:2, 1:4, and finally, 1:8. It is arbitrarily decided that the total volume for each dilution is to be 1 mL. Note that the least common denominator for these dilution factors is 2. Once the first dilution is made ($\frac{1}{2}$), a 1:2 (least common denominator between 2 and 4) dilution of it will yield

$$\frac{1}{2} \times \frac{1}{2} = \frac{1}{4}$$

(initial dilution factor)(next dilution factor)
= (final dilution factor) $\qquad$ **_Eq. 1–41_**

By making a 1:2 dilution of the first dilution, the second dilution factor of 1:4 is satisfied. Making a 1:2 dilution of the 1:4 dilution will result in the next dilution (1:8). To establish the dilution factor needed for subsequent dilutions, it is helpful to solve the following equation for (x):

Stock/preceding concentration $\times$ (x)
= final dilution factor needed $\qquad$ **_Eq. 1–42_**

To make up these dilutions, three test tubes are labeled 1:2, 1:4, and 1:8, respectively. One mL of diluent is added to each test tube. To make the primary dilution of 1:2, 1 mL of serum is added to test tube number 1. The solution is mixed, and 1 mL of the primary dilution is removed and added to test tube number 2. After mixing, this solution contains a 1:4 dilution. Then 1 mL of the 1:4 dilution from test tube number 2 is added to test tube number 3. After discarding 1 mL from test tube 3, the resultant dilution in this test tube is 1:8, and this satisfies all of the previously established criteria. Refer to Figure 1-14 for an illustration of this serial dilution.

Example 1–18

Another type of serial dilution combines several dilution factors that are not multiples of one another. In our previous example, 1:2, 1:4, and 1:8 dilutions are all related to one another by a factor of 2. Consider the situation when 1:10, 1:20, 1:100, and 1:200 dilution factors are required. There are several approaches to solving this type of dilution problem. One method is to treat the 1:10 and 1:20 dilutions as one serial dilution problem, the 1:20 and 1:100 dilutions as a second serial dilution, and the 1:100 and 1:200 dilutions as the last serial dilution. Another approach is to consider what dilution factor of the concentrate is needed to yield the final dilution. In this example, the initial dilution is 1:10, with subsequent dilutions of 1:20, 1:100, and 1:200. The first dilution may be accomplished by adding 1 mL of concentrate to 9 mL of diluent. The total volume of solution is 10 mL. Our initial dilution factor has been satisfied. In making the remaining dilutions, 2 mL of diluent is added to each test tube.

Initial/preceding dilution $\times$ (x) = dilution needed

Solve for x.

Using the dilution factors listed above and solving for (x), the equations become

1:10 $\times$ (x) = 1:20
where (x) = 2 (or 1 part stock to 1 part diluent)

1:20 $\times$ (x) = 1:100
where (x) = 5 (or 1 part stock to 4 parts diluent)

1:100 $\times$ (x) = 1:200
where (x) = 2 (or 1 part stock to 1 part diluent) $\qquad$ **_Eq. 1–43_**

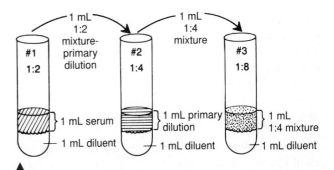

▲
Figure 1-14. Serial dilution.

In practice, the 1:10 dilution must be diluted by a factor of 2 in order to obtain a subsequent 1:20 dilution. Since the second tube already contains 2 mL of diluent, 2 mL of the 1:10 dilution should be added (1 part stock to 1 part diluent). In preparing the 1:100 dilution from this, a 1:5 dilution factor of the 1:20 mixture is required (1 part stock to 4 parts diluent). Since this tube already contains 2 mL, the volume of diluent in the tube is divided by its parts, which is 4; thus, 500 μL, or 0.500 mL, of stock should be added. The 1:200 dilution is prepared in the same manner using a 1:2 dilution factor (1 part stock to 1 part diluent) and adding 2 mL of the 1:100 to the 2 mL of diluent already in the tube.

Water of Hydration

Some compounds are available in a hydrated form. To obtain a correct weight for these chemicals, the attached water molecule(s) must be included.

Example 1–19

How much $CuSO_4 \cdot 5H_2O$ must be weighed to prepare 1 L of 0.5 M $CuSO_4$? When calculating the gmw of this substance the water weight must be considered so that the gmw is 250 g rather than gmw of $CuSO_4$ alone (160 g). Therefore,

$$\frac{250 \; (g) \; CuSO_4 \cdot 5H_2O}{mol} \times \frac{0.5 \; mol}{1 \; (L)} = 125 \text{ g/L} \qquad \textbf{\textit{Eq. 1–44}}$$

Cancel out "like" terms to obtain the result of 125 g/L.

A reagent protocol often designates the use of an anhydrous form of a chemical; frequently, however, all that is available is a hydrated form.

Example 1–20

A procedure requires 0.9 g of $CuSO_4$. All that is available is $CuSO_4 \cdot 5H_2O$. What weight of $CuSO_4 \cdot 5H_2O$ is needed?

Calculate the percentage of $CuSO_4$ present in $CuSO_4 \cdot 5H_2O$. The percentage is

$$\frac{160}{250} = 0.64, \quad \text{or} \quad 64\% \qquad \textbf{\textit{Eq. 1–45}}$$

Therefore, 1 g of $CuSO_4 \cdot 5H_2O$ contains 0.64 g of $CuSO_4$ so the equation becomes:

$$\frac{0.9 \text{ g } CuSO_4 \text{ needed}}{0.64 \; CuSO_4 \text{ in } CuSO_4 \cdot 5H_2O}$$

$$= 1.41 \text{ g } CuSO_4 \cdot 5H_2O \text{ required} \qquad \textbf{\textit{Eq. 1–46}}$$

Graphing Beer's Law

The Beer-Bouger/Lambert Law (Beer's Law) mathematically establishes the relationship between concentration and absorbance in many photometric determinations. Beer's Law is expressed as:

$$A = abc \qquad \textbf{\textit{Eq. 1–47}}$$

where A = absorbance; a = absorptivity constant for a particular compound at a given wavelength under specified conditions of temperature, pH, etc.; b = length of the light path; and c = concentration.

If a method follows Beer's Law, then absorbance is proportional to concentration as long as the length of the light path and the absorptivity of the absorbing species remains unaltered during the analysis. Even in automated systems, adherence to Beer's Law is often determined by checking the linearity of the test method over a wide concentration range. At the end of manual assays, the concentration results are obtained by using a Beer's Law graph, known as a standard graph. This graph is made by plotting absorbance versus the concentration of known standards. Since most photometric assays set the initial absorbance to zero (0) using a reagent blank, the initial data points are 0,0. Graphs should be labeled properly and the concentration units must be given. The horizontal axis is referred to as the x axis, while the vertical line is the y axis. It is not important which variable (absorbance or concentration) is assigned to an individual axis, but it is very important that the values assigned to them are uniformly distributed along the axis. By convention in the clinical laboratory, concentration is usually plotted on the x axis. On a standard graph, only the standard and the associated absorbances are plotted.

Once a standard graph has been established, it is permissible to run just one standard, or calibrator, as long as the system remains the same. *One point calculation* or *calibration* is a term that refers to the calculation of the comparison of the known standard/calibrator concentration and its corresponding absorbance to the absorbance of the unknown value according to the following ratio:

$$\frac{\text{Concentration of standard } (Cs)}{\text{Absorbance of standard } (As)}$$
$$= \frac{\text{Concentration of unknown } (Cu)}{\text{Absorbance of the unknown } (Au)} \qquad \textbf{\textit{Eq. 1–48}}$$

Solving for the concentration of the unknown, the equation becomes:

$$C_u = \frac{(A_u)(C_s)}{A_s} \qquad \textbf{\textit{Eq. 1–49}}$$

Example 1–21

The biuret protein assay is very stable and follows Beer's Law. Rather than make up a complete new standard graph, one standard (6 g/dL) was assayed. The absorbance of the standard was 0.400, and the absorbance of the unknown was 0.350. Determine the value of the unknown in g/dL.

$$Cu = \frac{(0.350)(6 \text{ g/dL})}{(0.400)} = 5.25 \text{ g/dL} \qquad \textbf{\textit{Eq. 1–50}}$$

This method of calculation is acceptable as long as everything, including the instrument and lot of reagents, remains the same. If anything in the system changes, a new standard graph should be done.

Enzyme Calculations

Another application of Beer's Law is the calculation of enzyme assay results. When calculating enzyme results, the *rate of absorbance change* is often monitored continuously during the reaction to give the difference in absorbance, known as the *delta absorbance*, or ΔA. Instead of using a standard graph or a one point calculation, the molar absorptivity of the product is used. If the absorptivity constant and absorbance, in this case ΔA, is given, Beer's Law can be used to calculate the enzyme concentration directly, as follows:

$$A = abc$$

$$C = \frac{A}{ab} \qquad \textit{Eq. 1–51}$$

When the absorptivity constant (a) is given in units of grams per liter (moles) through a 1 centimeter (cm) light path, the term *molar absorptivity* (ϵ) is used. Substitution of ϵ for a, and ΔA for A produces the following Beer's Law formula:

$$C = \frac{\Delta A}{\epsilon} \qquad \textit{Eq. 1–52}$$

Units used to report enzyme activity traditionally have included weight, time, and volume. In the early days of enzymology, method-specific units (*e.g.,* King-Armstrong, Caraway) were all different and very confusing. In 1961, the Enzyme Commission of the International Union of Biochemistry recommended using one unit, the international unit (IU), for reporting enzyme activity. The IU is defined as the amount of enzyme that will catalyze one micromole (μmol)/minute. These units were often expressed as units per liter (U/L). The designations IU, U, or IU/L were adopted by many clinical laboratories to represent the international unit. While the reporting unit is the same, unless the conditions of the analysis are identical, use of the IU does not standardize the actual enzyme activity. The SI recommended unit is the *katal*, which is expressed as mol/L/second. Whichever unit is used, calculation of the activity using Beer's Law required inclusion of the dilution and, depending upon the reporting unit, required conversion to the appropriate term (*e.g.,* μmol to mol, mL to L, minute to second, temperature factors, etc.). Beer's Law for the international unit now becomes:

$$C = \frac{(\Delta A)10^{-6}\,(TV)}{(\epsilon)(b)(SV)} \qquad \textit{Eq. 1–53}$$

where TV = total volume of sample plus reagents in mL and SV = sample volume used in mL. The 10^{-6} converts moles to μmol for the international unit. If another unit of activity is used, such as the katal, conversion into liters and seconds would be needed, but the conversion to and from micromoles would not be needed.

Example 1–22

The ΔA per minute for an enzyme reaction is 0.250. The product measured has a molar absorptivity of 12.2×10^3 at 425 nm at 30°C. The incubation and reaction temperature are also kept at 30°C. The assay calls for 1 mL of reagent and 0.050 mL of sample. Give the enzyme activity results in international units.

Applying Beer's Law and the necessary conversion information, the equation becomes:

$$c = \frac{(0.250)(10^{-6})(1.050 \text{ mL})}{(12.2 \times 10^3)(1)(0.050 \text{ mL})} = 430 \text{ U} \qquad \textit{Eq. 1–54}$$

Note: b is usually always given as 1 cm, and because it is a constant, may not be considered in the calculation.

SUMMARY

The primary goal of any clinical chemistry laboratory is to correctly perform analytical procedures that yield accurate and precise information to aid in patient diagnosis. In order to achieve accurate and reliable results, the clinical laboratorian must be familiar with the use of basic supplies and equipment. They also must have a fundamental understanding of the concepts critical to clinical chemistry testing. This chapter provides the information necessary for the important practice of clinical chemistry including units of measure, temperature, reagents (chemicals, standards, solutions, water specifications), laboratory supplies (glass and plasticware, pipets, burets, balances, and desiccants), separation techniques, and laboratory mathematics and calculations.

PRACTICE PROBLEMS

1. Determine each of the following for a solution containing 64 g of NaCl made up to 500 mL with distilled water.
   a. Molarity
   b. Normality
   c. Percent (w/v)
   d. Dilution factor

2. Determine the concentration values for each of the following for a solution containing 10 mg of $CaCl_2$ made with 200 mL of distilled water.
   a. mg/dL
   b. Molarity
   c. Normality

3. You must make 1 L of 0.1 M acetic acid (CH_3COOH). All you have available is concentrated glacial acetic acid whose assay value is 98% and whose specific gravity is 1.05 g/mL. How many mL are needed to make this solution?

4. What is the hydrogen ion concentration of an acetate buffer whose pH is 4.24?

5. Use the Henderson-Hasselbalch equation to solve each of the following buffer problems.
   a. What is the ratio of salt to weak acid for a Veronal buffer with a pH of 8.6 and a pK_a of 7.43?
   b. The pK_a for acetic acid is 4.76. What is the expected pH if the concentration of salt is 5 mmol/L and that of acetic acid is 10 mmol/L?

6. The hydrogen ion concentration of a solution is 0.0000937. What is the pH?

7. You are making up a standard curve for an albumin assay. The standard values needed are:

    1.0 g/dL
    2.0 g/dL
    4.0 g/dL
    6.0 g/dL
    8.0 g/dL
    10.0 g/dL

    You have a 30 g/dL stock solution.

    a. What is the dilution needed to obtain the standard values needed?

    b. You need a total volume of 15 mL per standard. List the amount of stock plus the amount of diluent needed to make up each standard.

8. Perform the following conversions.

    8×10^3 mg = _____ μg

    40 μg = _____ mg

    13 dL = _____ mL

    200 μl = _____ mL

    5 μl = _____ mL

    50 mL = _____ L

    4 cm = _____ mm

9. What volume of 14 N H_2SO_4 is needed to make 250 mL of 3.2 M H_2SO_4 solution?

10. A 24-hour urine has a total volume of 1200 mL. A 1:200 dilution of the urine specimen gives a creatinine result of 0.8 mg/dL.

    a. What is the creatinine concentration for the undiluted urine specimen?

    b. What is the concentration in terms of milligrams per milliliter?

    c. What is the result in terms of grams per 24 hours?

11. How much $CuSO_4 \cdot 5H_2O$ is needed to make up 500 mL of 2.5 M $CuSO_4$?

12. An enzyme reaction yields a ΔA of 0.031 per 30 seconds. The assay calls for 1.5 mL of reagent and 50 μL of sample. The ϵ of the product is 5.3×10^3. Give the enzyme results in international units.

REFERENCES

1. National Committee for Clinical Laboratory Standards. Quantities and units: SI. Committee Report C11-CR. Villanova, PA: NCCLS, 1979.

2. National Committee for Clinical Laboratory Standards. Temperature calibration of water baths, instruments, and temperature sensors. NCCLS Document 12-A2. Villanova, PA: NCCLS, 1990.

3. Bowers GN, Inman SR. The gallium melting-point standard. Clin Chem 1977; 23:733.

4. Bowie L, Esters F, Bolin J, Gochman N. Development of an aqueous temperature-indicated technique and its application to clinical laboratory instruments. Clin Chem 1976; 22:449.

5. Harris D. Quantitative chemical analysis. San Francisco: Freeman, 1982.

6. International Union of Pure and Allied Chemistry and International Federation of Clinical Chemistry. Quantities and units in clinical chemistry, bulletin No. 20. Oxford: IUPAC, 1972.

7. National Committee for Clinical Laboratory Standards. Development of certified reference materials for the national reference system for the clinical laboratory. NCCLS Document NRSCLS-3A. Villanova, PA: NCCLS, 1991.

8. National Institute of Standards and Technology. Standard reference materials. Publication 260. Washington, DC: US Department of Commerce, 1991.

9. Mears TW, Young DS. Measurement and standard reference materials in clinical chemistry. Am J Clin Pathol 1968; 50:411.

10. National Committee for Clinical Laboratory Standards. Preparation and testing of reagent water in the clinical laboratory. NCCLS Document C3-A2. Villanova: NCCLS, PA, 1991.

11. Glass engineering handbook. 2nd ed. New York: McGraw-Hill, 1958.

12. Morey GW. The properties of glass (American Chemical Society monograph). New York: Reinhold, 1954.

13. National Institute of Standards and Technology. The international system of units (SI). NIST Publication 330. Gaithersburg, MD: NIST, 1991.

14. Bio Rad Laboratories. Procedure for comparing precision of pipet tips. Product Information No. 81-0208, 1993.

15. Oxford Laboratories, Inc. Pipetting devices and their calibration. Fred G Williams Fluid Handling Systems, TIS:24

Chapter 2

Laboratory Safety

Carolyn Houghton, Mary Ruth Beckham, Barbara Smith Michael

Objectives

Upon completion of this chapter, the clinical laboratorian should be able to:

- *Discuss safety awareness for clinical laboratory personnel.*
- *List the responsibilities of employer and employee in providing a safe work place.*
- *Identify hazards related to handling chemicals, biological specimens, and radiological materials.*
- *Choose appropriate personal protective equipment when working in the clinical laboratory.*
- *Identify the classes of fires and the type of fire extinguishers to use for each.*
- *Describe steps used as precautionary measures when working with electrical equipment, cryogenic materials, and compressed gases, and avoiding mechanical hazards associated with laboratory equipment.*
- *Select the correct means for disposal of waste generated in the clinical laboratory.*
- *Outline the steps required in documentation of an accident in the workplace.*

KEY WORDS

Airborne Pathogen	Hazard Communi-	National Fire Pro-
Biohazard	cation	tection Associa-
Bloodborne	Hazardous Waste	tion (NFPA)
Pathogen	High Efficiency	Occupational Safety
Carcinogen	Particulate Air	and Health Act
Chemical Hygiene	(HEPA) Filter	(OSHA)
Corrosive Chemical	Mechanical Hazard	Radioactive Material
Cryogenic Material	Material Safety Data	Reactive Chemical
Fire Tetrahedron	Sheet (MSDS)	Teratogen

Michael L. Bishop, Janet L. Duben-Engelkirk, and Edward P. Fody.
CLINICAL CHEMISTRY. © 1996 Lippincott–Raven Publishers.

INTRODUCTION

General Information

All clinical laboratory personnel, by the nature of the work they perform, are exposed daily to a variety of real or potential hazards: electric shock, toxic vapors, compressed gases, flammable liquids, radioactive material, corrosive substances, mechanical trauma, poisons, and the risks of handling biologic materials, to name a few. As G. C. Roach (1975) stated, "A cautious attitude and an awareness of hazards are essential to . . . safety and productivity in the clinical laboratory. Maintaining a continuous concern for safety is an additional challenge to the already demanding field of clinical laboratory sciences."

Laboratory safety necessitates the effective control of all hazards that exist in the clinical laboratory at any time. Safety begins with the recognition of hazards and is achieved through the application of common sense, a safety-focused attitude, good personal behavior, good housekeeping in all laboratory work and storage areas, and above all, the continual practice of good laboratory technique. In most cases, accidents can be traced directly to two primary causes: unsafe acts (not always recognized by personnel) and unsafe environmental conditions.

Safety Responsibility

Who is responsible for safety? Both the employer and the employee share responsibility. The employer has the ultimate responsibility for safety and delegates authority for safe operation through all levels of supervision. Safety management in the laboratory should start with a written safety policy that is supported by all levels of management. The attitude of management toward safety is reflected in the attitudes of supervisors. Similarly, the worker's attitude usually reflects that of the supervisor. Thus, lab supervisors are essential members in any safety program because of their regular contact with the employees. Other safety responsibilities for the employer include:

- Establishing laboratory work methods and safety policies.
- Providing supervision and guidance to the employees.
- Providing safety information, training, personal protective equipment, and medical surveillance to employees.
- Providing and maintaining equipment and laboratory facilities that are adequate for the tasks required.

The employee also has a responsibility for his or her own safety and the safety of coworkers. Employee conduct in the laboratory is a vital factor in the achievement of a workplace without accidents or injuries. The employee's responsibilities include:

- Knowledge and compliance with the established laboratory work methods.

- Positive attitudes toward supervisors, coworkers, facilities, and training.
- Prompt notification of unsafe conditions to the immediate supervisor.
- Conduct of safe work practices and utilization of personal protective equipment.

SAFETY AWARENESS FOR CLINICAL LABORATORY PERSONNEL

Signage and Labeling

Appropriate signs to identify hazards are critical not only to alert laboratory personnel to potential hazards, but also to identify specific hazards that arise because of an emergency such as fire or explosion. The National Fire Protection Association (NFPA) has developed a standard hazards–identification system (diamond-shaped, color-coded symbol), which also has been adopted by many clinical laboratories. At a glance, emergency personnel can assess health hazards (blue quadrant), flammable hazards (red quadrant), reactivity/stability hazards (yellow quadrant), and other special information (white quadrant). In addition, each quadrant shows the magnitude of severity, graded from a low of 0 to a high of 4, of the hazards within the posted area. (Note the NFPA hazard-code symbol in Fig. 2-1).

Manufacturers of laboratory chemicals also provide precautionary labeling information for users. Indicated on the product label is information such as a statement of the hazard, precautionary measures, specific hazard class, first-aid instructions for internal/external contact, the storage code, the safety code, and personal protective gear and equipment needed. This information is in addition to specifications on the actual lot analysis of the chemical constituents and other product notes (see Fig. 2-1). All "in-house" prepared reagents and solutions should be labeled in a standard manner with the chemical identity, concentration, hazard warning, special handling, storage conditions, date prepared, expiration date if applicable, and preparer's initials.

OSHA

In 1970, Public Law 91-596, better known as the Occupational Safety and Health Act (OSHA), was enacted by Congress. The goal of this federal regulation was to provide all employees (clinical laboratory personnel included) with a safe work environment. Under this legislation, the Occupational Safety and Health Administration is authorized to conduct on-site inspections to determine whether an employer is complying with the mandatory standards. Safety is no longer only a moral obligation, but also a legal responsibility of the employer.

Cat. **C4324-5GL**

Qty. **5 gallons (18.9L)**

Baxter

Scientific Products

S|P™ Methyl Alcohol

(Methanol - Anhydrous)
CH_3OH FW 32.04

⑦

For Laboratory Use
Store at 68-86°F (20-30°C)

Flash Point: 11°C (52°F Closed Cup)
Maximum Limits and Specifications
Methyl Alcohol: 99.8% Minimum by volume
Residue after Evaporation: 0.001% Maximum
Water: 0.10% Maximum
CAS — (67-56-1);

Distributed by:
Baxter Healthcare Corporation
Scientific Products Division
McGaw Park, IL 60085-6787 USA
Rev. 11/87

Made in USA

Lot No. **KEAM** ⑧

② **DANGER! FLAMMABLE** ①

④ NFPA

☠ **DANGER! POISON**

When using, the following safety precautions are recommended:

SAFETY GLASSES & SHIELD

VENT HOOD

③

LAB COAT, APRON & GLOVES

FIRE EXTINGUISHER

⑤

FLAMMABLE • VAPOR HARMFUL • MAY BE FATAL OR CAUSE BLINDNESS IF SWALLOWED • CANNOT BE MADE NON-POISONOUS • HARMFUL IF INHALED • CAUSES IRRITATION • NON-PHOTOCHEMICALLY REACTIVE

Keep away from heat, sparks and flame. Keep container tightly closed and upright to prevent leakage. Avoid breathing vapor or spray mist. Avoid contact with eyes, skin and clothing. Use only with adequate ventilation. Wash thoroughly after handling. In case of fire, use water spray, alcohol foam, dry chemical or CO_2. In case of spillage, absorb and flush with large volumes of water immediately.

⑥

FIRST AID: In case of skin contact, flush with plenty of water; **for eyes,** flush with plenty of water for 15 minutes and get medical attention. **If swallowed,** if conscious, induce vomiting by giving two glasses of water and sticking finger down throat. Have patient lie down and keep warm. Cover eyes to exclude light. Never give anything by mouth to an unconscious person. **If inhaled,** remove to fresh air. If necessary, give oxygen or apply artificial respiration.

NOTE: It is unlawful to use this fluid in any food or drink or in any drug or cosmetic for internal or external use. Not for internal or external use on man or animal.

DOT Description: Methyl Alcohol, Flammable Liquid, UN1230

▲

Figure 2-1. Sample chemical label: (*1*) statement of hazard; (*2*) hazards class; (*3*) safety precautions; (*4*) NFPA hazard code; (*5*) fire extinguisher type; (*6*) safety instructions; (*7*) formula weight; (*8*) lot number.

Color of the label indicates hazard:

Red = flammable. Store in an area segregated for flammable reagents.

Blue = health hazard. Toxic if inhaled, ingested, or absorbed through the skin. Store in a secure area.

Yellow = reactive and oxidizing reagents. May react violently with air, water, or other substances. Store away from flammable and combustible materials.

White = corrosive. May harm skin, eyes, or mucous membranes. Store away from red-, blue-, and yellow-coded reagents.

Gray = presents no more than moderate hazard in any of categories. For general chemical storage.

Exception = reagent incompatible with other reagents of same color bar. Store separately.

Hazard code (*4*)—Following the National Fire Protection Association (NFPA) usage, each diamond shows a red segment (flammability), a blue segment (health, *i.e.,* toxicity), and yellow (reactivity). Printed over each color-coded segment is a black number showing the degree of hazard involved. The fourth segment, as stipulated by the NFPA, is left blank. It is reserved for special warnings, such as radioactivity. The numerical ratings indicate degree of hazard: 4 = extreme hazard; 3 = severe hazard; 2 = moderate hazard; 1 = slight hazard; 0 = none according to present data.

(Photo courtesy of Scientific Products Division, Baxter Healthcare Corporation.)

Other Regulations and Guidelines

There are also other federal regulations relating to laboratory safety, such as the Clean Water Act, the Resource Conservation and Recovery Act, and the Toxic Substances Control Act. In addition, clinical laboratories are required to comply with applicable local/state laws such as fire and building codes. Since laws, codes, and ordinances are updated on a frequent basis, current reference materials should be reviewed. Assistance can be obtained from local libraries and from federal, state, and local regulatory agencies.

Safety is also an important part of the requirements for accreditation and reaccreditation of health care institutions by voluntary accrediting bodies such as the Joint Commission on Accreditation of Health Care Organizations (JCAHO) and the Commission on Laboratory Accreditation of the College of American Pathologists (CAP). JCAHO publishes a yearly accreditation manual for hospitals and an "Accreditation Manual for Pathology and Clinical Labora-

tory Services," which includes a detailed section on safety requirements.

SAFETY EQUIPMENT

General

Safety equipment has been developed specifically for use in the clinical laboratory. The employer is required by law to have designated safety equipment available, but it is also the responsibility of the employee to comply with all safety rules and to use safety equipment.

All laboratories are required to have safety showers, eyewash stations, and fire extinguishers available and to periodically inspect the equipment for proper operation. Other items that must be available for personnel include fire blankets, spill kits, and first-aid supplies.

Mechanical pipetting devices must be used for manipulating all types of liquids in the laboratory, including water. Mouth pipetting is strictly prohibited.

Hoods

Fume hoods are required to expel noxious and hazardous fumes from chemical reagents. Biohazard hoods remove particles that may be harmful to the employee who is working with infective biological specimens. The Centers for Disease Control (CDC) and the National Institute of Health (NIH) have described four levels of biosafety, which consist of combinations of laboratory practices and techniques, safety equipment, and laboratory facilities. The biosafety level of a laboratory is based on the operations performed, the routes of transmission of the infectious agents, and the laboratory function or activity. Accordingly, biohazard hoods are designed to offer various levels of protection, depending on the biosafety level of your specific laboratory (Table 2-1).

Chemical Storage Equipment

Safety equipment is available for the storage and handling of chemicals and compressed gases. Safety carriers should always be used to transport gallon bottles of acids, alkalis, or other solvents, and approved safety cans should be used for storing, dispensing, or disposing of flammables. Safety cabinets are required for the storage of flammable liquids, and only specially designed explosion-proof refrigerators should be used to store flammable materials. Gas-cylinder supports or clamps must be used at all times, and large tanks should be transported using hand carts.

Personal Protection Equipment

The parts of the body most frequently subject to injury in the clinical laboratory are the eyes, skin, and respiratory/digestive tract. Hence, the use of personal protective equipment is very important. Safety glasses, goggles, visors, or protective work shields protect the eyes and face from splashes and impact. Contact lenses do not offer any eye protection. It is strongly recommended that they not be worn in the clinical chemistry laboratory. If any solution is accidentally splashed into the eye(s), thorough irrigation is required and will be hampered by the presence of the contact lens(es).

Gloves and rubberized sleeves protect the hands and arms. Latex gloves are recommended for laboratory use. Lab coats are to be full length, buttoned, and made of liquid resistant material. Proper footwear is required: shoes constructed of porous materials, open-toed shoes, or sandals are considered hazardous.

Respirators with high efficiency particulate air (HEPA) filters must be worn, for example, when working directly with tuberculosis (TB) patients or performing procedures that may aerosolize specimens of patients with suspected or confirmed cases of TB. Respirators are also required for chemical emergency use, and the correct type of respirator must be used for the specific hazard.

Each employer must provide (at no charge) lab coats, gloves, or other protective equipment to all employees who

TABLE 2-1. Comparison of Biological Safety Cabinets

			Applications		
Cabinets			**Applications**		
Type	*Face Velocity (Ifpm)*	*Airflow Pattern*	*Radionuclides/ Toxic Chemicals*	*Biosafety Level(s)*	*Product Protection*
Class I*, open front	75	In at front; out rear and top through HEPA filter	NO	2,3	NO
Class II: Type A	75	70% recirculated through HEPA; exhaust through HEPA	NO	2,3	YES
Type B1	100	30% recirculated through HEPA; exhaust via HEPA and hard-ducted	YES (Low levels/ volatility)	2,3	YES
Type B2	100	No recirculation; total exhaust via HEPA and hard-ducted	YES	2,3	YES
Type B3	100	Same as IIA, but plena under negative pressure to room and exhaust air is ducted	YES	2,3	YES
Class III	NA	Supply air inlets and exhaust through 2 HEPA filters	YES	3,4	YES

*Glove panels may be added and will increase face velocity to 150 Ifpm; gloves may be added with an inlet air pressure release that will allow work with chemicals/radionuclides

From Centers for Disease Control and Prevention and the National Institutes of Health. Table 3, Comparison of Biological Safety Cabinets. Biosafety in Microbiological and Biomedical Laboratories. U.S. Government Printing Office, 1993.

may be exposed to biological or chemical hazards. It is the employer's responsibility to clean and maintain all personal protective equipment. All contaminated personal protective equipment must be removed and properly disposed of before leaving the laboratory.

BIOLOGICAL SAFETY

General

All blood samples and other body fluids should be collected, transported, handled, and processed using strict precautions. Gloves, gowns, and face protection must be used if splash or splattering is likely to occur.

Centrifugation of biologic specimens produces finely dispersed aerosols that are a high-risk source of infection. Ideally, specimens should remain "capped" during centrifugation. Any blood spills must be cleaned up and the area or equipment disinfected immediately. As an additional precaution, the use of a centrifuge with an internal shield is recommended.

Bloodborne Pathogen Exposure Control Plan

In December 1991, OSHA issued the final rule for occupational exposure to bloodborne pathogens. To minimize employee exposure, each employer must have a written exposure control plan. The plan must be available to all employees whose reasonable anticipated duties may result in occupational exposure. The exposure control plan must be discussed with all employees and be available to them while they are working. The employee must be provided with adequate training of all techniques described in the exposure control plan. All necessary equipment and supplies must be readily available and inspected on a regular basis.

Clinical laboratory personnel are knowingly or unknowingly in frequent contact with potentially biohazardous materials. In recent years, new and serious occupational hazards to personnel have arisen, and this problem has been complicated because of the general lack of understanding of the epidemiology, mechanisms of transmission of the disease, or inactivation of the causative agent. Special precautions must be taken when handling all specimens due to the continual increase of infectious samples received in the laboratory. Therefore, in practice, specimens from patients with confirmed or suspected hepatitis, acquired immunodeficiency syndrome (AIDS), Creutzfeldt-Jakob disease (CJD), or other potentially infectious diseases should be handled no differently than other routine specimens. Adopting a Universal Precautions Policy, which considers blood and other body fluids from all patients as infective, is strongly recommended.

Airborne Pathogens

Due to the recent resurgence of TB, in 1993 OSHA issued a statement that the agency would enforce the CDC *Guidelines for Preventing the Transmission of Tuberculosis in Health Care Facilities.* The purpose of the guidelines is to encourage early detection, isolation, and treatment of active cases. A TB exposure control program must be established and risks to laboratory workers must be assessed. Those in high-risk areas may be required to wear a respirator for protection. All health care workers must be screened for TB infection.

CHEMICAL SAFETY

Hazard Communication

In the August 1987 issue of the *Federal Register*, OSHA published the new Hazard Communication Standard ("Right to Know Law"). This standard was expanded to include clinical laboratories in May of 1988. The Right to Know Law was developed for employees who may be exposed to hazardous chemicals. The employees must be informed of the health risks associated with those chemicals. The Law's intent is to ensure that the health hazards are evaluated for all chemicals that are produced and that this information is relayed to the employees.

To comply with the regulation, clinical laboratories must:

- Plan and implement a written hazard communication program.
- Obtain material safety data sheets (MSDS) for each hazardous compound present in the workplace and have the MSDS readily accessible to the employees.
- Educate all employees on an annual basis on how to interpret chemical labels, MSDS, and health hazards of the chemicals and how to work safely with these chemicals.

Material Safety Data Sheets

The MSDS is a major source of safety information for employees who may use hazardous materials in their occupations. Employers are responsible for obtaining from the chemical manufacturer or developing an MSDS for each hazardous agent used in the workplace. A standardized format is not mandatory, but all requirements listed in the Law must be addressed. A summary of the MSDS information requirements includes the following:

- Product name and identification.
- Hazardous ingredients.
- Permissible Exposure Limit (PEL).
- Physical and chemical data.
- Health hazard data and carcinogenic potential.
- Primary routes of entry.
- Fire and explosion hazards.
- Reactivity data.
- Spill and disposal procedures.
- Personal protective equipment recommendations.
- Handling.
- Emergency and first aid procedures.
- Storage and transportation precautions.

- Chemical manufacturer's name, address, and phone number.
- Special information section.

The MSDS must be printed in English and provide the specific compound identity along with all common names. All information sections must be completed, and the date that the MSDS was printed must be indicated. Copies of the MSDS will be readily accessible to employees during all shifts.

Chemical Hygiene Plan

OSHA requires each laboratory that uses hazardous chemicals to have a Chemical Hygiene Plan. This plan provides procedures and work practices for regulating exposure of laboratory personnel to hazardous chemicals. Procedures describing how to protect employees against teratogens, carcinogens, and other toxic chemicals must be described in the plan. The protocol must be reviewed annually and updated when regulations are modified or chemical inventory changes.

Toxic Effects from Hazardous Substances

Toxic substances have the potential of deleterious effects (local or systemic) by direct chemical action or interference with the function of body systems. They can cause acute or chronic effects related to the duration of exposure (*i.e.,* short-term or single contact versus long-term or prolonged repeated contact). Almost any substance, even the most harmless, can be toxic in excess. Moreover, some chemicals are toxic at very low concentrations. Exposure to toxic agents can be through direct contact, inhalation, or inoculation/injection.

In the clinical chemistry laboratory, personnel should be particularly aware of toxic vapors from chemical solvents, such as acetone, chloroform, methanol, or carbon tetrachloride, that do not give explicit sensory-irritation warnings, as do bromide, ammonia, and formaldehyde. Another source of poisonous vapors that is frequently disregarded is metallic mercury. It is highly volatile and toxic and is rapidly absorbed through the skin and respiratory tract. Mercury Spill Kits should be available in areas where mercury thermometers are used.

Storage and Handling of Chemicals

To avoid accidents when handling chemicals, it is important to develop respect for all chemicals and to have a complete knowledge of their properties. This is particularly important when transporting, dispensing, or using chemicals that, when in contact with certain other chemicals, could result in the formation of substances that are toxic, flammable, or explosive. For example, acetic acid is incompatible with other acids such as chromic and nitric, carbon tetrachloride is incom-

patible with sodium, and flammable liquids are incompatible with hydrogen peroxide and nitric acid.

Arrangements for the storage of chemicals will depend on the quantities of chemicals needed and the nature or type of chemicals. Proper storage is essential to prevent and control laboratory fires and accidents. Ideally, the storeroom should be organized so that each class of chemicals is isolated in an area that is not used for routine work. An up-to-date inventory should be kept indicating location of chemicals, minimum/maximum quantities required, shelf life, etc. Some chemicals deteriorate over time and become hazardous (*e.g.,* ether forms explosive peroxides). Storage should not be based solely on alphabetical order because incompatible chemicals might be stored next to each other and react chemically. (Refer to Appendix I, Examples of Incompatible Chemicals.)

Flammable/Combustible Chemicals

Flammable and combustible liquids, which are used in numerous routine procedures, are among the most hazardous materials in the clinical chemistry laboratory because of possible fire or explosion. They are classified according to flash point, which is the temperature at which sufficient vapor is given off to form an ignitable mixture with air. A flammable liquid has a flash point below 37.8°C (100°F), and combustible liquids, by definition, have a flash point at or above 37.8°C (100°F). Some of the commonly used flammable and combustible solvents are acetone, benzene, ethanol, heptane, isopropanol, methanol, toluene, and xylene. It is important to remember that flammable chemicals also include certain gases and solids such as paraffin.

Corrosive Chemicals

Corrosive chemicals are injurious to the skin or eyes by direct contact or to the tissues of the respiratory and gastrointestinal tracts if inhaled or ingested. Typical examples include acids (acetic, sulfuric, nitric, and hydrochloric) and bases (ammonium hydroxide, potassium hydroxide, and sodium hydroxide).

Reactive Chemicals

Reactive chemicals are substances that, under certain conditions, can spontaneously explode or ignite or that evolve heat and/or flammable or explosive gases. Some strong acids or bases react with water to generate heat (exothermic reactions). Hydrogen is liberated if alkali metals (sodium or potassium) are mixed with water or acids, and spontaneous combustion also may occur. The mixture of oxidizing agents, such as peroxides, and reducing agents, such as hydrogen, generate heat and may be explosive.

Carcinogenic Chemicals

Carcinogenic chemicals are substances that have been determined to be cancer-causing agents. OSHA has issued lists of confirmed carcinogens and suspected carcinogens and detailed standards for the handling of these substances. Benzidine is a common example of a known carcinogen. If possible, a substitute chemical or different procedure should be used to avoid exposure to carcinogenic agents.

Chemical Spills

Strict attention to good laboratory technique can help prevent chemical spills. However, emergency procedures should be established to handle any accidents. If a spill occurs, the first step should be to assist/evacuate personnel, then confinement and cleanup of the spill can begin. There are several commercial spill kits available for neutralizing and absorbing spilled chemical solutions (Fig. 2-2). However, no single kit is suitable for all types of spills. Emergency procedures for spills also should include a reporting system.

RADIATION SAFETY

Environmental Protection

A radiation-safety policy should include environmental and personnel protection. All areas where radioactive materials are used or stored must be posted with caution signs, and traffic in these areas should be restricted to essential personnel only. Regular and systematic monitoring must be emphasized, and decontamination of laboratory equipment, glassware, and work areas should be scheduled as part of routine procedures. Records must be maintained as to the quantity of radioactive material on hand as well as the quantity that is disposed.

Personal Protection

It is essential that only properly trained personnel work with radioisotopes and that users are monitored to ensure that the maximal permissible dose of radiation is not exceeded. Radiation monitors must be evaluated regularly to detect degree of exposure for the laboratory employee. Records must be maintained for the length of employment plus 30 years.

FIRE SAFETY

The Chemistry of Fire

Fire is basically a chemical reaction that involves the rapid oxidation of a combustible material or fuel, with the subsequent liberation of heat and light. In the clinical chemistry laboratory, all the elements essential for fire to begin are present—fuel, heat or ignition source, and oxygen (air). However, recent research suggests that a fourth factor is present. This factor has been classified as a reaction chain where burning continues and even accelerates. It is caused by the breakdown and recombination of the molecules that make up the material burning with the oxygen in the atmosphere.

In summary, the fire triangle has been modified into a three-dimensional pyramid known as the fire tetrahedron (Fig. 2-3). This modification does not eliminate established procedures in dealing with a fire but does provide additional means by which fires may be prevented or extinguished. A fire will extinguish if any of the three basic elements (heat, air, fuel) are removed.

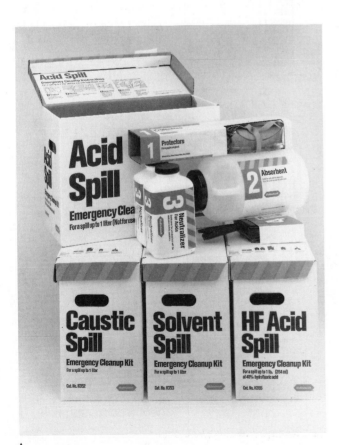

Figure 2-2. Spill cleanup kit. (Photo courtesy of Scientific Products Division, Baxter Healthcare Corporation)

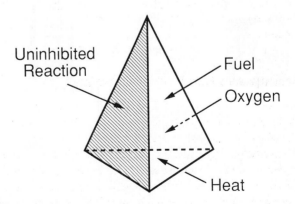

Figure 2-3. Fire tetrahedron.

Classification of Fires

Fires have been divided into four classes based on the nature of the combustible material and requirements for extinguishment:

Class A—ordinary combustible solid materials such as paper, wood, plastic, and fabric.
Class B—flammable liquids/gases and combustible petroleum products.
Class C—energized electrical equipment.
Class D—combustible/reactive metals such as magnesium, sodium, and potassium.

Types/Applications of Fire Extinguishers

Just as fires have been divided into classes, fire extinguishers are divided into classes that correspond to the type of fire to be extinguished. Be certain to choose the right type—using the wrong type of extinguisher may be dangerous. For ex-ample, do not use water on burning liquids or electrical equipment.

Pressurized-water extinguishers, as well as foam and multipurpose dry-chemical types, are used for class A fires. Multipurpose dry-chemical and carbon dioxide extinguishers are used for class B and C fires. Halogenated hydrocarbon extinguishers are recommended particularly for use with computer equipment. Class D fires present special problems, and extinguishment is left to trained fire-fighting personnel using special dry-chemical extinguishers (Fig. 2-4).

You should know the location and type of portable fire extinguisher near your work area and know how to use an extinguisher before a fire occurs. In the event of a fire, first evacuate all personnel, patients, and visitors who are in immediate danger and then activate the fire alarm, report the fire, and attempt to extinguish the fire, if possible. Personnel should work as a team to carry out emergency procedures. Fire drills must be conducted regularly and with appropriate documentation.

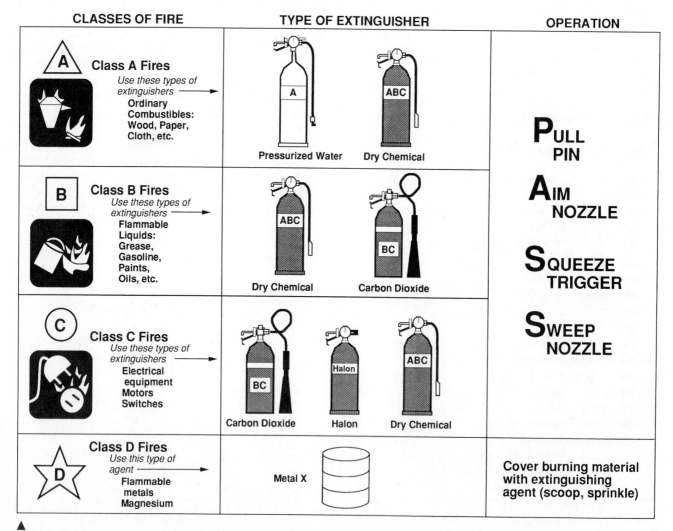

Figure 2-4. Proper use of fire extinguishers. (Adapted with permission from the Clinical and Laboratory Safety Department, The University of Texas Health Science Center at Houston)

CONTROL OF OTHER HAZARDS

Electrical

Most individuals are aware of the potential hazards associated with the use of electrical appliances and equipment. Hazards of electrical energy can be direct and result in death, shock, and/or burns. Indirect hazards can result in fire and/or explosion. Therefore, there are many precautionary procedures to follow when operating or working around electrical equipment:

- Use only explosion-proof equipment in hazardous atmospheres.
- Be particularly careful when operating high-voltage equipment, such as electrophoresis apparatus.
- Use only properly grounded equipment (three-prong plug).
- Check for "frayed" electrical cords.
- Promptly report any malfunctions or equipment producing a "tingle" for repair.
- Do not work on "live" electrical equipment.
- Never operate electrical equipment with wet hands.
- Know the exact location of the electrical control panel for the electricity to your work area.
- Use only approved extension cords and do not overload circuits. (Some local regulations prohibit the use of any extension cord.)
- Have ground checks and other periodic preventive maintenance performed on equipment.

Compressed Gases

Compressed gases, which serve a number of functions in the laboratory, present a unique combination of hazards in the clinical laboratory: danger of fire, explosion, asphyxiation, or mechanical injuries. There are several general requirements for safely handling compressed gases:

- Know the gas that you will use.
- Store tanks in a vertical position.
- Keep cylinders secured at all times.
- Never store flammable liquids and compressed gases in the same area.
- Use the proper regulator for the type of gas in use.
- Do not attempt to control or shut off gas flow with the pressure-relief regulatory.
- Keep removable protection caps in place until the cylinder is in use.
- Make certain that acetylene tanks are properly piped (the gas is incompatible with copper tubing).
- Do not force a "frozen" cylinder valve.
- Use a hand truck to transport large tanks.
- Always check tanks upon receipt and then periodically for any problems such as leaks.
- Make certain that the cylinder is properly labeled to identify the contents.
- Empty tanks should be marked "empty."

Cryogenic Materials

Liquid nitrogen is probably one of the most widely used cryogenic fluids (liquefied gases) used in the laboratory. There are, however, several hazards associated with the use of any cryogenic material: fire or explosion, asphyxiation, pressure buildup, embrittlement of materials, and tissue damage similar to that of thermal burns.

Only containers constructed of materials designed to withstand ultralow temperatures should be used for cryogenic work. In addition to the use of eye-safety glasses, hand protection to guard against the hazards of touching supercooled surfaces is recommended. The gloves, of impermeable material, should fit loosely so that they can be taken off quickly if liquid spills on them. Also, to minimize violent boiling/frothing and splashing, specimens to be frozen should always be inserted into the coolant very slowly. Cryogenic fluids should be stored in well-insulated but loosely stoppered containers that minimize loss of fluid due to evaporation by "boil-off" and that prevent "plugging" and pressure buildup.

Mechanical

In addition to physical hazards such as fire and electric shock, laboratory personnel should be aware of the mechanical hazards of equipment such as centrifuges, autoclaves, and homogenizers.

Centrifuges, for example, must be balanced to distribute the load equally. The operator should never open the lid until the rotor has come to a complete stop. Safety locks on equipment should never be rendered inoperable.

Laboratory glassware itself is another potential hazard. Agents such as glass beads should be added to help eliminate bumping/boil-over when liquids are heated. Tongs or gloves should be used to remove hot glassware from ovens, hot plates, or water baths. Glass pipettes should be handled with extra care, as should sharp instruments such as cork-borers, needles, scalpel blades, and other tools.

DISPOSAL OF HAZARDOUS MATERIALS

The safe handling and disposal of chemicals and other materials requires a thorough knowledge of their properties and hazards. Generators of hazardous wastes have a moral and legal responsibility, as defined in applicable local, state, and federal regulations, to protect both the individual and the environment when disposing of wastes. There are four basic waste-disposal techniques: flushing down the drain to the sewer system, incineration, landfill burial, and recycling.

Chemical Waste

In some cases, it is permissible to flush water-soluble substances down the drain with copious quantities of water. However, strong acids or bases should be neutralized before disposal. Foul-smelling chemicals should never be disposed of down the drain. Possible reaction of chemicals in the drain and potential toxicity must be considered when deciding if a particular chemical can be dissolved or diluted and then flushed down the drain. For example, sodium azide, which is used as a preservative, forms explosive salts with metals such as the copper in pipes.

Other liquid wastes, including flammable solvents, must be collected in approved containers and segregated into compatible classes. If practical, solvents such as xylene and acetone may be filtered or redistilled for reuse. If recycling is not feasible, disposal arrangements should be made by specifically trained personnel. Flammable material also can be burned in specially designed incinerators with afterburners and scrubbers to remove toxic products of combustion.

Also, before disposal, hazardous substances such as carcinogens and peroxides, which are explosive, should be transformed to less hazardous forms whenever feasible. Solid chemical wastes that are not suitable for incineration must be buried in a landfill. This practice, however, has created an environmental problem, and there is now a shortage of "safe" sites.

Radioactive Waste

The manner of use and disposal of isotopes is strictly regulated by the Nuclear Regulatory Commission (NRC) and depends on the type of waste (soluble or nonsoluble) and its level of radioactivity. The radiation safety officer should always be consulted about policies dealing with radioactive waste disposal. Many clinical laboratories transfer radioactive materials to a licensed receiver for disposal.

Biohazardous Waste

On November 2, 1988, President Reagan signed into law The Medical Waste Tracking Act of 1988. Its purpose was to: (1) charge the Environmental Protection Agency with the responsibility to establish a program to track medical waste from generation to disposal, (2) define medical waste, (3) establish acceptable techniques for treatment and disposal, and (4) establish a department with jurisdiction to enforce the new laws. Several states have implemented the federal guidelines and incorporated additional requirements. Some entities covered by the rules are any health care–related facility including but not limited to ambulatory surgical centers; blood banks and blood drawing centers; clinics, including medical, dental, and veterinary; clinical, diagnostic, pathologic, or biomedical research laboratories; emergency medical services; hospitals; long-term care facilities; minor emergency centers; occupational health clinics and clinical laboratories; and professional offices of physicians and dentists.

Medical waste is defined as "special waste from health care facilities" and is further defined as solid waste that if improperly treated or handled "may transmit infectious diseases." It comprises animal waste, bulk blood and blood products, microbiologic waste, pathologic waste, and sharps. The approved methods for treatment and disposition of medical waste are incineration, steam sterilization, burial, thermal inactivation, chemical disinfection, or encapsulation in a solid matrix.

Generators of medical waste must implement the following procedures:

1. Employers of health care workers must establish and implement an infectious waste program.
2. All biomedical waste should be placed into a bag that is marked with the biohazard symbol, and then placed into a leak-proof container that is puncture-resistant and equipped with a solid, tight-fitting lid. All containers must be clearly marked with the word *biohazard* or its symbol.
3. All sharp instruments, such as needles, blades, and glass objects, should be placed into special puncture-resistant containers prior to placing them inside the bag and container. Needles should not be transported, recapped, bent, or broken by hand.
4. All biomedical waste must then be disposed of by one of the recommended procedures.

Potentially biohazardous material, such as blood or blood products and contaminated laboratory waste, cannot be discarded directly. Contaminated combustible waste can be incinerated. Contaminated noncombustible waste, such as glassware, should be autoclaved before being discarded. Special attention should be given to the discarding of syringes, needles, and broken glass that also could inflict accidental cuts or punctures. Appropriate containers should be used for discarding these sharp objects.

ACCIDENT DOCUMENTATION AND INVESTIGATION

Any accidents involving personal injuries, even minor ones, should be reported immediately to a supervisor. Under OSHA regulations, employers are required to maintain records of occupational injuries and illnesses for the length of employment plus 30 years. The record-keeping requirements include a first report of injury, an accident investigation report, and an annual summary, which is recorded on an OSHA injury log (Form 200).

The first report of injury is used to notify the insurance company and the human resources/employee relations department that a workplace injury has occurred. The report is usually completed by the employee and the supervisor and contains information on the employer, injured person, time and place of the injury, cause of the injury, and nature of the injury. The report is signed and dated; then it is forwarded to the institution's risk manager/insurance representative.

The investigation report should include information on the injured person, a description of what happened, the cause of the accident (environmental/personal), other contributing factors, witnesses, the nature of the injury, and actions to be taken to prevent a recurrence. This report should be signed and dated by the person who conducted the investigation.

On an annual basis a log and summary of occupational injuries and illnesses should be completed and forwarded to the U.S. Department of Labor, Bureau of Labor Statistics (OSHA Injury Log No. 200). The standardized form requests information similar to the first report of injury and the accident investigation report. Information about every occupational death, nonfatal occupational illness, and nonfatal occupational injury that involved loss of consciousness, restriction of work or motion, transfer to another job, or medical treatment (other than first aid) must be reported.

Because it is important to determine why and how an accident occurred, an accident investigation should be conducted. Most accidents can be traced to two underlying causes: environmental (unsafe conditions) and/or personal (unsafe acts). Environmental factors include inadequate safeguards, use of improper or defective equipment, hazards associated with the location, or poor housekeeping. Personal factors include improper laboratory attire, lack of skills or knowledge, specific physical or mental conditions, and attitude. The employee's positive motivation is very important in all aspects of safety promotion and accident prevention.

It is particularly important that if any individual sustains a needle puncture during blood collection or a cut during subsequent specimen processing or handling, the appropriate authority should be notified immediately. For a summary of recommendations for the protection of laboratory workers refer to *Protection of Laboratory Workers from Infectious Disease Transmitted by Blood, Body Fluids and Tissue,* 2nd ed., Tentative Guideline M29-T2, (NCCLS).

SUMMARY

The cardinal safety rules of the clinical laboratory are to develop foresight and accident perception, use common sense, and develop and practice the following:

1. *Good personal behavior/habits.*
   Wear proper attire and protective clothing.
   Tie back long hair.
   Do not eat, drink, or smoke in the work area.
   Never mouth pipette.
   Wash hands frequently.
2. *Good housekeeping.*
   Keep work areas free of chemicals, dirty glassware, etc.
   Store chemicals properly.
   Label reagents and solutions.
   Post warning signs.
3. *Good laboratory technique.*
   Do not operate new or unfamiliar equipment until you have received instruction and authorization.
   Read all labels and instructions carefully.
   Use the personal safety equipment that is provided.
   For the safe handling, use, and disposal of chemicals, learn their properties and hazards.

Learn emergency procedures and become familiar with the location of fire exits, fire extinguishers, blankets, etc.
Be careful when transferring chemicals from container to container and always add acid to water slowly.

EXERCISES

Answer the following questions:

1. Who is responsible for safety in the laboratory?
2. In general, how should chemicals be stored?
3. List examples of personal protective equipment and its use.
4. Why is a Universal Precautions Policy important?
5. What information must be given to employees according to the Hazardous Communication Policy?
6. Where are material safety data sheets to be located?
7. Describe the different classifications of chemicals?
8. Name and give examples of the four classes of fires.
9. Explain the proper disposal of chemical, radioactive, and biohazardous waste.
10. Summarize the process for documentation and investigation of accidents.

SUGGESTED READINGS

ACS Committee on Chemical Safety, Smith GW, ed. Safety in academic chemistry laboratories. Washington: American Chemical Society, 1985.

Allocca JA, Levenson HE. Electrical and electronic safety. Reston, VA: Reston Publishing Company, 1985.

Boyle MP. Hazardous chemical waste disposal management. Clin Lab Sci 1992; 5:6.

Brown, JW. Fighting the TB dragon. Med Lab Obs 1994; 26:2.

Brown, JW. Tuberculosis alert: An old killer returns. Med Lab Obs 1993; 25:5.

Bryan RA. Recommendations for handling specimens from patients with confirmed or suspected Creutzfeldt-Jakob disease. Lab Med 1984; 15:50.

CDC. Update: Universal precautions for prevention of transmission of human immunodeficiency virus, hepatitis B virus and other bloodborne pathogens in health care settings. MMWR 1988; 37:229.

CDC, NIH. Biosafety in microbiological and biomedical laboratories. 3rd ed. Washington: US Government Printing Office, 1993.

Chaff LF. Safety guide for health care institutions. 4th ed. Chicago: American Hospital Publishing, 1989.

Chervinski D. Environmental awareness: It's time to close the loop on waste reduction in the health care industry. Adv Admin Lab 1994; 3:4.

DHHS Committee to Coordinate Environmental and Related Programs. NIH Guidelines for the laboratory use of chemical carcinogens. Washington: National Institutes of Health, 1981.

Favero MS. Biological hazards in the laboratory. Lab Management 1987; 18:10.

Furr AK. Handbook of laboratory safety. 3rd ed. Boca Raton, FL: CRC Press, 1990.

Gile TJ. Hazard-communication program for clinical laboratories. Clin Lab 1988; 1:2.

Halper HR, Foster HS. Laboratory regulation manual. Vol. 3. Germantown, MD: Aspen Systems, 1988.

Hartree E. Electrical hazards, fire and explosion. In: Hartree E, Booth V, eds. Safety in biological laboratories. London: The Biochemical Society, 1977.

Hayes DD. Safety considerations in the physician office laboratory. Lab Med 1994; 25:3.

Hazard communication. Federal Register 59:27, Feb 4, 1994.

Hicks RM. Basics of clinical laboratory safety. In: Snyder JR, Larsen AL, eds. Administration and supervision in laboratory medicine. Philadelphia: Harper & Row, 1983.

Karcher, RE. Is your chemical hygiene plan OSHA-proof? MLO 1993; 25:7.

Le Sueur CL. A three-pronged attack against AIDS infection in the lab. Med Lab Obs 1989; 21:37.

Miller SM. Clinical safety: Dangers and risk control. Clin Lab Sci 1992; 5:6.

NCCLS. Clinical laboratory hazardous waste (proposed guideline). Villanova, PA: National Committee for Clinical Laboratory Standards, 1986.

NCCLS. Protection of laboratory workers from infectious disease transmitted by blood, body fluids and tissue (tentative guideline). Villanova, PA: National Committee for Clinical Laboratory Standards, 1989.

NIH Radiation Safety Branch. Radiation: The National Institutes of Health safety guide. Washington: US Government Printing Office, 1979.

NRC Committee on the Hazardous Biological Substances in the Laboratory. Biosafety in the laboratory: Prudent practices for the handling and disposal of infectious materials. Washington: National Academy Press, 1989.

NRC Committee on Hazardous Substances in the Laboratory. Prudent practices for disposal of chemicals from laboratories. Washington: National Academy Press, 1983.

NRC Committee on Hazardous Substances in the Laboratory. Prudent practices for the handling of hazardous chemicals in laboratories. Washington: National Academy Press, 1981.

Occupational exposure to bloodborne pathogens; final rule. Federal Register 56:235, Dec 6, 1991.

Otto CH. Safety in health care: Prevention of bloodborne diseases. Clin Lab Sci 1992; 5:6.

Pipitone DA. Safe storage of laboratory chemicals. New York: Wiley, 1984.

Roach GC. Laboratory safety: Principles for personal practice. Washington: American Society for Medical Technology, 1975.

Rose SL. Clinical laboratory safety. Philadelphia: Lippincott, 1984.

Rudmann SV, Jarus C, Ward KM, Arnold DM. Safety in the student laboratory: A national survey of university-based programs. Lab Med 1993; 24:5.

Slockbower JM, Blumenfeld TA. Collection and handling of laboratory specimens: A practical guide. Philadelphia: Lippincott, 1983.

Stern A, Ries H, Flynn D, Vance A. Fire safety in the laboratory: Part I. Lab Med 1993; 24:5.

Stern A, Ries H, Flynn D, Vance A. Fire safety in the laboratory: Part II. Lab Med 1993; 24:6.

Chapter 3

Specimen Collection and Processing

Beverly S. Oxford, Susan Dovenbarger

Objectives

Upon completion of this chapter, the clinical laboratorian should be able to:

- State the role of the laboratory professional in specimen collection.

- Identify the types of blood specimens used for analysis in the clinical laboratory, including their characteristics and methods of collection.

- Identify the physiological factors that affect blood specimens.

- Relate the importance of isolation precautions and know the basic procedures to be followed by phlebotomists in isolation situations.

- Describe the venipuncture procedure to include the necessary equipment, supplies, technique, and precautions.

- Describe the skin puncture procedure to include the necessary equipment, supplies, technique, and precautions.

- Outline the special needs and precautions when collecting blood from infants and children whether by skin puncture or venipuncture.

- List the ways to monitor the quality of blood collections.

- Describe the special collection procedures for fasting specimens, timed-interval specimens, specimens for blood alcohol levels, and forensic specimens.

- Discuss the issues in proper specimen transport.

- State the importance of handling specimens as outlined in universal precautions procedures, and the proper disposal of infectious waste.

- State the importance of evaluating of each specimen for possible interfering substances such as lactescence, hemolysis, or icterus.

- Understand the role of individual laboratory instrumentation and analytical methods in determining specific laboratory specimen storage procedures.

- Describe the basic collection and handling requirements for cerebral spinal fluid, paracentesis fluid, and urine.

- List the ways in which specimen processing is monitored for quality in the laboratory.

Michael L. Bishop, Janet L. Duben-Engelkirk, and Edward P. Fody.
CLINICAL CHEMISTRY. © 1996 Lippincott–Raven Publishers.

KEY WORDS

Analyte	Glycolytic Inhibitor	Phalanx
Anticoagulant	Hemolysis	Phlebotomist
Antecubital Fossa	Heparin	Plasma
Arterial Bloods	Icterus	Random Urine
Basal State	Lactescence	Specimen
Capillary Blood	Lancet	Serum
Cerebral Spinal	Lateral	Skin Puncture
Fluid (CSF)	Medial	Timed Urine
Closed Collection	National Committee	Specimen
System	for Clinical	Venipuncture
Effusion	Laboratory	Venous Blood
Evacuated Tube	Standards	Whole Blood
Fasting Specimens	(NCCLS)	
Forensic Medicine	Paracentesis	

Specimen collection, handling, and processing are an integral part of the process of specimen analysis. In ensuring the quality and integrity of the patient sample from collection to the final result, it is important to be aware of, and follow, the guidelines for specimen collection and handling established by such agencies as the National Committee for Clinical Laboratory Standards (NCCLS).[29–38] Proper patient identification, correct use of equipment and supplies, and an adequate knowledge of the physiologic and environmental factors affecting specimen quality are all crucial in obtaining blood specimens for reliable laboratory testing. Additionally, timely and proper transport and processing of the collected specimen further ensure the accuracy of the final test result.

If the course of sample collection, transport, processing, and analysis is viewed as a chain, it is clear that each link is critical in providing the most accurate clinical picture of the patient. The chain is only as reliable as its weakest link; therefore, all four links must function consistently and correctly. If the integrity of the chain is compromised at any point along the way, whether as a result of poor collection technique or incorrect specimen handling, then, unfortunately, the final outcome of test results is also compromised.

ROLE OF THE LABORATORY PROFESSIONAL IN SPECIMEN COLLECTION

Appearance and Attitude

Specimen collection may be performed by the clinical laboratorian or by a trained phlebotomist. In most cases, the phlebotomist is the front-line representative of the hospital laboratory. Patients and hospital staff rarely glimpse the behind-the-scenes efforts in the laboratory, but rather form opinions of those operations based on the more visible efforts of the phlebotomist. Because of this, it is extremely important for the phlebotomist to maintain a professional manner when interacting with patients and staff.[4,18,20,43,47]

The phlebotomist should display an appearance and attitude that instills confidence in his or her role as a member of the health care team. A clean white coat and neat clothing not only identify the phlebotomist as a member of that team, but also represent the phlebotomist's professional commitment and reinforce the patient's confidence in the phlebotomist's ability to perform the procedure. The phlebotomist should enter the patient's room with a well-organized, clean tray of collection equipment. Exposed boxes of dirty needles and dried blood specimens on a tray not only create a negative impression in the patient, but constitute a safety hazard for both the phlebotomist and the patient.[16] Nervousness and absent-minded behaviors can lead to a negative impression of the phlebotomist's ability and performance.

Interpersonal Skills and Behavior

Blood collection is undoubtedly one of the most unpopular procedures that patients encounter during their hospital stay, and although no phlebotomist can make it a painless procedure, there are specific interpersonal skills that can be used for an optimal interaction.

Communication techniques should be focused on relieving patient anxiety over blood collection and continuing to reinforce patient confidence in the phlebotomist's skills. All patients should be approached in a considerate and respectful manner. Increased sensitivity to the sick and good listening skills will serve the phlebotomist well through the continued interaction.

The patient should always be the focus of the phlebotomist's undivided attention during the interaction. Blood-drawing procedures hurt and on occasion require more than one attempt. It is untruthful and unfair to tell a patient that the collection will not hurt. Patients with veins in such poor condition as to require more than two sticks should be referred to a laboratory supervisor or the physician for collection.

It is very likely that at some point the phlebotomist will encounter family and visitors when collecting specimens. Family members or visitors may be asked to leave the room at the phlebotomist's discretion. When working with pediatric patients, however, it is important to recognize that parents and family can provide a great source of comfort during uncomfortable procedures and as much as possible should be encouraged to remain with the child. Family members may even be able to assist the phlebotomist when working with children and should be encouraged to do so when appropriate.

Communicating with pediatric patients sometimes takes an extra amount of skill and patience. One bad blood-drawing episode can affect all subsequent blood-drawing procedures for that child. The added investment of time and effort will benefit patient care and management. The following are some specific tips on communicating with pediatric patients:

1. Use nonjudgmental language when talking with children.
2. As much as possible, offer choices to children and encourage their participation.

3. When it is age-appropriate, explain the procedure in non-threatening language.

4. Offer distractions such as holding someone's hand really tight, saying ouch very loudly, counting to 10, singing a song, or pretending to be somewhere else, and finally, give the okay to cry when needed.

Patient Confidentiality

Patient confidentiality is not only the phlebotomist's obligation, but also is guaranteed under the Patient's Bill of Rights adopted by the American Hospital Association in 1973. A patient's medical history or personal information should never be discussed with anyone unless it is a professional discussion pertaining to the patient's care. Moreover, any discussion of a patient's case should never be done in public areas.

Patient Identification

Proper patient identification is vital to quality patient care. Incorrectly identified samples can produce life-threatening consequences when results are subsequently reported on the wrong patient. The phlebotomist cannot give too much attention to this aspect of blood collection. The identification process must be done with total accuracy for every patient every time. In the hospital, the first check is to verify the card outside the patient's room. The card should coincide with the name and hospital number on the laboratory requisition. Next, the phlebotomist must ask the patient his or her name (*e.g.,* "What is your name?" but never "Are you Mr. Smith?"). In the case of pediatric patients or for adults who are not fully alert, family members or visitors may be asked to state the patient's name. Finally, check the patient's identification bracelet. The information on this bracelet *must* correspond *exactly* with the information on the test requisition. If there are any discrepancies, then the patient's nurse must be notified before blood is taken, and if necessary, action should be initiated to correct the information on the bracelet or requisition. In the event that the patient is not wearing an identification bracelet, as is sometimes the case with pediatric patients or recent surgical patients, then the nurse *must* provide positive identification of the patient.[17,33]

Outpatient areas will need to rely on different means of positive patient identification. In addition to asking the patient his or her name, if a clinic card is available, it can be used to help correctly identify the patient, as with the hospital identification bracelet. If such a card is not available, then another form of identification should be checked. Parents of pediatric patients should verify the child's identification before blood is taken.

BLOOD SPECIMEN TYPES

By far the most common specimen analyzed in the clinical chemistry laboratory is blood. Traditionally, blood is cen-trifuged, and the serum or plasma is used for analysis. With advances in instrumentation, heparinized whole blood is now also used, especially in acute-care (stat) environments. Blood can be obtained from arterial, venous, or capillary sites, each type with its own characteristics.

The most common use for arterial blood is blood gas analysis. Blood from arteries is considered uniform throughout the body, and the pH, PCO_2, and PO_2 values reflect the respiratory and metabolic status of the body. The venous blood content varies based on the metabolic activity of the tissue it drains; thus venous blood content will vary by collection site and time (depending on muscular activity). Any impairment to the arterial or venous circulation, whether local or general, can affect specific analytes. By far the largest difference between arterial and venous blood is the oxygen content. Other analytes that vary between arterial and venous blood are pH, CO_2 content, glucose and chloride content, packed-cell volume, lactic acid, and ammonia level.[37] Any impairment of general or local circulation will exaggerate the differences between venous and arterial samples.

Arterial blood sampling is usually limited to medical situations that require arterial puncture. Arterial sampling should be performed only by those who are instructed in the proper techniques, aware of the dangers of the procedures, and familiar with the precautions designed to prevent hazards to the patient.[37] Because of variations in venous samplings, each institution should standardize preferred sites and times for venous sampling. Location of venous sampling, positioning of the patient (sitting or supine), and use of a tourniquet can all influence the results obtained.

Specimens obtained by skin puncture reflect a combination of arterial and venous blood. Warming of the site prior to collection will cause "arterialization" of the capillary bed due to vasodilation and increased blood flow. For blood gas analysis, free-flowing blood obtained from an arterialized site is considered to be similar to arterial blood, except for PO_2 values, which will vary with the perfusion of the site.[30]

In reality, the skin-puncture sample is a mixture of capillary, venous, and arterial blood, and it also may contain intracellular and tissue fluids.[34] There may be significant differences between skin-puncture specimens and those collected by venipuncture.[7,8,21] For example, glucoses may be increased, while potassium, total protein, and calcium may be decreased. In addition, tissue-fluid contamination resulting from excessive squeezing may falsely elevate potassium levels. A number of chemistry analytes, however, show no clinically significant differences between venous and skin-puncture blood. It is therefore important to note on the requisition the method of specimen collection so that the physician can accurately interpret the results.[7,8]

Skin-puncture collection is indicated when venous sites are limited or not accessible, as is often the case with burn patients and patients receiving chemotherapy. It is also necessary when difficult or fragile veins are encountered, as might occur with obese or elderly patients. Because of the many risks associated with venipuncture in infants and neonates, skin puncture is a highly desirable alternative for specimen collection.[7,26,34]

PHYSIOLOGIC FACTORS AFFECTING SPECIMENS

The best time of the day to collect blood specimens is in the early morning before the patient has eaten or become physically active. At this point, known as the *basal state,* the body is at rest, and food has not been ingested during the night. Specimens collected during the basal state more often reflect normal values than those collected at other times of the day.[17]

Some blood chemistries can be significantly altered after the patient has had a meal. Glucose, triglyceride, and potassium values, for example, may be affected by dietary intake, causing abnormal test results.[22] For this reason, it is worthwhile for the phlebotomist to note whether the sample is a fasting or nonfasting sample to enable the physician to correctly interpret the test results. If a fasting specimen is required for a particular test, then the phlebotomist should make certain that the patient has been fasting for the required time.

Moderate exercise can effect the values for certain chemistries and enzymes. Although most of these levels return to normal shortly after stopping the exercise, some analytes may remain elevated for 24 hours or more.[17] Emotional stress may falsely elevate white blood cell counts and may be indicated by nervousness or anxiety. Further, the phlebotomist should keep in mind that mild to violent crying greatly elevates the number of circulating white blood cells. This, of course, is especially important to note when collecting blood specimens from infants and children.[16]

The body's diurnal rhythms and even patient posture also can alter the results for certain analytes. Patient posture (erect or supine) during or just prior to venipuncture has been found to be a variable factor in analyte values.[44] Some of the more commonly affected values are listed in Table 3-1. The posture variable should be considered when comparing inpatient and outpatient values. Erect posture is believed to promote movement of water from intravascular to interstitial compartments, thus increasing the concentration of some chemical analytes.[15,22]

ISOLATION PRECAUTIONS AND GUIDELINES

Any discussion of techniques used to prevent the continued spread of infection must always include good handwashing. Thorough handwashing in between contact with patients or contaminated items is the key to breaking this most common route of infection transmission. Additional measures also may be needed when certain infectious diseases are pre-

TABLE 3-1. Factors Affecting Chemical Analyte Values

Value	Hemolysis[23,24]	Prolonged Tourniquet Application[40]	Posture[40]
Albumin	×	ns	×
Calcium	×	ns	×
Cholesterol	×	↑	×
Enzymes	×	Some	Some
Iron	×	↑	ns
Lipids	×	↑	×
Potassium	×	↓	×
Sodium	ns	ns	ns
Total protein	×	↑	×

× = *significant abnormal effect*
ns = *no significant effect*

sent. Protective clothing such as gowns, gloves, masks, and shoe covers may be required depending on the specific isolation type encountered.

Not all isolation procedures are established to prevent the spread of infection from the patient to the staff. Reverse isolation may be necessary to protect patients who are extremely susceptible to infection as a result of deficient immune systems. The newborn and intensive care nursery areas both use a form of reverse isolation by requiring a 3-minute hand/arm scrub and gowning before coming into contact with the infants. Other areas, such as the adult intensive care and burn units, maintain protective guidelines that specifically address the needs of their patients.

Table 3-2 lists the different types of isolation precautions and the protective measures required when entering each patient's room. It is important to note that although not all isolation guidelines require the use of gloves when in contact with patients, gloves are always worn when collecting blood from patients on universal precautions and also when collecting blood from anyone by means of an open collection method (*e.g.,* skin puncture). Current Centers for Disease Control (CDC) and NCCLS guidelines state that gloves must be used for collecting blood by skin puncture (open collection system) and are suggested for use during venipuncture for all patients.[10,38] However, it is also recognized that phlebotomists may be more clumsy using gloves during venipuncture and therefore incur greater risk for accidental needlesticks. Under these circumstances, phlebotomists must thoroughly understand the hazards involved in order to make an informed decision regarding their glove use during routine venipuncture.

TABLE 3-2. Types of Isolation and the Protective Measures Required by Each

Isolation Type	Gowns	Masks	Gloves	Shoe Covers
Strict (*e.g.,* measles, chickenpox, strepto-coccal and staphylococcal pneumonia, diphtheria, rabies)	Yes	Yes	Yes	No
Respiratory (*e.g.,* tuberculosis, whooping cough, meningococcal meningitis, mumps, measles)	No	Yes	No	No
Contact (*e.g.,* wound and skin infections)	Yes	No	Yes	No
Universal precautions (*e.g.,* HIV, hepatitis)	Yes*	Yes*	Yes	No
Reverse (*e.g.,* severe combined immune deficiency syndrome?)	Yes	Yes	Yes	Yes
Neutropenic (*e.g.,* patients receiving certain chemotherapies, transplant patients)	No	Yes†	No	No

If procedure is likely to produce splatters or airborne droplets of blood or body fluids.
†*If health care provider has signs of cold or illness such as coughing and sneezing.*

SPECIMEN COLLECTION BY VENIPUNCTURE

Specimen collection by venipuncture involves a thorough understanding of the available equipment and supplies. Adherence to the procedural guidelines established by the NCCLS is encouraged.

There are three methods for collecting blood by venipuncture. The most common is the *evacuated-tube system,* also known as the *closed collection system.* Because blood is taken directly from the patient's vein into a stoppered tube, the sample is completely contained, thereby reducing the risk of outside contaminants to the sample and reducing the hazard of the collector's exposure to the blood. The system consists of a needle, plastic sleeve or tube holder, and the evacuated tube. The needle screws securely into the sleeve, and tubes slide easily into the opposite sleeve opening. The tube can be secured by pushing it up to the recessed guidelines inside the sleeve, taking care not to completely penetrate the stopper. Once the stopper is punctured completely by the needle, the tube's vacuum will draw blood into the tube. This evacuated system is shown in Figure 3-1.

The closed or evacuated system is usually preferable to *syringe collection.* However, use of syringes may be necessary when the patient has fragile or damaged veins that may easily collapse from the vacuum pressure introduced by the tube. When using a syringe, the phlebotomist creates a controlled vacuum by gently pulling the plunger back to collect blood into the syringe. The vacuum pressure thus created may be less forceful and less likely to damage fragile veins. Syringes are not screwed onto the needle base as in the first method, but rather are attached to the needle by means of a plastic fitting. Syringe collection is shown in Figure 3-2.

Butterfly infusion sets, although more expensive, are convenient, particularly when both a syringe and evacuated tubes must be collected during the same procedure. The set consists of a needle with a plastic butterfly-shaped holder for increased manipulation during needle insertion. A piece of plastic tubing, available in varying lengths, connects the needle to an adapter that may be attached to a syringe or to a Luer adapter for use with the evacuated-tube system. Some brands of butterfly infusion sets are now manufactured with an attached Luer adapter. Blood collection by means of a butterfly system is shown in Figure 3-3.

Equipment and Supplies

The equipment most commonly used for blood collection by venipuncture is as follows:

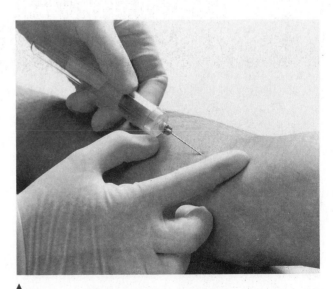

▲

Figure 3-1. Evacuated-system blood collection.

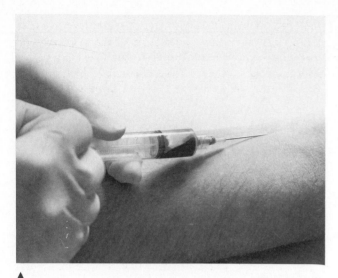

▲
Figure 3-2. Blood collection by syringe.

Tourniquet.
Sterile venipuncture needle (gauge to fit size and condition of vein).
Plastic tube holder or sleeve.
Evacuated tubes.
Sterile alcohol pads.
Gloves.
2 × 2 inch gauze pads.
Adhesive tape or bandages.

Other equipment may be needed for special collection problems, including butterfly infusion sets (used for pediatric patients or for adult patients with difficult veins), Luer adapters (to connect infusion sets to plastic sleeves or syringes), and syringes (to avoid collapsing fragile veins with evacuated tubes).

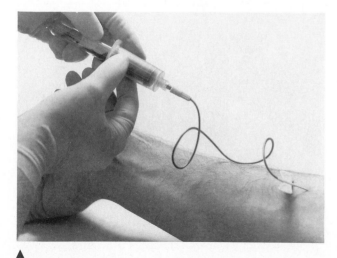

▲
Figure 3-3. Blood collection by modified evacuated system using a Luer adapter to attach tube holder to butterfly set.

Tourniquets

Several basic types of tourniquets are available, as shown in Figure 3-4. Most commonly, a piece of pliable rubber strapping or tubing is used. However, Velcro tourniquets are becoming more popular because of their convenience and comfort. A more sophisticated tourniquet type is made of cloth and clasps around the patient's arm using a seat belt-like apparatus. This style is advantageous because the venous pressure created by tightening the tourniquet may be partially released by a lever without completely releasing pressure from the patient's arm. Since the tourniquet remains on the patient's arm, it may be tightened again during collection if necessary (*e.g.*, when drawing multiple samples). This feature is particularly useful because prolonged venous pressure during venipuncture may cause erroneous laboratory values. The blood pressure cuff also offers an easy mechanism for increasing or decreasing venous pressure during multiple-tube collection and is readily available on most inpatient units. Table 3-1 lists some examples of chemical analytes affected by prolonged tourniquet application.

Venipuncture Needles

Needles are manufactured with a variety of bore sizes, referred to as the needle's *gauge*. The size of the needle used is determined by the condition of the patient's veins and by the volume of blood to be collected. Larger needles (18 gauge) are needed when collecting from blood donors, but a smaller bore (21 and 22 gauge) suffices for routine laboratory collection. It is important to note that the larger the gauge number, the smaller is the bore or opening of the needle. Butterfly infusion sets with 23- or 25-gauge needles may be needed when collecting from difficult hand or arm veins on

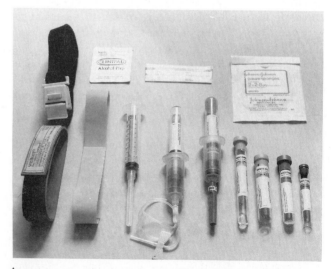

▲
Figure 3-4. Venipuncture supplies. (*Bottom, from left to right*) Tourniquets: Velket™ and rubber tubing types, syringe with Luer lock needle, evacuated tube in tube holder with Luer lock butterfly set, tube holder with evacuated tube and multisample needle, and other evacuated tubes. (*Top*) Alcohol prep pad, adhesive bandage, gauze sponge. (Photograph by Steven R. Conlon)

adults and should always be used when collecting routine specimens from children.

There has been concern over the relationship between needle size and the presence of hemolysis as a source of error in laboratory analysis. *Hemolysis* is defined as damage to erythrocyte membranes, causing release of cellular constituents (*e.g.,* hemoglobin) into the blood plasma or serum. Some reports indicate that using a higher-gauge needle (smaller bore) causes hemolysis, while others indicate that smaller-gauge needles (larger bores) cause more hemolysis because the larger opening allows increased vacuum pressure, drawing in the blood more quickly (Becton Dickenson: Data on file. Becton Dickenson Vacutetainer Systems, Rutherford NJ: 07070).[9] Hemolyzed samples are unacceptable for many laboratory tests (see Table 3-1).

Needles are manufactured for either single- or multiple-sample collection. Single-sample needles are normally used when collecting with a syringe rather than an evacuated tube. Multisample needles allow for tube changes without blood leakage within the plastic sleeve. This is accomplished with the aid of a rubber cover over the stopper-puncturing portion of the needle.

Plastic Sleeves and Evacuated Tubes

It is recommended that the phlebotomist use the plastic sleeve designed by the manufacturer of the evacuated tubes being used. The slight differences in sizes between manufacturers may make complete stopper puncture more difficult. The tubes and corresponding sleeves are manufactured in a wide range of sizes and volumes. The smaller-volume tubes (2 and 3 mL) are especially useful for pediatric collection.

Evacuated tubes come with a color-coded stopper indicating the additive present.[29] Additives may be needed to preserve specific analytes or to anticoagulate specimens for certain tests. Additives are also available to accelerate the blood's clotting action and thereby reduce laboratory turnaround time for determining test results. Anticoagulants inhibit the blood's clotting action and yield specimens containing intact clotting factors. These additives are used when whole blood or plasma is needed for analysis. Plain tubes without additives are used when serum is to be analyzed. Serum is the liquid portion of the blood without clotting factors, while the liquid plasma still contains these factors.

It is important to remember to use the appropriate tube when collecting samples for specific tests. Tubes needed for whole-blood analysis, such as hematology, will be of no value for tests requiring the analysis of serum. Table 3-3 lists the more commonly used evacuated tubes, their additives, and uses.

Because of the possibility of backflow or cross-contamination when collecting multiple tubes at one time, the phlebotomist should collect the tubes in a consistent sequence.

TABLE 3-3. Commonly Used Evacuated Tubes

Stopper Color	Additive	Resulting Sample	Additive Action	Possible Assays*	Comments
Red	None	Serum		Enzymes, electrolytes, iron, lipids, drug levels	Some tubes have silicone coating to accelerate clotting.
Red/gray	Inert polymer barrier	Serum	Cleaner separation of serum and cells	Same as red stopper	Required for CO_2 analysis
Orange	Thrombin	Serum	Accelerates clotting	Stat chemistries	
Gray	Iodoacetate	Serum	Inhibits glycolysis	Glucose, lactose	No interference with enzymatic glucose or BUN
Gray	Na fluoride and K oxalate	Plasma	Inhibits glycolysis	Glucose, lactose	Unacceptable for Na or K analysis
Green	Na/Li/NH₃ heparin salts	Plasma	Enhances antithrombin III†	Corticosteroids, electrolytes	Antithrombin III inactivates clotting factors.
Brown	Na Heparin	Plasma	Enhances antithrombin III†	Lead	Tube has only minute quantities of lead.
Royal blue	None	Serum		Trace elements	Tube has only minute quantities of trace elements.
Lavender	EDTA‡	Plasma	Binds calcium	CEA, hematology values	Calcium is needed for clotting.
Blue	Na Citrate	Plasma	Binds calcium	Coagulation values	

*Does not include all possible assays.
†Antithrombin III inhibits clotting factors.
‡Ethylenediaminetetraacetic acid.

This sequence is known as the *order of draw* (Table 3-4). The recommended order is to first fill blood culture tubes, then tubes with no additive (red or dark-blue top), then the tubes containing sodium citrate for coagulation studies (light-blue top), followed by tubes containing heparin (green top), ethylenediaminetetraacetic acid (EDTA) tubes (lavender top), and finally tubes containing oxalate or fluoride (gray top). This order must be followed because the consequences of contamination can severely affect results. For example, if a tube containing the potassium salt EDTA is collected prior to a heparinized tube for electrolyte analysis, the potassium values from the heparinized tube may be falsely increased by contamination from the EDTA in the tube. Since EDTA is a calcium chelator, calcium levels also may be falsely decreased in tubes subsequently drawn. Potassium oxalate or potassium fluoride tubes should be collected last for three reasons. First, the potassium additive may again falsely elevate subsequent electrolyte specimens. Second, oxalate damages cell membranes and some enzymes and therefore might cause abnormal erythrocyte morphology if an EDTA a tube for hematology is drawn following an oxalate tube. Third, fluoride has varying erroneous effects on other analytes.[9,22] When coagulation studies are the only tests ordered, then it is recommended that the phlebotomist first draw a prime tube (for discard) before the blue tube. This prevents the risk of contamination from tissue fluid, which may interfere with the coagulation test results.

The phlebotomist should be aware of the sample amount necessary for each test. A full tube is not necessary for some tests, particularly those which require a small amount of serum. This is especially important when working with pediatric patients, in whom the amount of blood removed must be minimized. However, the additive tubes must be full enough to achieve the correct blood-to-additive ratio for reliable results.

Alcohol and Gauze Pads, Tape, and Bandages

Sterile alcohol pads containing 70% isopropyl alcohol are necessary for cleansing the puncture site. Because alcohol is not bactericidal, povidone-iodine must be used when collecting samples for blood cultures or when collecting from severely neutropenic patients. Clean cotton 2 × 2 inch gauze is recommended to apply pressure to the site following the collection. Pressure bandages may be made using folded cotton gauze that is secured with tape, or an adhesive bandage may be used if the bleeding has stopped.

Determining the Site

The area of the arm known as the *antecubital fossa* is most commonly used for venipuncture. Other acceptable sites include the hand, ankle, or foot. However, when the only palpable site is an ankle or foot vein, the physician or nurse should be consulted, since this is not recommended for patients with certain conditions.[42] Some hospital and laboratory policies prohibit the phlebotomist from using any collection site other than the hand or arm. There are several special considerations to be taken into account when selecting the venipuncture site. Areas with hematomas or scarring present should be avoided. The phlebotomist should never insert a needle into a previously punctured site, since this heightens the risk of infection. The arm on the same side as a mastectomy also should be avoided. If intravenous (IV) fluids are being administered, then the opposite arm should be used. In the event that both arms are used for IV therapy, then a nurse should be called for assistance. The IV should be stopped for 2 minutes prior to blood being taken. The puncture should be made *below* the IV site, with a 5-mL prime tube drawn first for discard. This first tube is most likely contaminated with IV fluid and would not be a representative specimen for analysis. Once the specimens are collected, the nurse can restart the IV.

An anterior view of the antecubital fossa (Fig. 3-5) reveals several large veins, including the cephalic, median cubital, and basilic veins. The median cubital vein is the most preferable of the three because of its size and because venipuncture here is less painful for the patient. However, the other two veins are also suitable for specimen collection. Choosing the best vein is accomplished by palpating, or feeling, the vein with the tips of the fingers. The phlebotomist should use the index or middle finger for this task because they are the most sensitive to touch. Using the thumb for palpating is not recommended because the pulse in the thumb might be confused with an arterial pulse in the patient's arm. The selected vein should always feel pliable and spongy to touch, much like the feel of hollow tubing. The direction and length of the vein in the area chosen for puncture should be traced with the fingertips. Areas that feel hardened may be sclerosed or scarred veins or perhaps a tendon or muscle and should not be punctured. The detection of a pulse indicates the presence of an artery and disqualifies the site for venipuncture. After choosing the vein and assessing its condition, the phlebotomist decides which collection devices are necessary for successful venipuncture.

Venipuncture Procedure (Fig. 3-6)

1. Identify the patient, exactly matching the information on the hospital bracelet with the test requisition.

TABLE 3-4. Order of Tube Draw for Specimen Collection

1. Sterile blood-culture tubes
2. Nonadditive tubes
3. Coagulation tubes
4. Additive tubes

Recommended by Becton Dickinson, VACUTAINER™ tube manufacturer

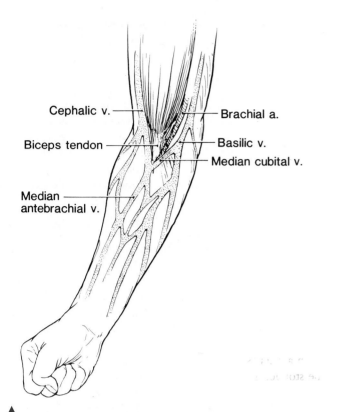

Figure 3-5. The venous structure; *a* = artery; *v* = vein. (Redrawn from Valaske MD, ed. So you're going to collect a blood specimen, rev ed., Skokie, IL: College of American Pathologists, 1982; 16.)

2. Select and prepare the supplies according to test requirements.
3. Assess the sites for venipuncture according to the criteria described earlier.
4. Position the patient comfortably. Patients may recline or sit with the arm supported and the site easily accessible to the phlebotomist.
5. Apply the tourniquet 3 to 4 inches above the site. The tourniquet should be tight but not painful to the patient. At this time the patient also may be asked to clench and unclench the fist to facilitate enlargement of the vein. It may be necessary for the patient to dangle the arm downward, or the phlebotomist may need to tap the area two or three times to help dilate the vein.
6. Palpate the vein, establishing surface length and direction. Choose the vein most easily palpated, not necessarily the one most easily seen.
7. Cleanse the site with alcohol, using outward or concentric circles. This step must be repeated if the site is touched again.
8. Inspect the needle for burrs or nicks. Allow the alcohol to air dry. Do not recontaminate the area by wiping with unsterile gauze. Fanning or blowing dry introduces airborne contaminants.

9. Fix the selected vein for puncture by placing the thumb approximately 1 inch below the site and the index finger the same distance above the site and pulling the skin taut. The other fingers may wrap around the elbow, aiding the support of the arm.
10. Align the needle with the vein, and with the bevel of the needle turned up, enter the vein at a 15° angle in a smooth, continuous motion. The phlebotomist will feel a "give" when the needle enters the vein.[45]
11. Anchor the needle by placing the index finger underneath at the hub and the thumb on top of the plastic sleeve. This reduces the risk of movement pulling the needle out of the vein.
12. Engage and fill collection tubes according to the order of draw. Release the tourniquet as soon as adequate blood flow is noted.
13. When collection is complete, disconnect the last tube from plastic sleeve. Place a 2 × 2 inch cotton gauze pad over the site and withdraw needle, being careful to apply pressure quickly after the needle is removed. Applying pressure and withdrawing the needle simultaneously can be quite painful to the patient.
14. Gently invert additive tubes 8 to 12 times for thorough mixing. Vigorous inversion may cause hemolysis.
15. Check the site for bleeding. Pressure should be applied for 3 minutes with the arm in a straight, extended position. A bandage may be applied when bleeding has stopped. Prolonged bleeding should be reported to the nurse.
16. Dispose of contaminated needles and supplies according to hospital procedure. Never recap or clip needles.
17. Label the tubes with patient name, hospital number, date, time, and phlebotomist's initials.
18. Package samples according to test specifications; for example, cold agglutinins should be placed in a warm pack, whereas ammonia must be placed on ice.
19. If patients logs are kept for specimen collection, these should be completed. If the patient had certain dietary restrictions enforced pending test collection, these should be removed.
20. Transport the specimens properly and promptly to the laboratory.

Collection using a syringe or butterfly set follows the same procedure just described, allowing for differences in the equipment. If transfer of blood from a syringe to an evacuated tube is necessary, then the phlebotomist should gently remove the stopper and allow the blood to flow slowly down the side of the tube. Fill the tube appropriately and replace the stopper. With a clean needle, puncture the stopper to prevent excess pressure from pushing the stopper out during transport. This method reduces the chances of accidental contaminated needle sticks. As previously mentioned, blood transfer should be done with care to avoid hemolysis, which may affect some analytes (see Table 3-1).

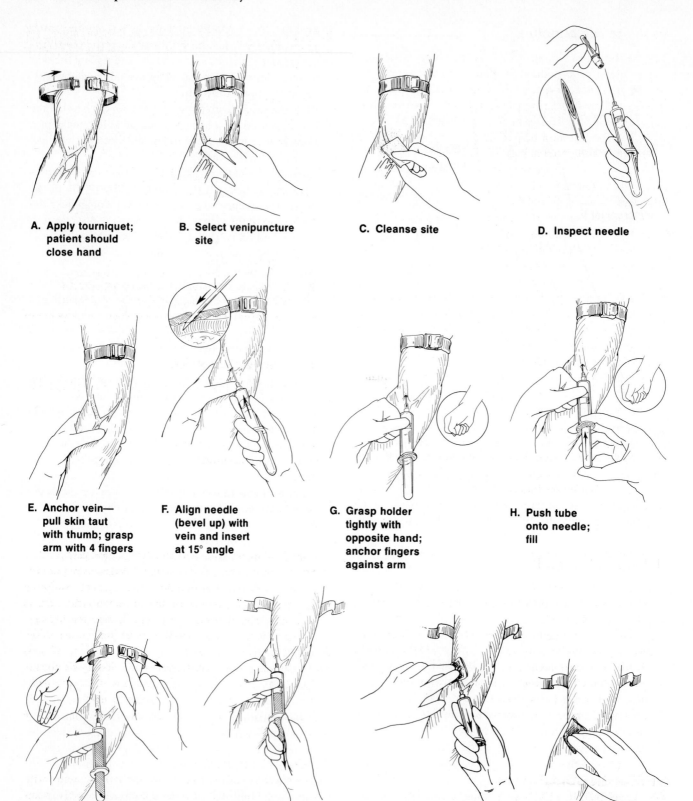

A. Apply tourniquet; patient should close hand

B. Select venipuncture site

C. Cleanse site

D. Inspect needle

E. Anchor vein— pull skin taut with thumb; grasp arm with 4 fingers

F. Align needle (bevel up) with vein and insert at 15° angle

G. Grasp holder tightly with opposite hand; anchor fingers against arm

H. Push tube onto needle; fill

I. Release tourniquet; patient should open hand

J. Gently remove tube

K. Place gauze; remove needle

L. Apply pressure

Figure 3-6. Venipuncture procedure.

Pediatric Venipuncture

The phlebotomist should use extreme caution when collecting blood from pediatric patients. Among the hazards present are cardiac arrest, hemorrhage, venous thrombosis, reflex arteriospasm, damage to surrounding tissue and organs, infection, and damage to the infant from holding.[5]

If venipuncture must be performed on patients less than 2 years of age, the NCCLS recommends using either a tuberculin or 3-mL syringe with a 21- or 23-gauge needle or a butterfly apparatus attached to an evacuated-tube system.[33] Whenever possible, the phlebotomist will want to choose 2- or 3-mL collection tubes for pediatric testing and thus reduce the volume of blood required while ensuring the correct additive dilution. The use of these smaller tubes also introduces a smaller amount of vacuum pressure to the vein.

Only superficial veins should be used, avoiding those needed for IV therapy. A child's arm should be well secured prior to venipuncture to avoid the possibility of injury. Following the venipuncture procedure, it is recommended that bandages not be used on children under 3 years of age because of the risk of accidental choking.

Special consideration should be given to the smaller blood volume of the pediatric patient. A newborn baby, for example, has approximately 80 to 110 mL of blood per kilogram of body weight; a young child has 75 to 100 mL/kg.[44] Many sites, especially intensive care nurseries, make it a practice to document each blood collection, including the amount of blood taken, the tests collected, the date and time, and the phlebotomist's initials. Phlebotomists should question requests to obtain large amounts of blood from small children or infants.

Notes and Precautions

Patients may experience fainting, nausea, or vomiting at any time during the venipuncture procedure. If the reaction should occur during collection, the phlebotomist should stop immediately. If fainting occurs when the patient is in a sitting position, the head should be lowered, cold compresses should be applied to the head and neck, and, if necessary, ammonia inhalant should be administered to revive the patient. A physician should be notified immediately if the patient fails to regain consciousness. Table 3-5 lists potential sources of error in venipuncture.

SPECIMEN COLLECTION BY SKIN PUNCTURE

Unlike venipuncture, skin puncture is an open collection system. Blood is brought to the surface of the skin by applying pressure to the site. The sample is dripped into a capillary blood collector (instead of vacuum pressure pulling the sample into the tube). For this reason, it is essential that gloves always be worn during skin puncture.

TABLE 3-5. Sources of Error in Venipuncture

1. Improper identification of patient
2. Failure to check patient adherence to dietary restrictions
3. Use of improper equipment/tubes
4. Prolonged tourniquet application
5. Failure to allow site to dry after cleansing
6. Inserting needle with bevel down
7. Wrong insertion depth of needle
8. Venipuncture above an intravenous line
9. Venipuncture in an unacceptable area
10. Wrong order of tube draw
11. Incomplete tube filling
12. Vigorous shaking/hemolysis of specimens
13. Neglecting to release tourniquet before needle withdrawal
14. Mislabeling of tubes
15. Failure to put initials, date, and time on requisition
16. Neglecting to chill specimens requiring refrigeration
17. Slow transport of specimens to laboratory

Equipment and Supplies

The following supplies are most often used for blood collection by skin puncture:

Sterile alcohol pads.
Sterile 2 × 2 inch gauze pads.
Capillary collection tubes.
Lancet.
Heelwarming device.
Tape or bandage.
Gloves.

Alcohol, Sterile Gauze, Tape, and Bandages

A sterile pad containing 70% isopropyl alcohol is required to decontaminate the skin-puncture site. Patients suffering from severe neutropenia may need a bactericidal such as povidone-iodine for sterilization of the site, followed by cleansing with an alcohol swab to remove the iodine residue. Sterile gauze pads are used to wipe away any trace of alcohol remaining on the skin surface, since alcohol hemolyzes blood. The phlebotomist should take care to use aseptic technique during this process, handling all sterile items carefully and making certain not to contaminate any items before they are used.

Capillary Collection Tubes

Microcollection tubes, like evacuated tubes, come with color-coded caps indicating the presence of specific additives. Instead of containing liquid additives, as is often the case with evacuated tubes, microcollection tubes contain a powdered or freeze-dried additive. The red caps indicate a plain tube with no additive. Tubes with green caps contain heparin for plasma studies. These tubes are available in both sodium and lithium heparin. In order to avoid falsely increased serum sodium levels, lithium heparin is the additive of choice when collecting for electrolyte monitoring. The lavender tubes contain EDTA for whole-blood or plasma testing. An additional feature of the microcollection tubes is the availability of silicon separator gel in the red- and green-capped tubes. This gel has a specific gravity intermediate to that of the liquid portion and cellular components of blood. Centrifugation pushes the cells down below the gel while allowing the serum or plasma to remain on top, thereby making processing these small tubes an easier task. Although design may vary slightly among manufacturers, the tubes come with an open collector cap to facilitate collection. After collection, this cap can be replaced with the color-coded one to seal the tube for transport. Figure 3-7 illustrates some of the available microcollection devices along with other supplies necessary for skin puncture. Additional information on skin-puncture devices can be found in the NCCLS publication H14–A: *Devices for Collection of Skin Puncture Specimens.*[36]

Lancets

It is imperative to use only those lancets designed for skin puncture. Other devices, such as surgical blades, can cause lacerations and can strike bone by puncturing too deeply. There are a number of skin-puncture devices designed to allow a controlled depth of puncture. This is of particular importance when collecting from infants and children who lack the fatty tissue necessary to prevent striking bone. Studies on neonates have shown that puncturing devices greater than 2.4 mm long can cause bone damage, resulting in abscesses or osteomyelitis, the most common complication of skin puncture.[5,6,25] Lancets are widely available in a variety of lengths for controlled puncture depth. Becton Dickinson

manufactures a lancet with a retractable blade in two sizes, known as the Microtainer Brand safety lancet, that not only provides a controlled depth but also provides protection against accidental sticks.

Heelwarmers

Heelwarming devices are commercially available or can be made by moistening a cloth or diaper with warm tap water. The temperature of either device should be no higher than 42°C to avoid the risk of burning the patient.[34] The selected site should be wrapped with the warm pack for 3 to 5 minutes prior to puncture. Warming the site increases blood flow sevenfold and can help avoid the need for multiple sticks and decrease the incidence of hemolyzed specimens. Microwave ovens should never be used to warm moistened diapers or cloths.

Gloves

Latex exam gloves are sufficient for blood collection. The collector should note that gloves must be changed after each patient. Washing gloves can cause wicking, or tiny holes to appear in the latex, leaving a pathway for potential contamination.

Determining the Site

Figures 3-8 and 3-9 show the acceptable areas for skin puncture on the finger and heel. Infants and toddlers less than 15 months should not have blood collected by fingerstick. Generally, tissue depth on the fingers of children at this age is not sufficient for safe puncture without the risk of striking bone. If the heel is to be punctured, then the stick should be made

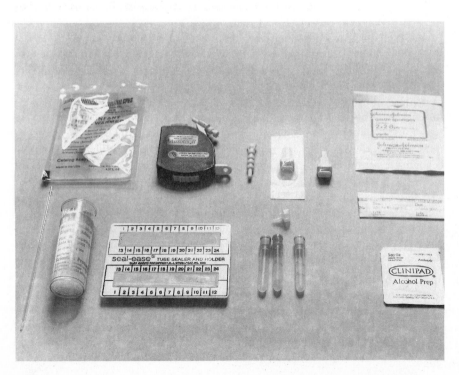

Figure 3-7. Skin puncture supplies. (*Top left*) Heel warmer, automatic lancets—Autolet™, lancet for autolet, Microtainer® Safety Lancets. (*Bottom*) Natelson™ heparinized capillary tube and micro capillary tubes, sealing clay for tubes, plain and anticoagulated Microtainers™ with Flo Top® Collector (above tubes). Also shown, gauze sponge, adhesive bandage, and alcohol prep pad. (Photograph by Steven R. Conlon)

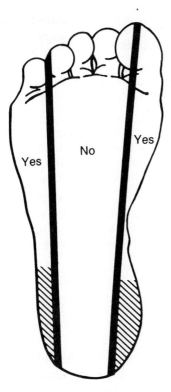

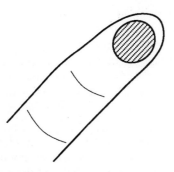

Figure 3-8. Acceptable areas for heel-skin punctures on newborns. (Blumenfeld TA, Turi GK, Blanc WA. Recommended site and depth of newborn heel skin punctures based on anatomical measurements and histopathology. Lancet 1979, 1:230–233.)

on the plantar surface medial to a line drawn from the middle of the large toe to the heel or lateral to a line starting between the fourth and fifth toes and drawn to the heel, as shown in Figures 3-8 and 3-9.[5,6,34] This area does not include the arch or middle area of the foot, where tendons and nerves are located. Fingers should be punctured on the fleshy portion of the palmar surface of the middle of the last phalanx, preferably using the third or fourth (ring) finger. The index finger may not be an optimal site because it is frequently used for touch, and the fifth or little finger offers decreased skin surface to bone depth. Skin puncture must not be made into previously punctured sites, since the possibility exists of pushing infection deeper into the tissue and thus

Figure 3-9. Acceptable area for finger-skin puncture.

causing serious abscesses. Bruised or edematous sites are also not appropriate for puncture because the presence of hemolyzed blood or tissue fluid can cause specimen contamination. Sites where IV infiltration or open sores are present should never be used for skin puncture.

Skin-Puncture Procedure

1. Identify the patient, exactly matching information on the hospital bracelet with the test requisition.
2. Select and prepare the supplies needed according to the test requirements.
3. Assess the sites available for skin puncture using the criteria described earlier.
4. Warm the site for 3 to 5 minutes.
5. Position the patient to maximize comfort and control during the procedure. At this time the phlebotomist should determine whether assistance may be needed from a parent or a nurse for pediatric patients. When performing fingersticks on children, the arm should be extended with the palmar surface of the fingers turned upward, using a steady surface for support during the puncture. This reduces the possibility of the patient jerking away during the puncture resulting in a laceration.
6. Cleanse the site with 70% isopropyl alcohol, taking care not to contaminate the alcohol pad prior to cleansing. Dry the site with a sterile 2 X 2 inch cotton gauze pad using sterile technique. This step removes any alcohol residue left on the skin that may contaminate the sample.
7. Grasp the heel or finger firmly but gently, and make the puncture in a smooth motion perpendicular to the skin surface. Puncture across the lines of the finger or heel to promote beading of blood into drops.
8. Firmly and slowly apply pressure toward the site. Do not "milk" or use excessive squeezing because this will hemolyze the specimen and also can cause blisters on sensitive skin.
9. Wipe away the first drop of blood because this may contain excess tissue fluid.
10. Position the site downward and touch the collector cap to the next drop of blood. The blood will flow into the tube by means of capillary action. A properly performed puncture can produce up to 0.5 mL of blood. If blood flow decreases, the site can be firmly wiped with gauze to remove platelet aggregation. If blood is being collected for pH or blood gas analysis, heparinized capillary tubes should be used, and air bubbles absolutely must be avoided. Air bubbles can cause erroneous pH and blood gas values after only a few seconds of air contact.[24,27]
11. Seal the collection tube with the accompanying caps for chemistry samples or with sealing clay for capillary pipets. Small magnetic beads can be placed in the capillary tube for mixing if blood gas or chemistry analysis will be delayed.

12. When collection is complete, apply pressure to the site with a clean gauze pad until bleeding stops. Newborn heels should be elevated above the body to discourage prolonged bleeding. Bandages may be made for heels by folding a gauze square lengthwise and securing with tape. Tape should never be placed directly against an infant's skin, since later removal of the tape also can remove or damage the skin. Regular or special fingertip bandages can be used following fingersticks. It is advisable, however, not to place bandages on toddlers less than 3 years of age to avoid accidental choking.
13. Dispose of contaminated needles and supplies according to hospital procedure.
14. Label tubes with patient's name, hospital number, date, time, and phlebotomist's initials, and indicate skin-puncture collection.
15. Package samples according to test specifications.
16. Record date and time of collection and amount of blood taken in the patient log if appropriate. Remove any dietary or drug restrictions.
17. Transport the specimens properly and promptly to the laboratory.

Notes and Precautions

1. A free-flowing puncture is critical to obtaining accurate test results. Do not use excessive squeezing because this will result in hemolysis and bruising.
2. Do not stick more than twice to obtain a specimen. If difficulty is encountered adhering to these limits, consult a supervisor or physician for follow-up.
3. Do not puncture in a previously punctured site because this greatly increases the risk of infection.
4. Use only lancets measuring less than 2.4 mm in depth.
5. Do not apply tape directly to an infant's skin; instead, wrap gauze around the ankle and secure with tape.
6. Do not use bandages that can be removed from the fingers or arms of toddlers and cause accidental choking.

Table 3-6 lists the potential sources of error in skin puncture.

QUALITY ASSURANCE IN SPECIMEN COLLECTION

High-quality sample collection is a crucial element in the process of patient diagnosis and treatment, and it should be monitored regularly in several areas. These areas include: (1) equipment and supply inspection, (2) procedure review, (3) procedure execution, including patient preparation, (4) specimen handling, and (5) specimen quality.

Blood-collection tubes have a limited shelf life because of additives, sterilization, and/or evacuation. Therefore, the phlebotomist should routinely check expiration dates on all supplies.[17] Additionally, new lot numbers of tubes should be

TABLE 3-6. Sources of Error in Skin Puncture
1. Misidentification of patient
2. Puncturing wrong area of infant heel (see Fig. 3-8)
3. Puncturing bone in infant heel
4. Puncturing fingers of infants
5. Puncturing wrong area of adult finger
6. Contaminating specimen with alcohol or Betadine
7. Failure to discard first blood drop
8. Excessive massaging of puncture site
9. Collecting air bubbles in pH or blood gas specimen
10. Hemolyzing specimen
11. Failure to seal specimens adequately
12. Failure to chill specimens requiring refrigeration
13. Erroneous specimen labeling
14. Failure to note skin puncture collection
15. Failure to warm site
16. Delaying specimen transport
17. Bruising site as a result of excessive squeezing

examined and verified using a standard procedure. This inspection should include an examination of the tubes for breakage and a visual inspection of the additives.

A regular review of collection procedures by the phlebotomy staff will improve the quality of collection by continually reinforcing correct methods. Procedure manuals also should be reviewed periodically by the supervisory staff for necessary updates and additions.

Each patient care area should possess a booklet containing all requirements for laboratory specimens (*e.g.,* size and type of container, known interferences, and patient preparation instructions). This will enable the patient care staff and the phlebotomist to work together as a quality team to ensure that the patient is properly prepared for collection. Laboratory supervisory staff should establish a system for evaluating all phlebotomists at regular intervals. Particular attention should be given to proper patient preparation, identification, and procedure execution.

Phlebotomists also should have access to a booklet containing specimen collection and handling requirements. This booklet should include requirements for minimum volume, tube color and size, and any special handling needs (*e.g.,* keep sample on ice, protect from light, and so on).

Pediatric units should maintain logs on each patient to record blood-collection times and the amount of blood taken. This is especially important for infants, whose small blood volume must be monitored carefully to avoid excessive loss of blood.

A phlebotomy problem log or record should be maintained by the laboratory to document and track uncollected specimens and unacceptable specimens. The log should be reviewed periodically to identify errors in the blood-collection process. Tables 3-5 and 3-6 list common errors made during blood collection that adversely affect the quality of the patient sample.

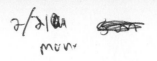

SPECIAL COLLECTION PROCEDURES

Fasting Specimens

If chemical analysis is desired on a fasting specimen, it is the technologist's responsibility to ascertain whether the patient has indeed been fasting for the required amount of time and has not ingested any restricted solids or fluids. The dietary effects on chemical-analyte values make it advisable to ask patients to fast for 12 hours before specimen collection. Although data collected on healthy subjects indicate that many chemical analytes are not affected when specimens are drawn 30 and 120 minutes postprandial, fasting may still be recommended.[1] Patient fasting may be indicated because patient food intake and metabolism may very well be abnormal and uncontrollable. Diets that are high in protein, saturated and unsaturated fatty acids, caffeine, serotonins, and many other substances have been reported to cause abnormal chemistry values.[22]

Timed-Interval Specimens

Timed specimens may be desired to observe certain metabolic processes. This is true for glucose tolerance tests and in some measurements of serum iron. Therapeutic-drug monitoring also may require this type of specimen. It is the laboratorian's responsibility to obtain the specimen at the required time and to note the time of collection on the test requisition or enter this information into the computer system.

Collection of Specimens for Blood Alcohol Level

When a specimen is being collected to measure blood alcohol level, alcohol must not be used to cleanse the site of venipuncture. Instead, soap and water may be used, and the site should be thoroughly dried before venipuncture. If this test is requested for legal reasons, the procedure is more involved and is often considered a forensic specimen collection.

Forensic Specimen Considerations

Forensic medicine refers to medical knowledge or results that apply to questions of law affecting life or property. A specimen result might be used as evidence in a court of law to prove cause of death, innocence or guilt of an accused individual, or possibly to prove alcohol or drug abuse. Specimens must be handled with great care and strict adherence to policies and procedures so that the results may serve as legal evidence. Any deviation from policy could lead to the dismissal of valuable evidence because of suspicions of tampering with or switching of the specimen.

Usually, a physician is required to draw forensic specimens in the presence of a law enforcement officer, who places the specimen in a special container and delivers it to the person who is to perform the analysis.

Forensic specimens require proper collection, appropriate containers, and the phlebotomist to seal, sign, date, and note the time of collection. The clinical laboratorian performing the assay must sign the container seal and note the hour and date received on the seal, which must be intact.[3] These steps are necessary to ensure knowledge of the specimen's whereabouts at all times so that results may be legally valid.

SPECIMEN TRANSPORT

As soon as the specimens are collected, they should be transported to the laboratory in a timely manner. All specimens should be in primary leak-proof containers placed in a secondary leak-proof container during transport to the laboratory. To promote clot formation and reduce the potential for hemolysis, specimens should be maintained in an upright position and agitated as little as possible during transport. The ambient temperature is also of concern during specimen transport, because analytes react differently to temperature changes. Maintaining temperatures in the 22 to 25°C range is preferable.[37,39] Ideally, serum/plasma should be separated from the blood cells within 2 hours of collection; therefore, specimens should be transported to the performing laboratory within this period of time.[37] Other fluid specimens also should be transported to the laboratory immediately upon collection to maintain their integrity. Collection and transport personnel should be aware of all special collection and handling requirements for each test.

Some specimens may require chilling after collection to inhibit metabolism by the blood cells or to stabilize labile analytes. Chilling should not be done unless there is a documented need to do so, because chilling may harm some analytes. Proper chilling is accomplished by placing the specimen in crushed ice or a mixture of ice and water. Be sure there is good contact between the specimen and the cooling medium to ensure rapid and even cooling.[37]

Certain tests require the specimen be protected from light to ensure recovery of the analyte (*e.g.,* bilirubin). Specially colored collection tubes or secondary containers should be provided to ensure this protection.

Several systems are available to transport samples. Mechanical distribution systems include cart and track systems, pneumatic tube systems, conveyor belts, and dumb waiters. Messenger services are also effective in transporting specimens. Pneumatic tube systems are becoming the most popular method of specimen transport, although there are concerns regarding specimen integrity. Concerns focus on the loss of integrity of red cell membrane. Tests that can be affected are the enzyme lactate dehydrogenase (LD), potassium, plasma hemoglobin, and acid phosphate determinations.[37] Similar concerns can be raised for any mechanical transportation system in which agitation of the blood tubes is a possibility. Each laboratory should evaluate automated transportation systems for their effects on the validity of test results, unless specific documentation already exists. When choosing transportation systems, each hospital should evaluate systems based on individual needs and cost-effectiveness.

Specimen transport concerns for specimens collected offsite are the same as those for specimens collected in the hospital setting. Blood specimens that are not received in the performing laboratory within 2 hours of collection should be centrifuged and the serum should be separated from the cells while awaiting transport. Couriers should be trained in proper specimen transport techniques. Care should be taken not to expose the specimens to temperature extremes, whether excessive heat (summer transportation) or excessive cold (winter transportation).[37] For specific details, see NCCLS publication H5-A3: *Procedures for the Handling and Transport of Domestic Diagnostic Specimens and Etiologic Agents: Approved Standard*.[35]

SPECIMEN PROCESSING AND RECORD KEEPING

Universal Precautions

It has been recommended by the NCCLS and the CDC that laboratory workers follow universal precaution procedures. These procedures are outlined in the NCCLS guideline, *Protection of Laboratory Workers from Infectious Disease Transmitted by Blood, Body Fluids, and Tissue*.[38] Precautions should be taken for all patients and for all specimens submitted to the laboratory. The potential for infectivity of any patient's blood and body fluids cannot be known with absolute certainty even when warning labels are used; therefore, treating all samples as possibly infective should be the rule.

According to universal precaution procedures, all laboratory workers should wear lab coats or aprons at all times to protect their clothing from contamination. Gloves should be worn when handling all specimens, and facial protection should be used when the possibility of specimen splashing exists. Gloves should be changed when contaminated, and hands should be washed after gloves are removed. A 1:10 dilution of bleach should be available to wipe lab surfaces routinely and to be used to clean up after spills. Special care also should be taken with biohazard waste. Infectious waste should be disposed of in special "red bags." Sharps should be disposed of in rigid containers. State and local ordinances should be followed for the proper storage, transport, and disposal of biohazard waste.

Specimen Receiving

Upon arrival of a specimen in the laboratory, the information on the specimen label (patient name, hospital number, date, and time of collection) should be matched with the same information on the specimen requisition. In the event that the information is incomplete or incorrect, verification should be obtained from the phlebotomist or nurse involved in taking the specimen.

After positive specimen identification has been established, the specimen acceptability should be determined. Each procedure performed in the laboratory should have established causes for rejection. These criteria and any special handling instructions should be available to the processing area, and specimens should be screened at this time. Table 3-7 lists some of the common criteria that would make a sample unacceptable.

Specimen Accession

Once accepted, specimens should be sorted as to type and, if necessary, distributed to various areas of the laboratory. Specimens also must be assigned accession numbers if this has not previously been performed by a laboratory computer system. When manually assigning accession numbers, care should be taken to ensure that each patient is given a unique accession number. A specimen log containing patient name, identification number, laboratory accession number, tests ordered, and any other pertinent information must be kept by the laboratory.

Specimen Processing

Whole-blood specimens are commonly used for blood gas analysis and occasionally for electrolyte, glucose, and lactic acid determinations. Blood gas specimens should be in a syringe that has been tightly sealed with caps specifically made for this purpose, because the sample must remain anaerobic. There should be no air bubbles in the specimen, since this would cause erroneous results. These collection requirements are extremely important, because significant amounts of CO_2 may be lost if the sample is exposed to air.[2,29]

Analysis of whole-blood specimens for blood gas and electrolytes should be performed as quickly as possible because there is usually no metabolic inhibitor present. The ice slurry necessary to maintain the integrity of the blood gases will begin to cause leakage of potassium from the red blood cells at a faster rate than at room temperature. Chilling beyond 2 hours is not recommended.[37] Reference ranges for tests performed on whole blood will vary slightly from those

TABLE 3-7. Common Criteria for Specimen Rejection
Inadequate specimen identification
Inadequate volume of blood collected in an additive tube
Inappropriate collection tube
Hemolysis
Improper transportation

used for plasma and serum, mostly owing to the dilutional effect of the red blood cells.

For tests requiring plasma or serum, the whole-blood specimens should be centrifuged to separate the serum/plasma from the red blood cells. NCCLS guidelines state that blood specimens should be centrifuged within 2 hours after collection and serum/plasma removed from contact with the red blood cells.[37] Serum specimens should be allowed to clot in an upright position for 30 to 60 minutes prior to centrifugation. Add time if the specimen is chilled.[37] If the sample is not given sufficient time to clot, latent fibrin formation may occur, which may cause problems during the analysis phase. The clotting time can be shortened by the addition of clotting agents such as thrombin or glass beads. Anticoagulated specimens can be centrifuged immediately upon arrival in the laboratory.

Specimens should be centrifuged for about 10 minutes at a relative centrifugal force (RCF) of 1000 to 1200 $\times$ g.[37] When using tubes with gel separators or other serum-separating devices, refer to the manufacturers guidelines for centrifugation speeds and time. Collection-tube stoppers should not be removed prior to centrifugation. Properly capped specimens eliminate changes in analyte concentration due to evaporation or pH changes and also eliminate aerosol formation during centrifugation. It is no longer necessary to open collection tubes to release clots prior to centrifugation, because technical improvements in collection tubes have eliminated clot hang-up.[37] This procedure also should be avoided because it increases the risk of hemolyzing the sample.

For most laboratory tests, maintaining centrifuge temperatures at 20 to 25°C should be adequate. Refrigerated centrifuges should be used only for those analytes that require them.

After centrifugation, care should be taken when removing the stopper from blood-collection tubes. Laboratory personnel should use a gauze pad when removing stoppers or snap caps to protect themselves from splashing. Alternatively, a face shield can be used. As a general rule, serum or plasma should be removed, by pipetting, from the collection tube within 2 hours of collection. The serum or plasma should be visibly free from red blood cells. Specimens collected in tubes with gel separators may remain in contact with the gel barrier for up to 48 hours. If a nongel device is used, remove the serum within 1 hour of centrifugation.[37] In general, a specimen should not be centrifuged more than once, because excessive handling will increase the potential for hemolysis.

Once the serum or plasma has been harvested, the specimens should be reviewed again to determine acceptability. Interfering substances or processes, undetectable at the time of sample accession, can now be dealt with. The most common problems with blood specimens are hemolysis, icterus, and lactescence. Caution should be taken when interpreting the amount of these interferences in a specimen. Studies have shown that visual estimation shows little agreement with actual interferent concentration.[19] Although no other practical

methods currently exist for quantitating interfering substances or processes, laboratories should be aware of the limitations.

Lactescent specimens, while normally unacceptable for many chemistry procedures, may be the only type available from patients with such conditions as diabetes mellitus, chronic renal failure, colon or liver carcinomas,[48] and familial hyperlipidemia, or occasionally from patients on intralipid therapy. Lactescent or lipemic specimens may be identified by a turbid appearance in the serum, and this should be noted on the test requisition. Ultracentrifugation at 10^5 g has been reported to remove most chylomicrons, including triglycerides, which, if not removed, may cause electrolyte-exclusion errors in certain methodologies.[48] Dialysis, dilution, and other specimen processing and testing methodologies also may be used to reduce analytical errors caused by lipemia.[17]

Icteric specimens also may be unacceptable because of the jaundiced or yellow color of the serum, which interferes with certain chemical analyses. Certain methodologies may have procedures to eliminate icteric interference.

Hemolysis can invalidate results for potassium, magnesium, phosphorus, lactate dehydrogenase, and acid phosphatase because the concentration of these analytes is higher in red cells than in plasma. The protein content of hemoglobin also will increase total protein and albumin concentrations, and the iron content of hemoglobin will increase serum iron concentrations. Other analytes affected by hemolysis are bilirubin, cholesterol, chloride, haptoglobin, most enzymes, and lipids.[23,24]

Processing with Serum/Plasma Separation Devices

The use of serum/plasma separators provides distinct advantages during sample processing and analysis and, therefore, has become very popular in the chemistry laboratory today. There are two major types of separation devices: those added prior to centrifugation and those added after. Of the former type, blood-collection tubes containing gel separators are probably the best and most convenient. The use of these tubes provides the following advantages: the tube does not need to be opened prior to centrifugation, processing time is shortened, aerosols are avoided, and the closed system prevents evaporation and creates a safe barrier between serum and cells.[11,37] Various studies have shown these gels to be inert, thus having no effect on the specific chemistries performed.[11,14,41,46] Unless documentation exists, however, each laboratory should verify the appropriate use of gel separators. Gel separators available for plasma, when appropriate, show the distinct advantages listed above and provide fibrin-free specimens. It has been reported that serum may be stored on these gel barriers for at least 48 hours.[28,37] These tube types are ideal for use with the new direct tube sampling analyzers.

Serum-separator materials other than the gel barrier may be added to the tube after collection but before centrifugation. Such materials include beads, crystals, or fiber. Upon centrifugation, these substances form a barrier between the serum or plasma and the cells. However, these devices are more permeable than gel barriers. It is recommended that the serum or plasma be removed within 1 hour of centrifugation. If serum is to remain longer, the laboratory should conduct appropriate studies to evaluate the barrier system for its effectiveness in preventing interaction between the serum and the cells.[37,40]

Postcentrifugation separator devices are usually plunger-type devices that are placed into the serum or plasma. The serum or plasma will flow through the filter into a protected tube. The filter should not be allowed to touch the cells. The device is placed into the sample such that an air gap is created. This gap is needed to avoid leakage of cellular constituents into the serum or plasma.[37]

Blood Specimen Storage and Stability

Each laboratory should determine the exact handling and storage requirements for each analyte tested. As a general rule, serum/plasma should be separated from the cells within 2 hours of collection. If analysis is not carried out within 5 hours after separation, storage at 2 to 8°C for up to 24 hours is acceptable for most analytes. Storage beyond 24 hours may require freezing (−20°C). Caution should be taken not to repeatedly freeze and thaw serum or plasma. Thaw only once.[37] There are exceptions to these processes; some analytes should be refrigerated or frozen immediately upon separation, and some (*e.g.,* LD) should be stored at room temperature.

With the advent of direct tube sampling and the common use of serum/plasma gel separators, the questions of sample stability are continually being reexamined. Use of serum-separating devices is highly recommended, with the use of gel separators becoming widely accepted, since specimens can be collected and processed with minimal handling. Studies on the stability of the serum/plasma left in gel separator tubes indicate that most analytes exhibit minimal clinically significant changes when stored 8 to 24 hours at room temperature. The exceptions are lactate dehydrogenase, glucose, bicarbonate, and potassium, which are also unstable under normal conditions.[11,14,40] Each laboratory should evaluate the system selected based on current analytical methods used within the laboratory and conduct appropriate studies if no documentation exists.

Specimens collected in sodium fluoride, a glycolytic inhibitor, will keep glucose stable for 24 hours at room temperature or 48 hours at refrigerated temperatures.[37] A recent study has shown, however, that the antiglycolytic properties of sodium fluoride take effect slowly, and during the first hour after collection, the decrease in glucose is at the same rate as in specimens collected without additives. Between 1 and 4 hours after collection, the glucose utilization rate decreases when compared with that of untreated specimens, with the antiglycolytic effect of sodium fluoride becoming totally effective at about 4 hours.[12] Therefore, for samples that will be processed within 1 hour of collection, additives to preserve glucose may not be necessary. This also increases the importance of separating serum and plasma quickly from newborn and pediatric specimens, in which blood cell glycolysis proceeds at an accelerated rate.

COLLECTION AND HANDLING OF OTHER BODY FLUIDS

Nonblood body fluids commonly analyzed in the laboratory are usually collected by the medical staff; however, proper transport and handling are important to maintain sample integrity. All specimens should be in leak-proof containers that are transported to the laboratory within a secondary leak-proof container.

Cerebral Spinal Fluid (CSF)

CSF is a selective ultrafiltrate of the plasma that surrounds the brain and spinal cord. Specimens of CSF are normally obtained by lumbar puncture. The appearance of normal CSF is clear and colorless. Any deviation from this appearance can be diagnostic. Laboratory analysis of the CSF is indicated in cases of suspected diseases that may involve the meninges or changes in the blood-brain barrier.

The most commonly requested chemistry tests are glucose and total and specific protein determinations. For proper evaluation of the results, a venous blood sample should be drawn at the same time. If possible, the patient should be fasting for 4 hours to allow equilibration of the glucose levels between the blood and CSF, since diagnostic decisions may be made based on the differences between the venous and CSF glucose levels.[42] Samples should be centrifuged and analyzed immediately upon receipt to minimize the effects of the metabolism of glucose by white cells or microorganisms.

If pH, PCO_2, or bicarbonate are to be measured, collect the sample in a heparinized glass syringe and keep the sample anaerobic and at 0°C. Analyze the sample within 20 minutes of collection. An arterial blood gas sample should be drawn simultaneously for comparison.[42]

Paracentesis Fluids

The pleural, pericardial, and peritoneal cavities contain small amounts of sterile, acellular fluids that are ultrafiltrates of plasma. Abnormal accumulations of these fluids are called *effusions.* Laboratory analysis of pericardial, pleural, and peritoneal fluids can help determine the cause of the accumulation. Physicians collect the samples by a procedure called *paracentesis* (aspiration of the fluid through the skin).

The most commonly ordered chemistry tests on paracentesis fluids are specific gravity, protein, and lactate dehydrogenase determinations. A simultaneously drawn venous blood sample should accompany the fluid to the laboratory to serve as a reference for the fluid results. Fluid samples should arrive in the laboratory within 1 hour of being obtained. If any delay is anticipated in the transport or analysis, the sample should be refrigerated. Samples should be allowed to clot and then should be centrifuged prior to analysis.[42]

Amniotic Fluid

Amniotic fluid bathes the fetus within the amniotic sac. Aspiration of this fluid by a process called *amniocentesis* is performed by the physician.

The most commonly ordered chemistry tests on amniotic fluid are performed to test for fetal lung maturity, congenital diseases, hemolytic diseases, neural tube defects, and gestational age. Upon collection, the specimen should be protected from light and transported to the laboratory immediately. Observation of the specimen for the presence of meconium, indicated by a green color, and for contamination with blood is necessary, because the presence of either of these will interfere with the bilirubin scan and the L/S ratio. The specimen should be centrifuged slowly before analysis for bilirubin or L/S ratio. Refer to the specific procedure to be used for centrifuge speed, because even slow speeds can remove phospholipid lamellar bodies, which contribute to the measurement of fetal lung maturity.[42]

Urine

Laboratory tests on urine can evaluate renal function, diagnose renal or urinary tract disease, and aid in the detection of metabolic or systemic disease. The method and timing of the specimen collection depend on the tests requested.

Random urine specimens

Random urine specimens are usually used for routine urinalysis (pH, glucose, protein, specific gravity, and osmolality). The first morning void is often requested because it is the most concentrated, but usually collection time and volume are not important.

The collection container should be chemically clean and hold approximately 50 mL. The lid should be tight fitting to prevent leakage. If the specimen is to be analyzed for bilirubin or porphyrin, a dark-colored plastic container should be used to protect the specimen from the light. The label should be on the container, not the lid, and it should adhere even when the sample is refrigerated and should not smear when wet.[31]

If transport to the laboratory is delayed, the specimen should be protected from direct light. If transport to the laboratory or analysis is delayed more than 2 hours, the specimen should be refrigerated. Refrigeration will slow changes in pH and chemical constituents such as true glucose, urobilinogen, and bilirubin, and will prevent the autolysis of formed elements.[31] For additional information refer to NCCLS guideline, *Collection and Transportation of Single-Collection Urine Specimens,* for procedures relating to random urine specimens.[31]

Timed Urine Collections

Timed urine collections may refer to specimens collected at specific intervals, such as before and after meals, or specimens to be collected over specific periods of time. Discrete samples collected over a period of time are used for tolerance tests (*e.g.,* glucose). The most common type of timed test is the 12- or 24-hour "pooled" urine collection. The 24-hour urine collection yields the total excretion of the analytes in question over a 24-hour period. It is the most effective because many solutes exhibit diurnal variations. For example, the lowest concentrations of electrolytes, catecholamines, and 17-hydroxysteroids occur in the morning, with the peak concentrations occurring around noon.[32]

Patient education and cooperation are necessary in order to obtain a complete collection of the timed urine specimen. It is most important that the entire amount of urine voided during the interval be obtained and that none is removed or spilled. In the hospital setting, it is also important to educate the patient care staff to ensure that the collection is complete and no aliquots are removed prior to completion of the collection. Refer to the NCCLS guideline, *Collection and Preservation of Timed Urine Specimens,*[32] for details on timed urine collection procedures. Some analytes will require that a preservative be added to the container at the beginning of the collection. These preservatives help the analytes maintain their integrity during specimen collection. Acids or bases may be added to maintain proper pH. In cases where no preservative is needed, refrigeration is preferred to reduce bacterial growth. Other analytes require no special treatment.

QUALITY CONTROL IN SPECIMEN PROCESSING

Just like any other area of the laboratory, the sample processing and handling area should be monitored for quality of work. All equipment used should be properly maintained. Centrifuges should have their timers, rotor speeds, and brushes checked at routine intervals. Temperatures of refrigerators and refrigerated centrifuges should be checked daily.

Each laboratory should evaluate its sample handling and processing procedures and look for specific areas where the potential for error is high. Once identified, ways should be found to eliminate these sources of error. The ideal sample processing area should have a steady work flow, with an appropriate and consistent time frame from time of specimen receipt to completion of processing. This time frame should be monitored for its consistency and appropriateness. Table 3-8 lists some of the common sources of errors in the processing and storage of blood specimens.

To monitor the sample processing area, laboratories should keep logs documenting sample collection, handling,

TABLE 3-8. Errors in Processing and Storage of Blood Specimens

1. Contamination of plasma/serum with RBCs
2. Leakage of RBC K^+ into serum (use serum-separator devices to avoid #1 and #2)
3. Centrifuging specimens requiring chilling at room temperature[14]
4. Exposure of affected analytes to air[14]
5. Repeated freezing, thawing, and centrifuging
6. Evaporation of serum H_2O because of
   a. ↑ Temperature
   b. ↑ Air flow
   c. ↑ Air exposure
   d. ↓ Humidity[12]
   e. ↑ Specimen surface-area air exposure

and processing errors. Logs should document specimen misidentification, breakage and spills necessitating recollects, sample mixups, inappropriate specimens, lost samples, and so on. These logs should be reviewed on a routine basis. When a recurring error is identified, the process involved should be examined to identify the cause of the error and then modified to eliminate it.

SUMMARY

Specimen collection, handling, and processing are an integral part of the process of specimen analysis.

The specimen collection process begins the chain of events in the testing process. Obtaining the best quality blood specimen begins with the appropriate equipment and a well-trained phlebotomist. Phlebotomist training should include proper phlebotomy procedures, interpersonal skills, and an understanding of the physiologic factors affecting blood specimens. The most common method used for obtaining blood specimens in adults is venipuncture, usually performed in the area of the arm known as the antecubital fossa. Special precautions should be taken when performing venipuncture on children. The special needs of children and elderly patients indicate the use of skin puncture for most routine blood specimen collection. In infants, the site of choice for skin puncture is the plantar surface of the heel; in older children and adults, the plantar surface on the finger tip of the third or fourth finger is used. Blood collection tubes come with color-coded stoppers indicating the additive present. It is important to to use the appropriate tube when collecting samples for specific tests, and to remember the correct order in which to draw when collecting multiple tubes. Special collection procedures should be followed for fasting specimens, timed interval specimens, specimens collected for blood alcohol levels, and for forensic specimens.

The second link in the chain is the process of specimen transport. All types of specimens should be transported to the laboratory as soon as possible after they are obtained to maintain their integrity. Specimens should be transported in primary leak-proof containers placed in secondary leak-proof containers. Collection and transport personnel should be aware of all special collection and handling requirements for each test.

Once specimens are received in the performing laboratory, they must be properly prepared for analysis. It is recommended that laboratory workers follow universal precaution procedures when handling blood and body fluid specimens. Upon arrival, the specimen identification and acceptability for analysis should be established. For tests requiring serum or plasma, the whole blood specimens should be centrifuged to separate the serum/plasma from the red cells. NCCLS guidelines state that serum/plasma should be separated from the cells within 2 hours of collection. Once centrifuged, the specimen should be revaluated for acceptability. The presence of interfering substances such as lactescence, hemolysis, or icterus should be evaluated based on the specific method of analysis. Each laboratory should determine the exact handling and storage requirements for each analyte tested. As a general rule, serum/plasma can be stored at 2 to 8°C for up to 24 hours, if not analyzed within 5 hours. For specimens collected using gel plasma/serum separated, specimens are stable for at least 48 hours after centrifugation. With minimal handling, these specimens are stable for 8 to 24 hours at room temperature.

Nonblood body fluids commonly analyzed in the laboratory are usually collected by the medical staff; however, proper transport and handling are important to maintain sample integrity. CSF, paracentesis fluid, amniotic fluid, and urine are typically sent to the laboratory for analysis and have individual processing and handling requirements. Timed urine collections require patient education and cooperation to obtain a complete collection. Some analytes will require that a preservative be added to the urine container at the beginning of the collection to maintain specimen integrity.

Quality assurance in the preanalytical phase involves the monitoring of equipment and supplies as well as processes. It is important that the processes of specimen collection, transportation, and processing be monitored constantly for process errors and specimen delays. When a recurring error is identified, the process involved should be examined to identify the cause of the error and then modified to eliminate it.

EXERCISES

1. State the primary sites for venous collection.
2. State the primary sites for skin puncture collection.
3. Identify the supplies and equipment required for a venous collection on an adult patient.
4. List the important factors for proper specimen transport.
5. List the steps of a venous collection.
6. List the steps of a skin puncture collection.

REFERENCES

1. Annino JS, Relman AS. Effect of eating on some of the clinically important chemical constituents of the blood. Am J Clin Pathol 1959;31:155.

2. Bandi ZL. Estimation, prevention, and quality control of carbon dioxide loss during aerobic sample processing. Clin Chem 1981;27:1676.

3. Baselt RC, Cravey RH. Forensic toxicology. In: Doull J, Klaassen CD, Amdur MO, eds. Casarett and Doull's toxicology: The basic science of poisons. 2nd ed. New York: Macmillan, 1980;661.

4. Ben-Sira Z. The function of the professional's affective behavior in client satisfaction: A revised approach to social interaction theory. Health Soc Behav 1976;17:3.

5. Blumenfeld TA. Blood collection by skin puncture in infants, Vol 2. Rutherford NJ: Becton Dickson, 1988;1.

6. Blumenfeld TA, Turi GK, Blanc WA. Recommended site and depth of newborn heel skin punctures based on anatomical measurements and histopathology. Lancet 979;1:230.

7. Blumenfeld TA. Infant blood specimen collection and chemical analysis. Diagn Med 1978;1:58.

8. Blumenfeld TA, Hertelendy WG, Ford SH. Simultaneously obtained skin puncture serum, skin puncture plasma and venous serum compared and effects of warming the skin before puncture. Clin Chem 1977;23:1705.

9. Calam RR, Cooper MH. Recommended "Order of Draw" for collecting blood specimens into additive containing tubes. Clin Chem 1982;28:1399.

10. Centers for Disease Control. Update: Universal precautions for prevention and transmission of human immunodeficiency virus, hepatitis B virus, and other blood borne pathogens in health-care settings. MMWR 1988;37:377.

11. Chan KM, Daft M, Koenig JW, Ladenson JH. Plasma separator tube of Becton Dickinson evaluated. Clin Chem 1988;34:2158.

12. Chan AYW, Swaminathan R, Cockran DS. Effectiveness of sodium fluoride as a preservative of glucose in blood. Clin Chem 1989;35:315.

13. Creer M, Ladenson J. Analytical errors due to lipemia. Lab Med 1983;14:351.

14. Doumas BT, Hause LL, Simuncak DM, Breitenfeld D. Differences between values for plasma and serum in tests performed in the Ektachem 700 XR analyzer and evaluation of "plasma separator tubes (PST)." Clin Chem 1989;35:151.

15. Fawcett JK, Wynn V. Effects of posture on plasma values and some blood constituents. J Clin Pathol 1960;13:304.

16. Fody EP, Schoen I, Bennett B, Duby M, Pappas AA, Richardson LO. So you're going to collect a blood specimen: An introduction to phlebotomy. 4th ed. Northfield, IL: College of American Pathologists, 1989.

17. Garza D, Becan-McBride K. Phlebotomy handbook. 3rd ed. Norwalk, CT: Appleton & Lange, 1993;78.

18. Gerard BA, Boniface WJ, Love BH. Interpersonal skills for health professionals. Reston, VA: Reston Publishing, 1980;2.

19. Glick MR, Ryder KW, Glick SJ, Woods JR. Unreliable visual estimation of the incidence and amount of turbidity, hemolysis, and icterus in serum from hospitalized patients. Clin Chem 1989;35:837.

20. Goldin P, Russell B. Therapeutic communication. Am J Nurs 1969;69:1928.

21. Haymond RE, Knight JA. Venous serum, capillary serum, and capillary plasma compared for use in determination of lactate dehydrogenase and aspartate aminotransferase activities. Clin Chem 1975;21:896.

22. Henry JB. Clinical diagnosis and management by laboratory methods. 16th ed. Philadelphia: Saunders, 1979;5.

23. Kaplan LA, Pesce A. Clinical chemistry theory, analysis, and correlation. St Louis: Mosby, 1984;46.

24. Ladenson JH. Nonanalytical sources of laboratory error in pH and blood gas analysis. In: Durst RA, ed. Blood, pH, gases and electrolytes. Washington, DC: National Bureau of Standards Special Publication 450, 1976;175.

25. Lilien LD, Harris VJ, Ramamurthy RS, Pildes RS. Neonatal osteomyelitis of the calcaneus: Complication of heel puncture. J Pediatr 1976;88:478.

26. McKay RJ Jr. Diagnosis and treatment: Risks of obtaining samples of venous blood in infants. Pediatrics 1966;38:906.

27. Meites S, Monte JL, Blumenfeld TA, et al. Skin-puncture and blood collecting techniques for infants. Clin Chem 1979;25:183.

28. Narayanan S, et al. Control of blood collection and processing variables to obtain extended (greater than 5 day) stability in serum constituents. Clin Chem 1979;25:1086.

29. National Committee for Clinical Laboratory Standards. Evacuated tubes for blood specimen collection. Approved Standard H1-A3. Villanova, PA:NCCLS, 1991;11(9);5.

30. National Committee for Clinical Laboratory Standards. Blood gas preanalytical considerations: Specimen collection, calibration, and controls. Proposed Guideline C27-P. Villanova, PA: NCCLS, 1985.

31. National Committee for Clinical Laboratory Standards. Collection and transportation of single-collection urine specimens. Proposed Guideline GP8-P. Villanova, PA: NCCLS, 1984.

32. National Committee for Clinical Laboratory Standards. Collection and preservation of timed urine specimens. Proposed Guideline GP13-P. Villanova, PA: NCCLS, 1987.

33. National Committee for Clinical Laboratory Standards. Procedures for the collection of diagnostic blood specimens by venipuncture. Approved Standard H3-A3. Villanova, PA: NCCLS, 1991;11(10);1.

34. National Committee for Clinical Laboratory Standards. Procedures for the collection of diagnostic blood specimens by skin puncture. Approved Standard H4-A3. Villanova, PA: NCCLS, 1991;11(11);1.

35. National Committee for Clinical Laboratory Standards. Percutaneous collection of arterial blood for laboratory analysis. Approved Standard H11-A2. Villanova, PA: NCCLS, 1992;12(8);1.

36. National Committee for Clinical Laboratory Standards. Devices for collection of skin puncture specimens. Approved Guideline H14-A2. Villanova, PA: NCCLS, 1990:10(11).

37. National Committee for Clinical Laboratory Standards. Procedures for handling and processing of blood specimens. Approved Guideline H18-A. Villanova, PA: NCCLS, 1990:10(12).

38. National Committee for Clinical Laboratory Standards. Tentative standard M29-T2: Protection of laboratory workers

from infectious disease transmitted by blood, body fluids, and tissue. Villanova, PA: NCCLS, 1991.

39. Rehak NN, Chiang BT. Storage of whole blood: Effect of temperature on the measured concentration of analytes in serum. Clin Chem 1988;34:2111.

40. Seckinger DL, Vasquez DA, Rosenthal PK, Heller ZH. Evaluation of a new serum separator. Clin Chem 1982;28:157.

41. Sibilia R, Lohff M, Bush D, Mahaffey R, Joseph R. Analyte stability in closed containers after 6 days storage. Clin Chem 1989;35:1158.

42. Slockbower JM, Blumenfeld TA. Collection and handling of laboratory specimens: A practical guide. Philadelphia: Lippincott, 1983;12.

43. Smiley O, Smiley C. Interviewing techniques for nurses. Can J Public Health 1974;65:281.

44. Statland BE, Winkle P, Bokelund H. Factors contributing to intraindividual variation of serum constituents: 4. Effects of posture and tourniquet application on variation of serum constituents in healthy subjects. Clin Chem 1974;20:1513.

45. Valaske MJ, ed. So you're going to collect a blood specimen. Danville, IL: College of American Pathologists, 1986;1.

46. Vaughn R, Lewis LM. Updated evaluation of the Becton Dickinson serum separator tube. Clin Chem 1989;35:1159.

47. Ware J, Snyder M. Dimensions of patient attitudes regarding doctors and medical care services. Med Care 1975;13:669.

48. Winkelman JW, Wybenga DR. Lipoproteins. In: Henry RJ, Winkelman JW, eds. Clinical chemistry: Principles and techniques. Hagerstown, MD: Harper & Row, 1974;1496.

Chapter 4

Quality Control and Statistics

George S. Cembrowski, Anne M. Sullivan

Objectives

Upon completion of this chapter, the clinical laboratorian should be able to:

- *Define the following terms: quality assurance, quality control, control, standard, accuracy, precision, descriptive statistics, inferential statistics, reference interval, random error, systematic error, dispersion, delta check, and confidence intervals.*

- *Calculate the following: sensitivity, specificity, efficiency, predictive value, mean, median, range, variance, and standard deviation.*

- *Evaluate laboratory data using the multirule system for quality control.*

- *Given laboratory data, graph the data and determine significant constant or proportional errors.*

- *Describe the preanalytic and postanalytic phases of quality assurance.*

- *Given laboratory data determine if there is a trend or a shift.*

- *Discuss the role of clinical laboratorians in point-of-care testing.*

- *Describe the important features/requirements of point-of-care analyzers.*

- *Discuss the processes involved in method selection and evaluation.*

- *Discuss proficiency-testing programs in the clinical laboratory.*

KEY WORDS

Accuracy	Quality Assurance
Analytical Variations	Quality Control
Control	Random Error
Delta Check	Reference Interval
Descriptive Statistics	Reference Method
Dispersion	Shift
F-test	Standard
Histogram	Systematic Error
Inferential Statistics	T-test
Precision	Trend
Predictive Value Theory	

STATISTICAL CONCEPTS

Statistics may be defined as the science of gathering, analyzing, interpreting, and presenting data. The volume of data generated by the clinical chemistry laboratory is enormous and must be summarized to be maximally useful to laboratorian and clinician. The introduction of a new test illustrates the extensive use of statistics in the laboratory. First, the laboratorian should introduce the test only after review of the data that document the usefulness of the test for diagnosing or monitoring a disease state. If several methods are available for performing the test, the laboratorian should study the published evaluations and select the most practical as well as the optimally accurate and precise method. During in-house evaluation of the method, precision and accuracy

Michael L. Bishop, Janet L. Duben-Engelkirk, and Edward P. Fody.
CLINICAL CHEMISTRY. © 1996 Lippincott–Raven Publishers.

must be evaluated. If the method's performance is acceptable, reference interval (normal range) data must be accumulated either to verify the manufacturer's recommended interval or else to set up a laboratory-specific reference interval. Patient data can then be properly interpreted by the clinician. Finally, once the method is in use, accuracy and precision must be continually assessed to ensure reliable analyses. The following sections will review some of the statistical concepts that must be understood by the laboratorian. For more comprehensive material, the reader is referred to several introductory textbooks.[3,15,69]

Descriptive Statistics

Descriptive statistics are used to summarize the important features of a group of data. Another type of statistics, *inferential statistics*, is used to compare the features of two or more groups of data. The descriptive statistics in this chapter will be applied to groups of single observations as well as to groups of paired observations.

Descriptive Statistics of Groups of Single Observations

One of the most useful ways of summarizing groups of data is by graphing them. Figure 4-1 shows the results obtained from the repeated analysis of a patient plasma specimen for the analyte antithrombin III (ATT), which is a potent inhibitor of many of the activated clotting factors. The results are plotted as a frequency diagram, or *frequency histogram,* with the number or frequency of each result plotted on the y axis and the value of the result plotted on the x axis. The frequency histogram of the repeated measurements has a bell shape, with most of the results falling close to the center of the distribution. Repeated measurements of the same specimen are nonidentical and are due to diverse causes, including instrument, reagent, and operator variations. The laboratorian often classifies these variations as being *analytical variations.*

For most assays, the analytical variation is usually much lower than the variation of samples obtained from different individuals (the *interindividual variations*). Table 4-1 shows the results of a reference interval study for ATT. The blood of fasting, healthy, ambulatory subjects was drawn in a standard manner, with the plasma analyzed for ATT. Frequency histograms of the ATT data are shown in Figure 4-2. The scale for the concentration is expressed in intervals of 2 units for Figure 4-2*A* and 5 units for Figure 4-2*B*. With a decrease in the number of intervals, the shape of the frequency histogram becomes more regular. Both histograms are almost symmetrical and bell-shaped. The bell shape of this distribution approximates the shape of a Gaussian distribution and allows analysis of the data by standard (parametric) statistical tests. Data that deviate greatly from the Gaussian distribution

TABLE 4-1. Antithrombin III Values from a Reference Interval

Value	Frequency	Cumulative Frequency
88	1	1
92	1	2
93	2	4
95	1	5
96	1	6
97	3	9
98	1	10
99	1	11
100	3	14
101	4	18
102	4	22
103	3	25
104	2	27
105	4	31
106	4	35
107	5	40
108	3	43
109	2	45
110	9	54
111	4	58
112	4	62
113	4	66
114	7	73
115	7	80
116	7	87
117	4	91
118	7	98
120	4	102
121	1	103
122	4	107
124	1	108
125	2	110
126	3	113
127	1	114
129	1	115
133	2	117
138	1	118
140	1	119

Mean = 111.6; median = 112; mode = 110; s = 9.5 units

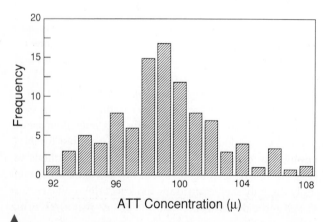

Figure 4-1. Frequency histogram of ATT results obtained from the repeated analysis of a single patient specimen ($x = 100$; $s = 3$ units).

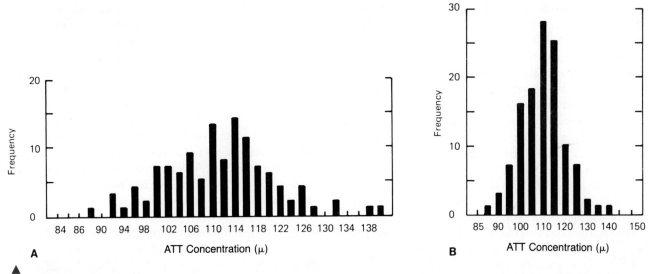

Figure 4-2. Frequency histograms of antithrombin III concentrations from a reference-interval study. The data in *A* have been grouped in intervals of 2 units. The data in *B* have been grouped by 5 units ($\bar{x}$ = 111.6; s = 9.5 units).

should be analyzed with distribution-free or nonparametric statistics. Small departures from Gaussian distributions do not seriously affect the results of parametric tests. The most common type of deviation from the Gaussian distribution in clinical laboratory observations is *skewness,* or the presence of increased numbers of observations in one of the tails of the distribution. (An example of a skewed distribution is shown in Figure 4-9.)

Another type of plot is the *cumulative-frequency histogram.* In this plot, the number of observations that are less than or equal to a certain observation are plotted against the value of that observation. Figure 4-3 shows a cumulative-frequency histogram for the normal-range ATT data. The cumulative frequencies in Table 4-1 have been divided by the total number of observations (*n*) and then multiplied by 100 to obtain relative cumulative frequencies that range from 0% to 100%.

Groups of Gaussian observations can be described by statistics that summarize their location and dispersion. The most common statistical test that summarizes location is the *mean,* which is calculated by summing the observations and dividing by the number of the observations. If the observations are $x_1, x_2, x_3, \ldots, x_n$, then the mean, or $\bar{x}$, is

$$\bar{x} = \frac{x_1 + x_2 + x_3 + \cdots + x_n}{n}$$

$$= \frac{\sum x_i}{n} \qquad \textbf{(Eq. 4-1)}$$

Three other measures of location are commonly used: median, mode, and percentiles. The *median* is the value of the observation that divides the observations into two groups, each containing equal numbers of observations. The values in the one group are smaller than the median, and those in the other are larger than the median. If the observations are arranged in increasing order and there is an odd number of observations, the median is the middle observation. If there

is an even number of observations, the median is the average of the two innermost observations.

The *mode* is the most frequent observation. For the ATT data, the mean is 111.6, the median is 112, and the mode is 110 units. For data with approximately a Gaussian distribu-

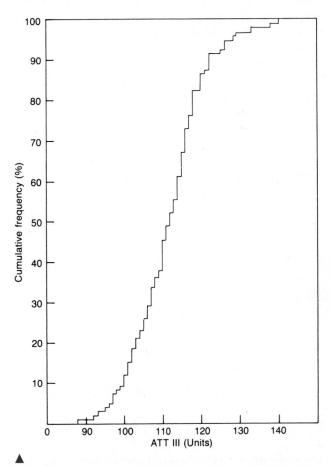

Figure 4-3. Cumulative frequency histogram for antithrombin III.

tion, the mean, mode, and median are approximately equal, and thus reporting of the mean is usually sufficient. Data that are significantly non-Gaussian should have the mode, median, and mean reported. The *percentile,* the only nonparametric statistic to be described in this chapter, is the value of an observation below which a certain proportion of the observations fall and which may be obtained from the cumulative-frequency histogram. The 5th percentile, or P_5, is the value below which 5% of the observations fall. For ATT, P_5 is 96. The 97.5 percentile, or $P_{97.5}$, is the value below which 97.5% of the observations fall. For ATT, $P_{97.5}$ is 133. The median is, of course, P_{50}. *Dispersion,* or the spread of data around its location, is most simply estimated by the range, which is the difference between the largest and smallest observations. The most commonly used statistic for describing the dispersion of groups of single observations is the *standard deviation,* which is usually represented by the symbol *s.* The standard deviation of the observations $x_1, x_2, x_3, \ldots, x_n$ is

$$s = \sqrt{\frac{\Sigma(\overline{x} - x_i)^2}{n - 1}} \qquad \textit{(Eq. 4-2)}$$

Figure 4-4 shows an idealized frequency histogram of glucose control values, which have a mean of 120 and a standard deviation of 5 mg/dL. For these data, as well as for other data with Gaussian distributions, approximately 68% of the observations will be between the limits of $\overline{x} - s$ and $\overline{x} + s$, 95.5% will be between $\overline{x} - 2s$ and $\overline{x} + 2s$, and 99.7% will be between $\overline{x} - 3s$ and $\overline{x} + 3s$. Limits can be constructed to include a specific proportion of the population. The usual limits for reference ranges include 95% of the population and correspond to $\overline{x} - 1.96s$ and $\overline{x} + 1.96s$. For ATT, the 95% or $\pm 1.96s$ limits would be 92.5 to 130.6 or 92 to 131 units (U). The percentile also may be used to express dispersion. The percentile limits that would enclose 95% of the population would be $P_{2.5}$ to $P_{97.5}$, or 93 to 133 U.

In Equation 4-2, calculation of *s* requires that the mean be calculated first. There is an alternate equation (Eq. 4-3) that does not need prior calculation of the mean:

$$s = \sqrt{\frac{n\Sigma x_i^2 - (\Sigma x_i)^2}{n(n - 1)}} \qquad \textit{(Eq. 4-3)}$$

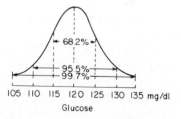

105 110 115 120 125 130 135 mg/dl
Glucose

▲
Figure 4-4. Idealized Gaussian frequency histogram of glucose control values with a mean of 120 and standard deviation of 5 mg/dL. The percentages indicate the area under the curve bounded by the ±1, ±2, and ±3, limits.

This equation is frequently used in computer programs to minimize computation time. Another way of expressing *s* is in terms of the coefficient of variation (CV), which is obtained by dividing *s* by the mean and multiplying by 100 to express it as a percent:

$$CV\ (\%) = \frac{100s}{\overline{x}} \qquad \textit{(Eq. 4-4)}$$

The CV, a unitless number, simplifies comparison of standard deviations of test results expressed in different units and concentrations. The CV of the glucose control data of Figure 4-4 is thus 100 times 5 mg/dL divided by 120 mg/dL, or 4.2%.

The *mean absolute deviation* (MAD), also known as the *average deviation,* is another measure of dispersion of groups of single observations and is calculated using the following equation:

$$MAD = \frac{\Sigma|x_i - \overline{x}|}{n} \qquad \textit{(Eq. 4-5)}$$

The *standard deviation of the mean,* also called the *standard error of the mean* (SEM), is calculated from the following equation, in which *n* represents the number of observations averaged to calculate the mean:

$$SEM = \frac{s}{\sqrt{n}} \qquad \textit{(Eq. 4-6)}$$

The SEM may be used to calculate confidence limits for the mean. The SEM can be interpreted as the average error encountered if the sample mean was used to estimate the population mean. The 95% confidence limits for the mean $\overline{x}$ would be $\overline{x} \pm 1.96s\sqrt{n}$. The SEM decreases as sample size increases, and the mean of a large sample is likely to be closer to the true mean than the mean of a small sample.

Descriptive Statistics of Groups of Paired Observations

Perhaps the most informative step in the evaluation of a new analytical method is the comparison-of-methods experiment, in which patient specimens are measured by both the new method and the old, or comparative, method.[70] The data obtained from this comparison consist of two values for each patient specimen. Graphing is the simplest way to visualize and summarize the paired-method comparison data. By convention, the values obtained by the old (comparative) method are plotted on the *x* axis, and the values obtained by the new (test) method are plotted on the *y* axis. Figure 4-5 shows a plot of serum creatinine determinations performed using two different instruments, the Technicon SMAC plotted on the *x* axis and Technicon RA 1000 plotted on the *y* axis. There appears to be a linear relationship between the two methods over the entire range of creatinine values.

The agreement between the two methods may be estimated from the straight line that best fits the points. Whereas visual estimation may be used to draw the line, the use of a statistical technique, *linear regression analysis,* will result in an

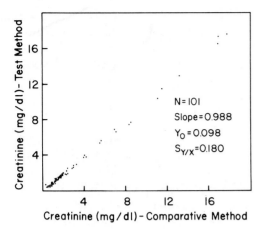

Figure 4-5. Graphic presentation of a comparison-of-methods experiment. Creatinine measured by the Technicon RA-1000 is compared with the creatinine measured by the Technicon SMAC.

impartial choice of line and will provide the laboratorian with measures of location and dispersion for the line. The straight line through the data will have the equation of

$$y = mx + y_0 \qquad \textbf{(Eq. 4-7)}$$

The slope of the line will be m; the value of the y intercept (the value of y at $x = 0$) will be y_0. If there is perfect agreement between the two methods, each value measured by the test method will be identical to that measured with the comparative method. The equation of the line would be $y = x$, with m being 1 and y_0 being 0. Figure 4-6A shows perfect agreement between the test and comparative methods. Figure 4-6B shows the situation in which values from the test method are consistently higher than those of the comparative method. The best line through the data still has a slope of 1 but a y intercept of 5.0. Fig-ure 4-6C shows the situation in which values from the test method are higher than values from the comparative method for nonzero concentrations; the slope is greater than 1 (1.1), but the y intercept is 0. For increasing concentrations, there are greater differences in the test values measured by the two methodologies. In Figure 4-6D, the y intercept is still 0 and the slope is less than 1 (0.9), showing that the test-method values are lower than the comparative-method values for all nonzero values.

Linear regression analysis usually provides unbiased estimates of the slope and y intercept. In linear regression, the line of best fit is one that minimizes the sum of the squares of the vertical distances of the observed points from the line. For the points (x_1, y_1), (x_2, y_2), . . . , (x_i, y_i), . . . , (x_n, y_n), the equation of the slope of the regression line is

$$m = \frac{\Sigma(x_i - \bar{x})(y_i - \bar{y})}{\Sigma(x_i - \bar{x})^2}$$

$$= \frac{n\Sigma x_i y_i - \Sigma x_i \, \Sigma y_i}{n\Sigma x_i^2 - (\Sigma x_i)^2} \qquad \textbf{(Eq. 4-8)}$$

The y intercept is calculated from m and the means of x_i and y_i:

$$y_0 = \frac{\Sigma y_i}{n} - m \, \frac{\Sigma x_i}{n}$$

$$= \bar{y} - m\bar{x} \qquad \textbf{(Eq. 4-9)}$$

Linear regression assumes that there is no measurement error in the comparative method and that the standard deviation of the regression line is due to random errors in the test method. The dispersion of the points about the regression line is referred to as the *standard deviation* of the regression line and is abbreviated as $s_{y/x}$. Another name for this dispersion is the *standard error* of the estimate. It is calculated using the following equation:

$$S_{y/x} = \sqrt{\frac{\Sigma(y_i - Y_i)^2}{n - 2}} \qquad \textbf{(Eq. 4-10)}$$

The method-comparison plots in Figure 4-6E and F show the influence of increased scatter of points about the regression line. In Figure 4-6E and F, the value of either 2 or 5, respectively, was alternately added to or subtracted from the values of y in Figure 4-6A. The slope and intercept did not change. Only $s_{y/x}$ increased to 2 or 5, respectively.

The *correlation coefficient*, r, is a measure of the strength of the relationship between the y and x variables. The correlation coefficient can have values from -1 to $+1$, with the sign indicating the direction of relationship between the two variables. A positive r indicates that both variables increase or decrease together, whereas a negative r indicates that as one variable increases, the other decreases. An r value of 0 indicates no relationship. The usual equation for the calculation of r is

$$r = \frac{n\Sigma x_i y_i - \Sigma x_i \Sigma y_i}{\sqrt{[(n\Sigma x_i^2 - (\Sigma x_i)^2] \times [(n\Sigma y_i)^2 - (\Sigma y_i)^2]}} \qquad \textbf{(Eq. 4-11)}$$

Whereas many laboratorians equate high positive values of r (0.95 or higher) with excellent agreement between the test and comparative methods, it should be noted that most clinical chemistry comparisons should have correlation coefficients greater than 0.98. The absolute value of the correlation coefficient can be significantly increased by widening the range of samples being compared. The correlation coefficient does have a use, however. When r is less than 0.99, use of the regression formula results in an estimate of the slope that is too small and a y intercept that is too large. Waakers and associates have recommended that if r is less than 0.99, alternate regression statistics should be used to derive more realistic estimates of the regression, slope, and y intercept.[68]

Error accounts for the difference between the test- and comparative-method results. Two kinds of error are measured in comparison-of-methods experiments: random and systematic. *Random error* is present in all measurements, is due to chance, and can be either positive or negative. The measure of dispersion $s_{y/x}$ provides an estimate of random

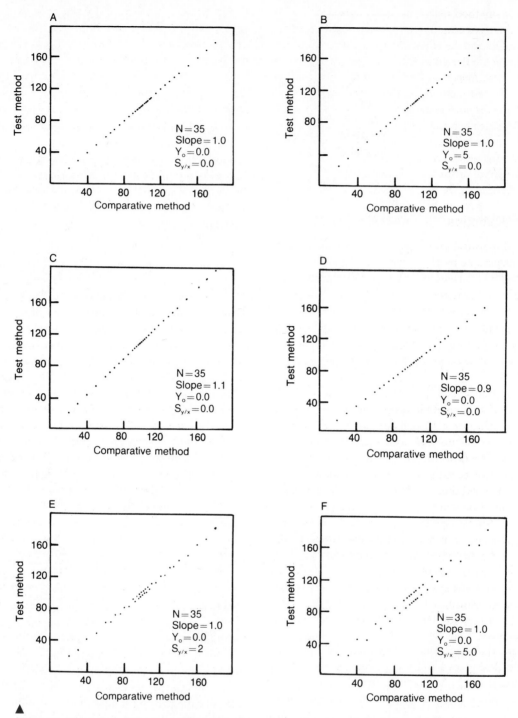

Figure 4-6. Comparison-of-methods experiments using simulated data. (*A*) shows no error; (*B*) shows constant error; (*C*) and (*D*) show proportional error; (*E*) and (*F*) show random error.

error. *Systematic error* is error that influences observations consistently in one direction. Unlike random error, systematic error should not be present in a method. The measures of location, the slope and *y* intercept, provide measures of the systematic error.

Because $s_{y/x}$ is an estimate of the standard deviation about the regression line, statistical limits can be calculated for any point on the line, just as for groups of single observa-

tions. The 95% limits of any *y* value on the regression line are $y \pm 1.96s_{y/x}$; that is, 95% of the *y* values for any *x* value will fall within the calculated value of *y*. If $mx + y_0$ is substituted for *y* (Eq. 4-7), then the 95% limits are $mx + y_0 \pm 1.96s_{y/x}$.

There are two types of systematic error: constant error and proportional error. *Constant systematic error* exists when there is a constant difference between the test method and the comparative method regardless of the concentration. In Figure 4-6*B*, there is a constant difference of 5 between the test-method values and the comparative-method values. This constant difference, reflected in the *y* intercept, is called *constant systematic error. Proportional error* exists when the difference between the test method and the comparative method is proportional to the analyte concentration. In Figures 4-6*C* and *D*, the difference between the test method and comparative method is proportional to the measured concentration. This difference, manifested by a slope different from unity, is therefore due to proportional systematic error.

Inferential Statistics

The inferential statistical tests described in this chapter will be used to compare the means or standard deviations of two groups of data. The *t* test, described by Gosset in 1908, is used to determine whether there is a statistically significant difference between the means of two groups of data. The *F* test is used to determine whether there is a statistically significant difference between the standard deviations of two groups of data. Both tests have limited usefulness in method-evaluation studies.

For both the *t* and *F* tests, a statistic is calculated and then compared with critical values found in the *t* and *F* tables of statistics books. The critical values are used to derive the significance level or the probability that the differences in the means or standard deviations are due to chance. By convention, if the probability of a difference occurring due to chance is less than 5%, then the difference evaluated by the *t* or *F* test is said to be *statistically significant*. If the probability of the difference occurring due to chance exceeds 5%, then the difference is usually said to be *not statistically significant*. The lower the probability, the more statistically significant is the difference: *e.g.,* a difference that occurs 1% of the time due to chance has greater significance than one that occurs 5% due to chance.

To apply the *t* test, the *t* statistic is calculated and compared with a table of critical *t* values for selected significance levels and degrees of freedom. Values of *t* should be obtained from the *t* table if a small number of observations (30 or fewer) is averaged. If more than 30 observations are averaged, the critical values are almost independent of the number of observations and depend primarily on the significance level. The critical values listed in Table 4-2 may be used if more than 30 observations are averaged.

TABLE 4-2. Critical *t* Values		
Significance Level	Two-Tailed	One-Tailed
5% (0.05)	1.96	1.64
1% (0.01)	2.58	2.33

If the *t* test is used to compare two means and the calculated *t* statistic exceeds the critical value, then a significant difference is said to exist. The larger the difference, the larger will be the *t* statistic and the lower will be the significance level or the probability that the difference is due to chance. The one-tailed critical values are used to test whether one mean is either significantly greater or less than the other mean. The two-tailed critical values are used to test whether the means are significantly different.

In its simplest application, the *t* test is used to determine whether the mean of a group of data ($\overline{x}$) is different from the true mean (abbreviated as M). The equation for the calculation of the *t* statistic is

$$t = \frac{|\overline{x} - \text{M}|}{s/\sqrt{n}}$$

Degrees of freedom $= n - 1$ *(Eq. 4-12)*

This *t* value is the absolute value of the arithmetic difference of the true mean and average divided by the standard error of the mean. For example, if the mean of a group of glucose determinations on one control product were tested and shown to be different from the usual mean value, the **two-tailed** critical *t* values would be used. If the calculated *t* value were less than 1.96, then the difference between the means would not be considered significant at the 5% level. A *t* value between 1.96 and 2.58 indicates a statistically significant difference; such a difference would occur due to chance with a probability between 1% and 5%. A difference with a *t* value greater than 2.58 is more statistically significant and has less than a 1% probability of occurring due to chance.

The **one-tailed** critical values would be used to determine whether the mean of a group of data is either significantly larger or smaller than the true mean. For example, a physician may have several cholesterol values from a patient and may wish to determine whether these values are significantly greater than the upper limit of acceptability. After calculating the *t* value, the physician should use the table of one-tailed critical values.

In clinical chemistry, the *t* test is usually applied to method-comparison data in which patient specimens are measured by both the test and the comparative methods. The measured values are averaged for each method, with the averages tested for statistically significant differences. If x_i and y_i are the values obtained by the comparative and test methods, respectively, the *t* value for a group of paired observations $(x_1, y_1), (x_2, y_2), \ldots, (x_i, y_i), \ldots, (x_n, y_n)$ is

$$t = \frac{\dfrac{\Sigma y_i}{n} - \dfrac{\Sigma x_i}{n}}{s_d/\sqrt{n}}$$

$$= \text{bias}/(s_d/\sqrt{n}) \qquad \textit{(Eq. 4-13)}$$

The numerator of the expression is the difference between the mean of the test method ($\Sigma y_i/n$) and the mean of the comparative method ($\Sigma x_i/n$). This difference between the means is called the *bias*. The symbol s_d stands for the standard deviation of the differences:

$$s_d = \sqrt{\frac{\Sigma[y_i - x_i - \text{bias}]^2}{n-1}} \qquad \textit{(Eq. 4-14)}$$

Equation 4-13 shows that the *t* value is simply the bias, or difference of the means, divided by a standard error. Westgard and Hunt have shown that the interpretation of the *t* test without regard for s_d and *n* may be misleading.[77,84] Statistically significant differences may exist between methods, but the size of the bias may not be clinically significant. The user is cautioned to determine whether the bias is clinically significant if the difference between the means is found to be statistically significant.[84] Also, the bias and s_d may be clinically large, but the resulting *t* value may be small. Thus all terms, bias, s_d, and *t* value, must be critically evaluated when interpreting the results of the *t* test.

The second inferential statistical test, the *F* test, has been used to compare the sizes of the standard deviations of two methods. To calculate the *F* statistic, the square of the larger standard deviation (s_L) is divided by the square of the smaller standard deviation (s_S):

$$F = \frac{(s_L)^2}{(s_S)^2} \qquad \textit{(Eq. 4-15)}$$

Like the *t* test, the *F* statistic is then compared with critical *F* values in statistical tables that are tabulated by degrees of freedom and significance level. Because the *F* test provides information about statistical significance but not clinical significance, Westgard and Hunt state that the *F* value should not be used as an indicator of acceptability of a test.[77,84] Rather, acceptability should depend on the size of random error.

REFERENCE INTERVALS (NORMAL RANGE)

Definition of Reference Interval

Physicians order laboratory tests for a variety of reasons. The most important of these are diagnosis of disease, screening for disease, and monitoring of levels of drugs and endogenous substances such as electrolytes. Other reasons for testing include determining prognosis, confirming a previously abnormal test, physician education, and medicolegal purposes. When a test is used for diagnosis, screening, or prognosis, the test result is usually compared with a reference interval (normal range) that is defined as the usual values for a healthy population. For example, if a patient appears to have signs and symptoms of hyperthyroidism, one of the follow-up tests ordered by the attending physician would be measurement of the serum thyroxine level. If the patient's thyroxine level exceeds the upper reference limit, the physician may require further testing to determine the cause of the hyperthyroidism. When a test is used for monitoring, the test result is usually compared with values that were previ-

ously obtained from the same patient. For example, in patients with surgically removed colonic carcinomas, the presence of carcinoembryonic antigen (CEA) is often used to detect recurrence of the carcinoma. In these patients, each new CEA value is compared with previous CEA values. The acceptable range for the CEA values should be derived from the previous test values of each patient.[32]

The International Federation of Clinical Chemistry (IFCC) has recommended use of the term *reference intervals* to denote the usual limits of laboratory data.[28] The presence of health is not implied in the definition, and thus reference intervals may be constructed for ill as well as healthy populations. The IFCC recommends[28] that the following factors be specified when reference intervals are established: (1) makeup of the reference population with respect to age, sex, and genetic and socioeconomic factors; (2) the criteria used for including or excluding individuals from the reference sample group; (3) the physiologic and environmental conditions under which the reference population was studied and sampled, including time and date of collection, intake of food and drugs, posture, smoking, degree of obesity, and stage of menstrual cycle; (4) the specimen-collection procedure, including preparation of the individual; and (5) the analytical method used, including details of its precision and accuracy. The IFCC considers the terms *normal values* and *normal range* to be specific reference intervals that correspond to the health-associated (central 95%) reference interval. In this chapter, we will discuss the health-associated reference interval almost exclusively and thus will use the terms *normal values, normal range*, and *reference range* and *reference interval* interchangeably.

In the past, many hospital laboratories have either used the reference interval recommended by the instrument or test manufacturer or the values published in medical or laboratory textbooks. Because of the diversity of instrumentation, methodologies, reagents, and populations, it is important that moderate- to large-sized hospital laboratories determine their own reference intervals. Smaller laboratories will lack the resources to conduct such work; instead they should analyze far fewer specimens (at least 20) and verify the reference interval specified in the method's package insert, which is provided by the manufacturer.

The selection of subjects for the reference interval study is very important. Many laboratory data are dependent on age and sex. For example, plasma testosterone level is low in both prepubertal boys and girls and increases during puberty, attaining higher levels in boys. If a physician obtains a testosterone level to rule out a testosterone-secreting tumor in a pubertal girl, he or she must be able to compare the girl's testosterone value with normal values for girls of her age. Similarly, alkaline phosphatase levels are elevated during growth (Fig. 4-7) as well as in men and women older than 60 years. Ideally, the laboratory should have age- and sex-stratified normal values for all populations tested. Thus, if a laboratory tests many specimens from older adults, proper reference intervals should be provided for this population. The University of Sherbrooke in Quebec, Canada, has

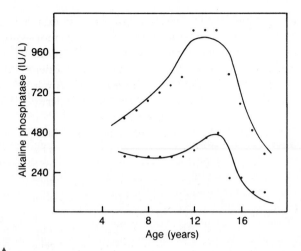

Figure 4-7. $P_{2.5}$–$P_{97.5}$ reference intervals for alkaline phosphatase in healthy boys determined by the Bowers-McComb method.[23]

published a monograph that graphically depicts age and sex reference intervals for 26 common analytes.[41]

To derive reliable estimates of reference intervals, at least 120 individuals should be tested in each age and sex category. However, it is often necessary to carry out reference range studies using fewer individuals. Sampling of 120 males and females would be adequate for determining the reference interval of an analyte that does not vary substantially with age and sex (*e.g.,* sodium). An analyte such as creatine kinase, in which there are substantial differences between males and females, would require sampling of at least 120 males and 120 females. If reference intervals were desired for both males and females from birth to age 70, and if each age category equaled a 10-year interval, then the minimum number of individuals to be tested would be 100 subjects times 2 sexes times 7 age classifications, or 1400 subjects. Systematic testing of such a large number of subjects is almost always prohibitive. Winsten[88] has suggested that four age categories be used: newborns, prepubertal, adult population (postpubertal and premenopausal), and older adult population (males after age 60 and postmenopausal females). While these divisions are not optimal, they reduce the number of categories that should be tested. The dependence of alkaline phosphatase on age (shown in Fig. 4-7) indicates that unless there is further stratification of alkaline phosphatase by age, the prepubertal reference interval may be limited in its usefulness.

Reference interval studies on healthy children are limited because of the psychological and physical pain of phlebotomy. Many such studies are done on pediatric patients for whom serum and plasma samples are already available. The diseases and treatments of these children usually will result in an erroneous reference interval. Because very few centers can undertake systematic reference-interval studies on healthy newborns and children, we recommend that published reference values be used and compared critically with the clinical laboratory's adult reference values. Meites[40] has compiled reference values for pediatric patients in a com-

prehensive monograph. The earlier edition of this monograph contains very useful information not included in the third edition and should not be discarded.

Collection of Data for Reference-Interval Studies

If reference-interval data are to be maximally useful, the reference population should consist of individuals in a state of good health. Hospital employees are the most readily available healthy individuals for reference interval studies, but unfortunately, there is a sampling bias, with the population consisting primarily of premenopausal females. Great effort often must be expended to recruit enough male adults and postmenopausal females. Once the desired reference population in identified, a consent form must be drafted and presented to the hospital's human subject experimentation committee or institutional review board so that volunteers can be canvased for donation of blood or other specimens.

All potential donors should be interviewed to gather the following data: age, sex, health status, activity level, height and weight, alcohol consumption, drug usage (including oral contraceptives), smoking history, and stage of menstrual cycle. If a normal-range or health-associated reference interval is desired, any individual with acute or chronic disease should be excluded. The donors should be instructed about preparation before specimen collection. Fasting specimens are usually required, in which case the donor should be instructed not to eat after 10 P.M. and to drink only noncaloric decaffeinated fluids before blood drawing. All donors should be sampled in a similar fashion, with care taken by the phlebotomist not to apply the tourniquet for longer than 1 minute. It should be noted that many analytes (*e.g.,* iron, ACTH, cortisol, and so on) exhibit significant diurnal variations. The time of sampling for these substances should be controlled.

Once the specimens are acquired, they must be labeled and handled in the same manner as the regular specimens. The instruments used to analyze these specimens should be in good running operation. Ideally, no more than 5 to 10 subjects should be sampled and analyzed daily. With this long-term analysis, the normal range will reflect the long-term state of analytical control. Analysis over a short period may introduce systematic differences or shifts in the reference range due to transient instrument or reagent differences.

Statistical Analysis of the Reference-Interval Data

The analysis of the large amount of data derived from reference-interval studies used to be very laborious. Today this task is simplified by the ready availability of microcomputer-based "spread sheet" or statistical programs. The test data are first entered into the computer along with donor demographic data, such as identifier code, sex, and age. Frequency histograms then are plotted for all the tests. Results for donors with outlying laboratory data should be forwarded

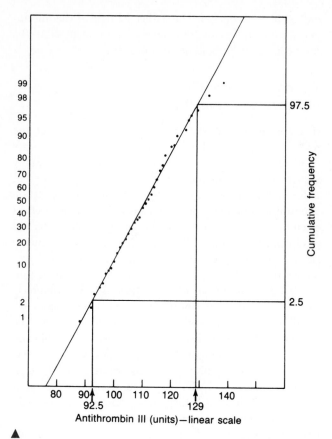

Figure 4-8. Probability plot for antithrombin III (linear scale).

for the y axis. Figure 4-8 shows a cumulative-frequency histogram for ATT plotted on probability paper. Cumulative frequency probability plots make Gaussian distributions linear and enables drawing of the best straight line through the points. Most of the reference-interval data are thus used to determine the reference interval, and the effect of outliers is diminished. The outer 2.5% and 97.5% limits of the population are determined by selecting values for ATT on the straight line that correspond to cumulative frequencies of 2.5% and 97.5%, respectively. The reference-intervals derived from Figure 4-8 are 92.5 to 129 U. If the distribution of the population is very smooth and Gaussian, then the 95% limits also can be calculated directly from $\bar{x} \pm 1.96s$. In the case of ATT, the limits calculated in this manner are $115.6 \pm 1.96 \times 9.5$ U, or 92.5 to 130.6 U. Alternately, the 95% limits can be determined from the 2.5 and 97.5 percentile limits, or 93 to 133 U (see Fig. 4-3).

If the data are not Gaussian, determination of the 95% limits is more difficult. Figure 4-9 shows a frequency histogram and Figure 4-10 presents a probability plot for gamma–glutamyltranspeptidase (GGT) in 118 females. The frequency histogram shows a skewing toward increased GGT values. The probability plot is nonlinear and indicates a non–Gaussian distribution. Percentiles do not depend on the shape of the distribution and may be used to determine the 2.5% and 97.5% limits for the population. The 2.5 and 97.5 percentile limits for GGT are 10 and 35 IU/L.

Since the early 1990s, there has been less enthusiasm in deriving reference intervals using probability plots and more in using percentile analysis. This change is due to at least three factors, including the fact that many sets of reference range data are not Gaussian, that percentile analysis is simpler than performing probability plots, and that the recent National Committee for Clinical Laboratory Standards (NCCLS) Document, *How to Define and Determine Reference Intervals in the Clinical Laboratory*,[45] promotes the percentile approach. Reference intervals are occasionally widened to include the lower limit of the analyte. For example,

to their physicians. If the reason for the outlying results is known (*e.g.,* elevated creatine phosphokinase levels after participation in a football game), the outlying results should be eliminated from further analysis.

Until approximately 1990, it was usual to plot cumulative-frequency histograms on probability paper and then derive reference intervals from the probability plot. Figure 4-2 shows the frequency histogram of ATT. Figure 4-3 shows the cumulative-frequency histogram of ATT using a linear scale

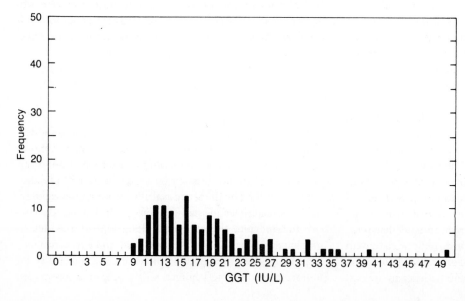

Figure 4-9. Frequency histogram of GGT in 118 females. The corresponding probability plots are shown in Figure 4-10.

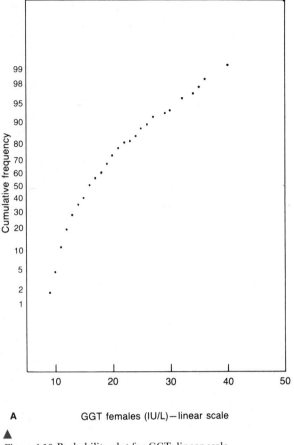

A   GGT females (IU/L)—linear scale

Figure 4-10. Probability plot for GGT: linear scale.

Figure 4–11 shows the frequency histogram and the probability plot for total bilirubin in 228 male and female subjects. The probability plot is nonlinear so the 2.5 and 97.5 percentile limits are used to define the reference interval: 0.3 to 1.4 mg/dL. Because the published lower limit of bilirubin is usually 0, and because there are few, if any, pathologic reasons for a bilirubin of 0, the reference interval derived from these data was set as 0 to 1.4 mg/dL.

Occasionally, the results of a new reference interval study may be quite different from the old reference intervals, even though the same instrument and methodology were used. In such a situation, careful investigation must be undertaken to discover whether there are differences in the analytical method or in the reference population.

The distribution of patient data has been analyzed by various computational methods in an attempt to derive health-associated reference intervals. Martin and associates have recommended that reference intervals be derived from the numerical analysis of distributions of patient data.[39] Problems exist in this approach to reference interval determination and account for its lack of acceptance.

A normal range or health-associated reference interval is adequate for most tests. There are a few tests, such as that for glycosylated hemoglobin (HbA$_{1c}$), for which an alternate reference interval is desirable. HbA$_{1c}$ is a measure of an individual's average blood glucose level over the last 6 to 12 weeks; it usually is measured to determine and improve compliance. The frequency histogram in Figure 4-12 compares the HbA$_{1c}$ values of normal subjects with those of patients with diabetes. There is little overlap between non-diabetic and diabetic patients. The upper range of normal has little significance to diabetic patients; many physicians use a somewhat arbitrary limit of less than 7% to designate good diabetic control. Probably the most useful reference interval to improve patient compliance is that based on the patient's previous values.

DIAGNOSTIC EFFICACY

Until recently, many years might elapse between the introduction of a new diagnostic test and its rigorous clinical evaluation. Now, in as little as 1 to 2 years, a new diagnostic test may gain sudden fame and then, just as quickly, disfavor. The analysis of prostatic acid phosphatase (PAP) by radioimmunoassay (RIA) is such an example. A 1977 *New England Journal of Medicine* article concluded that the RIA for PAP had the potential to detect well over half the cases of prostate cancer that had not yet penetrated the prostatic capsule.[24] Two and one-half years later, the same journal published an article which stated that

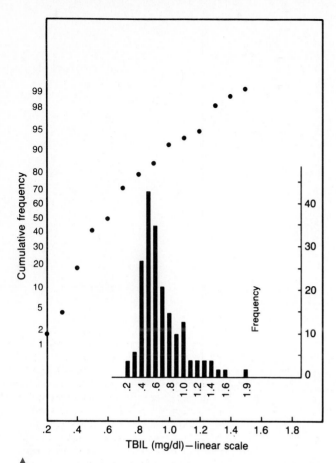

Figure 4-11. Frequency histogram of total bilirubin in 228 males and females (*inset*) and the corresponding probability plot (*linear scale*).

"the most economical and most reliable probe for the detection [of prostate cancer] remains not a needle in the patient's arm, but the gloved finger of the physician, engaged in a thorough rectal examination of the prostate[29]." The controversy associated with the PAP test probably caused the urologic community to be wary about other prostatic tumor markers. The subsequent introduction of a test for prostate-specific

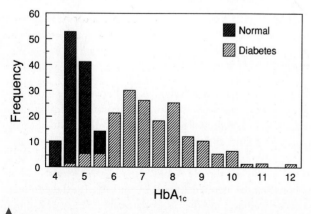

Figure 4-12. Comparison of frequency histograms of hemoglobin A_{1c} for subjects with normal fasting glucoses and patients with diabetes.

antigen (PSA), a sensitive but not specific test for prostate cancer, was not accompanied by the fanfare of the RIA test for PAP. The PSA test has turned out to be a more useful test in the management of prostate cancer.[7]

The material in this section will allow the reader to analyze frequency histograms of patient test data and to calculate a test's diagnostic sensitivity and specificity and predictive value. The reader also will be able to compare the diagnostic efficacies of various laboratory tests and to determine the most useful test.

Predictive Value Theory

Whereas the terms *diagnostic sensitivity, specificity,* and *predictive value* were used by radiologists beginning in the early 1950s, it was not until the 1970s that laboratorians became familiar with their meanings. The diagnostic sensitivity should not be confused with analytical sensitivity. For a test that is used to diagnose a certain disease, the *diagnostic sensitivity* is the proportion of individuals with that disease who test positively with the test. Sensitivity is usually expressed in percent:

$$\text{Sensitivity (\%)} = \frac{\substack{100 \times \text{the number of diseased} \\ \text{individuals with a positive test}}}{\substack{\text{total number of diseased} \\ \text{individuals tested}}} \quad \textbf{(Eq. 4-16)}$$

The specificity of a test is defined as the proportion of individuals without the disease who test negatively for the disease. The specificity (%) is defined as follows:

$$\text{Specificity (\%)} = \frac{\substack{100 \times \text{the number of individuals} \\ \text{without the disease with a} \\ \text{negative test}}}{\substack{\text{total number of individuals} \\ \text{tested without the disease}}} \quad \textbf{(Eq. 4-17)}$$

To be optimal, sensitivity and specificity should each be 100%. The sensitivity and specificity of a test depend on the distribution of test results for the diseased and nondiseased individuals and also on the value of the test that defines the abnormal levels. Figure 4-13 shows frequency histograms of PSA of two patient populations. These two populations are subsets of a group of 4962 male volunteers 50 years old or older who were evaluated for the presence of prostate cancer via digital rectal examination of the prostate and a PSA measurement.[9a] Of these volunteers, 770 had abnormal findings. The upper frequency histogram represents the subset of 578 patients with abnormal PSA values or abnormal digital rectal examinations who were biopsied for possible prostate cancer and found to be cancer free. The lower frequency histogram represents the subset of 192 patients with abnormal PSA values or abnormal digital rectal examinations who were biopsied for, and found to have, prostate cancer. It can be seen that while patients with prostate carcinoma tend to have higher values of PSA, there is a great deal of overlap between the two populations.

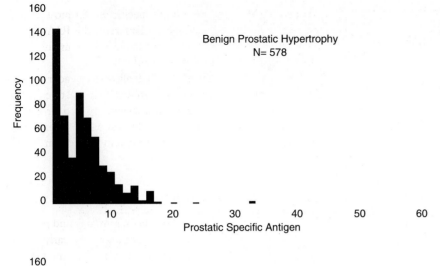

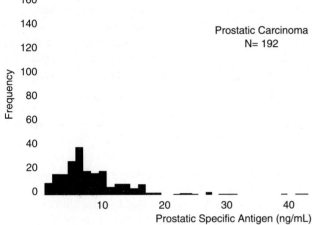

Figure 4-13. Frequency histograms of PSA results (Hybritech Tandem assay) of patients evaluated for possible prostate cancer. (*A*) The top plot shows the PSA of biopsied patients without prostate cancer; (*B*) the bottom shows the PSA of patients with biopsy-positive prostate cancer. (Adapted with permission from Catalona WJ, Richie JP, deKernion JB, et al. Comparison of prostate specific antigen concentration versus prostate specific antigen density in the early detection of prostate cancer: Receiver operating characteristic curves. J Urol. 1994;152:2031.)

The sensitivity and specificity can be calculated for any value of PSA. If high values of PSA are used to indicate the presence of disease (*e.g.,* if values exceeding 35 ng/mL are used to indicate prostate cancer), then the specificity of the test will be 100% (all the patients without prostate cancer are classified as negative using the test). The sensitivity, however, is quite low (2.6%), with only 5 of the 192 patients with carcinoma having PAP values greater than 35 ng/mL. The sensitivity can be increased by decreasing the test value. Thus, if values of PSA in excess of 4.0 ng/mL (the upper reference limit) are used to diagnose carcinoma, the sensitivity increases to 79%. The specificity, however, decreases to 46%. There are very few laboratory tests with sensitivities and specificities close to 100%. The sensitivity and specificity of the MB fraction of creatine kinase for the diagnosis of myocardial infarction are approximately 95% each. The urine hCG pregnancy tests used in the clinical laboratory have sensitivities and specificities approaching 100%. Many tests have sensitivities and specificities that are close to 50%. Galen and Gambino have stated that if the sum of the sensitivity and specificity of a test is approximately 100%, the test is no better than a coin toss.[25]

The sensitivity and specificity may be calculated from simple ratios. Patients with a disease who are correctly clas-

sified by a test to have the disease are called *true positives* (TP). Patients without the disease who are classified by the test not to have the disease are called *true negatives* (TN). Patients with the disease who are classified by the test as disease-free are called *false negatives* (FN). Patients without the disease who are incorrectly classified as having the disease are called *false positives* (FP). The sensitivity may be calculated from the formula 100TP/(TP + FN). Specificity may be calculated from 100TN/(TN + FP).

Three other ratios are important in evaluating diagnostic tests: predictive value of a positive test (PV⁺), predictive value of a negative test (PV⁻), and the efficiency. PV⁺ is that fraction of positive tests that are true positives: PV⁺ (%) = 100TP/(TP + FP). PV⁻ is that fraction of negative tests that are true negatives: PV⁻ (%) = 100TN/(TN + FN). The efficiency of a test is the fraction of all test results that are either true positives or true negatives: efficiency (%) = 100(TP + TN)/(TP + TN + FP + FN). The calculations for sensitivity, specificity, and predictive values for PSA exceeding 4.0 ng/mL are shown in Table 4-3. If a patient has a negative test result, he has an 87% probability of not having cancer. The probability that a patient has cancer if he has a positive test is rather low—approximately 33%.

TABLE 4-3. Sample Calculation of Sensitivity, Specificity, PV⁺, PV⁻, and Efficiency Calculated for Values of PSA > 4.0 ng/mL

	Number of Patients with Positive PSA (>4.0 ng/mL)	Number of Patients with Negative PSA (4.0 ng/mL)
Number of patients with prostate cancer	TP (151)	FN (41)
Number of patients without prostate cancer	FP (313)	TN (265)

$Sensitivity = 100 \times TP/(TP + FN) = 100 \times 151/192 = 79\%$
$Specificity = 100 \times TN/(TN + FP) = 100 \times 265/578 = 46\%$
$PV^+ = 100 \times TP/(TP + FP) = 100 \times 154/464 = 33\%$
$PV^- 100 \times TN/(TN + FN) = 100 \times 265/306 = 87\%$

The predictive value of a test can be expressed as a function of sensitivity, specificity, and disease prevalence, or the proportion of individuals in the population who have the disease:

$$PV^+ = \frac{\text{prevalence} \times \text{sensitivity}}{(\text{prevalence})(\text{sensitivity}) + (1 - \text{prevalence})(1 - \text{specificity})} \quad \textit{(Eq. 4-18)}$$

The PV⁺ for PSA for various prevalences and a cutoff of 4.0 ng/mL (sensitivity = 79%, specificity = 46%) are shown in Table 4-4. It can be seen that even in the situation in which disease has a high prevalence (0.2, or one-fifth of the population), the probability that the positive test truly indicates carcinoma is only 27%.

Because the selection of a cutoff level to define disease can be arbitrary, it is preferable to calculate and plot sensitivity and specificity for all values of a test. This allows comparison of sensitivities of two or more tests at defined specificities or comparison of specificities for certain sensitivities. Receiver operating characteristic (ROC) curves, which are plots of sensitivity (true-positive rate) versus 1 − specificity (false-positive rate), have been used to compare different laboratory tests.[60] Figure 4-14 illustrates two different ROC curves. Curve *A* is an ROC curve for a test in which there is wide separation between the test values of the diseased and nondiseased patients. Curve *B* illustrates the ROC curve obtained from a test for which there is little separation between diseased and nondiseased patients. The lower left part of the curve, where the false-positive rate is close to 0 (100% specificity) and the true-positive rate is close to 0 (0% sensitivity), corresponds to extreme test values in the diseased population. The upper right part of the curve, in which both the true-positive rate and the false-positive rate are high, corresponds to typical test values in the nondiseased population. For intermediate test values, a good test should have a high sensitivity (high true-positive rate) and a high specificity (low false-positive rate) and will form an ROC curve with its point close to the upper left-hand corner of the plot. Curve *A* corresponds to such a test. The test represented by curve *B,* in which the false-positive rate is equal to the true-positive rate, conveys no useful diagnostic information. Increasingly, clinical evaluations of diagnostic tests are being presented in the form of ROC curves. The NCCLS has published guidelines for clinically evaluating laboratory tests, including a comprehensive guide to ROC curves.[51]

TABLE 4-4. Dependence of Predictive Value on Disease Prevalence (Sensitivity = 79%, Specificity = 46%)

Prevalence of Disease	PV⁺
0.001	0.1%
0.01	1.5%
0.10	14.0%
0.2	26.8%
0.5	59.4%

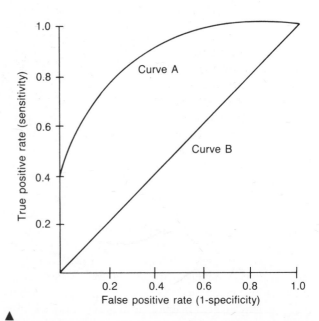

▲

Figure 4-14. ROC curves. Curve A corresponds to a test with wide separation between the diseased and nondiseased patients. Curve B corresponds to a test that provides no additional information.

Figure 4-15 shows the ROC curve of the PSA data of Figure 4-13. Also shown are the PSA concentrations at which the sensitivity and specificity were calculated. It can be seen that no PSA value offers both a high sensitivity and high specificity.

METHOD SELECTION AND EVALUATION

Method Selection

Before a new test or methodology is introduced into the laboratory, both managerial and technical information must be compiled and carefully considered. The information should be collected from many different sources, including manufacturers, sales representatives, colleagues, scientific presentations, and the scientific literature. The managerial information should include instrument cost, throughput, sample volume, personnel requirements, cost per test, specimen types, instrument size, and power and environment requirements. The technical information must include analytical sensitivity, analytical specificity, detection limit, linear range, interfering substances, and estimates of imprecision and inaccuracy.

The *analytical sensitivity* or *detection limit* refers to the smallest concentration that can be measured accurately. One professional group has defined the *detection limit* as equal to three times the standard deviation of the blank, or as located three standard deviations above the measured blank.[1] *Specificity* refers to a method's ability to measure only the analyte of interest. The *linear range* (sometimes called the *analytical* or *dynamic range*) is the concentration range over which the measured concentration is equal to the actual concentration without modification of the method. The wider the linear range, the less frequent will be specimen dilutions. Estimates of the inaccuracy of instruments may be obtained from the results of external proficiency-testing programs such as those offered by the College of American Pathologists (CAP). Estimates of imprecision are available from many vendors of quality-control products.

Method Evaluation

Once a method is brought in-house, the laboratorian must become proficient in using it. Then, in advance of the complete evaluation, a short initial evaluation should be carried out. This preliminary evaluation should include the analysis

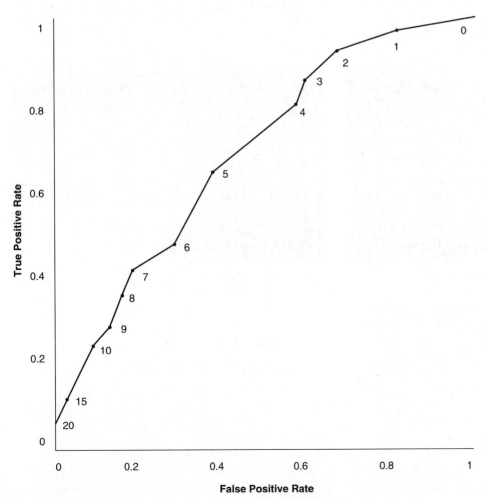

◄

Figure 4-15. ROC curve for PSA (Hybritech Tandem assay) for the diagnosis of prostate cancer. (Adapted with permission from Catalona WJ, Richie JP, deKernion JB, et al. Comparison of prostate specific antigen concentration versus prostate specific antigen density in the early detection of prostate cancer: Receiver operating characteristic curves. J Urol 1994;152:2031.)

of a series of standards to determine the linear range, the replicate analysis (at least 8) of two to four patient samples to obtain an estimate of short-term imprecision, and finally, preliminary interference and recovery studies. If any of these initial results fall short of the specifications published in the method's product information sheet (package insert), the method's manufacturer should be consulted. Without improvement in the method, more extensive evaluations would be pointless.[75]

In method evaluation, the imprecision and inaccuracy of a method are estimated and compared with the maximum allowable error, which usually is based on medical criteria. If the imprecision or inaccuracy exceeds this maximum allowable error, the method is judged as unacceptable and must be modified and reevaluated or rejected. *Imprecision,* the dispersion of repeated measurements about the mean, is due to the presence of random analytical error. Imprecision is estimated from studies in which aliquots of a specimen with a constant concentration of analyte are analyzed repetitively. *Inaccuracy,* the difference between a measured value and its true value, is due to the presence of systematic analytical error, which can be either constant or proportional. Inaccuracy can be estimated from three studies: recovery, interference, and a comparison-of-methods study.

Measurement of Imprecision

The first step in method evaluation is the precision study. This study estimates the random error associated with the test method and points out any problems affecting reproducibility. It is recommended that this study be done over a 20-day period, incorporating one or two analytical runs per day.[46,76] An *analytical run* is defined as a group of patient specimens and control materials that are analyzed, evaluated, and reported together. Imprecision should be measured at more than one concentration with control materials spanning the clinically meaningful range of concentrations. For example, glucose should be studied in the hypoglycemic (~50 mg/dL), hyperglycemic (~150 mg/dL), and normoglycemic (~100 mg/dL) ranges.

Once the precision data are collected, the mean, standard deviation, and coefficient of variation are calculated. The random error or imprecision associated with the test procedure is indicated by the standard deviation and the coefficient of variation. The within-run imprecision is indicated by the standard deviation of the controls analyzed within one run. The total imprecision may be obtained from the standard deviation of control data with one data point accumulated per day. A statistical technique, analysis of variance, may be used to analyze all the available precision data to provide estimates of the within-run, between-run, and total imprecision.[34]

Measurement of Inaccuracy

Once the short-term imprecision of the method is estimated and deemed adequate, the accuracy experiments[76] can begin.

Accuracy can be estimated in three ways: recovery, interference, and patient-sample comparison studies. Recovery and interference studies should be performed and documented by the manufacturer. As these two types of studies can be prohibitive both in terms of effort and materials, they are usually performed in larger clinical laboratories. Recovery experiments will show whether a method measures all the analyte or only part of it. In a recovery experiment, a test sample is prepared by adding a small aliquot of concentrated analyte in diluent to the patient sample. Another sample of the patient specimen is diluted by adding the same volume of diluent alone. Both diluted specimens are then analyzed by the test method. The amount recovered is the difference between the two measured values. Care should be taken to ensure that the original patient samples are diluted by no more than 10%; in this way, the solution matrix of the samples is minimally affected. If the comparative method is used to measure replicates of the diluted specimens,[76] preparation and procedural errors can be minimized. Table 4-5 illustrates a sample calculation of recovery, which is expressed as percentage recovered.

The interference experiment is used to measure systematic errors caused by substances other than the analyte. An interfering material can cause systematic errors in one of two ways. The material itself may react with the analytical reagents, or it may alter the reaction between the analyte and the analytical reagents. The interference experiment[48] is similar to the recovery experiment, except that the substance suspected of interference is added to the patient sample. A sample calculation of interference is shown in Table 4-6. The concentration of the potentially interfering material should be in the maximally elevated range. If an effect is observed, its concentration should be lowered to discover the concentration at which test results are first invalidated. Materials to be tested should be selected from literature reviews and recent method-specific references. Young[89] and others[63] have published extensive listings of the effects of drugs on laboratory tests. Common interferences (*e.g.,* hemoglobin, lipids, bilirubin, anticoagulants, preservatives, and so on) also should be tested. Glick and Ryder have presented "interferographs"[26,27] for various chemistry instruments—these are graphs relating analyte concentration measured versus interferent concentration. They have demonstrated that thousands of dollars of rework can be saved by the acquisition of instruments that minimize hemoglobin, triglyceride, and bilirubin interference.[62]

Comparison-of-Methods Experiment

Comparison-of-methods studies are done on the same patient samples using the method being evaluated (test method) and a comparative method. The quality of the comparative method will affect the interpretation of the experimental results. The best comparative method that can be used is the *reference method,* which is a method with negligible inaccuracy in comparison with its imprecision. Reference methods may

TABLE 4-5. Example of a Recovery Study

Sample Preparation

Sample 1: 2.0 mL serum + 0.1 ml H_2O
Sample 2: 2.0 mL serum + 0.1 ml 20 mg/dL calcium standard
Sample 3: 2.0 mL serum + 0.1 ml 50 mg/dL calcium standard

	Calcium Measured	Concentration		Recovery
		Added	*Recovered*	
Sample 1	7.50 mg/dL			
Sample 2	8.35 mg/dL	0.95 mg/dL	0.85 mg/dL	89%
Sample 3	9.79 mg/dL	2.38 mg/dL	2.29 mg/dL	96%

Calculation of Recovery

$$Concentration\ added = standard\ concentration \times \frac{mL\ standard}{mL\ standard + mL\ serum}$$

$$Concentration\ recovered = concentration\ (diluted\ test) - concentration\ (baseline)$$

$$Recovery = \frac{concentration\ recovered}{concentration\ added} \times 100\%$$

be laborious and time-consuming (*e.g.,* for cholesterol). Because most laboratories are not staffed and equipped to perform reference methods, the results of the test method are usually compared with those of the method routinely in use. The routine method will have certain inaccuracies, both known and unknown, depending on how well the laboratory has studied them or how well the method is documented in the literature. If the test method will be replacing the comparative method, the differences between the two methods should be well characterized.

Westgard et al[76] and the NCCLS[49] recommend that at least 40 samples, and preferably 100 samples, be run by both methods. They should span the clinical range and should represent many different pathologic conditions. Duplicate analyses of each sample by each method are recommended, with the duplicate samples analyzed in different runs and in a different order of analysis on the two runs. Analysis by both methods should be performed on the same day, preferably within 4 hours.

If duplicates are not analyzed, the validity of the experimental results must be checked by comparing the test and comparative-method results immediately after analysis, identifying those samples with large differences, and, if necessary, repeating the analyses. If 40 specimens are compared, two to

TABLE 4-6. Example of an Interference Study

Sample Preparation

Sample 1: 1.0 mL serum + 0.1 mL H_2O
Sample 2: 1.0 mL serum + 0.1 mL of 10 mg/dL magnesium standard
Sample 3: 1.0 mL serum + 0.1 mL of 20 mg/dL magnesium standard

	Calcium Measured	Magnesium Added	Interference
Sample 1	9.80 mg/dL		
Sample 2	10.53 mg/dL	0.91 mg/dL	0.73 mg/dl mg/dL
Sample 3	11.48 mg/dL	1.81 mg/dL	1.68 mg/dL

Calculation of Interference

$$Concentration\ added = standard\ concentration \times \frac{mL\ standard}{mL\ standard + mL\ serum}$$

$$Interference = concentration\ (diluted\ test) - concentration\ (baseline)$$

five patient specimens should be analyzed daily for a minimum of 8 days. If 100 specimens are compared, the comparison study should be carried out during the 20-day replication study.

A graph of the test-method data (plotted on the y axis) versus the comparative-method data (plotted on the x axis) helps to visualize the method comparison data (see Fig. 4-5). Data should be plotted on a daily basis and inspected for linearity and outliers. In this way, original samples can be available for reanalysis. Visual confirmation of linearity is usually adequate; in some cases, however, it may be necessary to evaluate linearity more quantitatively.[47] Many analysts have used the F test, the paired t test, and the correlation coefficient for the interpretation of the experimental data. The F test is used to compare the magnitude of the imprecision of the test method with that of the imprecision of the comparative method. The paired t test is used to compare the magnitude of the bias (difference between the means of the test and that of the comparative method) with that of the random error. Both the F test and t test indicate only whether a statistically significant difference exists between the two standard deviations or means, respectively. They provide no information about the magnitude of the existing error relative to the clinically allowable limits of error.[84]

Despite regular admonishments about the misuse of the correlation coefficient r, laboratorians continue to use it as an indicator of the acceptability of the test method. The value of r can be increased by increasing the range of the patient specimens. The main application of the correlation coefficient in method-evaluation studies should be in determining the type of regression analysis to be used. If the correlation coefficient is 0.99 or greater, the range of patient samples is adequate for the standard linear regression analysis described in the statistics section. If r is less than 0.99, then alternate regression analysis should be used.[16,22,68] Linear regression analysis is far more useful than the t test and F test for evaluating method-comparison data.[53] The constant systematic error can be estimated from the y intercept, proportional systematic error from the slope, and random error from the standard error of the estimate ($s_{y/x}$). If there is a nonlinear relationship between the test and comparative values, then linear regression can only be used over the linear range. Because outlying points are weighted more heavily in linear regression, it is important to make sure that these outliers are genuine and not the result of laboratory error.

Once all the estimates of the imprecision and inaccuracy are calculated, they are compared with predefined limits of medically allowable analytical error.[78] If the error is smaller than the allowable error, the performance is considered acceptable. If the error is larger than the allowable error, the errors must be reduced or the method rejected.

The allowable analytical error represents total error and includes components of both random and systematic error. Many sources, including Tonks,[66] Barnett,[2] and even the U.S. government have published estimates of medically allowable error for critical decision levels. Tonks' criteria are general

(smaller of one-fourth of the normal range or 10% of the measured value). Barnett's recommendations are widely used but are limited to common laboratory tests. Under the federal law, the Clinical Improvement Amendments of 1988 (CLIA-88), the Health Care Financing Administration (HCFA) has published allowable errors in a wide array of clinical laboratory tests.[67] While the HCFA's allowable errors are used to specify the maximum error allowable in federal-mandated proficiency testing, these limits are now being used as guidelines to determine the acceptability of clinical chemistry analyzers.[18,86] Table 4-7 compares, for certain critical concentrations, the allowable error levels specified by Barnett, Tonks, and the HCFA with the currently available analytical performance. Westgard et al[78] have transformed Barnett's limits to 95% limits of allowable error (*i.e.*, 95% of patient samples should have errors less than the limit, or only 1 of every 20 samples can have an error larger than the specified limit). Tonks' limits are already expressed as 95% limits.

Westgard et al[78] have recommended two different sets of criteria for the evaluation of error: confidence-interval criteria and single-value criteria. Because of the complexity of confidence-interval criteria, the reader is referred to the original description.[74] The single-value criteria are found in Table 4-8. In using the single-value criteria, estimates of the random error, proportional error, constant error, and systematic error are calculated and then compared to the medically allowable error. Several of the error estimates depend on the concentration of analyte measured and are usually calculated at critical concentrations of the analyte. In applying these performance criteria, all error criteria (random, proportional, constant, and systematic error) must be less than the allowable error for a method to be judged acceptable. Otherwise, the analytical method must be rejected or modified to reduce the error.

As an ultimate criterion, Westgard et al[78] suggest a total-error criterion, which combines random and systematic components of error, to estimate the magnitude of error that can be expected when a patient specimen is measured. The use of the single-value criteria is illustrated in Table 4-9, in which the creatinine method for the Technicon RA-1000 is evaluated and compared with the Technicon SMAC methodology. The method-evaluation experiment for this comparison is plotted in Figure 4-5.

Before a method can be put into routine use, the manufacturer's reference range must be validated or even re-established, or the procedure must be written or updated and personnel should be trained to use the method. A statistical quality-control program should be implemented for the procedure using the accumulated replication data to establish quality-control limits. It may be necessary to adjust the control limits once the procedure is in routine service.

Because of constraints of personnel, time, and budget, a laboratory may not be able to perform comprehensive comparison-of-methods experiments on every new method to be introduced. An understanding of the principles of method evaluation and a familiarity with statistical tests will allow the

TABLE 4-7. Performance Standards for Common Analytes

Test	Critical Concentration (X_c)	Medically Allowable Error (E_A)			
		Barnett	Tonks	CLIA[+]	State of the Art*
Calcium	11.0 mg/dL	0.5	1.8	1.0	0.3
Chloride	90 mmol/L	4.0	1.8	4.5	4.5
Chloride	110 mmol/L	4.0	2.2	5.5	4.5
Cholesterol	250 mg/dL	40	25	25	16
Glucose	100 mg/dL	10	10	10	5.2
Glucose	120 mg/dL	10	12	12	5.2
Phosphate	4.5 mg/dL	0.5	0.45		
Potassium	3.0 mmol/L	0.5	0.5	0.2	
Potassium	6.0 mmol/L	0.5	0.5	0.2	
Sodium	130 mmol/L	4.0	2.3	4.0	3.3
Sodium	150 mmol/L	4.0	2.7	4.0	3.3
Total Protein	7.0 g/dL	0.6	0.49	0.7	0.3
Urea Nitrogen	27 mg/dL	4.0	2.7	2.4	2.0
Uric Acid	6.0 mg/dL	1.0	1.0	0.2	

[+] *US Department of Health and Human Services. Medicare, Medicaid, and CLIA Programs. Regulations implementing the clinical laboratory improvement amendments of 1988 (CLIA) final rule. Federal Register 1992;57:7002.*

Arch Pathol Lab Med 116:781–787, 1992.

Source: Adapted from Westgard JO, de Vos DJ, Hunt MR, Quam EF, Carey RN, Garber CC. Concepts and practices in the evaluation of clinical chemistry methods. IV. Decisions of acceptability. Am J Med Technol 1978;44:727.

laboratory supervisor to choose a method that would probably fit the laboratory's performance criteria.[79] A series of abbreviated experiments could then be undertaken to estimate imprecision and inaccuracy. Figure 4-16 shows a simple data input form that can be used to simplify collection of method evaluation data.

QUALITY ASSURANCE AND QUALITY CONTROL

The *quality-control system* is the laboratory's system for recognizing and minimizing analytical errors.[8] Quality control is one component of the *quality-assurance system*, which has been defined as all systematic actions necessary to provide adequate confidence that laboratory services will satisfy given medical needs for patient care.[21] The quality-assurance system encompasses preanalytical, analytical, and postanalytical factors. The monographs *Laboratory quality management* by Cembrowski and Carey[10] and *Cost-effective quality control* by Westgard and Barry[71] provide detailed information about quality-control and quality-assurance practices in the clinical chemistry laboratory.

There are many preanalytical factors that can influence analytical results, including patient preparation, sample collection, sample handling, and storage. Ladenson has written probably the best review of the effects of preanalytical factors on chemistry tests.[36] Preanalytical factors are difficult to monitor and control because most occur outside the lab-

oratory. Health care professionals, especially physicians and nurses, should become more aware of the importance of patient preparation and how it can affect laboratory tests. Patient preparation for a test is critical. For example, nutritional status, a recent meal, alcohol, drugs,[89] smoking, exercise, stress, sleep, and posture all affect various laboratory tests. The laboratory must provide instructions, usually in the form of a procedure manual,[50] for proper patient preparation and specimen acquisition. This procedure manual should be found in all nursing units and thus be available to all medical personnel. Additionally, easy-to-understand patient handouts must be available for outpatients.

Specimen-collection procedures should follow specific guidelines such as those established by the NCCLS.[52,53,55,56] Blood-collection teams should be reminded periodically

TABLE 4-8. Single-Value Criteria of Westgard et al[60]

Analytical Error	Criterion
Random error (RE)	$1.96s < E_A$
Proportional error (PE)	$|(\text{Recovery} - 100) \times X_0/100)| < E_A$
Constant error (CE)	$|\text{Bias}| < E_A$
Systematic error (SE)	$|(y_0 + mX_c) - X_c| < E_A$
Total error (TE = RE + SE)	$1.96s + |(y_0 + mX_c) - X_c| < E_A$

E_A = medically allowable error
X_c = critical concentration

TABLE 4-9. Example of the Application of the Single-Value Criteria of Westgard et al to Creatinine Evaluation Data*

1. Random error (RE) = 1.96s
   Random error is estimated from analyzing a control product once daily for 20 days.

   $$X = 2.0 \text{ mg/dL}, \ s = 0.040 \text{ mg/dL}$$
   $$\text{RE} = 1.96 \times s$$
   $$= 1.96 \times 0.040 \text{ mg/dL}$$
   $$= 0.078 \text{ mg/dL}$$

   Because RE < E_A, RE is acceptable.

2. Proportional error (PE) = $|(\text{Recovery} - 100) \times (X_C/100)|$
   Proportional error is estimated from a series of recovery experiments, after which the average recovery is calculated.

   $$\text{Average recovery for creatinine} = 98\%$$
   $$\text{PE} = |(98 - 100) \times (2/100)|$$
   $$= 0.04 \text{ mg/dL}$$

   Because PE < E_A, PE is acceptable.

3. Constant error (CE) = bias
   Constant error is estimated from the bias $|\bar{Y} - \bar{X}|$, derived from the comparison-of-methods experiment.

   $$\text{CE} = |\bar{Y} - \bar{X}|$$
   $$= |2.90 - 3.05|$$
   $$= 0.15 \text{ mg/dL}$$

   Because CE < E_A, CE is acceptable.

4. Systematic error (SE) = $|(Y_O + mX_C) - X_C|$
   Systematic error is estimated from the above equation in which Y_O and m are derived from a comparison-of-methods experiments (Fig. 4-5).

   $$Y_O = -0.098, \ m = 0.988$$
   $$\text{SE} = |(-0.098 + 0.988 \times 2) - 2)|$$
   $$= |1.88 - 2|$$
   $$= 0.12 \text{ mg/dL}$$

   Because SE < E_A, SE is acceptable.

5. Total error (TE) = RE + SE
   Total error is estimated from the sum of RE and SE as calculated above

   $$\text{TE} = 0.078 + 0.12$$
   $$= 0.20 \text{ mg/dL}$$

   Because TE < E_A, the method is acceptable.

Assume a performance standard (E_A) of 0.4 mg/dL at a critical concentration (X_C) of 2 mg/dL.

about the guidelines for duration of application of tourniquets and the types of specimen-collection tubes and anticoagulants to be used. Methods of specimen transport, separation, aliquoting, and storage are critical.[54,57] The length of time elapsed between drawing and separation of the serum or plasma from the cells can be a factor in analytical testing. For example, leukocytes and erythrocytes metabolize glucose and cause a steady decrease in glucose concentration in clotted, uncentrifuged blood. The centrifuging and aliquoting of samples to secondary containers may be critical.[55,57] Contamination of the specimen may occur at this point, rendering a less than optimal specimen for analytical testing. For example, secondary containers for specimens submitted for lead analysis must be scrupulously clean because of the ubiquitous presence of lead and the low levels of the substance that must be measured.

Specimen storage also may lead to errors in the reported results. Guidelines on storage requirements for specimens should be established for each analyte. Specimens may be affected by evaporation (*e.g.,* electrolytes) and exposure to light (*e.g.,* bilirubin), refrigeration (*e.g.,* LD), freezing, and so on. Clerical errors may occur at any step in the processing of specimens. While the use of computers has simplified clerical tasks, one specimen may still be mistaken for another. Obviously, such mistakes should be minimized.

The laboratorian is more able to control the analytical factors, which depend heavily on instrumentation and reagents. A schedule of daily and monthly preventive maintenance for each piece of equipment is essential. Instrument function checks that are to be routinely performed should be detailed in the procedure manual and their performance should be documented. The NCCLS has developed standards for monitoring variables such as water quality,[42] calibration of analytical balances, calibration of volumetric glassware and pipettes, stability of electrical power,[59] and the temperature of thermostatically controlled instruments.[58] Reagents and kits should be dated when received and also when opened. New lots of reagents should be run in parallel with old reagent lots before being used for analysis.

If the reagents are to be used as standards or calibrators, the most highly purified chemicals, reagent grade or ACS (American Chemical Society) grade, should be used. Different types of standards are available. A *primary standard* is a stable, nonhygroscopic (does not absorb water), highly purified substance that can be dried, preferably at 104°C to 110°C without a change in composition. Primary standards can thus be dried and then weighed to prepare solutions of selected concentrations. When purchased, primary standards are supplied with a record of analysis for contaminating elements, which should not exceed 0.05% by weight. Some standards have been certified to be pure by various official bodies such as the National Bureau of Standards and the CAP.

A *primary standard* is defined as the most highly purified substance currently available that can be weighed analytically. A *secondary standard* is one whose concentration is usually determined by analysis by an acceptable reference method that is calibrated with a primary standard. Its concentration cannot be determined directly from the weight of solute and volume of solution. Calibrators are used in calibration processes to establish concentrations of patient specimens. Calibration materials should meet the identity, labeling, and performance requirements of NCCLS guideline C22, *Tentative Guideline for Calibration Materials in Clinical Chemistry.*[43] Whenever possible, calibrators should have their concentrations assigned through the use of either reference methods or other very specific methods. The comparative analytical response of the calibrator and the specimens provides the

PATIENT COMPARISONS:

COMPARATIVE METHOD: _____

TEST METHOD: _____

	DATE	COMP	TEST
1			
2			
3			
4			
5			
6			
7			
8			
9			
10			
11			
12			
13			
14			
15			
16			
17			
18			
19			
20			
21			
22			
23			
24			
25			
26			
27			
28			
29			
30			

(MID RANGE)

	DATE	COMP	TEST
31			
32			
33			
34			
35			
36			
37			
38			
39			
40			
41			
42			
43			
44			
45			
46			
47			
48			
49			
50			
51			
52			
53			
54			
55			
56			
57			
58			
59			
60			

(LO RANGE)

	DATE	COMP	TEST
61			
62			
63			
64			
65			
66			
67			
68			
69			
70			
71			
72			
73			
74			
75			
76			
77			
78			
79			
80			
81			
82			
83			
84			
85			
86			
87			
88			
89			
90			

(HI RANGE)

LINEARITY:

	A	B	C	D	E
TARGET					
1					
2					
3					
MEASURED					

PRECISION:

WITHIN RUN

	LO	MID	HI
1			
2			
3			
4			
5			
6			
7			
8			
9			
10			
$\bar{X}$			
SD			
CV			

CONTROL PRODUCT: _____
I: _____
II: _____
III: _____

BETWEEN RUN

	DATE	I	II	III
1				
2				
3				
4				
5				
6				
7				
8				
9				
10				
11				
12				
13				
14				
15				
16				
17				
18				
19				
20				
$\bar{X}$				
SD				
CV				

◄

Figure 4-16. Data input form for method evaluation experiment. (Courtesy of Kristen Lambrecht)

basis for calculating values for patient specimens. The same material should not serve as calibrator and control.

Laboratory error can be minimized if attention is paid to proper laboratory procedures and techniques. The quality-control system for individual test methodologies can focus on controlling the test-specific variables. The postanalytical factors consist of the recording and reporting of the patient data to the physician within the appropriate time interval. With automation and computer-generated patient reports, the incidence of errors in the postanalytical phase has decreased greatly.

Quality Control

The purpose of the quality-control system is to monitor analytical processes, detect analytical errors during analysis, and prevent the reporting of incorrect patient values. Analytical methods are usually monitored by analyzing stable control materials and then comparing the observed values with their expected value. The laboratory's budget must reflect the importance of quality control. The monetary commitment is important to ensure an adequate system for monitoring and improving the laboratory's performance. Many large laboratories have a full-time technologist responsible for quality control. This individual is in charge of reviewing quality-control data and keeping the staff informed about the status of the analytical methods, acting as liaison among the laboratory director, the chief technologist, and the bench technologists.

The statistical system used to interpret the measured concentrations of the controls is called the *statistical quality-control system*. The principles of statistical quality control were established by Shewhart[65] early in this century. In 1950,

Levey and Jennings[38] used these same basic principles when they introduced statistical quality control to the clinical laboratory. Since 1950, statistical quality-control systems in the laboratory have undergone many modifications.

Analytical error may be separated into random-error and systematic-error components (Fig. 4-17). *Random error* affects precision and is the basis for varying differences between repeated measurements. Increases in random error may be caused by factors such as technique and temperature fluctuations. *Systematic error* arises from factors that contribute a constant difference, either positive or negative, and directly affects the estimate of the mean. Increases in systematic error can be caused by poorly made standards or reagents, failing instrumentation, poorly written procedures, and so on.

Control materials should behave like real specimens, be available in sufficient quantity to last a minimum of 1 year, be stable over that period, be available in convenient vial volumes, and vary minimally in concentration and composition from vial to vial.[44] The control material should closely resemble the specimen it is simulating in both its physical and chemical characteristics. The control material should be tested in exactly the same manner as patient specimens. Control materials should span the clinically important range

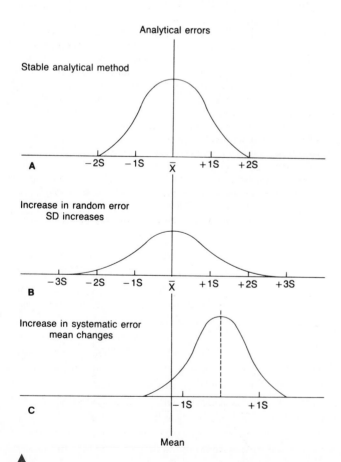

of the analyte's concentrations. Controls in the normal and low abnormal and high abnormal ranges are preferred. The manufacturer may assay its controls with various instruments and methodologies and then offer target ranges for its control product. These assayed controls are more expensive but can be used as external checks for accuracy.

Many commercially prepared control materials are lyophilized and require reconstitution before use. In reconstitution, care should be exercised to add and mix the proper amount of diluent. Incomplete mixing yields a partition of supernatant liquid and underlying sediment and will result in incorrect control values. Frequently, the reconstituted material will be more turbid (cloudy) than the actual patient specimen. Stabilized frozen controls do not require reconstitution but may behave differently from patient specimens in some analytical systems. It is important to carefully evaluate these stabilized controls with any new instrument system.

While some chemistry laboratories used to prepare all their control materials,[5] there are serious shortcomings in their preparation and use. It is difficult to minimize the infectious disease risk of "home-made" materials. Additionally, compared with commercial controls, these control materials are more susceptible to deterioration and contamination. On occasion, it is necessary to prepare control pools for selected analytes, such as drugs. Proper procedures should be followed,[5] but the task is more manageable because far smaller quantities are required than for a laboratory-wide pool.

Most control materials are produced from human serum. With greater emphasis placed on cost containment, more laboratories are using bovine-based control materials, which are lower in price than the human-based materials. The stability of bovine-based control materials is similar to that of human-based materials. For most analytes, bovine material meets the necessary requirements for monitoring imprecision.[9] Because bovine proteins differ greatly from human proteins, bovine material is inappropriate for immunochemistry assays of specific proteins. Similarly, bovine material is inappropriate for some dye-binding procedures for albumin and certain bilirubin methods. Bovine-based material can be used as a control in electrophoresis, but its electrophoretic pattern differs from that of human control serum and resembles a polyclonal gammopathy.

General Operation of a Statistical Quality-Control System

The statistical quality-control system in the clinical laboratory is used to monitor the analytical variations that occur during testing. In certain instances, these variations may be systematic and are caused by procedural errors due to technique, instrumentation, or failures of reagents or other materials. In other instances, however, random variations may appear despite tightly controlled, well-calibrated analytical methods.

The statistical quality-control program can be thought of as a three-stage process:

Figure 4-17. Schematic representation of distributions of control data. (*A*) represents the situation with no analytical error; (*B*) represents increased random error; (*C*) represents systematic shift.

1. Establishing allowable statistical limits of variation for each analytical method.
2. Using these criteria to evaluate the quality-control data generated for each test.
3. Taking remedial action when indicated (*i.e.,* finding causes of errors, rectifying them, and reanalyzing patient and control data).

To establish statistical quality control on a new instrument or on new lots of control material, the different levels of control material must be analyzed for at least 20 days. The means ($\bar{x}$) and standard deviations (s) of these control data are then calculated. Because of the small number of observations and possible outliers in the data, these initial estimates may not be totally reliable and should be revised as more data become available. When changing between different lots of similar material, many laboratorians use the newly obtained mean as the target mean but retain the standard deviation used for the previous control product. As more data are obtained, all of them should be averaged to derive the best estimates of the mean and standard deviation.[73]

Control values may be compared with statistical limits numerically or by display on a control chart. This chart is simply an extension of a Gaussian distribution curve (Fig. 4-18) with time expressed on the x axis. The y axis usually is scaled to provide a concentration made from $\bar{x} - 3s$ to $\bar{x} + 3s$. Horizontal lines corresponding to multiples of s are drawn around the x axis. The 2s lines correspond to the 95.5% limits for the control. If the analytical process is in control, approximately 95% of the points will be within these limits and approximately 5% of the points will be outside these limits. The 3s limits correspond to approximately the 99.7% limits. If the process is in control, no more than 0.3% of the points will be outside the 3s limits. An analytical method is considered in control when there is symmetrical distribution of control values about the mean and there are few control values outside the 2s control limits. Historically, many laboratories have defined an analytical method as out of control if a control value fell outside its 2s limits. In

laboratories that have used the 2s limits as warning limits and 3s limits as error limits, a control value between 2s and 3s would alert the technologist to a potential problem. A point outside the 3s limits would require corrective action.

Laboratories use different criteria for judging whether control results indicate out-of-control situations. Westgard and Groth have studied the error-detection capabilities of most of these criteria.[81] They used the term *control rule* to indicate the criterion for judging whether an analytical process is out of control. To simplify comparison of the various control rules, Westgard and associates used abbreviations for the different control rules. Table 4-10 lists most of the frequently used control rules and their abbreviations. The abbreviations have the form A_L, where A is a symbol for a statistic or is the number of control observations per analytical run and L is the

TABLE 4-10. Popular Control Rules

1_{2s}	Use as a rejection or warning when one control observation exceeds the $\bar{x} \pm 2s$ control limits; usually used as a warning.
1_{3s}	Reject a run when one control observation exceeds the $\bar{x} \pm 3s$ control limits.
2_{2s}	Reject a run when two consecutive control observations are on the same side of the mean and exceed the $\bar{x} + 2s$ or $\bar{x} - 2s$ control limits.
4_{1s}	Reject a run when four consecutive control observations are on the same side of the mean and exceed either the $\bar{x} + 1s$ or $\bar{x} - 1s$ control limits.
$10_{\bar{x}}$	Reject a run when ten consecutive control observations are on the same side of the mean.
R_{4s}	Reject a run if the range or difference between the maximum and minimum control observation out of the last 4 to 6 control observations exceeds 4s.
$\bar{x}_{0.01}$	Reject a run if the mean of the last N control observations exceeds the control limits that give a 1% frequency of false rejection ($P_{fr} = 0.01$).
$R_{0.01}$	Reject a run if the range of the last N control observations exceeds the control limits that give a 1% frequency of false rejection ($P_{fr} = 0.01$).

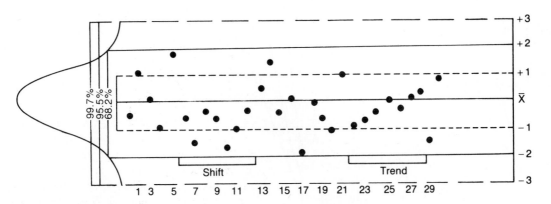

Figure 4-18. Control chart showing the relationship of control limits to the Gaussian distribution. Daily control values are graphed, and they show examples of a shift, an abrupt change in the analytical process, and a trend, a gradual change in the analytical process.

control limit.[83] For example, a 1_{2s} rule violation indicates the situation in which one control observation is outside the $\bar{x} \pm 2s$ limits. An *analytical run* is defined as a set of control and patient specimens assayed, evaluated, and reported together.

Ideally, if a method is in control, none of the control rules should be violated and there should be no rejection of the analytical run. Unfortunately, some analytical runs will be rejected as out of control even when there is no additional analytical error. For example, application of the 1_{2s} control rule when only one control is analyzed will result in 5% of the controls being outside the $2s$ limits when only the usual analytical variation is present. When more than one control is analyzed and no additional error is present, the probability of at least one control being outside the $2s$ limits becomes higher. When two controls are used, there is a 10% chance that at least one control will be outside the $2s$ limits. When four controls are used, there is a 17% chance of this occurring. For this reason, many analysts usually do not investigate the analytical method if a control exceeds the $2s$ limits when two or more controls are used. They merely reassay the controls or the entire analytical run.

Unfortunately, this intuitive approach to quality control achieves an unknown level of quality. What is needed is a control system that will reliably signal the presence of significant analytical error but will not respond to small errors. Defining such a control system requires an understanding of the response of control rules to analytical error.

Response of Control Rules to Error

Westgard and associates[81] have studied the response of control rules, either singly or in groups, to the presence of error, either systematic or random. Different control procedures (groups of control rules) have distinct responses, depending on the control rules and number of control observations (n) used. Using computers, Westgard and Groth[80] simulated the analysis of control materials by instruments with varying levels of error. A large number of simulations was done at each error level, with the proportion of out-of-control situations tabulated and then plotted against the size of the error. The resulting graphs, depicting probability of rejection versus size of analytical error, are called *power functions*. Ideally, a control rule should have a 0 probability of detecting no error and a 100% probability of detecting significant error. Figure 4-19A shows a graph of a family of power functions for the detection of systematic error by the 1_{2s}

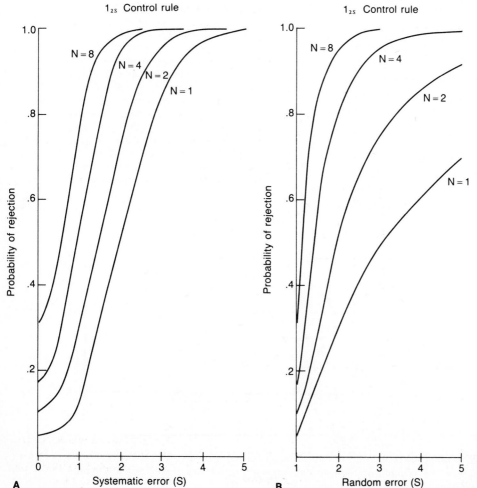

Figure 4-19. Power function curves for the 1_{2s} control rule for systematic error (*A*) and random error (*B*). (Provided by P. Douville)

control rule. The probability of rejection is plotted against the size of the systematic error. The size of the systematic error ranges from 0 to 5s, where s is the standard deviation. The different lines correspond to different numbers of controls analyzed. When there is no error, the probability of rejection is approximately 5% when one control is used. The probability of rejection when no error is present is called the *probability of false rejection* (P_{fr}). The probability of error detection (P_{ed}) is the probability of rejecting an analytical run as out of control when an error exists. The P_{ed} can be determined for any size error by locating the size of the error on the x axis and erecting a vertical line that intersects the power-function curve for the desired n. From this intersection, a horizontal line is drawn to the y axis. The value of the probability on the y axis is the P_{ed}. In Figure 4-19A, P_{ed} for the 1_{2s} control rule and a systematic error of 2s is 0.5 if one control is analyzed. The power-function graphs may be used to determine the effectiveness of various control procedures in detecting analytical errors. Two sets of power functions are necessary, one for systematic error (shift of the mean) and one for the random error (increase in imprecision). Figure 4-19B shows a family of power functions for the detection of random error by the 1_{2s} control rule. Be-

cause analytical processes have random error present, the x axis originates at 1s. In Figure 4-19B, P_{ed} for the 1_{2s} control rule and a doubling of the random error is 0.3 if one control is analyzed. The best control procedure is the one with the lowest P_{fr} and the highest P_{ed} for detecting analytical errors of a size that can compromise the quality of analytical results. Figures 4-20A and B show power functions for the 1_{3s} control rule for systematic and random error, respectively. While the 1_{2s} control rule results in a high detection of moderate-sized systematic errors, the P_{fr} is unacceptably high for *2 or more controls*. On the other hand, the 1_{3s} control rule is less responsive to moderate increases in systematic error but has a low P_{fr}.

Westgard and associates have suggested a manual implementation of a combination of control rules with at least two control observations per analytical run.[73] In addition to using the 1_{2s} rule as a warning rule and the 1_{3s} rule for rejection, this system also can include the 2_{2s}, R_{4s}, 4_{1s}, and $10_{\bar{x}}$ rules. This combination of rules allows for improved detection of both random and systematic error. The counting rules (the 2_{2s}, 4_{1s}, and $10_{\bar{x}}$ rules) are effective in detecting systematic error. The R_{4s} and 1_{3s} rules are effective in detecting random error. The 1_{3s} rule also can detect systematic error.

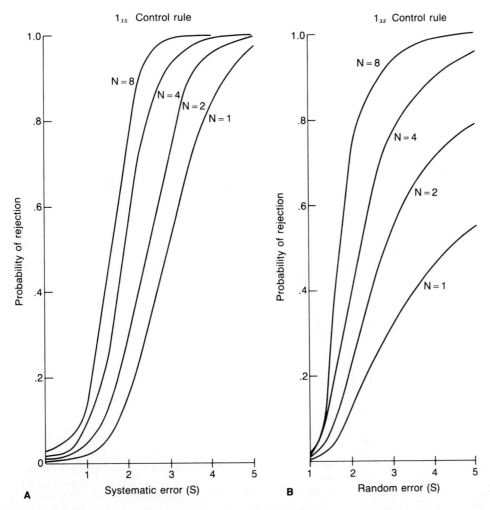

Figure 4-20. Power function curves for the 1_{3s} control rule for (A) systematic error and (B) random error. (Provided by P. Douville)

All new control results are evaluated to ensure that they are within their $\pm 2s$ limits. If they are, there is no further inspection. Otherwise, the other rules are applied in this order: 2_{2s}, 4_{1s}, $10_{\bar{x}}$, and R_{4s}. The 2_{2s} rule is invoked whenever two consecutive control observations on the same side of the mean exceed either the $\bar{x} + 2s$ or $\bar{x} - 2s$ control limits. This rule responds most often to systematic errors. The 2_{2s} rule is initially applied to the control observations within the most recent analytical run (across materials and within run). The rule can then be applied to the last two observations on the same control material but from consecutive runs (within materials and across runs), or it can be applied on the last two consecutive observations of the different control materials (across materials and across runs).

The 4_{1s} rule is violated when four consecutive control observations on the same side of the mean exceed the $\bar{x} + 1s$ or $\bar{x} - 1s$ control limits. It is most responsive to systematic errors. This rule is applied within and across materials and runs. The $10_{\bar{x}}$ rule is sensitive to systematic error and is violated when 10 consecutive control observations fall on one side of the mean. This rule is applied within materials and across runs as well as across materials and across runs.

The R_{4s} rule is violated when the range or difference between the highest and lowest control observations within the run exceeds 4_s. This rule is most responsive to random error or increased imprecision. The rule is invoked when the observation on one control material exceeds a $+2s$ limit and the observation on the other exceeds a $-2s$ limit. The two observations are outside $2s$ limits but in opposite directions, resulting in a range that exceeds $4s$. The range rule is intended for use within a single run with a maximum of four to six control observations, not across runs.

The combination of the 1_{2s}, 1_{3s}, 2_{2s}, 4_{1s}, $10_{\bar{x}}$, and R_{4s} control rules used in conjunction with a control chart has been called the *multirule Shewhart procedure*. The power functions for this multirule procedure are shown in Figure 4-21. This multirule procedure yields increased error detection over the use of the 1_{3s} rule alone. The inclusion of the 1_{3s} and R_{4s} rules improves the detection of random error, and the 2_{2s}, 4_{1s}, and $10_{\bar{x}}$ rules increase the detection of systematic error. This multirule procedure provides the laboratory with improved error detection and less false rejection of analytical runs. It is easily adapted to existing control procedures and has become extremely popular since its initial description in 1981.

Implementation of the multirule procedure involves the following:

1. Calculation of the means and standard deviations of the different concentrations of control materials.

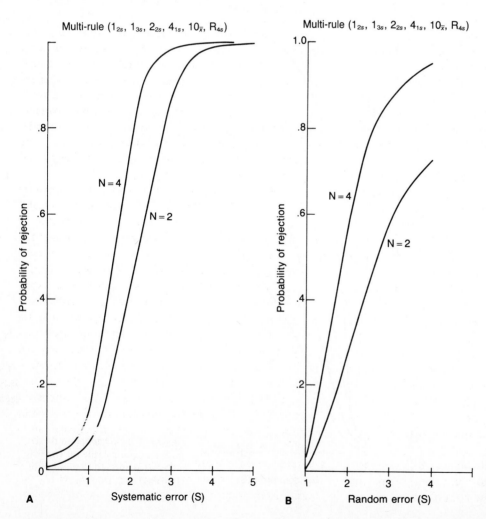

Multi-rule (1_{2s}, 1_{3s}, 2_{2s}, 4_{1s}, $10_{\bar{x}}$, R_{4s})

Multi-rule (1_{2s}, 1_{3s}, 2_{2s}, 4_{1s}, $10_{\bar{x}}$, R_{4s})

N = 4

N = 2

Probability of rejection

Systematic error (S)

A

N = 4

N = 2

Probability of rejection

Random error (S)

B

Figure 4-21. Power functions for the multirule control procedure (*A*) for systematic error and (*B*) random error. (Provided by P. Douville)

2. Construction of control charts, with lines indicating the $0, \pm 1, 2$, and 3 standard deviation limits.

3. Analysis of the different levels of control material in each analytical run, with plotting of the control data on the appropriate chart.

4. Accepting the analytical run if each observation falls within $2s$ limits.

5. Checking the $1_{3s}, 2_{2s}, R_{1s}, 4_{1s}$, and $10_{\bar{x}}$ rules for violations if one of the control materials is outside its $2s$ limits. If none of the rules is violated, the run is accepted. If a violation has occurred, the run is rejected, with the most likely type of error determined (random versus systematic). Once the error condition is identified and corrected, the patient and control specimens are reanalyzed.

Figure 4-22 shows a two-level control chart that simplifies implementation of the multirule procedure. Application of the multirule procedure to a two-level control system is illustrated in Figure 4-23. Interpretation of the control data is summarized in Table 4-11. Note that the R_{1s} rule is not applied across runs. The astute reader will note that the multirule procedure will detect $10_{\bar{x}}$ or 4_{1s} rule violations only if there is a 1_{2s} rule violation. On occasion, the $10_{\bar{x}}$ and 4_{1s} rules may be violated without the activation of the 1_{2s} warning rule. Such occurrences will be infrequent and would be indicative of small systematic errors that would eventually be detected by the multirule procedure.

The power functions in Figure 4-21 show that when the above multirule procedure is used with four control observations, it will detect moderate-sized errors (e.g., a $2s$ shift or a doubling of the random error) with a probability of approximately 50%. Smaller errors cannot be reliably detected. More sensitive control procedures are required for assays in which a $2s$ shift or a doubling of the standard deviation will cause misclassifications of clinical status. For example, the standard deviation of many calcium assays is 0.15 mg/dL for a normal range calcium. A positive shift of 0.30 mg/dL can easily place a normocalcemic patient into the hypercalcemic category. The sensitivity of the multirule procedure can be increased to detect smaller systematic errors by increasing the number of observations considered. Other control procedures can be used. The mean and range control procedure[30] has significantly greater power for the detection of error if each control material is analyzed in replicate in each analytical run. Practically, this procedure requires computer implementation, because calculations of the mean and range are required with each analytical run.

The *cumulative summation (cusum) technique* has been described for routine laboratory use[82]; it also requires computer implementation. The cusum procedure is very responsive to systematic error and can be used with the 1_{3s} rule. Cusum is very sensitive to small, persistent shifts that commonly occur in the modern, low-calibration-frequency analyzer. There are other control procedures that use exponentially smoothed averages and standard deviations.[13,80] These procedures have been implemented on several commercially available laboratory information quality-control systems.

Many of today's analyzers are capable of very high accuracy and precision. The magnitude of the analytical standard deviation may be very small when compared with that of the medically allowable error (e.g., the standard deviation of glucose on certain analyzers approaches 1 mg/dL, which is much smaller than the medically allowable error of 10 mg/dL). Only very large errors need be detected when these analyzers measure analytes such as glucose. The sensitivity of the control procedure must be reduced rather than increased. One way of doing this is by incorporating fewer observations, e.g., using the 1_{3s} control rule, or even expanding the control limits, e.g., using $\pm 3.5s$ control limits (i.e., the $1_{3.5s}$ control rule).[11]

Westgard et al recommend a novel approach for selecting control procedures based on the magnitude of the error to be detected and the frequency of errors.[85] They have constructed "selection grids" consisting of 3×3 tables that list optimal control procedures. Figure 4-24 shows such a grid. Multirule procedures with optimal performance characteristics form each of the nine entries in the grid.

The Use of Patient Data for Quality Control

Various algorithms have been proposed for manipulating patient data to determine whether processing or analytical errors have occurred. This section will focus on two different control procedures that use patient data: the average of patient data, and delta checks. Other quality-control procedures that use patient data include the review of individual outlying results to identify gross clerical errors (sometimes called *limit checks*) and the routine analysis of duplicate specimens, as is frequently done in endocrinology assays. Some laboratories use duplicate analyses of another type: patient-sample comparisons. These comparisons require the regular analysis of split samples on instruments that measure the same analyte. Differences between instruments that exceed predetermined limits are investigated and corrected.

The use of the averages of patient data was described by Hoffman and Waid in 1965.[33] In their "average of normals" method, an error condition was signaled when the average of consecutive centrally distributed patient data was beyond the control limits established for the average of the patient data. The assumption underlying the average of normals is that the patient population is stable. Any shift would thus be secondary to a systematic analytical error. Cembrowski, Chandler, and Westgard studied the use of average of patients with computer simulations and found that its error-detection capabilities depended on several factors.[12] The most important were the number of patient results averaged and the ratio of the standard deviation of the patient population (s_p) to the standard deviation of the analytical method (s_a). Other important factors included the limits for evaluating the mean (control limits), the limits for determining which patient data are averaged (truncation limits), and the magnitude of the population lying outside the truncation limits. Cembrowski et al stated that the technique could be used to supplement reference-sample quality control and recommended computer imple-

INSTRUMENT: ACA
ANALYTE: CSF Glucose
QC PRODUCT: Quantimetrix

LEVEL I LOT NUMBER: 45091 MEAN: 52 SD: 1.5
LEVEL II LOT NUMBER: 45092 MEAN: 93 SD: 1.5

DATE	REAGENT LOT #	EXP. DATE	TECH	COMMENTS	STOCK #	CAL CHECK	CALIBRATE
2-5-90	NQ188A	4/1/91					
2-6	"		KM	Return			
2-13	"						
2-16	"		BP				
2-19	"		BP				
2-22	"		BP	Return			X
2-28	"						
3-6	"		AO				

Level I (MEAN 52):
- 2-13: 52, 53
- 2-16: 52, 53
- 2-19: 52, 53
- 2-22: 52, 53

Level II (MEAN 93):
- 2-13: 93, 94
- 2-16: 95
- 2-19: 94
- 2-22: 95
- 2-28: 93
- 3-6: 94

Figure 4-22. Two-level control chart for implementing multirule procedures.

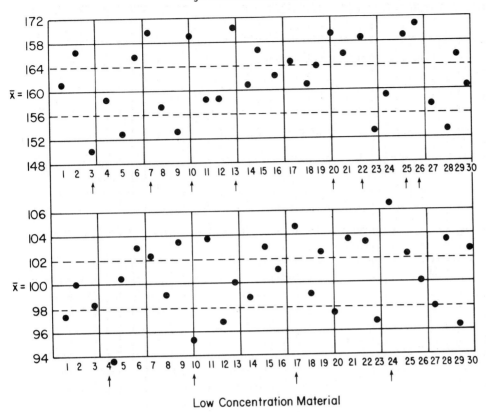

High Concentration Material

Low Concentration Material

◄ *Figure 4-23.* Control data to illustrate use of the multirule procedure. The arrows correspond to days in which there are violations of at least one rule.

mentation. Douville, Cembrowski, and Strauss evaluated the averages of patient endocrine data and demonstrated high error-detection capabilities for thyroid testing.[17]

Cembrowski et al investigated the use of the anion gap for quality control and found that the practice of reanalyzing single specimens with abnormally high or low anion gaps usually resulted in the needless repetition of analyzing patient specimens.[14] More often these single specimens had genuinely abnormal anion gaps. An improved control procedure consists of averaging eight consecutive patient anion

gaps and comparing the average with the control limits for the anion-gap average.[6]

The last algorithm to be discussed is the *delta check,* in which the most recent result of a patient is compared with the previously determined value. The differences between consecutive laboratory data (deltas) may be compared with limits that have been established by various investigators.[35,64] The differences are usually calculated in two ways: as numerical differences (current value minus last value) and as percentage differences (numerical differences times 100 di-

TABLE 4-11. Interpretation of the Multi-rule Shewhart Control Charts of Figure 4-20 for High and Low Concentration Materials

Control Material	Day	Rule Violation
High	Day 3	1_{2s} rule violation—warning—accept run
Low	Day 4	1_{3s} rule violation—reject run
High + low	Day 7	4_{1s} rule violation—reject run—across runs and materials
High + low	Day 10	R_{4s} rule violation—reject run—within run, across materials
High	Day 13	1_{2s} rule violation—warning—accept run
Low	Day 17	1_{2s} rule violation—warning—accept run
Low	Day 20	1_{2s} rule violation—warning—accept run
High	Day 22	$10_{\bar{x}}$ rule violation—reject run—within materials, across runs
Low	Day 24	1_{3s} rule violation—reject run
High	Day 25	1_{2s} rule violation—warning—accept run
High	Day 26	2_{2s} rule violation—reject run—within materials, across runs

Westgard Multi-rule QC Selection Grid	PROCESS STABILITY (Frequency of Errors, f)		
	>10%	2% – 10%	<2%
PROCESS CAPABILITY (Magnitude of errors, SEc) <2.0s	1_{3s} / 2_{2s} / R_{4s} / 4_{1s} / $12\bar{x}$ N =6	1_{3s} / 2_{2s} / R_{4s} / 4_{1s} / $8\bar{x}$ N = 4	1_{3s} / 2_{2s} / R_{4s} / 4_{1s} N = 2
2.0s to 3.0s	1_{3s} / 2_{2s} / R_{4s} / 4_{1s} / $8\bar{x}$ N = 4	1_{3s} / 2_{2s} / R_{4s} / 4_{1s} N = 2	1_{3s}/2_{2s}/R_{4s}/(4_{1s} W) N =2
>3.0s	1_{3s} / 2_{2s} / R_{4s} / 4_{1s} N = 2	1_{3s}/2_{2s}/R_{4s}/(4_{1s} W) N =2	1_{3s}/(4_{1s} W) N =2

© J. O. Westgard, E.F. Quam, P.L. Barry, University of Wisconsin, Madison, WI 1989

◀

Figure 4-24. QC selection grid for Westgard multirule algorithm. (Reproduced with permission)

vided by the current value). Wheeler and Sheiner evaluated the performance of several delta check methods and classified each delta check investigated as either a true or false positive.[87] They found that the percentage of true positives ranged from 5% to 29% and concluded that the delta check methods could detect errors otherwise overlooked but at the cost of investigating many false positives. In their tertiary-care hospital population, there were many false positives caused by large excursions in laboratory values secondary to disease or therapy.

External Quality Control

Proficiency-testing programs periodically provide samples of unknown concentrations of analytes to participating laboratories. Participation in these programs is concentrations of analytes to participating laboratories. Participation in these programs is mandated by the HCFA under the CLIA-88. The CAP and the American Association of Bioanalysts are probably the two largest providers of proficiency testing programs in the United States. They provide samples for all major qualitative and quantitative chemistry areas, including general chemistry, protein chemistry, urinalysis, toxicology, and endocrinology. Samples for quantitative analysis contain multiple analytes and are usually provided as lyophilized materials. Once a laboratory receives its samples, it must analyze and return its results within a specified time to a computer center for compilation and comparison with the results of other participating laboratories. The computer center establishes target values and ranges of acceptable results based on either the average of the participants' values or reference laboratories' values.

For the CAP proficiency program, the means and standard deviations of all results from peer (similar instrument and methodology) laboratories are computed. Then values beyond three standard deviations from the mean are discarded, and the mean and standard deviation are recomputed. A participant result is classified as acceptable if the difference between the result and the target answer (usually the peer mean) is less than the allowable error. CLIA-88 has defined allowable errors for a large number of regulated analytes[67]; some are shown in Table 4-7. These allowable errors are sometimes referred to as "fixed limits" and are expressed either in measurement units of the analyte (*e.g.,* ±0.5 mmol/L from the

mean for potassium) or as percentages (*e.g.,* ±10% for total cholesterol).

For a far smaller number of analytes (*e.g.,* thyroid stimulating hormone), CLIA-88 has statistically defined limits of acceptability. For these analytes the participant result is acceptable if it falls within ±3 SDI of the group mean. The comparison with statistical limits requires calculation of the deviation of the survey results from the mean, which is expressed in numbers of standard deviation indices (SDI) above or below the mean. The SDI is the numerical difference between an individual laboratory's results and the mean, divided by the standard deviation. A deviation outside ±3 SDIs usually is considered unacceptable because such deviations will occur only 0.3% of the time as a result of chance.

Compared with statistical limits, fixed-limits criteria will more often demonstrate the need for replacement of outmoded or unreliable instrumentation. Ehrmeyer and Laessig have used computer simulations to evaluate the ability of many different proficiency-testing schemes to detect poorly performing laboratories. The performance of all of these schemes is imperfect, with some good laboratories being judged as poor and some poor laboratories escaping detection.[19,20]

Cembrowski, Hackney, and Carey have proposed a multirule system to evaluate the HCFA-mandated proficiency test result.[12a,12b] When significant deviations are detected in a set of five survey results (one or more observations exceeding ±3 SDIs, or the range of the observations exceeding 4 SDIs or the mean of the 5 results exceeding ±1.5 SDIs), the laboratory records, including the internal quality-control results, should be reviewed. Mixups of proficiency specimens or of proficiency and clinical specimens should be ruled out. Whenever possible, an aliquot of the survey specimen should be saved and assayed again for the analytes yielding erroneous results. Results that still deviate significantly after retesting indicate a long-term bias. If the deviations are variable in magnitude and direction, there may be a problem with imprecision (random error). In the event that repeat analysis yields satisfactory results, the error probably represented a random error or transient bias encountered during the testing period.

External proficiency testing is also used to determine estimates of the state of the art of interlaboratory performance. The CAP regularly publishes summaries of its interlaboratory comparisons in the *Archives of Pathology and Laboratory Medicine.*

Toward Quality Patient Care

Much of this chapter has focused of the delivery of *accurate* laboratory test results. Accuracy in laboratory testing is only one quality characteristic that is required of the clinical chemistry laboratory.[10] Other equally important quality characteristics include effective test request forms; clear instructions for patient preparation and specimen handling; appropriate turnaround times for specimen processing, testing, and result reporting; appropriate reference ranges; and intelligible result reports. Most clinical laboratories provide accurate laboratory testing. We are just beginning to appreciate that most laboratory failures or mistakes occur in the preanalytical or postanalytical realm.

For example, Ross and Boone reviewed 363 incidents that occurred in a large tertiary-care hospital in 1987.[61] Of the 336 medical records investigated, they found that pre-analytical and postanalytical mistakes accounted for 46% and 47% of the total incidents, respectively. Preanalytical mistakes included missed or incorrectly interpreted laboratory orders, improper patient preparation, incorrect patient identification, wrong specimen container, and mislabeled or mishandled specimen. Postanalytical mistakes included delayed, unavailable, or incomplete results. Nonlaboratory personnel were responsible for 29% of the mistakes.

Most of these errors were interdepartmental; prevention of such errors requires a coordinated interdepartmental group approach. Committed representation of all relevant players is a prerequisite for success, whether they be physician, nurse, ward clerk, phlebotomist, or analyst. These individuals should form a *quality team* or *improvement team* and should meet regularly to characterize the problem and, eventually, to make recommendations for its solution. The group should be able to use various statistical and group techniques. In order for these *group processes* to succeed, there must be total commitment from management. Management must be trained in this "quality-improvement process"[31,72] and must support its growth throughout the institution. Such approaches are working extraordinarily well in Japanese industry and are now being successfully transferred to American businesses and even the hospital and clinic environment.[4,37] It is only through such quality-improvement efforts that we can significantly improve total patient care.

Point-of-Care Testing: The Newest Challenge

The development of portable, simple-to-use, clinical analyzers has made point-of-care (POC) or near patient testing a reality.[12c] Many different analytes can now be measured accurately and quickly at the patient's side, whether the patient is in a hospital, an ambulance, or even in an airliner. It is the thesis of POC testing that laboratory abnormalities can be rapidly detected and followed by effective therapy. Faster turnaround of results to the clinician should simplify patient care, improve outcomes, and thus reduce the total cost of patient care.

POC analyzers should be affordable, easily maintainable, and have a flexible test menu. They should be easily interfaced with available computer systems and calibration and quality control should be automated. The cost of consumables in a POC testing laboratory will never approach those in the high-volume central laboratory. However, the cost may be offset by savings in length of stay and greater staff efficiency. POC testing may prove cost beneficial if there is drastic reorganization of the laboratory, with acute testing done by POC instruments and all other tests sent to a cost-effective, low-overhead, high-quality off-site laboratory.

Clinical laboratorians must help define the evolving role of POC technology and testing. The implementation of POC testing will be successful when clinical laboratorians work with clinical users who require these services and with the manufacturers of the instrument and reagent systems. This collaboration sets the stage for communication and mutual resolution of POC testing problems. Work teams will make decisions related to the establishment of new POC testing programs based on review of the reliability of the method, training and staffing requirements, frequency of testing, necessity of near patient testing, and enhanced outcomes for patient care. One of the great advantages of POC testing is that it will be accomplished by nonlaboratorians. This advantage is a double-edged sword as these individuals will have little formal training in quality control. As such, the dominant issue in POC testing will be ensuring the quality of patient results.

PRACTICE PROBLEMS

Problem 4-1: Calculation of Sensitivity and Specificity

Alpha-fetoprotein (AFP) levels are used by obstetricians to help diagnose neural tube defects (NTD) in early pregnancy. For the following data, calculate the sensitivity, specificity, and efficiency of AFP for detecting NTD, as well as the predictive value of a positive AFP.

Number of Pregnancies
Interpretation of AFP Findings

Outcome of Pregnancy	Positive (NTD)	Negative (No NTD)	Total
NTD	5	3	8
No NTD	4	843	847
Total	9	846	855

Problem 4-2: A Management Decision in Quality Control

You are in charge of the clinical laboratory when a technologist presents you with his glucose worksheet. A 2_{2s} rule violation has occurred across runs and within materials on the high-concentration material. You ask to see the patient data and the previous control data. They follow:

Glucose Worksheet January 8

Samples	Results	January Glucose Control Values		
		Date	Low	High
Control—High	224	1/1	86	215
Patient	117	1/2	82	212
Patient	85	1/3	83	218
Patient	98	1/4	87	214
Patient	74	1/5	85	220
Patient	110	1/6	81	217
Control—Low	83	1/7	88	223
Patient	112	1/8	83	224
Patient	120			
Patient	97			
Patient	105			

1. Plot these control data.
2. What do you observe about these control data?
3. What might be a potential problem?
4. Should you report the patient data for today? Why or why not?

See Figure 4–25.

Problem 4-3: Interdepartmental Communication

You are having a problem with the medical intensive care unit (MICU) and the arterial blood gas specimens they submit to the laboratory. In the last 3 weeks you have refused to perform blood gas analyses on six different MICU specimens because of small clots found in the specimens. The MICU staff is furious with the rejection policy, yet you believe the analyses will be incorrect if these specimens are used.

1. Outline where the problem lies.
2. What can be done to remedy this problem?
3. Why would your present quality-control system not detect this sort of error?

The following problems represent the steps in a method-evaluation study. An abbreviated data set is used to encourage hand calculations by the student. Perform the calculations for the following experimental data, which were obtained from a glucose study. The test method is a coupled glucose oxidase procedure. The comparative method is the hexokinase method currently in use.

Problem 4-4: Precision (Replication)

For the following precision data, calculate the mean, standard deviation, and coefficient of variation for each of the two control solutions A and B. These control solutions were chosen because their concentrations were close to medical decision levels (X_c) for glucose: 120 mg/dL for control solution A

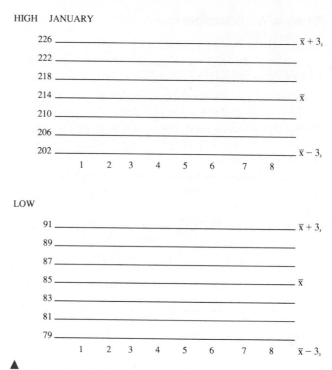

Figure 4-25. Blank control chart for Problem 4-2.

and 300 mg/dL for control solution B. Control solution A was analyzed daily, and the following values were obtained:

118, 120, 121, 119, 125, 118, 122, 116, 124, 123, 117, 117, 121, 120, 120, 119, 121, 123, 120, and 122 mg/dL

Control solution B was analyzed daily and gave the following results:

295, 308, 296, 298, 304, 294, 308, 310, 296, 300, 295, 303, 305, 300, 308, 297, 297, 305, 292, and 300 mg/dL

Problem 4-5: Recovery

For the recovery data below, calculate the percent recovery for each of the individual experiments and the average of all the recovery experiments. The experiments were performed by adding two levels of standard to each of five patient samples (A through E) with the following results:

Sample	0.9 mL Serum + 0.1 mL Water	0.9 mL Serum + 0.1 mL 500 mg/dL STD	0.9 mL Serum + 0.1 mL 1000 mg/dL STD
A	59	110	156
B	63	112	160
C	76	126	175
D	90	138	186
E	225	270	320

What do the results of this study indicate?

Problem 4-6: Interference

For the interference data below, calculate the concentration of ascorbic acid added, the interference for each individual sample, and the average interference for the group of patient samples. The experiments were performed by adding 0.1 mL of a 150-mg/dL ascorbic acid standard to 0.9 mL of five different patient samples (A through E). A similar dilution was prepared for each patient sample using water as the diluent. The results follow:

Sample	0.9 mL Serum + 0.1 mL Water	0.9 mL Serum + 0.1 mL 150 mg/dL STD
A	54	46
B	99	91
C	122	112
D	162	152
E	297	286

What do the results of this study indicate?

Problem 4-7: Linear Regression

The following method-comparison data were obtained (mg/dL):

Sample	By Hexokinase (mg/dL)	By Coupled Glucose Oxidase (mg/dL)
1	191	192
2	97	96
3	83	85
4	71	72
5	295	299
6	63	61
7	127	131
8	110	114
9	320	316
10	146	141

1. Graph these results. From inspection of the graph, determine whether there is significant constant or proportional error.
2. Calculate the linear regression statistics as follows:
   a. Set up a table with the following column headings:

   $$x_1, y_1, x^2i, y^2i, x_i\, y_i, Y_i$$

   b. Enter the x_i and y_i data into the table (remember that x is the comparative method and y the test method) and calculate $\Sigma x_i, \Sigma y_i, \Sigma x^2i, \Sigma y^2i,$ and $\Sigma x_i y_i$. Enter these data into the table.
   c. From the summations in (b), calculate $\bar{x}$ and $\bar{y}$. Use Equations 4-8 and 4-9 to calculate the slope m and the y intercept (y_0), respectively.
   d. Using the regression equation $Y = mx + y_0$, calculate y_i for each x_i, enter the Y_i values in the Y_i column, and then calculate:

$$(y_i - Y_i), (y_i - Y_i)^2, \text{and } \Sigma(y_i - Y_i)^2$$

   Enter these data into the table.
   e. From the summation in (d) and Equation 4-10, calculate $s_{y/x}$.
   f. From the summations in (b) and Equation 4-11, calculate r.
3. Report the following statistics for the comparison of methods experiment: $m, y_o, s_{y/x},$ and r.

Problem 4-8: Interpretation

Use the statistics calculated in Problem 4-7 to answer the following questions. Explain your answers by referring to the statistical value you used to arrive at your answer. Where appropriate, calculate errors at the two medical decision levels $X_{c1} = 120$ mg/dL and $X_{c2} = 300$ mg/dL.

1. What is the random error (RE) of the test method? What is the magnitude of the statistic that quantitates random error *between* the methods?
2. What are the constant (CE) and proportional (PE) errors?
3. Calculate the systematic error (SE) at both X_c. What is the predominant nature of the systematic error (refer to 2)?
4. What is the total error (TE = RE + SE) of the method? *Note:* Linear regression statistics should be used for error estimates only within the concentration range studied; we should have collected data up to 300 mg/dL in the comparison-of-methods experiment.
5. a. What is the statistic that quantitates random error between methods?
   b. What value did you calculate for this statistic?
6. Judge the acceptability of the performance of the test method. To reach your judgment, apply the following criteria:
   a. In order for the method to be accepted, all errors must be less than the allowable error (E_a) for a given X_c.
   b. The following are the E_a for glucose:

$$X_{c1} = 120 \text{ mg/dL} E_{a1} = 10 \text{ mg/dL}$$
$$X_{c2} = 300 \text{ mg/dL} E_{a2} = 25 \text{ mg/dL}$$

Problem 4-9

1. You are appointed as a clinical laboratorian to a point-of-care (POC) work team. Who else might serve on this team?
2. What needs to be in place before implementing a POC testing protocol to ensure accurate and precise patient results?
3. Once you decide to provide physicians and patients with POC testing, what are important features and/or requirements of POC analyzers?

REFERENCES

1. American Chemical Society, Committee on Environmental Improvement, Subcommittee on Environmental Analytical Chemistry. Guidelines for data acquisition and data quality evaluation in environmental chemistry. Anal Chem 1980;52:2242.

2. Barnett RN. Medical significance of laboratory results. Am J Clin Pathol 1968;50:671.

3. Barnett RN. Clinical laboratory statistics. 2nd ed. Boston: Little, Brown, 1979.

4. Berwick DM. Continuous improvement as an ideal in health care. N Engl J Med 1989;320:53.

5. Bowers GN, Burnett RW, McComb RB. Preparation and use of human serum control materials for monitoring precision in clinical chemistry. In: Selected methods for clinical chemistry. Vol 8. Washington: American Association of Clinical Chemists, 1977;21.

6. Bockelman HW, Cembrowski GS, Kurtycz DFI, Garber CC, Westgard JO, Weisberg HF. Quality control of electrolyte analyzers: Evaluation of the anion gap average. Am J Clin Pathol 1984;81:219.

7. Brawer MK, Lange PH. Prostate-specific antigen and premalignant change: Implications for early detection. CA Cancer J Clin 1989;39:361.

8. Buttner J, Borth R, Boutwell JH, Broughton PMG. Provisional recommendation on quality control in clinical chemistry. Clin Chem 1976;22:532.

9. Caputo MJ, Doumas BT, Gadsden RH, Laessig RH, Laessig RH. Bovine-based serum as quality control material: A comparative analytical study of bovine vs. human controls. Fullerton, CA: Hyland Diagnostics, 1981.

9a. Catalona WJ, Richie JP, deKernion JB, et al. Comparison of prostate specific antigen concentration versus prostate specific antigen density in the early detection of prostate cancer: Receiver operating characteristic curves. J Urol 1994;152:203.

10. Cembrowski GS, Carey RN. Laboratory quality management. Chicago: ASCP Press, 1989.

11. Cembrowski GS, Carey RN. Considerations for the implementation of clinically derived quality control procedures. Lab Med 1989;20:400.

12. Cembrowski GS, Chandler EP, Westgard J. Assessment of "average of normals" quality control procedures and guidelines for implementation. Am J Clin Pathol 1984;81:492.

12a. Cembrowski GS, Crampton C, Byrd J, Carey RN. Detection and classification of proficiency testing errors in HCFA-regulated analytes. Application to ligand assays. J Clin Immunoassay 1995;17:210.

12b. Cembrowski GS, Hackney JR, Carey N. The detection of problem analytes in a single proficiency test challenge in the absence of the Health Care Financing Administration rule violations. Arch Pathol Lab Med 1993;117:437.

12c. Cembrowski GS, Kiechle FL. Point of care testing: Critical analysis and practical application. Adv Pathol Lab Med 1994;7:3.

13. Cembrowski GS, Westgard JO, Eggert AA, Toren E. Trend detection in control data: Optimization and interpretation of Trigg's technique for trend analysis. Clin Chem 1975;21:139.

14. Cembrowski GS, Westgard JO, Kurtycz DFI. Use of anion gap for the quality control of electrolyte analyzers. Am J Clin Pathol 1983;79:688.

15. Chatfield C. Statistics for technology: A course in applied statistics. 3rd ed. New York: Chapman & Hall, 1983.

16. Cornbleet PJ, Gochman N. Incorrect least-squares regression coefficients in method-comparison analysis. Clin Chem 1979; 25:432.

17. Douville P, Cembrowski GS, Strauss J. Evaluation of the average of patients, application to endocrine assays. Clin Chim Acta 1987;167:173.

18. Ehrmeyer SS, Laessig RH, Leinweber JE, et al. Medicare/CLIA final rules for proficiency testing: Minimum intralaboratory performance characteristics (CV and bias) need to pass. Clin Chem 1990;36:1736.

19. Ehrmeyer SS, Laessig RH, Schell K. Use of alternate rules (other than the 1_{2s}) for evaluating interlaboratory performance data. Clin Chem 1988;34:250.

20. Ehrmeyer SS, Laessig RH. External proficiency testing. In: Cembrowski GS, Carey RN, eds. Laboratory quality management. Chicago: ASCP Press, 1989;227.

21. Elin RJ. Elements of cost management for quality assurance. Pathologist 1980;34:182.

22. Feldman U, Schmeider B, Klinkers H. A multivariate approach for the biometric comparison of analytical methods in clinical chemistry. Clin Chem Clin Biochem 1981;19:121.

23. Fleisher GA, Eickelberg ES, Elveback LR. Alkaline phosphatase activity in the plasma of children and adolescents. Clin Chem 1977;23:469.

24. Foti AG, Cooper JF, Herschman H, Malvaez RR. Detection of prostatic cancer by solid-phase radioimmunoassay of serum prostatic acid phosphatase. N Engl J Med 1977;297:1357.

25. Galen RS, Gambino SR. Beyond normality: The predictive value and efficacy of medical diagnoses. New York: Wiley, 1975.

26. Glick MR, Ryder KW, Jackson SA. Graphical comparisons of interferences in clinical chemistry instrumentation. Clin Chem 1986;32:470.

27. Glick MR, Ryder KW. Analytical systems ranked by freedom from interferences. Clin Chem 1987;33:1453.

28. Grasbeck R, Siest G, Wilding P, Williams GZ, Whitehead TP. Provisional recommendation on the theory of reference values: I. The concept of reference values. Clin Chem 1979;25:1506.

29. Guinan P, Bush I, Ray V, Vieth R, Rao R, Bhatti R. The accuracy of the rectal examination in the diagnosis of prostatic carcinoma. N Engl J Med 1980;303:499.

30. Hainline A Jr. Quality assurance: Theoretical and practical aspects. In: Selected methods for the small clinical chemistry laboratory. Washington: American Association of Clinical Chemistry, 1982;17.

31. Harrington HJ. The improvement process. New York: McGraw-Hill, 1987.

32. Harris EK, Cooil BK, Shakarji G, Williams GZ. On the use of statistical models of within person variation in long-term studies of healthy individuals. Clin Chem 1980;26:383.

33. Hoffman RG, Waid ME. The "average of normals" method of quality control. Am J Clin Pathol 1965;43:134.

34. Krouwer JS, Rabinowitz R. How to improve estimates of imprecision. Clin Chem 1984;30:290.

35. Ladenson JH. Patients as their own controls: Use of the computer to identify "laboratory error." Clin Chem 1975;21:1648.

36. Ladenson JH. Nonanalytical sources of variation in clinical chemistry results. In: Sonnenwirth AC, Jarett L, eds. Gradwohl's clinical laboratory methods and diagnosis. 8th ed. St. Louis: Mosby, 1980;149.

37. Laffel G, Blumenthal D. The case for using industrial quality management science in health care organizations. JAMA 1989;262:2869.

38. Levey S, Jennings ER. The use of control charts in the clinical laboratories. Am J Clin Pathol 1950;20:1059.

39. Martin HF, Gudzinowicz BJ, Driscoll JL. An algorithm for the selection of proper group intervals for histograms representing clinical laboratory data. Am J Clin Pathol 1975;64:327.

40. Meites S, ed. Pediatric clinical chemistry: Reference (normal) values. Washington: American Association of Clinical Chemistry, 1989.

41. Munan L, Kelly A, Petitclerc C, Billon B. Atlas of blood data. Canada: University of Sherbrooke, 1978.

42. NCCLS Document C3-A2. Approved guideline for preparation and testing of reagent water in the clinical laboratory. 2nd ed. Villanova, PA: NCCLS, 1991.

43. NCCLS Document C22-T. Tentative guideline for calibration materials in clinical chemistry. Villanova, PA: NCCLS, 1982.

44. NCCLS Document C23-T. Tentative guideline for control materials in clinical chemistry. Villanova, PA: NCCLS, 1982.

45. NCCLS Document C28-T. Tentative guideline for how to define and determine reference intervals in the clinical laboratory. Villanova, PA: NCCLS, 1994.

46. NCCLS Document EP5-T2. Tentative guideline for precision performance of clinical chemistry devices. Villanova, PA: NCCLS, 1992.

47. NCCLS Document EP6-P. Proposed guideline for evaluation of linearity of quantitative analytical methods. Villanova, PA: NCCLS, 1986.

48. NCCLS Document EP7-P. Proposed guideline for interference testing in clinical chemistry. Villanova, PA: NCCLS, 1986.

49. NCCLS Document EP9-T. Tentative guideline for method comparison and bias estimation using patient samples. Villanova, PA: NCCLS, 1993.

50. NCCLS Document GP2-A2: Approved Guideline for Clinical Laboratory Procedure Manuals. 2d ed. Villanova, Pa., NCCLS, 1992.

51. NCCLS Document GP10-T. Tentative guideline for assessment of clinical sensitivity and specificity of laboratory tests using receiver operating characteristic (ROC) plots. Villanova, PA: NCCLS, 1993.

52. NCCLS Document H3-A3. Approved standard for procedures for the collection of diagnostic blood specimens by venipuncture. 3rd ed. Villanova, PA: NCCLS, 1991.

53. NCCLS Document H4-A3. Approved standard for procedures for the collection of diagnostic blood specimens by skin puncture. 3rd ed. Villanova, PA: NCCLS, 1991.

54. NCCLS Document H5-A2. Approved standard for procedures for the handling and transport of domestic diagnostic specimens and etiologic agents. 2nd ed. Villanova, PA: NCCLS, 1985.

55. NCCLS Document H11-A2. Approved standard for the percutaneous collection of arterial blood for laboratory analysis. 2nd ed. Villanova, PA: NCCLS, 1992.

56. NCCLS Document H14-A2. Approved guideline for devices for collection of skin puncture specimens. 2nd ed. Villanova, PA: NCCLS, 1990.

57. NCCLS Document H18-A. Approved guideline for procedures for the handling and processing of blood specimens. Villanova, PA: NCCLS, 1990.

58. NCCLS Document 12-A2. Approved standard for temperature calibration of water baths, instruments, and temperature sensors. 2nd ed. Villanova, PA: NCCLS, 1990.

59. NCCLS Document 15-A. Approved standard for power requirements for clinical laboratory instruments and for laboratory power sources. Villanova, PA: NCCLS, 1980.

60. Robertson EA, Zweig MH, Van Steirteghem AC. Evaluating the clinical efficacy of laboratory tests. Am J Clin Pathol 1983;79:78.

61. Ross JW, Boone DJ. Assessing the effect of mistakes in the total testing process on the quality of patient care. [Abstract] Presented at the 1989 Institute on Critical Issues in Health Laboratory Practice, sponsored by the Centers for Disease Control and the University of Minnesota, April 9–12, 1989, Minneapolis, MN.

62. Ryder K, Glick M, Glick S. Calculation of charges for erroneous results caused by interfering substances in an outpatient chemistry laboratory. [Abstract]. Clin Chem 1989;35:1155.

63. Siest G, Galteau MM. Drug effects on laboratory test results. Littleton, MA: PSG Publishing, 1988.

64. Sheiner LB, Wheeler LA, Moore JK. The performance of delta check methods. Clin Chem 1979;25:2034.

65. Shewhart WA. Economic control of quality of the manufactured product. New York: Van Nostrand, 1931.

66. Tonks D. A study of the accuracy and precision of clinical chemistry determinations in 170 Canadian laboratories. Clin Chem 1963;9:217.

67. US Department of Health and Human Services. Medicare, Medicaid, and CLIA programs. Regulations implementing the clinical laboratory impovement amendments of 1988 (CLIA)k final rule. Federal Register 1992;57:7002.

68. Waakers PJM, Hellendoorn HBA, Op De Weegh GH, Heerspink W. Applications of statistics in clinical chemistry: A critical evaluation of regression lines. Clin Chim Acta 1975;64:173.

69. Weisbrot IM. Statistics for the clinical laboratory. Philadelphia: Lippincott, 1985.

70. Westgard JO. Precision and accuracy: Concepts and assessment by method evaluation testing. Crit Rev Clin Lab Sci 1981;13:28c.

71. Westgard JO, Barry Pd. Cost-effective quality control: Managing the quality and productivity of analytical processes. Washington: AACC Press 1986.

72. Westgard JO, Barry PL. Total quality control: Evolution of quality management systems. Lab Med 1989;20:377.

73. Westgard JO, Barry PL, Hunt MR, Groth T. A multi-rule Shewhart chart for quality control in clinical chemistry. Clin Chem 1981;27:493.

74. Westgard JO, Carey RN, Wold S. Criteria for judging precision and accuracy in method development and evaluation. Clin Chem 1974;20:825.

75. Westgard JO, deVos DJ, Hunt MR, Quam EF, Carey RN, Garber CC. Concepts and practices in the evaluation of clinical chemistry methods: I. Background and approach. Am J Med Tech 1978;44:290.

76. Westgard JO, deVos DJ, Hunt MR, Quam EF, Carey RN, Garber CC. Concepts and practices in the evaluation of clinical chemistry methods: II. Experimental procedures. Am J Med Tech 1978;44:420.

77. Westgard JO, deVos DJ, Hunt MR, Quam EF, Carey RN, Garber CC. Concepts and practices in the evaluation of clinical chemistry methods: III. Statistics. Am J Med Tech 1978;44:552.

78. Westgard JO, deVos DJ, Hunt MR, Quam EF, Carey RN, Garber CC. Concepts and practices in the evaluation of clinical chemistry methods: IV. Decisions of acceptability. Am J Med Tech 1978;44:727.

79. Westgard JO, deVos DJ, Hunt MR, Quam EF, Carey RN, Garber CC. Concepts and practices in the evaluation of clinical chemistry methods: V. Applications. Am J Med Tech 1978;44:803.

80. Westgard JO, Groth T. Design and evaluation of statistical control procedures: Applications of a computer "quality control simulator" program. Clin Chem 1981;27:1536.

81. Westgard JO, Groth T. Power functions for statistical control rules. Clin Chem 1979;25:863.

82. Westgard JO, Groth T, Aronsson T, de Verdier CH. Combined Shewhart-cusum control chart for improved quality control in clinical chemistry. Clin Chem 1977;23:1881.

83. Westgard JO, Groth T, Aronsson T, Falk H, de Verdier CH. Performance characteristics of rules for internal quality control: Probabilities for false rejection and error detection. Clin Chem 1977;23:1857.

84. Westgard JO, Hunt MR. Use and interpretation of common statistical tests in method-comparison studies. Clin Chem 1973;19:49.

85. Westgard JO, Quam BS, Barry PL. QC selection grids for planning QC procedures. Clin Lab Sci 1990;3:271.

86. Westgard JO, Seehafer JJ, Barry PL. European specifications for imprecision and inaccuracy compared with operating specifications that assure the quality required by US CLIA proficiency-testing criteria. Clin Chem 1994;40;1228.

87. Wheeler LA, Sheiner LB. A clinical evaluation of various delta check methods. Clin Chem 1981;27:5.

88. Winsten S. The ecology of normal values in clinical chemistry. Crit Rev Clin Lab Sci 1976;6:319.

89. Young DS. Effects of drugs on clinical laboratory tests. 3rd ed. Washington: American Association of Clinical Chemistry, 1990.

Chapter 5

Analytical Techniques and Instrumentation

Ioannis Laios, Alan H. B. Wu

Objectives

2/24   Tues.

Upon completion of this chapter, the clinical laboratorian should be able to:

- *Explain the general principles behind each of the analytical methods described.*
- *Discuss the limitations of each analytical technique.*
- *Compare and contrast the various analytical techniques.*
- *Discuss existing clinical applications for each analytical technique.*
- *Describe the operation and component parts of the following instruments: spectrophotometer, atomic absorption spectrometer, fluorometer, gas chromatograph, osmometer, ion selective electrode, and pH electrode.*
- *Outline the quality assurance and preventive maintenance procedures involved with the following instruments: spectrophotometer, atomic absorption spectrometer, fluorometer, gas chromatograph, osmometer, ion selective electrode, and pH electrode.*

KEY WORDS

Atomic Absorption	Fluorometry	Liquid Chromatography
Chemiluminescence	Gas Chromatography	Spectrophotometry
Electrochemistry	Ion Selective	
Electrophoresis	Electrodes	
Flame Photometry		

SPECTROPHOTOMETRY AND PHOTOMETRY

The instruments that measure electromagnetic radiation have several concepts and components in common. Shared instrumental components will be discussed in some detail in a later section. Photometric instruments measure light intensity without consideration of wavelength. Most instruments today either use filters (photometers), prisms, or gratings (spectrometers) to select (isolate) a narrow range of the incident wavelength. Radiant energy that passes through an object will be partially reflected, absorbed, and transmitted.

Electromagnetic radiation is described as photons of energy traveling in waves. The relationship between wavelength and energy E is described by Planck's formula.[1]

$$E = h\nu \qquad \textit{(Eq. 5-1)}$$

where h = a constant
ν = frequency

Since the frequency of a wave is inversely proportional to the wavelength, it follows that the energy of electromagnetic radiation is inversely proportional to wavelength. Figure 5-1A shows this relationship. Electromagnetic radiation includes a spectrum of energy from short-wavelength, highly energetic gamma and x-rays on the left in Figure 5-1B to long-wavelength radio frequencies on the right. Visible light falls in between, with the color violet at 400 nm and red at

Michael L. Bishop, Janet L. Duben-Engelkirk, and Edward P. Fody.
CLINICAL CHEMISTRY. © 1996 Lippincott–Raven Publishers.

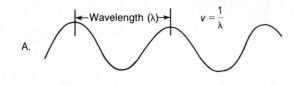

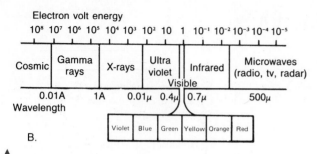

▲

Figure 5-1. Electromagnetic radiation—relationship of energy and wavelength.

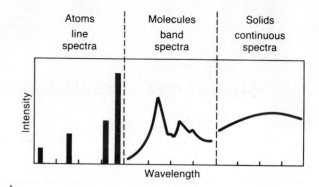

▲

Figure 5-2. Characteristic absorption or emission spectra. (From Coiner D. Basic concepts in laboratory instrumentation. ASMT Education and Research Fund, Inc. 1975–1979.)

700 nm wavelength being the approximate limits of the visible spectrum.

The instruments discussed in this section measure either absorption or emission of radiant energy to determine concentration of atoms or molecules. The two phenomena, absorption and emission, are closely related. In order for a ray of electromagnetic radiation to be absorbed, it must have the same frequency as a rotational or vibrational frequency in the atom or molecule that it strikes. Levels of energy that are absorbed move in discrete steps, and any particular type of molecule or atom will absorb only certain energies and not others. When energy is absorbed, valence electrons move to an orbital with a higher energy level. Following energy absorption, the excited electron will fall back to the ground state by emitting a discrete amount of energy in the form of a characteristic wavelength of radiant energy.

Absorption or emission of energy by atoms results in a line spectrum. Because of the relative complexity of molecules, they absorb or emit a bank of energy over a large region. Light emitted by incandescent solids is in a continuum. The three types of spectra are shown in Figure 5-2.[3,22]

Beer's Law

The relationship between absorption of light by a solution and the concentration of that solution has been described by Beer and others. Beer's law states that the concentration of a substance is directly proportional to the amount of light absorbed or inversely proportional to the logarithm of the transmitted light. Percent transmittance (% T) and absorbance A are related photometric terms that will be explained in this section.

Figure 5-3A shows a beam of monochromatic light entering a solution. Some of the light is absorbed, and the remainder passes through, strikes a light detector, and is converted to an electric signal. *Percent transmittance* is defined

as the ratio of the radiant energy transmitted T divided by the radiant energy incident on the sample I. All the light absorbed or blocked results in 0% T. A level of 100% T is obtained if no light is absorbed. In practice, the solvent without the constituent of interest is placed in the light path, as in Figure 5-3B. Most of the light is transmitted, but a small amount is absorbed by the solvent and cuvet or is reflected away from the detector. The electrical readout of the instrument is set arbitrarily at 100% T, while the light is passing through the "blank" or reference. The sample containing absorbing molecules to be measured is placed in the light path. The difference in amount of light transmitted by the blank and that transmitted by the sample is due only to the presence of the compound being measured. The % T measured by commercial spectrophotometers is the ratio of the sample transmitted beam divided by the blank transmitted beam.

Equal thicknesses of an absorbing material will absorb a constant fraction of the energy incident upon the layers. For example, in a tube containing layers of solution (Fig. 5-4A), the first layer transmits 70% of the light incident upon it.

▲

Figure 5-3. % Transmittance defined.

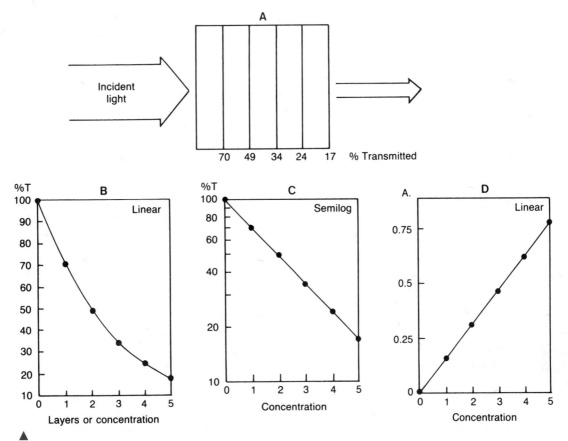

Figure 5-4. (A) % of original incident light transmitted by equal layers of light-absorbing solution; (B) % T versus concentration on linear graph paper; (C) % T versus concentration on semilog graph paper; (D) A versus concentration on linear graph paper.

The second layer will, in turn, transmit 70% of the light incident upon it. Thus, 70% of 70% (49%) is transmitted by the second layer. The third layer transmits 70% of 49%, or 34% of the original light. Continuing on, successive layers transmit 24% and 17%, respectively. The % T values, when plotted on linear graph paper, yield the curve shown in Figure 5-4B. Considering each equal layer as many monomolecular layers, we can translate layers of material to concentration. If semilog graph paper is used to plot the same figures, a straight line is obtained (Fig. 5-4C), indicating that as concentration increases, % T decreases in a logarithmic manner.

Absorbance A is the term used for the amount of light absorbed. It cannot be measured directly by a spectrophotometer, but rather is mathematically derived from % T as follows:

$$\% \ T = \frac{I}{I_0} \times 100 \qquad \textbf{(Eq. 5-2)}$$

where I_0 = incident light
I = transmitted light

Absorbance is defined as

$$A = -\log(I/I_0) = \log (100\%) - \log \% \ T = 2 - \log \% \ T$$

(Eq. 5-3)

According to Beer's law, absorbance is directly proportional to concentration:

$$A = \epsilon \times b \times c \qquad \textbf{(Eq. 5-4)}$$

where ϵ = molar absorptivity, the fraction of a specific wavelength of light absorbed by a given type of molecule
b = length of light path through the solution
c = concentration of absorbing molecules

Absorptivity depends on molecular structure and the way in which the absorbing molecules react with different energies. For any particular molecular type, absorptivity changes as wavelength of radiation changes. The amount of light absorbed at a particular wavelength depends on molecular and ion types present, and may vary with concentration, pH, or temperature.

Since the path length and molar absorptivity are constant for a given wavelength,

$$A \propto C \qquad \textbf{(Eq. 5-5)}$$

Unknown concentrations are determined from a calibration curve that plots absorbance at a specific wavelength versus concentration for standards of known concentration. For calibration curves that are linear and have a zero y intercept,

unknown concentrations can be determined from a single calibrator. Not all calibration curves result in straight lines. Deviations from linearity are typically observed at high absorbances. The stray light within an instrument will ultimately limit the maximum absorbance that a spectrophotometer can achieve. This is typically 2.0 absorbance units.

Spectrophotometric Instruments

A spectrophotometer is used to measure the light transmitted by a solution in order to determine the concentration of the light-absorbing substance in the solution. Figure 5-5 illustrates the basic components of a single-beam spectrophotometer, which are described in subsequent sections.

Components of a Spectrophotometer

Light Source

The most common source of light for work in the visible and near-infrared region is the incandescent tungsten or tungsten-iodide lamp. Only about 15% of the radiant energy emitted falls in the visible region, with most emitted as near-infrared.[3,22] Often, a heat-absorbing filter is inserted between the lamp and sample to absorb the infrared radiation.

The lamps most commonly used for ultraviolet work are the deuterium-discharge lamp or the mercury-arc lamp. Deuterium provides continuous emission down to 165 nm. Low-pressure mercury lamps emit a sharp-line spectrum, with both ultraviolet and visible lines. Medium- and high-pressure mercury lamps emit a continuum from ultraviolet to the midvisible region.[3,22] The most important factors for a light source are range, spectral distribution within the range, the source of radiant production, stability of the radiant energy, and temperature.

Monochromators

Isolation of individual wavelengths of light is an important and necessary function of a monochromator. The degree of wavelength isolation is a function of the type of device used and the width of entrance and exit slits. The bandpass of a monochromator defines the range of wavelengths transmitted and is calculated as width at over half the maximum transmittance (Fig. 5-6).

There are numerous devices used for obtaining monochromatic light. The least expensive are colored-glass filters. These usually pass a relatively wide band of radiant energy and have a low transmittance of the selected wavelength.

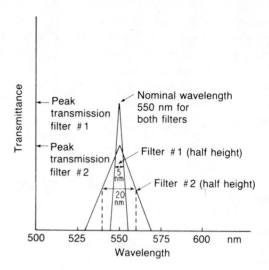

▲
Figure 5-6. Spectral transmittance of two monochromators with band pass at half height of 5 nm and 20 nm.

They are not precise, but they are simple, inexpensive, and useful.

Interference filters produce monochromatic light based on the principle of constructive interference of waves. Two pieces of glass, each mirrored on one side, are separated by a transparent spacer that is precisely one-half the desired wavelength. Light waves enter one side of the filter and are reflected at the second surface. Wavelengths that are twice the space between the two glass surfaces will reflect back and forth, reinforcing others of the same wavelengths, and finally pass on through. Other wavelengths will cancel out because of phase differences (destructive interference). Because interference filters also transmit multiples of the desired wavelengths, they require accessory filters to eliminate these harmonic wavelengths. Interference filters can be constructed to pass a very narrow range of wavelengths with good efficiency.

The prism is another type of monochromator. A narrow beam of light focused on a prism is refracted as it enters the more dense glass. Short wavelengths are refracted more than long wavelengths, resulting in dispersion of white light into a continuous spectrum. The prism can be rotated, allowing only the desired wavelength to pass through an exit slit.

Diffraction gratings are most commonly used as monochromators. This consists of many parallel grooves (15,000 or 30,000 per inch) etched onto a polished surface. Diffraction, the separation of light into component wavelengths, is based

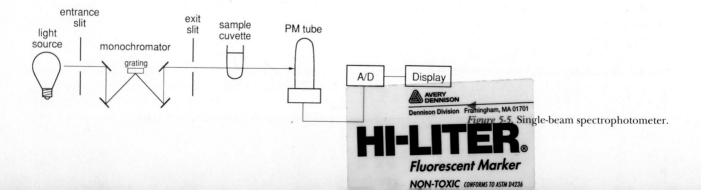

Figure 5-5. Single-beam spectrophotometer.

on the principle that wavelengths are bent as they pass a sharp corner. The degree of bending depends on the wavelength. As the wavelengths move past the corners, wavefronts are formed. Those which are in phase reinforce one another, while those which are not in phase cancel out and disappear. This results in complete spectra. Gratings with very fine line rulings produce a widely dispersed spectrum. They produce linear spectra, called *orders,* in both directions from the entrance slit. Since the multiple spectra have a tendency to cause stray light problems, accessory filters are used.

Sample Cell

The next component of the basic spectrophotometer is the sample cell or cuvet, which may be round or square. The light path must be kept constant in order to have absorbance proportional to concentration. This is easily checked by preparing a colored solution to read midscale when using the wavelength of maximum absorption. Fill each cuvet to be tested, take readings, and save those that match within an acceptable tolerance (*e.g.,* ± 0.25% T). Since it is difficult to manufacture round tubes with uniform diameters, they should be etched to indicate the position for use. Cuvets are sold in matched sets. Square cuvets have plane-parallel optical surfaces and a constant light path. They have an advantage over round cuvets in that there is less error from the lens effect, orientation in the spectrophotometer, and refraction. Cuvets with scratches on their optical surface scatter light and should be discarded. Inexpensive glass cuvets can be used for applications in the visible range, but they absorb light in the ultraviolet region. Quartz cuvets must therefore be used for applications requiring ultraviolet radiation.

Photodetectors

The purpose of the detector is to convert the transmitted radiant energy into an equivalent amount of electrical energy. The least expensive of the devices is known as a *barrier-layer cell,* or *photocell.* The photocell is composed of a film of light-sensitive material, frequently selenium, on a plate of iron. Over the light-sensitive material is a thin, transparent layer of silver. When exposed to light, electrons in the light-sensitive material are excited and released to flow to the highly conductive silver. In comparison with the silver, a moderate resistance opposes the electron flow toward the iron, forming a hypothetical barrier to flow in that direction. Consequently, this cell generates its own electromotive force, which can be measured. The produced current is proportional to incident radiation. Photocells require no external voltage source but rely on internal electron transfer to produce a current in an external circuit. Because of their low internal resistance, the output of electrical energy is not easily amplified. Consequently, this type of detector is used mainly in filter photometers with a wide bandpass producing a fairly high level of illumination so that there is no need to amplify the signal. The photocell is inexpensive and durable, but it is temperature-sensitive and is nonlinear at very low and very high levels of illumination.

Figure 5-7 shows a *phototube,* which is similar to a barrier-layer cell in that it has photosensitive material that gives off electrons when light energy strikes it. It differs in that an outside voltage is required for operation. Phototubes contain a negatively charged cathode and a positively charged anode enclosed in a glass case. The cathode is composed of a material such as rubidium or lithium that will act as a resistor in the dark but will emit electrons when exposed to light. The emitted electrons jump over to the positively charged anode, where they are collected and return through an external, measurable circuit. The cathode usually has a large surface area. Varying the cathode material changes the wavelength at which the phototube gives its highest response. The photocurrent is linear with the intensity of the light striking the cathode as long as voltage between the cathode and anode remain constant. A vacuum within the tubes avoids scattering of the photoelectrons by collision with gas molecules.

The third major type of light detector is the *photomultiplier* (PM) *tube,* which detects and amplifies radiant energy. As shown in Figure 5-8, incident light strikes the coated cathode, emitting electrons. The electrons are attracted to a series of anodes, known as *dynodes,* each having a successively higher positive voltage. These dynodes are of a material that will give off many secondary electrons when hit by single electrons. Initial electron emission at the cathode thus triggers a multiple cascade of electrons within the photomultiplier tube itself. Because of this amplification, the PM tube is 200 times more sensitive than the phototube. PM tubes are used in instruments designed to be extremely sensitive to very low light levels and light flashes of very short duration. The accumulation of electrons striking the anode produces a current signal, measured in amperes, that is proportional to the initial intensity of the light. The analog signal is converted first to a voltage and then to a digital signal through the use of an analog-to-digital (A/D) converter. Digital signals are processed electronically to produce absorbance readings.

In a *photodiode,* absorption of radiant energy by a reversed-biased *pn*-junction diode produces a photocurrent

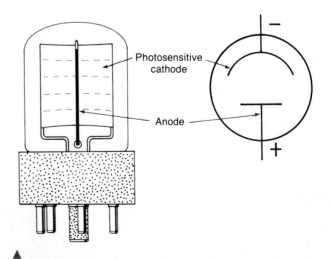

Figure 5-7. Phototube drawing and schematic.

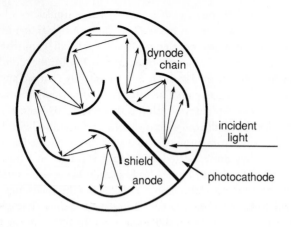

▲
Figure 5-8. Dynode chain in a photomultiplier.

that is proportional to the incident radiant power. Although photodiodes, because of the lack of internal amplification, are not as sensitive as PM tubes, their excellent linearity (six to seven decades of radiant power), speed, and small size make them useful in applications where light levels are adequate.[9] *Photodiode array* (PDA) detectors are available in integrated-circuits containing 256 to 2048 photodiodes in a linear arrangement. A linear array is shown in Figure 5-9. Each photodiode responds to a specific wavelength and as a result a complete UV/Vis spectrum can be obtained in less than a second. Resolution is 1 to 2 nm and depends on the number of discrete elements. In spectrophotometers using PDA detectors, the grating is positioned *after* the sample cuvet and disperses the *transmitted* radiation onto the PDA detector (Fig. 5-9).

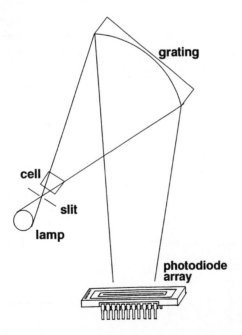

▲
Figure 5-9. Photodiode array spectrophotometer illustrating the placement of the sample cuvet before the monochromator.

For single-beam spectrophotometers, the absorbance reading from the sample must be blanked using an appropriate reference solution that does not contain the compound of interest. Double-beam spectrophotometers permit automatic correction of sample and reference absorbance, as shown in Figure 5-10. Because the intensities of light sources vary as a function of wavelength, double-beam spectrophotometers are necessary when the absorption spectrum for a sample is to be obtained.

Spectrophotometer Quality Assurance

Instrument function should be validated by performing at least the following checks: wavelength accuracy, stray light, and linearity.[23] *Wavelength accuracy* means that the wavelength indicated on the control dial is the actual wavelength of light passed by the monochromator. It is most commonly checked using standard absorbing solutions or filters with absorbance maxima of known wavelength. Didymium or holmium oxide in glass are stable and are frequently used as filters. The filter is placed in the light path, and the wavelength control is set at the wavelength at which maximal absorbance is expected. The wavelength control is then rotated in either direction to locate the actual wavelength that has maximal absorbance. If these two wavelengths do not match, the optics must be adjusted to calibrate the monochromator correctly.

Some instruments with narrow bandpass use a mercury-vapor lamp to verify wavelength accuracy. The mercury lamp is substituted for the usual light source, and the spectrum is scanned to locate mercury emission lines. The wavelength indicated on the control is compared with known mercury emission peaks to determine the accuracy of the wavelength indicator control.

Stray light refers to any wavelengths outside the band transmitted by the monochromator. The most common causes of stray light are reflection of light from scratches on optical surfaces or from dust particles anywhere in the light path, and higher-order spectra produced by diffraction gratings. The major effect is absorbance error, especially in the high-absorbance range. Stray light is detected by using cut-off filters, which eliminate all radiation at wavelengths beyond the one of interest. For example, to check for stray light in the near-ultraviolet region, insert a filter that does not transmit in the region of 200 to 400 nm. If the instrument reading is greater than 0% T, there is stray light present. Certain liquids, such as $NiSO_4$, $NaNO_2$, and acetone, absorb strongly at short wavelengths and can be used in the same way to detect stray light in the UV range.

Linearity is demonstrated when a change in concentration results in a straight-line calibration curve, as discussed under Beer's law. Colored solutions may be carefully diluted and used to check linearity, using the wavelength of maximal absorbance for that color. Sealed sets of different colors and concentrations are available commercially. They should be labeled with expected absorbance for a given bandpass instrument. Less than expected absorbance is an indication of stray

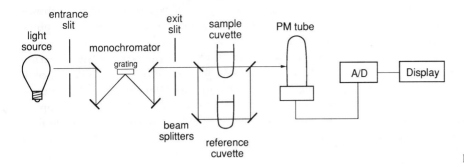

Figure 5-10. Double-beam spectro-photometer.

light or of a bandpass that is wider than specified. Sets of neutral-density filters are also available commercially to check linearity over a range of wavelengths.

A routine system should be devised for each instrument to check and record each of the parameters. The probable cause of a problem and the maintenance required to eliminate it are generally described in the individual instrument manuals.

Atomic Absorption Spectrophotometer

The atomic absorption spectrophotometer is used to measure concentration by detecting absorption of electromagnetic radiation by atoms rather than by molecules. Figure 5-11 shows the basic components. The usual light source, known as a *hollow-cathode lamp,* consists of an evacuated gas-tight chamber containing an anode, a cylindrical cathode, and an inert gas such as helium or argon. When voltage is applied, the filler gas is ionized. Ions attracted to the cathode collide with the metal, knock atoms off, and cause the metal atoms

to be excited. When they return to the ground state, light energy is emitted that is characteristic of the metal in the cathode. A separate lamp generally is required for each metal, *e.g.,* a copper hollow-cathode lamp is used to measure Cu.

Electrodeless discharge lamps are a relatively new light source for atomic absorption spectrophotometers. A bulb is filled with argon and the element to be tested. A radio-frequency generator around the bulb supplies the energy to excite the element, causing a characteristic emission spectrum of the element.[15]

The sample being analyzed must contain the reduced metal in the atomic vaporized state. This is commonly done by using the heat of a flame to break the chemical bonds and form free, unexcited atoms. The flame is the sample cell in this instrument, rather than a cuvet. There are various burner designs, but the most common one is known as a *premix long-path burner.* The sample, in solution, is aspirated as a spray into a chamber where it is mixed with air and fuel. This mixture passes through baffles, where large drops fall and are drained off. Only fine droplets reach the flame. The burner is a long

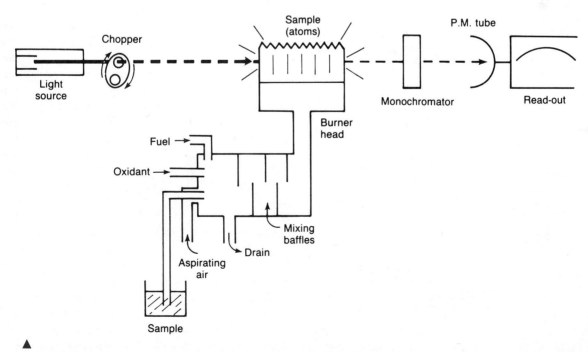

▲
Figure 5-11 Single-beam atomic absorption spectrophotometer—basic components.

and narrow slit, to permit a longer path length for absorption of incident radiation to occur. Light from the hollow-cathode lamp passes through the sample of ground-state atoms in the flame. The amount of light absorbed is proportional to the concentration. When a ground-state atom absorbs light energy, an excited atom is produced. The excited atom then returns to the ground state, emitting light of the same energy as it absorbed. The flame sample thus contains a dynamic population of ground-state and excited atoms, both absorbing and emitting radiant energy. The emitted energy from the flame will go in all directions, and it will be a steady emission. Since the purpose of the instrument is to measure the amount of light absorbed, the light detector must be able to distinguish between the light beam emitted by the hollow-cathode lamp and that emitted by excited atoms in the flame. To do this, the hollow-cathode light beam is modulated by inserting a mechanical rotating chopper between the light and the flame or by pulsing the electric supply to the lamp. Because the light beam being absorbed enters the sample in pulses, the transmitted light also will be in pulses. There will be less light in the transmitted pulses because part of it will be absorbed. There are thus two light signals from the flame: an alternating signal from the hollow-cathode lamp and a direct signal from the flame emission. The measuring circuit is tuned to the modulated frequency. Interference from the constant flame emission is eliminated electronically by accepting only the pulsed signal from the hollow cathode.

The monochromator is used to isolate the desired emission line from other lamp emission lines. In addition, it serves to protect the photodetector from excessive light emanating from flame emissions. A photomultiplier tube is the usual light detector.

Flameless atomic absorption requires an instrument modification that uses an electric furnace to break chemical bonds (electrothermal atomization). A tiny graphite cylinder holds the sample, either liquid or solid. An electric current passes through the cylinder walls, evaporates the solvent, ashes the sample, and finally heats the unit to incandescence to atomize the sample. This instrument, like the spectrophotometer, is used to determine the amount of light absorbed. Thus, again, Beer's Law is used for calculating concentration. A major problem is that background correction is considerably more necessary and critical for electrothermal techniques than for flame-based atomic absorption methods. Currently, the most common approach is to use a deuterium lamp as a secondary source and measure the difference between the two absorbance signals. However, there has also been extensive development of background correction techniques based on the Zeeman effect.[3] Atomic absorption spectrophotometry is very sensitive and precise. Its direct use, however, is limited to elements that exist in the atomic state.

Thurs. 2/26

Flame Photometer

The flame-emission photometer is used to measure light emitted by excited atoms. It is used primarily to determine concentration of Na^+, K^+, or Li^+. Figure 5-12 illustrates the basic components of a typical flame photometer. The sample solution is aspirated into the flame. The purpose of the

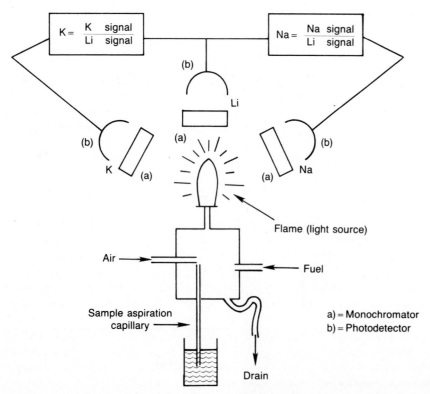

Figure 5-12. Flame-emission photometer—basic components.

flame is twofold. Chemical bonds are broken to produce atoms, and then atoms absorb energy from the flame and enter an excited electronic state. The excited atoms return to the ground state by emitting light energy that is characteristic for that atomic species. The emitted light, focused by lenses or mirrors, passes through monochromators that are selected to transmit radiant energy from the specific atoms. The monochromator for Na^+ is set at 589 nm, that for K^+ is set at 767 nm, and that for Li^+ is set at 671 nm. The emitted light then strikes a photodetector, which generates an electric signal proportional to the concentration of the atoms in the flame.

Fluctuation in the light source of the flame photometer is generally caused by changes in fuel or air pressure that affect flame temperature or rate of sample aspiration. Stability is achieved by using an internal-standard system. For sodium or potassium measurements, lithium can be added in equal amount to all standards and samples to act as an internal standard. The readout, as illustrated in Figure 5-12, is a ratio comparing the emission intensity for the internal standard with that of the element being analyzed. The same amount of lithium is present in the blank, sample, and standard. Any change in aspiration rate or flame temperature will affect the sodium, potassium, and lithium emissions proportionately. The ratio remains constant, thus compensating for possible fluctuations.

Several criteria are required for an internal standard. Concentration of the internal standard must be precisely the same in all samples and standards so that the reference beam is truly constant. The amount of energy required to excite the internal standard must be close to that required to excite the elements being measured. Emission lines must be far enough apart to be separated by available monochromators. Finally, the internal standard must not be normally found in the solution being analyzed. Lithium generally fulfills these requirements and is commonly used as an internal standard, except when lithium is already present in serum, as is sometimes used as an antidepressant agent. When lithium is to be measured, instrument modifications are necessary. Some instruments reverse the K/Li ratio and use a K^+ solution as internal reference so that

$$Li^+ \text{ concentration} = \frac{Li^+ \text{ signal}}{K^+ \text{ signal}} \qquad \textit{(Eq. 5-6)}$$

Other instruments use a fourth filter-photodetector with cesium as the internal standard. In this case, the readout of Na^+, K^+, or Li^+ is the ratio using cesium emission as the reference signal.

Comparison of Flame Methods

Flame-emission photometry is commonly used to measure sodium, potassium, and lithium because these alkali metals are fairly easy to excite. A flame using propane as fuel has sufficient energy to raise the valence electrons of approximately 1% to 5% of the atoms present in the flame to a higher energy level. Although this is relatively inefficient, it is sensitive

enough to measure the concentrations found in biologic specimens. Most other metals (*e.g.,* calcium) are less easily excited and are present in lower concentrations. In these cases, flame-emission photometry is not applicable.

Atomic absorption spectrophotometry is routinely used to measure concentration of trace metals that are not easily excited. It is generally more sensitive than flame emission, since the vast majority of atoms produced in the usual propane or air-acetylene flame remain in the ground state available for light absorption. It is accurate, precise, and very specific. One of the disadvantages, however, is the inability of the flame to dissociate samples into free atoms. For example, phosphate may interfere with calcium analysis by formation of calcium phosphate. This may be overcome by adding cations that compete with calcium for phosphate. Routinely, lanthanum or strontium is added to samples to form stable complexes with phosphate. Another possible problem is the ionization of atoms following dissociation by the flame. This can be decreased by reducing the flame temperature. Matrix interference is another source of error. This is due to the enhancement of light absorption by atoms in organic solvents and/or formation of solid droplets as the solvent evaporates in the flame. Such interference may be overcome by pretreatment of the sample by extraction.

Recently, inductively coupled plasma has been used to increase sensitivity for atomic emission. The torch is an argon plasma maintained by the interaction of a radio-frequency field and an ionized argon gas and is reported to have used temperatures between 5500° and 8000°K. At these temperatures, complete atomization of elements is thought to occur. Use of inductively coupled plasma as a source is recommended for determinations involving refractory elements such as uranium, zirconium, and boron.[15]

Fluorometry

As seen with the spectrophotometer, light entering a solution may pass mainly on through or may be absorbed partly or entirely, depending on the concentration and the wavelength entering that particular solution. Whenever absorption takes place, there is a transfer of energy to the medium. Each molecular type possesses a series of electronic energy levels and can pass from a lower energy level to a higher level only by absorbing an integral unit (quantum) of light that is equal in energy to the difference between the two energy states. There are additional energy levels owing to rotation or vibration of molecular parts. The excited state lasts about 10^{-8} seconds before the electron loses energy and returns to the ground state.[6] Energy is lost by collision, heat loss, transfer to other molecules, and emission of radiant energy. Since the molecules are excited by absorption of radiant energy and lose energy by multiple interactions, the radiant energy emitted is less than the absorbed energy. The difference between the maximum wavelengths, excitation, and emitted fluorescence is called *Stokes shift.* Both excitation (absorption) and fluorescence (emission) energies are characteristic for a given molecular type. For example, Figure 5-13 shows the absorption and fluorescence spectra of quinine in 0.1 N sulfuric acid. The dashed line on the left shows the short-wavelength excitation energy that is maximally absorbed, while the solid line on the right is the longer-wavelength (less energy) fluorescent spectrum.

Basic Instrumentation

Filter fluorometers are used to measure the concentrations of solutions that contain fluorescing molecules. A basic instrument is shown in Figure 5-14. The source emits short-wavelength high-energy excitation light. A mechanical attenuator controls the light intensity. The wavelength that is best absorbed by the solution to be measured is selected by the primary filter placed between the radiation source and the sample. The fluorescing sample in the cuvet emits radiant energy in all directions. The detector, placed at right angles to the sample cell, and a secondary filter that passes the longer wavelengths of fluorescent light prevent incident light from striking the photodetector. The electrical output of the photodetector is proportional to the intensity of fluorescent energy. In spectrofluorometers, filters are replaced by prisms or grating monochromators.

Gas-discharge lamps (mercury and xenon-arc) are the most frequently used sources of excitation radiant energy. Incandescent tungsten lamps are seldom used because little energy is released by them in the ultraviolet region. Mercury-vapor lamps are commonly used in filter fluorometers. Mercury emits a characteristic line spectrum. Resonance lines at 365 to 366 nm are commonly used. Energy at wavelengths other than the resonance lines is provided by coating the inner surface of the lamp with a material that absorbs the 254-nm mercury radiation and emits a broad band of longer wavelengths. Most spectrofluorometers use a high-pressure xenon lamp. Xenon has a good continuum, which is necessary for determining excitation spectra.

Monochromator fluorometers make use of grating, prisms, or filters for isolation of incident radiation. Light detectors are almost exclusively photomultiplier tubes because

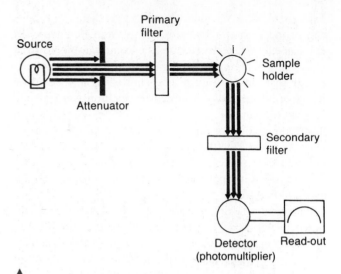

▲
Figure 5-14. Basic filter fluorometer. (From Coiner D. Basic concepts in laboratory instrumentation. ASMT Education and Research Fund, Inc, 1975–1979.)

of their higher sensitivity to low light intensities. Double-beam instruments are used to compensate for instability due to electric-power fluctuation.

Fluorescence concentration measurements are related to molar absorptivity of the compound, intensity of the incident radiation, quantum efficiency of the energy emitted per quantum absorbed, and length of the light path. In dilute solutions with instrument parameters held constant, fluorescence is directly proportional to concentration. Generally, a linear response will be obtained until the concentration of the fluorescent species is so high that the sample begins to absorb significant amounts of excitation light. Figure 5-15 shows a curve demonstrating nonlinearity as concentration increases. The solution must absorb less than 5% of the exciting radiation for a linear response to occur.[6] As with all quantitative measurements, a standard curve must be prepared to demonstrate that the concentration used falls in a linear range.

In fluorescence polarization, the radiant energy is polarized in a single plane. When the sample (fluorophor) is excited, it will emit polarized light along the same plane as the incident light if the fluorophor is attached to a large molecule. In contrast, a small molecule will emit depolarized light, since it will rotate out of the plane of polarization during its excitation lifetime. This technique is widely used for the detection of therapeutic and abused drugs. In the procedure, the sample analyte is allowed to compete with a fluorophor-labeled analyte for a limited antibody to the analyte. The lower the concentration of the sample analyte, the higher is the macromolecular antibody-analyte-fluorophor formed and the lower is the depolarization of the radiant light.

Advantages and Disadvantages of Fluorometry

The two advantages of fluorometry over conventional spectrophotometry are specificity and sensitivity. Fluorometry

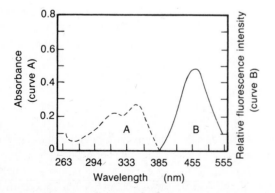

▲
Figure 5-13. Absorption and fluorescence spectra of quinine in 0.1 N sulfuric acid. (From Guilbault GG. Practical fluorescence, theory, methods and techniques. New York: Marcel Dekker, 1973.)

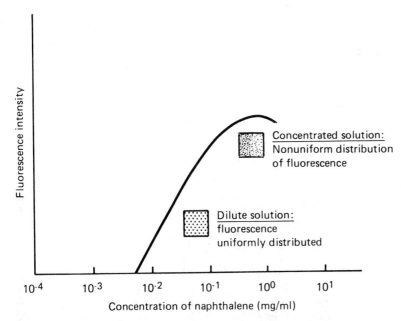

Figure 5-15. Dependence of fluorescence on the concentration of fluorophor. (From Guilbault GG. Practical fluorescence, theory, methods and techniques. New York: Marcel Dekker, 1973.)

increases specificity by selecting the optimal wavelength for both absorption and fluorescence, rather than just the absorption wavelength, as with spectrophotometry.

Fluorometry is about a thousand times more sensitive than most spectrophotometric methods.[6] One of the reasons is that because emitted radiation is measured directly, it can be increased simply by increasing the intensity of the exciting radiant energy. In addition, fluorescence measures the amount of light intensity present over a zero background. In absorbance, on the other hand, the quantity of absorbed light is measured indirectly as the difference between the transmitted beams. At low concentrations, the small difference between 100% *T* and the transmitted beam is difficult to measure accurately and precisely, thereby limiting the sensitivity.

The biggest disadvantage is that fluorescence is very sensitive to environmental changes. Changes in pH affect availability of electrons, and temperature changes the probability of loss of energy by collision rather than fluorescence. Contaminating chemicals or a change of solvents may change the structure. Ultraviolet light used for excitation can cause photochemical changes. Any decrease in fluorescence resulting from any of these possibilities is known as *quenching*. Since so many factors may change the intensity or spectra of fluorescence, extreme care is mandatory in analytical technique and instrument maintenance.[17]

Chemiluminescence

In chemiluminescence reactions, part of the chemical energy generated produces excited intermediates that decay to a ground state with the emission of photons.[13] The emitted radiation is measured with a photomultiplier tube, and the signal is related to analyte concentration. Chemiluminescence is different than fluorescence in that no excitation radiation is required and no monochromators are needed be-

cause the chemiluminescence arises from one species. Most important chemiluminescence reactions are oxidation reactions of luminol, acridinium esters, and dioxetanes and are characterized by a rapid increase in intensity of emitted light followed by a gradual decay. Usually, the signal is taken as the integral of the entire peak. Enhanced chemiluminescence techniques increase the chemiluminescence efficiency by including an enhancer system in the reaction of a chemiluminescent agent with an enzyme. The time course for the light intensity is much longer (60 minutes) than conventional chemiluminescent reactions, which last for about 30 seconds (Figure 5-16).

Advantages of chemiluminescence assays include subpicomolar detection limits, speed (with flash–type reactions light is measured for 10 seconds only), ease of use (most assays are one-step procedures), and simple instrumentation.[12]

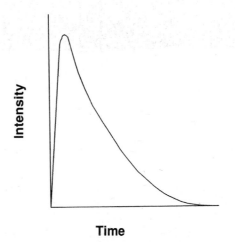

▲

Figure 5-16. Representative intensity versus time curve for a transient chemiluminescence signal.

The main disadvantage is that impurities can cause background signal that degrades sensitivity and specificity.

Turbidity and Nephelometry

Turbidimetric measurements are made with a spectrophotometer to determine concentration of particulate matter in a sample. The amount of light blocked by a suspension of particles depends on concentration, but also on size. Since particles tend to aggregate and settle out of suspension, sample handling becomes critical. Instrument operation is the same as for any spectrophotometer.

Nephelometry is similar, except that light scattered by the small particles is measured at an angle to the beam incident on the cuvet. Figure 5-17 demonstrates two possible optical arrangements for a nephelometer. Light scattering is dependent on wavelength and size of particle. For macromolecules whose size is close to or larger than the wavelength of incident light, sensitivity is increased by measuring forward light scatter.[16] Instruments are available with detectors placed at various forward angles, as well as at 90° to the incident light. Monochromatic light is used to obtain uniform scatter and to minimize sample heating. Some instruments use lasers as a source of monochromatic light, but any monochromator may be used.

Measuring light scatter at an angle other than at 180° in turbidimetry minimizes error from colored solutions and increases sensitivity. Since both methods are dependent on particle size, some instruments quantitate initial change in light scatter rather than total scatter. Reagents must be free of any particles, and cuvets must have no scratches.

Applications of Lasers

Light amplification by stimulated emission of radiation (LASER) is based on the interaction of radiant energy and suitably excited atoms or molecules. The interaction leads to stimulated emission of radiation. The wavelength, direction of propagation, phase, and plane of polarization of the emitted light are the same as those of the incident radiation. Laser light has narrow spectral width, small cross-sectional area with low divergence, and is polarized and coherent. The radiant emission can be very powerful and can be either continuous or pulsating.

Laser light can serve as the source of incident energy in a spectrometer or nephelometer. Some lasers can produce bandwidths of a few kilohertz in both the visible and infrared regions, thereby making these applications about three to six orders more sensitive than conventional spectrometers.[18]

Laser spectrometry also can be used for the determination of structure and identification of samples, as well as for diagnosis. Quantitation of samples is dependent on the spectrometer used. An example of the clinical application of the laser is the Coulter counter, which is used for differential analysis of white blood cells.[5]

ELECTROCHEMISTRY

Many types of electrochemical analyses are used in the clinical laboratory. Examples are potentiometry, amperometry, coulometry, and polarography. The two basic electrochemical cells involved in these analyses are galvanic and electrolytic cells.

Galvanic and Electrolytic Cells

An electrochemical cell can be set up as shown in Figure 5-18. It consists of two half-cells and a salt bridge, which can be a piece of filter paper saturated with electrolytes. The electrodes can be immersed in a single large beaker containing a salt solution instead of the two shown. In such a setup, the solution serves as the salt bridge.

In a galvanic cell, as the electrodes are connected, there is spontaneous flow of electrons from the electrode with the lower electron affinity (oxidation), *e.g.,* silver, and these electrons pass through the external meter to the cathode (reduction), where OH^- ions are liberated. This reaction continues until one of the chemical components is depleted, at which point the cell is "dead" and cannot produce electrical energy to the external meter.

Current may be forced to flow through the "dead" cell only by applying an external electromotive force E. This is

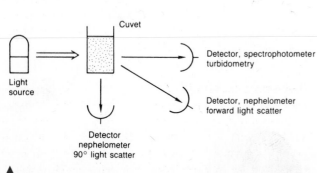

Figure 5-17. Nephelometer versus spectrophotometer—optical arrangements.

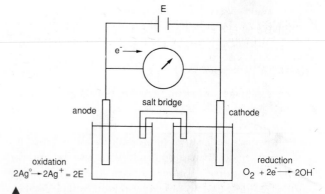

Figure 5-18. Electrochemical cell.

called an *electrolytic cell.* In short, a galvanic cell can be built from an electrolytic cell. When the external E is turned off, accumulated products at the electrodes will spontaneously produce current in the opposite direction of the electrolytic cell.

Half Cells

It is impossible to measure the electrochemical activity of one half-cell. Two reactions must be coupled, and one reaction compared with the other. In order to rate half-cell reactions, a specific electrode reaction is arbitrarily assigned 0.00 V. Every other reaction coupled with this arbitrary zero reaction is either positive or negative, depending on the relative affinity for electrons. The electrode defined as 0.00 V is the standard hydrogen electrode: H_2 gas at 1 atmosphere (atm). The hydrogen gas in contact with H^+ in solution develops a potential. The hydrogen electrode coupled with a zinc half-cell is cathodic, with the reaction $2H^+ + 2e^- \rightarrow H_2$, because H_2 has a greater affinity than Zn for electrons. Cu, however, has a greater affinity than H_2 for electrons, and thus the anodic reaction $H_2 \rightarrow 2H^+ + 2e^-$ occurs when coupled to the Cu-electrode half-cell.

The potential generated by the hydrogen-gas electrode is used to rate the electrode potential of metals in 1 mol/L solution. Reduction potentials for some metals are shown in Table 5-1.[21] A hydrogen electrode is used to determine the accuracy of reference and indicator electrodes, the stability of standard solutions, and the potentials of liquid junctions.

Ion-Selective Electrodes (ISE)

Potentiometric methods of analysis involve the direct measurement of electrical potential due to the activity of free ions. Ion-selective electrodes are designed to be sensitive toward individual ions.

pH Electrodes

An ion-selective electrode universally used in the clinical laboratory is the pH electrode. Figure 5-19 presents an illustration of the basic components of a pH meter.

TABLE 5-1. Standard Reduction Potentials

	Potential, V
$Zn^{2+} + 2e \leftrightarrow Z$	-0.7628
$Cr^{2+} + 2e \leftrightarrow Cr$	-0.557
$Ni^{2+} + 2e \leftrightarrow Ni$	-0.23
$2H^+ + 2e \leftrightarrow H_2$	0.000
$Cu^{2+} + 2e \leftrightarrow Cu$	0.3402
$Ag^+ + e \leftrightarrow Ag$	0.7996

(Data presented are examples from CRC Handbook of Chemistry and Physics, 61st ed. 1980–81.)

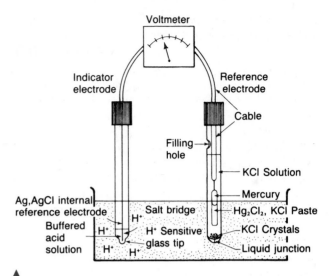

▲
Figure 5-19. Necessary components of a pH meter.

Indicator Electrode

The pH electrode consists of a silver wire coated with AgCl, immersed into an internal solution of 0.1 mmol/L HCl, and placed into a tube containing a special glass membrane tip. This membrane is sensitive only to hydrogen ions. Glass membranes that are selectively sensitive to H^+ consist of specific quantities of lithium, cesium, lanthanum, barium, or aluminum oxides in silicate. When the pH electrode is placed into the test solution, movement of H^+ ions near the tip of the electrode produces a potential difference between the internal solution and the test solution, which is measured as pH and read by a voltmeter. The combination pH electrode also contains a built-in reference electrode, either Ag/AgCl or calomel (Hg/Hg_2Cl_2) immersed in a solution of saturated KCl.

The specially formulated glass continually dissolves from the surface. The present concept of the selective mechanism that causes formation of electromotive force at the glass surface is that an ion-exchange process is involved. Cationic exchange occurs only in the gel layer. There is no penetration of H^+ through the glass. Although the glass is constantly dissolving, the process is slow, and the glass tip generally lasts for several years. pH electrodes are highly selective for hydrogen ions, but other cations in high concentration do interfere, the most common of which is sodium. Electrode manufacturers should list the concentration of interfering cations that may cause error in pH determinations.

Reference Electrode

The reference electrode commonly used is the calomel electrode. Calomel, a paste of mercurous chloride and potassium chloride, is in direct contact with metallic mercury in an electrolyte solution of potassium chloride. As long as the electrolyte stays at a constant concentration and the temperature remains constant, a stable voltage is generated at the interface of the mercury and its salt. A cable connected to the mercury leads to the voltmeter. The filling hole is needed for adding potassium chloride solution. A tiny opening at the bottom is required for completion of electric contact between the reference and indicator electrodes. The liquid junction consists of a fiber or ceramic plug that allows a small flow of electrolyte filling solution.

Construction varies, but all reference electrodes must generate a stable electrical potential. In general, reference electrodes consist of a metal and its salt in contact with a solution containing the same anion. Mercury/mercurous chloride, as in this example, is a frequently used reference electrode, but it has a disadvantage in that it is slow to reach a new stable voltage following temperature change and is unstable above 80°C.[3,22] Ag/AgCl is another common reference electrode. It can be used at high temperatures, up to 275°C, and the AgCl-coated Ag wire makes a more compact electrode than that of mercury. In measurements where chloride contamination must be avoided, a mercury sulfate and potassium sulfate reference electrode may be used.

Liquid Junctions

Electrical connection between the indicator and reference electrodes is achieved by allowing a slow flow of electrolyte from the tip of the reference electrode. A junction potential is always set up at the boundary between two dissimilar solutions because of positive and negative ions diffusing across the boundary at unequal rates. The resultant junction potential may increase or decrease the potential of the reference electrode. Therefore, it is important that the junction potential be kept to a minimum reproducible value when the reference electrode is in solution.

KCl is a commonly used filling solution because K^+ and Cl^- ions have nearly the same mobilities. When KCl is used as the filling solution for Ag/AgCl electrodes, addition of AgCl is required to prevent dissolution of the AgCl salt. A way of producing a lower junction potential is to mix K^+, Na^+, NO_3^-, and Cl^- in appropriate ratios.

Readout Meter

Electromotive force produced by the reference and indicator electrodes is in the millivolt range. Zero potential for the cell indicates that each electrode half-cell is generating the same voltage, assuming there is no liquid junction potential.

The isopotential is that potential at which a temperature change has no effect on the response of the electrical cell. Manufacturers generally achieve this by making midscale (pH 7.0) correspond to 0 V at all temperatures. They use an internal buffer whose pH changes due to temperature compensate for the changes in the internal and external reference electrodes.

Nernst Equation

The electromotive force generated because of H^+ at the glass tip is described by the Nernst equation, which is shown here in a simplified form:

$$\epsilon = \Delta pH \times \frac{RT \ln 10}{F} = \Delta pH \times 0.059\,V$$

(Eq. 5-7)

where ϵ = the electromotive force of the cell
F = Faraday's constant (96,500 C/mol)
R = the molar gas constant
T = temperature, in Kelvin

As temperature increases, hydrogen ion activity increases and the potential generated increases. Most pH meters have a temperature-compensation knob that amplifies the millivolt response when the meter is on pH function. pH units on the meter scale are usually printed for room-temperature use. On the voltmeter, 59.16 is read as 1 pH unit change. The temperature compensation changes millivolt response to compensate for changes due to temperature from 54.2 at 0°C to 66.10 at 60°C. However, most pH meters are manufactured for greatest accuracy in the 10° to 60°C range.

Calibration

The steps necessary to standardize a pH meter are fairly straightforward. First, balance the system with the electrodes in a buffer whose pH is 7.0. The balance or intercept control shifts the entire slope, as shown in Figure 5-20. Next, replace the buffer with one of a different pH. If the meter does not register the correct pH, amplification of the response changes the slope to match that predicted by the Nernst equation. If the instrument does not have a slope control, the temperature compensator performs the same function.

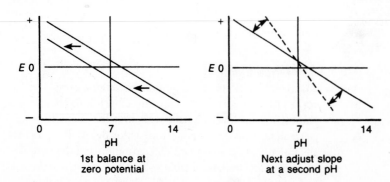

Figure 5-20. pH Meter calibration. (From Willard HH, Merritt LL, Dean JA, Settle FA. Instrumental methods of analysis. Belmont, CA: Wadsworth, 1981.)

pH Combination Electrode

The most common pH electrode used in laboratories today has both the indicator and reference electrodes combined in one small probe. This is convenient when small samples are tested. It consists of an Ag/AgCl internal reference electrode sealed into a narrow glass cylinder with a pH-sensitive glass tip. The reference electrode is an Ag/AgCl wire wrapped around the indicator electrode. The outer glass envelope is filled with KCl and has a tiny pore near the tip of the liquid junction. The solution to be measured must cover the glass tip completely. Examples of other ISE are shown in Figure 5-21. The reference electrode, electrometer, and calibration system described for pH measurements are applicable to all ISEs.

ISEs are of three major types: inert-metal electrodes in contact with a redox couple, metal electrodes that participate in a redox reaction, and membrane electrodes. The membrane can be of solid material (*e.g.,* glass), liquid (*e.g.,* ion-exchange electrodes), or special membrane (*e.g.,* compound electrodes) such as gas-sensing and enzyme electrodes.

The standard hydrogen electrode is an example of an inert-metal electrode. The Ag/AgCl electrode is an example of the second type. The electrode process $AgCl + e^- \leftrightarrow Ag^+ + Cl^-$ produces an electrical potential proportional to Cl^- ion activity. When the chloride ion is held constant, the electrode is used as a reference electrode. The electrode in contact with varying Cl^- concentrations is used as an indicator electrode to measure chloride concentration.

The H^+-sensitive gel layer of the glass pH electrode is considered a membrane. A change in the glass formulation makes the membrane more sensitive to sodium ions than to hydrogen ions, creating a sodium ion-selective electrode. Other solid-state membranes consist of either a single crystal or fine crystals immobilized in an inert matrix such as silicone rubber. Conduction depends on a vacancy defect mechanism, and the crystals are formulated to be selective for a particular size, shape, and change. Examples include F^--selective electrodes of LaF, Cl^--sensitive electrodes with AgCl crystals, and AgBr electrodes for the detection of Br^-.

The calcium ion–selective electrode is a liquid-membrane electrode. An ion-selective carrier, such as dioctyphenyl phosphate dissolved in an inert water-insoluble solvent, diffuses through a porous membrane. Since the solvent is insoluble in water, the test sample cannot cross the membrane, but Ca^{2+} ions are exchanged. The Ag/AgCl internal reference in a filling solution of $CaCl_2$ is in contact with the carrier by means of the membrane.

Potassium-selective liquid membranes use the antibiotic valinomycin as the ion-selective carrier. Valinomycin membranes show great selectivity for K^+. Liquid-membrane electrodes are recharged every few months to replace the liquid ion exchanger and the porous membrane.

Gas-Sensing Electrodes

Gas electrodes are similar to pH glass electrodes but are designed to detect specific gases (*e.g.,* CO_2 and NH_3) in solutions and are usually separated from the solution by a thin, gas-permeable hydrophobic membrane. Figure 5-22 shows a schematic illustration of the PCO_2 electrode. The membrane in contact with the solution is permeable only to CO_2, which diffuses into a thin film of sodium bicarbonate solution. The pH of the bicarbonate solution is changed as follows:

$$CO_2 + H_2O \leftrightarrow H_2CO_3 \leftrightarrow H^+ + HCO_3^- \quad \textit{(Eq. 5-8)}$$

The change in pH of the HCO_3^- is detected by a pH electrode. The PCO_2 electrode is widely used in clinical laboratories as a component of instruments for measuring serum electrolytes and blood gases.

In the NH_3 gas electrode, the bicarbonate solution is replaced by ammonium chloride solution, and the membrane is permeable only to NH_3 gas. As in the PCO_2 electrode, NH_3 changes the pH of NH_4Cl as follows:

$$NH_3 + H_2O \leftrightarrow NH_4^+ + OH^- \quad \textit{(Eq. 5-9)}$$

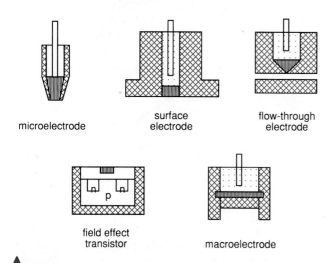

microelectrode

surface electrode

flow-through electrode

field effect transistor

macroelectrode

▲

Figure 5-21. Other examples of ion-selective electrodes.

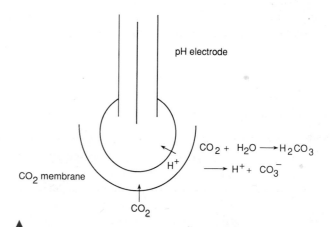

pH electrode

CO_2 membrane

$CO_2 + H_2O \longrightarrow H_2CO_3$

$\longrightarrow H^+ + CO_3^-$

H^+

CO_2

▲

Figure 5-22. The PCO_2 electrode.

The amount of OH⁻ ions produced varies linearly with the log of the partial pressure of NH_3 in the sample.

Other gas-sensing electrodes function on the basis of amperometric principle, *i.e.*, measurement of the current flowing through an electrochemical cell at a constant applied electrical potential to the electrodes. Examples are the determination of PO_2, glucose, and peroxidase.

The chemical reactions of the PO_2 electrode (Clark electrode) are illustrated in Figure 5-18. It is an electrochemical cell with a platinum cathode and an Ag/AgCl anode. The electrical potential at the cathode is set to $-0.65\,V$ and will not conduct current without oxygen in the sample. The membrane is permeable to oxygen, which diffuses through to the platinum cathode. Current passes through the cell and is proportional to the PO_2 in the test sample.

Glucose determination is based on the reduction in PO_2 during glucose oxidase reaction with glucose and oxygen. Unlike the PCO_2 electrode, the peroxidase electrode has a polarized platinum anode and its potential is set to $+0.6\,V$. Current flows through the system when peroxide is oxidized at the anode as follows:

$$H_2O_2 \rightarrow 2H^+ + 2e^- + O_2 \qquad \textit{(Eq. 5-10)}$$

Sat. 2128

Enzyme Electrodes

The various ion-selective electrodes may be covered by immobilized enzymes that can catalyze a specific chemical reaction. Selection of the ISE is determined by the reaction product of the immobilized enzyme. Examples include urease, which is used for the detection of urea, and glucose oxidase, which is used for glucose detection. A urea electrode must have an ISE that is selective for NH_4^+ or NH_3, while glucose oxidase is used in combination with a pH electrode.

Coulometric Chloridometers and Anodic Stripping Voltametry

Coulometric titrations for determination of chloride in body fluids has largely been replaced by chloride ion–selective electrodes.[2] Anodic stripping voltametry was widely used for analysis of lead and is best measured by electrothermal (graphite furnace) atomic absorption spectroscopy.

ELECTROPHORESIS

Theory and Principle

Electrophoresis is the migration of charged solutes or particles in an electrical field. *Iontophoresis* refers to the migration of small ions, whereas *zone electrophoresis* is the migration of charged macromolecules in a porous support medium such as paper, cellulose acetate, or agarose-gel film. An electrophoretogram is the result of zone electrophoresis and consists of sharply separated zones of a macromolecule. In a

clinical laboratory, the macromolecules of interest are proteins in serum, urine, cerebrospinal fluid, other biologic body fluids, erythrocytes, and tissues.

Electrophoresis consists of five components: the driving force (electrical power), the support medium, the buffer, the sample, and the detecting system. A typical electrophoretic apparatus is illustrated in Figure 5-23.

Charged particles migrate toward the opposite charged electrode. The velocity of migration is controlled by the net charge of the particle, the size and shape of the particle, the strength of the electric field, chemical and physical properties of the supporting medium, and the electrophoretic temperature. The rate of mobility[19] of the molecule μ is given by

$$\mu = \frac{Q}{k} \times r \times n \qquad \textit{(Eq. 5-11)}$$

where Q = net charge of particle
k = constant
r = ionic radius of the particle
n = viscosity of the buffer

From the equation, the rate of migration is directly proportional to the net charge of the particle and inversely proportional to its size and the viscosity of the buffer.

Procedure

The sample is soaked in hydrated support for approximately 5 minutes. The support is put into the electrophoresis chamber, which was previously filled with the buffer. Enough buffer must be added to the chamber to maintain contact with the support. Electrophoresis is carried out by applying a constant voltage or constant current for a specific length of time. The support is removed and placed in a fixative or is rapidly dried to prevent diffusion of the sample. This is followed by staining the zones with appropriate dye. The uptake of dye by the sample is proportional to sample concentration. After excess dye is washed away, the supporting medium may need to be placed in a clearing agent. Otherwise, it is completely dried.

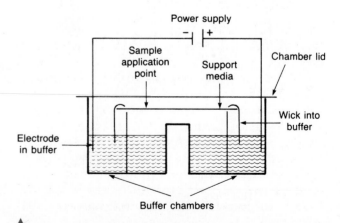

Figure 5-23. Electrophoresis apparatus—basic components..

Power Supply

Power supplies operating at either constant current or constant voltage are available commercially. In electrophoresis, when current flows through a medium that has resistance, heat is produced. This results in an increase in thermal agitation of the dissolved solute (ions), leading to a decrease in resistance and an increase in current. The latter increase leads to increases in the heat and evaporation of water from the buffer. This increases the ionic concentration of the buffer and subsequent further increases in the current. The migration rate can be kept constant by using a power supply with constant current. This is true because as electrophoresis progresses, a decrease in resistance as a result of heat produced also decreases the voltage.

Buffers

Two properties of a buffer that affect the charge of ampholytes are its pH and its ionic strength. The ions carry the applied electric current and also allow the buffer to maintain constant pH during electrophoresis. An ampholyte is a molecule, such as protein, whose net charge can be either positive or negative. If the buffer is more acidic than the isoelectric point (pI) of the ampholyte, it binds H^+ ions, becomes positively charged, and migrates toward the cathode. If the buffer is more basic than the pI, the ampholyte loses H^+ ions, becomes negatively charged, and migrates toward the anode. A particle without a net charge will not migrate, remaining at the point of application. During electrophoresis, ions cluster around a migrating particle. The higher the ionic concentration, the higher is the size of the ionic cloud and the lower is the mobility of the particle. Greater ionic strength produces sharper protein-band separation but leads to increased heat production. This may cause denaturation of heat-labile proteins. Consequently, for any electrophoretic system, the optimal buffer concentration should be determined. Generally, the most widely used buffers are made of monovalent ions because their ionic strength and molality are equal.

Support Materials

Cellulose Acetate

Paper electrophoresis is no longer widely used in clinical laboratories. It has been replaced by cellulose acetate or agarose gel. Cellulose is acetylated to form cellulose acetate by treating it with acetic anhydride. Cellulose acetate is produced commercially. It is a dry, brittle film composed of about 80% air space. When the film is soaked in buffer, the air spaces are filled with electrolyte, and the film becomes pliable. After electrophoresis and staining, cellulose acetate can be made transparent for densitometer quantitation. The dried transparent film can be stored for long periods. Cellulose acetate prepared to reduce electroendosmosis is available commercially. Cellulose acetate is also used in isoelectric focusing.

Agarose Gel

Agarose gel is another widely used supporting medium. It is used as a purified fraction of agar, is neutral, and therefore does not produce electroendosmosis. After electrophoresis and staining, it is detained (cleared), dried, and scanned with a densitometer. The dried gel can be stored indefinitely. Agarose-gel electrophoresis requires small amounts of sample (approximately 2 μl), it does not bind protein, and therefore, migration is not affected.

Polyacrylamide Gel

Polyacrylamide-gel electrophoresis involves separation of protein on the basis of charge and molecular size. Layers of gel with different pore sizes are used. The gel is prepared prior to electrophoresis in a tubular-shaped electrophoresis cell. At the bottom is the small-pore separation gel, followed by a large-pore spacer gel and finally another large-pore gel containing the sample. Each layer of gel is allowed to form a gelatin before the next gel is poured over it. At the start of electrophoresis, the protein molecules move freely through the spacer gel to its boundary with the separation gel, which slows their movement. This allows for concentration of the sample prior to separation by the small-pore gel. Polyacrylamide-gel electrophoresis separates serum proteins into 20 or more fractions rather than the usual 5 fractions separated by cellulose acetate or agarose. It is widely used to study individual proteins, *e.g.,* isoenzymes.

Starch Gel

Starch-gel electrophoresis separates proteins on the basis of surface charge and molecular size, as does polyacrylamide gel. The procedure is not widely used because of technical difficulty in preparing the gel.

Treatment and Application of Sample

Serum contains a high concentration of protein, especially albumin, and therefore, it is routine to dilute serum specimens with the buffer prior to electrophoresis. In contrast, urine and cerebrospinal fluid (CSF) are usually concentrated. moglobin hemolysate is used without further concentra-. Generally, preparation of a sample is done according to suggestion of the manufacturer, of the electrophoretic lies.

Cellulose-acetate and agarose-gel electrophoresis require approximately 2 to 5 μl of sample. These are the most commonly performed routine electrophoreses in clinical laboratories. Overloading of agarose gel with sample is not a frequent problem because most commercially manufactured plates come with a thin plastic template that has small slots through which samples are applied. After serum is allowed to diffuse into the gel for approximately 5 minutes, the template is blotted to remove excess serum before being removed from the gel surface. Sample is applied to cellulose acetate with a twin-wire applicator that is designed to transfer a small amount.

Detection and Quantitation

Separated protein fractions are stained to reveal their locations. Different stains come with different plates from different manufacturers. The simplest way to accomplish detection is visualization under UV light, whereas densitometry is the most common and reliable way for quantitation. Most densitometers will automatically integrate the area under a peak, and the result is printed as percentage of the total. A schematic illustration of a densitometer is shown in Figure 5-24.

Electroendosmosis

The movement of buffer ions and solvent relative to the fixed support is called *endosmosis* or *electroendosmosis*. Support media such as paper, cellulose acetate, and agar gel take on a negative charge from adsorption of hydroxyl ions. When current is applied to the electrophoresis system, the hydroxyl ions remain fixed, while the free positive ions move toward the cathode. The ions are highly hydrated, resulting in net cathodic movement of solvent. Molecules that are nearly neutral are swept toward the cathode with the solvent. Support media such as agarose and acrylamide gel are essentially neutral, thus eliminating electroendosmosis. The position of proteins in any electrophoresis separation depends on the nature of the protein, but also on all other technical variables.

Isoelectric Focusing

Isoelectric focusing is a modification of electrophoresis. An apparatus similar to that in Figure 5-24 is used. Charged proteins migrate through a support medium that has a continuous pH gradient. Individual proteins move in the electric field until they reach a pH equal to their isoelectric point, at which point they have no charge and cease to move.

Capillary Electrophoresis

In capillary electrophoresis (CE), separation is performed in narrow-bore fused silica capillaries (inner diameter 25–75 μm). Usually, the capillaries are filled only with buffer, although gel media can also be used. A schematic of CE instrumentation is shown in Figure 5-25. Initially, the capillary is filled with buffer, then the sample is loaded and the separation is performed by applying an electric field. Detection can be made near the other end of the capillary directly through the capillary wall.[7]

A fundamental concept in CE is the *electro-osmotic flow* (EOF). EOF is the bulk flow of liquid toward the cathode upon application of electric field and it is superimposed on electrophoretic migration. EOF controls the amount of time solutes remain in the capillary. Cations migrate fastest because both EOF and electrophoretic attraction are toward the cathode, neutral molecules are all carried by the EOF but are not separated from each other, and anions move slowest because although they are carried to the cathode by the EOF they are attracted to the anode and repelled by the cathode (Figure 5-26).

UV-visible detection is widely used for monitoring separated analytes and is performed directly on the capillary, but sensitivity is poor because of the small dimensions of the capillary, resulting in a short pathlength. Fluorescence, laser-induced fluorescence, and chemiluminescence detection can be used for higher sensitivity.

CE has been used for the separation, quantitation, and determination of molecular weights of proteins and peptides; for the analysis of polymerase chain reaction (PCR) products; and for the analysis of inorganic ions, organic acids, pharmaceuticals, optical isomers, and drugs of abuse in serum and urine.[13]

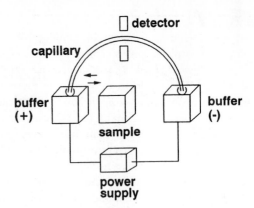

▲
Figure 5-25. Schematic of capillary electrophoresis instrumentation. Sample is loaded onto the capillary by replacing the anode buffer reservoir with the sample reservoir. (From Heiger DN. High performance capillary electrophoresis. Waldbronn, Germany: Hewlett-Packard, 1992.)

CHROMATOGRAPHY

Chromatography refers to a group of techniques used to separate complex mixtures on the basis of different physical interactions between the individual compounds and the stationary phase of the system. The basic components in any chromatographic technique are the mobile phase (gas or liquid), which carries the complex mixture (sample); the stationary phase (solid or liquid), through which the mobile

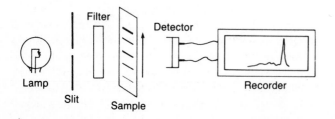

▲
Figure 5-24. Densitometer—basic components.

Figure 5-26. Differential solute migration superimposed on electro-osmotic flow in capillary zone electrophoresis. (From Heiger DN. High performance capillary electrophoresis. France: Hewlett-Packard, 1992.)

phase flows; the column holding the stationary phase; and the separated components (eluate).

Modes of Separation

Adsorption

Adsorption chromatography, also known as *liquid-solid chromatography,* is based on the competition between the sample and the mobile phase for adsorptive sites on the solid stationary phase. There is an equilibrium of solute molecules being adsorbed to the solid surface and desorbed and dissolved in the mobile phase. Those molecules which are most soluble in the mobile phase move fastest, while those which are least soluble move slowest. Thus a mixture is typically separated into classes according to polar functional groups. The stationary phase can be either acidic polar (*e.g.,* silica gel), basic polar (*e.g.,* alumina), or nonpolar (*e.g.,* charcoal). The mobile phase can be either a single solvent or a mixture of two or more solvents, depending on the analytes to be desorbed. Liquid-solid chromatography is not widely used in clinical laboratories because of technical problems with the preparation of a stationary phase that has homogeneous distribution of absorption sites.

Partition

Partition chromatography is also referred to as *liquid-liquid chromatography.* Separation of solute is based on relative solubility in an organic (nonpolar) solvent and an aqueous (polar) solvent. In its simplest form, partition (extraction) is performed in a separatory funnel. Molecules containing polar and nonpolar groups in an aqueous solution are added to an immiscible organic solvent. After vigorous shaking, the two phases are allowed to separate. Molecules that are polar remain in the aqueous solvent, whereas nonpolar molecules are extracted in the organic solvent. This results in the partitioning of the solute molecules into two separate phases.

The ratio of the concentration of the solute in the two liquids is known as the *partition coefficient:*

$$K = \frac{\text{solute in stationary phase}}{\text{solute in mobile phase}} \qquad \textbf{\textit{(Eq. 5-12)}}$$

Modern partition chromatography uses pseudo-liquid stationary phases that are chemically bonded to the support or high-molecular-weight polymers that are insoluble in the mobile phase.[14] Partition systems are called *normal phase* when the mobile solvent is less polar than the stationary solvent and are termed *reverse phase* when the mobile solvent is more polar.

Partition chromatography is applicable to any substance that may be distributed between two liquid phases. Since ionic compounds are generally soluble only in water, partition chromatography works best with nonionic compounds.

Steric Exclusion

Steric exclusion, a variation of liquid-solid chromatography, is used to separate solute molecules on the basis of size and shape. The chromatographic column is packed with porous material, as shown in Figure 5-27. A sample containing different-sized molecules moves down the column dissolved in the mobile solvent. Small molecules enter the pores in the packing and are momentarily trapped. Large molecules are excluded from the small pores and so move quickly between the particles. Intermediate-sized molecules are partially restricted from entering the pores and therefore move through the column at a rate that is intermediate between those of the large and small molecules.

Early methods used hydrophilic beads of cross-linked dextran, polyacrylamide, or agarose, which formed a gel when soaked in water. This method was termed *gel filtration.* A similar separation process using hydrophobic gel beads of polystyrene with a nonaqueous mobile phase was called *gel permeation chromatography.* Current porous packing uses rigid

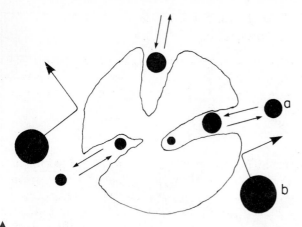

▲

Figure 5-27. Pictorial concept of steric exclusion chromatography. Separation of sample components by their ability to permeate pore structure of column-packing material. Smaller molecules (*a*) permeating the interstitial pores; large excluded molecules (*b*). (From Parris NA. Instrumental liquid chromatography: A practical manual on high performance liquid chromatographic methods. New York: Elsevier, 1976.)

inorganic materials such as silica or glass.[12] The term *steric exclusion* includes all these variations. The size of the pores is controlled by the manufacturer, and packing materials can be purchased with different pore sizes, depending on the size of the molecules being separated.

Sum 3/1

Ion-Exchange Chromatography

In ion-exchange chromatography, solute mixtures are separated by virtue of the magnitude and charge of ionic species. The stationary phase is a resin, consisting of large polymers of substituted benzene, silicates, or cellulose derivatives, with charge functional groups. The resin is very insoluble in water, and the functional groups are immobilized as side chains on resin beads that are used to fill the chromatographic column. Figure 5-28*A* shows resin with sulfonate functional groups. H^+ ions are held loosely and are free to react. This is an example of a cation-exchange resin. When a cation such as Na^+ comes into contact with these functional groups, an equilibrium is formed, following the law of mass action. Because there are many sulfonate groups, Na^+ ions are effectively and

completely removed from solution. The Na^+ ions that are concentrated on the resin column can be eluted from the resin by pouring acid through the column, driving the equilibrium to the left.

Anion-exchange resins are made with exchangeable hydroxyl ions such as the diethylamine functional group illustrated in Figure 5-28*B*. They are used like cation-exchange resins, except that hydroxyl ions are exchanged for anions. The example shows Cl^- ions in sample solution exchanged for OH^- ions from the resin functional group. Anion and cation resins mixed together (mixed-bed resin) are used to deionize water. The displaced protons and hydroxyl ions combine to form water. Other ionic functional groups besides the illustrated examples are used for specific analytical applications. Ion-exchange chromatography is used to remove interfering substances from a solution, to concentrate dilute ion solutions, and to separate mixtures of charged molecules, such as amino acids. Changing pH and ionic concentration of the mobile phase allows separation of mixtures of organic and inorganic ions.

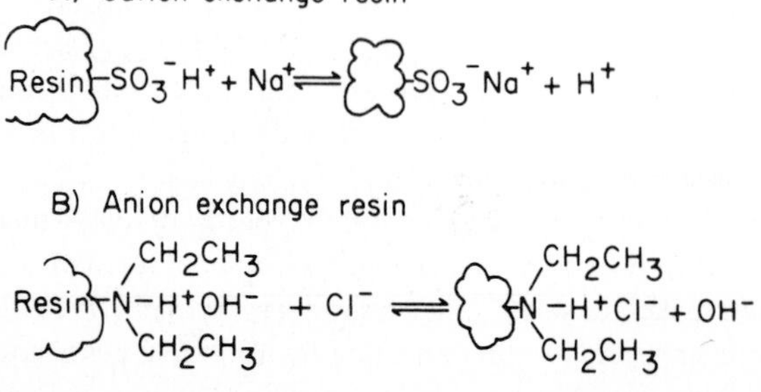

Chromatographic Procedures

Thin-Layer Chromatography (TLC)

Thin-layer chromatography (TLC) is a variant of column chromatography. A thin layer of **sorbent,** such as alumina, silica gel, cellulose, or cross-linked dextran, is uniformly coated on a glass or plastic plate. Each sample to be analyzed is applied as a spot near one edge of the plate, as shown in Figure 5-29. The mobile phase (solvent) is usually placed in a closed container until the atmosphere is saturated with solvent vapor. One edge of the plate is placed in the solvent, as shown. The solvent migrates up the thin layer by capillary action, dissolving and carrying sample molecules. Separation can be achieved by any of the four processes previously described, depending on the sorbent (thin layer) and solvent chosen. After the solvent reaches a predetermined height, the plate is removed and dried. Sample components are identified by comparison with standards on the same plate. The

A) Cation exchange resin

$$Resin-SO_3^-H^+ + Na^+ \rightleftharpoons SO_3^-Na^+ + H^+$$

B) Anion exchange resin

$$Resin-N \begin{matrix} CH_2CH_3 \\ -H^+OH^- \\ CH_2CH_3 \end{matrix} + Cl^- \rightleftharpoons N \begin{matrix} CH_2CH_3 \\ -H^+Cl^- \\ CH_2CH_3 \end{matrix} + OH^-$$

▲

Figure 5-28. Chemical equilibrium of ion–exchange resins.

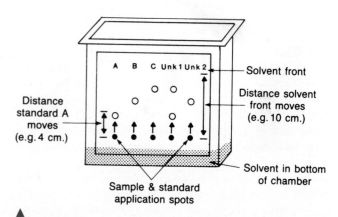

▲

Figure 5-29. TLC plate in chromatographic chamber.

distance a component migrates, compared with the distance the solvent front moves, is called the *retention factor* R_f

$$R_f = \frac{\text{distance leading edge of component moves}}{\text{total distance solvent front moves}}$$

(Eq. 5-13)

Each sample-component R_f is compared with the R_f of standards. Using Figure 5-29 as an example, standard A has an R_f value of 0.4, while standard B has an R_f value of 0.6 and standard C is 0.8. The first unknown contains A and C, since the R_f values are the same. This ratio is valid only for separations run under identical conditions. Since R_f values may overlap for some components, further identifying information is obtained by spraying different stains on the dried plate and comparing colors of the standards.

TLC is most commonly used as a semiquantitative screening test. Technique refinement has resulted in the development of semiautomated equipment and the ability to quantitate separated compounds. For example, sample applicators apply precise amounts of sample extracts in concise areas. Use of plates prepared with uniform sorbent thickness, finer particles, and new solvent systems has resulted in the technique of high-performance thin-layer chromatography (HPTLC).[20] Absorbance of each developed spot is measured using a densitometer, and the concentration is calculated by comparison with a reference standard chromatographed under identical conditions.

High-Performance Liquid Chromatography (HPLC)

Modern liquid chromatography uses pressure for fast separations, controlled temperature, in-line detectors, and gradient elution techniques. Figure 5-30 illustrates its basic components.

Pump

A pump forces the mobile phase through the column at a much greater velocity than is accomplished by gravity-flow columns. The pump can be pneumatic, syringe-type, reciprocating, or hydraulic amplifier. Pneumatic pumps are used for preoperative purposes, whereas hydraulic amplifier pumps are no longer commonly used. The most widely used pump today is the mechanical reciprocating pump, which is now used as a multihead pump with two or more reciprocating pistons. During pumping, the pistons operate out of phase (180° for two heads, 120° for three) to provide constant flow.

Column

The stationary phase is packed into long stainless steel columns. Usually, HPLC is run at ambient temperatures, although columns can be put in an oven and heated to enhance the rate of partition. Fine, uniform column packing results in much less band broadening but requires pressure to

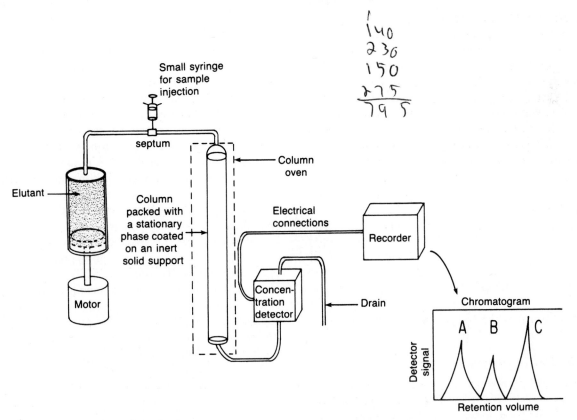

Figure 5-30. HPLC basic components. (From Bender GT. Chemical instrumentation: A laboratory manual based on clinical chemistry. Philadelphia: WB Saunders, 1972.)

force the mobile phase through. The packing also can be either pellicular (an inert core with a porous layer), inert and small particles, or macroporous particles. The most common material used for column packing is silica gel. It is very stable and can be used in different ways. It can be used as a solid packing in liquid-solid chromatography or coated with a solvent, which serves as the stationary phase (liquid–liquid). Owing to the short lifetime of coated particles, molecules of the mobile-phase liquid are now bonded to the surface of silica particles.

Reversed-phase HPLC is now very popular, and the stationary phase is nonpolar molecules (*e.g.,* octadecyl C–18 hydrocarbon) bonded to silica-gel particles. For this type of column packing, the mobile phase commonly used is acetonitrile, methanol, water, or any combination of solvents. A reversed-phase column can be used to separate ionic, nonionic, and ionizable samples. A buffer is used to produce the desired ionic characteristics and pH for separation of the analyte. Column packings vary in size (3 to 20 μm), with the smaller particles used mostly for analytical separations and the larger ones for preparative separations.

Sample Injector

As shown in Figure 5-30, a small syringe can be used to introduce the sample into the path of the mobile phase that carries it into the column. The best and most widely used method, however, is the loop injector. The sample is introduced into a fixed-volume loop. When the loop is switched, the sample is placed into the path of the flowing mobile phase and is flushed onto the column.

Loop injectors have high reproducibility and are used at high pressures. Many HPLC instruments now have loop injectors that can be programmed for automatic injection of samples. When the sample size is less than the volume of the loop, the syringe containing the sample is often filled with the mobile phase to the volume of the loop before filling the loop. This prevents the possibility of air being forced through the column, since such a practice may reduce the lifetime of the column packing material.

Detectors

Modern HPLC detectors monitor the eluate as it leaves the column and, ideally, produce an electronic signal proportional to the concentration of each separated component. Spectrophotometers that detect absorbances of visible or ultraviolet light are most commonly used. Photodiode array (PDA) and other rapid scanning detectors are also used for spectral comparisons and compound identification and purity. These detectors have been used for drug analyses in urine. Obtaining an ultraviolet scan of a compound as it elutes from a column can provide important information as to its identity. Unknowns can be compared against library spectra in a similar manner as mass spectrometry. Unlike gas chromatography/mass spectrometry, which requires volatilization of targeted compounds, however, LC-PDA enables direct injection of aqueous urine samples.

Fluorescence detectors are also used, since many biologic substances fluoresce strongly. The principles involved are the same as discussed in the section on spectrophotometric measurements. Another common HPLC detector is the amperometric or electrochemical detector. These devices measure current produced when the analyzer of interest is either oxidized or reduced at some fixed potential set between a pair of electrodes.

A mass spectrometer (MS) can also be used as a detector not only for the identification and quantitation of compounds, but also for structural information and molecular weight determination.[4] In an MS, the sample is first volatilized and then ionized to form charged molecular ions and fragments that are separated according to their mass-to-charge (m/z) ratio; the sample is then measured by a detector, which gives the intensity of the ion current for each species. Identification of molecules is based on the formation of characteristic fragments. Coupling a liquid chromatograph to a mass spectrometer is difficult because of the large amount of solvent in the eluate. Electrospray (ES) is a technique that allows ions to be transferred from solution to the gas phase.[11] The sample is passed through a metal capillary tip and converted, under the influence of a high electric field ($10^6 V/m$), to a fine mist of positively charged droplets, from which the solvent rapidly evaporates. The solute ions that remain are then transferred to the mass spectrometer to be analyzed.

Recorder

The recorder is used to record detector signal versus the time the mobile phase passed through the instrument starting from the time of sample injection. The graph is called a *chromatogram* (Fig. 5-31). The retention time is used to identify compounds when compared with standard retention times run under identical conditions. Peak area is proportional to concentration of the compounds that produced the peaks.

When the elution strength of the mobile phase is constant throughout the separation, it is called *isocratic elution*. For samples containing compounds of widely differing relative compositions, the choice of solvent is a compromise. Early eluting compounds may have retention times close to zero, thereby producing a poor separation (resolution), as shown in Figure 5-31*A*. Basic compounds often have low retention times because C-18 columns cannot tolerate high pH mobile phases. The addition of cation-pairing reagents to the mobile phase, *e.g.,* octane sulfonic acid, can result in better retention of negatively charged compounds onto the column.

The late-eluting compounds may have long retention times, producing broad bands resulting in decreased sensitivity. In some cases, certain components of a sample may have such a great affinity for the stationary phase that they do not elute at all. Gradient elution is an HPLC technique that can be used to overcome this problem. The composition of the mobile phase is varied to provide a continual increase in the solvent strength of the mobile phase entering the column

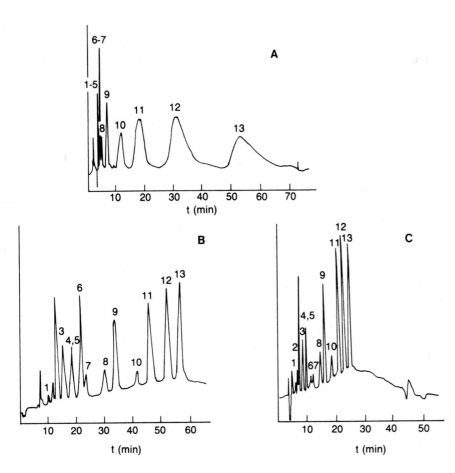

Figure 5-31. (*A*) Isocratic ion-exchange separation—mobile phase contains 0.055 M NaNO₃. (*B*) Gradient elution—mobile phase gradient from 0.01 to 0.1 MNaNO₃ at 2%/min. (*C*) Gradient elution—5%/min. (From Horváth C. High performance liquid chromatography, advances and perspectives. New York: Academic Press, 1980.)

(Fig. 5-31*B*). The same gradient elution can be performed with a faster change in concentration of the mobile phase (Fig. 5-31*C*).

Gas Chromatography

Gas chromatography is used to separate mixtures of compounds that are volatile or can be made volatile. Gas chromatography may be gas-solid chromatography (GSC), with a solid stationary phase, or gas-liquid chromatography (GLC), with a nonvolatile liquid stationary phase. Gas-liquid chromatography is commonly used in clinical laboratories. Figure 5-32 illustrates the basic components of a gas chromatographic system. The setup is almost the same as for HPLC, except that the mobile phase is a gas and samples are partitioned between a gaseous mobile phase and a liquid stationary phase. The carrier gas can be either nitrogen, helium, or argon. The selection of a carrier gas is determined by the detector used in the instrument. The instrument can be operated at a constant temperature or programmed to run at different temperatures if a sample has components with different volatilities. This is analogous to gradient elution described for HPLC.

The sample, which is injected through a septum, must be injected as a gas, or the temperature of the injection port must be above the boiling point of the components so that they vaporize upon injection. Sample vapor is swept through

the column partially as a gas and partially dissolved in the liquid phase. Volatile compounds that are present mainly in the gas phase will have a low partition coefficient and will move quickly through the column. Compounds with higher boiling points will move slowly through the column. The effluent passes through a detector that produces an electric signal proportional to the concentration of the volatile components. As in HPLC, the chromatogram is used both to

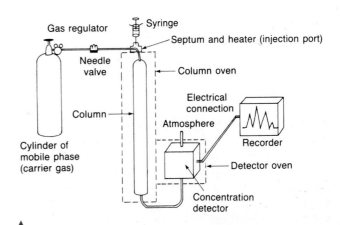

Figure 5-32. GLC basic components. (From Bender GT. Chemical instrumentation: A laboratory manual based on clinical chemistry. Philadelphia: WB Saunders, 1972.)

identify the compounds by the retention time and to determine their concentration by the area under the peak.

Columns

Sure Mon. 3/2

GLC columns are available in a variety of coil configurations and sizes. They are generally made of glass or stainless steel. Packed columns are filled with inert particles such as diatomaceous earth or porous polymer or glass beads coated with a nonvolatile liquid (stationary) phase. These columns usually have a diameter of $\frac{1}{8}$ to $\frac{1}{4}$ inch and a length of 3 to 12 feet. Capillary-wall coated open tubular columns have inside diameters in the range of 0.25 to 0.50 mm and are up to 60 m long. The liquid layer is coated on the walls of the column. A solid support coated with a liquid stationary phase may in turn be coated on column walls.

The liquid stationary phase must be nonvolatile at the temperatures used, must be thermally stable, and must not react chemically with the solutes to be separated. The stationary phase is termed *nonselective* when separation is primarily based on relative volatility of the compounds. Selective liquid phases are used to separate polar compounds based on relative polarity (as in liquid-liquid chromatography).

Detectors

There are many types of detectors, but only thermal conductivity and flame ionization detectors will be explained here because they are the most stable (Fig. 5-33). Thermal conductivity (TC) detectors contain wires (filaments) that change electrical resistance with change in temperature. The filaments form opposite arms of a Wheatstone bridge and are heated electrically to raise their temperature. Helium, which has a high thermal conductivity, is usually the carrier gas. Carrier gas from the reference column flows steadily across one filament, cooling it slightly. Carrier gas and separated compounds from the sample column flow across the other filament. The sample components usually have a lower thermal conductivity so that the temperature and resistance of the sample filament increase. The change in resistance results in an unbalanced bridge circuit. The electrical change is amplified and fed to the recorder. The electrical change is proportional to the concentration of the analyte.

Flame ionization detectors are widely used in the clinical laboratory. They are more sensitive than TC detectors. The column effluent is fed into a small hydrogen flame burning in excess air or atmospheric oxygen. The flame jet and a col-

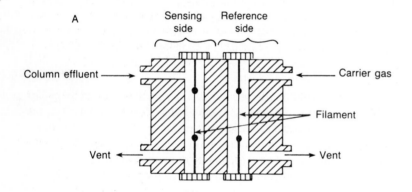

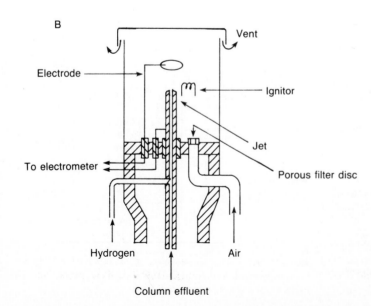

Figure 5-33. (*A*) Schematic diagram of a thermal conductivity detector. (*B*) Schematic diagram of a flame ionization detector. (From Tietz NW, ed. Fundamentals of clinical chemistry. Philadelphia: WB Saunders, 1987.)

lector electrode around the flame have opposite potentials. As the sample burns, ions form and move to the charged collector. Thus a current proportional to the concentration of the ions is formed and fed to the recorder.

Definitive determinations of samples eluting from gas chromatographic columns are possible when a mass spectrometer is used as a detector.[11] In an electron-impact mass spectrometer, samples are bombarded by electrons to form charged molecular ions and fragments that are filtered in terms of their mass-to-charge (m/z) ratio and measured by an electron multiplier. The characteristic fragmentation patterns produced by these ions are used for their identification. GC/MS systems are widely used for measuring unknown drugs in urine toxicology confirmations. Figure 5-35 illustrates the mass spectrum of carboxytetrahydrocannabinol, a metabolite of marijuana. Tandem mass spectrometers (GC/MS/MS), obtained by the addition of a second mass spectrometer to a GC/MS system, can be used for greater selectivity and lower detection limits.[23] The first mass spectrometer allows only ions of a specific m/z ratio to pass to the second mass spectrometer where they are further fragmented and analyzed (Fig. 5-34). Other gas chromatography detectors used for selective applications include electron capture and helium-ionization flame emission detectors.

Scintillation Counting

The scintillation counter is used for measuring gamma rays. Gamma rays are electromagnetic radiation of very high energy. Radioactivity will be explained briefly, and then the instrumentation measuring it will be discussed.

Gamma rays have extremely short wavelength and high energy. They are usually discussed in terms of energy rather than wavelength. Gamma rays with energy from 25,000 to over 1 million electron volts (eV) are commonly measured. Visible light, by contrast, has less than 10 eV. Another major difference between gamma rays and visible light is the source. Visible light comes from excited valence electrons as they emit energy and drop to a stable, lower energy orbital. Gamma rays, however, come from an unstable nucleus that rearranges to become more stable. In the process, the nucleus emits energy in the form of gamma rays or particles and, frequently, both gamma rays and particles.

All neutral atoms have an equal number of positively charged protons in the nucleus and negatively charged electrons in orbitals around the nucleus. The number of protons determines the atomic number (Z) and the identity of the element. Except for ordinary hydrogen, the nucleus also contains neutrons that have no charge. The mass of both protons and neutrons is slightly greater than one atomic mass unit. The mass of electrons is negligible in comparison; therefore, the atomic mass number (A) is equal to the sum of the protons and neutrons. The conventional symbol for nuclides is $^A_Z X$, where X is the chemical symbol of the element. The number of protons is always the same for any element, but the number of neutrons, and therefore the mass number, may differ. Atoms of the same element but with different atomic masses are called *isotopes*. For example, ordinary hydrogen has the symbol ^1_1H. It contains one proton and no neutrons, thus having a mass number and atomic number of 1. One isotope of hydrogen, ^3_1H, contains two neutrons, so its mass number is 3. This particular hydrogen isotope nucleus is unstable. Isotopes of any element that have an unstable ratio of neutrons and protons in the nucleus will spontaneously transform into other nuclear species with the emission of radiation. Such nuclei are said to be *radioactive* and are referred to as *radionuclides*. Thus, ^3_1H is a radionuclide of hydrogen, while $^{125}_{31}I$ is an iodine radionuclide.

Radionuclides emit three kinds of radiation: alpha particles, positively or negatively charged electrons, and gamma rays. The basic scintillation counter is used for the measurement of gamma-ray emission. Variations of the basic instrument are used to detect other emissions. A liquid scintilla-

GC / MS / MS

Tandem-in-Space

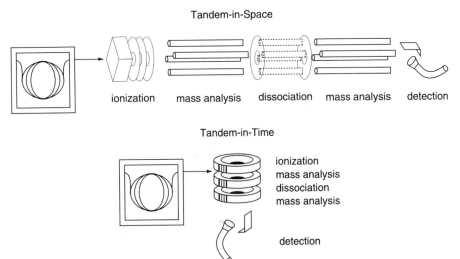

ionization   mass analysis   dissociation   mass analysis   detection

Tandem-in-Time

ionization
mass analysis
dissociation
mass analysis

detection

Figure 5-34. Triple quadruple mass spectrometer.

tion counter is used to measure negative electron (ß) emission, while a position scintillation counter measures positive electron emission.

Figure 5-35 shows the basic components of a scintillation counter. The radioactive sample is shown in a test tube, emitting gamma rays in all directions. Gamma rays that strike the detector of the scintillation counter cause electric pulses. The detector is a phosphor, usually an NaI crystal activated by trace amounts of thallium, optically joined to a photomultiplier (PM) tube. When gamma rays are absorbed by the phosphor, the material gives off a brief flash of visible or ultraviolet light. The brief flash, *scintillation,* is detected and amplified by the sensitive photomultiplier tube. High-energy gamma rays produce a brighter flash than do lower-energy gamma rays. The voltage pulse from the PM tube is proportional to the original gamma-ray energy.

The crystal frequently has a hole drilled into the center, as shown, so that a test tube can be inserted. This serves to increase the proportion of emitted radiation that strikes the phosphor, as well as ensuring that the sample is always in the same position. The crystal is covered with thin aluminum, except for the surface that interfaces with the PM tube. This prevents light from entering, while the high-energy gamma rays pass through easily to the crystal. In order to minimize detection of background radiation, the crystal/photomultiplier assembly is generally protected with a layer of lead.

Pulse amplitude may be increased by increasing the potential on the photomultiplier tube dynodes using the high-voltage control. Voltage pulses generated by the detector are further amplified and pass on to the spectrometer or pulse-height analyzer. Every type of radioactive atom has its own characteristic gamma-energy emission, just as atoms or molecules with excited electrons emit a characteristic light spectrum. By choosing the pulse size that will be detected, it is possible to count the gamma emissions from one type of atom and to exclude others.

The pulse-height analyzer contains two electronic circuits that will not conduct a current until a certain preset voltage is reached. The threshold, or base circuit, is set at less than the gamma peak energy to be counted, and the window, or upper circuit, is set slightly higher than the desired gamma energy. Voltage pulses from the amplifier that are less than the threshold will not be conducted. Any pulses passed simultaneously by both the window and the threshold circuits are blocked in the veto circuit. Only those pulses that fire the threshold and do not fire the window will pass on to the counting mechanism. The pulse-height analyzer controls may be calibrated in kiloelectron volts of gamma energy, and the levels may be set to selectively count the gamma-ray energy of any particular radionuclide. Many current scintillation counters have levels preset for commonly used radionuclides and allow for the selection of the energy to be counted by the push of a button.

The final component is the readout. Since the number of pulses is counted rather than light intensity, a timing mechanism is necessary. The timer is connected to the counting circuit. The readout is usually in counts per minute.

In order to determine concentration of radioactive atoms in an unknown sample, the count must be compared with the counts from a standard containing a known amount of radioactive material:

Concentration of unknown
$$= \frac{\text{concentration of standard}}{\text{counts of standard}} \times \text{counts of unknown}$$

(Eq. 5-14)

OSMOMETRY

An osmometer is used to measure the concentration of solute particles in a solution. The mathematical definition is

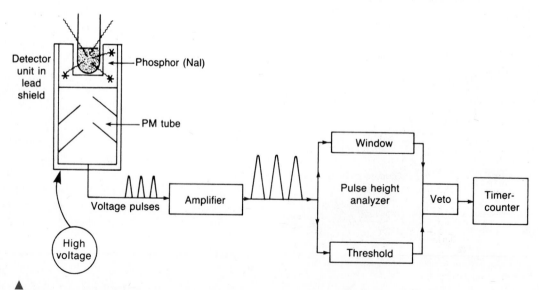

▲
Figure 5-35. Scintillation counter—basic components.

$$\text{Osmolality} = \phi \times n \times C \qquad \textit{(Eq. 5-15)}$$

where ϕ = osmotic coefficient

n = number of dissociable particles (ions) per molecule in the solution

C = concentration in moles per kilogram of solvent

The osmotic coefficient is an experimentally derived factor to correct for the fact that some of the molecules, even in a highly dissociated compound, exist as molecules rather than as ions.

The four physical properties of a solution that change with variations in the number of dissolved particles in the solvent are osmotic pressure, vapor pressure, boiling point, and freezing point. Osmometers measure osmolality indirectly by measuring one of these colligative properties, which change proportionally with osmotic pressure. Osmometers in clinical use measure either freezing-point depression or vapor-pressure depression, and results are expressed in milliosmolal per kilogram (mOsm/kg) units.

Freezing-Point Osmometer

Figure 5-36 illustrates the basic components of a freezing-point osmometer. The sample in a small tube is lowered into a chamber with cold refrigerant circulating from a cooling unit. A thermistor is immersed in the sample. To measure temperature, a wire is used to gently stir the sample until it is cooled to several degrees below its freezing point. It is possible to cool water to as low as $-40°C$ and still have liquid water, provided no crystals or particulate matter are present. This is referred to as a *supercooled solution*. Vigorous agitation when the sample is supercooled results in rapid freezing. Freezing also can be started by "seeding" a supercooled solution with crystals. When the supercooled solution starts to freeze because of the rapid stirring, a slush is formed, and the solution actually warms to its freezing-point temperature. The slush, an equilibrium of liquid and ice crystals, will stay

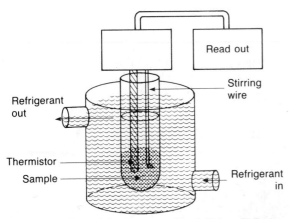

Figure 5-36. Freezing-point osmometer. (From Coiner D. Basic concepts in laboratory instrumentation. ASMT Education and Research Fund, Inc., 1975–1979.)

at the freezing-point temperature until the sample freezes solid and drops below its freezing point.

Impurities in a solvent will lower the temperature at which freezing or melting occurs by reducing the bonding forces between solvent molecules so that the molecules break away from each other and exist as a fluid at a lower temperature. The decrease in the freezing-point temperature is proportional to the number of dissolved particles present.

The thermistor is a material that has less resistance when temperature increases. The readout uses a Wheatstone bridge circuit that detects temperature change as proportional to change in thermistor resistance. Freezing point depression is proportional to the number of solute particles. Standards of known concentration are used to calibrate the instruments in mOsm/kg.

Vapor-Pressure Osmometer

Most osmometers used in the clinical laboratory use the freezing point depression method.

SUMMARY

The techniques and general principles used in a clinical chemistry laboratory are identical to those used in other analytical testing laboratories. Clinical laboratories do have special needs that require analyzers to have high throughput and sample turnaround times. The current generation of chemistry analyzers operates under the "random access" mode, *i.e.,* any combination of tests can be performed on a sample from an on-board menu of analytes. For general chemistry analytes, such as glucose or phosphorus, spectrophotometry is the most widely utilized technique. For electrolytes such as sodium and potassium, ion-specific electrodes are extensively used and have largely replaced flame photometers. Atomic absorption spectrophotometry is still used for metals such as zinc and copper, and remains the reference method for calcium and magnesium. However, colorimetric assays and ISEs are now used routinely for these latter two metals.

Electrophoresis is still an important technique in clinical laboratories, although the development of new assays have challenged its role. For example, lipoprotein electrophoresis largely has been replaced by direct measurement of high-density lipoprotein (HDL) cholesterol and calculation of low-density lipoprotein (LDL) cholesterol. The development of a specific assay for LDL cholesterol will further diminish the role of electrophoresis. With the development of specific immunoassays for creatine kinase isoenzyme form MB (CK-MB) and inhibition assays for lactate dehydrogenase form 1 (LD1), electrophoresis is not as popular today as before; however, an automated electrophoresis assay has recently been released for CK-MB isoforms. Electrophoresis will play a critical role in the area of molecular pathology for the identification of gene products and mutations. Capillary electrophoresis is an emerging technology that promises to have many clinical applications.

With the development of immunoassays, the role of liquid chromatography has shifted away from therapeutic drug monitoring toward toxicology. HPLCs with rapid scanning UV detectors are being used for comprehensive drug screens. GC/MS is the mainstay for confirmation analysis. The interface of LC to MS through electrospray atomization opens many new applications to the clinical laboratory, as this technique works well with aqueous materials and high molecular weight substances. However, HPLC/MS instrumentation is very expensive and will likely not be used in routine clinical laboratories in the near future.

EXERCISES

1. Which of the following is NOT necessary for obtaining the spectrum of a compound from 190–500 nm?
   A. deuterium light source
   B. double beam spectrophotometer
   C. quartz cuvets
   D. tungsten light source
   E. photomultiplier
2. Stray light in a spectrophotometer places limits on:
   A. sensitivity
   B. upper range of linearity
   C. photometric accuracy below 0.1 absorbance units
   D. ability to measure in the UV range
   E. use of a grating monochromator
3. Which of the following light sources is used in atomic absorption spectrophotometry?
   A. hollow cathode lamp
   B. xenon arc lamp
   C. tungsten light
   D. deuterium lamp
   E. laser
4. Which of the following is true concerning fluorometry?
   A. emission wavelengths are always set at lower wavelengths than excitation
   B. the detector is always placed at right angles to the excitation beam
   C. all compounds undergo fluorescence
   D. fluorescence is an inherently more sensitive technique than absorption
   E. fluorometers require special detectors
5. Which of the following techniques has the highest potential sensitivity?
   A. chemiluminescence
   B. fluorescence
   C. turbidimetry
   D. nephelometry
   E. phosphorescence
6. Which of the following is NOT a feature of lasers?
   A. no power supply needed
   B. monochromatic light is produced

   C. narrow spectral width
   D. emission of polarized radiation
   E. high intensity
7. Which electrochemical assay measures current at fixed potential?
   A. anodic stripping voltametry
   B. amperometry
   C. coulometry
   D. analysis with ion selective electrodes
   E. electrophoresis
8. Which of the following refers to the movement of buffer ions and solvent relative to the fixed support?
   A. isoelectric focusing
   B. iontophoresis
   C. zone electrophoresis
   D. electroendosmosis
   E. plasmapheresis
9. "Reverse-phase" liquid chromatography refers to:
   A. polar mobile phase and nonpolar stationary phase
   B. nonpolar mobile phase and polar stationary phase
   C. distribution between two liquid phases
   D. size is used to separate solutes instead of charge
   E. charge is used to separate solutes instead of size
10. Which of the following is NOT an advantage of capillary electrophoresis?
    A. very small sample size
    B. rapid analysis
    C. use of traditional detectors
    D. multiple samples can be assayed simultaneously on one injection
    E. cations, neutrals, and anions move in the same direction at different rates
11. Isocratic HPLC refers to:
    A. chromatography at a constant temperature
    B. use of a gradient
    C. use of ion pairing reagents
    D. chromatography at a constant elution strength
    E. use of a mobile phase with physiologic saline
12. Tandem mass spectrometers
    A. are two mass spectrometers placed in series with each other
    B. are two mass spectrometers placed in parallel with each other
    C. requires use of a gas chromatograph
    D. requires use of an electrospray interface
    E. does not require an ionization source

REFERENCES

1. Beiser, A Physics. 3rd ed. New York: Benjamin-Cummings, 1982.
2. Buchler Instruments, Inc. Instructions for the Buchler-Cotlove chloridometer, Models 4-2000, 4-2008. Fort Lee, New Jersey, 1976.
3. Christian GD, O'Reilly JE. Instrumental analysis, 2nd ed. Boston: Allyn and Bacon, 1986.

4. Constantin E, Schnell A. Mass spectrometry. New York: Ellis Horwood, 1990.

5. Coulter Hematology Analyzer: Multidimensional Leukocyte Differential Analysis, Vol. 11(1), 1989.

6. Guilbault GG. Practical fluorescence, theory, methods and techniques. New York: Marcel Dekker, 1973.

7. Heiger, DN. High performance capillary electrophoresis—An introduction. 2nd ed. Waldbronn, Germany: Hewlett-Packard Company, 1992.

8. Holland JF, Enke CG, Allison J, et al. Mass spectrometry on the chromatographic time scale: Realistic expectations. Anal Chem 1983;55:997A.

9. Ingle JD, Crouch SR. Spectrochemical analysis. New Jersey: Prentice-Hall, 1988.

10. Karasek FW, Clement RE. Basic gas chromatography—Mass spectrometry. New York: Elsevier, 1988.

11. Kebarle E, Liang T. From ions in solution to ions in the gas phase. Anal Chem 1993;65:972A.

12. Kricka LJ. Chemiluminescent and bioluminescent techniques. Clin Chem 1991;37:1472.

13. Monnig CA, Kennedy RT. Capillary electrophoresis. Anal Chem 1994;66:280R.

14. Parris NA. Instrumental liquid chromatography. A practical manual on high performance liquid chromatographic methods. New York: Elsevier, 1976.

15. Perkin-Elmer Corporate Bulletin. The guide to techniques and applications of atomic spectroscopy, 1981.

16. Ritchie RF, ed. Automated immunoanalysis, Part 1. New York: Marcel Dekker, 1978.

17. Sequoia-Turner Corporation Turner filter fluorometer, model III operating instructions and service manual, 1974–1977.

18. Svelto O, Hanna DC. Principles of lasers. 2nd ed. New York: Plenum, 1982.

19. Tietz NW, ed. Fundamentals of clinical chemistry. 3rd ed. Philadelphia: Saunders, 1987.

20. Touchstone JC, Rogers D. Thin layer chromatography: Quantitative environmental and clinical applications. New York: Wiley, 1980.

21. Weast, R.C. CRC handbook of chemistry and physics, 61st ed. 1980–1981. Cleveland: CRC Press, 1981.

22. Willard HH, Merritt LL, Dean JA, Settle FA. Instrumental methods of analysis. Belmont, CA: Wadsworth, 1981.

23. Winstead M. Instrument check systems. Philadelphia: Lea & Febiger, 1971.

Chapter 6

Thurs
3/5

Immunochemistry

Donald J. Cannon

Objectives

Upon completion of this chapter, the clinical laboratorian should be able to:

- *Describe the human immune system and its functions.*
- *Name the various techniques utilized in the evaluation of the immune system.*
- *Describe the diffusion techniques used to detect antigen-antibody reactions.*
- *List the various applications of electrophoresis to evaluate the humoral immune system.*
- *Describe the Western blot technique.*
- *Relate how DNA probes are used to detect specific structures, for example, viral DNA.*
- *Explain flow cell cytometry.*
- *Discuss the advantages and disadvantages of RIA as compared to other immunoassays.*
- *Explain the data reduction process for RIA.*

KEY WORDS

Antibody	Heterogeneous
Antigen	Assay
Antigenic	Homogeneous
Determinant	Assay
Autoimmune	Immune System
Disorder	Immunoassay
Chemiluminescence	Immunofixation
Competitive Protein	Indirect Immuno-
Binding	fluorescence
Diffusion	Nephelometry
DNA Probe	Titer
Electrophoresis	Turbidimetry
Flow Cell Cytometry	Western Blot
Fluorescence	
Polarization	

The immune system is a complex series of events that protects individuals from external, harmful agents. Individual survival depends on a properly functioning system. Disorders of the immune system can result in immunodeficient or hypersensitive states. Fortunately, the system can distinguish between self and non-self. The immune system consists of two parts: (1) a native series of defenses that range from intact skin to soluble factors (lysozyme, complement, etc.) to cell defenses (phagocytes) and (2) an adaptive defense that has humoral and cellular components (Table 6-1). Foreign agents that are recognized by the immune system are termed antigens. In many cases, an antigen contains a part of its structure that is recognized as foreign by the immune system. This structural domain is referred to as an antigenic determinant or epitope. The immune system can recognize large numbers of foreign antigens. In general, antigens are foreign, and must have a sufficient half-life to provoke a response; preferably have a high molecular weight (*e.g.*, 10,000; smaller molecules are referred to as haptens and may or may not bind to a larger carrier molecule); and must be structurally complex. Usually, antigens are large organic molecules—either proteins or large polysaccharides, or glycoproteins or glycolipids, but rarely lipids or nucleic acids.

The humoral system contributes complement, lysozyme, and interferon to natural immunity and antibodies and lymphokines to adaptive immunity. Antibodies are glycoproteins (immunoglobins) secreted by plasma cells, which in turn are under the control of many lymphocytes and their cytokines. When antibodies are produced in response to an antigen, the antibody recognizes (anti-binding site) and binds to the antigen (antigenic determinant). The extent of binding to a targeted (homologous) antigen is termed the *reaction specificity*, and the strength of the subsequent

Michael L. Bishop, Janet L. Duben-Engelkirk, and Edward P. Fody.
CLINICAL CHEMISTRY. © 1996 Lippincott–Raven Publishers.

TABLE 6-1. The Immune System

Organ Defense	Skin
	Respiratory
	Gastrointestinal
	Urogenital
	Eyes
Innate Immunity	Cells (neutrophils, macrophages, mast cells)
	Humoral (complement, lysozyme, interferon)
Acquired Immunity	Cells (lymphocytes, plasma cells)
	Humoral (antibodies, lymphokines)

TABLE 6-3. Serial Dilution to Determine Titer*

Tube	Dilution Solution (mL)	Sample (mL)	Titer (Final Dilution)
1	1.0	serum 1.0	2(1:2)
2	1.0	tube 1 1.0	4
3	1.0	tube 2 1.0	8
4	1.0	tube 3 1.0	16
5	1.0	tube 4 1.0	32
10	1.0	tube 9 1.0	1024

Titer expressed as lowest dilution at which reaction takes place

antigen–antibody bond is the *antibody affinity*. Antibody binding affinity can be expressed as an equilibrium constant, as in the equation for the law of mass action.

EVALUATION OF THE IMMUNE SYSTEM

When the immune system is functioning, protection is afforded. However, challenges to the system are continuous at times and can lead to a state of immunodeficiency. In addition, the system, in rare instances, can attack the host organism (autoimmune disorders) (Table 6-2). Immunodeficient conditions can be congenital as well as acquired. It is these and other clinical situations that necessitate evaluation of immune function. The simplest evaluation of the humoral system is serum protein electrophoresis, and the simplest evaluation of the cellular system is the total leukocyte count. However, in most clinical situations, a more thorough evaluation is both necessary and possible.

In general, the evaluation of the immune system began with the study of antibody-antigen reactions in bacterial infections. This type of study is termed *serology*. Measuring antibody production in response to an insult (*e.g.*, bacterial infection) is accomplished by defining antibody concentration as a *titer*, which is the highest dilution of serum that shows a positive reaction in the presence of antigen (*e.g.*, precipitation of antigen-antibody complex) (Table 6-3). Increased titers were indications of exposure or infection.

TABLE 6-2. Autoimmune Diseases*

Addison's disease
Chronic hepatitis
Hashimoto's thyroiditis
Juvenile diabetes
Lupus erythematosus
Multiple sclerosis
Rheumatoid arthritis

Selected examples

Diffusion

The ability of antigen-antibody mixtures to precipitate is the basis for one of the earliest and simplest identifications of soluble antigens—that of gel diffusion precipitation. Passive gel diffusion (single immunodiffusion) consists of placing a solution of antigen into a well that is cut into a clear, semi-solid matrix (*e.g.*, agar) that is impregnated with antibody. If initial concentrations are carefully selected, an opaque precipitin band may form. The shape of the band depends on molecular weights: if they are equal the band is straight, if not the band is concave toward the higher molecular weight species. This analysis is qualitative for the presence of antigen-antibody reaction. Another qualitative assessment is double immunodiffusion (Ouchterlony). In this technique, soluble antigen and antibody dilutions are placed in separate wells and allowed to diffuse toward each other. Patient sample is compared to standards for band location and precipitation pattern. The presence of rheumatoid arthritis (anti-RANA) and systemic lupus erythematosus (anti-SM) antibodies can be identified by this method.

Quantitative determination of protein antigens in serum can be assessed by the technique of radial immunodiffusion (RID). Very specific antibody is dispersed in a gel (agarose) and separate, increasing levels of antigen are allowed to diffuse from wells under controlled, constant conditions. In the presence of excess antibody, the precipitin zones (diameter) that are formed are proportional to the concentration of antigen in the wells. The RID plates can be read after a specific incubation time or after diffusion is complete. Interpolation of unknown data into a plot of ring diameter squared versus concentration will yield the concentration of an unknown antigen. The process can be reversed (antigen in the gel) for antibody determination. Complement components, alpha-1-antitrypsin, transferrin, and immunoglobins are some of the analytes measured by this method. The method is limited by the concentrations that can be measured and by the precision of the method, which is high.

Electrophoresis

Evaluation of the humoral immune system can also be done by electrophoresis. Serum proteins are screened by zone electrophoresis on a medium and are separated into five major regions. Details of this method can be found in Chapter 9, Amino Acids and Proteins. Overall, antibody deficiency (hypogammaglobulinemia) and possible autoimmune disorders or infection (hypergammaglobulinemia) may be observed. For more specific analysis, a hybrid of diffusion and electrophoresis (immunoelectrophoresis or IEP) is preferred. In this method, electrophoresis separates serum or spinal fluid proteins on cellulose acetate or agarose, and the separated constituents are allowed to diffuse toward troughs that contain specific antibodies to serum constituents. When the correct ratio of antigen to antibody exists (equivalence point), a precipitin line forms and the plate is then washed and stained. Proteins are identified by electrophoretic mobility, diffusion, and antibody specificity. Results of this technique are evaluated by observing precipitin arcs, one for each sample constituent. There are many uses for this technique. In clinical testing, it is used mainly to evaluate myeloma proteins. Other possible uses are to detect a structural abnormality, the absence of a protein antigen, or to screen for immune complexes. Urine can also be subjected to IEP, which is commonly used to detect the presence of Bence-Jones protein.

Immunofixation electrophoresis (IFE) has replaced IEP in many labs because it can be performed quicker and interpretation is easier. Aliquots of a single fresh serum (or urine or CSF) are electrophoresed in six tracks on an agarose gel (commercially available). Protein in one track is fixed. Each of the other five tracks is treated individually with different monospecific antisera. After brief incubation for the immunofixation step, the gels are washed and stained. The presence of an antigenic band will be visible and its location will be matched to the reference pattern and to the individual antisera.

Other Techniques

In addition to the basic techniques involving precipitation (following diffusion) and electrophoresis, there are a few other methods that should be mentioned. Electroimmunodiffusion (EID) is a variant of double diffusion, but antibody (placed near the anode) and antigen (near the cathode) are driven toward each other by an electric current. If the initial immunodiffusion procedure was double diffusion, then the technique is known as counter-current immunoelectrophoresis (CIE). The agar in this technique is formed at a pH that renders the antibodies positively charged and antigens negatively charged. Precipitin bands are formed following the electrophoresis, and the sensitivity is twenty times greater than simple Ouchterlony double diffusion.

If an electric charge is used in a radial diffusion setup as described for RID, the technique is referred to as rocket electrophoresis or electroimmunoassay. This technique will be accurate and sensitive and 10 to 15 times as fast to complete as RID.

Finally, if electrophoresis is used instead of diffusion as the second step in IEP, the technique is known as crossed immunoelectrophoresis (CRIE). This two-dimensional immunoelectrophoresis technique is powerful and sensitive and is used to evaluate structural changes in antigens. A summary of antigen antibody techniques is found in Table 6-4.

Western Blot

Under certain circumstances, the separation of antigens by electrophoresis may not allow for immunofixation and visualization. There may be too little antigen or the electrophoretic medium may not be suitable for immunofixation (e.g., polyacrylamide). In this case, the electrophoretically-separated antigens (e.g., proteins) are transferred to a new medium (e.g., nitrocellulose). The antigens will be fixed on the new medium by absorption or covalent bonding and then can be detected by a wide range of antibody probes. The probe could be labeled with a radioactive tag, an enzyme that can produce a visual product, or a fluorescent or chemiluminescent label. Detection would be accomplished with specialized instrumentation: spectrophotometer, fluorometer, and luminometer, respectively. This transfer technique is referred to as Western blotting, and is used to detect the presence of human immunodeficiency virus (HIV). Purified standard viral antigens are electrophoresed, transferred to nitrocellulose, and inoculated with patient serum. Serum antibody that reacts with certain viral antigens will be visualized by a second antibody reaction in which the second antibody is tagged. A positive reaction to core (p 24) and envelope (gp 41) antigenic proteins is indicative of a true positive HIV sample. This procedure is expensive, labor intensive, and technically demanding.

IMMUNOASSAYS

In addition to the use of immunodiffusion and electrophoresis for quantitative immunochemical analysis, there are other techniques that measure the rate of immune complex formation. These are generally classified as immunoassays (Table 6-5). Immunoassays can be labeled or nonlabeled (Table 6-6). Nonlabeled immunoassays measure immune complexes directly using either turbidimetry or nephelometry. These techniques measure the presence of particles in

TABLE 6-4. Antigen-Antibody Techniques

Diffusion	Single
	Double (Ouchterlony)
	Radial (RID)
Electrophoresis	Immuno-(IEP)
	Immunofixation-(IFE)
Hybrid	Electroimmunodiffusion (EID)
	Counter-current immunoelectrophoresis (CIE)
	Rocket electrophoresis
	Crossed immunoelectrophoresis (CRIE)
Other	Western blot

TABLE 6-5. Immunoassays

(A) Nonlabeled
    Labeled
(B) Competitive
    Non-Competitive
(C) Heterogeneous
    Homogeneous

---

solution, which are antigen-antibody complexes. These complexes are usually formed when increasing antigen is added to constant antibody amounts. This reaction proceeds through three stages: (a) antibody excess (soluble), (b) equivalence point (precipitate), and (c) antigen excess (soluble). The middle stage can be detected by turbidimetry on many automated clinical chemistry analyzers. A beam of light of single wavelength is focused on a cuvette containing the reaction, and the light scattered by the particles is measured. Both endpoint determinations and initial rate determinations can be used to measure antigen-antibody formation. They both are comparable but the latter (rate) has numerous advantages. In turbidimetry, measurements are made at 180° to the incident beam (unscattered light) and in nephelometry at 5° to 90° incident to the beam. The latter must be measured on dedicated instrumentation (high quality detection and low wavelength). Both methods are about equal in sensitivity; however, nephelometry may be more sensitive at low levels of antigen-antibody concentration. Complex formation is enhanced by the presence of polymers, which decrease reaction time. When antigen-antibody complexes are not large enough to scatter light, antigen can be linked to a carrier particle (*e.g.*, latex) to create a complex of light-scattering particles when antibody reacts with the antigen-particle. Alternatively, antibody attached to small particles reacts with specific antisera but not in the presence of free immunoglobulins. Serum, urine, and cerebrospinal fluid (CSF) are analyzed by turbidimetry and nephelometry for a wide range of analytes (plasma, proteins, hormones, drugs, rheumatoid factor, etc.)

The measurement of antigen-antibody reactions is widely used in clinical laboratory medicine to measure, with high specificity, and at very sensitive levels, a broad range of substances. This has been achieved by the use of labeled immunoassays and the detection of the resulting immunocomplexes under conditions of antibody excess where complex formation and antigen concentration are proportional.

One of the initial labels used in immunoassays was radioactivity. Radioactive iodine (^{125}I and ^{131}I) and tritium

---

TABLE 6-6. Immunoassay Labels

Radioactivity (RIA)
Enzymes (EIA, EMIT, CEDIA)
Fluorescence (FIA, FPIA)

(^{3}H) are used as labels for both antigens and antibodies. Labeled antigens need to be evaluated to determine whether antibody reacts equally with labeled and unlabeled antigen. Assays using labeled antibody are referred to as immunoradiometric assays (IRMA). Radioimmunoassays (RIA) using labeled antigen are very sensitive and resist interferences. RIA was a technological breakthrough for endocrine testing, and the technique was widely applied during the 1960s and 1970s. Although RIA is still the most popular technique in the clinical endocrine laboratory when high sensitivity is required, there are a number of drawbacks to its use. RIA is based on radioisotopes, and its use in the laboratory must be licensed and regulated by radiation safety committees. In addition, radioactive materials must be disposed of or stored under special conditions and usually must be separated from normal waste. If labeled hormones are not used within a short period of time (about 1 to 2 months) these substances must be discarded. The short shelf life of radioisotopes may result in waste and inefficiency in test-kit usage. Finally, standard curves must be run for every assay, with both patient and standards done in duplicate. It is also necessary to perform a separation step. Despite all these drawbacks, RIA is still used in the clinical laboratory because of its high sensitivity for a limited group of analytes.

Data reduction and analysis of RIA results is a complex process if done manually. In the past, the conversion of gamma-counter data to useful patient information has been a time-consuming step in RIA. Currently, with the availability of microcomputer linkage to counting instruments and the ready availability of computers and calculators in the laboratory, this task has been greatly simplified and can even be automated. It is important to construct a curve that spans the range of sensitivity of the method and clinical significance of the analyte. The first step in data reduction is to convert the radioactive counts to a number that can be related to concentration. Although a simple plot of counts versus concentration of standard will give a usable curve, it has been found to be more useful to calculate bound hormone (B) as a fraction of either the total counts added to the assay tube (T) or the counts obtained when no standard or sample competes with labeled hormone (represented as B_0 and referred to as the *zero standard*). Plots of the fraction or percent bound (B/B_0 or B/T) versus standard concentration or the logarithm of the standard concentration will generate a nonlinear standard curve. Although nonlinear curves can be plotted manually and used for interpretation, linear plots are more useful for detecting errors and for statistical treatment of data (Figure 6-1).

No simple mathematical equation exists that will always yield straight-line plots of binding data versus concentration. RIA binding may be described by a very complex mathematical expression, and a number of approximations and simplifying assumptions to this mathematical equation result in the *logit-log transformation*. A plot of logit of the bound fraction versus the log of the standard concentration is frequently linear. Logit is defined as given below:

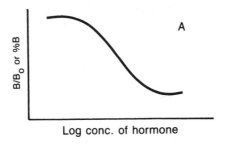

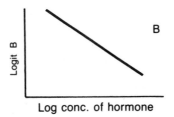

▲

Figure 6-1. (*A*) A plot of percent of labeled hormone bound to antibody on the y-axis, with log of the concentration of the hormone on the x-axis. (*B*) A logit plot of the bound labeled hormone on the y-axis and log of the concentration of the hormone on the x-axis.

Let B and B_0 be defined as described above, and let $y = B/B_0$:

and

$$\text{Logit } y = \ln_e [y/(1 - y)]$$

$$\text{Logit } (y) = A + B \ln x \qquad (\textit{Eq. 6-1})$$

where A = intercept
B = slope
x = hormone concentration

A special paper for plotting logit-log does exist, but microcomputers and calculator programs will convert radioactive counts into a logit number and mathematically calculate the best straight line through the values corresponding to the standard concentrations (linear regression). Patient values can then be calculated by the computer using this standard curve. Although logit-log plots are the most common RIA data transformation used, other smoothing and interpretation techniques are available for computer analysis.

It is important to remember that the best type of curve-plotting technique is determined by experiment, and there is no assurance that a logit-log plot of the data will always give a straight line for a given type of assay. Several different methods of data plotting should be tried when a new assay is being tested in order to determine which technique is most satisfactory. It is also best to examine a plot of the standard curve each time the assay is performed to check for out-of-place standard points. Finally, it should be pointed out that the relative error for all RIA standard curves is minimal when B/B_0 is 0.5 and increases at both the high- and low-concentration ends of the curve. This point can be recognized easily from a plot of B/B_0 versus log concentration of standard (see Fig. 6-1*A*), where a relatively large change

in concentration at either the low- or high-concentration end of the curve produces very little change in the B/B_0 value. Because the logit-log plot is linear, however, this point can be easily forgotten, and patient values derived from a B/B_0 value of greater than 0.9 or less than 0.10 should be interpreted with caution.

Initially, immunoassays used labeled antigens in a format known as competitive immunoassay. Labeled and unlabeled antigen compete for limited antibody sites. Labeled antigen bound to antibody will be inversely proportional to the concentration of antigen. This format requires excess antibody and a separation step. Competition for antibody can be performed following simultaneous or sequential mixing of components. Sensitivity is increased using sequential mixing, especially at low antigen concentration.

Later, when techniques to label antibodies were developed and the technology to produce monoclonal antibodies was perfected, labeled antibody immunoassays were used extensively. These assays were noncompetitive immunoassays in which, in the presence of excess labeled antibody, bound label was directly proportional to the antigen concentration. This assay employed a separation step to remove unbound antibody, usually by capturing unbound antibody on solid phase material (bead, tube, particle, well) coupled to the antigen followed by centrifugation. An extension of the above assay was to employ two labeled antibodies directed at two sites on the antigen of interest in a format known as sandwich immunoassay. In this latter scheme, sensitivity and specificity are much improved by initially coupling antigen to antibody attached to a solid phase then adding a second labeled antibody directed at a separate site on the antigen. After a wash step, the bound label is measured directly. A widely used example of this technique is referred to as enzyme-linked immunosorbent assay (ELISA).

In all of the previously described formats there is a separation step to remove bound from free label, and such assays are referred to as heterogenous immunoassays. Homogenous immunoassays do not require a separation step, and their introduction advanced the development of automated immunoassay systems. In the homogenous format, label attached to an antigen is changed by the act of antibody binding, and the magnitude of the change is proportional to the concentration of the antibody. An example of a homogenous immunoassay is a competitive immunoassay where an enzyme-labeled antigen (*e.g.*, a drug) competes with a sample antigen (drug) for an antibody. The activity of the enzyme that is used to label the antigen is blocked when the labeled antigen couples to the antibody. Therefore, enzyme activity is proportional to sample antigen concentration. Many enzymes are used as labels and one popular system uses glucose-6-phosphate dehydrogenase. In the presence of glucose-6-phosphate and nicotinamide adenine dinucleotide (NAD), enzyme activity is monitored via the production of NADH (absorbance at 340 nm). This system is known as enzyme multiplied immunoassay (EMIT). Another popular homogenous enzyme immunoassay uses the enzyme beta-galactosidase, in which two inactive fragments of this enzyme are conjugated to an antigen. When the labeled antigen reacts with antibody, the two fragments are prevented from assembling to form active enzyme. The enzyme fragments are produced by recombinant techniques in which two *E. coli* enzyme genes produce separate inactive fragments that must recombine to produce active enzyme. This system, known as cloned enzyme donor immunoassay (CEDIA), is also used in commercial automated clinical analyzers. Table 6-7 summarizes enzyme immunoassays.

In addition to the use of enzymes as nonisotopic labels, other substances can also provide amplification and high sensitivity. Fluorescent and chemiluminescent molecules and ligands are used. Fluorescent labels can be potentially very specific. Fluorescein isothiocyanate (FITC) is an example of a fluorescent label that is commonly used. Fluorescence is also used indirectly when, in the enzyme immunoassay format, a substrate (*e.g.*, methylumbelliferyl phosphate) is converted to a fluorophor (umbelliferone). Disadvantages in using fluorescence include excess background fluorescence and too narrow a difference between excitation and emission wavelengths. To minimize these effects, lanthanide chelates (*e.g.*, europium, etc.) are used as labels. These have a delayed fluorescence, a high specificity, and a large shift between excitation and emission wavelengths. Detection uses specialized equipment that measures time-resolved fluorescence.

A unique use of fluorescence is the basis for a homogenous immunoassay based on fluorescence polarization. A fluorescein-labeled antigen rotates very fast in solution and when excited does not emit polarized light. After binding to an antibody, the label rotates much slower and will emit polarized light. When polarized light is used as a source of excitation, emitted fluorescence is measured in two planes. The difference in fluorescence polarization (calculated) before and after the addition of labeled analyte is inversely proportional to the concentration of unknown analyte. This methodology (fluorescence polarization immunoassay or FPIA) is most useful for small antigens (drugs, hormones, etc.).

Chemiluminescent labels provide another very sensitive detection methodology for immunoassay analysis. Heterogeneous, competitive assays using labeled antigen and sandwich assays using labeled antibody are analyzed on automated equipment. The basic equipment is (automated or not) a luminometer, which is an instrument that measures light flashes in nanoseconds under temperature-controlled conditions. Specific labels (luminal, acridinium esters, etc.) coupled to antigen-antibody conjugates are exposed to alkali and hydrogen peroxide to activate a chemiluminescent reaction. Emitted light is proportional to unknown concentrations, which are calculated from standard curves. Alternatively, chemiluminescence substrates can be utilized in an enzyme immunoassay format with horseradish peroxidase and alkaline phosphatase and with substrates that are converted by enzymatic oxidation to unstable products that emit light (Table 6-8).

IMMUNOCYTOCHEMISTRY

Fluorescent labels as substitutes for radioisotopes or enzymes have been used extensively to identify organisms and tissue structure at the microscopic level. In one technique, fluorescent-labeled (*e.g.*, FITC) antibody is used to stain a tissue section or cell smear. With proper incident light and filters the antibody target is visualized. This technique, direct immunofluorescence, has been used to detect bacteria. Indirect immunofluorescence assay (IFA) is a popular technique in which serum antibody reacts with antigen fixed on a slide, and the antibody in turn reacts with conjugated antihuman

TABLE 6-7. Enzyme Immunoassays

Enzyme	Detection*
Alkaline phosphatase	P,F,C
Galactosidase	C,F
Horseradish peroxidase	P,C
Glucose-6-phosphate	C,P

P = photometry, F = fluorimetry, C = chemiluminescence

TABLE 6-8. Chemiluminescence

Label:	enzyme (peroxidase or alkaline phosphatase)
Substrate:	isoluminol or acridinium esters
Product:	light (long-lived)
Sensitivity:	highest
Usage:	antigens and antibodies

globulins. If the patient serum contains antibody of interest, this will be observed under a fluorescent microscope. Antinuclear antibodies (ANA) and antibodies to cytomegalovirus, rubella virus, and *Toxoplasma gondii* are but a few examples of the use of this technique. In addition to fluorescence, immunocytochemistry, such as described above, can be carried out with enzyme labels, such as horseradish peroxidase, followed by visualization of a colored compound (Table 6-9). This technique negates the problem of tissue autofluorescence, and can be used on paraffin-fixed tissue and is permanent.

Advances in molecular biology that have allowed for the production of large amounts of deoxyribonucleic acid (DNA) have made possible the production of small fragments of specific DNA, that can be labeled with biotin and that can be used to probe for the presence, for example, of certain viral DNA (*e.g.*, hepatitis). Following hybridization and washing, conjugate-labeled avidin complexes with the biotin and the resulting complex is detected, usually by using an enzyme (peroxidase or alkaline phosphatase) to produce a photometric or chemiluminescent signal. A format using fluorescent-labeled avidin is also used. Human papilloma virus DNA has been detected using DNA probes.

One of the most important advances in immunocytochemistry is the use of immunofluorescent labels to identify specific antigens on live cells in suspension (flow-cell cytometry). Stained cell suspensions are transported under pressure past a laser beam, and emitted fluorescence (at 90° relative to the beam) is measured and computer-analyzed. With this technique and using multiple labels, cells can be identified and either electronically or physically sorted. The technique has been used to analyze subpopulations of lymphocyte cells in various clinical diagnoses (Table 6-10).

SUMMARY

Immunochemistry consists of reactions between antibodies and antigens. This reaction can be very specific and sensitive and lends itself to qualitative and quantitative identification of either antigen or antibody. Part of the identification relies on the formation of a precipitate under correct conditions of concentration of the two interactants. This phenomenon can be measured by diffusion and electrophoretic tech-

niques. Tables 6-11 and 6-12 summarize the variables of immunoassay and some of the instrumentation used to perform immunoassay.

Further identification can be achieved in the absence of precipitation by labeling either antigen or antibody before or after complexation in solution. Using very sensitive labels that do not affect the antigen-antibody binding process, but which can be affected by the process, assays are achieved which identify various analytes. Furthermore, extending the above principles one can identify organisms and structures microscopically and separate live cell populations to assist in the diagnosis and prognosis of diseases.

TABLE 6-10. Cell Sorting

Fluorochrome Stain
  antibody label
  specific dye
    acridine orange
    thioflavin T
    pyronin Y
    fluorescein isothiocyanate (FITC)
    phycoerythrin
    tetramethyl rhodamine isothiocyanate (TRITC)
    tetra-m-cyclopropylrhodamine isothiocyanate (CXRITC)

TABLE 6-11. Immunoassay Variables

Format (homogeneous, competitive)
Detection mode (photometry, fluorescence)
Automation (yes, no, semi)
Calibration curve (proportional or inversely proportional)
Cost
Sensitivity (high-low)
Specificity (high-low)
Cross-reactivity (what, how much)
Menu (drugs, hormones, etc.)
Sample (urine, blood)

TABLE 6-9. Immunocytochemistry

Formats	Direct
	Indirect
Labels	Fluorescent
	Enzyme
	Peroxidase (color)*
	Biotin-avidin**

*Can vary with enzyme substrate used
**Enzyme or fluorescent labeled

TABLE 6-12. Instrumentation*

Nephelometer	Beckman Array
	Behring
Spectrophotometer	Boehringer (ES 300)
	Miles (Immuno-1)
Fluorometer	Tosoh (ES 300)
	Baxter (Stratus)
	Pharmacia (Delfia)
	Abbott (IMx)
Luminometer	Amersham (Amerlite)
	Ciba-Corning (ACS-180)

*All semi- or fully-automated, some batch or random access.

(From Donald J. Cannon, Professor of Pathology, Marshall University School of Medicine, Huntington, WV)

EXERCISES

1. Discuss the advantages and disadvantages of RIA.
2. Describe the use of standard curves and curve fit techniques in the calculation of immunoassay results.
3. Discuss the major functions of the human immune system.
4. Explain one application of electrophoresis to evaluate the humoral immune system.
5. Briefly explain the principle of the Western blot technique.
6. What is the major purpose for using a flow cell cytometry technique?
7. Describe two diffusion techniques to detect antigen-antibody reactions.

SUGGESTED READINGS

Burtis CA, Ashwood ER, eds. Tietz textbook of clinical chemistry. 2nd ed. Philadelphia: WB Saunders Co. 1994.

McClatchey KD, ed. Clinical laboratory medicine. Baltimore: Williams and Wilkins, 1994.

Noe DA, Rock RC, eds. Laboratory medicine: The selection and interpretation of clinical laboratory studies. Baltimore: Williams and Wilkins, 1994.

Rose NR, DeMacario EC, Fahey JL, Friedman H, Penn GM, eds. Manual of clinical laboratory immunology. 4th ed. Washington DC: American Society for Microbiology, 1992.

Turgeon ML. Immunology and serology in laboratory medicine. St. Louis: CV Mosby Co., 1990.

Ward KM, Lehmann CA, Leiken AM. Clinical laboratory instrumentation and automation. Philadelphia: WB Saunders Co., 1994.

Automated Techniques

Larry Schoeff

Objectives

Upon completion of this chapter, the clinical laboratorian should be able to:

- *Define the following terms: automation, continuous flow, discrete analysis, channel, flag, and random access.*

- *Discuss the history of the development of automated analyzers in the clinical chemistry laboratory.*

- *List three driving forces behind the development of more and better automated analyzers.*

- *Name the three basic approaches to sample analysis used by automated analyzers.*

- *Explain the major steps in automated analysis.*

- *State an example of a commercially available discrete analyzer and a centrifugal analyzer.*

- *Compare the different approaches to automated analysis used by instrument manufacturers.*

- *Discriminate between an open versus a closed reagent system.*

- *Relate three considerations in the selection of an automated analyzer.*

- *Discuss future trends in automated analyses.*

KEY WORDS

Automation	Dry Chemistry Slide
Bar-code	Flags
Centrifugal Analysis	Probe
Channel	Random Access
Continuous Flow	Robotics
Discrete Analysis	Rotor

AUTOMATED TECHNIQUES

Since the introduction of the first automated analyzer by Technicon in 1957, automated instruments have proliferated from many manufacturers.[4] This first "Auto Analyzer" (AA)© was a continuous flow, single channel, sequential batch analyzer capable of approximately 40 tests per hour. The next generation of Technicon instruments to be developed was the Simultaneous Multiple Analyzer (or SMA) series. As the name implies, SMA-6© and SMA-12© were analyzers with multiple channels (for different tests) working synchronously to produce 6 or 12 test results simultaneously at the rate of 360 or 720 tests per hour. It was not until the mid-1960s that these continuous flow analyzers had any significant competition in the marketplace.

In 1970, the first commercial centrifugal analyzer was introduced as a "spin-off" technology from NASA outer space research. Dr. Norman Anderson developed a prototype in 1967 at the Oak Ridge National Laboratory as an alternative to continuous flow technology, which had significant carry-over problems and costly reagent waste. He wanted to perform analyses in parallel and also take advantage of advances in computer technology. The second generation of these instruments in 1975 was more successful, as a result of miniaturization of computers and advances in the polymer industry for high grade optical plastic cuvets.

The next major development that revolutionized clinical chemistry instrumentation occurred in 1970 with the introduction of the DuPont Automated Chemical Analyzer (ACA©). It was the first noncontinuous flow, discrete analyzer, as well as the first instrument to have random access capabilities, whereby *STAT* specimens could be analyzed

Michael L. Bishop, Janet L. Duben-Engelkirk, and Edward P. Fody.
CLINICAL CHEMISTRY. © 1996 Lippincott–Raven Publishers.

out of sequence from the batch as needed. Plastic test packs, positive patient identification, and infrequent calibration were among the unique features of the ACA©.

The last major milestones were the introduction of thin film analysis technology in 1976, and the production of the Ektachem Analyzer© by Kodak in 1978. This instrument was the first to use micro-volumes of sample and reagents on slides for analysis, and it was the first to incorporate computer technology extensively into its design and use.

Since 1980, several primarily discrete analyzers have been developed that incorporate such characteristics as ion selective electrodes (ISE), fiber optics, polychromatic analysis, continually more sophisticated computer hardware and software for data handling, and larger test menus. Some of the popular and more successful analyzers utilizing these and other technologies since 1980 are the Beckman ASTRA© (now Synchron) analyzers extensively using ISEs; Baxter Paramax© using reagent tablet dispensing and primary tube sampling; the Hitachi analyzers with reusable reaction disks and fixed diode arrays for spectral mapping; and the Chem 1© by Technicon, which uses encapsulated oil segments of sample and reagents in a single continuous flow tube. Other automated systems that are commonly used in clinical chemistry are the Abbott Spectrum© and the Olympus and Roche series of analyzers.

Many of the manufacturers of these instrument systems have adopted the more successful features and technologies of other instruments, where possible, to make each generation of their product more competitive in the marketplace. The differences among the manufacturers' instruments, their operating principles, and their technologies are less distinct now than they were in the beginning years of laboratory automation.

Three specialized categories of clinical chemical instrumentation that are beyond the scope of discussion here are immunochemistry analyzers, physicians' office laboratory (POL) analyzers, and point of care, or portable analyzers. There are dozens of instrument systems on the market within each of these specialty classifications.

DRIVING FORCES TOWARD MORE AUTOMATION

There are many forces that currently drive clinical laboratories toward more and better automation. The sheer volume of testing is dictating how analyzers are designed for larger test menus and output of test results. Many laboratories around the country have consolidated into fewer and larger labs for comprehensive testing, which has vastly increased their workload volume. Coupled with this force is the expectation of faster turnaround time with test results, which means faster output is required by the analyzers. The quality of patients' test results is monitored continuously for improvement of the testing process. Wherever the human factor can be decreased or eliminated during the mechanical

and repetitive part of an assay, it is likely to result in improved quality.

Toward this goal, front-end automation (robotics) for specimen handling is beginning to evolve in some clinical laboratories. More sophisticated handling and monitoring of quality control data is available in most systems. Federal regulatory standards also play a part in demanding an increased standard of practice and level of quality with test results.

Escalating costs of health care in laboratory testing have forced labs into more productive and efficient ways of analysis, *i.e.*, automation. Point of care testing, whether at hospital bedside or at home, and physician office testing have created new markets for automated analyzers. Immunologic techniques for assaying drugs, specific proteins, hormones, etc., have been increasingly automated. Instruments that employ such techniques as fluorescence polarization (FPIA), nephelometry, and chemiluminescence have become popular in laboratories. Finally, intense competition among instrument manufacturers has driven automation into more sophisticated analyzers with creative technologies and unique features. Some anticipated future trends in laboratory automation are presented at the end of this chapter.

BASIC APPROACHES WITH AUTOMATED ANALYZERS

There are many advantages to automating procedures. One purpose is to increase the number of tests performed by one laboratorian in a given time period. Labor is a very expensive commodity in laboratories. Through mechanization, the labor component devoted to any single test is minimized, and this effectively lowers the cost per test. A second purpose is to minimize the variation in results from one laboratorian to another. By reproducing the components in a procedure as identically as possible, the coefficient of variation is lowered and reproducibility is increased. Accuracy is then not dependent on the skill or workload of a particular operator on a particular day. This allows better comparison of results from day to day and week to week. Automation, however, cannot correct for deficiencies inherent in methodology. A third advantage is gained because automation eliminates the potential errors of manual analyses such as volumetric pipetting steps, calculation of results, and transcription of results. A fourth advantage accrues because instruments can use very small amounts of samples and reagents. This allows less blood to be drawn from each patient. In addition, the use of small amounts of reagents decreases the cost of consumables.

There are three basic approaches with instruments: continuous flow, centrifugal analysis, and discrete analysis. All three can employ batch analysis (*i.e.*, large number of specimens in one run), but only discrete analyzers offer random access, or *STAT* capabilities.

In continuous flow, liquids (reagents, diluents, and samples) are pumped through a system of continuous tubing. Samples are introduced in a sequential manner, following each other through the same network. A series of air bubbles at regular intervals serve as separating and cleaning media. Continuous flow, therefore, resolves the major consideration of uniformity in performance of tests, because each sample follows the same reaction path. Continuous flow also assists the laboratory that needs to run many samples requiring the same procedure. The more sophisticated continuous flow analyzers use parallel single channels to run multiple tests on each sample, *e.g.*, SMA©, and SMAC©. The major drawbacks that contributed to the eventual demise of continuous flow analyzers (*i.e.*, AA, SMA, and SMAC) in the marketplace were significant carry-over problems and wasteful use of continuous flowing reagents. Technicon's answer to these problems was a noncontinuous flow discrete analyzer (the RA1000©) using random-access fluid, which is a hydrofluorocarbon liquid to reduce surface tension between samples/reagents and their tubing, and thereby reduce carryover. Later, the Chem 1© was developed by Technicon to utilize Teflon tubing and Teflon oil, virtually eliminating carryover problems. The Chem 1© is a continuous flow analyzer but only remotely comparable to the original continuous flow principle.

Centrifugal analysis uses the force generated by centrifugation to transfer and then contain liquids in separate cuvets for measurement at the perimeter of a spinning rotor. Centrifugal analyzers are most capable of running multiple samples, one test at a time, in a batch. Batch analysis is their major advantage, because reactions in all cuvets are read virtually simultaneously, taking no longer to run a full rotor of about 30 samples than it would take to run a few. Laboratories with a high workload of individual tests for routine batch analysis may use these instruments. Again, each cuvet must be uniformly matched to each other to maintain quality handling of each sample. The Roche Cobas-Bio©, with its xenon flash lamp and longitudinal cuvets[3], and the IL Monarch©, with its fully integrated walk away design, are two of the more successful centrifugal analyzers.

Discrete analysis is the separation of each sample and accompanying reagents in a separate container. Discrete analyzers have the capability of running multiple tests one sample at a time or multiple samples one test at a time. They are the most versatile analyzers. However, since each sample is in a separate reaction container, uniformity of quality must be maintained in each cuvet so that a particular sample's quality is not affected by the particular space that it occupies. The DuPont ACA©, Kodak Ektachem©, Baxter Paramax©, Beckman Synchron©, and Hitachi analyzers are all examples of discrete analyzers with random access capabilities.

Mon 3/9

STEPS IN AUTOMATED ANALYSIS

In clinical chemistry, *automation* is defined as the mechanization of the steps in a procedure. Manufacturers design their instruments to mimic manual techniques. The major steps in a procedure may be listed as follows:

Specimen preparation and identification.
Specimen measurement and delivery.
Reagent systems and delivery.
Chemical reaction phase.
Measurement phase.
Signal processing and data handling.

In this section, each step of automated analysis will be explained, and several different applications will be discussed. Several instruments have been chosen because they have components that represent either common features used in chemistry instrumentation or a unique method of automating a step in a procedure. None of the representative instruments are completely described, but rather the important components are described in the text as examples.

Specimen Preparation and Identification

Preparation of the sample for the analysis has been and remains a manual process in most laboratories. The clotting time (if using serum), centrifugation, and transferring the sample to an analyzer cup (unless using primary tube sampling) cause delay and expense in the testing process. One alternative to manual preparation is to automate this process by utilizing robotics, or front-end automation, to "handle" the specimen through these steps and load the specimen onto the analyzer. Another option is to bypass the specimen preparation altogether by using whole blood for analysis, *e.g.*, Abbott Vision. Robotics for specimen preparation has already become a reality in some clinical laboratories in the United States and other countries. In the future, automated analyzers may integrate this step into the automated instrument by adding components such as membrane filtration devices to separate serum or plasma.

The sample must be properly identified and its location in the analyzer must be monitored throughout the test. The simplest means of identifying a sample is by placing a manually labeled sample cup in a numbered analysis position on the analyzer, in accordance with a manually prepared worksheet or a computer generated load list. The most sophisticated approach utilizes a bar-code label affixed to the primary collection tube. This label contains patient demographics and also may include test requests.

The bar-code labeled tubes are then transferred to the loading zone of the analyzer where the bar-code is scanned and the information is stored in the computer's memory. The analyzer is then capable of monitoring all functions of identification, test orders and parameters, and sample position. Some analyzers may take test requests downloaded from the laboratory information system and run them when the appropriate sample is identified and ready to be pipetted.

Specimen Measurement and Delivery

Most instruments use either circular carousels or rectangular racks as specimen containers for holding disposable cups or primary sample tubes in the loading or pipetting zone of

▲
Figure 7-1. 2-mL sample cup with conical bottom. (Photograph courtesy of Scientific Products Division, Baxter Healthcare Corporation)

the analyzer. These cups or tubes hold standards, controls, and patient specimens to be pipetted into the reaction chambers of the analyzers. The slots in the trays or racks usually are numbered to aid in identification of the sample. The trays or racks move automatically in one-cup steps at preselected speeds. The speed determines the number of specimens to be analyzed per hour. As a convenience, the instrument can determine the slot number containing the last sample and terminate the analysis after that sample. The instrument's computer holds the number of cups in memory and aspirates only in slots containing sample cups.

To facilitate the loading of samples onto an automated sampling device, the necessity for exact measurement by the laboratorian should be eliminated. The configuration of

the sample cup should determine the minimum and maximum volumes required. Indentations or scorings molded into the plastic can indicate these volumes. The operator can then pour or pipet the samples into the containers easily and quickly.

One of the most commonly used containers is the 2-mL sample cup shown in Figure 7-1. These cups are available from many manufacturers and are relatively inexpensive.

On the Ektachem analyzer, sample cup trays are quadrants that hold 10 samples each in cups with conical bottoms. The four quadrants fit on a tray carrier (Fig. 7-2). Although the tray carrier accommodates only 40 samples, more trays of samples can be programmed and then loaded in place of completed trays while tests on other trays are in progress. A disposable sample tip is hand loaded adjacent to each sample cup on the tray.

In centrifugal analyzers, the samples and reagents are pipetted into a rotor with 20 or more positions. Each position contains a sample compartment, a reagent compartment, and a cuvet located at the periphery of the rotor (Fig. 7-3).

The DuPont ACA© has special sample cups to contain and identify each specimen (Fig. 7-4). Each sample kit consists of a sample cup and a cover. The operator must ensure that sufficient sample is in the cup for the number of tests to be run, plus a full well to correct for the dead volume. The ACA© cannot distinguish a short-sample error, consequently there will be no error code for insufficient sample.

The Baxter Paramax© allows sampling from primary collection tubes, or for limited samples there are microsample tubes. The tubes are placed in a circular tray that holds 96 specimens at one time. Bar-code labels for each sample, complete with patient name and identification number, are printed on demand by the operator (Fig. 7-5). This allows samples to be loaded in any order. The loading

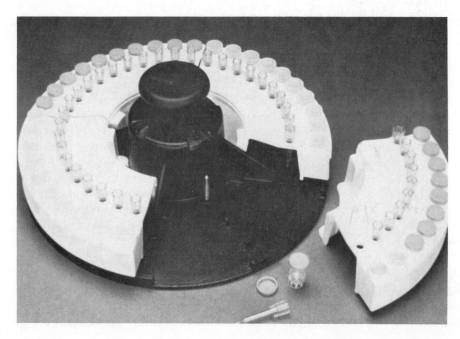

◄
Figure 7-2. The Ektachem 700. The four quadrant trays, each holding ten samples, fit on a tray carrier. (Photograph courtesy of Eastman-Kodak Co.)

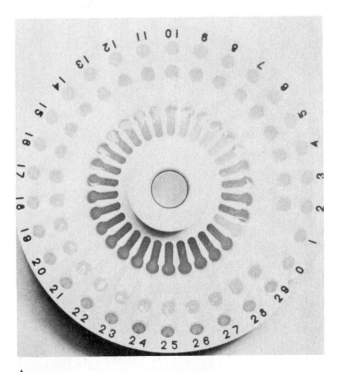

Figure 7-3. Centrifugal analyzer rotor. (Photograph courtesy of American Society for Medical Technology. Correlation of manual techniques to automated techniques. ©1978.)

the rate of one every 5 seconds. The cuvets are cut into sections or groups as required.

A problem with sample handling results from the exposure of the aliquot to the air, which can produce errors in analysis. Evaporation of the sample may be significant and may cause the concentration of the constituents being analyzed to rise in relation to the time of exposure. With instruments measuring electrolytes, the carbon dioxide present in the samples will be lost to the atmosphere, resulting in low carbon dioxide values. Manufacturers have devised a variety of mechanisms to minimize this effect, *e.g.*, lid covers for trays and individual caps that can be pierced.

The actual measurement of each aliquot for each test must be very accurate. This is generally done through aspiration of the sample into a probe. When the discrete instrument is in operation, the probe automatically dips into each sample cup and aspirates a portion of the liquid. After a preset, computer-controlled time interval, the probe quickly rises from the cup. Sampling probes on instruments using specific sampling cups are programmed or adjusted to reach a prescribed depth in those cups to maximize use of available sample. Those analyzers capable of aspirating sample from primary collection tubes usually have a parallel liquid level-sensing probe that will control entry of the sampling probe to a minimal depth below the surface of the serum, allowing full aliquot aspiration while avoiding clogging of the probe with serum separator gel or clot (Fig. 7-7).

In continuous-flow analyzers, when the sample probe rises from the cup, air is aspirated for a specified time to produce a bubble in between sample and reagent plugs of liquid. Then the probe descends into a container where wash solution is drawn into the probe and through the system. The wash solution is usually deionized water, possibly with a surfactant added. Remembering that all samples follow the same

carousel dispenses the tubes to a transfer carousel, from which the sampling occurs, and then the samples are transferred to an unloading carousel (Fig. 7-6). This analyzer makes use of a continuous belt of flexible, disposable plastic cuvets carried through the analyzer's water bath on a main drive track. The cuvets are loaded onto the Paramax from a continuous spool. These index through the instrument at

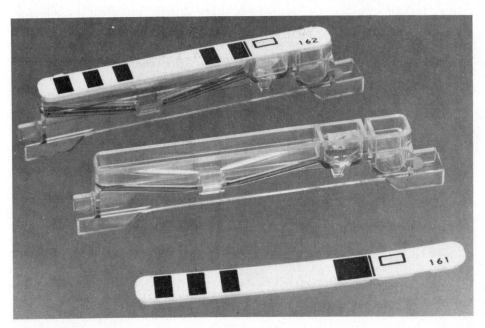

T ups.
3/10

Figure 7-4. The DuPont ACA. Sample cups. (Photograph courtesy of E.I. DuPont, Instrument Systems Division.)

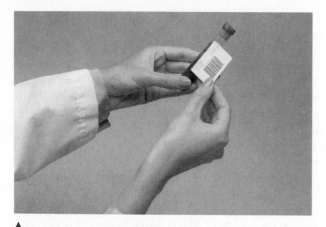

Figure 7-5. The Baxter Paramax. Sample collection tubes are identified with barcode labels. (Photograph courtesy of Baxter Diagnostics Inc.)

reaction path, the necessity for the wash solution between samples becomes obvious. Immersion of the probe into the wash reservoir cleanses the outside, while aspiration of an aliquot of solution cleanses the lumen. The reservoir is continually replenished with an excess of fresh solution. The wash aliquot, plus the previously mentioned air bubble, maintains sample integrity and minimizes sample carryover.

Some pipettors use a disposable tip and an air-displacement syringe to measure and deliver reagent. When this is used, the pipettor may be reprogrammed to measure sample and reagent for batches of different tests comparatively easily. Besides eliminating the effort of priming the reagent delivery system with the new solution, no reagent is wasted or contaminated, since nothing but the pipet tip contacts it.

The cleaning of the probe and tubing after each dispensing to minimize the carryover of one sample into the next is a concern for many instruments. In some systems, the reagent or diluent is also dispersed into the cuvet through

Figure 7-6. The Baxter Paramax. The loading carousel dispenses tubes to a transfer carousel, from which the sampling occurs, and then the samples are transferred to an unloading carousel. (Photograph courtesy of Baxter Diagnostics Inc.)

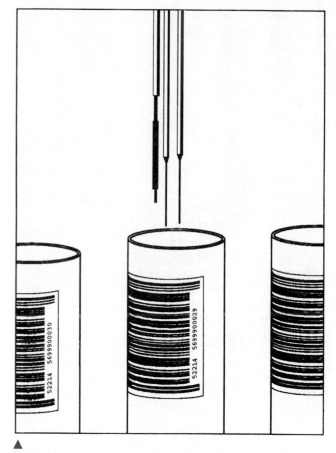

Figure 7-7. Dual sample probes of the Hitachi 736 analyzer. Note the liquid level sensor to the left of probes. (Photograph courtesy of Boehringer Mannheim Corp.)

the same tubing and probe. Deionized water may be dispensed into the cuvet after the sample to produce a specified dilution of the sample and also to rinse the dispensing system. In the Technicon RA1000©, a random-access fluid, a fluorocarbon, is the separation medium. The fluid is a viscous, inert, immiscible, non-wetting substance that coats the delivery system. The coating on the sides of the delivery system prevents carryover due to the wetting of the surfaces and, forming a plug of the solution between samples, prevents carryover by diffusion. A small amount, 10μL, of this fluid is dispensed into the cuvet with the sample. Surface tension leaves a coating of the fluid in the dispensing system.[6]

If a separate probe or tip is used for each sample and discarded after use, as in the Kodak Ektachem©, the issue of carryover is a moot point. The Ektachem© has a unique sample-dispensing system. A "proboscis" (Kodak's descriptive term) presses into a tip on the sample tray, picks it up, and moves over the specimen to aspirate the volume required for the tests programmed for that sample. The tip is then moved over to the slide-metering block. When a slide is in position to receive an aliquot, the proboscis is lowered so that a dispensed 10-μL drop touches the slide, where it is absorbed from the non-wettable tip. Aspiration and drop formation are controlled by a stepper motor-driven piston. The precision of dispensing is specified at ±5%.

In the centrifugal-analysis instrument, a loader is used for dispensing samples and reagents. Typically, it consists of two Hamilton syringes, a keyed turntable for holding the transfer disk or disposable rotor, a ring surrounding the disk to hold the sample cups, an autostop module, and a digital control panel. For each assay, the operator selects the appropriate volumes by pressing digital switches on the front panel of the loader. Sample and diluent, or a second reagent, are pipetted into one compartment on the disk, and the reagent and diluent are pipetted into a second compartment. The Hamilton syringes are driven by stepper motors for pipetting. After all the samples and reagents have been pipetted, the disk is removed from the loader and placed on the analyzer.

In several discrete systems, the probe is attached by means of non-wettable tubing to precision syringes. The syringes draw a specified amount of sample into the probe and tubing. Then the probe is positioned over a cuvet, into which the sample is dispensed. The Hitachi 736© utilizes two sample probes to simultaneously aspirate a double volume of sample in each probe immersed in one specimen container, and thereby delivering sample into four individual test channels, all in one operational step (Fig. 7-8). The loaded probes pass through a fine mist shower bath prior to delivery to wash off any sample residue adhering to the outer surface of the probes. After delivery the probes move to a rinse bath station for cleaning the inside and outside surfaces of the probes.

The ACA© filling station puts the appropriate amounts of sample and diluent into each analytical test pack. A filled sample cup is followed in the loading tray by the test packs for the tests required on the specimen. When the system is activated, a shuttle pushes the first sample cup left to the sampling position. While the specimen is moving left, a spring-loaded pack pusher pushes the first test pack onto the filling station rail below a decoder plate. The decoder senses the binary code on the top of the test pack. This code, when translated, provides the instrument with instructions determining the sample volume, type of diluent, and special handling characteristics. The pump flushes with diluent to be used for the particular method to purge the lines and needle prior to diluent intake. After the flush, the pump intakes diluent into the pump cylinder. The sample needle is positioned over the sample cup, dips down into the cup, and aspirates the proper volume of specimen. The needle rises out of the sample cup and moves right to a position over the test pack. The sample and diluent are injected into the test pack through a special pack-fill opening. After the pack is filled, the needle, tubing, and pump are flushed. The test pack is moved onto the transport system. The instrument also transfers the patient sample-identification number on the sample pack to the results-reporting system.

The Paramax© uses computer-controlled stepping motors to drive both the sampling and washout syringes. Every 5 seconds, the sampling probe enters a specimen container, withdraws the required volume, moves to the cuvet, and dispenses the aliquot with a volume of water to wash the probe. The washout volume is adjusted to yield the final reaction volume. If a procedure's range of linearity is exceeded, the system will retrieve the original sample tube, repeat the test using one-fourth the original sample volume for the repeat test, and calculate a new result, taking the dilution into consideration.

Economy of sample size is a major consideration in developing automated procedures, but methodologies have limitations to maintain proper levels of sensitivity and specificity. The factors governing sample and reagent measurement are interdependent. Generally, if sample size is reduced, then reagent concentration must be increased to ensure sufficient color development for accurate photometric readings.

Wed 3/11

Reagent Systems and Delivery

Reagents may be classified as liquid or dry systems for use with automated analyzers. Liquid reagents may be purchased

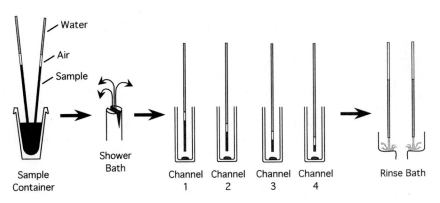

Sampling Operation

Figure 7-8. Sampling operation of the Hitachi 736 analyzer. (Courtesy of Boehringer Mannheim Corp.)

in bulk volume containers or in unit dose packaging as a convenience for *STAT* testing on some analyzers. Dry reagents are packaged in various forms. They may be bottled as lyophilized powder, which requires reconstitution with water or a buffer. Unless the manufacturer provides the diluent, water quality may be compromised by the laboratory. Other dry reagents may be in tablet form, such as those used by the ACA© and Paramax©. The ACA© crushes and dissolves reagent tablets in the plastic test pouch on board the instrument. The Paramax© uses an ultrasonic horn to break up and dissolve the tablet in a plastic cuvet filled with water. A third and unique type of dry reagent is the multilayered dry film slide for the Kodak Ektachem© analyzer. These slides have microscopically thin layers of dry reagents mounted on a plastic support. The slides are approximately the size of a postage stamp and not much thicker.

Reagent handling varies according to instrument capabilities and methodologies. Many test procedures use sensitive, short-lived working reagents, so contemporary analyzers use a variety of techniques to preserve them. One technique is to keep all reagents refrigerated until the moment of need then quickly pre-incubate them to reaction temperature or store them in a refrigerated compartment on the analyzer that feeds directly to the dispensing area. Another means of preservation is to provide reagents in a dried, tablet form and reconstitute them when the test is to be run. A third is to manufacture the reagent in two stable components that will be combined at the moment of reaction. If this approach is used, the first component also may be used as a diluent for the sample. Combinations of these reagent handling techniques are often used by the various manufacturers.

Reagents also must be dispensed and measured accurately. Many instruments use bulk reagents to decrease the preparation and changing of reagents. Instruments that do not use bulk reagents have unique reagent packaging.

In continuous-flow analyzers, reagents and diluents are supplied from bulk containers into which tubing is suspended. The inside diameter, or bore, of the tubing governs the amount of fluid that will be dispensed. A proportioning pump, along with a manifold, continuously and precisely introduces, proportions, and pumps liquids and air bubbles throughout the continuous-flow system.

To deliver reagents, many discrete analyzers use techniques similar to those used to measure and deliver the samples. A syringe or syringes driven by stepping motors pipet the reagents into reaction containers. Piston driven pumps connected by tubing may also dispense reagents. Another technique for delivering reagents to reaction containers uses pressurized reagent bottles connected by tubing to dispensing valves. The opening and closing of the valves are controlled by the computer. The fill volume of reagent into the reaction container is determined by the precise amount of time the valve remains open.

The Kodak Ektachem© analyzers use slides to contain their entire reagent chemistry system. Multiple layers on the slide are backed by a clear polyester support. The coating itself is sandwiched in a plastic mount. There are three or more layers: (1) a spreading layer, which accepts the sample; (2) one or more central layers, which can alter the aliquot; and (3) an indicator layer, where the analyte of interest may be quantitated (Fig. 7-9). The number of layers varies depending on the assay to be performed. The color developed in the indicator layer varies with the concentration of the analyte in the sample. Physical or chemical reactions can occur in one layer, with the product of these reactions proceeding to another layer where subsequent reactions can occur. Each layer may offer a unique environment and the possibility to carry out a reaction comparable to that offered in a solution assay, or it may promote an entirely different activity that does not occur in the liquid phase. The ability to create multiple reaction sites allows the possibility of manipulating and detecting compounds in ways not possible in solution chemistries. Interfering materials can be left behind or altered in upper layers.

All the reagents for the DuPont ACA© are contained in special test packs, which are compartmentalized plastic envelopes. A separate pack is used for each test performed on a specimen (Fig. 7-10). Compartments along the top of the

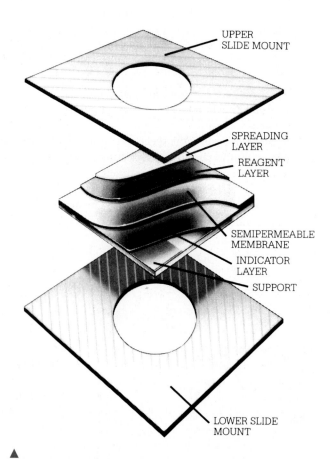

Figure 7-9. The Kodak Ektachem. Slides with multiple layers contain the entire reagent chemistry system. (Courtesy of Eastman-Kodak Co.)

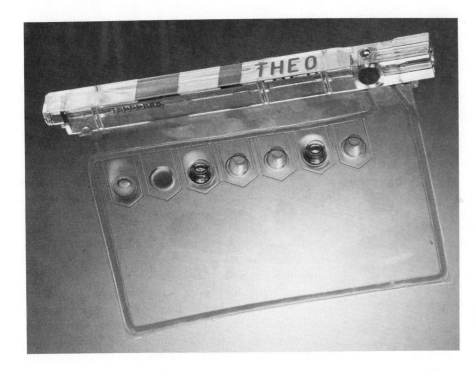

Figure 7-10. The DuPont ACA. A test pack with binary code on a bar at the top. (Photograph courtesy of E.I. DuPont, Instrument Systems Division)

envelope contain liquid or tablet reagents. The code name and a binary code depicting the test are printed on a plastic bar at the top of each test pack. All of the test packs require refrigerated storage.

The Paramax© has a reagent-dispensing carousel with positions for 32 disposable tablet dispensers, which contain a maximum of 300 tablets each. When activated by the instrument, the dispenser ejects a single tablet into a cuvet (Fig. 7-11). The reagent contained in each dispenser is identified to the instrument by an optical code, allowing dispensers to be loaded at random. The primary reagent for each test is contained in a tablet. Secondary reagents, which are liquid, are added after the tablet has been dissolved and the sample and diluent have been injected into the cuvet.

Chemical Reaction Phase

This phase consists of mixing, separation, incubation, and reaction time. In most discrete analyzers, the chemical reactants are held in individual moving containers that are either disposable or reusable. These reaction containers also function as the cuvets for optical analysis. If the cuvets are reusable, then wash stations are set up immediately after the read stations to clean and dry these containers (Fig. 7-12). This arrangement allows the analyzer to operate continuously without replacing cuvets. Examples of this type are the Hitachi and Synchron© analyzers. Alternatively, the reactants may be placed in a stationary reaction chamber (*e.g.*, ASTRA©) in which a flow-through process of the reaction

Figure 7-11. The Baxter Paramax. The primary reagents are contained in a tablet. (Photograph courtesy of Baxter Diagnostics Inc.)

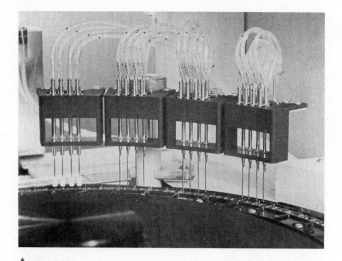

Figure 7-12. Wash stations on Hitachi 736 analyzer perform the following: (1) aspirate reaction waste and dispense water; (2) aspirate and dispense rinse water; (3) aspirate rinse water and dispense water for measurement of cell blank; (4) aspirate cell blank water to dryness. (Photograph courtesy of Boehringer Mannheim Corp.)

mixture occurs before and after the optical reading. In continuous flow systems, flow-through cuvets are utilized and optical readings are taken during the flow of reactant fluids. The Chem 1© uses this approach with its single analytical pathway Teflon tube (Fig. 7-13).

Mixing

A vital component of each procedure is the adequate mixing of the reagents and sample. Instrument manufacturers go to great lengths to ensure complete mixing. Non-uniform mixtures can result in noise in continuous-flow analysis and in poor precision in discrete analysis.

Mixing is accomplished in continuous flow analyzers (*e.g.*, the Chem 1©) through the use of coiled tubing. When the reagent and sample stream goes through coiled loops, the liquid rotates and tumbles in each loop. The differential rate of liquids falling through one another produces mixing in the coil.

The Technicon RA1000© uses a rapid start-stop action of the reaction tray. This causes a sloshing action against the walls of the cuvets, which mixes the components.

Centrifugal analyzers may use a start-stop sequence of rotation or bubbling of air through the sample and reagent to mix them while these solutions are moving from transfer disk to rotor. This process of transferring and mixing occurs in less than 3 seconds.[5] The centrifugal force is responsible for the mixing as it pushes sample from its compartment, over a partition into a reagent-filled compartment, and finally into the cuvet space at the perimeter of the rotor.

In slide technology (Kodak), the spreading layer provides a structure that permits a rapid and uniform spreading of the sample over the reagent layer(s) for even color development.

The ACA© has two components specially designed for mixing: the breaker-mixers. Platens press selectively against and collapse the reagent compartments along the top of the test pack, releasing the reagents into the interior environment of the pack. Meanwhile, a lower platen presses against the bottom portion of the test-pack envelope to force the fluids up into the ruptured reagent compartments. Then, with a patting motion, the breaker-mixer thoroughly mixes the reagents with the fluids.

The Paramax© uses ultrasonic sound waves for 45 seconds to dissolve the reagent tablets in the deionized water in each cuvet. Ultrasound is also used to mix the sample with the previously prepared reagents and to mix, if necessary, the contents of the cuvets after a second reagent addition.

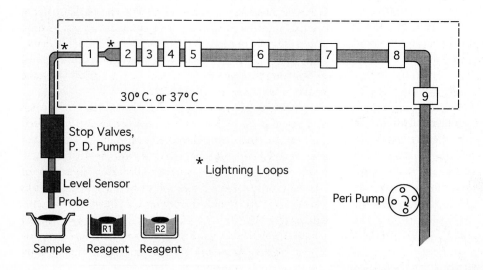

TECHNICON ™ **System Overview**

Figure 7-13. Technicon Chem 1 analytical pathway of single Teflon tube. (Courtesy of Technicon Instruments Corp.)

The Hitachi analyzers utilize spinning paddles that dip into the reaction container for a few seconds to stir sample and reagents, after which they return to a wash reservoir (Fig. 7-14). Other instruments, such as the ASTRA©, use magnetic stir bars lying in the bottom of the reaction container that, when activated, produce a whirling motion to mix. Still others may use forceful dispensing to accomplish mixing.

Separation

In chemical reactions, undesirable constituents that will interfere with an analysis may need to be separated from the sample before the other reagents are introduced into the system. Protein causes major interference in many analyses. One approach without separating protein is to use a very high reagent-to-sample ratio (the sample is highly diluted) so that any turbidity caused by precipitated protein is not sensed by the spectrophotometer. Another is to shorten reaction time to eliminate slower reacting interferents.

In the older continuous-flow systems, a dialyzer was the separation or filtering module. It performed the equivalent of the manual procedures of precipitation, centrifugation, and filtration, using a fine-pore cellophane membrane.

In slide technology (Kodak), the spreading layer of the slide traps cells, crystals, and other small particulate matter but also retains large molecules such as protein. In essence, what passes through the spreading layer is a protein-free filtrate.

Many discrete analyzers have no automated methodology by which to separate interfering compounds from the reaction mixture. Therefore, methods have been chosen that have few interferences or that have known interferences that can be compensated for by the instrument (*e.g.*, using correction formulas).

The use of automated column chromatography in some methods gives the ACA© the ability to remove from the sample substances that might adversely influence the reaction. Three types of columns can be used: gel-filtration, ion-exchange, and protein-removal. The column is located at the top of the test pack, immediately under the plastic pack header. If the test pack contains a chromatographic column, the sample and a prescribed quantity of diluent are injected into the column fill site opposite the usual sample and diluent fill site. The sample is moved through the column by the pressure of the diluent and into the test pack for analysis.

Incubation

A heating bath in discrete or continuous-flow systems maintains the required temperature of the reaction mixture and provides the delay necessary to allow complete color development. The principal components of the heating bath are the heat-transfer medium (*i.e.*, water or air), the heating element, and the thermoregulator. A thermometer is located in the heating compartment of an analyzer and is monitored by the system's computer. On many discrete analyzer systems, the multi-cuvets float in a water-bath incubator maintained at a constant temperature of usually 37°C.

Slide technology incubates colorimetric slides at 37°C. There is a precondition station to bring the temperature of each slide close to 37°C before it enters the incubator. The incubator moves the slides at 12-second intervals in such a manner that each slide is at the incubator exit four times during the 5-minute incubation time. This feature is used for two point-rate methods and enables the first point reading to be taken part way through the incubation time. Potentiometric slides are held at 25°C. The slides are kept at this temperature for 3 minutes to ensure stability before reading.

Reaction Time

Prior to the optical reading by the spectrophotometer, the reaction time may be dependent on the rate of transport through the system to the "read" station, timed reagent additions with moving or stationary reaction chambers, or a combination of both processes.

An environment conducive to the completion of the reaction must be maintained for a sufficient length of time before spectrophotometric analysis of the product is made. Time is a definite limitation. To sustain the advantage of speedy multiple analyses, the instrument must produce results in as short a time as possible.

It is possible to monitor not only completion of a reaction but also the rate at which the reaction is proceeding. The instrument may delay the measurement for a predetermined length of time or may present the reaction mixtures for measurement at constant intervals of time. Use of rate reactions may have two advantages: the total analysis time is shortened, and interfering chromogens that react slowly may be negated. Reaction rate is controlled by temperature; therefore, the reagent, timing, and spectrophotometric functions must be coordinated to work in harmony with the chosen temperature.

The ACA© has five delay stations located in the pack-processing area. At these points, the instrument performs no operations on the test packs. This interval allows enough

▲

Figure 7-14. Stirring paddles on Hitachi 736 analyzer. (Photograph courtesy of Boehringer Mannheim Corp.)

reaction time for endpoint or blank reactions to go to completion. The entire pack-processing area is maintained at 37°C through a closed environment with circulating fans.

The Paramax© has eight photometric stations located along the cuvet track. The addition of the sample initiates each reaction, which is monitored from 40 seconds to 10 minutes by the photometric stations for rate or endpoint reactions. The environment of the cuvets is maintained by a constant water bath in which the cuvets move.

Fri 3/13

Measurement Phase

After the reaction is completed, accommodation must be made to measure the products that are formed. Almost all available systems for measurement have been utilized, such as ultraviolet, fluorescent, and flame photometry; ion-specific electrodes; gamma counters; and luminometers. Still, the most common is visible and ultraviolet light spectrophotometry, although adaptations of traditional fluorescence measurement, such as fluorescence polarization, chemiluminescence, and bioluminescence have become popular. Abbotts' TDX©, for example, is a very popular instrument for drug analysis that employs fluorescence polarization to measure immunoassay reactions.

Analyzers that measure light require a monochromator to achieve the desired component wavelength. Traditionally, analyzers have used filters or filter wheels to separate light. The old Auto Analyzers used filters that were manually placed in position in the light path. Many instruments still employ rotating filter wheels, which the computer controls to position the appropriate filter into the light path. However, newer and more sophisticated systems offer the higher resolution afforded by diffraction gratings to achieve light separation into its component colors. Instruments that employ such monochromators include the Roche Cobas-Bio©, which has a mechanically rotating diffraction grating, and the Hitachi analyzers, which have a fixed grating that spreads its component wavelengths onto a fixed array of photo diodes (Fig. 7-15). This latter grating arrangement, as well as rotating filter wheels, easily accommodates polychromatic light analysis, which offers improved sensitivity and specificity over monochromatic measurement. By recording optical readings at different wavelengths the instrument's computer can then use these data to correct for reaction mixture interferences that may occur at adjacent, as well as desired, wavelengths.

Many newer instruments utilize fiber optics as a medium to transport light signals from remote read stations back to a central monochromator detector box for analysis of these signals. The Technicon SMAC© and the Baxter Paramax© have fiber optic cables, or "light pipes" as they are sometimes called, attached from multiple remote stations where the reaction mixtures reside, to a centralized filter wheel/detector unit that, in conjunction with the computer, sequences and analyzes a large volume of light signals from multiple reactions (Fig. 7-16).

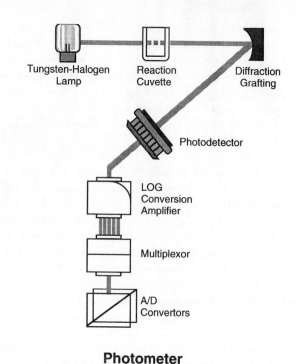

Photometer

Figure 7-15. Photometer for Hitachi 736 analyzer. Fixed diffraction grating separates light into specific wavelengths and reflects them onto a fixed array of eleven specific photodetectors. Photometer has no moving parts. (Courtesy of Boehringer Mannheim Corp.)

The containers holding the reaction mixture also play a vital role in the measurement phase. The reagent volume, and therefore sample size, speed of analysis, and sensitivity of measurement are some aspects influenced by the method of analysis.

A flow-through cuvet is used in continuous-flow analysis. The reagent stream under analysis flows continuously through the flowcell tubing. The SMAC© and Chem 1© "capture" absorbance signals in between air bubbles of the flowing stream (Fig. 7-17). This means that no debubbling is required as in the older continuous-flow analyzers. As the stream flows through the flowcell, a steady beam of light is focused through the stream. The amount of light that exits from the flowcell is dictated primarily by the absorbance of light by the stream. The exiting light strikes a photodetector, which converts the light into electrical energy. Filters and light-focusing components permit the desired light wavelength to reach the photodetector. The photometer continuously senses the sample photodetector output voltage and, as is the process in most analyzers, compares it with a reference output voltage. The electrical impulses are sent to a readout device, such as a printer or computer, for storage and retrieval.

In discrete analyzers, such as Hitachi or Synchron© or ACA© systems, the cuvet used for analysis is also the reaction vessel in which the entire procedure has taken place.

The ACA© photometer consists of a cuvet-forming device and a photometric system for making absorbance

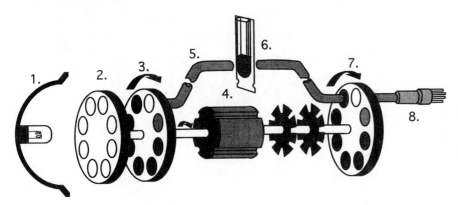

Figure 7-16. Paramax instrument photo optical system: (1) Source with reflector (2) Fixed focusing lens (3) Filter wheel (4) Double shafted motor (5) Fiber optic bundle (6) Cuvet (7) Filter wheel (8) Photomultiplier tube. (Courtesy of Baxter Diagnostics, Inc.)

measurements. When compressed by the photometer jaws, the test pack forms a cuvet between quartz windows. A wetting solution is introduced between the quartz windows and the test-pack walls to achieve a good optical interface.

Centrifugal analysis measurement occurs while the rotor is rotating at a constant speed of approximately 1000 rpm. Consecutive readings are taken of the sample, the dark current (readings between cuvets), and the reference cuvet. Each cuvet passes through the light source every 3 milliseconds (ms). After all the data points have been determined, centrifugation stops and the results are printed. The rotor is removed from the analyzer and discarded. For endpoint analyses, an initial absorbance is measured before the constituents have had time to react, usually at about 3 seconds, and is considered a blank measurement. After enough time has expired for the reaction to be completed, another absorbance reading is taken. For rate analyses, the initial absorbance is measured at 3 seconds, and then a lag time is allowed (preset into the instrument for each analysis). For each assay, several data points are determined at a programmed data time interval. The instrument monitors the absorbance measurements at each data point and calculates a result.

Slide technology depends on reflectance spectrophotometry, as opposed to traditional transmittance photometry, to provide a quantitative result. The amount of chromogen in the indicator layer is read after light passes through the indicator layer, is reflected from the bottom of a pigment-containing layer (usually the spreading layer), and is returned through the indicator layer to a light detector. For colorimetric determinations, the light source is a tungsten-halogen lamp. The beam focuses on a filter wheel holding up to eight interference filters, which are separated by a dark space. The beam is focused at a 45° angle to the bottom surface of the slide, and a silicon photodiode detects the portion of the beam that reflects down. Three readings are taken for the computer to derive reflectance density. The three recorded signals taken are: (1) the filter wheel blocking the beam, (2) reflectance of a reference white surface

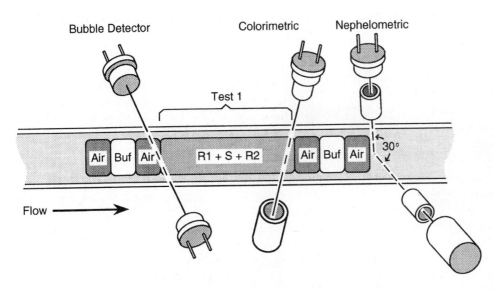

In-Line Reaction Detector

Figure 7-17. Flow-through cuvet for Technicon Chem 1 Analyzer. (Courtesy of Technicon Instruments Corp.)

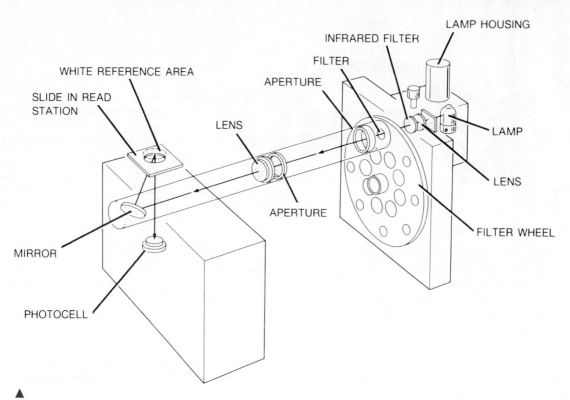

Figure 7-18. Components of the system for making colorimetric determinations in slide technology. (Courtesy of Eastman-Kodak Co.)

with the programmed filter in the beam, and (3) reflectance of the slide with the selected filter in the beam (Fig. 7-18).

After a slide is read, it is shuttled back in the direction from which it came, where a trap door allows it to drop into a waste bin. If the reading was the first for a two-point rate test, the trap door remains closed, and the slide re-enters the incubator.

Ciba Corning's ACS:180©, a fully automated, random access immunoassay system (Fig. 7-19), utilizes chemilumi-

nescence technology for reaction analysis. In chemiluminescence assays, quantitation of an analyte is based on emission of light resulting from a chemical reaction.[2] The principles of chemiluminescence immunoassays are similar to those of radioimmunoassay (RIA), except that an acridinium ester is used as the tracer and paramagnetic particles are used as the solid phase.[2] Sample, tracer, and paramagnetic particle reagent are added and incubated in disposable plastic cuvets depending on the assay protocol. After incubation,

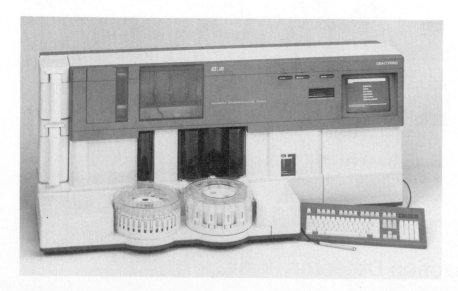

Figure 7-19. ACS:180© Automated Chemiluminescence System. (Photo courtesy Ciba-Corning Diagnostics Corp.)

magnetic separation and washing of the particles is performed automatically. The cuvets are then transported into a light-sealed luminometer chamber where appropriate reagents are added to initiate the chemiluminescent reaction.[1] Upon injection of the reagents into the sample cuvet, the system luminometer generates the chemiluminescent signal. Luminometers are similar to gamma counters in that they use a photomultiplier tube detector. However, unlike gamma counters, luminometers do not require a crystal to convert gamma rays to light photons. Light photons from the sample are detected directly, converted to electrical pulses, and then converted to counts.

Signal Processing and Data Handling

Since most automated instruments print the results in reportable form, accurate calibration is essential to obtaining accurate information. There are many variables that may enter into the use of calibration standards. The matrices of the standards and unknowns may be different. Depending on the methodology, this may or may not present problems. If secondary standards are used to calibrate an instrument, the methods used to derive the standard's constituent values should be known. Standards containing more than one assayable value per vial may cause interference problems. Since there are no primary standards available for enzymes, either secondary standards or calibration factors based on the molar extinction coefficients of the products of the reactions may be used.

Many times, a laboratory will have more than one instrument capable of measuring a constituent. Unless there are different normal ranges published for each method, the instruments should be calibrated so that the results are comparable.

The advantage of calibrating an automated instrument is the long-term stability of the standard curve, which requires only monitoring with controls on a daily basis. Some analyzers use low- and high-concentration standards at the beginning of each run and then use the absorbances of the reactions produced by the standards to produce a standard curve electronically for each run. Other instruments are self-calibrating after analyzing standard solutions.

The original continuous-flow analyzers used six standards assayed at the beginning of each run to produce a calibration curve for that particular batch. Now, continuous-flow analyzers use a single-level calibrator (secondary standard) to calibrate each run with water used to establish the baseline.

The centrifugal analyzer uses standards pipetted into designated cuvets in each run for endpoint analyses. After the delta absorbance for each sample has been obtained, the computer calculates the results by determining a constant for each standard. The constants are derived by dividing the concentration of the standard (pre-entered into the computer) by the delta absorbance and averaging the constants for all the standards to obtain a factor. The concentration of each control and unknown is determined by multiplying the delta absorbance of the unknown by the factor. If the concentration of an unknown exceeds the range of the standards, the result is printed in red as a "flag." Enzyme activity is derived by a linear regression fit of the delta absorbance versus time. The slope of the produced line is multiplied by the enzyme factor (pre-entered) to calculate the activity.[5]

Slide technology requires more sophisticated calculations to produce results. The calibration materials require a protein-based matrix because of the necessity for the calibrators to behave as serum when reacting with the various layers of the slides. Calibrator fluids are bovine serum-based, and the concentration of each analyte is determined by reference methods. Endpoint tests require three calibrator fluids, blank-requiring tests need four calibrator fluids, and enzyme methods require three calibrators. Colorimetric tests use spline fits to produce the standardization. In enzyme analysis, a curve-fitting algorithm estimates the change in reflection density per unit time. This is converted to either absorbance or transmission-density change per unit time. Then, a quadratic equation converts the change in transmission density to volume activity (U/L) for each assay.

The DuPont ACA© retains the calibrations for each lot of a particular method until the laboratorian programs the instrument for recalibration. Instrument calibration is initiated and/or verified by assaying a minimum of three levels of primary standards or, in the case of enzymes, reference samples. The values obtained are compared with the known concentrations by using linear regression with the x axis representing the expected values and the y axis representing the mean of the values obtained. The slope (scale factor) and y intercept (offset) are the adjustable parameters on the ACA©. On earlier models of this instrument, the parameters were determined and entered manually into the instrument computer by the operator, but in the more automated models, the calibration is performed by the instrument upon notification by the operator.

After calibration has been performed and the chemical or electrical analysis of the specimen is either in progress or completed, the instrument's computer goes into the data acquisition and calculation mode. The process may involve signal averaging, which may entail hundreds of data pulses per second, as with a centrifugal analyzer, and blanking and corrections formulas for interferents that are programmed into the computer for calculation of results.

All advanced automated instruments have some method of reporting printed results with a link to sample identification. In sophisticated systems, the demographic-sample information is entered in the instrument's computer along with the tests required. Then the sample identification is printed with the test results. The Paramax© prints bar-code labels for sample identification after the operator enters the patient information and tests requested into the computer terminal. Once the label is applied to the sample, the sample can be loaded on the analyzer. Microprocessors control the tests, reagents, and timing, while verifying the bar-code for each sample. This is the link between the results reported and the specimen identification. Even the simplest of systems sequentially number the test results to provide a connection with the samples.

Since most instruments now have either a built-in or attached video monitor, the sophisticated software programs that come with the instrument can be displayed for readily accessible inspection of different aspects of the testing process. Computerized monitoring is available for such parameters as reaction and instrument linearity, quality control data with various options for statistical display and interpretation, short sample sensing with "flags" on the print-out, abnormal patient results "flagged," clot detection, reaction vessel or test chamber temperature, and reagent inventories. The printer can also display patients' results as well as various warnings mentioned above. Many instrument manufacturers offer computer software for preventive maintenance schedules and algorithms for diagnostic trouble shooting. Some manufacturers also install phone modems on the analyzer for a direct communication link between the instrument and their service center for instant troubleshooting and diagnosis of problems.

SELECTION OF AUTOMATED ANALYZERS

Each manufacturer's approach to automation is unique. The instruments being evaluated should be rated according to previously identified needs. One laboratory may need a *STAT* analyzer, while another's need may be a batch analyzer for high test volumes. When considering cost, the price of the instrument, and even more important, the total cost of consumables, are significant. The high capital cost of an instrument may actually be small when divided by the large number of samples to be processed. It is also important to calculate the total cost per test for each instrument that is considered. Moreover, a break-even analysis to study the relationship of fixed costs, variable costs, and profits would be helpful in the financial justification and economic impact on a laboratory. Of course the mode of acquisition, *i.e.*, purchase, lease, rental, etc., has to be factored into this analysis also. The variable cost of consumables will increase as more tests are performed or samples are analyzed. The ability to use reagents produced by more than one supplier (open vs. closed reagent systems) can provide a laboratory with the ability to customize testing and, possibly, save money. The labor component also should be evaluated. For instruments that have the College of American Pathologists' workload recording units assigned, this provides an excellent basis for comparison, using time-motion studies. Unfortunately, the newest instruments on the market may not have been timed for labor, and thus this judgment may be more subjective than desired. With the large number of instruments available on the market, the goal is to find the right instrument for each situation.

Another major concern toward the selection of an instrument is its analytic capabilities. What is the instrument's performance characteristics for accuracy, precision, linearity, specificity, and sensitivity (which may be method dependent), calibration stability, and stability of reagents (both shelf life and on-board or reconstituted)? The best way to verify these performance characteristics of an analyzer before making a decision on an instrument is to see it in operation. Ideally, if a manufacturer will place an instrument in the prospective buyer's laboratory on a trial basis, then its analytic performance can be evaluated to the customer's satisfaction with studies to verify accuracy, precision, and linearity. At the same time, laboratory personnel can observe such design features as its test menus, true "walkaway" capability, "user friendliness," and the space that the instrument and its consumables occupy in their lab.

Clinical chemistry instrumentation provides speed and precision for assays that would otherwise be performed manually. The chosen methodologies and adherence to the requirements of the assay provide accuracy. No one may assume that the result produced is the correct value. Automated methods must be evaluated completely before being accepted as routine. It is important to understand how each instrument actually works.

FUTURE TRENDS IN AUTOMATION

Clinical chemistry automation will continue to evolve at a rapid pace in the next decade as it has in the last decade. System integration and miniaturization with more computer power will persist to accommodate more portable analyzers for increasingly popular point of care testing. Front-end automation (robotics) for specimen preparation prior to analysis will become common. There will be continuous improvement of analyzers, computer software for quality control, data management, diagnostics, etc. Artificial intelligence in computer systems will be developed, whereby the computer will "think" or make decisions if sufficiently programmed with infinite scenarios of data.

New technologies, such as polymerase chain reactions (PCR), and more immunologic techniques will drive automation in new directions. Analyzers will have larger and more varied test menus to include specific proteins and more drugs and hormones. Chemiluminescence analyzers, relatively new in the clinical laboratory, will become more popular. Spectral-mapping, or multiple wavelength monitoring, with high resolution photometers in analyzers will be routine with all specimens and tests as more instruments are designed with the monochromator device in the light path after the cuvet, not before. Spectral-mapping also has the potential for monitoring two or more chemical reactions (tests) simultaneously in the same cuvet. This will have a tremendous impact on throughput and turnaround time of test results. Mass spectrometry and capillary electrophoresis will be used more extensively in clinical laboratories for identification and quantitation of elements and compounds in extremely small concentrations. Finally, research and development of non-invasive, in vivo testing will continue. Transcutaneous monitoring is already available with some blood gasses. "True" or dynamic values from in vivo monitoring of constituents in blood and other body fluids will revolutionize laboratory medicine as we know it today. It all sounds futuristic, but then so did the first Auto Analyzer 50 years ago.

SUMMARY

Since the introduction of the first automated analyzer by Technicon, automated instruments have proliferated in the clinical chemistry laboratory. The driving forces behind the development of more and better automated analyzers include the increased volume of testing, faster turnaround time, and escalating costs. Automated analyzers utilize three basic approaches to sample analysis: continuous flow, centrifugal analysis, and discrete analysis. Manufacturers have designed their instruments to mimic the steps in a manual procedure to include: specimen preparation and identification, specimen measurement and delivery, reagent systems and delivery, chemical reaction phase, measurement phase, and signal processing and data handling. Each manufacturer's approach to automation is unique. When selecting an automated analyzer for the clinical chemistry laboratory, several factors need to be considered: needs, cost, mode of acquisition, consumables, analytic capabilities, space, and user friendliness. Clinical chemistry automation will continue to evolve. Future trends will undoubtedly include more computer power, more portability, increased use of robotics, continuous improvement in computer software, and the development of artificially intelligent computer systems. New technologies, such as polymerase chain reactions and non-invasive in vivo testing, will also have a large impact.

EXERCISES

1. Define the following terms:
   Automation
   Continuous flow
   Discrete analysis
   Random access
2. Describe three basic approaches to sample analysis used by automated analyzers.
3. What are the major steps in automated analysis?
4. Identify the instruments listed as either "A," a centrifugal analyzer, or "B," a discrete analyzer, or "C," a continuous flow analyzer.
   Kodak Ektachem©
   DuPont ACA©
   COBAS-BIO©
   Chem 1©
5. True or False. The ability to use reagents produced by more than one supplier is known as an "open reagent system."
6. Give three considerations in the purchase of a new automated analyzer.

REFERENCES

1. ACS:180 Product Backgrounder. The ACS:180™: First automated chemiluminescence system. East Walpole, MA: Ciba Corning Diagnostics Corporation, 1990.
2. Dudley RF. Chemiluminescence immunoassay: An alternative to RIA. Lab Med 1990;21(4):216.
3. Eisenwiener H, Keller M. Absorbance measurement in cuvets lying longitudinal to the light beam. Clin Chem 1979; 25(1)9:117.
4. Hodnett J. Automated analyzers have surpassed the test of time. Adv Med Lab Prof, January 10, 1994;8.
5. Pesce MA. Evaluation of the MULTISTAT III fluorescence/light scatter centrifugal analyzer. J Clin Lab Autom 1983;3(5):327.
6. Steindel SJ, Schoudt P. An assessment of random access fluid technology. Clin Lab Autom 1983;3(5):319.

SUGGESTED READINGS

Schoeff L, Williams R. Principles of laboratory instruments. St. Louis: Mosby-YearBook, Inc, 1993.

Ward K, Lehmann C, Leiken A. Clinical laboratory instrumentation and automation. Philidelphia: WB Saunders Co, 1994.

Karselis T. A pocket reference guide to laboratory instrumentation. Philidelphia: FA Davis Co, 1994.

Chapter 8

Computer Interfacing in the Clinical Chemistry Laboratory

Georgia A. Sehon

Computer Interfacing Standards
OSI Reference Model
HL7 Standard
ASTM E 1238 Standard

ASTM E 1381 Standard
ASTM E 1394 Standard
Exercises
References

Objectives

Upon completion of this chapter, the clinical laboratorian should be able to:

- *Compare and contrast the computer requirements of today's clinical laboratory vs. the laboratory of the future.*
- *Discuss the needs for a standardized format for medical data sharing.*
- *Identify the standard-based organizations for the communication of laboratory test results.*
- *Define the following terms: LIS, WAN, OSI, GUI, field, segment, separator, and network.*
- *Explain the Health Level 7 (HL7®)* standard.*
- *Discuss the ASTM E 1238, 1381, and 1394 standards.*
- *Describe the electronic medical record.*

KEY WORDS

ASTM E 1238 Standard	Open System
Data Types	Interconnection
Electronic Medical	Open Architecture
Record	RS-232
Field	Segment
Graphical User	Separator Character
Interface	Trigger Event
HL7® Standard	Unsolicited Update
Laboratory Information	Wide Area Network
System	

*Health Level 7 and HL7 are registered trademarks of Health Level Seven, Inc.

The modern day health care facility has most likely been computerized for several years. Some facilities chose an integrated system that was capable of accumulating patient demographics, sharing such demographics with ancillary services (*e.g.*, laboratory, pharmacy, dietetics), and even creating a financial statement suitable for billing patients and third party payers. Some facilities chose separate systems for many of these functions and may or may not have implemented various forms of data sharing among the separate systems. As these systems require updating, the decision to upgrade to or maintain an integrated system of hospital information becomes increasingly difficult. Typically, multiple vendors are involved and it becomes impractical to purchase an integrated solution from a single vendor. Reasonable alternatives have been to replace aging systems with solutions that provide highly sophisticated automated techniques to the individual services, but these alternatives leave the facility with many disparate systems.[13,16,20]

Today's health care delivery system places an ever increasing emphasis on the automation of total patient care information. The need to share critical data among the disparate automated systems has emerged. Health care delivery depends not only upon consolidated patient data but also upon consolidation of institutional data (provider practices, institutional policies, etc.). In the past, networking schemes to connect disparate systems have been implemented. However, these schemes failed to accommodate the different data structures that may have existed in each of the disparate systems. A consolidated view of the patient to accommodate all information is not practical, as each system may not contain all of the preferred variables. An integrated, automated solution that focuses on patient care will greatly enhance the quality of patient care given by health care facilities and

Michael L. Bishop, Janet L. Duben-Engelkirk, and Edward P. Fody.
CLINICAL CHEMISTRY. © 1996 Lippincott–Raven Publishers.

providers and will enable health care facilities to monitor approaches to medical care to maintain cost efficiency.[13,16,24]

How can a totally integrated, automated solution with disparate systems be achieved? A new generation of systems must be designed. The antiquated approach of simply networking disparate systems to link data must be abolished. Patient care information must be centralized into a total electronic medical record. Once the database design is in place, there will be few (if any) technical obstacles remaining to prevent data exchange across multiple facilities. The database for the future electronic record will need to be national or international in scope to promote data exchange and views for health care delivery across multiple institutions.[13,16,23]

As the electronic patient record is implemented, there will be little need for the clinical laboratory (and other ancillary services) to provide individual reports in the medical record. Electronic reports will be created from the electronic record and will give providers a total view of patient care.[7] The electronic record will enable the views to be customized for different care units and different providers.[1,25] The need to produce a standardized report for intra-institutional use will become minimal; however, when data are shared among institutions, a standardized format for medical data sharing will be needed.[17]

The new electronic health care systems must provide improved decision support systems for effective patient care. Laboratory personnel must be involved with other health care providers in the process of designing the laboratory reports of a central database.[17] It is important to note that the Laboratory Information System (LIS) should be viewed not only as a repository of information for internal use, but also as a repository of information that will need to be forwarded to a decision support system.[15] The LIS must be capable of an open architecture approach to allow the exchange of data between the central decision support system and the LIS. The exchange of data can be accomplished through the use of communication protocols that send data over a variety of high speed, high volume communication channels (*e.g.*, data or voice phone lines, fiber optics, Local Area Networks (LAN), Wide Area Networks (WAN), air waves, etc.). Most laboratory staff will need to consult with specialists in the communications field for guidance.[2,18,20,23,26]

Current health care trends suggest that clinical laboratory operations will continue to decentralize to include point-of-care and bedside testing.[3] The proliferation of remote laboratory testing facilities increases the need for developing a distributed computer architecture where processing tasks and data are shared across multiple hardware platforms attached to a network. The individual systems store data for local processing. To enable the composite storage of these data for regional viewing, the data must be transported to a centralized database. Several standards-based organizations have been working together to formulate standards for the communication of laboratory test results within and between health care delivery systems.[11] See Table 8-1 for a sample list of standards-based organizations. Several of these standards will be mentioned in this chapter and some of them will be explained.

As health care delivery systems move toward automation of total patient care, clinical laboratory automation must also be optimized. Human errors may be substantially reduced when many clerical, repetitive, and interpretive tasks are totally automated.[15,16] Bar-code label systems for identification of patients, samples, medications, and supplies show substantial improvements over manual input systems.[27] Computer interface software can be written to assist clinical instrument operators in the monitoring of the analyzers through window-based work stations. These object-oriented (OO) or graphical user interface (GUI) tools can greatly enhance the user's capability to interpret quality control results, monitor the instrument functions, and monitor patient results through the use of delta checks or other clinical monitors. All of these functions may take place before the patient results are actually transmitted to the LIS or central database.[5,6,12,21,28]

The discussion that follows presents information contained in several health care computer interfacing standards. It is not meant to be an all-inclusive discussion. Its purpose is to inform the reader of the intent of each standard and to provide a general overview of the specifications listed in each standard. As mentioned previously, consultation with experts in the communications field will be required for guidance in implementation. Although all of the presented standards have been published, some of the them are con-

TABLE 8-1. Sample List of Standards-based Organizations

ANSI	American National Standards Institute
ASO	Accredited Standards Organization
ASTM	American Society for Standards and Materials
CCITT	International Telegraph and Telephone Consultative Committee
DICOM	Working groups of the American College of Radiology and the National Electrical Manufacturers Association
HISPP	Health Information Systems Planning Panel
IEEE	Institute of Electrical and Electronics Engineers
ISO	International Standards Organization
NCDP	National Council of Prescription Drug Pharmacies

tinuing to be modified and updated and some are awaiting final approval status.

COMPUTER INTERFACING STANDARDS

OSI Reference Model

The International Standards Organization (ISO) has proposed the Open System Interconnection (OSI) Reference Model as a standard for all computer systems to use to communicate with each other. The model supports an open architecture approach, so that multiple disparate computer platforms used in the health care environment would have a common means of communicating with each other to transfer data. The OSI model contains seven layers (Table 8-2). Each layer can be implemented by the use of several different communication protocols. Each protocol also has a standard definition; use of one protocol by one manufacturer will be compatible with the same protocol used by another manufacturer. It is also possible for multiple protocols to be compatible with other protocols. It is beyond the scope of this publication to fully explain all of these protocols. Table 8-2 lists some examples of the protocols that conform to the definition of the various layers of the OSI Reference Model.[14,19,20,22]

HL7® Standard

One standard of particular interest to health care management is the Health Level 7® (HL7) Standard for electronic

	OSI Layers	Definition	Example Protocols/Standards
7	Application	Allows users access to the OSI environment and provides distributed information services	HL7® DNA DECnet File Transfer, Access & Management (FTAM) Virtual Terminal Protocol (VTP) CCITT X.400 & X.500
6	Presentation	Allows the application processes to be isolated from differences in data representation (syntax)	DNA DECnet File Transfer, Access & Management (FTAM) Virtual Terminal Protocol (VTP) ISO DIS 8822 & 8823 CCITT X.400 & X.409
5	Session	Supplies the control structure for communication between applications; institutes, controls, and terminates connections (sessions) between cooperating applications	DNA ISO EMM083 Standard CCITT X.215 & X.225 ISO DIS 8326 & 8327 Standard
4	Transport	Accommodates reliable, transparent transfer of data between end points; accountable for end-to-end error recovery and flow control	Transport Control Protocol (TCP) DNA DECnet CCITT T.70 & X.400
3	Network	Allows upper layers to be independent from the data transmission and switching technologies used to connect systems; accountable for establishment, maintenance, and termination of connections	ISO Internet Protocol (IP) DNA DECnet CCITT X.25 level 3
2	Data Link	Accommodates the reliable transfer of information across the physical link; transmits blocks of data (frames) with the necessary synchronization, error control, and flow control	IEEE 802.3/Ethernet DNA DECnet ANSI X3.28 HDLC Standard (ISO4335) ADCCP Standard (ANSI X3.66) CCITT X.25 level 2 CCITT X.75
1	Physical	Pertains to the transmission of unstructured bit stream over the physical medium; addresses the mechanical, electrical, functional, and procedural characteristics to access the physical medium	IEEE 802.3/Ethernet DNA DECnet RS-232-C & RS-449 Standards CCITT X.20, X.21, X.22, & V.35 CCITT X.25 level 1

TABLE 8-2. OSI Model

data exchange in health care environments, especially inpatient acute care facilities/hospitals. The term *Level 7* refers to the top level (7) of the OSI Reference Model. The HL7® Standard does not specify any particular protocols or standards to be used in layers 1–6 of the OSI model. The HL7® Standard does conform to the conceptual definition of an application-to-application interface placed in the seventh layer of the OSI model.[13]

The HL7® Standard currently addresses the interfaces among various systems that send or receive patient admission/registration, discharge, or transfer (ADT) data; queries; orders; results; clinical observations; billing data; and master file update information. The intention of the Standard is to build a robust framework to support many other interfaces. A common framework for data exchange among health care computer applications should eliminate or substantially reduce the need for custom interface programs and the subsequent maintenance of these programs.[4] The Standard even allows for site-specific variations of data to be exchanged.

Tables 8-3a–e display a partial listing of the HL7® data definition tables.[13] The encoding rules for HL7® message format specify data fields that are of variable length and are separated by a field separator character. Rules also specify how the various data types are encoded within a field and when an individual field may be repeated. Logical groupings composed of data fields are called segments. Segments are separated by segment separator characters. For proper identification, each segment begins with a three-character literal value. A segment is allowed to be repeated and may be required or optional.

The HL7® Standard assumes that an event in the health care environment creates a need for data exchange. This event is called the *trigger event*. For example, when a patient is admitted, information about both the patient and the admission needs to be shared by several other data systems. The same need is true for a clinical laboratory test order and a clinical laboratory test result. The trigger events for data exchange that are initiated by the application system are called *unsolicited update events*. These transactions will be sent to other systems by the application without any request from the other systems. Because not all systems are interested in receiving all unsolicited update events, a subscription system may be implemented in which an application would only subscribe to the trigger events that it is interested in receiving.

The person or service that requests an observation or test order is called the *requestor*. The person or service that produces the observation or test result is called the *producer* or *filler*. When one system is requesting data that are resident on another system, the requesting transaction is called a *query*. The query transaction is specifically requested, as opposed to the unsolicited update transaction, where the transaction is not specifically requested. A pharmacy system may need to review a clinical laboratory test result before dispensing a particular kind of medication. Because the clinical laboratory data reside on the clinical laboratory system and not on the pharmacy system, the pharmacy system queries the clinical laboratory system to view the specific clinical

TABLE 8-3a. HL7® Standard: Functional Area Definitions

Functional Area	Description
ADT	Registration/ADT
ANR	Ancillary report
BLN	Finance
CNT	Control/query
NMT	Network management
ORD	Order entry
QRY	Control/query

Reprinted with permission from Health Level Seven®, version 2.2 (Draft)

TABLE 8-3b. HL7® Standard: Message Types

Value	Description	Owner
ACK	General acknowledgment message	CNT
ADT	ADT message	ADT
ARD	Ancillary RPT (display)	ANR
BAR	Add/change billing account	BLN
DFT	Detail financial transaction	BLN
DSR	Display response	QRY
MCF	Delayed acknowledgment	CNT
ORF	Observational result/record resp	ANR
ORM	Order message	ORD
ORR	Order acknowledgment message	ORD
ORU	Observational result/unsolicited	ANR
OSQ	Order status query	ORD
RAR	Pharmacy administration information	ORD
RAS	Pharmacy administration message	ORD
RDE	Pharmacy encoded order message	ORD
RDR	Pharmacy dispense information	ORD
RDS	Pharmacy dispense message	ORD
RGV	Pharmacy give message	ORD
RGR	Pharmacy dose information	ORD
RER	Pharmacy encoded order information	ORD
ROR	Pharmacy prescription order response	ORD
RRA	Pharmacy administration acknowledgment	ORD
RRD	Pharmacy dispense acknowledgment	ORD
RRE	Pharmacy encoded order acknowledgment	ORD
RRG	Pharmacy give acknowledgment	ORD
QRY	Query	QRY
UDM	Unsolicited display message	QRY

Reprinted with permission from Health Level Seven®, version 2.2

laboratory test result. The clinical laboratory system then responds to the query by sending the requested data to the pharmacy system.

The lower levels (1–6) of the OSI Reference Model are responsible for delivering the encoded messages to the appropriate systems. The data message may be encapsulated several times, with each layer representing the necessary data required for the protocols to communicate with each other. A high degree of communication takes place between systems when the higher levels of the OSI Reference Model

TABLE 8-3c. HL7® Standard: Segment Identification (incomplete list)

Segment	Description
ADD	Addendum segment
BHS	Batch header
BLG	Billing segment
BTS	Batch trailer segment
DG1	Diagnosis
DSP	Display data segment
ERR	Error segment
EVN	Event type
FHS	File header segment
FT1	Financial transaction
FTS	File trailer segment
MFA	Master file acknowledgment segment
MFE	Master file entry segment
MFI	Master file identification segment
MRG	Merger patient information
MSA	Message acknowledgment segment
MSH	Message header segment
NTE	Notes and comments segment
OBR	Observation request segment
OBX	Observation segment
ORC	Common order segment
PID	Patient identification
PV1	Patient visit
QRD	Query definition segment
QRF	Query filter segment
STF	Staff identification segment
UB1	UB82 data
UB2	UB92 data
URD	Results/update definition segment
URS	Unsolicited selection segment

Reprinted with permission from Health Level Seven®, version 2.2 (Draft)

TABLE 8-3d. HL7® Standard: Data Types

Value	Description
ST	String data (used to transmit numeric)
TX	Text
FT	Formatted text
DT	Date
TM	Time
TS	Time stamp
PN	Person name
TN	Telephone number
AD	Address
CK	Composite ID with check digit
CN	Composite ID and name
CE	Coded entry
RP	Reference pointer
NM	Numeric

Reprinted with permission from Health Level Seven®, version 2.2 (Draft)

are used. The systems acknowledge the receipt of messages and return error messages whenever a message is not transmitted correctly. The lowest level protocols do the least amount of checking to ensure the integrity of the delivered message. It is probable that clinical laboratory instruments, which send data across physical connections using the RS-232-C Standard interface, do little to ensure the integrity of the transmitted message. It is only recently that standards have been published to include checksums in the lower level protocols (see ASTM E 1381 Standard). It is expected that instrument vendors will begin to comply with the newer standards, but there may be little incentive to do so with instruments that have been in use for several years.

TABLE 8-3e. HL7® Standard: Delimiter Values

Delimiter	Suggested Value	Encoding Character Position	Usage
Segment terminator	⟨cr⟩ hex 0D	-	Terminates a segment record. A subsequent line feed character is ignored. This value cannot be changed by implementors.
Field Separator	|	-	Separates two adjacent data fields within a segment. It also separates the segment ID from the first data field in each segment.
Component separator	∧	1	Separates adjacent components of data fields where allowed.
Subcomponent separator	&	4	Separates adjacent subcomponents of data fields where allowed. If there are no subcomponents, this character may be omitted.
Repetition separator	~	2	Separates multiple occurrences of a field where allowed.
Escape character	\	3	Escape character for TX and FT fields. If no escape characters are used in a message, this character may be omitted.

Reprinted with permission from Health Level Seven®, version 2.2 (Draft)

It is the responsibility of the application level to determine whether or not the content of an HL7® message contains all of the required data. For example, a pharmacy system may send a query transaction to the clinical laboratory system for specific clinical laboratory test data, but the message may not contain a date range for the inquiry. If the clinical laboratory system requires a date range to be specified, and the message does not contain the date range, then the clinical laboratory system must send a message back to the pharmacy system indicating that it could not process the request because of incomplete data. Agreements should be in place among the parties who expect to share data. These agreements should specify the data to be exchanged and the required elements.

The HL7® Standard defines specific components of an HL7® message. A *message* is the basic unit of data shared between systems. It consists of a group of segments in a defined sequence. The purpose of each message is defined by the message type. A *segment* is defined as a logical grouping of data fields. Segments may be required or optional and they may repeat. The *segment ID* is a unique three character code used to identify each segment. A *field* is defined as a string of characters and may contain formatted or unformatted data. Special characters are used as message delimiters. See Table 8-3e for a complete list of delimiters.

The HL7® Standard defines the message structure for multiple message types. Different message types are used to request and transfer specific data from one system to another system. When communication takes place between an analyzer and the clinical laboratory system, the analyzer takes the role of the producer (filler), and the clinical laboratory system takes the role of the requestor. When communication takes place between the central database system and the clinical laboratory system, the clinical laboratory system takes the role of the producer (filler), and the central database system takes the role of the requestor.

It is beyond the scope of this text to examine each of the message types. However, an example of two message types, Order Message (ORM) and Observation Result/Unsolicited (ORM) is found in Figure 8-1. Observations can be transmitted in a solicited (response to a query) or unsolicited mode as described earlier. These two message types are required to transmit observations and results of diagnostic studies from the producing system to the ordering system. The ORM message may be the result of a request (query) from an analyzer to the LIS for order information, or it may be the result of an LIS user sending order information to the analyzer without a request from the analyzer.

The definitions of the message segments used to transmit observations and results are defined in collaboration with the ASTM E1238 Standard. The OBR segment contains information that applies to all of the observations that follow. The OBX segment contains information about a single observation. An observation message requires at least one MSH, OBR, and OBX segment.

ASTM E 1238 Standard

The ASTM E 1238 Standard addresses standard specifications for transferring clinical observations between independent computer systems. It is under the jurisdiction of the ASTM Committee E-31 on Computerized Systems and is under the direct responsibility of Subcommittee E31.11.[8] The E 1238 Standard covers the two-way digital transmis-sion of requests for, and results of, tests, diagnostic studies, and care-provider observations between requestors and producers. It defines the logical format and encoding rules for messages intended for the interchange of any clinical information that can be re-

---

RESULTS MESSAGE

```
MSH|^~\&|LAB INTERFACE|Instrument Manager|LA AUTO INST|642|19941018095552||ORU|67320|P|2.2
PID|1||123-45-6789||DOE^JOHN^C|||M
OBR|1|107897||||||||||SER|||CX7|107897^CH 1018 195|||19941018094916|||||^^^^R
OBX|1|ST|82565^Creatinine; blood^AS4^173X4^CREATININE^99002||0.9|mg/dl||||F
OBX|2|ST|84520^Urea nitrogen; quantitative^AS4^174X3^UREA NITROGEN, BLOOD^99002||16|mg/dl||||F
OBX|3|ST|82948^Glucose; blood^AS4^175X2^GLUCOSE^99002||238|mg/dl|H|||F
OBX|4|ST|84295^Sodium; serum^AS4^176X5^SODIUM^99002||138|mmol/l||||F
OBX|5|ST|84132^Potassium; serum^AS4^177X6^POTASSIUM^99002||4.0|mmol/l||||F
OBX|6|ST|82435^Chloride; blood^AS4^178X7^CHLORIDE^99002||104|mmol/l||||F
OBX|7|ST|82374^Carbon Dioxide^AS4^179X8^CARBON DIOXIDE^99002||30|mmol/l||||F
```

DOWNLOAD REQUEST MESSAGE

```
MSH|^~\&|LA AUTO INST|642|LAB INTERFACE|Instrument Manager|19941024083551||ORM|2941024.083551|P|2.1
PID|||354^1^M11||SMITH^LARRY^S^J.||19281209|M|||||||012-34-5678
PV1|||4W
ORC|NW||100906|||||||^KILDARE^EDWARD^E
OBR|1|||82565^CREA^AS4^173X4^CREATININE^99002|||||||||PMX|113411^CH 1024 23^1^1
OBR|2|||84520^BUN^AS4^174X3^UREA NITROGEN, BLOOD^99002|||||||||PMX|113411^CH 1024 23^1^1
OBR|3|||82948^GLU^AS4^175X2^GLUCOSE^99002|||||||||PMX|113411^CH 1024 23^1^1
OBR|4|||84295^NA^AS4^176X5^SODIUM^99002|||||||||PMX|113411^CH 1024 23^1^1
OBR|5|||84132^K^AS4^177X6^POTASSIUM^99002|||||||||PMX|113411^CH 1024 23^1^1
OBR|6|||82435^CL^AS4^178X7^CHLORIDE^99002|||||||||PMX|113411^CH 1024 23^1^1
OBR|7|||82374^CARB^AS4^179X8^CARBON DIOXIDE^99002|||||||||PMX|113411^CH 1024 23^1^1
```

SUBSEQUENT UPLOAD OF RESULTS FOR THE DOWNLOAD REQUEST MESSAGE

```
MSH|^~\&|LAB INTERFACE|Instrument Manager|LA AUTO INST|642|19941024094329||ORU|70192|P|2.1
PID|1||012-34-5678
OBR|1|113411|||||||||PMX|113411^CH 1024 23^1^1
OBX|1|ST|84520^Urea nitrogen; quantitative^AS4^174X3^UREA NITROGEN, BLOOD^99002||27|mg/dl||||F
OBX|2|ST|82374^Carbon Dioxide^AS4^179X8^CARBON DIOXIDE^99002||30|mm/l||||F
OBX|3|ST|82948^Glucose; blood^AS4^175X2^GLUCOSE^99002||105|mg/dl||||F
OBX|4|ST|82565^Creatinine; blood^AS4^173X4^CREATININE^99002||1.6|mg/dl||||F
OBX|5|ST|84295^Sodium; serum^AS4^176X5^SODIUM^99002||137|mm/l||||F
OBX|6|ST|84132^Potassium; serum^AS4^177X6^POTASSIUM^99002||4.5|mm/l||||F
OBX|7|ST|82435^Chloride; blood^AS4^178X7^CHLORIDE^99002||100|mm/l||||F
```

◀

Figure 8-1. Sample Observation Transaction.

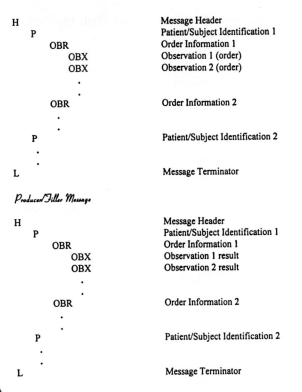

H			Message Header
	P		Patient/Subject Identification 1
		OBR	Order Information 1
		OBX	Observation 1 (order)
		OBX	Observation 2 (order)
		OBR	Order Information 2
	P		Patient/Subject Identification 2
L			Message Terminator

Producer/Filler Message

H			Message Header
	P		Patient/Subject Identification 1
	OBR		Order Information 1
		OBX	Observation 1 result
		OBX	Observation 2 result
	OBR		Order Information 2
	P		Patient/Subject Identification 2
L			Message Terminator

▲

Figure 8-2. Logical Relation of ASTM E 1238 Segments. Requestor Message.

ported in a textual form. This standard was initially developed for clinical laboratory results, but it has now been expanded to address general information about diagnostic testing as well as specific details which are useful for clinical practice, administration, and research in a broad, yet adaptable, form.

The terms used in the ASTM E 1238 Standard are similar to those used in the HL7® Standard. The requestor orders a procedure and the producer or filler provides the observa-

tion result in messages containing segments and fields. See Figure 8-2 for a logical relation of segments. The order segment is designated as OBR and the result segment is designated as OBX. Segment types are shown in Table 8-4a. Delimiters used in ASTM E 1238 are the same as those used in HL7®. See Table 8-4b for ASTM E 1238 Data Types. As indicated previously, the work of the ASTM computer standards committees is closely paralleling the work of the HL7® committee.

Although the ASTM E 1238 Standard defines many fields for data, not all data fields are transmitted with each message. For example, complete patient data may be sent only when the message is transmitted to the producer for the first time. Thereafter, only information specific to the current patient order may be sent. The producer is required to send a minimum amount of patient data and ordering information with its result message so that the result may be filed appropriately to match a particular order request. Although standards may address minimal amount of data necessary for a successful message transmission, it is the responsibility of the requestor and producer to agree upon the minimum amount of data required by the respective applications to maintain application data integrity.

Figure 8-3 presents an example of a valid message built according to the ASTM E 1238 specifications.

ASTM E 1381 Standard

The ASTM E 1381 Standard contains the specification for a low-level protocol to transfer messages between clinical laboratory instruments and computer systems. Like ASTM E 1238, it is under the jurisdiction of the ASTM Committee E-31 on Computerized Systems and is under the direct responsibility of Subcommittee E31.14.[9] This Standard de-

TABLE 8-4a. ASTM E 1238 Segment Types (incomplete list)		
Types	Description	Who Sends
H	Message header	Requestor and filler
P	Patient segment	Requestor and filler
OBR	Observation order segment	Requestor or filler
OBX	Result observation segment	Requestor or filler
E	Error checking segment	Requestor or filler
C	Comment segment	Requestor or filler
Q	Request results segment	Requestor or filler
L	Message terminator	Requestor and filler
S	Scientific segment	Filler
PV1	HL7 patient visit segment	
GT1	HL7 guarantor segment	
IN1	HL7 insurance segment	
OM1	General test/observation master segment	
OM2	Numeric test/observation segment	
OM3	Categorical test/observation master segment	
OM4	Observations that require specimens	
OM5	Test/observation segment for observation batteries	
OM6	Test/observation master segment for calculated observations	

TABLE 8-4b. ASTM Data Types

Code	Description
AD	Address
CE	Coded entry (*e.g.*, test IDs, Dx codes)
CK	Composite ID with check digit
CM	Composite miscellaneous
CNA	Composite ID and person name
CQ	Composite quantity with units ⟨number⟩^⟨units⟩
ID	Identifier
MO	Money
NM	Numeric
PN	Person name
RP	Reference pointer
ST	String for short text and numeric
TN	Telephone number
TS	Time stamp (date & time)
TX	Bulk text

Copyright ASTM. Reprinted with permission.

without the need for specific, customized programming efforts, which are costly and resource intensive.

ASTM E 1381 describes the mechanical and electrical connection for serial binary data bit transmission between instrument and computer system in the *physical layer*. It addresses signal levels, character structure, speed, interface connections, connectors, and cables. The *data link layer* addresses the link connection, contention (two systems transmitting data simultaneously), frames, frame number, checksum, acknowledgments, receiver interrupts, link release, error recovery, defective frames, time-outs, and restricted message characters. The data link layer interacts with higher layers with respect to send and receive messages, handles data link connection and release requests, and reports the data link status.

scribes the electronic transmission of digital information between clinical laboratory instruments and computer systems. The terminology for the OSI Reference Model is generally used in describing the c
ommunications protocol and services. This Standard defines the electrical parameters, cabling, data codes, transmission protocol, and error recovery for the information that passes between the instrument and the laboratory computer. As laboratory instrument manufacturers become familiar with this Standard, it is expected that they will produce instruments that conform to this Standard, thereby enabling the compatibility of clinical instruments and computer systems

ASTM E 1394 Standard

The ASTM E 1394 Standard also addresses the specification for transferring information between clinical instruments and computer systems. Like the previously described ASTM E Standards, it is under the jurisdiction of the ASTM Committee E-31 on Computerized Systems and is under the direct responsibility of Subcommittee E31.14.[10] This Standard specifies the message content for transferring information between a clinical instrument and a computer system through defined communications protocols. It applies to all text-oriented clinical instrumentation. This Standard is related to ASTM E 1238 in that both standards use positional convention to define the structure of messages that exchange information about clinical test requests and results. However, E 1238 contains many specifications that are not appropriate for use between clinical instrument and

```
H|^~\&|1234|LIPWS|Pickens Clinic|3414 Rosebud Rd SW^Suite 120^Cedar Rapids<CR>
A|^IA^52404||(319)363-4569||Mid-State Labs|(Example message)|T|A.2|198703311521< CR >
P|1|A9999|P10098 | 245-83-2033| HAMMOND^WILLLIM^EDWARD^II||19350109I|M|W|< CR >
   A|400 Eastern Ave^^Marion^IA^52332||393-9900 ~ 395-4533|321-6^SMITH&RICHARD&W.&&DR.|< CR >
   A|||165|63|401.9^Hypertension|Furosemide ~ Methyldopa||This is practice field<CR>
   A|#1.|This is practice field #2|19870420|0P<CR>

OBR......
   .
   .
   .

OBR|1|870930010|CM3562|80004^ELECTROLYTES|R|198703281530|198703290800|||401-0^INTERN&JOE&&&&MD^L|N| CR
>
   A||||SER|^SMITH&RICHARD&W.&&DR.|(319)377-4400|This is requestor field #1.| <CR>
   A|Requestor field #2|Diag.serv.field #1.|Diag.serv.field #2.|198703311400|||F<CR>
OBX|I|ST|84295^NA||150|mmol/l|136-148|H||A|F|19850301 < CR >
OBX|2|ST|84132^K+||4.5|mmol/l|3.5-5|N||N|F|19850301<CR>
OBX|3|ST|82435^CL||102|mmol/l|94-105|N||N|F|19850301< CR >
OBX|4|ST|82374^CO2||27|mmol/l|24-31|N||N|F|19850301<CR>
   .
   .
   .

**–End record–**
L|1||1|69|1014<CR>
```

▲

Figure 8-3. ASTM E 1238 Standard Message Example. (Reprinted with permission from ASTM, 1994 Annual Book of ASTM Standards)

computer system data exchange. Many of the requirements and limitations found in E 1238 are of little value to a clinical instrument. Additionally, clinical instruments contain test and instrument specific requirements which fall outside of the scope of E 1238.

The E 1394 Standard tries to identify and simplify all complex data structures and interface procedures and restrict multiple procedural options to single procedures appropriate for the clinical instrument setting. This Standard was developed independently of the E 1381 Standard, which addresses low-level data transfer and protocol specifications. The E 1394 Standard addresses message content and assumes that a protocol layer exists that will handle record blocking/deblocking, error detection and recovery, and other associated data transport tasks.

The ASTM E 1394 Standard specifies a message structure, which consists of a hierarchy of records of various types. The record levels range from zero to four. The smallest element in any record is the field, which contains a single item of information. The record types are related to one another in a definite hierarchy. Table 8-5 shows a hierarchical listing of record types and Figure 8-4 shows a logical relationship of a message structure. The comment and manufacturer information records have no assigned level. A sequence of records at one level is terminated by the appearance of another record type that has the same or higher level.

The hierarchy used in E 1394 is the same as the one used in E 1238. The record type IDs are different, and E 1394 has additional record types, which are pertinent to clinical instruments. The delimiter characters used in a given message are specified in the header message. Alphanumeric characters should not be used as delimiters. The five characters that immediately follow the "H" designation (the header ID) define the delimiters to be used throughout the following records of the message. Table 8-6 defines the delimiter character positions. It is the responsibility of the sender to ensure that the delimiter characters used in a message will not be

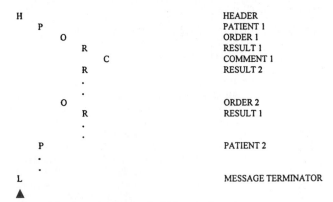

Figure 8-4. ASTM E 1394 Logical Message Structure.

included in the field contents of the message. Each message must define the delimiter characters in the header message. Different messages may define different delimiter characters.

To enable recovery of data from transmission errors, the transmitter and receiver must follow established guidelines for data storage and data retransmission when an error condition occurs. A decremental change in the message hierarchy must trigger a storage of all the data that preceded the change. Data may be stored more frequently but it is not a requirement. If the protocol level layer determines a transmission failure, the message must be retransmitted. The restart point may be based upon the amount of data stored before the actual error condition occurred, so that the entire message may not need to be retransmitted.

Field lengths and record lengths are considered variable. If an instrument has limits for these fields, it should be noted in the vendor's documentation for the instrument. Figure 8-5a–e shows a sample message formatted to the specifications of the ASTM E 1394 Standard.

SUMMARY

The modern day health care facility has most likely been computerized for several years. As these systems require updating, the decision to upgrade or to maintain an integrated system of hospital information becomes increasingly difficult. Typically, multiple vendors are involved and it becomes impractical to purchase an integrated solution from a single vendor. Reasonable alternatives have been to replace aging systems with solutions that provide highly sophisticated automated techniques to the individual services, but

TABLE 8-5. ASTM E 1394 Record Types

Hierarchy	Record Type	ID
Level 0	Message header	H
	Message terminator	L
	Manufacturer information	M
	Comment	C
Level 1	Patient information	P
	Request information	Q
	Scientific	S
	Manufacturer information	M
	Comment	C
Level 2	Test order	O
	Manufacturer information	M
	Comment	C
Level 3	Result	R
	Manufacturer information	M
	Comment	C
Level 4	Manufacturer information	M
	Comment	C

TABLE 8-6. ASTM E 1394 Delimiter Character Positions in Message Header

Character Position	Delimiter
1	H (Header)
2	Field
3	Repeat field
4	Component field
5	Escape character

H|\^&||PSWD|Harper Labs|2937 Southwestern Avenue^Buffalo^NY^73205||319 412-9722||||P|2.5|19890314<CR>
Q|1|^032989325|^032989327|ALL|||||||O<CR>

▲

Figure 8-5a. Sample Message in ASTM E 1394 Format. Request from Analyzer for Test Selections on Specimens 032989325-032989327. (Reprinted with permission from ASTM, 1994 Annual Book of ASTM Standards)

H|\^&||PSWD|Harper Labs|2937 Southwestern Avenue^Buffalo^NY^73205||319 412-9722||||P|2.5|19890314<CR>
P|1|2734|123|306-87-4587|BLAKE^LINDSEY^ANN^MISS<CR>
O|1|032989325||^^^BUN|R<CR>
O|2|032989325||^^^ISE|R<CR>
O|3|032989325||^^^HDL\^^^GLU|R<CR>
P|2|2462|158|287-17-2791|POHL^ALLEN^M.<CR>
O|1|032989326||^^^LIVER\^^^GLU|S<CR>
P|3|1583|250|151-37-6926|SIMPSON^ALBERT^^MR<CR>
O|1|032989327||^^^CHEM12\^^^LIVER|R<CR>

L|1|F<CR>

▲

Figure 8-5b. Sample Message in ASTM E 1394 Format. Response from Computer System from Previous Request. (Reprinted with permission from ASTM, 1994 Annual Book of ASTM Standards)

H|\^&||PSWD|Harper Labs|2937 Southwestern Avenue^Buffalo^NY^73205||319 412-9722||||P|2.5|19890314<CR>
P|1|2734|123|306-87-4587|BLAKE^LINDSEY^ANN^MISS<CR>
C|1|L|Notify IDC if tests positive|G<CR>
O|1|032989325||^^^BUN|R<CR>
R|1|^^^BUN|8.71<CR>
C|1|I|TGP^Test Growth Positive|P<CR>
C|2|I|colony count <10,000|P<CR>
O|2|032989325||^^^ISE|R<CR>
R|1|^^^ISE^NA|139|mEq/L<CR>
R|2|^^^ISE^K|4.2|mEq/L<CR>
R|3|^^^ISE^CL|111|mEq/L<CR>

O|3|032989325||^^^HDL|R<CR>
R|1|^^^HDL|70.29<CR>
O|4|032989325||^^^GLU|R<CR>
R|1|^^^GLU|92.98<CR>
C|1|I|Reading is Suspect|I<CR>
P|2|2462|158|287-17-2791|POHL^ALLEN^M.<CR>
O|1|032989326||^^^LIVER|S<CR >
R|1|^^^LIVER^AST|29<CR>
R|2|^^^LIVER^ALT|5O<CR>
R|3|^^^LIVER^TBILI|7.9<CR>
R|4|^^^LIVER^GGT|29<CR>
O|2|032989326||^^^GLU|S<CR>
R|1|^^^GLU|91.5<CR>
P|3|1583|250|151-37-6926|SIMPSON^ALBERT^^MR<CR>
O|1|032989327||LIVER|R<CR>
R|1|^^^AST|28<CR>   [Test ID field Implicitly Relates to LIVER order]
R|2|^^^ALT|49<CR>
R|3|^^^TBILI|7.3<CR>
R|4|^^^GGT|27<CR>
O|2|032989327||CHEM12|R<CR>
R|1|^^^CHEM12^ALB-G|28<CR>   [Test ID field Explicitly Relates to CHEM12 order]
R|2|^^^CHEM12^BUN|49<CR>
R|3|^^^CHEM12^CA|7.3<CR>
R|4|^^^CHEM12^CHOL|27<CR>
R|5|^^^CHEM12^CREAT|4.2<CR>
R|6|^^^CHEM12^PHOS|12<CR>
R|7|^^^CHEM12^GLUHK|9.7<CR>
R|8|^^^CHEM12^NA|138.7<CR>
R|9|^^^CHEM12^K|111.3<CR>
R|10|^^^CHEM12^CL|6.7<CR>
R|11|^^^CHEM12^UA|7.3<CR>
R|12|^^^CHEM12^TP|9.2<CR>

L|1<CR>

▲

Figure 8-5c. Sample Message in ASTM E 1394 Format. Results from Given Ordered Test Selections Shown in Various Formats. (Reprinted with permission from ASTM, 1994 Annual Book of ASTM Standards)

```
H|\^&||PSWD|Harper Labs|2937 Southwestern Avenue^Buffalo^NY^73205||319 412-9722||||P|2.5|19890314<CR>
    Q|1|032989326||ALL<CR>
L|1<CR>
```

▲

Figure 8-5d. Sample Message in ASTM E 1394 Format. Request from Computer System to Instrument for Previously Run Results. (Reprinted with permission from ASTM, 1994 Annual Book of ASTM Standards)

```
H|\^&||PSWD|Harper Labs|2937 Sou7thwestern Avenue^Buffalo^NY^73205||319 412-9722||||P|2.5|19890314<CR>
    P|1|2462|158|287-17-2791|POHL^ALLEN^M.<CR>
        O|1|032989326||^^^LIVER|S<CR>
            R|1|^^^AST|29<CR>
            R|2|^^^ALT|50<CR>
            R|3|^^^TBILI|7.9<CR>
            R|4|^^^GGT|29<CR>
        O|2|032989326||^^^GLU|S<CR>
            R|1|^^^GLU|91.5<CR>
L|1|F<CR>
```

▲

Figure 8-5e. Sample Message in ASTM E 1394 Format. Reply to Result Request. (Reprinted with permission from ASTM, 1994 Annual Book of ASTM Standards)

these alternatives leave the facility with many disparate systems.[13,16,20]. The need to share critical data among the disparate automated systems has emerged. Today's health care delivery system places an ever increasing emphasis on the automation of total patient care information. Health care delivery depends not only upon consolidated patient data but also on consolidation of institutional data (provider practices, institutional policies, etc.). An integrated, automated solution that focuses on patient care will greatly enhance the quality of patient care given by health care facilities and providers and will enable health care facilities to monitor approaches to medical care to maintain cost efficiency.[13,16,24]

A new generation of systems must be designed to accommodate an open architecture. All patient care information must be centralized into a consolidated electronic medical record. The new electronic health care systems must provide improved decision support systems for effective patient care.[15] Optimally, the database for the electronic record will need to be national or international in scope to promote data exchange and views for health care delivery across multiple institutions.[13,16,23] Electronic reports will be created from the electronic record and will give providers a total view of patient care, and that view will be customized for different care units and different providers.[1,7,25] However, when data are shared among disparate systems, a standardized format for medical data sharing is needed.[17]

Several standards-based organizations have been working together to formulate standards for the communication of laboratory test results within and among health care delivery systems.[11] As the health care delivery systems move toward automation of total patient care, clinical laboratory automation must be optimized. Software programs will be written to interface clinical instruments to work stations and to the LIS. Information must flow in both directions between the LIS and the clinical database system. Information will also be shared among various institutions. To inform the reader of the issues involved in computer interfacing, the specifications contained in the following standards are discussed: (1) the Open System Interconnection (OSI) Reference Model of the International Standards Organization (ISO), (2) the Health Level 7® (HL7) Standard, (3) the ASTM E 1238 Standard, (4) the ASTM E 1381 Standard, and (5) the ASTM E 1394 Standard.

EXERCISES

1. List some of the requirements of today's clinical laboratory with regard to computerization.
2. Name three standards-based organizations dealing with the communication of laboratory data.
3. Define the following terms:
   Network
   LIS
   WAN
   GUI
   Field
4. Briefly describe the HL7® standard.
5. Briefly describe the electronic medical record.

REFERENCES

1. Barrows RC Jr, Allen B, Fink DJ. An X Window system for statlab results reporting. Proceedings—The annual symposium on computer applications in medical care 1993;331.
2. Bergvin L, Johansson B, Borjesson U. Distribution of laboratory test results to primary health care centres with the EDI-FACT standard. Clin Chim Acta 1993;222(1–2):141.
3. Bond LW. Consideration of laboratory parameters in design and implementation of automated systems with bar-coding. Clin Chem 1990;36(9):1583.
4. Cahill BP, Holmen JR, Bartleson PL. Mayo foundation electronic results inquiry, the HL7 connection. Proceedings—

The annual symposium on computer applications in medical care 1991;516.

5. Chou D, Van Lente F, Castellani W. Considerations in the design and implementation of user interfaces between laboratory instrumentation and laboratory information systems. Clin Chem 1990;36(9):1586.

6. Connelly DP. Embedding expert systems in laboratory information systems. Am J Clin Pathol 1990;4 (suppl 1):57.

7. Dolin RH. Transfer of laboratory data from a mainframe to a microcomputer: An interim approach. MD Comput 1994; 11(1):49.

8. E 1238-94, Standard specification for transferring clinical observations between independent computer systems, 1994. American Society for Testing and Materials (ASTM), 1916 Race Street, Philadelphia, PA 19103.

9. E 1381-91, Specification for low-level protocol to transfer messages between clinical laboratory instruments and computer systems, 1991. American Society for Testing and Materials (ASTM), 1916 Race Street, Philadelphia, PA 19103.

10. E 1394-91, Standard specification for transferring information between clinical instruments and computer systems, 1991. American Society for Testing and Materials (ASTM), 1916 Race Street, Philadelphia, PA 19103.

11. Friedman BA, Mitchell W. Integrating information from decentralized laboratory testing sites. The creation of a value-added network. Am J Clin Pathol 1993;99(5):637.

12. Groth T, Moden H. A knowledge-based system for real-time quality control and fault diagnosis of multitest analyzers. Comput Methods Programs Biomed 1991;34(2–3):175.

13. Health Level Seven, Version 2.2, 1994. Health Level Seven, PO Box 66111, Chicago, IL 60666–9998.

14. Jean F-C, Jaulent M-C, Coignard J, Degoulet P. Distribution and communication in software engineering environments. Application to the HELIOS software bus. Proceedings—The annual symposium on computer applications in medical care 1991;506.

15. Johansson B, Bergvin L. Arden syntax as a standard for knowledge bases in the clinical chemistry laboratory. Clin Chim Acta 1993;222(1–2):123.

16. Korpman RA. Health care information systems. Patient-centered integration is the key. Clin Lab Med 1991;11(1):203.

17. Krieg AF. Laboratory information systems from a perspective of continuing evolution. Clin Lab Med 1991;11(1):73.

18. Peters M, Broughton PM. The role of expert systems in improving the test requesting patterns of clinicians. Ann Clin Biochem 1993;30(Pt 1):52.

19. Schatt S, Fox S. Voice/data telecommunications for business. Englewood Cliffs, NJ: Prentice-Hall, Inc, 1990;171.

20. Shafarman MJ, Meeks-Johnson J, Jones T, McCoy J, van Valkenburg T. Implementing a record-oriented clinical lab interface using HL7 version 2.1 at Indiana University Hospital. Proceedings—The annual symposium on computer applications in medical care 1991;511.

21. Solberg HE. Object-oriented methods. Clin Chim Acta 1993; 222(1–2):3.

22. Stallings W. Handbook of computer-communications standards. Indianapolis, IN: Howard W Sams & Co, 1987;1.

23. Stead WW. Systems for the year 2000: The case for an integrated database. MD Comput 1991;8(2):103.

24. Studnicki J, Bradham DD, Marshburn J, Foulis PR, Straumfjord, JV. A feedback system for reducing excessive laboratory tests. Arch Pathol Lab Med 1993;117(1):35.

25. Tate KE, Gardner RM. Computers, quality, and the clinical laboratory: A look at critical value reporting. Proceedings—The annual symposium on computer applications in medical care 1993;193.

26. Weilert M. Implementing information systems. Clin Lab Med 1991;11(1):41.

27. Weilert M, Tilzer LL. Putting bar-codes to work for improved patient care. Clin Lab Med 1991;11(1):227.

28. Wells IG, Cartwright RY, Farnan LP. Practical experience with graphical user interfaces and object-oriented design in the clinical laboratory. Clin Chim Acta 1993;222(1–2):13.

Part 2

Clinical Correlations and Analytical Procedures

Chapter 9

Amino Acids and Proteins

Barbara J. Lindsey

Objectives

Upon completion of this chapter, the clinical laboratorian should be able to:

- *Describe the structures and properties of amino acids and proteins.*

- *Discuss the general characteristics of the aminoacidopathies, including the metabolic defect in each and the procedure used for detection.*

- *Outline the process of protein synthesis and catabolism and list the site(s) of synthesis.*

- *Differentiate between simple and conjugated proteins.*

- *Briefly discuss the function and clinical significance of the following proteins:*

    pre-albumin
    albumin
    α_1-antitrypsin
    α_1-fetoprotein
    haptoglobin
    ceruloplasmin
    transferrin
    fibrinogen
    C-reactive protein
    immunoglobulins
    troponin
- *Discuss at least five general causes of abnormal serum protein concentrations.*
- *List the reference interval for total protein and albumin and discuss any non-pathological factors that influence the levels.*
- *Describe and compare methodologies used in the analysis of total protein and albumin, and include the structural characteristics and/or chemical properties that are relevant to each measurement.*

- *Briefly explain the principle and clinical usage of separation of proteins by:*
    electrophoresis
    high resolution electrophoresis
    immunodiffusion
    immunoelectrophoresis
    immunofixation electrophoresis
    turbidity/nephelometry
    isoelectric focusing
- *Given a densitometric scan of a serum protein electrophoresis utilizing the routine method (5 zones), recognize and name the fractions, interpret any abnormality in the pattern, and associate these patterns with common disease states.*

- *Differentiate the types of proteinuria on the basis of etiology and type of protein found in the urine.*

- *Describe the principle of the methods used for both qualitative and quantitative determination and identification of urine proteins.*

- *Describe the diseases associated with alterations in cerebrospinal fluid proteins.*

- *Describe the steps in Southern blot DNA analysis and PCR.*

KEY WORDS

Albumin	Hypoproteinemia
Amino Acid	Isoelectric Point (pI)
Aminoacidopathies	Nitrogen Balance
Amphoteric	Peptide Bond
Conjugated Protein	Proteinuria
Globulins	Simple Protein
Hyperproteinemia	

Michael L. Bishop, Janet L. Duben-Engelkirk, and Edward P. Fody.
CLINICAL CHEMISTRY. © 1996 Lippincott–Raven Publishers.

In 1839, the Dutch chemist G. J. Mulder was investigating the properties of a substance found in milk and egg whites that coagulated when heated. The Swedish scientist J. J. Berzelius suggested to Mulder that these substances should be called *proteins* from the Greek word *proteis,* meaning "first rank of importance," because he suspected that they might be the most important of all biologic substances. Mulder also thought that all protein substances were the same because they contained carbon, nitrogen, and sulfur. This was found to be untrue when Dumas discovered slightly varying nitrogen content in protein obtained from different sources.[65]

Today we know that proteins are macromolecules composed of polymers of covalently linked amino acids and that they are involved in every cellular process. This chapter will discuss the general properties of both amino acids and proteins, abnormalities related to each, and methods of analysis. New technologies involving proteins also will be introduced.

AMINO ACIDS

Basic Structure

α-Amino acids are small biomolecules containing at least one amino group (—NH$_2$) and one carboxyl group (—COOH) bonded to the α-carbon. They differ from one another by the chemical composition of their R groups (side chains). The general structure of an α-amino acid is depicted in Fig. 9-1. There are 20 different amino acids that are used as building blocks for protein. The R groups found on these α-amino acids are shown in Table 9-1.

Metabolism

About half of the 20 amino acids needed by humans cannot be synthesized at a rapid enough rate to support growth. These nine nutritionally essential amino acids must be supplied by the diet in the form of proteins. Under normal circumstances, proteolytic enzymes, such as pepsin and trypsin, completely digest dietary proteins into their constituent amino acids. Amino acids are then rapidly absorbed from the intestine into the portal blood and subsequently become part of the body pool of amino acids. Other contributors to the amino acid pool are newly synthesized amino acids and those released by the normal breakdown of body proteins.

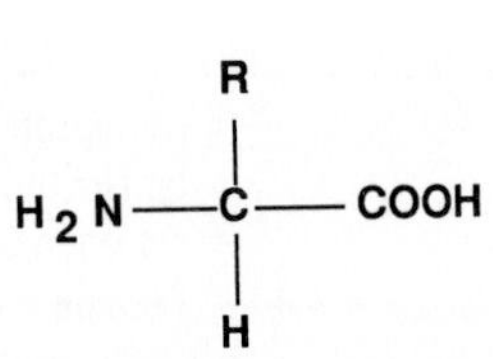

Figure 9-1.
General structure of an α amino acid.

TABLE 9-1. Amino Acids Required in the Synthesis of Proteins

Amino Acid	R
Glycine	—H
Alanine	—CH$_3$
Valine*	—CH—CH$_3$ (with —CH$_3$ above)
Leucine*	—CH$_2$—CH—CH$_3$ (with CH$_3$ above)
Isoleucine*	—CH—CH$_2$—CH$_3$ (with CH$_3$ above)
Cysteine	—CH$_2$—SH
Methionine*	—CH$_2$—CH$_2$—S—CH$_3$
Tryptophan*	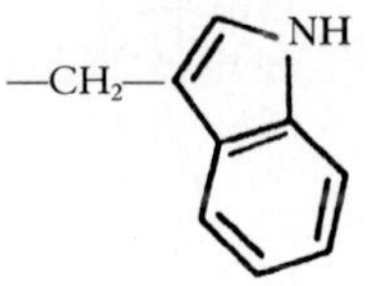
Phenylalanine*	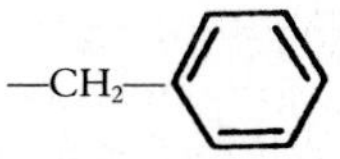
Asparagine	—CH$_2$—C—NH$_2$ (with O double bond)
Glutamine	—CH$_2$—CH$_2$—C—NH$_2$ (with O double bond)
Serine	—CH$_2$ (with OH above)
Threonine*	—CH—CH$_3$ (with OH above)
Tyrosine	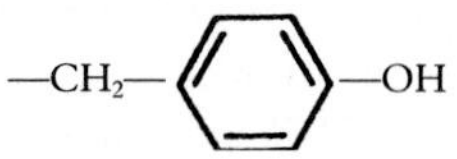 —CH$_2$—⬡—OH
Lysine*	—CH$_2$—CH$_2$—CH$_2$—CH$_2$—NH$_2$
Arginine	—CH$_2$—CH$_2$—CH$_2$—N—C—NH$_2$ (with NH$_2$ above, H below)
Histidine*	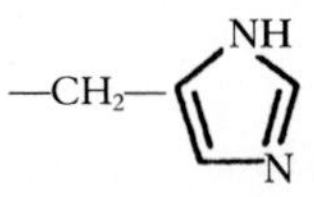
Aspartate	—CH$_2$—COOH
Glutamate	—CH$_2$—CH$_2$—COOH
Proline†	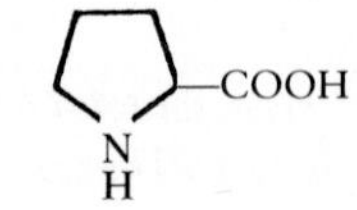 —COOH

The R group is the group attached to the alpha carbon.
*Nutritionally essential.
†Exception to alpha attachment of R group is proline.

The amino acid pool is drawn on primarily for the synthesis of body proteins, including plasma, intracellular, and structural proteins. Amino acids are also utilized for the synthesis of nonprotein nitrogen–containing compounds such as purines, pyrimidines, porphyrins, creatine, histamine, thyroxine, epinephrine, and the coenzyme NAD. In addition, protein provides 12% to 20% of the total daily body energy requirement. The amino group is removed from amino acids by either deamination or transamination. The resultant ketoacid can enter into a common metabolic pathway with carbohydrates and fats. Those amino acids that generate precursors of glucose, *e.g.,* pyruvate or a citric acid cycle intermediate, are referred to as *glucogenic*. Examples include alanine, which can be deaminated to pyruvate; arginine, which is converted to α-ketoglutarate; and aspartate, which is converted to oxaloacetate. Amino acids that are degraded to acetyl CoA or acetoacetyl CoA, such as leucine or lysine, are termed *ketogenic* because they give rise to ketone bodies. Some amino acids can be both ketogenic and glucogenic because some of their carbon atoms emerge in ketone precursors and others appear in potential precursors of glucose. Fig. 9-2 shows the points of entry of the carbon atoms of amino acids into the metabolic pathways of carbohydrates and fats. The ammonium ion that is produced during deamination of the amino acids is converted into urea by the urea cycle in the liver.

Aminoacidopathies

Aminoacidopathies are rare inherited disorders of amino acid metabolism. The abnormalities exist in either the activity of a specific enzyme in the metabolic pathway or in the membrane transport system for amino acids. Over a hundred diseases have been identified that result from inborn errors of amino acid metabolism.

Phenylketonuria

Phenylketonuria (PKU) is inherited as an autosomal recessive trait and occurs in approximately 1 in 14,000 births. Although there are several variants of this disease, the biochemical defect in the classic form of phenylketonuria is a deficiency of the enzyme phenylalanine hydroxylase (also called phenylalanine-4-mono-oxygenase), which catalyzes the conversion of phenylalanine to tyrosine (Fig. 9-3). In the absence of the enzyme, phenylalanine accumulates and is metabolized by an alternate degradative pathway. The catabolites include phenylpyruvic acid, which is the product of deamination of phenylalanine; phenyllactic acid, which is the reduction product of phenylpyruvic acid; phenylacetic acid, which is produced by decarboxylation and oxidation of phenylpyruvic acid; and phenylacetyl glutamine, which is the glutamine conjugate of phenylacetic acid. Although phenylpyruvic acid is the primary metabolite, all these compounds are seen in both the blood and the urine of a phenylketonuric patient, giving the urine a characteristic musty odor.

In infants and children with this inherited defect, retarded mental development occurs as a result of the toxic effects on the brain of phenylpyruvate or one of its metabolic by-products. The deterioration of brain function begins in the second or third week of life. Brain damage can be avoided

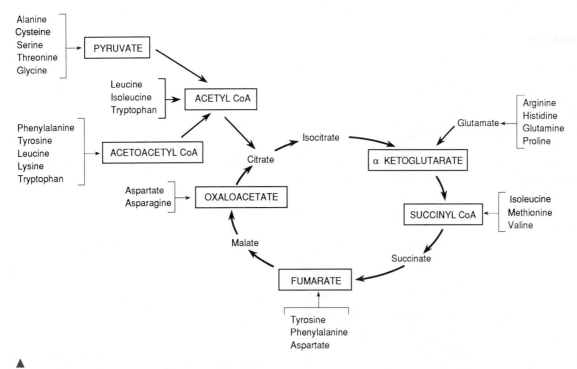

Figure 9-2. Conversion of the carbon skeletons of amino acids into pyruvate, acetyl CoA, or TCA cycle derivatives for further catabolism.

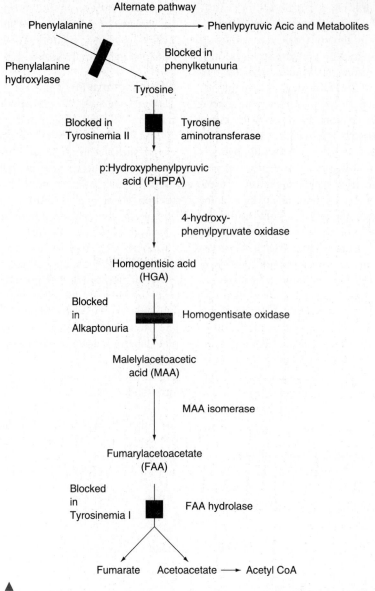

Figure 9-3. Metabolism of phenylanine and tyrosine.

if the disease is detected at birth and the infant is maintained on a diet containing very low levels of phenylalanine. Until recently, once the child reached age 5 or 6, the diet was terminated. This was based on the belief that at that age the brain was no longer vulnerable to damage from hyperphenylalaninemia. However, it has been reported that there is a slight reduction in IQ after discontinuation of the diet. Also, offspring of women with PKU that were untreated during pregnancy were almost always microcephalic and mentally retarded. The fetal effects of maternal PKU are preventable if the mother is maintained on a phenylalanine-restricted diet from before conception through term.[50] For both these reasons, it is now recommended that the PKU patient continue dietary treatment indefinitely.

There are cases of hyperphenylalaninemia, however, that are not responsive to dietary treatment. The defect in these cases is a deficiency in the enzymes needed for the regener-

ation and synthesis of tetrahydrobiopterin (BH_4). BH_4 is a cofactor required for the enzymatic hydroxylation of the aromatic amino acids phenylalanine, tyrosine, and tryptophan. A deficiency of BH_4 results in elevated blood levels of phenylalanine and deficient production of neurotransmitters from tyrosine and tryptophan. Examination of urinary pterins is helpful in diagnosis. While cofactor defects account for only 1% to 3% of all cases of elevated phenylalanine levels, they must be identified so that appropriate treatment can be initiated. Patients must be given the active cofactor along with the neurotransmitter precursors L-DOPA and 5-OH tryptophan.[59]

Today, all states have enacted legislation to enforce screening programs so that therapeutic measures can be taken early to prevent the resulting disability and mortality associated with hyperphenylalaninemia. The screening procedure used routinely is the Guthrie bacterial inhibition assay. In this test, spores of the organism *Bacillus subtilis* are incorporated into an agar plate that contains β-2-thienylalanine, a metabolic antagonist to *B. subtilis* growth. A filter paper disk impregnated with blood from the infant is placed on the agar. If the blood level of phenylalanine exceeds a range of 2 to 4 mg/dL, the phenylalanine counteracts the antagonist and bacterial growth occurs. In order to avoid false-negative results, the infant must be at least 24 hours of age to ensure adequate time for enzyme and amino acid levels to develop. Furthermore, the sample should be taken prior to administration of antibiotics or transfusion of blood or blood products. Premature infants can show false-positive results due to the immaturity of the liver's enzyme systems.

A more recent approach to the screening for PKU involves a microfluorometric assay for the direct measurement of phenylalanine in dried blood filter discs. This method yields quantitative rather than the semi-quantitative results of the Guthrie test, is more adaptable to automation, and is not affected by the presence of antibiotics. The procedure is based on the fluorescence of a complex formed of phenylalanine-ninhydrin-copper in the presence of a dipeptide (*i.e.*, L-leucyl-L-alanine).[60] The test requires pretreatment of the filter paper specimen with trichloroacetic acid (TCA). The extract is then reacted in a microtiter plate with a mixture of ninhydrin, succinate, and leucylalanine in the presence of copper tartrate. The fluorescence of the complex is measured using excitation/emission wavelengths of 360 nm and 530nm, respectively.[106]

Any positive results on the screening tests must be verified. The reference method for quantitative serum phenylalanine is HPLC; however, both fluorometric[60] and enzymatic[86] methods are available. The normal limits for serum phenylalanine levels for full term, normal weight, newborns range from 1.2 to 3.4 mg/dL (70 to 200 μmol/L).

Urine testing for phenylpyruvic acid, while not considered an initial screening procedure, can be used both for diagnosis in questionable cases and for monitoring of dietary therapy. The test, which may be performed by tube or reagent strip test (Phenistix, Miles Diagnostics, Elkhart, IN), involves the reaction of ferric chloride with phenylpyruvic acid in urine to produce a green color.

Prenatal diagnosis and detection of carrier status in families with PKU is now available using DNA analysis. Molecularly, PKU results from multiple independent mutations at the phenylalanine hydroxylase (PAH) locus. Analysis, using cloned human phenylalanine hydroxylase cDNA as a probe, has revealed the presence of numerous restriction fragment length polymorphisms (RFLPs) in the PAH gene. Since the restriction fragment patterns and the disease state have been proven to be tightly linked in PKU families, these polymorphisms can be used for diagnosis.[51]

Tyrosinemia and Related Disorders

A range of familial metabolic disorders of tyrosine catabolism are characterized by excretion of tyrosine and tyrosine catabolites in urine. Normally, the major path of tyrosine metabolism involves the removal of an amine group by tyrosine amino transferase forming p-hydroxyphenylpyruvic acid (PHPPA), which is oxidized to homogentisic acid (HGA). Homogentisic acid is further metabolized in a series of reactions to fumarate and acetoacetate, as shown in Figure 9-3. The defect in inherited tyrosine abnormalities is either a deficiency in tyrosine aminotransferase, resulting in *tyrosinemia II,* or, more commonly, a deficiency in fumarylacetoacetate (FAA) hydrolase, resulting in *tyrosinemia I.* The latter cleaves fumarylacetoacetic acid to fumaric and acetoacetic acids. The absence of these enzymes results in abnormally high levels of tyrosine and in some cases increases in PHPPA and methionine. The elevated tyrosine leads to liver damage, which may be fatal in infancy, or to cirrhosis and liver cancer later in life. The incidence of tyrosinemia I is approximately 1:100,000.

Alkaptonuria

This disorder is of considerable historical interest in that it was one of the original "inborn errors of metabolism" described. At the turn of the century, Archibald Garrod, a pediatrician, recognized that the syndrome which had been called *alkaptonuria* showed a pattern of familial inheritance. He proposed that alkaptonuria and certain other disorders are due to genetic defects each of which results in the lack of activity of a particular metabolic enzyme.[4] Forty-five years later it was confirmed that the biochemical defect in alkaptonuria is a lack of homogentisate oxidase in the tyrosine catabolic pathway (see Fig. 9-3). This disorder occurs in about 1 in 250,000 births. A predominant clinical manifestation of alkaptonuria is the darkening of urine upon standing exposed to the atmosphere. The phenomenon is due to an accumulation in the urine of homogentisic acid (HGA), which oxidizes to produce a dark polymer. Alkaptonuric patients have no immediate problems, but late in the disease the high level of HGA gradually accumulates in connective tissue causing generalized pigmentation of these tissues (ochronosis) and an arthritis-like degeneration.

Maple Syrup Urine Disease

As the name implies, the most striking feature of this hereditary disease is the characteristic maple syrup or burnt sugar odor of the urine, breath, and skin. *Maple syrup urine disease* (MSUD) results from an absence or greatly reduced activity of the enzyme, branched-chain keto acid decarboxylase, thereby blocking the normal metabolism of the three essential branched-chain amino acids leucine, isoleucine, and valine. Specifically, this enzyme is responsible for catalyzing the oxidative decarboxylation of all three branched-chain α-ketoacids to CO_2 and their corresponding acyl-CoA thioesters (Fig. 9-4). The result of this enzyme defect is an accumulation of the branched-chain amino acids and their corresponding ketoacids in the blood, urine, and cerebrospinal fluid (CSF).

If left untreated, the disease causes severe mental retardation, convulsions, acidosis, and hypoglycemia. In the classic form of the disease, death usually occurs during the first year; however, intermediate forms have been reported.

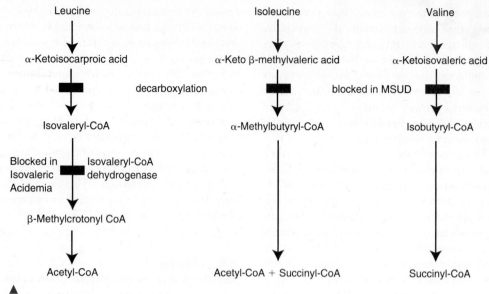

Figure 9-4. Metabolism of the branched-chain amino acids.

In these less severe variants, the activity of the decarboxylase is approximately 25% of normal. While this still results in a persistent elevation of the branched-chain amino acids, the levels frequently can be controlled by limiting dietary protein intake.[6]

Although the incidence of MSUD is low, occurring in only 1 in 216,000 births, early diagnosis is important in order to begin dietary treatment. Therefore, testing for MSUD is included in many of the metabolic screens required by law in certain states. A modified Guthrie test is commonly utilized for this neonatal screening. The metabolic inhibitor to *B. subtilis,* included in the growth media, is 4-azaleucine. In a positive test for MSUD, an elevated level of leucine from a filter paper disc impregnated with the infant's blood will overcome the inhibitor, and bacterial growth occurs.[25] Alternately, a microfluorometric assay for branched chain amino acids, utilizing leucine dehydrogenase (EC 1.4.1.9), can be used for mass screening. In the procedure, the filter paper specimen is treated with a solvent mixture of methanol and acetone to denature the hemoglobin. Leucine dehydrogenase is added to an aliquot of this sample extract. The fluorescence of the NADH produced in the subsequent reaction is measured at 450 nm using an excitation wavelength of 360 nm.[106]

A confirmed diagnosis is based on finding increased plasma and urinary levels of the three branched-chain amino acids and their ketoacids. The presence of alloisoleucine, an unusual metabolite of isoleucine, is characteristic. MSUD can be diagnosed prenatally by measuring the decarboxylase enzyme concentration in cells cultured from amniotic fluid.

Isovaleric Acidemia

Isovaleric acidemia results from a deficiency of the enzyme isovaleryl-CoA dehydrogenase in the degradative pathway of leucine (see Fig. 9-4). The resultant elevation of the glycine conjugate of isovaleric acid, isovalerylglycine, pro-

duces a characteristic "sweaty feet" odor. The abnormal organic acid levels can be identified by chromatography.

Homocystinuria

Homocysteine is an intermediate amino acid in the synthesis of cysteine from methionine. The synthesis of cysteine from methionine is outlined in Fig. 9-5. The usual cause of the hereditary disease, homocystinuria, is an impaired activity of the enzyme cystathionine β-synthase, which results in elevated

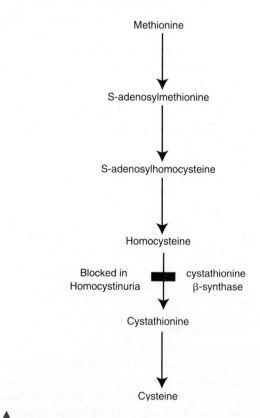

Figure 9-5. Cysteine synthesis.

plasma and urine levels of the precursors homocysteine and methionine. Newborns show no abnormalities, but gradually, physical defects develop. Associated clinical findings in late childhood include thrombosis due to toxicity of homocysteine to the vascular endothelium, osteoporosis, dislocated lenses in the eye due to the lack of cysteine synthesis essential for collagen formation, and frequently, mental retardation.

The enzyme cystathionine β-synthetase requires vitamin B_6 (pyridoxine) as its cofactor. Slightly different genetic defects lead to two forms of the disease: a vitamin B_6-responsive form, in which treatment consists of therapeutic doses of vitamin B_6, and a vitamin B_6-unresponsive form, in which the treatment is a diet low in methionine and high in cysteine.

The incidence of homocystinuria is approximately 1 in 200,000 births. Neonatal screening can be accomplished with a Guthrie test using L-methionine sulfoximine as the metabolic inhibitor. Increased plasma methionine levels from affected infants will result in bacterial growth. Elevations in urinary homocysteine can be detected by the presence of a red-purple color in the cyanide-nitroprusside spot test. Since cysteine also produces a positive result, the presence of homocysteine must be confirmed with a silver-nitroprusside test. Silver nitrate will reduce homocysteine but not cysteine, thus allowing only homocysteine to react with the nitroprusside and produce a reddish color.

Cystinuria

This inherited aminoacidopathy is caused by a defect in the amino acid transport system rather than a metabolic enzyme deficiency. Normally, amino acids are freely filtered by the glomerulus and then actively reabsorbed in the proximal renal tubules. In cystinuria, there is a 20- to 30-fold increase in the urinary excretion of cysteine due to a genetic defect in the renal resorptive mechanism. The transport mechanism is not specific for cysteine. Excretion of the other diamino acids, lysine, arginine, and ornithine, is also markedly elevated due to deficient resorption.

Of the four, cysteine is the amino acid that causes complications of the disease. Because cysteine is relatively insoluble, when it reaches these high levels in the urine it tends to precipitate in the kidney tubules and form urinary calculi. The formation of cysteine calculi can be minimized by a high fluid intake and alkalinizing the urine, which makes cysteine relatively more soluble. If this does not succeed, treatment with regular doses of penicillamine can be initiated.[90]

Cystinuria can be diagnosed by testing the urine for cysteine using cyanide-nitroprusside, which produces a red-purple color upon reaction with sulfhydryl groups. False-positive results due to homocysteine must be ruled out.

Amino Acid Analysis

Blood samples for amino acid analysis should be drawn after at least a 6 to 8 hour fast to avoid the effect of absorbed amino acids originating from dietary proteins. The sample is collected in heparin, and the plasma is promptly removed from the cells, taking care not to aspirate the layer of platelets and leukocytes. If this step is not carried out with caution, the plasma becomes contaminated with platelet or leukocyte amino acids, in which the contents of aspartic acid and glutamic acid, for example, are about 100-fold higher than those in plasma. Hemolysis should be avoided for the same reason. Deproteinization should be carried out within 30 minutes of sample collection, and analysis should be performed immediately or the sample should be stored at −20 to −40°C.

Urinary amino acid analysis can be performed on a random specimen for screening purposes, but for quantitation, a 24-hour urine preserved with thymol or organic solvents is required. Amniotic fluid also may be analyzed.

For preliminary screening, the method of choice is thin-layer chromatography. The application of either one- or two-dimensional separations depends on the purpose of the analysis. If one is searching for a particular category of amino acids, such as branched-chain amino acids, or even a single amino acid, usually one-dimensional separations are sufficient. Separation conditions can be selected in such a way as to offer good resolution of the amino acid in question.

For more general screening, two-dimensional maps are essential. In two-dimensional chromatography, the amino acids are allowed to migrate along one solvent front, and then the chromatogram is rotated 90° and a second solvent migration takes place. A variety of solvents have been used, including butanol-acetic acid-water and ethanol-ammonia-water mixtures. The chromatogram is visualized by staining with ninhydrin, which gives a blue color with most amino acids.

When the possibility of an amino acid disorder has been indicated in the preliminary steps, amino acids can be separated and quantitated by cation-exchange chromatography using a gradient buffer elution.[19] The other choice is to use an HPLC reversed-phase system equipped with fluorescence detection.[18]

PROTEINS

General Characteristics

Proteins are an essential class of compounds comprising 50% to 70% of the cell's dry weight. Proteins are found in all cells of the body, as well as in all fluids, secretions, and excretions.

Molecular Size

Biologically active proteins are macromolecules that range in molecular weight from approximately 6000 for insulin to several million for some structural proteins.

Structure

All proteins are composed of covalently linked polymers of amino acids. The amino acids are linked in a head-to-tail fashion; in other words, the carboxyl group of one amino acid combines with the amino group of another amino acid (Fig. 9-6). During this reaction, a water molecule is removed and the bond that is created is called a *peptide bond*. The amino acid that has the amino group free is the *N-terminal end*, and the amino acid that has the carboxyl group free is the *C-terminal end*. When two amino acids are joined together, the molecule is called a *dipeptide*; three amino acids are called a *tripeptide*; and four together are a *tetrapeptide*. As

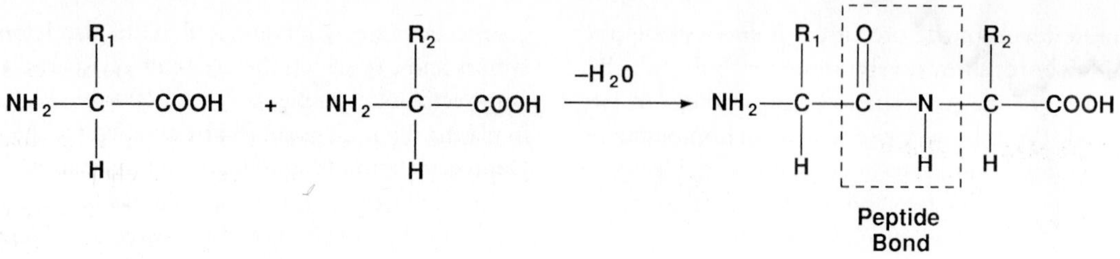

Figure 9-6. Formation of a dipeptide.

the chain increases further, it is called a *polypeptide*. In human serum, proteins average about 100 to 150 amino acids in the polypeptide chains.

The conformation of a protein is determined by interaction between a polypeptide and its aqueous environment in which the polypeptide attains a stable three-dimensional structure. There are four aspects of the structure of a protein that relate to its shape (conformation). The number and kinds of amino acids, as well as their sequence in the polypeptide chain, constitute the primary structure of a protein. The covalent peptide linkage is the only type of bonding involved at this level. The primary structure is crucial for the function and molecular characteristics of the protein. Any change in the amino acid composition can significantly alter the protein. For example, when the amino acid valine is substituted for glutamic acid in the β chain of hemoglobin A, hemoglobin S is formed, which results in sickle-cell anemia.

The secondary structure is the winding of the polypeptide chain. The usual pattern that is formed by the globular proteins (most serum proteins) is an α helix requiring 3.6 amino acids to make one turn in the coil. A β-pleated sheet or a stable but irregular pattern also can be seen. The secondary structure is maintained by hydrogen bonds between the NH and CO groups of the peptide bonds either within the same chain or between different chains within the same molecule. Figure 9-7 schematically pictures the secondary structure.

The next level of structure, the tertiary structure, refers to the way in which the twisted chain folds back upon itself to form the three-dimensional structure. The specific convolutions a polypeptide undergoes are determined by the in-

teraction of the R groups in the molecule. The reactions of the R groups include disulfide linkages, electrostatic attractions, hydrogen bonds, hydrophobic interactions, and van der Waals forces. The tertiary structure is responsible for many of the physical and chemical properties of the protein.

In addition to these first three levels that can exist in molecules composed of a single polypeptide chain, some proteins display a fourth level of organization called the *quaternary structure*. Quaternary structure is the arrangement of two or more polypeptide chains to form a functional protein molecule. Albumin, which is composed of a single polypeptide chain, would thus have no quaternary structure. On the other hand, hemoglobin is composed of four globin chains, lactate dehydrogenase consists of five peptide chains, and creatine kinase has two chains joined together. The polypeptide chains are united by non-covalent attractions such as hydrogen bonds and electrostatic interactions.

When the secondary, tertiary, or quaternary structure of a protein is disturbed, the protein may lose its functional and molecular characteristics. This loss of its native or original character is called *denaturation*. Denaturation can be caused by heat, hydrolysis by strong acid or alkali, enzymatic action, exposure to urea or other substances, or exposure to ultraviolet light.

Nitrogen Content

Proteins are composed of the elements C, O, H, N, and S. It is the nitrogen content that sets proteins apart from pure carbohydrates and lipids, which contain no nitrogen atoms. The nitrogen content of serum protein varies somewhat, but

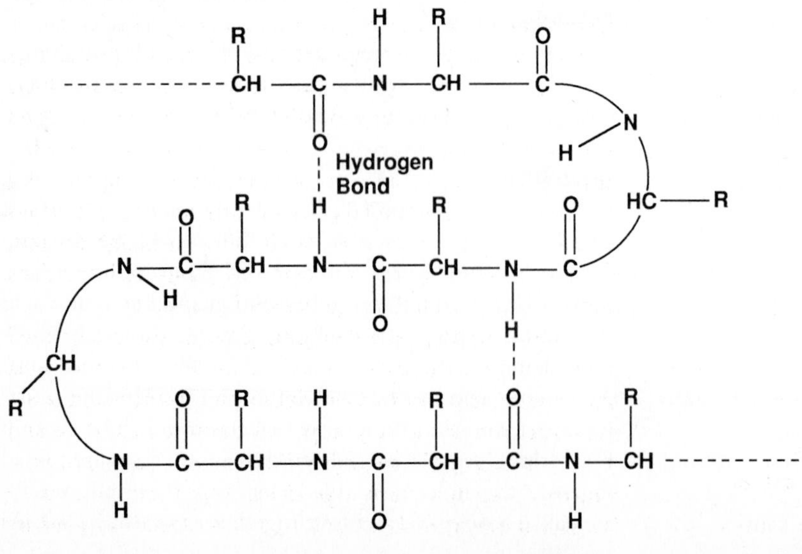

Figure 9-7. Secondary structure of proteins.

the average is approximately 16%. This characteristic is used in one method of total protein measurement.

Charge and Isoelectric Point

Because of their amino acid composition, proteins can bear positive and negative charges (*i.e.,* they are *amphoteric*). The

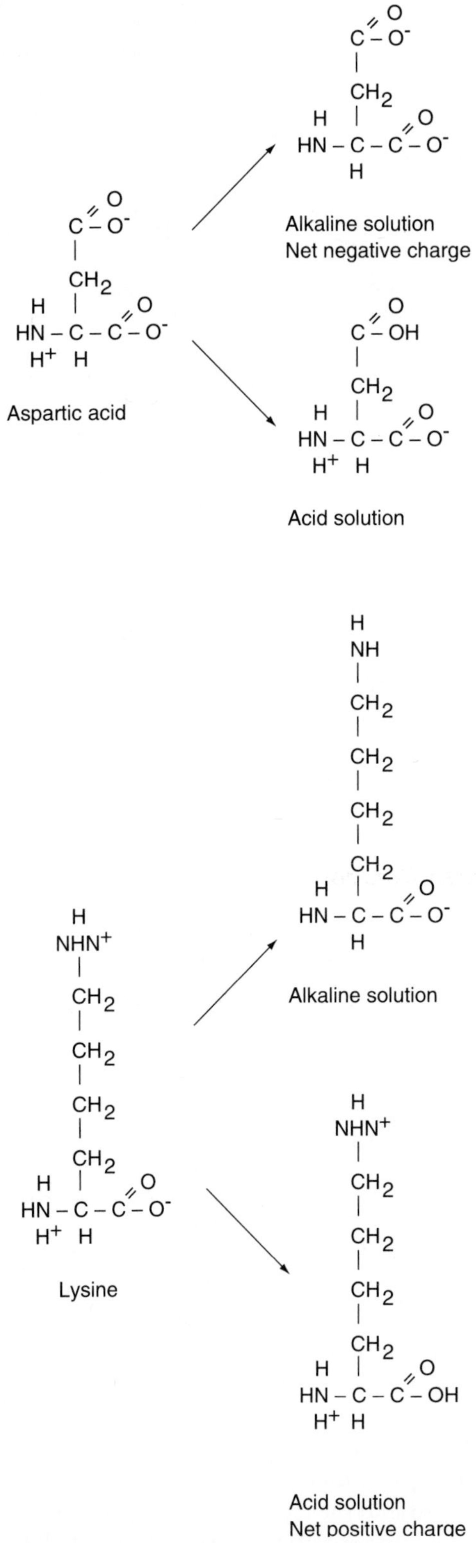

▲
Figure 9-8. Charged states of amino acids.

reactive acid or basic groups that are not involved in the peptide linkage can exist in three charged forms, depending on the pH of the surrounding environment. Aspartic acid and lysine are examples of amino acids that have free carboxyl or amino groups, respectively (see Table 9-1). Figure 9-8 shows these two amino acids and the charges they would have in solutions that are alkaline or acid.

At a pH of 2.98, L-aspartic acid has no net charge (equal or no ionization of the amino and carboxyl groups), whereas L-lysine has no net charge at a pH of 9.47. The pH at which an amino acid or protein has no net charge is known as its *isoelectric point* (pI). In other words, for a protein composed of several amino acids, the isoelectric point is the point at which the number of positively charged groups equals the number of negatively charged groups. If a protein is placed in a solution that has a pH greater than the pI, the protein will be negatively charged, whereas at a pH less than the pI, the protein will be positively charged. Since proteins differ in the number and type of constituent amino acids, they also differ in their pI values. For example, albumin has an isoelectric point of 4.9. The isoelectric points of some serum proteins are given in Table 9-2. Owing to the different pI values, proteins will carry different net charges at any given pH. This difference in magnitude of the charge is the basis for several procedures for separating and quantitating proteins such as electrophoresis. The procedure will be discussed later in this chapter.

Solubility

Proteins in aqueous solution swell and enclose water. Natelson and Natelson[65] advise that, for this reason, it is necessary when reconstituting lyophilized serum to mix the reconstituted vial gently for at least 30 minutes to complete the swelling process. Protein solutions are *colloidal emulsoids* or *micelles* because they are charged and because each molecule of protein has an envelope of water around it.

The solubility of proteins is promoted by a high dielectric character of the solvent and a high concentration of free water. Thus, low concentrations of salt (0.1 M) have been used to *salt in* globulins into solution. When the dielectric nature or the amount of free water is decreased, the charges on the protein promote aggregation. In a high salt concentration ($\sim$2 M), ionic salts compete with protein for water and thus decrease the amount of water available for hydration of the protein. The resultant precipitation of the globulins is called *salting-out.* By using a range of salt concentrations, a wide array of specific proteins can be separated from a mixture. Albumin remains in solution even in high salt concentrations because it has a higher dipole and holds water tighter, and, because of its smaller size in relation to the globulins, it does not require as much water to keep it hydrated.

Water-miscible, neutral organic solvents also can be used to separate proteins based on their solubility. These solvents have a dielectric constant less than water, which suppresses the ionization of the R groups on the protein surface and thereby decreases the solubility. The use of alcohol concentrations to separate serum proteins into fractions bearing Roman numeral designations was developed and used by

TABLE 9-2. Characteristics of Selected Plasma Proteins

	Reference Value (Adult, g/L)	Molecular Mass (D)	Isoelectric Point, pI	Electrophoretic Mobility, pH 8.6, I = 0.1	Comments
Prealbumin	0.1–0.4	55,000	4.7	7.6	Indicator of nutrition. Binds thyroid hormones and retinol binding protein.
Albumin	35–55	66,300	4.9	5.9	Binds bilirubin, steroids, fatty acids; major contributor to oncotic pressure.
α_1–Globulins					
α_1–Antitrypsin	2–4	53,000	4.0	5.4	Acute-phase reactant; protease inhibitor
α_1–Fetoprotein	1×10^5	76,000		6.1	Principle fetal protein
α_1–Acid glycoprotein (orosomucoid)	0.55–1.4	44,000	2.7	5.2	Acute-phase reactant
α_1–Lipoprotein(HDL)	2.5–3.9	200,000		4.4–5.4	Transports lipids
α_1–Antichymotrypsin	0.3–0.6	68,000		Inter α	Inhibits serine proteinases (*i.e.,* chymotrypsin)
Inter-α-Trypsin Inhibitor	0.2–0.7	160,000		Inter α	Inhibits proteinases (*i.e.,* trypsin)
Gc-Globulin	0.2–0.55	59,000		Inter α	Binds Vitamin D and actin
α_2–Globulins					
Haptoglobins					
Type 1-1	1.0–2.2	100,000		4.5	Acute-phase reactant; binds hemoglobin
Type 2-1	1.6–3.0	200,000	4.1	3.5–4.0	Binds hemoglobin
Type 2-2	1.2–1.6	400,000		3.5–4.0	Binds hemoglobin
Ceruloplasmin	0.15–0.60	134,000	4.4	4.6	Peroxidase activity; contains copper
α_2–Macroglobulin	1.5–4.2	725,000	5.4	4.2	Inhibits thrombin, trypsin, pepsin
β–Globulins					
Pre β–lipoproteins (VLDL)	1.5–2.3	250,000		3.4–4.3	Transports lipids (primarily triglyceride)
Transferrin	2.04–3.60	76,000	5.9	3.1	Transports iron
Hemopexin	0.5–1.0	57,000–80,000		3.1	Binds heme
β–Liproproteins (LDL)	2.5–4.4	3,000,000		3.1	Transports lipids (primarily cholesterol)
β_2–Microglobulin (B2M)	0.001–0.002	11,800		β_2	Component of human leucocyte antigen (HLA) molecules class 1
C_4 Complement	0.20–0.65	206,000		0.8–1.4	Immune response
C_3 Complement	0.55–1.80	180,000		0.8–1.4	Immune response
C_1 q Complement	0.15	400,000			Immune response
Fibrinogen	2.0–4.5	341,000	5.8	2.1	Precursor of fibrin clot
C-reactive Protein (CRP)	0.01	118,000	6.2		Acute-phase reactant; motivates phagocytosis in inflammatory disease
γ–Globulins					
Immunoglobulin G	8.0–12.0	150,000	5.8–7.3	0.5–2.6	Antibodies
Immunoglobulin A	0.7–3.12	180,000		2.1	Antibodies (in secretions)
Immunoglobulin M	0.5–2.80	900,000		2.1	Antibodies (early response)
Immunoglobulin D	0.005–0.2	170,000		1.9	Antibodies
Immunoglobulin E	6×10^4	190,000		2.3	Antibodies (reagins, allergy)

E. J. Cohn[16] during World War II, and human albumin, fraction V (Cohn), continues to be commercially available.

Immunogenicity

Most serum proteins are very effective antigens because of their molecular mass, their content of tyrosine, and their specificity by species. When injected into another species (*i.e.,* rabbit, goat, or chicken), human serum proteins elicit the formation of antibodies specific for each of the proteins present in the serum. When it became possible to obtain these antibodies for each of the serum proteins, methods using antigen–antibody reactions that are very specific were introduced into the clinical laboratory. Some of the methods commonly used are described later in this chapter.

Synthesis

Most of the plasma proteins are synthesized in the liver and secreted by the hepatocyte into the circulation. The immunoglobulins are exceptions because they are synthesized in plasma cells.

The amino acids of a polypeptide chain are placed in a sequence that is determined by a corresponding sequence of bases (guanine, cytosine, adenine, and thymine) in the deoxyribonucleic acid (DNA) that constitutes the appropriate gene. The double-stranded DNA unfolds in the nucleus, and one strand is used as a template for the formation of a complimentary strand of messenger RNA (mRNA). The mRNA, now carrying the genetic code from the DNA, moves to the cytoplasm, where it attaches to ribosomes. A sequence of three bases (*codon*) on the mRNA is required to specify the amino acid to be transcribed. The code on the mRNA also contains initiation and termination codons for the peptide chain.

The next step in protein synthesis is getting the amino acid to the ribosomes. First, the amino acid is activated in a reaction that requires energy and a specific enzyme for each amino acid. This activated amino acid complex is then attached to another kind of RNA, transfer RNA (tRNA), with the subsequent release of the activating enzyme and AMP. Transfer RNA is a short chain of RNA that occurs free in the cytoplasm. Each amino acid has a specific tRNA that contains three bases that correspond to the three bases in the mRNA. The tRNA carries its particular amino acid to the ribosome and attaches to the mRNA in accordance with the matching codon. In this manner the amino acids are aligned in sequence. As each new tRNA brings in the next amino acid, the preceding amino acid is transferred onto the amino group of the new amino acid and a peptide bond is formed by enzymes located in the ribosomes. The tRNA is released into the cytoplasm, where it can pick up another amino acid, and the cycle repeats itself. When the terminal codon is reached, the peptide chain is detached and the ribosome and mRNA dissociate. Figure 9-9 outlines protein synthesis. Intracellular proteins are generally synthesized on free ribosomes, whereas proteins made by the liver for secretion are made on ribosomes attached to the rough endoplasmic reticulum. Protein synthesis occurs at the rate of approximately 2 to 6 peptide bonds per second. Synthesis is controlled at two steps: selection of genes for transcription to mRNA and selection of mRNA for translation into proteins. Certain hormones are involved in controlling protein synthesis. Thyroxine, growth hormone, insulin, and testosterone promote synthesis, whereas glucagon and cortisol have a catabolic effect.

Proteins that are excreted after synthesis are thought to have a short sequence of amino acids, a secretory piece, that attracts the protein to the Golgi apparatus for the consequent secretion by exocytosis. The secretory piece is removed during the secretion process. Secreted proteins thus usually exist as precursor proteins within the cells from which they arise.

Catabolism and Nitrogen Balance

Most proteins in the body are constantly being repetitively synthesized and then degraded. It has been demonstrated that over a wide range of rates of synthesis, protein catabolism and synthesis are equal. Normally, this turnover totals about 125 to 220 g of protein each day. The rate of protein turnover, however, varies widely for individual proteins. For example, the plasma proteins and most intracellular proteins are rapidly degraded, having half-lives of hours or days, while some of the structural proteins such as collagen are metabolically stable and have half-lives of years.[12] The disintegration of protein takes place in the digestive tract, in the kidneys, and, particularly, in the liver. During catabolism, proteins are hydrolyzed to their constituent amino acids. The amino acids are deaminated, producing ammonia and ketoacids. The ammonia is converted to urea by the hepatocytes and is excreted in the urine. The ketoacids are oxidized by means of the citric acid cycle and converted to glucose or fat.

Ordinarily, a balance exists between protein anabolism (synthesis) and catabolism. In nutritional terms, this is called the *nitrogen balance*. When protein catabolism exceeds protein anabolism, the nitrogen excreted exceeds that ingested. This is "negative" nitrogen balance, and it occurs in conditions where there is excessive tissue destruction, such as burns, wasting diseases, continual high fevers, or starvation. The converse, "positive" nitrogen balance, is seen when anabolism is greater than catabolism. A positive balance is found during growth, pregnancy, and repair processes.

Classification

Proteins are generally classified into two major groups based on composition: simple and conjugated.

Simple Proteins

Simple proteins contain peptide chains which, upon hydrolysis, yield only amino acids. Simple proteins may be globular or fibrous in shape. Globular proteins are relatively symmetrical, with compactly folded and coiled polypeptide chains. Albumin is an example of a globular protein. Fibrous proteins are more elongated and asymmetrical and have a higher viscosity. Collagen and keratin are examples of fibrous proteins.

Conjugated Proteins

Conjugated proteins are composed of a protein (apoprotein) and a nonprotein moiety (prosthetic group). The prosthetic group may be lipid, carbohydrate, porphyrins, metals, etc. These groups impart certain characteristics to the proteins. Most of the names given to these conjugated proteins are self-descriptive. The *metalloproteins* have a metal ion attached to the protein, either directly, as in ferritin (which contains iron) and ceruloplasmin (which contains copper), or as complex metals (metal plus another prosthetic group), such as hemoglobin and flavoproteins. The metal (iron) in hemoglobin is first attached to protoporphyrin, which is then attached to the globin chains; in flavoprotein, the metal is first attached to FMN or FAD.

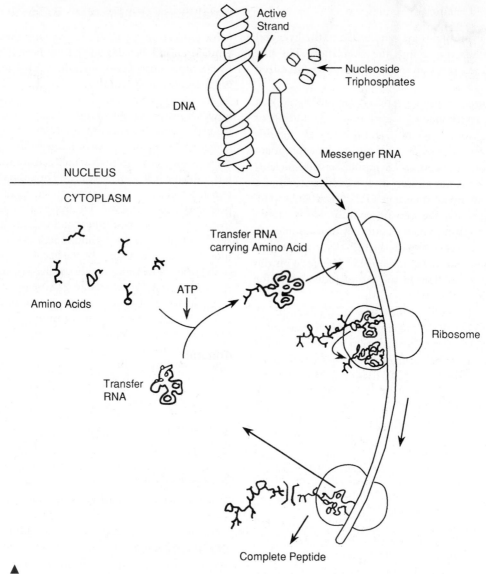

Figure 9-9. Schematic summary of protein synthesis.

When lipids such as cholesterol and triglyceride are linked to proteins, the molecules are called *lipoproteins*. When carbohydrates are joined to proteins, several terms may be used to describe the results. Generally, those molecules with 10% to 40% carbohydrate are called *glycoproteins*.[65] Examples of glycoproteins are haptoglobin and α_1-antitrypsin. When the percentage of carbohydrate linked to protein is higher, the proteins are often called *mucoproteins* or *proteoglycans* when heparan, keratan, or chondroitin sulfate are also present. An example of a mucoprotein is mucin, a compound that lubricates organ linings. *Nucleoproteins* are those proteins that are combined with nucleic acids (DNA or RNA). An example is chromatin.

General Function of Proteins

The broad array of molecules that make up proteins may function in general ways or, as individual molecules, may have some unique function. Plasma proteins and tissue proteins share the same amino acid pool, and thus alterations in one group will eventually affect the other. This is why the plasma proteins are important in tissue nutrition. The plasma proteins can be hydrolyzed to amino acids, which can be used for the production of energy by means of the citric acid cycle.

Another general function of plasma proteins is the distribution of water among the compartments of the body. Starling's law attributes a colloid osmotic force to plasma proteins, which, because of their size, cannot cross the capillary membranes. This osmotic force results in the absorption of water from the tissue into the venous end of the capillaries. When the concentration of plasma proteins is markedly decreased, the concomitant decrease in the plasma colloidal osmotic (oncotic) pressure results in increased levels of interstitial fluid and edema. This is often seen in renal disease when proteinuria results in a decreased plasma protein concentration and swelling of the face, hands, and feet occurs.

The amphoteric nature of proteins provides the mechanism for their participation as buffers within the plasma and interstitial tissues. The ionizable R groups can either bind or release excess hydrogen ions as needed.

Many plasma proteins function as specific transporters of metabolic substances. Examples include thyroxine-binding globulin, which carries thyroxine; haptoglobin, which binds free hemoglobin; and albumin, which transports free fatty acids, unconjugated bilirubin, calcium, sulfa drugs, and many other endogenous and exogenous compounds. In addition to transporting the molecules to the location where they can be used, proteins, through the binding process, serve to maintain the bound substance in a soluble state, provide a storage site for excess substance (transcobalamin II), and prevent the loss of small molecular mass compounds through the kidney (*i.e.,* transferrin in conserving iron).

Several proteins are glycoproteins. A major function of glycoproteins is in distinguishing which cells are native and which are foreign to the body. These are notably evident as histocompatibility antigens and erythrocyte blood groups. Antigens can stimulate the synthesis of antibodies, which are also proteins. Antibodies and components of the complement system help protect the body against infection.

Many cellular proteins act as receptors for hormones. The receptor binds to its specific hormone and allows the hormonal message to be transmitted to the cell. In addition, certain hormones (*e.g.,* growth hormone and ACTH) are themselves proteins.

Proteins also serve a structural role. Collagen is the most abundant protein in mammals, constituting a quarter of their total weight. Collagen is the major fibrous element of skin, bone, tendon, cartilage, blood vessels, and teeth. Elastin and proteoglycans are two other connective-tissue proteins.

Another important biological property of some proteins (enzymes) is their ability to catalyze biochemical reactions. Other proteins (clotting factors) aid in the maintenance of hemostasis. The vast array of functions attributed to various proteins is summarized in Table 9-3.

Plasma Proteins

While the proteins in all fluid compartments play a major physiological function, the plasma proteins are the most frequently analyzed. Over 500 plasma proteins have been identified. The properties of a few selected plasma proteins are discussed in the following section.

Prealbumin

Prealbumin is so named because it migrates ahead of albumin in the customary electrophoresis of serum or plasma proteins. The molecular characteristics of prealbumin can be seen in Table 9-2. Rarely is it observed as a distinct band on routine cellulose acetate electrophoretic patterns of serum, although it can be exhibited by high resolution electrophoresis or immunoelectrophoresis. It is rich in tryptophan and contains 0.5% carbohydrate. Prealbumin combines with thyroxine and triiodothyronine to serve as the transport mechanism for these thyroid hormones. Prealbumin also binds with retinol-binding protein to form a complex that transports retinol (vitamin A). Prealbumin is decreased in hepatic damage, acute phase inflammatory response, and tissue necrosis. A low prealbumin level is also a sensitive marker of poor protein nutritional status because of its short half-life (approximately 8 hours). Prealbumin is increased in patients receiving steroids. in alcoholism, and in chronic renal failure.

Albumin

Albumin is the protein present in highest concentration in the serum (see Table 9-2). It is synthesized in the liver. The earliest method for its determination involved the salting-out of the globulins with sodium sulfate, leaving the albumin in solution. The albumin was then determined by the Kjeldahl method and, later, by the biuret color development. The method commonly used today is one involving dye binding and the shift in color when a dye is bound by albumin. When more information about proteins is needed, an electrophoretic pattern is obtained, and the albumin is calculated as a percentage of the total protein (usually near 60%). At birth the reference value for serum albumin averages 39 g/L. The concentration falls to 28.4 g/L at about 9 months and then begins to increase slowly until adult values of 35 to 55 g/L are reached.[36] Serum albumin level after 60 years of age averages 38.3 g/L.[34]

Albumin has two well-known functions. One is the contribution albumin makes to the colloid osmotic pressure of the intravascular fluid. Due to its high concentration, albumin contributes nearly 80% of this pressure, which maintains the appropriate fluid in the tissues. The other prime function is its propensity to bind various substances in the blood. For example, albumin binds bilirubin, salicylic acid, fatty acids, calcium and magnesium ions, cortisol, and some drugs. This characteristic is also exhibited with certain dyes, providing a method for the quantitation of albumin.

Decreased concentrations of serum albumin may be caused by the following:

An inadequate source of amino acids, which is seen in malnutrition and muscle-wasting diseases.
Liver disease resulting in the inability of hepatocytes to synthesize albumin. The increase in globulins that occurs in early cirrhosis, however, will balance the loss in albumin

TABLE 9-3. Functions of Proteins

Tissue nutrition
Maintenance of water distribution between cells and tissues, interstitial compartments, and the vascular system of the body
Participation as buffers to maintain pH
Transportation of metabolic substances
Part of defense system (antibodies)
Hormones and receptors
Connective-tissue structure
Biocatalysts (enzymes)
Participation in the hemostasis and coagulation of blood

to give a total protein concentration within acceptable limits. The decline in serum albumin is insignificant in infectious hepatitis.

Gastrointestinal loss as interstitial fluid leaks out in inflammation and disease of the intestinal mucosa.

Loss in the urine in renal disease. Albumin is normally excreted in very small amounts. This excretion is increased when the glomerulus no longer functions to restrict the passage of proteins from the blood.

Abnormalities in serum albumin are also exhibited by the absence of albumin (*analbuminemia*), or the presence of albumin that has unusual molecular characteristics. This latter abnormality is called *bisalbuminemia* and is demonstrated by the presence of two albumin bands instead of the single band usually seen by electrophoresis. Analbuminemia is an abnormality of genetic origin resulting from an autosomal recessive trait. Increased serum albumin levels are seen in dehydration.

Globulins

The globulin group of proteins consist of α_1, α_2, β, and γ fractions. Each fraction consists of a number of different proteins with different functions. The following subsections describe selected examples of the globulins.

α_1-Antitrypsin. α_1-Antitrypsin is an acute-phase reactant. Its main function is to neutralize trypsin-like enzymes (*i.e.,* elastase) that can cause hydrolytic damage to structural protein. α_1-Antitrypsin is a major component (approximately 90%) of the fraction of serum proteins that migrates electrophoretically immediately following albumin. Molecular characteristics are seen in Table 9-2.

A deficiency of α_1-antitrypsin is associated with severe, degenerative, emphysematous pulmonary disease. The lung disease is attributed to the unchecked proteolytic activity of proteases from leukocytes in the lung during periods of inflammation. Juvenile hepatic cirrhosis is also a correlative disease in α_1-antitrypsin deficiency. The protein is synthesized but not released from the hepatocyte.

Several phenotypes of α_1-antitrypsin deficiency have been identified. The most common phenotype is *MM* (allele *Pi^M*) and is associated with normal antitrypsin activity. Other alleles are *Pi^S*, *Pi^Z*, *Pi^F*, and *Pi^-* (null). The homozygous phenotype *ZZ* individual is in serious jeopardy of liver and lung disease from a deficiency of α_1-antitrypsin. Individuals with the *MZ* or *MS* phenotype are usually not affected but should be counseled about having offspring who may be *ZZ*, *Z^-* and in danger. The *Pi^Z* allele occurs in 1 in 1500 Caucasians. Factors other than α_1-antitrypsin are also involved in disease, since some persons who have abnormal phenotypes and low concentrations of proteins do not develop overt disease. Increased levels of α_1-antitrypsin are seen in inflammatory reactions, pregnancy, and contraceptive use.

The discovery of abnormal α_1-antitrypsin levels is most often made by the lack of an α_1-globulin band on protein electrophoresis. This discovery is followed with one of the quantitative methods. A widely used method is radial immunodiffusion. Immunonephelometric assays by automated

instrumentation are also available. Phenotyping can be accomplished by immunofixation.

α_1-Fetoprotein (AFP). α_1-Fetoprotein (AFP) is synthesized initially by the fetal yolk sac and then by the parenchymal cells of the liver.[1] In 1956, it was first discovered, in fetal serum, to have an electrophoretic mobility between that of albumin and α_1-globulin. It peaks in the fetus at about 13 weeks' gestation (3 mg/mL) and recedes at 34 weeks' gestation.[43] At birth, it recedes rapidly to adult concentrations, which are normally very low (see Table 9-2).

The methods commonly used for AFP determinations are hemagglutination, radial immunodiffusion, and radioimmunoassay. Radioimmunoassay has been used for AFP determination because of the sensitivity of the method, which can detect the very low concentrations.

The function of AFP is not well established. It has been proposed that the protein protects the fetus from immunolytic attack by its mother.[43] AFP is detectable in the maternal blood up to the seventh or eighth month of pregnancy because it is transmitted across the placenta. Thus measurement of the level of AFP in maternal serum is a screening test for any fetal conditions in which there is increased passage of fetal proteins into the amniotic fluid. Conditions associated with an elevated AFP level include spina bifida and neural tube defects, atresia of the gastrointestinal tract, and fetal distress in general.[85] Its use in determining neural tube defects prior to term is an important reason for its assay. It is also increased in ataxia-telangiectasia, in tyrosinosis, and in hemolytic disease of the newborn. It is interesting to note that maternal serum AFP is also increased in the presence of twins. Low levels of maternal AFP indicate a three- to fourfold increased risk for Down's syndrome.[62]

The normal time for screening is between 15 to 20 weeks' gestational age. The maternal AFP increases gradually during this period; thus, interpretation requires accurate dating of the pregnancy. Maternal AFP levels are also affected by maternal weight, which reflects blood volume (inverse relationship), race (10% higher in African-Americans), and diabetes (lowered value). Thus, test results need to be adjusted for these variables. It has been found that reporting AFP in multiples of the median (MoM) leads to good correlation with infant risk. MoM are calculated by dividing the patient's AFP value by the median reference value for that gestational age. Most screening programs use 2.0 MoM as the upper limit, and 0.5 MoM as the lower limit for maternal serum AFP.

Serum levels of AFP can also be used as a tumor marker. Very high concentrations of AFP are found in many cases of hepatocellular carcinoma (approximately 80%) and certain gonadal tumors in adults[66] (see Chapter 24, Tumor Markers).

α_1-Acid Glycoprotein (Orosomucoid). α_1-Acid glycoprotein is composed of five carbohydrate units attached to a polypeptide chain. Owing to its low p*I* (2.7), it is negatively charged even in acid solutions, a fact that led to its name. Other molecular characteristics are seen in Table 9-2. Sequences of haptoglobin, immunoglobulins, and α_1-acid glycoprotein show considerable homology. This has led to speculation that these proteins share a precursor in the evolutionary past. α_1-Acid glycoprotein is implicated in the formation of certain membranes and fibers in association with collagen and may inactivate basic hormones such as progesterone.

The analytical method used most commonly for the determination of orosomucoid is radial immunodiffusion. Immunofixation has been used to study inherited variants.[39]

Increased concentration of this protein is the major cause of an increased glycoprotein level in the serum during inflammation. It is also increased in pregnancy, cancer, pneumonia, rheumatoid arthritis, and other conditions associated with cell proliferation.

α_1-Antichymotrypsin (α_1-ACT). α_1-antichymotrypsin is a serine proteinase with cathepsin G, pancreatic elastase, mast cell chymase, and chymotrypsin as target enzymes.[82] α_1-antichymotrypsin carries four oligosaccharide side chains. It migrates between the α_1 and α_2 zones on high resolution serum protein electrophoresis. Elevations are seen in inflammation, and it appears that the complex formed between plasma α_1-ACT and its target enzymes plays a major role in signaling for acute phase protein synthesis in response to injury. Hereditary deficiency of α_1-ACT is associated with asthma and liver disease.

Inter-α-Trypsin Inhibitor (ITI). Inter-α-trypsin inhibitor is composed of at least three distinct polypeptide subunits: two heavy chains designated H_1 and H_2 and a light chain designated L or *bikunin*. The light chain is responsible for the inhibition of the proteases trypsin, plasmin, and chymotrypsin.[5] On high resolution serum electrophoresis, ITI migrates in the zone between the α_1 and α_2 fractions. Elevations are seen in inflammatory disorders.

Gc-globulin (Group specific component; Vitamin D-binding protein). Gc-globulin is another protein that migrates in the α_1-α_2 interzone. This protein exhibits a high binding affinity for Vitamin D compounds and actin, the major constituent of the thin filaments of muscle. Due to genetic polymorphism, several phenotypes of Gc-globulin exist. Elevations of Gc-globulin are seen in the third trimester of pregnancy and in patients taking estrogen oral contraceptives. Severe liver disease and protein-losing syndromes are associated with low levels.

Haptoglobin. Haptoglobin (an α_2 glycoprotein) is synthesized in the hepatocytes and, to a very small extent, in cells of the reticuloendothelial system. Haptoglobin is composed of two kinds of polypeptide chains: two α chains and one β chain. There are three possible α chains and only one form of β chain. On starch-gel electrophoresis, haptoglobin exhibits three types of patterns, which serve to illustrate the polymorphism in the α chains. Homozygous haptoglobin 1-1 gives 1 band. The peptide chains form polymers with

each other and with haptoglobin 1 chains to provide the other two electrophoretic patterns, which have been designated as Hp 1-2 and Hp 2-2 phenotypes. Other characteristics are seen in Table 9-2.

Haptoglobin increases from a mean concentration of 0.02 g/L at birth to adult levels within the first year of life. As old age is approached, haptoglobin levels increase, with a more marked increase being seen in males. Radial immunodiffusion and immunonephelometric methods have been used for the quantitative determination of haptoglobin.

The function of haptoglobin is to bind free hemoglobin by its α chain. Abnormal hemoglobin such as Barts and hemoglobin H have no α chains and cannot be bound. The reticuloendothelial cells remove the haptoglobin-hemoglobin complex from circulation within minutes of its formation. Thus haptoglobin prevents the loss of hemoglobin and its constituent iron into the urine.

Serum haptoglobin concentration is increased in inflammatory conditions. It is one of the proteins used to evaluate the rheumatic diseases. Increases are also seen in conditions such as burns and nephrotic syndrome when large amounts of fluid and lower-molecular-weight proteins have been lost. Determination of a decreased free haptoglobin level (or haptoglobin-binding capacity) has been used to evaluate the degree of intravascular hemolysis that has occurred in transfusion reactions or hemolytic disease of the newborn. Mechanical breakdown of red cells during athletic trauma may result in a temporary lowering of haptoglobin levels.

Ceruloplasmin. Ceruloplasmin is a copper-containing α_2-glycoprotein that has enzyme activities (*i.e.,* copper oxidase, histaminase, and ferrous oxidase). It is synthesized in the liver, where six to eight atoms of copper, half as cuprous (Cu^+) and half as cupric (Cu^{2+}), are attached to an apoceruloplasmin. Ninety percent or more of the total serum copper is found in ceruloplasmin. Molecular characteristics are given in Table 9-2.

The early analytical method of ceruloplasmin determination was based on its copper oxidase activity. Most assays today use immunochemical methods, including radial immunodiffusion and nephelometry.[69]

Low concentrations of ceruloplasmin at birth gradually increase to adult levels and slowly continue to rise with age thereafter. Adult females have higher concentrations than do males, and pregnancy, inflammatory processes, malignancies, oral estrogen, and contraceptives cause an increased serum concentration.

Certain diseases or disorders are associated with low serum concentrations. In Wilson's disease (hepatolenticular degeneration), an autosomal recessive inherited disease, the levels may be low (0.1 g/L). Total serum copper is decreased, but the direct reacting fraction is elevated and the urinary excretion of copper is increased. The copper is deposited in the skin, liver, and brain, resulting in degenerative cirrhosis and neurologic damage. Copper also deposits in the cornea, producing the characteristic Kayser-Fleischer rings. Low ceruloplasmin is also seen in malnutrition, malabsorption, severe liver disease, nephrotic syndrome, and Menkes' kinky hair syndrome, in which a decreased absorption of copper results in a decrease in ceruloplasmin.

α_2-Macroglobulin. α_2-Macroglobulin, a dimeric, large protein (see Table 9-2), is synthesized by hepatocytes. It is found principally in the intravascular spaces because its movement is restricted due to its size. However, much lower concentrations of α_2-macroglobulin can be found in other body fluids, such as cerebrospinal fluid. Upon binding with and inhibiting proteases, it is removed by the reticuloendothelial tissues. The analytical methods that have been used for the satisfactory assay of this protein are radial immunodiffusion and immunonephelometry.

This protein reaches a maximum serum concentration at the age of 2 to 4 years[29] and then decreases to about one-third of that at about 45 years. Later in life, a moderate increase is seen. This change with age is more pronounced in males than in females.[54] There is a distinct difference in the reference values for males and females, adult females having higher values than males.

α_2-Macroglobulin inhibits proteases such as trypsin, pepsin, and plasmin. It also contributes more than one-fourth of the thrombin inhibition normally present in the blood. Very little is known of its correlation with disease or disorders, except in the case of renal disease.

In nephrosis, the levels of serum α_2-macroglobulin may increase as much as 10 times, because its large size aids in its retention. The protein is also increased in diabetes and liver disease. Use of contraceptive medications and pregnancy increase the serum levels by 20%.

Transferrin (Siderophilin). Transferrin, a glycoprotein, is synthesized primarily by the liver. Two molecules of ferric iron can bind to each molecule of transferrin. When the plasma is saturated with ferric iron, it assumes a pale pink color from the saturated molecules, whereas the ferrous iron complex is colorless.[65] Normally, only about 33% of the iron binding sites on transferrin are occupied. Transferrin is the major component of the β-globulin fraction and appears as a distinct band on high resolution serum protein electrophoresis. Genetic variation of transferrin has been demonstrated by electrophoresis on polyacrylamide gel. Other characteristics of transferrin are given in Table 9-2.

The analytical methods used for the quantitation of transferrin are immunodiffusion and immunonephelometry. Both are considered to give precise and accurate results.[76]

The major functions of transferrin are the transport of iron and the prevention of loss of iron through the kidney. Its binding of iron prevents iron deposition in the tissues during temporary increases in absorbed iron or free iron. Transferrin transports iron to its storage sites, where it is incorporated into another protein, apoferritin, to form ferritin. Transferrin also carries iron to cells such as bone marrow that synthesize hemoglobin and other iron-containing compounds.

The most common form of anemia is iron deficiency anemia, a hypochromic, microcytic anemia. In this type of anemia, transferrin in serum is normal or increased. A decreased transferrin level generally reflects an overall decrease in synthesis of protein, such as seen in liver disease or malnutrition, or it may be seen in protein-losing disorders such as nephrotic syndrome. A deficiency of plasma transferrin may result in the accumulation of iron in apoferritin or in histiocytes, or it may precipitate in tissues as hemosiderin. Patients with hereditary transferrin deficiencies have been shown to have marked hypochromic anemia. An increase of iron bound to transferrin is found in a hereditary disorder of iron metabolism, hemochromatosis, in which excess iron is deposited in the tissues, especially the liver and the pancreas. This disorder is associated with bronze skin, cirrhosis, diabetes mellitus, and low plasma transferrin levels.

Hemopexin. The parenchymal cells of the liver synthesize hemopexin, which migrates electrophoretically in the β-globulin region. Other characteristics are seen in Table 9-2. Hemopexin can be determined by radial immunodiffusion. The function of hemopexin is to remove circulating heme. When free heme (ferroprotoporphyrin IX) is formed during the breakdown of hemoglobin, myoglobin, or catalase, it binds to hemopexin in a 1:1 ratio. The heme-hemopexin complex is carried to the liver, where the complex is destroyed. Hemopexin also removes ferriheme and porphyrins.

The level of hemopexin is very low at birth but reaches adult values within the first year of life. Pregnant mothers have increased plasma hemopexin levels. Increased concentrations are also found in diabetes mellitus, Duchenne muscular dystrophy, and some malignancies, especially melanomas. In hemolytic disorders, serum hemopexin concentrations decrease. Administration of diphenylhydantoin also results in decreased concentrations. A more complete list of conditions that cause increased and decreased levels of hemopexin may be found in a thorough review by Natelson and Natelson.[65]

Lipoproteins. Lipoproteins are complexes of proteins and lipids whose function is to transport cholesterol, triglycerides, and phospholipids in the blood. Lipoproteins are subclassified according to the apoprotein and specific lipid content. Oh high resolution serum protein electrophoresis, high density lipoproteins (HDL) migrate between the albumin and α_1-zone; very low density lipoproteins (VLDL) migrate at the beginning of the β-fraction (pre-β); and the low density lipoproteins (LDL) appear as a separate band in the β region. For a more detailed discussion of structure and methods of analysis, refer to Chapter 15, Lipids and Lipoproteins.

β Microglobulin (B2M). β_2-microglobulin is the light chain component of the major histocompatibility complex (HLA). This protein is found on the surface of most nucleated cells and is present in high concentrations on lymphocytes. Due to its small size (MW 11,800), B2M is filtered by the renal glomerulus but most (>99%) is reabsorbed and catabolized in the proximal tubules. Elevated serum levels are the result of impaired clearance by the kidney. or overproduction of the protein, as occurs in a number of inflammatory diseases such as rheumatoid arthritis and systemic lupus erythematosus (SLE). In patients with HIV, a high B2M level in the absence of renal failure indicates a large lymphocyte turnover rate, which suggests the virus is killing lymphocytes. B2M may sometimes be seen on high resolution electrophoresis, but because of its low concentration, it is usually measured by immunoassay.

Complement. Complement is a collective term for several proteins that participate in the immune reaction and serve as a link to the inflammatory response. Molecular characteristics of selected complement components are listed in Table 9-2. These proteins circulate in the blood as nonfunctional precursors. In the classic pathway, activation of these proteins begins when the first complement factor, C1q, binds to an antigen-antibody complex. The binding takes place at the Fc, or constant, part of the IgG or IgM molecule. Each complement protein (C2 through C9) is then activated sequentially and can bind to the membrane of the cell to which the antigen-antibody complex is bound. The final result is lysis of the cell. In addition, complement is able to enlist the participation of other humoral and cellular effector systems in the process of inflammation. An alternate pathway (properdin pathway) for complement activation exists in which the early components are bypassed and the process begins with C3. This pathway is triggered by different substances (does not require the presence of an antibody), but the lytic attack on membranes is the same (sequence C5-C9).

Analytical methods have included a measurement of the titer of complement using complement activity in a hemolytic system and by immunochemical methods such as radial immunodiffusion and nephelometry.

Complement is increased in inflammatory states and decreased in malnutrition, lupus erythematosus, and disseminated intravascular coagulopathies. Inherited deficiencies of individual complement proteins also have been described. In most cases, the deficiencies are associated with recurrent infections.

Fibrinogen. Fibrinogen is one of the largest proteins in the blood plasma. It is synthesized in the liver, and because it has considerable carbohydrate content, it is classified as a glycoprotein. Other molecular characteristic are given in Table 9-2. On plasma electrophoresis, fibrinogen is seen as a distinct band between the β- and γ-globulins. The function of fibrinogen is to form a fibrin clot when activated by thrombin. Thus, fibrinogen is virtually all removed in the clotting process and is not seen in serum.

Fibrinogen customarily has been determined as clottable protein by a method to determine total protein (Biuret, Kjeldahl) performed on the clot that is formed when plasma is reacted with thrombin. The clot is washed, dried, dissolved, and analyzed. Fibrin split products (degradation products of fibrinogen and fibrin) are determined by immunoassay methods such as radial immunodiffusion, nephelometry,[40] and radioimmunoassay.

Fibrinogen is one of the acute-phase reactants, a term that refers to proteins that are markedly increased in plasma during the acute phase of the inflammatory process. Fibrinogen levels also rise with pregnancy and the use of birth control pills. Decreased values generally reflect extensive coagulation during which the fibrinogen is consumed.

C-Reactive Protein (CRP). C-reactive protein (CRP) appears in the blood of patients with diverse inflammatory diseases but is undetectable in healthy individuals. It is synthesized in the liver. Other characteristics are found in Table 9-2. C-reactive protein was so named because it precipitates with the C substance, a polysaccharide of pneumococci. However, it was found that CRP rises sharply whenever there is tissue necrosis, whether the damage originates from a pneumococcal infection or some other source. This led to the discovery that CRP recognizes and binds to molecular groups found on a wide variety of bacteria and fungi. C-reactive protein bound to bacteria promotes the binding of complement, which facilitates their uptake by phagocytes. This process of protein coating to enhance phagocytosis is known as *opsonization*.[79]

CRP is generally measured by its capacity to precipitate C substance or more commonly by immunologic methods including latex slide precipitation test, nephelometry, and enzyme immunoassay (EIA). C-reactive protein is elevated in acute rheumatic fever, bacterial infections, myocardial infarcts, rheumatoid arthritis, carcinomatosis, gout, and viral infections.

Immunoglobulins (Ig's). There are five major groups of immunoglobulins in the serum: IgA, IgG, IgM, IgD, and IgE. They are synthesized in plasma cells. Their synthesis is stimulated by an immune response to foreign particles and microorganisms. The immunoglobulins are not synthesized to any extent by the neonate. IgG crosses the placenta, and the IgG present in the newborn's serum is that synthesized by the mother. IgM does not cross the placenta and initially is 0.21 g/L, but this increases rapidly to adult levels by about 6 months.[29] IgA is virtually lacking at birth (0.003 g/L), increases slowly to reach adult values at puberty, and continues to increase during the lifetime. IgD and IgE levels are undetectable at birth by customary methods and increase slowly until adulthood. IgA is generally higher in males than in females; IgM and IgG levels are somewhat higher in females. IgE levels vary with the allergic condition of the individual.

The immunoglobulins are composed of two long polypeptide chains (heavy, or H, chains) and two short polypeptides (light, or L, chains), joined by disulfide bonds. An individual is capable of producing 10^6 different immunoglobulin molecules. The differences among these molecules are found in a region of the molecule called the *variable region*. This variable region is located on the end of the molecule that contains both the light and heavy chains and is the site at which the immunoglobulin (antibody) combines with the antigen. In this way, there are many different antibodies that are relatively specific for corresponding antigens.

The differences in the heavy chains (H) are called *idiotypes* and are designated IgG, IgA, IgM, IgD, and IgE. The heavy chains are called γ, α, μ, δ, and ϵ, respectively. The light chains (L) for all the immunoglobulin classes are of two kinds, either κ or λ. Each immunoglobulin or antibody molecule has two identical H chains and two identical L chains. For example, IgG has two γ type H chains and two identical L chains (either κ or λ).

When a foreign substance (antigen) is injected into an animal (*e.g.,* rabbit or goat), an antibody that will react with that antigen is synthesized. This reaction is relatively specific; that is, the antibodies will react specifically and selectively with the antigen used to raise them. Proteins and polysaccharides are strong antigens. Antibodies can be raised in rabbits and other animals to the human immunoglobulins as well as the other serum proteins. These rabbit anti-human immunoglobulin antibodies are used to detect and to quantitatively assay IgG, IgA, IgM, IgD, and IgE. The immunoglobulins have been determined using radial immunodiffusion and radioimmunoassay. Fluorescent immunoassay techniques and immunonephelometric assays also have been used. The automated immunonephelometric method is available for IgG, IgA, and IgM.

A molecule of immunoglobulin derived from the proliferation of one plasma cell (clone) is called a *monoclonal immunoglobulin* or *paraprotein*. A marked increase in such a monoclonal Ig is found in the serum of patients who have plasma cell malignancy (myeloma). Monoclonal increases are seen on electrophoretic patterns as spikes. Although these immunoglobulins are typically in the β or γ fractions, on occasion one may appear in the α_2 area. The monoclonal immunoglobulin is typed by the immunofixation method to determine it as IgG, IgA, IgM, IgD, and IgE, and the blood and urine are examined for free light chains.

Immunoglobulin G is increased in liver disease, infections, and collagen disease. A decrease in IgG is associated with an increased susceptibility to infections and monoclonal gammopathy, of one of the other idiotypes.

IgA in fluids other than serum has an additional secretory peptide, called a *J piece*. This piece enables the IgA to appear in secretions. Polyclonal increases in the serum IgA (without the J piece) are found in liver disease, infections, and autoimmune diseases. Decreased serum concentrations are found in depressed protein synthesis, ataxia-telangiectasia, and hereditary immunodeficiency disorders.

IgM is the first antibody that appears in response to antigenic stimulation. IgM also is the type of antibody that acts as anti-A and anti-B antibody to red cell antigens, rheumatoid factors, and heterophile antibodies. Increased IgM concentration is found in toxoplasmosis, cytomegalovirus, rubella, herpes, syphilis, and various bacterial and fungal diseases. A monoclonal increase is seen in Waldenström's macroglobulinemia. This increase is seen as a spike in the vicinity of the late β zone of a protein electrophoretic pattern. Decreases are seen in protein-losing conditions and immunodeficiency disorders.

IgD is the immunoglobulin with the fourth highest concentration in normal serum. Its concentration is increased in infections, liver disease, and connective-tissue disorders. IgD multiple myeloma also has been reported with a monoclonal spike in the late β zone of the electrophoretic pattern.

IgE is the idiotypic immunoglobulin present in the respiratory and gastrointestinal mucosa. These antibodies are associated with allergic and anaphylactic reactions. Polyclonal increases are seen in allergies, including asthma and hay fever. Monoclonal increases are seen in IgE myeloma, a rare disease.

Miscellaneous Proteins

Myoglobin

Myoglobin is an oxygen-binding protein found in skeletal and myocardial muscle. When muscle is damaged, myoglobin is released elevating the serum levels. In an acute myocardial infarction (AMI), this increase is seen within 3 to 4 hours of onset. For more detail, refer to Chapter 10, Enzymes.

Troponin

Troponin is a complex of three proteins that bind to the thin filaments of striated muscle (cardiac and skeletal) but are not present in smooth muscle. The complex consists of troponin T (TnT), troponin I (TnI), and troponin C (TnC). Together they function to regulate muscle contraction. Specifically, troponin T binds to tropomyosin and positions the complex along the actin filament (actin is the major constituent of muscle thin filaments); troponin I inhibits myosin ATPase and thereby blocks myosin movement (myosin forms the thick filaments of muscle); and troponin C binds to calcium to initiate muscle contraction.

Three genes code for TnT—one each in cardiac muscle and fast- and slow-skeletal muscle. TnI, also encoded by three genes, has similar isoforms while TnC, encoded by two genes, has only a cardiac and slow-skeletal muscle form. The isoforms are biochemically distinct and therefore can be differentiated from one another. Of particular interest is the cardiac isoform of troponin T. Cardiac troponin T levels in serum begin to rise within 4 hours following the onset of myocardial damage and remain elevated for 10 to 14 days following an AMI. Since cardiac TnT is specific for heart muscle, and even small amounts of cardiac necrosis cause release of discernible amounts of TnT into the serum, measurement of this protein is a valuable aid in the diagnosis of an acute myocardial infarction. Cardiac TnT is also useful in monitoring the effectiveness of thrombolytic therapy in myocardial infarction patients. The ratio of peak cardiac TnT concentration on day 1 to cardiac TnT concentration at day 4 discriminates between patients with successful (ratio greater than 1) and failed (ratio less than or equal to 1) reperfusion.[56]

Cardiac TnT is measured by an enzyme-linked immunosorbent assay using two monoclonal antibodies directed against different epitopes on the protein. The reference interval is 0 to 0.1 μg/L.

Total Protein Abnormalities

Measurement of total plasma protein content provides general information reflecting disease states in many organ systems.

Hypoproteinemia

A total protein level less than the reference interval occurs in any condition where a negative nitrogen balance exists. One cause of a low level of proteins in the plasma is excessive loss. Plasma proteins can be lost by excretion in the urine in renal disease (i.e., nephrotic syndrome), leakage into the gastrointestinal tract in inflammation of the digestive system, and the loss of blood in open wounds, internal bleeding, or extensive burns. Another circumstance producing hypoproteinemia is decreased intake either because of deficiency of protein in the diet (malnutrition) or through intestinal malabsorption due to structural damage (i.e., sprue). Without adequate dietary intake of proteins there is a deficiency of certain essential amino acids and protein synthesis is impaired. A decrease in serum proteins due to decreased synthesis is also seen in liver disease (site of all nonimmune protein synthesis) or in the inherited immunodeficiency disorders, in which antibody production is diminished. Additionally, hypoproteinemia may result from accelerated catabolism of proteins, such as occurs in burns, trauma, or other injuries.

Hyperproteinemia

An increase in total plasma proteins is not seen as commonly as hypoproteinemia. One condition in which an elevation of all the protein fractions is observed is dehydration. When excess water is lost from the vascular system, the proteins, because of their size, remain within the blood vessels. While the absolute quantity of proteins remains unchanged, the concentration is elevated due to a decreased volume of solvent water. Dehydration results from a variety of conditions, including vomiting, diarrhea, excessive sweating, diabetic acidosis, and hypoaldosteronism. In addition to dehydration, hyperproteinemia may be due to excessive production primarily of the gamma globulins.

Some disorders are characterized by the appearance of a monoclonal protein (MCP) in the serum and often in the urine as well. This protein is an intact immunoglobulin molecule, or occasionally, kappa or lambda light chains only. The most common disorder associated with MCP is multiple myeloma, in which the neoplastic plasma cells proliferate in the bone marrow. The MCP in this case is usually IgG, IgA, or kappa or lambda light chains. IgD and IgE MCPs occur very rarely. MCPs in multiple myeloma may reach a serum concentration of several g/dL.

Not all MCPs are associated with multiple myeloma. IgM MCP is often found in patients with Waldenström's macroglobulinemia, a more benign condition. A wide variety of disorders, including chronic inflammatory states, collagen vascular disorders, and other neoplasms may be associated with MCPs.

Polyclonal increases in immunoglobulins, which would be represented by increases in both kappa and lambda chains, are seen in the serum and urine in a variety of chronic diseases.

Methods of Analysis

Total Nitrogen

A total nitrogen determination measures all chemically bound nitrogen in the sample. The method can be applied to a variety of biologic samples, including plasma and urine.[98] In plasma, both the total protein and nonprotein nitrogenous compounds such as urea and creatinine are measured. The analysis of total nitrogen level is useful in assessing nitrogen balance. Monitoring the nitrogen nutritional status is particularly important in patients receiving total parenteral nutrition, such as individuals with neurologic injuries who are sustained on intravenous fluids for an extended period of time.

The method for total nitrogen analysis utilizes chemiluminescence. The sample, in the presence of oxygen, is heated to a very high temperature ($1100 \pm 20°C$). Any chemically bound nitrogen is oxidized to nitric oxide. The nitric oxide is then mixed with ozone (O_3) to form an excited nitrogen dioxide molecule ($NO_2^\star$). When this molecule decays to the ground state, it emits a photon of light. The amount of light emitted is proportional to the concentration of nitrogen in the sample. This chemiluminescence signal is compared with that of a standard for quantitation.[31] There are now available specialized instruments that perform the measurement automatically (Antek Instruments, Houston, Texas).

Total Proteins

The specimen most often used to determine the total protein is serum rather than plasma. The specimen need not be collected while the patient is fasting, although interferences in some of the determination methods occur with the presence of lipemia. Hemolysis will falsely elevate the total protein result because of the release of RBC proteins into the serum. Clear serum samples, tightly stoppered, are stable for a week or longer at room temperature, for a month at 2 to 4°C, and for at least 2 months at $-20°C$.[71]

The reference interval for serum total protein is 6.5 to 8.3 g/dL (65 to 83 g/L) for ambulatory adults. In the recumbent position, the serum total protein concentration is 6.0 to 7.8 g/dL (60 to 78 g/L). This lower normal range is due to shifts in water distribution in the extracellular compartments. The total protein concentration is lower at birth, reaching adult levels by 3 years of age. There is a slight decrease with age. In pregnancy, lower total protein levels are also seen. Methods for the determination of total protein are described below and summarized in Table 9-4.

Kjeldahl. The classical method for quantitation of total protein is the Kjeldahl method. Because of its precision and accuracy, it is used as a standard by which other methods are

TABLE 9-4. Total Protein Methods		
Method	Principle	Comment
Kjeldahl	Digestion of protein; measurement of nitrogen content	Reference method; assume average nitrogen content of 16%
Refractometry	Measurement of refractive index due to solutes in serum	Rapid and simple; assume nonprotein solids are present in same concentration as in the calibrating serum
Biuret	Formation of violet-colored chelate between Cu^{2+} ions and peptide bonds	Routine method; requires at least two peptide bonds and an alkaline medium
Dye-binding	Protein binds to dye and causes a spectral shift in the absorbance maximum of the dye	Research use

compared. In this method, the nitrogen is determined, and an average of 16% nitrogen mass in protein is assumed in order to calculate the protein concentration.

The serum proteins are precipitated with an organic acid such as trichloroacetic acid or tungstic acid. The nonprotein nitrogen is removed with the supernatant. The protein pellet is digested in H_2SO_4 with heat (340–360°C) and a catalyst, such as cupric sulfate, to speed the reaction. Potassium sulfate is also introduced to increase the boiling point to improve the efficiency of digestion. The H_2SO_4 oxidizes the C, H, and S in protein to CO_2, CO, H_2O, and SO_2. The nitrogen in the protein is converted to ammonium bisulfite (NH_4HSO_4), which is then measured by adding alkali and distilling the ammonia into a standard boric acid solution. The ammonium borate ($NH_4H_2BO_3$) formed is then titrated with a standard solution of HCL to determine the amount of nitrogen in the original protein solution.

This method is not used in the clinical laboratory because it is time-consuming and too tedious for routine use. It is important to note that the nitrogen content of each individual protein may differ from the 16% assumed in the Kjeldahl calculation described. The actual nitrogen content of serum proteins varies from 15.1% to 16.8%. Thus, if one uses a protein standard (calibrated with the Kjeldahl) that differs in composition from the serum specimen to be analyzed, an error is introduced because the percent of nitrogen will not be the same. It is also necessary to assume that no proteins of significant concentration in the unknown specimen are lost in the precipitation step. Despite these assumptions, the Kjeldahl method is still considered by some to be the reference method for proteins.

Refractometry. Refractometry is useful when a rapid method that requires a very small volume of serum is needed. The velocity of light is changed as it passes the boundary between two transparent layers (*i.e.,* air and water) causing the light to be bent (refracted). When a solute is added to the water, the refractive index at 20°C of 1.330 for water is increased by an amount proportional to the concentration of the solute in solution. This proportionality holds fairly well over a two- to threefold increase in concentration (*i.e.,* from 5–20 g/dL). Since the majority of the solids dissolved in serum are protein, the refractive index reflects the concentration of protein. However, in addition to protein, serum contains several nonprotein solids, such as electrolytes, urea, and glucose, that contribute to the refrac-

tive index of serum. Thus the built-in scale in the refractometer has to be calibrated with a serum of a known protein concentration that also has the nonprotein constituents present. An assumption is made that the test samples contain these other solutes in nearly the same concentration as in the calibrating serum. Error is introduced when these substances are increased or when the serum is pigmented (from bilirubin), lipemic, or hemolyzed. The refractive index is also temperature-dependent, and some refractometers incorporate a built-in temperature correction.

The total protein is commonly measured with a hand-held refractometer. A drop of serum is placed by capillary action between a coverglass and the prism. The refractometer is held so that light is refracted through the serum layer. The refracted rays cause part of the field of view to be light, producing a point at which there is a sharp line between light and dark. The number of grams per liter at this line on the internal scale is read. The temperature is corrected in the TS meter (American Optical Corp, Scientific Instruments Division, Buffalo, N.Y. 14215) by a liquid crystal system.

The measurement of total protein by refractometry is very easy and fast. The accuracy is acceptable,[52] with a reported agreement of ±3% with the biuret method,[81] but it is subject to false-positive interferences.

Biuret. The most widely used method, and the one recommended by the International Federation of Clinical Chemistry (IFCC) expert panel for the determination of total protein, is the biuret procedure. In this reaction, cupric ions (Cu^{2+}) complex with the groups involved in the peptide bond. In an alkaline medium and in the presence of at least two peptide bonds, a violet-colored chelate is formed. The reagent also contains sodium potassium tartrate to complex cupric ions to prevent their precipitation in the alkaline solution, and potassium iodide, which acts as an antioxidant. The absorbance of the colored chelate formed is measured at 540 nm. When small peptides react, the color of the chelate produced has a different shade than that seen with larger peptides. The color varies from a pink to a reddish violet. However, there is no discernible difference in the reaction given by the proteins normally seen in plasma. Thus, over a wide range of concentrations, the color that is formed is proportional to the number of peptide bonds present[91] and reflect the total protein level.

In addition to the—NHCO—group that occurs in the peptide bond, cupric ions will react with any compound that has two or more of the following groups:—NHCH$_2$— and—NHCS—. In fact, the method was given its name because a substance called biuret (NH$_2$CONHCONH$_2$) reacted with cupric ions in the same manner.

BIURET METHOD FOR DETERMINATION OF TOTAL SERUM PROTEINS[21]

Reagents

1. NaOH 6.0 mol/L. Dissolve 240 g of NaOH from a previously unopened bottle in freshly distilled or decarbonated, deionized water and dilute to 1 L.
2. Biuret
    CuSO$_4$—12 mmol/L
    KI—30 mmol/L
    C$_4$H$_4$KNaO$_6$—32.0 mmol/L
    NaOH—0.6 mol/L
    Dissolve 3.0 g of CuSO$_4$ in 500 mL of freshly distilled water. Add 9.0 g of potassium sodium tartrate (C$_4$H$_4$KNaO$_6$ · 4H$_2$O) and 5.0 g of KI. After these have dissolved, add 100 mL of 6 mol/L NaOH and dilute to 1 L with distilled H$_2$O. Store in a closed polyethylene bottle at room temperature. This is stable for 6 months.
3. Biuret Blank—Prepare the biuret reagent described previously, but omit the copper sulfate.
4. Protein Standard (bovine albumin 60 to 70 g/L)

Add 100 µL of the standard solution to 10.0 mL of H$_2$O and read the absorbance at 280 nm in a 1-cm^2 cuvet in a spectrophotometer that has a bandwidth of 2 nm or less. Divide the absorbance by 0.661 (the absorptivity of the bovine serum albumin at 280 nm) and multiply by 101 (dilution factor) to obtain the concentration of bovine serum albumin in g/L. The absorptivity of the bovine serum albumin solution in the reaction with biuret reagent described here is 0.2983 Lg^{-1} cm^{-1}.[20] This may be used to determine the concentration of the standard.

Procedure—For each serum unknown, standard, or control, label two tubes.

Mark one "test" and the other "test blank." Mark another single tube "reagent blank."

1. Measure 5.0 mL of biuret reagent into all tubes marked "test."
2. Measure 5.0 mL of biuret blank reagent into all tubes marked "test blank" and into the tube marked "reagent blank."
3. Add 100 µL of serum unknown, control, or standard into the appropriately labeled tubes of "test" and "test blank" series.
4. Add 100 µL of water to the tube marked "reagent blank."
5. Mix each by vortex or by inverting the tubes after covering them with parafilm.
6. Allow the tubes to stand at room temperature for 30 minutes for color development.
7. Set the spectrophotometer to 540 nm and zero absorbance with the biuret blank reagent.
8. Record the absorbances of each of the blank series, next the reagent blank, and finally each of the tubes of the test series.

Calculation—Calculate the concentration of each specimen in the following way:

Correct the absorbance of the test by subtracting the absorbance of the reagent blank and the test blank from the absorbance of the test. This is represented by the following formula:

$$A_{corrected} = A_{test} - A_{test\ blank} - A_{reagent\ blank}$$

Concentration of specimen (g/L) =

$$\frac{A_{corrected}\ of\ unknown}{A_{corrected}\ of\ standard} \times concentration\ of\ standard\ (g/L)$$

(Eq. 9-1)

Preparation of a Standard Curve—The reaction is linear from 10 to 120 g/L when the procedure is followed and the reagents are prepared according to instructions. A standard curve may be prepared by using 10.0 µL of dilutions in 0.15 M NaCl of the standard solution to give a concentration of 10 to 60 g/L. The reagent blank should read between 0.095 and 0.105 with blanks for the tests reading near 0.01 if clear. If lipemic sera must be analyzed, a method for overcoming this problem is available.[14] ■

Dye Binding. The dye-binding methods are based on the ability of most proteins in serum to bind dyes, although the affinity with which they bind may vary. The dyes bromphenol blue, Ponceau S, amido black 10B, lissamine green, and Coomassie blue, have been used to stain protein bands after electrophoresis. Bradford[7] has described a dye-binding method for the determination of total protein using Coomassie blue 250. The binding of Coomassie brilliant blue 250 to protein causes a shift in the absorbance maximum of the dye from 465 to 595 nm. The increase in absorbance at 595 is used to determine the protein concentration.

Bradford shows the dye-binding responses of 5 proteins to be similar; however, others have reported that for several other proteins this is not the case.[72,96] This drawback has prompted a recommendation for caution when applying this test to the complex mixture of protein that one finds in serum.

Ultraviolet Absorption. Serum proteins also have been estimated by the use of ultraviolet spectrophotometry. Proteins absorb light at 280 nm and at 210 nm. The absorptivity (absorbance of a 1 % solution in a 1-cm light path) at 280 nm is related to the absorbance of tyrosine, tryptophan, and phenylalanine amino acids in the protein. Human albumin has only one tryptophan residue in the molecule and has an absorptivity of 5.31, compared with that of fibrinogen which has 55 tryptophan residues and an absorptivity of 15.1.

The absorbance of proteins at 210 nm is due to the absorbance of the peptide bond at that wavelength. The wavelength at which maximal absorbance occurs depends to a small degree on the conformation of the protein. These methods have rarely been used in clinical laboratories but are used routinely in research laboratories to monitor eluates of protein separations from columns. In order to use these methods, the assumptions must remain that the composition of the unknown serum specimen is near that of the calibrating solution.

Fractionation, Identification, and Quantitation of Specific Proteins

In the assay of total serum proteins, useful diagnostic information can be obtained by determining the albumin fraction and the globulins. A reversal or significant change in the ratio of albumin and total globulin was first noticed in diseases of the kidney and liver. In order to determine the albumin-globulin (A/G) ratio, it is common to determine total protein and albumin. Globulins are calculated by subtracting the albumin from the total protein (the total protein − albumin = globulins).

When an abnormality is found in the total protein or albumin, an electrophoretic analysis is usually made. Serum proteins are separable into five or more fractions by the customary electrophoretic methods. If an abnormality is seen on the electrophoretic pattern, an analysis of the individual proteins within the area of abnormality is made.

The methods for measuring protein fractions are described in the following text. Table 9-5 lists the types of analyses used in quantitating albumin levels.

Salt Fractionation. Fractionation of proteins has been accomplished by various procedures using precipitation. Globulins can be separated from albumin by salting-out procedures using sodium salts.[57] Salts, by decreasing the water available for hydration of hydrophilic groups, will cause precipitation of the globulins. Several different concentrations (26% to 28% w/v) of different salts (i.e., Na_2SO_4, Na_2SO_3, etc.) have been used. The albumin that remains in solution in the supernatant can then be measured by any of the routine total protein methods.

The salting-out procedure is not used to separate the albumin fraction in most laboratories today because direct methods that react specifically with albumin in a mixture of proteins are available.

Dye Binding. The most widely used methods for determining albumin are the dye-binding procedures. The pH of the solution is adjusted so that albumin is positively charged. Then, by electrostatic forces the albumin is attracted to and binds to an anionic dye. When bound to albumin, the dye has a different absorption maximum than the free dye; thus, the amount of albumin can be quantitated. A variety of dyes have been used, including methyl orange, 2-(4′-hydroxyazobenzene)-benzoic acid (HABA), bromocresol green (BCG), and bromcresol purple (BCP). Methyl orange is nonspecific for albumin; β-lipoproteins and some α_1- and α_2-globulins also will bind to this dye. HABA, while more specific for albumin, has a low sensitivity. In addition, several compounds, such as salicylates, penicillin, conjugated bilirubin, and sulfonamides, interfere with the binding of albumin to the dye. BCG is not affected by interfering substances such as bilirubin and salicylates; however, hemoglobin can bind to the dye. For every 100 mg/dL of hemoglobin, the albumin is increased by 0.1 g/dL.[22] The measurement of albumin by BCG has also been reported to overestimate low albumin values. This was seen particularly in patients when the low albumin level was accompanied by an elevated alpha-globulin fraction, such as occurs in nephrotic syndrome or end-stage renal disease.[89,102] It was found that the alpha-globulins such as ceruloplasmin and α_1-acid glycoprotein[33] would react with BCG giving a color whose intensity is approximately one-third that of the reaction seen with albumin.[101] This reaction of the alpha-globulins contributed significantly to the absorbance of the test only after incubation times exceeded 5 minutes. Thus, the specificity of the reaction for albumin can be improved by taking absorbance readings within a standardized short interval after mixing.[33] The times used have varied from 0.5 to 30 seconds after mixing.

Bromcresol purple (BCP) is an alternate dye that may be used for albumin determinations.[68] BCP binds specifically to albumin, is not subject to most interferences, is precise, and exhibits excellent correlation with immunodiffusion reference methods. Linear regression analysis with RID gave an equation of $y = 0.95 + 0.93$.[68] Analysis of albumin by the BCP method, however, is not without its disadvantages. In patients with renal insufficiency, the BCP method underestimates the serum albumin.[55] It appears the serum of these patients contain either a substance tightly bound to albumin or a structurally altered albumin that effects the binding of BCP. Similarly, BCP binding to albumin is impaired in the presence of covalently bound bilirubin. BCG binding is unaffected in these situations. Today, both the BCG and BCP methods are commonly used to quantitate albumin.

TABLE 9-5. Albumin Methods

Method	Principle	Comment
Salt precipitation	Globulins are precipitated in high salt concentrations; albumin in supernatant is quantitated by biuret reaction	Labor-intensive
Dye binding		
Methyl orange	Albumin binds to dye, causes shift in absorption maximum	Nonspecific for albumin
HABA [2(4′-hydroxyazobenzene)-benzoic acid]	Albumin binds to dye, causes shift in absorption maximum	Many interferences (salicylates, bilirubin)
BCG (bromcresol green)	Albumin binds to dye, causes shift in absorption maximum	Sensitive; overestimates low albumin levels; most commonly used dye
BCP (bromcresol purple)	Albumin binds to dye, causes shift in absorption maximum	Specific, sensitive, precise
Electrophoresis	Proteins separated based on electric charge	Accurate; gives overview of relative changes in different protein fractions

Determination of Total Globulins. Another approach to fractionation of proteins is the measurement of total globulins. Albumin can then be calculated by subtraction of the globulin from total protein. The total globulin level in serum is determined by a direct colorimetric method employing glyoxylic acid.[30] Glyoxylic acid, in the presence of Cu^{2+} and in an acid medium (acetic acid and H_2SO_4), condenses with tryptophan found in globulins to produce a purple color. Albumin has approximately 0.2% tryptophan, compared with 2% to 3% for the serum globulins. When calibrated using a serum of known albumin and globulin concentrations, the total globulins can be determined.

The measurement of globulins based on their tryptophan content has never come into common use because of the ease and simplicity of the dye-binding methods for albumin.

Electrophoresis. Electrophoresis separates proteins on the basis of their electric charge properties. The charge characteristics of proteins were discussed earlier in this chapter. Proteins, when placed in an electric current, will move according to their net charge, which is determined by the pH of a surrounding buffer. At a pH greater than the pI, the protein is negatively charged, and *vice versa*. The direction of the movement is dependent on whether the charge is positive or negative; cations (positive net charge) migrate to the cathode (negative terminal), whereas anions (negative net charge) migrate to the anode (positive terminal). The speed of the migration is dependent to a large extent on the degree of ionization of the protein at the pH of the buffer. This can easily be estimated from the difference between the isoelectric point of the protein and the pH of the buffer. The more the pH of the buffer differs from the pI, the greater is the magnitude of the net charge of that protein and the faster it will move in the electric field. In addition to the net electric charge, the velocity of the movement also depends on the electric field strength, size and shape of the molecule, temperature and the characteristics of the buffer (*i.e.*, pH, qualitative composition, and ionic strength). The specific electrophoretic mobility μ of a protein can be calculated by:

$$\mu = \frac{(s/t)}{F} \qquad \textit{(Eq. 9-2)}$$

where s = distance traveled in cm
t = time of migration in seconds
F = field strength in V cm^{-1}

Tiselius developed electrophoresis using an aqueous medium. This is known as *moving-boundary* or *free electrophoresis*. Later, in the clinical laboratory, paper was used. The term given to the use of a solid medium is *zone electrophoresis*. Paper has been largely replaced by cellulose acetate or agarose gel as the support media used today.

Serum Protein Electrophoresis (SPE)

In the standard method for serum protein electrophoresis (SPE), serum samples are applied close the cathode end of a support medium strip that is saturated with an alkaline buffer (pH 8.6). The support strip is connected to two electrodes and a current is passed through the strip to separate the proteins. All major serum proteins carry a net negative charge at pH 8.6 and will migrate toward the anode. Utilizing the standard methods, the serum proteins arrange themselves into five bands: albumin travels farthest to the anode followed by α$_1$-globulins, α$_2$-globulins, β-globulins, and γ-globulins in that order. The width of the band of proteins in a fraction depends on the number of proteins with slightly different molecular characteristics that are present in that fraction. Homogenous protein gives a narrow band.

After separation, the protein fractions are fixed by immersing the support strip in an acid solution (*e.g.*, acetic acid) to denature the proteins and immobilize them on the support medium. The proteins are then stained. A variety of dyes have been used, including Ponceau S, Amido black, or Coomassie blue. The protein appears as bands on the support medium. Typical cellulose acetate electrophoretic patterns are shown in Figure 9-10B, while Figure 9-10A shows the patterns obtained using agarose gel as the support medium.

Visual inspection of the membrane is made, or the cleared transparent strip is placed in a scanning densitometer. Reflectance measurements also may be made on the uncleared membranes; however, scanning densitometry is used more commonly. The pattern on the membrane is made to move past a slit through which light is transmitted to a phototube to record the absorbance of the dye that is bound to each of the fractions. Usually this absorbance is recorded on a strip-chart recorder to obtain a pattern of the fractions (Fig. 9-11).

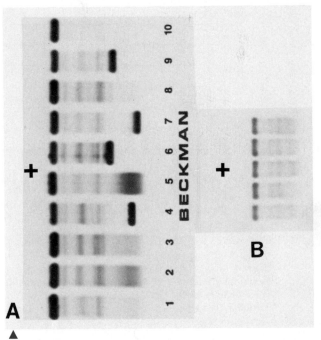

▲
Figure 9-10. Serum protein electrophoretic patterns on agarose and cellulose acetate. (*A*) Agarose gel—note the monoclonal γ-globulin. (*B*) Cellulose acetate. (Courtesy of Department of Laboratory Medicine, The University of Texas M.D. Anderson Hospital, Drs. Liu, Fritsche, and Jose Trujillo, Director, and Ms. McClure, Supervisor)

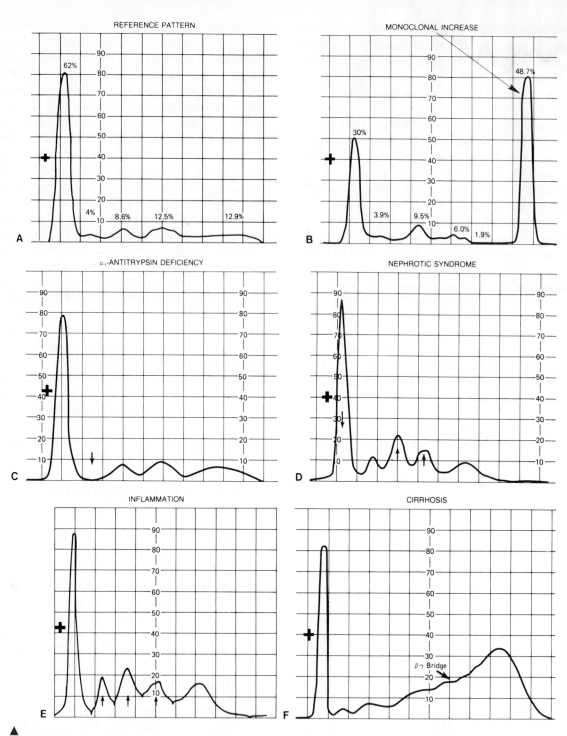

Figure 9-11. Selected densitometric patterns of protein electrophoresis. Albumin is at the anodal (+) end followed by α₁−, α₂−, β−, and γ-globulin fractions. Arrows indicate decrease or increase in fractions. (*A*) Reference pattern (agarose). (*B*) Monoclonal increase in γ area (agarose). (*C*) α₁−Antitrypsin deficiency (cellulose acetate). (*D*) Nephrotic syndrome (cellulose acetate). (*E*) Inflammation (cellulose acetate). (*F*) Cirrhosis (cellulose acetate). (*A* and *B* are courtesy of Drs. Liu, Fritsche, and Jose Trujillo, Director, and Ms. McClure of the Department of Laboratory Medicine, The University of Texas M.D. Anderson Hospital. Others are courtesy of Dr. Wu of the Hermann Hospital Laboratory/The University of Texas Medical School)

Many scanning densitometers compute the area under the absorbance curve for each band and the percentage of total dye that appears in each fraction. The concentration is then calculated as a percentage of the total protein that was determined by one of the protein methods, such as the biuret procedure.

The computation also may be made by cutting out the small bands from the membrane and eluting the dye from each band in 0.1 M NaOH. The absorbances are added to obtain the total absorbance, and the percentage of the total absorbance found in each fraction is then calculated.

A reference serum control should be run with each electrophoretic run (see Fig. 9-11A), and the results should be monitored to maintain 95% confidence limits for the fractions. Reference values for each fraction are as follows: albumin, 53% to 65% of the total protein (3.5–5.0 g/dL); α_1-globulin, 2.5% to 5% (0.1–0.3 g/dL); α_2-globulin, 7% to 13% (0.6–1.0 g/dL), β-globulin, 8% to 14% (0.7–1.1 g/dL); and γ-globulin, 12% to 22% (0.8–1.6 g/dL).

Inadvertent use of plasma will result in a narrow band in the β_2-globulin region because of the presence of fibrinogen. The presence of free hemoglobin will cause a blip in the pattern in the late α_2 or early β zone, and the presence of hemoglobin-haptoglobin complexes will cause a small blip in the α_2 zone.

Often the information obtained by quantitation of each fraction is approximately equal to that obtained by visual inspection. The great advantage of electrophoresis compared with the quantitation of specific proteins is the overview it provides. The electrophoretic pattern can give information about the relative increases and decreases within the protein population, as well as information about the homogeneity of a fraction.

Probably the most significant finding from an electrophoretic pattern is monoclonal immunoglobulin disease. The densitometric scan will show a sharp peak if the increase in immunoglobulins is due to a monoclonal increase (see Fig. 9-11B). A spike in the γ, β, or sometimes α_2 region signals the need for examination of the immunoglobulins and observation for clinical signs of myelomatosis. Likewise, a deficiency of the predominant immunoglobulin, IgG, is seen as a much paler stain in the γ area. Another significant finding is a decrease in α_1-antitrypsin (see Fig. 9-11C).

In nephrotic syndrome, the patient loses serum albumin and low-molecular-weight proteins in the urine. Some IgG is also lost. At the same time, an increase occurs in α_2-macroglobulin, β-lipoprotein, complement components, and haptoglobin.[41] These two events lead to a dramatic decrease in the relative amount of albumin and a marked increase in the relative amounts of α_2-globulin and β-globulin fractions (see Fig. 9-11D).

An inflammatory pattern indicating an inflammatory condition is seen when there is a decrease in albumin and an increase in the α_1-globulins (α_1-acid glycoprotein, α_1-antitrypsin), α_2-globulins (ceruloplasmin and haptoglobin), and β-globulin band (C-reactive protein) (see Fig. 9-11E). This type of pattern is also called an *acute-phase reactant pattern* and is seen in trauma, burns, infarction, malignancy, and liver disease. Acute-phase reactants are so named because they are found to be increased in the serum within days following trauma or exposure to inflammatory agents. Fibrinogen, haptoglobin, ceruloplasmin, and serum amyloid A increase several-fold, whereas C-reactive protein (CRP) and α_2-macrofetoprotein are increased several hundredfold. Interleukin-1, a protein factor from leukocytes, is recognized as an important mediator of the synthesis of acute-phase proteins by hepatocytes. These acute-phase reactants are thought to play some role in immunoregulatory mechanisms. Chronic infections also produce a decrease in the albumin, but the globulin increase is found in the γ fraction as well as the α_1, α_2, and β fractions.

The electrophoretic pattern of serum proteins in liver disease shows the decrease in serum albumin concentration and the increase in γ-globulin. The pattern in cirrhosis of the liver is rather characteristic, with the abnormalities cited previously, but in addition there are some fast moving γ-globulins that prevent resolution of the β- and γ-globulin bands. This is known as the β-γ bridge of cirrhosis[93] (see Fig. 9-11F).

In infective hepatitis, the γ-globulin fraction rises with increasing hepatocellular damage. In obstructive jaundice, there is an increase in the α_2- and β-globulins. Also noted in obstructive jaundice is an increased concentration of lipoproteins, which is an indicator of its biliary origin. This is especially the case when there is little or no decrease of the serum albumin.[87]

A computerized interpretive system for electrophoretic patterns has been introduced[104] and is available commercially (Helena Laboratories, Beaumont, Texas). The kinds of interrelationships of data used in the system are shown in Figure 9-12.

High Resolution Protein Electrophoreses (HRE)

Standard serum protein electrophoresis separates the protein into five distinct zones, which are composed of many individual proteins. By modifying the electrophoretic parameters these fractions may be further resolved into as many as 12 zones. This modification, known as *high resolution electrophoresis* (HRE), is accomplished by use of a higher voltage coupled with a cooling system in the electrophoretic apparatus and a more concentrated buffer. The support medium used most commonly is agarose gel. To obtain the HRE patterns, samples are applied on the agarose gel, electrophoresed in a chamber cooled by a gel block, stained, and then visually inspected. Each zone is compared to the same zone on a reference pattern for color density, appearance, migration rates, and appearance of abnormal bands or regions of density. A normal serum HRE pattern is shown in Figure 9-13. In addition, the patterns may be scanned with a densitometer to obtain semiquantitative estimates of the protein found in each zone. HRE is particularly useful in detecting small monoclonal bands and in differentiating unusual bands or prominent increases of normal bands that can masquerade as

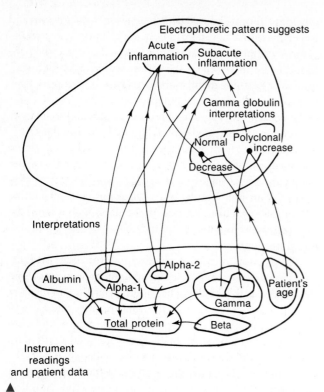

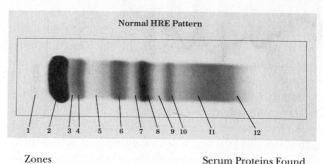

Normal HRE Pattern

Zones	Serum Proteins Found in Zones
1. **PREALBUMIN ZONE**	-Prealbumin
2. **ALBUMIN ZONE**	-Albumin
3. **ALBUMIN-ALPHA-1 INTERZONE**	-Alpha-lipoprotein (Alpha-fetoprotein)
4. **ALPHA-1 ZONE**	-Alpha 1-antitrypsin, Alpha-1-acid glycoprotein
5. **ALPHA-1-ALPHA-2 INTERZONE**	-Gc-globulin, Inter-alpha-trypsin inhibitor, Alpha-1-antichymotrypsin
6. **ALPHA-2 ZONE**	-Alpha-2-macroglobulin, Haptoglobin
7. **ALPHA-2-BETA-1 INTERZONE**	-Cold insoluble globulin, (Hemoglobin)
8. **BETA-1 ZONE**	-Transferrin
9. **BETA-1-BETA-2 INTERZONE**	-Beta-lipoprotein
10. **BETA-2 ZONE**	-C3
11. **GAMMA-1 ZONE**	-IgA, (Fibrinogen), IgM (Monoclonal Ig's, light chains)
12. **GAMMA-2 ZONE**	-IgG, (C-reactive protein) (Monoclonal Ig's, light chains)

Figure 9-12. Expert logic system for the interpretation of protein electrophoresis. (From Weiss SM, Kulikowski CA, Galen RS: Representing expertise in a computer program: The serum protein diagnostic program. Clin Lab Autom 1983;3(6). Reproduced with permission of Dr. Weiss and the publisher.)

Proteins listed in ( ) are normally found in too low a concentration to be visible in a normal pattern.

Figure 9-13. High-resolution electrophoretic pattern of serum. (Courtesy of Helena Laboratories, Beaumont, TX.)

a monoclonal gammopathy. For instance, in patients with nephrotic syndrome, an increased α_2–macroglobulin band in the α_2 region may be confused with a migrating monoclonal protein such as an IgA monoclonal protein gammopathy.[42]

Immunochemical Methods

Radial Immunodiffusion (RID)

Among the first immunochemical methods to be used in the clinical laboratory was radial immunodiffusion.[24,58,67] This precipitin reaction of antigen (protein) with its specific antibody is accomplished by incorporating the antibody in the agarose gel. The gel is then spread on a microscope slide or glass plate, and wells of 1 to 2 mm are cut from the agar. One well is filled with the unknown serum, and one well is filled with a standard solution of the protein to be determined. The antigen diffuses from the well into the agarose gel containing the antibody. The diffusion is the same throughout the 360° surrounding the well. When the proportion of antigen to the antibody in the agarose reaches optimal proportions, a precipitin line (circle) forms (Fig. 9-14).

The diameter of the precipitin ring is proportional to the concentration of the antigen. Standard solutions of varying concentrations are assayed, and a standard curve relating the square of the diameter of the circle to the concentration of the protein is prepared. The diameter (mm²) of the precipitin line formed by the unknown serum is then placed on

the standard plot to obtain the concentration of the protein (antigen) in the unknown serum. The Mancini method, which allowed the precipitin reactions to go to completion, was adopted by most laboratories because of its greater indifference to the effects of temperature. Drawbacks were the time that was required (longer than 24 h in some cases) and the difficulty in visualizing and measuring some of the more diffuse precipitin rings.

Immunoelectrophoresis (IEP)

In 1964, Grabar and Burtin[32] published methods for examining serum proteins using electrophoresis coupled with immunochemical reactions in agarose. The electrophoretic separation of the proteins in the serum is followed by the diffusion of these separated proteins into the agarose. Antiserum to all or one or more of the serum proteins, which has been placed in a trough parallel to the migration path of the separated proteins, also diffuses into the agarose. Where antigens and their corresponding antibodies meet in optimal proportions, precipitin lines form. These precipitin lines are

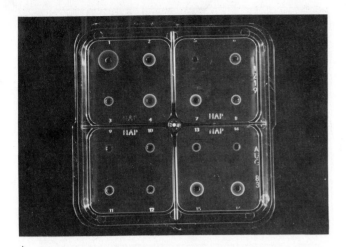

Figure 9-14. Radial Immunodiffusion (haptoglobin assay). (Upper left quadrant) Well #1—355 mg/dL standard; #2—115 mg/dL standard; #3—35 mg/dL standard; others are patient's unknowns.

arranged in characteristic sequence because of the electrophoretic mobility of the serum proteins and their diffusion characteristics (Fig. 9-15).

Immunoelectrophoresis is used primarily to determine if a monoclonal immunoglobulin is IgA, IgG, IgM, IgD, or IgE and if it is of kappa or lambda light-chain type. Successful interpretation of immunoelectrophoretic patterns, however, is difficult and requires experience.

Immunofixation

Immunofixation electrophoresis (IFE) was introduced by Alper and Johnson[2] for the study of microheterogeneity of protein antigens. A typical IFE plate, composed of agarose

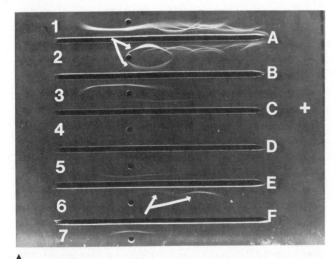

Figure 9-15. Immunoelectrophoresis (agarose). Wells—*1, 3, 5,* and *7* are reference human serum; *2, 4,* and *6* are patient's serum with 1:3 dilution. Troughs—(*A*) anti-human serum; (*B*) anti-IgG; (C) anti-IgA; (*D*) anti-IgM; (*E*) anti-kappa chains; (*F*) anti-lambda chains. Arrows at top—Monoclonal increase in fast γ area congruent with line developed with anti-IgG. Arrows at bottom—Line with monoclonal IgG mobility developed with anti-kappa and showing congruency with free kappa chains.

gel or cellulose acetate, has six lanes labeled: serum proteins, IgG, IgA, IgM, κ, and λ. The specimen is placed at the origin in each lane. The proteins are then separated by electrophoresis. After electrophoresis is completed, a protein fixative designed to denature and immobilize the serum proteins is added to the lane marked serum proteins. For each of the other lanes a monospecific antiserum is place directly over the portion of the electrophoretic pattern where the protein of interest is found. The antisera will precipitate the specific protein, and this immune precipitate is trapped (fixed) on the electrophoretic support medium. The excess protein is washed from the plate, and the remaining precipitation band is stained.[99] Interpretation is made by aligning the reference sample with the immunofixed patterns, and the common identities of the stained reference protein and the immunofixed proteins are made. The monoclonal protein fixation band will resemble the monoclonal band on the reference pattern in its position and configuration.[92] One class of heavy chains and one type of light chains usually will be present in a monoclonal immunoglobulin. IgG, IgA, IgM, and kappa and lambda chains can be identified utilizing IFE. If no identification of the monoclonal band can be made, consideration should be given to IgD, IgE, or heavy chain disease.

This method has superseded immunoelectrophoresis because it is faster and easier to perform, easier to interpret, and the volume of antiserum required is much smaller, making it more economical. Illustrations of immunofixation for the typing of gammopathies are provided in Figure 9-16 *A–D*.

Electroimmunodiffusion—"Rocket" Immunoelectrophoresis

The electroimmunodiffusion-immunoelectrophoresis method, described by Laurell,[47,48] allows the quantitative measurement of specific proteins after their electrophoresis in a gel that contains a monospecific antibody. In this procedure, the sample is added to a well in an agarose plate with antibody incorporated throughout the agarose. The plate is then placed in an electrophoresis chamber. As the protein moves through the gel, antigen-antibody complexes form and dissociate, until the point at which the concentrations of the antibody and antigen are at optimal proportions. At this point, the precipitate forms a "rocket." The distance of the precipitate from the point of sample application is related to the concentration of the antigen.

Turbidimetry and Nephelometry

Turbidimetry and nephelometry are used to measure the light scattering due to immune complexes that form when an antigen is mixed with its antibody in appropriate proportions. Turbidimetry is performed in a spectrophotometer or colorimeter. It is the measurement of light that remains after the incident light has been absorbed, reflected, or scattered by the particles in the solution in the cuvet. The photodetector is in direct alignment with the light source. Nephelometry is performed in an instrument that has the

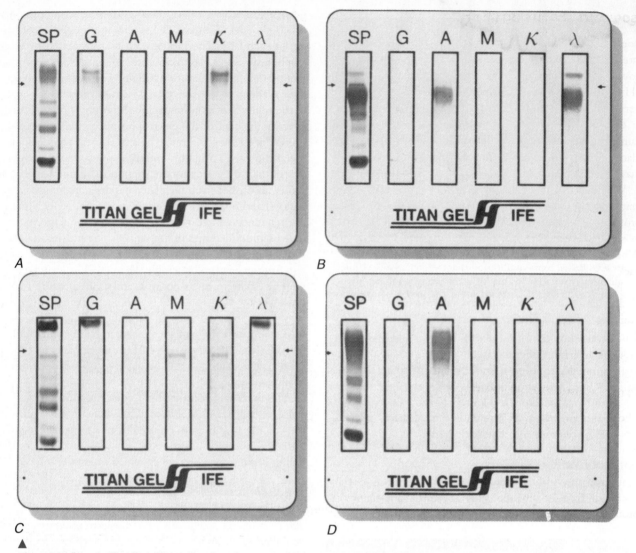

Figure 9-16. Immunofixation electrophoretic serum patterns on agarose gel. (*A*) IgG kappa monoclonal immunoglobulin. (*B*) Monoclonal free lambda light chain in addition to a typical IgA lambda monoclonal immunoglobulin. (*C*) Two monoclonal bands identified as IgG lambda and IgM kappa. (*D*) Abnormal IgA pattern with no corresponding kappa or lambda light chain, indicating heavy chain disease. (Courtesy of Helena Laboratories, Beaumont, TX)

photodetector placed at an angle to the incident light beam in a manner very much analogous to fluorometry. The nephelometer allows the measurement of the increase in light intensity when light has been scattered by the particles in solution. The total intensity of the scattered light is proportional to the number of particles present and thus can be used to measure the amount of antigen that was present to form antigen-antibody complexes.

Gel precipitin methods have been replaced to a large extent by nephelometric and turbidimetric measurements of protein-antibody complexes in solution. The precision of analysis is much improved, the time required is less, automation is possible,[75] and there have been continuing improvements in the immunochemical reagents and instrumentation for these light-scattering methods.[105]

Nephelometry requires dilute solutions in order to keep absorption and reflection at a minimum. The method shows linearity over a wide range of concentrations of antigen. On the other hand, turbidimetry requires higher concentrations of particles, and the relationship of concentration to the readings is linear over a limited range.

For both methods, all reagents and sera must be free of particles that would cause scattering of light. When the serum is pretreated to remove chylomicrons and other particles by mixing with polyethylene glycol, protamine sulfate, or dextran sulfate, an increase in sensitivity occurs. The selection of the incident light wavelength is also important in nephelometry. The wavelength chosen should not be absorbed by the compounds in solution, since any light absorption would decrease the emitted light. Turbidimetry and nephelometry can be done either as endpoint or kinetic measurements. In the endpoint mode, a measurement is taken at the beginning of the reaction and one at a set time later in the reaction. In kinetic mode, readings are taken as the complexes are forming and increasing in size.[73] The amount of scattered light is compared to a standard curve.

Isoelectric Focusing (IEF)

Isoelectric focusing (IEF) is zone electrophoresis that separates proteins on the basis of their pI. IEF employs constant power and polyacrylamide or agarose gel, which contains a pH gradient. The pH gradient is established by the incorporation of small polyanions and polycations (ampholytes) in the gel. The varying isoelectric points of the polyions cause them, in the presence of an electric field, to seek their place in the gradient and to remain there. The pH gradient may range from 3.5 to 10.

When a protein is electrophoresed in the gel, it will migrate to a place on the gel where the pH is that of its pI. The protein becomes focused there because, if it should diffuse in either direction, it leaves its pI and gains a net charge. When this occurs, the electric current once again carries it back to its point of no charge, or its pI.

The clinical applications of isoelectric focusing have included phenotyping of α_1-antitrypsin deficiencies,[38] determination of genetic variants of enzymes and hemoglobins,[97] detection of paraproteins in serum[88] and oligoclonal bands in CSF,[80] and isoenzyme determinations.

Proteins in Other Body Fluids

The kinds of body fluids being studied for their protein content have increased. This is due in part to the increased sensitivity of the test methods that are now available. Proteins in 16 body fluids are included in a contemporary publication by Ritzmann and Killingsworth.[77]

This section includes a discussion of the two fluids whose protein contents are studied most often: urine and cerebrospinal fluid (CSF).

Urinary Proteins

The majority of proteins found in the urine arise from the blood; however, urinary proteins can also originate from the kidney and urinary tract, and from extraneous sources such as the vagina and prostate. The proteins in the blood appear in the urine because they have passed through the renal glomerulus and have not been reabsorbed by the renal tubules. The qualitative tests for proteinuria are most commonly performed using a reagent test strip. These methods are based on the change in the response of an indicator dye in the presence of protein, known as protein error of indicators. A more detailed discussion of these strips may be found in the package inserts of the products and in the literature.[64]

Quantitative examination of urine protein requires considerable attention to the volume and/or time of the urine collection, since the concentration may vary with time and volume. Most quantitative assays are performed on urine specimens of 12 or 24 hours. The 24-hour timing allows for circadian rhythmic changes in excretion at certain times of day. The patient should void, completely emptying the bladder, and discard this urine. Urine is collected from that time for the next 24 hours, and at the end of that time, the urine is voided and the bladder is completely emptied. That urine is placed in the container as the last sample. The volume of the timed specimen is measured accurately and recorded. The results are reported generally in terms of weight of protein per 24 hours by calculating the amount of protein present in the total volume of urine collected during that time.

Several methods for the determination of total protein in urine and other body fluids have been proposed. These include the measurement of turbidity when urinary proteins are mixed with an anionic organic acid such as sulfosalicylic acid, trichloroacetic acid, or benzethonium chloride.[37] These methods are sensitive, but the reagent does not react equally with each protein fraction. This is particularly true of sulfosalicylic acid, which produces four times more turbidity with albumin than it does with γ globulin.[84]

Methods considered to give more accurate results consist of precipitation of the urine proteins, dissolution of the protein precipitate, and color formation with biuret reagent.[27] Another chemical procedure for urinary protein uses the Folin-Ciocalteu reagent, which is a phosphotungsto-molybdic acid solution, frequently called phenol reagent because it oxidizes phenolic compounds. The reagent changes color from yellow to blue during the reaction with tyrosine, tryptophan, and histidine residues in proteins. This method is about 10 times more sensitive than the biuret method. Lowry and associates[53] increased the sensitivity of the Folin-Ciocalteu reaction by incorporating a biuret reaction as the initial step. After the binding of the Cu^{2+} to the peptide bonds, the Folin-Ciocalteu reagent is added. As the Cu^{2+}-protein complex is oxidized, the reagent is reduced, forming the chromogens tungsten blue and molybdenum blue. This increased the sensitivity to 100 times greater than that of the biuret method alone. Another modification utilizes a pyrogallol red-molybdate complex that reacts with protein to produce a blue-purple complex.[28] This procedure is easily automated.

Dye-binding methods also have been used to determine the total protein content of body fluids. Methods using dyes such as Coomassie blue[35] and Ponceau S[70] are available in the literature. Table 9-6 summarizes the various methods for measurement of urinary total protein.[61]

The reference values or intervals for urinary proteins are highly method-dependent,[8] ranging from 100 to 250 mg per 24 hours. Because of their ease of use, speed, and sensitivity, the techniques that are used most frequently today are turbidimetric procedures. The method given below utilizes benzethonium chloride. This procedure is easily automated (ACA, DuPont Instruments, Wilmington, Del.).

TURBIDIMETRIC METHOD FOR URINARY TOTAL PROTEIN USING BENZETHONIUM CHLORIDE[37]

Principle—In an alkaline solution, protein reacts with the quaternary ammonium salt, benzyldimethyl {2-[2-(p-1,1,3,3-tetramethylbutylphenoxy) ethoxy] ethyl} ammonium chloride (benzethonium chloride) to produce a turbidity that is very stable.

Reagents

1. Benzethonium chloride—2.0 g/L
2. NaOH/EDTA mixture—0.5 mol/L NaOH in 33 mmol/L tetrasodium ethylenediamine-tetracetate (EDTA)
3. Standards—five levels of calibrators ranging from 60 to 2400 mg/L

Procedure

1. Mix 0.1 mL urine and 4.0 mL of the NaOH/EDTA mixture.
2. Immediately add 1.0 mL of 2.0 g/L benzethonium chloride.
3. Shake the mixture well (5 seconds on vortex).
4. After 40 to 60 minutes, read the absorbance at 660 nm against a distilled water blank.
5. Construct a calibration curve and calculate the protein content of each urine specimen.

Calculation

Protein in cuvet (mg/L) (obtained from calibration curve)

$$\times \; \frac{\text{total urine volume (mL/24 hr)}}{1000} = \text{mg protein/24 h}$$

(Eq. 9-3) ■

Physiologic Significance of Proteinuria. Proteinuria in renal disease may be classified as resulting from either glomerular or tubular dysfunction. Glomerular proteinuria is a consequence of loss of glomerular membrane integrity, which normally keeps proteins from passing through to the urine because of their large molecular weight. In early selective glomerular proteinuria, the proteins responsible for the increase are typically albumin (greater than 80%) and transferrin. As the glomerular lesion becomes more severe, the membrane becomes non-selective, and proteins of all sizes, including immunoglobulins, pass into the urine. Damage to the glomerular membrane occurs in diseases such as diabetes, amyloidosis, dysglobulinemia, and collagen disorders. The presence in the blood of toxic agents, such as mercury or heroin, may also result in loss of glomerular integrity. An early indicator of glomerular dysfunction is the presence of microalbuminuria. The term microalbuminuria is used to describe albumin concentrations in the urine that are greater than normal but not detectable with common urine dipstick assays. Recent studies[11] of patients with diabetes mellitus showed that microalbuminuria precedes the nephropathy associated with diabetes, particularly in Type I (IDDM) diabetics. Progression from microalbuminuria to clinical nephropathy can be delayed with intensive therapy to normalize blood glucose and blood pressure. Therefore, it has been recommended that all diabetics be tested annually for microalbuminuria. The normal albumin excretion rate is <20 μg/min or <30 mg/day. Microalbuminuria can be measured by radioimmunoassay, radial immunodiffusion, immunonephelometry, and enzyme immunoassay.

Tubular proteinuria is a consequence of the renal tubules either being unable to perform their usual function of reabsorption because of dysfunction, or because the amount of protein appearing in the tubular fluid exceeds the absorptive capability of a normal functioning tubule. In tubular proteinuria, small protein molecules that normally pass through the glomerulus and are reabsorbed, such as β_2-microglobulin (MW 11,800), retinol-binding protein (MW 21,000), and α_1-microglobulin (MW 30,000), appear in the urine.

Overflow proteinuria is an overabundance of proteins from the serum that appear in such high concentrations in the glomerular filtrate that the tubules are overwhelmed in their capacity to absorb them. This type of proteinuria is typified by the increased concentration of immunoglobulin light chains in multiple myeloma (Bence-Jones protein). The determination of specific proteins yields much more information than total urinary protein, particularly in the study of tubular proteinuria.

To determine the type of proteins that are being excreted, most methods require initial concentration of the urine. This is accomplished by precipitation, ultrafiltration, dialysis, or gel filtration techniques. The urine can then be electrophoresed by a method similar to that described for serum. Quantitative measurement of specific urine proteins can be made by use of immunochemical methods, such as radioimmunoassay, radial immunodiffusion, electroimmunoassay, and immunonephelometry.[13]

TABLE 9-6. Urine Protein Methods

Method	Principle	Comment
Turbidimetric methods (sulfosalicylic acid, trichloroacetic acid, or benzethonium chloride)	Proteins are precipitated as fine particles; turbidity is measured spectrophotometrically	Rapid, easy to use; unequal sensitivity for individual proteins
Biuret	Proteins are concentrated by precipitation, redissolved in alkali, then reacted with Cu^{2+}; Cu^{2+} form colored complex with peptide bonds	Accurate
Folin-Lowry	Initial biuret reaction; oxidation of tyrosine, tryptophan and histidine residues by Folin phenol reagent (mixture of phosphotungstic and phosphomolybdic acids): measurement of resultant blue color	Very sensitive
Dye-binding (Coomassie blue, Ponceau S)	Protein binds to dye, causes shift in absorption maximum	Limited linearity; unequal sensitivity for individual proteins

A urinary protein not found in the serum is Tamm-Horsfall protein, a mucoprotein produced in the renal tubules. Its concentration in urine is 40.0 mg per day. An IgA (secretory) is also produced in the kidney, and 1.1 mg per day is excreted.

Cerebrospinal Fluid Proteins

Cerebrospinal fluid (CSF) is formed in the choroid plexus of the ventricles of the brain by ultrafiltration of the blood plasma. Protein measurement is one of the tests that is usually requested, in addition to glucose level measurement and differential cell count, culture, and sensitivity. The accepted reference interval for patients between the ages of 10 to 40 years is 15 to 45 mg/dL of CSF protein.

The total CSF protein may be determined by several of the more sensitive chemical or optical methods referred to earlier in the discussion of urinary proteins. The most frequently used procedures are turbidimetric using trichloroacetic acid, sulfosalicylic acid with sodium sulfate,[63] or benzethonium chloride.[26] Also available are dye-binding methods (i.e., Coomassie brilliant blue), a kinetic biuret reaction, and the Lowry method using Folin phenol reagent.

Physiologic Significance of CSF Protein Analysis.

Abnormally increased total CSF proteins may be found in conditions where there is an increased permeability of the capillary endothelial barrier through which ultrafiltration occurs. Examples of such conditions include bacterial, viral, and fungal meningitis, traumatic tap, multiple sclerosis, obstruction, neoplasm, disk herniation, and cerebral infarction. The degree of permeability can be evaluated by measuring the CSF albumin and comparing it with the serum albumin. Albumin is usually employed as the reference protein for permeability because it is not synthesized to any degree in the CNS. The reference value for the CSF albumin-serum albumin ratio is less than 2.7 to 7.3.[94] A value greater than this indicates that the increase in the CSF albumin came from serum due to a damaged blood-brain barrier. Low CSF protein values are found in hyperthyroidism and when fluid is leaking from the CNS.

While total protein levels in the CSF are informative, diagnosis of specific disorders often require measurement of individual protein fractions. The pattern of types of proteins present can be seen by electrophoresis of CSF that has been concentrated about 100-fold. This may be performed on cellulose acetate or agarose gel. The pattern obtained normally from adult lumbar CSF shows prealbumin, a prominent albumin band, α_1-globulin composed predominantly of α_1-antitrypsin, an insignificant α_2 band, a β_1 band composed principally of transferrin, and a CSF-specific transferrin that is deficient in carbohydrate in the β_2 zone.[44]

Electrophoretic patterns of CSF from patients who have multiple sclerosis (MS) have multiple, distinct bands in the globulin zone.[46] This is called oligoclonal banding. Over 90% of patients with MS have oligoclonal bands, although the bands also have been found in inflammatory and infectious neurologic disease. These bands cannot be seen on routine cellulose acetate electrophoresis but require a high-resolution technique in which agarose is usually employed.

Another finding in demyelinating diseases, such as multiple sclerosis, is the production of IgG within the CNS. However, the increase in IgG may result from either intrathecal synthesis or increased permeability of the blood-brain barrier. In order to identify the source of the elevated CSF IgG levels, the IgG-albumin index can be calculated as follows:

$$\text{CSF IgG Index} = \frac{\text{CSF IgG (mg/dL)} \times \text{serum albumin (g/dL)}}{\text{serum IgG (g/dL)} \times \text{CSF albumin (mg/dL)}}$$

(Eq.9-4)

The CSF albumin concentration corrects for increased permeability. The reference range for the index is 0.25 to 0.80.[44] The index is elevated when there is increased CNS IgG production such as occurs in multiple sclerosis. IgG production is also increased in some bacterial infections and CNS inflammatory diseases. The IgG index is decreased when the integrity of the blood-brain barrier is compromised, as in some forms of meningitis and tumors.

Myelin basic proteins present in the CSF are also assayed because these proteins can provide an index of active demyelination. Myelin basic proteins are constituents of myelin, the sheath that surrounds many of the CNS axons. In very active demyelination, concentrations of myelin basic proteins of 17 to 100 ng/mL were found by radioimmunoassay. In slow demyelination, values of 6 to 16 ng/mL occurred, and in remission, the values were less than 4 ng/mL.[15]

DNA Probes

The detection of specific nucleic acid sequences using DNA probe technology has provided the clinical laboratory with a new approach to the diagnosis of disease. The greatest advances have been made in the area of genetic diseases, but the use of DNA probes in the detection and identification of infectious agents and the characterization of various types of cancer is growing rapidly. DNA probe tests are based on the principles of nucleic acid hybridization reactions. To understand these reactions, a review of basic DNA structure and characteristics will be presented.

DNA Structure

DNA consists of a double strand of nucleotides arranged in a helix. The backbone of DNA is a chain of deoxyribose molecules linked by phosphodiester bonds. The strands are antiparallel, meaning that the $3'$-$5'$-phosphodiester bonds are in opposite directions in the two strands. Attached to the sugar and protruding perpendicular into the center of the helix are the bases. For DNA, the bases are adenine (A), guanine (G), thymine (T), and cytosine (C). The double helix is maintained by hydrogen bonding between bases of the two DNA strands. Guanine specifically binds to cytosine with three hydrogen bonds, while two hydrogen bonds are formed between adenine and thymine.[23] The two DNA strands are thus complementary because every base of one strand is matched by its complementary hydrogen-bonding base on the other. Figure 9-17 shows the basic structure of a DNA molecule that was first described by Watson and Crick.[100] This specificity in the pairing of bases is a very important aspect of DNA probe technology.

Another element that plays a critical role is the fact that the hydrogen bonds holding the two DNA strands together are weak, as are the hydrophobic interactions between the stacked bases of the helix. Thus, it is possible to disrupt the dual strands but not break the bonds holding the backbone together. Increasing the temperature or adding acid or alkali will cause the double strand to separate much like a zipper being opened. This process of separation of DNA strands is called *denaturation*. In addition to the weak bonds, the negative charges on the phosphate groups of the backbone tend to repel the two strands. This repulsion is neutralized by the presence of ions such as Na^+ or Mg^{2+}. Diluting the solution will overcome this ion neutralization and cause denaturation.

Thus by increasing the temperature and pH and decreasing the ionic strength of a solution of DNA, the double strands of DNA can be separated. If the conditions are reversed, the single strands of DNA will re-form. The process is called *renaturation* or *annealing*. The annealing may not occur between the two original strands. As long as the bases are complementary, the two strands will join together. This is the principle in *hybridization*.

Hybridization also can occur between a DNA single strand and an RNA strand. RNA is similar to a single strand of DNA in that it has a sugar-phosphate backbone with purine and pyrimidine bases. However, the sugar is ribose, and the base uracil substitutes for the thymine.

DNA Probes: Production

DNA probes are simply short single strands of DNA whose bases are complementary to the sequence of bases on the target to be identified. Probes may be produced either by chemical synthesis of oligonucleotides or by molecular cloning. The bases on the DNA molecule carry the genetic information. A sequence of three bases code for an amino acid in a particular protein. Thus, to determine the sequence of bases needed in the probe, the protein product of the genetic transcription is examined. Once the amino acids of the protein are delineated, the bases that code for that amino acid are determined. When the nucleotide sequence is known, the bases can then be chemically linked in a process that has now been automated.[10]

The other approach to production of DNA probes, molecular cloning, usually results in a much larger section of DNA. In cloning DNA, the first step is to generate the DNA fragment of interest. Bacterial enzymes called *restriction endonucleases* recognize certain base sequences on the DNA and can cut the double-stranded DNA at that point. For example, Eco R_1 cleaves DNA at any point where the nucleotide sequence GAATTC occurs. When a long strand of DNA is treated with Eco R_1, several fragments of varying lengths will be cleaved wherever the bases GAATTC occurs. Other enzymes recognize other sequences. These fragments can be sorted by size using gel electrophoresis. By experimentation, the fragment of interest is isolated. This segment is then introduced into a vector. In most instances, the vector used is a bacterial plasmid, which is a circular double-stranded piece of DNA. The plasmid is cut, the foreign DNA is introduced, and the new ends are sealed with an enzyme called ligase. This process is illustrated in Figure 9-18. The vector is incorporated into a host cell, such as *Escherichia coli,* where it can replicate. After propagation, the "cloned" foreign DNA is isolated and purified.

The DNA probe also must be labeled so that it can be detected when it reacts with the sample. The label can be a radioactive molecule such as ^{32}P, a nonradioactive molecule such as biotin, which is incorporated into the probe, or a chemiluminescent compound such as acridinium esters.[103] When a radioactive label is used, the probes are detected by autoradiography, in which the radioactivity in the sample exposes x-ray film in the location of the probe. To detect the presence of the biotin-labeled DNA, avidin conjugated with an enzyme such as horseradish peroxidase or alkaline phosphatase is added.[49] The conjugated avidin binds to the biotin; then a substrate is added. The probe can be detected by observing a color change due to the activity of the

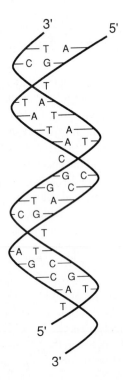

Figure 9-17. DNA structure.

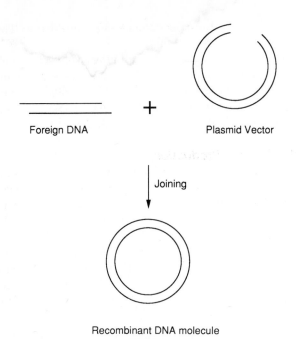

Figure 9-18. Construction of recombinant DNA.

electrical current. Following transfer of the DNA fragments, the membrane is then treated with a labeled probe. The probe will bind to DNA-containing base sequences that are complementary to the bases on the probe. Following this hybridization, excess probe is washed off, and the bound probe is visualized. In the Southern blot procedure, the probe is usually labeled with ^{32}P; thus, the position of labeled bound probe is detected by autoradiography. Figure 9-19 outlines the Southern blotting technique for DNA analysis.

In the past few years, a new technique known as polymerase chain reaction (PCR) has been developed. The primary advantage of PCR is the amplification of target DNA, thus increasing the sensitivity of DNA probe techniques. PCR requires only a knowledge of some base-sequence information on both sides of the desired locus. An oligonucleotide "primer" is constructed that consists of these specific sequences at the outside edge of the target region. The oligonucleotide primers are complementary to opposite DNA strands, and their 3′ ends are oriented toward each other. The general protocol for PCR is illustrated in Figure 9-20. The sample is appropriately pretreated by lysis or protease digestion, and the DNA is extracted. The DNA is

enzyme on the specific enzyme substrate. Alternately, the biotinylated probe can be detected by reaction with a fluorescein-conjugated antibody directed against biotin. The resultant fluorescence under ultraviolet light indicates the presence of the probe. Chemiluminescent probes can be detected utilizing wavelengths in the visible range. There have been several other detection systems described including ELISA techniques.[23,74]

DNA Probe Techniques

The traditional method for DNA analysis is the Southern blot procedure devised by E. M. Southern. In this technique, the first step is to extract the DNA from the specimen. The specimen may be whole blood, sputum, tissue, or cells such as lymphocytes or amniocytes. Cells or tissues are generally lysed and the DNA extracted with a phenol solution. Following extraction, the DNA is digested into varying-length fragments with restriction endonucleases. The fragments are separated according to size using agarose-gel electrophoresis. In many types of gels, the electrophoretic mobility of a DNA fragment is inversely proportional to the logarithm of the number of base pairs. The next step is to heat denature the DNA to separate the DNA strands. The single-stranded DNA is then transferred to a solid support, such as nitrocellulose or a charged nylon membrane, by blotting. The transfer is accomplished by placing the gel on a wick immersed in a concentrated salt solution. The solid support membrane is placed on top of the gel, followed with absorbent papers on top of the membrane. Capillary action draws the salt solution through the gel and forces the transfer of the DNA fragments from the gel onto the membrane at a position corresponding to their position in the gel. Alternately, this transfer can be accomplished by applying an

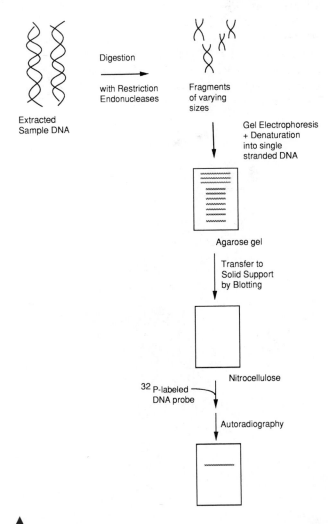

Figure 9-19. Southern blot technique.

then mixed with oligonucleotide primers, thermally stable DNA polymerase, and excess deoxyribonucleoside triphosphates (dNTP), and the mixture is heated. The increase in temperature causes the sample DNA strands to separate. Upon cooling, the primers anneal preferentially to the DNA template. These short pieces serve as "start" signals for DNA replication. The DNA polymerase enzyme, starting at the open 3′ end of the primer, adds a series of single nucleotides that are complementary to the original DNA molecule.[83] This process, using DNA polymerase, is the same series of actions that occurs in DNA replication *in vivo*. After one cycle of synthesis, the mixture is again heated to dissociate the strands, cooled to hybridize the primers to the DNA, and then the primers are extended again. In every cycle the total number of DNA strands is doubled. After 20 to 30 cycles, the target DNA has been amplified a million-fold. The target DNA can then be analyzed by blotting or even

simple staining. Instrumentation is now available to automate PCR.[45]

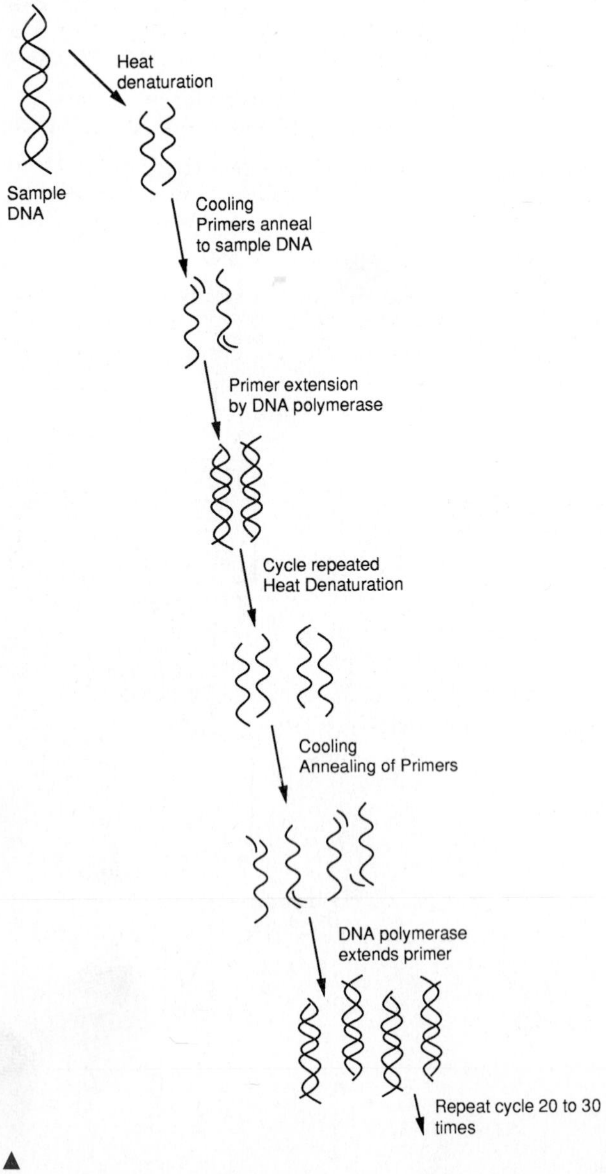

In addition to PCR, there are other approaches to increasing the sensitivity of DNA probe assays. One method uses the enzyme QB replicase to generate multiple copies of probes that have been hybridized to target RNA or DNA.[95] QB replicase is a polymerase found in the QB virus. Ligase chain reaction, another amplification technique, uses two pairs of probes that are complementary to two short target DNA sequences in close proximity. After denaturation of the target DNA into single strands at high temperature, the reaction temperature is lowered for the probes to anneal to the complementary target sequences. DNA ligase then fills in the gap between the probe pairs, generating ligation products that can serve as templates for another cycle.

Non-amplified probe assays require on the order of 500,000 to 1 million probe-target hybrids to be present in a sample before clear detection is possible. Amplification techniques increase the sensitivity of the assays at least 100-fold.

Applications of DNA Probes

The advantages of DNA probe analysis over other diagnostic methods is the affinity and specificity of nucleic acid probes for their target sequences. In genetic diseases, DNA probes have been used for prenatal diagnosis, detection of carriers, and discrimination of important genetic differences in patients with the same disease. The genetic diseases that have been diagnosed with DNA probes include sickle-cell anemia, Duchenne muscular dystrophy, and cystic fibrosis.[17] In sickle-cell anemia, the specific mutation in DNA sequence is known and a specific probe could be constructed. More often the specific gene mutation is undetermined, and an indirect approach using linkage analysis is used. For example, family studies of individuals with cystic fibrosis showed that the gene for cystic fibrosis is closely linked to identifiable markers. If these markers are present, it is sufficient evidence that the cystic fibrosis gene is also present, and in this way, carriers can be identified.

DNA probes are also advantageous in identifying pathogens because the organism can be detected before active replication or the appearance of an antibody response. Chlamydia, cytomegalovirus, and the human immunodeficiency virus (HIV) are examples of infectious organisms for which DNA analysis has aided detection.

Malignant disorders are another area where DNA probes have been found to be useful. For example, RNA from leukemic cells is isolated, and with DNA probes, the type of leukemia can be precisely identified and classified. DNA probes also have been developed for T- and B-cell lymphomas and oncogenes. Oncogenes are specific gene sequences that cause cancer when they are inappropriately expressed.

DNA fingerprinting in forensic medicine also makes use of DNA probes. The target area is a region of a highly polymorphic sequence of repeated bases that is unique to the individual. When sample DNA is digested with restriction endonucleases, separated by electrophoresis, blotted, and probed with the repeat DNA, the resultant pattern is unique to each

Figure 9-20. Polymerase chain reaction.

individual human being (identical twins excepted). With techniques such as PCR, even a tiny amount of DNA from dried blood, hair follicles, or semen—sources commonly found at the scenes of crime—can be amplified and analyzed.

The potential of DNA probe technology in the clinical laboratory has yet to be fully realized. As the research in molecular biology continues to identify DNA sequences associated with diseases and the technology improves to provide automated, less time-consuming techniques, DNA technology may find widespread use in clinical laboratory assays.

SUMMARY

Amino acids are the building blocks of proteins. Inherited abnormalities in amino acid metabolism results in a variety of conditions most of which are associated with mental retardation. The synthesis of most proteins occurs in the liver, with the exception of the immunoglobulins which are produced in the plasma cells. The linear sequence of amino acids in the protein, which comprises the primary structure, is defined by the genetic code in cellular DNA. Proteins may be "simple," composed only of amino acids, or "conjugated," where the peptide chain is attached to a non-protein moiety. With the large number of proteins in the plasma (over 500 identified) the functions are varied. For example, albumin is primarily responsible for the colloid osmotic pressure, haptoglobin and transferrin are transport proteins, binding free hemoglobin and iron respectively, immunoglobulins and components of the complement system help protect the body against infection, and other proteins such as fibrinogen aid in the maintenance of hemostasis. Low levels of total protein may be due to excessive loss, decreased synthesis, or accelerated catabolism, whereas elevated levels are associated with dehydration or excessive production.

The most widely used method for measuring total serum protein levels is the biuret reaction in which cupric ions complex with two or more peptide bonds. To determine the albumin fraction, dye-binding techniques are usually employed utilizing either BCG or BCP. The dye, when bound to albumin, is a different color than the free dye. Further fractionation of the serum proteins can be accomplished by electrophoresis, which utilizes the charge characteristics of the protein for separation. Routine serum protein electrophoresis arranges the proteins in five bands: albumin travels farthest to the anode followed by α_1-globulins, α_2-globulins, β globulins, and γ-globulins. High resolution electrophoresis separates protein into 12 zones. Specific proteins may be identified by immunochemical assays in which the reaction of the protein (antigen) and its antibody is measured. Methods utilizing various modifications of this principle include radial immunodiffusion (RID), immunoelectrophoresis (IEP), immunofixation electrophoresis (IFE), electroimmunodiffusion, immunoturbidimetry, and immunonephelometry.

Protein can also be measured in other body fluids. Either glomerular damage or tubular dysfunction will result in elevated urinary proteins. Abnormally increased CSF protein is found in conditions where there is an increased permeability of the capillary endothelial barrier.

Molecular biology is an area that is rapidly expanding. The detection of specific nucleic acid sequences using DNA probe technology has been employed in the investigation of genetic diseases, detection and identification of infectious agents, and the characterization of some cancers.

CASE STUDY 9-1

Immediately following the birth of a baby girl, the attending physician requested a protein electrophoretic examination of the mother's serum. This was done on a sample that was obtained upon the mother's admission to the hospital the previous day. An electrophoretic examination was also performed on the cord-blood specimen. Laboratory reports were as follows:

CASE STUDY TABLE 9-1. Electrophoresis (Values g/dL)

	Adult Reference Values	Mother's Serum	Cord Blood
Albumin	3.5–5.0	4.2	3.3
α_1-globulins	0.1–0.4	0.3	0.0
α_2-globulins	0.3–0.8	1.2	0.4
β-globulins	0.6–1.1	1.3	0.7
γ-globulins	0.5–1.7	1.3	1.0

The appearance of the mother's electrophoretic pattern was within that expected for a healthy person. The electrophoretic pattern of the cord-blood serum resembled the one shown in Figure 9-11C.

Questions

1. What protein fraction(s) is/are abnormal in the mother's serum and the cord-blood serum?
2. An abnormality in this/these fraction(s) is/are most often associated with what disease?
3. What other test(s) may be done to confirm this abnormality?

CASE STUDY
9-2

The patient is a 57-year-old Caucasian man who was admitted to the hospital on November 11. A battery of 12 screening tests was done on admission. The results showed the following values, which are abnormal. Other blood chemistries, including the total serum protein, were normal.

CASE STUDY TABLE 9-2A

	Reference Values	Patient's Values
Serum		
Albumin (g/dL)	3.5–5.0	2.7
Urea nitrogen (mg/dL)	10–20	34
Creatinine (mg/dL)	0–1.0	1.5
Alkaline phosphatase (U/L)	30–85	215
Aspartate aminotransferase (U/L)	20–50	85
Blood		
Platelet count (µL)	$150–440 \times 10^3$	113,000
White blood cell count (µL/cmm)	$4.5–11.0 \times 10^3$	15,900
Hemoglobin (g/dL)	13.5–18	7.9
Hematocrit (%)	40–54	22.7

The urine showed a 2+ protein and many white cells on microscopic examination. The quantitative urine protein determined the following day was 3.4 g/24 h. The creatinine clearance done several days later was 47 mL/min.

Later, a serum electrophoresis was ordered.

The densitometric tracing showed a slight peak in the early γ-globulin fraction of the pattern. The breadth of the band and its height caused the pathologist to disregard the peak as significant, although several discussions centered on this slightly unusual appearance.

CASE STUDY TABLE 9-2B

Electrophoresis	Reference Value	Patient's Value
Albumin (g/dL)	3.5–5.5	3.4
α_1-globulins (g/dL)	0.1–0.4	0.4
α_2-globulins (g/dL)	0.3–0.8	1.1
β-globulins (g/dL)	0.6–1.1	0.7
γ-globulins (g/dL)	0.5–1.7	1.1
(Total protein) (g/dL)	6–8	6.7

Quantitation of IgG, IgM, and IgA was performed. At the same time, a urine protein electrophoretic pattern showed a similar small peak in the gamma area.

The concentrations obtained for IgM, IgG, and IgA were 29 mg/dL, 370 mg/dL, and 54 mg/dL, respectively. All concentrations were below the lower limits of the reference ranges. A determination for IgD was then done by nephelometry. The IgD concentration was 750 mg/dL after calculation of the dilution.

At the same time, the radiographic examinations of this patient revealed punched-out areas in the bone. A bone-marrow aspirate revealed plasma cell infiltration, and following immunoelectrophoresis, the monoclonal protein was typed and found to be IgD lambda light-chain type.

Questions

1. What was the first laboratory result that indicated a possible protein abnormality?
2. Why was this first indication not taken seriously?
3. Why are the immunoglobulins IgG, IgM, IgA lower than their normal concentrations?

CASE STUDY
9-3[9]

A 13-month-old boy was admitted to a small rural hospital. He had been in a normal state of health until 10 days prior, at which time he had developed an upper respiratory tract infection. He experienced increasing problems with his balance and became lethargic. These symptoms prompted his mother to seek medical attention at the hospital where pertinent laboratory results at admission showed a serum glucose level of 23 mg/dL and moderate ketones in the urine and serum. An intravenous solution of 5% dextrose was initiated to correct the low glucose level. Because the clinical picture resembled Reye's syndrome, the child was transferred to a medical center hospital for a definitive diagnosis. Laboratory results on admission to the medical center hospital appear in Case Study Table 9-3A.

Questions

1. Which laboratory result can be useful in ruling out the diagnosis of Reye's syndrome?
2. What correlations can be made using the blood pH, pCO_2, serum glucose, and serum and urine ketone findings.
3. What other laboratory tests should be performed to verify a defect in protein metabolism?

CASE STUDY TABLE 9-3A. Admission Laboratory Results

Hematology		Urinalysis	
Hct: 37%		Specific gravity	1.022
WBC: 128×10^9/L		Protein	Trace
Bands	28%	Acetone	3+
Segmented	46%	Blood	1+
Lymphocytes	21%		
Monocytes	5%		

Chemistry (Reference Range)

Glucose	133 mg/dL (65–105)	Alk phos	129 U/L (20–70)
BUN	21 mg/dL (7–18)	AST	154 U/L (10–30)
Na	136 mmol/L (136–145)	ALT	133 U/L (8–20)
K	4.3 mmol/L (3.6–5.1)	CK	36 U/L (25–90)
TCO_2	10 mmol/L (23–29)	LD	119 U/L (45–90)
Cl	112 mmol/L (98–106)	Ketone	Moderate
NH_3	48 μmol/L (40–80)		

Arterial blood gases
pH: 7.17
PCO_2: 23 mmHg
PO_2: 90 mmHg

REFERENCES

1. Adinolfi A, Adinolfi M, Lessof MH. Alpha-fetoprotein during development in disease. J Med Genet 1975;12:138.
2. Alper CC, Johnson AM. Immunofixation electrophoresis: A technique for the study of protein polymorphism. Vox Sang 1969;17:445.
3. Anderson NG, Anderson NL. The human protein index. Clin Chem 1982;28:739.
4. Armstrong F. Biochemistry. 3rd ed. New York: Oxford University Press, 1989.
5. Balduyck M, Mizo J. Inter-alpha-trypsin inhibitor and its plasma and urine derivatives. Ann Biol Clin 1991;49(5):273.
6. Berry HK. Neonatal screening: III. The spectrum of metabolic disorders. Diagn Med 1984;7:39.
7. Bradford MM. A rapid and sensitive method for the quantitation of microgram quantities of protein utilizing the principle of protein-dye binding. Anal Biochem 1976;72:248.
8. Buffone GJ, Ellis D. Urinary proteins. In: Ritzmann SE, Killingsworth LM, eds. Proteins in body fluids, amino acids and tumor markers: Diagnostic and clinical aspects. New York: Alan Liss, 1983;117.
9. Campbell P. Case studies. J Med Tech 1985;2:9.
10. Caruthers M. Gene synthesis machines: DNA chemistry and its uses. Science 1985;230:281.
11. Cembrowski G. Testing for microalbuminuria: Promises and pitfalls. Lab Med 1990;21(8):491.
12. Champe P, Harvey R. Lippincott's illustrated reviews: Biochemistry. Philadelphia: JB Lippincott, 1987.
13. Christenson RH, Silverman LM. The determination of urinary protein adapted to the Beckman automated ICS system. [Abstract] Clin Chem 1982;23:1620.
14. Chromy V, Fischer J. Photometric determination of total protein in lipemic serum. Clin Chem 1977;23:754.
15. Cohen SR, Herndon RM, McKhann GM. RIA of myelin basic proteins in spinal fluid. N Engl J Med 1976;23:754.
16. Cohn EJ, Strong LE, Hughes WL Jr, et al. Preparation and properties of serum and plasma proteins: IV. A system for the separation into fractions of the protein and lipoprotein components of biological tissues and fluids. J Am Chem Soc 1946;68:459.
17. Dawson B. DNA diagnostic uses grow. Clin Chem News. September 1989;12.
18. Deyl Z, Hyanek J, Horokova M. Profiling of amino acids in body fluids and tissues by means of liquid chromatography. J Chromatogr 1986;379:177.

19. Dickinson JC, Rosenblum H, Hamilton PB. Ion exchange chromatography of the free amino acids in the plasma of the newborn infant. Pediatrics 1965;36:2.

20. Doumas BT, Bayse DD, Borner K, et al. A candidate reference method for determination of total protein in serum: II. Test for transferability. Clin Chem 1981;27:1651.

21. Doumas BT, Bayse D, Borner K, Carter RJ, Peters T Jr, Schaffer R. A candidate reference method for determination of total protein in serum: I. Development and validation. Clin Chem 1981;27:1642.

22. Doumas BT, Watson WA, Biggs HG. Albumin standards and the measurement of serum albumin with bromocresol green. Clin Chem Acta 1971;31:87.

23. Enns R. DNA probes: An overview and comparison with current methods. Lab Med 1988;19:295.

24. Fahey JL, McKelvey EM. Quantitative determination of serum immunoglobulins in antibodyagar plates. J Immunol 1965;98:84.

25. Feld R. Maple syrup urine disease. Clin Chem News. August 1986;8.

26. Flachaire E, Damour O, Bienvenu J, Aouiti T, Later R. Assessment of the benzethonium chloride method for routine determination of protein in cerebrospinal fluid and urine. Clin Chem 1983;29:343.

27. Frier E. Determination of urine protein with the biuret reaction cited by Kachmar. In: Tietz N, ed. Fundamentals of clinical chemistry. Philadelphia: Saunders, 1976;363.

28. Fujita Y, Mori I, Kitano S. Color reaction between pyrogallol red–molybdenum (VI) complex and protein. Bunseki Kajaku 1983;32:E379.

29. Geiger J, Hoffman P. Quantitative immunologische bestimimmung von 16 verschiedenen serum proteinen bei 260 normalen, 0–15 jahre alten kindern. Z Kinderheilk 1970;109:22.

30. Goldenberg J, Drewes PA. Direct photometric determination of globulin in serum. Clin Chem 1971;17:358.

31. Gorimar TS, Hernandez HA, Polczynsk MW. Rapid protein determination using pyro-chemiluminescence. Am Clin Prod Rev, Nov, 1984.

32. Grabar P, Burtin P. Immunoelectrophoresis. Amsterdam: Elsevier, 1964.

33. Gustafsson JEC. Improved specificity of serum albumin determination and estimation of "acute phase reactants" by use of the bromocresol green reaction. Clin Chem 1976;22:616.

34. Haferkamp O, Schlettwein-Gsell D, Schwick HG, Storiko K. Serum protein in an aging population with particular reference to evaluation of immune globulins and antibodies. Gerontologia 1966;12:30.

35. Heick HMC, Begin HN, Acharya C, Mohammed A. Automated determination of urine and cerebrospinal fluid proteins with Coomassie brilliant blue and the Abbott ABA-100. Clin Biochem 1980;13:81.

36. Hitzig WH, Joller PW. Developmental aspects of plasma proteins. In: Ritzmann SE, Killingsworth LM, eds. Body fluids, amino acids, and tumor markers: Diagnostic and clinical aspects. New York: Alan Liss, 1983;1.

37. Iwata J, Nishikaze O. New microturbidimetric method for determination of protein in cerebrospinal fluid and urine. Clin Chem 1979;25:1317.

38. Johnson AM. Isoelectric focusing. In: Ritzmann SE, Killingsworth LM, eds. Proteins in body fluids, amino acids, and tumor markers: Diagnostic and clinical aspects. New York: Alan Liss, 1983;101.

39. Johnson AM, Schmid K, Alper CA. Inheritance of human$_1$-acid glycoprotein variants. J Clin Invest 1969;48:2293.

40. Kang EP, Fitzpatrick MJ, O'Neill SP. Immunoturbidimetric assay of human fibrinogen using a centrifugal analyzer. [Abstract] Clin Chem 1982;28:1622.

41. Kawai T. Clinical aspects of the plasma proteins. Philadelphia: Lippincott, 1973.

42. Keren D. High resolution electrophoresis aids detection of gammopathies. Clin Chem News, May 11, 1989.

43. Keyser JW. Human plasma proteins. New York: Wiley, 1979;280.

44. Killingsworth LM. Cerebrospinal fluid proteins. In: Ritzmann SE, Killingsworth LM, eds. Proteins in body fluids, amino acids, and tumor markers: Diagnostic and clinical aspects. New York: Alan Liss, 1983;147.

45. Landegren U, Kaiser R, Caskey CT, Hood L. DNA diagnostics: Molecular techniques and automation. Science 1988;242:229.

46. Laterre EC, Callewaert A, Heremans JF, Sfaello A. Electrophoretic morphology of gamma globulins in cerebrospinal fluid of multiple sclerosis and other disease of the nervous system. Neurology 1970;20:982.

47. Laurell CB. Quantitative estimation of proteins by electrophoresis in agarose gel containing antibodies. Anal Biochem 1966;15:45.

48. Laurell CB. Electroimmunoassay. Scand J Clin Lab Invest 1972;29:21.

49. Leary JJ, Brigati DJ, Ward DC. Rapid and sensitive colorimetric methods for visualizing biotin-labeled DNA probes hybridized to DNA probes hybridized to DNA or RNA immobilized on nitrocellulose: Bioblots. Proc Natl Acad Sci USA 1983;80:4045.

50. Levy HL. Phenylketonuria—1986. Pediatr Rev 1986;7:269.

51. Lidsky A, Guttler F, Woo S. Prenatal diagnosis of classic phenylketonuria by DNA analysis. Lancet 1985;1:549.

52. Lines JG, Raines DN. Refractometric determination of serum concentration: 2. Comparison with biuret and Kjeldahl determination. Ann Clin Biochem 1970;7:6.

53. Lowry OH, Rosenbrough NH, Farr AL, Randall RJ. Protein measurement with the Folin-Phenol reagent. J Biol Chem 1951;193:265.

54. Lyngbye J, Kroll J. Quantitative immunoelectrophoresis of proteins in serum from a normal population: Season, age- and sex-related variations. Clin Chem 1971;17:495.

55. Maguire G, Price C. Bromcresol purple method for serum albumin gives falsely low values in patients with renal insufficiency. Clin Chim Acta 1986;155(1):83.

56. Mair J, Dienstl F, Puschendorf B. Cardiac troponin T in the diagnosis of myocardial injury. Crit Rev Clin Lab Sci 1992; 29(1):31.

57. Majoor CLH. The possibility of detecting individual proteins in blood serum by differentiation of solubility curves in concentrated sodium sulfate solutions. J Biol Chem 1947;169:583.

58. Mancini G, Carbonara AO, Heremans JF. Immunochemical quantitation of antigens by single radial immunodiffusion. Immunochemistry 1965;2:235.

59. Matalon R, Michals K. Phenylketonuria: Screening, treatment and maternal PKU. Clin Biochem 1991;24:337.

60. McCaman MW, Robins E. Fluorimetric method for the determination of phenylalanine in serum. J Lab Clin Med 1962; 59:885.

61. McElderry LA, Tarbit IF, Cassells-Smith AJ. Six methods for urinary protein compared. Clin Chem 1982;28:356.

62. Merkatz I, Nitowsky H, Macri J, Johnson W. An association between low maternal serum α fetoprotein and fetal chromosomal abnormalities. Am J Obstet Gynecol 1984;148:886.

63. Meulmans O. Determination of total protein in spinal fluid with sulfosalicylic acid and trichloroacetic acid. Clin Chem Acta 1960;5:757.

64. Michael BS. The routine examination of urine. In: Ross DL, Neely RA, eds. Textbook of urinalysis of body fluids. New York: Appleton-Century-Crofts, 1983;67.

65. Natelson S, Natelson E. Principles of applied clinical chemistry plasma proteins in nutrition and transport. Vol 3. New York: Plenum, 1980.

66. Nichols WS, Nakamura RM. Biologic serum markers associated with human neoplasia. In: Ritzmann SE, Killingsworth LM, eds. Proteins in body fluids, amino acids, and tumor markers: Diagnostic and clinical aspects. New York: Alan Liss, 1983;378.

67. Ouchterlony O. Antigen-antibody reactions in gels. Acta Pathol Microbiol Scand 949;26:507.

68. Parviainen M, Harmoinen A, Jokela H. Serum albumin assay with bromcresol purple dye. Scand J Clin Lab Invest 1985; 45(6):561.

69. Pesce MA, Bodourian SH. Nephelometric measurement of ceruloplasmin with a centrifugal analyzer. Clin Chem 1982; 28:516.

70. Pesce MA, Strande CS. A new micromethod for the determination of protein in cerebrospinal fluid and urine. Clin Chem 1973;19:1265.

71. Peters T Jr, Biamonte ET, Doumas BT. Protein (total) in serum, urine, cerebrospinal fluid, albumin in serum. In: Faulkner WR, Meites S, eds. Selected methods in clinical chemistry. Vol 9. Washington: American Association for Clinical Chemistry, 1982;317.

72. Pierce J, Svelter C. An evaluation of the Coomassie blue G250 dye binding method for quantitative protein determination. Anal Biochem 1977;81:478.

73. Reiber H. Kinetics of protein agglomeration. A nephelometric method for the determination of total protein in biological samples. J Biochem Biophys Methods 1983;7:153.

74. Renz M, Kruz C. A colorimetric method for DNA hybridization. Nucleic Acid Res 1984;12:3435.

75. Richie RF, Alper CA, Graves J, Pearson N, Larson C. Automated quantitation of proteins in serum and other biologic proteins. Am J Clin Pathol 1979;59:151.

76. Richie RF. Specific proteins. In: Henry JB, ed. Clinical diagnosis and management by laboratory methods. 6th ed. Vol 1. Philadelphia: Saunders, 1979;243.

77. Ritzmann SE, Killingsworth LM. Protein abnormalities, proteins in body fluids, amino acids, and tumor markers: Diagnostic and clinical aspects. 3rd ed. New York: Alan Liss, 1983.

78. Ritzmann SE, Finney MA. Proteins—Synopsis of characteristics and properties. In: Ritzmann SE, Killingsworth LM, eds. Proteins in body fluids, amino acids, and tumor markers: Diagnostic and clinical aspects. New York: Alan Liss, 1983;422.

79. Roitt I, Brostoff J, Male D. Immunology. St. Louis: Mosby, 1985;1.

80. Roos RP, Lichter M. Silver staining of cerebrospinal fluid IgG in isoelectric focusing gels. J Neurosci Methods 1983; 8:375.

81. Ross JN, Fraser MD. Analytical chemistry precision: State of the art for fourteen analytes. Am J Clin Pathol 1977;68:130.

82. Rubin H. The biology and biochemistry of antichymotrypsin and its potential role as a therapeutic agent. Biol Chem Hoppe Seyler 1992;373(7):497.

83. Saiki RK, Chang CA, Levenson CH, et al. Primer directed enzymatic amplification of DNA with a thermostable DNA polymerase. Science 1988;239:497.

84. Schriever H, Gambino SR. Protein turbidity produced by trichloroacetic acid and sulfosalicylic acid at varying temperatures and varying ratios of albumin and globulin. Am J Clin Pathol 1965;44:667.

85. Seppala M, Ruoslahti E. Alpha-fetoprotein in maternal serum: A new marker for detection of fetal distress and intrauterine death. Am J Obstet Gynecol 1973;115:48.

86. Shen RS, Abell CW. Phenylketonuria: A new method for the simultaneous determination of plasma phenylalanine and tyrosine. Science 1977;197:665.

87. Sherlock S. Diseases of the liver and biliary system. 4th ed. Oxford: Blackwell, 1971;38.

88. Sinclair D, Kumeraratre DS, Forrester JB, Lamont A, Stott DI. The application of isoelectric focusing to routine screening for paraproteinemia. J Immunol Methods 1983;64:147.

89. Speicher CE, Widish JR, Gaudot FJ, Helper BR. An evaluation of the overestimation of serum albumin by bromcresol green. Am J Clin Pathol 1978;69:347.

90. Stephens AD. Cystinuria and its treatment: 25 years experience at St. Bartholomew's Hospital. J Inherit Metab Dis 1989; 12:197.

91. Strickland RD, Freeman ML, Gurule FT. Copper binding by proteins in alkaline solution. Anal Chem 1961;33:545.

92. Sun T, Lien YY, Degnan T. Studies of gammopathies with immunofixation electrophoresis. Am J Clin Pathol 1979;72:5.

93. Sunderman FW. Studies of the serum proteins: VI. Recent advances in clinical interpretation of electrophoretic fractions. Am J Clin Pathol 1964;42:1.

94. Tibbling G, Link H, Ohman S. Principles of albumin and IgG analyses in neurological disorders: I. Establishment of reference values. Scand J Clin Lab Invest 1977;37:385.

95. Tune L. Method generates probe copies. Clin Chem News. June 1989;1.

96. Vankley J, Hales S. Assay for protein by dye binding. Anal Biochem 1977;81:485.

97. Vesterberg O. Isoelectric focusing of proteins. In:Colowick EJ and Kaplan, eds. Methods in enzymology. Vol 22. New York: Academic Press, 1971;389.

98. Ward MWN, Owens CWI, Rennie MJ. Nitrogen estimation in biological samples by use of chemiluminescence. Clin Chem 1980;26:1336.

99. Warren B. Immunofixation electrophoresis methodology and application. Beaumont, TX: Helena Laboratories, 77704-0752.

100. Watson JD, Crick FHC. Molecular structure of nucleic acid. A structure for deoxyribose nucleic acid. Nature 1953;171:737.

101. Webster D. A study of the interaction of bromcresol green with isolated serum globulin fractions. Clin Chem Acta 1974; 53:109.

102. Webster D, Bignell AHC, Attwood EC. An assessment of the suitability of bromcresol green for the determination of serum albumin. Clin Chem Acta 1974;53:101.

103. Weeks I, Beheshti I, Capra FM, Campbell AK, Woodhead JS. Acridinium esters as high-specific-activity labels in immunoassay. Clin Chem 1983;29:1474.

104. Weiss SM, Kulikowski CA, Galen RS. Representing expertise in a computer program: The serum protein diagnostic program. J Clin Lab Autom 1983;3:383.

105. Whicher JT, Warren C, Chambers RE. Immunochemical assays for immunoglobulins. Ann Clin Biochem 1984;21:78.

106. Yamaguchi A, Mizushima Y, Fukushi M, et al. Microassay system for newborn screening for phenylketonuria, maple syrup urine disease, homocystinuria, histidinemia and galactosemia with use of a fluoremetric microplate reader. Screening 1992;1:49.

Chapter 10

Enzymes

Bethany L. Hurtuk, Robin Gaynor Krefetz

Objectives

Upon completion of this chapter, the clinical laboratorian should be able to:

- *Define enzyme, including physical composition and structure.*
- *Classify enzymes according to the International Union of Biochemistry (IUB).*
- *Discuss the different factors affecting the rate of an enzymatic reaction.*
- *Describe zero-order kinetics.*
- *Discuss which enzymes are useful in the diagnosis of cardiac disorders, hepatic disorders, bone disorders, muscle disorders, malignancies, and acute pancreatitis.*
- *Discuss the tissue sources, diagnostic significance, and assays, including sources of error, for the following enzymes: CK, LD, AST, ALT, ALP, ACP, GGT, amylase, lipase, cholinesterase, and G-6-PD.*
- *Evaluate patient serum enzyme levels in relation to disease states.*
- *Explain why the measurement of serum enzyme levels is clinically useful.*

KEY WORDS

Activation Energy	Hydrolase	Oxidoreductase
Cofactor	International Unit	Transferase
Enzyme	Isoenzyme	Zero-Order Kinetics
Enzyme-Substrate	Kinetic Assay	
Complex	LD Flipped Pattern	

*E*nzymes are specific biologically synthesized proteins that catalyze biochemical reactions without altering the equilibrium point of the reaction or being consumed or undergoing changes in composition. Other reactants are converted to products. The catalyzed reactions are frequently specific and essential to physiologic functions, such as the hydration of carbon dioxide, nerve conduction, muscle contraction, nutrient degradation, and energy utilization.

Found in all body tissues, enzymes frequently appear in the serum following cellular injury or sometimes, in smaller amounts, from degraded cells or storage areas. Some enzymes, such as those which facilitate coagulation, are specific to plasma and are, therefore, present in significant concentrations in plasma. Hence plasma or serum enzyme levels are often useful in the diagnosis of particular diseases or physiologic abnormalities. This chapter discusses the general properties and principles of enzymes, aspects relating to the clinical diagnostic significance of specific physiologic enzymes, and assay methods for those enzymes.

GENERAL PROPERTIES/DEFINITIONS

As stated earlier, enzymes catalyze many specific physiologic reactions. The catalysis is facilitated by the structure of the enzymes themselves and by several other factors. As a protein, each enzyme is composed of a specific amino acid sequence (*primary structure*) with the resultant polypeptide chains in a steric arrangement (*secondary structure*), which is folded (*tertiary structure*), resulting in structural cavities. If an enzyme contains more than one polypeptide unit, the

Michael L. Bishop, Janet L. Duben-Engelkirk, and Edward P. Fody.
CLINICAL CHEMISTRY. © 1996 Lippincott–Raven Publishers.

quaternary structure refers to the spatial relationships between the subunits. Each enzyme contains an *active site,* often a water-free cavity, where the substance upon which the enzyme acts (the *substrate*) interacts with particular charged amino acid residues. An *allosteric site,* which is a cavity other than the active site, may bind regulator molecules and thereby be significant to the basic enzyme structure.

Even though a particular enzyme maintains the same catalytic function throughout the body, that enzyme may exist in different forms within the same individual. The different forms, which may originate from genetic or nongenetic causes, may be differentiated from each other based on certain physical properties, such as electrophoretic mobility, solubility, or resistance to inactivation. The term *isoenzyme* is generally used when discussing such enzymes, but the International Union of Biochemistry (IUB) suggests restricting this term to multiple forms of genetic origin.

In addition to the basic enzyme structure, a nonprotein molecule, a *cofactor,* may be necessary for enzyme activity. Inorganic cofactors, such as chloride or magnesium ions, are called *activators.* A *coenzyme* is an organic cofactor, such as nicotinamide adenine dinucleotide. When bound tightly to the enzyme, the coenzyme is called a *prosthetic group.* The enzyme portion (*apoenzyme*), with its respective coenzyme, forms a complete and active system, a *holoenzyme.*

Some enzymes, mostly digestive enzymes, are originally secreted from the organ of production in a structurally inactive form, called a *proenzyme* or *zymogen.* Other enzymes later alter the structure of the proenzyme to make active sites available by hydrolyzing specific amino acid residues. This mechanism prevents digestive enzymes from digesting their place of synthesis.

ENZYME CLASSIFICATION AND NOMENCLATURE

Enzyme nomenclature has historically been a source of confusion. Some enzymes have been named arbitrarily (ptyalin, trypsin), while others were named by attaching the suffix *-ase* to the name of the substrate upon which the enzyme acted (*e.g.,* esterase, urease) or to the type of reaction catalyzed (*e.g.,* aminotransferase, oxidase). Some enzymes have been identified by different names in different parts of the world. The continual discovery of more enzymes has complicated enzyme identification by name.

To standardize enzyme nomenclature, the Enzyme Commission (EC) of the IUB adopted a classification system in 1961 and revised the standards in 1972 and 1978. The IUB systems assigns a *systematic name* to each enzyme, defining the substrate acted on, the reaction catalyzed, and possibly, the name of any coenzyme involved in the reaction. Because many systematic names are lengthy, a more usable, trivial, *recommended name* is also assigned by the IUB system.[10]

In addition to naming enzymes, the IUB system identifies each enzyme by an Enzyme Commission (EC) numerical code containing four digits separated by decimal points.[10] The first digit places the enzyme in one of the following six classes:

1. Oxidoreductases—catalyze an oxidation–reduction reaction between two substrates.
2. Oxidoreductases—catalyze the transfer of a group other than hydrogen from one substrate to another.
3. Hydrolases—catalyze hydrolysis of ether, ester, acid–anhydride, glycosyl, C—C, C—halide, or P—N bonds.
4. Lyases—catalyze removal of groups from substrates without hydrolysis. The product contains double bonds.
5. Isomerases—catalyze the interconversion of geometric, optical, or positional isomers.
6. Ligases—catalyze the joining of two substrate molecules, coupled with breaking of the pyrophosphate bond in ATP or a similar compound.

The second and third figures represent the subclass and sub-subclass of the enzyme, respectively, divisions that are made according to criteria specific to the enzymes in the class. The final number is a serial number that is specific to each enzyme in a sub-subclass. Table 10-1 provides the EC code numbers, as well as the systematic and recommended names, for enzymes frequently quantitated in the clinical laboratory.

Table 10-1 also lists common and standard abbreviations for commonly analyzed enzymes. Without IUB recommendation, capital letters have been used as a convenience to identify enzymes. The common abbreviations, sometimes developed from previously accepted names for the enzymes, were used until the standard abbreviations listed in the table were developed.[2,3] These standard abbreviations are being used in the United States and may be used later in this chapter to indicate specific enzymes.

ENZYME KINETICS

Catalytic Mechanism of Enzymes

A chemical reaction may occur spontaneously if the free energy or available kinetic energy is higher for the reactants than for the products. The reaction then proceeds toward the lower energy if a sufficient number of the reactant molecules possess enough excess energy to break their chemical bonds and collide to form new bonds. The excess energy, called *activation energy,* is defined as the energy required to raise all molecules in one mole of a compound at a certain temperature to the transition state at the peak of the energy barrier. At the transition state, each molecule is equally likely either to participate in product formation or to remain an unreacted molecule. Reactants possessing enough energy to overcome the energy barrier participate in product formation.

One way to provide more energy for a reaction is to increase the temperature and thus increase intermolecular collisions, but this does not normally occur physiologically. Enzymes catalyze physiologic reactions by lowering the

TABLE 10-1. Classification of Frequently Quantitated Enzymes

Class	Recommended Name	Common Abbreviation	Standard Abbreviation	EC Code Number	Systematic Name
Oxidoreductases	Lactate dehydrogenase	LDH	LD	1.1.1.27	L-Lactate: NAD⁺ oxidoreductase
	Glucose-6-phosphate dehydrogenase	G–6–PDH	G6PD	1.1.1.49	D-Glucose-6-phosphate: NADP⁺ 1-oxidoreductase
Transferases	Aspartate aminotransferase	GOT (glutamate oxaloacetate transaminase)	AST	2.6.1.1	L-Aspartate: 2-oxaloglutarate aminotransferase
	Alanine aminotransferase	GPT (glutamate transaminase)	ALT	2.6.1.2	L-Alanine: 2-oxaloglutarate aminotransferase
	Creatine kinase	CPK (creatine phosphokinase)	CK	2.7.3.2	ATP: creatine N-phospho-transferase
	Gamma glutamyltransferase	GGTP	GGT	2.3.2.2	(5-Glutamyl) peptide: amino acid-5-glutamyltransferase
Hydrolases	Alkaline phosphatase	ALP	ALP	3.1.3.1	Orthophosphoric monoester phosphohydrolase (alkaline optimum)
	Acid phosphatase	ACP	ACP	3.1.3.2	Orthophosphoric monoester phosphohydrolase (acid optimum)
	α-Amylase	AMY	AMS	3.2.1.1	1,4-D-glucan gluconhydrolase
	Triacylglycerol lipase		LPS	3.1.1.3	Triacylglycerol acylhydrolase
	Cholinesterase	CHS	CHS	3.1.1.8	Acylcholine acylhydrolase

Adapted from Competence Assurance, ASMT, Enzymology, An Educational Program, RMI Corporation, 1980, with permission.

activation energy level that the reactants (*substrates*) must reach for the reaction to occur (Fig. 10-1). The reaction may then occur more readily to a state of equilibrium in which there is no net forward or reverse reaction, even though the equilibrium constant of the reaction is not altered. The extent to which the reaction progresses depends on the number of substrate molecules that pass the energy barrier.

The general relationship among enzyme, substrate, and product may be represented as follows:

$$E + S \rightleftharpoons ES \rightleftharpoons E + P \qquad \textit{(Eq. 10-1)}$$

where E = enzyme
S = substrate
ES = enzyme-substrate complex
P = product

The ES complex is a physical binding of a substrate to the active site of an enzyme. The structural arrangement of amino acid residues within the enzyme makes the three-dimensional active site available. The transition state for the ES complex has a lower energy of activation than has the transition state of S alone so that the reaction proceeds after the complex is formed. An actual reaction may involve two substrates and two products.

Different enzymes are specific to substrates to different extents or in different respects. Some enzymes exhibit *absolute specificity,* meaning that the enzyme combines with only one substrate and catalyzes only the one corresponding reaction. Other enzymes are *group-specific* because they com-

bine with all substrates containing a particular chemical group, such as a phosphate ester. Still other enzymes are specific to chemical bonds and thereby exhibit *bond specificity.* *Stereoisometric specificity* refers to enzymes that combine with only one optical isomer of a certain compound.

Factors that Influence Enzymatic Reactions

Substrate Concentration

The rate at which an enzymatic reaction proceeds and whether the forward or reverse reaction occurs depend on several reaction conditions. One major influence on enzymatic reactions is substrate concentration. In 1913, Michaelis and Menten hypothesized the role of substrate concentration in formation of the enzyme-substrate complex. According to the hypothesis, which is represented in Figure 10-2, the substrate readily binds to free enzyme at a low substrate concentration. With the amount of enzyme exceeding the amount of substrate, the reaction rate steadily increases as more substrate is added. The reaction is following *first-order kinetics,* because the reaction rate is directly proportional to substrate concentration. Eventually, however, the substrate concentration is high enough to saturate all available enzyme, and the reaction velocity reaches its maximum. When product is formed, the resultant free enzyme immediately combines with excess free substrate. The reaction is in *zero-order kinetics,* and the reaction rate is dependent on enzyme concentration only.

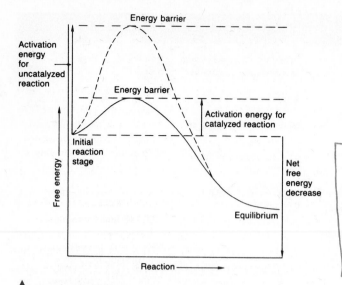

Figure 10-1. Energy versus progression of reaction, indicating the energy barrier that the substrate must surpass to react with and without enzyme catalysis. The enzyme considerably reduces the free energy needed to activate the reaction.

From their theory, Michaelis and Menten derived the *Michaelis-Menten constant* (K_m), which is a constant for a specific enzyme and substrate under defined reaction conditions and is an expression of the relationship between the velocity of an enzymatic reaction and substrate concentration. The assumptions are made that an equilibrium among E, S, ES, and P is established rapidly and that the E+P → ES reaction is negligible. The rate-limiting step is the formation of product and enzyme from the enzyme-substrate complex. Then, maximum velocity is fixed, and the reaction rate is a function of only the enzyme concentration. As designated in Figure 10-2, K_m is specifically the substrate concentration at which the enzyme yields half the possible maximum velocity. Hence K_m is an indication of the amount of substrate needed for a particular enzymatic reaction.

The Michaelis–Menten hypothesis of the relationship between reaction velocity and substrate concentration can be represented mathematically as follows:

$$V = \frac{V_{max}\,[S]}{K_m + [S]} \qquad \textbf{(Eq. 10-2)}$$

where V = measured velocity of reaction
V_{max} = maximum velocity
$[S]$ = substrate concentration
K_m = Michaelis–Menten constant of enzyme for specific substrate

Theoretically, V_{max} and then K_m could be determined from the plot in Figure 10-2. However, V_{max} is difficult to determine from the hyperbolic plot and often is not actually achieved in enzymatic reactions, because enzymes may not function optimally in the presence of excessive substrate. A more accurate determination of K_{min} may be made through a *Lineweaver-Burk plot,* a double-reciprocal plot of the Michaelis–Menten constant, which yields a straight line (Fig. 10-3). The reciprocal is taken of both the substrate concentration and the velocity of an enzymatic reaction. The equation becomes

$$\frac{1}{V} = \frac{K_m}{V_{max}}\,\frac{1}{[S]} + \frac{1}{V_{max}} \qquad \textbf{(Eq. 10-3)}$$

As indicated in Figure 10-3, both V_{max} and K_m may conveniently be determined from the plot.

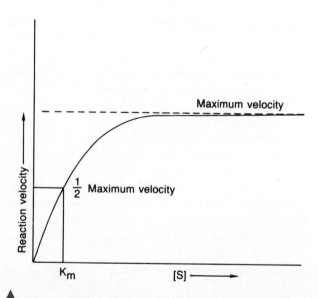

Figure 10-2. Michaelis-Menten curve of velocity versus substrate concentration for enzymatic reaction. K_m is the substrate concentration at which the reaction velocity is half of the maximum level.

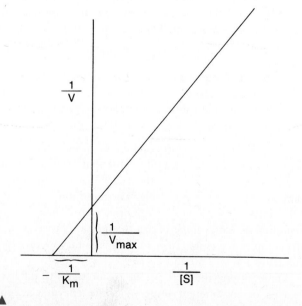

Figure 10-3. Lineweaver-Burk transformation of Michaelis-Menten curve. V_{max} is the reciprocal of the γ intercept of the straight line. K_m is the negative reciprocal of the x intercept of the same line.

Enzyme Concentration

Since enzymes catalyze physiologic reactions, the enzyme concentration affects the rate of the catalyzed reaction. As long as the substrate concentration exceeds the enzyme concentration, the velocity of the reaction is proportional to the enzyme concentration. The higher the enzyme level, the faster the reaction will proceed, because more enzyme is present to bind with the substrate.

pH

Enzymes are proteins that carry net molecular charges. Extreme pH levels may denature an enzyme or influence its ionic state, resulting in structural changes or a change in the charge on an amino acid residue in the active site. Hence, each enzyme operates within a specific pH range and maximally at a specific pH. Most physiologic enzymatic reactions occur in the pH range of 7 to 8, but some enzymes are active in wider pH ranges than others. In the laboratory, the pH for a reaction is carefully controlled at the optimal pH by means of appropriate buffer solutions.

Temperature

Increasing temperature usually increases the rate of a chemical reaction by increasing the movement of molecules, the rate at which intermolecular collisions occur, and the energy available for the reaction. Such is the case with enzymatic reactions, until the temperature is high enough to denature the protein composition of the enzyme. The Q_{10} value, the increase in reaction rate for every 10°C increase in temperature, is approximately 2 for enzymes, indicating that this increase in temperature usually doubles the reaction rate.

Each enzyme functions optimally at a particular temperature, which is influenced by other reaction variables, especially the total time for the reaction. The optimal temperature is usually close to that of the physiologic environment of the enzyme. However, some denaturation may occur at the human physiologic temperature of 37°C. The rate of denaturation increases as the temperature increases and is usually significant at 40 to 50°C.

Because low temperatures render enzymes reversibly inactive, many serum or plasma specimens for enzyme quantitations are refrigerated or frozen to prevent activity loss until analysis. Storage procedures may vary from enzyme to enzyme because of individual stability characteristics. However, repeated freezing and thawing tends to denature protein and should be avoided.

Because of their temperature sensitivity, enzymes should be analyzed under strictly controlled temperature conditions. Incubation temperatures should be accurate within ± 0.1°C. Laboratories usually attempt to establish an analysis temperature for routine enzyme quantitations of 25, 30, or 37°C. Attempts to establish a universal temperature for enzyme analysis have been futile, and thus reference ranges for enzyme levels may vary significantly from laboratory to laboratory. In the United States, however, 37°C is most commonly used.

Cofactors

Cofactors, as defined earlier, are nonprotein entities that must bind to particular enzymes before catalysis occurs. Common *activators,* inorganic cofactors, are metallic (Ca^{2+}, Fe^{2+}, Mg^{2+}, Mn^2 Zn^{2+}, and K^+) and nonmetallic (Br^- and Cl^-). The activator may be essential for the reaction or may only enhance the reaction rate in proportion with concentration to the point at which the excess activator begins to inhibit the reaction. Activators function by alternating the spatial configuration of the enzyme for proper substrate binding, linking substrate to the enzyme or a coenzyme, or undergoing oxidation or reduction.

Some common *coenzymes,* organic cofactors, are nucleotide phosphates and vitamins. Coenzymes serve as second substrates for enzymatic reactions, and when bound tightly to the enzyme, they are called *prosthetic groups.* For example, nicotinamide adenine dinucleotide (NAD) as a cofactor may be reduced to nicotinamide adenine dinucleotide phosphate (NADP) in a reaction in which the primary substrate is oxidized. Hence, increasing coenzyme concentration will increase the velocity of an enzymatic reaction in a manner synonymous with increasing substrate concentration. When quantitating an enzyme that requires a particular cofactor, that cofactor should always be provided in excess so that the extent of the reaction is not dependent on the concentration of the cofactor.

Inhibitors

Enzymatic reactions may not progress normally if a particular substance, an *inhibitor,* interferes with the reaction. *Competitive inhibitors* physically bind to the active site of an enzyme and thereby compete with the substrate for the active site. With a substrate concentration significantly higher than the concentration of the inhibitor, the inhibition is reversible, because the substrate is more likely than the inhibitor to bind the active site, and the enzyme has not been destroyed.

A *noncompetitive inhibitor* binds an enzyme at a place other than the active site and may be reversible in the respect that some naturally present metabolic substances combine reversibly with certain enzymes. Noncompetitive inhibition also may be *irreversible* if the inhibitor destroys part of the enzyme involved in catalytic activity. Since the inhibitor binds the enzyme independently from the substrate, increasing substrate concentration does not reverse the inhibition.

In still another kind of inhibition, *uncompetitive inhibition,* the inhibitor binds to the enzyme-substrate complex so that increasing substrate concentration results in more enzyme-substrate complexes to which the inhibitor binds and thereby increases the inhibition. The enzyme-substrate-inhibitor complex does not yield product.

Each of the three kinds of inhibition is unique with respect to effects on the V_{max} and K_m of enzymatic reactions (Fig. 10-4). In competitive inhibition, the effect of the inhibitor can be counteracted by adding excess substrate to bind the enzyme. The amount of the inhibitor is then negligible by comparison, and the reaction will proceed at a

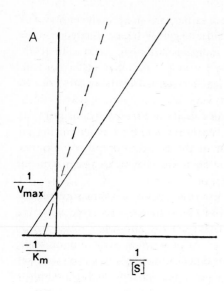

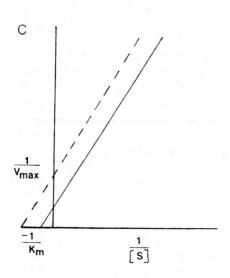

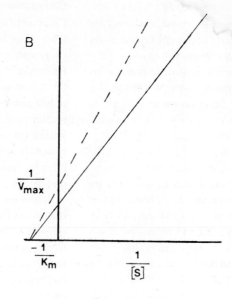

◄

Figure 10-4. Normal Lineweaver-Burk plot (solid line) compared with each type of enzyme inhibition (dotted line). (*A*) Competitive inhibition—V_{max} unaltered; K_m appears increased. (*B*) Noncompetitive inhibition—V_{max} decreased; K_m unchanged. (*C*) Uncompetitive inhibition—V_{max} decreased; K_m appears decreased.

slower rate but to the same maximum velocity as an uninhibited reaction. The K_m is a constant for each enzyme and cannot be altered. However, since the amount of substrate needed to achieve a particular velocity is higher in the presence of a competing inhibitor, the K_m appears to increase when exhibiting the effect of the inhibitor.

The substrate and inhibitor, commonly a metallic ion, may bind an enzyme simultaneously in noncompetitive inhibition. The inhibitor may inactivate either an enzyme-substrate complex or just the enzyme by causing structural changes in the enzyme. Even if the inhibitor binds reversibly and does not inactivate the enzyme, the presence of the inhibitor when it is bound to the enzyme slows the rate of the reaction. Thus, for noncompetitive inhibition, the maximum reaction velocity cannot be achieved. Increasing substrate levels has no influence on the binding of a noncompetitive inhibitor, so the K_m is unchanged.

Since uncompetitive inhibition requires the formation of an enzyme-substrate complex, increasing substrate concentration increases inhibition. Thus, maximum velocity equal to that of an uninhibited reaction cannot be achieved, and the K_m appears to be decreased.

Measuring Enzyme Activity

Since enzymes are often present in very small quantities in biologic fluids and are often difficult to isolate from similar compounds, the most convenient quantitations of enzymes measure catalytic activity and relate this activity to concentration. A given method might photometrically measure: (1) an increase in product concentration, (2) a decrease in substrate concentration, (3) a decrease in coenzyme concentration, or (4) an increase in the concentration of an altered coenzyme.

If the amount of substrate and any coenzyme is in excess in an enzymatic reaction, the amount of substrate or coenzyme used or product or altered coenzyme formed will depend only on the amount of enzyme present to catalyze the reaction. Hence, enzyme concentrations are performed in zero-order kinetics, with the substrate in sufficient excess to ensure that not more than 20% of the available substrate is converted to product. Any coenzymes also must be in excess. NADH is a coenzyme frequently measured in the laboratory. NADH absorbs light at 340 nm, whereas NAD does not, and a change in absorbance at 340 nm is readily measurable.

In specific laboratory methodologies, substances other than substrate or coenzyme are necessary and must be present in excess. NAD or NADH is often convenient as a reagent for a *coupled-enzyme* assay when neither NAD nor NADH is a coenzyme for the reaction. In other coupled-enzyme assays, more than one enzyme is added in excess as a reagent, and multiple reactions are catalyzed. After the enzyme under analysis catalyzes its specific reaction, a product of that reaction becomes the substrate upon which an intermediate *auxiliary enzyme* acts. A product of the intermediate reaction becomes the substrate for the final reaction, which is catalyzed by an *indicator enzyme* and commonly involves the conversion of NAD to NADH or *vice versa.*

When performing an enzyme quantitation in zero-order kinetics, inhibitors must be lacking, and other variables that may influence the rate of the reaction must be carefully controlled. A constant pH should be maintained by means of an appropriate buffer solution. The temperature should be constant within $\pm 0.1°C$ throughout the assay at a temperature at which the enzyme is active (usually 25, 30, or 37°C).

During the progress of the reaction, the time period for the analysis also must be carefully selected. When the enzyme is initially introduced to the reactants and the excess substrate is steadily combining with available enzyme, the reaction rate rises. After the enzyme is saturated, the rates of product formation, release of enzyme, and recombination with more substrate proceed linearly. After a length of time, usually 6 to 8 minutes after reaction initiation, the reaction rate decreases as the substrate is depleted, the reverse reaction is occurring appreciably, and the product begins to inhibit the reaction. Hence, enzyme quantitations must be performed during the linear phase of the reaction.

One of two general methods may be used to measure the extent of an enzymatic reaction: (1) fixed-time and (2) continuous-monitoring. For the *fixed-time method,* the reactants are combined, the reaction proceeds for a designated length of time, the reaction is stopped (usually by inactivating the enzyme), and a measurement is made of the amount of reaction that has occurred. The reaction is assumed to be linear over the reaction time, and so the larger the reaction, the more enzyme is present.

In *continuous-monitoring assays,* multiple measurements, usually of absorbance change, are made during the reaction, either at specific time intervals, usually every 30 or 60 seconds, or continuously by a continuous-recording spectrophotometer. Such assays are advantageous over fixed-time methods because the linearity of the reaction may be more adequately verified. If absorbance is measured at intervals, several data points are necessary to increase the accuracy of linearity assessment. Continuous measurements are preferred because any deviation from linearity is readily observable.

The most common cause of deviation from linearity occurs when the enzyme is so elevated that all substrate is used early in the reaction time. For the remainder of the reaction the rate change is minimal, and the implication is that the coenzyme concentration is very low. With continuous monitoring, the analyst may observe a sudden decrease in the reaction rate (deviation from zero-order kinetics) of a particular determination and may repeat the determination using less patient sample. The decrease in the amount of patient sample operates as a dilution, and the answer obtained may be multiplied by the dilution factor to obtain the final answer. The sample itself is not diluted so that the diluent cannot interfere with the reaction. (Sample dilution with saline may be necessary to minimize negative effects in analysis caused by hemolysis or lipemia.)

Immunoassay methodologies are also available for some enzymes. Since enzymes are proteins, they usually cause antibody production when injected into animals to which they are foreign. The antisera formed is very specific for the enzyme and can be used in conjunction with the sensitivity of immunoassay to quantitate an enzyme or isoenzyme that is present in serum in small amounts. Also, since the method is independent of the catalytic activity of the enzyme, inactive enzyme protein may be measured.

Calculation of Enzyme Activity

Since enzymes are quantitated relative to their activity rather than direct concentration, the units used to report enzyme levels are *activity units.* The definition for the activity unit must consider variables that may alter results (pH, temperature, substrate, etc.). Historically, specific method developers frequently established their own units for reporting results with respect to their method conditions. To standardize the system of reporting quantitative results, the CE defined the *international unit* (U) as the amount of enzyme that will catalyze the reaction of one micromole of substrate per minute under specified conditions of temperature, pH, substrates, and activators. Because the specified conditions may vary from laboratory to laboratory, reference values are still often laboratory-specific. Enzyme concentration is usually expressed in units/liter (U/L).

Another unit recommended for expressing an enzyme activity unit is the Katal (mol/s). The mole is the unit for substrate concentration, and the unit of time is the second. Enzyme concentration is then expressed as Katals per liter (Kat/L). $1.0 \text{ U} = 60 \text{ } \mu\text{Kat}.$

When enzymes are quantitated by measuring the increase or decrease of NADH at 340 nm, the molar absorptivity (6.22×10^3) of NADH is used to calculate enzyme activity. Because an NADH solution with a concentration of 1.0 μmol/mL has an absorbance of 6.22, a 0.001 absorbance decrease represents a concentration change of 0.001/6.22 (1.6×10^{-4}) μmol/mL.

Enzymes as Reagents

Enzymes may be used as reagents to measure nonenzymatic physiologic constituents. For example, glucose, cholesterol, and uric acid are frequently quantitated by means of enzymatic reactions. Such methods measure the "true" level of the analyte due to the specificity of the enzyme. Enzymes also might serve as reagents for methods of quantitating analytes that are substrates for corresponding enzyme quantitations. For example, lactate dehydrogenase may be a reagent when lactic acid levels are evaluated. For such methods, the enzyme is added in excess in a quantity sufficient to provide a complete reaction in a short time period.

Immobilized enzymes are chemically bonded to adsorbents such as agarose or some types of cellulose by azide groups, diazo, and triazine. The enzymes act as recoverable reagents. When substrate is passed through the preparation, the product is retrieved and analyzed, and the enzyme is present and free to react with more substrate. Immobilized enzymes are convenient for continuous-flow or batch analyses and are more stable than enzymes in a solution.

ENZYMES OF CLINICAL SIGNIFICANCE

Table 10-2 lists the commonly analyzed enzymes, including their systematic names and clinical significance. Each enzyme will be discussed in this chapter with respect to tissue sources, diagnostic significance, assay methods, sources of error, and reference range.

Creatine Kinase

Creatine kinase (CK) is an enzyme that is generally associated with ATP regeneration in contractile or transport systems. Its predominate physiologic function occurs in muscle cells, where it is involved in the storage of high-energy creatine phosphate. Every contraction cycle of muscle results in creatine phosphate utilization, with the production of ATP. This results in relatively constant levels of muscle ATP. The reversible reaction catalyzed by creatine kinase is shown in Equation 10-4.

$$\text{Creatine} + \text{ATP} \overset{CK}{\rightleftharpoons} \text{Creatine phosphate} + \text{ADP} \qquad (Eq.\ 10\text{-}4)$$

Tissue Sources

CK is widely distributed in tissues, with highest activities found in skeletal muscle, heart muscle, and brain tissue.

TABLE 10-2. Major Enzymes of Clinical Significance

Enzyme	Clinical Significance
Acid phosphatase (ACP)	Prostatic carcinoma
Alanine aminotransferase (ALT)	Hepatic disorders
Alkaline phosphatase (ALP)	Hepatic disorders
	Bone disorders
Amylase (AMS)	Acute pancreatitis
Aspartate aminotransferase (AST)	Myocardial infarction
	Hepatic disorders
	Skeletal muscle disorders
Creatine kinase (CK)	Myocardial infarction
	Skeletal muscle disorders
γ-Glutamyltransferase (GGT)	Hepatic disorders
Glucose-6-phosphate dehydrogenase (G–6–PD)	Drug-induced hemolytic anemia
Lactate dehydrogenase (LD)	Myocardial infarction
	Hepatic disorders
	Carcinomas
Lipase (LPS)	Acute pancreatitis
Pseudocholinesterase (PChE)	Organophosphate poisoning
	Genetic variants

Other tissue sources in which CK is present in much smaller quantities include the bladder, placenta, gastrointestinal tract, thyroid, uterus, kidney, lung, prostate, spleen, liver, and pancreas.

Diagnostic Significance

Because of the high concentrations of CK in muscle tissue, CK levels are frequently elevated in disorders of cardiac and skeletal muscle. The CK level is considered a very sensitive indicator of acute myocardial infarction (AMI) and muscular dystrophy, particularly the Duchenne type. In fact, very striking elevations of CK occur in Duchenne muscular dystrophy, with values reaching 50 to 100 times the upper limit of normal (ULN). Although total CK levels are sensitive indicators of these disorders, they are not entirely specific indicators inasmuch as CK elevation is found in various other abnormalities of cardiac and skeletal muscle.

Elevated CK levels also are seen occasionally in central nervous system disorders such as cerebral vascular accident, seizures, nerve degeneration, and central nervous system shock. Damage to the blood-brain barrier must occur to allow enzyme release to the peripheral circulation.

Other pathophysiologic conditions in which elevated CK levels occur are hypothyroidism, malignant hyperpyrexia, and Reye's syndrome. Table 10-3 lists the major disorders associated with abnormal levels of CK.

Because enzyme elevation is found in numerous disorders, the separation of total CK into its various isoenzyme fractions is considered a more specific indicator of various disorders than are total levels. Typically, the clinical relevance of CK activity depends more on isoenzyme fractionation than on total levels.

TABLE 10-3. Creatine Kinase Isoenzymes— Tissue Localization and Sources of Elevation

Isoenzyme	Tissue	Condition
CK-MM	Heart	Myocardial infarction
	Skeletal muscle	Skeletal muscle disorders
		Muscular dystrophy
		Polymyositis
		Hypothyroidism
		Malignant hyperthermia
		Physical activity
		Intramuscular injections
CK-MB	Heart	Myocardial infarction
	Skeletal muscle	Myocardial injury
		Ischemia
		Angina
		Inflammatory heart disease
		Cardiac surgery
		Duchenne muscular dystrophy
		Polymyositis
		Malignant hyperthermia
		Reye's syndrome
		Rocky Mountain spotted fever
		Carbon monoxide poisoning
CK-BB	Brain	CNS shock
	Bladder	Anoxic encephalopathy
	Lung	Cerebrovascular accident
	Prostate	Seizures
	Uterus	Placental or uterine trauma
	Colon	Carcinoma
	Stomach	Reye's syndrome
	Thyroid	Carbon monoxide poisoning
		Malignant hyperthermia
		Acute and chronic renal failure

1st Rise / Full

CK occurs as a dimer consisting of two subunits, which can be separated readily into three distinct molecular forms. The three isoenzymes have been designated as CK-BB (brain type), CK-MB (hybrid type), and CK-MM (muscle type). On electrophoretic separation, CK-BB will migrate fastest toward the anode and hence is called *CK-1*. CK-BB is followed by CK-MB (*CK-2*) and finally by CK-MM (*CK-3*), exhibiting the slowest mobility (Fig. 10-5). Table 10-3 indicates the tissue localization of the isoenzymes and the major conditions associated with elevated levels.

In the sera of healthy persons, the major isoenzyme is the MM form. Values for the MB isoenzyme range from undetectable to trace (less than 6% of total CK). It also appears

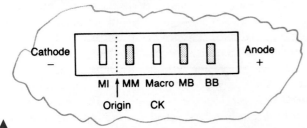

Figure 10-5. Electrophoretic migration pattern of normal and atypical CK isoenzymes.

that CK-BB is present in small quantities in the sera of healthy persons. However, the presence of CK-BB in serum depends on the method of detection. Most techniques cannot detect CK-BB in normal serum.

CK-MM is the major isoenzyme fraction found in striated muscle and normal serum. Skeletal muscle contains almost entirely CK-MM with a small amount of CK-MB. The majority of CK activity in heart muscle is also attributed to CK-MM, with approximately 20% due to CK-MB.[13] Normal serum consists of approximately 94% to 100% CK-MM. Injury to both cardiac and skeletal muscle accounts for the majority of cases of CK-MM elevations (see Table 10-3). Hypothyroidism results in CK-MM elevations because of the involvement of muscle tissue (increased membrane permeability), the effect of thyroid hormone in enzyme activity, and possibly, the slower clearance of CK as a result of hypometabolism.

Mild to strenuous activity may contribute to elevated CK levels, as may intramuscular injections of certain drugs. In physical activity, the extent of elevation is variable. However, the degree of exercise in relation to the exercise capacity of the individual is the most important factor in determining the degree of elevation.[39] Physically well-conditioned subjects show lesser degrees of elevation than do unconditioned subjects. Levels may be elevated for as long as 48 hours following exercise.

CK elevations seen following intramuscular injections are generally less than $5 \times$ ULN and are usually not apparent after 48 hours, although elevations may persist for 1 week. The predominant isoenzyme is CK-MM.

The quantity of CK-BB in the tissues listed in Table 10-3 is usually small. The small quantity, coupled with its relatively short half-life (1–5 hr), results in CK-BB activities that are generally low and transient and not usually measurable when tissue damage occurs. Highest concentrations are found in the central nervous system, the gastrointestinal tract, and the uterus during pregnancy.

Even though brain tissue has high concentrations of CK, serum rarely contains CK-BB of brain origin. Because of its molecular size (80,000), its passage across the blood-brain barrier is hindered. However, when extensive damage to the brain has occurred, significant amounts of CK-BB can sometimes be detected in the serum.

Recently it has been observed that CK-BB may be significantly elevated in patients with carcinoma of various organs. It has been found in association with untreated prostatic carcinoma and other adenocarcinomas. These findings indicate that CK-BB may be useful as a tumor-associated marker.[37]

The most common causes of CK-BB elevations are central nervous system damage, tumors, childbirth, and the presence of macro-CK, an enzyme-immunoglobulin complex. In the majority of these cases, the CK-BB level is greater than 5 U/L, usually in the range of 10 to 50 U/L. Other conditions listed in Table 10-3 usually show CK-BB activity below 10 U/L.[19]

The value of CK isoenzyme separation can be found principally in detection of myocardial damage. Cardiac tissue contains significant quantities of CK-MB, approximately 20% of all CK-MB. Whereas CK-MB is found in small quantities in other tissues, myocardium is essentially the only tissue from which CK-MB enters the serum in significant quantities. *Demonstration of elevated levels of CK-MB, ≥6% of the total CK, is considered the most specific indicator of myocardial damage, particularly acute myocardial infarction (AMI).* Following myocardial infarction, the CK-MB levels begin to rise within 4 to 8 hours, peak at 12 to 24 hours, and return to normal levels within 48 to 72 hours. This time frame must be taken into account when interpreting CK-MB levels.

CK-MB activity has been observed in other cardiac disorders (see Table 10-3). Therefore, increased quantities are not entirely specific for AMI but probably reflect some degree of ischemic heart damage. The specificity of CK-MB levels in the diagnosis of AMI can be increased if interpreted in conjunction with LD isoenzymes and if measured sequentially over a 48-hour period to detect the typical rise and fall of enzyme activity seen in AMI (Fig. 10-6).

Exchange of the subunits of CK isoenzymes resulting in transformation into another isoenzyme form has been observed both *in vitro* and *in vivo*.[19] The clinical significance of this finding relates to the fact that CK-BB can be transformed into CK-MB by the following reaction:

$$CK\text{-}BB + CK\text{-}MM \rightleftharpoons CK\text{-}MB + CK\text{-}BB + CK\text{-}MM$$

(Eq. 10-5)

This transformation could account for unexplained CK-MB levels in patients with lung cancer, acute cerebral disorders, and other disorders in which CK-MB activity would not normally be expected. It is also possible that this transformation of subunits results in the formation of CK-BB in sera with high CK-MB activity.

The MB isoenzyme also has been detected in the sera of patients with noncardiac disorders. CK-MB levels found in these conditions probably represent leakage from skeletal muscle, although in Duchenne muscular dystrophy there may be some cardiac involvement as well. CK-MB levels in Reye's syndrome also may reflect myocardial damage.

Despite the findings of CK-MB levels in disorders other than myocardial infarction, its presence still remains a highly significant indicator of AMI.[18] The typical time course of CK-MB elevation following AMI is not found in other conditions.

In recent years there have been numerous reports describing the appearance of unusual CK isoenzyme bands displaying electrophoretic properties that differ from the three major isoenzyme fractions (see Fig. 10-5).[1,4,20,25,30] These atypical forms are generally of two types and are referred to as *macro-CK* and *mitochondrial CK*.

Macro-CK appears to migrate to a position midway between that of CK-MM and CK-MB. This type of macro-CK is largely composed of CK-BB complexed with immunoglobulin. In most instances, the associated immunoglobulin is IgG, although a complex with IgA also has been described. The term *macro-CK* has been used to describe complexes of lipoproteins with CK-MM as well.

The incidence of macro-CK in sera ranges from 0.8% to 1.6%. Currently, no specific disorder has been associated with its presence, although it appears to be age- and sex-related, occurring most frequently in women over the age of 50.

The mitochondrial form of CK (CK-Mi) is bound to the exterior surface of the inner mitochondrial membranes of

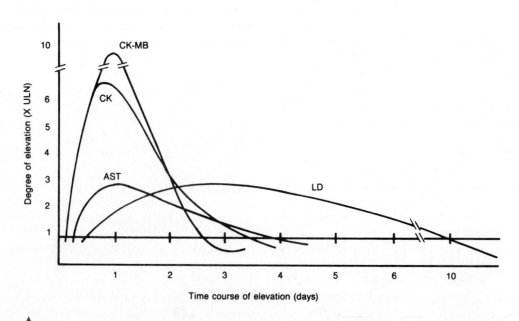

Figure 10-6. Time activity curves of enzymes in myocardial infarction for AST, CK, CK-MB, and LD. CK, specifically the MB fraction, increases initially, followed by AST and LD. LD is elevated the longest. All usually return to normal within 10 days.

muscle, brain, and liver. It migrates to a point cathodal to CK-MM and exists as a dimeric molecule of two identical subunits. It occurs in serum in both the dimeric state and in the form of oligomeric aggregates of high molecular weight (350,000). CK-Mi is not present in normal serum, and it is typically not present following myocardial infarct. The incidence of CK-Mi ranges from 0.8% to 1.7%. Extensive tissue damage must occur, causing breakdown of the mitochondrion and cell wall, for it to be detected in serum. Its presence does not correlate with any specific disease state but appears to be an indicator of severe illness. It has been detected in cases of malignant tumor and cardiac abnormalities.

In view of the lack of a definite correlation between these atypical CK forms and a specific disease state, it appears that their significance relates primarily to the methods used for detecting CK-MB. In certain analytic procedures, these atypical forms may be measured as CK-MB, resulting in erroneously high CK-MB levels.

Methods that are currently used for measurement of CK isoenzymes include electrophoresis, ion-exchange chromatography, RIA, and immunoinhibition methods. Electrophoresis has become a major method for detecting the CK isoenzymes and is the reference method. The electrophoretic properties of the CK isoenzymes are shown in Figure 10-5. Generally, the technique consists of performing electrophoresis on the sample, measuring the reaction using an overlay technique, and then visualizing the bands under ultraviolet light. With electrophoresis, the atypical bands can be separated, allowing their detection apart from the three major bands. Often a strongly fluorescent band appears, which migrates in close proximity to the CK-BB form. The exact nature of this fluorescence is not known, but it has been attributed to the binding of fluorescent drugs or bilirubin by albumin.

In addition to visualizing atypical CK bands, other advantages of electrophoresis methods include detecting an unsatisfactory separation and allowing visualization of adenylate kinase (AK). AK is an enzyme released from erythrocytes in hemolyzed samples and appearing as a band cathodal to CK-MM. AK may interfere with chemical or immunoinhibition methods, causing a falsely elevated CK or CK-MB value.

Ion-exchange chromatography has the potential for being more sensitive and precise than electrophoretic procedures performed with good technique. On an unsatisfactory column, however, CK-MM may merge into CK-MB and CK-BB may be eluted with CK-MB. Also, macro-CK will elute with CK-MB.

Antibodies against both the M and B subunits have been used to determine CK-MB activity. Anti-M inhibits all M activity but not B activity. CK activity is measured before and after inhibition. Activity remaining after M inhibition is due to the B subunit of both MB and BB activity. The residual activity after inhibition is multiplied by 2 to account for MB activity (50% inhibited). The major disadvantage of this method is that it detects BB activity, which, though not normally detectable, will cause falsely elevated MB results when BB is present. In addition, the atypical forms of CK-Mi and macro-CK are not inhibited by anti-M antibodies and also may cause erroneous results for MB activity.[33]

Recently, a double-antibody immunoinhibition technique has become available. This technique allows differentiation of MB activity due to adenylate kinase and the atypical isoenzymes, resulting in a more specific analytical procedure for CK-MB.[33]

RIA methods detect CK-MB reliably with no cross-reactivity. RIA methods measure the concentration of enzyme protein rather than enzymatic activity and can, therefore, detect enzymatically inactive CK-MB. This leads to the possibility of permitting detection of infarction earlier than other methods. RIA methods employing a double-antibody system also have been described. The technique is time-consuming and may not be suitable for routine use, but the sensitivity is greater, which may be helpful in borderline cases.[4]

Physicians sometimes want to know a patient's CK-MB level rather quickly and do not feel they can wait the several hours needed to perform an electrophoresis or RIA procedure. For these reasons, often a screening method is performed using a double-antibody immunodiffusion method or other simple, quick method. The specimen is then saved, and a more time-consuming and more specific procedure, such as RIA or electrophoresis, is performed.

A new rapid assay for the subforms of CK-MB has been tested to help diagnose acute myocardial infarction in the first 6 hours after chest pain. The assay is performed on plasma samples for MB_2 and MB_1, activity using rapid high voltage electrophoresis on an automated analyzer. The assay takes approximately 25 minutes and allows ER physicians to better decide which patients need the costly care in the Cardiac Care Unit of the hospital. This test may prove to be extremely helpful in the cost containment horizon of health care reform.[32]

Assay

As indicated by Equation 10-4, CK catalyzes both forward and reverse reactions involving phosphorylation of creatine or ADP. Typically, for analysis of CK activity, this reaction is coupled with other enzyme systems, and a change in absorbance at 340 nm is determined. The forward reaction is coupled with the pyruvate kinase–lactate dehydrogenase–NADH system and proceeds according to Equation 10-6:

$$\text{Creatine} + \text{ATP} \overset{\text{CK}}{\rightleftharpoons} \text{creatine phosphate} + \text{ADP}$$

$$\text{ADP} + \text{phosphoenolpyruvate} \overset{\text{PK}}{\rightleftharpoons} \text{pyruvate} + \text{ATP}$$

$$\text{Pyruvate} + \text{NADH} + \text{H}^+ \overset{\text{LD}}{\rightleftharpoons} \text{lactate} + \text{NAD}^+$$

(Eq. 10-6)

The reverse reaction is coupled with the hexokinase–glucose-6-phosphate dehydrogenase–NADP system, as indicated in Equation 10-7:

$$\text{Creatine phosphate} + \text{ADP} \underset{}{\overset{\text{CK}}{\rightleftharpoons}} \text{creatine} + \text{ATP}$$

$$\text{ATP} + \text{glucose} \overset{\text{HK}}{\rightleftharpoons} \text{ADP} + \text{glucose-6-phosphate}$$

$$\text{Glucose-6-phosphate} + \text{NADP}^+ \overset{\text{G-6-PD}}{\rightleftharpoons}$$

$$\text{6-phosphogluconate} + \text{NADPH}$$

(Eq. 10-7)

The reverse reaction proposed by Oliver and modified by Rosalki is the most commonly performed method in the clinical laboratory. The reaction proceeds two to six times faster than the forward reaction, depending on the assay conditions, and there is less interference from side reactions. The optimal pH for the reverse reaction is 6.8, while that for the forward reaction is 9.0.

CK activity in serum is very unstable, being rapidly inactivated because of oxidation of sulfhydryl groups. Inactivation can be partially reversed by the addition of sulfhydryl compounds to the assay reagent. Compounds such as N-acetylcysteine, mercaptoethanol, thioglycerol, and dithiothreitol are among those used.

Sources of Error

Hemolysis of serum samples may be a source of elevated CK activity. Erythrocytes are virtually devoid of CK; however, they are rich in AK activity. AK reacts with ADP to produce ATP, which is then available to participate in the assay reaction, causing falsely elevated CK levels. This interference can occur with hemolysis of greater than 320 mg/L of hemoglobin, which releases sufficient AK to exhaust the AK inhibitors in the reagent. Trace hemolysis causes very little, if any, CK elevation. Serum should be stored in the dark, because CK is inactivated by daylight. Activity can be restored after storage in the dark at 4°C for 7 days or at −20°C for 1 month when the assay is conducted using a sulfhydryl activator.[11] Because of the effect of muscular activity on CK levels, it should be noted that physically well-trained persons tend to have elevated baseline levels, while patients who are bedridden for prolonged periods of time may have decreased CK activity.

Reference Range

See Table 10-4. The higher values in males are attributed to increased muscle mass. Note that the reference ranges listed for all enzymes are subject to variation depending on the method used and the assay conditions.

TABLE 10-4. Reference Ranges for CK

Total CK	
Male	15–160 U/L (37°)
Female	15–130 U/L (37°)
CK-MB	<6% of total CK

Lactate Dehydrogenase

Lactate dehydrogenase (LD) is an enzyme that catalyzes the interconversion of lactic and pyruvic acids. It is a hydrogen-transfer enzyme that utilizes the coenzyme NAD^+ according to Equation 10-8:

$$\begin{array}{ccc} CH_3 & & CH_3 \\ | & & | \\ HC-OH + NAD^+ \overset{LD}{\rightleftharpoons} & C=O + NADH + H^+ \\ | & & | \\ COOH & & COOH \\ \text{Lactate} & & \text{Pyruvate} \end{array}$$

(Eq. 10-8)

Tissue Source

LD is widely distributed in the body. Very high activities are found in the heart, the liver, skeletal muscle, the kidney, and erythrocytes. It is present in lesser amounts in the lung, smooth muscle, and the brain.

Diagnostic Significance

Because of its widespread activity in numerous body tissues, LD is elevated in a variety of disorders. Increased levels are found in cardiac, hepatic, skeletal muscle, and renal diseases, as well as in several hematologic and neoplastic disorders. The highest levels of LD are seen in pernicious anemia. Intramedullary destruction of erythroblasts causes elevations due to the high concentration of LD in erythrocytes. Liver disorders, such as viral hepatitis and cirrhosis, show slight elevations of two to three times ULN. AMI and pulmonary infarct also show slight elevations of approximately the same degree (2–3 × ULN). In AMI, LD levels begin to rise within 12 to 24 hours, reach peak levels within 48 to 72 hours, and may remain elevated for 10 days. Skeletal muscle disorders and some leukemias contribute to increased LD levels. Marked elevations can be observed in most patients with acute lymphoblastic leukemia in particular (see Enzymes in Malignancy later in this chapter).

Because of the many conditions that contribute to its increased activity, an elevated total LD value is a rather nonspecific finding. LD assays, therefore, assume more clinical significance when separated into isoenzyme fractions. The enzyme can be separated into five major fractions, each composed of four subunits. It has a molecular weight of 128,000 daltons. Each isoenzyme is composed of four polypeptide chains with a molecular weight of 32,000 daltons each. Two different polypeptide chains, designated H (heart) and M (muscle), combine in five arrangements to yield the five major isoenzyme fractions.

Table 10-5 indicates the tissue localization of the LD isoenzymes and the major disorders associated with elevated levels. LD-1 migrates most quickly toward the anode, followed in sequence by the other fractions, with LD-5 migrating the slowest.

In the sera of healthy individuals, the major isoenzyme fraction is LD-2, followed by LD-1, LD-3, LD-4, and

TABLE 10-5. Lactate Dehydrogenase Isoenzymes—Tissue Localization and Sources of Elevation

Isoenzyme	Tissue	Disorder
LD-1 (HHHH) (M.I.) and LD-2 (HHHM)	Heart Red blood cells Renal cortex	Myocardial infarct Hemolytic anemia Megaloblastic anemia Acute renal infarct Hemolyzed specimen
LD-3 (HHMM)	Lung Lymphocytes Spleen Pancreas	Pulmonary embolism Extensive pulmonary pneumonia Lymphocytosis Acute pancreatitis Carcinoma
LD-4 (HMMM) and LD-5 (MMMM)	Liver Skeletal muscle	Hepatic injury or inflammation Skeletal muscle injury

LD-5 (see Table 10-6 for the isoenzyme ranges). LD-1 and LD-2 are present to approximately the same extent in the tissues listed in Table 10-5. However, cardiac tissue and red blood cells contain a higher concentration of LD-1. Therefore, in conditions involving cardiac necrosis (AMI) and intravascular hemolysis, the serum levels of LD-1 will increase to a point at which they are present in greater concentration than is LD-2, resulting in a condition known as the *LD flipped pattern* (LD-1 > LD-2).[8] This flipped pattern is of greatest clinical significance when used with CK isoenzyme analysis in the diagnosis of AMI. LD-1/LD-2 ratios greater than 1 also may be observed in hemolyzed serum samples.[24] Elevations of LD-3 occur most frequently with pulmonary involvement and are also observed in patients with various carcinomas. The LD-4 and LD-5 isoenzymes are found primarily in liver and skeletal muscle tissue, with LD-5 being the predominant fraction in these tissues. LD-5 levels have greatest clinical significance in the detection of hepatic disorders, particularly intrahepatic disorders. Disorders of skeletal muscle will reveal elevated LD-5 levels, as depicted in the muscular dystrophies.

Recent reports have identified a sixth LD isoenzyme, which migrates cathodic to LD-5.[7,9,16] LD-6 is neither an artifact nor an immunoglobulin complex but a distinct isoenzyme fraction. In the studies reporting its existence, it has been present in patients with arteriosclerotic cardiovascular failure. It is believed that its appearance signifies a grave prognosis and impending death. LD-5 is elevated concurrently with the appearance of LD-6, probably representing hepatic congestion due to cardiovascular disease. It is suggested, therefore, that LD-6 may reflect liver injury secondary to severe circulatory insufficiency.

LD has been shown to complex with immunoglobulins and to reveal atypical bands on electrophoresis. LD complexed with IgA and IgG usually migrates between LD-3 and LD-4. This macromolecular complex is not associated with any specific clinical abnormality.

Analysis of LD isoenzymes can be accomplished by electrophoretic or immunoinhibition methods or by differences in substrate affinity. The electrophoretic procedure is the most widely used method. After electrophoretic separation, the isoenzymes can be detected either fluorometrically or colorimetrically. LD can utilize other substrates in addition to lactate, such as α-hydroxybutyrate. The H subunits have a greater affinity for α-hydroxybutyrate than to the M subunits. This has led to the use of this substrate in an attempt to measure the LD-1 activity, which consists entirely of H subunits. The chemical assay, known as the measurement of α-hydroxybutyrate dehydrogenase activity (α-HBD), is outlined in Equation 10-9.

TABLE 10-6. LD Isoenzymes as a Percentage of Total LD[24]

Isoenzyme	Percentage
LD-1	14–26
LD-2	29–39
LD-3	20–26
LD-4	8–16
LD-5	6–16

$$\begin{array}{ccc} CH_3 & & CH_3 \\ | & & | \\ CH_2 + NADH + H^+ \xrightarrow{\alpha-HBD} & CH_2 + NAD^+ \\ | & & | \\ HC=O & & HC-OH \\ | & & | \\ COOH & & COOH \\ \alpha\text{-Ketobutyrate} & & \alpha\text{-Hydroxybutyrate} \end{array}$$

(Eq. 10-9)

α-HBD is not a separate and distinct enzyme but is considered to represent the LD-1 activity of total LD. However, α-HBD activity is not entirely specific for the LD-1 fraction, since LD-2, LD-3, and LD-4 also contain varying amounts of the H subunit. HBD activity is increased in those conditions in which the LD-1 and LD-2 fractions are increased. In AMI, α-HBD activity is very similar to the time course of total LD activity and can be useful in the diagnosis when other LD isoenzyme procedures are not readily available.

Immunoinhibition methods that selectively measure LD-1 activity have been described for LD isoenzymes. One such method involves the use of antibodies that react with all M subunit-containing isoenzymes. LD-1 activity can then be quantitated following removal of the M subunit-inhibited fractions.[34] Several reagent manufacturers are releasing similar procedures to quantitate LD-1 levels quickly.

Assay

LD catalyzes the interconversion of lactic and pyruvic acids utilizing the coenzyme NAD^+. The reaction sequence is outlined in Equation 10-10:

$$Lactate + NAD^+ \underset{}{\overset{LD}{\rightleftharpoons}} pyruvate + NADH + H^+$$

(Eq. 10-10)

The reaction can proceed in either the forward, lactate (L), or reverse, pyruvate (P), direction. Both reactions have been used in clinical assays. The rate of the reverse reaction is approximately three times faster, allowing smaller sample volumes and shorter reaction times. However, the reverse reaction is more susceptible to substrate exhaustion and loss of linearity. The optimal pH for the forward reaction is 8.3 to 8.9, while that for the reverse reaction is 7.1 to 7.4.

Sources of Error

Erythrocytes contain an LD concentration approximately 100 to 150 times that found in serum. Therefore, any degree of hemolysis should render a sample unacceptable for analysis. LD activity is unstable in serum regardless of the temperature at which it is stored. If the sample cannot be analyzed immediately, it should be stored at 25°C and analyzed within 48 hours. LD-5 is the most labile isoenzyme. Loss of activity occurs more quickly at 4°C than at 25°C. Serum samples for LD isoenzyme analysis should be stored at 25°C and analyzed within 24 hours of collection.

Reference Range

See Table 10-7.

Aspartate Aminotransferase

Aspartate aminotransferase (AST) is an enzyme belonging to the class of transferases. It is commonly referred to as a *transaminase* and is involved in the transfer of an amino group

between aspartate and α-keto acids. The older terminology, *glutamic-oxaloacetic transaminase (GOT)*, is also still used. Pyridoxal phosphate functions as a coenzyme. The reaction proceeds according to equation 10-11:

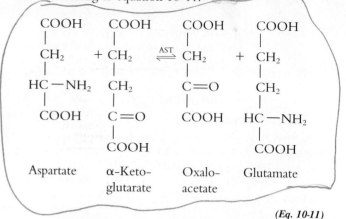

(Eq. 10-11)

The transamination reaction is important in intermediary metabolism because of its function in the synthesis and degradation of amino acids. The ketoacids formed by the reaction are ultimately oxidized by the tricarboxylic acid cycle to provide a source of energy.

Tissue Sources

AST is widely distributed in human tissues. The highest concentrations are found in cardiac tissue, the liver, and skeletal muscle, with smaller amounts found in the kidney, the pancreas, and erythrocytes.

Diagnostic Significance

The clinical use of AST is confined mainly to the evaluation of myocardial infarct, hepatocellular disorders, and skeletal muscle involvement. In AMI, AST levels begin to rise within 6 to 8 hours, peak at 24 hours, and generally return to normal within 5 days. However, due to the wide tissue distribution, AST levels in the diagnosis of AMI are not as useful as CK and LD isoenzyme analysis.

AST elevations are frequently seen in pulmonary embolism. Following congestive heart failure, AST levels also may be increased, probably reflecting liver involvement due to inadequate blood supply to that organ. AST levels are highest in acute hepatocellular disorders. In viral hepatitis, levels may reach 100 times ULN. In cirrhosis, only moderate levels, approximately four times ULN are detected (see Chapter 18, Liver Function). Skeletal muscle disorders, such as the muscular dystrophies, and inflammatory conditions also cause increases in AST levels (4–8 × ULN).

AST exists as two isoenzyme fractions located in the cell cytoplasm and mitochondria. The cytoplasmic isoenzyme is the predominant form occurring in serum. In disorders producing cellular necrosis, such as liver cirrhosis, the mitochondrial form may be markedly increased. Isoenzyme analysis of AST is not routinely performed in the clinical laboratory.

Assay

Assay methods for AST are generally based on the principle of the Karmen method, which incorporates a coupled en-

TABLE 10-7. Reference Ranges for LD	
Forward (L→P)	100–225 U/L (37°)
Reverse (P→L)	80–280 U/L (37°)

zymatic reaction using malate dehydrogenase (MD) as the indicator reaction and monitors the change in absorbance at 340 nm continuously as NADH is oxidized to NAD⁺ (see Eq. 10-12). The pH optimum is 7.3 to 7.8.

$$\text{Aspartate} + \alpha\text{-ketoglutarate} \xrightleftharpoons{\text{AST}}$$
$$\text{oxaloacetate} + \text{glutamate}$$
$$\text{Oxaloacetate} + \text{NADH} + \text{H}^+ \xrightleftharpoons{\text{MD}} \text{malate} + \text{NAD}^+$$

(Eq. 10-12)

Sources of Error

Hemolysis should be avoided because it can increase AST activity as much as tenfold. AST activity is stable in serum for 3 to 4 days at refrigerated temperatures.

Reference Range

AST   5–30 U/L (37°C)

Alanine Aminotransferase

Alanine aminotransferase (ALT) is a transferase with enzymatic activity similar to that of AST. Specifically, it catalyzes the transfer of an amino group from alanine to α-ketoglutarate with the formation of glutamate and pyruvate. The older terminology still in use is *glutamic-pyruvic transaminase (GPT)*. Equation 10-13 indicates the transferase reaction. Pyridoxal phosphate acts as the coenzyme.

(Eq. 10-13)

Tissue Sources

ALT is distributed in many tissues and is found in comparatively high concentrations in the liver. It is considered to be the more liver-specific enzyme of the transferases.

Diagnostic Significance

Clinical applications of ALT assays are confined mainly to evaluation of hepatic disorders. Higher elevations are found in hepatocellular disorders than in extra- or intrahepatic obstructive disorders. In acute inflammatory conditions of the liver, ALT elevations will frequently be higher than those of AST and will tend to remain elevated longer due to the longer half-life of ALT in serum.

Cardiac tissue contains a small amount of ALT activity, but the serum level usually remains normal in AMI unless subsequent liver damage has occurred. Assay of ALT levels

will often be requested along with that of AST to help determine the source of an elevated AST level and to detect liver involvement concurrent with myocardial injury.

Assay

The typical assay procedure for ALT consists of a coupled enzymatic reaction employing LD as the indicator enzyme, which catalyzes the reduction of pyruvate to lactate with the simultaneous oxidation of NADH. The change in absorbance at 340 nm measured continuously is directly proportional to ALT activity. The reaction proceeds according to Equation 10-14. The pH optimum is 7.3 to 7.8.

$$\text{Alanine} + \alpha\text{-ketoglutarate} \xrightleftharpoons{\text{ALT}} \text{pyruvate} + \text{glutamate}$$
$$\text{Pyruvate} + \text{NADH} + \text{H}^+ \xrightleftharpoons{\text{LD}} \text{lactate} + \text{NAD}^+$$

(Eq. 10-14)

Sources of Error

ALT is stable for 3 to 4 days at 4°C. It is relatively unaffected by hemolysis.

Reference Range

ALT   6-37 U/L (37°C)

Alkaline Phosphatase

Alkaline phosphatase (ALP) belongs to a group of enzymes that catalyze the hydrolysis of a wide variety of phosphomonoesters at an alkaline pH. Consequently, ALP is a nonspecific enzyme capable of reacting with many different substrates. Specifically, ALP functions to liberate inorganic phosphate from an organic phosphate ester with the concomitant production of an alcohol. The reaction proceeds according to Equation 10-15:

$$\begin{array}{ccc} & & \text{O} \\ & & \parallel \\ \text{R}-\overset{\displaystyle\text{O}}{\underset{\displaystyle\text{O}^-}{\overset{\parallel}{\text{P}}}}-\text{O}^- + \text{H}_2\text{O} & \xrightleftharpoons[\text{pH 9–10}]{\text{ALP}} & \text{R}-\text{OH} + \text{HO}-\overset{\displaystyle}{\underset{\displaystyle\text{O}^-}{\overset{\parallel}{\text{P}}}}-\text{O}^- \end{array}$$

Phosphomonoester   Alcohol   Phosphate ion

(Eq. 10-15)

The optimal pH for the reaction is 9.0 to 10.0, but this depends on the substrate used. The enzyme requires Mg^{2+} as an activator.

Tissue Sources

ALP activity is present in most human tissues. The highest concentrations are found in the intestines, liver, bone, spleen, placenta, and kidney. In the liver, the enzyme is located on both sinusoidal and bile canalicular membranes, while in bone, activity is confined to the osteoblasts, those cells involved in production of bone matrix.

The specific location of the enzyme within these tissues accounts for the more predominant elevations in certain disorders.

Diagnostic Significance

Elevations of ALP are of most diagnostic significance in the evaluation of hepatobiliary and bone disorders. In hepatobiliary disorders, elevations are seen more predominantly in obstructive conditions than in hepatocellular disorders. In disorders of the bone, elevations are seen when there is involvement of the osteoblasts.

In biliary tract obstruction, ALP levels range from three to ten times ULN. Increases are due primarily to increased synthesis of the enzyme induced by cholestasis. In contrast, hepatocellular disorders such as hepatitis and cirrhosis show only slight increases, usually less than three times ULN. Owing to the degree of overlap of ALP elevations that occurs in the various liver disorders, a single elevated ALP level is difficult to interpret. It assumes more diagnostic significance when evaluated along with other tests of hepatic function (see Chapter 18, Liver Function).

Elevated ALP levels may be seen in a variety of bone disorders. Perhaps the highest elevations of ALP activity occur in Paget's disease (osteitis deformans). Other bone disorders include osteomalacia, rickets, hyperparathyroidism, and osteogenic sarcoma. In addition, increased levels are seen in the case of healing bone fractures and during periods of physiologic bone growth (see Enzymes in Bone Disorders later in this chapter).

In normal pregnancy, increased ALP activity, averaging approximately one and one-half times ULN, can be detected between the 16th and 20th week of pregnancy. ALP activity increases and persists until the onset of labor. Activity then returns to normal within 3 to 6 days.[27,31] Elevations also may be seen in complications of pregnancy such as hypertension, preeclampsia, and eclampsia, as well as in threatened abortion.

ALP levels are significantly decreased in the inherited condition of hypophosphatasia. Subnormal activity is due to the absence of the bone isoenzyme and results in inadequate bone calcification.

ALP exists as a number of isoenzymes, which have been studied by a variety of techniques. The major isoenzymes that are found in the serum and that have been most extensively studied are those derived from the liver, bone, intestine, and placenta.[38]

Electrophoresis is considered the most useful single technique for ALP isoenzyme analysis. However, because there may still be some degree of overlap between the fractions, electrophoresis in combination with another separation technique may provide the most reliable information. A direct immunochemical method for the measurement of bone-related ALP is now available; this has made ALP electrophoresis unnecessary in most cases.

The liver fraction migrates the fastest, followed by bone, placental, and intestinal fractions. Because of the similarity between the liver and bone phosphatases, there often is not a clear separation between them. Quantitation by a densitometer is sometimes difficult because of the overlap between the two peaks. The liver isoenzyme can actually be divided into two fractions, the major liver band and a smaller fraction called *fast liver,* or α_1 liver, which migrates anodal to the major band and corresponds to the α_1 fraction of protein electrophoresis. When total ALP levels are increased, it is the major liver fraction that is most frequently elevated. Many hepatobiliary conditions cause elevations of this fraction, usually early in the course of the disease. The fast-liver fraction has been reported in metastatic carcinoma of the liver, as well as in other hepatobiliary diseases. Its presence is regarded as a valuable indicator of obstructive liver disease.[27] However, it is even present occasionally in the absence of any detectable disease state.[38]

The bone isoenzyme increases due to osteoblastic activity and is normally elevated in children during periods of growth and in adults over the age of 50. In these cases, an elevated ALP level may be difficult to interpret.[12]

The presence of intestinal ALP isoenzyme in serum depends on the blood group and secretor status of the individual. Individuals who have B or O blood group and are secretors are more likely to have this fraction. Apparently, intestinal ALP is bound by erythrocytes of group A.[27] Furthermore, in these individuals, increases in intestinal ALP occur after consumption of a fatty meal. Intestinal ALP may increase in several disorders, such as diseases of the digestive tract and cirrhosis, and increased levels are also found in patients undergoing chronic hemodialysis.

Difference in heat stability is the basis of another approach used to identify the isoenzyme source of an elevated ALP. Typically, ALP activity is measured before and after heating the serum at 56°C for 10 minutes. If the residual activity after heating is less than 20% of the total activity before heating, then the ALP elevation is assumed to be due to bone phosphatase. If greater than 20% of the activity remains, the elevation is probably due to liver phosphatase. These results are based on the finding that placental ALP is the most heat-stable of the four major fractions, followed by intestinal, liver, and bone fractions in decreasing order of heat stability. Placental ALP will resist heat denaturation at 65°C for 30 minutes.

Heat inactivation is an imprecise method for differentiation, since inactivation depends on many factors, such as correct temperature control, timing, and analytic methods sensitive enough to detect small amounts of residual ALP activity. In addition, there is some degree of overlap between heat inactivation of liver and bone fractions in both liver and bone diseases.

A third approach to identification of ALP isoenzymes is based on selective chemical inhibition. Phenylalanine is one of several inhibitors that have been used. Phenylalanine inhibits intestinal and placental ALP to a much greater extent than liver and bone ALP. With the use of phenylalanine, however, it is not possible to differentiate placental from intestinal ALP or liver from bone ALP.

In addition to the four major ALP isoenzyme fractions, there exist some abnormal fractions that are associated with neoplasms. The most frequently seen of these are the Regan and the Nagao isoenzymes. They have been referred to as *carcinoplacental alkaline phosphatases* because of their similarities to the placental isoenzyme. The frequency of occurrence ranges from 3% to 15% in cancer patients. The Regan isoenzyme has been characterized as an example of an ectopic production of an enzyme by malignant tissue. It has been detected in various carcinomas, such as lung, breast, ovarian, and colon, with highest incidences in ovarian and gynecologic cancers. Because of its low incidence in cancer patients, diagnosis of malignancy is rarely based on its presence. It is, however, useful in monitoring the effects of therapy because it will disappear upon successful treatment.

The Regan isoenzyme migrates to the same position as does the bone fraction and is the most heat-stable of all ALP isoenzymes, resisting denaturation at 65°C for 30 minutes. Its activity is inhibited by phenylalanine.

The Nagao isoenzyme may be considered a variant of the Regan isoenzyme. Its electrophoretic, heat-stability, and phenylalanine-inhibition properties are identical to those of the Regan fraction. However, Nagao also can be inhibited by L-leucine. Its presence has been detected in metastatic carcinoma of pleural surfaces and in adenocarcinoma of the pancreas and bile duct.

Assay

Because of the relative nonspecificity of ALP with regard to substrates, a variety of methodologies for its analysis have been proposed and are still in use today. The major differences between these relate to the concentration and types of substrate and buffer used and the pH of the reaction. A continuous-monitoring technique based on a method devised by Bowers and McComb allows calculation of ALP activity based on the molar absorptivity of p-nitrophenol. The reaction proceeds according to Equation 10-16:

p-Nitrophenyl-phosphate   *p*-Nitro-phenol   Phosphate ion

(Eq. 10-16)

p-Nitrophenylphosphate (colorless) is hydrolyzed to p-nitrophenol (yellow), and the increase in absorbance at 405 nm, which is directly proportional to ALP activity, is measured.

Sources of Error

Hemolysis may cause slight elevations, because ALP is approximately six times more concentrated in erythrocytes than in serum. ALP assays should be run as soon as possible after collection. Activity in serum increases approximately 3% to 10% upon standing at 25° or 4°C for several hours. Diet may induce elevations in ALP activity of blood group B and O individuals who are secretors. Values may be 25% higher following ingestion of a high-fat meal.

Reference Range

ALP   30 to 90 U/L (30°C)

Acid Phosphatase

Acid phosphatase (ACP) belongs to the same group of phosphatase enzymes as ALP and is a hydrolase that catalyzes the same type of reactions. The major difference between ACP and ALP is the pH of the reaction. ACP functions at an optimal pH of approximately 5.0. Equation 10-17 outlines the reaction sequence:

Phosphomonoester   Alcohol   Phosphate ion

(Eq. 10-17)

Tissue Source

ACP activity is found in the prostate, bone, liver, spleen, kidney, erythrocytes, and platelets. The prostate is the richest source, with many times the activity found in other tissues.

Diagnostic Significance

The major diagnostic significance of ACP measurement is its use as an aid in the detection of prostatic carcinoma, particularly metastatic carcinoma of the prostate. Prostatic cancer is the third most prevalent fatal male cancer in the United States. The incidence is more common in men over the age of 50 and increases with age.

Prostatic carcinoma is commonly classified into four stages. In stages A and B, the malignant tumor is confined to the prostate. Stage C tumors have spread beyond the prostate but are confined to the pelvic area, whereas stage D tumors have metastasized to various regions of the body. Total ACP determinations are relatively insensitive techniques and can detect elevated ACP levels due to prostatic carcinoma in the majority of cases only when the tumor has metastasized.

Because of the importance of detecting prostatic carcinoma before it has metastasized, when it may still be curable, attempts have been made to increase both the sensitivity and specificity of ACP measurements, with an emphasis on detection of the prostatic contribution only. Assay methods for this purpose include the use of differential substrates, specific chemical inhibitors, and immunologic techniques. One of the most specific substrates for the prostatic portion is thymolphthalein monophosphate. Other tissue sources of ACP can react, but to a much lesser degree, with this substrate.

Chemical-inhibition methods used to differentiate the prostatic portion most frequently employ tartrate as the inhibitor. The prostatic fraction is inhibited by tartrate. Serum and substrate are incubated both with and without the addition of L-tartrate. ACP activity remaining after inhibition with L-tartrate is subtracted from total ACP activity determined without inhibition, and the difference represents the prostatic portion:

$$\text{Total ACP} - \text{ACP after tartrate inhibition} = \text{prostatic ACP} \qquad \textit{(Eq. 10-18)}$$

The reaction is not entirely specific for prostatic acid phosphatase, but other tissue sources are largely uninhibited.

Neither of these methods of ACP determinations is sensitive to prostatic carcinoma that has not metastasized. Values are usually normal in the majority of cases and, in fact, may be elevated only in as many as 50% of cases of prostatic carcinoma that has metastasized.[41]

A technique with much improved sensitivity over conventional ACP assays is the immunologic approach using antibodies that are specific for the prostatic portion. Immunochemical techniques, however, are not of value as screening tests for prostatic carcinoma.

Because of improved sensitivity and specificity of immunochemical assays, these methods are also of value in monitoring the course of the disease following therapy. After successful surgical removal or radiotherapy, the elevated enzyme levels usually return to the normal range.

In addition to using prostatic ACP to monitor prostatic carcinoma, there is an increasing trend to monitor prostatic-specific antigen (PSA) (see Chapter 24 Tumor Markers). PSA is more likely than ACP to be elevated at each stage of prostatic carcinoma, even though a normal PSA level may be found in stage D tumors. PSA is particularly useful to monitor the success of treatment. PSA is controversial as a screening test for prostatic malignancy, because PSA elevation may occur in conditions other than prostatic carcinoma.[14,15,29]

Other prostatic conditions in which elevations of ACP activity have been reported include hyperplasia of the prostate and prostatic surgery. There are conflicting reports of elevations following rectal examination and prostate massage. Some studies report ACP elevations, while others indicate no detectable change. When elevations are found, levels usually return to normal within 24 hours.[17]

ACP assays have proven useful in forensic clinical chemistry, particularly in the investigation of rape. Vaginal washings are examined for seminal fluid–ACP activity, which can persist for up to 4 days.[21] Elevated activity is presumptive evidence of rape in such cases.

Serum ACP activity may frequently be elevated in bone diseases. Activity has been shown to be associated with the osteoclasts.[40] Elevations have been noted in Paget's disease, breast cancer with bone metastases, and Gaucher's disease, in which there is an infiltration of bone marrow and other tissues by Gaucher cells rich in ACP activity. Because of

ACP activity in platelets, elevations are seen when damage to platelets occurs, as in thrombocytopenia due to excessive platelet destruction.

Assay

Assay procedures for total ACP use the same techniques as in ALP assays but are performed at an acid pH (see Eq. 10-19):

$$p\text{-Nitrophenylphosphate} \underset{\text{pH 5}}{\overset{\text{ACP}}{\rightleftharpoons}} p\text{-nitrophenol} + \text{phosphate ion} \qquad \textit{(Eq. 10-19)}$$

The reaction products are colorless at the acid pH of the reaction, but the addition of alkali stops the reaction and transforms the products into chromogens, which can be measured spectrophotometrically.

Some substrate specificities and chemical inhibitors for prostatic ACP measurements have been discussed previously. Thymophthalein monophosphate is the substrate of choice for quantitative end-point reactions. For continuous monitoring methods, α-naphthyl phosphate is preferred.

Immunochemical techniques for prostatic ACP employ several approaches, including RIA, counter immunoelectrophoresis (CEIP), and immunoprecipitation. Also, an immunoenzymatic assay (Tandem E) includes incubation with an antibody to prostatic ACP followed by washing and an incubation with p-NPP. The p-NP formed, measured photometrically, is proportional to the prostatic ACP in the sample.

Sources of Error

Serum should be separated from the red cells as soon as the blood has clotted to prevent leakage of erythrocyte and platelet ACP. Serum activity decreases within 1 to 2 hours if the sample is left at room temperature without a preservative being added. Decreased activity is due to a loss of carbon dioxide from the serum, with a resultant increase in pH. If not assayed immediately, serum should be frozen or acidified to a pH below 6.5. With acidification, ACP is stable for 2 days at room temperature. Hemolysis should be avoided because of contamination from erythrocyte ACP.

RIA procedures for measurement of prostatic acid phosphatase require nonacidified serum samples. Activity is stable for 2 days at 4°C.

Reference Range

See Table 10-8.

TABLE 10-8. Reference Ranges for ACP	
Total ACP	
Males	2.5–11.7 U/L
Females	0.3–9.2 U/L
Prostatic ACP	
Males	0.2–5.0 U/L
Females	0.0–0.8 U/L

Gamma-Glutamyltransferase

Gamma-glutamyltransferase (GGT) is an enzyme involved in the transfer of the γ-glutamyl residue from γ-glutamyl peptides to amino acids, H_2O, and other small peptides. In most biological systems, glutathione serves as the γ-glutamyl donor. Equation 10-20 outlines the reaction sequence:

$$\text{Glutathione} + \text{amino acid} \xrightarrow{\text{GGT}}$$
$$\text{glutamyl} - \text{peptide} + \text{L-cysteinylglycine}$$

(Eq. 10-20)

The specific physiologic function of GGT has not been clearly established, but it has been suggested that GGT is involved in peptide and protein synthesis, regulation of tissue glutathione levels, and the transport of amino acids across cell membranes.[35]

Tissue Sources

GGT activity is found primarily in tissues of the kidney, brain, prostate, pancreas, and liver. Clinical applications of assay, however, are confined mainly to evaluation of liver and biliary system disorders.

Diagnostic Significance

In the liver, GGT is located in the canaliculi of the hepatic cells and particularly in the epithelial cells lining the biliary ductules. Because of these locations, GGT is elevated in virtually all hepatobiliary disorders, making it one of the most sensitive of enzyme assays in these conditions (see Chapter 18, Liver Function). Higher elevations are generally observed in biliary tract obstruction.

Within the hepatic parenchyma, GGT exists to a large extent in the smooth endoplasmic reticulum and is, therefore, subject to hepatic microsomal induction. Thus, GGT levels will be increased in patients receiving enzyme-inducing drugs such as warfarin, phenobarbital, and phenytoin. Enzyme elevations may reach levels four times ULN.[27]

Because of the effects of alcohol on GGT activity, GGT assays are considered sensitive indicators of alcoholism, particularly occult alcoholism, when other indications are lacking. Generally, enzyme elevations in alcoholics and heavy drinkers range from two to three times ULN, although higher levels have been observed. In contrast, studies of the effects of normal social drinking reveal no significant enzyme elevations.[35] GGT assays are also useful in monitoring the effects of abstention from alcohol and are used as such by alcohol treatment centers. Levels usually return to normal within 2 to 3 weeks after cessation but can rise again if alcohol consumption is resumed. In view of the susceptibility to enzyme induction, any interpretation of GGT levels must be done with consideration of the consequent effects of drugs and alcohol.

GGT levels are elevated in other conditions such as acute pancreatitis, diabetes mellitus, and myocardial infarction. The source of elevation in pancreatitis and diabetes is probably the pancreas, but the source of GGT in myocardial infarction is unknown. GGT assays are of limited value in the diagnosis of these conditions and are not routinely requested.

GGT activity is useful in differentiating the source of an elevated ALP level, because GGT levels are normal in skeletal disorders and during pregnancy. It is particularly useful in evaluating hepatobiliary involvement in adolescents, since ALP activity will invariably be elevated owing to bone growth.

Assay

The substrate most widely accepted for use in the analysis of GGT is γ-glutamyl-p-nitroanilide. The γ-glutamyl residue is transferred to glycylglycine, releasing p-nitroaniline, a chromogenic product with a strong absorbance at 405 to 420 nm. The reaction, which can be used as a continuous-monitoring or a fixed-point method, is outlined in Equation 10-21:

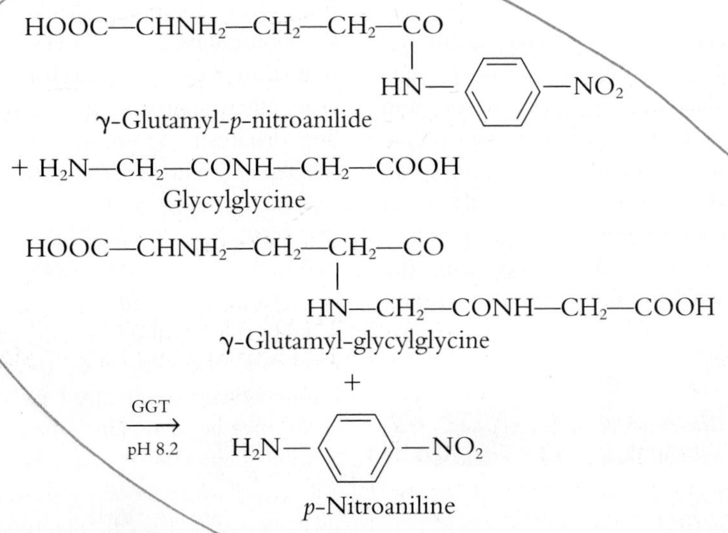

$$HOOC-CHNH_2-CH_2-CH_2-CO$$

γ-Glutamyl-p-nitroanilide

$$+ H_2N-CH_2-CONH-CH_2-COOH$$
Glycylglycine

$$HOOC-CHNH_2-CH_2-CH_2-CO$$

γ-Glutamyl-glycylglycine

$$+$$

$$\xrightarrow[\text{pH 8.2}]{\text{GGT}} \quad H_2N-\langle \rangle -NO_2$$

p-Nitroaniline

(Eq. 10-21)

Sources of Error

GGT activity is very stable, with no loss of activity for 1 week at 4°C. Hemolysis does not interfere with GGT levels, since the enzyme is lacking in erythrocytes.

Reference Range

See Table 10-9. Values are lower in females, presumably because of suppression of enzyme activity due to estrogenic or progestational hormones.

Amylase

Amylase (AMS) is an enzyme belonging to the class of hydrolases. It catalyzes the breakdown of starch and glycogen. Starch consists of both amylose and amylopectin. Amylose is a long, unbranched chain of glucose molecules, linked by $\alpha,1$–4 glycosidic bonds, whereas amylopectin is a branched-chain polysaccharide with $\alpha,1$–6 linkages at the branch points. The structure of glycogen is similar to that of amylopectin but is more highly branched. α-AMS attacks only the $\alpha,1$–4 glycosidic bonds to produce degradation products consisting of glucose, maltose, and intermediate chains, called *dextrins,* which contain $\alpha,1$–6 branching linkages. Cellulose and other structural polysaccharides consisting of α linkages are not attacked by α-AMS. AMS is thus an important enzyme in the physiologic digestion of starches. The reaction proceeds according to Equation 10-22:

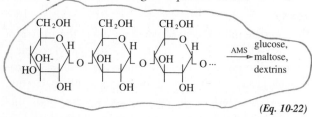

(Eq. 10-22)

AMS requires calcium and chloride ions for its activation.

Tissue Sources

The acinar cells of the pancreas and the salivary glands are the major tissue sources of serum amylase. Lesser concentrations are found in skeletal muscle, the small intestine, and the fallopian tubes. Amylase is the smallest in size of the enzymes, with a molecular weight of 50,000 to 55,000. Because of its small size, it is readily filtered by the renal glomerulus and also appears in the urine.

Digestion of starches begins in the mouth with the hydrolytic action of salivary AMS. Salivary amylase activity,

however, is of short duration because, upon swallowing, it is inactivated by the acidity of the gastric contents. Pancreatic amylase then performs the major digestive action of starches once the polysaccharides reach the intestine.

Diagnostic Significance

The diagnostic significance of serum and urine amylase measurements is in the diagnosis of acute pancreatitis.[36] Disorders of tissues other than the pancreas can also produce elevations in amylase levels. Thus, an elevated amylase level is a nonspecific finding. However, the degree of elevation of AMS is helpful, to some extent, in the differential diagnosis of acute pancreatitis. In addition, other laboratory tests, such as measurements of urinary amylase levels, amylase clearance studies, amylase isoenzyme studies, and measurements of serum lipase levels, when used in conjunction with serum amylase measurement, increase the specificity of amylase measurements in the diagnosis of acute pancreatitis (see Enzymes in Acute Pancreatitis later in this chapter).

In acute pancreatitis, serum amylase levels begin to rise 2 to 12 hours after onset of an attack, peak at 24 hours, and return to normal levels within 3 to 5 days. Values generally range from 250 to 1000 Somogyi units/dL (2.5–5 × ULN). Values can reach much higher levels.

Other disorders causing an elevated serum amylase level include salivary gland lesions such as mumps and parotitis and other intraabdominal diseases such as perforated peptic ulcer, intestinal obstruction, cholecystitis, ruptured ectopic pregnancy, mesenteric infarction, and acute appendicitis. In addition, elevations have been reported in renal insufficiency and diabetic ketoacidosis. Serum amylase levels in intraabdominal conditions other than acute pancreatitis are usually less than 500 SU/dL.

An apparently asymptomatic condition of hyperamylasemia has been noted in approximately 1% to 2% of the population. This condition, called *macroamylasemia,* results when the amylase molecule combines with immunoglobulins to form a complex that is too large to be filtered across the glomerulus. Serum AMS levels increase because of the reduction in normal renal clearance of the enzyme, and consequently, the urinary excretion of AMS is abnormally low. The diagnostic significance of macroamylasemia lies in the need to differentiate it from other causes of hyperamylasemia.

Much interest has been focused recently on the possible diagnostic use of amylase isoenzyme measurements.[23,36] Serum AMS is a mixture of a number of isoenzymes that can be separated by differences in physical properties, most notably electrophoresis, although chromatography and isoelectric focusing also have been applied. In normal human serum, two major bands and as many as four minor bands may be seen. The bands are designated as P-type and S-type isoamylase. P isoamylases are derived from pancreatic tissue, while S isoamylase is derived from salivary gland tissue, as well as the fallopian tube and lung. The isoenzymes of salivary origin migrate most quickly (S1, S2, S3), while

TABLE 10-9. Reference Ranges for GGT	
Male	6–45 U/L
Female	5–30 U/L

those of pancreatic origin are the slower ones (P1, P2, P3). In normal human serum, the isoamylases migrate in regions corresponding to the β- to γ-globulin regions of protein electrophoresis. The most commonly observed fractions are P2, S1, and S2. P1 is rare, occurring in less than 2% of patients studied.

In acute pancreatitis, there is typically an increase in P-type activity, with P3 being the most predominant isoenzyme. However, P3 also has been detected in cases of renal failure and, therefore, is not entirely specific for acute pancreatitis. S-type isoamylase represents approximately two-thirds of amylase activity of normal serum, while P-type predominates in normal urine. Currently, amylase isoenzyme analysis is not readily adaptable as a routine laboratory procedure. However, with further refinement, it would appear to hold considerable promise as an aid in the diagnosis of pancreatic disease.

Recent studies have shown a marked increase in serum amylase levels in patients with AIDS. These levels ranged from 282 to 2340 U/L (reference range 92–278). Demographics of these AIDS patients indicate that African-Americans and Hispanics as well as IV drug abusers are more likely to have elevated serum amylase levels.[28]

Assay

Amylase can be assayed by a variety of different methods, which are summarized in Table 10-10. The four main approaches are categorized as amyloclastic, saccharogenic, chromogenic, and continuous-monitoring.

In the amyloclastic method, AMS is allowed to act on a starch substrate to which iodine has been attached. As AMS hydrolyzes the starch molecule into smaller units, the iodine is released, and a decrease in the initial dark-blue color intensity of the starch-iodine complex occurs. The decrease in color is proportional to the AMS concentration.

The saccharogenic method uses a starch substrate that is hydrolyzed by the action of AMS to its constituent carbohydrate molecules that have reducing properties. The amount of reducing sugars is then measured where the concentration is proportional to amylase activity. The saccharogenic method is the classic reference method for determining amylase activity and is reported in Somogyi units. Somogyi units are an ex-

pression of the number of milligrams of glucose released in 30 minutes at 37°C under specific assay conditions.

Chromogenic methods employ a starch substrate to which a chromogenic dye has been attached, forming an insoluble dye-substrate complex. As AMS hydrolyzes the starch substrate, smaller dye-substrate fragments are produced, and these are water-soluble. The increase in color intensity of the soluble dye-substrate solution is proportional to AMS activity.

Recently, coupled-enzyme systems have been employed to determine AMS activity by a continuous-monitoring technique in which the change in absorbance of NAD^+ at 340 nm is measured. Equation 10-23 is an example of a continuous-monitoring method. The optimal pH for amylase activity is 6.9.

$$Maltopentose \xrightarrow{AMS} maltotriose + maltose$$

$$Maltotriose + maltose \xrightarrow{\alpha\text{-Glucosidase}} 5\text{-glucose}$$

$$5\text{-Glucose} + 5\ ATP \xrightarrow{Hexokinase}$$

$$5\text{-glucose-6-phosphate} + 5\ ADP$$

$$5\text{-Glucose-6-phosphate} + 5\ NAD^+ \xrightarrow{G\text{-}6\text{-}PD}$$

$$5,6\text{-phosphogluconolactone} + 5\ NADH$$

(Eq. 10-23)

Sources of Error

Amylase in serum and urine is quite stable. Little loss of activity occurs at room temperature for 1 week or at 4°C for 2 months. Because plasma triglycerides suppress or inhibit serum AMS activity, AMS values may be normal in acute pancreatitis with hyperlipemia.

The administration of morphine and other opiates for the relief of pain prior to sampling of blood will lead to falsely elevated serum AMS levels. The drugs presumably cause constriction of the sphincter of Oddi and pancreatic ducts, with consequent elevation of intraductal pressure causing regurgitation of amylase into the serum.

Reference Range

Because of the variety of amylase procedures currently in use, activity is expressed according to each procedure. There is no uniform expression of AMS activity, although Somogyi units are frequently used (Table 10-11). The approximate conversion factor between Somogyi units and international units is 1.85.

TABLE 10-10. Amylase Methodologies

Amyloclastic	Measures the disappearance of starch substrate
Saccharogenic	Measures the appearance of the product
Chromogenic	Measures the increasing color from production of product coupled with a chromogenic dye
Continuous monitoring	Coupling of several enzyme systems to monitor amylase activity

TABLE 10-11. Reference Ranges for AMS

Serum	60–180 SU*/dL
	95–290 U/L
Urine	35–400 U/h

*Somogyi units

Lipase

Lipase (LPS) is an enzyme that hydrolyzes the ester linkages of fats to produce alcohols and fatty acids. Specifically, LPS catalyzes the partial hydrolysis of dietary triglycerides in the intestine to the 2-monoglyceride intermediate, with the production of long-chain fatty acids. The reaction proceeds according to Equation 10-24:

$$CH_2-O-\overset{\overset{\displaystyle O}{\|}}{C}-R_1$$
$$CH-O-\overset{\overset{\displaystyle O}{\|}}{C}-R_2 + 2H_2O \underset{}{\overset{LPS}{\rightleftharpoons}} CH-O-\overset{\overset{\displaystyle O}{\|}}{C}-R_2 + 2 \text{ fatty}$$
$$CH_2-O-\overset{\overset{\displaystyle O}{\|}}{C}-R_3$$
$$\text{Triacylglycerol} \qquad \text{2-Monoglyceride}$$

$$CH_2OH \qquad CH_2OH \qquad \text{acids}$$

(Eq. 10-24)

The enzymatic activity of pancreatic lipase is specific for the fatty acid residues at positions 1 and 3 of the triglyceride molecule.

Tissue Sources

Lipase concentration is found primarily in the pancreas, although it is also present in the stomach and small intestine.

Diagnostic Significance

Clinical assays of serum lipase measurements are confined almost exclusively to the diagnosis of acute pancreatitis. In this respect, it is similar to amylase measurements but is considered more specific for pancreatic disorders than is amylase measurement, although somewhat less sensitive. Elevated lipase levels also may be found in other intra-abdominal conditions but with less frequency than elevations of serum amylase. Elevations have been reported in cases of penetrating duodenal ulcers and perforated peptic ulcers, intestinal obstruction, and acute cholecystitis. In contrast to amylase levels, lipase levels are normal in conditions of salivary gland involvement. Thus, LPS levels are useful in differentiating a serum amylase elevation due to pancreatic versus salivary involvement.

A new latex test for serum lipase levels has been reported as a possible future test to help evaluate acute abdominal pain. This may also be useful as a near patient testing procedure to be used as a screening test in the Emergency Department.[6]

Assay

Procedures used to measure LPS activity include estimation of liberated fatty acids and turbidimetric methods. The reaction is outlined in Equation 10-25:

$$\text{Triglyceride} + 2 H_2O \underset{\text{pH } 8.6-9.0}{\overset{LPS}{\rightleftharpoons}}$$
$$\text{2-monoglyceride} + 2 \text{ fatty acids} \qquad \textbf{(Eq. 10-25)}$$

Early methods were hampered by long incubation times, but numerous modifications have produced current methods with moderately short incubation times.

The classic method of Cherry and Crandall used an olive oil substrate and measured the liberated fatty acids by titration after a 24-hour incubation. Modifications of the Cherry-Crandall method have been complicated by the lack of stable and uniform substrates. However, triolein is one substrate that now is used as a more pure form of triglyceride.

Turbidimetric methods are simpler and more rapid than titrimetric assays. Fats in solution create a cloudy emulsion. As the fats are hydrolyzed by LPS, the particles disperse, and the rate of clearing can be measured as an estimation of LPS activity.

Sources of Error

Lipase is very stable in serum, with negligible loss in activity at room temperature for 1 week or for 3 weeks at 4°C. Hemolysis should be avoided, since hemoglobin inhibits the activity of serum lipase, causing falsely low values.

Reference Range

Lipase   0–1.0 U/mL

Cholinesterase

The cholinesterases are a group of enzymes that hydrolyze the esters of choline. Two distinct types of cholinesterase occur in human tissue. Acetylcholinesterase (AChE) is also known as *true cholinesterase,* while the other form, cholinesterase, is referred to as *pseudocholinesterase* (PChE). The substrate for AChE is acetylcholine, which is hydrolyzed to choline. Acetylcholine is important as a neurotransmitter and, if not hydrolyzed adequately, will result in excessive quantities causing a block in neuromuscular activity. PChE has a greater affinity for butyryl rather than acetyl esters but also has the ability to hydrolyze many different substrates.

Tissue Sources

Acetylcholinesterase is found primarily in the erythrocytes, brain, and nerve cells. Pseudocholinesterase is found mostly in the serum, liver, pancreas, heart, and white matter of the central nervous system.

Diagnostic Significance

It is decreased activity of PChE that is of clinical significance. PChE concentration in plasma is dependent on the rate of protein synthesis. When cellular impairment exists, decreases in PChE levels are sensitive indicators of changes in the rate of protein synthesis and can be used as an indication for the development of complications in hepatocellular disease and as a prognostic indicator in cancer patients.[40] A fall in PChE levels is also seen in conditions of starvation and after injuries from burns. However, PChE is rarely used as a liver function test in the United States.

The most important application of PChE is its reflection of the presence of organophosphate and insecticide poisoning. These chemicals are potent inhibitors of both forms of the enzyme, causing almost complete inactivation. Inhibition is characterized by tremors and paralysis.

It is also useful to characterize PChE in terms of the genetic type of the enzyme. Some individuals have low PChE activity because of an atypical variant that renders them susceptible to prolonged apnea when exposed to certain types of anesthetics, most importantly succinylcholine. Hence, testing for the presence of the atypical variant may be indicated in patients who have experienced unusual anesthetic reactions.

Assay

Cholinesterase activity can be determined by different techniques, the most important of which include manometric measurements, electrometric measurements, and photometric measurements. The photometric method is most widely used and can incorporate the use of different substrates. One approach is outlined in Equation 10-26:

$$\text{Butyrylthiocholine} \xrightleftharpoons{\text{PChE}} \text{thiocholine} + \text{butyric acid}$$

$$\text{Thiocholine} + 5,5'\text{-dithiobis-(2-nitrobenzoic acid)} \rightarrow \text{5-thio-nitrobenzoic acid}$$

(Eq. 10-26)

The final product is a yellow chromogen whose formation is proportional to enzyme activity.

Sources of Error

Plasma cholinesterase is very stable at 4°C for several days. Serum samples should not be hemolyzed because of the high concentration of AChE in erythrocytes.

Reference Range

AChE   4.7–11.8 U/mL (37°C)

Glucose-6-Phosphate Dehydrogenase

Glucose-6-phosphate dehydrogenase (G-6-PD) is an oxidoreductase that catalyzes the oxidation of glucose-6-phosphate to 6-phosphogluconate or the corresponding lactone. The reaction is important as the first step in the pentose-phosphate shunt of glucose metabolism with the ultimate production of NADPH. The reaction is outlined in Equation 10-27:

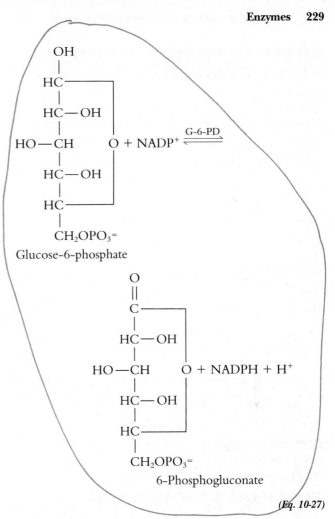

Glucose-6-phosphate

6-Phosphogluconate

(Eq. 10-27)

Tissue Sources

Sources of G-6-PD include the adrenal cortex, spleen, thymus, lymph nodes, lactating mammary gland, and erythrocytes. Very little activity is found in normal serum.

Diagnostic Significance

Most of the interest of G-6-PD focuses on its role in the erythrocyte. Here, it functions to maintain NADPH in its reduced form. An adequate concentration of NADPH is required to regenerate sulfhydryl-containing proteins, such as glutathione, from the oxidized to the reduced state. Glutathione in the reduced form, in turn, protects hemoglobin from oxidation by agents that may be present in the cell. A deficiency of G-6-PD consequently results in an inadequate supply of NADPH and, ultimately, in the inability to maintain reduced glutathione levels. When erythrocytes are exposed to oxidizing agents, hemolysis occurs because of oxidation of hemoglobin and damage of the cell membrane.

G-6-PD deficiency is an inherited sex-linked trait. The disorder can result in several different clinical manifestations, one of which is drug-induced hemolytic anemia. When exposed to an oxidant drug such as primaquine, an antimalarial drug, affected individuals experience a hemolytic episode. The severity of the hemolysis is related to the drug concentration. G-6-PD deficiency is most common in African-Americans but has been reported in virtually every ethnic group.

TABLE 10-12. Reference Ranges for G-6-PD	
Serum	0–0.18 U/L
Erythrocytes	0.117–0.143 U/10⁹ cells

Increased levels of G-6-PD in the serum have been reported in myocardial infarction and megaloblastic anemias. No elevations are seen in hepatic disorders. G-6-PD levels, however, are not routinely performed as diagnostic aids in these conditions.

Assay

The assay procedure for G-6-PD activity is outlined in Equation 10-28:

$$\text{Glucose-6-phosphate} + NADP^+ \xrightleftharpoons{\text{G-6-PD}}$$
$$\text{6-phosphogluconate} + NADPH + H^+$$

(Eq. 10-28)

A red cell hemolysate is used to assay for deficiency of the enzyme, while serum is used for evaluation of enzyme elevations.

Reference Range

See Table 10-12.

PATHOPHYSIOLOGIC CONDITIONS

Enzymes in Cardiac Disorders

Acute myocardial infarction (AMI) occurs when there is an abrupt reduction in blood circulation to a region of myocardial tissue. Atherosclerosis of the coronary arteries is a common precipitating event in many patients with AMI. Because of the cessation of blood flow, the myocardium becomes ischemic and undergoes necrosis, and ultimately, its cellular contents are released into the circulation, reflecting the myocardial damage.

Increased activity of serum enzymes has been used in diagnostic testing for AMI for several years. Enzymes reach their peak levels of serum activity at different times following a myocardial infarct. Figure 10-6 shows the relationship of these enzymes to the occurrence of the infarct. The total enzyme levels that are used most routinely in the diagnosis of AMI are CK, AST, and LD. Of these, CK elevations are the first to appear, followed by AST and LD. Table 10-13 summarizes the time course of enzyme activity in AMI.

Despite the delayed rise of LD levels relative to the occurrence of the infarct, the main advantage to performing total LD levels is to aid in the detection of those infarctions that have occurred more than 5 days prior to enzyme analysis. The probability of CK or AST levels remaining elevated beyond 5 days is very small unless reinfarction has occurred.

The degree of enzyme elevation is roughly proportional to the size of the infarct. CK levels greater than eight times the upper limit of normal (ULN) or AST levels greater than five times ULN are associated with large infarcts with an increased risk of mortality. However, lower values do not necessarily guarantee a more favorable prognosis.

Total serum enzyme levels are sensitive indices of AMI, but they are not specific for AMI, since elevations also occur in other cardiac conditions and following injury to non-cardiac tissues (see Table 10-2). The development of assay methods for the isoenzymes of CK and LD, however, has produced diagnostic measures that are considered both highly sensitive and specific for AMI when interpreted according to specific criteria. The following four isoenzyme patterns represent the major criteria used for interpreting CK and LD isoenzyme results that are obtained during the 48-hour period following onset of symptoms.

Pattern 1: CK-MB ≥ 6% of Total CK; LD-1 > LD-2

Pattern 1 is considered to provide the most reliable diagnostic criteria for detecting AMI. Such a finding confirms that AMI has occurred. Typically, the CK-MB band begins rising 4 to 8 hours after infarct and reaches peak levels 12 to 38 hours after infarct. Peak values range from 11% to 31% of total CK. CK-MB levels usually return to normal within 2 to 3 days. The LD flipped pattern occurs 12 to 24 hours after infarct. It can persist for 5 days, being present after CK-MB has returned to normal. This isoenzyme pattern is found in 80% to 85% of patients with AMI.

Pattern 2: CK-MB ≥ ×−6% of Total CK; LD-2 > LD-1

The presence of CK-MB in quantities ≥6% of total CK with a normal LD isoenzyme pattern reflects some degree of myocardial damage but not necessarily AMI. It does not

TABLE 10-13. Time Course of Enzyme Activity in Myocardial Infarction				
Enzyme	Onset of Elevation (h)	Peak Activity (h)	Degree of Elevation (× ULN)	Duration of Elevation (days)
CK	4–8	12–24	5–10	3–4
CK-MB	4–8	24–38	5–15	2–3
AST	8–12	24	2–3	5
LD	12–24	72	2–3	10
LD-1>LD-2	12–24			5

provide biochemical confirmation of AMI. Approximately 15% to 20% of suspected AMI patients present with this pattern. However, this pattern may also occur in patients who have not sustained myocardial infarction.

Pattern 3: CK-MB < 6% of Total CK

CK-MB values that are less than 6% of total CK indicate that no myocardial infarction has occurred, regardless of the LD isoenzyme data. The lack of elevated CK-MB levels reveals the high degree of sensitivity and specificity of this isoenzyme in ruling out AMI.

Pattern 4: CK-MB < 6% of Total CK; LD-1 > LD-2

Pattern 4 can be found in a variety of noncardiac disorders. It is most frequently seen in intravascular hemolysis, renal cortex infarct, and megaloblastic anemia and has been reported in hemolyzed serum samples. The lack of significant elevations of CK-MB, however, rules out myocardial damage, even though the LD pattern is flipped.

Other conditions resembling AMI can usually be differentiated by the use of combined isoenzyme data. CK-MB has been noted in cases of angina pectoris and coronary insufficiency without evidence of infarction. However, in these cases, LD isoenzyme analysis does not reveal the flipped pattern.

Pulmonary embolism (PE) causes clinical symptoms similar to those of AMI. Isoenzyme analysis does not show abnormal CK-MB levels, since lung tissue contains the CK-BB isoenzyme. Total CK levels are usually normal because of the low activity of CK in lung tissue and the instability of CK-BB in blood. In addition, PE does not produce a flipped LD pattern unless the embolism is associated with a hemorrhagic pulmonary infarct. The typical LD isoenzyme pattern in PE is the appearance of elevated LD-2 and LD-3 fractions.

Congestive heart failure produces enzyme abnormalities suggestive of liver involvement due to inadequate blood perfusion. Increased levels of AST, ALT, LD, and LD-5 are seen, rather than the typical cardiac enzyme profile.

Commonly performed diagnostic procedures, such as cardiac catheterization and coronary arteriography, do not usually produce elevated CK-MB levels unless acute myocardial infarction occurs. However, patients undergoing cardiac surgery, such as coronary bypass surgery, usually have increased CK-MB levels.

It is recommended that blood specimens for enzyme and isoenzyme analysis in the diagnosis of AMI be collected initially and every 12 hours over a 48-hour period until the diagnostic changes are observed. If the CK-MB fraction appears significantly elevated, LD isoenzymes should be determined for at least 48 hours, unless the flipped pattern appears earlier. It is further recommended that, in the event that total CK levels are not elevated, CK isoenzyme analysis be performed whenever the total CK levels fall within the upper third of the normal range of CK reference values. Elevated CK-MB levels may occur in the presence of normal total CK levels. If the laboratory is capable of performing these isoenzyme

studies, routine assay of AST levels is unnecessary because it provides no further diagnostic information.

α-HBD assays are occasionally requested in the diagnosis of AMI. Because α-HBD represents primarily the LD-1 isoenzyme fraction, it may provide some useful information when other methods of LD isoenzyme evaluation are not available. α-HBD follows the same general time course as total LD levels. α-HBD analysis is unnecessary if LD isoenzyme fractionation is available.

Enzymes in Hepatic Disorders

Enzyme assays perform a vital function as laboratory aids in the diagnosis of hepatic disorders. Specific enzyme profiles can be adapted to increase the specificity of enzyme assays in the differential diagnosis of liver disease. Those enzymes of most clinical significance in the diagnosis of hepatocellular disorders include AST, ALT, LD, and LD-5. Generally, in inflammatory hepatocellular disorders, ALT activity is higher than AST activity, whereas in liver cell necrosis AST shows greater elevations.

The enzymes most predominantly elevated in biliary tract obstruction include ALP and GGT. GGT is considered the most sensitive of all enzyme assays for all types of liver disorders, with highest elevations observed in obstructive disorders. Despite this presumptive classification of enzymes, there is a considerable overlap of enzyme elevations in hepatic disorders. However, the degree of elevation, as well as the combined use of enzyme assays and nonenzymatic liver function tests, is helpful, to some extent, in differential diagnosis. Refer to Chapter 18, Liver Function, for a more detailed discussion.

Enzymes in Skeletal Muscle Disorders

Serum enzyme levels play an important role in the diagnosis and assessment of primary muscle disorders. The enzymes most commonly assayed are CK, AST, LD, and aldolase. CK is considered the most sensitive enzyme index of primary muscle involvement and is the enzyme of choice in the evaluation of muscle disorders. Muscle conditions in which enzyme elevations are seen include myopathies, muscle trauma, physical exertion, intramuscular injections, malignant hyperthermia, and muscle hypoxia. In contrast to the enzyme elevations seen in these primary muscle disorders, muscle weakness due to neurogenic muscular disorders, such as multiple sclerosis, myasthenia gravis, and poliomyelitis, is characterized by enzyme levels that are normal or only slightly elevated.

Perhaps the highest elevations of CK activity are seen in the muscular dystrophies, particularly the most severe sex-linked form of Duchenne muscular dystrophy. CK elevations occur early in the course of the disease, usually in the first year of life, and reach values 50 to 100 times ULN. CK elevations are attributed to leakage of intracellular enzymes due to genetically induced alterations in cell membrane function. As the disease progresses, enzyme levels decrease, reflecting the inability of diseased muscle fibers to function normally. In the terminal stages of the disease, enzyme activity may be within normal limits. Other forms of the muscular dystrophies, such as limb-girdle and facioscapulohumeral muscular dystrophies, reveal slight to moderate elevations of CK that do not tend to decrease as the disease progresses. In addition to abnormal CK-MM levels, approximately 90% of Duchenne muscular dystrophy patients also will have elevated CK-MB levels. These levels are attributed to both cardiac involvement and abnormal muscle composition resembling fetal muscle with CK-MB production.

LD-5 is the predominant LD isoenzyme present in skeletal muscle. However, in the muscular dystrophies, there is an increase in levels of the LD-1 and LD-2 forms. Levels of the other skeletal muscle enzymes are abnormal in over 90% of affected muscular dystrophy patients but show lesser elevations than does CK.

Polymyositis, considered a connective-tissue disorder characterized by inflammatory and degenerative muscle changes, also reveals elevated enzyme levels. Serum CK and aldolase levels are elevated in approximately two-thirds of patients. Because CK levels in childhood polymyositis approximate the levels observed in Duchenne muscular dystrophy, it may be difficult to differentiate the two diseases on the basis of enzyme levels. However, patients with polymyositis may respond to steroid therapy with a reduction in enzyme levels, which does not occur in muscular dystrophy.

Muscle trauma such as that occurring from surgery or intramuscular injections will cause serum elevations of muscle enzymes. CK elevations following surgical procedures may be similar to those occurring in AMI. However, the predominant isoenzyme is CK-MM. All muscle enzymes may be elevated following muscular exercise, although CK is most frequently and predominantly elevated.

Malignant hyperthermia is a serious complication of general anesthesia that occurs following the use of halothane and succinylcholine. It occurs usually in genetically susceptible persons and is characterized by impaired respiration, muscle spasm, shock, and increased body temperature. CK elevations can rise to extremely high elevations (300 × ULN) in the early postanesthetic period. The CK isoenzyme pattern is abnormal, with the appearance of all three isoenzyme forms. The majority of affected individuals show evidence of subclinical myopathy. The presence of CK enzyme elevations are also used to detect other affected relatives, the majority showing elevations less than two times ULN.

Enzymes in Bone Disorders

The major enzyme that is increased in disorders of the skeletal system is ALP, particularly the bone isoenzyme, which is associated with osteoblastic activity. In addition, acid phosphatase is elevated in certain bone disorders.

Paget's disease of bone (osteitis deformans) is one of the more common chronic skeletal diseases. It is characterized by an excessive resorption of bone caused by osteoclastic activity. Bone resorption is followed by an osteoblastic phase in which new bone is formed and deposited in a random fashion, causing an irregular pattern of bone deposition. ALP activity is elevated because of the stimulation of the osteoblasts and can reach extreme elevations ranging from 10 to 100 times ULN. The incidence of occurrence is about 3% in individuals over the age of 40 and increases with advancing age.

Bone disorders resulting from vitamin D deficiency are known as *rickets* in children and *osteomalacia* in adults. The two diseases are not synonymous, however, owing to differences in growing bone in children and calcified bone in adults. Deposition of uncalcified bone matrix occurs in both conditions. Both conditions are characterized by an increase in osteoblastic activity with resultant elevations in serum ALP activity.

Hyperparathyroidism results from excessive secretion of parathyroid hormone. Because of the effects of parathyroid hormone on calcium and phosphorous metabolism, there is usually a consequent abnormality in bone metabolism. However, bone involvement is primarily due to abnormal bone resorption with little osteoblastic activity. Therefore, the majority of patients have normal serum ALP levels, with elevations seen only when there is significant bone involvement, in approximately 30% of cases.

Osteogenic sarcoma is one of the most common of primary bone tumors and occurs most frequently in the 10- to 20-year-old age group. ALP values are variable, but considerably higher values are observed in patients with osteoblastic lesions than in those with lytic lesions. Patients whose bone disorder is primarily confined to osteoclastic activity, such as those with multiple myeloma, usually have normal ALP values.

During the course of healing of bone fractions, ALP elevations of approximately two times ULN may be seen. Values return to normal within 2 months.

ALP levels can be expected to be increased during periods of physiologic bone growth. In children younger than 10 years, the average ALP level is almost four times that of adult activity. Peak activity occurs in girls between the ages of 11 and 12 and in boys between 13 and 14. Elevations are more pronounced in boys than in girls and can reach levels 50% higher than those of the younger age groups. Adult levels are usually reached in girls by the age of 15, while boys still have values one and one-half times adult levels at age 18.[12]

Elevations in acid phosphatase activity in bone disorders appear to be associated more with osteoblastic activity. Therefore, those disorders characterized by excessive bone resorption will more likely reveal increased ACP levels. Additionally, metastases to bone will show elevations of both ACP and ALP.

Enzymes in Malignancy

Several of the enzymes and isoenzymes of clinical interest have some value in the diagnosis and management of various malignancies and, therefore, serve as tumor markers. A discussion of specific enzymes which may indicate particular malignancies is included in Chapter 24, Tumor Markers.

Sensitive and specific immunologic methods are usually employed for the serum quantitation.

The use of enzymes to detect, monitor, and follow treatment for malignancies is a rapidly changing field. Researchers are continually searching for serum levels of cellular enzymes that will be elevated during specific malignancies. Specificity is one of the main concerns—there are not many enzymes that are elevated in the serum of patients only with certain malignancies. Many unusual procedures for serum levels of enzymes, such as glutamate dehydrogenase (GD), 5′ nucleotidase 5′ NT, ribonuclease, and polyhexose isomerase, are not performed in most clinical laboratories. Their limited diagnostic use and low test volume result in their being sent to a reference laboratory for analysis.

Enzymes in Acute Pancreatitis

Acute pancreatitis should be considered in the differential diagnosis of any patient presenting with persistent, severe abdominal pain. Acute pancreatitis occurs when inactive proteolytic and lipolytic enzymes of the pancreas become activated and initiate autodigestion of pancreatic tissue and blood vessels. The severity of the condition ranges from pancreatic edema, with a mortality of 5% to 10%, to hemorrhagic pancreatic necrosis, with a mortality of 50% to 80%. Acute pancreatitis is most frequently associated with alcoholism and biliary tract disease, although approximately 20% of cases are of an unknown etiology.[36]

The laboratory diagnosis of acute pancreatitis includes evaluation of serum and urine amylase levels and serum lipase levels and calculation of the amylase-creatinine clearance ratio (C_{am}/C_{cr}). Measurement of serum amylase isoenzymes may become an important diagnostic tool in the future.

Serum amylase levels generally become elevated within 2 to 12 hours of an attack, reach peak levels at 24 hours, and return to normal within 3 to 5 days. There is a lack of good correlation between the degree of serum amylase elevation and the severity of the pancreatitis. In fact, normal levels have been observed in cases of severe, fulminating pancreatitis. Generally, amylase levels range between 250 and 1000 Somogyi units/dL. Higher elevations tend to have a greater specificity for acute pancreatitis. Values greater than 500 Somogyi units/dL are highly suggestive of acute pancreatitis. Other causes of acute abdominal pain, such as acute appendicitis and intestinal obstruction, usually show serum AMS levels less than 500 Somogyi units/dL. Considering that amylase is readily filtered by the glomerulus, urine amylase is an expected normal finding, with elevations being detected in acute pancreatitis.

Urine levels may rise to relatively higher values than those of serum amylase and usually remain elevated longer, for 7 to 10 days. Urine determinations are generally performed on a timed urine collection. If renal function is impaired, levels may be falsely low.

An attempt to enhance the specificity of amylase measurement for the diagnosis of acute pancreatitis is calculation of the C_{am}/C_{cr} ratio. The ratio, although not entirely specific for acute pancreatitis, has some bearing on the differ-

ential diagnosis. In acute pancreatitis, the renal clearance of amylase is generally much greater than that of creatinine, causing an elevated ratio. The mechanism accounting for this increase in renal clearance is, in part, attributed to a disturbance in tubular reabsorption of the enzyme. The ratio is derived from the creatinine clearance formula, and its calculation requires both amylase and creatinine values obtained from simultaneously collected samples of serum and urine. The calculation is according to the following formula:

$$C_{am}/C_{cr} \% = \frac{[AMS]\ urine}{[AMS]\ serum} \times \frac{[creatinine]\ serum}{[creatinine]\ urine} \times 100$$

(Eq. 10-29)

Because the ratio is independent of urine volume, a random urine sample is adequate. Comparison of amylase clearance to creatinine clearance is corrected for abnormal amylase clearance because of changes in the glomerular filtration rate, although an elevated ratio may be found in severe renal insufficiency.[41]

Normally, the ratio ranges from 2% to 5%. In acute pancreatitis, the ratio is greater than 5%, frequently ranging from 7% to 15%. A normal ratio is not commonly seen early in the course of acute pancreatitis. An elevated ratio also may occur in diabetic ketoacidosis, in extensive burns, and in duodenal perforation, thus decreasing the specificity of the test for acute pancreatitis. In acute pancreatitis, the elevated C_{am}/C_{cr} ratio returns to normal after serum and urine amylase levels become normal.

Another advantage of calculating the ratio is its ability to differentiate macroamylasemia from other causes of hyperamylasemia. Because of the large complex of macroamylase, its renal clearance is decreased, resulting in a low ratio. In macroamylasemia the C_{am}/C_{cr} ratio is usually less than 2%.

In conjunction with other laboratory measures, serum lipase measurements serve as a useful measure of acute pancreatitis. Serum lipase analysis has not achieved a degree of popularity equal to that of serum amylase primarily because of the lengthy procedures used in the past to measure lipase. Lipase activity in acute pancreatitis generally parallels that of serum amylase. However, peak levels probably occur a bit later, at 12 to 24 hours, and levels may remain elevated longer than those of amylase. The combined sensitivity of both serum amylase and lipase lends greater confidence to exclusion of a diagnosis of acute pancreatitis when neither enzyme is elevated.

Measurement of amylase levels in pleural or peritoneal fluid also may be requested. In acute pancreatitis, pleural fluid invariably has an elevated amylase level, so that a normal pleural fluid amylase rules out acute pancreatitis.[36] However, elevations can also occur in primary or metastatic carcinoma of the lung and esophageal perforation.

SUMMARY

Enzymes are proteins that are found in all body tissues and catalyze biochemical reactions. The catalyzed reactions are specific and essential to physiologic well-being. In injury or in many disease states certain enzymes are released from their normal locations and appear in increased amounts in the general circulation. Hence, an understanding of biologically significant enzymes and the reactions they catalyze is useful in the diagnosis and treatment of certain disease states.

This chapter has reviewed the general properties of enzymes and their classification and catalytic mechanisms. Several factors that affect the rate of enzymatic reactions and general methods for measuring enzyme activity were also discussed.

Many enzymes are clinically important. Quantitating serum levels of certain enzymes or isoenzymes can assist in

the diagnosis and prognosis of hepatic disorders, skeletal muscle disorders, bone disorders, cardiac disorders, malignancy, or acute pancreatitis. This chapter has discussed the tissue sources, diagnostic significance, preferred assay methodologies, and reference ranges of several enzymes of major clinical significance.

Methodologies for identifying and quantitating enzymes are continuously becoming more sensitive and specific. As the methodologies improve, the importance of evaluating enzymes to assist in diagnosis and treatment of disease will continue to increase.

CASE STUDY
10-1

A 52-year-old man was admitted to the hospital with complaints of substernal, oppressive chest pain, difficult and rapid breathing, and anxiety. Symptoms had begun 1 day prior to admission. The patient was con-valescing from abdominal surgery performed 2 weeks earlier. There was no previous history of heart disease. Refer to Case Study Table 10-1.

CASE STUDY TABLE 10-1. Laboratory Data

	CK (30–160)	CK-MB (<6%)	AST (5–30)	ALT (6–37)	LD (100–225)	LD Isoenzymes
Day 1—6 PM	210	3%	35	70	370	LD1>LD2, ↑LD3, ↑LD1
Day 2—6 AM	190	4%	35	62	280	LD1>LD2, ↑LD3, ↑LD1
Day 2—6 PM	160	3%	30	58	262	↑LD5
Day 3—6 AM	145	3%	27	45	240	↑LD5

Questions

1. Based on the laboratory data, would the working diagnosis be myocardial infarction?

2. Which laboratory data support the working diagnosis?

3. Suggest an explanation for the increases in CK and ALT values and the LD isoenzyme data.

CASE STUDY
10-2

A 32-year-old man decided to train for an amateur boxing match. He trained on his own and discovered, during the match, that he was not very well prepared. He was knocked out late in the fight and, upon regaining consciousness, experienced some tightness in his chest. He was taken to the hospital. Upon admission, the following abnormal laboratory results were obtained.

CASE STUDY TABLE 10-2

Total CK	210 U/L
LD	336 U/L
AST	45 U/L
ALT	47 U/L

Questions

1. What could be causing these enzyme elevations?

2. How do you know from the laboratory data whether the man suffered a myocardial infarction?

3. When should these abnormal results return to normal?

CASE STUDY
10-3

A 20-year-old woman arrived at the emergency room complaining that she thought she had "the flu" but had been ill for 3 days and was feeling worse instead of better. Her symptoms were fever, intense abdominal pain, nausea, and vomiting. She stated that her pain worsened when she moved. Physical examination revealed tenderness in the upper middle abdomen.

Questions

1. Given the symptoms and physical findings, what is the most probable diagnosis?

2. What enzyme(s) discussed in this chapter might be elevated with this disease? Why?

REFERENCES

1. Bark CJ. Mitochondrial creatine kinase—A poor prognostic sign. JAMA 1980;243:2058.

2. Baron DN, Moss DW, Walker PG, Wilkinson JH. Abbreviations for names of enzymes of diagnostic importance. J Clin Pathol 1971;24:656.

3. Baron DN, Moss DW, Walker PG, Wilkinson JH. Revised list of abbreviations for names of enzymes of diagnostic importance. J Clin Pathol 1975;28:592.

4. Batsakis J, Savory J, eds. Creatine kinase. Crit Rev Clin Lab Sci 1982;16:291.

5. Batsakis J, Savory J, eds. Creatine kinase isoenzymes: Electrophoretic and quantitative measurements. Crit Rev Clin Lab Sci 1981;15:187.

6. Benner, et al. Lipase latex test for acute abdominal pain. Ital J Gastroenterol February 1994;24(2).

7. Bhagavan NV, Darm JR, Scottolini AG. A sixth lactate dehydrogenase isoenzyme (LD6) and its significance. Arch Pathol Lab Med 1982;106:521.

8. Bruns DE, Emerson JC, Intemann S, Bertholf R, Hill KE, Savory J. Lactate dehydrogenase isoenzyme-1: Changes during the first day after acute myocardial infarction. Clin Chem 1981;27:1821.

9. Cabello B, Lubin J, Rywlin AM, Frenkel R. Significance of a sixth lactate dehydrogenase isoenzyme (LDH₆). Am J Clin Pathol 1980;73:253.

10. Enzyme nomenclature 1978, recommendations of the nomenclature committee of the international union of biochemistry on the nomenclature and classification of enzymes. New York: Academic Press, 1979.

11. Faulkner WR, Meites S, eds. Selected methods for the small clinical chemistry laboratory, selected methods of clinical chemistry. Vol 9. Washington, D.C. American Association for Clinical Chemistry, 1982.

12. Fleisher GA, Eickelberg ES, Elveback LR. Alkaline phosphatase activity in the plasma of children and adolescents. Clin Chem 1977;23:469.

13. Galen RS. The enzyme diagnosis of myocardial infarction. Human Pathol 1975;6:141.

14. Gittes RF. Carcinoma of the prostate. N Engl J Med 1991; 324:236.

15. Gittes RF. Prostate-specific antigen. N Engl J Med 1987; 317:954.

16. Goldberg DM, Werner M, eds. LD6, A sign of impending death from heart failure, selected topics in clinical enzymology. New York: Walter de Greyter, 1983.

17. Griffiths JC. The laboratory diagnosis of prostatic adenocarcinoma. Crit Rev Clin Lab Sci 1983;19:187.

18. Irvin RG, Cobb FR, Roe CR. Acute myocardial infarction and MB creatine phosphokinase—Relationship between onset of symptoms of infarction and appearance and disappearance of enzyme. Arch Intern Med 1980;140:329.

19. Lang, H, ed. Creatine kinase isoenzymes—Pathophysiology and clinical application. Berlin: Springer-Verlag, 1981.

20. Lang H, Wurzburg U. Creatine kinase, an enzyme of many forms. Clin Chem 1982;28:1439.

21. Lantz RK, Berg MJ. From clinic to court: Acid phosphatase testing. Diagn Med, March/April 1981;55.

22. Latner AL, Schwartz MK, eds. The plasma cholinesterases: A new perspective. Adv Clin Chem 1981;22:2.

23. Legaz ME, Kenny MA. Electrophoretic amylase fractionation as an aid in diagnosis of pancreatic disease. Clin Chem 1976; 22:57.

24. Leung FY, Henderson AR. Influence of hemolysis on the serum lactate dehydrogenase–1/Lactate dehydrogenase–2 ratio as determined by an accurate thin-layer agarose electrophoresis procedure. Clin Chem 1981;27:1708.

25. Lott JA. Electrophoretic CK and LD isoenzyme assays in myocardial infarction. Lab Management, February 1983;23.

26. Lott JA, Stang JM. Serum enzymes and isoenzymes in the diagnosis and differential diagnosis of myocardial ischemia and necrosis. Clin Chem 1980;26:1241.

27. Mass DW. Alkaline phosphatase isoenzymes. Clin Chem 1982;28:2007.

28. Murthy, Uma et al. Hyperamylasemia in patients with acquired immunodeficiency syndrome. Am J Gastroenterol 1992;87(3):332.

29. Oesterling JE. Prostate specific antigen: A critical assessment of the most useful tumor marker for adenocarcinoma of the prostate. J Urol 1991;145:907.

30. Pesce MA. The CK isoenzymes: Findings and their meaning. Lab Management, October 1982;25.

31. Posen S, Doherty E. The measurement of serum alkaline phosphatase in clinical medicine. Adv Clin Chem 1981; 22:165.

32. Puleo, P, et al. Use of a rapid assay of subforms of creatine kinase MB to rule out acute myocardial infarction. N Engl J Med 1994;331:561.

33. Roche Diagnostics. Isomune-CK package insert. Nutley, NJ: Hoffman–La Roche, Inc, 1979.

34. Roche Diagnostics. Isomune-LD package insert. Nutley, NJ: Hoffman–La Roche, Inc, 1979.

35. Rosalki SB. Gamma-glutamyl transpeptidase. Adv Clin Chem 1975;17:53.

36. Salt WB II, Schenker S. Amylase—Its clinical significance. A review of the literature. Medicine 1976;4:269.

37. Silverman LM, Dermer GB, Zweig MH, Van Steirteghem AC, Tokes ZA. Creatine kinase BB: A new tumor-associated marker. Clin Chem 1979;25:1432.

38. Warren BM. The isoenzymes of alkaline phosphatase. Beaumont, Tx: Helena Laboratories, 1981.

39. Wilkinson JH. The principles and practice of diagnostic enzymology. Chicago: Year Book Medical Publishers, 1976.

40. Yam LT. Clinical significance of the human acid phosphatases, a review. Am J Med 1974;56:604.

41. Yoshito B, Murphy GP. Antigen marker assays in prostate cancer. Lab Management, April 1983;19.

SUGGESTED READINGS

Cornish-Bowden A. Fundamentals of enzyme kinetics. London: Butterworth, 1979.

Dixon M. Enzymes. New York: Academic Press, 1979.

Foster RL. The nature of enzymology. New York: Wiley, 1980.

Holley JW. Enzymology: Basic text. Houston: CASMT, sponsored by BMC, 1980.

Pincus MR, Zimmerman HJ, Henry JB. Clinical enzymology. In: JB Henry, ed. Clinical diagnosis and management by laboratory methods. Philadelphia: WB Saunders Co, 1991;250.

Remaley AL, Wilding P. Macroenzymes: Biochemical characterization, clinical significance, and laboratory detection. Clin Chem 1989;35(2):2261.

Roger GP. Fundamentals of enzymology. New York: Wiley, 1982.

Segel IH. Enzyme kinetics. New York: Wiley, 1975.

Wilkinson JH. Isoenzymes. 2nd ed. Philadelphia: Lippincott, 1970.

Wilkinson JH. The principles and practice of diagnostic enzymology. Chicago: Year Book Medical Publishers, 1976.

Chapter 11

Blood Gases, pH, and Buffer Systems

Sharon S. Ehrmeyer, Joan B. Shrout

— Test #4

Objectives

Upon completion of this chapter, the clinical laboratorian should be able to:

- *Describe the principles involved in the measurement of pH, PCO_2, PO_2, and the various hemoglobin species.*

- *Outline the interrelationship of the buffering mechanisms of bicarbonate-carbonic acid and hemoglobin.*

- *Explain the significance of and, where possible, calculate values for the following pH and blood gas parameters: pH, PCO_2, PO_2, actual bicarbonate, carbonic acid, base excess, oxygen saturation, fractional oxyhemoglobin, hemoglobin oxygen (binding) capacity, oxygen content, $P50$, and total CO_2.*

- *Using the Henderson-Hasselbalch equation and blood gas data, determine whether data are normal or represent metabolic or respiratory acidosis or metabolic or respiratory alkalosis. Identify whether the data represent uncompensated or compensated conditions.*

- *Identify some common causes of metabolic acidosis and alkalosis and of respiratory acidosis and alkalosis. State how the body attempts to compensate (kidney and lungs) for the various conditions.*

- *Describe the significance of the hemoglobin-oxygen dissociation curve and the impact of pH, 2,3-DPG, temperature, and PCO_2 on its shape, $P50$ values, and release of oxygen to the tissues.*

- *Discuss problems and precautions in collecting and handling samples for pH and blood gas analysis. Include*

syringes, anticoagulants, mixing, icing, and capillary and venous samples as well as arterial samples in the discussion.

- *Describe the approaches to quality control including commercial controls, tonometry, external (proficiency testing), and delta checks.*

- *Given oxygen saturation data from both the blood gas analyzer and oximeter, discuss reasons for possible discrepancies.*

- *Calculate partial pressures for PCO_2 and PO_2 for various percentages of carbon dioxide and oxygen. In doing the calculations, account for the barometric pressure and vapor pressure of water.*

KEY WORDS

Acid	Fractional	Oxygen Saturation
Acidemia	Oxyhemoglobin	$P50$
Acidosis	Hemoglobin Oxygen	Partial Pressure
Alkalemia	(binding) Capacity	PCO_2
Alkalosis	Hemoglobin-Oxygen	pH
Analytical Error	Dissociation	pK
Base	Curve	PO_2
Base Excess	Henderson-	Pre-analytical Error
Bicarbonate	Hasselbalch	Quality Control
Buffers	Equation	Respiratory
Calibration	Metabolic	Acidosis and
Carbonic Acid	(nonrespiratory)	Alkalosis
Compensation	Acidosis and	
Electrodes	Alkalosis	
FIO$_2$	Oxygen Content	

Michael L. Bishop, Janet L. Duben–Engelkirk, and Edward P. Fody.
CLINICAL CHEMISTRY. © 1996 Lippincott–Raven Publishers.

An important component of clinical biochemistry is information on acid–base/blood gas homeostasis. It forms an extensive knowledge base for handling data in life-threatening situations. Many principles and facts must be interrelated; focusing on only one result can be misleading.

This chapter presents information about the buffer systems the body employs for the maintenance of pH in the blood. The exchange of gases, specifically carbon dioxide and oxygen, is discussed. Techniques and instrumentation used in the measurement of pH and blood gases are described. Since the sample collection and handling can greatly affect the quality of the laboratory results, preanalytical considerations are discussed. Quality control approaches also are presented.

DEFINITIONS: ACID, BASE, BUFFER

A discussion of acid–base balance requires a review of several basic definitions, such as acid, base, buffer, pH, and pK. In addition, the applicability of general biochemical principles of equilibrium and the laws of mass action are relevant.

An *acid* is a substance that can yield a hydrogen ion or hydronium ion when dissolved in water. A *base* is a substance that can yield hydroxyl ions (OH^-). The relative strengths of acids and bases as well as their ability to dissociate in water can be ranked. Dissociation constant (K-value) tables can be found in most biochemistry texts. For an acid, the larger the K, the greater the tendency to dissociate into ions. The pK, defined as the negative log of the ionization constant, is that pH where the protonated and unprotonated forms are present in equal concentration. Strong acids have pK values of less than 3.0, while strong bases have pK values greater than 9.0. For acids, raising the pH above the pK will cause the acid to dissociate and yield a hydrogen ion (H^+). For bases, lowering the pH below the pK will cause the base to release OH^-. Many species have more than one pK, meaning they can accept or donate more than one H^+.

A *buffer,* the combination of a weak acid or weak base and its salt, is a system that resists changes in pH. The effectiveness of a buffer is dependent on the pK of the buffering system and the pH of the environment in which it is placed. In plasma, the bicarbonate–carbonic acid system, having a pK of 6.1, is one of the principal buffers.

$$CO_2 \leftrightarrow CO_2 + H_2O \leftrightarrow H_2CO_3 \leftrightarrow HCO_3^- + H^+$$

| Gas | CO₂ dissolved in plasma | Carbonic acid | Bicarbonate |

(Eq. 11-1)

When the blood plasma pH is 7.4, this buffer is more effective than the lactic acid–lactate system (pK=3.9) or the ammonium–ammonia system, with a pK of 9.4.

Weisberg cited an example to demonstrate the effectiveness of the blood buffers.[17] If 100 mL of distilled water is at a pH of 7.35, and one drop of 0.05 N HCl is added, the pH will change to 7.00. In order to change 100 mL of "normal"

blood from a pH of 7.35 to 7.00, approximately 25 mL of 0.05 N HCl is needed. With 5.5 L of blood in the average body, more than 1300 mL of HCl would be required to make this same change in pH.

ACID–BASE BALANCE

Maintenance of H^+

The body produces 15 to 20 mol of H^+ per day; however, the normal concentration of H^+ in the extracellular body fluids ranges from only 36 to 44 nmol/L. Deviations in H^+ concentration from this range will cause alterations in the rates of chemical reactions within the cell and affect the many metabolic processes of the body. Concentrations greater than 44 nmol/L can alter consciousness and lead to coma and death, whereas H^+ concentrations less than 36 nmol/L can cause neuromuscular irritability, tetany, loss of consciousness, and death.

Hydrogen ion concentration is frequently expressed by the logarithmic pH scale. The two quantities are related:

$$pH = \log \frac{1}{cH^+} = -\log cH^+ \qquad \textbf{(Eq. 11-2)}$$

where c is concentration. The normal pH of arterial blood is 7.40 and is equivalent to an H^+ concentration of 40 nmol/L. Because of the reciprocal relationship between cH^+ and pH, an increase in H^+ decreases the pH, while a decrease in H^+ increases the pH. A pH below the reference range is referred to as *acidosis,* whereas a pH above the reference range is referred to as *alkalosis.* Technically, the suffix *-osis* refers to a process in the body, whereas the suffix *-emia* refers to the corresponding state in the blood.

The arterial pH is controlled by systems that regulate the production and retention of acids and bases. These include buffers, the respiratory center and lungs, and the kidneys.

Buffer Systems: Regulation of H^+

The body's first line of defense against changes in H^+ concentration is the buffer systems present in all body fluids. All buffers consist of a weak acid, such as H_2CO_3, and its salt or conjugate base, HCO_3^-, for the bicarbonate–carbonic acid buffer system. H_2CO_3 is a weak acid because it does not completely dissociate into H^+ and HCO_3^-. (In contrast, a strong acid such as HCl completely dissociates into H^+ and Cl^- in solution.) When an acid is added to the bicarbonate–carbonic acid buffer system, the HCO_3^- will combine with the H^+ from the acid to form H_2CO_3. When a base is added, H_2CO_3 will combine with the OH^- group to form H_2O and HCO_3^-. In both cases, there will be a smaller change in pH than there would be from adding the acid or base directly to an unbuffered solution.

The rapid exchange of CO_2 between the tissues and blood and then the blood and lungs, plus the effect of pH on the respiration rate and the fact that the bicarbonate level can be altered in the renal tubules, makes the bicarbonate–carbonic acid system an important buffer despite its low buffering capacity. In addition, this buffering system immediately counters the effects of fixed nonvolatile acids (H^+A^-) by binding the dissociated hydrogen ion ($H^+A^- + HCO_3^- = H_2CO_3 + A^-$). The resultant H_2CO_3 then dissociates, and the H^+ is neutralized by the buffering ability of hemoglobin. A large percentage of the buffering of whole blood is done by hemoglobin, since hemoglobin, once oxygen is released, binds hydrogen ion. Figure 11-1 shows the interrelationship of the hemoglobin and bicarbonate buffering system.

Other buffers also are important. The phosphate buffer system ($cHPO_2^{2-}$–$cH_2PO_4^-$) plays a role in plasma and erythrocytes and is involved in the exchange of sodium ion for hydrogen ion in the urine filtrate. Plasma proteins, especially the imidazole groups of histidine, also form an important buffer system in plasma. In general, since most circulating proteins have a net negative charge, they are capable of binding hydrogen ion.

Respiratory System: Regulation of Carbonic Acid

The end product of many aerobic metabolic processes is carbon dioxide. Since dissolved CO_2 (dCO_2) is more concentrated in the tissues, it easily diffuses out of the tissue into the plasma and erythrocytes (RBC). In plasma, a small amount remains as dCO_2, while some combines with water to form carbonic acid (H_2CO_3), which quickly dissociates into H^+ and HCO_3^- (see Fig. 11-1). The freed H^+ is buffered by plasma proteins and plasma buffers. Additional CO_2 combines with proteins to form carbamino compounds.

Although some of the CO_2 entering the RBCs remains as dCO_2 and some combines with hemoglobin to form carbamino hemoglobin, most of the CO_2 combines with water

to form H_2CO_3. This reaction is accelerated by the enzyme carbonic anhydrase found in the RBC membrane. Again, the H_2CO_3 quickly dissociates to H^+ and HCO_3^-. The H^+ would alter the pH if left unbound. However, once the oxygen picked up in the lungs is unloaded from oxyhemoglobin (O_2Hb) at the tissues, hemoglobin readily accepts the H^+, forming deoxyhemoglobin (HHb). As the HCO_3^- concentration rises in the RBC, it diffuses out into the plasma. In order to maintain electroneutrality (the same number of positively and negatively charged ions on each side of the RBC membrane), chloride diffuses into the cell. This is known as the *chloride shift*.

In the lungs, the H^+ carried on the deoxyhemoglobin in the venous blood is released to recombine with HCO_3^- to form H_2CO_3, which dissociates into H_2O and CO_2. The CO_2 diffuses into the alveoli and is eliminated through respiration. The inspired oxygen diffuses from the alveoli into the blood and is loaded onto the hemoglobin, forming oxyhemoglobin (O_2Hb) in the arterial blood. The net effect of the interaction of these two buffering systems is minimal change in H^+ concentration between the venous and arterial circulations. However, when CO_2 is not removed by the lungs at the rate of its production, it will accumulate in the blood, causing an increase in H^+ concentration. If, on the other hand, CO_2 removal is faster than production, the H^+ concentration will be decreased. Consequently, ventilation affects the pH of the blood. In addition, a change in the H^+ concentration of blood due to nonrespiratory disturbances causes the respiratory center to respond by altering the rate of ventilation in an effort to restore the blood pH to normal.

The Kidney System: Regulation of Bicarbonate

When the hydrogen ion concentration deviates from normal, the kidneys also respond by selectively excreting or reabsorbing hydrogen, sodium, chloride, phosphate, potassium, ammonia, and bicarbonate ions to restore the equilibrium between the production and removal of hydrogen ions (Fig. 11-2). An abnormal bicarbonate level may be the direct result of kidney malfunction, or it may be due to another disease process that secondarily affects the kidney.

The process of reabsorption of bicarbonate actually comprises several reactions in the renal tubule cells, lumen, and blood vessels supplying the tubule cells. The overall reactions result in the reabsorption of sodium and bicarbonate and the "loss" into the filtrate (which will become urine) of carbon dioxide, water, sodium, potassium, chloride, dihydrogen phosphate, and ammonium sulfate.[17]

In the proximal and distal kidney tubules, sodium and bicarbonate travel from the filtrate—in the lumen—into the tubule cell. Sodium is exchanged for hydrogen. The hydrogen combines with bicarbonate still in the lumen, forming carbonic acid, which dissociates into water and carbon dioxide. The carbon dioxide easily diffuses into the tubule cell and reacts with available hydroxyl ions to re-form bicarbonate, which is then reabsorbed into the

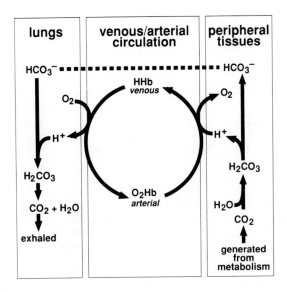

Figure 11-1. Interrelationship of the bicarbonate and hemoglobin buffering systems.

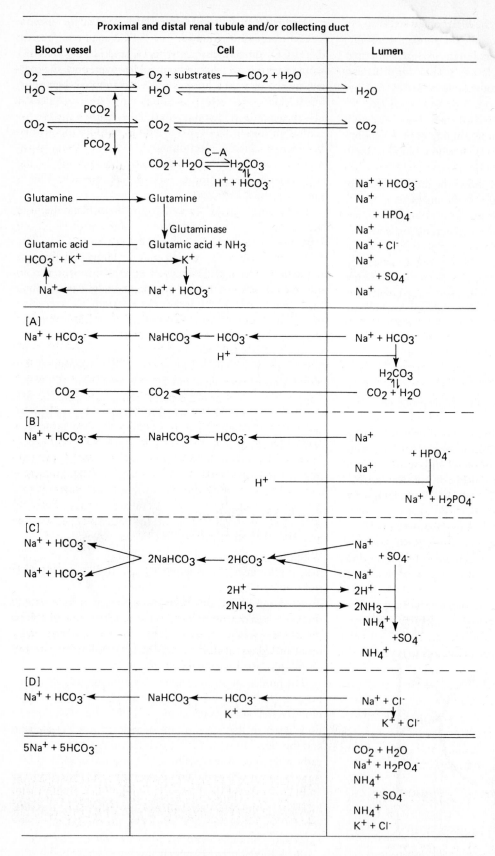

Figure 11-2. Sodium and hydrogen exchange in the kidney. (Weisberg HF. Water, electrolyte, and acid-base balance. 2nd ed. Baltimore: Williams & Wilkins, 1962; used with permission of Dr. HF Weisberg and the Williams & Wilkins Co.)

bloodstream. These two movements of CO_2 are referred to as the *reclamation* of bicarbonate.

The sodium ion also exchanges for a hydrogen ion, which subsequently reacts with a molecule of disodium hydrogen phosphate (Na_2HPO_4) to form sodium dihydrogen phosphate (NaH_2PO_4) in the filtrate. Sodium is further conserved or reabsorbed in the overall reaction, in which hydrogen ions again pass from the tubule cell into the filtrate and combine with ammonia, forming ammonium ions, which cannot easily diffuse into the tubule cells. The ammonium ions complex with sulfate ions in the filtrate to form ammonium sulfate. Sodium further exchanges with potassium. The potassium then combines with chloride in the filtrate.

In each of the sodium reabsorption reactions, bicarbonate is reabsorbed from the filtrate into the tubule cell and then into the blood. (*Reabsorption* involves the process of reentering the blood. *Secretion* and/or *excretion* by the tubule cells puts substances into the filtrate.) These reactions determine the pH of the urine, as well as the pH of the blood.

There are various factors that affect the reabsorption of bicarbonate. When the blood or plasma bicarbonate level is higher than 26 to 30 mmol/L, bicarbonate will be excreted.[17] While reabsorption apparently continues, excretion also occurs. Therefore, it is unlikely that the plasma bicarbonate will exceed 30 mmol/L unless these excretory capabilities fail.

The bicarbonate level may become increased if an excessive amount of lactate, acetate, or bicarbonate is intravenously infused. It also may increase if there is an excessive loss of chloride without replacement (as from vomiting or prolonged nasogastric suction), since the bicarbonate anion will be retained to preserve electroneutrality.

Several factors may result in decreased bicarbonate levels. Most diuretics, regardless of mechanism of action, favor the excretion of bicarbonate. Reduced bicarbonate reabsorption also occurs in conditions such as diarrhea, enteric fistulas, or Addison's disease, in which there is an excessive loss of cations. In kidney dysfunction (such as chronic nephritis or infections), bicarbonate reabsorption may be impaired. Failure to exchange sodium and hydrogen may lead to an accumulation of sulfates, phosphates, and chloride in the blood and to the loss of bicarbonate in the filtrate. Increased organic acid production (*i.e.,* ketones and/or lactic acid) also will result in decreased bicarbonate reabsorption. Sodium is excreted in complex with the anionic portion of the ketoacid. Bicarbonate is subsequently lost, binding with the "extra" load of hydrogen ions—the cationic portion of the acetoacetic acid, betahydroxybutyric acid, and acetone.[18]

ASSESSING ACID–BASE HOMEOSTASIS

The Bicarbonate Buffering System and the Henderson-Hasselbalch Equation

In assessing acid–base homeostasis, the bicarbonate buffering system is used. From the data, inferences can be made pertaining to the other buffers and the systems that regulate the production, retention, and excretion of acids and bases.

In the bicarbonate buffering system, the dCO_2 is in equilibrium with CO_2 gas which can be expelled by the lungs. Therefore, the bicarbonate buffering system is referred to as an *open* system, and the dCO_2, which is controlled by the lungs, is the *respiratory component*. The lungs can rapidly participate in the regulation of blood pH through hypo- or hyperventilation. The bicarbonate concentration is controlled mainly by the kidneys, making it the *nonrespiratory,* or *metabolic, component* in evaluating acid–base homeostasis.

The Henderson-Hasselbalch equation expresses acid–base relationships in a mathematical formula:

$$pH = pK' + \log \frac{cA^-}{cHA} \qquad \textbf{\textit{(Eq. 11-3)}}$$

where A = proton acceptor (*e.g.*, HCO_3^-)
HA = proton donor, or weak acid (*e.g.*, H_2CO_3)
pK' = pH at which there is an equal concentration of protonated and unprotonated species

Knowing three of the variables allows for the calculation of the fourth.

In plasma and at body temperature, the pK' of the bicarbonate buffering system is 6.1. Since the equilibrium between carbonic acid and carbon dioxide in plasma is approximately 1:800, the concentration of carbonic acid is proportional to the partial pressure exerted by the dissolved CO_2. In plasma at 37°C, the value for the combination of the solubility constant for PCO_2 and the factor to convert to millimoles per liter of H_2CO_3 is 0.0307 mmol/L per millimeter of mercury (mmol $L^{-1}mmHg^{-1}$). The constant is affected by the temperature and the solvent. If either of these changes, the solubility constant also will change. Both pH and PCO_2 are measured in the blood gas analysis, and the pK' is a constant; therefore, HCO_3^- can be calculated:

$$pH = pK' + \log \frac{cHCO_3^-}{0.031 \times PCO_2} \qquad \textbf{\textit{(Eq. 11-4)}}$$

In health, when the kidneys and lungs are functioning properly, a 20:1 ratio of bicarbonate to carbonic acid will be maintained (resulting in a pH of 7.40). This is illustrated by substituting normal values (Table 11-1) for bicarbonate and PCO_2 into the preceding equation:

$$\frac{24 \text{ mmol/L}}{(0.031 \text{ mmol/L-mmHg}) \times 40 \text{ mmHg}} = \frac{24}{1.2} = \frac{20}{1}$$

$$\textbf{\textit{(Eq. 11-5)}}$$

TABLE 11-1. Arterial Blood Gas Reference Range at 37°C

pH	7.35–7.45
PCO_2 (mmHg)	35–45
HCO_3^- (mmol/L)	22–26
Total CO_2 content (mmol/L)	23–27
PO_2 (mmol/L)	80–110
SO_2 (%)	>95

Adding the log of 20 (1.3) to the pK' of the bicarbonate system yields a normal pH of 7.40 (7.40 = 6.1 + 1.3).

Acid–Base Disorders

Acid–base disorders result from a variety of pathologic conditions. A pH of blood less than the reference range is termed *acidemia.* A pH greater than the reference range is termed *alkalemia.* A disorder due initially to ventilatory dysfunction (a change in the PCO_2, the respiratory component) is termed a *primary respiratory acidosis* or *alkalosis.* A disorder due to a change in the bicarbonate level (a renal or metabolic function) is termed a *nonrespiratory (metabolic) disorder.* Mixed respiratory and nonrespiratory disorders due to more than one pathologic process are also quite common.

Because the body's cellular and metabolic activities are so pH-dependent, the body tries to return the pH toward normal whenever an imbalance occurs. This action by the body is termed *compensation,* and the body accomplishes this by altering the factor not primarily affected by the pathologic process. For example, if the imbalance is of a nonrespiratory origin, the body will compensate by altering the degree of ventilation. For disturbances of the respiratory component, the kidneys will compensate by selectively excreting or reabsorbing anions and cations. The lungs' compensating response is immediate, but short term and often incomplete, whereas the kidneys' response is slower (2 to 4 days), but long term and potentially complete. *Fully compensated* implies that the pH has returned to the normal range (the 20:1 ratio has been restored), and *partially compensated* implies that the pH is approaching normal. Compensation is only an effort by the body to return the blood pH to normal; the primary abnormality is not corrected.

Acidosis

Acidosis can be caused by a primary nonrespiratory (metabolic) abnormality or a primary respiratory problem. In primary *nonrespiratory* acidosis, there is a decrease in bicarbonate (< 24 mmol/L) resulting in a decreased pH as a result of the ratio for the nonrespiratory to respiratory component in the Henderson-Hasselbalch equation being less than 20:1:

$$pH \propto \frac{\downarrow c HCO_3^-}{N\,(0.0307 \times PCO_2)} < \frac{20}{1} \quad \textbf{\textit{(Eq. 11-6)}}$$

where N = normal value, and ↓ indicates a decreased level.

Nonrespiratory acidosis may be caused by the direct administration of an acid-producing substance, such as ammonium chloride or calcium chloride, or by excessive formation of organic acids seen with diabetic ketoacidosis and starvation. Nonrespiratory acidosis is also seen with reduced excretion of acids, as in renal tubular acidosis, and with excessive loss of bicarbonate from diarrhea or drainage from a biliary, pancreatic, or intestinal fistula.

The body compensates for nonrespiratory acidosis through *hyperventilation,* which is an increase in the rate or depth of breathing. By "blowing off" CO_2, the base-to-acid ratio will return toward normal. Secondary compensation occurs when the "original" organ (the kidney, in this case) begins to correct the ratio by retaining bicarbonate.

Primary *respiratory acidosis* results from a decrease in the rate of alveolar ventilation, *hypoventilation,* causing a decreased elimination of CO_2 by the lungs:

$$pH \propto \frac{N c HCO_3^-}{\uparrow (0.0307 \times PCO_2)} < \frac{20}{1} \quad \textbf{\textit{(Eq. 11-7)}}$$

Respiration is regulated in the medulla of the brain. Chemoreceptors present in the aortic arch and the carotid sinus respond to levels of oxygen, carbon dioxide, and pH in the blood and cerebrospinal fluid (CSF). There are several situations in which the CO_2 will not be removed in the usual manner by the lung. Most lung diseases will decrease the effective removal of CO_2. For example, in emphysema, a disease in which there is an abnormal, permanent increase in the size of the alveolar air spaces and destructive changes in the alveolar walls, the surface area available for gas exchange is reduced, and CO_2 will be retained in the blood. In bronchopneumonia, the alveoli contain secretions, white blood cells, bacteria, and fibrin, all of which impede gas exchange. Hypoventilation caused by drugs such as barbiturates, morphine, or alcohol will increase PCO_2 levels, as will mechanical obstruction or asphyxiation (strangulation or aspiration). Decreased cardiac output, such as that seen with congestive heart failure, also will result in less blood presented to the lungs for gas exchange and, therefore, an elevated PCO_2.

For primary respiratory acidosis, the compensation occurs through nonrespiratory processes. The kidneys increase the excretion of H^+ and increase the reabsorption of HCO_3^-. Although the renal compensation begins immediately, it takes days to weeks for maximal compensation to occur. With an increase in $c HCO_3^-$, the base-to-acid ratio will be altered and the pH will return toward normal.

Alkalosis

Primary *nonrespiratory alkalosis* is due to a gain in bicarbonate causing an increase in the nonrespiratory component and an increase in the pH:

$$pH \propto \frac{\uparrow c HCO_3^-}{N(0.0307 \times PCO_2)} > \frac{20}{1} \quad \textbf{\textit{(Eq. 11-8)}}$$

This condition may result from the excess administration of sodium bicarbonate or through ingestion of bicarbonate-producing salts such as sodium lactate, citrate, or acetate. Excessive loss of acid through vomiting or nasogastric suctioning or prolonged use of diuretics that augment renal excretion of H^+ produces an apparent increase in $c HCO_3^-$. The body responds by depressing the respiratory center. The resulting hypoventilation increases the retention of CO_2.

Primary *respiratory alkalosis* from an increased rate of alveolar ventilation causes excessive elimination of CO_2 by the lungs:

$$pH \propto \frac{Nc(HCO_3^-)}{\downarrow (0.0307 \times PCO_2)} > \frac{20}{1} \qquad \textbf{\textit{(Eq. 11-9)}}$$

The causes of respiratory alkalosis include chemical stimulation of the respiratory center by drugs, such as salicylates, an increase in the environmental temperature, fever, hysteria, pulmonary emboli, and pulmonary fibrosis. The kidneys compensate by excreting bicarbonate and retaining H^+.

OXYGEN AND GAS EXCHANGE

Oxygen and Carbon Dioxide

Oxygen's role in metabolism is crucial to all life, maintaining the integrity of cells and tissues. In cell mitochondria, electron pairs from the oxidation of NADH and $FADH_2$ are transferred to molecular oxygen, causing release of the energy used to synthesize ATP from the phosphorylation of ADP. Although the measurement of intracellular oxygen is not feasible with current technology, evaluation of a patient's oxygen status is possible using the partial pressure of oxygen (PO_2) measured along with pH and PCO_2 in the blood gas analysis.

For adequate tissue oxygenation, the following conditions are necessary: (1) available atmospheric oxygen, (2) gas exchange between the lungs and arterial blood, (3) loading of oxygen onto hemoglobin, and (4) transport (cardiac output) and release of oxygen to the tissues. Poor tissue oxygenation can result from disturbances in any of the preceding.

The amount of oxygen available in atmospheric air is dependent on the barometric pressure (BP). At sea level, the BP is 760 mmHg. (In the International System of units, 1 mmHg is equal to 0.133 kPa, where 1 Pa is equal to 1 N/m^2.) Dalton's law states that the total atmospheric pressure is the sum of the individual gas pressures. One atmosphere exerts 760 mmHg pressure and is made up of oxygen (20.93%), carbon dioxide (0.03%), nitrogen (78.1%), and inert gases (approximately 1%). While the percentage for each of the gases is the same at all altitudes, the *partial pressure* for each of the gases in the atmosphere is equal to the BP at a particular altitude times the appropriate percentage for each gas. Because the gases are humidified by the water present in the earth's atmosphere, the vapor pressure of water (47 mmHg at 37°C) must be accounted for in calculating the partial pressure for the individual gases. For example,

Partial pressure of O_2 at sea level = (760 mmHg − 47 mmHg) × 20.93% = 149 mmHg

Partial pressure of CO_2 at sea level = (760 mmHg − 47 mmHg) × 0.03% = 0.2 mmHg

Thus, 149 mmHg of O_2 is available to the alveoli of a person who is at sea level (Fig. 11-3). Since the oxygen tension in the lungs is less than that in the atmosphere, this partial-pressure gradient causes oxygen to diffuse into the lungs.

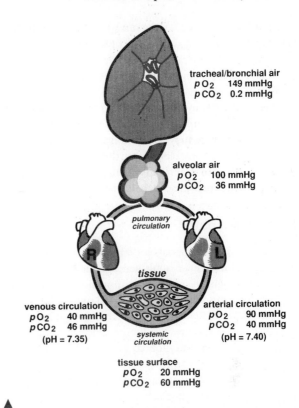

Figure 11-3. Gas content in lungs, pulmonary, and systemic circulation.

Exactly how much oxygen diffuses depends on the makeup or composition of inspired air. The FIO_2 (the fraction of inspired oxygen) can be as much as 100% when oxygen is being supplied. The quantity of oxygen diffusion is dependent also on the rate of cellular CO_2 production, the rate of breathing, and the volume of air exchanged. The amount of oxygen inspired is reduced in the lung owing to the presence of CO_2 and the loss of O_2 to the alveolar membrane, interstitial fluid, plasma, capillary wall, and red cell membrane. Therefore, the tension of O_2 in the alveoli and available to the arterial blood in health and at sea level is reduced to approximately 100 mmHg.

While there are several factors that affect the amount of oxygen entering the arterial blood, the effective alveolar surface area exposed to the blood is one of the most important. Anything that decreases this surface area results in decreased oxygen intake. For example, fibrotic or edematous tissue occurring in tuberculosis or certain carcinomas may cause less oxygen to cross into the blood. Diseases such as asthma or diphtheria lead to increased obstruction of the airways. Others reduce *compliance*—the expandability of the lungs and thorax—resulting in less uptake of oxygen from the atmosphere. The diffusion rate of oxygen across the alveolar membrane to the arterial blood also determines the PO_2 level. The distance through which the oxygen must diffuse is inversely proportional to the rate of diffusion. An example of an increase in distance might occur in pneumonia, where accumulated fluid reduces the exchange of oxygen and carbon dioxide. The adequacy of blood perfusion to the alveoli is another factor. A decreased cardiac output, for whatever reason, as well as shunting of blood away from the lungs' alveoli, will generally result in decreased PO_2.[8]

While oxygen is being transported to the tissues, carbon dioxide is being eliminated by the lungs in the exhaled air. The tension of CO_2 in venous blood ($PvCO_2$) is approximately 46 mmHg, while the tension of CO_2 in alveolar air ($PACO_2$) is approximately 36 mmHg. Carbon dioxide is driven from the venous blood into the alveoli due to this 10 mmHg difference in tension and the increased diffusion rate, which is 20 times greater than that of oxygen.

Oxygen Transport

Most of the oxygen in arterial blood is transported to the tissues by hemoglobin. Each adult hemoglobin (A_1) molecule can combine reversibly with up to four molecules of oxygen. The actual amount of oxygen loaded onto hemoglobin depends on the availability of oxygen, the concentration and type(s) of hemoglobin present, the presence of nonoxygen substances, such as carbon monoxide, the pH and the temperature of the blood, and levels of PCO_2 and 2,3-diphosphoglycerate. With adequate atmospheric and alveolar oxygen available, and with normal diffusion of oxygen to the arterial blood, more than 95% of "functional" hemoglobin (hemoglobin capable of *reversibly* binding oxygen) will bind oxygen. Increasing the availability of oxygen to the blood further saturates the hemoglobin. However, once the hemoglobin is 100% saturated, an increase in oxygen to the alveoli serves only to increase the concentration of dissolved oxygen in the arterial blood and may cause oxygen toxicity as well as decreased ventilation and an increased PCO_2.

Quantities Associated with Assessing the Oxygen Status of a Patient

Oxygen saturation (SO_2) represents the ratio of oxygen that is bound to the carrier protein—hemoglobin—compared with the total amount that the hemoglobin could bind.[8]

$$SO_2 = \frac{cO_2Hb}{(cO_2Hb + cHHb)} \times 100 \quad \textit{(Eq. 11-10)}$$

The symbol cO_2Hb represents the concentration of oxyhemoglobin, the oxygen-bound derivative of hemoglobin. The symbol $cHHb$ represents the concentration of deoxyhemoglobin, the hemoglobin that is *capable* of binding oxygen but *currently* is not binding any oxygen. While SO_2 can be determined directly on arterial blood samples and on mixed-venous (drawn from the pulmonary artery) and venous samples using an oximeter, SO_2 is also calculated from the PO_2 and total hemoglobin by the blood gas instrument. However, the calculated results can differ significantly from those determined by direct measurement due to algorithms in the blood gas instruments, assuming an oxyhemoglobin dissociation curve, and not accounting for the presence of other hemoglobin species.[6]

Fractional oxyhemoglobin (FO_2Hb) is the ratio of the concentration of oxyhemoglobin to the concentration of total hemoglobin ($ctHb$),

$$FO_2Hb = \frac{cO_2Hb}{ctHb} = \frac{cO_2Hb}{cO_2Hb + cHHb + cdysHb}$$

$$\textit{(Eq. 11-11)}$$

where the $cdysHb$ represents the hemoglobin derivatives, such as carboxyhemoglobin ($COHb$), that cannot reversibly bind with oxygen but are still part of the "total" hemoglobin measurement.[7]

These two terms, SO_2 and FO_2Hb, can be confused because, in most healthy individuals (and even with some disease states), the numerical values for SO_2 are very close to those for FO_2Hb. However, the values for FO_2Hb and SO_2 will deviate in several conditions (such as when the subject is a smoker) owing to the preferential binding of carbon monoxide (CO) to hemoglobin, forming carboxyhemoglobin, and the resultant loss of hemoglobin to bind oxygen.

The maximum amount of oxygen that can be carried by hemoglobin in a given quantity of blood is the *hemoglobin oxygen (binding) capacity*.[6,13] The molecular weight of the tetramer hemoglobin is 64,458 g/mol. One mole of a perfect gas occupies 22,414 mL. Therefore, each gram of hemoglobin carries 1.39 mL of oxygen:

$$\frac{22,414 \text{ mL/mol}_4}{64,458 \text{ g/mol}} = 1.39 \text{ mL/g} \quad \textit{(Eq. 11-12)}$$

When the total hemoglobin (tHb) is 15 g/100 mL and the hemoglobin is 100% saturated with oxygen, the oxygen capacity is

$$15 \text{ g/100mL} \times 1.39 \text{ mL/g} = 20.8 \text{ mL O}_2/100 \text{ mL of blood}$$

$$\textit{(Eq. 11-13)}$$

Oxygen content is the sum of the oxygen bound to hemoglobin as O_2Hb and the amount dissolved in the blood.[6] (Since PO_2 and PCO_2 are only indices of gas-exchange efficiency in the lungs, they do not reveal the *content* of either gas in the blood.) For every mmHg PO_2, 0.00314 mL of O_2 will be dissolved in 100 mL of plasma at 37°C. For example, if the PO_2 is 100 mmHg, 0.3 mL of O_2 will be dissolved in every 100 mL of blood plasma. The amount of dissolved oxygen is usually not clinically significant. However, with low tHb or at hyperbaric conditions, it may become a significant source of oxygen to the tissues. "Normally," approximately 97% of the hemoglobin is saturated with oxygen. Assuming a tHb of 15 g/100 mL, the oxygen content for every 100 mL of blood plasma becomes

$$0.3 \text{ mL} + (20.8 \text{ mL} \times 0.97) = 20.5 \text{ mL}$$

$$\textit{(Eq. 11-14)}$$

Hemoglobin–Oxygen Dissociation

Once the oxygen is transported to the tissues, some hemoglobin usually relinquishes its oxygen. The increased H^+

concentration and PCO_2 levels at the tissues due to cellular metabolism change the molecular configuration of O_2Hb, facilitating this release of oxygen.

Oxygen dissociates from adult hemoglobin (A_1) in a characteristic fashion. If this dissociation is graphed (Fig. 11-4) with the PO_2 on the x axis and the percent oxygen saturation on the y axis, the resulting curve is sigmoid, or slightly S-shaped. Hemoglobin "holds on" to the oxygen until the oxygen tension is reduced to about 60 mmHg. Below this tension, the oxygen is released rapidly. The *affinity* that hemoglobin has for oxygen affects the rate of this dissociation.

The P_{50} represents the partial pressure of oxygen at which the hemoglobin oxygen saturation (SO_2) is 50%. The P_{50} is a measure of the O_2Hb binding characteristics and identifies the position of the oxygen–hemoglobin dissociation curve at half saturation. A low P_{50} indicates an increased O_2Hb affinity; a high P_{50} reflects a decreased O_2Hb affinity. Terms for changes in the affinity between hemoglobin and oxygen include *shift to the right* (increased P_{50}) and *shift to the left* (decreased P_{50}) with reference to the dissociation curve. In cases in which the hemoglobin releases the oxygen more easily or at a faster-than-normal rate, the patient benefits, provided sufficient O_2 is taken up by the lungs. When this occurs, the tissues receive oxygen, even though the PO_2 or the hemoglobin level may not be within "normal" limits.

Although it is rarely done, the position of the oxygen dissociation curve can be determined by the $P50$ measurement. An arterial blood gas sample is equilibrated (tonometered) with a humidified gas mixture having known concentrations of carbon dioxide and oxygen for a few minutes prior to measurement of SO_2 for the calculation of the $P50$ value. A "normal" $P50$ is 26 to 27 mmHg of PO_2.

Several factors affect the affinity of hemoglobin for oxygen. The pH affects dissociation and an increase in hydrogen

ions in the tissues leads to a decrease in affinity of hemoglobin for oxygen and a shift to the right of the dissociation curve. As PCO_2 decreases, the pH becomes more alkaline in the lung, and the curve shifts slightly to the left, enhancing oxygen binding to hemoglobin. An elevated *temperature* results in a shift to the right and oxygen leaving the hemoglobin. Conversely, hypothermia shifts the curve to the left, and the hemoglobin binds the oxygen more tightly. An increase in *carbon dioxide* causes a shift to the right. An elevation in *carbon monoxide* causes the curve to shift to the left. As the percentage of carboxyhemoglobin increases, the shape of the curve loses some of its sigmoid characteristics, and the oxygen bound to hemoglobin will not be released to the tissues as easily.

2,3-Diphosphoglycerate (2,3-DPG) is a phosphate compound in erythrocytes that exerts an effect on the oxyhemoglobin dissociation curve. When the 2,3-DPG is bound by the β chains of the hemoglobin molecule, a shift to the right occurs and oxygen is unloaded. DPG stimulation occurs with a decrease in intracellular pH, a decrease in oxygen level in the tissues, and adaptation to high altitude. Many patients with slow onset of anemia demonstrate elevated levels of 2,3-DPG. This may partially explain the failure of patients with extremely low hemoglobin values to lose consciousness.

All the preceding discussion refers to normal adult (A_1) hemoglobin. In hemoglobinopathies and in newborns, the pattern of dissociation may differ. For example, fetal hemoglobin causes a shift to the left, but with little change in the sigmoid shape.

MEASUREMENT

Oximetry: Determination of Oxygen Saturation

The *actual percent oxygen saturation* (SO_2) is determined spectrophotometrically using an oximeter, an instrument designed to measure the various hemoglobin species directly. Each species of hemoglobin has a characteristic absorbance curve (Fig. 11-5). The number of hemoglobin species measured will depend on the number and specific wavelengths incorporated into the instrumentation. For example, two-wavelength instrument systems can measure only two hemoglobin species, *i.e.*, oxyhemoglobin and deoxygenated hemoglobin.

Clinically useful oximeters utilize a minimum of four wavelengths for measurements of HHb, O_2Hb, and the two most common dyshemoglobins, COHb and MetHb. Oximeters with more than four wavelengths enable the instrument to recognize the presence of dyes, turbidity, SulfHb, and abnormal proteins as well as other pigments in concentrations sufficient to significantly affect the values of the reported analytes.[9] Microprocessors control the sequencing of multiple wavelengths of light through the sample and

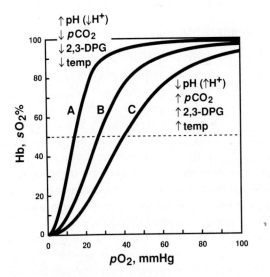

Figure 11-4. Oxygen-dissociation curves. Curve *B* is the normal human curve (p50 of 27 mmHg at pH of 7.40). Curves *A* and *C* are from blood with increased affinity (reduced p50) and decreased affinity (increased p50), respectively.

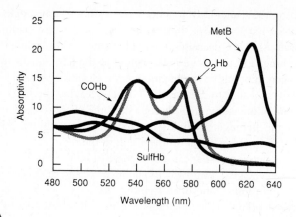

Figure 11-5. Optical absorption of hemoglobin fractions. (Reproduced with permission of Clin Chem News, January 1990)

apply the necessary "matrix" equations after absorbance readings are made to calculate the percent saturation for oxygen and carboxyhemoglobin or other hemoglobin species:

$$O_2HB = a_1A_1 + a_2A_2 + \cdots + a_nA_n$$

$$HHb = b_1A_1 + b_2A_2 + \cdots + b_nA_n$$

$$COHb = c_1A_1 + c_2A_2 + \cdots + c_nA_n$$

$$MetHb = d_1A_1 + d_2A_2 + \cdots + d_nA_n$$

(Eq. 11-15)

where a_1, a_n, b_n, etc. are coefficients that are analogues of the absorption constant a that are derived from established methods, and A_1, A_2, etc. are the absorbances of the sample. The matrix equations will change depending on the number of wavelengths of light passed through the sample (which is manufacturer-specific).[14] (The "calculation" made by these instruments should not be confused with a calculated SO_2 from a blood gas analyzer, which, in reality, *estimates* the value from a measured PO_2 and an empirical equation for the oxyhemoglobin dissociation curve. The "estimated" values for SO_2 from blood gas analyzers should not be used because of the possibility of significant clinical error.)

As with any spectrophotometric measurement, there exist potential sources of error. These include faulty calibration of the instrument, random measurement problems, and the presence of spectral-interfering substances. The presence of any substances absorbing light at the wavelengths used in the measurement of any hemoglobin pigment has the potential of being a source of error. Product claims for specific instruments must be consulted for specific interferences.

Since the primary purpose of determining SO_2 is to assess oxygen transport from the lungs, it is best if the patient's ventilation status be stabilized prior to blood-sample collection. Changes in supplemental oxygen or mechanical ventilation should be followed by an appropriate waiting period before the sample is redrawn. Ideally, the patient should not smoke for at least 4 hours before blood is collected. All blood samples should be collected under anaerobic conditions, mixed immediately with the heparin anticoagulant, immediately iced, and promptly analyzed to avoid changes

in saturation resulting from the utilization of oxygen by metabolizing cells.[9]

Laboratory determination of *in vitro* SO_2 is simple because of relatively sophisticated spectrophotometry application in oximeters. *In vivo,* noninvasive measurements for following "trends" in oxygen saturation also are possible using *pulse oximetry*. By passing light of two different wavelengths through the tissue of the toe, finger, or ear, the pulse oximeter differentiates between the absorption of light due to oxyhemoglobin and deoxyhemoglobin in the capillary bed. The accuracy of pulse oximetry can be compromised by many factors. Dysfunctional hemoglobin (since only two wavelengths of light are used), diminished pulse (due to poor perfusion), and severe anemia are but a few.

Blood Gas Analyzers: pH, PCO_2, and PO_2

Blood gas analyzers use electrodes as sensing devices to measure PO_2, PCO_2, and pH. The PO_2 measurement is amperometric, meaning that the amount of current flow is an indication of the oxygen present. The PCO_2 and pH measurements are potentiometric, where a change in voltage indicates the activity of each analyte.

For review, a *cathode* can be defined in at least three ways: (1) the negative electrode, (2) a site to which cations tend to travel, or (3) a site at which reduction occurs. *Reduction* is the gain of electrons by a particle (atom, molecule, or ion). The *anode* is: (1) the positive electrode, (2) the site to which anions migrate, or (3) the site at which oxidation occurs. *Oxidation* is the loss of electrons by a particle. An *electrochemical cell* is formed when two opposite electrodes are immersed in a liquid that will conduct the current.

Several additional parameters can be calculated by the blood gas analyzer: bicarbonate, total carbon dioxide, base excess (BE), and calculated oxygen saturation.

Measurement of PO_2

In the measurement of PO_2, the amount of current flow in a circuit is related to the amount of O_2 being reduced at the cathode. A gas-permeable membrane covering the tip of the electrode allows the O_2 to diffuse into the electrolyte solution while inhibiting other reducible substances in the blood from reaching the cathode. Electrons are drawn from the anode surface to the cathode surface in order to reduce the O_2. A small, constant polarizing potential (typically $-0.65V$) is applied between the anode and cathode to overcome the nonspontaneity of the reaction. A *microammeter* placed in the circuit between anode and cathode measures the movement of electrons (current). Four electrons are drawn for every mole of oxygen reduced, making it possible to determine the PO_2. (*Note:* The semipermeable membrane also will pass other gases, such as CO_2 and N_2, but these will not be reduced at the cathode.)

There are several possible sources of error in the measurement of PO_2. Any debris such as bacteria, mold, or yeast in the cuvet will impede diffusion of the oxygen and may even consume the oxygen. The membrane covering the tip of the electrode must be wrinkle- and bubble-free. Oxygen in the sample can be lost by diffusion into rubber gaskets, tubing, cracks, the cuvet material, or the electrolyte solution in the main body of the electrode itself. Any change in temperature of the cuvet or the electrode will produce an error in measurement.

Noninstrumental concerns, including sample collection and handling, are addressed in another section of this chapter. However, it is particularly important not to expose the sample to room air when collecting, transporting, and making oxygen measurements. Equilibration between the sample and room air ($PO_2 \cong 150$ mmHg) can result in significant error. If the exposure to room air occurs prior to placing the sample in ice, the magnitude of error will be even greater, since the solubility of oxygen in blood increases as the temperature of the sample is lowered. Patients with elevated white blood cell counts pose another potential source of error. Leukocytes will metabolize oxygen, producing a falsely low PO_2 level unless the sample is analyzed immediately after drawing.

Continuous measurements for PO_2 also are possible using *transcutaneous*(TC) electrodes placed directly on the skin. Measurement is dependent on oxygen diffusing from the capillary bed through the tissue to the electrode. While most commonly used with neonates and infants, this noninvasive approach is not without problems.[1] Skin thickness is a major concern, since the diffusion of oxygen to the electrode is more difficult when the skin is thicker. Diffusion of oxygen to the electrode can be enhanced by heating the electrode placed on the skin, but burns can result unless the electrodes are moved regularly. Although PO_2 measured by TC electrodes may *reflect* the arterial PO_2, the two values are not equivalent because of the consumption of oxygen by the tissue where the electrode is applied and the effects of heating the tissue.

Measurement of pH and PCO_2

In order to understand potentiometric measurements, it is helpful to think of atoms and ions as having a chemical energy. An increased concentration or *activity* of the ions leads to an increase in force exerted by those ions.

To measure how much force—energy or potential—a given ion possesses, certain elements in the measuring device are required; namely, two electrodes (the measuring electrode responsive to the ion of interest and the reference electrode) and a voltmeter, which measures the potential difference (ΔE) between the two electrodes. The potential difference is related to the concentration of the ion of interest by the Nernst equation:

$$\Delta E = \Delta E^\circ + \frac{0.05916}{n} \log a_i \quad \text{at } 25°C$$

(Eq. 11-16)

where ΔE° = standard potential of the electrochemical cell
n = charge of the analyte ion i
a_i = activity of the analyte ion i

To measure pH, a glass membrane sensitive to H^+ is placed around an internal Ag–AgCl electrode to form a glass measuring electrode. The potential that develops at the glass membrane as a result of H^+ from the unknown solution diffusing into the membrane's surface is proportional to the difference in cH^+ between the unknown sample and the buffer solution inside the electrode. For the potential developed at the glass membrane of the glass electrode to be measured, a reference electrode must be introduced into the solution, and both electrodes must be connected to a pH (volt) meter. The reference electrode [commonly either a calomel (Hg–HgCl) or an Ag–AgCl half-cell] provides a steady reference voltage against which voltage changes from the measuring (glass electrode) are compared. The pH meter reflects the potential difference between the two electrodes.

For the cell described, it can be calculated from the Nernst equation that a change of +59.16 mV, at 25°C, is the result of a tenfold increase in hydrogen ion activity or a decrease of an entire pH unit (*e.g.*, pH 7 to pH 6). Changing the temperature affects the response; therefore, the electrodes must be kept at a constant temperature. At 37°, a change of 1 pH unit elicits a 61.5-mV change.

Problems such as the following must be minimized if determinations of pH are to be made with adequate precision and accuracy. Creeping KCl crystals from the salt bridge (junction) connecting the measuring and reference electrodes may lead to a short circuit or an inappropriate path to ground. Dilution of the chloride by blood or water in this junction may decrease conductivity and increase response time. The KCl junction can entrap bubbles and cause instability of the response. The glass membrane of the measuring electrode must be kept free from protein, lipid, and platelet deposits, because coating of the membrane causes sluggish and/or erratic responses. The membrane must be inspected periodically for cracks in the glass, which will lead to inaccuracies or to a break in the circuit. In addition, any changes in the temperature of the electrode and the sample chamber will affect the voltage output.

The PCO_2 is determined with a modified pH electrode. The glass pH membrane is covered by an outer semipermeable membrane of material (similar to those materials used for measuring oxygen) that allows nonionic diffusion of particles. Molecules of gases, such as carbon dioxide, diffuse across the membrane. Between the outer membrane and the glass membrane of the pH electrode is a layer of electrolyte, usually a bicarbonate buffer. The carbon dioxide that diffuses across the membrane reacts with the buffer, forming carbonic acid, which then dissociates into bicarbonate plus *hydrogen ion*. The change in activity of the hydrogen ion is measured by the pH electrode and related to PCO_2.

Sources of error include each of those mentioned for pH. In addition, the outer membrane must be wrinkle- and bubble-free, and the buffer solution must not become contaminated or depleted.

Calibration

As stated previously, temperature is an important factor in the measurement of pH and blood gases. The Nernst equation specifies the expected voltage output of an electrochemical cell at a given temperature. If the temperature of the measurement system changes, the output (voltage) will change. The solubility of gases in a liquid medium also is dependent on the temperature: as the temperature goes down, the solubility of the gas increases. Because pH and blood gas measurements are extremely sensitive to temperature, it is critical that the electrodes' sample chamber and waterbath be maintained at constant temperature for all measurements.[14] Because of this requirement, all of today's blood gas analyzers have electrode chambers thermostatically controlled to $37 \pm 0.1°C$.

The pH electrode is calibrated with two phosphate buffer solutions traceable to standards prepared by the National Institute of Standards and Technology. The calibrators have a pH between 6.84 and 7.39 at 37°C. These buffers must be stored at the stated temperature and not exposed to room air, because pH changes with the absorption of CO_2.

Calibration of any blood gas analyzer will vary depending on the manufacturer. In general, for PCO_2 and PO_2, two gas mixtures are used. Typically, one gas mixture has PCO_2 and PO_2 values in the physiologic normal range, while the other gas mixture contains no oxygen (zero PO_2) and sufficient CO_2 to produce an elevated PCO_2 reading.

Most instruments are self-calibrating (calibrate automatically at specified time intervals) and are programmed to reject data when particular criteria (*e.g.*, voltage or ampere readings) are not met. For example, if the value(s) obtained during calibration exceed(s) a programmed tolerance limit, flagging of a *drift* error will occur at the time of calibration, and corrective action will need to be taken before patient samples can be analyzed.

Calculated Parameters

Several parameters can be calculated from the measured pH and PCO_2 values. Algorithms, tables, nomograms, charts, and diagrams have been developed to assist in the calculation of these parameters. There is controversy over the application of some of the factors, coefficients, and derived factors used. Presented here are the basics. The references cited will provide more detailed derivations and explanations.

The calculation of *bicarbonate* is based on the Henderson-Hasselbalch equation. One basic assumption is that the pK' of the bicarbonate buffer system in plasma at 37°C is 6.1.

Carbonic acid concentration can be calculated using the solubility coefficient of carbon dioxide in plasma at 37°C.

The solubility constant to convert PCO_2 to millimoles per liter of H_2CO_3 is 0.0307. If the temperature or the composition of plasma changes (*e.g.*, an increase in lipids, in which gases are more soluble), the constant will change.

Total carbon dioxide content ($ctCO_2$) is defined as the bicarbonate plus dissolved carbon dioxide (carbonic acid) plus carbon dioxide associated with proteins (carbamates). A blood gas analyzer approximates total carbon dioxide content by adding the bicarbonate and carbonic acid values $[ctCO_2 = cHCO_3^- + (0.0307 \times PCO_2)]$.[5] Actual total carbon dioxide content can be measured directly by other chemical methods from a sample that has been maintained anaerobically. Since these measurements are not made routinely on whole blood, the results are more correctly considered to be bicarbonate levels due to the loss of the dissolved fraction on exposure to air.

Base excess (BE) is used by some clinicians to assess the metabolic component of a patient's acid–base disorder and is calculated from the patient's pH, PCO_2, and hemoglobin. Base excess of blood, an *in vitro* measurement, is defined as the theoretical amount of titratable acid or base required to return the plasma pH to 7.40 at a PCO_2 of 40 mmHg at 37°C.[5] A positive value (base excess) indicates an excess of bicarbonate or relative deficit of noncarbonic acid and suggests metabolic alkalosis. A negative value (base deficit) indicates a deficit of bicarbonate or relative excess of noncarbonic acids and suggests metabolic acidosis. The indicated metabolic alkalosis or acidosis may be due to primary disturbances or compensatory mechanisms. Blood gas analyzers are programmed with algorithms that calculate this parameter. Care needs to be taken not to confuse base excess of blood and base excess of extracellular fluid. These numbers will not be the same with severe acidosis or alkalosis.[5]

Correction for Temperature

The patient's temperature is sometimes a factor in interpreting blood gas data. However, if the pH, PCO_2, and PO_2 values are "corrected" for the temperature of the blood (the actual temperature of the patient) by changing the values measured at 37°C, then the appropriate reference range *for that temperature* must be used to interpret the data and identify the acid–base status. For example, a pH of 7.20 at 37°C (reference range is 7.35 to 7.45) "corrected" to the patient's temperature of 30°C becomes 7.30 (reference range 7.45 to 7.55).[19] The pH indicates acidemia at 37°C or 30°C. Correcting for temperature has the potential for the misinterpretation of the data unless the new reference ranges are included and the physician is familiar with using non-37°C data. This problem is eliminated when all results are reported for 37°C and only one set of reference ranges—those at 37°C—are needed. Blood gas analyzers with microprocessors have the capability of correcting for temperature if desired. Specific manufacturers' algorithms may vary.

QUALITY ASSURANCE

Pre-analytical Considerations

Blood gas measurements, like all measurements, are subject to *analytical errors*—errors due to the instrument's handling of the specimen. Few other measurements, however, are as affected by *pre-analytical errors*—those introduced during the collection and transport of samples prior to analysis—as are blood gas measurements.[10]

Figure 11-6 depicts the quality-assurance cycle for blood gas analyses. The steps included in the analytical area are under the direct control of the laboratory. Much of the quality-assurance cycle lies outside the laboratory. Therefore, the laboratorian must take an active role in educating *all* people involved in the cycle to ensure quality.

Because the collection and handling of the sample can affect blood gas results, the steps in these processes must be controlled. Drawing samples for pH and blood gas analyses should be done only by personnel who have experience with the drawing equipment and have knowledge of the possible sources of error. Since collection may be painful and result in patient hyperventilation, which lowers the PCO_2 and increases the pH, the ability to reassure the patient is essential. The choice of site—radial, brachial, femoral, or temporal artery—is usually customary within an institution, depending on the predominant patient population (*e.g.,* pe-

diatric patients, burn patients, outpatients). The National Committee for Clinical Laboratory Standards (NCCLS) publication *Standard for the Percutaneous Collection of Arterial Blood for Laboratory Analysis* is an excellent reference.[20] Cited in this document are the hazards of arterial puncture: hematoma, arteriospasm, thrombosis, and anaphylaxis (if a local anesthesia is used).

The use of venous samples for pH and blood gas studies is controversial. Reference ranges are less familiar and cannot be called to memory as readily. There is nothing wrong with venous samples as long as pulmonary function or O_2 transport is not being assessed, but the sample must be clearly identified as venous. Capillary blood may need to be used to measure pH and PCO_2. Capillary PO_2 values, even with warming of the skin prior to drawing the sample, do not correlate well with arterial PO_2 values, primarily due to the exposure of the sample to room air.[2]

Sources of error in the collection and handling of blood gas specimens include the collection device, the form and concentration of heparin, speed of syringe filling, maintenance of the anaerobic environment, mixing, transport, and storage due to delay in analysis. For proper interpretation of blood gas results, the patient's status in terms of ventilation (on room air or supplemental oxygen), temperature, and posture must be documented at the time the sample is collected.

Recommended collection devices for arterial blood samples include glass and plastic syringes. Glass syringes have been preferred because the plunger responds to arterial pressure and the glass is impermeable to gases that can alter PCO_2 and PO_2 values over time. With the new compositions of plastics, plastic syringes have become more acceptable. However, plastic syringes can alter PO_2, and to some extent PCO_2, values due to room air dissolved in the syringe barrel and plunger tip.[12,15] Due to the increased solubility of oxygen with decreasing temperature, the error is magnified by icing the specimen. If plastic syringes are used, it is recommended that the sample not be iced and should be analyzed within 20 minutes.

Both dry and liquid heparin are acceptable anticoagulants. Because of the potential dilution and equilibration errors, liquid heparin is less preferable.[8] The amount of liquid heparin needed to lubricate the syringe and the needle depends on the size and type of syringe and the size of the needle. An excess of heparin that is in equilibrium with room air ($PCO_2 \cong 0$ mmHg) will cause a decrease in PCO_2 and the bicarbonate level as well as dilute the sample. Excess liquid heparin also falsely elevates PO_2 values (room air $PO_2 \cong 150$ mmHg) of less than 150 mmHg.

The use of dry heparin eliminates the dilution and equilibration problems. However, care must be taken to ensure that the heparin is dissolved and no clot forms. While sodium and lithium heparin are recommended for pH and blood gas analysis, other salt forms are available: ammonium, zinc, electrolyte-balanced, and calcium titrated. Selection of

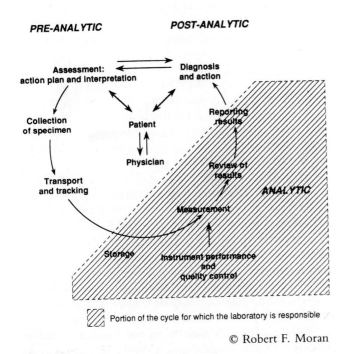

BLOOD GAS ANALYSIS QUALITY ASSURANCE CYCLE

PRE-ANALYTIC   **POST-ANALYTIC**

Assessment: action plan and interpretation ⟷ Diagnosis and action

Collection of specimen — Patient — Reporting results

Physician

Review of results

Transport and tracking

ANALYTIC

Measurement

Storage — Instrument performance and quality control

© Robert F. Moran

Portion of the cycle for which the laboratory is responsible

▲

Figure 11-6. Blood gas analysis quality assurance cycle. (Reproduced with permission of Robert F. Moran)

the proper type of heparin is particularly important with instruments combining electrolyte analyses.

Slow filling of the syringe may be caused by a mismatch of syringe and needle sizes. Too small a needle, while reducing the pain and therefore the likelihood of arteriospasm and hematoma, may produce bubbles as well as hemolysis of the sample, something of particular importance when potassium is measured along with pH/blood gases. Maintenance of an anaerobic environment is critical to correct results, since the gases in the sample will equilibrate with gases in room air, in bubbles, or in the syringe material itself. Cracks between the barrel and plunger of the syringe, the syringe material, and bubbles in the specimen—due to aspirating with the plunger or using a needle of incorrect size—may result in loss of anaerobic conditions and thus may be a significant source of error.

The blood in the syringe must be mixed with the heparin anticoagulant in order to prevent clotting and injection/aspiration of the sample into the blood gas analyzer. Adequate mixing immediately prior to analysis is essential, especially with respect to the parameter that will be measured spectrophotometrically—hemoglobin.

Transport time of the sample should be minimal. Immersing the specimen in an ice water slurry prevents changes in PCO_2, pH, and PO_2. If the alveolar PO_2 is above 100 mmHg, which could be anticipated with anyone receiving oxygen therapy, the NCCLS guidelines recommend rapid analysis of PO_2.[8]

Since sample procurement and handling are the source of many possible errors in blood gas analysis, it is necessary that procedures and policies be constructed carefully and adherence monitored to ensure quality. No quality-control product can monitor the preanalytical aspects of blood gas analysis, however.

Quality Control

An internal quality-control (QC) system monitors primarily random errors or imprecision problems in the measurement of blood gases. There are three basic approaches to monitoring instrument performance by an internal QC program.[8,9] For blood gases, these include commercial controls, tonometry, and duplicate analysis made on separate instruments. All these have limitations. The "ideal" approach would encompass some combination of the three.

Commercial control materials, which are used to monitor the acceptability of patient results, provide the basis for most QC practices. Ideally, such materials are stable, have minimal vial-to-vial variance, and are available in large lots that can be used for at least several months. While a laboratory tries to choose a control material that closely mimics actual patient samples, this is not possible for blood gasses.

At present, three general classes of commercial controls are available for blood gas analysis: equilibrated aqueous solutions, blood-based (solutions containing free hemoglobin or tanned red blood cells), and perfluorocarbon emulsions. Controls can include additional analytes, such as sodium, potassium, chloride, and calcium. All vary in their stability and are susceptible to temperature variation in storage and handling. Each must be handled as described by the manufacturer to eliminate precision errors caused by the handling itself. Improper handling does result in large errors.

Blood gas controls are available in at least three levels corresponding to values observed with acidosis, normal, and alkalosis pH conditions. These three levels of controls are analyzed to ascertain proper function of the analyzer across the range of patient values. For each of these levels, the manufacturer specifies target values. However, each laboratory should establish its own ranges. The ranges should be based on what goes on in the individual laboratory rather than on what goes on in the manufacturer's laboratory on a limited number of instruments.

The main problem with commercial QC materials is that the matrix is not fresh whole blood. Consequently, they may not detect problems that affect patient samples, or they may be detecting errors induced by improper handling of the commercial controls. *Tonometry,* using fresh whole blood, is the ideal approach.[3,8] Equilibrating, or *tonometering,* a blood sample with gases of known concentration over a period of time and at a constant temperature is a relatively inexpensive way to check the precision and the accuracy of the PCO_2 and PO_2 measurements.[8]

While tonometry is the ideal QC approach for blood gases, many problems have been documented.[3,4,8,16] Variations in technique from one person to another may be significant. Premixed gas samples, the ability to maintain intactness of the sample over a period of time (or at least to allow for not analyzing it immediately), and the ability to keep records for a mean value (in millimeters of mercury) rather than having to correct for the daily barometric pressure will simplify as well as improve precision of the documentation itself. Incorporation of a reliable pH standard also would help, since a tonometered sample does not provide for pH assessment. Although this is the basis for the reference method for PO_2 and PCO_2, it is perceived to be cumbersome and time consuming. Consequently, few laboratories utilize this recommended approach.

A laboratory may choose to perform *duplicate assays* using two or more instruments for simultaneous analysis of a patient sample. *Delta checks,* or the difference in values obtained on the two instruments, often pick up problems that might be missed in routine QC.[9] The allowable difference in duplicates run on split patient samples should be tighter than those observed on commercial controls. However, discrepancies between results provide no clue regarding which data point is wrong or which instrument is malfunctioning. While two instruments are unlikely to have the same error at the same time, this is not always true. Therefore, the duplicate-assay approach should not be used as the sole method of QC. Used in conjunction with commercial controls and/or tonometry, it can be a useful technique for detecting errors and also for troubleshooting the instruments.

Whatever the QC approach, the QC needs of the blood gas laboratory contrast sharply with those of the general laboratory, which analyzes many patient samples as a group and includes multiple control specimens with each run. In the blood gas laboratory, time or patient sample volume do not always allow for repeat analyses if problems exist. Consequently, the blood gas laboratory must perform *prospective* QC, since instruments must be *prequalified* to ensure proper performance *before* the patient sample arrives for analysis.[9]

External (Proficiency Testing) Quality Control

Participating in external QC through interlaboratory surveys or proficiency-testing (PT) programs will lend considerable help in the identification and monitoring of accuracy problems resulting from systematic errors or biases.[11] Ongoing comparisons of results through PT are necessary to ensure that systematic (accuracy) errors do not slowly increase and go undetected by internal QC procedures. A rigorous internal QC program ensures internal consistency. Good performance in a PT program ensures the absence of significant bias relative to other laboratories and reconfirms the validity of a laboratory's patient results. If an individual analyzer does not produce PT results consistent with its peer laboratories, or if the differences between values change over time, suspicion of the analyzer's performance is warranted.

Interpretation of Results

Laboratory professionals need certain knowledge, attitudes, and skills for obtaining and analyzing specimens for pH and blood gases. Although it is certainly the patient's physician who assimilates all results—laboratory, radiology, nuclear medicine, and surgical pathology findings, along with the clinical history of the patient—laboratory personnel are involved in the assessment and interpretation of groups of results to make preliminary judgments about the "fit," *i.e.*, do the results make sense? Simple analysis of the data may reveal an instrument problem (possible bubble in the sample chamber or fibrin plug) or a possible sample-handling problem [PO_2 out of line with previous results and current inspired oxygen (FIO_2) levels]. The application of knowledge saves time. The ability to correlate data quickly reduces turnaround time and prevents mistakes.

SUMMARY - Final

Arterial pH and blood gas measurements are ordered to facilitate the care and treatment of critically ill patients. The body maintains acid–base balance through various buffering systems. In the laboratory, the bicarbonate–carbonic acid buffer system, which works in conjunction with and reflects the status of the body's other buffering systems, is used to evaluate acid–base status. pH and PCO_2 measurements assess the patient's acid–base status. Other calculated parameters, such as HCO_3^-, base excess, and total CO_2 content, help to differentiate metabolic (nonrespiratory) from respiratory conditions.

The PO_2 measurement also is important. In arterial blood, it assesses the ability of the lungs to oxygenate the blood and it is used as an indirect measurement of the body's tissue oxygenation status. However, this measurement alone can be misleading. For example, an anemic patient will have a decreased O_2 content and capacity and a normal PO_2, provided the cardiovascular and pulmonary systems are intact. Consequently, other parameters are used in conjunction with PO_2. These include SO_2 (preferably measured by oximetry), identification of the presence of dyshemoglobins, $P50$, and the calculated parameters of O_2 content and capacity.

CASE STUDY 11-1

A 51-year-old man with symptoms of respiratory distress was admitted to a hospital. Until the previous year, the man had smoked three packs of cigarettes a day. Shortly after the man was admitted, blood drawn for blood gas studies revealed the following:

pH = 7.41
PCO_2 = 40.8 mmHg
HCO_3 = 26.0 mmol/L
PO_2 = 50.8 mmHg
SO_2 = 76.9% (actual)
cCOHb = 11.5% (reference range <2.0%)
ctHb = 20.1 g/dL

Further testing indicated that
$P50$ = 32.5 mmHg (reference range 25–29 mmHg)
2,3-DPG = 5.3 units (reference range 3.5–5.3 units)

Questions

1. What is the acid–base status of the patient at the time this sample was obtained?
2. Was the oxyhemoglobin dissociation curve shifted? If so, was it shifted to the left or to the right, and why?
3. What other result is especially significant?
4. Would a calculated SO_2 generated from the blood gas results be valid?

CASE STUDY
11-2

A 24-year-old graduate student was brought to the emergency room comatose. He was found in his room unconscious with an empty bottle of secobarbital by his bedside. He did not respond to painful stimuli, his respiration was barely perceptible, and his pulse was weak. Blood gases were drawn, and the results were as follows:

pH = 7.10
PCO_2 = 70 mmHg
PO_2 = 58 mmHg
HCO_3^- = 20 mmol/L
BE = −3.5 (reference range −2 to + 2)
FO_2Hb = 80%

Questions

1. What caused the profound hypoventilation and hypoxemia?
2. What is the acid–base status of the patient?
3. What does the base excess (BE/D) value indicate? Is this consistent with the initial diagnosis?
4. Once the respiratory component is restored to normal, what will be the expected acid–base status of the patient?

CASE STUDY
11-3

An 80-year-old woman fell on the ice, fracturing her femur. After several hours, she arrived in the emergency room anxious, panting, and complaining of a pain in her chest and not being able to breathe. Her pulse indicated a rapid heart rate (tachycardia); her respiration also was rapid (tachypnea). Her blood gases were as follows:

pH = 7.31
PCO_2 = 27 mmHg
PO_2 = 62 mmHg
HCO_3^- = 12 mmol/L
SO_2 = 78% (calculated)
BE = −5 mmol/L (reference range −2 to + 2)

Questions

1. What was the acid–base status of the patient at the time the blood was drawn?
2. Is the decreased HCO_3^- level (the metabolic factor) the primary or compensatory component?
3. What clinically caused the acid–base imbalance?

REFERENCES

1. Beyerl D. Non-invasive measurement of blood oxygen levels. Am J Med Tech 1982;48(5):355.
2. Bruck E, et al. Procedure for the collection of diagnostic blood specimens by skin puncture (H4-A2). Villanova, PA: National Committee for Clinical Laboratory Standards, 1991.
3. Burnett RW Quality control in blood pH and gas analysis by use of a tonometered bicarbonate solution and duplicate blood analysis. Clin Chem 1981;27:1761.
4. Burnett RW, Covington AK, Maas AHJ, et al. IFCC method (1988) for tonometry of blood: Reference materials for PCO_2 and PO_2. J IFCC 1989;1(2):78.
5. Ehrmeyer S, et al. Definitions of quantities and conventions related to blood pH and gas analysis (C12-T2). Villanova, PA: National Committee for Clinical Laboratory Standards, 1991.
6. Ehrmeyer S, et al. Fractional oxyhemoglobin, oxygen content and saturation, and related quantities in blood: Terminology, measurement and reporting (C-25T). Villanova, PA: National Committee for Clinical Laboratory Standards, 1992.
7. Ehrmeyer S, et al. Performance characteristics for devices measuring PO_2 and PCO_2 in blood samples (C21-A). Villanova, PA: National Committee for Clinical Laboratory Standards, 1992.
8. Ehrmeyer S, et al. Blood gas pre-analytical considerations: Specimen collection, calibration, and controls (C27-A). Villanova, PA: National Committee for Clinical Laboratory Standards, 1993.
9. Ehrmeyer SS, Laessig RH. Stat labs need prospective quality control. Clin Chem News 1988;14(4):11.
10. Fallon KD. Demand for speed challenges quality control. Clin Chem News 1988;14(4):6.

11. Laessig RH, Ehrmeyer SS. Proficiency testing in clinical chemistry: A new look at an old problem. Lab Med 1989;20(6):422.

12. Mahoney JJ, et al. Changes in oxygen measurements when whole blood is stored in iced plastic or glass syringes. Clin Chem 1991;37(7):1244.

13. Moran RF, Cormier AD. The blood gases: pH, PO_2, PCO_2, part I. Clin Chem News 1988;14(4):12.

14. Moran RF, Fallon KD. Oxygen saturation, content, and the dyshemoglobins, part II. Clin Chem News 1990;16(2):8.

15. Müller-Plathe O, Heyduck S. Stability of blood gases, electrolytes and haemoglobin in heparinized whole blood samples influence of the type of syringe. Eur J Clin Chem Biochem 1992;30:349.

16. Shrout JB. Controlling the quality of blood gas results. Am J Med Tech 1982;48(5):347.

17. Weisberg HF. Water, electrolyte, and acid–base balance. 2nd ed. Chapters 4, 8. Baltimore: Williams & Wilkins, 1962.

18. Weisberg HF. Water, electrolytes, acid–base, and oxygen. In: Davidsohn I, Henry JB, eds. Clinical diagnosis by laboratory methods. 15th ed. Chapter 12. Philadelphia: Saunders, 1974.

19. Weisberg HF. Tri-Slide calculator for Henderson-Hasselbalch equation and CO_2rrec°t-O_2 slide for temperature corrections of pH, PCO_2 and PO_2. [Monograph]. 1978.

20. Wiseman JD, et al. Percutaneous collection of arterial blood for laboratory analysis (H11-A2). Villanova, PA: National Committee for Clinical Laboratory Standards, 1992.

Chapter 12

Electrolytes — Test - 4

John G. Toffaletti

Objectives

Upon completion of this chapter, the clinical laboratorian should be able to:

- *Define the following terms: electrolyte, osmolality, anion gap, anion, cation.*
- *State the clinical significance of each of the electrolytes mentioned in the chapter.*
- *Discuss the physiology of each electrolyte described in the chapter.*
- *Calculate osmolality and an anion gap.*
- *Discuss the analytical techniques used to assess electrolyte concentrations.*
- *Discuss the clinical significance of osmolality.*
- *Given patient data, correlate the information with disease state.*
- *Identify the reference ranges for sodium, potassium, chloride, bicarbonate, and calcium.*
- *State the specimen of choice for the major electrolytes.*
- *Discuss the role of the kidney in electrolyte excretion and conservation in a healthy individual.*

KEY WORDS

Active Transport	Hyperkalemia	Hypovolemia
Anion	Hypermagnesemia	Intracellular Fluid
Anion Gap	Hypernatremia	(ICF)
Cation	Hyperphosphatemia	Osmolal Gap
Diffusion	Hypochloremia	Osmolality
Electrolyte	Hypokalemia	Osmolarity
Extracellular Fluid	Hypomagnesemia	Osmometer
Hypercalcemia	Hyponatremia	Polydipsia
Hyperchloremia	Hypophosphatemia	Tetany

Electrolytes are ions capable of carrying an electric charge. They are classified as anions or cations based on the type of charge they carry. These names were determined years ago based on how the ion migrates in an electric field. Anions have a negative charge and move toward the anode, whereas cations migrate in the direction of the cathode because of their positive charge.

The numerous processes in which electrolytes are an essential component are: volume and osmotic regulation (Na, Cl, K), myocardial rhythm and contractility (K, Mg, Ca), cofactors in enzyme activation (Mg, Ca, Zn, etc.), regulation of ATPase ion pumps (Mg), acid/base balance (HCO_3, K, Cl), blood coagulation (Ca, Mg), neuromuscular excitability (K, Ca, Mg), and the production and utilization of ATP from glucose (Mg, PO_4, etc.). Because many of the functions listed here require electrolyte concentrations to be held within narrow ranges, the body has complex systems for monitoring and maintaining the concentrations of electrolytes.

This chapter will explore both the metabolic physiology and regulation of each electrolyte and relate these factors to the clinical significance of electrolyte measurements. In addition, methodologies used in determining concentrations of the individual analytes will be discussed.

WATER

The average water content of the human body varies from 40% to 75% of the total body weight, with values declining with age and especially with obesity. Water is the solvent for all processes in the human body. It transports nutrients to cells, determines cell volume by its transport into and out of cells, removes waste products via urine, and acts as the

Michael L. Bishop, Janet L. Duben-Engelkirk, and Edward P. Fody.
CLINICAL CHEMISTRY. © 1996 Lippincott–Raven Publishers.

body's coolant via sweating. Water is located in both intracellular and extracellular compartments. The *intracellular fluid (ICF)* is fluid inside the cells. *Extracellular fluid (ECF)* can be subdivided in the *intravascular extracellular fluid (plasma)* and the *interstitial cell fluid (ISCF)* that surrounds the cells in the tissues. Normal plasma is about 93% water, with the remaining volume occupied by lipids and proteins. The concentrations of ions within cells and in plasma are maintained both by energy-consuming active transport processes and by diffusion or passive transport processes.

Active transport is a mechanism that requires energy in order to move ions across cellular membranes. For example, maintaining a high intracellular concentration of potassium and a high extracellular (plasma) concentration of sodium requires utilization of energy from adenosine triphosphate (ATP) in ATPase-dependent ion pumps. *Diffusion* is the passive movement of ions across a membrane. It is dependent on the size and charge of the ion being transported and on the nature of the membrane through which it is passing. The rate of diffusion of various ions also may be altered by physiologic and hormonal processes.

By maintaining the concentration of proteins and electrolytes in a controlled yet somewhat flexible environment, the distribution of water in these compartments also can be controlled. Since water is free to pass through most biologic membranes, the osmotic effects of sodium and other ions, proteins, and blood pressure influence the flow of water across a membrane.

Osmolality

Osmolality is a physical property of a solution, which is based on the concentration of solutes (expressed as millimoles) per kilogram of solvent. Osmolality is related to several changes in the properties of a solution relative to pure water, such as freezing point depression and vapor pressure decrease. These colligative properties are the basis for routine measurements of osmolality in the laboratory. The term *osmolarity* is still occasionally used, with results reported in milliosmoles per liter, but it is inaccurate in cases of hyperlipidemia or hyperproteinemia, for urine specimens, or in the presence of certain osmotically active substances, such as alcohol or mannitol. Both the sensation of thirst and antidiuretic hormone (ADH) secretion are stimulated by the hypothalamus in response to an increased osmolality of blood. The natural response to the thirst sensation is to consume more fluids, thus increasing the water content of the ECF, diluting out the elevated sodium levels, and decreasing the osmolality of the plasma. Thirst is, therefore, important in mediating fluid intake. The other means of controlling osmolality is by secretion of antidiuretic hormone (ADH, vasopressin). This hormone is secreted by the posterior pituitary gland and acts on the cells of the collecting ducts in the kidneys to increase water reabsorption. As water is conserved, osmolality decreases, turning off ADH secretion.[26]

Clinical Significance of Osmolality

Osmolality in plasma is important because it is the parameter to which the hypothalamus responds. The regulation of osmolality also affects the sodium concentration in plasma, largely because sodium and its associated anions account for approximately 90% of the osmotic activity in plasma. Another important process affecting the sodium concentration in blood is the regulation of blood volume. As we will see later, although osmolality and volume are regulated by separate mechanisms (except for ADH and thirst), they are related because osmolality (sodium) is regulated by changes in water balance, whereas volume is regulated by changes in sodium balance.[26,32]

To maintain a normal plasma osmolality ($\sim$275–290 mOsm/kg of plasma H_2O), osmoreceptors in the hypothalamus respond quickly to small changes in osmolality: A 1% to 2% increase in osmolality causes a fourfold increase in the circulating concentration of ADH, and a 1% to 2% decrease in osmolality shuts off ADH production. ADH acts by increasing the reabsorption of water in the cortical and medullary collecting tubules. ADH has a half-life in the circulation of only 15 to 20 minutes.

Renal water regulation by ADH and thirst each play important roles in regulating plasma osmolality. Renal water excretion is more important in controlling water excess, whereas thirst is more important in preventing water deficit or dehydration. Consider what happens in several conditions.

Water Load

As excess intake of water (for example, in polydipsia) begins to lower plasma osmolality, both ADH and thirst are suppressed. In the absence of ADH, a large volume of dilute urine will be excreted, as much as 10 to 20L of water daily, well above any normal intake of water. Therefore, hypoosmolality and hyponatremia usually occurs only in patients with impaired renal excretion of water.[26]

Water Deficit

As a deficit of water begins to increase plasma osmolality, both ADH secretion and thirst are activated. Although ADH contributes by minimizing renal water loss, thirst is the major defense against hyperosmolality and hypernatremia. Although hypernatremia rarely occurs in a person with a normal thirst mechanism and access to water, it becomes a concern in infants, unconscious patients, or anyone who is unable to either drink or ask for water. Osmotic stimulation of thirst progressively diminishes in people who are more than 60 years old. Particularly in the older patient with illness and diminished mental status, dehydration becomes increasingly likely. As an example of the effectiveness of thirst in preventing dehydration, a patient with diabetes insipidus (no ADH) may excrete 10L of urine per day. However, because thirst persists, water intake matches output and plasma sodium remains normal.[26]

Regulation of Blood Volume

Adequate blood volume is essential to maintain blood pressure and ensure good perfusion to all tissues and organs. Regulation of both sodium and water are interrelated in controlling blood volume. The renin-angiotensin-aldosterone system responds primarily to a decreased blood volume. Renin is secreted near the renal glomeruli in response to decreases renal blood flow (decreased blood volume and/or blood pressure). Renin converts angiotensinogen to angiotensin I, which then becomes angiotensin II. Angiotensin II causes both vasoconstriction, which quickly increases blood pressure, and secretion of aldosterone, which increases retention of sodium and the water that accompanies the sodium. The effects of blood volume and osmolality on sodium and water metabolism are shown in Figure 12-1. Changes in blood volume (actually pressure) are initially detected by a series of stretch receptors located in areas such as the cardiopulmonary circulation, the carotid sinus, the aortic arch, and the glomerular arterioles. These receptors then activate a series of responses (effectors) that restore volume by appropriately varying vascular resistance, cardiac output, and renal sodium and water retention.[26]

Other factors that affect blood volume are: (1) atrial natriuretic peptide (ANP), released from the myocardial atria in response to volume expansion, promotes sodium excretion in the kidney; (2) volume receptors independent of osmolality stimulate the release of ADH, which conserves water by renal reabsorption; (3) glomerular filtration rate (GFR) increases with volume expansion and decreases with volume depletion; (4) all other things equal, an increased plasma sodium will increase urinary sodium excretion, and

vice versa. The normal reabsorption of 98% to 99% of filtered sodium by the tubules conserves nearly all of the 150L of glomerular filtrate produced daily. A 1% to 2% reduction in tubular reabsorption of Na can increase water loss by several liters per day.

Urine osmolality values may vary widely depending on water intake and the circumstances of collection. However, it is generally decreased in diabetes insipidus (inadequate ADH) and polydipsia (excess H_2O intake due to chronic thirst), and increased in conditions such as the syndrome of inappropriate ADH secretion (SIADH) and hypovolemia (although urinary Na is usually decreased).

Determination of Osmolality

Specimen

Osmolality may be measured in serum or urine. The use of plasma is not recommended because osmotically active substances may be introduced into the specimen from the anticoagulant.

Discussion

The methods for determining osmolality are based on properties of a solution that are related to the number of molecules of solute per kilogram of solvent, such as changes in freezing point and vapor pressure. An increase in osmolality decreases the freezing-point temperature and the vapor pressure. Measurement of freezing-point depression and vapor pressure decrease (actually the "dew-point") are the two most frequently used methods of analysis.[9] There are several different osmometers now on the laboratory market.

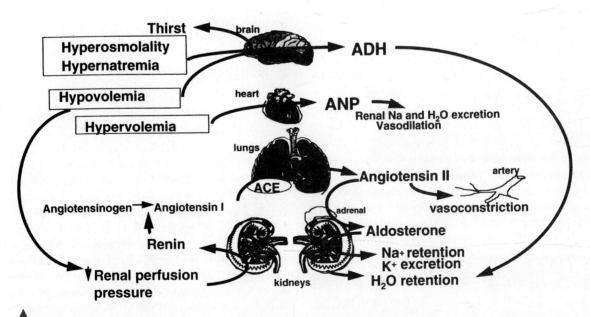

Figure 12-1. Responses to changes in blood osmolality and blood volume. ADH = antidiuretic hormone, ANP = atrial natriuretic peptide. The primary stimuli are shown in boxes (*e.g.,* hypovolemia).

For a detailed operating procedure, consult the operator's manual of the instrument in question.

Samples must be free of particulate matter to obtain accurate results. Serum and urine samples that are turbid should be centrifuged before analysis to remove any extraneous particles. If reusable sample cups are used, they should be thoroughly cleaned and dried between each use to prevent contamination.

Osmometers that operate by freezing-point depression are standardized using sodium chloride reference solutions. After calibration, the appropriate amount of sample is pipetted into the required cuvet or sample cup and placed in the analyzer. The sample is then supercooled to −7°C and seeded to initiate the freezing process. Once temperature equilibrium has been reached, the freezing point is measured, with results for serum and urine osmolality reported as milliosmoles per kilogram.

Many formulas have been proposed for calculating osmolality.[38] Calculation of osmolality has some usefulness either as an estimate of the true osmolality or to determine the *osmolal gap,* which is the difference between the measured osmolality and the calculated osmolality. The osmolal gap indirectly indicates the presence of osmotically active substances other than sodium, urea, or glucose, such as ethanol, methanol, ethylene glycol, lactate, or β–hydroxybutyrate.

Two formulas are presented here, each having theoretical advantages and disadvantages. Both are adequate for the purpose described above. For more discussion, the reader should consult other references.[38]

$$2\,Na + \frac{glucose}{20} + \frac{BUN}{3}$$

$$1.86\,Na + \frac{glucose}{18} + \frac{BUN}{2.8} + 9 \quad \textit{(Eq. 12-1)}$$

Reference Ranges[31]

See Table 12-1.

THE ELECTROLYTES

Sodium

Sodium is the most abundant cation in the extracellular fluid, representing 90% of all extracellular cations, and largely determines the osmolality of the plasma. A normal plasma osmolality is approximately 295 mmol/L, with 270 mmol/L being the result of sodium and associated anions.

TABLE 12-1. Reference Ranges for Osmolality

Serum	275–295 mOsmol/kg
Urine: (24h-specimen)	300–900 mOsmol/kg
Urine: Serum ratio	1.0–3.0

Sodium concentration in the ECF is much larger than inside cells. Since a small amount of sodium can diffuse through the cell membrane, the two sides would eventually reach equilibrium. To prevent equilibrium from occurring, active transport systems such as ATPase ion pumps are present in all cells. Potassium, as discussed in the next section, is the major intracellular cation. Like sodium, potassium would eventually diffuse across the cell membrane until equilibrium is reached. The Na/K ATPase ion pump moves three sodium ions out of the cell in exchange for two potassium ions moving into the cell as ATP is converted to ADP. Since water follows electrolytes across cell membranes, the continual removal of sodium from the cell prevents osmotic rupture of the cell by also drawing water from the cell.

Regulation

The plasma sodium concentration depends greatly on the intake and excretion of water and, to a somewhat lesser degree, the renal regulation of sodium. Three processes are of primary importance: (1) the intake of water in response to thirst, as stimulated or suppressed by plasma osmolality; (2) the excretion of water, largely affected by ADH release in response to changes in either blood volume or osmolality; and (3) the blood volume status, which affects sodium excretion through aldosterone, angiotensin II, and ANP. The kidneys have the ability to conserve or excrete large amounts of sodium, depending on the sodium content of the ECF and the blood volume. Normally, 60% to 75% of filtered Na is reabsorbed in the proximal tubule. Some sodium is also reabsorbed in the loop and distal tubule and (under the control of aldosterone) is exchanged for K in the connecting segment and cortical collecting tubule. The regulation of osmolality and volume have been summarized in Figure 12-1.

Clinical Applications

Hyponatremia. The pathogenesis of hyponatremia is perhaps most easily understood by categorizing the causes of hyponatremia with the associated blood volume status, as shown in Table 12-2. Volume status may be assessed by skin turgor (fullness), jugular venous pressure, and urine Na concentration, with a low urine Na indicating hypovolemia.[26] Hypovolemic hyponatremia results from Na loss in excess of water loss. There are several common causes of this condition: (1) use of thiazide diuretics (but not loop diuretics) induces Na and K loss without interfering with ADH–mediated water retention; (2) loss of hypotonic fluid by prolonged vomiting, diarrhea, sweating, or burns can occur, with replacement by relatively more hypotonic fluid as thirst is stimulated by hypovolemia; (3) potassium depletion, which favors intracellular loss of K to the blood, may also cause Na loss, and cellular loss of K promotes Na movement into the cell with an associated decrease in plasma Na and volume; (4) aldosterone deficiency increases renal loss of Na and water, with Na loss in excess of water loss; and (5) salt-wasting nephropathy may infrequently develop in renal tubular and interstitial diseases,

TABLE 12-2. Hyponatremia Related to Blood Volume Status

Hypovolemia (Na loss in excess of H_2O loss)
    Loss of fluid (GI, burns) with hypotonic fluid replacement
    Thiazide diuretics
    Potassium depletion
    Aldosterone deficiency
Normovolemia [defective (excess) water balance]
    Inappropriate ADH secretion (SIADH)
    Artifactual due to severe hyperlipidemia
    Severe hyperglycemia
    Polydipsia
    Diminished cortisol (glucocorticoid)
    Reset osmostat
Hypervolemic (water overload)
    CHF, hepatic cirrhosis
    Advanced renal disease—decreased GFR with excess water intake.

such as medullary cystic and polycystic kidney disease, usually as renal insufficiency becomes severe [serum creatinine >610 μmol/L (>8 mg/dL)].

Normovolemic hyponatremia typically indicates a problem with water balance and can have the following causes. (a) Inappropriate ADH secretion initiates mild hypervolemia, which then leads to excretion of Na and water by release of atrial natriuretic peptide (ANP). (b) Artifactual hyponatremia can occur in cases of severe hyperlipidemia or hyperproteinemia. Methods that dilute plasma before Na analysis by flame photometry or ion-selective electrode (ISE) give erroneously low Na results on these samples because they measure mmol/L Na per L of plasma. "Direct" methods by ISE, which do not dilute plasma or whole blood, give accurate Na results because they detect the Na concentration only in the plasma water. (c) Severe hyperglycemia induces water movement into plasma to normalize plasma osmolality, which results in hyponatremia. (d) Chronic excess intake of water, as with polydipsia (chronic thirst), can eventually lead to hyponatremia that is usually mild but occasionally severe. (e) Adrenal insufficiency can decrease cortisol, and aldosterone which contributes to the development of hyponatremia. Because cortisol usually inhibits ADH release, a deficiency of cortisol promotes ADH release and water retention. Although the initial phase of this process results in hypovolemia, the ADH-induced water retention typically restores volume status to normal.[26] (f) In pregnancy, the hypothalamic set-point for osmolality is offset such that the plasma Na concentration is regulated approximately 5 mmol/L lower than normal. This may be initiated by vasodilation, which leads to an ADH and possibly an aldosterone response to apparent hypovolemia.

Hypervolemic hyponatremia is nearly always a problem of water overload, which usually causes edema. The usual therapy is water restriction. Sodium should not be given, because it could increase the severity of edema. For example, congestive heart failure or hepatic cirrhosis increase venous back-pressure in the circulation, which promotes movement of fluid from the blood to the interstitium, causing edema. Volume receptors sense hypovolemia, which leads to secretion of ADH and eventual hypervolemia and hyponatremia. Also, in advanced renal disease, the inability to excrete water promotes hypervolemia whenever fluid intake is excessive.

Hypernatremia. Hypernatremia (increased serum sodium concentration) usually results from excess loss of water relative to sodium loss. Loss of hypotonic fluid may occur either by the kidney or through profuse sweating or diarrhea. The measurement of urine osmolality is necessary to evaluate the cause of hypernatremia. In renal loss, the urine osmolality is low or normal. In extrarenal losses, the urine osmolality is increased. Interpretation of the urine osmolality in hypernatremia is shown in Table 12-3.

There are several causes of hypernatremia. Water loss through the skin and by breathing (insensible loss) accounts for about 1L of water loss per day in adults. Any condition that increases water loss, such as fever, burns, or exposure to heat, will increase the likelihood of developing hypernatremia. Very commonly, hypernatremia occurs in adults with altered mental status and in infants, both of whom may be thirsty but who are unable to ask for or obtain water.

Note: Persons who cannot fully concentrate their urine, such as neonates, young children, the elderly, and some patients with renal insufficiency, may show a relatively lower urine osmolality.

Chronic hypernatremia in an alert patient is indicative of hypothalamic disease, usually with a defect in the osmoreceptors rather than from a true resetting of the osmostat. A reset osmostat may occur in primary hyperaldosteronism, in which excess aldosterone induces mild hypervolemia, which retards ADH release, shifting plasma sodium upwards by ~3 to 5 mmol/L.[26]

Hypernatremia may be from excess ingestion of salt or administration of hypertonic solutions of sodium. Neonates are especially susceptible to hypernatremia from this cause. In these cases, ADH response is appropriate, resulting in urine osmolality >800 mOsm/kg (see Table 12-3).

TABLE 12-3. Interpretation of Urine Osmolality in Evaluation of Hypernatremia (plasma Na> 150 mmol/L)

Urine osmolality < 300 mOsmol/kg
    Diabetes insipidus (impaired secretion of ADH or kidneys cannot respond to ADH)
Urine osmolality 300–700 mOsmol/kg
    Partial defect in ADH release or response to ADH
    Osmotic diuresis
Urine osmolality over 700 mOsmol/kg
    Loss of thirst
    Insensible loss of water (breathing, skin)
    GI loss of hypotonic fluid
    Excess intake of sodium

Hypernatremia may result from loss of water in diabetes insipidus, either because the kidney cannot respond to ADH (nephrogenic diabetes inspidus), or ADH secretion is impaired (central diabetes insipidus). Diabetes insipidus is characterized by copious production of dilute urine (3–20 L/d). Since people with diabetes insipidus drink large volumes of water, hypernatremia usually does not occur in diabetes insipidus unless the thirst mechanism is also impaired. Partial defects of either ADH release or the response to ADH may also occur. In such cases, urine is concentrated to a lesser extent than appropriate to correct the hypernatremia. Excess water loss may also occur in renal tubular disease, such as acute tubular necrosis, in which the tubules become unable to fully concentrate the urine.

Treatment of hypernatremia is directed at correction of the condition that caused either water depletion or sodium retention. Because too rapid correction of serious hypernatremia (>160 mmol/L) can induce cerebal edema and death, hypernatremia must be corrected gradually: the maximal rate should be 0.5 mmol/L per hour.[26]

Determination of Sodium

Specimen. Serum, plasma, and urine are all acceptable for sodium measurements. When plasma is used, lithium heparin, ammonium heparin, and lithium oxalate are suitable anticoagulants. The specimen of choice in urine sodium analyses is a 24-hour collection.

Methods. Through the years, sodium has been measured in a variety of ways, including chemical methods, flame-emission spectrophotometry (FES), atomic absorption spectrophotometry (AAS), and, most recently, ion-selective electrode (ISE). Chemical methods are now outdated because of their large sample volume requirements and lack of precision. Of the methods now readily available, ISE is by far the most routinely used in clinical laboratories.

Ion-selective electrodes use a semipermeable membrane to develop a potential produced by having different ion concentrations on either side of the membrane. In this type of system, two electrodes are used. One electrode has a constant potential, making it the reference electrode. The difference in potential between the reference and measuring electrodes can be used to calculate the "concentration" of the ion in solution. However, it should be noted that it is the activity of the ion, not the concentration, that is being measured.

The CX3 (Beckman Instruments, Inc.) uses a lithium aluminum silicate glass membrane in its ion-selective electrode system. The module in which the sodium system is contained also houses the potassium electrode. The reference electrode is used for both the sodium and potassium methods.[19]

The Ektachem (Eastman Kodak Company) uses a single-use ISE system. Each disposable slide contains a reference and measuring electrode (Fig. 12-2). A drop of sample fluid and a drop of reference fluid are simultaneously applied to the slide, and the potential difference between the two is

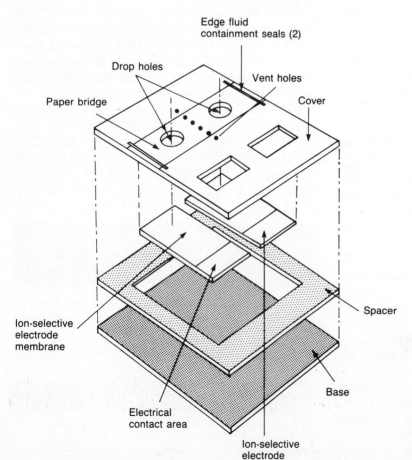

Figure 12-2. Schematic diagram of the ion-selective electrode system for the potentiometric slide on the Ektachem. (Courtesy of Eastman-Kodak Co., Rochester, NY)

measured with an electrometer. Because the sample is not diluted, this procedure is a direct ISE method.[21]

In flame-emission spectrophotometry, a sample is aspirated into a flame, producing atoms in an excited state, which are capable of emitting light of a specific wavelength, depending on the element of interest. The intensity of the emitted light can then be correlated to the quantity of ion present in the sample.[3]

SODIUM AND POTASSIUM DETERMINATION BY FLAME-EMISSION SPECTROPHOTOMETER

Principle–Na^+ and K^+ emit light of specific wavelengths when aspirated into a propane flame. This light can be measured by comparison with a known standard after it is isolated from interfering light. Lithium is used as an internal standard to eliminate environmental interferences.

Reagents

1. Stock lithium nitrate (1 mol/L). Add 68.95 g of $LiNO_3$ to a 1-L volumetric flask containing 500 mL of deionized water. Mix and dilute to volume with deionized water. Mix thoroughly and store in a polyethylene container at room temperature.
2. Working lithium nitrate. In a 4-L volumetric flask, dilute 60 mL of the stock $LiNO_3$ to 4000 mL with deionized water. Mix and store in a polyethylene container at room temperature.
3. Stock sodium chloride standard (1 mol/L). Add 58.45 g of NaCl to a 1-L flask containing 500 mL of deionized water. Mix and dilute to volume. Mix again and store in a polyethylene container at room temperature.
4. Stock potassium chloride standard (100 mmol/L). Add 7.45g of KCl to 500 mL of deionized water in a 1-L volumetric flask. Mix and dilute to volume. Mix again and store in a polyethylene container at room temperature.
5. Working serum standard. Add 140 mL of NaCl stock and 5 mL of KCl stock to a 1-L volumetric flask and dilute to volume with deionized water. Mix thoroughly and store in a polyethylene container. (Na^+ = 140 mmol/L; K^+ = 5 mmol/L)
6. Working urine standards. (1) Add 50 mL of NaCl stock and 50 mL KCl stock standards to a 1-L volumetric flask. Dilute to volume with deionized water. Mix and store in a polyethylene container. (Na^+ = 50 mmol/L; K^+ = 50 mmol/L) (2) Add 100 mL of NaCl stock and 100 mL of KCl stock standards to a 1-L volumetric flask and dilute to volume with deionized water. Mix and store in a polyethylene container at room temperature. (Na^+ = 100 mmol/L; K^+ = 100 mmol/L)

Procedure

1. Either by hand or using an automatic dilutor, prepare all standards and samples using a 1:201 dilution with working $LiNO_3$ (*e.g.*, 25 μL of specimen and 5 mL of $LiNO_3$).
2. The procedure for operating the analyzer will vary from one instrument to another. Consult the operator's manual for detailed instructions.
3. For serum samples, set the standard values into the instrument (140^+ Na; 5^+ K), run controls and samples, and recheck the standard after every 5 samples.
4. When running urine samples, set the 50 Na/50 K values into the analyzer. Then check the linearity by aspirating the 100 Na/100 K standard. Run patient samples and controls, checking standards after every 5 samples.

Comments

1. Results are reported in millimoles per liter.

2. For calculating urine results when running a 24-h urine collection, use this formula:

$$\text{(Concentration in mmol/L} \times \text{total volume in L per 24 h) } 1000 = \text{mmol/24 h}$$

(Eq. 12-2)

3. The spectral lines for sodium and potassium are 589 and 768 nm, respectively. However, most commercial analyzers use 676 nm for K^+.[19]
4. The dilution required for biologic assays will vary according to the analyzer used and the specimen to be analyzed. Biologic samples require diluting in order to decrease the protein concentration and viscosity (which otherwise might lead to plugs in the atomizer) and also to dilute the ion concentration of the sample so that the intensity of the emitted light will be linear in relation to the concentration.
5. The presence of hyperglycemia, hyperlipidemia, or hyperproteinemia may cause an apparent decrease in serum sodium. This is apparent rather than *pseudohyponatremia* because it is due to the displacement of plasma water by proteins or lipids or by the transfer of water from the intracellular to the extracellular space in response to hyperglycemia. In the case of hyperproteinemia and hyperlipidemia, a true sodium concentration may be obtained by running the specimen undiluted on a direct-reading ISE. These examples do not constitute true electrolyte disorders. ∎

Reference Ranges[31]

See Table 12-4.

Potassium

Potassium is the major intracellular cation in the body, with a 20 times greater concentration inside the cells than outside. Many cellular functions require that the body maintain a low ECF concentration of K^+. As a result, only 2% of the body's total potassium circulates in the plasma.

Functions of potassium in the body include regulation of neuromuscular excitability, contraction of the heart, intracellular fluid volume, and hydrogen ion concentration.[26]

The potassium ion concentration has a major effect on the contraction of cardiac muscles. An elevated plasma potassium slows the heart rate by decreasing the resting membrane potential of the cell, which increases the net difference between the cell's resting potential and threshold (action) potential. A decrease in the extracellular potassium concentration increases myocardial excitability and often causes an arrhythmia. The heart may cease to contract in extreme cases of either hyper- or hypokalemia.

The potassium concentration also affects the hydrogen ion concentration in the blood. For example, in *hypokalemia* (low serum potassium), as potassium ions are lost from the

TABLE 12-4. Reference Ranges for Sodium

Serum, plasma	135–145 mmol/L
Urine (24h collection)	40–220 mmol/L
CSF	138–150 mmol/L

body, sodium and hydrogen ions move into the cell. The hydrogen ion concentration is therefore decreased in the ECF, resulting in alkalosis.

Regulation

The kidneys are important in the regulation of potassium balance. Initially, the proximal tubules reabsorb nearly all the potassium. Then, under the influence of aldosterone, additional potassium is secreted into the urine in exchange for sodium in both the distal tubules and the collecting ducts. Thus, the distal nephron is the principal determinant of urinary potassium excretion. Most individuals consume far more potassium than is needed; the excess is excreted in the urine but may accumulate to toxic levels if renal failure occurs.

Potassium uptake from the ECF into the cells is important in normalizing an acute rise in plasma K concentration due to an increased K intake. Excess plasma K rapidly enters cells to normalize plasma K. As the cellular K then gradually returns to the plasma, it is removed by urinary excretion. Note that chronic loss of cellular K may result in cellular depletion before there is an appreciable change in the plasma K concentration, since excess K is normally excreted in the urine.

There are several important factors that influence the distribution of potassium between cells and extracellular fluid: (1) potassium loss frequently occurs whenever the Na-K ATPase pump is inhibited by conditions such as hypoxia, hypomagnesemia, or digoxin overdose; (2) insulin promotes acute entry of K ions into skeletal muscle and liver by increasing Na-K ATPase activity; and (3) catecholamines, such as epinephrine (β2 stimulator), promote cellular entry of K, whereas propranolol (β-blocker) impairs cellular entry of K. Dietary deficiency or excess is rarely a primary cause of hypokalemia or hyperkalemia. However, with a pre-existing condition, dietary deficiency (or excess) can enhance the degree of hypokalemia (or hyperkalemia).

Exercise. Potassium is released from cells during exercise, which may increase plasma K by 0.3 to 1.2 mmol/L with mild to moderate exercise and by as much as 2 to 3 mmol/L with exhaustive exercise. These changes are usually reversed after several minutes of rest. Note that forearm exercise during venipuncture can cause erroneously high plasma K concentrations.[32]

Hyperosmolality. Hyperosmolality causes diffusion of water out of cells, carrying K ions with the water, which leads to gradual depletion of potassium.

Cellular Breakdown. Cellular breakdown releases K into the ECF. Examples are severe trauma, tumor lysis syndrome, and massive blood transfusions.

Causes of Hypokalemia. Common causes of hypokalemia are shown in Table 12-5. Of these, therapy with thiazide-type diuretics is by far the most common. Hypokalemia is defined as a plasma potassium concentration below the

TABLE 12-5. Factors that Cause Hypokalemia	
• Diuretics:	Thiazides, furosemide
	Enhance distal K+ secretion by increasing tubular Na and H_2O flow
• Insulin:	Enhances cellular uptake of K+
• Alkalemia:	Promotes cellular H+ loss/K+ gain
• Decreased Mg:	Promotes cellular and renal loss of K+
• Excess aldosterone:	Enhances K+ excretion in urine
• Inadequate intake:	Rarely a cause, frequently a contributor

lower limit of the reference range. Hypokalemia can occur with GI or urinary loss of potassium or with increased cellular uptake of potassium.

GI loss occurs when GI fluid is lost through vomiting, diarrhea, gastric suction, or discharge from an intestinal fistula. Renal loss of potassium can result from kidney disorders such as renal tubular acidosis and potassium-losing nephritis. Since aldosterone promotes Na retention and K loss, excess aldosterone can lead to hypokalemia and metabolic alkalosis.[26] Hypomagnesemia can lead to hypokalemia by promoting urinary loss of potassium. Magnesium deficiency also diminishes the activity of Na-K ATPase and enhances the secretion of aldosterone. Effective treatment requires supplementation with both Mg and K.[26] While reduced dietary intake of K rarely causes hypokalemia in healthy persons, decreased intake may intensify hypokalemia caused by use of diuretics, for example.

Both alkalemia and insulin increase the cellular uptake of potassium. Since alkalemia promotes intracellular loss of H^+ to minimize elevation of intracellular pH, both K and Na enter cells to preserve electroneutrality. Plasma K decreases by about 0.4 mmol/L per 0.1 unit rise in pH.[26] Insulin promotes the entry of K into skeletal muscle and liver cells. Since insulin therapy can sometimes uncover an underlying hypokalemic state, plasma K should be monitored carefully whenever insulin is administered to susceptible patients.[26]

The symptoms of hypokalemia are muscle weakness and cardiac arrhythmias, which often become apparent as plasma potassium decreases below 3 mmol/L. The dangers of hypokalemia are of concern in all patients, but especially so in those with cardiovascular disorders. Hypokalemia can lead to muscle weakness or paralysis, which can interfere with breathing. A variety of cardiac arrhythmias can be induced by hypokalemia: premature atrial and ventricular beats, sinus bradycardia, atrioventricular block, and ventricular tachycardia and fibrillation. These may cause sudden death in certain patients with cardiac disorders. Mild hypokalemia is usually asymptomatic.

Treatment of hypokalemia includes oral or I.V. replacement of potassium. In some cases, chronic mild hypokalemia may be corrected simply by including food with a high potassium content, such as bananas and orange juice, in the diet.

Causes of Hyperkalemia. The most common causes of hyperkalemia are shown in Table 12-6. Patients with hyperkalemia often have an underlying disorder, such as renal insufficiency, diabetes mellitus, or metabolic acidosis, that contributes to hyperkalemia.[5] For example, during administration of KCl, a person with renal insufficiency is far more likely to develop hyperkalemia than is a person with normal renal function.

In healthy persons, an acute oral load of potassium will increase plasma K briefly, because most of the absorbed K rapidly moves intracellularly. Normal cellular processes gradually release this excess K back into the plasma, where it is normally removed by renal excretion. Impairment of urinary K excretion is almost always associated with chronic hyperkalemia.[26]

If a shift of K from cells into plasma occurs too rapidly to be removed by renal excretion, acute hyperkalemia develops. In diabetes mellitus, insulin deficiency promotes cellular loss of K. Hyperglycemia also contributes by producing a hyperosmolar plasma that pulls water and K from cells, promoting further loss of K into the plasma.[26]

In metabolic acidosis, as excess H^+ moves intracellularly to be buffered, K leaves the cell to maintain electroneutrality. Plasma K increases by 0.2 to 1.7 mmol/L for each 0.1 unit reduction of pH.[26] Because cellular K often becomes depleted in cases of acidosis with hyperkalemia (including diabetic ketoacidosis), treatment with agents such as insulin and bicarbonate can cause a rapid intracellular movement of K, producing severe hypokalemia.

A variety of drugs may cause hyperkalemia, especially in patients with either renal insufficiency or diabetes mellitus. These drugs include captopril (inhibits angiotensin converting enzyme), nonsteroidal antiinflammatory agents (inhibit aldosterone), spironolactone (K-sparing diuretic), digoxin (inhibits Na-K pump), cyclosporine (inhibits renal response to aldosterone), and heparin therapy (inhibits aldosterone secretion).

Hyperkalemia may result when potassium is released into the ECF during enhanced tissue breakdown or catabolism,

especially if renal insufficiency is present. Increased cellular breakdown may be caused by trauma, administration of cytotoxic agents, massive hemolysis, tumor lysis syndrome, and blood transfusions. In the last example, K is gradually released from erythrocytes during storage, often resulting in markedly elevated plasma K concentrations in blood stored for several weeks.

Patients on cardiac bypass may develop mild elevations in plasma potassium during warming after surgery, since warming causes cellular release of potassium. Note that hypothermia causes movement of potassium into cells.

Symptoms of Hyperkalemia. Hyperkalemia can cause muscle weakness by altering neuromuscular conduction. Muscle weakness does not usually develop until plasma potassium reaches 8 mmol/L.[26]

Hyperkalemia disturbs cardiac conduction, which can lead to cardiac arrhythmias and possible cardiac arrest. Plasma potassium concentrations of 6 to 7 mmol/L may alter the electrocardiogram, and concentrations > 10 mmol/L may cause cardiac arrest.[26]

Treatment of Hyperkalemia. To offset the effect of potassium, which lowers the resting potential of myocardial cells, calcium may be given to reduce the threshold potential of myocardial cells. Therefore, calcium provides immediate but short-lived protection to the myocardium against the effects of hyperkalemia. Substances that acutely shift potassium back into cells, such as sodium bicarbonate, glucose, and/or insulin, may also be administered. Patients treated with these agents must be monitored carefully to prevent hypokalemia as K moves back into cells. A cation-exchange resin, such as kayexelate, lowers plasma K by binding K ions in the gut. If the preceding measures do not prove successful or if renal failure occurs, peritoneal dialysis or hemodialysis may be necessary.[26]

Collection of Samples. Proper collection and handling of samples for K analysis is extremely important, since there are many causes of artifactual hyperkalemia. First, the coagulation process releases K from platelets, so that serum K may be 0.1 to 0.5 mmol/L higher than plasma K concentrations.[30] If the patient's platelet count is elevated (thrombocytosis), serum potassium may be further elevated. Second, if a tourniquet is left on the arm too long during blood collection, or if the patient excessively clenches their fist or otherwise exercises their forearm before venipuncture, cells may release potassium into the plasma. The first situation may be avoided by using a heparinized tube to prevent clotting of the specimen, and the second, by using proper care in the drawing of blood. Third, because storing blood on ice promotes the release of potassium from cells,[11] whole blood samples for potassium determinations should be stored at room temperature (never iced) and analyzed promptly or centrifuged to remove the cells. Fourth, if hemolysis occurs after the blood is drawn, potassium may be falsely elevated.

TABLE 12-6. Causes of Hyperkalemia	
• Excess intake:	Oral (diet, salt substitute)
• Cellular loss:	Acidemia (cellular H+ gain promotes K+ loss)
	Insulin deficiency
	Drugs (captopril, digoxin, cyclosporine, heparin)
	Transfusions
	Crush injuries
	Moderate to severe exercise
• Decreased excretion:	Renal failure
	Hypoaldosteronism
	Hypovolemia
• Factitious:	*In vitro* hemolysis, ice storage, clotting

Determination of Potassium

Specimen. Serum, plasma, and urine may be acceptable for analysis. Hemolysis must be avoided because of the high K^+ content of erythrocytes. Heparin is the anticoagulant of choice. Whereas serum and plasma generally give similar potassium levels, markedly elevated platelet counts may result in the release of potassium during clotting from rupture of these cells, causing a spurious hyperkalemia. For these cases, plasma is preferred. Urine specimens should be collected over a 24-hour period to eliminate the influence of diurnal variation.

Methods. As with sodium, the current method of choice is ISE. For ISE measurements, a valinomycin membrane is used to selectively bind K^+, causing an impedance change that can be correlated to K^+ concentration.

Sodium and potassium are measured simultaneously by flame emission. (Refer to the previous section on sodium determination by flame emission for a detailed procedure.)

Reference Ranges[31]

See Table 12-7.

Chloride

Chloride (Cl^-) is the major extracellular anion. Its precise function in the body is not well understood; however, it is involved in maintaining osmolality, blood volume, and electric neutrality. In most processes, chloride ions shift secondarily to a movement of Na^+ or bicarbonate ions.

Chloride ingested in the diet is almost completely absorbed by the intestinal tract. Chloride ions are then filtered out by the glomerulus and passively reabsorbed, in conjunction with sodium, by the proximal tubules. Excess chloride is excreted in the urine and sweat. Excessive sweating stimulates aldosterone secretion, which acts on the sweat glands to conserve sodium and chloride.

Chloride maintains electric neutrality in two ways. First, Na^+ is reabsorbed along with Cl^- in the proximal tubules. In effect, Cl^- acts as the rate-limiting component in that Na^+ reabsorption is limited by the amount of Cl^- available. Electroneutrality is also maintained by chloride through the chloride shift. In this process, carbon dioxide (CO_2) generated by cellular metabolism within the tissue diffuses out into both the plasma and the red cell. In the red cell, CO_2 forms carbonic acid (H_2CO_3), which splits into H^+ and HCO_3^- (bicarbonate). Deoxyhemoglobin buffers H^+, while the HCO_3^- eventually diffuses out into both the plasma and the red cell in order to maintain the electric balance of the cell.

TABLE 12-7. Reference Ranges for Potassium	
Plasma, serum	3.4–5.0 mmol/L
Urine, 24 h	25–125 mmol/24hr

Clinical Applications

Hyperchloremia may occur when there is an excess loss of bicarbonate ion due to either gastrointestinal losses, renal tubular acidosis, or aldosterone deficiency. *Hypochloremia* may occur with excessive loss of chloride from prolonged vomiting, diabetic ketoacidosis, aldosterone excess, or salt-losing renal diseases such as pyelonephritis. A low serum level of chloride may also be encountered in conditions associated with high serum bicarbonate concentrations, such as compensated respiratory acidosis or metabolic alkalosis.

Determination of Chloride

Specimen. Serum or plasma may be used, with lithium heparin being the anticoagulant of choice. For urine specimens, a 24-hour collection is preferred because of the large diurnal variation. Sweat is also suitable for analysis. Sweat collection and analysis is discussed in Chapter 23, Body Fluids.

Methods. There are several methodologies available for measuring chloride. The most common ones currently in use are amperometric-coulometric titration, mercurimetric titration, ISE, and colorimetry.

Amperometric-coulometric titration is a method using coulometric generation of silver ions (Ag^+), which combine with Cl^-. Amperometric indication of the endpoint is used at the first sign of free Ag^+. The elapsed time is used to calculate the concentration of Cl^- in the sample. Both the Cotlove Chloridometer (Buchler Instruments, Inc.) and the Astra 8 (Beckman Instruments, Inc.) use this principle in chloride analysis.

Mercurimetric titration is also a common method for chloride analysis, particularly the Schales–Schales method. This technique involves titrating Cl^- with a solution of mercury (Hg^{2+}), forming soluble but nonionized mercuric chloride.

$$2Cl^- + Hg^{2+} \rightarrow HgCl_2 \qquad \textit{(Eq. 12-3)}$$

The endpoint is reached when excess Hg^{2+} forms a complex with an indicator, such as diphenylcarbazone, producing a violet-blue color. Both this method and amperometric-coulometric titration suffer from halogen interferences. Other ions, such as Br^-, I^-, CN^-, CNS^-, and $-SH$, will react, creating positive errors.

Another technique for measuring chloride is the ion-selective electrode. It has found limited use in general clinical laboratories, although it is important in measuring sweat chloride for cystic fibrosis. There is also a colorimetric method for chloride that uses mercuric thiocyanate and ferric nitrate to form a reddish colored complex with a peak at 480 nm. This method is commonly used in AutoAnalyzers (Technicon, Inc.) but is not suitable for urine or sweat analysis.

CHLORIDE DETERMINATION BY CHLORIDOMETER

Principle–The chloridometer uses amperometric-coulometric titration to quantitate chloride. Free silver ions in the system are used to indicate the endpoint of the reaction:

$$Ag^{+2} + 2Cl^- \rightarrow AgCl_2 \qquad \textit{(Eq. 12-4)}$$

Reagents

1. Acid reagent

   6.4 mL of concentrated nitric acid

   100 mL of glacial acetic acid

   Add both the nitric acid and glacial acetic acid to 900 mL of deionized water and mix. This reagent is stable at room temperature for 6 months.

2. Gelatin reagent—6.2 g of Gelatin Reagent (obtained from Buchler Instruments, part no. 4–2028)

   Heat 1 L of deionized water. Add the gelatin and mix with a stirring bar over heat until gelatin is dissolved. Aliquot into 13 × 100-mm test tubes, cover, and store at 2 to 6°C. This reagent is stable for 6 months. Use a fresh tube of gelatin daily.

3. Stock standard—5.845 g of sodium chloride (NaCl)

   Add the NaCl to a 100-mL volumetric flask containing approximately 50 mL of deionized water. Mix to dissolve NaCl and then dilute to 100-mL volume with deionized water. Mix in a glass bottle at 26°C. Standard is stable for 6 months.

4. Working standard (100 mmol/L)

   Using a volumetric pipet, pipet 10 mL of stock standard into a 100-mL volumetric flask. Dilute to volume with deionized water and mix well. Standard is stable for 2 months.

Procedure

1. Daily maintenance of any chloridometer should include cleaning the silver electrode. Consult the instrument manual for the appropriate cleanser.

2. Perform other maintenance as required by the instrument manual.

3. Run a blank, using 4 mL of acid reagent and 4 drops of the gelatin reagent in the appropriate vial. Set the analyzer to 0 mEq/L on the digital readout.

4. Check the calibration of the instrument by pipetting 100 μL of the working standard into a vial containing 4 mL of acid reagent. Add 4 drops of gelatin. Standard should read 100 ± 3 mEq/L. (The Digital Chloridometer by Buchler Instruments is internally calibrated to the Faraday constant and normally requires no external calibration. For adjustment, see the operator's instruction manual.)[13]

5. Run the patients and controls in duplicate, using 100 μL of sample, 4 mL of acid reagent, and 4 drops of gelatin reagent.

Comments

1. Rinse the electrodes with deionized water between each determination and blot dry.

2. Leave the electrode in a vial of deionized water when the analyzer is not in use.

3. Lipemia, icteremia, and hemolysis do not affect this test, but other halogens, such as Br^- and I^-, will interfere.

4. This method is suitable for serum, urine, or spinal fluid and is linear to 150 mmol/L. ■

Reference Ranges

See Table 12-8.

TABLE 12-8. Reference Ranges for Chloride

Plasma, serum	98–106 mmol/L
Urine, 24h	110–250 mmol/L

Bicarbonate

Bicarbonate is the second most abundant anion in the ECF. Total CO_2 is composed of the bicarbonate ion (HCO_3^-), carbonic acid (H_2CO_3), and dissolved CO_2, with bicarbonate accounting for over 90% of the total CO_2 at physiologic pH.

Bicarbonate is the major component of the buffering system in the blood. Carbonic anhydrase in red blood cells converts CO_2 and H_2O to carbonic acid, which dissociates into H^+ and HCO_3^-. Bicarbonate diffuses out of the cell in exchange for chloride to maintain ionic charge neutrality within the cell. This process converts potentially toxic CO_2 in the plasma to an effective buffer, bicarbonate. Bicarbonate buffers excess hydrogen ion by combining with acid, then eventually dissociating into H_2O and CO_2 in the lungs where the acidic gas CO_2 is eliminated.

Regulation

In the kidneys, most (85%) of the bicarbonate ion is reabsorbed by the proximal tubules, with 15% being reabsorbed by the distal tubules. Because tubules are only slightly permeable to bicarbonate, it is usually reabsorbed as CO_2. This happens as bicarbonate, after filtering into the tubules, combines with hydrogen ions to form carbonic acid, which then dissociates into H_2O and CO_2. The CO_2 readily diffuses back into the ECF. Normally, nearly all the bicarbonate ions are reabsorbed from the tubules, with little lost in the urine. When bicarbonate ions are filtered in excess of hydrogen ions available, almost all excess HCO_3^- flows into the urine.

In alkalosis, with a relative increase in bicarbonate ion compared to CO_2, the kidneys increase excretion of HCO_3^- into the urine, carrying along a cation such as sodium. This loss of HCO_3^- from the body helps correct pH.

Among the responses of the body to acidosis is an increased excretion of H^+ into the urine. In addition, HCO_3 reabsorption is virtually complete, with 90% of the filtered bicarbonate reabsorbed in the proximal tubule, and the remainder in the distal tubule.[26]

Clinical Applications

A decreased bicarbonate may occur from metabolic acidosis as bicarbonate combines with H^+ to produce CO_2, which is exhaled by the lungs. The typical response to metabolic acidosis is compensation by hyperventilation, which lowers PCO_2. Elevated total CO_2 concentrations occur in metabolic alkalosis as bicarbonate is retained, often with an increased PCO_2 due to compensation by hypoventilation. Typical causes of metabolic alkalosis include severe vomiting, hypokalemia, and excessive alkali intake.

Determination of Carbon Dioxide

Specimen. This chapter deals specifically with venous serum or plasma determinations. For discussion of arterial and whole blood PCO_2 measurements, refer to Chapter 11, Blood Gases, pH, and Buffer Systems.

Serum or lithium heparin plasma is suitable for analysis. While specimens should be anaerobic for the highest accuracy, many current analyzers (excluding blood gas analyzers) do not permit anaerobic sample handling. In most instances, the sample is loosely capped until the analysis is performed.

Carbon dioxide measurements may be obtained in several ways, but the actual portion of the total CO_2 being measured may vary with the method used. Two common methods, ISE and a colorimetric method adapted to continuous-flow analyzers, measure all fractions of CO_2.

Beckman Instruments, Inc., has an ISE for measuring total CO_2. It uses an acid reagent to convert all the forms of CO_2 to CO_2 gas, which diffuses through a membrane into an alkaline reagent. The pH change of this reagent is proportional to the amount of CO_2 gas present. A second, reference electrode is also used, and the difference in electric impulses from these two electrodes is used to calculate the total CO_2 of the sample.

Reference Range[31]

Carbon dioxide   22–29 mmol/L (plasma, serum)

Magnesium

Magnesium Physiology

In 1695, from well water in Epsom, England, Dr. Nehemiah Grew prepared "Epsom salts," a name still given to magnesium sulfate. Following Black's recognition of magnesium as an element in 1755, Sir Humphrey Davy isolated magnesium in 1808.[13]

The biologic significance of magnesium as a constituent of plants has been known since the 18th century. Specifically, magnesium ion is an essential component of chlorophyll, as iron is an essential component of hemoglobin. Chlorophyll and hemoglobin share a similar chemical structure.

The clinical symptoms of hypomagnesemia were not reported until 1934 by Hirschfelder and Haury. In 1969, Shils reported that magnesium depletion was associated with hypomagnesemia, hypokalemia, and hypocalcemia. While hypomagnesemia was often overlooked as a clinical diagnosis, over the last 10 years, physicians have developed a renewed interest in the clinical importance of magnesium. Many recent articles have described the effects of magnesium on the heart,[2] the role of magnesium as a calcium-channel blocking agent,[17] the clinical consequences of hypomagnesemia,[15] and the value of magnesium supplementation in both acute myocardial infarction[27] and critical care.[6]

The importance of magnesium should not be surprising. Magnesium is an essential activator of over 300 enzymes, including those important in glycolysis, transcellular ion transport, muscle contraction, and oxidative phosphorylation. It is sufficient to say that magnesium is second in biochemical importance to no other electrolyte or compound.

Regulation

The average dietary intake of magnesium should be 10 to 15 mmol/d, with rich sources being green vegetables, meat, grains, and seafood. The small intestine may absorb from 20% to 60% of the dietary magnesium, depending on the need. Secretions from the lower GI tract are rich in Mg, while secretions of the upper GI tract are relatively poor in Mg. This difference is why prolonged diarrhea often causes Mg depletion, while vomiting has little effect. The overall regulation of body magnesium is controlled largely by the kidney, which can avidly reabsorb magnesium in deficiency states or readily excrete excess magnesium in overload states. The loop of Henle appears to be the major renal regulatory site, where approximately 65% of filtered Mg may be reabsorbed in the thick ascending limb. The renal threshold for magnesium is approximately 0.60 to 0.85 mmol/L (~1.46–2.07 mg/dL). Because this is close to the normal serum concentration, slight excesses of magnesium in serum are rapidly excreted by the kidneys.[32]

Although the regulation of magnesium appears to be related to that of calcium, it is not as well characterized. Parathyroid hormone (PTH) increases the renal reabsorption of magnesium and enhances the absorption of magnesium in the intestine. However, for equivalent decreases in ionized calcium and ultrafiltrable magnesium in blood, changes in ionized calcium have a far greater effect on PTH secretion.[35] Aldosterone apparently has the opposite effect of PTH in the kidney, increasing the renal excretion of magnesium.

Distribution of Magnesium in Blood and Cells

About 53% of magnesium in the body is in the skeleton, and about 46% is intracellular magnesium in the skeletal muscle, liver, and myocardium. Only about 1% is in the blood and ECF. In a manner similar to calcium, magnesium in serum exists as protein-bound (24%), complex-bound (10%), and ionized (66%) forms.[35] Of course, these proportions vary among individuals, especially in disease or during medical interventions. As with calcium, the pH of blood apparently affects the magnesium binding by proteins in the blood.

Clinical Applications

Because the reported incidence of hypomagnesemia in hospitalized patients varies from 5% to 50%, there is debate both for and against the need for routine measurement of magnesium in serum. However, it is clear that hypomagnesemia contributes to a poorer prognosis in many clinical settings, especially in patients with cardiac disorders.

Although the measurement of total magnesium concentrations in serum remains the usual diagnostic test for detection of magnesium abnormalities, it has limitations. First, because approximately 25% of magnesium is protein-bound, total magnesium may not reflect the physiologically active free ionized magnesium. Second, because magnesium is primarily an intracellular ion, serum concentrations will not necessarily reflect the status of intracellular magnesium. Even when tissue and cellular magnesium is depleted by as much as 20%, serum magnesium concentrations may remain normal.

A magnesium load test may detect body depletion of magnesium in patients with a normal concentration of magnesium in serum. However, a magnesium load test requires >48 hours to complete. After collection of a baseline 24-h urine, 30 mmol (729 mg) of Mg in a 5% dextrose solution is administered intravenously, as another 24-h urine is collected. While individuals with adequate body stores of magnesium will excrete 60% to 80% of the magnesium load within 24 h, magnesium-deficient patients will excrete 50% or less.[5]

Cardiac Disorders. The detection of hypomagnesemia is probably most critical in cardiovascular disease, because magnesium deficiency is associated with coronary vasospasm, acute infarction, and sudden death.[16] These conditions may arise because hypomagnesemia both contributes to the loss of myocardial potassium and increases the ratio of calcium to magnesium, all of which alter myocardial electrophysiology.[28]

Congestive heart failure (CHF) causes a state of diminished cardiac output inadequate for the metabolic needs of tissues. Since deficiencies of magnesium and potassium lead to high blood pressure, causing increased vascular resistance, the problems of diminished cardiac output in CHF are further intensified.

Drugs. Several drugs, including diuretics, gentamicin, cisplatin, and cyclosporine, increase renal loss of magnesium and frequently result in hypomagnesemia. The loop diuretics, such as furosemide, are especially effective in increasing renal loss of Mg. Thiazide diuretics require a longer period of use to cause hypomagnesemia. Cisplastin has a nephrotoxic effect that inhibits the ability of the renal tubule to conserve magnesium. Cyclosporine, an immunosupressant, severely inhibits the renal tubular reabsorption of magnesium and has many side effects, including nephrotoxicity, hypertension, hepatotoxicity, and neurologic symptoms such as seizures and tremors.[18]

Diabetes. Hypomagnesemia is found in many patients with diabetes. Hypomagnesemia can aggravate the neuromuscular and vascular complications commonly found in this disease. While not well understood, the mechanism of magnesium deficiency appears to be related to the hormones insulin, PTH, and vitamin D. Osmotic diuresis due to hyperglycemia is another likely factor in the development of hypomagnesemia. The American Diabetic Association has made a statement regarding dietary intake of magnesium and measurement of serum Mg in diabetic patients.[1]

Alcoholism. Hypomagnesemia in alcoholic patients apparently results from a combination of dietary magnesium deficiency, ketosis, vomiting, diarrhea, and hyperaldosteronism. Because alcoholic patients usually become hypomagnesemic during hospitalization, normalization of serum magnesium is beneficial in many cases.

Hypomagnesemic patients with Paget's disease tend to have more active disease, as measured by increased serum alkaline phosphatase and urinary hydroxyproline. Since urinary magnesium is not significantly increased in Paget's disease, the low serum magnesium is apparently caused by increased uptake of magnesium into bone. For these reasons, patients with Paget's disease should increase their intake of magnesium to about 10 mmol per day.[29]

Because magnesium requirements increase during pregnancy, hypomagnesemia may develop if intake is not also increased. Premature labor or eclampsia are clearly associated with the development of hypomagnesemia. Treatment with magnesium salts is a long-accepted practice for these conditions with extreme hypermagnesemia sometimes resulting from excessive doses.

DETERMINATION OF MAGNESIUM

Specimen–Nonhemolyzed serum or lithium heparin plasma may be analyzed. Since the Mg^{2+} concentration inside erythrocytes is 10 times greater than that in the ECF, hemolysis should be avoided, and the serum should be separated from the cells as soon as possible. Urine is also suitable for analysis, with an acidified, 24-hour urine collection being the preferred specimen.

Methods. The two primary methods for measuring magnesium are atomic absorption spectrophotometry (AAS) and colorimetry using titan yellow dye. Other methods that can be used are complexometric titration (as with 8-hydroxyquinoline), EDTA titration, flame-emission spectrophotometry, and fluorometry.

Magnesium methods used in the clinical laboratory have been reviewed.[40] Atomic absorption is an accurate method that is also practical because of its ease of use, sensitivity, and specificity. By aspirating a diluted sample into a flame, the magnesium present is dissociated into free magnesium atoms, which will absorb light at a characteristic wavelength (285.2 nm). The titan yellow procedure for magnesium is simple and fast but lacks accuracy and has several interferences. The Automatic Clinical Analyzer (ACA, E. I. DuPont Company) uses complexometric titration with methylthymol blue and a chelating agent to remove calcium interference in its magnesium analysis. The Kodak Ektachem uses a thin-film reflectance method based on the Mg chelator BAPTA.[34] ∎

Reference Ranges[31]
See Table 12-9.

Calcium

Calcium Physiology
In 1883, Ringer showed that calcium was essential for myocardial contraction.[25] While attempting to study how bound and free forms of calcium affected frog heart contraction, McLean and Hastings showed that the ionized calcium concentration was proportional to the amplitude of frog heart contraction, while protein-bound and citrate-bound calcium had no effect.[23] From this observation, they developed the first assay for ionized calcium using isolated frog hearts. While the method had poor precision by today's standards, the authors were able to show that blood ionized calcium was both closely regulated and had a mean concentration in humans of about 1.18 mmol/L.

TABLE 12-9. Reference Ranges for Magnesium	
Plasma, serum	0.65–1.0 mmol/L (1.6–2.4 mg/dL)

More recent studies have shown that since decreased ionized calcium impairs myocardial function, it is important to maintain ionized calcium at a near normal concentration during surgery and in critically ill patients.[8]

Decreased ionized calcium concentrations in blood can cause neuromuscular irritability, which may become clinically apparent as irregular muscle spasms, called *tetany*. Studies have shown that the rate of fall in ionized calcium initiates tetany as much as the absolute concentration of ionized calcium.[14]

Regulation

Three hormones are known to regulate serum calcium by altering their secretion rate in response to changes in ionized calcium. These hormones are parathyroid hormone (PTH), vitamin D, and calcitonin. The actions of these hormones are shown in Figure 12-3.

PTH secretion into blood is stimulated by a decrease in ionized calcium, and conversely, PTH secretion is stopped by an increase in ionized calcium. PTH exerts major effects on both bone and kidney. In the bone, PTH activates osteoclasts, which break down bone and subsequently release calcium into the ECF. In the kidneys, PTH conserves calcium by increasing tubular reabsorption of calcium ions. PTH also stimulates renal production of active vitamin D.

Vitamin D_3, a cholecalciferol, is obtained either in the diet or from exposure of skin to sunlight. Vitamin D_3 is then converted in the liver to 25-hydroxycholecalciferol (25-OH-D_3), still an inactive form of vitamin D. In the kidney 25-OH-D_3 is specifically hydroxylated to form 1,25-dihydroxycholecalciferol [$1,25$-$(OH)_2$-D_3], the biologically active form. This active form of vitamin D increases calcium absorption in the intestine and enhances the effect of PTH on bone resorption.

Calcitonin, which originates in the medullary cells of the thyroid gland, is secreted when the concentration of calcium in blood increases. Calcitonin apparently exerts its calcium-lowering effect by inhibiting the actions of both parathyroid hormone and vitamin D. Although calcitonin is apparently not secreted during normal regulation of the ionized calcium concentration in blood, it is secreted in response to a hypercalcemic stimulus. A study on hemodialysis patients indicated that a 10% increase in ionized calcium caused serum calcitonin to increase by about 35%.[10]

Distribution

Over 99% of calcium in the body is part of bone. The remaining 1% is mostly in the blood and other ECF. Very little is in the cytosol of most cells. In fact, the concentration of ionized calcium in blood is 5,000 to 10,000 times higher than in the cytosol of cardiac or smooth-muscle cells. Maintenance of this large gradient is vital to maintain the essential rapid inward flux of calcium ions.

Calcium in blood is distributed among several forms. About 45% circulates as free calcium ions, 40% is bound to protein, mostly albumin, and 15% is bound to anions such as bicarbonate, citrate, phosphate, and lactate.[36] Clearly, this distribution can change in disease. It is especially noteworthy that concentrations of citrate, bicarbonate, lactate, phosphate, and albumin can change dramatically during surgery or critical care. This is the principal reason why ionized calcium cannot be reliably calculated from total calcium measurements, especially in acutely ill individuals.

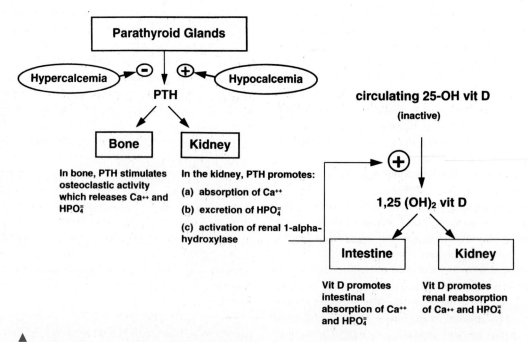

Figure 12-3. Hormonal response to hypercalcemia and hypocalcemia. PTH = parathyroid hormone; 25-OH vit D = 25 hodroxy vitamin D; $1,25(OH)_2$ vit D = dihydroxy vitamin D.

Clinical Applications

Calcium measurements appear to be most useful in the following four clinical areas: (1) in the differential diagnosis of hypercalcemia, (2) in renal diseases, (3) in monitoring of neonates, and (4) during surgery and intensive care. While both total calcium and ionized calcium measurements are available in many laboratories, ionized calcium is usually a more sensitive and specific marker for calcium disorders.

Hypercalcemic Disorders. Hyperparathyroidism, or excess secretion of parathyroid hormone, may show obvious clinical signs or may be asymptomatic. While either total or ionized calcium measurements are elevated in serious cases, ionized calcium is more frequently elevated in subtle or asymptomatic hyperparathyroidism. In general, ionized calcium measurements are elevated in 90% to 95% of cases of hyperparathyroidism, while total calcium is elevated in 80% to 85% of cases.

Hypercalcemia may be present in various types of malignancy, with hypercalcemia sometimes being the sole biochemical marker for disease. In general, while ionized calcium may be slightly more sensitive, either ionized or total calcium measurements have about equal utility in the detection of malignancy.

Renal Diseases. Patients with renal disease caused by glomerular failure often have altered concentrations of calcium, phosphate, albumin, magnesium, and hydrogen ion (pH). Very frequently in chronic renal disease, secondary hyperparathyroidism develops as the body tries to compensate for hypocalcemia caused either by hyperphosphatemia (phosphate binds and lowers ionized calcium) or altered vitamin D metabolism. Monitoring and controlling ionized calcium concentrations may avoid problems due to hypocalcemia, such as osteodystrophy, unstable cardiac output or blood pressure, or problems arising from hypercalcemia, such as renal stones and other calcifications.

Neonatal Monitoring. Typically, blood ionized calcium concentrations in neonates are high at birth and then rapidly decline by 10% to 20% after 1 to 3 days. After about a week, ionized calcium concentrations in the neonate stabilize at levels slightly higher than in adults.[37]

The concentration of ionized calcium may decrease rapidly in the early neonatal period, because the infant may lose calcium rapidly and not readily reabsorb it. Several possible etiologies have been suggested: abnormal PTH and vitamin D metabolism, hypercholesterolemia, hyperphosphatemia, and hypomagnesemia.

Surgery and Intensive Care. Because appropriate calcium concentrations promote good cardiac output and maintain adequate blood pressure, the maintenance of a normal ionized calcium concentration in blood is beneficial to the patient in either surgery or intensive care. Controlling calcium concentrations may be critical in open heart surgery when the heart is restarted and during liver transplantation because large volumes of citrated blood are given.

Because these patients may receive large amounts of citrate, bicarbonate, calcium salts, and/or fluids, the greatest discrepancies between total calcium and ionized calcium concentrations may be seen during major surgical operations. Consequently, ionized calcium measurements are the only calcium measurement of clinical value.

Hypocalcemia occurs commonly in critically ill patients, *i.e.*, those with sepsis, thermal burns, renal failure, and/or cardiopulmonary insufficiency. These patients frequently have abnormalities of acid–base regulation and losses of protein and albumin, which are best suited to monitoring calcium status by ionized calcium measurements. Normalization of ionized calcium may have beneficial effects on cardiac output and blood pressure.

Hypomagnesemia. As hypomagnesemia in hospitalized patients has become more frequent, chronic hypomagnesemia has also become recognized as a frequent cause of hypocalcemia. Hypomagnesemia may cause hypocalcemia by three mechanisms: (1) it inhibits the glandular secretion of PTH across the parathyroid gland membrane, (2) it impairs PTH action at its receptor site on bone, and (3) it causes vitamin D resistance.[32]

Pancreatitis. About half of the patients with acute pancreatitis develop hypocalcemia. The most consistent cause appears to be diminished secretion of PTH, which may either be low or inappropriately normal for the degree of hypocalcemia.[24]

Determination

Specimen. The preferred specimen for total calcium determinations is either serum or lithium heparin plasma collected without venous stasis. Because anticoagulants such as EDTA or oxalate bind calcium very tightly and interfere with its measurement, they are unacceptable for use.

The proper collection of samples for ionized calcium measurements requires greater care. Since loss of CO_2 will increase pH, samples must be collected anaerobically. While heparinized whole blood is the preferred sample, serum from sealed evacuated blood-collection tubes may be used if clotting and centrifugation are done quickly (<30 minutes) and at room temperature. No liquid heparin products should be used. Most heparin anticoagulants (sodium, lithium) partially bind to calcium and lower ionized calcium concentrations. A heparin concentration of 25 IU/mL, for example, decreases ionized calcium by about 3%. Dry heparin products are available either titrated with small amounts of Ca or Zn ions, or have small amounts of heparin dispersed in an inert "puff" that essentially eliminates the interference by heparin.

For analysis of calcium in urine, an accurately timed urine collection is preferred. The urine should be acidified with 6 mol/L HCl, with approximately 1 mL of the acid added for each 100 mL of urine.

Methods. While atomic absorption remains the reference method for total calcium,[4] accurate measurements require careful maintenance, calibration, and sample preparation. In routine testing of serum, many automated analyzers give results that are comparable with and probably somewhat more reliable than atomic absorption results. However, for analysis of total calcium in urine and other fluids, atomic absorption is clearly preferable.

Atomic absorption involves the introduction of diluted sample into an air-acetylene flame, where calcium ions are atomized. Since these atoms absorb light of exact wavelengths, corresponding to the absorption of discrete quanta of energy, the 422.7-nm wavelength is monitored.

Many automated methods for total calcium are based on the complexometric reaction between calcium and ortho-cresolphthalein complexone, often with 8-hydroxyquinoline added to prevent magnesium interference. This methodology is used on the DuPont ACA, Beckman Astra 8, and Technicon SMAC. Some automated analyzers, including the Kodak Ektachem and Beckman CX3, use the dye arsenazo III to determine calcium.

All current commercial analyzers that measure ionized calcium use ion-selective electrodes for this measurement. These systems utilize membranes impregnated with special molecules that selectively, but reversibly, bind calcium ions. As calcium ions bind to these membranes, an electric potential develops across the membrane that is proportional to the ionized calcium concentration. A diagram of one such electrode is shown in Figure 12-4.

DETERMINATION OF TOTAL SERUM CALCIUM BY ATOMIC ABSORPTION SPECTROPHOTOMETRY

(Refer to the serum magnesium procedure earlier in this chapter for a discussion of the principle and comments concerning this procedure.)

Reagents

1. Lanthanum chloride stock reagent—refer to magnesium procedure.
2. Working lanthanum diluent—refer to magnesium procedure.
3. Stock calcium standard (100 mg/dL)—To a 1-L volumetric flask containing 500 mL of deionized water, add 2.497 g of calcium carbonate (NBS 915) and 5 mL of concentrated hydrochloric acid. Dilute to volume with deionized water.
4. Working calcium standards—Three concentrations of working standards are used. Add approximately 50 mL of deionized water to each of three 100-mL volumetric flasks. Pipet the appropriate amount of stock standard, as shown in Table 12-4, for each concentration of working standard and dilute to volume with deionized water.

Procedure

After performing preliminary maintenance, turn on the lamp and ignite the flame, allowing a warm-up period of approximately 15 minutes. Prepare 1:50 dilutions of standards, controls, and samples, using 0.1 mL of specimen and 4.9 mL of working lanthanum diluent. Prepare specimens in duplicate and cover all dilutions to prevent evaporation.

Set the wavelength to 422.7 nm. Aspirate analyzer concentration to zero. Aspirate the no. 2 standard (10 mg/dL), and set the analyzer readout to a concentration of 10 mg/dL. To ensure linearity, aspirate the 5 mg/dL and 15 mg/dL standards. Their values should read 0.2 of their assigned concentrations. Run patients and controls in duplicate, aspirating deionized water between each sample. All duplicates should agree within 0.2 mg/dL.

For urine analyses, dilute samples, standards, and controls using the urine diluent described in the magnesium procedure. Dilute all specimens 1:50 with the urine diluent and analyze following serum procedure for calcium.

Calculations—For 24-h urine collections, the concentration can be repeated as follows:

$$Ca^{2+} \text{ mg/24-h} = (\text{mg/dL of } Ca^{2+} \times \text{total volume of urine in milliliters}) \times 100$$

(Eq. 12-5)

Comments—This procedure is linear to at least 15 mg/dL. ∎

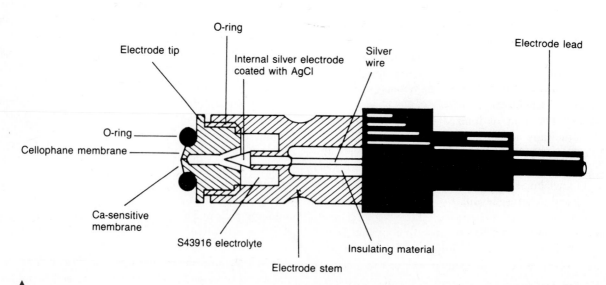

Figure 12-4. Diagram of ionized calcium electrode for the ICA ionized calcium analyzer. (Courtesy of Radiometer America, Westlake, OH)

Reference Ranges

For total calcium, the reference range varies slightly with age. In general, calcium concentrations are higher up through adolescence, while bone growth is most active. Ionized calcium concentrations can change rapidly during the first 1 to 3 days of life. Following this, they stabilize at relatively high levels with a gradual decline through adolescence; see Table 12-10.

Phosphate

Phosphate Physiology

Compounds of phosphate are everywhere in living cells and participate in many of the most important biochemical processes.[39] The genetic materials DNA and RNA are complex phosphodiesters. Most coenzymes are esters of phosphoric or pyrophosphoric acid. The most important reservoirs of biochemical energy are adenosine triphosphate (ATP), creatine phosphate, and phosphoenol pyruvate. Phosphate deficiency can lead to ATP depletion, which is ultimately responsible for many of the clinical symptoms observed in hypophosphatemia.

Alterations in the concentration of 2,3-disphosphoglycerate (2,3-DPG) in the red blood cells affect the affinity of hemoglobin for oxygen, with an increase facilitating the release of oxygen in the tissues and a decrease making oxygen bound to hemoglobin less available.[7] By affecting the formation of 2,3-DPG, the concentration of inorganic phosphate indirectly affects the release of oxygen from hemoglobin.

Understanding the cause of an altered phosphate concentration in the blood is often difficult because transcellular shifts of phosphate are a major cause of hypophosphatemia in blood. That is, an increased shift of phosphate into cells can deplete phosphate in the blood. Once phosphate is taken up by the cell, it remains there to be used in the synthesis of phosphorylated compounds. As these phosphate compounds are metabolized, inorganic phosphate slowly leaks out of the cell into the blood, where it is regulated principally by the kidney.

Regulation

Phosphate in blood may be absorbed in the intestine from dietary sources, released from cells into blood, and lost from bone. In healthy individuals, all these processes are relatively constant and easily regulated by renal excretion or reabsorption of phosphate.

While disturbances to any of these processes can alter phosphate concentrations in the blood, the loss of regulation by the kidneys will have the most profound effect. While other factors such as vitamin D, calcitonin, growth hormone, and acid–base status can affect renal regulation of phosphate, the most important factor is parathyroid hormone (PTH), which overall lowers blood concentrations by increasing renal excretion.

Vitamin D acts to increase phosphate in the blood. Vitamin D increases both phosphate absorption in the intestine and phosphate reabsorption in the kidney.

Growth hormone, which helps regulate skeletal growth, can affect circulating concentrations of phosphate. In cases of excessive secretion or administration of growth hormone, phosphate concentrations in the blood may increase because of decreased renal excretion of phosphate.

Distribution

While the concentration of all phosphate compounds in blood is about 12 mg/dL (3.9 mmol/L), most of that is organic phosphate and only about 3 to 4 mg/dL is inorganic phosphate. Phosphate is the predominant intracellular anion, with intracellular concentrations varying depending on the type of cell. About 80% of the total body pool of phosphate is contained in bone.

Clinical Applications

Both hyperphosphatemia and hypophosphatemia are seen frequently in clinical medicine. Both conditions are important to recognize and treat appropriately.

While hyperphosphatemia may be caused by either an increased intake of phosphate or increased release of cellular phosphate, hyperphosphatemia is most commonly caused by a decrease in renal phosphate excretion due to renal failure. Because they may not yet have developed mature PTH and vitamin D metabolism, neonates are especially susceptible to hyperphosphatemia caused by increased intake, such as from cow's milk or laxatives. Increased breakdown of cells can sometimes lead to hyperphosphatemia. This may be caused by severe infections, intensive exercise, or neoplastic diseases. Since immature lymphoblasts have about four times the phosphate content of mature lymphocytes, patients with lymphoblastic leukemia are especially susceptible to hyperphosphatemia.

Hypophosphatemia is very common in hospitalized patients. While most such cases are moderate and seldom cause problems, severe hypophosphatemia (< 1.0 g/dL or 0.3 mmol/L) requires monitoring and possible replacement therapy. A recent study[20] found that the most frequent causes of severe hypophosphatemia in hospitalized patients were as follows:

1. Infusion of dextrose solution. The flow of glucose into cells is often accompanied by an influx of phosphate. This cellular uptake of phosphate can be substantially enhanced by either insulin administration or respiratory alkalosis.

TABLE 12-10. Reference Changes for Calcium		
Total Calcium (serum, plasma)		
Child:	2.15–2.65 mmol/L	(8.6–10.6 mg/dL)
Adult:	2.18–2.55 mmol/L	(8.7–10.2 mg/dL)
Ionized Calcium (whole blood)		
Neonate:	1.20–1.48 mmol/L	(4.8–5.9 mg/dL)
Child:	1.20–1.38 mmol/L	(4.8–5.5 mg/dL)
Adult:	1.16–1.32 mmol/L	(4.6–5.3 mg/dL)
Urine reference ranges vary with diet. Patients on an average diet should excrete 50–300 mg/day.		

2. Nutritional recovery syndrome. This syndrome occurs as people who are severely malnourished are re-fed. Severe hypophosphatemia is often seen, and death sometimes occurs.

3. Use of antacids that bind phosphate. The divalent cations in antacids, such as aluminum hydroxide, magnesium hydroxide, or aluminum carbonate, will bind phosphate and prevent its absorption in the gut. Hypophosphatemia usually requires chronic excessive use of antacids.

4. Alcohol withdrawal. Since alcohol itself has little effect on phosphate excretion, the cause of hypophosphatemia in this condition is complex. Several factors may partially contribute. These are: (a) poor diet, (b) use of antacids, (c) vomiting, (d) alcohol-induced magnesium deficiency, and (e) ketoacidosis.

Determination of Inorganic Phosphorus

Specimen. Circulating phosphate levels are subject to circadian rhythm, with highest levels in late morning and lowest in the evening. Serum or lithium heparin plasma is acceptable for analysis. Hemolysis should be avoided because of the higher concentrations inside the red cells. Urine analysis for phosphate requires a 24-hour sample collection because of significant diurnal variations.

Methods. Most of the current methods for phosphorous determination involve the photometric determination of molybdenum blue formed by the reduction of phosphomolybdenum. Most of these techniques vary in the reducing agent used.

MANUAL DETERMINATION OF INORGANIC PHOSPHORUS IN SERUM

Principle—The sample is deproteinized using an iron–trichloroacetic acid reagent. After centrifugation, the supernate is mixed with molybdic acid, forming phosphomolybdate. This is reduced to molybdenum blue by Fe^{2+}, and the absorbance is read at 660 nm.

Reagents

1. Iron–trichloroacetic acid—Add 50 g of trichloroacetic acid to a 500-mL volumetric flask containing 300 mL of deionized water. Dissolve 5 g of thiourea and 15 g of Mohr's soft (ferrous ammonium sulfate hexahydrate) in the solution and dilute to volume with water. Store in an amber bottle up to 6 to 12 months. (Discard when absorbance changes more than 20%.)
2. Ammonium molybdate (0.0355 *M*)—Slowly add 45 mL of concentrated sulfuric acid to a 500-mL volumetric flask containing 200 mL of cold deionized water. (Cool solution in an ice bath while adding acid.) Dissolve 22 g of $(NH_4)_6Mo_7O_{24} \cdot 4H_2O$ in 200 mL of deionized water. This reagent has an extremely long shelf life.
3. Phosphorus standard (5 mg/dL)—Dry 0.220 g of KH_2PO_4 for 1 h at 110°C. Transfer to a 1-L volumetric flask and dissolve in deionized water. Dilute to volume with water. Add a few drops of chloroform to preserve the solution and store it in a polyethylene container.
4. Hydrochloric acid (6*N*)—Add 10 mL of concentrated HCl to 10 mL of water.

Procedure

1. Pipet 0.2 mL of serum or plasma into a test tube and add 5 mL of iron-reagent. Mix and let stand 10 minutes; centrifuge until supernate is clear.

2. In 13 × 100-mm test tubes, prepare a blank (5 mL of iron–TCA reagent + 0.2 mL of deionized water), the standard (5 mL of iron–TCA + 0.2 mL of standard), and unknowns (carefully decant supernate from step A into tube).
3. To each tube add 0.5 mL of the molybdate reagent and mix. Let stand 20 minutes and read at 660 nm. The color is stable for 2 hours.

Calculation

$$\text{I Phos, mg/dL} = \frac{\text{abs of unknown}}{\text{abs of standard}} \times 5$$

(Eq. 12-6)

Comments

1. This procedure is linear to at least 200 mg/dL.
2. This procedure can be used for urine analyses by adding the following steps:

   Acidify alkaline urines to pH 6 using concentrated HCl. Dilute samples 1:10 with deionized water and mix.

   Pipet 0.2 mL of the diluted urine to a test tube and add 5 mL of the iron–TCA reagent and 0.5 mL of the molybdate reagent. Mix well. (Urine specimens containing protein may develop turbidity when the iron–TCA reagent is added. Allow the protein to settle 15 minutes before centrifuging and decant supernate into a clean 13 × 100-mm test tube.)

   Set up blank and standard tubes as in step 2 of serum procedure.
3. Calculation:

$$\text{I Phos g/24 h} = \frac{\text{abs unk}}{\text{abs std}} \times \frac{0.01}{1000} \times \frac{10}{0.2}$$

$$\times \text{ urine volume in mL}$$

(Eq. 12-7) ∎

Reference Ranges

Phosphate values vary with the age of the subject. Divided into age groups, the ranges are shown in Table 12-11.

Lactate

Lactate Biochemistry and Physiology

Lactate is a by-product of an emergency mechanism that produces a small amount of ATP when oxygen delivery is severely diminished.[33] Pyruvate is the normal endproduct of glucose metabolism (glycolysis). The conversion of pyruvate to lactate is activated when a deficiency of oxygen leads to an accumulation of excess NADH (Fig. 12-5). Normally, sufficient oxygen maintains a favorably high ratio of NAD to NADH. Under these conditions, pyruvate is converted

TABLE 12-11. Reference Ranges for Phosphate		
Serum, Plasma		
Neonate:	1.13–2.78 mmol/L	3.5–8.6 mg/dL
Child:	1.45–1.78 mmol/L	4.5–5.5 mg/dL
Adult:	0.87–1.45 mmol/L	2.7–4.5 mg/dL
Urine (average, nonrestricted diet)		
Urine, 24 h	13–42 mmol/day	0.4–1.3 g/day

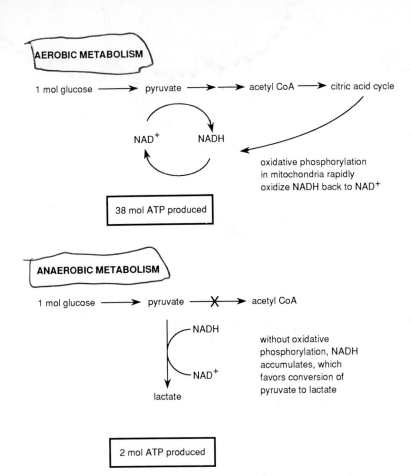

AEROBIC METABOLISM

1 mol glucose ⟶ pyruvate ⟶ ⟶ acetyl CoA ⟶ citric acid cycle

NAD⁺   NADH

oxidative phosphorylation
in mitochondria rapidly
oxidize NADH back to NAD⁺

38 mol ATP produced

ANAEROBIC METABOLISM

1 mol glucose ⟶ pyruvate ⟶ X ⟶ acetyl CoA

NADH

without oxidative
phosphorylation, NADH
accumulates, which
favors conversion of
pyruvate to lactate

NAD⁺

lactate

2 mol ATP produced

Figure 12-5. Aerobic vs. anaerobic metabolism of glucose.

to acetyl coenzyme A (CoA), which enters the citric acid cycle and produces 38 moles of ATP for each mole of glucose oxidized. However, under hypoxic conditions NADH accumulates, inhibiting the formation of acetyl CoA and favoring the conversion of pyruvate to lactate through anaerobic metabolism. As a result, only 2 moles ATP are produced for each mole of glucose metabolized to lactate, with the excess lactate released into the blood. This release of lactate into blood has clinical importance because the accumulation of excess lactate in blood is an early, sensitive, and quantitative indicator of the severity of oxygen deprivation (Fig. 12-6).

Regulation

Since lactate is a by-product of anaerobic metabolism, lactate is not specifically regulated, as is potassium or calcium, for example. As oxygen delivery decreases below a critical level, blood lactate concentrations rise very rapidly and indicate tissue hypoxia earlier than does pH. The liver is the major organ for removing lactate by converting lactate back to glucose by a process called gluconeogenesis.

Clinical Applications

Studies performed since the early 1970s have shown that measurements of blood lactate are useful for metabolic monitoring in critically ill patients, for indicating the severity of the illness, and for objectively determining the prognosis of these patients.[33] In general, the following are indicators that a greater level of therapeutic intervention will be needed in a patient:

1. A very high lactate (over 5–6 mmol/L) following surgery.
2. A lactate that remains elevated despite medical treatment.
3. A definite rise in blood lactate during treatment.

The above conditions indicate that the patient is experiencing a deficit of oxygen delivery or oxygen consumption in the tissues. Therapeutic procedures that may be utilized in these conditions include: vasodilators, cardiac stimulants, fluid administration (volume support), ventilation, and transfusion of red cells (for anemia).

Lactate concentrations also have been used to monitor administration of nitroprusside, a fast-acting, potent vasodilator that is sometimes used following surgery to increase peripheral blood flow. Since cyanide is a metabolite of nitroprusside, excess nitroprusside can block oxidative metabolism, resulting in the production of lactate. Therefore, blood lactate measurements sensitively indicate when excess nitroprusside is causing cyanide toxicity.

Determination of Lactate

Methods. Although lactate may sensitively indicate inadequate tissue oxygenation, the use of blood lactate measurements has been hindered because older methods were slow and laborious and required relatively large volumes of blood. Thus, other means of following perfusion or oxygenation have been used, such as indwelling catheters that measure blood flow, pulse oximeters, base-excess determinations, and measurements of oxygen consumption (VO₂). However, as more suitable methods for lactate become available, lactate may become routinely used in critical care.

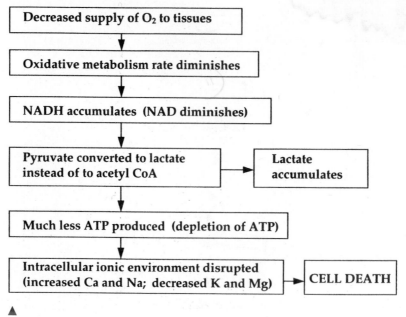

Figure 12-6. Metabolic effects of hypoxia leading to cell death.

A review of methods for lactate shows a progression from those using chemical oxidation to those using specific enzyme reactions or enzyme electrodes. The early methods, based on chemical oxidation, used permanganate along with either H_2SO_4 or oxygen to produce acetaldehyde and carbon monoxide or carbon dioxide. These methods were fairly accurate but also were laborious, time-consuming, and not applicable to *STAT* determinations.

Enzymatic methods use lactate dehydrogenase with the cofactor NAD^+ to convert lactate to pyruvate, with the formation of NADH proportional to the concentration of lactate. This method is used in the most widely available procedure for lactate—that done on the DuPont ACA. Because it is readily available as a *STAT* procedure all day and requires only about 40 to 50 μL of serum or plasma, this enzymatic method has increased awareness of the potential uses of lactate measurements.

A further development in the evolution of lactate methods has been the development of a small lactate analyzer by Yellow Springs Instruments. This analyzer uses the enzyme lactate oxidase bound to a membrane to catalyze the reaction

$$\text{Lactate} + O_2 \xrightarrow{\text{Lactate oxidase}} \text{pyruvate} + H_2O_2$$

(Eq. 12-8)

The H_2O_2 is oxidized back to O_2 by a platinum anode. Because the analyzer requires only 30 μL whole blood, the *STAT* and micro-capabilities have been further enhanced.

Specimen Handling. Special care should be practiced when handling specimens for lactate analysis to prevent *in vitro* accumulation of lactate as glucose is anaerobically metabolized. Heparinized blood may be used if it is both kept on ice and the plasma is quickly separated. Iodoacetate, which inhibits glycolysis without affecting coagulation, is another satisfactory additive.

Reference Ranges
See Table 12-12.

ANION GAP

Routine measurement of electrolytes usually involves only Na^+, K^+, Cl^-, and HCO_3^- (as total CO_2). These values may be used to approximate the *anion gap (AG)*, which is the difference between unmeasured anions and unmeasured cations. Note that there is never a "gap" between total cationic charges and anionic charges. The AG is created by the concentration difference between commonly measured cations (Na + K) and commonly measured anions (Cl + HCO_3), as shown in Figure 12-7. AG is useful in indicating an increase in one or more of the unmeasured anions in the serum. It is also helpful as a form of quality control for the analyzer used to measure these electrolytes. Consistently abnormal anion gaps in serum from healthy persons may indicate an instrument problem.[22]

There are two commonly used methods for calculating the anion gap. The first equation is

$$AG = Na^+ - (Cl^- + HCO_3^-)$$ *(Eq. 12-9)*

It is equivalent to unmeasured anions minus the unmeasured cations in this way:

TABLE 12-12. Reference Ranges for Lactate[38]

Venous blood:	0.5–2.2 mmol/L
Arterial blood:	0.5–1.6 mmol/L
(alert value 5 mmol/L or higher)	
CSF:	0.6–2.2 mmol/L

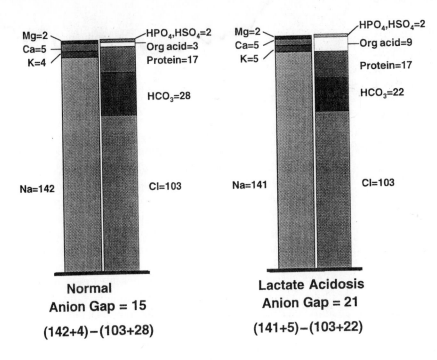

Figure 12-7. Demonstration of anion gap from concentrations of anions and cations in normal state and in lactate acidosis.

$$AG = (\text{protein} + \text{organic acids*} + PO_4^- + 2SO_4^{2-}) - (K^+ + 2Ca^{2+} + 2Mg^{2+}) \qquad \textit{(Eq. 12-10)}$$

The reference range for the AG using this calculation is 7 to 14 mmol/L.[31] The second calculation method is

$$AG = (Na^+ + K^+) - (Cl^- + HCO_3^-)$$

$$\textit{(Eq. 12-11)}$$

It has a reference range of 10 to 18 mmol/L.[31]

Low anion gap values are rare but may be seen in multiple myeloma (due to abnormal proteins) or in cases of instrument error. An elevated AG may be caused by uremia, which leads to PO_4^- and SO_4^{2-} retention; ketoacidosis, as seen in cases of starvation or diabetes; poisoning due to ingestion of methanol, ethylene glycol, salicylate, or paraldehyde; lactic acidosis; and severe dehydration, which causes increased plasma proteins.

ELECTROLYTES AND RENAL FUNCTION

The kidney is central to the regulation and conservation of electrolytes in the body. For a review of kidney structure, refer to Figure 12-8 and Chapter 17, Renal Function. The following is a summary of electrolyte excretion and conservation in a healthy individual:

1. Glomerulus—This portion of the nephron acts as a filter, retaining large proteins and protein-bound constituents, while most of the other plasma constituents pass into the filtrate. Their concentrations in the filtered plasma should be approximately equal to the ECF without protein.
2. Renal tubules
   a. Phosphate reabsorption is inhibited by PTH and increased by 1,25-dihydroxycholecalciferol. Excretion of PO_4 is stimulated by calcitonin.

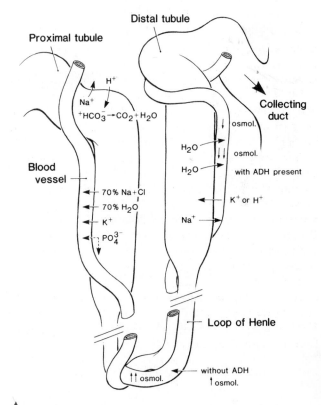

Figure 12-8. Summary of electrolyte movements in the renal tubules.

   b. Calcium is reabsorbed under the influence of PTH and 1,25-dihydroxycholecalciferol. Calcitonin stimulates excretion of calcium.
   c. Magnesium reabsorption occurs largely in the thick ascending limb of the loop of Henle.
   d. Sodium reabsorption can occur through three mechanisms:

(1) Approximately 70% of the sodium in the filtrate is re-absorbed in the proximal tubules by iso-osmotic re-absorption. It is limited, however, by the availability of chloride in order to maintain electrical neutrality.

(2) Sodium is reabsorbed in exchange for H^+. This reaction is linked with HCO_3^- and is dependent on carbonic anhydrase.

(3) Stimulated by aldosterone, Na^+ is reabsorbed in exchange for K^+ in the distal tubules. (H^+ competes with K^+ for this exchange.)

e. Chloride is reabsorbed, in part, by passive transport in the proximal tubule along the concentration gradient created by Na^+.

f. Potassium is reabsorbed by two mechanisms:

(1) Active reabsorption in the proximal tubule almost completely conserves K^+.

(2) Exchange with Na^+ is stimulated by aldosterone. H^+ competes with K^+ for this exchange.

g. Bicarbonate is recovered from the glomerular filtrate and is converted to CO_2 when H^+ is excreted in the urine.

3. Loop of Henle—With normal ADH function, it creates an osmotic gradient that enables water reabsorption to be increased or decreased in response to body fluid changes in osmolality.

4. Collecting ducts—Also under ADH influence, final adjustment of water excretion is made here.

SUMMARY

Electrolytes are ions capable of carrying an electric charge. They are classified as anions or cations based on the type of charge they carry. Anions have a negative charge and move toward the anode; cations have a positive charge and migrate toward the cathode. Electrolytes are an essential component in numerous processes in the body including: volume and osmotic regulation, myocardial rhythm and contractility, enzyme activation, regulation of ATPase ion pumps, acid-base balance, blood coagulation, neuromuscular excitability, and the production and utilization of ATP from glucose. Electrolytes discussed in this chapter include: sodium, potassium, chloride, bicarbonate, magnesium, calcium, phosphate, and lactate. This chapter also discussed the metabolic physiology and regulation of each electrolyte, as well as the commonly used methods of assessment. Regulation of electrolyte concentrations in rather narrow ranges is primarily accomplished by the kidneys.

CASE STUDY 12-1

A 15-year-old girl in a coma is brought to the emergency room by her parents. She has been an insulin-requiring diabetic for 7 years. Her parents state that there have been several episodes of hypoglycemia and ketoacidosis in the past, and that she has often been "too busy" to take her insulin injections. The laboratory results obtained upon admission are shown in Case Study Table 12-1.

CASE STUDY TABLE 12-1. Laboratory Results

	Result	Reference Range
VENOUS BLOOD		
Sodium	148 mmol/L	135–145 mmol/L
Potassium	5.8 mmol/L	3.4–5.0 mmol/L
Urea nitrogen	35 mg/dL	7–18 mg/dL
Creatinine	1.3 mg/dL	0.5–1.3 mg/dL
Chloride	87 mmol/L	98–106 mmol/L
Bicarbonate	8 mmol/L	22–29 mmol/L
Lactate	5 mmol/L	0.5–2.2 mmol/L
Osmolality	385 m Osmol/kg	275–295 m Osmol/kg
Glucose	1050 mg/dL	70–120 mg/dL
ARTERIAL BLOOD		
pH	7.11	7.35–7.45
PO_2	95 mmHg	83–100 mmHg
PCO_2	20 mmHg	35–45 mmHg
URINE		NORMAL
Ketones	4+	Negative
Glucose	4+	Negative

Questions

1. What is the diagnosis?
2. Calculate the anion gap. What is the cause of an increased anion gap in this patient?
3. What is the significance of the plasma osmolality?
4. Why are the chloride and bicarbonate decreased? What is the significance of the "normal" sodium and elevated potassium values?

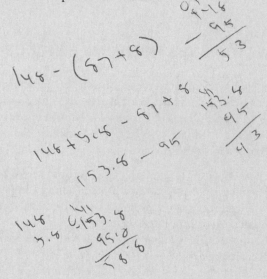

CASE STUDY 12-2

An 80-year-old female with partial senility was admitted to the hospital following a viral illness with several days of diarrhea that led to an increased state of confusion. Her laboratory results were as follows:

plasma Na: 165 mmol/L
urine Na: 15 mmol/L
urine Osmolality: 550 mOsm/kg

Questions

1. What are the factors that have contributed to the hypernatremia?
2. What is most likely the major factor?
3. How can the urine Na be low while the urine osmolality is in the upper normal range?

CASE STUDY[1] 12-3

A 36-year-old woman takes diuretics for mild hypertension. The following laboratory results are noted on a routine office visit:

plasma Na: 136 mmol/L
plasma K: 3.0 mmol/L
plasma Cl: 98 mmol/L
plasma HCO$_3^-$:   29 mmol/L

pH: 7.47
plasma aldosterone: 500 ng/L (high)
urine K: 45 mmol/L (high)

Question

What are the factors that may contribute to the hypokalemia?

CASE STUDY 12-4

Consider the following sets of laboratory results.

Case	Ion Ca (1.16–1.32)	Total Mg (0.7–0.9)	PO4 (0.7–1.4)	Hct (35–45%)	Intact PTH (13–64 ng/L)
A	1.44	0.90	0.7	42	100
B	1.08	0.50	0.8	40	25
C	1.70	0.90	1.4	30	12

Reference ranges shown in parentheses; units are mmol/L unless otherwise indicated.

Questions

1. Which set of laboratory results (A, B, or C) is most likely associated with each of the following diagnoses?
   - Primary hyperparathyroidism
   - Malignancy
   - Hypomagnesemic hypocalcemia

REFERENCES

1. American Diabetes Association. Magnesium supplementation in the treatment of diabetes. Diabetes Care 1992;15:1065.
2. Aresenian M. Magnesium and cardiovascular disease. Prog Cardiovasc Dis 1993;35:271.
3. Beaty RD. Concepts, instrumentation, and techniques in atomic absorption spectrophotometry. Norwalk, CT: Perkin-Elmer Corp, 1978.
4. Cali JP, Bowers GN Jr, Young DS. A referee method for the determination of total calcium in serum. Clin Chem 1973; 19:1208.
5. Chernow B, ed. The pharmacologic approach to the critically ill patient. 2nd ed. Baltimore: Williams and Wilkins, 1988.
6. Chernow B, Bamberger S, et al. Hypomagnesemia in patients in postoperative intensive care. Chest 1989;95:391.
7. Ditzel J. Effect of plasma inorganic phosphate on tissue oxygenation during recovery from diabetic ketoacidosis. Adv Exp Med Biol 1973;37A:163.
8. Drop LJ. Ionized calcium, the heart, and hemodynamic function. Anaesth Analg 1985;64:432.
9. Dufour, DR. Osmometry: The rational basis for use of an underappreciated diagnostic tool. Workshop at AACC National Meeting, 1993.

10. Felsenfeld AJ, Machado L, Rodriquez M. The relationship between serum calcitonin and calcium in the hemodialysis patient. Am J Kidney Dis 1993;21:293.

11. Fleisher M, Gladstone M, Crystal D, Schwartz MK. Two whole-blood multianalyte analyzers evaluated. Clin Chem 1989;35:1532.

12. Gambino SR, Schreiber H. The measurement of CO_2 content with the autoanalyzer. Am J Clin Pathol 1966;45:406.

13. Gambling DR, Birmingham CL, Jenkins LC. Magnesium and the anaesthetist. Can J Anaesth 1988;35:644.

14. Gray TA, Paterson CR. The clinical value of ionized calcium assays. Ann Clin Biochem 1988;25:210.

15. Gums JG. Clinical significance of magnesium: A review. Drug Intell Clin Pharmacol 1987;21:240.

16. Hanline M. Hypomagnesemia causes coronary artery spasm. JAMA 1985;253:342.

17. Iseri LT, French JH. Magnesium: Nature's physiologic calcium blocker. [Editorial]. Am Heart J 1984;108:188.

18. June CH, et al. Profound hypomagnesemia and renal magnesium wasting associated with the use of cyclosporine for marrow transplantation. Transplantation 1985;39:620.

19. Karselis T. Electrolyte instrumentation then and now. Am J Med Technol 1982;48:329.

20. King AL, Sica DA, Miller G, Pierpaoli S. Severe hypophosphatemia in a general hospital population. South Med J 1987; 80:831.

21. Kodak Ektachem 400 Operator's Manual. Rochester, NY: Eastman Kodak Co, 1993.

22. Lolekha PH, Lolekha S. Value of the anion gap in clinical diagnosis and laboratory evaluation. Clin Chem 1983;29:279.

23. McLean FC, Hastings AB. A biological method for estimation of calcium ion concentration. J Biol Chem 1934;107:337.

24. Maxwell MH, Kleeman CR, Narins RG. Clinical disorders of fluid and electrolyte metabolism. 4th ed. New York: McGraw-Hill, 1987;774.

25. Ringer S. A further contribution regarding the influence of different constituents of blood on contractions of the heart. J Physiol 1883;4:29.

26. Rose BD, ed. Clinical Physiology of acid-base and electrolyte disorders 3rd ed. New York: McGraw-Hill, 1989.

27. Shechter M, Kaplinsky E, Rabinowitz B. The rationale of magnesium supplementation in acute myocardial infarction. Arch Intern Med 1992;152:2189.

28. Sheehan JP, Seelig MS. Interactions of Mg and K in the pathogenesis of cardiovascular disease. Magnesium 1984;3:301.

29. Taylor WM. Low serum magnesium concentration in Paget's disease of bone (osteitis deformans). Ann Clin Biochem 1985; 22:591.

30. Tietz NW. Textbook of clinical chemistry. Philadelphia: Saunders, 1986;1178.

31. Tietz NW, ed. Clinical guide to laboratory tests. Philadelphia: Saunders, 1983.

32. Toffaletti, J. Self study course: Understanding the clinical uses of blood gases and electrolytes. Washington DC: AACC Press, 1993.

33. Toffaletti JG. Blood lactate: Biochemistry, laboratory methods, and clinical interpretation. Crit Rev Clin Lab Science 1991;28:253.

34. Toffaletti J, Abrams B, Bird C, Schwing M. Clinical validation of an automated thin-film reflectance method for measurement of Mg in serum and urine. Magnesium 1988;7:84.

35. Toffaletti J, Cooper D, Lobaugh B. The secretion of parathyroid hormone in response to changes in either ionized calcium, protein-bound calcium, or magnesium in normal adults. Metabolism 1991;40:814.

36. Toffaletti J, Gitelman HJ, Savory J. Separation and quantitation of serum constituents associated with calcium by gel filtration. Clin Chem 1976;22:1968.

37. Wandrup J. Critical analytical and clinical aspects of ionized calcium in neonates. Clin Chem 1989;35:2027.

38. Weisberg HF. Osmolality-calculated, "delta," and more formulas. Clin Chem 1975;21:1182.

39. Westheimer FM. Why nature chose phosphates. Science 1987;235:1173.

40. Wills MR, Sunderman FW, Savory J. Methods for estimation of serum magnesium in clinical laboratories. Magnesium 1986;5:317.

Chapter 13

Trace Elements in Clinical Chemistry

Kenneth R. Copeland

Objectives

Upon completion of this chapter, the clinical laboratorian should be able to:

- *Define the following terms: trace element, ultratrace element, and essential element.*

- *Discuss the clinical significance of essential and non-essential elements.*

- *Discuss the issues related to the laboratory's role in trace and ultratrace element assessment.*

- *Explain the body's mechanism for maintaining iron homeostasis.*

- *Relate the tests that can be used to evaluate iron status.*

- *Correlate disease state or patient status with a trace element excess or deficit.*

- *Compare and contrast the daily dietary requirements, biochemical functions, deficiency characteristics, and toxicity characteristics of the major trace elements discussed in this chapter.*

- *Relate serum iron, transferrin, %saturation, and ferritin values, in various disease states.*

KEY WORDS

Ceruloplasmin
Essential Element
Ferritin
Total Iron Binding
   Capacity (TIBC)

Trace Element
Transferrin Saturation
Ultratrace Element

Trace elements (trace metals) have an important role in both normal biological function and toxicity in humans. The control of the homeostasis of trace elements is under rigid control mechanisms in order to optimize their physiological functions. Any alteration due to a deficit or an excess of the element often produces significant biochemical alterations. In this chapter, the importance of trace metals as nutritional factors and the consequences associated with altered levels resulting from altered dietary intake, several hereditary factors, and environmental exposure will be examined.

A *trace element* is an element that occurs in biological systems at concentrations of mg/kg amounts, or less (parts per million). Typically, the daily requirement of such an element is a few milligrams per day. An *ultratrace element* is present in tissues at concentrations of µg/kg amounts, or less (parts per billion), and has extremely low daily requirements (usually less than 1 mg).[52,54] Elements can also be defined as being essential or non-essential. An *essential element* is an element that is absolutely necessary for life. Deficiency or absence of the element will cause a severe alteration of function and/or eventually lead to death. Three criteria must be met before an element is considered to be essential.[44,52,54,62] First, a predictable deficient state that produces impaired function must be well defined. Second, the disease state must be ameliorated by the administration of the element. And third, there must be a recognized biochemical basis for the element's essential function. Because of these rigid criteria, only a few trace or ultratrace elements have been deemed essential to date; however, others may be deemed essential in the future. A summary of the elements commonly encountered in clinical chemistry is presented in Table 13-1.

Michael L. Bishop, Janet L. Duben-Engelkirk, and Edward P. Fody.
CLINICAL CHEMISTRY. © 1996 Lippincott–Raven Publishers.

TABLE 13-1. Essential Trace Elements in Clinical Chemistry

	Proven Essential	Probably Essential	Non-Essential To Date
Trace:	Iron		
	Zinc		
	Copper		
Ultratrace:	Manganese	Nickel	Aluminum
	Cobalt	Vanadium	Arsenic
	Selenium	Tin	Cadmium
	Molybdenum		Fluoride
	Chromium		Gold
	Iodine		Lead
			Mercury
			Silicon

Trace elements have many diverse and crucial biochemical functions.[16,44,54,62] Commonly, they are involved directly in electron transport, as a cofactor for a wide variety of enzymes, or as a vital component of several metalloproteins. Trace elements act as cofactors for many diverse enzymes involved in a wide variety of biochemical functions including DNA synthesis, protein synthesis, bone and collagen production, and several cellular metabolism pathways. The biochemical functions and daily requirements of several essential trace elements are summarized in Table 13-2.

The clinical interest in the determination of trace metals for diagnosis and prognosis of different diseases is widespread.[3,15,16,24,33,34,36,40,44,49,52,54,62,64] Trace element deficiencies can occur as a result of many genetically determined enzyme defects. These defects result in altered absorption, transport, storage, or excretion of the particular element. More commonly, a trace element deficiency can be the result of nutritional deficiency, insufficient intestinal absorption, or increased loss or utilization of the element. Several biochemical, physiological, and morphological alterations have been associated with trace element deficiencies, some of which are summarized in Table 13-2.

LABORATORY ISSUES

Several issues regarding the clinical laboratory's role in trace and ultratrace element assessment pose considerable challenges.[16,52,54] First, the concentrations of these elements in biological fluids are extremely low (ng/L to mg/L, or less). For this reason, complex analytical techniques utilizing a high degree of expertise are required to achieve the required sensitivity, precision, and accuracy. Second, a major concern and common problem is that of specimen contamination during sample processing and analysis. And third, the significance of the data obtained can be difficult to interpret in light of underlying clinical conditions.

The analytical methods used for the determination of trace metals in biological specimens must be sensitive, pre-

TABLE 13-2. Trace Element Function and Abnormalities

Metals	Daily Dietary Requirements	Biochemical Functions	Deficiency Characteristics	Toxicity Characteristics
Iron	10–20 mg	Oxygen transport		Hepatic failure
		Respiration	Anemia	Cardiomyopathy
		Oxidative processes		Peripheral neuropathy
Zinc	10–15 mg	Hemoglobin synthesis	Impaired wound healing	Ataxia
		Collagen metabolism	Retarded growth	Pancreatitis
		Bone development	Skeletal abnormalities	Anemia, fever
		Growth & reproduction	Male impotence	Nausea, vomiting
Copper	2–6 mg	Pigmentation	Disorders in pigmentation	Nausea, vomiting
		Bone development	Retarded growth	Liver necrosis
		Oxygen transport	Anemia in children	Hemolytic anemia
		Nucleic acid synthesis	Wilson's disease	Renal dysfunction
		Protein synthesis	Menkes kinky hair disease	Neurological dysfunction
Manganese	20 µg	Growth & reproduction	Disorders in spermatogenesis	Neurological alterations
		Oxidative phosphorylation	Bone abnormalities	Psychosis
		Cholesterol metabolism	Bleeding disorders	Speech disturbances
Cobalt	0.1 µg	Vitamin B$_{12}$ function	Megaloblastic anemia	Gastrointestinal function
		Methionine metabolism		Cardiomyopathy
Selenium	50–200 µg	Oxygen metabolism	Cardiovascular disease	Neurotoxicity
		Free radical protection	Muscle degeneration	Hepatotoxicity
			Carcinogenesis	
Molybdenum	150–500 µg	Xanthine metabolism	Mental disturbances	Hyperuricemia
			Esophageal cancer	
Chromium	5–20 µg	Insulin uptake	Reduced glucose tolerance	Renal failure
		Glucose metabolism		Pulmonary cancer

cise, accurate, and relatively fast. Several different methods have been used to determine metals in biological fluids.[1,17,20,21,23,26,29,38,39,46,48,51,52,54–56] Chemical methods were once the most common means of metal determination. However, these methods were non-specific, lacked sensitivity, and involved complex, cumbersome experimental procedures. Today, the most common method of analysis is atomic absorption spectroscopy (AAS), either with flame or electrothermal (graphite furnace) atomization. Graphite furnace AAS is generally preferred for the determination of ultratrace elements, because its detection limit is 10 to 100-fold better than conventional flame AAS. The most serious limitation of this method is due to interferences caused by the biological sample matrix; however, the use of improved methods of background correction and improved matrix modifiers have helped to minimize this problem. More recently, technologies such as neutron activation analysis, mass spectrometric methods with thermal ionization, and, especially, inductively couple plasma coupled to mass spectrometry (ICP-MS) are being used to determine ultratrace elements in biological fluids and tissues.[26,46,51,54] The high capital cost of these methods, along with the high degree of technical expertise required, make them prohibitive for most laboratories.

Several different biological matrices can be used to assess trace element status including plasma, serum, blood, urine, sweat, intracellular fluids, fingernails, and hair. The most commonly encountered in the clinical laboratory are serum/plasma, whole blood, and urine. Regardless of the specimen type, it is imperative that specimens be collected, processed, and analyzed without contamination. Given the extremely low concentration of metals in biological matrices, it is not surprising that even small amounts of exogenous metal contamination can drastically and erroneously elevate results.[43,52,54,63] The most common sources of contamination come from rubber, plastics, paper products, metal surfaces, dust, skin, hair, and water. In order to minimize contamination, all glassware should all be washed in acid and well rinsed with ultra pure water. The work station should have limited access to personnel to help maintain a clean working environment, and powder-free gloves should be worn at all times. If possible, an environmental fume hood should be used to provide a clean working environment. Finally, good laboratory practice is essential to minimize contamination from utensils and vessels used to prepare samples and to contain water and reagents.

IRON

Biochemical Significance

Iron is the most abundant of all essential elements. Iron participates in many biochemical processes including cellular oxidative mechanisms and oxygen transport.[17,54] The major heme proteins (*i.e.,* hemoglobin and myoglobin) are the most abundant iron-containing compounds in the body. Hemoglobin transports oxygen from the lungs to the tissues, where it is used in respiration, and myoglobin stores oxygen in the skeletal muscles. Iron itself is also essential for normal function of the electron transport chain and, hence, respiration. It is a major component of several cytochromes, where it acts as an electron acceptor or donor during oxidative phosphorylation. Additionally, several enzymes require iron as a cofactor including cytochrome oxidase, xanthine oxidase, peroxidase, catalase, and NADH dehydrogenase.

Normal Iron Homeostasis

Iron homeostasis is maintained under tight control to ensure that appropriate requirements for normal physiological and biochemical function are maintained (see Table 13-2). Unlike most elements, iron homeostasis is regulated primarily by absorption (by the small intestine) and not by excretion.[13,22] To meet the body's high metabolic demands for iron, various protein ligands are utilized to ensure that iron is in a readily available, soluble, non-toxic form. The major iron transport/storage proteins are transferrin, lactoferrin, and ferritin.[4,58,65] A fourth iron complex, hemosiderin, is an insoluble iron storage complex derived from ferritin.

The daily requirement for iron varies according to age, sex, weight, and state of health.[18,22,28,32] In adult males, the daily requirement is approximately 10 mg/day, which is easily obtained by a well-balanced diet (see Table 13-2). Foods high in iron include animal viscera, especially liver and kidney, egg yolks, fish, and dried legumes. Children and adult females, particularly due to increased blood loss associated with menstruation or pregnancy, may require more than 20 mg/day. Whole blood contains a large proportion of the body's iron (as hemoglobin) (Table 13-3) and even a small amount of blood loss can cause a significant amount of iron loss from the body. Because the body is able to conserve iron extremely well, normally only 6% to 12% of the dietary iron intake (1–2 mg/day) is actually absorbed to meet daily requirements. During conditions of iron deficiency the gastrointestinal tract is able to absorb a greater percentage of dietary iron (30–40%) in order to meet the increased requirements. The vast majority of ingested iron is excreted in the feces (greater than 10 mg/day). In addition smaller amounts of iron are lost in the feces by way of the GI tract (0.6 mg/day). Other sources of normal iron loss include sweat, the exfoliation of squamous cells, and the urinary tract (0.9 mg/day). As described above, iron loss due to menstruation can be significant. Since the normal iron storage capacity of the body is large, it may take an extended period of time before iron stores are depleted, unless there is a large sudden loss of blood (*i.e.,* due to trauma injuries).

Transferrin is a glycoprotein that is synthesized by the liver and which contains two iron binding sites per molecule. It has a molecular weight of approximately 90 kDa and a half-life in the body of 10 days. Transferrin is an integral component of iron homeostasis and normal iron metabo-

TABLE 13-3. Distribution of Iron in Various Body Compartments

Compartment	Iron Content (mg)	Percent
Red blood cells (Hemoglobin)	2500–2800	68–70%
Plasma (Transferrin)	3–5	0.1–1%
Plasma (Ferritin)	0.1	0.002
Myoglobin	135–150	4%
Reticulo-endothelial (spleen, liver, bone marrow) as ferritin, hemosiderin	500–1000	25–28%
Cytochromes, enzymes	8–20	1–2%
Total body iron	3000–4000	100%

lism.[13,18,32,58] Its primary function is to transport intestinally absorbed iron in the plasma to the sites of erythropoiesis and to the liver, where excess iron is stored in hepatocytes as ferritin or hemosiderin. Under normal conditions, approximately one third of serum transferrin binding sites are bound with iron. Besides carrying iron absorbed by the intestinal tract, transferrin also carries iron taken up by macrophages subsequent to its release as hemoglobin from the breakdown of old erythrocytes. Most of the iron required for hemoglobin synthesis (20–25 mg/day) is provided by iron stored in the reticulo-endothelial cells of the liver, spleen, and bone marrow. This iron, when released, is transported bound to transferrin to its required sites. The balance of iron, if required, is obtained from daily intake.

Ferritin consists of a spherical protein shell composed of 24 subunits with a molecular mass of 500 kDa. The protein can bind up to 4,000 iron molecules, making it a large potential source of iron. Not surprisingly, the major function of ferritin is to provide a store of iron that may be used for heme synthesis.[4,32,54,65] Ferritin is present in all tissues but is most abundant in liver parenchymal cells and macrophages found in the liver, spleen, and bone marrow. Plasma normally contains small quantities of ferritin that is derived by normal tissue release. In healthy individuals, serum ferritin levels are a sensitive and precise reflection of body iron stores. Serum ferritin levels are elevated when there is an excess amount of iron in the body and are decreased when iron stores are depleted.

Clinical Significance of Alterations in Iron Status

There are many situations where the determination of iron status is required to make an accurate clinical diagnosis.[8,27,32,53,54,58] The two most common conditions that warrant further discussion are iron deficiency and iron overload or toxicity.

Iron Deficiency[8,27,28,32,53]

Iron deficiency is defined as diminished body iron content. This condition develops as a result of inadequate iron intake and/or excessive iron loss. If it is not corrected, an anemic

state, characterized by impaired erythrocyte synthesis, will develop. Iron deficiency anemia is the most prevalent nutritional deficiency in humans, with an estimated 750 million people worldwide being afflicted.

The early stage of iron deficiency is manifested by a depletion of body iron stores.[32,54] Iron depletion occurs as a result of prolonged inadequate intake of iron, inadequate iron absorption, or significant iron loss (mainly due to blood loss). Iron depletion is generally not associated with any of abnormalities associated with anemia (such as altered red cell indices or morphology). The most sensitive and early indicator of iron depletion is decreased serum ferritin. This can be observed before a significant decrease in serum iron is seen. If iron loss continues and/or iron stores are not replenished, a more significant iron deficient state will soon develop. Iron deficiency is most common during periods of rapid growth (particularly in children or in pregnancy) and in menstrual-age females. Clinically, it is characterized by mild anemia, resulting in decreased serum ferritin and iron levels, mild microcytosis (small erythrocytes), a moderately low hemoglobin (< 12 g/dL), and elevated serum transferrin levels. Iron deficiency in adult males or post-menopausal females is most often due to gastrointestinal bleeding, often resulting from peptic ulcers, gastritis, or neoplasm.

Clinically significant anemia is the late-stage manifestation of iron deficiency and does not present until iron stores are essentially nonexistent. Microscopic examination of blood reveals microcytic and hypochromic erythrocytes. Laboratory alterations include decreased serum ferritin and iron levels, elevated transferrin levels, and hemoglobin concentrations less than 8 g/dL. In addition, iron stains of bone marrow and liver cells reveal virtually nonexistent storage iron.

Once iron deficiency has been diagnosed as the cause of anemia, it is important to institute treatment. Initially, corrective measures should be undertaken to prevent ongoing blood loss. This is followed by improvement in dietary habits and the administration of oral iron supplementation using ferrous sulfate.

Iron Overload[18,27,32,53,54,58]

Toxicity from excessive iron intake or excessive tissue storage of iron presents a considerable risk, given the extreme toxic nature of iron. There are several causes of iron overload (hematochromatosis). Elevations in serum iron occur in several pathological conditions including hemolytic anemia, lead toxicity, pyridoxine deficiency, pernicious anemia, acute liver necrosis, and iron poisoning. The intake of large amounts of iron, particularly by children as a result of accidental ingestion of iron-containing vitamin supplements, is a common cause of iron poisonings. In addition, several genetic causes have been identified, including hereditary hematochromatosis, that cause abnormally increased production of erythrocytes and increased storage of iron. Treatment of these individuals includes therapeutic phlebotomy in order to decrease erythrocyte volume. Acute iron overdoses are typically treated by chelation therapy using desferrioxamine.

Laboratory Determination

The evaluation of iron status is best performed using a battery of specific iron related parameters.[8,27,28,32,53] While many tests can be used, the most useful and commonly employed ones include hematocrit and other red cell indices, hemoglobin, serum iron, transferrin and ferritin, percent saturation, transferrin iron binding capacity, and, in some cases, erythrocyte protoporphyrin. The alterations in these parameters for certain disease states are summarized in Table 13-4.

Serum/Plasma Iron

One of the oldest tests to assess iron status has been the quantitation of serum or plasma iron.[19,22,30,59] Several methods, including a wide variety of colorimetric and spectrophotometric ones, have been used. Today colorimetric assays are commonly used because of their suitability to automation. Regardless of the chromogenic agent employed, the basic principle of all colorimetric assays is the same. The first step involves dissociating iron from protein (mainly transferrin) by denaturation with strong acid. Second, a reducing agent, commonly ascorbic acid, is added to convert Fe(III) to Fe(II). Third, Fe(II) is complexed with the chromogenic agent, commonly ferrozine or bathophenanthroline. Finally, the colored complex is quantitated spectrophotometrically at an appropriate wavelength.

Only serum or heparinized plasma is suitable for analysis. Other anticoagulants, such as EDTA, oxalate, and citrate, are not acceptable as they chelate iron. One of the most common causes of erroneous iron results is caused by the analysis of hemolyzed specimens. Given the high concentration of iron in erythrocytes (from hemoglobin), even minimal hemolysis will give falsely elevated results. To minimize this effect, serum or plasma should be separated from erythrocytes within 1 hour of collection, and hemolyzed (even slightly) specimens should not be analyzed.

Serum iron levels are very labile in individuals. For this reason, several exogenous factors must be taken into account to interpret results properly. The amount of iron in the serum is regulated both by its rate of absorption from the GI tract and by its exchange with ferritin in the storage sites. Thus, increased absorption, due to dietary modification and/or increased release from tissues, can lead to transient elevations in serum iron levels. In addition, serum iron shows marked diurnal variation, with morning values approximately 30% higher than values later in the day. Finally, the use of iron-containing medications can lead to elevated serum iron levels, which may not mimic tissue levels.

The reference range for serum iron in adult males is 65 to 170 μg/dL. Females and children tend to have slightly lower values (Table 13-5). There are many disease states characterized by decreased serum iron. Most commonly it is seen in association with a generalized iron deficiency due to lack of dietary intake, inadequate absorption, or chronic loss as a result of bleeding or renal disease. Another significant cause is impaired release of iron from its storage sites, as observed in infection and malignancy (see Table 13-4). Elevated levels are much less common but are observed in both acute and chronic iron overdose situations.

Serum Ferritin

Since the concentration of circulating ferritin is proportional to the iron stores of an individual, serum ferritin is a very sensitive indicator of depleted iron stores.[4,65] In adult males, serum ferritin levels less than 20 μg/L (less than 10 μg/L in adult females and children) indicate the presence of signifi-

TABLE 13-4. Laboratory Markers of Iron Status in Several Disease States

Condition	Serum Iron (65–170 μg/dL)	Transferrin (200–400 mg/dL)	% Saturation (20–55)	Ferritin (20–250 μg/L)
Iron deficiency	Decreased	Increased	Decreased	Decreased
Iron poisoning/Overdose	Increased	Decreased	Increased	Increased
Hematochromatosis	Increased	Decreased	Increased	Increased
Malnutrition	Decreased	Decreased	Variable	Decreased
Malignancy	Decreased	Decreased	Decreased	Increased
Chronic infection	Decreased	Decreased	Decreased	Increased
Viral hepatitis	Increased	Increased	Normal/Incr.	Increased
Anemia of chronic disease	Decreased	Normal/Decr.	Decreased	Normal/Incr.
Sideroblastic anemia	Increased	Normal/Decr.	Increased	Increased

TABLE 13-5. Reference Ranges for Parameters Used to Assess Iron Status

Patient Population	Serum Iron (µg/dL)	Transferrin (mg/dL)	Ferritin (µg/L)	% Saturation	TIBC (µg/dL)
Adult, male	65–170	200–400	20–250	20–55	250–450
Adult, female	50–170	200–400	10–120	15–50	250–450
0–5 months	40–250	130–275	50–600	12–50	100–400
6 months–6 years	50–120	150–400	7–140	12–60	100–450
7–17 years	50–150	200–400	10–150	15–55	250–450

cantly diminished iron stores. Several conditions, including chronic disorders such as inflammation, infection, viral hepatitis, and malignancy, and conditions of iron overload, such as hematochromatosis or thalassemia, are characterized by elevated ferritin levels (see Table 13-4). It is important to realize that individuals with iron deficiency in conjunction with any of these disorders can present with normal or even elevated serum ferritin levels, despite the fact that they may actually have clinically significant iron deficiency. However, in general, a decreased serum ferritin is almost always indicative of iron deficiency. Serum ferritin is routinely determined by radioimmunoassay, enzyme immunoassay, or chemoluminescence immunoassay methods.

Transferrin

Since serum iron principally reflects the amount of iron bound to transferrin, determination of transferrin can provide important information regarding the maximal iron binding capacity, and hence the available iron, in an individual. In the past, measurement of transferrin was analytically difficult and was determined most easily by an indirect method (see later discussion). However, with the advent of improved technology, serum transferrin is now routinely determined directly using a variety of immunological techniques in most clinical laboratories.[11,18,28,32,58]

In adults and older children, serum transferrin levels range from 200 to 400 mg/dL. However, lower values are commonly seen in young children and pregnant females (see Table 13-5). Elevated transferrin levels are seen in iron deficiency as the liver synthesizes more protein in an attempt to sequester iron more efficiently from a diminished pool. Transferrin is also useful in assessing nutritional status and the ability of the liver to synthesize proteins. Since transferrin has a smaller body pool and half-life (8–10 days) than albumin, it is more sensitive to changes in nutritional status. In malnourished states, transferrin levels can decrease to less than 100 mg/dL. Additionally, transferrin is a "negative" acute phase reactant, with decreased levels observed in inflammation and malignancy, as well as in chronic liver disease, infection, and nephritic diseases.

Other Tests of Iron Status Assessment

Total iron binding capacity (TIBC) is has seen limited use since the advent of improved transferrin assays. TIBC is an estimate of serum transferrin levels that is obtained by measuring the total iron binding capability of a patient's serum. Since transferrin represents most of the iron-binding capacity of serum, TIBC is generally a good estimate of serum transferrin levels. Generally, TIBC is determined as follows: All iron binding sites are saturated (mainly transferrin) by the addition of exogenous iron. The excess iron is removed using an adsorbent such as charcoal or magnesium carbonate. The total amount of iron in the serum (the iron bound to transferrin) is then determined colorimetrically. It is important to realize that while transferrin represents most of the iron-binding capacity of the serum, other proteins also bind iron. However, a good correlation is observed between TIBC estimation of transferrin and the direct determination of transferrin, particularly when evaluating iron status in anemic and overdose cases. Typically, the TIBC is 10% to 15% higher than actual transferrin levels (see Table 13-5 for reference values).

The measurement of serum iron and transferrin (or TIBC) allows for the calculation of the percent of transferrin molecules that have iron bound to them. This calculation is a ratio of serum iron (actual iron in the serum) to serum transferrin or TIBC (potential quantity of iron that can be bound). This term is called the *transferrin saturation* (or *percent saturation*). Normally, only one-third (20–55%) of the transferrin iron binding sites are occupied by iron. Several disease states (see Table 13-4) result in decreased transferrin saturation. Iron deficient states are characterized by a low transferrin saturation. In contrast, conditions of iron overload (such as iron poisoning or hematochromatosis) have increased levels of transferrin saturation. Because various disease states may alter transferrin (and therefore TIBC) levels, serum ferritin is the single best test to diagnose iron deficiency.

ZINC[2,34,49,60]

Zinc (Zn) is the second most abundant essential trace element. The majority of zinc in the human body (2.5 gm) is present in muscle (60%) and bone (30%), with the remaining being unequally distributed in other tissues and organs. The daily requirement of zinc is 10 to 15 mg. This is obtained mainly from dietary intake of foods rich in zinc (*e.g.*, meats, fish, and seafood). Only 20% to 30% of dietary zinc is actually absorbed, mainly by the small intestine. Once absorbed, it is transported in the blood to its sites of utilization and storage (mainly tissue), where it is complexed to various proteins. In blood, zinc is found mainly bound to erythrocytes (75–85%), with the remainder in the plasma bound to albumin, transferrin, and immunoglobulins. Most zinc is excreted by the gastrointestinal tract and eliminated in feces, urine, and other fluids.

Zinc functions primarily as an enzyme cofactor. To date, over 300 diverse zinc-dependent enzymes have been identified, including carboxypeptidase A, carbonic anhydrase, alkaline phosphatase, DNA polymerase, and thymidine kinase. These enzymes are key components in the regulation of normal growth and cell function, wound healing, collagen synthesis, bone metabolism, reproduction, taste, sight, integrity of the immune system, anti–inflammatory actions, and protection against free radical damage (see Table 13-2).

Clinical Significance

Zinc Deficiency[5,33,34,47,49,60]

Zinc deficiency can occur as a result of prolonged inadequate dietary intake, increased body metabolic requirements, or abnormal loss of zinc-containing fluids.[5,47,60] The most prevalent cause of zinc deficiency is due to malnutrition and inadequate intake. This commonly occurs in conjunction with long-term parenteral feeding or increased metabolic requirements such as childhood growth, pregnancy, and lactation. In addition, zinc deficiency also occurs as a result of several diseases that cause excessive excretion of zinc (chronic liver diseases, protein loosing renal diseases, chelation therapy) or decreased absorption of dietary zinc (Crohn's disease). Decreased levels of zinc often accompany high catabolic states (tissue destruction from burns, surgery, hemolytic anemia), acute myocardial infarction, and the administration of certain drugs. The symptoms of zinc deficiency are widespread and include growth retardation, skeletal and skin abnormalities, loss of appetite, impotence, and impaired sensory perception. The supplemental administration of zinc corrects most of these alterations.

Zinc Toxicity[25,34,37,54,60]

Zinc is a relatively non-toxic metal; however, there have been reported cases of toxicity caused by excessive intake of zinc salts or industrial occupational exposure. Zinc toxicity is marked by GI disturbances (diarrhea, vomiting, fever), loss of reflexes, paralysis, and ataxia.

Laboratory Determination[1,17,23,39,46,48,51,55,56]

Zinc status is assessed most easily by measuring plasma zinc levels. In some cases, urine levels may sometimes be useful, particularly in the assessment of zinc toxicity. Atomic absorption is the most common and most reliable method of analysis. The reference range for serum zinc in adults is 75 to 125 mg/dL.

SERUM ZINC DETERMINATION BY ATOMIC ABSORPTION SPECTROPHOTOMETRY[11,39]

Principle

When atomized in a flame, zinc produces a characteristic absorption band at 213.9 nm with an intensity relative to the concentration of the element in the sample. Comparing the absorption of the sample to the absorption of standard zinc solutions at this wavelength allows quantitation of the analyte. The serum samples are diluted 1:4 with deionized H_2O to bring the zinc content into optimum analytical range. Glycerol is added to the standards to approximately match the viscosity of the diluted sera.

Specimen Collection and Preparation

The specimen should be collected via venipuncture in a metal-free blood collection tube containing no additives. Since zinc is used in the manufacture of most rubber, other tubes are unacceptable. In addition, hemolyzed specimens are *not* acceptable.

Apparatus

Atomic Absorption Spectrophotometer (Perkin Elmer 5100 or other)
Parameters:

- Wavelength: 213.9 nm
- Lamp source: Zinc Hollow Cathode Lamp
- Burner: Single Slot with Flow Spoiler
- Flame: Oxidizing
- Air flow: 10 L/min.
- Acetylene flow: 2 L/min.
- Slit width: 0.7 nm
- Detection limit: 20 µg/L

Reagents

1. *Sample Diluent (1% HNO_3):*
   Ten mL of concentrated nitric acid is added to a 1-L volumetric flask. Bring to volume using deionized water.
2. *5% Glycerol Solution in Acid:*
   Fifty mL of glycerol and 10 mL of concentrated nitric acid is added to a 1-L volumetric flask. Bring to volume using deionized water.
3. *Stock Zinc Standard (1 µg/mL):*
   One mL of a 1000 ppm certified zinc standard is added to a 1-L flask. Bring to volume with the 5% glycerol/1% nitric acid solution.
4. *Working Zinc Standards:*
   Prepare the working zinc standards in 100 mL volumetric flasks as described below. The standards must be stored in polypropylene containers in order to avoid possible leaching of zinc from the glass.

Zinc Concentration (µg/L)	Stock Standard (mL)	5%Glycerol/ Acid Solution (mL)
50	5.0	95.0
150	15.0	85.0
300	30.0	70.0

Procedure

1. *Instrument Setup:*
   Set up instrument according to the manufacturer's recommendations and operating procedures.
2. *Sample Setup:*
   A. Dilute each patient specimen and quality control specimen (1:4) as follows: Add 1 mL of serum and 4.0 mL of sample diluent into a 13 × 100 polystyrene tube.
   B. Cover the tubes with parafilm and mix thoroughly by vortexing. (Do not use polyethylene stoppers, as they leach zinc into the sample.)

C. Standards are not diluted, but analyzed as directly as prepared.
3. *Sample Analysis:*
   A. Analyze the working zinc standards. Use the results obtained to calibrate the instrument.
   B. Analyze all quality control and patient samples.
4. *Calculation of Results:*

$$\text{Zinc (µg/dL)} = \frac{\text{Zinc (µg/L)} \times 5 \text{ (dilution factor)}}{10 \text{ (dL/L)}}$$

(Eq. 13-1)

Technical Notes
1. If concentrations exceed dynamic range, the prepared (X5 dilution) sample should be diluted further with the 5% glycerol solution to obtain a concentration within the method linearity. Additional dilution factors must be included in the calculation of the result.
2. The linear range of the assay is 20 µg/L to 300 µg/L.
3. The dynamic range of the assay is 10 µg/dL to 150 µg/dL.
4. Samples having concentrations below the dynamic range are reported as less than 10 µg/dL.
5. All re-usable labware must be acid washed with 1N HCl prior to use. ■

COPPER[15,16,54,56,57]

Copper is the third most abundant trace element in the human body. The daily requirement of copper is 2 to 6 mg (see Table 13-2), most of which is obtained from dietary sources. Foods high in copper content include grains, shellfish, legumes, organ meats, cocoa, and water, especially when the water comes from copper plumbing. In some cases, environmental and occupational exposure can significantly contribute to the daily intake and often causes copper toxicity.

Copper is absorbed mainly by the duodenum and complexed to sulfhydryl-groups of proteins (mainly albumin and metallothionine). Typically, only 40% to 60% of orally ingested copper is actually absorbed. Several factors can affect the extent of copper absorption, including gender (females absorb more than males), the quantity ingested, the chemical form of copper, diet, age, and health. The complexed copper is transported to the liver where it is stored in hepatocytes. When required, copper is released from the liver and transported bound to *ceruloplasmin* to its required sites. More than 90% to 95% of the circulating copper in the plasma is bound to ceruloplasmin. Copper is primarily excreted through the biliary tract and GI secretions. Urinary excretion is low due to efficient reabsorption of copper in the renal tubules.

Copper is an essential cofactor required for function of many diverse metalloenzymes and proteins. It is required for enzymes involved in hemoglobin synthesis, iron metabolism (*e.g.*, ceruloplasmin), normal bone and joint development (*e.g.*, lysyl oxidase, which is required for collagen cross-linking), and pigmentation (*e.g.*, tyrosinase, which is required for melanin production). Copper is also a component of several enzymes responsible for scavenging oxygen free radicals (*e.g.*, superoxide dismutase) and electron transport (*e.g.*,

cytochrome c oxidase). Deficiency of copper results in decreased activity of these enzymes resulting in anemia, impaired iron metabolism, cardiomyopathy, muscle weakness, growth failure, and brain degeneration. Recently, much interest has focused on the importance of copper in relation to ischemic heart disease, bone and joint diseases like osteoporosis, and its role in oxygen free radical tissue damage.

Clinical Significance

Copper Deficiency[16,35,42,54,57]
The incidence of copper deficiency in the general population is low; however, it is quite common in certain population sub-groups. Copper deficiency is commonly observed in premature infants and in children suffering from malnutrition and malabsorption syndromes. These children are susceptible to copper deficiency because of their small storage amounts of copper in the liver. In addition, increased nutritional amounts place a burden on the limited copper supplies. Commonly, copper depletion is manifested by anemia. In cases of sustained copper deficiency, bone and joint abnormalities, and often osteoporosis, defects in pigmentation, and neurological abnormalities can be observed. Dietary copper supplementation can prevent and reverse many of these abnormalities. For this reason, the monitoring of copper levels in susceptible children is of great value. In older children and adults, copper deficiency is observed in several disease states. In particular, chronic GI disease, malabsorption syndromes, renal disease, long-term total parenteral feeding, and administration of chelation agents can lead to copper depletion.

A rare and severe form of copper deficiency is caused by the disease *Menkes' kinky hair syndrome*.[35,57] This is an X-linked genetic disease that results in a defect in copper transport and storage, and ultimately leads to progressive brain disease and eventual death. Early symptoms include depigmentation of the skin and hair, growth retardation, seizures, and cerebral and cerebellar degeneration. The inability to incorporate copper into copper-requiring enzymes produces these symptoms. They usually appear early in life and lead to death by 3 years of age.

Copper Toxicity[15,40,54,57,66]
Copper toxicity, characterized by increased tissue and serum levels, can occur as a result of environmental exposure, ingestion of excess copper, and as a result of several diseases. Acute copper poisoning may be caused by ingestion of excess copper (often when acidic food or drink is stored in metal containers made of copper, or leached from copper pipes), ingestion of fungicides containing copper sulfate, or exposure from industrial sources. Symptoms of acute copper toxicity often include nausea, vomiting, diarrhea, GI bleeding, hemoglobinuria, hematuria, hypotension, and, in severe conditions, liver necrosis, convulsions, and death. Elevated serum copper levels (*hypercupremia*) are associated with a number of conditions including pregnancy, lym-

phoma, leukemia, thalassemia, Hodgkin's disease, rheumatoid arthritis, anemia, and ingestion of some drugs and steroids (*i.e.,* estrogen).

Wilson's disease is an autosomal recessive disorder that results in excessive accumulation of copper in tissues, particularly the liver and basal ganglia of the brain. The biochemical defect in Wilson's disease has yet to be elucidated. The disease, if not treated, leads to progressive hepatolenticular degeneration and neurological dysfunction.[35] As tissue copper concentrations increase (due to the inability to use the copper), the brain and liver become particularly susceptible. Early in the disease, weakness, anorexia, jaundice, changes in mental status, and Kayser-Fleisher rings develop in the cornea (caused by copper deposition). As the disease progresses, liver necrosis and cirrhosis, and central nervous system manifestations that resemble Parkinson's disease or multiple sclerosis (incoordination, tremors, excessive salivation), and psychiatric disorders occur. Laboratory abnormalities of the disease include decreased serum ceruloplasmin and serum copper and increased urine and tissue copper levels. Prompt diagnosis of Wilson's disease is important so that therapy can be instituted. If therapy is implemented early in the course of the disease, the hepatic and brain damage can be prevented. Therapy consists of dietary modifications and chelation therapy using penicillamine, which chelates copper, thus preventing its deposition in tissues and allowing it to be excreted from the system. Wilson's disease may be diagnosed by measurement of serum and urine copper and ceruloplasmin.

Laboratory Determination

Routine laboratory tests for the evaluation of copper status include serum and urine copper and serum ceruloplasmin, and sometimes include the determination of tissue copper levels.[21,38,46,51,56,63]

Serum copper provides a relatively easy means for assessing long-term copper status. Low serum copper levels (hypocupremia) are consistent with several of the clinical disorders described above. However, serum copper levels are a relatively poor indicator of short-term and marginal copper deficiency, because hypocupremia may not be evident until tissue levels are severely depleted. Several physiological factors can effect serum copper levels. Women have higher levels than men, and concentrations increase with age. Pregnancy, oral contraceptives, infection, and inflammation also tend to increase levels, while corticosteroids reduce levels. In addition, serum copper exhibits diurnal variation with peak levels observed in the morning. *Urine copper* determination is generally of limited clinical use, although it is sometimes used to aid in the diagnosis of Wilson's disease or to follow the effectiveness of chelation therapy in patients who are diagnosed with the disease.

Serum and urine copper is most commonly determined using atomic absorption methodologies (see below). The reference ranges for serum copper are both age- and gender-specific:

Adult Males	50–150 µg/dL
Adult Females	60–260 µg/dL
Children (< 5 yr)	27–153 µg/dL

Ceruloplasmin accounts for >95% of copper found in serum or plasma, thus it is affected by factors that influence serum copper levels. The reference range in adults is 20 to 45 mg/dL. Ceruloplasmin levels are decreased in Wilson's disease and Menke's kinky hair syndrome. In addition, decreased levels are seen during most types of copper deficiency, protein-loosing nephrotic syndromes, and patients receiving long-term total parenteral nutrition. Elevated levels can be seen during pregnancy, biliary cirrhosis, and certain infections. Serum ceruloplasmin levels are most commonly determined photometrically or by immuno-chemical methods such as nephelometry.

SERUM COPPER DETERMINATION BY ATOMIC ABSORPTION SPECTROPHOTOMETRY[11,21]

Principle
When atomized in a flame, copper produces a characteristic absorption band at 324.8 nm with an intensity relative to the concentration of the element in the sample. Comparing the absorption of the sample to the absorption of standard copper solutions at this wavelength allows quantitation of the analyte. The serum or urine sample is diluted (1:1) with deionized water to bring the copper content into the optimum analytical range and to reduce protein clogging of the burner slot. Glycerol is added to the standards to approximately match the viscosity of the diluted sera or urine.

Apparatus
Atomic Absorption Spectrophotometer (Perkin Elmer 5100 or other)
Parameters:
- Wavelength: 324.8 nm
- Lamp source: Copper Hollow Cathode Lamp
- Burner: Single slot with Flow Spoiler
- Air flow: 10 L/min.
- Acetylene flow: 2 L/min.
- Slit: 0.7
- Detection limit: 50 µg/L

Reagents
1. *10% Glycerol Solution in 1% Nitric acid:*
   One hundred mL of glycerol and 10 mL of concentrated nitric acid is added to a 1-L volumetric Flask. Bring to volume using deionized water.
2. *Stock Copper Standard (10 µg/mL):*
   Ten mL of 1000 ppm certified atomic absorption copper solution is added to a 100-mL flask. Ten mL of glycerol is added, and the mixture prepared to volume using deionized water.
3. *Working Copper Standards:*
   Prepare the working copper standards in 100-mL volumetric flasks, as described below.

Copper Concentration (µg/L)	Stock Standard (mL)	10% Glycerol/ Acid Solution (mL)
200	2.0	98.0
600	6.0	94.0
1200	12.0	88.0

Procedure

1. *Instrument Setup:*

   Set up instrument according to the manufacturer's recommendations and operating procedures.

2. *Sample Setup:*

   A. Dilute each patient specimen and quality control specimen as follows: Add 1 mL of serum and 1 mL of deionized water to a polystyrene culture tube.

   B. Stopper and mix thoroughly by vortexing.

3. *Sample Analysis:*

   A. Analyze the working copper standard solutions. Use the values obtained to calibrate the instrument.

   B. Analyze all quality control and patient samples.

4. *Calculation of Results:*

$$\text{Copper } (\mu g/dL) = \frac{\text{Copper } (\mu g/dL) \times 2 \text{ (dilution factor)}}{10 \text{ (dL/L)}}$$

(Eq. 13-2)

Technical Notes

1. If concentrations exceed dynamic range, the prepared sample should be diluted further with the 10% glycerol solution to obtain a concentration within the method linearity. Additional dilution factors must be included in calculation of the result.

2. The linear range of the assay is 50 $\mu g/L$ to 1200 $\mu g/L$.

3. The dynamic range of the assay is 10 $\mu g/dL$ to 240 $\mu g/dL$.

4. Samples having concentrations below the dynamic range should be reported as less than 10 $\mu g/dL$.

5. All reusable lab ware must be acid washed with 1N HCl prior to use. ■

ESSENTIAL ULTRATRACE ELEMENTS

Manganese[6,7,44,45,54,64]

Manganese is an essential cofactor of several metabolic processes including lipid and carbohydrate metabolism, bone and tissue formation, skeletal growth, and reproduction. To date, over 50 manganese-dependent enzymes have been identified including catalase, peroxidase, superoxide dismutase, 5′-nucleotidase, RNase, and glucosyl- and galactosyltransferases. The daily intake of manganese is 20 $\mu g/day$, which easily exceeds the daily requirement of 2 to 5 $\mu g/day$. For this reason, manganese deficiency is extremely rare. Manganese is absorbed by the small intestine in a process closely linked to the absorption of iron. Once absorbed, manganese is transported bound to β-microglobulin to mitochondrial rich organs such as the liver, pancreas, kidney, and pituitary gland. The major route of excretion is through bile and pancreatic excretions.

Since manganese deficiency is extremely rare, the etiology of it has not been well characterized. Regardless, several defects have been directly attributed to it. These include alterations in hair color and growth, loss of weight, sterility, nausea, inhibition of vitamin K-dependent blood clotting, and altered cholesterol transport and utilization. Manganese toxicity, particularly due to occupational exposure, results in both psychological and neurological alterations. Inhalation of manganese dust by miners often results in the develop-

ment of headaches, impotence, leg cramps, apathy, loss of appetite, and speech disturbances.

Manganese is not routinely measured in the clinical laboratory except in rare cases of suspected toxicity. Clinically, serum manganese increases following myocardial infarction and a decrease is observed in children with epilepsy and other convulsive disorders. The normal reference range of manganese in serum is 0.5 to 1.0 $\mu g/L$.

Cobalt[9,14,31]

Despite being widely distributed in nature, the biological function of cobalt is limited. To date, vitamin B_{12} (cobalamin) is the only known cobalt containing compound in humans. The majority of the daily intake of cobalt is obtained from dietary sources high in vitamin B_{12}, especially meat and dairy products. Inorganic sources of cobalt minimally contribute to daily intake. As with manganese, the daily requirement of cobalt (0.1 μg) far exceeds the daily intake (5–45 $\mu g/day$). For this reason, cobalt deficiency is limited to conditions with severely diminished vitamin B_{12} absorption. The GI absorption of vitamin B_{12} is dependent on the presence of intrinsic factor. When intrinsic factor is present, more than 70% of dietary vitamin B_{12} is absorbed. When intrinsic factor is absent, less than 5% is taken up. The majority of cobalt absorbed is reincorporated into newly synthesized cobalamin and stored in the liver.

Cobalt deficiency *per se* has not been reported, since deficiency states are always the result of a deficiency of vitamin B_{12}. This is most often due to the absence of intrinsic factor (and hence the inability to absorb vitamin B_{12}) as a result of disorders affecting intestinal absorption. Vitamin B_{12} deficiency is usually manifested as severe megaloblastic anemia. Cobalt toxicity is rare and limited to occupational inhalation or accidental ingestion of cobalt. It is characterized by pulmonary edema, nausea, cardiomyopathy, hematological disorders, and GI dysfunction. The direct measurement of cobalt is not clinically relevant and is rarely performed. Typical plasma concentrations are 0.5 to 4.0 $\mu g/L$.

Selenium[10,24,61]

Selenium is one of the least abundant and most toxic of all the essential elements. The intake of selenium is highly variable depending on the location. In the United States, the average daily intake is 50 to 200 $\mu g/day$. However, in certain regions of the world, particularly those with low soil selenium levels, supplementation may be required. Selenium is widely distributed, but the highest concentrations are found in the liver, kidney, heart, spleen, pancreas, and lungs. The functions of selenium are widespread but not well characterized. It is an integral component of several antioxidant enzymes including glutathione peroxidase, which is required for normal immune system function, and plays a critical role in the control of oxygen metabolism, particularly in the breakdown of H_2O_2.

The concentration of selenium in human serum is 0.2 to 1.0 µg/L. Selenium deficiency is rare in humans; however, it is the cause of Keshan disease, a syndrome that is endemic to Keshan Province in China where the soil is deficient in selenium. This disease is characterized by severe arrhythmias, cardiogenic shock, and muscle degeneration. Children and women of child-bearing age are particularly susceptible. Selenium deficiency has been linked to cardiovascular disease, myocardial infarction, and carcinogenesis; however, the evidence of a direct causal relationship is far from conclusive. In animals, it is characterized by hepatic necrosis, retarded growth, muscular degeneration, and infertility. Selenium toxicity is rare and only occurs as a result of industrial exposure. Its symptoms are poorly defined. In animals, selenium is both neurotoxic and hepatotoxic.

Molybdenum[44,50]

Although molybdenum has a recognized essential role in life, its precise function is not well understood. It is an essential cofactor for several metalloenzymes, including xanthine oxidase and other oxidoreductases. The absence of molybdenum as a cofactor for sulfite oxidase and xanthine dehydrogenase produces mental retardation, neurological seizures, and bilateral lens dislocation.

Chromium[3,12,41]

Chromium is an essential factor in the metabolism of glucose and lipids. It is required for the stimulation of fatty acid and cholesterol synthesis, and it improves glucose tolerance by facilitating insulin release and tissue binding. Chromium deficiency results in retarded growth and impaired glucose tolerance and has been demonstrated to be a casual factor in atherosclerosis.

SUMMARY

A trace element is defined as a constituent that occurs in biological systems at concentrations of mg/kg or less. Ultratrace elements are present in even smaller quantities. Elements may also be defined as essential or non-essential. An essential element is absolutely necessary for life. A deficiency or absence of an essential element can cause severe impairment and/or lead to death. Trace elements or trace metals have an important role in both normal biological function and toxicity in humans. The control of trace element homeostasis is under rigid control mechanisms to optimize their physiological functions. Any alteration (*i.e.,* a deficit or an excess) of a trace element can produce a significant biochemical alteration. Essential trace elements include, iron, zinc, and copper. Essential ultratrace elements include, manganese, cobalt, selenium molybdenum, chromium, and iodine. Nickel, vanadium, and tin are also probably essential. The biochemical functions of trace elements are diverse and crucial. They are commonly involved directly in electron transport, as a cofactor for a wide variety of enzymes, or as a vital component of several metalloproteins. Trace element deficiencies can occur as a result of many genetically determined enzyme defects. More commonly, however, a deficiency can be the result of nutritional deficiency, insufficient intestinal absorption, or increased loss or utilization of the element.

CASE STUDY 13-1

A 33-year-old male with cerebral palsy who is confined to a wheel chair presents with recurring chronic gastritis (hemorrhagic gastrointestinal bleeding); recent bouts of diminished strength, weakness, and fatigue; and depression. He has been taking Zantac© required intermittently for treatment of his gastritis. He admits to a loss of 20 pounds over the previous 3 months. A complete cell count was performed. Of note were the following results:

Hemoglobin	8.7 g/dL	(13.3–17.7 g/dL)
Hematocrit	29%	(40–52%)
Mean cell volume	67 fL	(81–100 fL)

Subsequent laboratory tests were performed:

Iron	10 µg/dL	(55–190 µg/dL)
Transferrin	407 mg/dL	(200–400 mg/dL)
% Saturation	< 2%	(20–555%)
TIBC	484 µg/dL	(250–450 mg/dL)
Ferritin	< 5 µg/L	(25–250 µg/L)

Questions

1. What condition is consistent with these laboratory results?
2. What factors may have contributed to the development of this condition?
3. What medical attention should be provided?
4. Besides the correct diagnosis, what other conditions can produce microcytic anemia? How could you differentiate between them?

CASE STUDY
13-2

A 26-year-old female of Vietnamese ancestry presents to her family physician for a routine physical examination. She reports that she and her husband are interested in starting a family. A complete blood count (CBC) revealed the presence of a microcytic anemia. For this reason further studies were performed.

Erythrocyte count	5.70×10^{12}/L	$(3.8–5.2 \times 10^{12}$/L)
Hemoglobin	9.8 g/dL	(11.7–15.7 g/dL)
Hematocrit	31%	(35–47%)
MCV	64 fL	(81–100 fL)
Serum iron	75 µg/dL	(75–190 µg/dL)
Ferritin	88 µg/L	(20–200 µg/L)
Transferrin	288 mg/dL	(200–400 mg/dL)

Questions

1. What is the cause of this woman's microcytic anemia?
2. What further analysis could be performed to confirm your suspicion?
3. What medical treatment is indicated?
4. Would iron supplementation be advised in this patient?

CASE STUDY
13-3

A 36-year-old male had undergone multiple partial small bowel resections because of Crohn's disease. After recovery from surgery, he was placed on total parenteral nutrition (TPN), and after adequate rehabilitation he was discharged under the care of a home health-care nurse who visited him three times a week to follow his overall progress and check his physical and vital conditions. Periodic routine laboratory tests were performed to assess nutritional status. His first few months post-surgery were unremarkable. Eventually he experienced symptoms of weakness, diarrhea, general malaise, and hair loss. He visited his physician, who did a complete work-up of his problem and obtained the following laboratory values:

Sodium	147 mmol/L	(136–146 mmol/L)
Potassium	4.9 mmol/L	(3.5–5.1 mmol/L)
Chloride	118 mmol/L	(98–106 mmol/L)
Glucose	102 mg/dL	(80–100 mg/dL)
Creatinine	0.6 mg/dL	(0.6–1.2 mg/dL)
Calcium	7.9 mg/dL	(8.4–10.2 mg/dL)
Hemoglobin	10.4 g/dL	(13.3–17.7 g/dL)
Hematocrit	31%	(40–52%)
Albumin	2.5 g/dL	(3.5–5.5 g/dL)
Transferrin	112 mg/dL	(200–400 mg/dL)
Copper	38 µg/dL	(50–150 µg/dL)
Zinc	46 µg/dL	(75–125 µg/dL)

Questions

1. What do the laboratory studies suggest as a possible cause of this patient's problem?
2. What is the significance of the low copper and zinc levels? What is the underlying cause of their discrepancy?
3. What medical treatment would be instituted in this patient?

REFERENCES

1. Accominotti M, Pegon Y, Vallon JJ. Determination of zinc in blood serum by electrothermal atomic absorption spectrometry with matrix modification. Clin Chim Acta 1988;173:99.
2. Aggett PJ, Favier A. Zinc. Int J Vitam Nutr Res 1993;63:301.
3. Anderson RA. Recent advances in the clinical and biochemical effects of chromium deficiency. Prog Clin Biol Res 1993; 380:221.
4. Andrews SC, Arosio P, Bottke W, et al. Structure, function, and evolution of ferritins. J Inorg Biochem 1992;47:161.
5. Apgar J. Zinc and reproduction. Ann Rev Nutr 1985;5:43.
6. Aschner M, Aschner JL. Manganese neurotoxicity: Cellular effects and blood-brain barrier transport. Neurosci Biobehav Rev 1991;15:333.
7. Baruthio F, Guillard O, Arnaud J, Pierre F, Zawislak R. Determination of manganese in biological materials by electrothermal atomic absorption spectrometry: A review. Clin Chem 1988;34:227.
8. Beutler E. The common anemias. J Am Med Assoc 1988; 259:2433.
9. Beyersmann D, Hartwig A. The genetic toxicology of cobalt. Toxicol Appl Pharmacol 1992;115:137.
10. Burk RF. Clinical effects of selenium deficiency. Prog Clin Biol Res 1993;380:181.
11. Burtis CA, Ashwood ER, eds. Tietz textbook of clinical chemistry. 2nd ed. Philadelphia: WB Saunders Co, 1994.
12. Cohen MD, Kargacin B, Klein CB, Costa M. Mechanisms of chromium carcinogenicity and toxicity. Crit Rev Toxicol 1993;23:255.
13. Conrad ME. Regulation of iron absorption. Prog Clin Biol Res 1993;380:203.
14. Cugell DW. The hard metal diseases. Clin Chest Med 1992; 13:269.

15. Danks DM. Copper and liver disease. Eur J Pediatr 1991; 150:142.

16. Delves HT. Assessment of trace element status. Clin Endocrinol Metab 1985;14:725.

17. D'Haese PC, Lambers LV, Vanheule AO, DeBroe ME. Direct determination of zinc in serum by Zeeman atomic absorption spectrometry with a graphite furnace. Clin Chem 1992;38:2439.

18. de Jong G, van Eijk JP, van Eijk HG. The biology of transferrin. Clin Chim Acta 1990;190:1.

19. Eckfeldt JH, Witte DL. Serum iron: Would analytical improvement enhance patient outcome? Clin Chem 1994; 40:505.

20. English JL, Hambidge KM. Plasma and serum zinc concentrations: Effect of time between collection and separation. Clin Chim Acta 1988;175:211.

21. Evanson MA. Measurement of copper in biological samples by flame or electrothermal atomic absorption spectrometry. Methods Enzymol 1988;158:351.

22. Fairweather-Tait S. Iron. Int J Vitam Nutr Res 1993;63:296.

23. Falchuk KH, Hilt KL, Vallee BL. Determination of zinc in biological samples by atomic absorption spectrophotometry. Methods Enzymol 1988;158:422.

24. Fan AM, Kizer KW. Selenium: Nutritional, toxicologic, and clinical aspects. West J Med 1990;153:160.

25. Fosmire GJ. Zinc toxicity. Am J Clin Nutr 1990;51:225.

26. Gercken B, Barnes RM. Determination of lead and other trace element species in blood by size exclusion chromatography and inductively coupled plasma/mass spectrometry. Anal Chem 1991;63:283.

27. Giardina PJ, Hilgartner MW. Update on thalassemia. Pediatr Rev 1992;13:55.

28. Guyatt GH, Oxman AD, Ali M, Willan A, McIlroy W, Patterson C. Laboratory diagnosis of iron-deficiency anemia: An overview. J Gen Intern Med 1992;7:145.

29. Holcombe JA, Hassell DC. Atomic absorption, atomic fluorescence and flame emission spectrometry. Anal Chem 1990; 62:169R.

30. Iron Panel of the International Committee for Standardization in Hematology. Revised recommendations for the measurements of serum iron in human blood. Br J Heamatol 1990;75:615.

31. Jensen AA, Tuchsen F. Cobalt exposure and cancer risk. Crit Rev Toxicol 1990;20:427.

32. Johnson MA. Iron: Nutrition monitoring and nutrition status assessment. J Nutr 1990;120(Suppl 11):1486.

33. Keen CL, Gershwin ME. Zinc deficiency and immune function. Ann Rev Nutr 1990;10:415.

34. King JC. Assessment of zinc status. J Nutr 1990;120(Suppl 11):1479.

35. Kodama H. Recent developments in Menkes disease. J Inherit Metab Dis 1993;16:791.

36. Lauwerys R, Bernard A, Cardenas A. Monitoring of early nephrotoxic effects of industrial chemicals. Toxicol Lett 1985; 64–65:33.

37. Leonard A, Gerber GB, Leonard F. Mutagenicity, carcinogenicity and teratogenicity of zinc. Mutat Res 1986;168:343.

38. Liska SK, Kerkay J, Pearson KH. Determination of copper in whole blood, plasma, and serum using Zeeman effect atomic absorption spectroscopy. Clin Chim Acta 1985;150:11.

39. Liska SK, Kerkay J, Pearson KH. Determination of zinc and copper in urine using Zeeman effect flame atomic absorption spectrometry. Clin Chem Acta 1985;151:231.

40. McClain CJ, Marsano L, Burk RF, Bacon B. Trace metals in liver disease. Semin Liver Dis 1991;11:321.

41. Mertz W. Chromium in human nutrition: A review. J Nutr 1993;123:626.

42. Milne DB. Assessment of copper nutritional status. Clin Chem 1994;40:1479.

43. Moyer TP, Mussmann GV, Nixon DE. Blood-collection device for trace and ultra-trace metal specimens evaluated. Clin Chem 1991;37:709.

44. Neilson FH. Ultratrace elements of possible importance for human health: An update. Prog Clin Biol Res 1993;380:355.

45. Neve J, Leclercq N. Factors affecting determinations of manganese in serum by atomic absorption spectroscopy. Clin Chem 1991;37:723.

46. Nixon DE, Moyer TP, Johnson P, et al. Routine measurement of calcium, magnesium, copper, zinc, and iron in urine and serum by inductively coupled plasma emission spectroscopy. Clin Chem 1986;32:1660.

47. O'Dell BL. Roles of zinc and copper in the nervous system. Prog Clin Biol Res 1993;380:147.

48. Passey RB, Maluf KC, Fuller R. Quantitation of zinc in nitric acid digested plasma by atomic absorption spectrophotometry. Anal Biochem 1985;151:462.

49. Prassad AS. Clinical manifestations of zinc deficiency. Ann Rev Nutr 1985;5:341.

50. Rajagopalan KV, Johnson JL, Wuebbens MM, et al. Chemistry and biology of the molybdenum cofactors. Adv Exp Med Biol 1993;338:355.

51. Roberts NB, Fairlough D, McLoughlin S, Taylor WH. Measurement of copper, zinc and magnesium in serum and urine by DC plasma emission spectrometry. Ann Clin Biochem 1985;22:533.

52. Savory J, Wills MR. Trace metals: Essential nutrients or toxins. Clin Chem 1992;38:1565.

53. Sears DA. Anemia of chronic disease. Med Clin North Am 1992;76:567.

54. Seiler HG, Sigel A, Sigel H, eds. Handbook on metals in clinical and analytical chemistry. New York: Marcel Dekker, Inc, 1994.

55. Smith JC, Butrimovitz GP. Direct measurement of zinc in plasma by atomic absorption spectroscopy. Clin Chem 1979;25:1487.

56. Smith JC, Holbrook JT, Danford DE. Analysis and evaluation of zinc and copper in human plasma and serum. J Am Coll Nutr 1985;4:627.

57. Solomons NW. Biochemical, metabolic, and clinical role of copper in human nutrition. J Am Coll Nutr 1985;4:83.

58. Spiekerman AM. Proteins used in nutritional assessment. [Review]. Clin Lab Med 1993;13(2):353.

59. Valcour AA, Kryzymowski G, Onoroski M, Bowers GN, McComb RB. Proposed reference method for iron in serum used to evaluate two automated iron methods. Clin Chem 1990;36:1789.

60. Vallee BL, Fachuk KH. The biochemical basis of zinc physiology. Physiol Rev 1993;73:78.

61. Van Dael P, Deelstra H. Selenium. Int J Vitam Nutr Res 1993;63:312.

62. Versieck J. Trace elements in human body fluids and tissues. Crit Rev Clin Lab Sci 1985;22:97.

63. Want ST, Demshar HP. Rapid Zeeman atomic absorption determination of copper in serum and urine. Clin Chem 1993;39:1907.

64. Wedler FC. Biological significance of manganese in mammalian systems. Prog Med Chem 1993;30:89.

65. Worwood M. Ferritin. Blood Rev 1990;4:259.

66. Yarze JC, Martin P, Munoz SJ, Friedman LS. Wilson's disease: Current status. Am J Med 1992;92:643.

Chapter 14

First exam

Carbohydrates and Alterations in Glucose Metabolism

James G. Donnelly

Objectives

Upon completion of this chapter, the clinical laboratorian should be able to:

- *Identify the major classes of carbohydrates and give examples of each.*

- *Identify different isomers of carbohydrates with the same molecular formula.*

- *State the mechanism for glucose and other carbohydrates to act as reducing substances and explain mutarotation.*

- *Explain the fate of glucose and other carbohydrates in the body including the pathways of anaerobic glycolysis, glycogenesis, and the hexose monophosphate shunt.*

- *Discuss the fate of other substrates such as pyruvate, lactate, glycerol, ketones, and acetyl coA in the cell.*

- *Relate the importance of gluconeogenesis and glycogenolysis for maintaining blood sugar concentrations.*

- *Describe the roles of insulin, glucagon, epinephrine, cortisol, and growth hormone in regulating extracellular glucose concentrations, and the role of the liver, skeletal muscle, and adipose tissue in regulating extracellular glucose concentrations.*

- *Explain how eating, starvation, adrenergic stimulation, severe stress or injury, diabetes, and hyperinsulinism affect glycemic control.*

- *Understand that diabetes is a complex group of diseases and how these diseases have multifactorial causes including genetic, environmental, and immune contributions. Evaluate the different (expected) biochemical profiles of patients with uncompensated IDDM or NIDDM.*

- *Identify the different forms of hypoglycemia.*

- *Discuss the roles of each of the assays used to assist in the diagnosis of diabetes including glucose measurements, the fasting glucose, two-hour postprandial, and OGTT measurements and the utility of assays used in the management of diabetes, including HbA$_{1c}$ fructosamine, and urinary albumin.*

KEY WORDS

Carbohydrate	Gluconeogenesis	Hypoglycemic
Diabetes Mellitus	Glycogen	Insulin
Disaccharide	Glycogenolysis	Ketone
Embden-Myerhof Pathway	Glycolysis	Microalbuminuria
Fisher Projection	Haworth Projection	Monosaccharide
Fructosamine	HbA$_{1c}$	Polysaccharide
Glucagon	Hyperglycemic	Triose

Michael L. Bishop, Janet L. Duben-Engelkirk, and Edward P. Fody.
CLINICAL CHEMISTRY. © 1996 Lippincott–Raven Publishers.

Organisms rely on the oxidation of complex organic compounds to obtain energy. Three general types of such compounds are carbohydrates, amino acids, and lipids. Although all three are used as a source of energy, carbohydrates are the primary source for brain, erythrocytes, and retinal cells in humans. Consequently, the concentration of glucose in the extracellular water (including plasma) requires stringent regulation. This is accomplished by many complex and interrelated endocrine and metabolic processes facilitated by the normal functioning of several glands and organs. If one or more of these systems is not functioning properly, a loss of glycemic control occurs, which ultimately results in the formation of a pathologic state. However, disorders of glucose metabolism, such as diabetes mellitus, are more than a simple dysfunction of glucose regulation. These diseases often have multifactorial causes, which can include genetic, immune, autoimmune, and other molecular mechanisms.

CLASSIFICATION AND CHEMISTRY

Monosaccharides

Carbohydrates are either polyhydroxy aldehydes or polyhydroxy ketones, or multimeric units of such compounds. The general formula of a carbohydrate is $(CH_2O)_n$. There are some deviations from this basic formula, because carbohydrate derivatives can be formed by the addition of other chemical groups such as phosphates, sulfates, and amines. The smallest molecule that conforms to the general formula is formaldehyde. However, this compound usually is not considered to be a carbohydrate and is highly reactive and toxic. Glyceraldehyde, a three-carbon compound, is normally considered to be the smallest carbohydrate. This compound is optically active; that is, it rotates the plane of polarized light. The central carbon of glyceraldehyde is chiral; in other words, all of the chemical bonds originating at it are unique. For each chiral atom there are 2^n possible isomers; therefore, there are 2^1, or two, forms of glyceraldehyde. These isomers are mirror images of each other and are called *stereoisomers*.

Carbohydrates are grouped into generic classifications based on the number of carbons in the molecule. Trioses contain three carbons, tetroses contain four, pentoses contain five, hexoses contain six carbons, and so on. If the carbohydrate is an aldehyde, it is called an aldose. If the compound is a ketone, it is called a ketose. Thus glucose (Fig. 14-1), a six-carbon aldehyde, is classified as an aldohexose. Fructose, a six-carbon ketone, is a ketohexose (Fig. 14-1).

Several models can be used to represent carbohydrates (Fig. 14-1). The Fisher projection of a carbohydrate has the aldehyde or ketone at the top of the drawing. The carbons

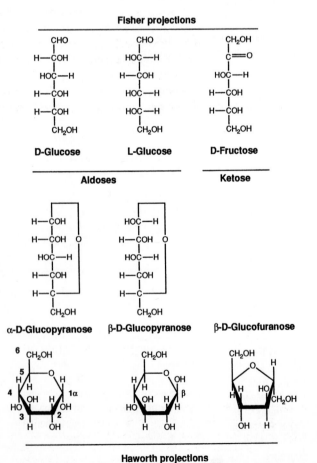

◀ *Figure 14-1.* Fisher and Haworth projections of glucose and fructose. D-glucose, L-glucose, and D-fructose are represented as open chain Fisher projections (top of figure). α-D-glucopyranose and β-D-glucopyranose are represented as cyclic Fisher projections (middle) and Haworth projections (bottom of figure). β-D-glucofuranose is also represented at the bottom.

are numbered starting at the aldehyde or ketone end, and the compound can be represented as either a straight chain or a cyclic, hemiacetal form.

D- and *L-* are terms used to describe some of the possible optical isomers of glucose and other compounds that exist as stereoisomers. Because there are four chiral carbons in an aldohexose, there are 2^4 or 16 possible isomers, two of which are stereoisomers named D-glucose and L-glucose. In Figure 14-1, D-glucose is represented in the Fisher projection with the hydroxy group on carbon number five positioned on the right. L-glucose has the hydroxy group of carbon number five positioned on the left.

Many carbohydrates favor the formation of hemiacetal ring structures. In fact, only a small fraction of glucose (less than 1%) in solution is in the open chain form. The formation of a hemiacetal ring structure introduces a new chiral carbon at position one. This increases the number of different cyclic aldohexoses to 2^5, or 32. Thus, the D- or L-forms of glucose each have two possible representations in the ring form (Fig. 14-1). The α form is represented by drawing the hydroxy group of carbon one on the right. The β form of glucose is represented by drawing the hydroxy group of carbon one on the left.

The Haworth projection represents the compound in a cyclic form, which is more representative of the actual structure. When glucose is drawn in a Haworth projection, the α form of D-glucopyranose is represented by the hydroxy group of carbon one oriented downward or below the plane of the paper (Fig. 14-1). β-D-glucopyranose has the hydroxy group of carbon one oriented upward to figuratively represent it to be above the plane of the paper (Fig. 14-1).

Most sugars in humans are of the D form. The proportion of α-D-glucopyranose and β-D-glucopyranose in solution is determined by the nature of the solvent and its temperature.

Thus, if a pure solution of α-D-glucose is allowed to sit at 40°C, it forms an equilibrium mixture of 36% α-D-glucose and 64% β-D-glucose. This reaction is called *mutarotation.* Theoretically, it is also possible for the six-carbon hemiacetal to form two different rings. One ring is similar to a furan, which contains four carbons and one oxygen; the other ring contains five carbons and one oxygen and is similar to a pyran ring (Fig. 14-1). Glucose thermodynamically favors the formation of six-member rings (five carbons with one oxygen); therefore it is a pyranose. Fructose, on the other hand, favors the formation of furanose rings.

Disaccharides and Polysaccharides

Separate carbohydrates, or *monosaccharides,* can be joined together to form *disaccharides* or multimeric structures called *oligosaccharides* and *polysaccharides.* When two carbohydrate molecules join together, a water molecule is generated. When they split, one molecule of water is lost. This reaction is called *hydrolysis.* Polysaccharides are the most abundant organic molecules in nature. Cellulose, for example, is a highly insoluble polysaccharide occurring widely in plants; starches are the principal carbohydrate (polysaccharide) storage products of higher plants; and glycogen is the principal carbohydrate storage product in animals.

The glycoside linkages of carbohydrates can involve any of the carbons; however, certain carbons are favored depending on the carbohydrate. A common disaccharide used by humans is sucrose, often known as table sugar. Sucrose is α-D-glucopyranosyl (1→2) β-D-fructofuranoside (Fig. 14-2). The oxygen from carbon one of the α form of D-glucose is bound to the second carbon on β-D-fructose. Two other common disaccharides are maltose and lactose (Fig. 14-2). Maltose is composed two α-D-glucose units in

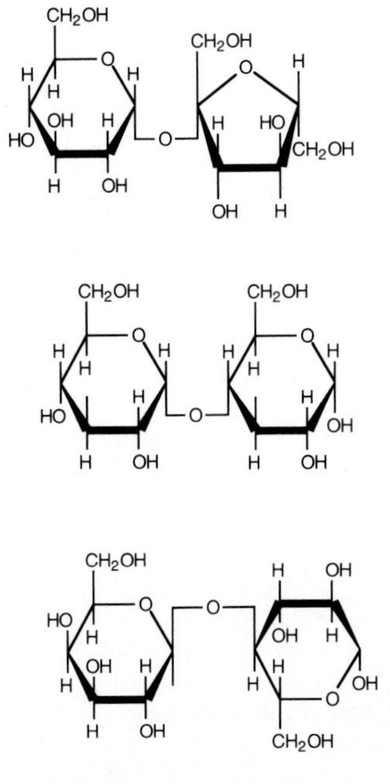

Sucrose
α-D-glucopyranosyl
(1 →2) β-D-fructofuranoside

Maltose
α-D-glucopyranosyl
(1 →4) α-D-glucopyranoside

Lactose
β-D-glucopyranosyl
(1 →4) α-D-glucopyranoside

Figure 14-2. Haworth projections of sucrose, maltose, and lactose—three common disaccharides. The carbons involved in each glycoside linkage are designated in parentheses. Sucrose is not a reducing sugar because fructofuranoside will not furnish a reactive carbonyl group during mutarotation. Maltose and lactose are both reducing sugars because glucopyranoside has an available carbonyl group.

a (1→4) linkage. Maltose is a disaccharide, which is formed by the digestion of starch by the enzyme amylase. Lactose is the disaccharide found in milk. Lactose is composed of β-D-galactose and α-D-glucose in a (1→4) linkage.

All monosaccharides and many disaccharides are reducing agents. This is because a free aldehyde or ketone (the open chain form) can be oxidized under the proper conditions. As a disaccharide, one of the aldehydes or ketones is usually α to the glycosidic linkage. The other aldehyde or ketone is still free to function as a reducing agent. However, when both aldehydes or ketones are α to the glycosidic linkage, such as in sucrose, then the disaccharide is incapable of undergoing mutarotation and is therefore incapable of functioning as a reducing agent. Both maltose and lactose are reducing agents, whereas sucrose is not.

Single units of carbohydrates cannot be stored easily in cells. All carbohydrates are powerful osmotic agents due to the number of hydroxy groups that can contribute to hydrogen bonding and organization of water. Additionally, intracellular enzymes readily recognize glucose and its congeners and quickly shunt them into metabolic pathways. Therefore, to store glucose molecules efficiently until they are needed for energy, cells convert the carbohydrate into the polymers starch in plants and glycogen in animals (Fig. 14-3). Starch and glycogen polymers can range in size from less than 100 to more than 2000 linked glucose units. Starch and glycogen are similar because their main chains are composed of 1,4-glycosidic linkages (Fig. 14-3). Branches occur off the main chain at 1,6-glycosidic linkages (Fig. 14-3). Glycogen is more branched than starch, and its branches are somewhat shorter. Presumably, branching permits the organism to store a larger amount of carbohydrate in a smaller, more compact volume than would otherwise

be possible for linear polymers. The two tissues that synthesize and store glycogen are liver and muscle.

Reactivity of Carbohydrates

The ability of glucose to function as a reducing agent has been useful in the detection and quantitation of carbohydrates in body fluids. Glucose and other carbohydrates are capable of converting cupric ions in alkaline solution to cuprous ions. The solution loses its deep blue color and a red precipitate of cuprous oxide forms. Benedict's and Fehling's reagents, which contain an alkaline solution of cupric ions stabilized by citrate or tartrate, respectively, have been used to detect reducing agents in urine and other body fluids. Another chemical characteristic that used to be exploited to quantitate carbohydrates in the past is the ability of these molecules to form Schiff bases with aromatic amines. O-toluidine in a hot acidic solution will yield a colored compound with an absorbance maxima at 630 nm. Galactose, an aldohexose, and mannose, an aldopentose, will also react with o-toluidine and produce a colored compound that can interfere with the reaction. The Schiff base reaction with o-toluidine is of historical interest only and has been replaced by more specific enzymatic methods, which will be discussed later.

CARBOHYDRATE METABOLISM AND FUEL HOMEOSTASIS

As mentioned, glucose is a primary source of energy for humans. The nervous system, including the brain, is totally dependent on glucose from the surrounding extracellular

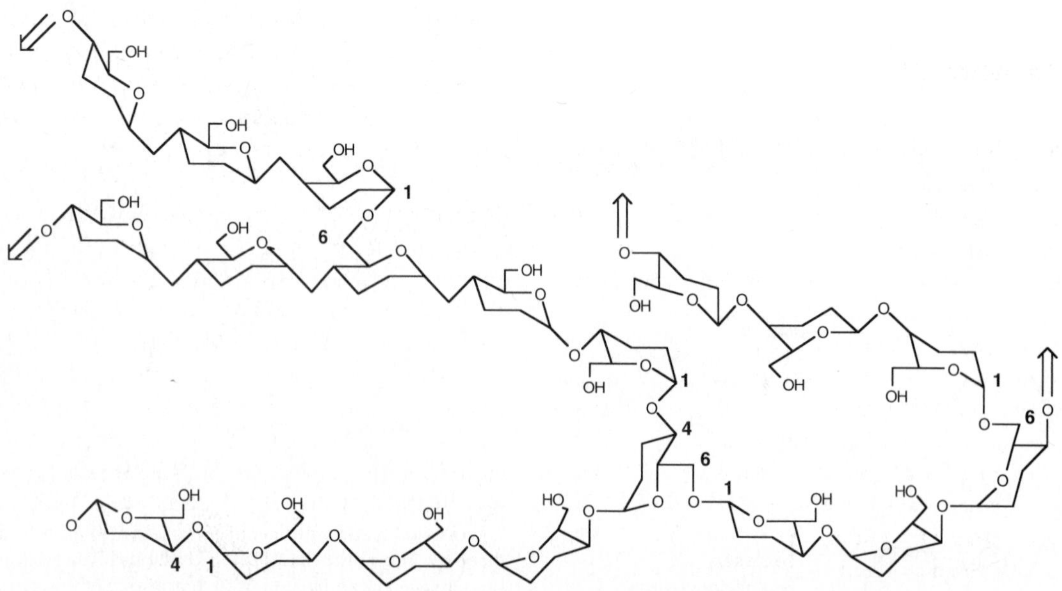

Figure 14-3. Representation of a growing glycogen molecule. Numbers indicate the main chain (1→4) and branched (1→6) glycosidic linkages. Open arrows (⇒) designate potential extension points for the chains.

fluid (ECF) for energy. Nervous tissue cannot concentrate carbohydrates nor can it store them; therefore it is critical to maintain a steady supply to the tissues. For this reason, the concentration of glucose in the ECF must be maintained in a narrow range. When the concentration falls below a certain level, the nervous tissues lose their primary energy source and are incapable of maintaining normal function.

Fate of Glucose

Most of our ingested carbohydrates are polymers such as starch and glycogen. Salivary amylase and pancreatic amylase are responsible for the digestion of these non-absorbable polymers to dextrins and disaccharides, which are further hydrolyzed to monosaccharides by maltase, an enzyme released by the intestinal mucosa. Sucrase and lactase are two other important gut-derived enzymes that hydrolyze sucrose to glucose and fructose and lactose to glucose and galactose.

Many individuals are lactose intolerant. This can be either congenital or acquired in adulthood, and is caused by a decrease in the expression of lactase on the intestinal lumen brush border. The consumption of milk products results in an osmotic imbalance due to the retention of lactose in the gut. Intestinal flora are capable of utilizing lactose. The result is an accumulation of waste products, such as lactic acid, along with lactose. Water accumulates in the intestine and distention, pain, flatulence, diarrhea, and dehydration leading to electrolyte imbalances can occur. Adults will normally self-regulate the ingestion of lactose to avoid the discomfort that they have experienced in the past. Congenital lactase deficiency is more serious because lactose is a major component of milk. Severe dehydration in infants can result in life-threatening electrolyte imbalances. The treatment of lactose intolerance in infants is the substitution of milk for a lactose deficient brand.

Glucose Metabolism

Once disaccharides are converted into monosaccharides, they are absorbed by the gut and transported to the liver by the hepatic portal venous blood supply. Glucose is the only carbohydrate to be either directly utilized for energy or stored as glycogen. Galactose and fructose must be converted to glucose before they can be utilized. After glucose enters the cell, it is quickly shunted into one of three possible metabolic pathways (described below) depending on the availability of substrates or the nutritional status of the cell. The ultimate goal of the cell is to convert glucose to carbon dioxide and water, during which process the cell obtains the high energy molecule adenosine triphosphate (ATP) from inorganic phosphate and adenosine diphosphate (ADP). The cell requires oxygen for the final steps in the electron transport chain. Nicotinamide adenine dinucleotide in its reduced form (NADH) will act as an intermediate to couple glucose oxidation to the electron transport chain (ETC) in the mitochondria where much of the ATP is gained.

The first step for all three pathways requires glucose to be converted to glucose-6-phosphate using the high-energy molecule ATP. This reaction is catalyzed by the enzyme hexokinase (Fig. 14-4). Glucose-6-phosphate can enter the Embden-Meyerhof pathway or the hexose monophosphate pathway, or can be converted to glycogen (Fig. 14-4). The first two pathways are important for the generation of energy from glucose; the latter pathway is important for the storage of glucose.

The Embden-Meyerhof (EM) pathway occurs in the cytosolic compartment of the cell. In this pathway, glucose is broken down into two three-carbon molecules of pyruvic acid, which can enter the tricarboxylic acid cycle (TCA cycle) or be further oxidized to two molecules of lactic acid (Fig. 14-4). The EM pathway requires no oxygen. Anaerobic glycolysis is important for tissues such as muscle, which often have important energy requirements without an adequate oxygen supply. These tissues can derive ATP from glucose in oxygen deficit and will accumulate lactic acid. The lactic acid diffuses from the muscle cell, enters the systemic circulation, and is then taken up and utilized by the liver. This process is called the Cori, or glucose-lactate, cycle.

In order for anaerobic glycolysis to take place, two moles of ATP must be consumed for each mole of glucose; however, four moles of ATP are directly produced resulting in a net gain of two moles of ATP. Further gains of ATP result from the introduction of pyruvate into the TCA cycle and NADH into the ETC. Other substrates have the opportunity to enter the pathway at several points. Glycerol released from the hydrolysis of triglycerides can enter at 3-phosphoglycerate, and fatty acids and ketones and some amino acids are converted or catabolized to acetyl coenzyme A (Acetyl CoA), which is part of the TCA cycle. Other amino acids enter the pathway as pyruvate or as deaminated α keto- and oxoacids. The conversion of fatty acids to acetyl CoA is called β-*Knoops oxidation*. The conversion of amino acids by the liver, and other specialized tissues such as the kidney, to substrates that can be converted to glucose is called *gluconeogenesis*. Gluconeogenesis also encompasses the conversion of glycerol, lactate, and pyruvate to glucose.

The second energy pathway is the hexose monophosphate shunt (HMP shunt), which is actually a detour of glucose-6-phosphate from the glycolytic pathway to become 6-phosphogluconic acid. This oxidized product permits the formation of ribose-5-phosphate and nicotinamide dinucleotide phosphate in its reduced form (NADPH). NADPH is important for erythrocytes, which lack mitochondria and are therefore incapable of the TCA cycle. The reducing power of NADPH is required for the protection of the cell from oxidative and free radical damage. Without NADPH, the lipid bilayer membrane of the cell and critical enzymes would eventually be destroyed, resulting in the death of the cell. The HMP shunt also permits pentoses, such as ribose, to enter the glycolytic pathway.

When the cell's energy requirements are being met, glucose can be stored as glycogen. This third pathway, which is

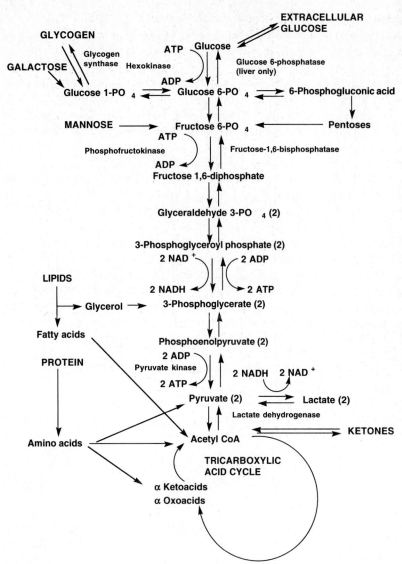

Figure 14-4. The Embden-Meyerhof pathway for anaerobic glycolysis. Entry points for glycogen, galactose, mannose, pentoses, lipids, ketones, and amino acids are illustrated. Glucose yields two molecules of pyruvate and two moles of ATP in this pathway. The entry point for gluconeogenesis (ketones, amino acids, and lactate) is at pyruvate or acetyl CoA. The entry point for glycerol is at 3–phosphoglycerate. Glucose can be synthesized and released to the ECF only from the liver because other tissues do not synthesize glucose-6-phosphatase. Glucose-6-phosphate cannot pass from the cell. The liver isoform of lactate dehydrogenase catalytically favors the synthesis of pyruvate, whereas muscle lactate dehydrogenase favors the synthesis of lactate. Therefore, muscles and other tissues can generate lactate in times of need, and the liver is capable of transforming this lactate back into glucose.

called *glycogenesis,* is relatively straightforward. Glucose-6-phosphate is converted to glucose-1-phosphate, which is then converted to uridine diphosphoglucose and then to glycogen by glycogen synthase. Several tissues are capable of the synthesis of glycogen, especially the liver and muscles. Hepatocytes are capable of releasing glucose from glycogen or other sources to maintain the blood glucose concentration. This is because the liver synthesizes the enzyme glucose-6-phosphatase. Without this enzyme, glucose is trapped in the glycolytic pathway. Muscle cells do not synthesize glucose-6-phosphatase, and therefore they are incapable of dephosphorylating glucose. Once glucose enters a muscle cell it remains as glycogen unless it is catabolized. *Glycogenolysis* is the process by which glycogen is converted back to glucose-6-phosphate for entry into the glycolytic pathway. Table 14-1 outlines the major energy pathways involved either directly or indirectly with glucose metabolism.

In summary, dietary glucose and other carbohydrates can either be utilized by the liver and other cells for energy or can be stored as glycogen for later use. When the supply of

glucose is low, the liver will use glycogen and other substrates to elevate the blood glucose concentration. These substrates include glycerol from triglycerides, lactic acid from skin and muscles, and amino acids. If the lipolysis of triglycerides is upregulated, it results in the formation of ketone bodies, which the brain can use as a source of energy through the TCA cycle. The synthesis of glucose from amino acids is gluconeogenesis. This process is used in conjunction with the formation of ketone bodies when glycogen stores are depleted—conditions normally associated with starvation. The principle pathway for glucose oxidation is through the Embden-Meyerhof pathway. NADPH can be synthesized through the HMP shunt, which is a side pathway from the anaerobic glycolytic pathway.

Regulation of Glucose Metabolism

The liver, pancreas, and other endocrine glands do a remarkable job of keeping blood glucose concentrations within a narrow range. During a brief fast, glucose is sup-

TABLE 14-1. Important Energy Pathways Associated with Glucose Metabolism

Metabolic Pathway	Substrate	Product	Use
Anaerobic Glycolysis	Glucose	Pyruvate/Lactate	Energy (ATP)
Glycogenesis	Glucose	Glycogen	Storage of glucose in liver/muscle
Glycogenolysis	Glycogen	Glucose-6-phosphate	Reclaim glucose to increase blood sugar (liver) or obtain energy (muscle)
Gluconeogenesis	Glycerol, Lactate, Amino Acids	Glucose	Increase blood sugar (liver)
TCA/ETC	Pyruvate	Carbon dioxide	Oxidative phosphorylation pathway of the mitochondria
β-Knoops Oxidation	Fatty acids	Acetyl CoA/Acetone/ Acetoacetic acid/ β-hydroxybutyric acid	Shuttles lipids into energy pathway. Can supply brain with ketones for energy

plied to the ECF from the liver through glycogenolysis. When the fasting period is longer than one day, glucose is synthesized from other sources through gluconeogenesis. Control of blood glucose is under two major hormones: insulin and glucagon (Figs. 14-5 and 14-6a). Other hormones and neuroendocrine substances also exert some control over blood glucose concentrations. This permits the body to respond to increased demands for glucose or to survive prolonged fasts. It also permits the conservation of energy as lipids when excess substrates are ingested.

Hormonal Regulation

Hormones function at several levels and in many different tissues. Their actions can affect the entry of glucose into the cell and can also influence the fate of glucose once it has en-

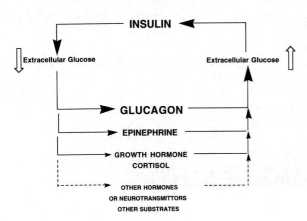

Figure 14-5. The role of hormones and other substances in glycemic control. Insulin is a hypoglycemic agent, glucagon is the predominant hyperglycemic agent. The open arrows (⇑) indicate either an increase or decrease in extracellular glucose concentration brought about by the various hormones. The size of the lettering emphasizes the relative importance of one hormone versus another. Epinephrine, growth hormone, and cortisol have a lesser role in glycemic control; however, the additive effects of each hormone can greatly enhance the activity of glucagon. (Adapted from Cryer PE. Glucose homeostasis and hypoglycemia. In: Wilson JD, Foster DW, eds. Williams Textbook of Endocrinology. Philadelphia: WB Saunders Co., 1992.)

tered the cell. To understand the action of hormones on the cell, first consider what would occur if we did not have regulatory mechanisms for glucose control. After a meal is ingested and as digestion occurs, the simple sugars will be absorbed first. This will be followed by the digestion and absorption of more complex sugars. The absorptive state is referred to as the *postprandial state.* The concentration of glucose in blood increases soon after a meal, and without endocrine control this concentration would keep increasing for a few hours. *Hyperglycemia,* an elevation of glucose in the ECF, including blood, would pull water from cells. If the concentration of glucose rose above 180 mg/dL, the tubular cells of the renal epithelium would not be able to completely reabsorb it from the filtrate, and some glucose would pass into the urine. This is called *glucosuria* and would result in the loss of water and electrolytes from the body. Remember that the brain requires a steady supply of glucose. Other tissues, such as muscle, vary their glucose requirements depending on their energy needs. After the carbohydrates from the meal are completely absorbed, the body's needs must still be met. This is the postabsorptive state. Unless another meal is eaten, the concentration of ECF glucose will decrease and the body's nervous tissue will lose function, resulting in hypoglycemic shock and, eventually, coma. Therefore, the endocrine system must assist in the meeting of three requirements: a steady supply of glucose must be available under all normal circumstances, excess glucose must be safely stored to prevent it from causing alterations in fluid and electrolyte balance or being lost, and the stored glucose must be used to supply the ECF when glucose is not being absorbed by the gut.

Insulin is a peptide hormone that is synthesized in the β cells of the islets of Langerhans in the pancreas. Insulin is the primary hormone responsible for the entry of glucose into cells; therefore insulin production and release increases after a meal, when ECF glucose concentrations are increasing (see Fig. 14-5). This protein is synthesized and stored in vesicles in the cytosol of the β cells until it is needed. As with most proteins that are synthesized with the specific aim of release into the blood stream, insulin is synthesized with a short leader peptide sequence. Once the pre-proinsulin leader

a. NORMAL

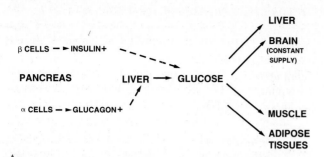

Figure 14-6a. Normal glycemic control. Insulin, released from the β cells of the pancreas, has a positive effect on the delivery of glucose to tissues. Glucagon, released from the α cells of the pancreas, stimulates the synthesis and release of glucose from the liver.

b. ADRENERGIC RESPONSE

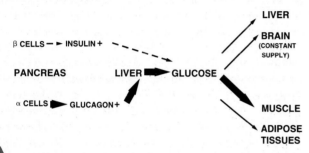

Figure 14-6b. The effect of epinephrine (adrenergic) activity on glucose. Glucagon release is increased and the effects of glucagon are also potentiated. Blood flow to the muscles increases, and this facilitates the delivery of glucose.

c. STARVATION

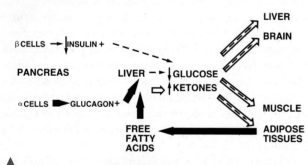

Figure 14-6c. The effect of prolonged starvation on the body. Starvation exhausts the supply of glycogen in the liver. In response to hypoglycemia, glucagon is greatly increased and insulin is suppressed. Adipose tissues increase the release of fatty acids, which in turn are converted to ketones by the liver. The ketones are used as energy by the brain and other tissues. Ketones are also converted to glucose by the liver (gluconeogenesis).

d. SEVERE STRESS OR INJURY

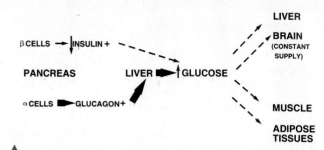

Figure 14-6d. Severe stress or injury. Stress or injury including surgery, causes an increase in glucagon which, in turn, causes blood glucose to become elevated. Epinephrine and cortisol are also increased in this case. The liver mobilizes glucose even though insulin secretion is suppressed. The decrease in insulin decreases the delivery of glucose to tissues, and therefore the patient becomes hyperglycemic.

e. POST PRANDIAL ABSORPTIVE STATE

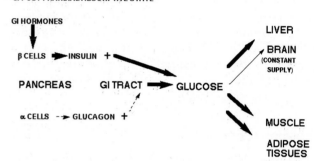

Figure 14-6e. Postprandial absorptive state. Insulin release increases in response to neurologic stimulus, hormone release, and increases in circulating substrates. The increase in insulin causes increased uptake by the liver, muscle, and adipose tissues.

f. IDDM

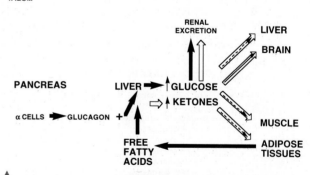

Figure 14-6f. IDDM. In IDDM, there is an absence of insulin due to β cell destruction. Glucagon is usually upregulated in response to the hyperglycemic state. Glucose uptake to the tissues is decreased due to the lack of insulin. Ketones are synthesized and released due to the effects of glycogen. Both ketones and glucose are lost to the urine.

Figure 14-6a–h. The relationship of the pancreas and the liver in regulating the glucose concentrations delivered to the muscles and adipose tissues. Note that under most circumstances, the brain receives a constant supply of glucose. Dashed arrows (- - →) indicate normal hormonal control and diminished glucose delivery where appropriate. Solid arrows (→) indicate normal or increased delivery of substrates to tissues. Solid arrows can also indicate increased release of hormones where appropriate. Open arrows (⇑) indicate the delivery of ketones from the liver to tissues. (From Unger RH, Orci L. Glucagon and the A cell. Physiology and Pathophysiology. The New England Journal of Medicine 304:1518–1524, Fig 3, 1981.)

g. NIDDM

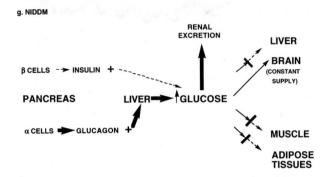

▲
Figure 14-6g. NIDDM. In NIDDM, glucose is increased; however, the concentration of glucagon does not increase to the same extent as that observed in IDDM. Insulin resistance is present and less glucose is delivered to the tissues. Ketones are usually not observed.

h. INSULIN SECRETING TUMOR

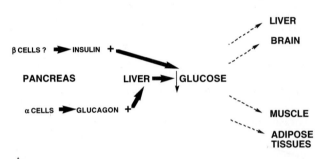

▲
Figure 14-6h. Insulinoma. Insulin secreting tumors are autonomous. Hypoglycemia will not decrease the release of insulin from an insulinoma. As a result glucose decreases even if glucagon release is stimulated.

peptide is cleaved, the proinsulin peptide, which is 82 amino acids long, is further cleaved by specific proteases. A 31-amino acid stretch is removed from the middle of the protein leaving a dipeptide with an α and β chain of 21 and 30 amino acids long, respectively. The dipeptide is held together by disulfide bonds between cysteines on the α and β chains. The internal peptide that was cleaved by the proteases is called the *C-peptide.* It has no known function in glucose regulation and is released from the vesicle along with insulin.

Insulin release in response to an increase in ECF glucose concentration is biphasic. There is an initial small rapid increase in circulating insulin that is additional to the prolonged release of insulin from the β cell. Insulin release is terminated when ECF glucose concentrations begin to decrease. Insulin binds to a receptor on the surface of most cells except neurons, erythrocytes, and retinal epithelium. These specialized cells obtain glucose without endocrine control. On all other cells the insulin-insulin receptor complex initiates a chain of events in the cell, which first increases the number of glucose transporters on the cell surface and then increases glycogenesis. The glucose transporters provide for the passage of glucose into the cell.

Glucagon is a peptide hormone synthesized by α cells in the pancreas. This hormone indirectly antagonizes the action of insulin. Both insulin and glucagon function in a "push and pull" type of antagonism to maintain glucose within its expected range.

Glucagon is secreted in response to a decrease in glucose concentration in the ECF (Figs. 14-5 and 14-6a). The overall function of glucagon is to increase hepatic glycogenolysis, inhibit glycolysis, and increase gluconeogenesis. Glucagon inhibits glycolysis and increases gluconeogenesis by increasing cyclic 3'5'-adenosine monophosphate (cAMP) concentrations within the cell. In conditions where glucagon is absent and glucose is present, the cell contains a small amount of fructose-2,6-bisphosphate, which is formed from fructose 6-phosphate and ATP under the catalytic control of the dienzyme complex phosphofructokinase 2/fructose-2,6-bisphosphatase. Fructose-2,6-bisphosphate acts as a potent inhibitor of fructose-1,6-bisphosphatase and as a stimulator of phosphofructokinase.[11] This serves to promote glycolysis and block the reverse reaction. When cAMP is increased, a number of different events occur. A protein kinase is activated and, in turn, further activates and deactivates other enzymes. Fructose-2,6-bisphosphatase is activated while the phosphofructokinase 2 end of the protein is inhibited. This permits the destruction of fructose-2,6-bisphosphate. Pyruvate kinase is deactivated, which allows for an accumulation of phosphoenolpyruvate in the cell. This permits an accumulation of fructose 1,6-bisphosphate and other precursors. These two steps inhibit glycolysis and increase gluconeogenesis.

Epinephrine is a neuroendocrine catecholamine that is synthesized by the adrenal medulla. Epinephrine is released in response to a decrease in glucose below a certain concentration (Fig. 14-6b). As an individual becomes even slightly hypoglycemic, epinephrine is released from the adrenal glands and an increase in activity in the sympathetic nervous system can be observed. Epinephrine is the "fight or flight" hormone. It antagonizes parasympathetic activity such as digestion. The blood supply to liver and muscle increases in the presence of epinephrine. Glycogenolysis and gluconeogenesis are also stimulated in the liver. In addition, glucose utilization is limited by the partial inhibition of insulin release.

Cortisol is another hormone that can influence ECF glucose (Fig. 14-5). Cortisol, a glucocorticoid hormone of the adrenal cortex, is under synthetic control by ACTH. Extended elevations of cortisol result in an increase in gluconeogenesis and insulin antagonism. This, in turn, results in hyperglycemia.

Growth hormone (GH) or *somatotropin* is an anterior pituitary hormone that antagonizes insulin action (Fig. 14-5). Prolonged elevations of growth hormone will result in a hyperglycemic state. Somatostatins are a group of peptides synthesized by the δ-cells of the pancreas and other tissues. Among their many functions, somatostatins inhibit the release of both insulin and glucagon, thereby blunting their effects and preventing rapid or unnecessary fluctuations of glucose.

Somatomedins are a group of polypeptides synthesized by the liver and released in response to growth hormone. They include the insulin-like growth factors that have both growth-promoting and insulin-like effects.

Thyroxine (T_4) is an amino acid derivative that is secreted by the thyroid gland. This hormone contains four iodine molecules on tyrosine residues. There is hardly any metabolic function that is not somehow affected by this hormone. Its effect on glucose metabolism and control is minimal; however, it is capable of promoting glycogenolysis and depleting the liver of glycogen. Thyroxine is also capable of increasing the rate of absorption of glucose from the gut.

Other hormones that increase serum glucose concentrations include β-endorphins, vasopressin (antidiuretic hormone), and angiotensin. Individually, they have only minor roles in glycemic control; however, the additive effects of several glycemic hormones should not be underestimated. This is especially true in patients with surgical stress or long-term clinical depression.

Glucose has some autoregulatory functions. When glucose concentrations are high and insulin is present, glucose will enter hepatocytes and muscle and increase glycogen synthesis and storage. Lactate also appears to function as a promoter of glycogen synthesis through gluconeogenesis.[7] It is of interest to note that the central role of the liver in glucose metabolism correlates with both its location and blood supply. First, all carbohydrates and other substrates including amino acids and lipids absorbed by the gut proceed to the liver through the hepatic portal system. Also, the pancreas secretes its hormones to the blood supply of the liver, and therefore this organ is exposed to more insulin and glucagon than the systemic tissues. Furthermore, the metabolic products of the mesenteric adipose tissues drain to the portal vessel as well. These metabolic products are some of the substrates required for gluconeogenesis.

ALTERATIONS IN CARBOHYDRATE METABOLISM

Humans can have both hyper- or hypoglycemic states. Hyperglycemic states are more prevalent. The expected glucose concentration in plasma is dependent on the time and contents of the subject's last meal. After an overnight fast, the plasma glucose concentration is normally between 70 to 110 mg/dL, or 3.9 to 6 mmol/L. The expected overall range for random glucose in non-fasting individuals is 70 to 150 mg/dL, or 3.9 to 8.3 mmol/L. The distribution curve of plasma glucose concentrations is unimodal in most races. This indicates that there is no clear-cut boundary for hyperglycemia. Diagnosis of hyperglycemia depends on the observation of glucose concentrations beyond a certain recommended concentration and under defined conditions. It must be stressed that transient hyperglycemic states can be caused by a number of physiologic and endocrine changes and that the diagnosis of diabetes mellitus is not always absolute. Figure 14-6 summarizes the hormonal regulation of glucose and other substrates in response to a number of situations.

Diabetes Mellitus

Diabetes mellitus is a diverse group of hyperglycemic disorders with different etiologies and clinical pictures. Common to diabetes mellitus is a relative or absolute deficiency of insulin. In other words, the hyperglycemia observed in diabetes mellitus is caused by either a lack of insulin or insulin insensitivity by the tissues. Complications of diabetes include diseases of the eyes, kidneys, nervous system, and blood vessels.

Diabetes mellitus can be divided into two major types (Table 14-2). Type I diabetes is characterized by an absolute lack of insulin, while Type II is a relative deficiency of insulin without loss of production. Historically, Type I diabetes mellitus has been called *juvenile onset diabetes*. A name for Type I diabetes that is used more often is *insulin-dependent diabetes mellitus (IDDM)*. These descriptions inferred the early age of onset and dependence of these individuals on exogenous insulin for survival. Type II diabetes mellitus, with its relative lack of insulin, has been termed *adult onset diabetes* and *non-insulin-dependent diabetes mellitus (NIDDM)*. These classifications are somewhat misleading and may eventually need to be modified. The onset of Type I diabetes can occasionally occur in adults, and NIDDM individuals sometimes require exogenous insulin.

Type I, Insulin-Dependent Diabetes Mellitus (IDDM)

The risk factors for IDDM and the etiology of this disease are beginning to be understood (Table 14-2). IDDM is common among Caucasians and African-Americans. It is uncommon in Asians, American Indians, African Blacks, Inuits, and other certain races.[10] However, genetics only play a partial role in the formation of IDDM. Within a country, there can be geographical and seasonal differences in disease prevalence.[10] Additionally, the concordance rate of IDDM in genetically identical twins is less than 50%.

There is much evidence to suggest that the genetic association to IDDM is through the major histocompatibility complex (MHC) human leukocyte antigen (HLA) alleles. The products of these genes extrude from the surface of cells. The role of the products in IDDM may be due in part to their antigenic composition and in part to their role in the immune response. The MHC genes are divided into two classes. Class I antigens are found on all nucleated cells. These antigens are polymorphic and belong to either the A, B, or C series. The class I antigens associate with β$_2$-microglobulin on the cell membrane. The class I antigens play a role in recognition of self and the destruction of foreign cells and cells that are altered by viruses or malignant diseases. To prevent self-destruction of cells bearing these class

TABLE 14-2. Classes of Diabetes Mellitus

Type			Age of Onset	Pathogenesis	Risk Factors
Type I	IDDM	Juvenile Onset	≤ 20 years	β-cell destruction Absence of insulin	Genetic: HLA DR3/4, DQw3.2 Environmental: Toxins, Viruses Immune: IL-1 and TNFγ
Type II	NIDDM	Adult Onset	≥ 20 years	Relative lack of insulin or insulin resistance	Genetic: Unknown Environmental: Diet
	GDM			Likely insulin resistance	Possible pre-NIDDM

I antigens, cytotoxic or killer T lymphocytes will normally lyse only cells bearing class I antigens that have a foreign antigen associated with them. Class II antigens are normally found only on the surfaces of monocytes, macrophages, dendritic cells, and B lymphocytes. Interferon-γ (IFN-γ) is capable of inducing the class II antigens on the surface of other cells.[9] The class II antigens encompass the polymorphic series of antigens called DR, DQ, and DP. Genetic studies on the HLA types of IDDM individuals has revealed that 95% of IDDM patients are positive for HLA-DR3 and/or DR4, compared to 65% of the general population. Heterozygotes for DR3,4 are at greater risk than homozygotes. The HLA DQ series appears to be even more closely associated with IDDM. When an individual has an aspartate residue at position 57 on the β peptide of the DQ chain, IDDM does not occur. However, individuals homozygous for the expression of DQ β chains that contain another amino acid residue at position 57 are very susceptible to IDDM. Another close HLA association with the occurrence of IDDM is the HLA DQw3.2 series. On the other hand, inheritance of the HLA DR2 gene appears to be protective because individuals with this gene are 10 times less likely to develop IDDM.

The environmental factors that may contribute to, or precipitate, IDDM in a susceptible individual include environmental toxins, such as nitroso compounds or Vacor™, a rodenticide for warfarin-resistant rats, and a wide range of viruses including mumps, rubella, cytomegalovirus, and the Epstein-Barr virus. Recent viral illnesses are associated with the occurrence of IDDM; however, there is still no cause-and-effect relationship that explains how viruses play a role in IDDM.

IDDM also appears to be an autoimmune disease. Some of the evidence to that extent includes the following: the HLA class II D series is associated with other autoimmune disease, lymphocyte infiltration of the islets of Langerhans occurs in the early stages of IDDM, and islet cell antibodies for specific membrane and cytoplasmic antigens are present in IDDM. However, what remains to be answered is whether these observed phenomena are the cause of IDDM or the result of prior injury to β cells. A plausible hypothesis that explains the development of β cell destruction is as follows:[10] the β cells of the pancreas are known to be

extremely sensitive to free radical damage of the type produced by tumor necrosis factor α (TNF α) and interleukin 1 (IL 1). Nonspecific virus or bacterial infections induce the macrophage-mediated release of IL 1 and TNF α. These agents are then responsible for the destruction of a small number of β cells. Normally, an individual would not be affected by this loss; however, susceptible individuals produce autoantibodies to β cell components, the destruction of β cells progresses rapidly, and IDDM presents itself.

Patients presenting with a clinical picture of IDDM are typically below the age of 20 years and have polyphagia, polyuria, and weight loss with a concomitant elevated blood glucose. Islet cell destruction is present in these individuals. Symptoms usually occur abruptly, and ketonemia and ketonuria can be present with the hyperglycemia. Therapy consists of administration of exogenous insulin and diet counseling. The early treatment of diabetic ketoacidosis may require intensive therapy to correct the hyperglycemia and electrolyte disturbance as well as the concomitant metabolic acidosis.

Type II, Non-Insulin Dependent Diabetes Mellitus (NIDDM)

NIDDM is the most common form of diabetes. Typically, this disease is characterized by an underproduction of insulin or insulin insensitivity at target tissues. Genetic factors may play a greater role in this disease than in IDDM. Concordance rates for NIDDM in identical twins are 100%. Currently, a genetic marker has not been discovered for NIDDM. The etiology of this disease is unknown, although some observed correlations may eventually assist in determining its causes. Environmental factors, such as diet, may play an important role in the development of NIDDM. Obesity, especially of the abdominal viscera, is common in individuals with NIDDM. Changes in the normal substrate concentrations delivered to the liver, adipose tissue, and skeletal muscle are thought to affect the normal functioning of the insulin receptor.[2] This, in turn, may decrease the number of glucose transporters that are delivered to the cell surface, thereby decreasing the delivery of glucose into the cell. Although this model oversimplifies the actual molecular mechanisms, it is known that restriction of caloric intake and weight loss are often sufficient to control NIDDM.

Otherwise, the patient is usually treated with sulfonylureas or other oral hypoglycemic drugs. These drugs appear to facilitate the release of insulin from β cells and may also increase the target tissues' sensitivity to insulin. Interestingly, NIDDM may be, in part, a disease caused by an imbalance between insulin and glucagon concentrations. It has been observed in true NIDDM individuals that the number of β cells do not decrease; however, there is a significant increase in the number of α cells in the islets.[10]

Pathophysiology of Diabetes Mellitus

In both IDDM and NIDDM, the individual will be hyperglycemic, which can be severe. Glucosuria can also occur after the renal tubular transporter system for glucose becomes saturated. This happens when the glucose concentration of plasma exceeds roughly 180 mg/dL in an individual with normal renal function and urine output. As hepatic glucose overproduction continues, the plasma glucose concentration reaches a plateau around 300 to 500 mg/dL (17 to 28 mmol/L). Provided renal output is maintained, glucose excretion will match the overproduction causing the plateau.

The individual with IDDM has a higher tendency to produce ketones. NIDDM patients seldom generate ketones, but instead have a greater tendency to develop hyperosmolar nonketotic states. The difference in glucagon and insulin concentrations in these two groups appears to be responsible for the generation of ketones through increased β-Knoops oxidation. In IDDM, there is an absence of insulin with an excess of glucagon. This permits gluconeogenesis and lipolysis to occur. In NIDDM, insulin is present as is (sometimes) hyperinsulinemia, therefore glucagon is attenuated. Fatty acid oxidation is inhibited in NIDDM. This causes fatty acids to be incorporated into triglycerides for release as very low density lipoproteins (VLDL).

The laboratory findings of a diabetic with ketoacidosis tend to reflect dehydration, electrolyte disturbances, and acidosis. Acetoacetate, β-hydroxybutyrate, and acetone are produced from the oxidation of fatty acids. The two former ketone bodies contribute to the acidosis. Lactate, fatty acids, and other organic acids can also contribute to a lesser degree. Bicarbonate and total carbon dioxide are usually decreased due to Kussmaul breathing (deep respirations). This is a compensatory mechanism to blow off carbon dioxide and remove hydrogen ions in the process. The anion gap in this acidosis can exceed 16 mmol/L. Serum osmolality is high due to hyperglycemia; sodium concentrations tend to be lower due in part to losses (polyuria) and in part to a shift of water from cells because of the hyperglycemia. Care must be taken not to falsely underestimate the sodium value because of hypertriglyceridemia. Grossly elevated triglycerides will displace plasma volume and give the appearance of decreased electrolytes when flame photometry or prediluted ion specific electrodes are used for sodium determinations. Hyperkalemia

is almost always present due to the displacement of potassium from cells in acidosis. This is somewhat misleading because the patient's total body potassium is usually decreased.

More typical of the untreated NIDDM patient is the nonketotic hyperosmolar state. The individual presenting with this syndrome has an overproduction of glucose; however, there appears to be an imbalance between production and elimination in urine. Often, this state is precipitated by heart disease, stroke, or pancreatitis. Glucose concentrations exceed 300 to 500 mg/dL (17 to 28 mmol/L) and severe dehydration is present. The severe dehydration contributes to the inability to excrete glucose in the urine. Mortality is high with this condition. Ketones are not observed, because the severe hyperosmolar state inhibits the ability of glucagon to stimulate lipolysis. The laboratory findings of nonketotic hyperosmolar coma include: plasma glucose values exceeding 1000 mg/dL (55 mmol/L), normal or elevated plasma sodium and potassium, slightly decreased bicarbonate, elevated blood urea nitrogen (BUN) and creatinine, and an elevated osmolality. The gross elevation in glucose and osmolality, the elevation in BUN, and the absence of ketones distinguish this condition from diabetic ketoacidosis.

Gestational Diabetes Mellitus (GDM)

GDM is a form of diabetes mellitus that occurs in a woman during pregnancy. In half the cases, the woman reverts back to a nondiabetic state after delivery. Infant mortality rates in untreated GDM are 7%. Insulin can control the hyperglycemia; however, it is not known whether this improves infant mortality.[10] Although this form of diabetes appears to be related to NIDDM, other factors may be etiologically important. Some gestational diabetic women may be progressing to IDDM and have marginal β cell function. GDM is diagnosed using the oral glucose tolerance test (OGTT). Results greater than 190 mg/dL (10.6 mmol/L) at 1 h, 165 mg/dL (9.2 mmol/L) at 2 h, and 145 mg/dL (8 mmol/L) at 3 h after a 100 g glucose load are diagnostic. Screening at the initial visit and at around 24 weeks is recommended.[10]

Pregnancy and Infants of Diabetic Mothers

An increase in insulin resistance is normal in the mother as pregnancy progresses.[4] This serves to increase the delivery of substrates to the fetus. The supply of substrates to the fetus is dependent on the maternal blood substrate supply and placental blood flow. Because the mother's lipolysis pathways are upregulated, ketosis may occur when just one meal is missed. For this reason, fasting blood tests are not recommended in pregnancy.[10]

In pregnant, diabetic women, the increase in substrates is exaggerated and the infant receives more glucose, lipids, and amino acids than its tissues would normally encounter if the mother was nondiabetic. These substrates trigger the fetus to produce insulin in its own β cells. The fetus of the diabetic woman overproduces insulin to compensate for the elevated substrates, which, in turn, increases fetal growth. Diabetic mothers often have macrosomic infants. These infants are larger by weight and length. The internal organs are enlarged, with the exception of the brain. Adipose tissues are also increased. Because the infant has hyperplastic β cells and insulinemia, it is important to monitor these infants continuously for hyperglycemia shortly after birth. Complications for the infant of a diabetic mother are numerous and can include the aforementioned hypoglycemia, respiratory distress syndrome, hypocalcemia, polycythemia, and hyperbilirubinemia.

Maturity Onset Diabetes of Youth (MODY)

MODY is a very rare form of type II diabetes or NIDDM that is inherited in an autosomal dominant fashion.[3] This is the only form of diabetes having a known genetic inheritance pattern that can be traced in family studies. However, the actual gene responsible for this form of diabetes is unknown.

Glucose Intolerance in Nondiabetic States and Secondary Diabetes

Impaired glucose tolerance and so-called secondary diabetes can occur for many reasons. Recall that glycemic control requires normal hepatic and pancreatic function and an intact endocrine system. Skeletal muscle and adipose tissue, in conjunction with the liver, also play pivotal roles in glucose usage and production. Any factor that can influence the aforementioned systems can ultimately affect glycemic control.

Impaired glucose intolerance very often refers to the abnormal outcome of an oral glucose tolerance test (OGTT). It can, however, imply a loss of glycemic control resulting in an elevation in the fasting blood glucose and an exaggeration of the postprandial glucose values. Increasing age is the most common factor that will ultimately affect the OGTT in the general population. Other factors that can affect the OGTT include patient preparation, such as fasting compliance and duration of the fast prior to the test. In addition, patients that are bedridden may develop glucose intolerance. Other important factors that should be considered include the time of day for the test, the nutritional status of the patient, and the carbohydrate load that the patient has received during the days preceding the OGTT. During the test, beverages containing caffeine must be avoided. If carbohydrate intake in the days prior to the test is low, then the OGTT will be increased. The list of drugs that can elevate or decrease blood glucose is extensive, and the physician must be aware of these when ordering an OGTT on a patient.

Secondary diabetes is a syndrome of diabetes that can be defined by a known cause other than the development of true IDDM, NIDDM, or GDM. Some causes include pancreatic disease, elevations in glucocorticoids or administration of exogenous glucocorticoids, abnormalities in the synthesis of insulin or the insulin receptor, anti-insulin antibodies, and

drug-induced diabetes. Pentamidine, which is used in the treatment of *Pneumocystis carinii* pneumonia in HIV-infected patients, has been reported to cause drug-induced islet cell destruction in about one fifth of the patients treated.[6]

Diseases and Conditions That Cause Hypoglycemia

Hypoglycemia can have many causes; some are transient and relatively insignificant, others can be life-threatening (Table 14-3). The plasma glucose concentration at which glucagon and other glycemic factors are released is between 65 to 70 mg/dL (3.6 to 3.9 mmol/L), and at about 50 to 55 mg/dL (2.8 to 3.0 mmol/L) observable symptoms of hypoglycemia appear. The warning signs and symptoms of hypoglycemia are all related to the central nervous system. The release of epinephrine into systemic circulation and norepinephrine at nerve endings of specific neurons act in unison with glucagon to increase plasma glucose. A sense of anxiety or anxiousness ensues. Heart rate may increase and the individual may feel dizzy, cold, and sweaty. Tremors may be noticed in the hands. This is the typical response of the sympathetic nervous system along with the sustained release of epinephrine from the adrenal medulla. If the individual is incapable of correcting the hypoglycemia, then the symptoms will include loss of neuronal function including slurred speech; loss of motor coordination, unconsciousness, and eventually coma; irreversible neuronal damage; and death.

Hypoglycemia can be classified as postabsorptive and postprandial. Postabsorptive hypoglycemia is referred to as fasting hypoglycemia. Postprandial hypoglycemia is often called reactive hypoglycemia. Postabsorptive hypoglycemia infers a loss of glycemic control because, during a fast, the normal individual can rely on gluconeogenesis to maintain the extracellular glucose concentration. Fasting hypoglycemia therefore is a serious form of the disease. Reactive hypoglycemia, on the other hand, is usually not serious. Postabsorptive (fasting) hypoglycemia can be due to insulin, sulfonylureas, or alcohol. That insulin and sulfonylureas cause hypoglycemia is self-explanatory. Alcohol causes hypoglycemia by preventing glycogen storage and inhibiting glycolysis. The oxidation of alcohol depletes NAD^+ from cells, and without this substrate glycolysis is inhibited. Furthermore, the end product of ethanol metabolism is acetic acid, which enters the TCA cycle as acetyl coA or is shunted into lipid synthesis. Organ failure, including liver, heart, or kidneys, can cause hypoglycemia. Deficiencies of cortisol (Addison's disease), glucagon, or epinephrine can also cause hypoglycemia. Insulin-secreting tumors (insulinoma) can release insulin autonomously without counter-regulation.

In children, there are three important types of hypoglycemia. These are neonatal, congenital, and ketotic hypoglycemia. Neonatal hypoglycemia usually occurs as a result of hyperinsulinism in the neonate. This is normally transient; however, a diagnosis of an insulinoma must be ruled out if the condition persists.

TABLE 14-3. Causes of Hypoglycemia	
Type and Cause	**Mechanism**
Postabsorptive (Fasting)	
Drugs/alcohol	Hypoglycemic agents: reduce substrates or inhibit gluconeogensis
Organ Failure	Diminished metabolic capacity
Insulinoma	Autonomous insulin secreting tumor
Neonatal	Transient hyperinsulinism, impaired gluconeogenesis
Congenital	Glucose-6-phosphatase deficiency (von Gierke disease)
	Glycogen synthase deficiency
	Fructose-1,6-bisphosphatase deficiency
Ketotic Hypoglycemia	Idiopathic, usually self-correcting
Postprandial (Reactive)	
Congenital	Galactosemia
Idiopathic	Fructose Intolerance

The most common congenital form of hypoglycemia is glucose-6-phosphatase deficiency type I, which is also called von Gierke disease. This disease is characterized by a severe postabsorptive hypoglycemia that coincides with metabolic acidosis ketonemia and elevated lactate and alanine. Glycogen accumulation in the liver causes hepatomegaly. The patients usually have hyperlipidemia, uricemia, and growth retardation. Although the glycogen accumulation is irreversible, the disease can be kept in control by avoiding the development of hypoglycemia. Liver transplantation corrects the hypoglycemic condition.

Other enzyme defects or deficiencies that cause hypoglycemia include glycogen synthase, fructose-1,6-bisphosphatase, phosphoenolpyruvate carboxy kinase, and pyruvate carboxylase. Glycogen debrancher enzyme deficiency does not cause hypoglycemia but does cause hepatomegaly.

Ketotic hypoglycemia of childhood is a mild form of hypoglycemia that usually corrects itself. The development of a hypoglycemic state occurs due to fasting because of illnesses such as the flu. The formation of ketones appears to be a compensatory condition to the hypoglycemia. The mechanism of ketotic hypoglycemia of childhood is unknown; however, the amino acid alanine, a primary substrate for gluconeogenesis, is decreased in the plasma of these children.

Postprandial (reactive) hypoglycemia causes include the congenital deficiencies of enzymes of carbohydrate metabolism, including galactosemia and fructose intolerance. Galactosemia is characterized by a defect of the enzyme galactose-1-phosphate uridyltransferase. Postprandial hypoglycemia is accompanied by diarrhea and vomiting, and occurs because of the inhibition of glycogenolysis. Galactose must be removed from the diet to prevent the development of irreversible complications. The disorder can be identified by measuring erythrocyte galactose-1-phosphate uridyltransferase activity. On the other hand, fructose-1-phosphate aldolase deficiency causes nausea and hypoglycemia after fructose ingestion.

Specific inborn errors of amino acid metabolism and long chain fatty acid oxidation are also responsible for hypoglycemia. There are also alimentary and idiopathic hypoglycemias in the postprandial category. Alimentary hypoglycemia appears to be due to an increase in the release of insulin in response to rapid absorption of nutrients after a meal or the rapid secretion of insulin-releasing gastric factors. Idiopathic postprandial hypoglycemia is a controversial diagnosis that may be overused.[1]

ROLE OF THE LABORATORY IN DIFFERENTIAL DIAGNOSIS AND MANAGEMENT OF PATIENTS WITH GLUCOSE METABOLIC ALTERATIONS

The demonstration of hyperglycemia or hypoglycemia under specific conditions is used to diagnose diabetes mellitus and hypoglycemic conditions. Other laboratory tests have been developed to identify insulinomas and to monitor glycemic control and the development of renal complications.

Glucose Measurements

Glucose can be measured from serum or whole blood. Today, most glucose measurements are performed on serum or plasma. The glucose concentration in whole blood is approximately 15% lower than the glucose concentration in serum or plasma. Serum or plasma must be refrigerated and separated from the cells within 1 hour to prevent substantial losses of glucose by the cellular fraction, particularly if the white blood cell count is elevated. Fluoride ions are often used as an anticoagulant and preservative of whole blood, particularly if the analysis is delayed. Fluoride inhibits glycolytic enzymes.

The most used methods of glucose analysis employ the enzymes glucose oxidase or hexokinase (Fig. 14-7). Glucose oxidase converts β-D-glucose to gluconic acid. Mutarotase may be added to the reaction to facilitate the conversion of α-D-glucose to β-D-glucose. Oxygen is consumed and hydrogen peroxide is produced. The reaction can be monitored polarographically either by measuring the rate of disappearance of oxygen using an oxygen electrode or by consuming hydrogen peroxide in a side reaction. Horseradish peroxidase is used to catalyze the second reaction, and the hydrogen peroxide is used to oxidize a dye compound. Two commonly used chromagens are 3-methyl-2-benzothiazolinone hydrazone and N,N-dimethylaniline. The shift in absorbance can be monitored spectrophotometrically. This coupled reaction is known as the *Trinder reaction*. One disadvantage of the glucose oxidase procedure is that uric acid, bilirubin, and ascorbic acid can also be oxidized by peroxidase preventing the oxidation and detection of the chromagen. The direct measurement of oxygen using the polarographic technique avoids this interference.

The hexokinase procedure allows less interference than the coupled glucose oxidase procedure. Hexokinase in the presence of ATP converts glucose to glucose-6-phosphate. Glucose-6-phosphate and the cofactor $NADP^+$ are converted to 6-phosphogluconate and NADPH by glucose-6-phosphate dehydrogenase. NADPH has a strong absorbance maxima at 340 nm, and the rate of appearance of NADPH can be monitored spectrophotometrically.

Fasting Plasma Glucose

The random glucose measurement is of little diagnostic value unless it is grossly elevated or decreased. A more important measurement is the fasting glucose concentration. This is drawn after an overnight fast (10–16 h). A fasting glucose concentration greater than 140 mg/dL (7.8 mmol/L) demonstrated on more than one occasion is considered diagnostic for diabetes mellitus by the National Diabetes Data Group.[5]

Two-Hour Postprandial Plasma Glucose

The two-hour postprandial glucose measurement is often used in conjunction with the fasting plasma glucose. The patient is advised to consume a meal that contains approximately 100 grams of carbohydrates. Two hours after eating, a blood sample is drawn for plasma glucose measurement. A

glucose value greater than 200 mg/dL (11.1 mmol/L) indicates diabetes mellitus. A variation of this test is to use a standardized load of glucose. A solution containing 75 grams of a glucose is administered, and a specimen for plasma glucose measurement is drawn 2 hours later.

Oral Glucose Tolerance Test (OGTT)

The OGTT is the most sensitive test for the diagnosis of diabetes. The National Diabetes Data Group recommends that the test be performed on ambulatory individuals who are not on restricted diets. Approximately 150 g of carbohydrates should be consumed on each of the preceding 3 days. A sample of the patient's blood is drawn after an overnight fast. The patient then consumes 75 g of a glucose solution and blood is drawn every 30 minutes for two hours. For children, glucose is administered at 1.75 g glucose/kg body weight to a 75 g maximum. A plasma glucose greater than or equal to 200 mg/dL (11.1 mmol/L) at the 2-hour time point and at one other time point indicates diabetes mellitus. Impaired glucose tolerance is diagnosed with a plasma glucose between 140 and 200 mg/dL (7.8 and 11.1 mmol/L) at 2 hours and greater than or equal to 200 mg/dL (11.1 mmol/L) at one time point in the test. Gestational diabetes is considered present if two or more of the values in the OGTT are greater than the following: fasting, 105 mg/dL (5.8 mmol/L); 1 h, 190 mg/dL (10.6 mmol/L); and 2 h, 165 mg/dL (9.2 mmol/L). A variation of the OGTT is the intravenous glucose tolerance test. This test is relatively insensitive and is not useful for the diagnosis of diabetes mellitus. This assay provides a measurement of how fast glucose is cleared from blood. A value ($K = 0.69/T_{1/2} \times 100$) is derived from the half-life of glucose clearance ($T_{1/2}$). The half-life of glucose is obtained from a plot of glucose concentrations obtained after the infusion of glucose.

Urinary Glucose

Glucose can be detected readily in urine using the specific test strips that contain glucose oxidase, peroxidase, and a

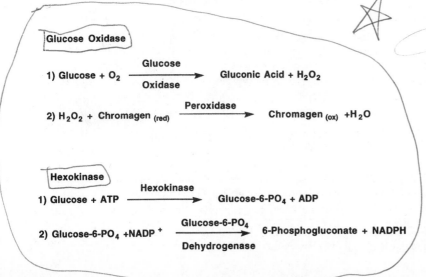

Glucose Oxidase

1) Glucose + O_2 $\xrightarrow{\text{Glucose Oxidase}}$ Gluconic Acid + H_2O_2

2) H_2O_2 + Chromagen (red) $\xrightarrow{\text{Peroxidase}}$ Chromagen (ox) + H_2O

Hexokinase

1) Glucose + ATP $\xrightarrow{\text{Hexokinase}}$ Glucose-6-PO$_4$ + ADP

2) Glucose-6-PO$_4$ + $NADP^+$ $\xrightarrow{\text{Glucose-6-PO}_4 \text{ Dehydrogenase}}$ 6-Phosphogluconate + NADPH

Figure 14-7. Reaction schemes for glucose oxidase and hexokinase reactions. (*A*) In the presence of oxygen, glucose oxidase converts glucose to gluconinc acid and hydrogen peroxide. The rate of oxygen consumption can be measured polarographically. Hydrogen peroxide can also be used to oxidize a chromagen to a colored product using peroxidase. The development of the chromagen is related to the concentration of glucose in the specimen. (*B*) Hexokinase phosphorylates glucose in the presence of ATP. Glucose-6-phosphate can then be converted to 6-phosphogluconate by glucose-6-phosphate dehydrogenase in the presence of $NADP^+$. The reaction is monitored spectrophotometrically because reduced NADPH absorbs light at 340 nm.

chromagen. Reducing agents, such as uric acid, ascorbic acid, levodopa, methyldopa and salicylates, can all interfere with the oxidation of the chromogen, resulting in negative bias. Other carbohydrates containing reducing groups, along with glucose, can be detected using Benedict's and Fehling's reagents, as mentioned earlier.

Urinary Ketones

Acetone and acetoacetic acid can be detected in urine using the Acetes™ or Ketostix™ systems. These tablets or strips use nitroprusside (sodium nitroferricyanide) to detect ketones. Note that β hydroxybutyric acid lacks a ketone group and is not detected by this assay. However, this does not present a problem, because all three fatty acid breakdown products are produced simultaneously, albeit not equally. β-hydroxybutyric acid may be the predominant ketone formed in diabetic ketoacidosis; therefore a weak positive result may not reflect the extent of the ketoacidosis. This is further confused by the appearance of acetoacetate later, often while the patient is being treated.

Quantitative assays for acetoacetate and β-hydroxybutyric acid are available using β-hydroxybutyrate dehydrogenase and either NADH or NAD. If NAD is used as the cofactor and the reaction is buffered at around pH 9.0, then β-hydroxybutyric acid is measured. On the other hand, a separate reaction using NADH and buffered around pH 7.0 would measure acetoacetic acid.

Glycosylated Proteins and HbA$_{1C}$

The aim of diabetic management is to maintain the blood glucose concentration within or near the non-diabetic range with a minimal amount of fluctuations. Serum or plasma glucose concentrations can be measured by laboratories in addition to self-monitoring of whole blood glucose concentrations by the patient. Long-term blood glucose regula-

tion can be followed by measurement of glycosylated hemoglobins. This provides the clinician with a time-averaged picture of the patient's blood glucose concentration.

Many proteins are known to react with carbohydrates at the peptide N-terminus forming glycosylated peptides. Glucose can rapidly react with hemoglobin to form a labile aldimine (Schiff base, Fig. 14-8). This equilibrium product can then undergo an Amadori rearrangement to form glycosylated hemoglobin (ketoamine). The ketoamine product is stable and cannot revert back to hemoglobin and glucose. HbA$_{1c}$ is the largest subfraction of normal HbA in both diabetic and non-diabetic subjects. This fraction is formed by the reaction of the β chain of HbA with glucose.

The ketoamine (HbA$_{1c}$) fraction reflects the concentration of glucose present in the body over a prolonged time period related to the 60-day half-life of erythrocytes. Normal values are method-dependent, although a concentration of around 5% to 7% is within the normal range of most methods. HbA$_{1c}$ values of 12% or greater can be expected in diabetics with poor control. Methods employed in the measurement of HbA$_{1c}$ must not measure the ketoamine intermediate; otherwise the results will be influenced by short-term fluctuations in glucose concentration. Table 14-4 outlines the various methods for glycosylated hemoglobin analysis.

Cation exchange has been the most accepted technique for the measurement of HbA$_{1c}$. High-pressure liquid chromatography (HPLC) or minicolumns can be used to determine HbA$_{1c}$. Hemoglobin F interferes with this procedure because it is co-eluted with HbA$_{1c}$. The interference of hemoglobin F in this assay is especially problematic in the first 6 months of life, and more commonly in sickle cell patients that express both HbS and HbF. In uremic patients, carbamylated hemoglobins can also elute in the cation exchange methods. The interference of HbF and carbamylated hemoglobin is eliminated in the affinity chromatography methods that employ a phenyl boronate affinity support.

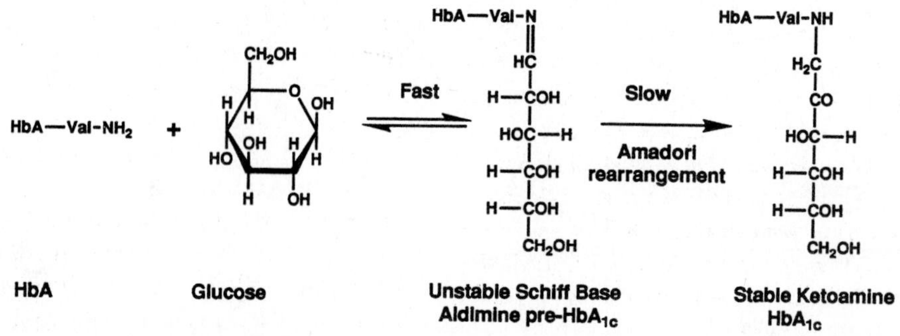

Figure 14-8. The nonenzymatic glycosylation of HbA to HbA$_{1c}$. Glucose initially forms an unstable aldimine (Schiff base) with the amino group of valine, the terminal amino acid on the β chain of HbA. The unstable base can rapidly reverse or stabilize itself through an Amadori rearrangement. The stable ketoamine does not readily reverse itself, and HbA$_{1c}$ is then present until the erythrocyte is destroyed. (From Higgis PJ, Bunn HF. Kinetic analysis of the nonenzymatic glycosylation of hemoglobin. J Biol Chem 1981; 256:5204–5208.)

TABLE 14-4. Selected Methods for the Measurement of Glycated Hemoglobins

Method	Measurement	Interference
Cation exchange	HbA_{1c}	Carbamyl Hb HbF Temperature-sensitive
Affinity chromatography Phenyl boronate matrix	Glycated Hb	
Monoclonal antibody Latex agglutination	HbA_{1c}	
Phenylboronate capture Fluorescence quenching	Glycated Hb	

One such procedure has been automated on the Abbott Im_x™ system, which employs fluorescence quenching for hemoglobin measurement. Individuals with accelerated hemoglobin turnover, such as hereditary spherocytosis or sickle cell, will have considerably low HbA_{1c} values irrespective of glycemic control. Some procedures do not separate the other glycated fractions (A_{1a} and A_{1b}) and a total glycated hemoglobin is reported.

Miles has introduced a new method based on an immunochemical technique for the measurement of HbA_{1c}. A monoclonal antibody is used in the assay and only HbA_{1c} is measured. The antibody employed in the assay has been selected on the basis of its ability to recognize unique amino acids around the N-terminal valine and also to recognize glucose. No other glycated hemoglobins are recognized, nor is the labile Schiff base recognized. A proprietary agglutinator (that most likely resembles polyvalent glycated HbA_{1c} peptides) is present in the reagent pack and causes the agglutination of specific anti-HbA_{1c}-coated latex particles. The agglutination is measured by an increase in turbidity in the test cartridge. HbA_{1c}, present in the specimen or control, will cause an inhibition of the agglutination reaction. The degree of inhibition is proportional to the concentration of HbA_{1c} present in the specimen. The final concentration of HbA_{1c} is calculated as a ratio of HbA_{1c} to total hemoglobin. Serum proteins are also non-enzymatically glycosylated. Fructosamine refers to any glycated serum protein. Typically, the predominant glycated serum protein is albumin, and the degree of glycosylation of albumin is then a measure of the glucose control of the patient over a period of time related to the half-life of albumin. The principle for the analysis of fructosamine is the reduction of the dye nitroblue tetrazolium. The half-life of albumin is approximately 2½ weeks, and thus fructosamine reflects short-term glucose control.

Microalbuminuria

Diabetes mellitus causes progressive changes to the kidneys and ultimately results in diabetic renal nephropathy. This complication progresses over a period of years and may be delayed by aggressive glycemic control. An early sign that nephropathy is occurring is an increase in urinary albumin. Microalbumin measurements are useful to assist in diagnosis at an early stage and prior to the development of proteinuria. Microalbumin concentrations are between 20 to 300 mg/d. Proteinuria is typically greater than 0.5 g/d.[8]

Insulin and C-Peptide

Insulin measurements are not required for the diagnosis of diabetes mellitus. However, in certain hypoglycemic states, it is important to know the concentration of insulin in relation to the plasma glucose concentration. To investigate an insulinoma, the patient is required to fast under controlled conditions. Males and females have different metabolic patterns in prolonged fasts. The normal male will maintain a plasma glucose of 55 to 60 mg/dL (3.1 to 3.3 mmol/L) for several days. Females will produce ketones more readily and permit plasma glucose to decrease to approximately 35 mg/dL (1.9 mmol/L) after around 36 hours. A decreased glucose coincident with a nonsuppressed insulin of greater than 8 units/mL is indicative of insulinoma. An insulinoma can also be diagnosed if serial measurements are taken and the glucose concentration decreases faster than the insulin concentration. In the male, insulinoma may be ruled out if hypoglycemia is not observed after a 3-day fast.

Proinsulin assays may be utilized in the future because proinsulin represents less than 20% of the total circulating insulin in a normal individual. In contrast, in an individual with an insulinoma, 30% to 90% of the total circulating insulin is in the proinsulin form. This assay, which uses monoclonal antibodies, may prove useful and convenient, allowing the clinician to avoid hospitalizing the patient for the fasting period and sparing the patient the discomfort and cost of the hospital stay.

C-peptide assays are of little diagnostic use, except in instances where inappropriate insulin administration must be distinguished from endogenous insulin production. Hypoglycemia in a diabetic with access to insulin may be due to excessive use or due to endogenous insulin. Exogenous insulin has been purified and C-peptide has been removed. If insulin concentrations are greater than the C-peptide concentration, the plasma insulin is from exogenous administration.

SUMMARY

Carbohydrates have the general formula $(CH_2O)_n$. Glucose is a six-carbon aldohexose. There are 32 different possible isomers of aldohexoses with the glucose chemical formula. Glucose and other sugars can exist in the open chain or ring form. The open chain form permits the free carbonyl to reduce Benedict's or Feeling's reagents. β-D-glucose is a primary source of energy for humans. Energy in the form of ATP can be obtained from glucose through the anaerobic pathway. Additional energy is then obtained from the product pyruvate as it passes through the TCA cycle. The nervous system relies solely on glucose for energy in normal circumstances. We maintain our glucose concentration within a narrow range for this reason.

Insulin, which is produced in the β cells of the pancreas, is responsible for the uptake of glucose into cells and the reduction of plasma glucose postprandially. Insulin also promotes glycogenolysis and triglyceride synthesis. Glucagon, which is produced in the β cells of the pancreas, opposes the action of insulin. Both glucagon and epinephrine increase plasma glucose by activating gluconeogenesis and glycogenolysis in the liver. Gluconeogenesis is the formation of glucose from lactate, amino acids, pyruvate, and glycerol.

Diabetes mellitus can be classified as insulin-dependent (IDDM) and non-insulin-dependent (NIDDM). The classifications may sometimes be confusing, because NIDDM patients sometimes require insulin administration for glycemic control. The development of IDDM appears to be partly re-lated to the HLA genotype of an individual. HLA-DR3/4 and DQw3.2 genotypes have a higher tendency to developing IDDM than does the general population. IDDM may also have an environmental component, which is thought to trigger an immune reaction leading to an autoimmune response that causes β cell destruction. Untreated hyperglycemia in diabetes usually is no higher than 500 mg/dL (28 mmol/L) when normal renal function is present. Ketoacidosis is more common in IDDM, osmolality is increased, plasma potassium is increased, and plasma sodium is slightly decreased. Bicarbonate is decreased in response to the acidosis.

NIDDM is also thought to have a genetic factor. NIDDM individuals have no β cell destruction and may have decreased, normal, or increased insulin concentrations—but they are insulin resistant at the tissues. In NIDDM, there is a greater tendency toward hyperosmolar nonketotic coma. This is characterized by a glucose concentration of greater than 600 mg/dL (33 mmol/L) and an absence of ketones. The BUN and osmolality are increased and urine output is decreased. GDM may be related to NIDDM. The definitive tests for diabetes are fasting glucose concentrations greater than 140 mg/dL (7.8 mmol/L) on two occasions, or a value of 200 mg/dL (11.1 mmol/L) at the 2-hour and at one other time point during a two-hour OGTT.

Hypoglycemia can occur as one of two types: reactive (postprandial) or (fasting) postabsorptive. Common causes of fasting hypoglycemia include insulin excess, oral hypoglycemic drugs, and prolonged alcohol ingestion. Neonatal, congenital, and ketotic hypoglycemia occur in children. Congenital forms of hypoglycemia include von Gierke disease. Galactosemia is another relatively common congenital variety of hypoglycemia.

CASE STUDY 14-1

A 19-year-old woman complained of recent onset blurred vision, nausea, constant fatigue, and dizziness. The patient was polydipsic and polyuric and woke several times through the night to urinate.

Chemistry test results:

random serum glucose:	210 mg/dL
serum sodium:	134 mmol/L
serum potassium:	5.4 mmol/L
serum chloride:	102 mmol/L
serum carbon dioxide:	16 mmol/L
serum BUN:	17 mg/dL
Urinalysis:	Positive for glucose and ketones.

Questions

1. Based on the above information can this patient be diagnosed as a diabetic? Justify your answer. If further tests are required, which one(s) would you suggest and what criteria would be required for the diagnosis of diabetes using these tests?

2. Assuming that this person is diabetic, would her history lead you to believe that she was an IDDM or NIDDM diabetic? Give reasons to substantiate your choice.

3. What is the reason for her increased anion gap (16 mmol/L)? What other compounds could contribute to an increased anion gap?

CASE STUDY 14-2

A 70-year-old comatose male was admitted through the emergency room. The patient was unresponsive to pain and had diminished muscle tone on his right side. His respirations were shallow, and the patient was hypotensive. The patient's medical history included a mild stroke 2 years before admission, NIDDM first diagnosed when the patient was 55-years-old, and reduced renal function as a consequence of the diabetes. The patient's diabetes is managed with oral hypoglycemic agents.

Chemistry test results:

admission serum glucose:	1121 mg/dL
serum sodium:	146 mmol/L
serum potassium:	5 mmol/L
serum chloride:	112 mmol/L
serum carbon dioxide:	21 mmol/L
serum BUN:	65 mg/dL
serum creatinine:	2.5 mg/dL

serum osmolality:	380
serum ketones:	negative
Urinalysis:	not available.

Questions

1. What chemistry results distinguish non-ketotic hyperosmolar coma from diabetic ketoacidosis? Why is the glucose concentration so elevated? What is the glucose concentration in SI units (mmol/L)?
2. What chemistry results would indicate that renal function is impaired in this patient? Could this patient have other complications?
3. If the serum ketones result was unavailable, how could acidosis be tentatively ruled out? What other chemistry tests would be required to absolutely rule out acidosis?

CASE STUDY 14-3

A 28-year-old woman delivered, by cesarean section, a 4.3 kg (9.5 lb.) infant. The infant was above the 95th percentile for weight and length. The infant exhibited a generous amount of subcutaneous fat. The mother's history was incomplete: she claimed to have had no medical care through this, her second pregnancy. She claimed that her first infant was large as well. Shortly after birth, the infant became lethargic and flaccid. A whole blood glucose and ionized calcium were determined on a specimen of blood drawn from the infant. These tests were performed *STAT* in the nursery. The ionized calcium was 4.9 mg/dL (normal) and the whole blood glucose was 25 mg/dL (1.4 mmol/L). A plasma glucose was drawn and analyzed in the main laboratory to confirm the whole blood findings. The plasma glucose results were 33 mg/dL (1.8 mmol/L). An intravenous glucose solution was started and whole blood glucose was measured hourly. The infant subsequently developed respiratory distress and required a 2-week hospital stay including 7 days in the neonatal intensive care unit prior to discharge.

Questions

1. Give a possible reason for the infants large birth weight and size.
2. If the mother was a gestational diabetic, why was her baby hypoglycemic?
3. Why is there a discrepancy between the whole blood glucose concentration and the plasma glucose concentration?
4. Would it be worthwhile to perform a diabetic screen on the mother now?
5. When is the appropriate time to perform an OGTT during pregnancy? Why?

REFERENCES

1. Cryer PE. Glucose homeostasis and hypoglycemia. In: Wilson JD, Foster DW, eds. Williams textbook of endocrinology. Philadelphia: WB Saunders Co, 1992.
2. Häring HU. The insulin receptor: Signalling mechanism and contribution to the pathogenesis of insulin resistance. Diabetologia 1991;34:848.
3. Malchoff CD. Diagnosis and classification of diabetes mellitus. Conn Med 1991;55(11)625.

4. Mazze RS. Detection of glucose intolerance in pregnancy. Int J Clin Pharmacol Ther Toxicol 1993;31(9):440.

5. National Diabetes Data Group. Classification and diagnosis of diabetes mellitus and other categories of glucose intolerance. Diabetes 1979;28:1039.

6. Perronne C, Cricaire F, Leport C, Assari D, Vilde JL, Assan R. Hypoglycemia and diabetes mellitus following parenteral pentamidine mesylate treatment in AIDS patients. Diabet Med 1990;7:585.

7. Radziuk J, Pye S, Zhang Zi. Substrates and the regulation of hepatic glycogen metabolism. Adv Exp Med Biol 1993;334:235.

8. Stehouwer CDA, Donker AJM. Clinical usefulness of measurement of urinary albumin excretion in diabetes mellitus. Neth J Med 1993;42:175.

9. Thorsby E, Kjersti RS. Role of HLA genes in predisposition to develop insulin-dependent diabetes mellitus. An Med 1992;24(6):523.

10. Unger RH, Foster DW. Diabetes mellitus. In: Wilson JD, Foster DW, eds. Williams textbook of Endocrinology. Philadelphia: WB Saunders Co, 1992.

11. Van Schaftingen E. Glycolysis revisited. Diabetologia 1993; 36:581.

Chapter 15

Lipids and Lipoproteins

G. Russell Warnick, Judith R. McNamara, Lily L. Wu

Objectives

Upon completion of this chapter, the clinical laboratorian should be able to:

- *Explain lipid/lipoprotein physiology and metabolism.*

- *Define the following terms: lipoprotein, exogenous, endogenous, chylomicrons, fatty acids, phospholipids, triglycerides, cholesterol, VLDL, LDL, HDL, Lp(a).*

- *Describe the clinical tests used to assess lipids and lipoproteins, including the principles and procedures.*

- *Given clinical data, evaluate the patient's lipid or lipoprotein status.*

- *Identify the reference ranges for the major lipids discussed.*

- *Discuss the interaction in the body between the lipids and lipoproteins and various hormones.*

- *Relate the clinical significance of lipid and lipoprotein values in the assessment of coronary heart disease.*

- *Discuss the incidence and types of lipid and lipoprotein abnormalities.*

KEY WORDS

Arteriosclerosis	HDL
Cholesterol	LDL
Chylomicrons	Lipoprotein
Dyslipidemias	Lp(a)
Endogenous	Phospholipids
Exogenous	Triglycerides
Fatty Acids	VLDL

The *lipoproteins* constitute the body's "petroleum industry." Like the great tankers that travel the world's oceans transporting petroleum for fuel needs, the large *chylomicrons* carry dietary triglycerides throughout the circulatory system to cells, finally docking at the liver to deposit the chylomicron remnants. The *very low density lipoproteins* (VLDL) are like tanker trucks, carrying triglycerides assembled in the liver out to cells for energy needs or storage as fat. The *low-density lipoproteins* (LDL), rich in cholesterol, are the almost empty tankers that deliver cholesterol to the peripheral cells after the triglycerides have been off-loaded. The *high-density lipoproteins* (HDL) are the clean-up crew, gathering up extra cholesterol for transport back to the liver. Thus, cholesterol, which in excess contributes to heart disease, is used by the body for such useful functions as the maintenance of cell membranes and as a precursor for hormone synthesis, and facilitates triglyceride transport in serving the fuel needs of the body.

The *lipids* and lipoproteins, which are central to the metabolism of the body, have become increasingly important in clinical practice, primarily because of their association with coronary heart disease (CHD). Many national and international epidemiological studies have demonstrated that, especially in affluent countries with high fat consumption, there is a clear association with the development of atherosclerosis. Decades of basic research have contributed to knowledge about the nature of the lipoproteins and their lipid and protein constituents, as well as their role in the pathogenesis of the atherosclerotic process. International efforts to reduce the impact of CHD on public health have

Michael L. Bishop, Janet L. Duben-Engelkirk, and Edward P. Fody.
CLINICAL CHEMISTRY. © 1996 Lippincott–Raven Publishers.

focused attention on improving the reliability and convenience of the lipid and lipoprotein assays. Expert panels have developed guidelines for the detection and treatment of high cholesterol, as well as laboratory performance goals and detailed recommendations for reliable measurement of the lipid and lipoprotein analytes.[130–133,143] The lipids and lipoproteins are reviewed here primarily in the context of clinical and laboratory guidelines from the National Cholesterol Education Program (NCEP).

LIPID CHEMISTRY

The major lipids of the body triglycerides, cholesterol, phospholipids, and glycolipids play a variety of biological roles. They serve as a primary source of fuel and are important components of cell membranes and many cell structures, they provide stability to the cell membrane and allow for transmembrane transport, and they are transported through the blood stream in the form of lipoproteins.

Lipid and Lipoprotein Constituents

Fatty Acids

Fatty acids are the major constituents of triglycerides and phospholipids. There are short-chain (4–6 carbon atoms), medium-chain (8–12 carbon atoms), and long-chain fatty acids (>12 carbon atoms). Many dietary fatty acids are straight-chain compounds with even numbers of carbon atoms (4–24 carbon atoms). Depending on the number of double-bonds in the molecule, the fatty acids may be saturated (no double-bonds), monounsaturated (one double-bond), or polyunsaturated (two or more double-bonds). The double-bonds of unsaturated fatty acids are usually arranged in *cis* form, which causes the fatty acids to bend. These bends increase the space that the fatty acids require and, as a result, increase the fluidity. Double-bonds occurring in the *trans* form do not bend, and have properties more closely resembling saturated fatty acids.

Triglycerides

The triglyceride molecule is composed of one molecule of glycerol with three fatty acid molecules attached (usually three different fatty acids including both saturated and unsaturated molecules). Triglycerides containing saturated fatty acids without bends pack together closely and tend to be solids at room temperature. Unsaturated fatty acids do not pack together closely and tend to be liquids at room temperature. Most triglycerides from plants (*e.g.*, corn, sunflower seed, and safflower oil) are composed primarily of polyunsaturated fatty acids, oriented in the *cis* formation, and are in liquid form even at 4°C. Triglycerides from animal sources, which contain largely saturated fats, are typically solid at room temperature. Vegetable oils that have been hydrogenated to solidify them for use as butter and lard replacements have had the *cis* double-bonds changed to *trans* double-bonds.

The source of triglycerides in the body can be either exogenous (dietary) or endogenous (synthesized in liver and other tissues). Triglyceride molecules allow the body to compactly store long carbon chains (fatty acids) for energy that can be used during fasting states between meals. The high-energy triglyceride molecules, which constitute 95% of fats stored in tissues, are transported in plasma mostly in the form of large triglyceride-rich particles called chylomicrons and VLDL. When the triglycerides are metabolized, their fatty acids are released to the cells and converted into energy. The glycerol of the triglyceride is recycled into additional triglycerides. Triglycerides also provide insulation to vital organs in the form of fat deposits in adipose cells. The breakdown of triglycerides is facilitated by hormone-sensitive lipase, lipoprotein lipase (LPL), epinephrine, and cortisol.[184] The LPL molecules are attached to heparan sulfate stalks in the capillaries. As the triglyceride-rich lipoproteins (chylomicrons and VLDL) are carried through the circulation, the triglycerides are hydrolyzed as they come in contact with LPL.[83] The supply of free fatty acids available to the cells to be used in the TCA cycle for energy production is dependent on LPL interaction with chylomicrons and VLDL. Hormone-sensitive lipase acts inside adipose (fat) cells to release free fatty acids from triglyceride stores for energy when dietary sources are not available or are insufficient for the body's energy needs. Epinephrine and cortisol promote triglyceride breakdown when the cells need energy and glucose stores have been depleted.

Cholesterol

Cholesterol is an unsaturated steroid alcohol of high molecular weight, consisting of a perhydrocyclopentanthroline ring and a side chain of eight carbon atoms (Fig. 15-1). In its esterified form, it contains one fatty acid molecule. Cholesterol is found almost exclusively in animals. Virtually all cells and body fluids contain some cholesterol. Cholesterol is used for the manufacture and repair of cell membranes, for synthesis of bile acids and vitamin D, and is the precursor of five major classes of steroid hormones: progestins, glucocorticoids, mineralocorticoids, androgens, and estrogens. As with triglycerides, there are both exogenous (dietary) and endogenous (primarily hepatic) sources of cholesterol. Diet contributes 100 to 700 mg cholesterol per day (<300 mg/day is recommended for most adults) (Table 15-1). Another 500 to 1000 mg is produced in the liver and other tissues, and an additional 600 to 1000 mg of biliary cholesterol is secreted into the intestine daily, about 50% of which is reabsorbed into the blood. In the body, about 70% of cholesterol is located in stationary pools located in the skin, adipose tissue, and muscle cells; the remaining 30%, or so, forms a mobile pool that is transported in the form of lipoproteins and circulates through the liver. In the blood circulation two thirds of the cholesterol is esterified, and one third is in free form.

Phospholipids

Phospholipid, glycolipid, and cholesterol are the three major types of membrane lipids. Phospholipids are amphipathic, branes and the outer shells of lipoprotein particles. Most phospholipids are formed by the conjugation of two fatty acids and a phosphorylated glycerol. The two fatty acids normally are 14 to 24 carbon atoms long, with one being saturated and one unsaturated. The phosphorus group can be complexed to choline to form phosphatidylcholine (lecithin), or to ethanolamine, inositol, or serine to form cephalins.

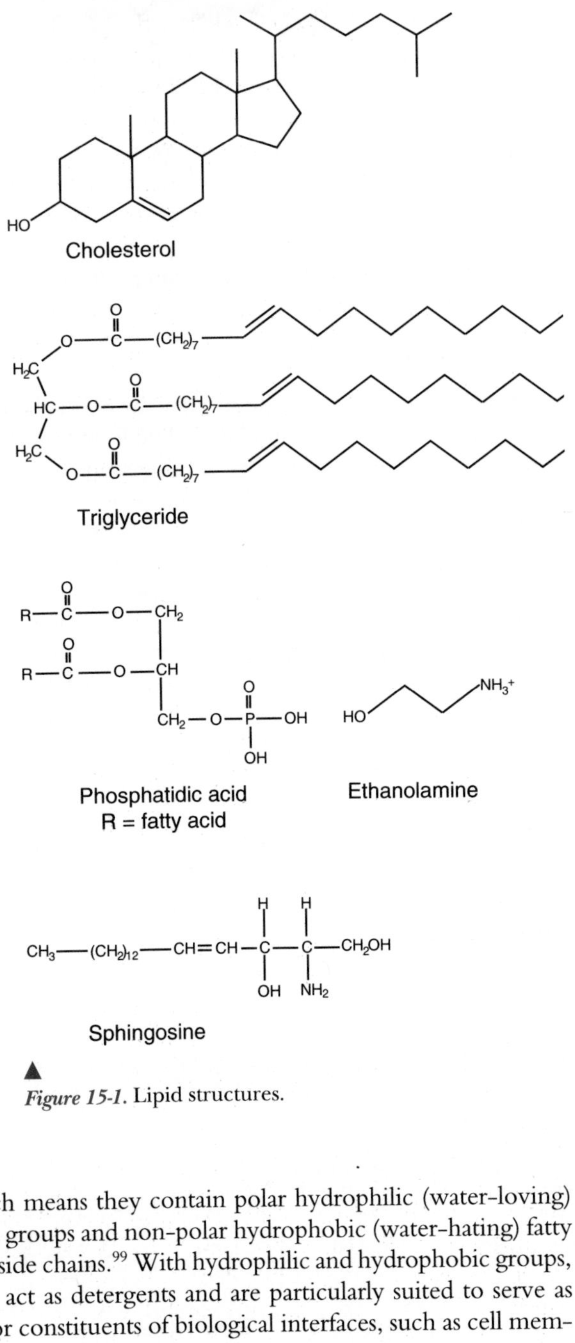

Different fatty acid residues result in several different lecithins and cephalins. Sphingomyelin is the only phospholipid in membranes that is not derived from glycerol but rather from an amino alcohol called *sphingosine* by affixing a fatty acid to its amino group and a phosphoryl choline to its hydroxy group.

Glycolipids

Glycolipids are sugar-containing lipids that consist of a sphingosine molecule that has a fatty acid attached to its amino group and a sugar linked to the primary alcohol group. The whole molecule is often referred to as a *ceramide*. The simplest glycolipids are galactosylceramide and glucosylceramide. More complex glycolipids, such as gangliosides, may contain branched chains with as many as seven sugar residues. Gangliosides are the major lipids of cell membranes of the brain and central nervous system, and are synthesized from ceramides. Membrane glucosphingolipids also play an important role in cell recognition and blood typing.

Prostaglandins (PG)

Prostaglandins (PG) are long-chain polyunsaturated fatty acids, called *eicosanoids,* which contain 20 carbon atoms, including a five-carbon (cyclopentane) ring, and affect a wide variety of physiological processes. PG are synthesized in almost all tissues from arachidonate and other polyunsaturated fatty acids. The major classes of PG are designated PGA through PGI, with a subscript number to indicate the number of carbon-carbon double bonds outside the ring. Other PG-related compounds that are also derived from arachidonate include thromboxanes, with a six-member ether ring, and leukotrienes. PG, thromboxanes, and leukotrienes are short-lived and can alter the activities of the cells in which they are synthesized as well as those of adjoining cells.

Figure 15-1. Lipid structures.

which means they contain polar hydrophilic (water-loving) head groups and non-polar hydrophobic (water-hating) fatty acid side chains.[99] With hydrophilic and hydrophobic groups, they act as detergents and are particularly suited to serve as major constituents of biological interfaces, such as cell membranes and the outer shells of lipoprotein particles. Most

TABLE 15-1. Dietary Guidelines Developed by the American Heart Association and Recommended by the Adult Treatment Panel II of the National Cholesterol Education Program (as compared to the average American diet)

Dietary Nutrient	Step I Diet	Step II Diet	Average American Diet
Total Fat (% of calories)	≤30%	≤30%	36%
Saturated	<10%	<7%	15%
Monounsaturated	<15%	<15%	15%
Polyunsaturated	<10%	<10%	6%
Cholesterol	<300 mg/day	<200 mg/day	>400 mg/day

Apolipoproteins (apo)[114]

Apolipoproteins (apo) are the protein moieties associated with the plasma lipoproteins (Table 15-2). Apos are structural elements in the amphipathic shell of lipoprotein particles and help to keep the lipids in solution during circulation through the blood stream. Apos also regulate plasma lipid metabolism by activating and inhibiting enzymes that are involved in the process. They interact with specific cell-surface receptors and direct the lipids to the correct target organs and tissues in the body.

Apo A-I is the major apo of HDL. Apo B, which is responsible for the binding of LDL to LDL receptors, is a large protein with a molecular weight of approximately 500 kDa and is the functional protein for transporting cholesterol to cells. Over 95% of the protein in LDL is apo B. Apo B is synthesized in two forms: apo B-100 in the liver and apo B-48 in the intestine. Apo B-48 consists of the first 48% of the amino acid sequence found in apo B-100. Apo B-100 is found in VLDL, IDL, and LDL; apo B-48 is found in chylomicrons. Apo E, which promotes binding of lipoproteins (LDL, VLDL, and apo E-HDL) to the LDL receptor and a specific chylomicron remnant receptor, is also associated both with transport of cholesterol ester in plasma and with the redistribution of cholesterol in tissues.[196] There are three major isoforms of apo E: apo E-2, E-3, and E-4. Individuals have any one (homozygous) or a combination of any two (heterozygous) isoforms. Apos E-3 and E-4 both bind well to the LDL-receptor, whereas E-2 is defective in binding. The apo E isoform pattern influences circulating cholesterol levels. For example, individuals with apo E-4 (4/2, 4/3, 4/4) average about 10% higher plasma cholesterol levels than ones having apo E-3 (3/3).[46] The apo E2/2 phenotype is present in Type III hyperlipoproteinemia.[81] Apo (a) is a highly glycosylated apolipoprotein with approximately 70% structural homology with plasminogen. There are several isoforms of apo (a) ranging in molecular weight from 300 to 700 kDa. The size heterogeneity is under genetic control and is due to the variability in the length of the polypeptide chain.

Lipoproteins[36,37,145,176]

Lipoproteins (Fig. 15-2) may be classified by either their density or their electrophoretic mobility. The four major classes of lipoproteins separated by ultracentrifugation according to their differences in density are chylomicrons, VLDL, LDL, and HDL. These lipoproteins also differ in their chemical composition, size, and potential atherogenicity (Table 15-3). Because of the need to transport water-soluble lipids through the bloodstream to various tissues, cholesterol and triglyceride are packaged in spherical lipoproteins. Each lipoprotein contains a nonpolar, hydrophobic lipid core consisting primarily of cholesteryl esters and triglycerides. This core is surrounded by a water-soluble surface composed of apolipoproteins, phospholipids, and free (or unesterified) cholesterol. The polar surface components confer the solubility, which makes possible the transport of the highly insoluble cholesterol esters and triglycerides in the blood circulation. Within each of the major lipoprotein classes, there are subpopulations with subtle variations in size and relative composition.

TABLE 15-2. Characteristics of Human Apolipoproteins and Their Variants

Apolipoproteins	Functions	Major Source	Normal Plasma Conc. (mg/dL)	Relative Mass (Mr)	Isoelectric Point
Apo A-I	Major structural protein in HDL Activates LCAT Ligand for HDL binding	Liver & Intestine	100–200	27,000	5.3–5.4
Apo A-II	Structural protein in HDL Activates LCAT Enhances hepatic triglyceride lipase activity	Liver	20–50	17,400 (dimer)	5.0
Apo A-IV	Component of intestinal lipoproteins	Intestine	10–20	38,000	—
Apo B-100	Major structural protein in VLDL & LDL Ligand for the LDL receptor	Liver	70–125	5.4×10^5	—
Apo B-48	Primarily structural protein in chylomicrons	Intestine	<5	2.6×10^5	—
Apo C-I	Activates lipoprotein lipase	Liver	5–8	6630	6.5
Apo C-II	Activates lipoprotein lipase Activates LCAT	Liver	3–7	8835	4.8
Apo C-III	Inhibits lipoprotein lipase Inhibits receptor recognition of apo E	Liver	10–12	9960	4.5–4.9
Apo E2,3,4	Binds to LDL-receptor and Remnant-receptor	Liver	3–15	34,145	5.4–6.1
Apo (a)	Structural protein for Lp(a) May inhibit plasminogen binding	Liver	<30	$3–7 \times 10^5$	—

*There are many minor apolipoproteins, such as apo D, apo J, apo H, apo F, and apo G.

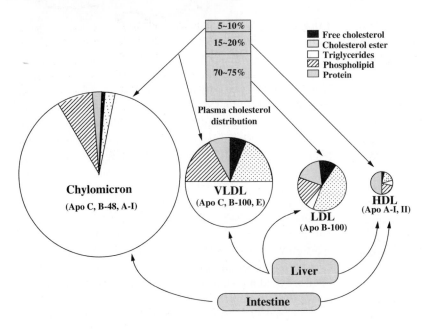

Figure 15-2. Classification of lipoproteins.

Chylomicrons

Chylomicrons are the largest of the lipoprotein particles with diameters ranging from 80 to 1200 nm and with density <0.95 g/mL. They are the major carriers of exogenous triglycerides.

Chylomicrons are composed of 90% to 95% (by weight) triglyceride, 2% to 6% phospholipid, 2% to 4% cholesteryl ester, 1% free cholesterol, and 1% to 2% apolipoprotein. Apo C and apo B-48 are the most abundant apolipoproteins in chylomicrons, but small amounts of others, including apo E and apos A-I, II, and IV, can also be found. Following absorption of dietary lipids through the microvilli of the intestine, chylomicrons are responsible for transporting dietary triglycerides and some cholesterol to the rest of the body. The clearance time from the formation of chylomicrons after a meal and the removal of remnants by the liver is about 6 hours. Normally, chylomicrons are not found in 12- to 14-hour fasting blood specimens. A creamy layer rising to the top of a fasting serum specimen that has been cooled overnight indicates the presence of chylomicrons and signals a defect in their clearance.

Very Low Density Lipoproteins (VLDL)

VLDL, like chylomicrons, are also rich in triglycerides. They are the major carriers of endogenous (liver synthesized) triglycerides and have a diameter ranging from 40 to 80 nm, with a density range of 0.95 to 1.006 g/mL. VLDL are composed of 50% to 65% (by weight) triglyceride, 8% to 14% cholesteryl ester, 12% to 16% phospholipid, 4% to 7% free cholesterol, and 5% to 10% apolipoprotein. The protein composition includes 40% to 50% apo C, 30% to 40% apo B-100, and 10% to 15% apo E. Excess dietary intake of carbohydrate enhances hepatic synthesis of triglycerides, which, in turn, increases VLDL production. Elevations in

TABLE 15-3. Characteristics of Human Lipoproteins

Lipoprotein	Density (g/mL)	Electrophoretic Mobility	Source	Apolipoproteins Major & Minor (Italic)	Molecular Weight (10⁶)	Diameter (nm)
Chylomicron	~0.93	Origin	Intestine	CI, II, III; B-48 *A-I, II, IV & E*	50–1,000	80–1200
Very low density lipoprotein (VLDL)	0.93–1.006	Pre-beta	Liver	B-100 *CI, II, III; E*	10–80	40–80
Intermediate-density lipoprotein (IDL)	1.006–1.019	Slow pre-beta	Catabolism of VLDL	B-100 *CIII, E*	5–15	30–40
Low-density lipoprotein (LDL)	1.019–1.063	Beta	Catabolism of IDL	B-100 *C*	~3	18–30
High-density lipoprotein (HDL)	1.063–1.21	Alpha	Liver, intestine	A-I, II *CI, II, III; E, D*	0.36–0.20	5–12
Lipoprotein (a) [Lp(a)]	1.050–1.100 (sinking pre-beta)	Pre-beta	Liver	B-100 & apo(a)	~5	25–35

VLDL are evidenced by increased serum triglyceride concentrations, and significant elevations can produce turbid serum because VLDL and chylomicrons are large enough particles to scatter light.

Low-Density Lipoproteins (LDL)

LDL contain 50% cholesterol by weight and are the most cholesterol-rich of the lipoproteins. They are synthesized in the liver and are responsible for transporting cholesterol from the liver to the peripheral tissues. LDL have diameters ranging from 18 to 30 nm, with a density range of 1.019 to 1.063 g/mL, and contain 35% to 45% cholesteryl esters, 6% to 15% free cholesterol, 22% to 26% phospholipid, 4% triglyceride, and 22% to 26% apolipoprotein in the form of apo B-100. LDL are the most atherogenic lipoproteins, and high serum levels are regarded as a major CHD risk factor.[7,89,91]

LDL particles of different size and composition have been separated into as many as eight subclasses by density ultracentrifugation or by gradient electrophoresis gels.[8,122] LDL size is inversely related to serum triglyceride levels.[125] Smaller, more dense LDL are associated with a higher risk of CHD. Very small LDL, however, are not associated with a higher risk of CHD but are observed in severe elevations of triglycerides and a disease called pancreatitis, which will be discussed later in the chapter.

High-Density Lipoprotein (HDL)

HDL are the smallest lipoproteins with diameters ranging from 5 to 12 nm and density from 1.063 to 1.21 g/mL. HDL particles are synthesized by both the liver and intestine, and consist of 25% to 30% phospholipid, 15% to 20% cholesteryl esters, 5% free cholesterol, 3% triglyceride, and 45% to 59% apolipoprotein. The major apolipoproteins are apo A-I (70%) and apo A-II (10–23%); minor proteins include apo C and apo E. HDL typically carry 20% to 35% of total plasma cholesterol, but unlike LDL, which carry cholesterol to the tissues, HDL take excess cholesterol from the tissues and return it to the liver (reverse transport). Based on density differences, there are two major groups of HDL subclasses: HDL_2 and HDL_3. HDL_2 are larger in size and richer in lipid than HDL_3 and may be the more efficient vehicles for the transfer of cholesterol from peripheral tissue to the liver.[141] In addition to the two major classes of HDL, there are as many as fourteen subclasses that have been identified by various methods.[15,34,105]

Lipoprotein (a) [Lp(a)][6,79,116]

Lp(a) are LDL-like lipoprotein particles with one additional molecule of apo (a) linked to apo B-100 by a disulfide bond. Lp(a) are heterogeneous both in size and density due to differing numbers of peptide sequences called *kringles,* which are named for their resemblance to Danish pastries. Lp(a) have a density of approximately 1.050 to 1.13 g/mL and migrate at the pre-β position on plasma lipoprotein electrophoresis. The concentration of Lp(a) is inversely related to the size of the isoform.

The presence of elevated serum levels of Lp(a) is thought to be an independent risk factor for the development of premature CHD, myocardial infarction, and cerebrovascular disease. Because the kringle domains of Lp(a) have a high level of homology with plasminogen, which is an element of the blood coagulation cascade, Lp(a) may compete with plasminogen for binding sites. This interference with plasminogen binding may then retard the rate of clot lysis, thereby increasing the likelihood of myocardial events. Plasma levels of Lp(a) vary widely among individuals in a population, but the levels tend to remain relatively constant within the individual. Elevated plasma levels are found in approximately 10% to 15% of Caucasians, and levels average two times higher in African-Americans. Plasma levels of Lp(a) are predominately genetically determined, but a number of other factors, such as age, race, hormone status, and liver and renal function, moderate the values.

LIPOPROTEIN PHYSIOLOGY AND METABOLISM

Lipid Absorption[97,175]

Lipids constitute an important part of the diet. An average person ingests, absorbs, resynthesizes, and transports about 60 to 130 g of fat daily in the body, mostly in the form of triglycerides. Since fats are insoluble in an aqueous system, special mechanisms are required for their absorption by the intestine. In the intestinal lumen, the process of digestion converts lipid into more polar compounds with amphophilic properties. Thus, triglycerides are transformed into monoglycerides, diglycerides, and free fatty acids; cholesterol esters are transformed into free cholesterol; and phospholipids are transformed into their lyso-derivatives. These compounds and bile salts form mixed micelles, in which the amphipathic lipids orient themselves with the hydrophobic regions on the inside of the micelles and the polar groups exposed to the aqueous environment. Lipid absorption occurs when the micellar solutions of lipids come in contact with the microvillus membranes of the mucosal cells. The smaller free fatty acids with 10 or fewer carbon atoms can pass directly into the portal circulation and be carried by albumin to the liver. The longer-chain fatty acids, monoglycerides, and diglycerides can be absorbed into the intestinal mucosal cells from bile acid micelles. Inside the intestinal cells, the long-chain free fatty acids are immediately re-esterified to form triglycerides and cholesteryl esters. The triglycerides and cholesteryl esters are packaged into the core of chylomicrons for delivery to the blood circulation. Approximately 90% to 95% of triglycerides in the diet are absorbed from the intestinal lumen, whereas only about 50% of the cholesterol is absorbed.

Lipid Synthesis

Triglyceride[99]

Fatty acids are stored in cells as triglycerides. In the liver, adipose tissue, and many other organs and tissues, the synthesis of triglycerides begins with the activation of fatty acids to acyl-CoA ester. This process requires the input of energy from ATP. There are different routes involved in the formation of triglycerides. Two acyl-CoA molecules may either condense with glycerol-3-phosphate to form phosphatidate (diacylglycerol-3-phosphate) or be added to dihydroxyacetone phosphate through an esterification process. The final stage involves removal of the phosphoryl group followed by acylation to form the triglyceride.

Cholesterol[149]

The liver is the major site of cholesterol synthesis, although cholesterol is also produced in many other organs and tissues. Cholesterol is formed from acetyl-CoA. Three molecules of acetyl-CoA are condensed to produce 3-hydroxy-3-methylglutaryl coenzyme A (HMG CoA), which, in turn, is converted to mevalonic acid through the action of HMG-CoA reductase. Mevalonic acid is converted into squalene after a series of condensation and rearrangement steps. Squalene then cyclizes to form lanosterol, which is further modified to yield cholesterol.

The synthesis of cholesterol in the liver is regulated by intracellular cholesterol concentration and the activity of HMG-CoA reductase, which is the rate-limiting enzyme in cholesterol biosynthesis. Consequently, an important class of drugs target this enzyme as a means of lowering plasma cholesterol levels. In the liver, some of the cholesterol is degraded to bile acids, primarily cholic acid and chenodeoxycholic acid. Formation of bile acids is essential for removing cholesterol from the body.

Lipoprotein Metabolism

The mechanisms by which lipids are utilized, transported, and removed from the body are complex. Several specialized receptors, enzymes (Table 15-4), transfer proteins, and transport pathways are involved, with each playing an important role or roles. Several defects in cholesterol transport and removal have been shown to lead to the development and progression of atherosclerosis.[20,175]

Lipoproteins transport lipids in three separate but interacting pathways through the body: the exogenous, endogenous, and reverse cholesterol transport pathways. The exogenous pathway transports dietary lipids, through the formation of chylomicrons in the small intestine, to the liver; the endogenous pathway is responsible for transporting hepatic lipids by way of VLDL and LDL to the peripheral tissues; and the reverse cholesterol pathway utilizes HDL to transport cholesterol from the peripheral tissues back to the liver for excretion or reutilization.

The Exogenous Pathway (Fig. 15-3)

Lipids of dietary origin, mainly dietary triglycerides, absorbed in the small intestine are packaged into chylomicrons. The newly synthesized chylomicrons, collected in the intestinal lymphatics, are released into the blood stream via the thoracic duct. After being released into the blood circulation, chylomicrons travel to the capillaries of skeletal muscle, heart, and adipose tissues. On the surface of capillary endothelial cells, they are cleaved by the membrane-bound LPL, and free fatty acids and glycerol are released from the core triglycerides. Some of the free fatty acids are used as an immediate energy source but most are taken up by the tissue and re-esterified into triglycerides for storage. During this metabolic process, phospholipids, free cholesterol, and apos A-I, II, and IV are transferred from chylomicrons to HDL in exchange

TABLE 15-4. Enzymes of Plasma Lipoprotein Metabolism

Enzyme	Major Tissue Source	Substrates	Function	Location	Cofactors
Lipoprotein lipase (LPL)	Adipose tissue (adipocytes) Muscle	Triglycerides and phospholipids of chylomicrons and VLDL	Hydrolyzes triglycerides	Muscle, adipose capillary	Apo-CII Apo CIII inhibits
Hepatic lipase (HTLP)	Liver (hepatocytes)	Triglycerides and phospholipids of VLDL, LDL, and HDL	Hydrolyzes triglycerides	Liver	Apo C-II
Lecithin-cholesterol acyltransferase (LCAT)	Liver	Cholesterol and phosphotidylcholine of HDL	Esterifies free cholesterol	Plasma	Apo A-I, C-I

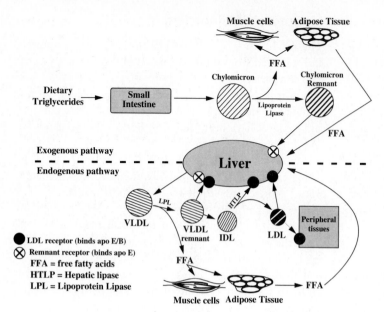

Figure 15-3. Overall exogenous pathway of triglyceride metabolism.

for apo C and apo E from HDL and VLDL. These processes result in the transformation of chylomicrons into cholesteryl ester-rich chylomicron remnants. The chylomicron remnants are rapidly taken up by the liver through interaction of apo E on the remnant particles with remnant receptors on the surface of liver cells. Once in the liver, lysosomal enzymes break down the remnants to release free fatty acids, free cholesterol, and amino acids. Some of the cholesterol is converted to bile acids. The bile acids and free cholesterol are excreted into the bile and hence into the intestine. However, as mentioned above, approximately 50% of the excreted cholesterol in the intestine is reabsorbed and returns to the liver; only about 50% of the total excreted cholesterol ends up in the stool as fecal neutral steroids. On the other hand, about 97% to 98% of the bile acids are reabsorbed in the lower intestine, with the remainder excreted in stool as fecal acidic steroids.

The Endogenous Pathway (Fig. 15-4)

Most triglycerides in the liver that are destined to be incorporated in VLDL are derived from the diet after recirculation from adipose tissue. Only a relatively small fraction appear to be synthesized *de novo* from dietary carbohydrate. Cholesterol utilized in VLDL formation is derived from both dietary cholesterol and that synthesized directly in the liver.

VLDL particles, once secreted into the bloodstream, undergo a similar degradation process as chylomicrons. Through the action of LPL and apo C-II, VLDL lose their core lipids, mainly triglycerides, and also lose some of the apolipoproteins and phospholipids from the surface.[176] During this process, VLDL become VLDL remnants, which can then be converted to intermediate-density lipoproteins (IDL) and then further processed to LDL. However, only about 50% of VLDL are converted to LDL; the other 50%

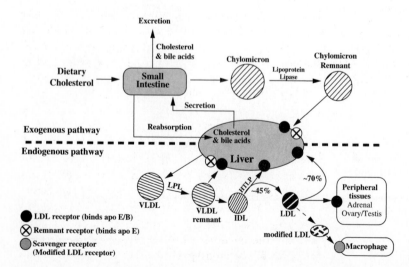

Figure 15-4. Overall pathway of cholesterol metabolism.

are taken up by either the remnant receptor or the LDL receptor of the liver through interaction with apo B or apo E.

LDL are the major cholesterol-carrying lipoproteins and function primarily to transport cholesterol to peripheral cells. LDL bind to high-affinity receptors of plasma membranes and, through internalization, deliver cholesterol to peripheral tissues. Endocytosis of an LDL particle into the cell is followed by its degradation into component parts by cellular lysozymes and hydrolases. The triglycerides are saponified into free fatty acids and glycerol to be used by the cell for energy or to be re-esterified for later use. The apolipoproteins are degraded to their constituent amino acids for use by the cell. Free cholesterol derived from degraded LDL can be utilized in cell membranes, for hormone synthesis, or can be re-esterified for storage. In cells of the adrenal and other steroid-synthesizing tissues, cholesterol is used to synthesize the steroids cortisol, testosterone, estrogen, and androgen. Regulation of intracellular cholesterol concentration is controlled by a metabolite of free cholesterol, which turns off cellular cholesterol synthesis and then down-regulates LDL receptors as levels rise. This mechanism allows some control of the routing of cholesterol around the body. Abnormalities in the LDL receptor mechanism can result in elevations of LDL in the circulation and thus lead to hypercholesterolemia and premature atherosclerosis.[21,66]

The Reverse Cholesterol Transport Pathway

A major function of lipoproteins is the transport of cholesterol from the liver through the blood vessels to peripheral tissues. However, when supply exceeds need, cholesterol accumulates in tissues. Since cholesterol can only be excreted from the body through the liver as bile acid or free cholesterol, excess cholesterol must be transported back to the liver in order to be removed from peripheral tissues. The major role of HDL is in the reverse cholesterol transport pathway,[166] a process which removes cholesterol from peripheral tissues (Fig. 15-5). In this process, nascent HDL produced in the liver and intestine absorb free cholesterol from peripheral cells (even from macrophages)[161] and convert it to cholesteryl ester for storage inside the HDL core during transport. Both lecithin cholesterol acetyltransferase (LCAT) and apos A-I and D are essential for the esterification of cholesterol. There are several routes by which HDL can deliver esterified cholesterol to the liver:[85] direct uptake of HDL by the liver can be mediated by specific receptors capable of binding apo A; apo E-containing HDL may enter the liver through the binding of apo E to the LDL receptor or the remnant receptor on the liver; and, finally, with the help of cholesteryl ester transfer protein (CETP), the cholesteryl ester in HDL may be indirectly delivered to the liver by transferring it to triglyceride-rich lipoproteins, such as chylomicrons and VLDL, in exchange for triglycerides.

Lipoprotein Receptors

Perhaps the most important mechanism involving lipoprotein metabolism is the interaction between apolipoproteins on the lipoprotein surfaces and the receptors on various cell surfaces. This critical interaction is necessary not only for delivering lipoproteins to the cells, but also for efficient removal of potentially atherogenic lipoproteins from the blood and peripheral tissues. Lipoprotein receptors are plasma membrane proteins that are capable of binding with high affinity to circulating lipoprotein particles through their interaction with apolipoproteins. The following paragraphs discuss three of the major receptors.

LDL Receptors[20,21]

LDL receptors recognize both apo E and apo B-100 and mediate cellular binding, uptake, and degradation of LDL and the other apo B-100-containing lipoproteins (VLDL and IDL). They are widely expressed and, therefore, play an important role in cellular and systemic cholesterol homeostasis.[88]

The receptors are synthesized inside the cell and then migrate to particular regions of the cell membrane called *coated pits*. The coated pits act like protective harbors. Once an LDL particle binds to a receptor, it is drawn into the cell for degradation. Receptors that migrate to areas of the cell membrane outside the coated pits are unable to bind and internalize LDL. Synthesis of LDL receptors is inhibited by high intracellular cholesterol levels, thus allowing the amount of cholesterol entering the cell to be controlled. Conversely, insufficient or defective receptors will stimulate intracellular cholesterol synthesis. In patients that are heterozygous for a disease called familial hypercholesterolemia (approximately 1 in 500 persons in many populations), half of the LDL receptors are defective, causing insufficient LDL to be taken up by the cells. Consequently, LDL cholesterol removal rate from the circulation is about 40%, and the circulating LDL cholesterol level approximately doubles. Homozygotes without any normally func-

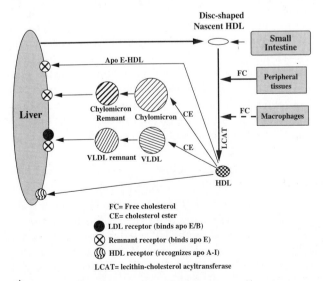

Figure 15-5. Reverse cholesterol transport pathway.

tioning LDL receptors have LDL elevations that are even more severe.

Remnant Receptors[54,157]

Remnant receptors recognize apo E and are the major receptors for the clearance of chylomicron remnants and beta-VLDL from blood circulation. They also bind apo E-containing HDL.

Scavenger Receptors[67]

Scavenger receptors can be found on the surfaces of macrophages and on a subset of other cells, such as muscle cells. These receptors mediate the removal of modified LDL, including oxidized LDL and β-VLDL (collective name for chylomicron remnants and VLDL remnants) from blood circulation. The scavenger receptors are biochemically and genetically distinct from native LDL receptors. Unlike LDL receptors, scavenger receptor expression is not regulated by intracellular cholesterol concentration. Macrophages can continuously take up cholesterol from modified LDL through scavenger receptors, resulting in cholesterol accumulation and the formation of foam cells, which is the hallmark of early atherosclerotic lesions.

Effect of Hormones

Insulin

One of insulin's major functions is the inactivation of hormone-sensitive lipase. In adipose tissues, the result is suppression of the release of free fatty acid, causing a reduction of free fatty acid delivery to the liver. However, insulin and glucose also promote esterification of free fatty acids into triglycerides. Thus, in the liver, insulin action tends to regulate VLDL synthesis. Insulin is also required for expression of LPL activity. This enzyme is essential for triglyceride-rich lipoprotein clearance. Genetic deficiency of the enzyme leads to chylomicronemia.[32] In insulin-dependent diabetes, severe insulin deficiency can lead to fatty liver and marked hypertriglyceridemia, as free fatty acids are released unchecked from adipose tissue, transported to the liver, re-esterified into triglyceride, and transported out as VLDL. The loss of LPL activity leads to defects in the clearance of chylomicrons and VLDL, greatly exacerbating the hypertriglyceridemia frequently found in this condition.

Insulin deficiency or insulin resistance (a diminution of the biological response to a given concentration of insulin),[14,53] which is associated with diabetes, often leads to elevation of plasma triglycerides, reduction of HDL cholesterol concentrations, and production of atherogenic, small, dense LDL. These unfavorable changes in lipid metabolism can increase the risk for CHD in non-insulin-dependent diabetic patients.[53]

Growth Hormone[5]

Growth hormone can increase the production of VLDL by increasing the production of apo E and apo B-48 and by stimulating lipolysis in adipose tissues and triglyceride synthesis in the liver. It also stimulates the synthesis of LDL receptors and enhances the clearance of lipoproteins by the liver. In patients with growth hormone deficiency, administration of growth hormone results in a normalization of dyslipidemia.

Sex Hormones[58,150]

Estrogens appear to raise VLDL levels by accelerating the production rate of large, triglyceride-rich VLDL. On the other hand, an increase in the number of LDL receptors by estrogen results in a lowering of LDL levels, and an increase in apo A-I production results in an increase in HDL levels.[68] Progesterone balances these effects by lowering VLDL and triglyceride levels by increasing their rate of catabolism.

Thyroid Hormone

Thyroid hormone exerts multiple effects on the metabolism of fatty acids and glycerol. Thyroid hormone stimulates both LPL and hepatic triglyceride lipase synthesis, and also increases LDL receptor production. As a result, patients with hypothyroidism will exhibit elevated levels of plasma LDL-cholesterol.[76]

LIPID AND LIPOPROTEIN ANALYSES

Lipid Measurement

Accuracy is especially important in the analysis of the lipids and lipoproteins. Since these analytes are important indicators of CHD risk and require results from large population studies in order to assign the levels associated with that risk, decision cut-points cannot be easily established by an individual laboratory or manufacturer as is done for many other diagnostic analytes. Decision cut-points have been set by NCEP expert panels based on population distributions and coronary disease risk relationships established in large long-term epidemiological studies. Standardization of the lipid analytical methods in the participating research laboratories involved in these epidemiological studies made results comparable among laboratories and over time. For other research laboratories to generate comparable results and for routine clinical laboratories to obtain reliable classification of patients using the national decision cut-points, the methods must be standardized to the same accuracy base used for the national studies, a process which will be described subsequently.

Cholesterol Measurement

The lipid work-up traditionally has begun with measurement of total serum cholesterol. The lipoproteins are also generally quantitated based on their cholesterol content. The early analytical methods[197] used strong acids (such as sulfuric and acetic) and chemicals (such as acetic anhydride or ferric chloride), which produce a measurable color with cholesterol. Since the strong acid reactions are relatively nonspecific, partial or full extraction by organic solvents was used to improve specificity. The current reference method for cholesterol employs hexane extraction after hydrolysis with alcoholic KOH followed by reaction with Liebermann-Burchard color reagent, which is composed of sulfuric and acetic acids and acetic anhydride.[1,48] The multi-step manual method is complicated but gives good agreement with the "gold standard" method developed and applied at the U. S. National Institute for Standards and Technology—the so-called Definitive Method employing isotope-dilution mass spectrometry.[35]

In recent years, virtually all routine clinical laboratories have progressed to using enzymic reagents. Enzymes, selected for specificity to the analyte of interest, provide reasonably accurate quantitation without the necessity for extraction or other pre-treatment. The reagents are mild compared to earlier strong acid reagents and suitable for use in modern automated microprocessor-controlled robotic instruments. Cholesterol, and often triglycerides and HDL cholesterol, are included in routine test panels on automated batch and discrete chemistry analyzers. Measurement of the lipoproteins, HDL and LDL, requires separation steps which are generally performed manually, but the cholesterol content is quantified using automated analyzers.

Although several enzymic reaction sequences have been described, only one sequence is in common use for measuring cholesterol.[3,144] About two thirds of circulating cholesterol occurs as cholesteryl ester, so the enzyme cholesteryl ester hydrolase is used to cleave the fatty acid residue and thus convert cholesteryl esters to unesterified or free cholesterol. The free cholesterol is reacted by cholesterol oxidase, producing hydrogen peroxide, which is substrate for a common enzymic color reaction using horseradish peroxidase to couple two colorless chemicals into a colored compound. The intensity of the color is proportional to the amount of cholesterol in the samples and can be measured by a spectrophotometer at a wavelength of about 500 nm. The enzymes and reagents have improved over the years so most commercial reagents, when calibrated appropriately, can be expected to give reliable results. This reaction sequence is generally used on serum without an extraction step but can be subject to interference. For example, vitamin C and bilirubin are reducing agents that can interfere with the peroxidase-catalyzed color reaction.[119]

Triglycerides

Measurement of triglycerides in conjunction with cholesterol is useful in detecting certain genetic and other types of metabolic disorders. However, in many cases the triglyceride measurement is made primarily for estimation of LDL cholesterol by the Friedewald equation (discussed subsequently). Several enzymic reaction sequences are available for triglyceride measurement, all having in common the use of lipases to cleave fatty acids from glycerol.[92] The freed glycerol participates in any one of several enzymic sequences. One of the more common earlier reactions, ending in a product measured in the ultraviolet region, used glycerol kinase and pyruvate kinase, culminating in the conversion of NADH to NAD⁺ with an associated decrease in absorbency.[23] This reaction is quite susceptible to interference and side reactions. The UV endpoint is also less convenient for modern analyz-

ers, so this and other UV sequences are gradually being replaced by a second sequence involving glycerol kinase and glycerol-phosphate oxidase, which feeds into the same peroxidase color reaction described for cholesterol.[118]

All enzymic triglyceride reaction sequences also react with endogenous free glycerol, which is universally present in serum and constitutes a significant source of interference.[39,178] In most specimens, the endogenous free glycerol contributes to a 10 to 20 mg/dL overestimation of triglycerides. A few specimens, about 20%, will have higher glycerol, with levels increased in certain conditions such as diabetes and liver disease or from glycerol-based medications. Most research laboratories incorporate some type of correction for endogenous free glycerol, but this practice is uncommon in clinical laboratories. The most common correction, designated double cuvette blanking, results from the performance of a second parallel measurement using reagent without lipase to quantitate only the free glycerol blank. This measurement is subtracted from the total glycerol measurement of the complete reaction sequence to determine a net or blank-corrected triglyceride result.[188] Another approach, designated the single cuvet blank, begins with the lipase-free reagent. After a brief incubation, a blank reading is taken, which measures only the concentration of free glycerol. The lipase is then added as a second separate reagent and after additional incubation, a total reading is taken which, corrected for the blank by the instrument, represents net triglycerides.[180] Since these blank corrections add to the cost of the measurement and are perhaps unjustified by the accuracy requirements in routine practice, a convenient and easily implemented alternative, designated calibration blanking, involves adjustment of the calibrator set points to compensate for the average free glycerol content of specimens. This approach is usually reasonably accurate because the free glycerol levels are relatively low and consistent in most specimens. Thus, only a few specimens will be undercorrected but will still be more accurate than uncorrected.

The triglyceride reference method involves alkaline hydrolysis, solvent extraction, and a color reaction with chromotropic acid.[111] The assay is tedious, poorly characterized, and not used except in the standardization laboratory at the Centers for Disease Control and Prevention (CDC). A more suitable reference method with solvent extraction followed by an optimized enzymic assay is in development. Accuracy in the triglyceride measurement is not as critical as it is for cholesterol, because the cut-points are widely spaced and because the physiological variation is so large, with CVs for biological variation in the 25% to 30% range, that analytical variation is relatively insignificant.

Phospholipids

Quantitative measurement of phospholipids is rare in routine clinical practice. At one time, there was a suggestion that HDL be quantified in terms of phospholipids because it is the predominant lipid constituent, but this practice has never been adopted.[181] Phospholipids are sometimes measured in research (e.g., in studies of dietary influences). An enzymic reaction sequence utilizing phospholipase D, choline-oxidase, and horseradish peroxidase measures the choline-containing phospholipids lecithin, lysolecithin, and sphingomyelin, which account for at least 95% of total phospholipids in serum.[120,181] Commercial kit methods with this enzymic sequence can be purchased. Prior to the availability of enzymic reagents, the common quantitative method involved extraction and acid digestion with analysis of the total lipid-bound phosphorus.[11]

Certain choline-containing phospholipids within this class, or their ratios, have been determined in the clinical laboratory. For example, fetal lung maturity has been evaluated from characteristic patterns of phospholipids in amniotic fluid. In this instance, phospholipids may be recovered by solvent extraction, applied to a silica gel plate for separation by thin-layer chromatography, and quantified after visualization by treating with iodine vapor. The ratio between lecithin and sphingomyelin is predictive of the stage of development of the fetal lung.

Apolipoprotein Measurement

There are a number of apolipoproteins that are incorporated into the lipoproteins (Table 15-3) Three apolipoproteins in particular have been of interest for clinical diagnostic purposes. Apo B, the major protein of LDL and VLDL, is an indicator of combined LDL and VLDL concentration and can be measured directly in serum by immunoassay methods. This characteristic has provided some researchers with a rationale for replacing the alternative separation and measurement of LDL cholesterol.[167] Apo A-I, as the major protein of HDL, could be measured directly in serum in place of separation and analysis of HDL cholesterol, and Lp(a) could be measured as an indicator of CHD risk. These apolipoproteins are commonly measured in research laboratories and in some clinical laboratories supporting cardiovascular practices in addition to the conventional lipoprotein measurements. The apolipoproteins are useful in patient management by skilled practitioners, but they have not yet been accepted or recommended by any organization, such as NCEP, for use in routine practice. Prospective studies have not consistently shown them to be independent CHD risk factors, after adjusting for LDL and HDL cholesterol.

The apolipoproteins are measured by immunoassays of various types with several commercial kit methods available. Most common in routine laboratories are turbidimetric assays for chemistry analyzers or nephelometric assays for dedicated nephelometers. Especially for apo B and Lp(a), these light-scattering assays are often subject to interference from the larger triglyceride-rich lipoproteins, chylomicrons, and VLDL. ELISA, RID, and RIA methods are also available but, especially for the latter two types, are becoming less common. Antibodies used in the immunoassays may be polyclonal or monoclonal. International efforts to develop reference materials and standardization programs for the assays are in progress. Since Lp(a) is genetically heterogeneous and the levels and CHD risk correlate with the isoform size, qualitative assessment of isoform distribution may also be important.

Lipoprotein Methods

Several methods have been used for the separation and quantitation of the serum lipoproteins. Lipoprotein separations take advantage of physical properties such as density, size, charge, and apolipoprotein content. The range in density observed among the lipoprotein classes is a function of the relative lipid and protein content and enables fractionation by ultracentrifugation. Electrophoretic separations take advantage of differences in charge and size. Chemical precipitation methods, which are most common in clinical laboratories, depend on particle size, charge, and differences in the apolipoprotein content. Antibodies specific to apolipoproteins can be used to bind and separate lipoprotein classes. Chromatographic methods take advantage of size differences in molecular sieving methods or composition in affinity methods using, for example, heparin sepharose.

Many ultracentrifugation methods have been used in the research laboratory, but ultracentrifugation is uncommon in the clinical laboratory.[73] The most common approach, called preparative ultracentrifugation, uses sequential density adjustments of serum to fractionate major and minor lipoprotein classes.[74] Density gradient methods[33,96,140,142,147]—either non-equilibrium techniques, in which separations are based on the rate of flotation, or equilibrium techniques in which separations run to completion—permit fractionation of several or all classes in a single run. The available methods use different types of ultracentrifuge rotors: swinging bucket, fixed angle, vertical, and zonal. Newer methods have trended toward smaller scale separations in small rotors using table-top ultracentrifuges.[17,195] Ultracentrifugation, even though tedious, expensive, and technically demanding, remains a workhorse for separation of lipoproteins both for quantitative purposes and for preparative isolations. Ultracentrifugation is also used in the reference methods for lipoprotein quantitation, which is appropriate because the lipoproteins are classically defined in terms of hydrated density.

Electrophoretic methods[103,104] allow separation and quantitation of major lipoprotein classes and provide a visual display useful in detecting unusual or variant patterns. Agarose gel has been the most common medium for separation of intact lipoproteins, providing a clear background and convenient use.[41,107,137,139,194] Electrophoretic methods have, in general, been considered useful for qualitative analysis but less than desirable for lipoprotein quantitation because of poor precision and large systematic biases compared to other methods.[146] Evaluations of one of the newer commercial automated electrophoretic systems demonstrate, however, that electrophoresis can be precise and accurate.[193]

Electrophoresis in polyacrylamide gels is used for separation of lipoprotein classes,[128] of subclasses, and of the apolipoproteins. Of particular recent interest have been methods that fractionate LDL and HDL subclasses in order to characterize the more atherogenic,[15,93,105,122] heavier, lipid-depleted, and smaller fractions versus the larger, lighter subclasses.

Separation methods involving chemical precipitation, usually with polyanions such as heparin and divalent cations such as manganese, can be used to separate any of the lipoprotein classes. Because of their convenience, such methods have become common in clinical laboratories in the separation of HDL.[24,102,190] Apo B in VLDL and LDL is rich in positively charged amino acids, which preferentially form complexes with the polyanions; the divalent cations neutralize the charged groups on the lipoproteins making them aggregate and become insoluble. The insoluble lipoproteins aggregate and precipitate, leaving HDL in the solution. By varying the concentration of polyanion and divalent cation any of the lipoprotein classes can be separated with reasonable specificity.

Separation of lipoproteins by immunochemical means using antibodies that are specific to epitopes on the apolipoproteins has potential for both research and routine use.[90,164] Antibodies can be immobilized on a solid support, such as a column matrix or latex beads, to facilitate separations. The apo B–containing lipoproteins as a group can be bound by antibodies to apo B. Selectivity within the apo B–containing lipoproteins, such as removing VLDL while retaining LDL, can be obtained by using antibodies to the minor apolipoproteins. HDL can be selectively removed using antibodies to apo A-I, the major protein of HDL. The use of latex-immobilized antibodies is the basis for a recently introduced commercial method for direct quantitation of LDL cholesterol.[124]

High-Density Lipoproteins

The measurement of HDL cholesterol has assumed greater importance in the latest NCEP treatment guidelines. Previously, HDL cholesterol was measured as a risk factor but otherwise was not considered in treatment decisions. Following recommendations of an NIH-sponsored consensus panel,[136] the NCEP guidelines now include HDL cholesterol measurement with total cholesterol in the first medical work-up. Since the risk associated with HDL cholesterol is expressed over a relatively small concentration range, accuracy in the measurement is especially important.

For routine diagnostic purposes, HDL is separated almost exclusively by chemical precipitation. In practice, the HDL cholesterol measurement is a two-step procedure. The precipitation reagent is added to serum or plasma to aggregate non-HDL lipoproteins, which are sedimented by centrifugation. Early methods used centrifugation forces of approximately 1500 × gravity requiring lengthy centrifugation times of 10 to 30 minutes. Newer methods use high-speed centrifuges with forces of 10,000 to 15,000 × gravity, decreasing centrifugation times to 3 to 5 minutes. The HDL is then quantified as the cholesterol in the supernate with analysis commonly by one of the enzymic assays modified for the lower HDL cholesterol range.

The earliest common precipitation method used heparin in combination with manganese to precipitate the apo B–containing lipoproteins.[25,55,115] Because manganese produced interference with enzymic assays, alternative reagents were developed.[174] Sodium phosphotungstate[26,112] with magnesium became common in routine use but, because of its sensitivity to reaction conditions and greater variability, is being replaced by dextran sulfate (a synthetic heparin) with magnesium.[191] The earliest dextran sulfate methods used material

of 500 kDa, but 50 kDa, considered more specific, is now becoming more common.[190] Polyethylene glycol also precipitates lipoproteins, but 100-fold higher concentrations of reagent are required than for polyanions. The consequent larger dilutions and highly viscous reagents, which are difficult to pipet precisely, make this reagent less common.[4,47,186] Numerous commercial versions of these precipitation reagents are available. In the past, the various methods have sometimes given quite different results, since there has been a lack of standardization, but recently a protocol has become available to reagent manufacturers that should reduce the amount of variability in the future.

A significant problem with HDL precipitation methods is interference from elevated triglyceride levels, which prevents sedimentation of the precipitate.[189] When the triglyceride-rich VLDL and chylomicrons are present, the low density of the aggregated lipoproteins may prevent them from sedimenting or may even cause floating during centrifugation. This incomplete sedimentation, indicated by cloudiness, turbidity, or particulate matter floating in the supernate, results in overestimation of HDL cholesterol. High-speed centrifugation will reduce the proportion of turbid supernates. Pre-dilution of the specimen also promotes clearing but may lead to errors in the cholesterol analysis. Turbid supernates may also be cleared by ultrafiltration, a method that works well but is tedious and inefficient.

New methods currently in development will streamline and possibly automate the preliminary HDL separation step. Improvement must occur in HDL measurement to provide results that are sufficiently reliable for diagnostic purposes.

The accepted reference method for HDL cholesterol has been a three-step procedure developed at the CDC. This method involves ultracentrifugation to remove VLDL, heparin-manganese precipitation from the 1.006 g/mL infranate to remove LDL, and analysis of supernatant cholesterol by the Abell-Kendall assay.[48] Since this method is tedious and expensive, a simpler direct precipitation method has been validated by the CDC Network laboratory group as a designated comparison method—it uses direct dextran sulfate (50 kDa) precipitation with Abell-Kendall cholesterol analysis.[129]

Low-Density Lipoproteins

LDL cholesterol, as the proven atherogenic lipoprotein, is the primary basis for treatment decisions in the NCEP clinical guidelines.[51,52] The common research method for accurate LDL cholesterol quantitation and the basis for the reference method is designated *beta-quantification,* where beta refers to the electrophoretic term for LDL. The beta-quantification technique involves a combination of ultracentrifugation and chemical precipitation.[12,115] Ultracentrifugation of serum at the native density of 1.006 g/L is used to float VLDL and any chylomicrons for separation. The fractions are recovered by slicing the tube between the fractions and pipetting. Ultracentrifugation is preferred for VLDL separation because other methods, such as precipitation, are not as specific for VLDL and are subject to interference from chylomicrons. Ultracentrifugation is a robust but tedious method that can give reliable results provided the technique is meticulous.

In a separate step, chemical precipitation (as described above) is used to separate HDL from either the whole serum or the infranate obtained from ultracentrifugation. Compared to ultracentrifugation, the precipitation step is efficient and convenient, as well as relatively robust for HDL separation. Cholesterol is quantitated in serum, in the 1.006 g/mL infranatant, and in the HDL supernate by enzymic or other assay methods. LDL cholesterol is calculated as the difference between cholesterol measured in the infranate and in the HDL fraction. VLDL cholesterol is usually calculated as the difference between that in whole serum and the amount in the infranate fraction. The ultracentrifugation step makes beta-quantification tedious and inaccessible for routine diagnostic purposes.

A simpler technique for LDL cholesterol quantitation that is common in both routine and research laboratories and which bypasses ultracentrifugation is the so called "*Friedewald calculation*" or *derived beta-quantification*.[57] HDL is separated by precipitation, its cholesterol is assayed, and cholesterol and triglycerides are measured in the serum. VLDL cholesterol is estimated as triglycerides divided by 5 (when using mg/dL units), an approximation which works reasonably well in most normolipemic specimens. The presence of elevated triglycerides, 400 mg/dL is the accepted limit, chylomicrons, and a beta-VLDL characteristic of the rare type III hyperlipoproteinemia preclude this estimation. The estimated VLDL cholesterol and measured HDL cholesterol are subtracted from total serum cholesterol to estimate or derive LDL cholesterol.

The Friedewald estimation has been used almost universally in estimating LDL cholesterol in routine clinical practice. Investigations in lipid specialty laboratories have suggested the method is reliable for patient classification, provided the underlying measurements are made with appropriate accuracy and precision.[123,192] There is considerable concern about the reliability in routine laboratories, however, since the error in estimated LDL cholesterol includes the cumulative error in each of the underlying measurements used in the calculation of cholesterol, triglycerides, and HDL cholesterol. The NCEP laboratory expert panel reviewed performance data and concluded that the level of analytical performance required to derive LDL cholesterol accurately enough to meet clinical needs was beyond the capability of most routine laboratories. In order to meet the requisite NCEP precision goal of 4% CV for LDL cholesterol (Table 15-5), a laboratory would be required to achieve half the NCEP goals for each of the underlying measurements. The NCEP panel concluded that alternative methods are needed for routine diagnostic use, preferably methods which directly separate LDL for cholesterol quantitation.[132]

In response to the NCEP request, several direct LDL cholesterol methods have recently been developed or refined for general use.[146] One commercial method that is compatible as a pretreatment step with a variety of analytical systems employs immunochemical separation.[124] A mixture of antibodies specific to epitopes on the apolipoproteins of VLDL and HDL is immobilized on latex beads. Specimen is added to the beads in a micro-filtration device, which is mixed and subjected to centrifugation. VLDL and HDL are retained by the filter while LDL passes through. Cholesterol in the LDL filtrate is assayed by enzymic reagent using an automated chemistry analyzer. This convenient method has been found to correlate with beta-quantification on both fasting and non-fasting specimens.[124]

Fatty Acids

Analysis of fatty acids is used in the research laboratory for studies of dietary factors,[106] but is used less commonly in the routine laboratory for diagnosis of rare genetic conditions. Fatty acids are commonly analyzed by gas-liquid chromatography.[184] The fatty acids are extracted, undergo alkaline hydrolysis, and then are converted to methyl esters of diazomethane. Application of the sample to a gas chromatograph allows separation of fatty acids. A reference standard typically contains laurate, myristate, palmitate, palmitoleate, phytanate, stearate, oleate, linoleate, linolenate, arachidate, and arachidonate.[100]

Stool Fat

Fat in stool or urine can have significant diagnostic implications. An adult with a normal diet will not have more than 6 g per day of fat in the feces. Higher values can indicate the presence of malabsorption in children and pancreatic insufficiency in adults. The common method in clinical laboratories requires a 72-hour collection of feces, followed by an analysis of lipid content and calculation of lipid as a percentage of the entire mass.[184]

Compact Analyzers

A major area of development in recent years has focused on compact analysis systems for use in point of care testing at the patient's bedside, in the physician's office, in wellness centers, and even in the home.[187] First-generation systems, introduced in the 1980s, were relatively large and measured cholesterol and triglycerides as well as other common analytes, usually separately and sequentially. HDL separations involving off-line pretreatment steps were subsequently developed. Second-generation systems became smaller and more sophisticated, offering separation of HDL and analysis of cholesterol and triglycerides simultaneously from fingerstick blood in an integrated system. A third-generation system measures cholesterol, triglycerides, HDL cholesterol, and glucose simultaneously from a fingerstick sample. Noninstrumented systems with a thermometer-like reading are available for total cholesterol and are in development for HDL cholesterol. These new technologies offer the capability of measuring lipids and lipoproteins reliably outside the conventional laboratory.

Standardization

Precision

Precision is a prerequisite for accuracy; a method may have no overall systematic error or bias, but if it is imprecise it will still be inaccurate on individual measurements. With the shift to modern automated analyzers, analytical variation has, in general, become less of a concern than biological and other sources of pre-analytical variation. Cholesterol levels are affected by many factors that can be categorized into biological, clinical, and sampling sources.[43,44] Changes in lifestyle that affect usual diet, exercise, weight, and smoking patterns can result in fluctuations in the observed cholesterol and triglyceride values and the distribution of the lipoproteins. Similarly, the presence of clinical conditions, various diseases, or the medications used in their treatment, affect the circulating lipoproteins. Conditions during blood collection, such as fasting status, posture, the choice of anticoagulant in the collection tube, and storage conditions, can alter the measurements. Typical observed biological variation over 1 year for total cholesterol averages approximately 6.1% CV. Thus, in the average patient, measurements made over the course of a year would fall 66% of the time within ±6.1% of the mean cholesterol concentration and 95% of the time within twice this range. Some patients may exhibit substantially more biological variation. Thus, pre-analytical variation generally is relatively large in relation to the usual analytical variation, which is typically less than 3% CV, and must be considered in interpreting cholesterol results. Some factors, such as posture and blood collection, can be standardized to minimize the variation. The NCEP guidelines[52] recommend making decisions based on the average of two or three measurements to factor out the effect of both pre-analytical and analytical sources. The use of stepped cutpoints in the work-up also reduces the practical effect of variation.

Accuracy

It is essential for accuracy that a method be calibrated or traceable to its respective reference method. In the case of cholesterol, the reference system is quite advanced and complete, having served as a model for standardization of laboratory analytes.[129,143] The Definitive Method at the National Institute for Standards and Technology provides the accuracy target but is too expensive and complicated for frequent use.[35] The Reference Method developed and applied at the CDC, and calibrated by an approved primary reference standard to the Definitive Method, provides a transferable, practical reference link.[48] The Reference Method is now accessible in a network of standardized laboratories, which compose the Cholesterol Reference Method Laboratory Network. This network was established in the U.S. and some European and Asian countries to extend standardization to manufacturers and clinical laboratories.[129] The network was established primarily to provide direct accuracy comparisons using fresh native serum specimens, which are necessary for reliable accuracy transfer because of analyte-matrix interaction problems on processed reference materials.[49,69,94]

Matrix Interactions

In the early stages of cholesterol standardization, which were directed toward diagnostic manufacturers and routine laboratories, commercial lyophilized or freeze-dried materials were used. These materials, made in large quantities, often with spiking or artificial addition of analytes, were assayed by the Definitive and/or Reference Methods and distrib-

uted widely for accuracy transfer. Subsequently, biases were observed in enzymic assays on fresh patient specimens. Even though such manufactured reference materials are convenient, stable, and amenable to shipment at ambient temperatures, the manufacturing process, especially spiking and lyophilization, altered the measurement properties in enzymic assays. These enzymic assays, sensitive to the nature of the analyte and specimen matrix, were altered such that results were not representative of results on patient specimens. In order to achieve reliable feedback on accuracy and transfer of the accuracy base, direct comparisons with the Reference Method on actual patient specimens were determined to be necessary.[129]

CDC Cholesterol Reference Method Laboratory Network

In response, the CDC Cholesterol Network program was organized. The Network offers reliable accuracy comparisons and has developed a formal certification program whereby laboratories and manufacturers can document traceability to the National Reference System for Cholesterol.[129] Clinical laboratories performing total and HDL cholesterol analyses can select a commercial method which has been certified by the Network. Certification does not ensure all aspects of quality in a reagent system but primarily ensures that the accuracy is traceable to the National Reference System for Cholesterol within accepted limits and that the precision is acceptable. Certification is most efficient through manufacturers, but individual laboratories desiring to confirm the performance of their system by completing a certification protocol can contact the CDC at 770-488-4126 for information about the network.

Analytical Performance Goals

The NCEP laboratory panels[131-133,143] have established requisite analytical performance goals based on clinical needs for routine measurements (Table 15-5). For analysis of total cholesterol the performance goal for total error is 8.9%. That is, the overall error should be such that each individual cholesterol measurement falls within ±8.9% of the Reference Method value. Actually, since the goals are based on 95% certainty, 95 of 100 measurements should fall within the total error limit. One can assay a specimen many times and calculate the mean to determine the usual value or the cen-

TABLE 15-5. NCEP Analytical Performance Goals

Cholesterol	Precision CV 3%	Bias ±3%	Total Error ±8.9%
HDL			
1993			
≥42 mg/dL	6%	±10%	±21.8%
<42 mg/dL	SD ≤ 2.5 mg/dL		
1998			
≥42 mg/dL	4%	±5%	±12.8%
<42 mg/dL	SD <1.7 mg/dL		
LDL	4%	±4%	±11.8%
Triglycerides	5%	±5%	±14.8%

tral tendency. The scatter or random variation around the mean is described by the standard deviation—an interval of plus and minus one standard deviation around the mean includes by definition two thirds of the observations. In the laboratory, since the scatter or imprecision is often proportional to the concentration, random variation is usually specified in relative terms as CV, the coefficient of variation or relative standard deviation, which equals the standard deviation divided by the mean. Overall accuracy or systematic error is described as bias, the difference between the mean and the true value. Bias is primarily a function of the method's calibration and may vary by concentration. Of greatest concern in this context is bias at the NCEP decision cutpoints. The bias and CV targets presented in Table 15-5 are representative of performance which will meet the NCEP goals for total error.

Quality Control

Achieving acceptable analytical performance requires the use of reliable quality control materials, which should preferably closely emulate actual patient specimens. Currently, the most suitable are quality control pools that are prepared from freshly collected patient serum, aliquoted into securely sealed vials, quick frozen, and stored at −70°C. Pools of fresh frozen serum are essential for reliability when monitoring accuracy in lipoprotein separation and analysis, and are preferable for monitoring cholesterol and other lipid measurements. Since commercial pools often undergo matrix alterations that change their analysis characteristics, results may not represent results on patients. At least two pools should be analyzed, preferably with levels at or near decision points for each analyte.

Serum, usually collected in serum separator vacuum tubes with clotting enhancers, has been the fluid of choice for lipoprotein measurement in the routine clinical laboratory. EDTA plasma has traditionally been the choice in lipid research laboratories, especially for lipoprotein separations, because the anticoagulant is thought to enhance stability. EDTA does have potential disadvantages which discourage routine use. Micro-clots, which form during storage, plug the sampling probes on the modern chemistry analyzers. EDTA osmotically draws water from red cells, diluting the plasma constituents, and the dilution effect can vary depending on such factors as fill volume, the analyte being measured, and the extent of mixing. Since the NCEP cutpoints are based on serum values, cholesterol measurements made on EDTA plasma require correction by the factor of 1.03.

LIPID AND LIPOPROTEIN DISTRIBUTION IN THE POPULATION

In the general adult population, serum lipoprotein concentrations differ between men and women, the result, primarily, of differences in sex hormone levels.[77,108] Women generally have higher levels of HDL cholesterol and lower levels of total cholesterol and triglycerides than men due to higher estrogen levels.[154] The difference in total cholesterol values

disappears after menopause as estrogen decreases.[27,58] Men and women both, however, show a tendency toward increases in concentrations of total cholesterol, LDL cholesterol, and triglycerides with age.[36,108,122] HDL cholesterol concentrations generally remain stable after the onset of puberty, and do not drop in women with the onset of menopause.

Circulating levels of total cholesterol, LDL cholesterol, and triglycerides in young children are generally much lower than those seen in adults.[50,59,108] In addition, concentrations do not differ significantly between boys and girls. HDL cholesterol levels for both boys and girls are comparable to those of adult women. At the onset of puberty, however, HDL cholesterol concentrations in boys fall to adult male levels, a drop of approximately 20%, while those of girls remain the same. It is the lower concentration of HDL cholesterol in men that is associated with their increased risk of premature heart disease.

Epidemiological studies have shown that the incidence of heart disease is strongly associated with serum cholesterol concentration,[30,91] and comparisons performed in various societies have shown that those that traditionally eat less animal fat and more grains, fruits, and vegetables, such as many Asian countries, for example, have lower levels of LDL cholesterol and lower rates of heart disease than societies that have more fat, particularly animal fat, in their diet.[170,183] These differences can be attributed to both genetic and dietary inputs. The importance of dietary factors was shown clearly in a study that compared the dietary patterns and heart disease rates in Japanese men living in Japan, Hawaii, and California.[89] In this study, as dietary intake became more westernized, with increased consumption of fat and cholesterol, the LDL cholesterol concentrations increased significantly, as did the rates of heart disease, so that Japanese men living in California had much higher rates of heart disease than Japanese men living in Japan; those in Hawaii were intermediate. Within societies where diet tends to be more homogeneous, LDL cholesterol levels become somewhat less discriminatory as a risk factor, and HDL cholesterol levels become more discriminatory because of the ability of HDL to remove excess cholesterol from the circulation.[87]

A few years ago, as part of a campaign to alert the American population to the risk factors associated with heart disease and to try to reduce the rate of its morbidity and mortality, NCEP was formed. NCEP has utilized panels of experts, including the Adult Treatment Panel, the Children and Adolescents Treatment Panel, and the Laboratory Standardization Panel, to produce recommendations within the scope of each panel's activities.[50–52,130–133]

In 1988, the first Adult Treatment Panel (ATP I) developed a set of heart disease risk factors,[51] which were revised and further refined by ATP II in 1993.[52] The current list of risk factors is shown in Table 15-6. The ATP II also has determined that all adults be screened for concentrations of both total and HDL cholesterol, and has developed guidelines for the diagnosis and follow-up treatment of individuals with abnormal levels (Table 15-7). The Children and Adolescents Treatment Panel has developed similar criteria for the pediatric population.[50]

TABLE 15-6. Positive Risk Factors Associated with Coronary Heart Disease, as Determined by the (Adult) Treatment Panel of the National Cholesterol Education Program

- Age: ≥45 y for men; ≥55 y or premature menopause for women
- Family history of premature CHD
- Current cigarette smoking
- Hypertension (BP ≥140/90 mm Hg or taking antihypertensive medication)
- LDL cholesterol concentration ≥160 mg/dL (≥4.1 mmol/L), with <2 risk factors
- LDL cholesterol concentration 130–159 mg/dL (3.4–4.1 mmol/L), with ≥2 risk factors
- HDL cholesterol concentration <35 mg/dL (<0.9 mmol/L)
- Diabetes mellitus

Negative Risk Factor Associated with Coronary Heart Disease

- HDL cholesterol concentration ≥60 mg/dL (≥1.6 mmol/L)

TABLE 15-7. Treatment Guidelines Established by the Adult Treatment Panel of the National Cholesterol Education Program

Initial Testing (can be nonfasting):	
Risk Category	*Action*
TC <200 mg/dL (5.2 mmol/L) & HDLC ≥35 mg/dL (0.9 mmol/L)	Repeat within 5 yrs Provide risk reduction information
TC <200 mg/dL (5.2 mmol/L) & HDLC <35 mg/dL (0.9 mmol/L)	Perform lipoprotein analysis (see below)
TC 200–239 mg/dL (5.2–6.2 mmol/L), HDLC ≥35 mg/dL (0.9 mmol/L) & <2 risk factors	Repeat in 1–2 yrs Provide risk reduction information
TC 200–239 mg/dL (5.2–6.2 mmol/L), HDLC <35 mg/dL (0.9 mmol/L) or ≥2 risk factors	Perform lipoprotein analysis (see below)
TC ≥240 mg/dL (6.2 mmol/L)	Perform lipoprotein analysis (see below)

Follow-up Testing (lipoprotein analysis after 12-hr fast):	
Risk Category	*Action*
LDLC <130 mg/dl (3.4 mmol/L)	Repeat TC and HDLC within 5 yrs Provide risk reduction information
LDLC 130–159 mg/dL (3.4–4.1 mmol/L) & <2 risk factors	Provide Step 1 diet & physical activity information, and reevaluate in 1 yr
LDLC 130–159 mg/dL (3.4–4.1 mmol/L) & ≥2 risk factors	Do clinical evaluation, incl. family history Start dietary therapy (see below)
LDLC ≥160 mg/dL (4.1 mmol/L)	Do clinical evaluation, incl. family history Start dietary therapy (see below)

Treatment Decisions:		
Risk Category	*Action Level*	*Goal*
Dietary Therapy:		
no CHD; <2 risk factors	≥160 mg/dL (4.1 mmol/L)	<160 mg/dL (4.1 mmol/L)
no CHD; ≥2 risk factors	≥130 mg/dL (3.4 mmol/L)	<130 mg/dL (3.4 mmol/L)
CHD	>100 mg/dL (2.6 mmol/L)	100 mg/dL (2.6 mmol/L)
Drug Therapy:		
no CHD; <2 risk factors	≥190 mg/dL (4.9 mmol/L)	<160 mg/dL (4.1 mmol/L)
no CHD; ≥2 risk factors	≥160 mg/dL (4.1 mmol/L)	<130 mg/dL (3.4 mmol/L)
CHD	≥130 mg/dL (3.4 mmol/L)	≤100 mg/dL (2.6 mmol/L)

The NCEP Laboratory Standardization Panel[130,143] and its successor, the Lipoprotein Measurement Working Group,[130–133] have set laboratory guidelines for the precision and accuracy associated with the measurement of total cholesterol, triglycerides, and lipoprotein cholesterol (HDL and LDL cholesterol), as previously discussed (Table 15-5).

Obviously, the best way to reduce the prevalence of heart disease in the general population is through prevention. Learning and practicing good dietary patterns early in life, maintaining these patterns throughout life,[168] and refraining from smoking[78,80] and controlling blood pressure are important means for reducing the incidence of CHD. Screening programs to measure total and HDL cholesterol, with follow-up testing for triglycerides and LDL cholesterol where indicated, is a method of identifying individuals who may have levels that put them at risk, so that they can receive treatment to reduce the level of risk. Treatment of other diseases that may affect lipoproteins is also important.

A prudent diet, low in fat and cholesterol, with a caloric intake adjusted to meet and maintain ideal body weight, along with regular exercise, can reduce the risk of heart disease and cancer.[75,138,163,165] Dietary intake of fat and cholesterol have been shown to have a synergistic effect, *i.e.,* dietary cholesterol is more efficiently absorbed when in the presence of fat,[42] and saturated fat has been found to be more atherogenic than unsaturated fat.[106,134] The American Heart Association has recommended dietary guidelines for the intake of fat and cholesterol for most adult Americans. These guidelines are shown in Table 15-1.

DISEASE: PREVENTION, DIAGNOSIS, AND TREATMENT

Diseases associated with abnormal lipid concentrations are referred to as dyslipidemias. They can be caused directly by genetic abnormalities or through environmental/lifestyle imbalances, or they can develop secondarily, as a consequence of other diseases.[40,63,65] Dyslipidemias are generally defined by clinical characteristics of the patients and the results of blood tests and do not necessarily define the specific defect associated with the abnormality. Many dyslipidemias, however, regardless of etiology, are associated with CHD, or arteriosclerosis.

Arteriosclerosis
In the United States and many other developed countries arteriosclerosis is the single leading cause of death and disability. The mortality rate has decreased in the U.S. in the last few years, partly due to advances in diagnosis and treatment, but also due to changes in lifestyle in the American population, resulting from increased awareness of the relationship between cholesterol and heart disease. This increased awareness has resulted in an overall decrease in the average serum cholesterol concentration and in a lower prevalence of heart disease, but it still exceeds all other causes of death combined. As many women as men develop arteriosclerosis, but women develop it ten years later than men, on average.

The relationship between heart disease and lipid abnormalities stems from the deposition of lipids, mainly in the form of esterified cholesterol, in the walls of the arteries. This lipid deposition starts with thin layers called fatty streaks. In studies examining blood vessels at autopsy, fatty streaks have been seen in almost everyone over the age of 15 years, regardless of cause of death.[172,173] Under certain conditions, the fatty streaks develop over time into plaques which partially block (occlude) the flow of blood. When the plaque formation develops in arteries of the arms or legs, it is called peripheral vascular disease (PVD); when it develops in the heart, it is referred to as coronary artery disease (CAD); and when it develops in the vessels of the brain, it is called cerebrovascular disease (CVD). CAD is associated with angina and myocardial infarction, and CVD is associated with stroke. Many genetic and acquired abnormalities may also lead to deposits of lipid in the liver and kidney, resulting in impaired function of these vital organs. Lipid deposits in skin are called *xanthomas* and are a clue to genetic abnormalities. They are nodules that disfigure and can indicate further lipid complications.

Plaque formation is a process of cell injury followed by infiltration and cell proliferation to repair the site. As blood travels through blood vessels, small injuries occur that signal macrophages and platelets to heal the break. LDL bring cholesterol to the site so that new cell membranes can be formed and macrophages can repair the area. LDL that have been modified by oxidative processes and chemical alterations can be taken up by the macrophages, producing foam cells.[22,84,177] These foam cells accumulate beneath the endothelial layer of the arterial wall. Future injury leads to more deposits, and eventually a plaque is formed. Continual injury and repair will lead to additional narrowing of the opening, or lumen, causing the blood to pass through under greater and greater pressure.

Deposits in the vessel walls are frequently associated with increased serum concentrations of LDL cholesterol and/or decreased HDL cholesterol.[30,64] Lowering the LDL cholesterol concentration is an important step in preventing and treating CHD.[28,29,56,101,108,109,126,135,136] It has been estimated that for every 1% decrease in LDL cholesterol concentration there is a 2% decrease in a person's of risk of developing arteriosclerosis.[109,110] For patients with established heart disease, recent studies have shown that aggressive treatment to reduce LDL cholesterol levels below 100 mg/dL (2.6 mmol/L) is effective in the stabilization and sometimes regression of plaques.[2,16,18,19,86] In some individuals, high levels of blood cholesterol or triglycerides are caused by genetic abnormalities whereby either too much is synthesized or too little is removed.[65] High levels of cholesterol or triglycerides in most people, however, are due to increased consumption of foods rich in fat and cholesterol, smoking, and lack of exercise, or to other disorders or disease states which have effects on lipid metabolism, such as diabetes, hypertension, hypothyroidism, obesity, other hormonal imbalances, liver and kidney diseases, and alcoholism. While low levels of HDL cholesterol have been associated with increased risk of heart disease,[30] so far, due to a lack of well-tolerated drug therapies to raise HDL cholesterol levels, there are not enough studies to document whether raising HDL cholesterol decreases risk.

Such studies are currently underway and in the future should provide valuable information in this regard.[148]

Laboratory analyses can aid in the diagnosis of arteriosclerosis. Accurate determinations of total, HDL, and LDL cholesterol levels can indicate the need for diet or diet and drug therapy. As shown in Table 15-7, individuals on a low-fat diet who continue to have LDL cholesterol levels ≥190 mg/dL (≥4.9 mmol/L) on repeated measurement will benefit from drug intervention. If they have ≥2 CAD risk factors and continue to have LDL cholesterol levels of ≥160 mg/dL (≥4.1 mmol/L), they also would benefit from drug therapy. And if they have already been diagnosed with heart disease, drug therapy should be considered when the LDL cholesterol level is ≥130 mg/dL (≥3.4 mmol/L). The average of at least two measurements, taken 1 to 8 weeks apart, should be used to determine treatment.[51,52]

Classic drug treatments with bile-acid sequestrants, such as cholestyramine or cholestipol, work by sequestering cholesterol in the gut to keep it from being absorbed, and are still considered to be the only safe drug for use in children since the drug is not also absorbed.[109,110] They have uncomfortable side effects, however, such as bloating and constipation. Niacin is a very potent reducer of LDL cholesterol, which also raises HDL cholesterol; however, it can be hepatotoxic and can aggravate glucose intolerance and hyperuricemia.[28] The newest class of drugs includes the HMG-CoA reductase inhibitors lovastatin, simvastatin, and pravastatin. These drugs are effective in reducing LDL cholesterol from 20% to 40%[16,18] and are generally well tolerated. They block intracellular cholesterol synthesis by inhibiting the enzyme, HMG CoA reductase. The major safety issues are hepatotoxic effects and the lack of data on possible long-term side effects. Short-term patient monitoring in clinical trials has shown that <2% of patients have sustained increases in liver enzymes. A big drawback to these drugs is their cost. Other drugs being considered are probucol, which prevents lipid oxidation and macrophage uptake, and fibric acid derivatives, such as clofibrate, gemfibrozil, fenofibrate, and etiofibrate, which reduce triglyceride and VLDL cholesterol levels and increase HDL cholesterol.

Hyperlipoproteinemias

As stated earlier, lipoproteins are complex transport vehicles for moving cholesterol, cholesteryl esters, and triglycerides in the blood. Disease states associated with abnormal serum lipids are generally caused by malfunctions in the synthesis, transport, or catabolism of the lipoproteins.[152,158] Dyslipidemias can be subdivided into two major categories: *hyper*lipoproteinemias, which are diseases associated with elevated lipoprotein levels, and *hypo*lipoproteinemias, which are associated with decreased lipoprotein levels.[157] The hyperlipoproteinemias can be subdivided into hypercholesterolemia, hypertriglyceridemia, and combined hyperlipidemia, in which there are elevations of both cholesterol and triglycerides.

Hypercholesterolemia

Hypercholesterolemia is the lipid abnormality most closely linked to heart disease.[30] One form of the disease, which is associated with genetic abnormalities that predispose affected individuals to elevated cholesterol levels, is called familial hypercholesterolemia, or FH. Homozygotes for FH are fortunately rare (1:1,000,000 in the population) but can have total cholesterol concentrations as high as 800 to 1000 mg/dL (20–26 mmol/L), and frequently have their first heart attack while still in their teens.[169] Heterozygotes for the disease are seen much more frequently, since the disease is caused by an autosomal codominant disorder. They tend to have total cholesterol concentrations in the range of 300 to 600 mg/dL (8–15 mmol/L) and, if not treated, will become symptomatic for heart disease in their 20s to 50s. Approximately 5% of patients with CAD under 50 years of age are FH heterozygotes. Other symptoms associated with FH include tendinous and tuberous xanthomas, which are cholesterol deposits under the skin, and arcus, which indicates cholesterol deposition in the vessels of the eye.

In both homozygotes and heterozygotes, the cholesterol elevation is primarily associated with an increase in LDL cholesterol. These individuals synthesize intracellular cholesterol normally, but lack, or are deficient in, active LDL receptors. Consequently, cholesterol derived through absorption and incorporated into LDL builds up in the circulation because there are no receptors to bind the LDL and transfer the cholesterol into the cells. The cells, on the other hand, which require cholesterol for use in cell membrane and hormone production, synthesize cholesterol intracellularly at an increased rate to compensate for the lack of cholesterol from the receptor-mediated mechanism.

In FH heterozygotes, and other forms of hypercholesterolemia in which there is insufficient LDL receptor activity, reduction in the rate of internal cholesterol synthesis, through inhibition of HMG CoA reductase activity by use of HMG CoA reductase inhibitors, stimulates the production of additional receptors, thereby increasing cell internalization of cholesterol from LDL, which in turn lowers the serum levels. As previously stated, these drugs have been associated with 20% to 40% decreases in serum cholesterol concentrations in these patients. Homozygotes, however, cannot benefit from this type of therapy, because they have no functional receptors to stimulate. Homozygotes rely primarily on a technique called LDL pheresis, in which blood is periodically drawn from the patient, processed to remove LDL, and returned to the patient in a similar fashion to the dialysis treatment used for kidney patients.

Most individuals with elevated LDL cholesterol levels do not have FH, but they are all at increased risk for premature CHD[179] and should be maintained on a low-fat, low-cholesterol diet, with the caloric intake adjusted to attain or maintain ideal body weight.[70,82,95] Regular physical activity should also be incorporated. Drug therapy is incorporated when necessary (see Table 15-7).

Hypertriglyceridemia

The NCEP Adult Treatment Panel identifies borderline high triglycerides as levels of 200 to 400 mg/dL (2.3–4.5 mmol/L), high as 400 to 1000 mg/dL (4.5–11.3 mmol/L), and very high as >1000 mg/dL (11.3 mmol/L). Hypertriglyceridemia can derive from a genetic abnormality, and is then called familial hypertriglyceridemia, or from secondary causes, such as hormonal abnormalities associated with the pancreas, adrenal glands, and pituitary, or from diabetes mellitus or nephrosis.[9] Diabetes mellitus, with its lack of insulin, leads to increased shunting of glucose into the pentose pathway, causing increased fatty acid synthesis. Nephrosis depresses the removal of large-molecular-weight constituents like triglycerides, causing increased serum levels. Hypertriglyceridemia is generally due to an imbalance between synthesis and clearance of VLDL in the circulation.[31,37,38] In most studies, hypertriglyceridemia has not been statistically implicated as an independent risk factor for CHD, but many CHD patients have moderately elevated triglyceride levels in conjunction with decreased HDL cholesterol levels.[159] It is difficult to separate the risk associated with increased triglycerides from that of decreased HDL cholesterol since the two are linked and serum concentrations are usually inversely related.

Triglycerides are influenced by a number of hormones, such as pancreatic insulin and glucagon, pituitary growth hormone, ACTH and thyrotropin, and adrenal medulla epinephrine and norepinephrine from the nervous system. Epinephrine and norepinephrine influence serum triglyceride levels by triggering production of hormone-sensitive lipase, which is located in adipose tissue.[162] Other body processes that trigger hormone-sensitive lipase activity are cell growth (growth hormone), adrenal stimulation (ACTH), thyroid stimulation (thyrotropin), and fasting (glucagon). Each of these processes, through its action on hormone-sensitive lipase, will result in an increase in serum triglyceride values.

Severe hypertriglyceridemia (>1000 mg/dL or 11 mmol/L), although it does not put individuals at risk for CHD, is a potentially life-threatening abnormality because it can cause pancreatitis, or inflammation of the pancreas. It is therefore imperative that these patients be diagnosed and treated with triglyceride-lowering medication and that they be monitored closely. Severe hypertriglyceridemia is generally caused by a deficiency of LPL or by a deficiency in apolipoprotein C-II, which is a necessary cofactor for LPL activity.[83] Normally, LPL hydrolyzes triglycerides carried in chylomicrons and VLDL to provide cells with free fatty acids for energy from exogenous and endogenous triglyceride sources. A deficiency in LPL or apo C-II activity keeps chylomicrons from being cleared, and serum triglycerides remain extremely elevated, even when the patient has fasted for 12 to 14 hours.

Treatment of hypertriglyceridemia consists of diet, triglyceride-lowering drugs (primarily fibric acid derivatives) in cases of severe hypertriglyceridemia or where an accompanying low HDL cholesterol value indicates risk of CHD, and sometimes oils with specific fatty acid compositions.[72,117]

Combined Hyperlipoproteinemia

Combined hyperlipoproteinemia is generally defined as the presence of elevated levels of both serum total cholesterol and triglycerides. Individuals presenting with this syndrome are considered to be at increased risk for CHD. In the ge-

netically derived form, called familial combined hyperlipoproteinemia, or FCH, some individuals of an affected kindred may have only elevated cholesterol; others, only elevated triglycerides; and others, elevations of both.

Another rare genetic form of combined hyperlipoproteinemia is called familial dysbetalipoproteinemia, or Type III hyperlipoproteinemia. The latter name is a holdover from a former lipoprotein typing system developed by Fredrickson et al,[54] but which is otherwise generally no longer used. The disease stems from an accumulation of cholesterol-rich VLDL and chylomicron remnants due to defective catabolism of those particles. The disease is also associated with the presence of a relatively rare form of apo E, called apo E2/2. Individuals with Type III will frequently have total cholesterol values of 200 to 300 mg/dL (5–8 mmol/L) and triglycerides of 300 to 600 mg/dL (3–7 mmol/L). To distinguish them from other combined hyperlipoproteinemics it is first necessary to isolate the VLDL fraction of their serum by ultracentrifugation. A ratio derived from the cholesterol concentration in VLDL to total serum triglycerides will be >0.30 in the presence of Type III hyperlipoproteinemia. If the VLDL fraction is subjected to agarose electrophoresis, the particles will migrate in the beta region, rather than in the normal pre-beta region. Definitive diagnosis requires a determination of apo E isoforms by isoelectric focusing or DNA typing, resulting in either apo E2/2 homozygosity or very rarely, apo E deficiency. Treatment, however, does not rely on a diagnosis, as these patients are treated with standard therapy of diet, niacin, gemfibrozil, and HMG CoA reductase inhibitors, as the clinical results dictate.

Lp(a) Elevation

Elevations in the serum concentration of Lp(a) are currently thought to confer increased risk of CHD[45,62,79,151,156] and CVD.[6,198] Higher levels of Lp(a) have been seen in patients with CHD than in normal control subjects, although prospective studies have not yet conclusively determined the positive association. As described earlier, Lp(a) are LDL that contain an extra apolipoprotein, called apo(a); the size and serum concentration of Lp(a) are genetically determined. Since apo(a) has a high degree of homology with plasminogen, a coagulation factor,[121] the general hypothesis involves a supposed competition between plasminogen and apo(a) for fibrin binding sites.[71,113,127] Under this hypothesis, if the apo(a) competes successfully, blocking plasminogen, clots that form along the arterial wall will not be dissolved. Most LDL-lowering drugs have no effect on Lp(a) concentration, even when LDL becomes markedly reduced. The one drug that has been shown to have some effect is niacin. Until prospective studies confirm Lp(a)'s atherogenicity, however, treatment with niacin is not advised except in conjunction with other dyslipidemic conditions where niacin is also indicated.

Hypolipoproteinemias

Hypolipoproteinemias are abnormalities marked by decreased lipoprotein concentrations. They fall into two major categories: hypo*alpha*lipoproteinemia and hypo*beta*lipoproteinemia. Hypobetalipoproteinemia is associated with isolated low levels of LDL cholesterol (*i.e.*, without other accompanying lipoprotein disorders), but since it is not generally associated with CHD, it will not be further discussed.

Hypoalphalipoproteinemia

Hypoalphalipoproteinemia indicates an isolated decrease in circulating HDL, usually defined as an HDL cholesterol concentration <35 mg/dL (0.9 mmol/L) without the presence of hypertriglyceridemia. The use of the term, alpha, denotes the region in which HDL migrate on agarose electrophoresis. There are several defects, often genetically determined, that are associated with hypoalphalipoproteinemia.[60,153,155,160,182,185] Virtually all of these defects are associated with increased risk of premature CHD. In the defect called *Tangier disease,* HDL cholesterol concentrations can be as low as 1 to 2 mg/dL (0.03–0.05 mmol/L) accompanied by total cholesterol concentrations of 50 to 80 mg/dL (1.3–2.1 mmol/L). Treatment of individuals with isolated decreases of HDL cholesterol is limited. Niacin is somewhat effective but has side effects, such as hepatotoxicity, that sometimes precludes its use; estrogen replacement in postmenopausal women is also effective.[10,68,171]

Acute, transitory hypoalphalipoproteinemia can be seen in cases of severe physiological stress, such as acute infections (primarily viral), other acute illnesses, surgical procedures, etc.[61] HDL cholesterol, as well as total cholesterol concentrations, can be markedly reduced under these conditions but will return to normal levels as recovery proceeds. For this reason, lipoprotein concentrations drawn during hospitalization or with a known disease state should be reassessed in the healthy, non-hospitalized state before intervention is considered.

SUMMARY

In summary, lipoproteins are complex particles that interact with many other metabolic pathways of the body. Their homeostasis can be affected by perturbations of other pathways, such as hormone imbalances and diabetes, and likewise, abnormal lipoprotein metabolism can disrupt the body's system and cause disease, as seen in pancreatitis and coronary disease. Some aspects of lipoprotein balance are genetically determined (gender, enzyme, or receptor defects), while others can be controlled (diet, smoking, exercise, blood pressure). Careful measurements are required for proper diagnosis of disease, and education is important to help prevent disease.

ACKNOWLEDGMENTS

The authors wish to acknowledge production support provided by Sherry Hamman and suggestions by the laboratory staff of Pacific Biometrics, Inc.

CASE STUDY 15-1

A 52-year-old male went to his doctor for a physical. The patient was 24 pounds overweight and had been a district manager for an automobile insurance company for the last 10 years. He had missed his last two appointments to the physician because of business. The urinalysis dipstick was not remarkable. His blood pressure was elevated. The blood chemistry results listed in Case Study Table 15-1 were obtained.

CASE STUDY TABLE 15–1. Laboratory Results

Analyte	Patient Value	Reference Range
Sodium	151	135–143 mEq/L
Potassium	4.5	3.0–5.0 mEq/L
Chloride	106	98–103 mEq/L
CO_2 content	13	22–27 mmol/L
Total protein	5.7	6.5–8.0 g/dL
Albumin	1.6	3.5–5.0 g/dL
Calcium	7.9	9.0–10.5 mg/dL
Cholesterol	210	140–200 mg/dL
Uric acid	6.2	3.5–7.9 mg/dL
Creatinine	2.5	0.5–1.2 mg/dL
BUN	95	7–25 mg/dL
Glucose	88	75–105 mg/dL
Total bilirubin	1.2	0.2–1.0 mg/dL
Alkaline phosphatase	27	7–59 IU/L
Lactate dehydrogenase	202	90–190 IU/L
Aspartate transaminase	39	8–40 IU/L
Amylase	152	76–375 IU/L

Questions

1. Given the abnormal tests, what additional information would you like to have?

2. If this patient had triglycerides of 100 mg/dL (1.1 mmol/L) and an HDL cholesterol of 23 mg/dL (0.6 mmol/L), what would be his calculated LDL-cholesterol value?

3. If, on the other hand, his triglycerides were 476 mg/dL (5.4 mmol/L), with an HDL cholesterol of 23 mg/dL (0.6 mmol/L), what would be his calculated LDL-cholesterol value?

**CASE STUDY
15-2**

A 30-year-old male was brought to the ER with chest pain after a softball game. He was placed in the coronary care unit when his ECG showed erratic waves in the ST region. A family history revealed that his father died of a heart attack when he was 45 years old. The patient had always been athletic in high school and college, so he had not concerned himself with a routine physical. The laboratory tests listed in Case Study Table 15-2 were run.

CASE STUDY TABLE 15–2. Laboratory Results

Analyte	Patient Values	Reference Range
Sodium	139	135–143 mEq/L
Potassium	4.1	3.0–5.0 mEq/L
Chloride	101	98–103 mEq/L
CO_2 content	29	22–27 mmol/L
Total protein	6.9	6.5–8.0 g/dL
Albumin	3.2	3.5–5.0 g/dL
Calcium	9.3	9.0–10.5 mg/dL
Cholesterol	278	140–200 mg/dL
Uric acid	5.9	3.5–7.9 mg/dL
Creatinine	1.1	0.5–1.2 mg/dL
BUN	20	7–25 mg/dL
Glucose	97	75–105 mg/dL
Total bilirubin	0.8	0.2–1.0 mg/dL
Alkaline phosphatase	20	7–59 IU/L
Lactate dehydrogenase	175	90–190 IU/L
Aspartate transaminase	35	8–40 IU/L
Amylase	98	76–375 IU/L

Questions

1. Given the symptoms and the family history, what additional tests should be recommended?
2. If his follow-up total cholesterol remains in the same range after he has been released from the hospital, and his triglycerides and HDL cholesterol are within the normal range, what course of treatment should be recommended?

**CASE STUDY
15-3**

A 60-year-old female came to her physician because she was having urination troubles. Her previous history included hypertension and episodes of edema. The physician ordered various laboratory tests on blood drawn in his office. The results are shown in Case Study Table 15-3.

CASE STUDY TABLE 15–3. Laboratory Results

Analyte	Patient Values	Reference Range
Sodium	149	135–143 mEq/L
Potassium	4.5	3.0–5.0 mEq/L
Chloride	120	98–103 mEq/L
CO_2 content	12	22–27 mmol/L
Total protein	5.7	6.5–8.0 g/dL
Albumin	2.3	3.5–5.0 g/dL
Calcium	7.6	9.0–10.5 mg/dL
Cholesterol	201	140–200 mg/dL
Uric acid	15.4	3.5–7.9 mg/dL
Creatinine	4.5	0.5–1.2 mg/dL
BUN	87	7–25 mg/dL
Glucose	88	75–105 mg/dL
Total bilirubin	1.3	0.2–1.0 mg/dL
Triglycerides	327	65–157 mg/dL
Lactate dehydrogenase	200	90–190 IU/L
Aspartate transaminase	45	8–40 IU/L
Amylase	380	76–375 IU/L

Questions

1. What are the abnormal results in this case?
2. Why do you think the triglycerides are abnormal?
3. What is the primary disease exhibited by this patient's laboratory data?

REFERENCES

1. Abell LL, Levy BB, Brody BB, Kendall FC. A simplified method for the estimation of total cholesterol in serum and demonstration of its specificity. J Biol Chem 1952;195:357.
2. Alderman E, Haskell WL, Fain JM, et al. Beneficial angiographic and clinical response to multifactor modification in the Stanford Coronary Risk Intervention Project (SCRIP). Circulation 1991;84(II):140.
3. Allain CC, Poon LS, Chan CS, et al. Enzymatic determination of total serum cholesterol. Clin Chem 1974;20:470.
4. Allen JK, Hensley WJ, Nichols AV, Whitfield JB. An enzymic and centrifugal method for estimating high-density lipoprotein cholesterol. Clin Chem 1979;25:325.
5. Angelin B, Rudling M. Growth hormone and hepatic lipoprotein metabolism. Curr Opin Lipidol 1994;5:160.
6. Armstrong VW, Cremer P, Eberle E, et al. The association between serum Lp(a) concentrations and angiographically

assessed coronary atherosclerosis. Dependence on serum LDL levels. Atherosclerosis 1986;62:249.

7. Assmann G, Funke H, Schmitz G. Low density lipoproteins and hypercholesterolemia. Drug Res 1989;39:996.

8. Austin MA, Breslow JL, Hennekens CH, Buring JE, Willett WC, Krauss RM. Low density lipoprotein subclass patterns and risk of myocardial infarction. JAMA 1988;260:1917.

9. Avogaro P, Cazzolato G. Changes in the composition and physico-chemical characteristics of serum lipoproteins during ethanol-induced lipemia in alcoholic subjects. Metabolism 1975;24:1231.

10. Barrett-Connor E, Bush TL. Estrogen and coronary heart disease in women. JAMA 1991;265:1861.

11. Bartlett GR. Phosphorus assay in column chromatography. J Biol Chem 1959;234:466.

12. Belcher JD, McNamara JR, Grinstead GF, et al. Measurement of low density lipoprotein cholesterol concentration. In: Rifai N, Warnick GR, eds. Laboratory measurement of lipids, lipoproteins, and apolipoproteins. Washington DC: AACC Press, 1994;107.

13. Bilheimer DW, Grundy SM. Role of LDL receptor in lipoprotein metabolism. In: Malmendier CL, Alaupovic P, eds. Advances in experimental medicine and biology. Vol 210. Lipoproteins and atherosclerosis. New York: Plenum Press, 1987;123.

14. Bjorntorp. Fatty acid, hyperinsulinemia, and insulin resistance: Which come first. Curr Opin Lipidol 1994;5:166.

15. Blanche PJ, Gong EL, Forte TM, Nichols AV. Characterization of human high-density lipoproteins by gradient gel electrophoresis. Biochem Biophys Acta 1981;665:408.

16. Blankenhorn DH, Azen SP, Kramsch DM, et al. Coronary angiographic changes with lovastatin therapy. The monitored atherosclerosis regression study (MARS). Ann Intern Med 1993;119:969.

17. Brousseau T, Clavey V, Bard JM, Fruchart JC. Sequential ultracentrifugation micromethod for separation of serum lipoproteins and assays of lipids, apolipoproteins, and lipoprotein particles. Clin Chem 1993;39:960.

18. Brown BG, Albers JJ, Fisher LD, et al. Regression of coronary artery disease as a result of intensive lipid-lowering therapy in men with high levels of apolipoprotein B. N Engl J Med 1990;323:1289.

19. Brown BG, Zhao XQ, Sacco DE, Albers JJ. Lipid lowering and plaque regression. New insights into prevention of plaque disruption and clinical events in coronary disease. Circulation 1993;87:1781.

20. Brown MS, Goldstein JL. Lipoprotein receptors, cholesterol metabolism, and atherosclerosis. Arch Pathol Lab Med 1975;99:181.

21. Brown MS, Goldstein JL. A receptor mediated pathway for cholesterol homeostasis. Science 1986;232:34.

22. Brown MS, Goldstein JL. The LDL receptor concept: Clinical and therapeutic implications. In: Stokes J III, Mancini M, eds. Hypercholesterolemia: Clinical and therapeutic implications. Vol 18. Atherosclerosis reviews. New York: Raven Press, 1988;85.

23. Bucolo G, Yabut J, Chang TY. Mechanized enzymatic determination of triglycerides in serum. Clin Chem 1975;21:420.

24. Burstein M, Legmann P. Lipoprotein precipitation. In: Clarkson TB, Kritchevsky D, Pollak OJ, eds. Monographs on atherosclerosis. Vol II. New York: S Karger, 1982;1.

25. Burstein M, Samaille J. Sur un dosage rapide du cholesterol lie aux α- et aux β-lipoproteines du serum. Clin Chim Acta 1960;5:609.

26. Burstein M, Scholnick HR. Lipoprotein-polyanion-metal interactions. Adv Lipid Res 1973;11:68.

27. Campos H, McNamara JR, Wilson PWF, Ordovas JM, Schaefer EJ. Differences in low density lipoprotein subfractions and apolipoproteins in premenopausal and postmenopausal women. J Clin Endocrinol Metab 1988;67:30.

28. Canner PL, Berge KG, Wenger NK, et al. Fifteen-year mortality in Coronary Drug Project patients: Long-term benefit with niacin. J Am Coll Cardiol 1986;8:1245.

29. Carlson LA, Rosenhamer G. Reduction of mortality in the Stockholm Ischaemic Heart Disease Secondary Prevention Study by combined treatment with clofibrate and nicotinic acid. Acta Med Scand 1988;223:405.

30. Castelli WP, Garrison RJ, Wilson PWF, Abbott RD, Kalousdian S, Kannel WB. Incidence of coronary heart disease and lipoprotein cholesterol levels: The Framingham Study. JAMA 1986;256:2835.

31. Chait A, Albers JJ, Brunzell JD. Very low density lipoprotein overproduction in genetic forms of hypertriglyceridemia. Eur J Clin Invest 1980;10:17.

32. Chait A, Robertson HT, Brunzell JD. Chylomicronemia syndrome in diabetes mellitus. Diabetes Care 1981;4:343.

33. Chapman MJ, Goldstein S, Lagrange D, Laplaud PM. A density gradient ultracentrifugal procedure for the isolation of the major lipoprotein classes from human serum. J Lipid Res 1981;22:339.

34. Cheung MC, Albers JJ. Characterization of lipoprotein particles isolated by immunoaffinity chromatography. Particles containing A-I and A-II and particles containing A-I but no A-II. J Biol Chem 1984;259:12201.

35. Cohen A, Hertz HS, Mandel J, et al. Total serum cholesterol by isotope dilution/mass spectrometry: A candidate definitive method. Clin Chem 1980;26:854.

36. Cohn JS, McNamara JR, Cohn SD, Ordovas JM, Schaefer EJ. Postprandial plasma lipoprotein changes in human subjects of different ages. J Lipid Res 1988;29:469.

37. Cohn JS, McNamara JR, Krasinski SD, Russell RM, Schaefer EJ. Role of triglyceride-rich lipoproteins from the liver and intestine in the etiology of postprandial peaks in plasma triglyceride concentration. Metabolism 1989;38:484.

38. Cohn JS, McNamara JR, Schaefer EJ. Lipoprotein cholesterol concentrations in the plasma of human subjects as measured in the fed and fasted states. Clin Chem 1988;34:2456.

39. Cole TG. Glycerol blanking in triglyceride assays: Is it necessary? Clin Chem 1990;36:1267.

40. Coleman MP, Key TJ, Wang DY, et al. A prospective study of obesity, lipids, apolipoproteins, and ischaemic heart disease in women. Atherosclerosis 1992;92:177.

41. Conlon D, Blankstein LA, Pasakarnis PA, et al. Quantitative determination of high-density lipoprotein cholesterol by agarose gel electrophoresis updated. Clin Chem 1979;24:227.

42. Connor WE, Stone DB, Hodges RE. The interrelated effects of dietary cholesterol and fat upon human serum lipid levels. J Clin Invest 1964;43:1691.

43. Cooper GR, Myers GL, Smith SJ, Schlant RC. Blood lipid measurements. Variations and practical utility. JAMA 1992; 267:1652.

44. Cooper GR, Smith SJ, Myers GL, Sampson EJ, Magid E. Estimating and minimizing effects of biologic sources of variation by relative range when measuring the mean of serum lipids and lipoproteins. Clin Chem 1994;40:227.

45. Dahlen GH, Guyton JR, Attar M, Farmer JA, Kautz JA, Gotto AM Jr. Association of levels of lipoprotein Lp(a), plasma lipids, and other lipoproteins with coronary artery disease documented by angiography. Circulation 1986;74:758.

46. Davignon J, Gregg RE, Sing CF. Apolipoprotein E polymorphism and atherosclerosis. Arteriosclerosis 1988;8:1.

47. Demacker PNM, Vos-Janssen HE, Hijmans AGM, Van't Laar A, Jansen AP. Measurement of high-density lipoprotein cholesterol in serum: Comparison of six isolation methods combined with enzymatic cholesterol analysis. Clin Chem 1980;26:1780.

48. Duncan IW, Mather A, Cooper GR. The procedure for the proposed cholesterol reference method. Division of Environmental Health Laboratory Sciences, CEH, Centers for Disease Control, Atlanta, GA, 1982.

49. Eckfeldt JH, Copeland KR. Accuracy verification and identification of matrix effects. The college of American pathologist's protocol. Arch Pathol Lab Med 1993;117:381.

50. Expert panel. Blood cholesterol levels in children and adolescents. NIH Publication No. 91-2732. Washington, DC: NIH, 1991.

51. Expert panel: Report of the National Cholesterol Education Program expert panel on detection, evaluation, and treatment of high blood cholesterol in adults. Arch Intern Med 1988;148:36.

52. Expert panel: Summary of the second report of the National Cholesterol Education Program (NCEP) expert panel on detection, evaluation, and treatment of high blood cholesterol in adults (Adult Treatment Panel II). JAMA 1993;269:3015.

53. Frayn KN. Insulin resistance and lipid metabolism. Curr Opin Lipidol 1993;4:197.

54. Fredrickson DS, Levy RI, Lees RS. Fat transport in lipoproteins: An integrated approach to mechanisms and disorders. N Engl J Med 1967;276:34, 94, 148, 215, 273.

55. Fredrickson DS, Levy RI, Lindgren FT. A comparison of heritable abnormal lipoprotein patterns as defined by two different techniques. J Clin Invest 1968;47:2446.

56. Frick MH, Elo O, Haapa K, et al. Helsinki heart study: Primary-prevention trial with gemfibrozil in middle-aged men with dyslipidemia. Safety of treatment, changes in risk factors, and incidence of coronary heart disease. N Engl J Med 1987;317:1237.

57. Friedewald WT, Levy RI, Fredrickson DS. Estimation of the concentration of low-density lipoprotein cholesterol in plasma, without use of the preparative ultracentrifuge. Clin Chem 1972;18:499.

58. Furman RH, et al. Effects of androgens and estrogens on serum lipids and the composition and concentration of serum lipoproteins in normolipemic and hyperlipidemic states. Prog Biochem Pharmacol 1967;2:215.

59. Garcia RE, Moodie DS. Routine cholesterol surveillance in childhood. Pediatrics 1989;84:751.

60. Genest JJ Jr, Bard JM, Fruchart JC, Ordovas JM, Schaefer EJ. Familial hypoalphalipoproteinemia in premature coronary artery disease. Arterioscler Thromb 1993;13:1728.

61. Genest JJ, Corbett H, McNamara JR, Schaefer MM, Salem DN, Schaefer EJ. Effect of hospitalization on high-density lipoprotein cholesterol in patients undergoing elective coronary angiography. Am J Cardiol 1988;61:998.

62. Genest J Jr, Jenner JL, McNamara JR, et al. Prevalence of lipoprotein (a) [Lp(a)] excess in coronary artery disease. Am J Cardiol 1991;67:1039.

63. Genest JJ Jr, Martin-Munley SS, McNamara JR, et al. Familial lipoprotein disorders in patients with premature coronary artery disease. Circulation 1992;85:2025.

64. Genest JJ, McNamara JR, Salem DN, Schaefer EJ. Prevalence of risk factors in men with premature coronary artery disease. Am J Cardiol 1991;67:1185.

65. Genest JJ Jr, Ordovas JM, McNamara JR, et al. DNA polymorphisms of the apolipoprotein B gene in patients with premature coronary artery disease. Atherosclerosis 1990;82:7.

66. Goldstein JL, Brown MS. Familial hypercholesterolemia. In: Scriver CR, Beaudet AL, Sly WS, Valle D, eds. The metabolic basis of inherited disease. 6th ed. New York: McGraw Hill, 1989.

67. Goldstein JL, Brown MS. Lipoprotein metabolism in the macrophage: Implications for cholesterol deposition in atherosclerosis. Ann Rev Biochem 1983;52:223.

68. Granfone A, Campos H, McNamara JR, et al. Effects of estrogen replacement on plasma lipoproteins and apolipoproteins in postmenopausal dyslipidemic women. Metabolism 1992;41:1193.

69. Greenberg N, Li ZM, Bower GN. National reference system for cholesterol (NRS-CHOL): Problems with transfer of accuracy with matrix materials. Clin Chem 1988;24:1230.

70. Grundy SM, Denke MA. Dietary influences on serum lipids and lipoproteins. J Lipid Res 1990;31:1149.

71. Hajjar KA, Gavish D, Breslow JL, Nachman RL. Lipoprotein(a) modulation of endothelial cell surface fibrinolysis and its potential role in atherosclerosis. Nature 1989;339:303.

72. Harris WS, Connor WE, Alam N, Illingworth DR. Reduction of postprandial triglyceridemia in humans by dietary n-3 fatty acids. J Lipid Res 1988;29:1451.

73. Hatch FT, Lees RS. Practical methods for plasma lipoprotein analysis. In: Pachetti R, ed. Advances in Lipid research. Orlando, FL: Academic Press, 1968;6:1.

74. Havel RJ, Eder HA, Bragdon JH. The distribution and chemical composition of ultracentrifugally separated lipoproteins in human serum. J Clin Invest 1955;34:1345.

75. Hegsted DM, Ausman LM, Johnson JA, Dallal GE. Dietary fat and serum lipids: An evaluation of the experimental data. Am J Clin Nutr 1993;57:875.

76. Heimberg M, Olubadewo JO, Wilcox HG. Plasma lipoproteins and regulation of hepatic metabolism of fatty acids in altered thyroid states. Endocr Rev 1985;6:590.

77. Heiss G, Tamir I, Davis CE, et al. Lipoprotein cholesterol distributions in selected North American populations: The lipid research clinics program prevalence study. Circulation 1980;61:302.

78. Hjermann I, Velve Byre K, Holme I, Leren P. Effect of diet and smoking intervention on the incidence of coronary heart disease. Report from the Oslo Study Group of a randomized trial in healthy men. Lancet 1981;2:1303.

79. Hoefler G, Harnoncourt F, Faschke E, Mitrl W, Pfeiffer KH, Kostner GM. Lipoprotein Lp(a). A risk factor for myocardial infarction. Arteriosclerosis 1988;8:398.

80. Holme I, Hjermann I, Helgelend A, Leren P. The Oslo Study: Diet and antismoking advice. Additional results from

a 5-year primary preventive trial in middle-aged men. Prev Med 1985;14:279.

81. Hopkins PN, Wu LL, Schumacher MC, et al. Type III dyslipidemia in patients heterozygous for familial hypercholesterolemia and apolipoprotein E2. Evidence for a gene-gene interaction. Arterioscler Thromb 1991;11:1137.

82. Hunninghake DB, Stein EA, Dujovne CA, et al. The efficacy of intensive dietary therapy alone or combined with lovastatin in outpatients with hypercholesterolemia. New Engl J Med 1993;328:1213.

83. Jackson RL, McLean LR, Ponce E, Rechtin A, Demel RA. Mechanism of action on LPL and hepatic triglyceride lipase. In: Malmendier CL, Alaupovic P, eds. Advances in experimental medicine and biology. Vol 210: Lipoproteins and atherosclerosis. New York: Plenum Press, 1987;73.

84. Jialal I, Chait A. Differences in the metabolism of oxidatively modified low density lipoprotein and acetylated low density lipoprotein by human endothelial cells: Inhibition of cholesterol esterification by oxidatively modified low density lipoprotein. J Lipid Res 1989;30:1561.

85. Johnson WJ, Mahlberg FH, Rothblat GH, Philips MC. Cholesterol transport between cells and high-density lipoproteins Biochim Biophys Acta 1991;1085:273.

86. Kane JP, Malloy MJ, Ports TA, Phillips NR, Diehl JC, Havel RJ. Regression of coronary atherosclerosis during treatment of familial hypercholesterolemia with combined drug regimens. JAMA 1990;264:3007.

87. Kannel WB. High-density lipoproteins: Epidemiologic profile and risks of coronary artery disease. Am J Cardiol 1983;52:9B.

88. Kasaniemei YA, Grungy SM. Significance of low density lipoprotein production in the regulation of plasma cholesterol level in man. J Clin Invest 1982;70:13.

89. Kato H, Tillotson J, Nichaman MZ, Rhoads GG, Hamilton HB. Epidemiologic studies of coronary heart disease and stroke in Japanese men living in Japan, Hawaii, and California. Am J Epidemiol 1973;97:372.

90. Kerscher L, Schiefer S, Draeger B, Maier J, Ziegenhorn J. Precipitation methods for the determination of LDL-cholesterol. Clin Biochem 1985;18:118.

91. Keys A, ed. Coronary heart disease in seven countries. Circulation 1970;41(Suppl I):1.

92. Klotzsch SG, McNamara JR. Triglyceride measurements: A review of methods and interferences. Clin Chem 1990;36:1605.

93. Krauss RM, Lindgren FT, Ray RM. Interrelationships among subgroups of serum lipoproteins in normal human subjects. Clin Chem Acta 1980;104:275.

94. Kroll MH, Chesler R, Elin RJ. Effect of lyophilization on results of five enzymatic methods for cholesterol. Clin Chem 1989;35:1523.

95. Kromhout D, de Lezenne Coulander C. Diet, prevalence and 10-year mortality from coronary heart disease in 871 middle-aged men. The Zutphen Study. Am J Epidemiol 1984;119:733.

96. Kulkarni KR, Garber DW, Schmidt CF, et al. Analysis of cholesterol in all lipoprotein classes by single vertical ultracentrifugation of fingerstick blood and controlled-dispersion flow analysis. Clin Chem 1992;38:1898.

97. Kuksis A, ed. Fat absorption. Vol 1. Boca Raton, Fl: CRC Press, 1986.

98. Labeur C, Shepherd J, Rosseneu M. Immunological assays of apolipoproteins in plasma: Methods and instrumentation. Clin Chem 1990;36:591.

99. Lehninger AL. Biochemistry: The molecular basis of cell structure and function. 2nd ed. New York: Worth, 1975.

100. LePage G, Roy C. Direct transesterification of all classes of lipids in a one-step reaction. J Lipid Res 1986;27:114.

101. Leren P. The effect of plasma cholesterol lowering diet in male survivors of myocardial infarction. Acta Med Scand 1966;466(Suppl):92.

102. Levin SJ. High-density lipoprotein cholesterol: Review of methods. The American Society of Clinical Pathologists. Check Sample, Core Chemistry, No. PTS 89-2(PTS-36), 1989;5(2).

103. Lewis LA, Opplt JJ. CRC Handbook of electrophoresis. Vol I. Boca Raton: CRC Press, Inc, 1980.

104. Lewis LA, Opplt JJ. CRC Handbook of electrophoresis. Vol II. Boca Raton: CRC Press, Inc, 1980.

105. Li Z, McNamara JR, Ordovas JM, Schaefer EJ. Analysis of high density lipoproteins by a modified gradient gel electrophoresis method. J Lipid Res 1994;35:1698.

106. Lichtenstein AH, Ausman LM, Carrasco W, Jenner JL, Ordovas JM, Schaefer EJ. Hydrogenation impairs the hypolipidemic effect of corn oil in humans: Hydrogenation, trans fatty acids, and plasma lipids. Arterioscler Thromb 1993;13:154.

107. Lindgren FT, Silvers A, Jutaglr R, et al. A comparison of simplified methods for lipoprotein quantitation using the analytic ultracentrifuge as a standard. Lipids 1977;12:278.

108. Lipid research clinics population studies data book. Vol 1. The prevalence study. Washington, DC, Dept of Health and Human Services, Public Health Service. NIH Publication No. 80-1527, 1980.

109. Lipid research clinics program: The Lipid research clinic coronary primary prevention trial results. II. The relationship of reduction in incidence of coronary heart disease to cholesterol lowering. JAMA 1984;251:365.

110. Lipid research clinics program: The Lipid research clinics coronary primary prevention trial. I. Reduction in incidence of coronary heart disease. JAMA 1984;251:351.

111. Lofland HB Jr. A semiautomated procedure for the determination of triglycerides in serum. Anal Biochem 1964;9:393.

112. Lopes-Virella MF, Stone P, Ellis S, Colwell JA. Cholesterol determination in high-density lipoproteins separated by three different methods. Clin Chem 1977;23:882.

113. Loscalzo J, Weinfield M, Fless GM, Scanu AM. Lipoprotein(a), fibrin binding, and plasminogen activation. Arteriosclerosis 1990;10:240.

114. Mahley WR, Innerarity TL, Rall SC Jr, et al. Plasma lipoproteins: Apolipoprotein structure and function. J Lipid Res 1984;25:1277.

115. Manual of laboratory operations, lipid research clinics program, lipid and lipoprotein analysis. Washington DC. Revised 1983. NIH, U.S. Dept. of Health and Human Services.

116. Marcovina SM, Levine DM, Lippi G. Lipoprotein (a): Structure, measurement, and clinical significance. In: Rifai N, Warnick GR, eds. Methods for clinical laboratory measurement of lipid and lipoprotein risk factors. Washington, DC: AACC Press, 1994;235.

117. Mascioli EA, Lopes S, Randall S, et al. Serum fatty acid profiles after intravenous medium chain triglyceride administration. Lipids 1989;24:793.

118. McGowan M, Artiss J, Strandbergh DR, Zak, B. A peroxidase-coupled method for the colorimetric determination of serum triglycerides. Clin Chem 1983;29:538.

119. McGowan MW, Artiss JD, Zak B. Spectrophotometric study on minimizing bilirubin interference in an enzyme reagent mediated cholesterol reaction. Microchem J 1983;27:564.

120. McGowan, MW, Artiss JD, Zak B. A procedure for the determination of high-density lipoprotein choline-containing phospholipids. J Clin Chem Clin Biochem 1982;20:807.

121. McLean JW, Tomlinson JE, Kuang WJ, et al. cDNA sequence of human apolipoprotein(a) is homologous to plasminogen. Nature 1987;330:132.

122. McNamara JR, Campos H, Ordovas JM, Peterson J, Wilson PWF, Schaefer EJ. Effect of gender, age, and lipid status on low density lipoprotein subfraction distribution: Results of the Framingham Offspring Study. Arteriosclerosis 1987;7:483.

123. McNamara JR, Cohn JS, Wilson PWF, Schaefer EJ. Calculated values for low-density lipoprotein cholesterol in the assessment of lipid abnormalities and coronary disease risk. Clin Chem 1990;36:36.

124. McNamara JR, Cole TG, Centois JH, Ferguson CA, Ordovas JM, Schaefer EJ. Immunoseparation method for measuring low-density lipoprotein cholesterol directly from serum evaluated. Clin Chem 1995;41:232.

125. McNamara JR, Jenner JL, Li Z, Wilson PWF, Schaefer EJ. Change in low density lipoprotein particle size is associated with change in plasma triglyceride concentration. Arteriosclerosis 1992;12:1284.

126. Miettinen M, Turpeinen O, Karvonen MJ, Elosuo R, Paavilainen E. Effect of cholesterol-lowering diet on mortality from coronary heart-disease and other causes. A twelve-year clinical trial in men and women. Lancet 1972;2:835.

127. Miles LA, Fless GM, Levin EG, Scanu AM, Plow EF. A potential basis for the thrombotic risks associated with lipoprotein(a). Nature 1989;339:301.

128. Muniz N. Measurement of plasma lipoproteins by electrophoresis on polyacrylamide gel. Clin Chem 1977;23:1826.

129. Myers GL, Cooper GR, Henderson LO, Hassemer DJ, Kimberly M.M. Standardization of lipid and lipoprotein measurements. In: Rifai N, Warnick GR, eds. laboratory measurement of lipids, lipoproteins and Apolipoproteins. Washington DC: AACC Press, 1994;177.

130. National Cholesterol Education Program Laboratory Standardization Panel. Current status of blood cholesterol measurements in clinical laboratories in the United States. Clin Chem 1988;34:193.

131. National Cholesterol Education Program Lipoprotein Measurement Working Group. Recommendations for measurement of high density lipoprotein cholesterol. NIH Publication. In Press.

132. National Cholesterol Education Program Lipoprotein Measurement Working Group. Recommendations for measurement of low density lipoprotein cholesterol. NIH Publication. In Press.

133. National Cholesterol Education Program Lipoprotein Measurement Working Group. Recommendations for triglyceride measurement. NIH Publication. In Press.

134. Nicolosi RJ, Stucchi AF, Kowala MC, Hennessy LK, Hegsted DM, Schaefer EJ. Effect of dietary fat saturation and cholesterol on LDL composition and metabolism. *In vivo* studies of receptor and nonreceptor-mediated catabolism of LDL in cebus monkeys. Arteriosclerosis 1990;10:119.

135. NIH Consensus Conference. Lowering blood cholesterol to prevent heart disease. JAMA 1985;253:2080.

136. NIH Consensus Conference. Triglyceride, HDL cholesterol and coronary heart disease. JAMA 1993;269:505.

137. Noble RP. Electrophoretic separation of plasma lipoproteins in agarose gel. J Lipid Res 1968;9:963.

138. Ornish D, Brown SE, Scherwitz LW, et al. Can lifestyle changes reverse coronary heart disease? The lifestyle heart trial. Lancet 1990;336:129.

139. Papadopoulos NM. Hyperlipoproteinemia phenotype determination by agarose gel electrophoresis updated. Clin Chem 1978;24:227.

140. Patsch JR, Patsch W. Zonal ultracentrifugation. In: Albers JJ, Segrest JP, eds. Methods of enzymology. Orlando FL: Academic Press, 1986;129:3.

141. Patsch JR, Prasad S, Gotto AM Jr, et al. Postprandial lipemia: A key for the conversion of HDL 2 into HDL 3 by hepatic lipase. J Clin Invest 1984;74:2017.

142. Redgrave TG, Roberts DCK, West CE. Separation of plasma lipoproteins by density-gradient ultracentrifugation. Anal Biochem 1975;65:42.

143. Report from the Laboratory Standardization Panel of the National Cholesterol Education Program. Recommendations for improving cholesterol measurement. Bethesda, MD: National Institutes of Health. NIH publication no. 90-2964, 1990.

144. Richmond W. Preparation and properties of a cholesterol oxidase from *Nocardia sp.* and its application to the enzymatic assay of total cholesterol in serum. Clin Chem 1973;19:1350.

145. Rifai N. Lipoproteins and apolipoproteins composition, metabolism and association with coronary heart disease. Arch Pathol Lab Med 1986;110:694.

146. Rifai N, Warnick GR, McNamara JR, Belcher JD, Grinstead GF, Frantz ID Jr. Measurement of low-density-lipoprotein cholesterol in serum: A status report. Clin Chem 1992;38:150.

147. Rosseneu M, Van Diervliet JP, Bury J, Vinaimont N. Isolation and characterization of lipoprotein profiles in newborns by density gradient ultracentrifugation. Pediatr Res 1983;17(10):788.

148. Rubins HB, Robins SJ, Iwane MK, et al. Rationale and design of the Department of Veterans Affairs High-Density Lipoprotein Cholesterol Intervention Trial (HIT) for secondary prevention of coronary artery disease in men with low high-density lipoprotein cholesterol and desirable low-density lipoprotein cholesterol. Am J Cardiol 1993;71:45.

149. Russell DW. Cholesterol biosynthesis and metabolism. Cardiovasc Drug Ther 1992;6:103.

150. Sacks FM, Walsh BW. Sex hormones and lipoprotein metabolism. Curr Opin Lipidol 1994;5:236.

151. Sandkamp M, Funke H, Schulte H, Kohler E, Assmann G. Lipoprotein(a) is an independent risk factor for myocardial infarction at a young age. Clin Chem 1990;36:20.

152. Schaefer EJ. Hyperlipoproteinemia. In: Rakel RE, ed. Conn's current therapy. Philadelphia: WB Saunders, 1991;515.

153. Schaefer EJ, Blum CB, Zech LA, et al. Metabolism of high density apolipoproteins in Tangier disease. New Engl J Med 1978;299:905.

154. Schaefer EJ, Lichtenstein AH, Lamon-Fava S, McNamara JR, Ordovas JM. Lipoproteins, nutrition, aging, and atherosclerosis. Am J Clin Nutr 1995;61(suppl):7265.

155. Schaefer EJ, Heaton WH, Wetzel MG, Brewer HB Jr. Plasma apolipoprotein A-I absence associated with a marked reduction of high density lipoproteins and premature coronary artery disease. Arteriosclerosis 1982;2:16.

156. Schaefer EJ, Lamon-Fava S, Jenner JL, et al. Lipoprotein(a) levels and risk of coronary heart disease in men. The Lipid Research Clinics Coronary Primary Prevention Trial. JAMA 1994;271:999.

157. Schaefer EJ, Levy RI. Pathogenesis and management of lipoprotein disorders. N Engl J Med 1985;312:1300.

158. Schaefer EJ, McNamara JR. Overview of the diagnosis and treatment of lipid disorders. In: Rifai N, Warnick GR, eds. Methods for clinical laboratory measurement of lipid and lipoprotein risk factors. 2nd ed. Washington, DC: AACC Press, 1994;21.

159. Schaefer EJ, McNamara JR, Genest J Jr, Ordovas JM. Clinical significance of hypertriglyceridemia. Semin Thromb Hemost 1988;14:142.

160. Schaefer EJ, Ordovas JM, Law SW, et al. Familial apolipoprotein A-I and C-III deficiency, variant II. J Lipid Res 1985;26:1089.

161. Schmitz G, Robene H, Assmann G. Role of the high density lipoprotein receptor cycle in macrophage cholesterol metabolism. Atherosclerosis Rev 1987;16:95.

162. Schonfeld G. Lipoproteins in atherogenesis. Artery 1975; 5:305.

163. Schuler G, Hambrecht R, Schlierf G, et al. Regular physical exercise and low fat diet. Effects on progression of coronary artery disease. Circulation 1992;86:1.

164. Schumaker VN, Robinson MT, Curtiss LK, Butler R, Sparkes RS. Anti-apoprotein B monoclonal antibodies detect human low density lipoprotein polymorphism. J Biol Chem 1984;259:6423.

165. Shekelle RB, Shryock AM, Paul O, et al. Diet, serum cholesterol, and death from coronary heart disease. The Western Electric study. N Engl J Med 1981;304:65.

166. Small DM. The HDL system: A short review of structure and metabolism. Atherosclerosis Rev 1987;16:1.

167. Sniderman AD. Is it time to measure apolipoprotein B? Arteriosclerosis 1990;10:665.

168. Spady DK, Dietschy JM. Interaction of aging and dietary fat in the regulation of low-density lipoprotein transport in the hamster. J Lipid Res 1989;30:559.

169. Sprecher DL, Schaefer EJ, Kent KM, et al. Cardiovascular features of homozygous familial hypercholesterolemia: Analysis of 16 patients. Am J Cardiol 1984;54:20.

170. Stamler J, Stamler R, Shekelle RB. Regional differences in prevalence, incidence, and mortality from atherosclerotic coronary heart disease. In: deHaas JH, Hemker HC Snellen HA, eds. Ischaemic heart disease. Leiden, The Netherlands: Leiden University Press, 1970;84.

171. Stampfer MJ, Colditz GA. Estrogen replacement therapy and coronary heart disease: A quantitative assessment of the epidemiologic evidence. Prev Med 1991;20:47.

172. Stary HC. Evolution and progression of atherosclerotic lesions in coronary arteries of children and young adults. Arteriosclerosis 1989;9(Suppl I):I-19.

173. Stary HC. Macrophages, macrophage foam cells, and eccentric intimal thickening in the coronary arteries of young children. Atherosclerosis 1987;64:91.

174. Steele BW, Koehler DF, Azar MM, Blaszkowski TP, Kuba K, Dempsey ME. Enzymatic determinations of cholesterol in high-density-lipoprotein fractions prepared by a precipitation technique. Clin Chem 1976;22:98.

175. Stein EA, Gary LM. Lipids lipoproteins and apolipoproteins. In: Tietz N, ed. Textbook of clinical chemistry. 2nd Ed. Burtis & Ashwood ER.(ed) Philadelphia: WB Saunders Co, 1994;1002.

176. Stein T, Stein O. Catabolism of VLDL and removal of cholesterol from intact cells. In: Schettler G, Strange E, Wissler RW, eds. Atherosclerosis is it reversible? New York: Springer Verlag, 1978.

177. Steinberg D, Parthasarathy S, Carew TE, Khoo JC, Witztum JL. Beyond cholesterol. Modifications of low-density lipoprotein that increase its atherogenicity. New Engl J Med 1989;320:915.

178. Stinshoff K, Weisshaar D, Staehler F, Hesse D, Gruber W, Steier E. Relation between concentrations of free glycerol and triglycerides in human sera. Clin Chem 1977;23:1029.

179. Stone NJ, Levy RI, Fredrickson DS, Verter J. Coronary artery disease in 116 kindred with familial type II hyperlipoproteinemia. Circulation 1974;49:476.

180. Sullivan DR, Kruijswijk Z, West CE, Kohlmeier M, Katan MB. Determination of serum triglycerides by an accurate enzymatic method not affected by free glycerol. Clin Chem 1985;31:1227.

181. Takayama M, Itoh S, Nagasaki T, Tanimizu I. A new enzymatic method for determination of serum choline-containing phospholipids. Clin Chim Acta 1977;79:93.

182. Third JL, Montag J, Flynn M, Freidel J, Laskarzewski P, Glueck CJ. Primary and familial hypoalphalipoproteinemia. Metabolism 1984;33:136.

183. Thom TJ, Epstein FH, Feldman JJ, Leaverton PE, Wolz M. Total mortality and mortality from heart disease, cancer, and stroke from 1950 to 1987 in 27 countries. NIH Pub. No. 92-3088, Washington, DC, 1992.

184. Tietz N. Fundamentals of clinical chemistry. 3rd ed. Philadelphia: Saunders, 1987.

185. Vergani C, Bettale G. Familial hypo-alpha-lipoproteinemia. Clin Chim Acta 1981;114:45.

186. Viikari J. Precipitation of plasma lipoproteins by PEG-6000 and its evaluation with electrophoresis and ultracentrifugation. Scand J Clin Lab Invest 1976;36:265.

187. Warnick GR. Compact analysis systems for cholesterol, triglycerides and high-density lipoprotein cholesterol. Curr Opin Lipidol 1991;2:343.

188. Warnick GR. Enzymatic methods for quantification of lipoprotein lipids. In: Albers JJ, Segrest JP, eds. Methods in enzymology. Vol 129. Orlando: Academic Press, 1986;101.

189. Warnick GR, Albers JJ, Bachorik PS, et al. Multi-laboratory evaluation of an ultrafiltration procedure for high density lipoprotein cholesterol quantification in turbid heparin-manganese supernates. J Lipid Res 1981;22:1015.

190. Warnick GR, Benderson J, Albers JJ, et al. Dextran sulfate-Mg^{2+} precipitation procedure for quantitation of high-density-lipoprotein cholesterol. Clin Chem 1982;28:1379.

191. Warnick GR, Cheung MC, Albers JJ. Comparison of current methods for high-density lipoprotein cholesterol quantitation. Clin Chem 1979;25:596.

192. Warnick GR, Knopp RH, Fitzpatrick V, Branson L. Estimating low-density lipoprotein cholesterol by the Friedewald equation is adequate for classifying patients on the basis of nationally recommended cutpoints. Clin Chem 1990;36:15.

193. Warnick GR, Leary ET, Goetsch J. Electrophoretic quantification of LDL-cholesterol using the Helena REP. [Abstract] Clin Chem 1993;39:1122.

194. Warnick GR, Nguyen T, Bergelin RO, Wahl PW, Albers JJ. Lipoprotein quantification: An electrophoretic method compared with the Lipid Research Clinics method. Clin Chem 1982;28:2116.

195. Wu LL, Warnick GR, Wu JT, Williams RR, Lalouel JM. A rapid micro-scale procedure for determination of the total lipid profile. Clin Chem 1989;35(7):1486.

196. Wu LH, Wu JT, Hopkins PN. Apolipoprotein E. Laboratory determination and clinical significance. In: Rifai N, Warnick GR, eds. Methods for clinical laboratory measurement of lipid and lipoprotein risk factors. 2nd ed. Washington, DC: AACC Press, 1994;279.

197. Zak B. Cholesterol methodologies: A review. Clin Chem 1977;23:1201.

198. Zenker G, Költringer P, Bone G, Niederkorn K, Pfeiffer K, Jurgens G. Lipoprotein(a) as a strong indicator for cerebrovascular disease. Stroke 1986;17:942.

Chapter 16

First Exam

Nonprotein Nitrogen

Susan T. Smith

Objectives

Upon completion of this chapter, the clinical laboratorian should be able to:

- List the nonprotein nitrogen (NPN) substances in the blood and recognize their relative concentration.
- Recognize the chemical structures of the following non-protein nitrogen substances: urea, creatinine, uric acid, and ammonia.
- Describe the biosynthesis and excretion of urea, uric acid, creatine, creatinine, and ammonia.
- Relate the level of protein in the diet, protein metabolism, renal blood flow, and renal function to the plasma level of urea.
- Discuss the BUN/creatinine ratio and its use in distinguishing between pre-renal and renal causes of an elevated blood urea.
- Recognize the major clinical conditions associated with increased and decreased levels of urea, creatine, and creatinine in the plasma.
- Describe specimen requirements and storage conditions for determinations of urea, uric acid, creatine, creatinine, and ammonia.
- Recognize the reference interval for plasma urea, creatinine, uric acid, and ammonia.
- Relate plasma levels of creatinine to diet, muscle mass and turnover, and renal function.
- Discuss commonly used methods for the determination of creatine, creatinine, urea, uric acid, and ammonia in plasma and urine.

- Discuss sources of error in the Jaffe method for creatinine and methods that have been utilized to minimize these problems.
- Recognize the impact of specific methods for creatinine on the estimation of GFR with creatinine clearance.
- Recognize reference intervals for plasma and urine creatinine and the effect of age and sex on these values.
- Relate plasma levels of uric acid to diet, purine metabolism, and renal disease.
- Relate the solubility of uric acid to the pathological consequences of increased plasma uric acid.
- Recognize reference intervals for plasma uric acid and the effect of age and sex on these values.
- Discuss clinical conditions and toxic effects related to increased plasma level of ammonia.
- Describe specimen collection and handling requirements for measurement of ammonia in plasma.
- Given patient values for urea, creatinine, uric acid, and supporting clinical history, suggest possible clinical conditions associated with these results.

KEY WORDS

Allantoin	Coupled Enzymatic	Distal Tubule
Ammonia	Method	Encephalophy
Azotemia	Creatine	GFR Glomerular
BUN/Creatinine	Creatinine	Filtration Rate
Ratio	Creatinine	Glomerular Filtrate
Clearance	Clearance	Glomerulus

Michael L. Bishop, Janet L. Duben-Engelkirk, and Edward P. Fody.
CLINICAL CHEMISTRY. © 1996 Lippincott–Raven Publishers.

Gout
Hyperuricemia
Hypouricemia
Kinetic Method of
  Measurement
NPN (Nonprotein
  Nitrogen)

Pre-renal
Post-renal
Protein-free Filtrate
Proximal Tubule
Renal Tubular
  Secretion

Specificity
Urea
Uric Acid
Uremia or Uremic
  Syndrome

The determination of nonprotein nitrogenous substances in the blood has traditionally been used to monitor renal function. The term *nonprotein nitrogen* (*NPN*) originated in the early days of clinical chemistry when analytical methodology required that protein be removed from the sample before analysis. After removal, the concentration of nitrogen-containing compounds in this "protein-free filtrate" was quantitated in a manner very similar to the method (Kjeldahl) used to determine protein nitrogen.[30] Kjeldahl digestion, although technically very difficult, is still the most accurate method for the determination of the total NPN concentration. However, much more valid clinical information is obtained by analyzing a patient's sample for the individual components of the nonprotein nitrogen fraction.

The NPN fraction is made up of about 15 compounds of clinical interest (Table 16-1). The majority of these compounds arise from the catabolism of proteins and nucleic acids.

UREA

Biochemistry

Urea constitutes nearly half the nonprotein nitrogen substances in the blood (Fig. 16-1). It is synthesized in the liver from CO_2 and the ammonia arising from the deamination of amino acids by means of the ornithine or Krebs-Henseleit cycle. Urea constitutes the major excretory product of protein metabolism. Following synthesis in the liver, urea is transported by the plasma to the kidney, where it is readily filtered from the plasma by the glomerulus. Most of the urea in the glomerular filtrate is excreted in the urine, although up to 40% is reabsorbed by passive diffusion during passage of the filtrate through the renal tubules. The

amount reabsorbed is dependent on urine flow rate and level of hydration. Small amounts of urea ($<10\%$ of the total) are excreted through the GI tract and skin.

The level of urea in the plasma is markedly affected by renal function and perfusion, the protein content of the diet, and the level of protein catabolism occurring.

Disease Correlations

The conditions causing abnormal levels of urea in the plasma can be divided into three main categories: prerenal, renal, and postrenal. One major cause of prerenal changes in plasma urea concentration is related to renal circulation. Urea cannot be filtered from the blood if the blood does not flow through the kidney. Factors that cause decreased renal blood flow and thus an elevated level of urea include congestive heart failure, shock, hemorrhage, dehydration, and a marked decrease in blood volume. In a hospital population, congestive heart failure is three times more common than any other cause of an elevated urea.[50]

An elevated level of urea in the blood is called *azotemia*. Very high levels of plasma urea accompanied by renal failure is called *uremia* or the *uremic syndrome*. This is eventually fatal if not treated by dialysis.

Another factor that will cause prerenal changes in blood urea concentration is the level of protein metabolism. Increased or decreased protein intake or metabolism over a period of time will increase or decrease, respectively, the urea level, although this change is more readily reflected when urinary excretion of urea is measured. Thus, a high-protein diet, or the increased protein catabolism such as occurs in fever, major illness, stress, corticosteroid therapy, and gastrointestinal hemorrhage, may increase urea levels. Levels will be decreased during periods of low protein intake or increased protein synthesis, such as late pregnancy and infancy.

Because of its major role in the excretion of urea, decreased renal function, in general, causes an increase in plasma urea concentration. Renal causes of an elevated urea include acute and chronic renal failure, glomerular nephritis, tubular necrosis, and other intrinsic renal disease (see Chapter 17, Renal Function). Postrenal effects that elevate the urea level are due to obstruction to the flow of urine anywhere in the urinary tract. Obstruction may be caused by renal stones, tumors of the bladder or prostate, or severe infection.

Differentiation between the various causes of an increased plasma urea concentration may be aided by calculation of the

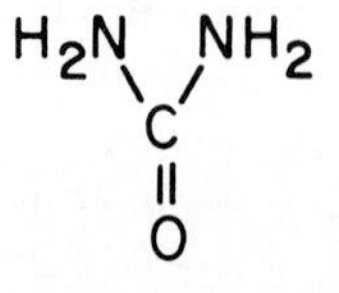

TABLE 16-1. Components of the Nonprotein Nitrogen Fraction

Compound	Approximate Plasma Concentration (% of Total NPN)
Urea	45
Amino acids	20
Uric acid	20
Creatinine	5
Creatine	1–2
Ammonia	0.2

Figure 16-1. Structure of urea.

12 now

BUN/CREAT ratio. This ratio is normally 10:1 to 15:1.[2] Non-renal conditions such as congestive heart failure or increased protein breakdown tend to elevate the plasma level of urea to a greater extent than the plasma level of creatinine. As a result, in these situations, the BUN/CREAT ratio will be elevated. An elevated BUN/CREAT ratio will indicate to the physician that the patient's elevated UN is probably due to causes other than renal disease.

The major causes of decreased plasma urea levels include decreased protein intake and severe liver disease.[2] The conditions affecting plasma urea levels are summarized in Table 16-2.

Analytical Methods

Because measurements of urea were initially done on a protein-free filtrate of whole blood, and early analytical methods were based on measuring the amount of nitrogen, this assay is commonly called a _BUN_ or _blood urea nitrogen_ determination. Current analytical methods use serum or plasma instead of whole blood. However, the term _BUN_ is still often used to denote this assay, and the results are reported in terms of nitrogen concentration rather than the more correct urea concentration.

Urea's nitrogen concentration can be converted to urea concentration by multiplying by 2.14, as shown in Figure 16-2. Concentration of urea nitrogen in milligrams per deciliter may be converted to millimoles per liter of urea by multiplying by 0.36.

Two analytical approaches have been used to assay for urea. The oldest and most often used involves the hydrolysis of urea by the enzyme urease (urea aminohydrolase EC 3.5.1.5) and quantitation of the NH_4^+ that is produced by a variety of techniques. Early colorimetric methods were based on using Nesslers reagent or the Berthelot reaction to detect the NH_4^+ produced.[18,31] Current practice has replaced these methods with a coupled enzymatic method utilizing L-glutamate dehydrogenase (EC 1.4.1.3), 2-oxoglu-

$$\frac{\text{Mol wt. urea}}{\text{at wt. N} \times 2} = \frac{60}{28} = 2.14$$

$$(\text{urea N mg/dL})\ (2.14) = \text{urea mg/dL}$$

▲ _Figure 16-2._ Conversion of UN concentration to urea.

tarate, and NADH, as shown in Figure 16-3.[27,78] The coupled enzymatic method demonstrates good specificity and sensitivity and is currently the most popular method for the measurement of plasma urea on automated systems. The major disadvantage of the coupled enzymatic method is that endogenous enzymes, such as lactic dehydrogenase, compete for the NADH in the reaction mixture and the equilibrium for the indication reaction (GLDH) lies to the left. As a result, this method is most accurate if done kinetically rather than in an endpoint mode.[27,78] A modification of this method using bacterial urease has recently been described for use in measuring urea in urine specimens.[66]

An alternative coupled enzymatic method has been described that quantitates the NH_4^+ produced from urease hydrolysis by a series of coupled reactions that produce H_2O_2. The H_2O_2 is then quantitated by the formation of a quinone-imine dye in the presence of phenol and 4-aminophenazone.[44] The major advantage of this method is that it can be readily performed manually using a spectrophotometer in the visible range as well as on automated systems.

Another approach that has been used to measure the NH_4^+ produced by urease hydrolysis is to measure the color change generated by a pH indicator. This approach has been incorporated into several instruments using dry reagents in a multilayer film (Kodak EKTACHEM Fuji Dri Chem™), using reagent strips (BMC Reflotron),[30,54] or using a liquid reagent system.[55]

An alternate approach is utilized by Beckman Instruments on their BUN, ASTRA, and Synchron analyzers.[58] These instruments use an electrode to measure the rate of increase in conductivity as the (nonionic) urea in the sample is hydrolyzed by urease to the ionic species NH_4^+ and HCO_3^-. Because the rate of change in conductivity is measured, NH_3 contamination is not a problem as it is in other methods. A similar conductimetric method has recently been introduced by NOVA on the NOVA™ 12.[14] Potentiometric methods using an ammonia ISE have also been developed.[21]

TABLE 16-2. Causes of Abnormal Plasma Urea Concentration

Increased Levels	
Prerenal	Congestive heart failure
	Shock, hemorrhage
	Dehydration
	Increased protein catabolism
	Corticosteroid therapy
Renal	Acute and chronic renal failure
	Glomerular nephritis
Postrenal	Urinary tract obstruction
Decreased Levels	
	Decreased protein intake
	Severe liver disease
	Severe vomiting & diarrhea
	Normal pregnancy

$$\text{Urea} \xrightarrow{\text{urease}} 2NH_4^+ + HCO_3^-$$

$$NH_4^+ + 2\text{-oxoglutarate} + NADH$$
$$\searrow \text{GLDH}$$
$$\text{Glutamate} + NAD^+ + H_2O$$

▲ _Figure 16-3._ Enzymatic assay for urea.

Urea may be measured by direct condensation with diacetyl monoxime in the presence of strong acid and an oxidizing agent to form a yellow diazine derivative. Thiosemicarbazide is added to the reaction mixture to stabilize the color.[32,47] The major advantage of this method is that it measures urea directly and ammonia does not interfere. This method has been successfully automated on Technicon and some other discrete analyzers, but it has the disadvantage that it uses caustic chemicals, requires elevated temperature, and has limited linearity. As a result, it is seldom used in modern clinical laboratories.

Another direct approach has recently been used by one instrument manufacturer (Ames) and involves the reaction of urea with o-phthalaldehyde and naphthylethylanediamine to form a colored product.[30,60]

A method utilizing isotope-dilution mass spectrometry has been proposed as a definitive method for urea.[87] The current reference method is the coupled urease/GLDH enzymatic method, which was described earlier. When used as a reference method, it is recommended that the assay be done on a protein-free filtrate.[65] Analytical methods are summarized in Table 16-3.

Specimen Requirements and Interfering Substances

Urea concentration may be readily measured in serum, plasma, or urine. When collecting plasma, NH_4^+ ions and high concentrations of sodium fluoride or sodium citrate must be avoided. Fluoride or citrate will inhibit the urease

TABLE 16-3. Summary of Analytical Methods—Urea

Urease Methods

All methods use similar first step	$Urea + H_2O \xrightarrow{Urease} 2NH_4^+ + CO_3^{-2}$	
GLDH *coupled enzymatic*	$NH_4^+ + 2\text{-oxoglutarate} + NADH \xrightarrow{GLDH} NAD^+ + glutamate + H_2O$	Most popular on common automated instruments; best done kinetically; candidate reference method
Conductimetric	Conversion of unionized urea to NH_4^+, HCO_3^-; results in a decrease in conductivity	Used on ASTRA and NOVA; very specific and rapid
Nesslerization	$NH_4^+ + 2HgI_2 + 4KI \longrightarrow NH_2Hg_2I_3$ (yellow-orange colloid) $+ H_3O^+ + 4KI + NaI$	Nonspecific, historical interest only
Berthelot	$NH_4^+ + NaOCl + phenol \xrightarrow{Nitroprusside} indophenol$	Nonspecific; very sensitive to interference from endogenous NH_3
Indicator dye	$NH_4^+ + pH$ indicator dye $\longrightarrow$ color change	Used in multilayer film reagents, dry reagent strips, and automated systems
H_2O_2 *coupled*	$NH_4^+ + 2\text{-glutamate} + 2 ATP \xrightarrow{Glutamine\ synthetase} 2 ADP + 2$ L-glutamine	Recently introduced; not commercially available
	$ADP + 2$ phosphoenolpyruvate $\xrightarrow{Pyruvate\ kinase} 2$ pyruvate $+ 2 ATP$	
	$Pyruvate + 2H_2PO_4^- + 2O_2 + 2H_2O \xrightarrow{Pyruvate\ oxidase} 2$ acetyl phosphate $+ CO_2 + 2H_2O_2$	
	$H_2O_2 + phenol + 4$ aminophenazone $\xrightarrow{Peroxidase}$ quinonemonoimine dye	

Direct Methods

Diacetyl monoxime	Diacetyl monoxime $+$ urea $\xrightarrow{Str\ acid}$ diazine (yellow)	Not specific; uses toxic reagents
O-Phthaldehyde	Urea $+$ O-phthaladehyde $\xrightarrow{H^+}$ isoindoline $\xrightarrow{H^+}$ Isoindoline $+$ N-(I-naphthyl)ethylenediame $\longrightarrow$ colored product	Used on some automated instruments; no NH_3 interference, positive interference from sulfa drugs

utilized in the coupled enzymatic methods. Although the overall protein content of the diet will influence the level of urea, the effect of a single protein-containing meal will cause only about a 10% increase in the blood level,[5] and a fasting sample is usually not required. A non-hemolyzed sample is recommended. Urea is quite susceptible to bacterial decomposition, so samples (especially urine) that cannot be analyzed within a few hours should be refrigerated. A few crystals of thymol also may be added to urine samples to improve preservation. Methods used for serum or plasma must be modified for use with urine because of the high urea concentration and presence of endogenous ammonia.

Reference Interval[76]

Urea nitrogen 7 to 18 mg/dL   (2.5–6.4 mmol/L urea)

CREATININE/CREATINE

Biochemistry

Creatine is synthesized mainly in the liver from arginine, glycine, and methionine. It is then transported to other tissues, such as muscle, where it is converted to phosphocreatine, which serves as a high energy source. Either creatine phosphate or creatine, under physiologic conditions, spontaneously loses phosphoric acid or water, respectively, to form its anhydride, creatinine, in which form it is then excreted into the plasma. The structures of these compounds are shown in Figure 16-4.

Creatinine is excreted into the circulation at a relatively constant rate that has been shown to be proportional to the muscle mass of the individual. It is removed from the circulation by glomerular filtration and is excreted in the urine. Additional amounts of creatinine are secreted by the proximal tubule such that, in a normal individual, creatinine clearance exceeds the glomerular filtration rate by 10% to 40%.[68] Small amounts may also be reabsorbed by the renal tubules, especially at low flow rates.[9]

Plasma levels of creatinine are related to the relative muscle mass, the rate of creatine turnover, and renal function. Although it has been accepted for many years that the plasma level of creatinine is relatively unaffected by diet, recent reports indicate that the protein content of the diet may indeed affect the plasma level of creatinine if it affects the muscle mass of the individual.[6] Because of the observed constancy of endogenous production, the measurement of creatinine excretion has been used as a measure of the completeness of 24-hour urine collections in a given individual, although a number of studies have shown this method to be unreliable.[53]

Disease Correlations

Creatinine

Elevated levels of creatinine are mainly associated with abnormal renal function, especially as it relates to the glomerular filtration rate. Unfortunately, it is a relatively insensitive monitor and may not be measurably increased until renal function has deteriorated more than 50%.[63] The rate of endogenous creatinine clearance has been shown to be a relatively accurate reflection of the glomerular filtration rate (GFR),[51] although this observation may be a coincidence that has occurred due to the overestimation of plasma creatinine by the nonspecific Jaffee method (see subsequent discussion under Analytical Methods) and the fact that creatinine clearance is actually greater than inulin clearance in normal individuals.[59] The observed relationship between plasma creatinine and the GFR and the observation that plasma creatinine levels are relatively constant and unaffected by diet should make it an excellent analyte for the assessment of renal function. However, because of the difficulties encountered in analyzing the small amount of creatinine normally present (see discussion under Analytical Methods), several authors have suggested that this use of

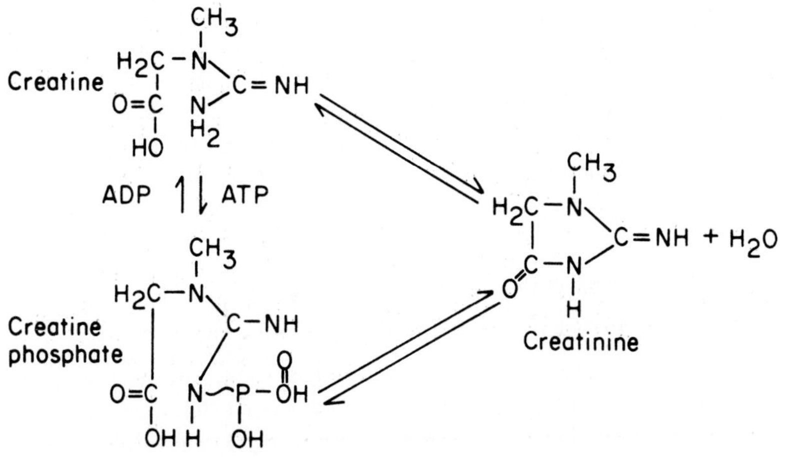

Figure 16-4. The structures of creatine, creatine phosphate, and creatinine.

plasma creatinine does not provide needed sensitivity in the detection of mild renal dysfunction, and it has been suggested that it is better to use other methods for monitoring GFR.[29,31,59,81] Despite these problems, plasma creatinine level is the most commonly used monitor of renal function.

As noted earlier in the chapter, the serum creatinine level has been used in combination with the BUN to differentiate between prerenal and renal causes of azotemia.[2] The normal BUN/creatinine ratio is approximately 10:1. Values markedly greater than this suggest that the cause of an elevated BUN is prerenal rather than renal. A low BUN/CREAT ratio may suggest interference in the analysis of creatinine (see Case Study 16–2) or a marked increase in creatine turnover.

Creatine

In muscle disease such as muscular dystrophy or poliomyelitis, hyperthyroidism, and trauma, both plasma creatine and urinary creatinine are often elevated. Plasma creatinine levels are usually normal in these patients. Because analytical methods for creatine are not readily available in most clinical laboratories, its measurement has been replaced by the measurement of creatine kinase levels for the diagnosis of muscle disease.[75] Plasma creatine levels are not elevated in renal disease.

Analytical Methods for Creatinine

The methods most frequently used to measure creatinine are based on the Jaffe reaction first described in 1886.[36] In this reaction, creatinine reacts with picric acid in alkaline solution to form a red-orange chromogen, which is a tautomer of creatinine and picric acid.[53] This reaction was first adopted for the measurement of blood creatinine by Folin and Wu in 1919.[20] It has subsequently been shown to be rather nonspecific and subject to interference by a large number of compounds present in blood, such as ascorbate, pyruvate, acetone, and glucose.[27,32,53,85] These compounds will also form a color in the presence of the reagent and cause a significant positive error. Because of this nonspecificity, the most accurate results are obtained when the creatinine in a protein-free filtrate of the sample is adsorbed onto an adsorbent, such as Fuller's earth or Lloyd's reagent, and then eluted and reacted with alkaline picrate.[27,32] Since this method is time-consuming and not readily automated, it is not used routinely.

Two approaches have been used to increase the specificity of the assay methods for creatinine: a kinetic Jaffe method and the use of various enzymes. In the kinetic Jaffe method, serum is mixed with alkaline picrate, and the rate of change in absorbance at 520 nm is measured between 20 and 80 seconds.[11,18,37,43,46,52] Although this method eliminates some of the nonspecific reactants, it is still subject to erroneously elevated results in patients with elevated levels of alpha-ketoacids[71,85] and in those on cephalosporin therapy.[72] The presence of significant levels of bilirubin and hemoglobin may cause a negative bias, probably due to their destruction in the strong base used.[32,86]

Despite these problems, the kinetic Jaffee method is quite popular. It is inexpensive, rapid, and easy to do. It is used on a number of automated systems such as the Beckman ASTRA™, DuPont ACA™, and many discrete analyzers.

In an effort to eliminate the nonspecificity of the Jaffee reaction, several coupled enzymatic methods have been developed.[4,13,24,37,42,52,74,82] Of these, the methods employing the reaction sequence shown below using creatininase (creatinine aminohydrolase [EC 3.5.2.10]), creatinase (creatine amidinohydrolase [EC 3.5.3.3]), sarcosine oxidase (EC 1.5.3.1), and peroxidase (EC 1.11.1.7) seem to show the most promise for an accurate and precise assay method.[13,25,86]

$$Creatinine + H_2O \xrightarrow{\text{Creatinine aminohydrolase}} creatine$$

$$Creatine + H_2O \xrightarrow{\text{Creatine amidinohydrolase}} sarcosine + urea$$

$$Sarcosine + H_2O + O_2 \xrightarrow{\text{Sarcosine oxidase}} formaldehyde\ glycine + H_2O_2$$

$$H_2O_2 + phenol\ derivative + dye \xrightarrow{\text{POD}} colored\ product + H_2O$$

(Eq. 16-1)

This method is used in the recently developed single-slide method on the Kodak Ektachem[4,24,61] and the Boehringer Mannheim PAP,[13] and can be adapted for use on many automated systems.

Another approach to the analysis of creatinine has been the measurement of the color formed when creatinine reacts with 3,5-dinitrobenzenoic acid or its derivatives.[41,53] This method has been successfully adapted to a reagent strip method. However, the color produced is less stable than that of the Jaffee chromogen.[60]

Several HPLC methods have been developed.[7,64,70] The most popular of these uses pretreatment of the sample with trichloroacetic acid to remove protein, cation-exchange chromatography, and UV detection of the creatinine.[64] This HPLC method has been recommended as a reference method to replace the Fuller's earth–Jaffe method.

It is important to note that the development and use of more accurate methods for plasma creatinine that tend to give lower values will have a significant effect on results obtained for creatinine clearance, since results for urinary creatinine are not subject to as many interferences as plasma levels. Use of more specific methods for plasma will thus result in apparently higher clearance rates. Such results may mask mild renal dysfunction using current reference intervals.[81] Analytical methods for creatinine are summarized in Table 16-4.

Specimen Requirements and Interfering Substances

Creatinine can be measured in serum, plasma, or urine. Hemolysis should be avoided, since there are a significant number of nonspecific chromogens in the red cells. Lipemic

TABLE 16-4. Summary of Analytical Methods—Creatinine

Methods Based on Jaffe Reaction

$$\text{Creatinine} + \text{picrate} \xrightarrow{\text{OH}^-} \text{Janovski complex (red)}$$

Jaffe-Fuller's earth	Creatinine in protein-free filtrate adsorbed onto Fuller's earth; released with alkaline picrate to yield colored Janovski	Previously considered reference method
Jaffe without adsorbent	Creatinine in protein-free filtrate reacts with alkaline picrate to form colored complex	Subject to positive bias from glucose, uric acid, glutathione, ascorbic acid, alpha-ketoacids, and cephalosporins
Jaffe-kinetic	Jaffe reaction done directly on serum; rate of color formation determined early in reaction	Subject to positive bias from alpha-ketoacids and cephalosporins; requires automated equipment for precision

Enzymatic

Creatininase-CK

$$\text{Creatinine} + H_2O \xrightarrow{\text{Creatinine aminohydrolase}} \text{creatine}$$

$$\text{Creatine} + \text{ATP} \xrightarrow{\text{CK}} \text{creatine phosphate} + \text{ADP}$$

$$\text{ADP} + \text{phosphoendpyruvate} \xrightarrow{\text{PK}} \text{ATP} + \text{pyruvate}$$

$$\text{Pyruvate} + \text{NADH} \xrightarrow{\text{LD}} \text{lactate} + \text{NAD}^+$$

Requires large sample of preincubation; not widely used

Creatininase-H_2O_2

$$\text{Creatinine} + H_2O \xrightarrow{\text{Creatininase}} \text{creatine}$$

$$\text{Creatine} + H_2O \xrightarrow{\text{Creatinase}} \text{sarcosine} + \text{urea}$$

$$\text{Sarcosine} + H_2O + O_2 \xrightarrow{\text{Sarcosine oxidase}} \text{glycine} + \text{formaldehyde} + H_2O_2$$

$$H_2O_2 + \text{phenol derivative} + 4\text{ aminophenazone} \xrightarrow{\text{POD}} \text{benzoquinone-immine dye}$$

Automated on Kodak Ektachem and some other instruments; potential to replace Jaffee; some positive bias due to lidocaine; no interference from acetoacetate or cephalosporins

Other Methods

3,5-Dinitrobenzoic acid (DNBA)	$\text{Creatinine} + \text{DNBA} \xrightarrow{\text{OH}^-} \text{purple color}$	Used on reagent strips; colored product unstable
HPLC	Reversed-phase or cation-exchange chromatography	Highly specific; requires pretreatment step; proposed reference method

and icteric samples will give erroneous results. Fasting is not required. A recent meal will not affect the creatinine level although heavy protein ingestion may transiently elevate levels. Urine should be refrigerated after collection or frozen if storage longer than four days is required.[77]

Sources of Error

When creatinine is measured by the Jaffe reaction, glucose, uric acid, alpha-ketoacids, and ascorbate may increase results, especially at temperatures above 30°C. As discussed above, this interference is markedly decreased when the reaction is measured kinetically. Depending on the concentration of reactants and measuring time, some interference from alpha-keto acids may still exist.[86] These substances do not interfere in the enzymatic methods described above. Bilirubin causes a negative bias in both Jaffe and enzymatic methods.[86]

Patients on cephalosporin antibiotics may show erroneously elevated results when the Jaffee reaction is used.[72] Other drugs also have been shown to increase the results, particularly dopamine, which affects both enzymatic and

Jaffee methods.[86,89] Ascorbate will interfere in any of the enzymatic methods that utilize peroxidase.[86] Lidocaine has been shown to cause a positive bias in some enzymatic methods.[24,67] The antifungal agent, 5-fluorocytosine, will also cause errorenous results.[59]

Reference Interval[77]

See Table 16-5. Reference intervals vary with age and sex and may decrease in the elderly.[77]

Analytical Methods for Creatine

The traditional method for the analysis of creatine is based on analyzing the sample for creatinine before and after heating in acid solution using an endpoint Jaffe method.[31] Heating converts all the creatine to creatinine, and the difference between the two samples will be the creatine concentration. Unfortunately, heating may result in the formation of additional nonspecific chromogens, and the precision of this

TABLE 16-5. Reference Ranges for Creatinine		
Cord blood	0.6–1.2 mg/dL	(53–106 mmol/L)
Children	0.3–0.7 mg/dL	(27–62 mmol/L)
Adult		
Male	0.6–1.2 mg/dL	(53–106 mmol/L)
Female	0.5–1.1 mg/dL	(44–97 mmol/L)

method is poor. In recent years, several enzymatic methods have been developed including the use of the creatininase assay described above[13] with omission of the initial enzyme and the utilization of CK(EC 2.7.3.2), PK(EC 2.1.7.40), and LD(EC 1.1.1.27) in a coupled sequence.[3,48,56,83] At least one of these methods has been adapted for automated systems.

URIC ACID

Biochemistry

In the higher primates, such as humans and apes, uric acid is the final breakdown product of purine metabolism. Most other mammals have the ability to catabolize purines one step further to allantoin, a much more water-soluble end-product. This reaction sequence is shown in Figure 16-5.

Purines such as adenosine and guanine, resulting from the breakdown of nucleic acids that are ingested or come from the destruction of tissue cells, are converted into uric acid, mainly in the liver. Uric acid is transported by the plasma from the liver to the kidney, where it is filtered by the glomerulus. Of the uric acid in the glomerular filtrate, 98% to 100% is reabsorbed in the proximal tubules. Small amounts of uric acid are then secreted by the distal tubules and ultimately appear in the urine. Triglycerides, ketone bodies, and lactic acid have been shown to compete with urate for excretory sites in the renal tubules. This route accounts for two-thirds to three-quarters of the daily uric acid excretion. The remainder is excreted into the GI tract and degraded by bacterial enzymes.

Nearly all (96.8%) of the uric acid in plasma is present as monosodium urate. At the pH of plasma, urate in this form is relatively insoluble, and, at levels above 6.4 mg/dL, the plasma is saturated and urate crystals may form and precip-

itate in the tissues.[61] In the urine at pH levels <5.75, uric acid is the predominant form and frequently appears as uric acid crystals.

Disease Correlations

There are three major disease states associated with elevated levels of uric acid in the plasma. They include gout, increased nuclear breakdown, and renal disease. Gout is a disease found primarily in males and usually first diagnosed between the ages of 30 and 50. Patients have pain and inflammation of the joints caused by precipitation of sodium urates in the joint resulting from the high levels of uric acid found in extracellular fluids. In 25% to 30% of these patients, hyperuricemia has been shown to be due to overproduction of uric acid, although other common causes include ingestion of some drugs or alcohol. The plasma uric acid levels in these patients is usually above 6.0 mg/dL. Because of their high uric acid level, patients with gout are also highly susceptible to the development of renal calculi, although not all patients with elevated serum urate levels develop these complications. In women, urate levels rise after menopause; some postmenopausal women may develop hyperuricemia and gout.

Another common cause of elevated plasma uric acid levels is increased breakdown of cell nuclei, such as that which occurs in patients on chemotherapy for such proliferative diseases as leukemia, lymphoma, multiple myeloma, or polycythemia. Levels of uric acid should be monitored in these patients to avoid excessively high levels, which may be toxic to the kidneys. To prevent hyperuricemia in these patients, they are often treated with allopurinol, a drug that inhibits the synthesis of urate by replacing it with more soluble metabolites. Patients with hemolytic and megaloblastic anemia may also exhibit elevated levels of uric acid.

Because uric acid is filtered by the glomerulus and secreted by the distal tubules, chronic renal disease will also cause elevated levels of uric acid, although this assay is not very useful in the evaluation of renal function, because so many other factors affect its plasma level (Table 16-6).

High levels of uric acid are also found secondary to a variety of diseases, such as glycogen storage disease and other congenital enzyme deficiencies that result in the production of excess metabolites, such as lactate and triglycerides, which compete with urate for excretion.[29]

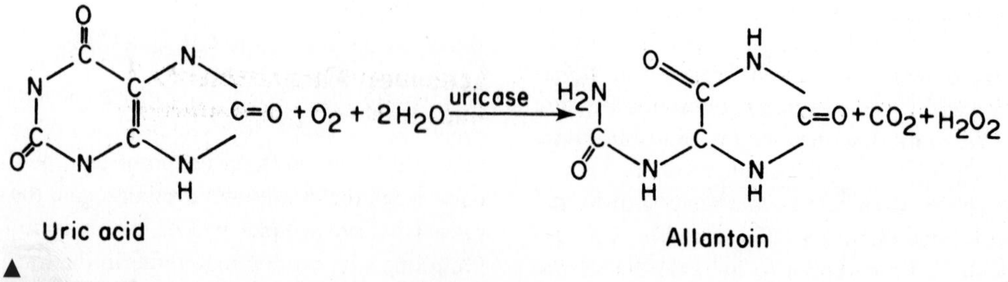

▲
Figure 16-5. The catabolism of uric acid to allantoin.

TABLE 16-6. Causes of Increased Plasma Uric Acid Concentration

Increased dietary intake
Increased urate production
    Gout
    Treatment of myeloproliferative disease with cytotoxic drugs
Decreased excretion
    Lactic acidosis
    Toxemia of pregnancy
    Glycogen storage disease type I
    Drug therapy
    Poisons: lead, alcohol
    Renal disease
Specific enzyme defects
    Lesch-Nyhan syndrome

Lesch-Nyhan syndrome, a sex-linked genetic disorder characterized by a deficiency of hypoxanthine-guanine phosphoribosyl transferase (HGPRT), a major enzyme in the biosynthesis of purines, also results in increased levels of uric acid in plasma and urine.

Hyperuricemia is also a common feature of toxemia of pregnancy and lactic acidosis. The cause of increased plasma levels here is apparently competition for binding sites in the renal tubules.[29] Elevated levels also may be found after ingestion of a diet rich in purines (liver, kidney, sweetbreads, shellfish, etc.) or a marked decrease in total dietary intake, resulting in increased tissue breakdown.

Decreased levels of uric acid are much less common than increased levels and are usually secondary to severe liver disease or defective tubular reabsorption as in Falconi's syndrome.[77] Overtreatment with allopurinol or cancer chemotherapy with 6-mercaptopurine or azathioprine may also be a cause of hypouricemia.[29]

Analytical Methods

As shown in Figure 16-5, uric acid is readily oxidized to allantoin and thus can function as a reducing agent in many chemical reactions. This property formed the basis for nearly all the early analytical methods for uric acid. The most popular method of this type is the Caraway method, which is based on the oxidation of the uric acid in a protein-free filtrate, with subsequent reduction of phosphotungstic acid to tungsten blue.[8] The Caraway method uses sodium carbonate to provide the alkaline pH necessary for color development. The blue color produced can be intensified by the addition of cyanide to the reaction mixture, but adequate color can be obtained without cyanide by proper adjustment of the pH.

Because this method is relatively nonspecific, attempts have been made to find more specific alternatives. Several methods using the very specific enzyme uricase (urate oxidase, EC 1.7.3.3) have been described.[16,38,45,79] As noted previously, this enzyme catalyzes the oxidation of uric acid to allantoin with subsequent production of H_2O_2.

The simplest of these methods is based on the fact that uric acid has a significant UV absorbance band with a peak at 293 nm, whereas allantoin does not. The difference in absorbance before and after incubation with uricase will be proportional to the uric acid concentration.[45,79] It has been shown that the only substance to cause significant interference in this assay is xanthine at a concentration much higher than normally present in plasma.[15] The major disadvantages of this very simple method are: (1) it requires a spectrophotometer capable of measuring absorbance at 293 nm, and (2) the presence of proteins in the serum can cause a high background absorbance. Common laboratory spectrophotometers are not capable of measuring at this wavelength with needed sensitivity. This method is currently used on several automated instruments, including the DuPont ACA™. It also has been proposed as a candidate reference method by the American Association for Clinical Chemistry.[16]

A more common approach that takes advantage of the specificity of uricase but avoids the need for UV spectrophotometry uses a coupled enzymatic reaction that measures the amount of peroxide formed in the primary reaction. In these methods, either peroxidase or catalase is used to couple the peroxide produced to a color-producing reaction. The methods using peroxidase use either 4-amino antipyrine and a phenol derivative[38] or 3-methyl-2-benzothiazolinone hydrazone (MBTH) and dimethylaniline (DMA)[23] to produce a colored product that absorbs in the visible range. The sequence of reactions using catalase is as follows.[26]

$$H_2O_2 + ethanol \xrightarrow{Catalase} acetaldehyde + 2H_2O$$

$$Acetaldehyde + NAD^+ \xrightarrow{ADH^*} acetate$$

$$+ NADH + 2H^+ \qquad (Eq.\ 16\text{-}2)$$

Both methods are subject to interference from substances such as bilirubin and ascorbic acid, which destroy peroxide. Commercial reagent preparations often include potassium ferricyanide and ascorbate acid oxidase to minimize these interferences. These methods have been adapted for use on a variety of automated systems and will continue to be the most popular methods until more automated systems develop the capability of measuring absorbance at 293 nm, as required by the selected reference method.

HPLC methods using several types of columns also have been developed.[57,73] These methods show good specificity and sensitivity but may require pretreatment of the sample to remove protein. The National Bureau of Standards is currently developing a definitive method for uric acid based on isotope dilution/mass spectrometry.[17] Analytical methods are summarized in Table 16-7.

Specimen Requirements and Interfering Substances

Uric acid may be determined in serum, urine, or heparinized plasma. Serum should be removed from cells as quickly as possible. Although diet in general may affect the uric acid

*ADH = alcohol dehydrogenase.

TABLE 16-7. Summary of Analytical Methods—Uric Acid

PHOSPHOTUNGSTIC ACID	Uric acid + phosphotungstate $\xrightarrow{Na_2CO_3/OH^-}$ CO_2 + allantoin + tungsten blue	Nonspecific; requires protein removal; seldom used anymore
URICASE	All based on same initial reaction with subsequent determination of H_2O_2 Uric acid + O_2 $\xrightarrow{Uricase}$ allantoin + CO_2 + H_2O_2	Enzyme highly specific
Spectrophotometric	Measurement of decrease in absorbance at 290–293nm due to destruction of uric acid	Basis for candidate reference method; only interference from xanthine
Coupled enzymatic (I)	H_2O_2 + phenol derivative + 4 aminoantipyrine $\xrightarrow{POD}$ quinone-immine dye	Subject to negative bias from bilirubin, ascorbate; used on many automated instruments
Coupled enzymatic (II)	$H_2O_2^+$ ethanol $\xrightarrow{Catalase}$ H_2O + acetaldehyde Acetaldehyde + NAD $\xrightarrow{ADH}$ acetate + NADH + $2H^+$	Popular; readily automated
HPLC	Several types of ion-exchange or reversed-phase columns used	Highly specific; may require sample pretreatment
Isotope Dilution/ Mass Spectrometry	Sample diluted with known amount of labeled uric acid; record ratio of two isotopes	Being developed by NBS as definitive method

levels, a recent meal has no significant effect, and therefore, the patient need not be fasting. Gross lipemia should be avoided. Increased levels of bilirubin may give decreased results in the analytical methods using peroxidase. Significant hemolysis, which releases glutathione, also may give low values; a number of drugs, such as thiazides and salicylates, have been shown to cause elevated values for uric acid.[89]

Uric acid itself is quite stable in serum or plasma once they have been separated from the red cells. Serum samples may be stored at room temperature for up to 3 days if protected from bacterial contamination.

Reference Interval[77]

Values are slightly lower in children and premenopausal females (Table 16-8).

AMMONIA

Biochemistry

The level of ammonia in the circulation is extremely low (11–35 µmol/L).[77] It arises from the deamination of amino acids, which occurs mainly through the action of digestive and bacterial enzymes on proteins in the intestinal tract.

TABLE 16-8. Reference Ranges for Uric Acid*

Children	2.0–5.5 mg/dL	(0.12–0.32 mmol/L)
Adult males	3.5–7.2 mg/dL	(0.21–0.42 mmol/L)
Adult females	2.6–6.0 mg/dL	(0.15–0.35 mmol/L)

Cross-reference ranges based on uricase methods

Ammonia is also released from metabolic reactions that occur in skeletal muscle during exercise. It is used by the parenchymal cells of the liver in the production of urea, as described earlier in this chapter. In severe liver disease, when there is significant collateral circulation, or when parenchymal liver cell function is severely impaired, ammonia is not removed from the circulation and blood levels rise. Unlike the other NPN substances, plasma levels of ammonia are not dependent on renal function but on liver function and thus are not useful in the study of kidney disease.

Markedly elevated levels of ammonia are quite neurotoxic and are often associated with encephalopathy. It is thought that this toxicity may be due in part to depletion of alpha-ketoglutarate and subsequent impairment of the cerebral citric acid cycle,[69] but other mechanisms, such as altered neurotransmitter function, probably play a role.[2] Because of its important role in monitoring the progress of several severe clinical conditions and the extreme liability of the sample, this assay should be available in every laboratory.

Disease Correlations

The two major clinical conditions in which blood ammonia levels offer useful information are hepatic failure and Reye's syndrome. Severe liver disease represents the most common cause of disturbed ammonia metabolism, and monitoring of the blood ammonia level is often used to determine the prognosis of patients with such disease. Unfortunately, the correlation between the degree of hepatic encephalopathy and plasma NH_3 is not always consistent, and some patients may have encephalopathy with normal plasma ammonia levels.

Reye's syndrome is an often fatal disease that occurs most commonly in children. It is frequently preceded by a viral infection. It is now known that this is an acute metabolic disorder of the liver, and autopsy findings show severe fatty infiltration of that organ. Blood ammonia levels have shown good correlation with both the severity of the disease and prognosis. Fitzgerald and associates have shown that when plasma ammonia levels are less than five times normal, the survival rate in Reye's syndrome is 100%.[19]

Blood ammonia levels are also elevated in the inherited deficiency of any of the urea-cycle enzymes except argininosuccinase. Patients with these disorders are mentally retarded and suffer from behavior problems. They often exhibit elevated levels of glutamine and protein intolerance, as well as elevated ammonia levels.[22]

Because of the importance of portal circulation in the removal of ammonia from the circulation, portal-systemic shunting of blood also will lead to elevated ammonia levels. The patency of a portocaval shunt may be evaluated by measuring the plasma ammonia before and after a standard dose of NH_4^+ salts.[10]

Analytical Methods

The accurate measurement of ammonia in plasma is complicated by its low concentration, the extreme instability of the sample, and the ubiquitous nature of NH_3 contamination. There are two distinct approaches that have been used for the measurement of plasma ammonia. One is a two-step approach in which ammonia is first isolated from the sample and then assayed. The second involves direct measurement of ammonia by an enzymatic method or ion-selective electrode.

One of the first analytical methods for ammonia, developed by Conway in 1935, involved use of the Conway microdiffusion chamber.[12] In this method, ammonia was liberated from the sample by alkali, absorbed into acid, and then quantitated by titration or colorimetry. This method is generally not very accurate and is seldom used today.

A second, more successful approach for isolating NH_3 is the use of a cation-exchange resin (Dowex 50) followed by elution of the NH_3 with NaCl and quantitation by the Berthelot reaction.[39] This method is available in kit form and gives acceptable, although slightly higher, results. Because it requires an isolation step, it is time-consuming and not readily automated.

A more specific and convenient approach is a coupled enzymatic assay using the following reaction sequence.[35,40,49]

$$NH_4^+ + \text{2-oxoglutarate} + NADPH \xrightarrow{GLDH^*}$$

$$\text{L-glutamate} + NADP^+ + H_2O \qquad \textbf{(Eq. 16-3)}$$

Under appropriate reaction conditions, the decrease in absorbance at 340 nm is proportional to the ammonia concentration. NADPH is the preferred coenzyme because it is used specifically by the glutamic dehydrogenase and is not consumed by side reactions with endogenous substrates such as pyruvate. ADP is added to the reaction mixture to stabilize the GLDH. This method is used on many automated systems and is available in kit form from numerous manufacturers.

Another approach is the use of an ammonia electrode.[1] This electrode measures the changes in pH of a solution of ammonium chloride as ammonia diffuses across a semipermeable membrane. Because the electrode membrane used is not very stable, this method has not gained much popularity. A recent report described successful use of a microflow-through electrode in a semiautomated system.[80] Analytical methods for ammonia are summarized in Table 16-9.

Specimen Requirements and Interfering Substances

Proper specimen handling is of the utmost importance in plasma ammonia assays. Ammonia levels rise rapidly in whole blood after drawing because of the deamination of amino acids.

Venous blood should be obtained without trauma and placed on ice immediately. EDTA is the preferred anticoagulant. It had previously been shown that use of heparin as an anticoagulant helped to stabilize the sample and decrease endogenous ammonia production.[33] However, use of heparinized blood has recently been shown to give erroneously low results in some samples.[15] The anticoagulant and evaluated tube used should be tested for ammonia contamination. Samples should be centrifuged at 0 to 4°C within 20 minutes of collection, and the plasma removed. Some authors report that separated plasma samples are then stable for up to 3½ h in an ice bath.[49] However, for best results, it is recommended that the assay be completed as soon as possible to prevent an erroneous increase due to *in vitro* deamination. Frozen plasma is reportedly stable for several days at −20°C. Because red cells contain 2 to 3 times as much ammonia as plasma, hemolysis should be avoided.

*GLDH = glutamic dehydrogenase (EC 1.4.1.3).

TABLE 16-9. Summary of Analytical Methods—Ammonia

Conway microdiffusion	NH_3 released from alkaline solution determined by back titration with acid	Time-consuming; poor accuracy and precision; no longer used
Ion-exchange	NH_3 absorbed onto Dowex 50 resin, eluted, and quantitated with Berthelot reaction	Time-consuming manual method; good accuracy
Coupled enzymatic	$NH_4^+ + \text{2-oxoglutarate} + NADPH \xrightarrow{GLDH} \text{glutamate} + NADP^+ + H_2O$	Most common on automated instruments; precise and accurate
Ion-selective electrode	Diffusion of NH_3 through selective membrane into NH_4Cl causing pH change which is measured potentiometrically	Good accuracy and precision; some problem with membrane stability

In normal patients, blood ammonia levels do not appear to be significantly affected by eating, but they are affected by cigarette smoking. It is recommended that patients not smoke for several hours before the sample is drawn.[22]

It is also recommended that all glassware and reagents to be used be specially cleaned to eliminate ammonia contamination and that the assay be performed in an isolated laboratory, free from ammonia contamination.

Sources of Error

The two biggest pitfalls in accurate ammonia analysis are sample handling and ammonia contamination. Ammonia levels rise rapidly in shed blood and these samples must be handled as described previously. It has been shown that elimination of sources of ammonia contamination in the laboratory can significantly increase the accuracy of ammonia assay results.[33] Sources of contamination include tobacco smoke and urine and the presence of ammonia in detergents, glassware, reagents, and water. Although enzymatic methods are affected less than resin methods, ammonia contamination is still a potential problem in all assay methods for ammonia.

Another problem with ammonia assays is that serum-based control material cannot be used for quality control because the ammonia content is unstable. Frozen aliquots of human serum albumin containing known amounts of ammonium chloride or ammonium sulfate may be used. Solutions containing known amounts of ammonium sulfate are available from the College of American Pathologists.

A number of substances have been shown to influence the ammonia level *in vivo*.[77,89] Ammonium salts, asparaginase, barbiturates, ethanol, hyperalimentation, narcotic analgesics, and diuretics all cause increased levels of ammonia in plasma. Levodopa, kanamycin, diphenhydramine, neomycin, *Lactobacillus acidophilus,* and lactulose all cause decreased values.

Reference Interval[77]

Ammonia 14 to 49 µg/dL (11–35 µmol/L)

The values obtained will vary somewhat with the method used. Enzymatic methods tend to give slightly lower values than do ion-exchange methods. Higher values are usually seen in newborns.

SUMMARY

The nonprotein nitrogen compounds found in the plasma include urea, creatinine, creatine, uric acid, and ammonia. Urea, the major excretory product of protein metabolism, represents nearly 50% of the total NPN compounds. Its plasma level is related to the protein content of the diet, renal blood flow, level of protein catabolism, and renal function. Clinical conditions causing elevated urea levels can be divided into prerenal, renal, and postrenal conditions. The BUN/CREAT ratio can be used to distinguish between these conditions. The most common prerenal cause of an elevated plasma urea level is congestive heart failure. Urea is usually measured by a coupled enzymatic method that quantitates the amount of ammonium produced by urease hydrolysis of the urea in the sample.

Creatinine is formed as creatine and phosphocreatine in muscle spontaneously lose water or phosphoric acid. In a given individual, it is excreted into the plasma at a constant rate that is related to the individual's muscle mass. Plasma levels are related to rate of creatine turnover, muscle mass, and renal function. It is the most commonly used monitor of renal function and has been shown to be inversely related to glomerular filtration rate. Measurement of creatinine in plasma and urine has traditionally been done using the Jaffee reaction. This reaction is relatively nonspecific and subject to interference from many commonly occurring substances such as glucose, protein, and alpha-keto acids. Specificity of this reaction can be improved by utilizing a kinetic rather than an endpoint method; however, the kinetic Jaffe method is still subject to interference from some keto acids, bilirubin, and some drugs. A specific enzymatic method has been developed but has not been widely accepted. Use of a more specific method for measurement of creatinine may require readjustment of reference intervals for both plasma creatinine and creatinine clearance. Without such adjustment, mild levels of renal dysfunction may be overlooked.

Uric acid is the final breakdown product of the purines from nucleic acids in humans. It is filtered by the glomerulus, reabsorbed in the proximal tubules, and secreted by the distal tubules into the urine. Additional uric acid is excreted into the GI tract and degraded by bacterial enzymes. Uric acid demonstrates low solubility in the plasma, and elevated levels of urate will be deposited in the joints and tissues causing painful inflammation. Elevated levels of plasma urate are due to gout, increased nuclear breakdown, and renal disease. Increased uric acid may also be seen secondary to conditions that result in excess production of metabolites, such as lactate and triglycerides, which compete for urate for secretion in the distal tubule. Uric acid is usually quantitated by a coupled enzymatic method in which uric acid is oxidized by uricase to allantoin and peroxide, and the peroxide produced is quantitated with a variety of dyes. Increased levels of bilirubin and other substances that destroy peroxide in the sample will give erroneously decreased results.

Ammonia is present in the plasma in extremely low concentrations. It arises from the deamination of amino acids resulting from the breakdown of protein, and it is normally rapidly removed from the circulation and converted to urea in the liver. Elevated levels of ammonia are neurotoxic. Elevated levels of ammonia in the plasma are usually associated with hepatic failure or Reyes' syndrome. Ammonia is usually measured by a coupled enzymatic method using glutamic dehydrogenase and measuring change in absorbance due to conversion of NADPH to NADP. The major source of error in ammonia determinations is improper sample handling. The ammonia level of freshly drawn blood rises rapidly on standing. The sample should be placed on ice and the plasma separated and analyzed as rapidly as possible. The presence of ammonia contamination and cigarette smoke may also affect the results.

CASE STUDY 16-1

A 65-year-old man was first admitted for treatment of chronic obstructive lung disease, renal insufficiency, and marked cardiomegaly. Pertinent laboratory data on admission (5/31) is shown in Case Study Table 16–1.

Because of severe respiratory distress, the patient was transferred to the ICU, placed on a respirator, and given diuretics and I.V. fluids to promote diuresis. This treatment brought about a marked improvement in both cardiac output and renal function, as shown by laboratory results several days later (6/3). After 2 additional days on a respirator with I.V. therapy, the patient's renal function had returned to normal, and at discharge his laboratory results were normal (6/7).

The patient was readmitted 6 months later because of the increasing inability of his family to arouse him. On admission, he was shown to have a tremendously enlarged heart with severe pulmonary disease, heart failure, and probable renal failure. Laboratory studies on admission were as shown in Case Study Table 16-2. Numerous attempts were made to improve the patient's cardiac and pulmonary function, all to no avail, and the patient died 4 days later.

CASE STUDY TABLE 16-1. Laboratory Results— First Admission

Test	5/31	6/3	6/7
BUN, mg/dL	45	24	11
Creatinine, mg/dL	1.8	1.3	0.9
BUN/Creatinine	25	18.5	12.2
pH	7.22	7.5	
PCO_2, mmHg	74.4	48.7	
PO_2, mmHg	32.8	57.6	
O_2 sat, %	51.3	91.0	

CASE STUDY TABLE 16-2. Laboratory Results— Second Admission

BUN, mg/dL	90
Creatinine, mg/dL	3.9
UA, mg/dL	12.0
BUN/CREAT	23
pH	7.351
PCO_2, mmHg	59.9
PO_2, mmHg	34.6
O_2 sat, %	63.7

Question

1. What is the most likely cause of the patient's elevated BUN? What data support your conclusion?

CASE STUDY 16-2

An 80-year-old woman was admitted with a diagnosis of hypertension, congestive heart failure, anemia, possible diabetes, and chronic renal failure. She was treated with diuretics and I.V. fluids and released 4 days later. Her laboratory results are shown in Case Study Table 16-3.

CASE STUDY TABLE 16-3. Laboratory Results

	First Admission		Second Admission	
	2/11	2/15	7/26	7/28
BUN, mg/dL	58	★	★	61.0
Creatinine, mg/dL	6.2	6.2	6.4	6.0
Uric acid, mg/dL	10.0	★	★	9.2
BUN-creatine	9.4	★	★	10.1
Glucose, mg/dL	86	80	113	★

★Not done

Five months later, she was readmitted for treatment of repeated bouts of dyspnea. She was placed on a special diet and drugs to control her hypertension and was discharged. It was felt that she had not been taking her medication as prescribed, and she was counseled regarding the importance of regular doses.

Questions

1. What is the most probable cause of the patient's elevated BUN? What data support your conclusion?
2. Note that this patient's admitting diagnosis is "possible diabetes." If this patient had truly been a diabetic with an elevated blood glucose and a positive acetone, what effect would this have had on the measured values for creatinine? Explain.

CASE STUDY
16-3

A 3-year-old girl was admitted with a diagnosis of acute lymphocytic leukemia. Her admitting laboratory data are shown in Case Study Table 16-4. After admission, she was treated with packed cells, 2 units of platelets, I.V. fluids, and allopurinol. On the second hospital day, chemotherapy was begun, using I.V. vincristine and prednisone and intrathecal injections of methotrexate, prednisone, and cytosine arabinoside. She was discharged for home care 5 days later. She was continued on prednisone and allopurinol at home. She received additional chemotherapy 1 month later (11/1) and again on 11/14. On 12/6, she was readmitted because she had painful sores in her mouth and was unable to eat.

Questions

1. How would you explain the marked elevations of uric acid on admission?
2. What two factors are responsible for the normal levels of uric acid seen in subsequent admissions?
3. What is the most likely cause of the abnormally low level of BUN seen on 12/6? What other laboratory result would be useful to confirm your suspicions?

CASE STUDY TABLE 16-4. Laboratory Results

	10/1	10/2	10/3	10/4	11/14	12/6	6/20
BUN, mg/dL	12.0	★	★	15	4.0	2.0	★
Creatinine, mg/dL	0.7	★	★	1.0	0.7	★	0.7
Uric Acid, mg/dL	12.0	9.2	4.0	1.9	2.3	★	3.1
WBC mm³	56,300				3,700	2,800	3,700

★Not done.

REFERENCES

1. Attili AF, Autizi D, Capocaccia L. Rapid determination of plasma ammonia, using an ion specific electrode. Biochem Med 1975;14:109.
2. Bakerman S. ABC's of interpretive laboratory data. Greenville, NC: S Bakerman, 1983.
3. Beyer, C. Creatine measurement in serum and urine with an automated enzymatic method. Clin Chem 1993;39:1613.
4. Bissell MG, Ward E, Sanghavi P, et al. Multilaboratory evaluation of the new single slide enzymatic creatine method for the Kodak Ektachem analyzer. [Abstract]. Clin Chem 1987; 33:951.
5. Black D, Cameron S. Renal function. In: Brown S, Mitchell F, Young D, eds. Chemical diagnosis of disease. New York: Elsevier, North Holland Biomedical Press, 1979.
6. Bleiler RE, Schedl HP. Creatinine excretion: Variability and relationships to diet and body size. J Lab Clin Med 1972; 59:945.
7. Brown ND, Sing HC, Neeley WE, Koetitz ES. Determination of "true" serum creatinine by high-performance liquid chromatography combined with a continuous-flow microanalyzer. Clin Chem 1977;23:128.
8. Caraway WT. Standard methods of clinical chemistry. Vol 4. New York: Academic Press, 1963;239.
9. Chesley LC. Renal excretion at low urine volumes and the mechanism of oliguria. J Clin Invest 1935;17:591.
10. Conn H. Studies of the source and significance of blood ammonia: IV. Early ammonia peaks after ingestion of ammonium salts. Yale J Biol Med 1972;45:543.
11. Cook JGH. Factors influencing the assay of creatinine. Ann Clin Biochem 1975;12:219.
12. Conway EJ. Apparatus for the microdetermination of certain volatile substances: The blood ammonia with observations on normal human blood. Biochem J 1935;29:2755.
13. Crocker H, Shephard MDS, White GH. Evaluation of an enzymatic method for determining creatinine in plasma. J Clin Pathol 1988;41:576.
14. Demko PR, Welch R, Bellino L, Griben D, Young CC, Handani W. Simultaneous determination of urea, glucose, sodium, potassium, chloride, and total carbon dioxide in undiluted human serum using and electrode based automated analyzer. [Abstract]. Clin Chem 1989;35:1091.
15. Dowart W, Saner M. "Heparinized plasma is an unacceptable specimen for ammonia determination." Clin Chem 1992; 38:161.
16. Duncan PH, Gochman N, Cooper T, et al. A candidate reference method for uric acid in serum: I. Optimization and evaluation. Clin Chem 1982;28:284.
17. Ellerbe P, Cohen A, Welch MJ, White EV. Determination of serum uric acid by isotope dilution mass spectrometry as a new candidate reference method. Anal Chem 1990;62:2173.
18. Faulkner WR, Meites S, eds. Selected methods of clinical chemistry. Vol 9. Washington DC: AACC, 1982;357.

19. Fitzgerald JF, et al. The prognostic significance of peak ammonia levels in Reye's syndrome. Pediatrics 1982;70:997.

20. Folin O, Wu H. System of blood analysis. J Biol Chem 1919;31:81.

21. Georges J. Determination of ammonia and urea in urine and of urea in blood by use of an ammonia-selective electrodes. Clin Chem 1979;25:1888.

22. Glasgow M. Clinical application of blood ammonia determinations. Lab Med 1981;12:151.

23. Gochman N, Schmitz JM. Automated determination of uric acid with use of a uricase-peroxidase system. Clin Chem 1971;17:1154.

24. Gosney K, Adachi-Kirkland J, Schiller HS. Evaluation of lidocaine interference in the Kodak Ektachem 700 analyzer single-slide method for creatinine. Clin Chem 1987;33:2311.

25. Gruder WG, Hoffman GE, Hubbuch A, Poppe WA, Siedel J, Price CP. Multicenter evaluation of an enzymatic method for creatinine using a sensitive color reagent. J Clin Chem Biochem 1986;24:889.

26. Haeckel R. The use of aldehyde dehydrogenase to determine H₂O₂ producing reactions: The determination of uric acid concentration. J Clin Chem Clin Biochem 1976;14:101.

27. Haeckel R, Godsden RH, Sherwin JE, et al. Assay of creatinine in serum with use of Fuller's earth to remove interferents. In: Cooper GR, ed. Selected methods of clinical chemistry. Vol 10. Washington, DC: American Association for Clinical Chemistry, 1983;225.

28. Hallett CJ, Cook JGH. Reduced nicotinamide adenine dinucleotide-coupled reaction for emergency blood urea estimation. Clin Chem Acta 1971;35:33.

29. Halstead JA, Halstead CH. The laboratory in clinical medicine: Interpretation and application. 2nd ed. Philadelphia: Saunders, 1981.

30. Hammond BR, Lester E. Evaluation of a reflectance photometric method for determination of urea in blood, plasma, or serum. [Letter]. Clin Chem 1984;30:596.

31. Henry RJ. Clinical chemistry: Principles and technics. Chap 17. New York: Harper & Row, 1968.

32. Henry RJ, Cannon DC, Winkelman JW. Clinical chemistry: Principles and technics. New York: Harper & Row, 1964;287.

33. Howanitz JH, Howanitz PJ, Skrodzki CA, Iwanski JA. Influences of specimen processing and storage conditions on results for plasma ammonia. Clin Chem 1984;30:906.

34. Humphries BA, Melnychuk M, Donegan EJ, et al. Automated enzymatic assay for plasma ammonia. Clin Chem 1979;25:26.

35. Ishihard A, Kurahasi K, Uihard H. Enzymatic determination of ammonia in blood and plasma. Clin Chem Acta 1972;41:255.

36. Jaffe M. Uber den niederschlag welchen pikrinsaure in normalen harn erzurgt und uber eine neue reaction des kreatinins. Z Physiol Chem 1886;10:391.

37. Jaynes PK, Feld RD, Johnson GF. An enzymatic reaction rate assay for serum creatinine with a centrifugal analyzer. Clin Chem 1982;28:114.

38. Kabasakalian P, Kalliney S, Wescott A. Determination of uric acid in serum, with the use of uricase and a tribromophenolaminoantipyrine chromogen. Clin Chem 1973;19:522.

39. Kingsley GR, Tager HS. Ion-exchange method for the determination of plasma ammonia nitrogen with the Berthelot reaction. In: MacDonald RP, ed. Standard methods of clinical chemistry. Vol 6. New York: Academic Press, 1970;115.

40. Kristen E, Gerez C, Kristen R. Eine enzymatische microbesimmung du ammoniasks geeignet fur extrakte tierischer bewebe und flussigkeiten. Biochem J 1963;337:312.

41. Langley WD, Evans M. The determination of creatinine with sodium 3,5-dinitrobenzoate. J Biol Chem 1936;115:333.

42. Lanser A, Blijenberg BG, Jeijnse B. Evaluation of a new diagnostic kit for the enzymatic determination of creatinine. J Clin Chem Clin Biochem 1979;17:633.

43. Larsen K. Creatinine assay by a reaction kinetic principle. Clin Chem Acta 1972;41:209.

44. Lespinas F, Dupuy G, Revol F, Aubry C. Enzymic urea assay: A new colorimetric method based on hydrogen peroxide measurement. Clin Chem 1989;35:654.

45. Liddle L, Seegmillu JE, Loster L. The enzymatic spectrophotometric method for the determination of uric acid. J Lab Clin Med 1959;54:903.

46. Lustgarten JA, Wenk RA. Simple, rapid kinetic method for serum creatinine measurement. Clin Chem 1972;18:1419.

47. Marsh WH, Fingerhut B, Miller H. Automated and manual direct methods for determination of blood urea. Clin Chem 1965;11:624.

48. Marymount JH, Smith JN, Klotsch S. A simple method for urine creatine. Am J Clin Pathol 1968;49:289.

49. Mondzac A, Erlich GE, Seegmiller JE. An enzymatic determination of ammonia in biological fluids. J Clin Lab Med 1965;66:526.

50. Morgan DB, Carver ME, Payne RB. Plasma creatinine and urea: Creatinine ratio in patients with raised plasma urea. Br J Med 1977;2:929.

51. Morgan DB, Dillon S, Payne RB. The assessment of glomerular function: Creatinine clearance or plasma creatinine? Postgrad Med 1978;54:302.

52. Moss G, Bondar RJL, Buzelli D. Kinetic enzymatic method for determining serum creatinine. Clin Chem 1975;21:1422.

53. Narayanan S, Appelton H. Creatinine: A review. Clin Chem 1980;26:1119.

54. Ohkubo A, Kanel S, Yamanaka M, Katsuyama H, Iwata Y, Sekikawa N. Multilayer-film analysis for urea nitrogen in blood, serum, or plasma. Clin Chem 1984;30:1222.

55. Orsonneau J, Massoubre C, Cabanes M, Lustenberger P. Simple and sensitive determination of urea in serum and urine. Clin Chem 1992;38:619.

56. Oversteegen HJ, DeBoer J, Bakker AJ, Van Leeuwen C. A simpler enzymatic determination of creatine in serum with a commercial creatinine kit. [Letter]. Clin Chem 1984;33:720.

57. Pachla LA, Kissinger PT. Measurement of serum uric acid by liquid chromatography. Clin Chem 1979;25:1847.

58. Paulson G, Ray R, Sternberg J. A rate sensing approach to urea measurement. Clin Chem 1971;17:644.

59. Perrone RD, Madias NE, Levey AS. Serum creatinine as an index of renal function: New insight into old concepts Clin Chem 1992;38:1933.

60. Pesce AJ, Kaplan LA. Methods in clinical chemistry. St. Louis: CV Mosby Co, 1987.

61. Powers D, Kringle R, et al. Specificity of the Ektachem analyzer enzyme-based creatinine assay compared with Jaffee and HPLC methods. [Abstract]. Clin Chem 1987;33:951.

62. Raphael S, et al. Lynch's medical laboratory technology. 4th ed. Chap 9. Philadelphia: Saunders, 1983.

63. Renkin EM, Robinson RR. Glomerular filtration. N Engl J Med 1974;290:785.

64. Rosano TG, Ambrose RT, Wu AHB, Swift TA, Yadejari P. Candidate reference method for determining creatinine in serum: Method development and interlaboratory validation. Clin Chem 1990;36:1951.

65. Sampson EJ, Baird MA, Burtis CA, et al. A coupled-enzyme equilibrium method for measuring urea in serum: Optimization and evaluation of the AACC Study Group on urea candidate reference method. Clin Chem 1980;26:816.

66. Scott P, Maguire GA. A kinetic assay for urea in undiluted urine specimens. Clin Chem 1990;36:1830.

67. Sena SF, Syed D, Romeo R, Krzymowski GA, McCoumbs RB. Lidocaine metabolite and creatinine measurements in the Ektachem 700: Steps to minimize its impact on patient care. Clin Chem 1988;34:2144.

68. Shemesh O, Golbetz H, Kriss JP, Myers BD. Limitations of creatinine as a filtration marker in glomerulopathic patients. Kidney Int 1985;28:830.

69. Shorey J, McCandless MW, Schenker S. Cerebral alpha-ketoglutarate in ammonia intoxication. Gastroenterology 1967;53:706.

70. Soldin SJ, Hill GJ. Micromethod for determination of creatinine in biological fluids by high-performance liquid chromatography. Clin Chem 1978;24:747.

71. Soldin SJ, Henderson L, Hill GJ. The effect of bilirubin and ketones on reaction rate methods for the measurement of creatinine. Clin Biochem 1978;11:82.

72. Swain RR, Briggs SL. Positive interference with the Jaffe reaction by cephalosporin antibiotics. Clin Chem 1977;23:1340.

73. Tanaka M, Hama M. Improved rapid assay of uric acid in serum by liquid chromatography. Clin Chem 1988;34:2567.

74. Tanganelli E, Prencipe L, Bassi D, et al. Enzymatic assay of creatinine in serum and urine with creatinine iminohydrolase and glutamate dehydrogenase. Clin Chem 1982;28:1461.

75. Thompson H, Rechnitz GA. Ion electrode based enzymated analysis of creatinine. Anal Chem 1974;46:246.

76. Tietz N. Fundamentals of clinical chemistry. 3rd ed. Philadelphia: Saunders, 1987.

77. Tietz N. Clinical guide to laboratory tests. Philadelphia: Saunders, 1983;152.

78. Tiffany TO, Jansen JM, Burtis CA, et al. Enzymatic kinetic rate and endpoint analysis of substrate by use of GEMSAEC fast analyzer. Clin Chem 1972;18:829.

79. Trivedi R, Berta E, Rebar L, Desai K, Stong LJ. New ultraviolet method for assay of uric acid in serum or plasma. Clin Chem 1978;24:562.

80. van Anken HC, Schiphorst ME. A kinetic determination of ammonia in plasma. Clin Chem Acta 1974;56:151.

81. vanLente F, Suit P. Assessment of renal functions by serum creatinine and creatinine clearance: Glomerular filtration rate estimated by four procedures. Clin Chem 1989;35:2326.

82. vanLente F. Creatinine. Clin Chem News Oct 1990;8.

83. Wahlefield AW, Herz G, Bergmeyer HU. A completely enzymatic determination of creatinine in human sera or urine. [Abstract]. Scand J Clin Lab Invest 1972;29(Suppl 126).

84. Wahlefeld AW, Siedel J. Creatine and creatinine. In: Bergmeyer HU. Methods of enzymatic analysis. 3rd ed. Vol VIII. New York: Academic Press, 1985;488.

85. Watkins PJ. The effects of ketone bodies on the determination of creatinine. Clin Chem Acta 1967;18:191.

86. Weber JA, van Zanten AP. Interferences in current methods for measurements of creatinine. Clin Chem 1991;37:695.

87. Welch MJ, Cohen A, Hertz HS, et al. Determination of serum urea by isotope dilution mass spectrometry as a candidate definitive method. Anal Chem 1984;56:713.

88. Willems D, Steenssens W. Ammonia determination in plasma with a selective electrode. Clin Chem 1988;34:2372.

89. Young DS, Pestaner LC, Gibberman V. Effects of drugs on clinical laboratory tests. Clin Chem 1975;21:5.

Part 3

Clinical Correlations of Organ Systems

Chapter 17

Renal Function

Suzanne Elle-Cartledge, Alan H. B. Wu

Renal Anatomy
Renal Physiology
Glomerular Function
Tubular Function
Nonprotein Nitrogenous (NPN)
    Metabolite Elimination
Water, Electrolyte, and Acid-Base
    Homeostasis

Endocrine Function
Analytical Procedures
Clearance Measurement
Urine Electrophoresis
Beta$_2$-Microglobulin
Myoglobin
Pathophysiology
Glomerular Diseases

Tubular Diseases
Urinary Tract
    Infection/Obstruction
Renal Calculi
Renal Failure
Summary
Case Studies
References

Objectives

Upon completion of this chapter, the clinical laboratorian should be able to:

- *Diagram the anatomy of the nephron.*
- *Describe the physiologic role of each of the following: nephron, glomerulus, proximal tubule, loop of Henle, distal tubule, and collecting duct.*
- *In conjunction with hormones, describe the mechanisms by which the kidney maintains fluid and electrolyte balance.*
- *Discuss the concept of renal clearance and how it is measured.*
- *Distinguish between the glomerular filtration rate from the plasma renal flow.*
- *List the tests in a urinalysis and microscopy profile and understand the clinical significance of each.*
- *Relate the pathophysiologic role of total urine proteins, urine albumin, serum β_2-microglobulin, and the myoglobin clearance rate.*
- *Describe various diseases of the glomerulus and tubules and how laboratory tests are used in these disorders.*
- *Distinguish between acute and chronic renal failure.*
- *Explain the complications of renal failure and the goals for laboratory monitoring.*
- *Discuss the therapy of chronic renal failure with regard to renal dialysis and transplantation.*

KEY WORDS

Acute Renal Failure	Diabetes Mellitus	Hemofiltration
Allograft	Glomerular	Inulin
Chronic Renal	Filtration Rate	Nephrotic
Failure	Glomerulonephritis	Syndrome
Creatinine	Glomerulus	Plasma Renal Flow
Clearance	Graft	Tubules
Cyclosporin	Hemodialysis	Urinalysis

RENAL ANATOMY

The kidneys are paired, bean-shaped organs located retroperitoneally on either side of the spinal column. Macroscopically, each kidney is enclosed by a fibrous capsule of connective tissue. When dissected longitudinally two regions can be clearly discerned—an outer region called the cortex and an inner region called the medulla (Fig. 17-1*A*). The pelvis can also be seen. It is a basin-like cavity at the upper end of the ureter into which newly formed urine passes. The bilateral ureters are thick-walled canals connecting the kidneys to the urinary bladder. Urine is temporarily stored in the bladder until voided from the body via the urethra. Figures 17-1*B–E* show the arrangement of nephrons in the kidney. These are the functional units of the kidney and can only be seen microscopically. Each kidney contains approximately 1 million nephrons. Each nephron is a complex apparatus composed of six basic parts expressed diagrammatically in Figure 17-2:

1. The glomerulus—located in the cortex and derived from the Latin word *glomus* meaning ball of yarn. It is formed by a pair of afferent and efferent arterioles almost totally surrounded by the expanded end of a renal tubule, known as *Bowman's capsule.*
2. The proximal convoluted tubule—located in the cortex.
3. The thin descending limb of the loop of Henle—spans the medulla from the corticomedullary junction to the inner medulla.
4. The ascending limb of the loop of Henle—located in both the medulla and the cortex and composed of a thin, then thick, region.
5. The distal convoluted tubule—located in the cortex.
6. The collecting duct—formed by two or more distal convoluted tubules as they pass back down through the

Michael L. Bishop, Janet L. Duben-Engelkirk, and Edward P. Fody.
CLINICAL CHEMISTRY. © 1996 Lippincott-Raven Publishers.

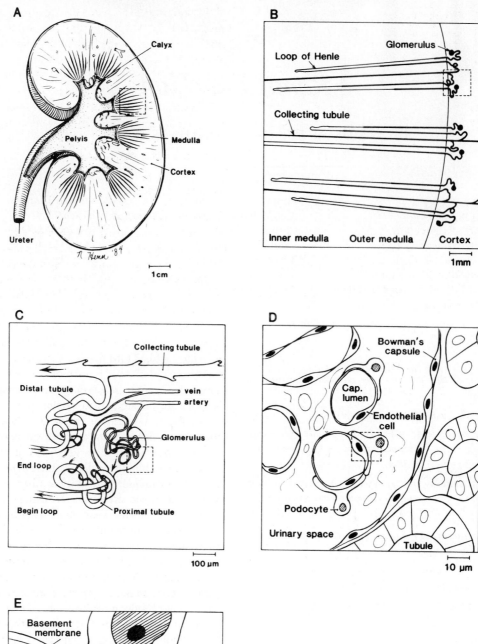

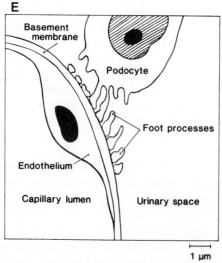

Figure 17-1. Successive magnifications of kidney by powers of ten.

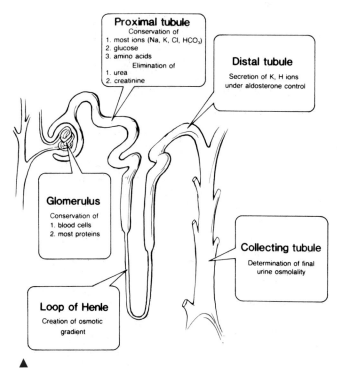

Figure 17-2. The nephron shown in diagrammatic form.

cortex and the medulla to collect the urine that drains from each nephron. Collecting ducts eventually merge and empty their contents into the renal pelvis.

RENAL PHYSIOLOGY

The kidneys perform a variety of functions for the body, the most prominent probably being the removal from plasma of unwanted substances (both waste and surplus), homeostasis (the maintenance of equilibrium) of the body's water, electrolyte and acid/base status, and participation in endocrine regulation. The following sections will consider how each of the five parts of the nephron normally functions.

Glomerular Function

The glomerulus is the first part of the nephron to receive incoming blood and functions to filter this blood. Every substance except cells and large molecules continues into further sections of the nephron. Several factors facilitate filtration. One is the unusually high pressure in the glomerular capillaries, which is due to their position between two arterioles. This sets up a steep pressure difference across the walls. Another factor is the semipermeable glomerular basement membrane, which has a molecular size cut-off value of approximately 66,000 Daltons, about the molecular size of albumin. This means that water, electrolytes, and small dissolved solutes such as glucose, amino acids, urea, and creatinine pass freely through and enter the proximal convoluted tubule but albumin, many plasma proteins, cellular elements, and protein-bound substances, such as lipids and bilirubin, are

stopped. A third factor is that the basement membrane is negatively charged and this contributes to the filtration by repelling negatively charged molecules, such as proteins. Thus, of the 1200 to 1500 mL of blood that the kidneys receive each minute (approximately one-quarter of the total cardiac output), the glomerulus filters out 125 to 130 mL of essentially protein-free, cell-free fluid. The volume of blood filtered per minute is known as the *glomerular filtration rate* (*GFR*) and its determination is essential in evaluating renal function, as discussed in the section on Analytical Procedures.

Tubular Function

Proximal Convoluted Tubule

The proximal tubule is the next part of the nephron to receive the now cell- and essentially protein-free blood. This filtrate contains both waste products, which are toxic to the body above a certain concentration, and substances that are valuable to the body. One of the functions of the proximal tubule is to return the bulk of each valuable substance back to the blood circulation. Thus, three-quarters of the water, sodium, and chloride, all of the glucose (up to the renal threshold), almost all of the amino acids, vitamins, and proteins, and varying amounts of ions such as magnesium, calcium, potassium, and bicarbonate are reabsorbed. With the exception of water and chloride ions the process is active; that is, the tubular epithelial cells utilize energy to transport the substances across their plasma membranes thence to the blood. The transport processes that are involved normally have sufficient reserve for efficient reabsorption, but they are saturable, *i.e.,* there is a concentration for each substance in the filtrate above which the relevant transport system cannot function fast enough to bind and therefore remove the substance from the filtrate. The substance is excreted in the urine. The plasma concentration above which the substance appears in urine is known as the *renal threshold* and its determination is useful in assessing both tubular function and non-renal disease states. A renal threshold for water does not exist, since it is always transported passively through diffusion down a concentration gradient. Chloride ion in this instance diffuses in the wake of sodium.

A second function of the proximal tubule is to secrete products of kidney tubular cell metabolism, such as hydrogen ions, and drugs such as penicillin. Transport across the membrane of the cell is again either active or passive.

Almost all (98–100%) of uric acid, a waste product, is reabsorbed actively, only to be secreted at the distal end of the proximal tubule. Urea, another waste product, is a highly diffusible molecule and passively passes out of the renal tubule and into the interstitium where it is contributes to the osmolality gradient found in the medulla.

Loop of Henle

In this portion of the nephron, the osmolality in the medulla, increasing steadily from the corticomedullary junction inwards, facilitates the reabsorption of water, sodium, and chloride. Thus, the thin descending limb is increasingly hypo-osmolal compared to the surrounding interstitium. Water diffuses out of the lumen to relieve the osmolal difference. A concentrated filtrate results. The situation is reversed in the ascending limb. The filtrate is hyperosmolal compared to the interstitium and the tubular cells are impermeable to water. So, in order to redress the imbalance, sodium and chloride diffuse out of the tubular lumen in the thin region and chloride is actively reabsorbed from the lumen in the thick portion of the limb with sodium passively following. This accounts for about 20% of the sodium and chloride and 15% of water. Urea readily diffuses into the lumen of the thin descending limb and out of the lumen of the thin ascending limb of Henle.

Distal Convoluted Tubule

The filtrate entering this penultimate section of the nephron is close to its final composition. About 95% of the sodium chloride and 90% of water have already been reabsorbed from the original glomerular filtrate. The function of the distal tubule is to effect small adjustments to achieve electrolyte and acid/base homeostasis. It is under the hormonal control of aldosterone. This hormone is produced by the adrenal cortex and its secretion is triggered by decreased blood flow in the afferent renal arteriole. It is regulated by the renin-angiotensin mechanism and to a lesser extent by ACTH. Aldosterone stimulates sodium reabsorption by the distal tubule and potassium and hydrogen ion secretion. Hydrogen ion secretion is linked to bicarbonate regeneration and ammonia secretion, which occur here. In addition to these, small amounts of chloride are reabsorbed.

Collecting Duct

The upper portion of the collecting duct is still under the control of aldosterone, so, again, sodium is reabsorbed. Chloride is also reabsorbed here, as is urea. In addition, the collecting duct is under the control of antidiuretic hormone (ADH). This peptide hormone is secreted by the posterior pituitary in response to neural impulses triggered mainly by increased blood osmolality or decreased intravascular volume. It has an extremely short half-life (on the order of minutes). ADH stimulates water reabsorption. The walls of the collecting duct are normally impermeable to water (like the ascending loop of Henle), but in the presence of ADH they become permeable to water, which diffuses passively from the lumen of the collecting duct to the medulla, resulting in a more concentrated urine.

Nonprotein Nitrogenous (NPN) Metabolite Elimination

Nonprotein nitrogenous (NPN) waste products are formed in the body as a result of the degradative metabolism of nucleic and amino acids and proteins. Excretion of these compounds is an important function of the kidneys. The three principal substances are urea, creatinine, and uric acid.[9,10] For a more detailed treatment of their biochemistry and disease correlations, see Chapter 16, Nonprotein Nitrogen.

Urea

Urea makes up the majority (over 75%) of the NPN wastes excreted daily as a result of the oxidative catabolism of proteins. Proteins are broken down to amino acids, which are "detoxified" by removal of the nitrogen atoms. Ammonia is formed and readily converted to urea, thereby avoiding toxicity. The kidney is the only significant route of excretion for urea. It has a molecular weight of 60 and, therefore, is readily filtered by the glomerulus. Instead of being excreted wholly, 40% to 60% is reabsorbed by the collecting duct. The reabsorbed urea contributes to the high osmolality in the medulla, which is one of the processes of urinary concentration mentioned earlier (see Loop of Henle). The amount of reabsorption depends on the GFR, the plasma renal flow, and the urine flow rate.

Creatinine

Muscle contains creatine phosphate, which is a reservoir of high-energy phosphoryl groups for the rapid formation of ATP as catalyzed by creatine kinase. It is the first source of metabolic fuel that the muscle uses. It is formed from creatine as shown below.

$$creatine\ phosphate + ADP + H^+ \longrightarrow$$
$$creatine + ATP \xrightarrow{nonenzymatic} creatinine$$

(Eq. 17-1)

Every day, up to 20% of total muscle creatine (and its phosphate) spontaneously dehydrates and cyclizes to form the waste product creatinine. Thus, creatinine levels are a function of muscle mass and remain approximately the same from day to day unless muscle mass changes. Creatinine is inert and has a molecular weight of 113. It is thus readily filtered by the glomerulus and unlike urea is not reabsorbed by the tubules. However, a small amount is secreted by the kidney tubules at high serum concentrations.

Uric Acid

Uric acid is the primary waste product of purine metabolism. The purines adenine and guanine are precursors of nucleic acids ATP and GTP, respectively. Uric acid has a molecular weight of 168. Thus, like creatinine, it is readily filtered by the glomerulus but then it goes through a complex cycle of reabsorption and secretion as it courses through the nephron. Only 6% to 12% from the original filtrate is finally excreted. Uric acid exists in its ionized and more soluble form, usually sodium urate, at urinary pH values above 5.75 (the first pK_a of uric acid). At pH values below 5.75, it is undissociated. This fact has great clinical significance in the development of urolithiasis and gout.

Water, Electrolyte, and Acid-Base Homeostasis

Water Balance

The kidney's contribution to water balance in the body is through water loss. It is the dominant organ in this regard. The water loss is under the hormonal control of ADH.

ADH responds primarily to changes in osmolality and intravascular volume. Therefore, increased osmolality stimulates secretion of ADH, which increases the permeability of the collecting ducts to water resulting in water reabsorption and the excretion of a more concentrated urine. Decreased intravascular volume will have a similar effect. In contrast, the major system regulating water intake is thirst, which appears to be triggered by the same stimuli that trigger ADH.

In states of dehydration, the renal tubules reabsorb water at their maximal rate, resulting in production of a small amount of maximally concentrated urine (high urine osmolality,[3] 1200 mOsmol/kg). In states of water surplus, the tubules reabsorb water at only a minimal rate, resulting in excretion of a large volume of extremely dilute urine (low urine osmolality, down to 50 mOsmol/kg).[2,7,8] The continuous fine-tuning possible between these two extreme states results in the precise control of fluid balance in the body (Fig. 17-3).

Ionic Equilibria

The following description is a brief overview of the notable ions involved in maintenance of ionic equilibria within the body. For a more comprehensive treatment of this subject, refer to Chapter 12, Electrolytes.

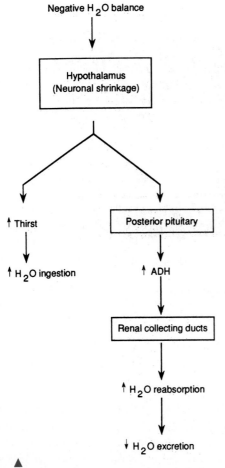

Figure 17-3. ADH control of thirst mechanism.

Sodium. Sodium is the primary extracellular cation in the human body and is excreted principally through the kidneys. Sodium balance in the body is only controlled via excretion. There appears to be no control over intake. The renin-angiotensin-aldosterone hormonal system is the major mechanism for control of sodium balance.

Potassium. Potassium is the main intracellular cation in the body. The precise regulation of its concentration is of overriding importance to cellular metabolism and is controlled chiefly by renal means. Like sodium, it is freely filtered by the glomerulus and then actively reabsorbed throughout the entire nephron (except for the descending limb of the loop of Henle). Both the distal convoluted tubule and the collecting ducts can reabsorb and excrete potassium, and this excretion is controlled by aldosterone. Potassium ions can compete with hydrogen ions in their exchange with sodium (in the proximal convoluted tubule), and this process is utilized by the body to conserve hydrogen ions and thereby compensate in states of metabolic alkalosis (Fig. 17-4).

Chloride. Chloride is the principal extracellular anion and is involved in the maintenance of extracellular fluid balance. It is readily filtered by the glomerulus and is passively reabsorbed as a counter ion when sodium is reabsorbed in the proximal convoluted tubule. In the ascending limb of the loop of Henle, potassium is actively reabsorbed by a distinct chloride "pump," which also reabsorbs sodium. This pump can be inhibited by loop diuretics such as furosemide. As expected, the regulation of chloride is controlled by the same forces that regulate sodium.[3,8]

Phosphate, Calcium, and Magnesium. The phosphate ion occurs in approximately equal concentrations in both the intracellular and extracellular fluid environments. It exists as either a protein-bound or a non-protein-bound form; homeostatic balance is chiefly determined by proximal tubular reabsorption under the control of parathyroid hormone (PTH). Calcium, the second-most predominant intracellular cation, is the most important inorganic messenger in the cell. It also exists in both protein-bound and non-protein-bound states. The latter form is either ionized and physiologically active or nonionized and complexed to small, diffusible ligands (*e.g.,* phosphate and bicarbonate). The ionized form is freely filtered by the glomerulus and reabsorbed in the proximal tubule under the control of PTH. However, renal control of calcium concentration is not the major means of regulation. PTH- and calcitonin-controlled regulation of calcium absorption from the gut and bone stores is more important than renal secretion. Magnesium, a major intracellular cation, is important as an enzymatic cofactor. Like phosphate and calcium, it exists in both protein-bound and non-protein-bound states. The latter fraction is easily filtered by the glomerulus and reabsorbed in the proximal convoluted tubule under the influence of PTH.

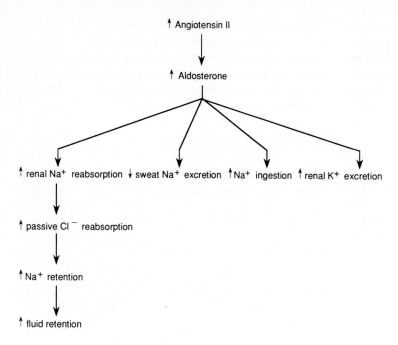

Figure 17-4. Electrolyte regulation by aldosterone.

Acid-Base Equilibria

Many nonvolatile acidic waste products are formed by normal body metabolism each day. Carbonic acid, lactic acid, ketoacids, and others must be continually transported in the plasma and excreted from the body while causing only minor alterations in the physiologic pH. The renal system constitutes one of three means by which constant control of overall body pH is accomplished. The other two strategies involved in this regulation are the respiratory system and the acid-base buffering system.[12]

The kidneys manage their share of the responsibility for controlling body pH by dual means: conserving bicarbonate ions and removing metabolic acids. For a more in-depth examination of these processes, refer to Chapter 11, Blood Gases and Buffer Systems.

Regeneration of Bicarbonate Ions. In a complicated process, bicarbonate ions are first filtered out of the plasma by the glomerulus. Once in the lumen of the renal tubules, this bicarbonate combines with hydrogen ions to form carbonic acid, which subsequently degrades to carbon dioxide (CO_2) and water. This CO_2 then diffuses into the brush border of the proximal tubular cells, where it is reconverted by carbonic anhydrase to carbonic acid and then degrades back to hydrogen ions and regenerated bicarbonate ions. This reaction is detailed below:

$$H_2O + CO_2 \underset{CA}{\longleftrightarrow} H_2CO_3 \longleftrightarrow H^+ \; HCO_3^-$$

(Eq. 17-2)

This regenerated bicarbonate is transported into the blood to replace that which was depleted by metabolism;

the accompanying hydrogen ions are secreted back into the tubular lumen and from there enter the urine. Thus, filtered bicarbonate is "reabsorbed" into the circulation, thereby helping to return blood pH to its optimal level and effectively functioning as another buffering system.

Excretion of Metabolic Acids. Hydrogen ions are manufactured in the renal tubules as part of the regeneration mechanism for bicarbonate. These hydrogen ions, as well as others that are dissociated from nonvolatile organic acids, are disposed of by means of several different reactions with buffer bases.

Reaction with Ammonia (NH_3). The glomerulus does not filter NH_3. However, this substance is formed in the renal tubules when the amino acid glutamine, which serves to transport ammonia in a nontoxic form throughout the circulation to the kidneys, is deaminated by glutaminase. This NH_3 then reacts with secreted hydrogen ions to form ammonium (NH_4^+), which is unable to readily diffuse out of the tubular lumen and is therefore excreted into the urine.

$$\text{Glutamine} \xrightarrow{\text{Glutaminase}} \text{glutamic acid} + NH_3$$

$$NH_3 + H^+ + NaCl \longrightarrow NH_4Cl + Na^+$$

(Eq. 17-3)

This mode of acid excretion is the primary means by which the kidneys compensate for states of metabolic acidosis.

Reaction with Monohydrogen Phosphate (HPO_4^{2-}). Phosphate ions filtered by the glomerulus can exist in the tubular fluid as disodium hydrogen phosphate. This moiety can react with extant hydrogen ions to yield dihydrogen phosphate, which is then excreted and is responsible for the "titratable acidity" of the urine. The released sodium then combines with bicarbonate to yield sodium bicarbonate and is reabsorbed.

$$Na_2HPO_4 + H^+ \longleftrightarrow NaH_2PO_4 + Na^+$$

(Eq. 17-4)

These mechanisms can excrete increasing amounts of metabolic acid until a maximum urine pH of approximately 4.4 is reached. After this, the renal compensatory apparatus is unable to adjust to any further decreases in blood pH, and metabolic acidosis must ensue. Very few free hydrogen ions are excreted directly.

Endocrine Function

In addition to its numerous excretory and regulatory functions, the kidney has endocrine responsibilities as well. It is both a primary endocrine site, as the producer of its own hormones, and a secondary site, as the target locus for hormones manufactured by other endocrine organs.

Primary Endocrine Function

The kidneys synthesize renin, the prostaglandins, and erythropoietin.

Renin. Renin is the initial component member of the renin-angiotensin-aldosterone feedback system. It serves to catalyze the synthesis of angiotensin by means of cleavage of the circulating plasma precursor angiotensinogen. Renin is produced by the juxtaglomerular cells of the renal medulla whenever extracellular fluid volume decreases. It serves as a vasoconstrictor to increase blood pressure and is responsive to changes in both the sodium and potassium levels in the blood.[2,8] For a more rigorous look at the complexities of this feedback loop, see Chapter 19, Endocrinology.

Prostaglandins. The prostaglandins are a group of potent cyclic fatty acids formed from essential (dietary) fatty acids, primarily the unsaturated arachidonic acid. They are formed in almost all tissues and their actions are diverse. They behave like hormones but differ in that they are synthesized at the site of action. Once formed, prostaglandins exert a very short-lived effect and are rapidly catabolized. Prostaglandins can be either PG1 or PG2. The latter type is more common. The prostaglandins produced by the kidneys such as PGA2, PGE2, PGD2, and PG12 increase renal blood flow, sodium and water excretion, and renin release. They act to oppose renal vasoconstriction due to angiotensin and norepinephrine. Angiotensin II is believed to stimulate PGE2. Prostaglandins have been used in antihypertensive therapy. Arachidonic acid also gives rise to *leukotrienes* and *thromboxanes* whose actions, although less well documented, have also been associated with diverse biological activity such as inflammation and bronchoconstriction.

Erythropoietin. Erythropoietin is a single-chain polypeptide. It is produced by cells close to the proximal tubules and its production is regulated by blood oxygen levels. Thus, hypoxia produces increased serum concentrations within 2 hours. Erythropoietin acts on the erythroid progenitor cells in the bone marrow causing their maturation and increasing the number of red blood cells. In chronic renal insufficiency, erythropoietin production is significantly reduced. Recently, recombinant human erythropoietin has been developed and is now in routine use in chronic renal failure patients. Prior to this therapy, anemia in these patients was a clinical reality.[2,9] Erythropoietin concentrations in blood can be measured by an enzyme-linked immunoassay.

Secondary Endocrine Function

The kidneys are the target locus for the action of aldosterone; for the catabolism of insulin, glucagon, and aldosterone; and as the point of activation for vitamin D metabolism.[3,9] Vitamin D is one of the three major hormones that determine phosphate and calcium balance and bone calcification in the human body. For a more detailed treatment of this bioactivation, refer to Chapter 29, Vitamins. Chronic renal insufficiency is therefore often associated with *osteomalacia* (inadequate bone calcification, the adult form of rickets) owing to the continual distortion of normal vitamin D metabolism.

ANALYTICAL PROCEDURES

There are several tests available that assess the various aspects of nephron function including glomerular filtration, renal plasma flow, and proximal and distal tubular secretion and reabsorption.

Clearance Measurement

All laboratory methods used for evaluation of renal function rely on the measurement in blood of a waste product, usually the nitrogenous substances urea and creatinine, that accumulates once the kidneys begin to fail. However, before the concentration of either of these substances begins to rise in the blood, renal failure must be quite advanced, with only about 20% to 30% of the nephrons still functioning. Measurement of clearance affords the clinical chemist the opportunity to detect much earlier stages of the disease. A summary of these clearance assays is shown in Table 17-1.

Creatinine

Creatinine is a nearly ideal substance for the measurement of clearance for various reasons: it is an endogenous metabolic product synthesized at a constant rate for a given individual, it is cleared essentially only by glomerular filtration (it is not reabsorbed and is only slightly secreted by the proximal tubule), and it can be analyzed inexpensively by readily available colorimetric methods. As a result, creatinine clearance has become the standard laboratory assay for de-

TABLE 17-1. Summary of Clearance Assays

Clearance	Renal Handling	Source
1. Creatinine	Filtered by glomerulus; no reabsorption; slight secretion	Endogenous
2. Inulin	Filtered by glomerulus; no reabsorption; no secretion	Exogenous
3. Na[^{125}I]iothalamate	Filtered by glomerulus; no reabsorption; no secretion	Exogenous

termination of early renal failure. This value is derived by arithmetically relating the total urine creatinine concentration to the total serum creatinine concentration excreted during a 24-hour period.

Specimen collection, therefore, must include both a 24-hour urine specimen and a serum creatinine value (ideally taken at the midpoint of the 24-hour urine collection and, realistically, no later than 24 hours *before* or *after* the urine collection). The urine container (clean, dry, and free of contaminants or preservatives) must be kept refrigerated throughout the duration of both the collection procedure and the subsequent storage period until laboratory analyses can be performed. The patient must maintain an adequate fluid intake throughout the procedure in order to ensure a minimum urine flow rate of at least 2 mL/min. This cancels out the possibility of negative error due to urine retention in the bladder. Additionally, coffee, tea, heavy exercise, medications, and recreational drugs should be avoided in order to minimize possible interferences.

Once the specimens have been collected, the concentration of creatinine in both serum and urine is measured by any of the applicable methods discussed in Chapter 16, Nonprotein Nitrogen. The total volume of urine is carefully measured, and the clearance is calculated by the following formula:

$$C_{Cr} \text{ (mL/min)} = \frac{U_{Cr} \text{ (mg/dL)} \times V_{Ur} \text{ (mL/24 h)}}{P_{Cr} \text{ (mg/dL)} \times 1440 \text{ min/24 h}} \times \frac{1.73}{A}$$

(Eq. 17-5)

where as C_{Cr} = creatinine clearance
U_{Cr} = urine creatinine concentration
V_{Ur} = urine volume excreted in 24 hours
P_{Cr} = serum creatinine concentration
$1.73/A$ = normalization factor for body surface area

Normally, the urine output of a healthy individual averages 1500 mL/24 h; since this total volume is only about 1% of the glomerular filtrate formed each day, a large volume is reabsorbed elsewhere along the nephron.[2,9]

Most clinical chemists use the shorter, more easily remembered notation for the general clearance formula: UV/P. Correction for body surface area must be included in the formula, since creatinine excretion varies with regard to lean body mass. In this factor, 1.73 is the generally accepted average body surface in square meters. If the patient's body surface area varies greatly from the average (*e.g.,* in obese or pediatric

patients), this correction prevents the accumulation of up to several hundred percent error. Nomograms for the more exact determination of body surface area from weight and height values can be found in Appendix J, Nomogram for the Determination of Body Surface Area. Reference ranges for creatinine clearance are 97 to 137 mL/min/1.73 m^2 and 88 to 128 mL/min/1.73 m^2 for males and females, respectively. Creatinine clearance normally decreases with age, such that each decade of life accounts for a decrease of about 6.5 mL/min/1.73 m^2. Specific methodologies for the measurement of creatinine in serum and urine are given in Chapter 16, Nonprotein Nitrogen.

Inulin

Inulin is an exogenous plant polysaccharide derived from artichokes and dahlias. It is completely filtered by the glomeruli and neither secreted nor reabsorbed by the tubules. Inulin is the most accurate of all GFR assays, but as a nonendogenous substance, it requires intravenous infusion under constant medical surveillance. It is therefore unwieldy, time-consuming, expensive, and uncomfortable to the patient and so is not clinically practical to use. In addition, there is only a limited availability of methods for its quantitation. Of the GFR assays, creatinine behaves most closely like inulin.[2,9]

Sodium [^{125}I]Iothalamate

This substance is a radioactively labeled filtration marker that is simpler to administer than inulin. On the night before the test, the patient ingests an iodine solution to saturate the thyroid gland. The patient is then fully hydrated and kept that way throughout the procedure. Baseline samples of blood and urine are collected, and then the [^{125}I]iothalamate and epinephrine are injected subcutaneously. After an equilibration period of 1 hour, blood and urine specimens are collected again, and the urine is discarded. Subsequently, three consecutive matched 2-hour urine and blood specimens are obtained. A scintillation counter is then used to measure radioactive counts in all the samples. The same formula, UV/P, is used for the clearance calculations, except that the U value is minus the control urine counts and the P value is minus the control serum counts.[14]

Para-Aminohippurate (PAH)

Para-Aminohippurate (PAH) is a nontoxic, nonprotein-bound exogenous compound. It is eliminated with over

90% efficiency by glomerular filtration *and* tubular secretion in one pass through the renal system. It is therefore used to estimate renal secretory capability or plasma renal flow (PRF). However, as a nonendogenous substance, like inulin, it is time-consuming and difficult to administer. Usually, renal biopsies are performed to assess tubular damage rather than PAH clearance.

Tubular secretory capacity can be estimated by measuring the clearance of a substance that is freely filtered by the glomerulus and readily secreted from the tubules on the very first pass through the kidneys. *Para*-Aminohippurate (PAH) is such a substance; over 90% is removed in the first pass, and it is therefore used to estimate PRF. The secretory function of the kidneys is determined by the difference between the amount of PAH excreted and the amount filtered by the glomerulus. As will be discussed later, the GFR itself is best measured by inulin clearance.

Urinalysis (UA)

Urinalysis (UA) is an underutilized tool for the overall evaluation of renal function. UA permits a detailed, in-depth assessment of renal status with an easily obtained specimen. Urinalysis also serves as a quick indicator of an individual's glucose status and hepatic-biliary function. Routine UA consists of a group of screening tests generally performed as part of a patient's admission work-up or physical examination. It includes assessment of physical characteristics, chemical analyses, and a microscopic examination of the sediment from a (random) urine specimen.

Physical Characteristics

Specimen Collection. The importance of a properly collected and stored specimen for UA cannot be overemphasized. Initial morning specimens are preferred, particularly for protein analyses, since they are more concentrated from overnight retention in the bladder. The specimen should be obtained by a clean midstream catch or catheterization. The urine should be freshly collected into a clean, dry container with a tight-fitting cover. It must be analyzed within 1 hour of collection if held at room temperature or else refrigerated at 2° to 8°C for not more than 8 hours before analysis. If not assayed within these time limits, several changes will occur. Bacterial multiplication will cause false-positive nitrite tests, and urease-producing organisms will degrade urea to ammonia and alkalinize the pH. Loss of CO_2 by diffusion into the air adds to this pH elevation, which in turn causes cast degeneration and red-cell lysis.

The urine container must be sterile if the urine is to be cultured. For children who are not yet toilet trained, specially designed polyethylene plastic bags with adhesive backing are available to fit around the perineum so that the urine is voided directly into the bag. Specimens for routine UA are usually random, or spot, collections. Fasting and postprandial specimens are used for glucose determinations. Timed urine specimens (for clearances, etc.) should always be stored in refrigerated containers.[11]

Visual Appearance. Color intensity of urine correlates with concentration: the darker the color, the more concentrated is the specimen. The various colors observed in urine are due to different excreted pigments. Yellow and amber are generally due to urochromes (derivatives of urobilin, the end product of bilirubin degradation), while yellowish brown to green are due to bile pigment oxidation. Red and brown after standing are due to porphyrins, while reddish brown in fresh specimens comes from hemoglobin/red cells. Brownish black after standing is seen in *alkaptonuria* (due to excreted homogentisic acid) and in malignant melanoma (in which the precursor melanogen oxidizes in the air to melanin). Drugs also may affect urine color; methylene blue can turn urine blue, while the analgesic dye pyridium stains urine a bright orange. Some foods, such as beets, also may alter urine color.

Odor. Odor ordinarily has little diagnostic significance. The characteristic pungent odor of fresh urine is due to volatile aromatic acids, in contrast to the typical ammoniacal odor of urine that has been allowed to stand. Urinary tract infections impart a noxious, fecal smell to urine, while the urine of diabetics often smells fruity due to the presence of ketones.

Turbidity. The cloudiness of a urine specimen depends on both its pH and its dissolved solids composition. Turbidity in general may be due to gross bacteriuria, while a smoky appearance is seen in hematuria. Threadlike cloudiness is observed when the specimen is full of mucus. In alkaline urine, suspended precipitates of amorphous phosphates and carbonates may be responsible for turbidity, while in acidic urine, amorphous urates may be the cause.[11]

Volume. The volume of urine excreted is indicative of the balance between fluid ingestion and water lost from the lungs, sweat, and intestines. Most adults produce from 750 to 2000 mL/24 h, averaging about 1.5 L per person. (For a routine UA, a 10- to 12-mL aliquot from a well-mixed sample is optimal for accurate analysis of sedimentary constituents.) Polyuria is observed in cases of diabetes mellitus and insipidus (in the latter case due to the lack of ADH), as well as in chronic renal disease, *acromegaly* (overproduction of the growth hormone somatostatin), and *myxedema* (hypothyroid edema). Anuria and/or *oliguria* (<200 mL/day) is found in nephritis, end-stage renal disease, urinary tract obstruction, and acute renal failure.

Specific Gravity (SG). The specific gravity of urine is defined as the weight of the urine divided by the weight of the water standard (1.000). SG gives an indication of the density of a fluid, depending on the concentration of dissolved total solids, and therefore measures a property similar to that measured by osmolality. For a more detailed discussion of urine osmolality, refer to Chapter 12, Electrolytes. SG varies with the solute load to be excreted (consisting primarily of NaCl and urea), as well as with the urine volume. As such, it can be used as a marker for the amount of hydration/dehydration of an individual.

Laboratory Methods. The analytical method most commonly encountered today consists of a refractometer, or total solids meter. This operates on the principle that the refractive index (RI) of a urine specimen will vary directly with the total amount of dissolved solids in the sample. Thus, this instrument, as typified by the Goldberg refractometer, measures the RI of the urine as compared with water on a scale that is calibrated directly into the ocular and viewed while held up to a light source. Correct calibration is vital for accuracy. Most recently, an indirect colorimetric reagent strip method for assaying SG has been added to the dipstick screen by one manufacturer, but is not yet more widely used than the refractometry method.

Disease Correlation. Normal range for urinary SG is 1.005 to 1.030. Dilute specimens are classified in the range of 1.000 to 1.010, while concentrated samples fall between 1.025 and 1.030. SG can vary in pathologic states, as discussed below. Low SG can occur in diabetes insipidus, where it may never exceed the range of 1.001 to 1.003, and in pyelonephritis and glomerulonephritis, in which the renal concentrating ability has become dysfunctional. High SG can be seen in diabetes mellitus, congestive heart failure, dehydration, adrenal insufficiency, liver disease, and nephrosis. SG will increase about 0.004 units for every 1% change in glucose concentration and about 0.003 units for every 1% change in protein. Fixed SG (*isothenuria*) around 1.010 is observed in severe renal damage, in which the kidney excretes a urine that is iso-osmotic with the plasma. This generally occurs after an initial period of anuria, because the damaged tubules are unable to concentrate/dilute the glomerular filtrate.[7,11]

pH. Determinations of urinary pH *must* be done on fresh specimens, owing to the marked tendency of urine to alkalinize upon standing. Normal urine pH falls within the range of 4.5 to 8.0. Acidity in urine (pH <7.0) is primarily caused by phosphates, which are excreted as salts conjugated to Na^+, K^+, Ca^{2+}, and N_4^+. Acidity also reflects the excretion of the nonvolatile metabolic acids pyruvate, lactate, and citrate. Owing to the Na^+/H^+ exchange pump mechanism of the renal tubules, pH (H^+ ion concentration) increases as sodium is retained. Pathologic states in which increased acidity is observed include systemic acidosis, as seen in diabetes mellitus, and renal tubular acidosis. In the latter condition, the tubules are unable to excrete excess H^+ even though the body is in metabolic acidosis, and urinary pH remains around 6. Some medications such as ammonium chloride and mandelic acid, will also acidify the urine.

Alkaline urine (pH >7.0) is observed postprandially as a normal reaction to the acidity of gastric HCl dumped into the duodenum and thence into the circulation. Urinary tract infections and bacterial contamination also will alkalinize pH, as has been mentioned. Medications such as potassium citrate and sodium bicarbonate, will reduce urine pH. Alkaline urine is also found in *Fanconi's syndrome,* a congenital generalized aminoaciduria due to defective proximal tubular function.

Chemical Analyses

Today, routine urine chemical analysis is rapid and easily performed with commercially available reagent strips or dipsticks. These strips are plastic coated with different reagent bands directed toward different analytes. When dipped into urine a color change signals a deviation from normality. Colors on the dipstick bands are matched against a color chart provided with the reagents. Automated and semiautomated instruments that detect by reflectance photometry provide an alternative to the color chart and offer better precision and standardization. Abnormal results are followed up by specific quantitative or confirmatory urine assays. The analytes routinely tested are glucose, protein, ketones, nitrite, and bilirubin/urobilinogen.

Glucose and Ketones. These constituents are normally absent in urine. The testing for and pathologic significance of detecting these analytes are discussed in Chapter 14, Carbohydrates.

Protein. Reagent strips for UA are used as a general qualitative screen for proteinuria. They are primarily specific for albumin, and they may give false-positive results in specimens that are alkaline and highly buffered. Therefore, positive dipstick results should be confirmed by a more specific acid precipitation test, such as the trichloroacetic acid and precipitation method described in Chapter 9, Amino Acids and Proteins.

Nitrite. This assay semiquantitates the amount of urinary reduction of nitrate (on the reagent strip pad) to nitrite by the enzymes of gram-negative bacteria. This scheme is shown in the following reaction:

$$\text{Nitrite} + p\text{-arsanilic acid} \longleftrightarrow \text{diazonium compound}$$

$$+ \ N\text{-1-naphthylethylenediamine} \longleftrightarrow \text{pink color}$$

(Eq. 17-6)

A negative result does not mean that no bacteriuria is present, however. A gram-positive pathogen, such as the *Staphylo-, Entero-,* or *Streptococci,* may not produce nitrate-reducing enzymes; alternatively, a spot urine sample may not have been retained in the bladder long enough to pick up a sufficient number of organisms to register on the reagent strip.[11]

Bilirubin/Urobilinogen. Hemoglobin degradation ultimately results in the formation of the waste product bilirubin, which is then converted to urobilinogen in the gut through bacterial action. Ordinarily, the majority of this urobilinogen is excreted as urobilin in the feces. A minority, however, is excreted in urine as a colorless waste product. This amount is not sufficient to be detected as a positive dipstick reaction in normal patients. If, however, hepatic or posthepatic jaundice is present (such as occurs in biliary atresia, Crigler–Najjar syndrome type II, and hepatitis), conjugated (water-soluble) bilirubin levels will rise in serum. Consequently, the kidneys will aid in excretion of this excess waste, and urinary bilirubin and urobilinogen levels will increase. A more in-depth view of this metabolic process and these assay methods is given in Chapter 18, Liver Function. Reagent strip tests for bilirubin involve its diazotization to form a brown azo dye. Dipstick methods for urobilinogen differ, but most rely on a modification of the Ehrlich reaction with *p*-dimethylaminobenzaldehyde.[11]

Sediment Examination

A centrifuged, decanted urine aliquot leaves behind a sediment of formed elements that is used for microscopic examination.

Cells. For cellular elements, evaluation is best accomplished by counting and then taking the average of at least 10 microscopic fields.

Red blood cells. Erythrocytes greater in number than 0 to 2/high-powered field (hpf) are considered abnormal. Such hematuria may result simply from severe exercise or menstrual blood contamination. However, it also may be indicative of trauma, particularly vascular injury, renal/urinary calculi obstruction, pyelonephritis, or cystitis; in conjunction with leukocytes, it is diagnostic for infection.

White blood cells. Leukocytes greater in number than 0 to 1/hpf are considered abnormal. These cells are usually polymorphonuclear phagocytes, commonly known as *segmented neutrophils*. They are observed whenever there is acute glomerulonephritis, urinary tract infection, or inflammation of any type. In hypotonic urine (low osmotic concentration), white blood cells can become enlarged, exhibiting a sparkling effect in their cytoplasmic granules. These cells possess a noticeable Brownian motion and are called *glitter cells,* but they have no pathologic significance.

Epithelial cells. Several types of epithelial cells are frequently encountered in normal urine as they are continuously sloughed off the lining of the nephrons and urinary tract. Large, flat, squamous vaginal epithelia are often seen in urine specimens from female patients; samples heavily contaminated with vaginal discharge may show clumps or sheets of these cells. Renal epithelia are round, uninucleate cells; their presence in urine in numbers >2/hpf is a clinically significant marker for active tubular injury/degeneration. Transitional bladder epithelial cells (urothelial cells) may be flat, cuboidal, or columnar and also can be observed in urine on occasion. Large numbers will be seen only in cases of urinary catheterization or bladder inflammation or neoplasia.

Miscellaneous elements. Spermatozoa are often encountered in the urine of both males and females; however, they are usually not reported because they are of no pathologic significance. Yeast cells are also frequently found in urine specimens. Since they are extremely refractile and of a similar size to red blood cells, they can easily be mistaken for them under low magnification. Higher-power examination for the presence of budding or mycelial forms serves to differentiate these fungal elements from erythrocytes. Most yeasts are *Candida albicans* species, and they commonly infect women who are diabetic or on oral contraceptives. They are an indication of the presence of either a urinary or vaginal moniliasis. Parasites found in urine are generally contaminants from fecal or vaginal material. In the former category, the most commonly encountered organism is *Enterobius vermicularis* (pinworm) infestation in children. In the latter category, the most common offender is the intensely motile flagellate *Trichomonas vaginalis*. A true urinary parasite, sometimes seen in patients from endemic areas of the world, is the ova of the trematode *Schistosoma hematobium*. This condition will usually occur in conjunction with a significant hematuria.[11]

Bacteria. Normal urine is sterile and contains no bacteria. Small numbers of organisms seen in a fresh urine specimen usually represent skin or air contamination. In fresh specimens, however, large numbers of organisms, or small numbers accompanied by white blood cells and the symptoms of urinary tract infection, are highly diagnostic for true infection. Clinically significant bacteriuria is considered to be more than 20 organisms/hpf or, alternatively, 10^5 or greater registered on a microbiologic colony count. Most pathogens seen in urine are gram-negative coliforms (microscopic "rods") such as *E. coli* and *Proteus sp.* Asymptomatic bacteriuria, in which there are significant numbers of bacteria without appreciable clinical symptoms, occurs somewhat commonly in young girls, pregnant women, and diabetics. This condition must be taken seriously; if left untreated, it may result in pyelonephritis and, eventually, in permanent renal damage.

Casts. Casts are precipitated, cylindrical impressions of the nephrons. They are composed of Tamm-Horsfall mucoprotein (*uromucoid*) from the tubular epithelia in the ascending limb of the loop of Henle. Casts form whenever there is sufficient renal stasis, increased urine salt/protein concentration, and decreased urine pH. In patients with severe renal disease, truly accurate classification of casts may require use of "cytospin" centrifugation and Papanicolaou staining for adequate differentiation. Unlike cells, casts should be examined under low power and are most often located around the edges of the coverslip.

Hyaline. The matrix of these casts is clear and gelatinous, without cellular or particulate matter embedded in them. They may be quite difficult to visualize unless a high intensity lamp is used. Their presence indicates glomerular leakage of protein. This leakage may be temporary (due to fever, upright posture, dehydration, or emotional stress) or may be permanent; however, their occasional presence is not considered pathologic.

Granular. These casts are classified descriptively as either coarse or finely granular. The type of particulate matter embedded in them is simply a matter of the amount of degeneration that the epithelial-cell inclusions have undergone. Their occasional presence is not pathologic; however, large numbers may be found in cases of chronic lead toxicity and pyelonephritis.

Cellular. Several different types of casts are included in this category. *Red blood cell* or *erythrocytic casts* are always considered pathologic because they are diagnostic for glomerular inflammation that results in renal hematuria. They are seen in subacute bacterial endocarditis, kidney infarcts, collagen diseases, and acute glomerulonephritis. *White blood cell* or *leukocytic casts* are also always considered pathologic because they are diagnostic for inflammation of the nephrons. They are observed in pyelonephritis, nephrotic syndrome, and acute glomerulonephritis. In cases of asymptomatic pyelonephritis, these casts may be the only clue to detection. Epithelial-cell casts are sometimes formed by fusion of renal tubular epithelia after desquamation. Their occasional presence is not abnormal. Many, however, are observed in severe desquamative processes and renal stases such as occur in heavy metal poisoning, renal toxicity, eclampsia, nephrotic syndrome, and amyloidosis. *Waxy casts* are uniformly yellowish, refractile, and brittle-appearing, with sharply defined, often broken edges. They are almost always pathologic because they are indicative of tubular inflammation or deterioration. They are formed by renal stasis in the collecting ducts and are therefore found in the chronic renal diseases. *Fatty casts* are abnormal, coarse, granular casts with lipid inclusions that appear as refractile globules of different sizes. *Broad (renal failure) casts* may be up to two to six times wider than "regular" casts and may be cellular, waxy, or granular in composition. Like waxy casts, they are derived from the collecting ducts in cases of severe renal stasis.

Crystals

Acid environment. Crystals seen in urines with pH values of <7 include calcium oxalate, which are normal, colorless octahedrons or "envelopes"; they may have an almost starlike appearance. Also seen are amorphous urates, which are normal yellow-red masses with a sand grain-like appearance. Uric acid crystals found in this environment are normal, yellow to red-brown crystals that appear in extremely irregular shapes such as rosettes, prisms, or rhomboids. Cystine crystals are also sometimes observed in acid urine; they are highly pathologic and appear as colorless, refractile, nearly flat hexagons, somewhat similar to uric acid. These are observed in *cystinuria* (an inherited aminoaciduria resulting in mental retardation) and *homocyst-*

inuria (a rare defect of cystine reabsorption resulting in renal calculi).

Alkaline environment. Crystals seen in urines with pH values >7 include amorphous phosphates, which are normal crystals that appear as fine, colorless masses resembling sand. Also seen are calcium carbonate crystals, which are normal forms that appear as small, colorless dumbbells or spheres. Triple phosphate crystals are also observed in alkaline urines; they are colorless prisms of three to six sides resembling "coffin lids." Ammonium biurate crystals are normal forms occasionally found in this environment; they appear as spiny yellow-brown spheres, or "thornapples."

Other. Sulfonamide crystals are abnormal precipitates shaped like yellow-brown sheaves, clusters, or needles, formed in patients undergoing antimicrobial therapy with sulfa drugs. These drugs are seldom used today. Tyrosine/leucine crystals are abnormal types shaped like clusters of smooth, yellow needles or spheres. These are sometimes seen in patients with severe liver disease.[11]

Urine Electrophoresis

Owing to the efficiency of renal glomerular sieving and tubular reabsorption, normal urinary protein excretion is only about 50 to 150 mg/24 hours. Proteinuria may develop when there are defects in renal reabsorption or glomerular capillary permeability or when there is a marked increase in serum immunoglobulins. As a result, urine electrophoresis in the clinical laboratory is primarily used to aid in distinguishing between acute glomerular nephropathy and tubular proteinuria and to screen for monoclonal or polyclonal dysglobulinemias. In the latter category, urine electrophoresis can help in the differential diagnosis of multiple myeloma and Waldenstrom's macroglobulinemia. Positive identification and subtyping of the urinary paraproteins (*Bence Jones proteins*) detected there can be done by immunoelectrophoresis and immunofixation electrophoresis.

Beta₂-Microglobulin

Recently, interest has arisen concerning the use of clearance of β_2-microglobulin (β_2-M) as an indicator of GFR. β_2-M is a small, nonglycosylated peptide of molecular weight 11,800 daltons that constitutes the constant portion of the light chain of class I major histocompatibility complex antigens on the surface of most nucleated cells. The plasma membrane sheds β_2-M as a relatively intact moiety into the surrounding extracellular fluid. This process is fairly constant in adults; thus, levels of β_2-M remain stable in normal patients. Additionally, these values do not differ between males and females. As a small endogenous peptide, β_2-M is easily filtered by the glomerulus. About 99.9% is then reabsorbed by the proximal tubules through pinocytosis. Elevated levels in serum are indicative of increased cellular turnover. This elevation can be seen in the myelo- and lymphoproliferative disorders, such as AIDS and multiple myeloma. Increased urine levels are then observed when such concentrations exceed the renal threshold. In the absence of increased synthesis, however, increased urinary β_2-M is due to the inability of the proximal tubules to reabsorb it. Thus, this peptide can be used to monitor the GFR as a marker for glomerular disease, particularly tubular proteinuria. Elevations are found in the intermediate and late stages of diabetic nephropathy, and thus measurements are also useful for monitoring renal transplant status, since it is not removed from the circulation by dialysis. Indeed, it has been found in some studies to be a more efficient marker of renal graft rejection than serum creatinine values, since it is not dependent on lean muscle mass or daily variation in excretion. However, β_2-M is unstable in acidic urine specimens, and this poses a problem for reliable analyses. β_2-M are assayed by nephelometric techniques.[4]

Myoglobin

Recently, myoglobin clearance has been proposed as an effective early indicator of myoglobinuric acute renal failure—a high clearance or a low clearance and low serum concentration indicating low risk and a low clearance and high serum concentration indicative of high risk.[15] Myoglobin is a low molecular weight protein (16,900 daltons) that is associated with acute skeletal and cardiac muscle injury. It can induce acute renal failure if its release from muscle is sufficient to overload the renal proximal tubules and cause myoglobinuria. For this to occur, about 200 grams of muscle must be damaged. Thus, insignificant myoglobinuria is seen in an acute myocardial infarction but significant in skeletal myoglobinuria muscle injury. Acute renal failure caused by elevated myoglobin levels is a complication whose severity may be significantly reduced or even prevented if recognized early and treated aggressively. Myoglobin functions to bind and transports oxygen from the plasma membrane to the mitochondria in muscle cells. The normal range in serum is <90 ng/mL. To a large extent, it is bound to plasma proteins and therefore is filtered to a much smaller degree than would be predicted from its size. Its molecular sieving coefficient (glomerular filtrate to plasma ratio) is 0.75. Serum and urine myoglobin can be measured easily and rapidly by nonisotopic immunoassays. Urine myoglobin can also be measured qualitatively by dipstick methods after removing hemoglobin, but this method suffers from a lack of sensitivity and specificity. Further discussion is in Chapter 25, Porphyrins, Hemoglobin, and Myoglobin.

PATHOPHYSIOLOGY

Glomerular Diseases

Disorders or diseases that directly damage the renal glomeruli may, at least initially, exhibit normal tubular function. With time, however, disease progression involves the renal tubules as well. The following syndromes have discrete symptomologies, which are recognizable by their patterns of clinical laboratory findings.[2,9]

Acute Glomerulonephritis

Pathologic lesions in acute glomerulonephritis primarily involve the glomerulus. Histologic examination shows large, inflamed glomeruli with a decreased capillary lumen. Abnormal laboratory findings virtually always include rapid onset of hematuria and proteinuria (usually albumin, and generally of <3 g/day). There is usually the rapid development of a decreased GFR, anemia, elevated BUN and serum creatinine, oliguria, sodium and water retention (with consequent hypertension and some localized edema), and sometimes congestive heart failure (CHF). Numerous hyaline and granular casts are generally seen upon urinalysis. The presence of actual red blood cell casts is regarded as highly suggestive of this syndrome.

Acute glomerulonephritis is often related to recent infection by group A β-hemolytic streptococci. It is theorized that circulating immune complexes trigger a strong inflammatory response in the glomerular basement membrane, resulting in a direct injury to the glomerulus itself. Other possible causes include drug-related exposures, acute kidney infections due to other bacterial (and possibly viral) agents, and other systemic immune complex diseases, such as systemic lupus erythematosus (SLE) and subacute bacterial endocarditis (SBE). For unknown reasons, the course of the syndrome becomes fulminant in some cases, quickly progressing to renal failure; this is termed *rapidly progressive glomerulonephritis.*

Chronic Glomerulonephritis

Lengthy glomerular inflammation, whether from a renal disease or from an idiopathic cause, may lead to glomerular scarring and the eventual loss of operational nephrons. This process is often undetected for lengthy periods, since only minor decreases in renal function occur at first, and only slight proteinuria and hematuria are observed. Gradual development of uremia (or *azotemia,* excess nitrogen compounds in the blood) may sometimes be the first sign of this process.

Nephrotic Syndrome

In nephrotic syndrome, many different etiologies can result in a specifically demonstrable type of glomerular injury: an abnormally increased permeability of the glomerular basement membrane. This defect almost always yields such abnormal findings as extremely massive proteinuria (>3.5 g/day and even up to 20 g/day), resultant hypoalbuminemia, and subsequent decreased plasma oncotic pressure, culminating in a generalized edema due to the movement of body fluids out of the vascular and into the interstitial spaces. Other hallmarks of this syndrome are hyperlipidemia (possibly secondary to lipoprotein alterations) and lipiduria. The latter takes the form of oval fat bodies in the urine; these bodies are degenerated renal tubular cells containing reabsorbed lipoproteins. Owing to both the extreme proteinuria and the osmotically induced hypovolemia in these cases, disorders of several different co-agulation factors are observed, and intravascular thrombus formation becomes a danger. A schematic diagram of the pathophysiologic sequence in nephrotic syndrome is shown in Figure 17-5.

Primary causes are associated directly with glomerular disease states. Secondary causes are associated with infections (SBE, syphilis), mechanical derangements of the circulatory system (constrictive pericarditis, renal vein thromboses), allergic/toxic reactions (bee venom, penicillamine), generalized disease states (carcinomas, SLE, amyloidosis), or miscellaneous disorders (transplant rejections, severe preeclampsia).

Tubular Diseases

Tubular defects occur to a certain extent in the progression of all renal diseases as the GFR falls. In some instances, however, this aspect of the overall dysfunction becomes predominant. The result is decreased excretion/reabsorption of certain substances and/or reduced urinary concentrating capability. Clinically, the most important defect is the primary tubular disorder affecting acid—base balance: renal tubular acidosis (RTA). This disease can be classified into two types, depending on the nature of the tubular defect: *distal RTA*, in which the renal tubules are unable to keep up the vital pH gradient between the blood and tubular fluid; and *proximal RTA*, in which there is decreased bicarbonate reabsorption, resulting in hyperchloremic acidosis. In general, reduced reabsorption in the proximal tubule is manifested by findings of abnormally low serum values for phosphorus and uric acid and by the presence of glucose and amino acids in the urine. In addition, there may be some proteinuria (usually <2 g/day).

Acute inflammation of the tubules and surrounding interstitium also may occur as a result of analgesic drug or radiation toxicity, methicillin hypersensitivity reactions, renal transplant rejection, and viral-fungal-bacterial infections. Characteristic clinical findings in these cases are decreases in GFR, urinary concentrating ability, and metabolic acid excretion; the presence of leukocyte casts in the urine; and inappropriate control of sodium balance.[2,9]

Urinary Tract Infection/Obstruction

Infection

The site of infection may be either in the kidneys themselves (*pyelonephritis*) or in the urinary bladder (*cystitis*). In general, a microbiologic colony count of $>10^5$ colonies/mL is considered diagnostic for infection in either locale. *Bacteriuria* (as evidenced by positive nitrite dipstick findings for some organisms), hematuria, and *pyuria* (presence of leukocytes in the urine, as shown by positive leukocyte esterase dipstick) are all frequently encountered abnormal laboratory results in these cases. In particular, the presence of white blood cell (leukocyte) casts in the urine is considered diagnostic for pyelonephritis.[2,9,11]

Obstruction

Renal obstructions can cause disease in one of two ways. They may either gradually raise the intratubular pressure until nephrons necrose and chronic renal failure ensues, or they may predispose the urinary tract to repeated infections.

Obstructions may be located in either the proximal or distal urinary tract. Blockages in the upper tract are characterized by histologic evidence of a constricting lesion below a dilated collecting duct. Obstructions of the lower tract are evidenced by the presence of residual urine in the bladder after cessation of *micturition* (urination); symptoms include

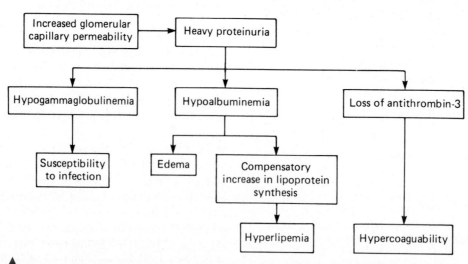

Figure 17-5. Pathophysiologic sequence in nephrotic syndrome.

slowness of voiding, both initially and throughout urination. Causes of obstructions are quite varied. They can include neoplasias (such as prostate/bladder carcinoma or lymph node tumors constricting ureters), acquired diseases (such as urethral strictures or renal calculi), and congenital deformities of the lower urinary tract. The clinical symptoms of advancing obstructive disease follow the course of its progression, with decreased urinary concentrating capability seen first, succeeded by diminished metabolic acid excretion, decreased GFR, and reduced renal blood flow. Laboratory tests useful in elucidating the nature of the blockage are the urinalysis, urine culture, BUN, serum creatinine, and the complete blood count. Final diagnosis is usually made by radiologic imaging techniques.[2,9,11]

Renal Calculi

Renal calculi, commonly termed kidney stones, are formed by the combination of a variety of crystallized substances. These are listed in Table 17-2. Of these, calcium oxalate stones are by far the most commonly encountered, particularly in the tropics and subtropics.

It is currently believed that recurrence of calculi in susceptible individuals is due to a complex mixture of causes. Chief among these are a reduced urine flow rate (related to a decreased fluid intake) and saturation of the urine with large amounts of essentially insoluble substances (such as those mentioned above). Qualitative chemical analyses of these stones permit differentiation of mixed etiologies (producing "mixed stones") in complicated diagnostic cases. Recently, specialized x-ray diffraction and infrared spectroscopy techniques have come into more widespread use for this purpose. Clinical symptoms are, of course, similar to those encountered in other obstructive processes: hematuria, urinary tract infections, and "renal colic" (characteristic abdominal pain).[2,9,11]

TABLE 17-2. Types of Renal Calculi

Calculi Type	Etiology
1. Calcium oxalate	Hypercalciuria
	Hyperparathyroidism
	Vitamin D toxicity
	Sarcoidosis
	Osteoporosis
	Burnett's (milk-alkali) syndrome
	RTA
2. Magnesium ammonium phosphate	Infectious processes
3. Calcium phosphate	Alkali consumption
	RTA
	Infection by urease producers
4. Uric acid	Hyperuricaciduria
	Gout
	Hyperuricemia
5. Cystine	Primary inherited cystinuria

Renal Failure

Acute Renal Failure

Acute renal failure is defined as a sudden, sharp decline in renal operation due to an acute toxic or hypoxic insult to the kidneys. This has been defined as occurring when the GFR is reduced to <10 mL/min. This syndrome is subdivided into three types, depending on the location of the precipitating defect. In prerenal failure, the defect lies in the blood supply before it reaches the kidney. Causes can include cardiovascular system failure and consequent hypovolemia. In primary renal failure, the defect involves the kidney itself. The most common cause is acute tubular necrosis; other etiologies include vascular obstructions/inflammations and glomerulonephritis. In postrenal failure, the defect lies in the urinary tract after it exits the kidney. Generally, acute renal failure occurs as the sequelae to a lower urinary tract obstruction or to rupture of the urinary bladder.

Toxic insults to the kidney severe enough to initiate acute renal failure include hemolytic transfusion reactions, heavy metal/solvent poisonings, antifreeze ingestion, and analgesic and aminoglycoside toxicities. These conditions directly damage the renal tubules. Hypoxic insults include conditions that severely compromise renal blood flow, such as septic/hemorrhagic shock, burns, and cardiac failure.

The most commonly observed symptoms of acute renal failure are oliguria and anuria (<400 mL/day). The diminished ability to excrete electrolytes and water results in a marked increase in extracellular fluid volume, leading to peripheral edema, hypertension (HTN), and CHF. If more water is retained than sodium, hyponatremia may develop; in extreme cases, central nervous system effects are seen, progressing from profound drowsiness to seizures, coma, and death. The hyperkalemia also may become severe enough to cause dangerous cardiac arrhythmias. In addition, variable amounts of erythrocytic casts, hematuria, proteinuria, and metabolic acidosis (with resulting respiratory attempts at compensation) are seen. Generally, systemic bone disease and anemia are not as marked as they are in chronic renal failure. Most prominent, however, is the onset of the *uremic syndrome,* or *end-stage renal disease* (*ESRD*), in which increased BUN and serum creatinine values (azotemia) are observed in conjunction with the preceding symptoms. The outcome of this disease is either recovery or, in the case of irreversible renal damage, progression to chronic renal failure.[2,9]

Chronic Renal Failure

Chronic renal failure is a clinical syndrome that occurs when there is a gradual decline in renal operation over time. The interrelated pathophysiology of this disease process is depicted in Figure 17-6. Chronic renal failure is classified into four progressive stages. The first stage is marked by a period of silent deterioration in renal status. Here kidney function decreases, but BUN and creatinine values stay within normal limits. The second stage is characterized by develop-

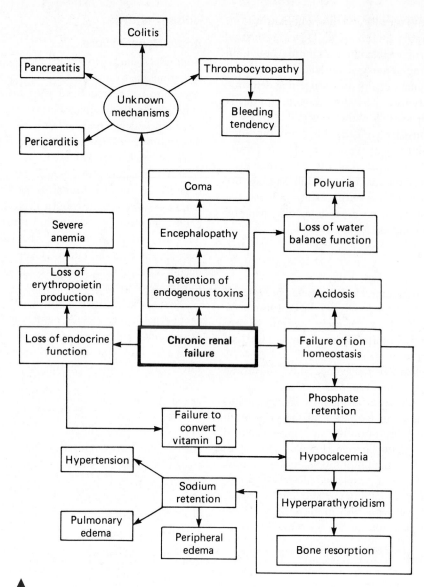

Figure 17-6. Pathophysiology of chronic renal failure.

ment of a slight renal insufficiency. A 50% reduction in normal functioning is necessary before the BUN and creatinine values reflect the pathologic changes by increasing above reference ranges. The third stage is typified by impending renal failure. Anemia begins to develop (due to the constant deficit in erythropoietin production), and systemic acidosis commences (due to the faulty clearance of endogenous metabolic acids). The fourth and last stage commences with the onset of the classic symptoms of the uremic syndrome (see preceding section). The conditions that can precipitate acute renal failure also may lead to chronic renal failure.[2,9] In addition, there are several other etiologies for this syndrome, as shown in Table 17-3.

Diabetes Mellitus

Diabetes mellitus can have profound effects on the renal system. In insulin-dependent diabetes mellitus (IDDM, type I),

patients suffer from a deficit of insulin activity. Approximately 40% to 50% of these patients will develop progressive deterioration of kidney function (*diabetic nephropathy*) within 15 to 20 years after their diagnosis. A small number of non-insulin-dependent (NIDDM, type II) diabetics also will go on to develop this condition. The lesions are primarily glomerular, but they may affect all other kidney structures as well; they are theorized to be caused by the abnormally hyperglycemic environment that constantly bathes the vascular system.[2,9]

Typically, diabetes affects the kidneys by causing them to become glucosuric, polyuric, and nocturic. These states are caused by the heavy demands (*nephromegaly*) made on the kidneys in order to diurese a hyperosmotic urine. In addition, a mild proteinuria (*microalbuminuria*) often develops between 10 and 15 years after the original diagnosis is made. The normal albumin concentration in urine ranges from 4 to

TABLE 17-3. Etiology of Chronic Renal Failure

Etiology	Examples
1. Renal circulatory diseases	Renal vein thrombosis, malignant HTN
2. Primary glomerular diseases	SLE, chronic glomerulonephritis
3. Renal sequelae to metabolic disease	Gout, DM, amyloidosis
4. Inflammatory diseases	Tuberculosis, chronic pyelonephritis
5. Renal obstructions	Prostatic enlargement, calculi
6. Congenital renal deformity	Polycystic kidneys, renal hypoplasia
7. Miscellaneous conditions	Radiation nephritis

25 mg/24 h. In patients with microalbuminuria, urine concentrations are in the range of 30 to 300 mg/24 h, which is not detectable by conventional urinalysis, dipsticks, which have a detection limit of about 550 mg/24 h. During the next 3 to 5 years, this situation becomes macroproteinuric, and thereafter, a steady decrease in GFR is usually seen from month to month. Indeed, plotting of the reciprocal of the serum creatinine value against time often results in a straight-line approximation sufficiently sensitive to allow prediction of the deterioration rate. Hypertension often manifests itself next, further exacerbating the renal damage. Eventually, chronic renal insufficiency or nephrotic syndrome may evolve, and each may be identified by their characteristic symptoms and laboratory findings (see preceding section). The early appearance of this proteinuria in the progression of the disease correlates significantly with the likelihood of developing diabetic nephropathy in later life. Microalbuminuria is apparently caused by the creation of increasing numbers of large pores at the glomerulus.[14] Interventional therapy of microalbuminuric patients, such as tight control of blood glucose, may prolong the onset of chronic renal failure.

Renal Hypertension

Etiology. Renal disease-induced hypertension can be caused by either decreased perfusion to all or part of an area of the kidney or by local tissue accumulation of sodium. Chronic ischemia of any sort results in nephron dysfunction and eventual necrosis; thus, the lack of perfusion may be due to trauma damage or stenosis to a main or branching artery or to sclerosis of intrarenal arterioles. In either case, the resulting changes in blood and body fluid volumes within the kidney trigger the activation of the renin-angiotensin-aldosterone system, setting off the vasoconstrictive responses, which are manifested as persistent hypertension.

Laboratory Tests. Renal hypertension can be evaluated by monitoring for elevations in serum aldosterone, Na^+, and plasma renin levels (particularly as renal vein differential samples). As a consequence of the increased Na^+ retention, urine K^+ levels will be increased (due to the accelerated excretion rate), and therefore, serum K^+ levels will be decreased.

Therapy of Acute Renal Failure

In patients with acute renal failure, uremic symptoms, uncontrolled hyperkalemia, and acidosis have traditionally been indications that the kidneys are unable to excrete the body's waste products and a substitute method in the form of dialysis was necessary. Dialysis is often instituted before this stage, however. There are several forms of dialysis available but they all employ a semipermeable membrane surrounded by a dialysate bath in common. In traditional hemodialysis, the membrane is synthetic and outside the body. Arterial blood and dialysate are pumped at high rates (150–250 and 500 mL/min, respectively) in opposite directions. The blood is returned to the venous circulation and the dialysate discarded. The diffusion of low molecular weight solutes (<500 daltons) into the dialysate is favored by this process, but mid-molecular-weight solutes (500–2000 daltons) are inadequately cleared. Creatinine clearance is roughly 150 to 160 mL/min. In peritoneal dialysis, the peritoneal wall acts as the dialysate membrane and gravity is used to introduce and remove the dialysate. Two variations of this form are available, continuous ambulatory peritoneal dialysis (CAPD) and continuous cycling peritoneal dialysis (CCPD), but in both the process is continuous, being performed 24 hours a day, 7 days a week. This method is not as rigorous as the traditional method. Small solutes (*e.g.,* potassium) have significantly lower clearance rates compared to the traditional method, but more large solutes are cleared and steady-state levels of blood analytes are maintained. Continuous arteriovenous hemofiltration (CAVH), continuous venovenous hemofiltration (CVVH), continuous arteriovenous hemodialysis (CAVHD), and continuous venovenous hemodialysis (CVVHD) together make up the slow continuous renal replacement therapies developed to treat acute renal failure in critically ill patients in intensive care settings. In these methods, the semipermeable membrane is again outside the body. Solutes up to 5000 daltons (the pore size of the membranes) and water are slowly (10 mL/min) and continuously filtered from the blood in the first two methods, causing minimal changes in plasma osmolality. The ultrafiltrate is lost to the body and the volume loss can be replaced in the form of parenteral nutrition and obligatory intravenous medications. The final two methods are similar to the filtration methods but a continuous trickle of dialysis fluid is pumped past the dialysis membrane resulting in continuous diffusion. This doubles the urea clearance.

Therapy of End-Stage Renal Disease

For patients with irreversible renal failure, dialysis and transplantation are the two therapeutic options. Initiation of either treatment occurs when the GFR falls to 5 mL/min (10–15 mL/min in patients with diabetic nephropathy).

Dialysis. Traditional hemodialysis or its more recent high efficiency form as well as peritoneal dialysis are the available methods. The clinical laboratory utilized in conjunction with a hemodialysis facility must be able to adequately monitor procedural efficiency in a wide variety of areas. Renal dialysis has basic goals, and specific laboratory tests should be performed to evaluate the achievement of each.

Monitoring of nitrogen balance. Rapid removal of sodium, urea, creatinine, and several other osmotically active compounds can cause *dialysis disequilibrium syndrome*. This may result in headaches, lassitude, seizures, and torpor and should be monitored by determinations of BUN, creatinine, uric acid, total protein, and albumin.

Monitoring of hydration and electrolyte balance. Swift changes in fluid and electrolyte distribution throughout the body can derange albumin concentrations and skew buffering capabilities, causing hypotension. Additionally, rapid shifts in potassium may precipitate cardiac toxicity. Important laboratory tests include Na^+, K^+, Cl^-, HCO_3^-, osmolality, and arterial pH.

Monitoring of generalized osteomalacia. Chronic alterations in bone deposition are caused by continued derangement of Ca^{2+}, phosphate, and vitamin D metabolism. A condition known as *renal osteodystrophy* can result. Laboratory analyses should assess serum phosphorus, Ca^{2+}, alkaline phosphatase, magnesium, and (at regular intervals) parathyroid hormone levels. The last assay is indicated to monitor the secondary hyperparathyroidism that often ensues due to upregulation of parathyroid gland activity.

Viral hepatitis. Dialysis patients suffer from a variety of acute and chronic liver diseases including the many forms of hepatitis. The common forms of dialysis-associated hepatitis include hepatitis A, hepatitis B, hepatitis C, cytomegalovirus hepatitis, and drug-induced hepatitis. At least 50% of patients with hepatitis are asymptomatic. The type of liver disease must be established for both prognosis and therapy. Essential tests are alanine aminotransferase (ALT), aspartate aminotransferase (AST), alkaline phosphatase (ALP), bilirubin, and hepatitis B surface antigen (HBsAg). Antibody to hepatitis C virus (anti-HCV) testing has also been proposed for all patients entering dialysis programs because of the high incidence of anti-HCV positives among hemodialysis patients.[3]

Hyperlipidemia. There is a 50% incidence of hyperlipidemia in the dialysis population and the most prevalent pattern is type IV (see Chapter 15, Lipids and Lipoproteins). Patients with this classification have an increased incidence of ischemic heart disease. Essential tests are a lipid panel,

calcium, and phosphorus. Calcium and phosphorus control and subsequent avoidance of hyperparathyroidism may lower the risk of ESRD-related atherosclerosis.[6]

Aluminum neurotoxicity. Aluminum neurotoxicity is well recognized in dialysis patients. Two syndromes have been defined—dialysis encephalopathy brought on by chronic oral or parenteral aluminum exposure, and acute aluminum neurotoxicity brought on by the ingestion of a combination of citrate and aluminum compounds after deferoxamine therapy or after dialysis contaminated with aluminum. Aluminum concentrations range between 100 to 150 mg/L (normal is <6 mg/L) and >500 mg/L in these disorders, respectively.[1]

Assessment of residual function. Evaluation of renal status after a dialysis session should be made by determinations of 24-hour urine volume and GFR. The GFR, however, usually drops temporarily in the first day after dialysis, so measurements of clearance (of creatinine or of [^{125}I]iothalamate) should be made after the first 24 hours.

Therapeutic drug monitoring (TDM). Inadequate renal function, coupled with dialysis-induced changes, can dramatically derange pharmacokinetics of therapeutic agents. TDM must be undertaken in order to ensure the maintenance of appropriate therapeutic drug levels. In particular, the aminoglycoside antibiotics are highly nephrotoxic and must be followed using regular serum drug levels.

Transplantation. The most efficient hemodialysis techniques provide only 10% to 12% of the small solute removal of two normal kidneys and considerably less removal of larger solutes. Hence, even well-dialyzed patients have physical disabilities and a decrease in the quality of life. Kidney transplantation offers the greatest chance for a full return of a healthy, productive life. However, this option is limited by the significant shortage of donor organs. For ESRD patients, the length of time waiting for an organ can vary from several months to several years.

Renal transplantation is from a compatible donor to a recipient suffering from irreversible renal failure. The organ can be from a cadaver or a live individual (80% and 20%, respectively, of all kidney transplants in the United States). For this procedure to be successful, the body's immune response to the transplanted organ must be suppressed. Thus, the donor and recipient are carefully screened for ABO blood group, HLA compatibility, and performed HLA antibodies. The HLA (human leukocyte antigen) system is the major inhibitor to transplantation. In addition, cytokine production is stopped by using immunosuppressive agents. Cytokines are peptide products of immune cells, which stimulate an immune response. There are four major immunosuppressive drugs or drug groups available: cyclosporine, corticosteroids, azathioprine, and the monoclonal and polyclonal antibodies. Cyclosporine has enjoyed widespread use since the 1980s, its use coinciding with improved graft survival. The transplantation procedure itself requires considerable skill to ensure suitable positioning of the kidney(s) and vascular connections as well as attention to detail, aseptic techniques, and hemostasis, because many of these patients are anemic or malnourished at the time of surgery and the use of immunosuppressive drugs potentially compromises the healing process.

Cyclosporine is a cyclic polypeptide of fungal origin. It consists of 11 amino acids and has a molecular weight of 1203. It is neutral and insoluble in water but is soluble in organic solvents and lipids. Cyclosporine inhibits the immune response selectively by inhibiting the interleukin-2 dependent proliferation of activated T cells, which destroy the allograft. The immune response is frozen and unresponsive. Oral cyclosporine is incompletely and variably absorbed from the GI tract. Peak concentration is on average 3 hours and the absolute bioavailability, which increases with time, is about 30%. Thus, the amount of cyclosporine required to achieve a given blood level tends to fall with time and typically reaches a steady level within 4 to 8 weeks. In the blood, one third of cyclosporine is found in plasma, bound primarily to lipoproteins with the rest bound to erythrocytes. Consequently, whole blood drug levels will typically be threefold higher than plasma levels. Cyclosporine has a half-life of approximately 8 hours and is metabolized to at least 15 metabolites by the liver, some of which may have immunosuppressive and nephrotoxic potential. Cyclosporine is also excreted in the bile. The measurement of cyclosporine levels is an intrinsic part of the management of transplant patients because of the variation in its metabolism in individual patients and from patient to patient. There is also some correlation between drug blood levels and episodes of rejection and toxicity. Cyclosporine monitoring can be a source of much confusion because of the various assays available and the option of using different matrixes (i.e., plasma or whole blood) for its measurement.

Laboratory support in patient management is aimed at assessing renal function and, hence, rejection status as well as toxicity due to the immunosuppressant drugs. During the early post-transplant period (usually 2 months), tests include urine volume; urinalysis for cells, protein, and bacteria; serum creatinine and urea levels; electrolytes panel including phosphorus, calcium, and hepatic enzyme levels; complete blood count; and the blood immunosuppressive drug levels. These tests monitor for occurrence of early rejection, immunosuppressant drug toxicity, or other underlying pathological events in the allograft. In the long-term, other tests are added to monitor for potential complications. One example is a lipid profile, because approximately 50% of patients treated with cyclosporine are hypercholesterolemic by the third month after transplant. Individuals with evidence of prior hepatitis B or C infection who are clinically quiescent are eligible for transplantation, and there is some

indication that immunosuppressive therapy appears to stimulate progression from benign to aggressive liver disease. Although kidney transplants have the capacity to function for decades, the mean half-life of a cadaveric transplant is approximately 7 years. The mortality rate is not significantly different from hemodialysis. Three-year graft survival figures vary from 65% to 85%, with live grafts doing better. It has been reported that there is no difference in patient survival among hemodialysis, CAPD, and cadaveric kidney transplantation. Live related donor transplantation is associated with a better patient survival than other ESRD therapeutic options.

SUMMARY

The kidney plays a vital role in the maintenance of water and electrolyte balance, homeostasis, and removal of waste products. The kidney produces several important hormones (such as renin and the prostaglandins) needed to perform these tasks. Other renal hormones (such as erythropoietin) are important for other physiologic functions. The kidneys themselves are the target organ for catabolism of insulin, glucagon, aldosterone, and vitamin D, among others.

Renal function tests focus largely on glomerular clearances, as assessed by creatinine and inulin measurements, and tubular functions, as assessed by protein measurements (e.g., urine electrophoresis). The analysis of urine for analytes, such as pH, glucose, ketones, and bilirubin, continue to be important screening tests for many non-renal diseases such as diabetes mellitus, ketoacidosis (and other acid-base imbalances), hemolysis, and liver diseases. Newer protein assays, such as for urine albumin, serum β_2-microglobulin, and serum and urine myoglobin, can provide important prognostic information useful for patient management. Microalbuminuria is useful for early detection of diabetic nephropathy, β_2-M is useful for early renal transplant rejection, and myoglobin clearance rates are helpful in predicting rhabdomyolysis-induced acute renal failure.

Common renal diseases include infectious and inflammatory processes to the glomerulus, tubules, and urinary tract, obstructions to normal kidney function, and acute and chronic renal failure. Common glomerular diseases include nephritis and nephrotic syndrome. Tubular diseases include renal tubular acidosis and Fanconi's syndrome. In situations of chronic renal failure, aggressive therapeutic approaches based on dialysis and transplantation have enabled prolonged survival of what was once a terminal condition. Variations in dialysis techniques have made this process more available and convenient, and with the implementation of powerful immunosuppressive drugs such as cyclosporin, widespread renal transplantation is now limited only by the availability of appropriate donor organs.

A 45-year-old male with end-stage renal disease (ESRD) secondary to hypertension was admitted for a living related donor kidney transplant. He was on peritoneal dialysis for 1½ years prior to transplant surgery. Transplantation was uneventful with no complications. During surgery when the kidney was connected, urine formation occurred within a few seconds. A renal scan the next morning showed good perfusion and function with no outflow obstruction. The patient was discharged 7 days later. Applicable lab data are shown in Case Study Table 17-1.

Questions

1. What tests did the surgeon use to assess whether the transplanted kidney was functioning?
2. How soon after transplant was kidney function observed?
3. Were the pre-transplant BUN/CR values expected for a patient on peritoneal dialysis?
4. Why were cholesterol and triglyceride levels requested prior to surgery?
5. Is a high PTH level expected in an ESRD patient?
6. What results in the first week indicated the transplant was a success?

CASE STUDY TABLE 17-1. Laboratory Results[a]

Day	Na⁺	K⁺	Cl⁻	HCO₃⁻	Cr	BUN	β₂-M Urine	24-h	Gluc	Chol	Trig	Ca²⁺	Phos	TP	AST	ALT	ALK	Bili	PTH	CyA
1	138	5.2↑	96↓	25	27.4↑	104↑	—	—	—	132	0.6	5.0	6.8↑	6.1↓	29	21	64	0.4	281↑	—
2	141	4.6	103	17↓	22.4↑	94↑	—	—	130↑	—	—	—	—	—	—	—	—	—	—	—
3	139	5.1↑	106	22	10.3↑	59↑	6.0↑	—	141↑	—	—	—	—	—	—	—	—	—	—	—
4	138	5.3↑	107	22	4.2↑	31↑	2.9↑	2350	119↑	—	—	5.8↑	3.7	—	22	23	75	0.4	—	163
5	135	5.7↑	105	22	2.5↑	23	2.2	2100	122↑	—	—	—	—	—	—	—	—	—	—	96
6	137	4.5	105	24	2.0↑	27↑	2.2	4880	94	—	—	—	—	—	—	—	—	—	—	69
7	136	4.3	102	23	1.8↑	31↑	2.0	2500	83	—	—	—	—	—	—	—	—	—	—	108
8	139	4.8	100	26	1.7	34↑	2.2	1900	84	—	—	—	—	—	—	—	—	—	—	190
9	137	4.3	98	27	1.7	32↑	2.0	—	86	—	—	5.76↑	2.8	—	46	102↑	76	1.1	—	900

[a]Transplant surgery was performed on day 2, before these reported results. Ca²⁺ is ionized calcium. Serum protein electrophoresis (SPEP) was ordered on day 1, which was normal for albumin and globulins.

A 27-year-old man was admitted for observation after complaining of back pain radiating to the upper chest for 3 days and a rash on his trunk and upper extremities for 1 day. The patient reported being exposed to a child with chicken pox. The patient is a post renal transplant candidate whose graft, done 1 year previously, was secondary to idiopathic chronic renal failure. No problems were reported until this incident. On the day 11 post admission, varicella was confirmed.

Questions

1. Why was this patient admitted for observation?
2. Did patient suffer acute renal failure?
3. What did the β₂-M results indicate?
4. Why were a serum and urine myoglobin requested?
5. Was the patient at high or low risk of acute renal failure due to myoglobin?
6. What other organs seemed to be involved from the result?
7. Were the BUN/CR results helpful in indicating whether the patient was improving?

CASE STUDY TABLE 17-2

Day	Cr	BUN	β₂-M	Myo Ser	Myo Urine	Phos	Amy	Lip	AST	ALT	ALP	Bili	CyA
1	2.3↑	25	5.0↑	—	—	5.0↑	—	—	1218↑	697↑	105	1.7↑	<28
3	3.1↑	38↑	—	—	—	—	136↑	356↑	5856↑	2963↑	147↑	—	<28
5	3.4↑	47↑	10.4↑	299↑	178↑	—	—	—	4416↑	993↑	—	—	—
7	—	—	—	1450↑	15600↑	—	—	—	—	—	—	—	—
9	2.6↑	37↑	9.4↑	448↑	3920↑	—	563↑	5332↑	—	—	—	—	—
10	2.5↑	35↑	6.0↑	—	—	—	555↑	5928↑	—	—	—	—	—
11	2.4↑	41↑	6.5↑	—	—	—	479↑	4462↑	—	—	—	—	—

CASE STUDY
17-3

A 19-year-old man was admitted to the emergency room in a coma with no overt signs of trauma or seizure. Physical examination was unremarkable, except for quick, gasping respirations and a weak, rapid pulse rate. Applicable laboratory data are shown in Case Study Table 17-3.

Questions

1. What is the most obvious diagnosis?
2. Does this patient have azotemia?
3. Does this patient have uremia?

CASE STUDY TABLE 17-3. Applicable Laboratory Data

Laboratory Test	Patient Results	Reference Range
Urine drug screen	Negative	—
Urine pH	5.0	4.5–8.0
Urine specific gravity	1.038	1.003–1.035
Urine glucose	4+	Negative
Urine ketones	3+	Negative
Urine osmolality	1288 mOsmol/kg	300–1300 mOsmol/kg
Urine protein	2+	Negative
Arterial blood pH	7.15	7.35–7.45
Arterial PCO_2	25 mmHg	35–45 mmHg
Serum acetone	3+	Negative
Blood urea nitrogen (BUN)	40 mg/dL	10–20 mg/dL
Serum creatinine	1.7 mg/dL	0.5–1.4 mg/dL
Serum glucose	723 mg/dL	65–110 mg/dL
Serum osmolality	355 mOsmol/kg	280–300 mOsmol/kg

EXERCISES

1. Calculate the creatinine clearance given the following information: serum creatinine 1.5 mg/dL, urine creatinine 150 mg/dL, urine volume 2000 mL/day, body surface area 1.90 m².
2. Which of the following is most commonly used to prevent renal transplant rejection?
    A. interleukin-2
    B. cyclosporine A
    C. FK506
    D. prednisone
    E. interferon
3. Which procedure is most widely used for chronic renal failure patients?
    A. hemodialysis
    B. peritoneal dialysis
    C. continuous arteriovenous hemofiltration
    D. continuous venovenous hemofiltration
    E. all are used equally
4. Which kidney disease is characterized by excessive secretion of proteins (>3.5 g/day) into the urine?
    A. glomerulonephritis
    B. acute renal failure
    C. renal calculi
    D. urinary tract infection
    E. nephrotic syndrome
5. Which renal hormone is responsible for the producing red blood cells?
    A. antidiuretic hormone
    B. prostaglandins
    C. renin
    D. erythropoietin
    E. aldosterone
6. Which of the following is NOT an ideal charcteristic for substance used to estimating the glomerular filtration rate?
    A. freely filterable
    B. nontoxic to the kidney
    C. completely removed on first pass
    D. is not reabsorbed by the tubules
    E. is not secreted by the tubules
7. What is the mechanism by which bicarbonate is excreted into the urine?
    A. direct excretion by the tubules
    B. filtration through the glomerulus
    C. filtration, conversion to CO_2, diffusion through proximal tubules, conversion to carbonic acid, dissociation of H^+, and excretion of bicarbonate
    D. conversion to CO_2, excretion by the tubules
    E. conversion to carbonic acid by carbonic anhydrase, and filtration through the glomerulus

8. Which of the following is FALSE concerning creatinine clearance rates?
   A. values are influenced by diet and exercise
   B. because of tubular secretion, clearance values are falsely high when the serum creatinine concentrations are high
   C. shorter urine collection times can be used although the accuracy will be affected
   D. creatinine excretion varies with regard to lean body mass
   E. rates remain constant throughout life

9. Which of the following is true concerning the urinary infections?
   A. a negative nitrite indicates no bacteria
   B. gram positive bacteria all convert nitrate to nitrite
   C. positive result indicates presence of gram negative bacteria
   D. small numbers of organisms seen by microscopy are likely due to contaminations
   E. the presence of white cells indicates a bacterial infection

10. Increases in β_2-microglobulin are seen in
    A. acquired immune deficiency syndrome
    B. multiple myeloma
    C. diabetic nephropathy
    D. renal transplant rejection
    E. all of the above

REFERENCES

1. Alfrey AC. Neurological aspects of uremia. In: Nissenson AR, Fine RN, eds. Dialysis therapy. 2nd ed. Philadelphia: Hanley and Belfus, 1993;275.

2. First MR. Renal function. In: Kaplan LA, Pesce AJ, eds. Clinical chemistry: Theory, analysis, and correlation. 2nd ed. St. Louis: Mosby, 1989;346.

3. Fraser D, Jones G, Kooh SW, Radde IC. Calcium and phosphate metabolism. In: Tietz NW, ed. Fundamentals of clinical chemistry. 3rd ed. Philadelphia: Saunders, 1987;705.

4. Frauenhoffer E, Demers LM. Beta$_2$-microglobulin. ASCP Check Sample Continuing Education Program, Clinical Chemistry, No. CC 86-5 (CC-173), 1986.

5. Gitnick G. Gastrointestinal disease. In: Nissenson AR, Fine RN, eds. Dialysis therapy. 2nd ed. Philadelphia: Hanley and Belfus, 1993;209.

6. Golper TA. Metabolic abnormalities. In: Nissenson AR, Fine RN, eds. Dialysis therapy. 2nd ed. Philadelphia: Hanley and Belfus, 1993;261.

7. Kaplan LA. Measurement of colligative properties. In: Kaplan LA, Pesce AJ, eds. Clinical chemistry: Theory, analysis, and correlation. 2nd ed. St. Louis: Mosby, 1989;207.

8. Kleinman LI, Lorenz JM. Physiology and pathophysiology of body water and electrolytes. In: Kaplan LA, Pesce AJ, eds. Clinical chemistry: Theory, analysis, and correlation. 2nd ed. St. Louis: Mosby, 1989;313.

9. Rock RC, Walker WG, Jennings CD. Nitrogen metabolites and renal function. In: Tietz NW, ed. Fundamentals of clinical chemistry. 3rd ed. Philadelphia: Saunders, 1987;669.

10. Russell PT, Sherwin JE, Obernolte R, Murray RL, Kaplan LA, Schultz AL. Nonprotein nitrogenous compounds. In: Kaplan LA, Pesce AJ, eds. Clinical chemistry: Theory, analysis, and correlation. 2nd ed. St. Louis: Mosby, 1989;1005.

11. Schumann GB, Schweitzer SC. Examination of urine. In: Kaplan LA, Pesce AJ, eds. Clinical chemistry: Theory, analysis, and correlation. 2nd ed. St. Louis: Mosby, 1989;820.

12. Sherwin JE, Bruegger BB. Acid–base control and acid–base disorders. In: Kaplan LA, Pesce AJ, eds. Clinical chemistry: Theory, analysis, and correlation. 2nd ed. St. Louis: Mosby, 1989;332.

13. Steffes MW, Mauer SM. Diabetic nephropathy: A disease causing and complicated by hypertension. Clin Chem 1991; 37:1838.

14. Van Lente F, Suit P. Assessment of renal function by serum creatinine and creatinine clearance: Glomerular filtration rate estimated by four procedures. Clin Chem 1989;35:2326.

15. Wu AHB, Laios I, Green S, et al. Immunoassays for serum and urine myoglobin: Myoglobin clearance assessed as a risk factor for acute renal failure. Clin Chem 1994;40:796.

Chapter 18

Liver Function

Edward P. Fody

Objectives

Upon completion of this chapter, the clinical laboratorian should be able to:

- *Diagram the anatomy of the liver.*

- *Explain the physiologic functions of the liver to include bile secretion, synthetic activity, and detoxification.*

- *Discuss the basic disorders of the liver and what laboratory tests may be performed to diagnose them.*

- *Given a patient's clinical data, evaluate the information to determine any disorder.*

- *Classify the three types of jaundice and discuss their causes.*

- *Explain the principles of the tests for bilirubin.*

- *Identify the enzymes most commonly used in the assessment of hepatobiliary disease.*

- *Differentiate the various types of hepatitis to include cause (i.e., bacteria, viruses), transmission, occurrence, alternate name, physiology, diagnosis, and treatment.*

KEY WORDS

Bile	Kupffer's Cells
Bilirubin	Lobule
Cirrhosis	Posthepatic
Conjugated Bilirubin	Prehepatic
Hepatitis	Sinusoids
Hepatoma	Urobilinogen
Icterus	

In the past, the liver has been referred to as the center of courage, passion, temper, and love and even as the center of the soul. It was once believed to produce "yellow bile" necessary for good health.

Today, we recognize the liver to be a very complex organ responsible for many major metabolic functions in the body. More than 100 tests measuring these diverse functions have existed in the clinical laboratory at one time or another. However, many were abandoned early on in favor of those which have proven to be most clinically useful. The purpose of this chapter is to discuss these more commonly used liver function tests with particular emphasis on current methodology.

ANATOMY

The liver is the largest and most versatile organ in the body (Fig. 18-1). It consists of two main lobes that together weigh from 1400 to 1600 g in the normal adult. This organ is reddish brown in color and is located under the diaphragm in the right upper quadrant of the abdomen. It has an abundant blood supply, receiving approximately 15 mL per minute from two major vessels: the hepatic artery and the portal vein. The hepatic artery, a branch of the aorta, contributes 20% of the blood supply and provides most of the oxygen requirement. The portal vein, which drains the gastrointestinal (GI) tract, transports the most recently absorbed material from the intestines to the liver. Within the connective tissue of the liver, these vessels give off numerous small branches that form a vascular network around so-called lobules.[49,52]

Michael L. Bishop, Janet L. Duben-Engelkirk, and Edward P. Fody.
CLINICAL CHEMISTRY. © 1996 Lippincott–Raven Publishers.

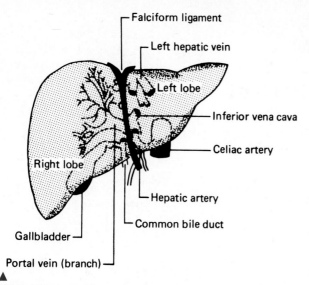

Figure 18-1. Gross anatomy of the liver showing major blood vessels and bile channels. (Adapted from Tietz NW. Fundamentals of clinical chemistry. Philadelphia: WB Saunders, 1976.)

Structural Unit

The "lobule," which measures 1 to 2 mm in diameter, forms the structural unit of the liver (Fig. 18-2). It is composed of cords of liver cells (hepatocytes) radiating from a central vein. The boundary of each lobule is formed by a portal tract made up of connective tissue containing a branch of the hepatic artery, portal vein, and bile duct. Between the cords of liver cells are vascular spaces, called *sinusoids,* that are lined by endothelial cells and Kupffer's cells. These spaces receive blood from the small branches of the hepatic artery and portal vein, located in the portal tracts. The Kupffer's cells are phagocytic macrophages capable of ingesting bacteria or other foreign material from the blood that flows through the sinusoids. The blood from the sinusoids drains into the central veins (hepatic venule) and then to the hepatic veins and inferior vena cava. Primary bile canaliculi are conduits, 1 to 2 μm in diameter, located between the hepatocytes. The bile canaliculi interconnect extensively and increase in size until they connect with the larger bile ducts in the portal tracts.[49,52]

PHYSIOLOGY

There are several hundred known functions performed by the liver each day, including numerous metabolic as well as secretory and excretory functions. The total loss of the liver usually results in death from hypoglycemia within 24 hours. Of the many liver functions, it is only possible in this chapter to discuss those which have prime significance in liver disease.

Excretory and Secretory Function

One of the more important liver functions, and one that is disturbed in a large number of hepatic disorders, is that of the excretion of bile.

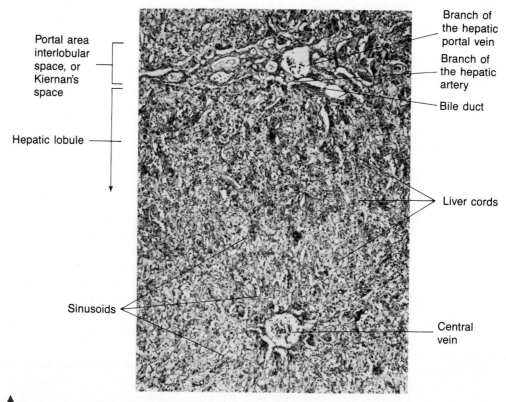

Figure 18-2. Liver lobule (panoramic view). (Photograph courtesy of James Furlong, MD)

Bile is composed of bile acids or salts, bile pigments (primarily bilirubin esters), cholesterol, and other substances extracted from the blood.[54] Total bile production averages about 3 L per day, although only 1 L is excreted. The primary bile acids cholic acid and chenodeoxycholic acid are formed in the liver from cholesterol. The bile acids are conjugated with the amino acids glycine or taurine, forming bile salts. Bile salts (conjugated bile acids) are excreted into the bile canaliculi by means of a carrier-mediated active transport system. During fasting and between meals, a major portion of the bile acid pool is concentrated up to tenfold in the gallbladder. Bile acids reach the intestines when the gallbladder contracts after each meal. Approximately 500 to 600 mL of bile enter the duodenum each day. Here it is intimately involved with the digestion and absorption of lipids. When the conjugated bile acids (salts) come into contact with bacteria in the terminal ileum and colon, dehydration to secondary bile acids (deoxycholic and lithocolic) takes place, and these are subsequently absorbed. The absorbed bile acids enter the portal circulation and return to the liver, where they are reconjugated and reexcreted. The enterohepatic circulation of bile takes place two to five times daily.[25,49]

Bilirubin, the principal pigment in bile, is derived from the breakdown of hemoglobin when aged red blood cells are phagocytized by the reticuloendothelial system, primarily in the spleen, liver, and bone marrow. About 80% of the bilirubin formed daily comes from the degradation of hemoglobin. The remainder comes from the destruction of heme-containing proteins (myoglobin, cytochromes, catalase) and from the catabolism of heme (Fig. 18-3).

When hemoglobin is destroyed, the protein portion, globin, is reused by the body. The iron enters the body's iron stores and is also reused. The porphyrin is broken down as a waste product and excreted. This action of splitting the porphyrin ring and releasing the iron and globin forms biliverdin, which is easily reduced to bilirubin.[12,55]

Bilirubin is transported to the liver in the bloodstream bound to proteins, chiefly albumin. It is then separated from the albumin and taken up by the hepatic cells. Two non-albumin proteins, isolated from liver cell cytoplasm and designated Y and Z, account for the intracellular binding and transport of bilirubin. The conjugation (esterification) of bilirubin takes place in the endoplasmic reticulum of the hepatocyte. An enzyme, uridyldiphosphate glucuronyl transferase (UDPG-T), transfers a glucuronic acid molecule to each of the two proprionic acid side chains in bilirubin, converting bilirubin into a diglucuronide ester. This product, bilirubin diglucuronide, is referred to as *conjugated bilirubin*. Conjugated bilirubin, which is water-soluble, is secreted from the hepatic cell into the bile canaliculi and then passes along with the rest of the bile into larger bile ducts and eventually into the intestines. In the lower portion of the intestinal tract, especially the colon, the bile pigments are acted on by enzymes present in the intestinal bacteria. The first product of this reaction is mesobilirubin, which is reduced to form mesobilirubinogen and then urobilinogen, which is a

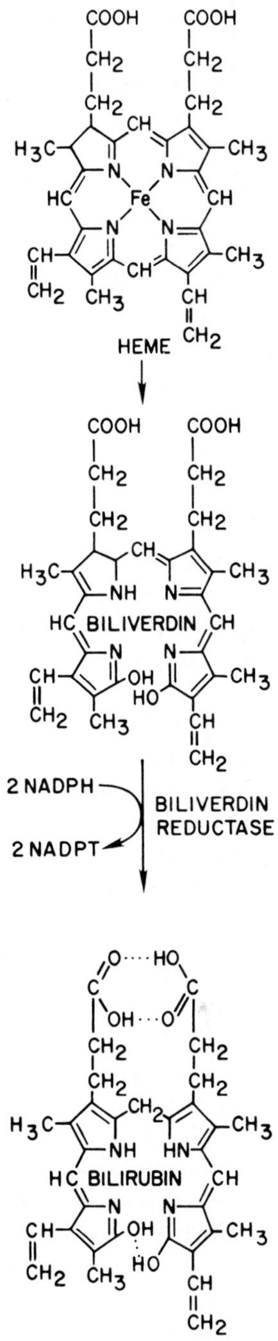

Figure 18-3. The catabolism of heme leading to the formation of bilirubin.

colorless product. The oxidation of urobilinogen produces the red-brown pigment urobilin, which is excreted in the stool. A small portion of the urobilinogen is reabsorbed into the portal circulation and returned to the liver, where it is again excreted into the bile. There is, however, a small quantity that remains in the blood. This urobilinogen is ultimately filtered by the kidney and excreted in the urine (Fig. 18-4).

A total of 200 to 300 mg of bilirubin is produced daily in the healthy adult. A normally functioning liver is required to eliminate this amount of bilirubin from the body. This

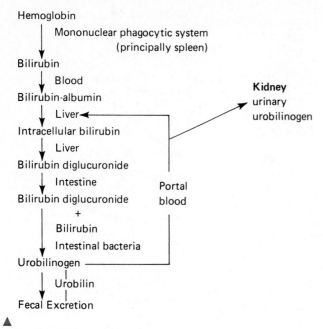

Hemoglobin
↓   Mononuclear phagocytic system
           (principally spleen)
Bilirubin
↓   Blood
Bilirubin-albumin
↓   Liver ←   **Kidney**
Intracellular bilirubin   urinary
↓   Liver   urobilinogen
Bilirubin diglucuronide
↓   Intestine   Portal
Bilirubin diglucuronide   blood
+
Bilirubin
↓   Intestinal bacteria
Urobilinogen
↓
Urobilin
↓
Fecal Excretion

▲

Figure 18-4. Metabolism of bilirubin.

excretory function requires that bilirubin be in the conjugated form, that is, the water-soluble diglucuronide. As mentioned previously, almost all the bilirubin formed is eliminated in the feces, and a small amount of the colorless product urobilinogen is excreted in the urine. Under these normal circumstances, a low concentration of bilirubin (0.2–1.0 mg/dL) is found in the serum, the majority of which is in the unconjugated form. A small percentage (0.2 mg/dL) of this total bilirubin exists in normal serum as the conjugated form.[58]

When the bilirubin concentration in the blood rises, the pigment begins to be deposited in the sclera of the eyes and in the skin. This yellowish pigmentation in the skin or sclera is known as *jaundice* or *icterus.*[12,39]

Jaundice may be caused by a variety of pathophysiologic mechanisms. For example, there may be an increased bilirubin load on the liver cell or a disturbance in uptake and transport of bilirubin within the liver cell. In addition, there may be defects in conjugation or excretion of bilirubin into the bile. Further difficulties may be due to obstruction of the large bile ducts before the bilirubin reaches the intestines. Several classifications of jaundice are found in the literature. One of the more frequently used classifications is based on the presumed site of physiologic or anatomic abnormality. In this classification, there are predominately three types of jaundice: prehepatic, hepatic, and posthepatic.

Prehepatic jaundice results when an excessive amount of bilirubin is presented to the liver for metabolism, such as in hemolytic anemia. This type of jaundice is characterized by unconjugated hyperbilirubinemia. However, the serum bilirubin levels rarely exceed 5 mg/dL because the normal liver is capable of handling most of the overload. As mentioned previously, unconjugated bilirubin is not water-soluble and is bound to albumin so that it is not filtered out

of the blood by the kidney. Therefore, bilirubin will not appear in the urine in this type of jaundice.

The largest percentage of patients with jaundice have the hepatic type. Hepatic jaundice may result from impaired cellular uptake, from defective conjugation, or from abnormal secretion of bilirubin by the liver cell.

Gilbert's syndrome is a relatively common disorder characterized by impaired cellular uptake of bilirubin. Affected individuals have no symptoms but may have mild icterus. The elevated level of bilirubin is less than 3 mg/100 mL and is unconjugated. Crigler-Najjar syndrome is a more serious disorder caused by a deficiency of the enzyme UDPG-T. Two types have been described. Type I, in which there is a complete absence of the enzyme in the liver, is very rare. No conjugated bilirubin is formed, and the bile is colorless. This type is uniformly fatal. In type II Crigler-Najjar syndrome, there is a less severe deficiency of the enzyme, and some conjugated bilirubin is formed. Dubin-Johnson syndrome and Rotor's syndrome are two hereditary disorders characterized by conjugated hyperbilirubinemia from defective excretion by the liver cell. Any cause of severe hepatocellular damage also may interfere with uptake, conjugation, or secretion of bilirubin. This will lead to unconjugated as well as conjugated hyperbilirubinemia.

Posthepatic jaundice results from the impaired excretion of bilirubin caused by mechanical obstruction of the flow of bile into the intestines. This may be due to gallstones or a tumor. When bile ceases to flow into the intestines, there is a rise in the serum level of conjugated bilirubin, and the stool loses its source of normal pigmentation and becomes clay-colored. Conjugated bilirubin appears in the urine, and urine urobilinogen levels decrease.[39,48,52] The various laboratory tests that assist in making the distinction between the different causes of jaundice will be discussed later in this chapter.

Major Synthetic Activity

Among the many diverse metabolic functions carried out by the liver is the synthesis of many major biologic compounds to include proteins, carbohydrates, and lipids. The liver plays an important role in plasma protein production, synthesizing albumin and the majority of the α- and β-globulins. All the blood-clotting factors (except VIII) are also synthesized in the liver. In addition, the deamination of glutamate in the liver is the primary source of ammonia, which is then converted to urea.[18,40,49]

The synthesis and metabolism of carbohydrates is also centered in the liver.[51] Glucose is converted to glycogen, a portion of which is stored in the liver and later reconverted to glucose as necessary. An additional important liver function is gluconeogenesis from amino acids.

Fat is formed from carbohydrates in the liver when nutrition is adequate and the demand for glucose is being met from dietary sources. The liver also plays a key role in the metabolism of fat. It is the major site for the removal of chylomicron remnants and for the conversion of acetyl CoA to fatty acids, triglycerides, and cholesterol. As mentioned earlier, further metabolism of cholesterol into bile acids also occurs in the liver. Very-low-density lipoproteins, which are responsible for transporting triglycerides into the tissues, are synthesized primarily in the liver. High-density lipoproteins are also made in the liver, as are phospholipids.

The formation of ketone bodies occurs almost exclusively in the liver. When the demand for gluconeogenesis depletes oxaloacetate and acetyl CoA cannot be converted rapidly enough to citrate, acetyl CoA accumulates, and a decyclase in the liver liberates ketone bodies into the blood.[52]

The liver is the storage site for all the fat-soluble vitamins (A,D,E, and K) and several water-soluble vitamins, such as B_{12}. Another vitamin-related function is the conversion of carotene into vitamin A.

The liver is the source of somatomedin (an insulin-like factor that mediates the activity of growth hormone) and angiotensinogen and is a major site of metabolic clearance of many other hormones. As the source of transferrin, ceruloplasmin, and metallothionein, the liver plays a key role in the transport, storage, and metabolism of iron, copper, and other metals.[49,52]

Many enzymes are synthesized by liver cells, but not all of them have been found useful in the diagnosis of hepatobiliary disorders. Those enzymes that have been used often include aspartate aminotransferase (AST)(SGOT) and alanine aminotransferase (ALT)(SGPT), which escape into the plasma from damaged liver cells; alkaline phosphatase (ALP), and 5′ nucleotidase (5NT), which are induced or released when the canalicular membrane is damaged and biliary obstruction occurs; and α-glutamyl transferase (GGT), which is increased in both hepatocellular and obstructive disorders.[3,68] These will be discussed in greater detail later in this chapter.

Detoxification and Drug Metabolism

Because the liver is interposed between the splanchnic circulation and the systemic blood, it serves to protect the body from potentially injurious substances absorbed from the intestinal tract and toxic by-products of metabolism. The most important mechanism in this detoxification activity is the microsomal drug-metabolizing system of the liver. This system is induced by many drugs (*e.g.,* phenobarbital) and other foreign compounds and is responsible for many of the detoxification mechanisms to include oxidation, reduction, hydrolysis, hydroxylation, carboxylation, and demethylation.[23] These various mechanisms serve to convert many noxious or comparatively insoluble compounds into other forms that are either less toxic or more water-soluble and therefore excretable by the kidney. For example, ammonia, a very toxic substance arising in the large intestines through bacterial action on amino acids, is carried to the liver by the portal vein and converted into the innocuous compound urea by the hepatocytes.

Conjugation with moieties such as glycine, glucuronic acid, sulfuric acid, glutamine, acetate, cysteine, and glutathione occurs mainly in the cytosol or smooth endoplasmic reticulum. This mechanism was discussed earlier in the chapter as the mode of bilirubin and bile acid excretion.

DISORDERS OF THE LIVER

Jaundice

Jaundice, or *icterus,* refers to the yellowish discoloration of the skin and sclerae resulting from hyperbilirubinemia. Although the upper limit of normal for total serum bilirubin is 1 mg/dL, jaundice is not clinically apparent until the bilirubin level exceeds 2 to 3 mg/dL. In African-American or Asian patients, yellowing of the sclerae may be the only clinical evidence of jaundice.[62]

Jaundice is one of the oldest medical disorders known, having been described in ancient Greek, Roman, Chinese, and Hebrew texts. Hippocrates related jaundice to dysfunction of the liver.

Except in infants, hyperbilirubinemia is generally well tolerated and does not produce serious clinical side effects. However, in infants, hyperbilirubinemia (levels exceeding 15–20 mg/dL) may be associated with kernicterus, a potentially very serious disorder of the central nervous system resulting from the increased bilirubin levels. This occurs only in infants because the immature central nervous system does not have a well-developed blood-brain barrier.[31,64]

While all cases of jaundice result from hyperbilirubinemia, not all are caused by hepatic dysfunction. While the majority of cases of jaundice are associated with liver disorders, hyperbilirubinemia may also result from erythrocyte destruction, or hemolysis, in patients with normal liver function. Thus, distinction between hepatic and hemolytic disease in the patient presenting with jaundice is an important task for which the attending physician will rely heavily on laboratory results. This will be discussed in detail later in this chapter.[35,40]

It is worth noting that hypercarotenemia, a disorder caused by the excessive ingestion of vitamin A, may produce skin discoloration indistinguishable from that of hyperbilirubinemia. In hypercarotenemia, however, the sclerae are usually not discolored.[39,60]

Cirrhosis

Cirrhosis is derived from the Greek word that means "yellow." However, in current usage, *cirrhosis* refers to the irreversible scarring process by which normal liver architecture is transformed into abnormal nodular architecture. One way to classify cirrhosis is by the appearance of the liver, *i.e.,* by the size of the nodules. These conditions are referred to as *macronodular* and *micronodular cirrhosis,* although mixed forms occur.[49,50]

Another way to classify cirrhosis is by etiology. In the United States, Canada, and western Europe, the leading cause of cirrhosis is alcohol abuse. This leads to a micronodular type of cirrhosis.

Other causes of cirrhosis include hemochromatosis, post-necrotic cirrhosis (which occurs as a late consequence of hepatitis), and primary biliary cirrhosis (which is an auto-immune disorder). Other uncommon etiologies of cirrhosis exist. About 10% to 20% of cases cannot be classified as to etiology. Cirrhosis is a serious disorder and is one of the 10 leading causes of death in the United States. It causes many complications.[21]

Portal hypertension results when the blood flow through the portal vein is obstructed by the cirrhotic liver. This may result in splenomegaly, which may not be clinically significant, and esophageal varices, which may rupture and lead to fatal hemorrhage.[39] The synthetic ability of the liver is reduced, causing hypoalbuminemia and deficiency of the clotting factors, which may lead to hemorrhage. Ascitic fluid may accumulate in the abdomen. Although some patients with cirrhosis are capable of prolonged survival, generally this diagnosis is an ominous one.[9,39,49]

Tumors of the Liver

On a worldwide basis, primary malignant tumors of the liver, known as *heptocellular carcinoma, hepatocarcinoma,* or *hepatoma,* are an important cause of cancer mortality. In the United States, these tumors are relatively uncommon. Most cases of hepatocellular carcinoma can be related to previous infection with a hepatitis virus. These tumors are especially common in parts of Africa and Asia and are infrequent in North America and western Europe. However, the liver is frequently involved secondarily by tumors arising in other organs. Metastatic tumors to the liver from primary sites such as the lung, pancreas, gastrointestinal tract, or ovary are common. Benign tumors of the liver are relatively uncommon.

Whether primary or secondary, any malignant tumor in the liver is a serious finding with a poor prognosis. Generally, the only hope for cure relies on surgical resection, which is not usually possible. Patients with malignancies of the liver usually have a survival measured in months.[14,34,59]

Reye's Syndrome

Reye's syndrome is a disorder of unknown cause involving the liver and arising primarily in children, although cases in adults have been reported. It is a form of hepatic destruction that usually occurs following recovery from a viral infection, such as varicella (chickenpox) or influenza. Shortly after the infection, the patient develops neurologic abnormalities, which may include seizures or coma. Liver functions are always abnormal, but the bilirubin level is not usually elevated. Without treatment, rapid clinical deterioration leading to death may occur.[20,22,28]

Drug- and Alcohol-Related Disorders

Many drugs and chemicals are toxic to the liver. This toxicity may take the form of overwhelming hepatic necrosis

leading to coma and death, or it may be subclinical and pass entirely unnoticed. Of all hepatic toxins, probably the most important is ethanol. In small amounts, alcohol may cause mild, inapparent injury. Heavier consumption leads to more serious damage, and prolonged heavy use may lead to cirrhosis. The exact amount of alcohol needed to cause cirrhosis is not known, and only a minority of alcoholics develop this condition. It is, however, a leading cause of morbidity and mortality in the United States.

Certain drugs, including tranquilizers such as phenothiazines, certain antibiotics, antineoplastic agents, and anti-inflammatory drugs, may cause liver injury. Usually this is mild and manifested only by elevation of liver function tests, which return to normal when the offending agent is discontinued. However, occasionally, this may lead to massive hepatic failure or cirrhosis.[23,38]

One of the most common drugs associated with serious hepatic injury is acetaminophen. This drug, when taken in massive overdose, is virtually certain to produce fatal hepatic necrosis unless rapid treatment is initiated.[23,37]

ASSESSMENT OF LIVER FUNCTION

Analysis of Bilirubin

Brief Review of Classical Methodology

Because its yellow color is detectable by the human eye, concentrations of serum bilirubin have been estimated for centuries. In 1883, Ehrlich first described a reaction in urine samples of the formation of a red or blue pigment when bilirubin was coupled with a diazotized sulfanilic acid solution. In 1913, Van den Bergh applied the Ehrlich reaction to serum bilirubins. He also used alcohol as an accelerator for the coupling of bilirubin to diazotized sulfanilic acid. Malloy and Evelyn developed the first useful quantitative technique for bilirubin in 1937 by accelerating the reaction with a 50% methanol solution, a technique that avoided the precipitation of proteins that was a source of error in the Van den Bergh method. In 1938, Jendrassik and Grof used a procedure containing caffeine-benzoate-acetate as an accelerator for the azo-coupling reaction.[61]

Bilirubin also has been quantified by methods other than coupling to sulfanilic acid. These include a direct measurement of its natural color. This principle was successfully employed in the development of the icterus index, which was introduced in 1919.[23] The test involves diluting serum with saline until it visually matches the color of a 0.01% potassium dichromate solution. The number of times the serum must be diluted is called the *icterus index*. However, substances in the serum other than bilirubin, such as carotene, xanthophyll, and hemoglobin, also may contribute to the icterus index, limiting its clinical usefulness. This test is now considered obsolete. In recent years, bilirubin has been quantitated by dilution in a buffer, followed by direct measurement of the absorption, using a well-calibrated spectrophotometer. This method is employed in the pediatric laboratory on newborn infants whose serum does not yet contain the interfering yellow lipochromes. The hemolysis that is so often a problem with pediatric specimens is "blanked out" by measuring a second wavelength.

With several of these methods, it was determined that two types of bilirubin existed.[4] The fraction that produced a color in the Van den Bergh method in aqueous solution was described as *direct bilirubin,* whereas the fraction that produced a color only after alcohol was added was called *indirect bilirubin.* For many years, results of bilirubin determinations were reported as direct or indirect. This terminology is now outdated. Since 1956, it has been known that the direct reaction is given by the diglucuronide of bilirubin or conjugated bilirubin, which is water-soluble. The indirect reaction, on the other hand, is given by unconjugated bilirubin, which is water-insoluble but dissolves in alcohol to couple with the diazo reagent. Direct and indirect bilirubin should be reported as conjugated and unconjugated, respectively. Most commonly, conjugated and total bilirubin are reported.

Method Selection

Unfortunately, no single method for the determination of bilirubin will meet all the requirements of the clinical laboratory. For the evaluation of jaundice in the newborn, the direct spectrophotometric method is satisfactory.[43] The sources of error in this technique are turbidity, hemolysis, and yellow lipochrome pigments. Hemolysis and turbidity can be "blanked out" by measuring a second wavelength, but the yellow lipochromes cannot be "blanked out." This method is, therefore, only valid for newborn infants whose serum does not contain lipochromes. In patients older than 1 month, a diazo colorimetric procedure is necessary. Most investigators who have compared the different methods for the measurement of bilirubin agree that the majority of techniques available today will give accurate results for total bilirubin, provided that good standards are available. The choice of method, then, should be based on whether one prefers a manual versus automated technique and whether direct bilirubin is to be determined. The choice of a direct bilirubin method poses the greatest problem because there is no reference method or adequate standardization available.

If a manual procedure is desired, then either the Evelyn-Malloy or Jendrassik-Grof method is suitable. The Jendrassik-Grof method is slightly more complex but has the following advantages over the Evelyn-Malloy:[16]

1. It is insensitive to sample pH changes.
2. It is insensitive to a 50-fold variation in protein concentration of the sample.
3. It has adequate optical sensitivity even for low bilirubin concentrations.
4. It has minimal turbidity and a relatively constant serum blank.
5. It is not affected by hemoglobin up to 750 mg/dL.

Since the development of the original procedure by Jendrassik-Grof, a number of modifications have been made

to speed up the reaction, reduce interference, and so on. There are now several commercial bilirubin procedures that employ a modified Jendrassik-Grof method.[17] It seems to be a particularly popular technique for the discrete sampler analyzers currently on the market.

The recommended methods for total bilirubin determination using all automated machines will generally give equivalent results. The techniques for direct bilirubin are unfortunately not as reliable. The best method for measuring small amounts of conjugated serum bilirubin is a research method that uses high-performance liquid chromatography, but this method is too difficult for routine laboratory use. Most clinical laboratories use either the Evelyn-Malloy or the Jendrassik-Grof method. Because space in this chapter does not allow for a detailed description of all the bilirubin test methodologies mentioned previously, we will emphasize only the most widely used principles for measuring adult and pediatric bilirubin.

JENDRASSIK-GROF METHOD FOR TOTAL AND CONJUGATED BILIRUBIN DETERMINATION[29]

Principle—Serum or plasma is added to a solution of sodium acetate and caffeine–sodium benzoate, which is then added to diazotized sulfanilic acid to form a purple azobilirubin. The sodium acetate buffers the pH of the diazotization reaction, while the caffeine–sodium benzoate accelerates the coupling of bilirubin with diazotized sulfanilic acid. This reaction is terminated by the addition of ascorbic acid, which destroys the excess diazo reagent. A strongly alkaline tartrate solution is then added to convert the purple azobilirubin to blue azobilirubin, and the intensity of the color is read at 600 nm.

Specimen Collection and Storage—A fasting serum specimen that is neither hemolyzed nor lipemic in nature is preferred. Prior to testing, serum should be stored in the dark and measured as soon as possible (within 2–3 h) after collection. Serum may be stored in the dark in a refrigerator for up to 1 week and in the freezer for 3 months without appreciable change in the bilirubin concentration.

Reagents
1. 0.05 Molar hydrochloric acid
2. Caffeine reagent—Add 50 g of caffeine, 75 g of sodium benzoate, and 125 g of sodium acetate to 450 mL of warm (50–60°C) distilled water. Cool, then add 1 g of Na_2 EDTA and dilute to 1 L. Stable for 6 months at room temperature.
3. Diazo 1—0.05% sulfanilic acid
4. Diazo 2—0.05% sodium nitrate
5. Working diazo (0.25 m diazo 1 plus 10 mL diazo 2)—Use within 30 minutes.
6. 4% Ascorbic acid solution
7. Tartrate solution—Add 100 g of sodium hydroxide and 350 g of sodium tartrate to 450 mL of distilled water. Cool; then dilute to 1 L. Stable for 6 months at room temperature.

Procedure for Total Bilirubin Determination
1. Add 1.0 mL of caffeine reagent to each of 2 cuvets (1.0-cm rectangular), one labeled "test" and one labeled "blank."
2. Add 100 µL of serum or plasma to each cuvet, followed by 150 µL of water.
3. Add 250 µL of sulfanilic acid to the blank cuvet and mix.
4. Add 250 µL of fresh diazo reagent to the test cuvet and mix.
5. Exactly 10 minutes after the addition of diazo reagent in step 4, add 0.5 mL of tartrate solution to the test cuvet and mix.

6. Read the absorbance of the test cuvet at 600 nm against the blank cuvet set at zero absorbance.

Procedure for Conjugated Bilirubin Determination

1. Add 1.0 mL of 0.05 M HCl to a cuvet labeled "test" and 1.0 mL of caffeine reagent to a cuvette labeled "blank."
2. Add 100 μL of serum or plasma to each cuvet followed by 150 μL of water.
3. Add 250 μL of sulfanilic acid to the blank cuvet and mix.
4. Add 250 μL of fresh diazo reagent to the test cuvet and mix.
5. Exactly 1 minute later, add 50 μL of ascorbic acid solution to both cuvets; then immediately add 0.5 mL of tartrate solution to each cuvet and mix.
6. Read the absorbance of the test cuvet at 600 nm against the blank cuvet at zero absorbance.

*Calculations–*Values for total and conjugated bilirubin are obtained by comparing the absorbance read to that of a calibration curve prepared with acceptable bilirubin standards. Values for unconjugated (indirect) bilirubin are obtained by subtracting the conjugated (direct) bilirubin from the total bilirubin.

*Comments and Sources of Error–*Normal blood contains no conjugated bilirubin. The reason some conjugated bilirubin is reported as normal is because current available methodology picks up some of the total bilirubin as a false-positive. The best routine methods, however, keep this technical error to a minimum and report upper limits or normal of less than 0.2 mg/dL for conjugated serum bilirubin. This method compensates for lipochrome pigments, which are present in the serum of adults and children greater than a few months of age. However, a hemolyzed specimen will cause a decrease in serum bilirubin by this method. In addition, because lipemia causes interference, fasting blood specimens are preferable. Serious loss of bilirubin occurs after exposure to fluorescent and indirect and direct sunlight. It is, therefore, imperative that exposure of samples and standards to light be kept to a minimum and that specimens and standards be refrigerated in the dark until the tests are performed. ■

Reference Range

Reference ranges for infants after 1 month and adults are shown in Table 18-1.

DIRECT SPECTROPHOTOMETRIC METHOD FOR DETERMINATION OF TOTAL BILIRUBIN IN SERUM[43]

*Principle–*The absorbance of bilirubin in the serum at 455 nm is proportional to its concentration. The serum of newborn infants does not contain lipochromes, such as carotene, that would increase the absorbance at 455 nm. The absorbance of hemoglobin at 455 nm is corrected by subtracting the absorbance at 575 nm.

*Specimen–*Serum is collected and stored with the same precautions as mentioned previously for adult specimens.

Reagents

1. 15 M phosphate buffer, pH 7.4.
2. To obtain accurate values, a spectrophotometer capable of transmitting bandwidths of 10 nm or less should be used.

TABLE 18-1. Reference Ranges for Bilirubin

Conjugated	0–0.2 mg/dL (0–3 μmol/L)
Unconjugated	0.2–0.8 mg/dL (3–14 μmol/L)
Total	0.2–1.0 mg/dL (3–17 μmol/L)

Procedure

1. With a micropipet, add 20 μL of serum to a microcuvet (1.0-cm light path) containing 1 mL of phosphate buffer.
2. Add 1 mL of phosphate buffer to a microcuvet as a blank and set at zero absorbance.
3. Read the absorbances of the diluted serum at 455 and 575 nm.

Calculation

$$(\text{Absorbance}_{450\,nm} - \text{absorbance}_{575\,nm}) \times 79 = \text{bilirubin in mg/dL}$$

*Comments and Sources of Error–*Error will be introduced if the buffer is turbid. Since the method depends on the extinction coefficient of bilirubin, all volumes must be accurate, and cuvets must be flat-surfaced with a path length of exactly 1 cm. A control should be employed with a level near 20 mg/dL, which is a critical decision point for the clinician, since exchange transfusion is necessary if this level is exceeded.

Precautions such as those mentioned in the previous method should also be followed for collection and storage of specimens. This method is relatively insensitive to hemolysis, which is often present in specimens obtained from infants, due to difficulty in skin puncture technique. However, it is markedly effected by the presence of lipochromes, and therefore cannot be used in infants more than a few months of age. ■

Reference Range

See Table 18-2.

Urobilinogen in Urine and Feces

As mentioned earlier, urobilinogen is a colorless end product of bilirubin metabolism that is oxidized by intestinal bacteria to the brown pigment urobilin. In the normal individual, part of the urobilinogen is excreted in the feces, and the remainder is reabsorbed into the portal blood and returned to the liver. A small portion is not taken up by the hepatocytes and is excreted by the kidney as urobilinogen. Increased levels of urinary urobilinogen are found in hemolytic disease and in defective liver-cell function such as that seen in hepatitis. Absence of urobilinogen from the urine and stool is most often seen with complete biliary obstruction. Fecal urobilinogen is also decreased in biliary obstruction, as well as in hepatocellular disease.[5]

The majority of the quantitative methods for urobilinogen are based on the reaction of this substance with *p*-dimethyl aminobenzaldehyde to form a red color. This reaction was first described by Ehrlich in 1901. Many modifications of this procedure have been made over the years to improve specificity. Major improvements were made in 1925 by Terwen, who used alkaline ferrous hydroxide to reduce urobilin to

TABLE 18-2. Reference Ranges for Infant Total Bilirubin

Infants	Premature, Total	Full-Term, Total
24 h	1–6 mg/dL	2–6 mg/dL
48 h	6–8 mg/dL	6–7 mg/dL
3–5 days	10–12 mg/dL	4–6 mg/dL

urobilinogen and added sodium acetate to eliminate interference from such compounds as indole. The use of petroleum ether rather than diethyl ether for the extraction of urobilinogen was introduced in 1936 by Watson to help in the removal of other interfering substances.[63] However, because studies indicate that the quantitative methods described do not completely recover urobilinogen from the urine, most laboratories use the less laborious, more rapid, semiquantitative method described next.[5,61]

DETERMINATION OF URINE UROBILINOGEN (SEMIQUANTITATIVE)[66]

Principle—Urobilinogen reacts with *p*-dimethyl aminobenzaldehyde to form a red color, which is then measured spectrophotometrically. Ascorbic acid is added as a reducing agent to maintain urobilinogen in the reduced state. The use of saturated sodium acetate stops the reaction and minimizes the combination of other chromogens with the Ehrlich's reagent.

Specimen
A *fresh* 2-h urine is collected. This specimen should be kept cool and protected from light.

Reagents
1. Ehrlich's reagent (*p*-dimethylaminobenzaldehyde in HCl)
2. Sodium acetate, saturated
3. Ascorbic acid powder
4. Stock standard (20 mg of phenolsulfonaphthalene)
5. Working standard (0.2 mg/dL with absorbance equivalency of a 0.346 mg/dL solution or urobilinogen)

Procedure
1. Measure the volume of the 2-hr urine sample.
2. Test the urine for the presence of bilirubin. If more than a trace is present, mix 2.0 mL of 10 g/dL $BaCl_2$ solution with 8.0 mL of urine and filter.
3. Dissolve 100 mg of ascorbic acid in 10 mL of urine (centrifuge if cloudy). Then place 1.5-mL aliquots in each of 2 tubes labeled "blank" and "unknown."
4. To the blank tube, add 4.5 mL of a freshly prepared mixture of 1 volume of Ehrlich's reagent and 2 volumes of saturated sodium acetate, and mix.
5. To the unknown tube add 1.5 mL of Ehrlich's reagent, mix well, and immediately add 3.0 mL of saturated sodium acetate and mix.
6. Within 5 minutes, measure the absorbance of the tubes marked blank and unknown at 562 nm against water set at zero. Measure the absorbance of the working standard against water at the same wavelength.

Calculations
See Table 18-3.

Comments and Sources of Error
1. The results of this test are reported in Ehrlich units rather than in milligrams of urobilinogen because substances other than urobilinogen account for some of the final color development.
2. Compounds other than urobilinogen that may be present in the urine and react with Ehrlich's reagent include porphobilinogen, sulfonamides, procaine, and 5-hydroxyindoleacetic acid. Bilirubin will form a green color and therefore must be removed, as described above.
3. Fresh urine is necessary, and the test must be performed without delay to prevent oxidation of urobilinogen to urobilin. Similarly, the spectrophotometric readings should be made within 5 minutes after color production because the urobilinogen-aldehyde color slowly decreases in intensity. ■

Reference Range
Urine urobilinogen   0.1–1.0 Ehrlich units/2 h or
0.5–4.0 Ehrlich units/day
(0.8–6.8 µ mol/day)

(1 Ehrlich unit is equivalent to approximately 1 mg of urobilinogen.)

Fecal Urobilinogen
Visual inspection of the feces usually suffices to detect decreased urobilinogen. However, the semiquantitative determination of fecal urobilinogen is available and involves the same principle described earlier for the urine.[66] It is carried out in an aqueous extract of fresh feces, and any urobilin present is reduced to urobilinogen by treatment with alkaline ferrous hydroxide before Ehrlich's reagent is added. A range of 75 to 275 Ehrlich units/100 g of fresh feces or 75 to 400 Ehrlich units per 24-hour specimen is considered a normal reference range.[5]

Measurement of Serum Bile Acids

Unfortunately, complex methods are required for the analysis of bile acids in serum. These involve extraction with organic solvents, partition chromatography, gas chromatography–mass spectroscopy, spectrophotometry, ultraviolet light absorption, fluorescence, radioimmunoassay, and enzyme immunoassay methods.[1,6,57] Despite the fact that serum bile acid levels are elevated in liver disease, the total con-

---

TABLE 18-3. Calculations

$$\text{Ehrlich units/100 mL urine} = \frac{\text{Abs (unk)} - \text{Abs (blk)}}{\text{Abs (std)}} \times 0.346 \times \frac{6.0}{1.5}$$

$$= \frac{\text{Abs (unk)} - \text{Abs (blk)} \times 1.38}{\text{Abs (std)}}$$

$$\text{Ehrlich units/2 h} = \frac{\text{Abs (unk)} - \text{Abs (blk)}}{\text{Abs (std)}} \times 0.0138 \times \text{urine vol in mL}$$

Note: Multiply answer by 1.25 if bilirubin was removed by $BaCl_2$.

centration is extremely variable and adds no diagnostic value to other tests of liver function. The variability of the type of bile acids present in serum, together with their existence in different conjugated forms, suggests that more relevant information of liver dysfunction may be gained by examining patterns of individual bile acids and their state of conjugation. For example, it has been suggested that the ratio of the trihydroxy to dihydroxy bile acids in serum will differentiate patients with obstructive jaundice from those with hepatocellular injury and that the diagnosis of primary biliary cirrhosis and extrahepatic cholestasis can be made on the basis of the ratio of the cholic to chenodeoxycholic acids. However, the high cost of these tests, the time required to do them, and the current controversy concerning their clinical usefulness render this approach unsatisfactory for routine use.

Enzyme Tests in Liver Disease

Any injury to the liver that results in cytolysis and necrosis causes the liberation of various enzymes. The measurement of these hepatic enzymes in the serum is used to assess the extent of liver damage and to differentiate hepatocellular (functional) from obstructive (mechanical) disease. The methods used to measure these enzymes, their normal reference ranges, and other general aspects of enzymology are considered in another chapter. Our discussion will be limited to the characteristic changes in serum enzyme levels seen in various hepatic disorders.

The most common enzymes assayed in hepatobiliary disease include alkaline phosphatase and the aminotransferases. Employed less often are γ-glutamyl transpeptidase, lactate dehydrogenase and its isoenzymes, 5′-nucleotidase, ornithine carbamyl transferase, and leucine aminopeptidase.[11,44]

Alkaline Phosphatase (ALP)

Alkaline phosphatase (ALP) is found in a number of tissues but is used most often in the clinical diagnosis of bone and liver disease. Slight to moderate increases in alkaline phosphatase activity occur in many patients with hepatocellular disorders, such as hepatitis and cirrhosis, and transient increases may occur in all types of liver disease. The most striking elevations occur in extrahepatic biliary obstruction, such as a stone in the common bile duct, or in intrahepatic cholestasis, such as drug cholestasis or primary biliary cirrhosis. This enzyme is almost always increased in metastatic liver disease and may be the only abnormality on routine liver function tests. Since bone is a source of the enzyme, Paget's disease, bony metastases, and other diseases associated with increased osteoblastic activity may produce high levels of alkaline phosphatase in the absence of liver disease. The enzyme is found in placenta, and pregnant women also have elevated levels.[11,30,39]

Aminotransferases (Transaminases)

Aspartate aminotransferase (AST), or glutamate oxaloacetic transaminase (SGOT), and alanine aminotransferase (ALT), or glutamate pyruvate transaminase (SGPT), are two enzymes widely used to assess hepatocellular damage. AST (SGOT) is found in all tissues, especially the heart, liver, and skeletal muscle. ALT (SGPT) is present primarily in the liver and, to a lesser extent, in kidney and skeletal muscle, making it more "liver-specific."

In the absence of acute necrosis or ischemia of other organs, elevated aminotransferase levels suggest hepatocellular damage. In severe viral hepatitis causing extensive acute necrosis, markedly elevated serum aminotransferase levels may be found, whereas only moderate increases are found in less severe cases. Mild chronic or focal liver diseases, such as subclinical or anicteric viral hepatitis, alcoholic cirrhosis, granulomatous infiltration, and tumor invasion, may be associated with only mild abnormalities. Minimal elevations occur in biliary obstruction.[66]

It is often helpful to conduct serial determinations of aminotransferases when following the course of a patient with acute or chronic hepatitis. However, one should exercise caution in interpreting these abnormal levels, since serum transaminases may actually decrease in some patients with severe acute hepatitis owing to the exhaustive release of hepatocellular enzymes.[11,44,60]

5′-Nucleotidase

5′-Nucleotidase is another phosphatase, which originates largely in the liver and is used clinically to determine whether an elevation of the alkaline phosphatase is caused by liver or bone disease. Levels of both 5′-nucleotidase and alkaline phosphatase are elevated in liver disease, whereas in primary bone disease alkaline phosphatase level is elevated, but the 5′-nucleotidase level is usually normal or only slightly elevated. This enzyme is much more sensitive to metastatic liver disease than is alkaline phosphatase because, unlike ALP, its level is not markedly elevated in other conditions such as pregnancy or childhood. In addition, some increase in enzyme activity may be noted after abdominal surgery.[3,11,39]

γ-Glutamyl Transpeptidase (GGT)

γ-Glutamyl transpeptidase (GGT) is found in high concentrations in the kidney and the liver and is elevated in the serum of almost all patients with hepatobiliary disorders. It is not specific for any type of liver disease but is frequently the first abnormal liver function test demonstrated in the serum of heavy drinkers. The highest levels are seen in biliary obstruction. It is, therefore, a sensitive test for alcoholic liver disease. Measurement of this enzyme is also useful if jaundice is absent for the confirmation of hepatic neoplasms and is a useful test to confirm hepatic disease in patients with elevated alkaline phosphates.[3,11,39,60]

Ornithine Carbamyl Transferase (OCT)

Ornithine carbamyl transferase (OCT) is a urea cycle enzyme present only in liver and intestinal tissue. Elevated levels of this enzyme occur primarily in liver disease but are not specific for any particular types of liver disease. Its diagnostic usefulness is therefore limited.[11,44]

Leucine Aminopeptidase

Leucine aminopeptidase is widely distributed in human tissues and is found in the pancreas, gastric mucosa, liver, spleen, large intestine, brain, small intestine, and kidney. Most investigators now believe that the serum activity of leucine aminopeptidase cannot be used to differentiate hepatocellular from obstructive jaundice. Furthermore, the measurement of this enzyme does not provide any useful information that cannot be obtained by other tests, such as the determination of 5'-nucleotidase or γ-glutamyl transferase.[3,11,24]

Lactate Dehydrogenase (LD)

Measurement of total serum lactate dehydrogenase (LD) is usually not helpful diagnostically because it is present in all organs and released into the serum from a variety of tissue injuries. However, fractionation of LD into its five tissue-specific isoenzymes may give useful information about the site of origin of the LD elevation. LD-5 is present, for the most part, in liver and skeletal muscle. An interpretation of isoenzyme patterns may be simple with an elevated LD-5 in a patient with jaundice. However, the similarity of isoenzyme patterns between different tissue-damaging states may necessitate the use of additional laboratory tests for interpretation.[11,15,39]

Moderate elevations of total serum LD levels are common in acute viral hepatitis and in cirrhosis, whereas biliary tract disease may produce only slight elevations. High serum levels may be found in metastatic carcinoma of the liver.[66]

Tests Measuring Hepatic Synthetic Ability

The measurement of the end products of hepatic synthetic activity can be used to assess liver disease. Although these tests are not very sensitive to minimal liver damage, they are useful in quantitating the severity of hepatic dysfunction.

Almost all serum proteins are produced by the liver. A decreased serum albumin may be due to decreased liver protein synthesis. The albumin level correlates well with the severity of functional impairment and is found more often in chronic rather than acute liver disease. The serum α-globulins also tend to decrease with chronic liver disease. However, a very low or absent α-globulin suggests α-antitrypsin deficiency as the cause of the chronic liver disease. Serum γ-globulin levels are transiently increased in acute liver disease and remain elevated in chronic liver disease. The highest elevations are found in chronic active hepatitis and postnecrotic cirrhosis. In particular, IgG and IgM levels are more consistently elevated in chronic active hepatitis, IgM in primary biliary cirrhosis, and IgA in alcoholic cirrhosis.[49]

The prothrombin time is commonly increased in liver disease because the liver is unable to manufacture adequate amounts of clotting factors or because the disruption of bile flow results in inadequate absorption of vitamin K from the intestine. Response of the prothrombin time to the administration of vitamin K is therefore of some value in differentiating intrahepatic disease with decreased synthesizing capacity from extrahepatic obstruction with decreased absorption of fat-soluble vitamins. A marked prolongation of the prothrombin time indicates severe diffuse liver disease and a poor prognosis.[11,21,33]

Tests Measuring Nitrogen Metabolism

The liver plays a major role in removing ammonia from the bloodstream and converting it to urea so that it can be removed by the kidneys. In liver failure, ammonia and other toxins increase in the bloodstream and may ultimately cause hepatic coma.[42] In this condition, the patient becomes increasingly disoriented and gradually lapses into unconsciousness. The cause of hepatic coma is not fully defined, although ammonia is presumed to play a major role. However, the correlation between blood ammonia levels and the severity of the hepatic coma is very poor. Therefore, the ammonia level is useful only when the patient serves as his own control and multiple measurements are made over a period of time.[40]

The production of glutamine by an enzymatic intracellular reaction between ammonia and glutamic acid provides a mechanism for removal of ammonia from the central nervous system. Elevations of CSF glutamine have been described in hepatic encephalopathy and in some cases of Reye's syndrome. Glutamine levels can be measured by a simple method involving hot acid hydrolysis, releasing ammonia, which is, in turn, measured by a colorimetric reaction.

Hepatitis

Hepatitis means "inflammation of the liver," which may be caused by virus, bacteria, parasites, radiation, drugs, chemicals, or toxins. Among the viruses causing hepatitis are hepatitis types A, B, C, D (or delta), and E, cytomegalovirus, Epstein-Barr virus, and probably several others. (Table 18-4)

Hepatitis A

Hepatitis A, also known as *infectious hepatitis* and *short-incubation hepatitis,* is usually transmitted by contaminated food or water. Epidemiologic data suggest that a carrier state in hepatitis A is unlikely or is of a short duration. The virus of hepatitis A has been identified by electron microscopy as a spherical particle, 27 nm in diameter containing KNA. The official designation for the hepatitis A virus is HAV. HAV has recently been cultured *in vitro*. However, this tissue-culture system is more a research tool than a diagnos-

TABLE 18-4. The Hepatitis Viruses

	Nucleotide	Incubation Period	Primary Mode of Transmission	Vaccine	Chronic Infection	Serologic Diagnosis Available
Hepatitis A	RNA	2–6 weeks	Fecal-oral	Yes	No	Yes
Hepatitis B	DNA	8–26 weeks	Parenteral, sexual	Yes	Yes	Yes
Hepatitis C	RNA	2–15 weeks	Parenteral, ? sexual	No	Yes	Yes
Hepatitis D	RNA	—	Parenteral, sexual	Yes	Yes	Yes
Hepatitis E	RNA	3–6 weeks	Fecal-oral	No	?	Yes*

At the time of publication, hepatitis E tests were available on a research basis only.

tic aid. The fecal shedding of the hepatitis A antigen is transient, and the antigen disappears from the stool after peak elevations of liver enzymes. The diagnosis of hepatitis A is made by the serologic detection of hepatitis A antibody. The production of hepatitis A antibody (anti-HAV) represents a specific host response to the hepatitis A virus and is believed to confer immunity from reinfection. Use of immune electron microscopy of the stool and radioimmunoassays for the detection of this antibody suggests that different types of antibody (both IgM and IgG) may appear at different times during the course of the illness. The IgM⁻ specific antibody peaks in 1 week and disappears by 8 weeks, whereas the IgG antibody peaks within 1 to 2 months and persists for years.

The documentation of anti-HAV positivity indicates exposure to HAV, absence of infectivity, and presence of immunity to recurrent infection. Anti-HAV positivity neither implies previous clinically apparent hepatitis nor establishes an etiologic relationship between HAV and acute or chronic liver disease. The demonstration of seroconversion or fecal shedding or antigen during the acute phase of illness is the only reliable method of establishing an HAV etiology.[26,27,47]

Hepatitis B

Hepatitis B is also known as *serum hepatitis* or *long-incubation hepatitis.* There are three major routes of transmission: parenteral, perinatal, and sexual. An additional fecal-oral route of transmission also exists, although this is not important from an epidemiologic basis. Infected patients manifest hepatitis B in virtually all body fluids, including blood, feces, urine, saliva, semen, tears, and milk.[8]

The hepatitis B virus consists of a 42-nm double-shelved spherical particle with a central core of DNA surrounded by a protein coat. This particle, which is present in low concentrations in the serum of patients with active viral hepatitis, was originally called the *Dane particle,* but it is now named *HBV.* Following an infection with HBV, the core of the antigen is synthesized in the nuclei of hepatocytes and then passes into the cytoplasm of the liver cell, where it is surrounded by the protein coat. An antigen present in the core of the virus (designated HBcAg) and a surface antigen present on the surface protein (designated HBsAg) have been identified by serologic studies. Another antigen, called the e antigen or HBeAG, also has been identified.[27]

The clinical course of hepatitis B is extremely variable. Approximately two-thirds of the cases may be asymptomatic or may produce only a mild flu-like illness. In the remaining third of the cases, a patient may develop a hepatitis-like syndrome of malaise, irregular fevers, right upper quadrant tenderness, jaundice, and dark urine.[47]

In approximately 1% of infected patients, the syndrome of fulminant hepatitis may develop. This is an overwhelming clinical illness with a high mortality.[41]

Approximately 90% of patients infected with the hepatitis B virus recover within 6 months. This recovery is manifested by the development of antibody to hepatitis B surface antigen. Approximately 10% of the infected patients will go on to develop chronic hepatitis, which is discussed below.[19]

On a worldwide basis, infection with hepatitis B virus is fairly common and usually occurs at the time of birth. In some areas, such as parts of Africa, Asia, and the Pacific islands, up to 80% of the general population may show serologic evidence of past hepatitis infection. The chronic carrier rate is high. Such persons are at high risk for development of cirrhosis or hepatocellular carcinoma (hepatoma).

In the United States, less than 10% of the population show serologic evidence of past infection with the hepatitis B virus. The chronic carrier rate is less than 1% of the general population. Persons at high risk for infection in this country include male homosexuals, intravenous drug abusers, children born to mothers who are hepatitis B surface antigen positive at the time of delivery, immigrants from endemic areas, and sexual partners and household contacts of patients who have hepatitis B. Although transmission of hepatitis B by blood products still occurs, effective screening tests now make this relatively uncommon. Health care workers, including laboratory personnel, may be at increased risk for developing hepatitis B depending on the degree of their exposure to blood and body fluids.

It is important to note that an effective vaccine and also immunoglobulin exist for hepatitis B. The vaccine is highly effective in stimulating the production of hepatitis B surface antibody and, therefore, in rendering the recipient immune. Following an acute exposure, such as a needle-stick injury, the patient should receive both the hepatitis vaccine and immunoglobulin. A similar treatment is highly effective in preventing the development of hepatitis in infants born to infected mothers.[2,36]

It is especially important that all health care workers exposed to blood and body fluids take the hepatitis vaccine. Each year thousands of cases of hepatitis B occur among health care workers, and several hundred die from its complications. This is entirely preventable.[32] Everyone who works in the clinical laboratory should take the hepatitis vaccine.

Hepatitis BsAg

Hepatitis BsAg, previously known as the *Australia antigen* and *hepatitis-associated antigen (HAA),* is the antigen for which routine testing is performed on all donated units of blood. HBsAg is the first serologic marker to appear during the prodrome of acute hepatitis B, and it identifies infected patients before the onset of clinical illness more reliably than any of the others. Although it has been shown that the isolated viral protein coat is not infectious, persons who chronically carry HBsAg in their serum must be considered potentially infectious, since the presence of the intact virus cannot be excluded. Patients who recover from hepatitis B develop anti-HBs following the disappearance of the HBsAg at about the time of clinical recovery (Fig. 18-5). This anti-body is common in the general population and is believed to confer immunity to future reinfection with HBV.

Hepatitis BcAg

The core antigen, HBcAg, has not been demonstrated in the plasma of hepatitis victims or blood donors. This antigen is seen only in the nuclei of hepatocytes during an acute infection with hepatitis B. The antibody to the core antigen, anti-HBc, usually develops earlier in the course than the antibody to the surface antigen (see Fig. 18-5). A recent serologic marker assay developed for general use is a test for IgM antibody to hepatitis B core antigen. In the proper clinical setting, this IgM assay has been shown to be specific for acute hepatitis B. Closely associated with the core antigen is a viral DNA⁻ dependent DNA polymerase. This viral enzyme is required for viral replication and is detectable in the serum early in the course of viral hepatitis, during the phase of active viral replication.

Hepatitis BeAg

Another marker of HBV infection is the e antigen. This antigen appears to be more closely associated with the core than the surface of the viral particle. The presence of the e antigen appears to correlate well with both the number of infectious virus particles and the degree of infectivity of HBsAg-positive sera. In other words, the presence of HBeAg in HBsAg carriers is an unfavorable prognostic sign and predicts a severe course and chronic liver disease. Conversely, the presence of anti-HBe antibody in carriers indicates a low infectivity of the serum (Fig. 18-6). The e antigen is detected in serum only when surface antigen is present (Fig. 18-7).

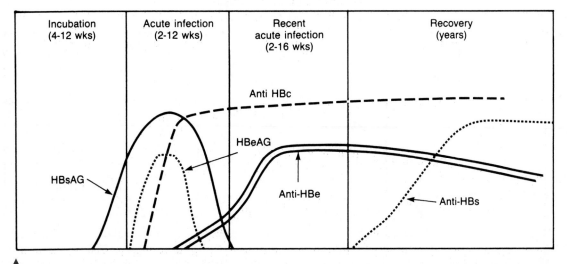

Sequence of HBV markers

Figure 18-5. Serology of hepatitis B infection with recovery.

Sequence of HBV surface markers
chronic hepatitis

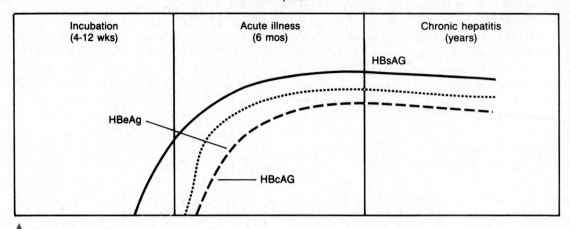

Figure 18-6. No antibody is formed against HBsAG. The persistence of HBeAG implies high infectivity and a generally poor prognosis. This patient would be likely to develop cirrhosis unless seroconversion occurs or treatment is given.

Hepatitis C

Until fairly recently, cases of viral hepatitis that could not be identified as type A or type B by serologic markers were logged in the general category of non-A or non-B hepatitis. Recently, the agent responsible for approximately 80% of these cases was identified as the RNA containing the *hepatitis C virus (HCV)*.[7,56]

Tests for detecting those at risk for transmitting HCV are now available. They detect antibody to hepatitis C (HCV) and identify most infectious carriers.[67]

Although much remains to be learned about hepatitis C, it appears most often to be transmitted parenterally. Sexual and fecal-oral routes also exist, and it is the most common form of hepatitis to be transmitted by blood transfusion.

Approximately 1% of blood donors in the United States are positive for HCV. Most of these patients are presumed to be infectious. The HCV antibody test will detect most infectious patients, although false-positive results do occur.

Clinically, acute hepatitis C is usually mild and may be entirely inapparent. However, infection has a high rate of progression to chronic hepatitis and cirrhosis. Thus, hepatitis C appears to be a major cause of chronic hepatitis in this country.[65]

Delta Hepatitis

Delta hepatitis, or hepatitis D (HDV), constitutes a unique example of satellite virus infection in human disease. The hepatitis delta virus causes disease only in patients who are currently infected with hepatitis B virus. It is incapable of causing any illness at all in patients who do not have hepatitis B.

Sequence of HBV surface markers
chronic hepatitis

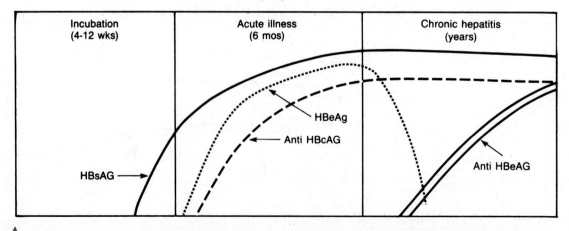

Figure 18-7. Serology of chronic hepatitis with formation of antibody to HBeAG. This is a favorable sign and suggests that the chronic hepatitis may resolve. Complete recovery would be heralded by the disappearance of HBsAG and formation of its corresonding antibody.

Hepatitis delta virus is an RNA virus with a high degree of base-pair homology with hepatitis B virus. When delta virus infection occurs in a patient who already has hepatitis B virus infection, the delta virus uses the hepatitis B virus for replication. Then, both HBV and HDV are produced. Unlike the hepatitis B virus, hepatitis delta virus appears directly toxic to human hepatocytes.

Two basic types of HDV infection exist, but both have the effect of worsening the prognosis of the hepatitis B infection. In coinfection, the patient acquires HBV and HDV at the same time. A patient with this coinfection is more likely to develop serious complications and has a higher rate of progression to chronic hepatitis. However, the majority of the patients still recover.

The other possibility is superinfection, in which a patient who is a chronic HBV carrier is superinfected with the HDV. Fulminant hepatitis may develop, or the progression to cirrhosis may be accelerated.

Epidemologically, HDV infection appears concentrated in countries surrounding the Mediterranean, Black, and Red Seas. In the United States, between 10% and 20% of patients who are chronic HBV carriers will be serologically positive for HDV. The risk factors for HDV infection are generally the same as for HBV infection.[45,46,53]

Hepatitis E

The RNA-containing hepatitis E virus is transmitted primarily by the fecal-oral route. This illness is found mainly in underdeveloped countries, although sporadic cases have been reported, primarily among travelers, in the United States and Western Europe. The incubation period is short, generally between 21 and 45 days. The virus may be detected in feces and bile by about 7 days after infection.

Hepatitis E infection is generally mild, except in pregnant women, where it may be a devastating illness. Serologic tests are available for diagnosis.[8,17,63]

Chronic Hepatitis

When evidence of hepatitis, such as elevated serum transaminase levels, is present for more than 6 months, *chronic hepatitis* is said to exist. Many different agents, including viruses, drugs, and alcohol, may cause chronic hepatitis. However, this discussion will center on the viral causes.[26]

Approximately 10% of cases of HBV will progress to chronic hepatitis. The severity of the initial illness has nothing to do with the risk of developing chronicity. Therefore, many patients may develop chronic hepatitis without even having been aware that they had hepatitis B infection. The serologic findings in patients with chronic hepatitis B are shown in Figure 18-7. These patients may be clinically quite sick, or they may be apparently entirely healthy. However, as long as HBsAg is present, they are infectious and at risk for developing complications, including cirrhosis and hepatocellular carcinoma. Patients who manifest the e antigen are highly infectious and are said to have the worst prognosis. The appearance of antibody to e antigen may herald recovery. Complete recovery occurs when hepatitis B surface antigen disappears and the corresponding antibody is then detected. These patients are then immune to further infection.[10]

While many patients with chronic hepatitis may recover spontaneously, others may require aggressive treatment.

On a worldwide basis, hepatitis B infection constitutes a major cause of morbidity and mortality. In underdeveloped countries, infection often occurs at the time of birth, and these children are at high risk for the development of cirrhosis or hepatocellular carcinoma.[19]

Another form of hepatitis with a high degree of chronicity is hepatitis C. Although patients with chronic hepatitis C infection appear to be at high risk for cirrhosis, the role of hepatitis C in the development of hepatocellular carcinoma is much less clear.[65] Interferon is currently used to treat chronic hepatitis.[13]

Hepatitis A is rarely, if ever, associated with chronic disease.[19]

Other Forms of Hepatitis

The test for hepatitis C will detect approximately 80% of patients who have viral hepatitis but do not have hepatitis A, hepatitis B, or other forms of noticeable viral hepatitis. What, then, of the other 20%? Other forms of viral hepatitis may exist.[7]

CASE STUDY
18-1

The following laboratory test results were obtained in a patient with severe jaundice, right upper quadrant abdominal pain, fever, and chills (Case Study Table 18-1).

Question

1. What is the most likely cause of jaundice in this patient?

CASE STUDY TABLE 18-1. Laboratory Results

Serum alkaline phosphatase	4 times normal
Serum cholesterol	Increased
AST (SGOT)	Normal or slightly increased
5′-Nucleotidase	Increased
Total bilirubin	25 mg/dL
Conjugated bilirubin	19 mg/dL
Prothrombin time	Prolonged but improves with a vitamin K injection

CASE STUDY
18-2

The following laboratory test results were found in a patient with a mild weight loss and nausea and vomiting, who later developed jaundice and an enlarged liver (Case Study Table 18-2).

Question

1.   What disease process is most likely in this patient?

CASE STUDY TABLE 18-2. Laboratory Results

Total serum bilirubin	20 mg/dL
Conjugated bilirubin	10 mg/dL
Alkaline phosphatase	Mildly elevated
AST	Markedly elevated
ALT	Moderately elevated
Albumin	Decreased
γ-Globulin	Increased

CASE STUDY
18-3

The following results were obtained in the patient from Case Study 18-2 (Case Study Table 18-3).

Questions

1.   What is the most likely diagnosis?
2.   What is the prognosis?
3.   What complications may develop?

CASE STUDY TABLE 18-3. Laboratory Results

Hepatitis A antibody (IgG)	Positive
Hepatitis A antibody (IgM)	Negative
Hepatitis B surface antigen	Positive
Hepatitis B surface antibody	Negative
Hepatitis Core antibody (IgM)	Positive
Hepatitis C antibody	Negative

CASE STUDY
18-4

The following laboratory results are obtained from a 19-year-old college student who consulted the Student Health Service because of fatigue and lack of appetite. She adds that recently she has noted that her sclera appears somewhat yellowish and that her urine has become dark.

Laboratory Results

Alanine aminotransferase (ALT)(SGPT)—elevated
Aspartate aminotransferase (AST)(SGOT)—elevated
Alkaline phosphatase—minimally elevated
LD—elevated
Serum bilirubin—5 mg/dL

Urine bilirubin—increased
Hepatitis A antibody (IgG)—negative
Hepatitis A antibody (IgM)—positive
Hepatitis B surface antigen—negative
Hepatitis B surface antibody—negative
Hepatitis C antibody—negative

Questions

1.   What is the most likely diagnosis?
2.   What additional factors in the patient's history should be sought?
3.   What is the prognosis?

CASE STUDY 18-5

A 36-year-old man consults his family physician because of liver function abnormalities, which had been noted initially during a pre-insurance physical examination 6 months ago. The following laboratory results are obtained, which are identical to those obtained 6 months previous.

Hepatitis A antibody (IgG)—positive
Hepatitis A antibody (IgM)—negative
Hepatitis B surface antigen—positive
Hepatitis B surface antibody—negative

Hepatitis B core antibody (IgM)—positive
Hepatitis C antibody—negative

Questions

1. What is the most likely diagnosis?
2. What is the prognosis?
3. What complications may develop?
4. What additional tests should be done?

REFERENCES

1. Adadi Y, et al. Determination of individual serum bile acids in chronic liver disease. Gastroenterology Jpn 1988;23:401.
2. Arevalo JA, Washington AE. Cost-effectiveness of prenatal screening and immunization for hepatitis B virus. JAMA 1988;259:365.
3. Batsakis JG, Soderman TH, Deegan MJ. Enzymatic evaluation of hepatobiliary disease. Lab Med 1974;5:33.
4. Billings BH. 25 years of progress in bilirubin metabolism (1952–1977). Gut 1978;19:481.
5. Binder L. Urobilinogen and urine bilirubin assays. South Med J 1988;81(10):1229.
6. Bouchier I, Pennington CR. Serum bile acids in hepatobiliary disease. Gut 1978;19:492.
7. Bradley DW. The agents of non-A, non-B viral hepatitis. J Virol Methods 1985;10:307.
8. Bradley DW, Beach MJ, Purday MA. Molecular characterization of hepatitis C and E viruses. Arch Virol Suppl 1993;7:1.
9. Bredfeldt JE. Hepatitis B virus update on the spectrum of clinical infection. Postgrad Med 1985;78(6):71.
10. Brunt PW, Losowsky MS, Read AE. The Liver and biliary system. London: William Heinemann, 1984;48.
11. Chang Y. Serological markers of viral hepatitis. Diag Med 1983;6:28.
12. Chapra S, Griffin PH. Laboratory tests and diagnostic procedures in the evaluation of liver disease. Am J Med 1985;79:221.
13. Davidsohn I, Henry JB. Clinical diagnoses by laboratory methods. 18th ed. Philadelphia: Saunders, 1991.
14. Dusheiko G, Di Bisceglie A, Bowyer S, Sachs E, Ritchie M, Schoub B, Kew M. Recombinant leukocyte interferon treatment of chronic hepatitis B. Hepatology 1985;5:556.
15. Ellis JC. Hepatocellular carcinoma. West J Med 1988;149(2):183.
16. Ericson RJ, Morales DRF. Clinical use of lactate dehydrogenase. N Engl J Med 1961;265:478.
17. Favorov MO, Khudyakov YE, Fields HA, et al. Enzyme immunoassay for the detection of antibody to hepatitis E virus based on synthetic peptides. J Virol Methods, Feb 1994; 46(2):237.
18. Gambino SR, Dire J. Manual on bilirubin assay. Chicago:ASCP Committee on Continuing Education, 1963.
19. Gambino SR, Schreirer H. Technicon Symposium Paper No. 54, 1964.
20. Gammal SH. Hepatic encephalopathy. Med Clin North Am 1989;73(4):793.
21. Garcia G, Gentry KR. Chronic viral hepatitis. Med Clin North Am 1989;73:971.
22. Gauthier M, Guay J, Lacroix J, Lortie A. Reyes syndrome: A reappraisal. Am J Dis Child 1989;143(10):1181.
23. Gitnick G. Principles and practices of gastroenterology and hepatology. Amsterdam: Elsevier, 1988.
24. Henry RJ, Golub OJ, Berman S, Segalove M. Critique on the icterus index determination. Am J Clin Pathol 1953;23:841.
25. Haddad LM, Winchester JF. Clinical management of poisoning and drug overdose. Philadelphia: Saunders, 1983.
26. Hoffman E, Nachlos MN, Gadby SD, et al. Limitations of the diagnostic value of serum leucine aminopeptidase. N Engl J Med 1960;263:541.
27. Hofmann AF. The enterohepatic circulation of bile acids. In: Sleisenger MH, Fordhan JS, eds. Gastrointestinal disease: Pathophysiology, diagnosis, and management. 4th ed. Philadelphia: Saunders, 1989.
28. Hollinger FB, Melnick JL, Robinson WS. Viral hepatitis. New York: Raven Press, 1985.
29. Hoofnagle JH. Type A and type B hepatitis. Lab Med 1983;14(11):705.
30. Hurvitz ES. Reye's syndrome. Epidemiol Rev 1989;11:249.
31. Jendrassik L, Grof P. Quantitative methods of measuring urobilinogen. Biochemistry 1938;297:81.
32. Kaplan MM. Serum alkaline phosphatase: Another piece is added to the puzzle. Hepatology 1986;6:526.
33. Kiviahan C, James E. The natural history of neonatal jaundice. Pediatrics 1984;74(3):365.
34. Krugman S. Hepatitis B immunoprophylaxis. Lab Med 1983;14(11):727.
35. Kumar R, Deykin DF. Pathogenesis and practical management of coagulopathy of liver disease. In: Davidsohn CS, ed. Problems in liver disease. New York: Stratton Intercontinental, 1979;66.
36. Luna G, Florence L, Johansen K. Hepatocellular carcinoma: A 5-year institutional experience. Am J Surg 1985;149:591.

37. Matzen P, et al. Differential diagnosis of jaundice. Liver 1984;4(6):360.

38. McLean AA. Hepatitis B vaccine: A review of the clinical data. J Am Dent Assoc 1984;110(4):624.

39. Mitchell JR. Acetaminophen toxicity. N Engl J Med 1988; 319(24):1601.

40. Maruyama H, et al. Hepatotoxins. Arch Toxicol 1988; 62(6):465.

41. Petersdorf RG, et al, eds. Harrison's principles of internal medicine. 10th ed. New York: McGraw-Hill, 1983.

42. Peltz S, et al. Ammonia encephalopathy. N Engl J Med 1989; 86(7):554.

43. Rakela J, Lange SM, Ludwig J. Fulminant hepatitis: Mayo clinic experience. Mayo Clin Proc 1985;60(5):289.

44. Reiner M, Thomas JL. Bilirubin studies on the serum of newborn infants. Clin Chem 1962;8:278.

45. Reichling JJ, Kaplan MM. Clinical use of serum enzymes in liver disease. Dig Dis Sci 1988;33(12):1601.

46. Rizzetto M. The delta agent. Hepatology 1983;3:729.

47. Rizzetto M, Verme G, Geris JL, Purcell RH. Hepatitis delta virus disease. Prog Liver Dis 1986;8:417.

48. Sass MA, Cianflacco AJ. The diagnosis of acute viral hepatitis. Resident and Staff Physician 1985;31(4):17.

49. Schiff L. Jaundice: Five and a half decades of historic perspective. Selective aspects. Gastroenterology 1980;78:831.

50. Schiff L, Schiff ER. Diseases of the liver. 6th ed. Philadelphia: Lippincott, 1987.

51. Schmid R. Bilirubin metabolism: State of the art. Gastroenterology 1978;76:1307.

52. Sherlock S. Carbohydrate changes in liver disease. Am J Clin Nutr 1970;23:462.

53. Sherlock S. Diseases of the liver and biliary system. 6th ed. Oxford: Blackwell Scientific Publications, 1981.

54. Shiels MT, et al. Frequency and significance of hepatitis D in acute and chronic hepatitis B: A U.S. experience. Gastroenterology 1985;89(6):1230.

55. Sleisenger MH, Fordtran JS. Gastrointestinal diseases: Pathophysiology diagnosis and management. 4th ed. Philadelphia: Saunders Co, 1989.

56. Stemmel W. What's new in the hepatocellular uptake mechanism of bilirubin and fatty acids? Pathol Res Pract 1988; 183(4):524.

57. Stevens CE, et al. Epidemiology of hepatitis C virus. JAMA 1990;263(1):49.

58. Street JM, et al. Chromatographic methods for bile acid analysis. Biol Med Chromatogr 1988;2(6):624.

59. Sorrentino D, et al. Mechanistic aspects of bilirubin uptake. Semin Liver Dis 1988;8(2):119.

60. Tabibian N. Hepatocellular carcinoma in the U.S. Am Fam Physician 1988;38(4):115.

61. Tietz NW. Fundamentals of clinical chemistry. Philadelphia: Saunders, 1986.

62. Tovill AS. The synthesis and degradation of liver produced proteins. Gut 1972;13:225.

63. Tsarev SA, Tsareva TS, Emerson SU, Kapikian AZ, Ticehurst J, London W, Purcell RH. ELISA for antibody to hepatitis E virus (HEV) based on complete open-reading frame-2 protein expressed in insect cells: Identification of HEV infection in primates. J Infect Dis Aug 1993;168(2):369.

64. Warnes TW. Investigation of the jaundiced patient. Br J Hosp Med 1982;28(4):385.

65. Watson CJ, Howkinson V. Studies of urobilinogen. Am J Clin Pathol 1947;17:108.

66. Wennberg RP, Goetzman BW. Neonatal intensive care manual. Chicago: Year Book Medical Publishers, 1985;189.

67. Weisigner RA. Non-A, non-B hepatitis: Promise and implications of the "hepatitis C" assay. Mod Med 1990;58:58.

68. Zimmerman HJ, Henry JB. Clinical enzymology. In: Henry JB, ed. Clinical diagnosis and management by laboratory methods. Philadelphia:Saunders, 1979;365.

Chapter 19

Endocrinology

Betsy D. Bennett, David J. Wells

<table>
<tr><td>

Hormones
Definition
Chemical Nature
Mechanism of Action
Mechanism of Control
Methodology
Classical Assays
Immunologic Assays
High-Performance Liquid
   Chromatography (HPLC)

</td><td>

Colorimetry
**Components of the Endocrine
   System**
Hypothalamus
Anterior Pituitary
Posterior Pituitary (Neurohypophysis)
Thyroid
Parathyroid Glands

</td><td>

Adrenal Glands
Islets of Langerhans
Reproductive System
Other Hormones of Clinical Interest
Summary
Case Studies
References

</td></tr>
</table>

Objectives

Upon completion of this chapter, the clinical laboratorian should be able to:

- *Define the following terms: hormone, prohormone, steroid, peptide, and amine.*
- *Describe the location and major hormones produced by the following endocrine glands: hypothalamus, anterior pituitary, posterior pituitary, parathyroids, adrenal medulla, adrenal cortex, pancreas, ovaries, and testes.*
- *Relate the mechanisms involved in hormone regulation.*
- *Given a list of hormones, categorize them as steroid, protein, or amine.*
- *Discuss the difference between primary and secondary hormone disorders.*
- *Explain the mechanisms of hormone action and control.*
- *Compare and contrast the various methodologies used to assay hormone levels.*
- *Relate the expected laboratory results associated with the following disease states: Cushing's syndrome, Addison's disease, diabetes mellitus, Conn's syndrome, pheochromocytoma, hyperparathyroidism, and hypoparathyroidism..*
- *State the importance of multiple measurements of some hormones levels.*
- *Discuss the importance of stimulation and suppression tests in diagnosing hormone disorders.*
- *State the functions of the endocrine system.*
- *List two problems in assaying hormone levels.*
- *Discuss the laboratory evaluation of infertility.*
- *Describe any special procedures or precautions in the collection of samples for hormone assay.*

KEY WORDS

ACTH	FSH	Mineralocorticoid
ADH	Gastrin	Neuroblastoma
Adrenal Glands	GH	Norepinephrine
Aldosterone	Glucagon	Oxytocin
Amenorrhea	Glucocorticoid	Panhypopituitarism
Amine	HCG	Peptides
Androgen	Hormone	Pheochromocytoma
Angiotensin	HPL	Placenta
Anterior Pituitary	Hyperaldosteronism	Posterior Pituitary
CAH	Hyperglucagonemia	Precocious Puberty
Conn's Syndrome	Hyperinsulinemia	Progesterone
Cortisol	Hyperpara-	Prohormone
CRH	thyroidism	Proinsulin
Cytokines	Hyperprolactinemia	Prolactin
DHEA	Hypoaldosteronism	PTH
Ectopic	Hypoinsulinemia	Receptor
Ectopic Hormone	Hypopara-	Sella Turcica
Production	thyroidism	Serotonin
Endocrine Gland	Hypothalamus	SIADH
Epinephrine	Islets of	Steroids
Estriol	Langerhans	Testosterone
Estrogen	LH	TRH
Exocrine Gland	MEN	TSH
Follicular Phase	Metanephrine	VMA

The endocrine system and the manifestations of disease in endocrine glands are complex and involve the interaction of many other body systems. Because hormones produced by the endocrine glands circulate in the blood, they are readily accessible to laboratory evaluation. However, evaluation of endocrine function differs in two major ways from most chemistry analyses. First, the substances to be

Michael L. Bishop, Janet L. Duben-Engelkirk, and Edward P. Fody.
CLINICAL CHEMISTRY. © 1996 Lippincott–Raven Publishers.

measured circulate in extremely small amounts (nanograms per milliliter in many cases). Detection of such low levels of hormones has required development of very sensitive analytical methods, which use complex techniques and sophisticated equipment. Consequently, these tests are relatively expensive to both the laboratory and the patient. In many cases, it is not economically feasible for small hospital laboratories to perform these tests, and consequently, they are sent to reference laboratories. Second, instead of receiving only one or two requests for a particular test on a patient in one day, the laboratory will frequently receive requests for the same test on several blood samples drawn from the same patient in a relatively short time interval. These requests will be part of a stimulation or suppression test designed to evaluate an endocrine gland's response to the usual control mechanisms. Such tests are necessary because: (1) the normal secretion of some hormones is erratic, and thus a single measurement under uncontrolled conditions is of no use; (2) destruction of as much as 90% of the functioning mass of a gland is necessary before clinical evidence of deficiency occurs; therefore, stimulation tests designed to evaluate glandular reserve may be the only way to detect early disease; and (3) some hormones circulate in such low levels that one unstimulated measurement cannot distinguish the normal from the hypofunctioning gland. In addition, stimulation tests allow measurement of these hormones at higher values where the technical reliability is greater.[19,97,98]

HORMONES

Definition

A *hormone* is a chemical substance that is produced and secreted into the blood by an organ or tissue and has a specific effect on a target tissue. This target tissue is usually, but not always, located at some distance from the site of hormone production.

Hormones act in conjunction with the nervous system to maintain the internal chemical conditions necessary for cellular function and to allow the body to respond to emergency demands. Interactions occur between different portions of the endocrine system, with multiple hormones affecting one physiologic function. (For example, insulin, glucagon, growth hormone, cortisol, and epinephrine all affect carbohydrate metabolism.) The opposite situation is also true. That is, a single hormone (cortisol, for example) may affect many different organ systems, producing a variety of physiologic effects. Normal growth and development (physical and mental) and reproduction also require the correct degree of endocrine function.[97,98]

Chemical Nature

Three chemical types of hormones have been identified— *steroids, peptides,* and *amines.* These differences in chemical structure are accompanied by differences in target tissue, mechanism of action, and metabolism.

Steroids

Steroids are produced by the adrenal glands (cortisol, aldosterone, sex steroids) and the gonads and placenta (testosterone, estrogen, progesterone). Regardless of the site of production, they are all synthesized from cholesterol, using the same initial biochemical pathways (Fig. 19-1). The end result depends on the enzymatic machinery that is predominant in a particular organ. Interconnections between enzyme systems and hormone precursors may result in production of large amounts of a hormone inappropriate for a particular organ if the usual enzymatic pathway is blocked. The adrenogenital syndrome results from such an enzymatic abnormality (see below).

In general, steroids are synthesized as they are needed and are not stored to any great extent. Because of their derivation from cholesterol, they are lipid-soluble and circulate bound to a carrier protein. They have a relatively long plasma half-life, ranging from 60 to 100 minutes.[97,98]

Protein

Protein hormones may be either peptides or glycoproteins (FSH, LH, TSH, HCG). The glycoprotein hormones are composed of α and β chains. The α chains of all four of these hormones are immunologically identical. The func-

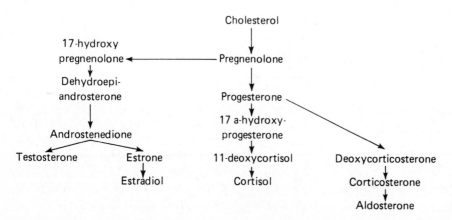

Figure 19-1. Simplified scheme of steroid synthesis from cholesterol.

tional and immunologic specificity of each are derived from the β chain. The peptides include insulin, glucagon, parathyroid hormone, growth hormone, and prolactin. These hormones are synthesized as a *prohormone,* which is cleaved to produce the usual circulating form. As a rule, protein hormones are synthesized and stored within the cell in the form of secretory granules and are released as needed. These hormones are water-soluble and are not bound to carrier proteins in the circulation. Their half-life is short, ranging from 5 to 60 minutes.[97,98]

Amines

The amines include epinephrine, norepinephrine, thyroxine, and triiodothyronine. These compounds have properties intermediate between those of the steroids and protein hormones. Epinephrine and norepinephrine tend to behave like the protein hormones—that is, they circulate unbound to protein and have a short half-life and a similar mechanism of action (see the following). In contrast, thyroxine and triiodothyronine have a long half-life, circulate bound to a carrier protein, and have a mechanism of action that is similar to the steroids.[97,98]

Mechanism of Action

The production of the appropriate response to a hormone by its target tissue depends on the interaction of the hormone and a specific receptor. The presence of high-affinity receptors accounts for the ability of the target cell to accumulate relatively high concentrations of a particular hormone. These receptors also provide for target-organ specificity, since not all cells have receptors for all hormones.

Protein Hormones

The protein hormones and catecholamines (epinephrine and norepinephrine) produce their effects by interaction with a receptor on the outer surface of the cell membrane. The hormone-receptor complex then activates an enzyme, adenyl cyclase, located on the inner cell membrane. This enzyme converts ATP into cyclic AMP, which acts as a "second messenger" and translates the hormone binding into cellular action. In general, cAMP activates protein kinases, resulting in phosphorylation of certain enzymes. Such phosphorylation results in either activation or inhibition of the enzyme, depending on the particular system involved. Hormones that act through cAMP produce rapid results (within minutes), which cease rapidly when the stimulus stops. Other mechanisms of action for these hormones, such as entry of the hormone-receptor complex into the cell with subsequent modification of cellular action, also may be important but have not been completely explained.[97,98]

Steroids

In contrast to the protein hormones, which cannot enter the cell, steroids pass through the cell membrane and interact with an intracellular receptor. The hormone-receptor complex eventually binds to a particular segment of chromatin, inducing the formation of messenger RNA (mRNA). The mRNA enters the cytoplasm and initiates the synthesis of proteins or peptides that carry out the action attributed to the hormone. These peptides may represent enzymes, structural proteins, secretory products, or receptors for other hormones. Because this type of hormone action depends on actual synthesis of an effector protein rather than phosphorylation of an existing enzyme system, the response of steroid hormones takes longer to occur initially and longer to stop when the stimulus for release of steroids ends. Thyroxine and triiodothyronine behave in a manner similar to the steroid hormones, except that they enter the nucleus directly without the necessity of binding to a receptor.[97,98]

Mechanism of Control

The endocrine system is regulated primarily by means of control of hormone synthesis rather than by the rate of hormone degradation. Synthesis of a particular hormone may be stimulated by another "tropic" hormone or by a change in the serum level of a substance that the hormone controls. For example, the release of cortisol is stimulated by ACTH, and insulin release is initiated by an increase in serum glucose levels. Cessation of hormone release is accomplished either directly or indirectly by a system of negative feedback. An example of negative feedback by a direct mechanism would be the inhibition of parathyroid hormone secretion by elevated serum calcium levels. Indirect mechanisms of negative feedback involve cessation of production of tropic hormones in response to elevations of levels of the effector hormone. Lack of tropic hormone then stops production of the effector hormone. For example, increased levels of cortisol act on the hypothalamus and pituitary to decrease secretion of corticotropin-releasing hormone (CRH) and ACTH, respectively. With decreased levels of ACTH, cortisol production by the adrenal ceases and cortisol levels drop. When plasma cortisol levels decrease below the desired range, low levels of this hormone stimulate the hypothalamus and pituitary to release CRH and ACTH, and adrenal production of cortisol begins again.[97,98]

METHODOLOGY

Classical Assays

Bioassays

Before the development of modern assay methods, the detection and quantitation of hormones were based on cumbersome, expensive assays that were dependent on a biologic response from a live animal or tissue/organ culture. Such assays required a large amount of technical time and experience. Interpretation was often based on a subjective evaluation of results that might be influenced by nonspecific factors having little or nothing to do with the desired hormonal effect.

An example of a bioassay previously used in the clinical evaluation of endocrine status is the detection of pregnancy based on corpus luteum formation in prepubertal female mice or ovulation in mature female rabbits. The assay involved injection of human urine samples into virgin animals and subsequent examination of the ovaries for corpus luteum formation. Bioassays are now used almost exclusively for research purposes and have been replaced in the hospital laboratory by modern hormone assays using competitive protein-binding methods and immunoassay techniques.

Competitive Protein Binding (CPB)

Competitive protein-binding (CPB) assays are based on competition for a limited number of protein-binding sites between a known added amount of "tagged" or "labeled" hormone and the unlabeled hormone present in the patient's sample. These binding proteins may be specific antisera (immunoassay), specific receptor sites (radioreceptor assay), or serum-binding proteins (such as thyroid-binding globulin or cortisol-binding protein). Assays using serum-binding proteins have, for the most part, been replaced by immunoassay techniques. Radioreceptor assays have not been particularly useful in the clinical laboratory because the binding protein is an intact receptor in the membrane of a live cell, and it is difficult to maintain the cells or receptor membrane preparations in a biologically active state.

Immunologic Assays

The majority of endocrine assays in current use depend on the binding of an antibody prepared against the specific hormone being tested. Radioimmunoassay (RIA) is currently a popular method for clinical endocrine determinations. The development of a radioimmunoassay for insulin by Yalow and Berson in 1960[101] was a major technologic breakthrough for hormone analysis and was rapidly followed by development of a number of endocrine assays based on these principles. In general, it appears that RIA has reached its peak in terms of clinical usage and will probably be replaced by immunofluorescence or other immunoassay techniques. The relative advantages and disadvantages of the various immunologic techniques are discussed in Chapter 6, Immunochemistry.

Radioimmunoassay

As mentioned previously, RIA is a competitive protein-binding technique that uses radiolabeled hormone as the tagged hormone and antisera prepared against the specific hormone as a binding site. Competition between unlabeled hormone in the patient sample and the added labeled hormone for a limited number of antibody-binding sites forms the basis of the assay. An understanding of RIA is important because it is the most common competitive protein-binding assay used clinically and because it illustrates many of the techniques and principles common to all assays of this type.

Advantages and Disadvantages of RIA. As mentioned previously, RIA was a technological breakthrough for en-

docrine testing, and the technique was widely applied during the 1960s and 1970s. Although RIA is still the most popular technique in the clinical endocrine laboratory when high sensitivity is required, there are a number of drawbacks to its use. RIA is based on radioisotopes, and its use in the laboratory must be licensed and regulated by radiation safety committees. In addition, radioactive materials must be disposed of or stored under special conditions and usually must be separated from normal waste. If labeled hormones are not used in a short period of time (about 1–2 months) these substances must be discarded. The short shelf-life of radioisotopes may result in waste and inefficiency in test-kit usage. Finally, standard curves involving complex mathematical manipulations must be run for every assay, with both patient and standards done in duplicate.[17,64,91]

A number of new testing techniques have been developed that are alternatives to RIA and eliminate the problems associated with radioisotopes. Such techniques have found greater use at present in therapeutic drug monitoring, but already are replacing RIA in some endocrine testing. These techniques, as well as others similar to RIA, are discussed in the following sections.

Immunoradiometric Assays (IRMA)

Immunoradiometric assay (IRMA) is similar to RIA in that a radiolabeled substance is used in an antibody-antigen reaction. However, the radioactive label is attached to the antibody instead of the hormone. In addition, an excess of antibody, rather than a limited quantity, is present in the assay. Since all the unknown antigen becomes bound in IRMA rather than just a portion, as in RIA, IRMA assays are more sensitive. As shown in Figure 19-2, both *one-site* and *two-site* IRMAs exist. In the one-site assay, the excess antibody that is not bound to the patient sample is removed by addition of a precipitating binder. In many cases, this binder is antigen (hormone) bound to some solid support. In the two-site assay ("sandwich" technique), a hormone with at least two antibody-binding sites is adsorbed onto a solid phase, to which one of the antibodies is firmly attached (either the walls of the assay tube itself or beads that are added to the patient sample in assay buffer). After binding to this antibody is completed, a second antibody labeled with ^{125}I is added to the assay. This antibody reacts with the second antibody-binding site to form the "sandwich," composed of antibody-hormone–labeled antibody. In contrast to RIA and similar competitive protein-binding assays, the amount of hormone present is *directly* proportional to the amount of radioactivity measured in the assay. Patient data are derived from a standard curve.

Enzyme-Linked Immunosorbent Assay (ELISA)

Enzyme-linked immunosorbent assay (ELISA) is similar to the immunoradiometric assay except that an enzyme tag is attached to the antibody instead of a radioactive label. This assay has the advantage of avoiding radioactive materials and produces an end product that can be measured in a spectrophotometer. The hormone is bound to the enzyme-

Immunoradiometric assays (IRMA)

A One-site

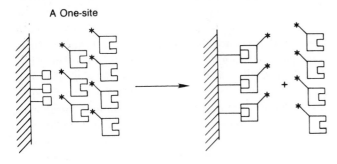

B Two-site

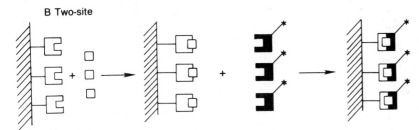

◀

Figure 19-2. In part *A*, antigen or hormone (□) is attached to solid support. Labeled antibody (◌) is added, which binds to the antigen with subsequent removal of the unbound antibody. In part *B*, an antibody is attached to solid support. Antigen or hormone is added to give an initial antigen-antibody complex, which can be reacted with a second labeled antibody (◼).

labeled antibody, and the excess antibody is removed as in immunoradiometric assays. Again, both one-site and two-site systems are possible. Once the excess antibody has been removed or the second antibody containing the enzyme has been added (two-site assay), the substrate and cofactors necessary for visualization of enzyme activity are added. The amount of hormone present is directly related to the amount of enzymatic activity (substrate formed) during a detection incubation. The sensitivity of the assay may be increased by increasing the incubation time for producing substrate. In some cases, substrate formed may give an optical color change so that detection of the hormone being measured can be determined by visual inspection. This type of assay has been used in pregnancy testing kits designed for home use.

Enzyme-Multiplied Immunoassay Technique (EMIT)

In enzyme-multiplied immunoassay technique (EMIT), enzyme tags are used instead of radiolabels, but the antibody binding alters the enzyme characteristics, allowing measurement of hormone without separating the bound and free components (homogeneous assay). This technique has wide applicability in drug monitoring but has not been popular as a technique for endocrine testing because of problems with sensitivity. The enzyme is attached to the hormone or drug being tested. This enzyme-labeled antigen is incubated with the patient sample and with antibody to the hormone or drug. Binding of the antibody to the enzyme-linked hormone either physically blocks the active site of the enzyme or changes the protein conformation so that the enzyme is no longer active. After antibody binding occurs, the enzyme substrate and cofactor are added, and enzyme activity is measured. If the patient sample contains the hormone of interest, it will compete with enzyme-linked hormones for

antibody binding, enzyme will not be blocked by antibody, and more enzyme activity will be measurable. Thus the greater the hormone concentration in the patient sample, the greater will be the enzyme activity (Fig. 19-3).

Fluorescent Immunoassay

Fluorescence as an assay technique has been used in endocrine testing for the determination of catecholamines and their metabolites.[45,97] Although capable of very high sensitivity, classical fluorescence techniques have not been very

Enzyme multiplied immunoassay technique (EMIT)

A.

+ →

+ △ → No reaction

B.

+ + → +

+ △ → Reaction

▲

Figure 19-3. In part *A*, or antigen or hormone (□) is attached to enzyme (◌). The antibody (◲) is then added, which inhibits enzyme activity if it binds to the enzyme-tagged hormone. Free patient hormone, shown in part *B*, prevents antibody binding to enzyme-labeled hormone so that substrate (△) can be hydrolyzed and monitored spectrophotometrically.

useful because many endogenous serum compounds, drugs, and dietary factors interfere by quenching or contributing to the fluorescence. In fact, fluorescence assays for catecholamines are being replaced by high-performance liquid chromatography (HPLC) techniques. Immunofluorescence techniques are, however, a relatively recent development. Ideally, fluorescence immunoassay represents a combination of high sensitivity (fluorescence) with high specificity (antigen-antibody reactions). In addition, microprocessors and computers have automated sample handling and data collection, permitting ease of operation and decreased technologist time. Because these assays measure relative change in fluorescence rather than absolute fluorescence, background and quenching problems can be minimized or corrected. Although most applications of fluorescence immunoassay have been in the area of drug monitoring, techniques for endocrine testing are rapidly being developed. Unlike EMIT procedures, in which sensitivity appears to be limited to the microgram per milliliter range, fluorescence should permit measurements at the lower detection levels needed for hormone measurement.

Substrate-Labeled Fluorescent Immunoassay. Substrate-labeled fluorescent immunoassay was one of the first methods developed using fluorescence immunoassay. A potentially fluorescent compound is attached to a drug or hormone, which is then added to the patient sample along with antibody to the hormone. If the patient sample contains the substance being tested, competition for antibody binding occurs between labeled and unlabeled hormone. After incubation, an enzyme is added that releases the fluorescent tag from the unbound labeled hormone, and fluorescence is measured. When antibody binds to the tagged substance, enzymatic hydrolysis of the tagging compound is prevented, and fluorescence cannot occur.

Fluorescence Polarization. Fluorescence polarization measures the change in the angle of polarized fluorescent light emitted by a fluorescent molecule. The hormone or drug, tagged with a fluorescent molecule, rotates rapidly in solution so that there is very little polarization of the fluorescence. When this tagged hormone binds to a large antibody, the rate of rotation of the tagged hormone-antibody complex becomes much slower relative to the unbound tagged hormone, and polarization increases. In the assay, patient sample is added to the tagged hormone and an initial fluorescent reading is made by the polarization fluorometer to compensate for sample quenching or background variation. Antibody to the hormone is then added. To the extent that patient material contains the substance of interest, competition occurs between unlabeled and labeled hormone for antibody-binding sites, and the amount of binding of labeled hormone determines the amount of fluorescence polarization. The fluorometer can be standardized by using a series of calibrators containing different concentrations of the substance of interest. The fluorescence polarization is measured

for these calibrators, and patient results are calculated from this information.

Other Fluorescence Techniques. Other fluorescence techniques are being developed or are available. These include fluorescent excitation transfer immunoassay and direct quenching fluorescent immunoassay. The former technique involves fluorescent tagging of both the antigen (hormone) and antibody with energy transfer from the antigen fluorescent compound to the antibody fluorescent compound when binding occurs. The latter involves direct quenching of antigen-tagged fluorescence when unlabeled antibody binding occurs.

High-Performance Liquid Chromatography (HPLC)

HPLC was developed during the 1960s primarily as a research tool. It involves pumping a liquid phase at high pressure through a solid-phase column. The separation is based on differential partitioning of the compounds between the mobile liquid phase and the stationary solid phase. The solid-phase column is composed of very small particles, and this results in much more efficient and faster separation than do other forms of chromatography. Partitioning is based primarily on the polarity of the solvent and the solubility in the liquid phase of the compounds to be separated. The technique offers versatility and control of the method development in that a great variety of column materials is available, and the investigator can vary the solvent composition over a wide range of polarities from extremely hydrophobic to purely hydrophilic.

HPLC does not lend itself to high throughput and requires a relatively large amount of technologist time and training. The technique is currently used primarily for toxicology and drug assays and not much for endocrine testing. However, one endocrine-testing function that is being performed by HPLC is catecholamine analysis. In the past, measurement of catecholamines was done by classical fluorescence methodology and was subject to a large number of background and quenching problems, as discussed earlier. The use of HPLC for separation of catecholamines and metabolites, combined with an electrochemical detection method, has resulted in more reliable measurement of these compounds.[21,97]

Colorimetry

Colorimetry refers to the measurement of light absorption at a specific wavelength or color. This light absorption may be a property of the substance being measured or may be possible only after the addition of reagents that form light-absorbing compounds with the hormone. As a technique for endocrine testing, colorimetry is not widely applicable because sensitivity is not usually sufficient in the physiologic concentration range of hormones. However, the urinary concentrations of cortisol and other adrenal steroids and their metabolites, as well as catecholamine metabolites, are high enough to allow their colorimetric determination. Measurement of the steroid compounds by these methods is not specific for a single adrenal steroid or metabolite but measures the quantity of those steroids that possess a common functional group.

Porter-Silber Method

The Porter-Silber method is based on the yellow pigment produced by a reaction of 17,21-dihydroxy-20-ketone (dihydroxyacetone side chain) structures with phenylhydrazine in the presence of alcohol and sulfuric acid.[97] Corticosteroids with this configuration include cortisol, cortisone, 11-deoxycortisol, and the tetrahydro derivatives (metabolites). The procedure, in outline form, consists of hydrolysis of glucuronide conjugates, extraction with organic solvents, removal of interfering compounds with alkali, and color reaction with the previously mentioned reagents. Absorption occurs at 410 nm.

Zimmerman Reaction

The Zimmerman reaction procedure measures those steroids possessing a 17-keto structure or those which can be oxidized to 17-keto structures.[97] The oxidation procedure used to produce 17-ketosteroids, called the *Norymberski procedure,* uses sodium bismuthate as the oxidizer. Compounds that can be oxidized are called *17-ketogenic steroids.* These include cortisol, cortisone, and their tetrahydro derivatives; 11-deoxycortisol and its tetrahydros; cortols; cortolones; and pregnanetriol and its 11-oxygenated derivative.

A later modification of the oxidation procedure involves reduction of C-20 ketones and C-21 methyl groups using sodium borohydride treatment before the oxidation step. This modification permits the measurement of the metabolites of the corticosteroid precursors and is of significance in testing for the adrenogenital syndrome.

The Zimmerman color-forming reaction employs *m*-dinitrobenzene to produce a reddish purple color with maximal absorption at 520 nm. The procedure, like the Porter-Silber method, requires a preliminary hydrolysis, extraction, and washing before the color-forming reaction occurs.

Both the Porter-Silber and the Zimmerman reactions require a large amount of technologist time and experience in solvent extractions and processing. They are not easily adapted to high sample numbers and require a 24-hour urine sample for accurate interpretation of results. An additional drawback is the fact that the color reaction occurs with a number of drugs and dietary by-products. These disadvantages have resulted in a shift to the use of immunoassay or chromatographic methods for measuring corticosteroids or their metabolites.

Analysis of Catecholamine Metabolites

Although sophisticated gas chromatography and HPLC techniques exist for measuring catecholamines and their metabolites, the classical procedure for quantitating metanephrines and normetanephrines is the Pisano procedure. Twenty-four-hour urine samples are subjected to acid hydrolysis and absorbed on an ion-exchange resin (amberlite CG-50). After elution with ammonium hydroxide, the compounds are converted to vanillin, reacted with periodate, and assayed spectrophotometrically at 360 nm.[62] A similar conversion to vanillin and spectrophotometric measurement are used in the analysis of another clinically important catecholamine metabolite, vanillylmandelic acid (VMA).[63]

COMPONENTS OF THE ENDOCRINE SYSTEM

Hypothalamus

The hypothalamus is the portion of the brain located in the walls and floor of the third ventricle. It is directly above the pituitary gland and is connected to the posterior pituitary by the pituitary stalk. Neurons located in the supraoptic and paraventricular nuclei of the hypothalamus send processes through the pituitary stalk, which terminate in the posterior pituitary (neurohypophysis). In addition to this direct neural connection with the posterior pituitary, other portions of the hypothalamus are in contact with the anterior pituitary (adenohypophysis) through the hypophyseal portal system.[67,97]

Neurons in the anterior portion of the hypothalamus release a number of hormones that are carried by means of portal veins to the pituitary sinusoids, where they either stimulate or inhibit the release of anterior pituitary hormones. At least six such substances [thyrotropin-releasing hormone (TRH), gonadotropin-releasing hormone (GnRH), somatostatin (somatotropin release-inhibiting hormone, SRIH), corticotropin-releasing hormone (CRH), prolactin-inhibiting factor (PIF), and growth hormone–releasing hormone (GHRH)] have been recognized, and the chemical structures of most of these have been determined. Three of the hypophyseal hormones (TRH, GnRH, somatostatin) whose structures are known are peptides composed of 3 to 14 amino acids. A fourth releasing factor (PIF) has been shown to be dopamine.[86,97] CRH is a larger peptide composed of 41 amino acids.[40,85] The tropic hormones of the anterior pituitary are mediated by negative feedback, which involves interaction of the effector hormones with the hypothalamus, as well as with cells of the anterior pituitary.[86,97]

The supraoptic and paraventricular nuclei of the hypothalamus produce antidiuretic hormone (ADH, vasopressin) and oxytocin, which pass down the pituitary stalk and are stored in the neurohypophysis. The release of ADH is mediated by neural impulses from hypothalamic osmoreceptors, as well as from pressure receptors in the right atrium and carotid sinus. Oxytocin is released in response to suckling and also may play a role in initiation of labor. These hormones are discussed below.[67,84,86]

The releasing and inhibiting factors of the hypothalamus are not present in the systemic circulation in amounts that are measurable by current techniques. Therefore, it is impossible to assess hypothalamic function directly. Instead, hypothalamic function or malfunction is inferred from evaluation of the anterior or posterior pituitary and the response of the anterior pituitary to synthetic releasing factors. In some instances, it may be virtually impossible to distinguish a hypothalamic from a pituitary problem by laboratory tests.

Diseases involving the hypothalamus that may result in endocrine dysfunction include tumors, inflammatory or degenerative processes, and congenital problems. These will be manifested in excesses or, more commonly, deficiencies of anterior pituitary hormones. In addition, complex interrelationships between the hypothalamus and other portions of the brain may produce hormone irregularities in association with a variety of psychological problems such as growth failure in emotionally deprived children and the complex endocrine abnormalities seen in patients with anorexia nervosa.[18]

Evaluation of hypothalamic function with regard to specific hormones will be included in discussions of anterior pituitary, posterior pituitary, and other effector hormones (see following sections).

Anterior Pituitary

The anterior pituitary is derived from a portion of the oral ectoderm (Rathke's pouch), which evaginates and eventually fuses with a projection of the hypothalamus. The latter becomes the posterior pituitary (neurohypophysis). The pituitary is located in a small cavity in the sphenoid bone of the skull called the *sella turcica*. Fibrous tissue (diaphragma sella) derived from the meninges covers the sella and is pierced by the pituitary stalk.

The anterior pituitary has classically been said to be composed of three types of cells, which are classified according to their histologic staining characteristics as acidophils, basophils, and chromophobes. More sophisticated studies have now shown the pituitary to be composed of at least five cell types, with each major hormone except the gonadotropins secreted by a different cell. The gonadotropins (LH and FSH) appear to be secreted by the same cell type.[86,93,97]

Hormones secreted by the anterior pituitary are either peptides (ACTH, GH, prolactin) or glycoproteins (TSH, LH, FSH). The glycoprotein hormones are composed of α and β subunits. The α chains are immunologically identical for all three pituitary glycoproteins, as well as for human chorionic gonadotropin. Functional and immunologic specificity resides in the β subunit. ACTH is derived from cleavage of a large prohormone (pro-opiomelanocortin) into ACTH and β-lipotropin. The latter is, in turn, cleaved to yield a variety of peptides, including α-MSH and a group of peptides (endorphins) that function as endogenous opiates.[67,93,97] The glycoprotein hormones and ACTH will be discussed in detail in the section of this chapter dealing with the effector hormones that they stimulate.

Excess secretion of pituitary hormones is usually due to a pituitary tumor and almost always involves only one hormone. Isolated deficiencies of pituitary hormones occur but

are rare, with GH being most common. Pituitary hormone deficiencies usually involve more than one and eventually all the anterior pituitary hormones (panhypopituitarism). These hormones are lost in a characteristic order, with growth hormone and gonadotropins disappearing first, followed by TSH, ACTH, and prolactin. Consequently, patients with this problem develop generalized endocrine dysfunction. Because these symptoms do not develop until approximately 75% of the gland has been destroyed, and because the symptoms tend to develop slowly or incompletely, diagnosis is often delayed.

Panhypopituitarism may be due to a variety of processes but most commonly results from a pituitary tumor (adenoma or craniopharyngioma) or from ischemia. In the case of pituitary tumors, the expansile growth of the lesion compresses the normal gland, resulting in loss of function. In addition, surgical treatment or radiation may result in destruction of the remaining normal pituitary as well as the tumor and often leads to generalized pituitary deficiency.

The usual cause of pituitary ischemia is hemorrhage or shock in a pregnant female at the time of delivery (Sheehan's syndrome). The pituitary is enlarged at this time and is more susceptible than usual to ischemic change. The extent of damage and, consequently, the resultant symptoms are variable. In many cases, the major symptom is loss of sexual function and secondary sex characteristics. These symptoms may be followed much later by symptoms of thyroid or adrenal insufficiency. However, in some cases, severe panhypopituitarism is present from the onset, and the patients may develop Addisonian crisis (see the following).[86,93,97]

Growth Hormone

Growth hormone (GH) is a peptide composed of 191 amino acids. In contrast to most of the other protein hormones, GH does not act through cAMP. Binding of GH to its membrane receptor leads to glucose uptake, amino acid transport, and lipolysis. Its effects are mediated through a group of second-messenger compounds called insulin-like growth factors (IGFs). These compounds are synthesized primarily by the liver under the influence of GH and have strong metabolic and mitogenic effects, especially on cartilage, adipose tissue, and striated muscle.[6]

Release of GH is controlled by two hypothalamic hormones: GHRH and somatostatin. GHRH appears to control the amount of GH released while somatostatin governs the frequency and duration of secretory pulses.[6] GH secretion is relatively erratic and occurs primarily in short bursts. Most GH secretion occurs during sleep, but secretion also increases transiently after exercise and in response to hypoglycemia. GH secretion and release are also affected by obesity, sex steroid levels, and renal and hepatic function.[10] A typical reference range for GH is given in Table 19-1. A single determination of GH provides little clinical information, however, because of the erratic nature of its release. Basal levels of GH in children are the same as for adults, but children show more frequent secretory bursts. GH appears to act in conjunction with sex steroids to mediate the adolescent growth spurt. GH

TABLE 19-1. Typical Reference Ranges for Selected Endocrinology Tests (Serum/Plasma)*[60,97]

Hormone	Reference Range	Stimulated	Suppressed
ACTH	AM 8–79 pg/mL PM 7–30 pg/mL	80 ± 7 pg/mL	
Aldosterone	Male 6–22 pg/dL Female 5–30 pg/dL		
Catecholamines urine plasma	<100 µg/m²/d norepinephrine <1700 pg/mL epinephrine <140 pg/mL		
Cortisol	AM 5–23 µg/dL PM 3–16 µg/dL	>20 µg/dL	<3 µg/dL
Gastrin	<100 pg/mL		
Glucagon	30–210 pg/mL		
GH	Male <2 ng/mL Female <10 ng/mL	>7 ng/mL	<2 ng/mL
5-HIAA (urine)	2–6 mg/d		
Insulin	2–25 µU/mL		
Metanephrines (urine)	0.05–1.20 µg/mg creatinine		
Prolactin	Male 0–20 ng/mL Female 5–40 ng/mL		
PTH (midmolecule)	0.29–0.85 ng/mL		

*Results are for ambulatory adult patients. A different reference range may apply to children or elderly individuals or to pregnant women. These reference ranges are guidelines only and will differ among laboratories.

production decreases with age and responses to some but not all secretory stimuli are also diminished.[6,10]

Excess GH. Elevated levels of GH can be seen in patients with severe malnutrition, chronic liver or kidney disease, uncontrolled diabetes mellitus, and certain types of growth failure. However, symptomatic excess GH production is usually due to the presence of a functioning pituitary adenoma.

The consequences of excess GH secretion depend on the age of onset. If the tumor develops before the epiphyses fuse, the patient becomes a pituitary giant. Patients with this problem have grown as tall as 8 feet, 11 inches. These patients retain normal proportions but have massive generalized organomegaly. They die at a relatively early age, usually of heart failure.

In the more common situation, GH-secreting pituitary tumors develop in adults whose epiphyses are closed. Under these circumstances, continued bony proliferation results in an increase in width rather than length. These changes are particularly apparent in the small bones of the hands and feet and in the jaw and result in the clinical picture of acromegaly. In addition to the bony changes, the patients develop generalized organomegaly and coarsening of the facial features. Metabolic effects of GH are manifested by development of impaired glucose tolerance (due to the anti–insulin effects of GH), hypertension, and occasionally, galactorrhea. The last complication may be due to lactogenic effects of GH itself or to an associated increase of prolactin that occurs in 25% to 40% of patients with GH producing pituitary adenomas.[51] As

the pituitary tumor grows, these patients usually develop signs of hypopituitarism owing to compression of normal pituitary by the expanding tumor.[24,51,86,97]

Laboratory confirmation of the diagnosis of GH excess can be made by demonstrating an elevated GH level [>7 ng/mL (>310 pmol/L)] that does not suppress after glucose administration. The latter is classically tested by measuring GH during a glucose tolerance test. After administration of 100 g of glucose, GH should decrease to <2 ng/mL (220 pmol/L) by 120 minutes.[97] Serum levels of growth hormone greater than 50 ng/mL (2210 pmol/L) may be considered diagnostic of acromegaly, while undetectable growth hormone levels rule out this diagnosis.[1,97] Acromegaly may rarely result from ectopic secretion of GH or GHRH by a carcinoma, usually arising in the lung or pancreas. Measurement of GHRH, available in research and specialty laboratories, will be useful if the ectopic GHRH production is suspected.[8,15,41]

Treatment of GH-secreting tumors is through surgical removal of the tumor or elimination of functioning tumor by radiation. Ectopic hormone production is also treated by removal of the neoplasm.

GH Deficiency. GH deficiency may occur as an isolated problem or as part of the general picture of panhypopituitarism. Until recently, it was thought that GH deficiency in adults had no significant consequences. Recent evidence, however, indicates that some changes in body composition thought to be associated with "normal aging" may actually be reflections of GH deficiency. These changes include decreased lean body mass, increased body fat, decreased bone density, and loss of skeletal muscle mass and may be at least partially reversible with GH administration. The significance of these findings and the appropriate use of GH treatment in adults remains to be determined.[13,27,29,30]

In infants, GH deficiency may be manifested as hypoglycemia. Since there are other more common causes of hypoglycemia in infancy, however, diagnosis is rarely made at this time. More often in children, absence of GH results in pituitary dwarfism. Children with pituitary dwarfism retain normal proportions and show no abnormalities of intellectual development. These patients appear much younger than their age, not only because they are short, but because their facial features and body proportions are immature. A similar clinical appearance occurs in Laron's dwarfism. In this situation, the patients lack IGFs and GH levels are elevated, allowing for laboratory distinction from pituitary dwarfism.

When GH deficiency occurs as part of a generalized hypopituitarism in children, the symptoms of pituitary dwarfism are accompanied by lack of sexual development, hypothyroidism, and hypoadrenocorticism. In an adult with hypopituitarism, the GH deficiency produces no symptoms, and problems result from loss of other pituitary hormones.[86,97]

Laboratory diagnosis of GH deficiency requires stimulation testing, since normal basal levels of GH may be essentially zero. Exercise may be a sufficient stimulus to detect normal GH production in some children.[97] The stimulation tests most often used involve infusion of arginine, L-DOPA, or insulin. GH levels should increase to >7 ng/mL in response to infusion of an appropriate dose of one of these agents. If no response is seen, repeat testing with another agent is indicated, since up to 30% of normal children may show a decreased response to a single agent. The insulin tolerance test results in elevations of plasma cortisol (via ACTH stimulation) as well as GH and allows for evaluation of more than one pituitary function. Consequently, in cases of suspected panhypopituitarism, it may be the test of choice. Patients must be watched very carefully during the insulin tolerance test to prevent complications of hypoglycemia.[6,10,97]

If panhypopituitarism is suspected, insulin infusion may be combined with administration of TRH and GnRH for a generalized test of pituitary function. In the normal patient, such a test would result in elevations of GH and cortisol (insulin), TSH and prolactin (TRH), and FSH (GnRH). If properly performed, this test may give a 1-day evaluation of anterior pituitary function. It is very expensive, however, both for the laboratory and the patient, and results are frequently inconclusive.[24,42,73,86,97]

Isolated GH deficiency in a child is treated by administration of exogenous GH. Panhypopituitarism requires replacement of hormones stimulated by the missing pituitary tropic hormones (*i.e.,* thyroxine, cortisol, and estrogens or androgens).

Prolactin

Prolactin is a protein hormone whose amino acid composition is similar to that of GH. In humans, it appears to function solely in the initiation and maintenance of lactation. In this regard, prolactin acts in conjunction with estrogens, progesterone, corticosteroids, and insulin to promote full development of breast tissue. These additional hormones, among other functions, increase the number of prolactin receptors in mammary tissue. It is thought that the high levels of estrogen present during pregnancy inhibit prolactin-induced lactation. When estrogen levels fall after delivery, prolactin then combines with mammary receptors and initiates lactation. The intracellular mechanism by which prolactin results in milk production is unclear but does not appear to involve cAMP as a second messenger.[52,86,97]

Reference ranges for prolactin are given in Table 19-1. Slightly lower levels are present in children. Prolactin secretion is controlled primarily by prolactin-inhibiting factor (PIF), which is released from the hypothalamus. This factor has been shown to be dopamine. Other inhibitory factors have been identified but their physiological significance is not clear. Several prolactin-releasing factors have been identified including thyroid-releasing hormone (TRH), vasoactive intestinal peptide (VIP), and peptide histidine methionine (PHM). These compounds are found in the hypothalamus and/or anterior pituitary but their physiologic role is uncertain.[52] Levels of prolactin increase during sleep and decline during waking hours. There also appears to be a gradual increase in prolactin secretion in the first half of the menstrual cycle, with a decline immediately after ovulation. The level then increases again until menstruation occurs. Prolactin secretion increases during pregnancy, reaching levels as high as 200 to 300 ng/mL (8,510–12,760 pmol/L) in the third trimester. In the postpartum period, stimulation of the breast produces a rapid rise in prolactin concentration, resulting in lactation. Mechanical stimulation of the breast also may produce an increase in prolactin concentration in some normal nonpregnant, nonlactating women.[86,97]

Hyperprolactinemia. Excessive release of prolactin from the anterior pituitary may come about through a decrease in PIF, an increase in PRF (not yet demonstrated), or autonomous production of prolactin by a pituitary tumor.

Alterations of dopamine production by the hypothalamus can be brought about by sectioning of the pituitary stalk, irritation of the C8–T5 segments of the spinal cord, or by pharmacologic manipulation. In particular, aldomet (which interferes with dopamine synthesis) and the phenothiazines and reserpine (which interfere with dopamine release) may produce hyperprolactinemia. Hyperprolactinemia is also common in patients with renal failure and other endocrine diseases such as hypothyroidism and adrenal insufficiency. The pathogenesis of the hyperprolactinemia in these situations involves increased production as well as delayed metabolic breakdown in the case of renal failure.[52,86,93,97]

Pituitary tumors producing prolactin are the most common type of pituitary adenoma. Many of these tumors (especially in women) are microadenomas.

In females, the presenting symptom of hyperprolactinemia is usually amenorrhea (cessation of menstrual periods). Galactorrhea (production of milk in any situation other than the postpartum period) is present in 50% to 90% of cases. Males usually have larger tumors at the time of presentation and may present with symptoms related to tumor size (headache, visual symptoms, etc.). Impotence is also a common complaint. Galactorrhea in males may occur but is uncommon. The mechanism for amenorrhea and impotence in association with hyperprolactinemia is not clear. However, it may relate to suppression of gonadotropin release by prolactin or to inhibited gonadotropin-releasing factor activity.[1,52,86]

The diagnosis of prolactin-secreting pituitary adenoma requires the exclusion of other possible causes of hyperprolactinemia, as mentioned previously. In the presence of an enlarged sella, hyperprolactinemia is almost certainly due to a pituitary neoplasm. If radiologic evidence of a pituitary tumor is not present, the diagnosis is more difficult. Findings suggestive of a pituitary tumor in these circumstances include: (1) serum prolactin levels >200 ng/mL (8510 pmol/L), (2) lack of response to TRH infusion (may also occur with drug-induced hyperprolactinemia), and (3) lack of response of serum prolactin levels to pharmacologic manipulation of dopamine. The last category is not very helpful with methods currently available, since both tumorous and nontumorous conditions may respond to dopamine manipulation.[1,24,86]

Treatment of hyperprolactinemia due to a pituitary adenoma is usually surgical. Pharmacologic treatment with a dopamine antagonist (bromocriptine) is also effective.

Prolactin Deficiency. Prolactin deficiency is usually seen as part of the general picture of panhypopituitarism. It is particularly noticeable in acute hypopituitarism secondary to postpartum hemorrhage (Sheehan's syndrome), since lactation does not develop in these patients. Isolated prolactin deficiency has been reported but is very rare.[86]

Posterior Pituitary (Neurohypophysis)

As noted in the previous discussion of the hypothalamus, the hormones released by the posterior pituitary are actually synthesized in the supraoptic and paraventricular nuclei of the hypothalamus. These hormones are transported in membrane-bound vesicles through the neuronal axons in the pituitary stalk to the neurohypophysis, where they are stored. Release of these hormones occurs in response to stimulation of the hypothalamus by changes in serum osmolality or by suckling.[67,84]

Antidiuretic Hormone (ADH, Vasopressin)

Antidiuretic hormone (ADH) is an octapeptide that is synthesized primarily in the supraoptic nucleus of the hypothalamus and is stored in the posterior pituitary until a stimulus is received for its release. It acts on the distal convoluted tubule and collecting tubule of the kidney, making them permeable to water. This action is mediated through cAMP and results in the production of a concentrated urine. The hormone is metabolized primarily in the liver.[36,67,84]

Release of ADH is mediated primarily through changes in plasma osmolality, which are detected by osmoreceptors in the hypothalamus. Normally, osmolality is maintained within a very narrow range, and changes of as little as 1% to 2% will stimulate or inhibit ADH release. A decrease in blood volume or blood pressure will also stimulate ADH release. The effect of these changes is to alter the set-points for osmoregulation to a lower level. This alteration increases blood volume at the expense of "normal" osmolality. This stimulus is mediated by baroreceptors in the thorax and requires changes in the extracellular volume of 5% to 10% before it is activated.[36,84,92]

In addition to tonicity and volume, ADH release may also be stimulated by nausea, pain, stress, various infectious, and vascular and neoplastic disorders. Such conditions may result in the syndrome of inappropriate ADH (see the following). ADH release may be inhibited by cold and by numerous drugs, most commonly ethanol.[84,92]

Hypersecretion of ADH. Hypersecretion of ADH may be relative or absolute. Relative hypersecretion occurs in situations in which hypovolemia takes precedence over osmolality as a stimulus for ADH release. In this situation, ADH acts to retain water in the presence of low serum osmolality. This condition is associated with a low urinary excretion of sodium, which distinguishes it from the syndrome of inappropriate ADH.[66]

The syndrome of inappropriate ADH (SIADH) results when ADH is released despite low serum osmolality in association with a normal or increased blood volume. This situation develops in patients with tumors producing ectopic ADH (most commonly oat-cell carcinoma of the lung) or in association with CNS trauma or infections, administration of certain drugs (opiates, barbiturates, clofibrate, chlorpropamide), and some pulmonary diseases, especially tuberculosis and sarcoidosis.[24,84]

SIADH can be diagnosed by measuring serum ADH and determining if the level is too high for the serum osmolality. Diagnosis of this syndrome requires that other conditions which could result in hypo-osmolality and/or generalized edema (hypovolemia, hypotension, congestive heart failure, renal insufficiency, etc.) be absent.[36] ADH assays are not readily available, however, and SIADH is usually diagnosed by the presence of three factors: (1) hypo-osmolality (hyponatremia), (2) urine osmolality greater than serum osmolality, and (3) urine Na$^+$ greater than 20 to 25 mEq/L. The latter serves to distinguish SIADH from relative excess of ADH due to volume depletion. In SIADH, expansion of extracellular volume stimulates sodium excretion even in the face of hyponatremia.[24,36,84,97]

The most appropriate therapy of SIADH is removal of the stimulus. If this is not possible, treatment by water restriction or drugs that block the renal response to ADH may be effective.

Hyposecretion of ADH. Deficiency of ADH [diabetes insipidus (DI)] results in severe polyuria (10–12 L/day). As long as the thirst mechanism is intact, the patient can usually compensate adequately by drinking large quantities of water. If anything occurs to disrupt the thirst mechanism or prevent the patient from drinking water, rapid development of hypernatremia will occur.

DI results from destruction of the posterior pituitary or hypothalamus secondary to neurosurgical procedures, tumor, trauma, or a degenerative or infiltrative process. These patients present with polyuria and polydipsia. In order to establish the diagnosis, the patient's ability to produce a concentrated urine must be tested. This is accomplished by water deprivation accompanied by hourly measurements of urine osmolality. When a constant urine osmolality is obtained or the patient loses 3% to 5% of body weight, serum osmolality is measured and a subcutaneous injection of ADH is administered. Urine osmolality is again measured hourly for 2 to 3 hours. Normal individuals or those with psychogenic polydipsia (excessive drinking of water because of a psychiatric disturbance) will show a urine/plasma osmolality ratio greater than 1 with water deprivation, and urine osmolality will increase greater than 5% after ADH administration. Patients with DI will show a urine/plasma osmolality less than 1 even after loss of 3% to 5% of body weight and will show a much greater response to ADH. A third category of patients who have an end-organ unresponsiveness to ADH (nephrogenic DI) will show urine and plasma osmolalities similar to DI but will not respond to ADH.[4,75,97]

Treatment of DI is replacement of ADH by injection of arginine vasopressin or administration of the vasopressin analog 1-desamino 8-D-arginine vasopressin (DDAVP, desmopressin) by nasal spray or injection. If only a partial deficiency is present, the patient may respond to therapy with drugs such as chlorpropamide or clofibrate, which increase ADH release. Nephrogenic DI is treated with diuretics or reduction of the osmotic load. The latter lowers extracellular volume and/or reduces the glomerular filtration rate, allowing increased reabsorption of water from the proximal tubules.[4,24,84,92]

Oxytocin

Oxytocin is a nonapeptide produced in the paraventricular nucleus of the hypothalamus. It is very similar in composition to ADH, and like ADH, it is secreted in association with a carrier protein and stored in the posterior pituitary. Oxytocin stimulates contraction of the gravid uterus at term and also results in contraction of myoepithelial cells in the breast, causing ejection of milk. It is released in response to neural stimulation of receptors in the birth canal and uterus and of touch receptors in the breast. In lactating females, emotional stimuli also may affect this pathway, resulting in stimulation or inhibition of oxytocin release and lactation. There is some evidence for a negative feedback effect of oxytocin on the posterior pituitary to inhibit further secretion.

Oxytocin exerts its physiologic effect by binding to receptors on myometrial or myoepithelial cell membranes and initiating formation of cAMP or cGMP. The hormone appears to be metabolized by the liver and kidneys and/or the lactating mammary gland. Radioimmunoassays have been developed to measure oxytocin but are not in routine use.[97]

No disorders associated with excess oxytocin production have been described. Deficiency states may exist but have not been identified conclusively. Synthetic preparations of oxytocin are used to increase weak uterine contractions during labor and to aid lactation in women who have difficulties in the ejection (not production) of milk. These situations may represent deficiencies in oxytocin production.[24,97]

Thyroid

The thyroid gland and its two principal hormones (thyroxine and triiodothyronine) are discussed in Chapter 20, Thyroid Function.

Calcitonin

A third thyroid hormone, calcitonin, is produced by the parafollicular or C cells of the thyroid. In pharmacologic doses, calcitonin has a calcium-lowering effect, and it is thought that this hormone plays a role in calcium metabolism in lower animals. No such physiologic role for calcitonin has been proved in humans.

Calcitonin measurement plays an important role in detection and diagnosis of thyroid cancers arising from the parafollicular (C) cells (medullary carcinoma of the thyroid). Patients with these tumors have elevated calcitonin levels and show an abnormally large release of calcitonin in response to infusion of substances known to stimulate its secretion (calcium, pentagastrin). Since many tumors of this type tend to be familial and/or to be associated with other endocrine neoplasms, stimulation of calcitonin release by pentagastrin or calcium in patients at high risk may allow for early diagnosis of medullary thyroid cancer.[14,81] (See Multiple Endocrine Neoplasia in this chapter.)

Parathyroid Glands

The parathyroid glands are usually located in the neck adjacent to the middle and lower portions of the thyroid gland. Most people have four parathyroid glands, but more are not uncommon. The glands may be found in other locations in the neck and also in the mediastinum.

Parathyroid hormone (PTH) is synthesized as a prohormone containing 115 amino acids. The active form of the hormone contains 84 amino acids and is secreted by the chief cells of the parathyroid glands. The intact molecule is quickly cleaved into N-terminal and C-terminal fragments. The N-terminal portion is the active portion of the molecule and has a short serum half-life. The inactive C-terminal fragment is more stable. PTH differs from many of the other hormones in that its release is regulated by direct negative feedback from the substance it controls (calcium) rather than by negative feedback on a tropic hormone from the pituitary.

PTH acts to increase serum calcium concentration by: (1) increasing resorption of calcium from bone, (2) stimulating calcium retention by the renal tubules, and (3) promoting 1-hydroxylation of 25-hydroxy vitamin D in the kidney. In turn, $1,25(OH)_2$ vitamin D increases calcium absorption from the intestine. PTH also increases renal excretion of phosphate and, to a lesser degree, that of bicarbonate. These actions are mediated by cAMP.

PTH is measured by immunoassay. Serum PTH levels were formerly determined by measurement of the N- or C-terminal fragment, but now a two-site immunoassay for intact PTH is rapidly replacing fragment-based assays and has provided a means for direct measurement of parathyroid function independent of peripheral metabolism and renal clearance of hormone fragments.[14,20,100] A typical reference range is given in Table 19-1.

Hyperparathyroidism

Hyperparathyroidism may be due to: (1) autonomous release of PTH from a benign or malignant parathyroid tumor or from diffuse hyperplasia of all four glands (primary hyperparathyroidism), (2) release of increased amounts of PTH from hyperplastic glands in an attempt to maintain normal serum calcium in the face of vitamin D deficiency or renal failure (secondary hyperparathyroidism), or (3) autonomous release of PTH developing in the setting of secondary hyperparathyroidism (tertiary hyperparathyroidism). All forms are associated with elevated serum PTH levels.[34,14,99]

Primary hyperparathyroidism is most commonly (80%) due to the presence of a functioning parathyroid adenoma. Hyperplasia accounts for virtually all the remaining cases of primary disease, with parathyroid carcinoma contributing less than 1% of cases. Since use of chemistry profiles has become widespread as a part of a general physical examination or admission to a hospital, most cases of primary hyperparathyroidism are discovered incidentally during a routine examination or evaluation for another disease. Prior to chemical screening, patients commonly presented with complications due to hypercalcemia or PTH excess, such as renal stones or other renal disease, peptic ulcers, pancreatitis, pathologic fractures, and so on.

The major complications of hyperparathyroidism occur in kidney and bone. High levels of serum calcium result in excess calcium excretion in the urine. Phosphaturia is also present due to the action of PTH on renal tubular cells. Consequently, calcium-containing renal stones form, and renal colic is a presenting symptom in approximately 25% of patients with hyperparathyroidism. Precipitation of calcium phosphate may also occur within the renal parenchyma, resulting in decreased renal function and, occasionally, hypertension. PTH also results in activation of osteoclasts and resorption of calcium from bony trabeculae. If hyperparathyroidism goes undetected for longer periods of time, severe demineralization of bone may occur, with replacement by fibrous tissue and formation of "cysts" (osteitis fibrosa cystica). The bone is very weak in these areas and may break with minimal stress (pathologic fracture).[14,99]

Other complications of hyperparathyroidism are related to increased calcium levels and include peptic ulcer disease (stimulation of hydrochloric acid and gastrin by calcium) and central nervous system symptoms.[14,99]

Laboratory diagnosis of primary hyperparathyroidism is based on the demonstration of increased serum calcium, decreased serum phosphorus, and increased PTH. If significant bone disease is present, alkaline phosphatase activity will be elevated because of increased osteoblastic activity. Chemical examination of urine shows increased levels of calcium and phosphorus and increased urinary hydroxyproline, reflecting increased bone turnover.[14,99]

Secondary hyperparathyroidism develops in response to decreased serum calcium and is reflected morphologically as diffuse hyperplasia of all four glands. The most common causes of secondary hyperparathyroidism are vitamin D deficiency and chronic renal failure. Vitamin D deficiency causes decreased intestinal absorption of calcium and a compensatory increase in PTH is required to maintain normal calcium levels. Development of increased PTH secretion in association with chronic renal failure is more complex and relates primarily to three factors: (1) phosphate retention by the failing kidney; (2) decreased functioning renal mass, resulting in decreased levels of $1,25\text{-}(OH)_2$ vitamin D; and (3) relative resistance of bone to the effects of PTH. All of these factors result in a tendency toward hypocalcemia, and excess PTH is required to maintain a normal calcium concentration.[14,99]

Patients with secondary hyperparathyroidism frequently develop severe bone disease identical to that seen in primary hyperparathyroidism. Laboratory findings in these patients include low or normal serum calcium levels, elevated serum phosphorus levels, increased alkaline phosphatase activity (if bone disease is present), increased urinary calcium levels, decreased urinary phosphorus levels, and increased urinary hydroxyproline levels.[14,99]

Tertiary hyperparathyroidism occurs when patients with secondary hyperparathyroidism develop autonomous function of the hyperplastic parathyroid glands or of a parathyroid adenoma. Laboratory findings are similar to those in primary hyperparathyroidism, except that phosphate levels are normal to high. Because of elevated serum phosphorus levels, the solubility product of calcium and phosphorus is exceeded, and calcium phosphate precipitates in soft tissues.[14,99]

Hyperparathyroidism may rarely occur in association with ectopic secretion of PTH by tumors, most commonly squamous cell carcinoma of the lung. The presentation may relate to the primary tumor or may be identical to that of patients with primary hyperparathyroidism. Laboratory findings in this situation are identical to those in primary hyperparathyroidism.[14,99] In addition, a hypercalcemic factor (called parathyroid hormone-related protein) has been isolated and characterized as a widely distributed protein in both normal and malignant tissues with a common amino acid sequence to the N-terminal region of PTH. Its level in serum is low or undetectable in normal persons and in patients with primary hyperparathyroidism but elevated in certain malignancies. Several immunoassays for the protein have been developed and should prove to be useful in the diagnosis of malignancy-associated hypercalcemia.[5,14,26,43]

Treatment of hyperparathyroidism is surgical removal of the source of PTH. If the problem is due to parathyroid hyperplasia, three and one-half glands are removed. The remaining half is usually sufficient to provide normal parathyroid function.

Hypoparathyroidism

Decreased parathyroid function is most commonly due to accidental injury to the parathyroid glands during thyroid or other neck surgery or to idiopathic atrophy. Patients with hypoparathyroidism are unable to maintain their serum calcium concentration without calcium supplementation and develop hypocalcemia. If serum calcium levels drop below 8 mg/dL (2.0 mmol/dL), patients develop tetany and other manifestations of altered neuromuscular activity. Latent tetany may be demonstrated in patients with borderline low calcium by tapping on the facial nerve, producing jerking movements of the jaw (Chvostek's sign), or by inflating a blood pressure cuff on the arm to produce ischemia and tetany in the arm (Trousseau's sign). Calcium levels below 6 mg/dL (1.5 mmol/dL) may result in laryngeal stridor and grand mal seizures.[14,99]

Laboratory evaluation of patients with hypoparathyroidism shows hypocalcemia, hyperphosphatemia, and decreased PTH levels. Similar clinical signs and laboratory findings occur in patients with pseudohypoparathyroidism—a hereditary disorder characterized by end-organ unresponsiveness to PTH. These patients have characteristic round faces and deformities of the bones of the hand. The defect responsible for this disorder is a reduction in a regulatory protein in the adenyl cyclase enzyme complex.[14]

Adrenal Glands

The adrenal glands are paired organs located one at the upper pole of each kidney. Each gland consists of an outer cortex and an inner medulla, which have different embryological origins, different mechanisms of control, and different products.

Adrenal Cortex

The adrenal cortex is composed of three layers. The outermost layer (zona glomerulosa) is responsible for secretion of mineralocorticoids. The second layer (zona fasciculata) secretes glucocorticoids, and the third layer (zona reticularis) secretes sex steroids, principally androgens.

All the hormones produced by the adrenal cortex are steroids derived from cholesterol. Although these hormones are under different control mechanisms, the fact that they share a common precursor and some common enzymatic pathways results in a complex interrelationship among them. Abnormalities in production of one hormone may be accompanied by increased or decreased amounts of other adrenocortical hormones, depending on the specific lesion involved. Thus, the clinical presentation of these patients results from a combination of hormone abnormalities. The individual hormones, their mechanisms of control, and abnormalities of production are discussed below.

Mineralocorticoids. Aldosterone is the principal mineralocorticoid (electrolyte-regulating hormone) produced by the zona glomerulosa. The steps in the synthesis of aldosterone and the enzymes required are outlined in Figure 19-1.

Normally, aldosterone production is primarily controlled by the renin-angiotensin system (Fig. 19-4). ACTH has a slight stimulatory effect on aldosterone synthesis, but this is usually of no significance.

Renin is a protein produced by the juxtaglomerular apparatus of the kidney in response to decreased renal perfusion pressure and/or decreased serum sodium levels. Renin acts on angiotensinogen, a plasma protein produced by the liver, to produce angiotensin I. In turn, angiotensin I is converted to angiotensin II by angiotensin-converting enzyme present mainly in the lungs. Angiotensin II is a potent vasoconstrictor and also stimulates secretion of aldosterone by the zona glomerulosa. Aldosterone acts on renal tubular epithelium to increase retention of Na^+ (and, secondarily, chloride and water) and excretion of K^+ and H^+. Thus the effect of the renin-angiotensin system is to increase blood pressure, both by vasoconstriction and increased plasma volume. When renal perfusion pressure increases, renin secretion and aldosterone production are suppressed.[31,94,103]

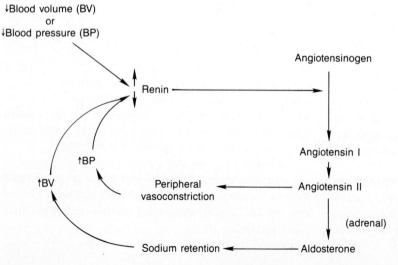

Renin-angiotensin system

Figure 19-4. Simplified scheme of the renin-angiotensin system and its relationship to blood volume and blood pressure.

Aldosterone production also varies with dietary sodium intake. With a "normal" intake of 100 to 150 mEq of Na per day, 40 to 160 μg (111–444 nmol) of aldosterone is secreted daily. Typical reference ranges for aldosterone concentration are found in Table 19-1. Serum concentration is higher in the morning than in the afternoon. Only approximately 30% is bound to plasma protein. The hormone is metabolized by the liver, and metabolites are excreted in the urine.[31,94]

Hyperaldosteronism. Hyperaldosteronism may be primary (due to an adrenal lesion) or secondary to abnormalities in the renin-angiotensin system.

Primary hyperaldosteronism is a relatively uncommon condition and is most often due to the presence of an aldosterone-secreting adrenal adenoma (Conn's syndrome). Cases of hyperaldosteronism due to bilateral hyperplasia of the adrenal cortex may occur, and rare cases due to primary adrenal hyperplasia or aldosterone-secreting adrenal carcinomas have been reported.[49] Clinical symptoms result from hypertension and/or hypokalemia.

Hypertension results from volume expansion and sodium retention produced by aldosterone. Although less than 1% of hypertensive patients are found to have Conn's syndrome, this condition is surgically curable and should always be considered in the differential diagnosis of hypertension. From the laboratory standpoint, the first step in evaluation of the hypertensive patient is determination of serum electrolytes. Patients with primary hyperaldosteronism usually show hypokalemia associated with mild hypernatremia and metabolic alkalosis. If renal potassium wasting (>30 mEq/day) is demonstrated in a hypertensive patient who has hypokalemia and who is not on diuretic therapy, serum aldosterone levels should be determined. More than 50% of such patients are found to have primary hyperaldosteronism.[31,49]

Samples for aldosterone determination should be drawn in the morning before the patient has gotten out of bed and must be interpreted with consideration of sodium intake and posture. If serum aldosterone is above the reference range, the response of the patient to manipulation of dietary sodium should be evaluated. Normally, sodium restriction results in increased plasma renin activity and aldosterone production. In the patient with an aldosterone-producing tumor, however, plasma renin activity will remain suppressed under these circumstances, and serum aldosterone levels will show no response. Sodium loading, which normally produces a decrease in plasma aldosterone levels, has no effect in patients with an aldosterone-producing tumor.[31,94,103]

If laboratory evidence suggests Conn's syndrome, the tumor can usually be localized by radiographic procedures, and surgical therapy is curative. If primary hyperaldosteronism is due to adrenal hyperplasia, medical therapy with angiotensin converting enzyme inhibitors or diuretics that blocks aldosterone receptors in the kidney is indicated.[49]

Secondary hyperaldosteronism results from excess production of renin. This condition may be due to renal artery stenosis (narrowing), which results in decreased renal perfusion, to malignant hypertension, or to the presence of a renin-secreting renal tumor (rare). In this situation, elevated aldosterone levels are associated with increased plasma renin activity that does not suppress with sodium loading. A vascular lesion in the main renal artery may be demonstrated by angiographic x-ray examination and is often surgically correctable. Increased plasma renin activity in patients with malignant hypertension is secondary to hypertensive damage to small vessels deep in the kidney and is not surgically correctable.[31,103]

Secondary hyperaldosteronism also may result from sodium depletion or circulatory abnormalities such as congestive heart failure. In this situation, increased aldosterone represents an attempt to maintain appropriate plasma volume and is not usually associated with hypertension.[103]

Hypoaldosteronism. Aldosterone deficiency is most often due to destruction of the adrenal glands and, as such, is associated with glucocorticoid deficiency (see the following on Addison's disease). Patients with congenital deficiency of the enzyme 21-hydroxylase may also have deficient aldosterone production, resulting in inability to conserve sodium. If sodium intake is not adequate, the patient will develop severe water and electrolyte abnormalities and may die from vascular collapse and/or hyperkalemia (see the following on congenital adrenal hyperplasia).[24,58]

Glucocorticoids. Cortisol, the principal glucocorticoid hormone, is synthesized in the zona fasciculata of the adrenal cortex as depicted in Figure 19-1. The actions of cortisol are widespread and include anti-insulin effects on carbohydrate, fat, and protein metabolism, effects on water and electrolyte balance, stabilization of lysosomal membranes, and suppression of inflammatory and allergic reactions.[48,58,97]

Cortisol is regulated through a feedback loop to pituitary ACTH and hypothalamic corticotropin releasing hormone (CRH) secretion. ACTH is a 39-amino acid polypeptide that is synthesized as part of the prohormone pro-opiomelanocortin. Release of ACTH from the anterior pituitary is stimulated by CRH from the hypothalamus and results in increased synthesis of cortisol by the adrenal cortex. Elevated cortisol levels then act on the pituitary and hypothalamus to shut off ACTH release, allowing serum cortisol to fall. When cortisol drops below a certain level, this inhibitory effect diminishes, and ACTH is again released (negative feedback). A variety of factors other than serum cortisol levels, including pain, fever, hypoglycemia, and fear, also may stimulate release of ACTH.[48,58]

ACTH and cortisol normally show a diurnal variation, with the highest levels present in the morning around 8 A.M. and lowest levels present in the late afternoon. Typical reference ranges for serum cortisol and ACTH are found in Table 19-1.

In the plasma, approximately 90% of serum cortisol is bound to a carrier protein (cortisol-binding globulin). Because the carrier sites on this protein are essentially saturated at normal levels of plasma cortisol, increases in free cortisol occur rapidly with increased cortisol production. Since free cortisol is readily filtered and excreted by the kidney, increased urine free cortisol levels are a sensitive indicator of adrenal hyperfunction (see the following). Determination of the level of cortisol metabolites excreted in the urine (17-hydroxycorticosteroids; 17-OHCS) also may be helpful in evaluating adrenal function.[48,58,97]

Hypercortisolism. The syndrome of hypercortisolism (Cushing's syndrome) may result from: (1) excessive production of cortisol by an adrenal adenoma or carcinoma; (2) excessive production of ACTH by a pituitary adenoma, leading to adrenal hyperplasia; (3) production of "ectopic" ACTH by a tumor of nonendocrine tissue, most commonly a lung cancer; and (4) exogenous administration of cortisol. Because of its widespread use in the treatment of a variety of systemic inflammatory diseases, exogenous administration is the most common cause of the group of symptoms known as *Cushing's syndrome*. In this situation, production of cortisol by the patient's adrenals is suppressed, and when exogenous cortisol is stopped abruptly, adrenal insufficiency may develop. Consequently, the usual approach to discontinuation of cortisol therapy is to decrease the dose gradually (taper the dose) over several days to allow the patient's adrenals to begin to produce cortisol again.[58,97]

The symptoms of hypercortisolism are the same regardless of the etiology and include the following:[58,97]

Truncal obesity—weight gain predominantly in the face, neck, shoulders, and abdomen with relatively thin extremities; many patients develop a characteristic "buffalo hump," a fat pad on the upper back at the base of the neck
Abnormal glucose metabolism (hyperglycemia)
Protein wasting, with thinning of the skin and development of striae, easy bruising, muscle wasting, and poor wound healing
Hypertension
Decreased ability to limit infection, with loss of delayed hypersensitivity response, decreased lymphoid tissue, and decreased circulating lymphocytes

Laboratory evaluation of the patient with apparent Cushing's syndrome is crucial in establishing the diagnosis of hypercortisolism and in distinguishing among the adrenal, pituitary, and ectopic causes of the syndrome. (Hypercortisolism due to ACTH production by a pituitary adenoma is referred to as *Cushing's disease* to distinguish it from the other causes of Cushing's syndrome.)

The first step must be to establish a diagnosis of hypercortisolism. This is usually done by measuring levels of cortisol at 8 A.M. and 8 P.M. to determine if the normal diurnal variation is present and by measuring the patient's response to exogenous suppression of the pituitary-adrenal axis. The latter is done by giving the patient a dose of dexamethasone at 11 P.M. and measuring plasma cortisol at 8 A.M. the next day. Dexamethasone is a potent glucocorticoid, which will normally inhibit the pituitary-adrenal axis at levels too low to interfere with measurement of plasma cortisol. If plasma cortisol does not show a diurnal pattern and does not suppress with dexamethasone administration, a diagnosis of hypercor-

tisolism is highly likely. Measurement of urine free cortisol in a 24-hour urine collection may aid in the diagnosis.[7,48,58,97]

Once a diagnosis of endogenous hypercortisolism is made, a variety of tests may be used in conjunction with other clinical and radiographic data to help establish where the abnormality lies.

1. Measurement of plasma ACTH—Since cortisol and ACTH interact in a feedback loop, increased plasma cortisol levels will be associated with suppression of ACTH if the syndrome is due to a primary adrenal lesion. In contrast, increased cortisol due to an ACTH-producing pituitary adenoma (Cushing's disease) or to ectopic production of ACTH would be associated with increased plasma ACTH levels. Measurement of plasma ACTH is not readily available in most hospital laboratories and must be sent to a reference lab. Consequently, turnaround time is relatively slow.[7,56]

2. Dexamethasone suppression test—Administration of 2 mg dexamethasone every 6 hours in a normal individual will result in suppression of plasma cortisol and 17-OHCS excretion. Patients with Cushing's syndrome due to endogenous cortisol production will not suppress with this regimen. Administration of high doses of dexamethasone (8 mg every 6 h) will, however, frequently result in suppression of pituitary Cushing's. In this situation, the set point for suppression of ACTH production by the pituitary tumor seems to be higher than normal, but the feedback loop is still intact. Patients with adrenal tumors or ectopic ACTH will not respond even to high doses of dexamethasone, since hormone production in these settings has totally escaped feedback control.[7,48,56,97]

3. Metyrapone test—Metyrapone blocks the action of 11-β-hydroxylase, the enzyme responsible for the final step in the formation of cortisol. Administration of this substance in the normal person results in a decreased production of cortisol, stimulation of ACTH, and an increase in the metabolic precursors of cortisol. Like cortisol, these metabolic precursors are excreted in the urine as 17-OHCS, and a normal response to metyrapone is, therefore, an increase in 17-OHCS excretion in the urine. A similar response is seen in patients with pituitary Cushing's who are still subject to feedback control, even though at a higher level. Patients whose cortisol production is completely autonomous, that is, those with adrenal tumors or ectopic ACTH, are already under maximum stimulation and show no response to metyrapone.[7,48,56,58]

4. CRH stimulation test—Administration of CRH to patients with Cushing's disease (pituitary adenoma) has been reported to increase their already elevated levels of ACTH. In contrast, no increase in ACTH was seen when CRH was given to patients with elevated ACTH due to ectopic ACTH production or to patients with suppressed ACTH due to primary hypercortisolism. This test may thus be useful in distinguishing between causes of hypercortisolism with an elevated plasma ACTH.[7,40,48,56,97]

Treatment of hypercortisolism depends on removal of the adrenal or pituitary tumor or of the tumor producing ectopic ACTH. If surgical removal cannot be accomplished, radiation therapy to the tumor or chemotherapy with an agent that inhibits cortisol formation may be helpful.

Hypocortisolism. Decreased production of serum cortisol may be associated with primary adrenal disease or may be secondary to pituitary abnormalities resulting in decreased ACTH. Hypocortisolism due to primary adrenal disease (Addison's disease) is most often due to idiopathic atrophy of the gland. This atrophy is thought to represent an autoimmune process, and Addison's disease may be associated with other autoimmune phenomena.[32] Other causes of primary adrenal hypofunction include destruction of the gland by tuberculosis or histoplasmosis, replacement of both glands by metastatic tumor, and DIC with adrenal hemorrhage. When tuberculosis was less well controlled, tuberculous destruction of the adrenals was very common, but with better therapy for tuberculosis, this condition is now relatively rare. Similarly, although invasion of the adrenals by metastatic tumor is common, this process rarely results in sufficient destruction of both adrenals to produce hypocortisolism. When hypocortisolism is due to Addison's disease, it is usually accompanied by deficiency of mineralocorticoids, since all three layers of the adrenal cortex are damaged.[58,95]

Secondary hypocortisolism is due to pituitary insufficiency with loss of ACTH. In this situation, mineralocorticoid deficiency is not a problem, since ACTH plays only a minor role in stimulation of the zona glomerulosa.

Patients with chronically decreased cortisol levels may present complaining of weight loss, weakness, and a variety of gastrointestinal complaints. Any stress may precipitate an acute syndrome of vascular collapse and shock. Patients with primary adrenal destruction and decreased mineralocorticoid levels have a more severe clinical presentation than those with secondary disease, because of the fluid and electrolyte abnormalities associated with aldosterone deficiency.[58,95]

Laboratory findings in the patient with adrenal insufficiency include low serum sodium, bicarbonate, and glucose levels and high serum potassium and BUN levels. Low plasma cortisol levels in a setting of vascular collapse is diagnostic of Addison's disease, since cortisol would normally be elevated in response to the stress associated with such circumstances. Acute adrenal insufficiency is a life-threatening medical emergency, and the patient must be treated promptly with cortisol replacement and fluids.[58,95]

Once a patient has survived an episode of acute adrenal insufficiency, or if the diagnosis of chronic adrenal insufficiency is suspected, laboratory evaluation is useful in distinguishing between primary and secondary disease. Laboratory tests appropriate in these circumstances include the following:

1. Plasma ACTH—With primary adrenal disease, ACTH levels will be elevated because the cortisol produced is insufficient to inhibit the feedback loop. (The presence of excess ACTH may be suggested clinically by the occurrence of increased pigmentation in these patients as well as in those patients with excess release of ACTH by a pituitary tumor. The increased pigmentation results from the occurrence of

an amino acid sequence identical to MSH within the ACTH molecule.) If adrenal production of cortisol is decreased because of pituitary disease, plasma ACTH will be low.[58,48,97]

2. ACTH stimulation test—Administration of synthetic ACTH will normally cause a prompt increase in plasma cortisol levels. Protocols for this test may look at incremental increases or peak values of cortisol release to determine if adrenal dysfunction is present. In patients with normal adrenal function, a peak value of 18 to 20 $\mu g/dL$ should be reached by 30 minutes after injection of synthetic ACTH. No significant response occurs in patients with adrenal insufficiency. A lengthier test, with infusion of ACTH over 2 to 3 days and daily measurement of plasma cortisol levels, is necessary to distinguish between adrenal and pituitary causes of insufficient cortisol production. In the latter, a stepwise increase in cortisol production will occur during the period of the test as the atrophic adrenals again become responsive to ACTH. In the former, no increase in cortisol production will occur.[48,58,97,95]

3. Metyrapone test—In patients who show signs of panhypopituitarism or in whom pituitary insufficiency might be expected to develop, the metyrapone test may provide a useful estimate of pituitary reserve. As noted previously, metyrapone blocks 11-β-hydroxylation of 11-deoxycortisol, the last step in the synthesis of cortisol. Decreased amounts of cortisol stimulate the pituitary-adrenal axis, resulting in an increased production of cortisol precursors, which are excreted in the urine as 17-OHCS. Failure of such an increase to occur indicates lack of pituitary reserve. This test must be undertaken with caution because the decrease in cortisol production in a patient with borderline adrenal function may precipitate acute adrenal insufficiency.[48,58]

Sex Steroids. Congenital adrenal hyperplasia (CAH) results from the lack of an enzyme necessary for the production of cortisol. This lack causes both a deficiency of cortisol and an accumulation of metabolites prior to the enzymatic block. Because of the interconnecting pathways of synthesis of aldosterone, cortisol, and the sex steroids, these excess metabolites are shunted into the formation of sex steroids, especially androgens. Excessive production of androgens results in virilization, with formation of ambiguous genitalia in female infants. (Male infants generally have no genital abnormalities at birth; although, if the syndrome goes undetected, they may have early development of secondary sex characteristics.) Symptoms of hypocortisolism and/or hypertension may be present, depending on the degree of the enzyme deficiency and the mineralocorticoid activity of the accumulating metabolite.

The most common (95%) enzyme deficiency producing the syndrome of CAH is 21-hydroxylase deficiency. This enzyme is involved in the synthesis of both aldosterone and cortisol. In some variants of the disease, the enzyme is missing from both the zona glomerulosa and the zona fasciculata, resulting in severe salt wasting in association with hypocortisolism. In other variants, enzyme loss appears to be confined to the zona fasciculata, and aldosterone production is unaffected. This type of enzyme deficiency is diagnosed by increased urinary excretion of 17-ketosteroids (17-KS, products of sex steroid metabolism), decreased excretion of 17-OHCS, and elevated urinary pregnanetriol. The only other congenital enzyme deficiency that occurs in significant numbers is deficiency of 11-hydroxylase (5% of cases of CAH). Shunting of cortisol precursors into other pathways results in increased synthesis of mineralocorticoids (deoxycorticosterone, DOC), as well as of androgens, so that these patients often have hypertension as well as the abnormalities of sexual development described previously. Diagnosis of this condition rests on markedly increased 17-KS levels in the urine, only a modest increase in urinary pregnanetriol levels, and increased plasma levels of 11-deoxycortisol and DOC. Both forms may be diagnosed prenatally by measurement of pregnanetriol and DOC levels in amniotic fluid.[24,54,72]

Other enzyme deficiencies producing similar symptoms and signs have been reported, but they are rare. All forms of CAH are treated by replacement of cortisol.

Adrenal Medulla

Cells composing the adrenal medulla and sympathetic nervous system are derived from the embryonic neural crest. Cellular migration from the neural crest to the paraspinal area to form the sympathetic chain and from there to the adrenal medulla occurs during the 5th and 6th weeks of gestation. These cells produce catecholamines (dopamine, norepinephrine, epinephrine) from tyrosine. Sympathetic ganglia produce norepinephrine, which is released from nerve endings and acts as a neurotransmitter. The adrenal medulla primarily produces epinephrine, which is released into the circulation and acts as a hormone. Some adrenal norepinephrine production and release also occur. As hormones, both norepinephrine and epinephrine serve to mobilize energy stores and prepare the body for muscular activity (increase heart rate and blood pressure, increase blood sugar, etc.). They are secreted in increased amounts with stress (pain, fear, etc.).[31,97]

Epinephrine and norepinephrine are metabolized by the enzymes monoamine oxidase and catechol-O-methyl transferase to form metanephrines and vanillylmandelic acid (VMA). These metabolites are excreted by the kidney (Fig. 19-5).

Excess Catecholamine Production.
Pheochromocytomas are tumors of the adrenal medulla or sympathetic ganglia that produce and release large quantities of catecholamines. The majority (90%) are located in the adrenal and produce both epinephrine and norepinephrine. Tumors in extra-adrenal locations produce only norepinephrine. Excessive production of these catecholamines increases cardiac output and causes peripheral vasoconstriction leading to hypertension. Classically, the patient with a pheochromocytoma

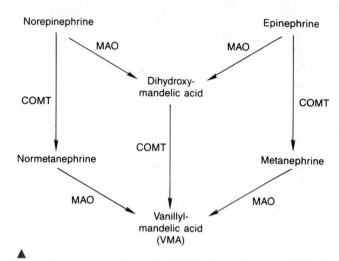

Metabolism of catecholamines

Figure 19-5. Production of metanephrines and VMA from epinephrine and norepinephrine through the action of monoamine oxidase (MAO) and catechol-*o*-methyl transferase (COMT).

complains of "spells" of increased heart rate, headache, and a sensation of tightness in the chest, accompanied by sweating and pallor. These episodes correspond to the release of large amounts of catecholamines by the tumor. Only a small percentage of patients with these tumors actually have the classic presentation with no signs or symptoms of hypertension between attacks. These classic "spells" are seen in some form in about 45% of patients with pheochromocytomas. Fifty percent of patients with this tumor have sustained hypertension and the remaining 5% are asymptomatic. Pheochromocytomas may occur in all age groups but are most common in patients in their third to fifth decades.[31,44,97]

Although pheochromocytomas account for less than 1% of cases of hypertension, diagnosis of these tumors is important because they are surgically curable and approximately 10% are malignant. Bilateral tumors are present in about 10% of isolated cases and in a higher percentage (70%) of cases associated with a multiple endocrine neoplasia syndrome (see the following).[31,44,97]

Diagnosis of pheochromocytoma should be considered in the following:

All patients who complain of attacks similar to those described previously, especially if these attacks are associated with exercise or stress. (Patients with extra-adrenal tumors may present with unusual or even bizarre symptoms. For example, patients with pheochromocytomas of the urinary bladder may complain of hypertensive attacks every time they urinate.)

Children with hypertension

Severe hypertension unresponsive to therapy

Hypertension with diabetes mellitus or hypermetabolic states (due to the anti-insulin activity of the catecholamines)

Patients with other endocrine tumors or with relatives who have pheochromocytomas

In the past, diagnosis depended on the precipitation of hypertensive attacks with provocative agents. Such tests were dangerous and neither very sensitive nor very specific. These provocative tests have been replaced by measurement of urinary metanephrines and VMA and of plasma catecholamines.[31,44,97]

Urinary metanephrines and VMA are traditionally measured in a 24-hour urine collection by colorimetric or chromatographic means.[31,44,97] Determination of urinary metanephrines is considered to be the best screening test for the presence of a pheochromocytoma (reference range—Table 19-1). Metanephrine measurement is favored over VMA as a primary screen because more substances (drugs, food, etc.) interfere with the measurement of VMA. Repetition of metanephrine analysis resulting in a second abnormal value or follow-up of abnormal metanephrine levels with the finding of elevated VMA excretion has a diagnostic accuracy of approximately 99%.[31,44,97]

Theoretically, measurement of plasma catecholamines would be preferable to analysis of urinary metabolites, since such measurement would eliminate the necessity for a 24-hour urine collection and allow sampling during an attack. Measurement of plasma catecholamines is helpful in patients with sustained hypertension but is less reliable in patients who are normotensive between attacks. Even in patients with sustained hypertension, increased plasma catecholamines may be due to neurogenic or idiopathic causes. The response of these patients to administration of clonidine is necessary to separate them from those with pheochromocytoma. Patients with pheochromocytomas show no response to clonidine, while those with neurogenic or essential hypertension show a 50% decrease in plasma catecholamine levels. Other problems with measurement of plasma catecholamine are:

The range of normal catecholamine concentration is so wide that a level that is diagnostic of pheochromocytoma is difficult to establish.

Blood must be obtained under carefully standardized conditions to avoid stimulation of catecholamine release.

Secretion of catecholamines may be episodic, and finding plasma levels within the reference range does not rule out the diagnosis.

Elevated levels of catecholamines may occur in a variety of other conditions.

Some authors suggest that epinephrine is more useful than nonepinephrine in diagnosis of pheochromocytoma, since epinephrine comes only from the adrenal and not from the sympathetic nervous system. Measurement of plasma catecholamines may be very useful in patients with extra-adrenal tumors, because blood samples can be drawn at different levels along the inferior vena cava and help to localize the tumor.[24,31,44,97] (See Table 19-1 for typical reference ranges.)

Treatment is surgical removal. The prognosis for patients with benign tumors is excellent.

Neuroblastomas. Neuroblastomas are malignant tumors of the adrenal medulla that occur in children. They produce catecholamines and may occasionally be associated with hypertension. Detection of elevated levels of VMA and/or homovanillic acid (HVA; product of dopamine metabolism) may be helpful both in diagnosis of the tumor and in detection of tumor recurrence or metastases in treated patients.[31]

Islets of Langerhans

Insulin

Insulin is a peptide hormone that is synthesized by the beta cells of the pancreatic islets. It consists of a total of 51 amino acids in two chains connected by two disulfide bridges. Insulin is synthesized as a large single chain *preproinsulin* that is cleaved to a more immediate precursor *proinsulin* in the rough endoplasmic reticulum. Proinsulin is then packaged into secretory granules, where it is broken down into equimolar amounts of insulin and an inactive C-peptide.[25,74,76]

Release of insulin and C-peptide occurs in response to a variety of stimuli, the most important of which is serum glucose. In the fasting state, release of insulin is minimal, resulting in serum levels of insulin of 4 to 25 µU/mL. Increasing levels of glucose stimulate secretion of insulin by a mechanism that probably involves both glucose metabolism within the islet cells and interaction of glucose with membrane receptors on the cell surface. Other factors that increase insulin secretion in humans are: (1) direct effect of glucose within the gastrointestinal tract, probably mediated by gastrointestinal hormones; (2) ingestion or infusion of amino acids; and (3) vagal stimulation. Inhibition of insulin secretion occurs with release or administration of epinephrine or norepinephrine and with certain drugs (thiazides, Dilantin, diazoxide). Administration of exogenous somatostatin also inhibits insulin release, but the role of physiologic amounts of this peptide in insulin regulation is unknown. The response of peripheral tissues to insulin may be influenced by a number of other factors. These include hormones that antagonize the action of insulin (growth hormone, human placental lactogen, corticosteroids, estrogens), obesity, physical inactivity, and low carbohydrate diet.[11,25,74,76]

Insulin functions to enhance anabolic activities of the body—that is, it stimulates incorporation of carbohydrate, fats, and amino acids into body tissues. Its major effects involve primarily the liver, adipose tissue, and muscle and include: (1) promotion of glucose uptake by adipose tissue and muscle, thus decreasing plasma glucose concentration; (2) promotion of glycogen synthesis in liver and muscle; (3) stimulation of fatty acid synthesis by liver and adipose tissue; and (4) stimulation of protein synthesis. The mechanism by which insulin induces all these events is not clear. Since insulin is not under the control of the hypothalamic-pituitary axis, abnormalities in insulin function reflect either primary abnormalities in production or release of insulin by the pancreatic islets or abnormalities of insulin receptors in peripheral tissues.[25,76,74]

Hyperinsulinemia. Excessive and/or inappropriate release of insulin causes a marked decrease in plasma glucose levels. If hypoglycemia [glucose <40 mg/dL (2.2 mmol/L)] occurs, patients develop symptoms related to the effects of lack of glucose on the brain (weakness, headache, blurred vision, convulsions, coma) and to the release of catecholamines (sweating, pallor, palpitations).

Hypoglycemia is a relatively common condition and is most often "reactive." Episodes of reactive hypoglycemia occur within minutes to hours after a meal and are not usually associated with excessive, unstimulated insulin release. Fasting blood glucose and insulin levels are normal. Insulin and glucose determinations made during an attack reveal an insulin level that is appropriate for the low serum glucose level (insulin-glucose ratio <0.3) (mmol/L:mg/dL). Most of these patients have a characteristic history of attacks occurring in relation to meals, and insulin measurement is not usually required for diagnosis. Many of these patients have a family history of diabetes mellitus or have had a surgical procedure that affects glucose absorption (usually removal of part of the stomach).[25,27,78]

In contrast to episodes of reactive hypoglycemia, hypoglycemic attacks due to insulin-producing tumors of the pancreas (insulinomas) occur while the patient is fasting. Simultaneous measurement of glucose and insulin during an attack reveals an insulin-glucose ratio of >0.3, (mmol/L:mg/dL) suggesting that insulin release is occurring independent of the serum glucose concentration. [Occasionally in this situation, insulin levels may fall within the reference range (see Table 19-1). However, it is the insulin-glucose ratio, not the absolute insulin value, that makes the diagnosis.] Hypoglycemic attacks due to insulin-producing tumors may be precipitated by fasting, and the usual provocative test to allow diagnosis of this disorder is placing the patient on a 72-hour fast. A similar clinical picture may occur in patients with tumors that produce an insulin-like substance immunologically distinct from true insulin. Tumors producing such a substance are usually large, bulky tumors of connective tissue arising in the chest or abdomen. Production of true insulin by tumors outside the pancreas is very rare.[11,78,74,76]

In some cases, patients may induce attacks of hypoglycemia in themselves by administration of commercial insulin or ingestion of oral hypoglycemic agents. These patients present with symptoms identical to that of an insulinoma. Laboratory analysis of the serum level of C-peptide may be helpful in this situation. In the patient who is injecting insulin, C-peptide levels will be low, since this substance is not present in commercial insulin preparations. In contrast, patients with insulinomas are releasing equimolar amounts of insulin and C-peptide, and thus C-peptide concentration will be elevated. If a patient is taking oral hypoglycemic agents, direct measurement of the drug in the serum may be

the only means of detection, since these drugs act by increasing release of endogenous insulin.[11,74,76]

Most insulinomas are solitary tumors of the pancreas, and most are benign. They are treated by surgical removal.

Hypoinsulinemia. Lack of insulin results in development of diabetes mellitus, which is characterized by fasting hyperglycemia accompanied by abnormalities of fat and protein metabolism. Depending on the severity of the insulin deficiency, hyperglycemia may or may not be accompanied by ketoacidosis. Laboratory determination of insulin does not play a role in diagnosis or management of this disease (see Chapter 14, Carbohydrates).

Glucagon

Like insulin, glucagon is a protein hormone that is involved in the regulation of carbohydrate, fat, and protein metabolism. It is synthesized by the α cells of the pancreatic islet and is composed of 29 amino acids. A larger proglucagon molecule is synthesized first, and this precursor is cleaved within secretory granules before being released.

Glucagon release is relatively constant throughout the day (reference range—Table 19-1). Serum levels increase in response to high-protein meals, acute hypoglycemia, exercise, and stimulation of the adrenergic system and are suppressed by high carbohydrate intake.

The physiologic role of glucagon is basically the opposite of that of insulin. It stimulates hepatic glycogenolysis and gluconeogenesis and lipolysis in adipose tissue, thus serving to maintain or increase fasting glucose levels. Glucagon appears to produce these effects through stimulation of adenyl cyclase, resulting in an increase in intracellular cAMP.[11,74,76,25]

Hyperglucagonemia. Although there is some speculation that increased levels of glucagon may play a role in the development of diabetes mellitus, and particularly diabetic ketoacidosis, current evidence would suggest that such a role is only ancillary and that hyperglucagonemia in most diabetics is not sufficient to produce ketosis. In any case, measurement of serum glucagon has no role in evaluation of patients with diabetes.

The major clinical use of serum glucagon measurements is in evaluation of the patient suspected of having a glucagon-secreting tumor of the pancreatic islets (glucagonoma). The clinical syndrome that these rare tumors produce consists of mild diabetes, anemia, anorexia and weight loss, and a skin rash. The rarity of the tumor plus the nonspecific nature of the symptoms usually delays recognition of this syndrome until relatively late in its course. Once the diagnosis is suspected, it can be confirmed by analysis of serum glucagon levels, which, in most cases, are greater than 1000 pg/mL (287 pmol/L). Hyperglucagonemia due to tumors must be differentiated from other causes of increased glucagon, which include diabetes mellitus, pancreatitis, and trauma. Levels of glucagon in these situations are usually considerably less than those found in patients with glucagonomas, and the clinical history will usually allow differentiation of these conditions.

Unlike insulinomas, most glucagonomas are malignant, and greater than 50% have metastasized at the time of diagnosis. The prognosis is poor.[11,74,76]

Other Hormones

Excessive production of a variety of other hormones has been described in association with islet-cell pancreatic tumors. These include gastrin, somatostatin, pancreatic polypeptide, and serotonin. The carcinoid syndrome produced by excess serotonin secretion is discussed in a later section. The clinical syndromes produced by excessive production of somatostatin and pancreatic polypeptide are non-specific, and the diagnosis is rarely made preoperatively except in patients with multiple endocrine tumors (see the following). Measurement of somatostatin and pancreatic polypeptide in plasma can make the diagnosis of an islet-cell tumor producing one of these substances, but assays for these compounds are performed primarily in reference laboratories and are not generally available.

Reproductive System

The embryonic gonads, both ovaries and testes, are derived from the germinal epithelium, which begins to develop on the urogenital ridge around the fourth week of gestation. Primitive germ cells migrate into this area from the yolk sac and, by the sixth week, are present in the epithelial tissue of the primitive gonad. At this point, the gonad is bipotential—that is, it may develop into either a testis or an ovary. The direction of further differentiation depends on the presence or absence of a Y chromosome. If a Y chromosome is present, the gonad develops into a testis, and the presence of human chorionic gonadotropin (HCG) from the placenta stimulates Leydig cells within the gonad to secrete testosterone. Testosterone then causes the male internal and external genitalia to develop and the testes to descend into the scrotum. In contrast, if the embryo has no Y chromosome, the gonad develops into an ovary. In the absence of testosterone production, the internal and external genitalia develop along female lines.[97]

Ovaries

The ovarian follicle (layer of cells surrounding the ovum) secretes estrogens, progesterone, and small amounts of androgens (see Fig. 19-1). The principal estrogenic hormone produced by the ovary is estradiol. Smaller amounts of estrone are also secreted. Estrogen is responsible for growth of the uterus, fallopian tubes, and vagina, promotion of breast development, maturation of the external genitalia, deposition of body fat into the female distribution, and termination of linear growth. Progesterone serves to prepare the uterus for pregnancy and the lobules of the breast for lactation. Development of the follicle with appropriate release of estrogen and progesterone is under the control of the gonadotropic hormones, follicle-stimulating hormones (FSH), and luteinizing hormone (LH) secreted by the anterior pituitary. The low levels of adrenal estrogens produced in the prepubertal child are sufficient to suppress pituitary gonadotropin release. At puberty, however, the pituitary becomes less sensitive to low levels of estrogen, and the levels necessary for inhibition are reset. Release of pituitary gonadotropin then initiates the cyclic changes characteristic of the female reproductive system. FSH results in rapid growth of several ovarian follicles and an associated increase in plasma estrogen levels during the first half of the menstrual cycle. Rising estrogen levels suppress FSH but stimulate LH release, resulting in a midcycle LH surge. This LH surge triggers ovulation (release of an ovum by one or more developing ovarian follicles) and is followed by a rapid drop in estrogen and LH concentration. The follicle that released the ovum is then transformed into a corpus luteum, which synthesizes both estrogen and progesterone, with peak levels occurring approximately 8 to 9 days after ovulation. In the absence of HCG from an implanted ovum, the corpus luteum degenerates, and estrogen and progesterone levels drop, resulting in menstruation. Reference ranges for estrogen, progesterone, FSH, and LH at different ages and during different phases of the menstrual cycle are given in Table 19-2.[39,97]

Because estrogen stimulates the proliferation of the uterine lining, the first half of the menstrual cycle, when estrogen effect is unopposed by progesterone, is called the *proliferative* or *follicular phase*. After ovulation, progesterone blocks estrogen-induced proliferation and stimulates secretory activity by the uterine glands, preparing the endometrium to receive a fertilized ovum. Because progesterone is synthesized by the corpus luteum, this portion of the cycle is called the *luteal* or *secretory phase*.

TABLE 19-2. Reproductive Hormones—Reference Ranges[60]

	TOTAL TESTOSTERONE ng/dL	TOTAL ESTROGEN pg/mL	PROGESTERONE ng/dL	FSH mU/mL	LH mU/mL
Children less than 10	<3–10	<25	7–52	<1–3	<1–5
Adult Females	20–75				
Proliferative phase		60–200	15–70	1–9	1–12
Midcycle				6–26	16–104
Luteal phase		160–400	200–2500	1–9	1–12
Postmenopausal	8–35	<130	(1–10)	30–118	16–66
Adult Males	300–1000	20–80	13–97	1–7	1–8

Estrogen. Abnormalities of estrogen secretion may occur because of primary ovarian abnormalities, secondary to irregularities in production of FSH, or both. These abnormalities are reflected as disturbances in the menstrual cycle, accelerated or delayed onset of puberty, and/or infertility.

Hyperestrinism. The clinical presentation of hyperestrinism in the female depends on the age of the patient at the time of presentation—that is, whether she is prepubertal, in her reproductive years, or postmenopausal. The clinical presentation in these situations would be early onset of puberty, development of excessive or irregular menses and/or infertility, and resumption of uterine bleeding, respectively.[39,97]

Precocious Puberty. Onset of puberty in females in the United States normally occurs between the ages of 10 and 14. These changes occur in a predictable order: (1) breast development, (2) development of pubic hair, and (3) onset of menses. Onset of these changes before 8 to 9 years of age is considered *precocious puberty* or *sexual precocity*. Ninety percent of cases of precocious puberty in females are not associated with a definable etiology (idiopathic). Levels of plasma estrogens, gonadotropins, and urinary 17-ketosteroids (17-KS, products of androgen metabolism) are appropriate for the patient's developmental (not chronological) age. The remaining 10% of cases usually involve a hypothalamic or ovarian tumor. Feminizing adrenal tumors occasionally produce precocious puberty, but these are rare.[9]

Menstrual Irregularity and Infertility. A woman in the reproductive age group with excessive estrogen production will most commonly present with irregular menstrual periods and abnormalities in the amount of menstrual flow.[28] Estrogen excess in this age group is most commonly associated with failure of ovulation (anovulatory cycles). If ovulation does not take place, the corpus luteum cannot develop, and consequently, progesterone synthesis does not occur. Estrogenic effect on the uterine lining (endometrium) continues unopposed, and the latter becomes excessively thick. Eventually, the endometrium is shed, usually in an irregular fashion, resulting in prolonged and excessive menstrual flow. In an extreme case, a woman may develop secondary amenorrhea (absence of menses for 6 months in a woman who has previously menstruated) followed by severe, prolonged bleeding.

Anovulatory cycles occur commonly at the beginning and end of a woman's reproductive life and may occur occasionally in almost all women. Chronic anovulation results, of course, in infertility. Endocrine abnormalities associated with anovulation become very complex, and the initial site of the problem in the hypothalamic-pituitary-ovarian axis cannot be determined in most cases. In the usual case, gonadotropins and estrogens do not cycle. That is, serum measurement of these hormones reveal relatively constant levels over a period of several weeks. Levels of estrogen produced by developing ovarian follicles are high enough to stimulate some LH production (although no LH surge) but not high

enough to completely suppress FSH secretion by the pituitary. Consequently, continued FSH stimulation produces large numbers of developing follicles that never go on to ovulation. LH-stimulated secretion of androgens by the outer portion of the ovarian follicle occurs and results in excessive androgen effect, predominantly hirsutism (excess growth of androgen-dependent hair, typically on the upper lip, chin, thighs, chest, and abdomen).[103] This condition is frequently associated with obesity, and a classic syndrome of amenorrhea, hirsutism, and obesity was described in 1935 by Stein and Levanthal.[39] This syndrome was noted to be associated with polycystic ovaries, and it was originally thought that the ovarian abnormality was the initiating event in this disturbance. It is now recognized that polycystic ovaries are most likely a result of chronic anovulation and do not imply a primary ovarian disorder.

Laboratory evaluation of the woman with chronic anovulation will reveal a slightly elevated LH level and a normal or low FSH level, resulting in an increased LH-FSH ratio. Estrogen levels may be within the reference range or may be elevated. Normal fluctuation of these hormones does not occur, and it is the demonstration of this loss of fluctuation that establishes the diagnosis of anovulation. Thus, hormone levels should be measured weekly for 3 to 4 weeks. These laboratory abnormalities are accompanied by increased levels of plasma androgens (androstenedione and/or testosterone) and urinary 17-KS. Since serum and urine levels of these compounds also vary at different stages of the menstrual cycle, multiple measurements must be taken to establish a diagnosis of excess androgen production.[9,28,88,102]

Estrogen-producing ovarian tumors, although uncommon in women in the reproductive age group, may occur and may produce symptoms similar to those described previously. In this case, physical examination should reveal an ovarian mass. Laboratory findings would include elevated estrogen and suppressed gonadotropin levels. Androgen levels and accompanying hirsutism may or may not be present, depending on the degree of androgen production by the tumor.

Postmenopausal Bleeding. Postmenopausal bleeding is a very serious symptom, and patients with this problem must always be evaluated to rule out the presence of an endometrial or cervical carcinoma. If postmenopausal bleeding is secondary to increased estrogen stimulation of the endometrium, the most common causes are exogenous administration of estrogen and estrogen-secreting ovarian tumors. Occasionally, no source of estrogen can be identified, and the process is self-limited.

Hyperestrinism in Males. Excessive production or administration of estrogen in males results in testicular atrophy and gynecomastia (enlargement of the male breasts). Estrogen-producing neoplasms are extremely rare in males. It is thought that most cases of hyperestrinism in men are relative—that is, the ratio of estrogen to testosterone is increased because of decreased testosterone production rather than elevated estrogen production. This situation is commonly seen in alcoholics, in whom the development of hyperestrinism probably reflects a combination of decreased testosterone production due to a toxic effect of ethanol on the testis and decreased estrogen metabolism by a damaged liver. These symptoms also may be seen in men who are being treated for prostate carcinoma with exogenous estrogens.

Hypoestrinism. Hypoestrinism, or ovarian insufficiency, may be primary or secondary to disturbances in gonadotropin secretion. Consequences of hypoestrinism depend on the age of the patient and may be reflected as delayed puberty or amenorrhea.

Delayed Puberty. This condition is defined as failure of puberty to occur by age 16. If breast and pubic hair development have occurred, but menses have not begun, the disorder is called *primary amenorrhea*. The condition may be primary in the ovary because of failure of ovarian development or an inborn error of steroid biosynthesis (rare) or may be secondary to a defect in gonadotropin production (pituitary or hypothalamic tumor, idiopathic).[9,28] Failure of ovarian development is most commonly due to Turner's syndrome—a genetic defect in which only one X chromosome is present. This defect occurs in 1 of 3000 to 6000 females. These patients have a characteristic physical appearance (short stature, webbed neck, broad chest), and other congenital anomalies are frequently present. Laboratory evaluation reveals increased gonadotropin and decreased estrogen levels. Administration of exogenous estrogen can induce development of secondary sex characteristics, but these patients remain sterile.

Another genetic defect resulting in delayed onset of puberty in apparently female children is the testicular feminization syndrome. These patients are genetic males (XY) who are unresponsive to testosterone. That is, although testes are present and producing androgens, the male reproductive system does not develop appropriately, and the external genitalia develop along female lines. Consequently, these children are identified as females at birth and are raised as such, coming to medical attention usually because of delayed puberty. At this point, physical examination usually reveals a tall female-appearing child who frequently has bilateral inguinal hernias that are actually small testes in the inguinal canals. The uterus is lacking, and the vaginal canal is a short, blind pouch. Estrogen levels are low, and androgen levels are usually in the reference range for males. Gonadotropin levels are within the reference range. Since the patient is phenotypically and psychologically female, treatment is removal of the testes (which, in this disorder, have an increased propensity to develop tumors) and administration of exogenous estrogen to stimulate the development of female secondary sex characteristics. These patients are also sterile.

Amenorrhea. Amenorrhea due to ovarian failure occurs in all women at the end of their reproductive years (menopause).

In the United States, menopause usually occurs between 48 and 55 years of age. Decline in ovarian function is gradual, and a period of menstrual irregularity may last for several months before actual amenorrhea develops. The signs and symptoms of menopause are due to decreased estrogen production by the failing ovaries. Lack of inhibition of the pituitary-gonadal axis results in elevated blood levels of LH and FSH. Occasionally, the ovaries fail prematurely, resulting in amenorrhea, hypergonadotropinism, and hypoestrinism in women less than 40 years of age. Although the syndrome may develop in women who have undergone radiation or chemotherapy for malignancy, in most cases the etiology of this problem is unknown. Diagnosis rests on demonstration of low estrogen and high gonadotropin levels in weekly blood samples obtained over a period of 1 month. Cases of temporary ovarian failure with these laboratory findings have been reported but are rare.[65]

A wide variety of conditions involving the pituitary and hypothalamus can result in primary or secondary amenorrhea. The most obvious would be hypopituitarism secondary to pituitary infarction or tumor. Other situations that result in disturbances of hypothalamic function sufficient to produce amenorrhea include marked weight loss, anorexia nervosa, severe emotional distress, intense participation in athletics (probably related to weight loss and stress), and chronic disease of any sort (for example, diabetes mellitus).[9,39]

Progesterone. As noted previously, progesterone is released from the corpus luteum after ovulation and functions in preparing the endometrium for pregnancy and the breasts for lactation. Like estrogen, it is a steroid that is inactivated in the liver.

Hyperprogesteronemia. Excessive progesterone production occurs when a corpus luteum fails to regress in the absence of fertilization. Blood levels of progesterone remain elevated, and menstruation does not occur. This situation is uncommon and is usually self-limited.

Hypoprogesteronemia. Inadequate production of progesterone by the corpus luteum ("inadequate luteal phase") is a relatively common condition that produces variations in duration and amount of menstrual flow, although the cycle is regular. The most common symptom of this problem is infertility. Laboratory determination of progesterone reveals levels below those anticipated for the time in the cycle. If pregnancy does occur, insufficient production of progesterone by the corpus luteum may lead to habitual abortion in the first trimester. Treatment is progesterone supplementation.[88]

Placenta

The placenta synthesizes and secretes a variety of protein hormones, as well as the steroids estrogen and progesterone. Evaluation of maternal serum and urine concentration of these hormones may be of value not only in diagnosing pregnancy but also in monitoring placental development and fetal well-being.

Human Chorionic Gonadotropin. Human chorionic gonadotropin (HCG) is a protein hormone produced by the placenta and consisting of α and β subunits. The α subunit is identical to the α subunits of FSH, LH, and TSH. The β subunit, however, is sufficiently different that antibodies to this subunit show essentially no cross-reactivity with anterior pituitary hormones. Immunoassays detecting the β subunit have served as the primary means of measuring HCG since their introduction in the early 1970s. More recently, assays of the intact molecule have become available and are felt by many investigators to be the method of choice for pregnancy testing. Measurement of the β subunit, however, is still recommended for monitoring patients with HCG-producing neoplasms.[59]

HCG secretion serves to maintain progesterone production by the corpus luteum in the early part of pregnancy. By the time that HCG levels drop at the beginning of the second trimester, the placenta has developed sufficiently to produce enough progesterone to maintain the endometrium and to allow the pregnancy to continue. In addition, HCG stimulates development of fetal gonads and synthesis of androgen by the fetal testes.[50]

Measurement of HCG is used primarily for the diagnosis of pregnancy. With the development of monoclonal antibodies, HCG can now be detected in serum and urine as early as 1 to 2 days after fertilization—even before the woman has missed a menstrual period. In addition, HCG determinations may be useful in the following:

The diagnosis of ectopic pregnancy—Implantation of the fertilized ovum in a site other than the uterus (usually the fallopian tube).

Prediction of spontaneous abortion—Lower than expected HCG values and failure of HCG to increase appropriately in the first few weeks of pregnancy indicate that the placenta is not developing correctly and have been shown to be associated with an increased incidence of spontaneous abortions.

Detection of multiple pregnancies—HCG levels in women carrying multiple fetuses are higher than in women carrying only one.

Detection and follow-up of HCG-producing tumors—Neoplasms of trophoblastic (placental) tissue produce HCG in amounts greater than those associated with normal pregnancies, and higher than expected HCG levels may be the first indication of the presence of such a tumor. After the neoplasm is diagnosed and treated, continued monitoring of HCG levels allows for the early diagnosis of recurrent or metastatic disease. HCG also may be produced by tumors arising in other sites (see Ectopic Hormones). Although not useful in the diagnosis of a particular type of tumor, follow-up measurements of HCG in a patient who has been treated for a tumor with

demonstrated HCG production is useful in the detection of recurrence and/or metastases.[55,59,70]

Human Placental Lactogen. Human placental lactogen (HPL) is a protein hormone that is structurally, immunologically, and functionally very similar to growth hormone and prolactin. Like HCG, it is produced by the placenta and can be measured in maternal urine and serum as well as amniotic fluid. Levels rise until approximately 37 weeks of gestation, paralleling increasing placental weight, and decline slightly in the last few weeks of pregnancy.[70,89,90]

HPL appears to act in concert with HCG to stimulate estrogen and progesterone synthesis by the corpus luteum. In addition, it stimulates development of the mammary gland (similar to prolactin) and has somatotropic actions similar to those of growth hormone. That is, it increases maternal plasma glucose levels and mobilization of free fatty acids and promotes positive nitrogen balance.

HPL is measured by immunologic assays and may show significant cross-reactivity with growth hormone and prolactin. Measurement is useful in monitoring conditions associated with a decrease in functioning placental tissue (most commonly in maternal hypertension) and in diagnosis of intrauterine growth retardation. In these situations, HPL levels are lower than expected for the length of gestation. HPL might be expected to be increased in conditions associated with larger than normal placental mass (diabetes mellitus, erythroblastosis fetalis, multiple pregnancies). However, most complications related to these conditions are not due to placental insufficiency, and therefore, serial measurement of HPL in patients with these problems has not proved to be clinically useful.[70,89,90]

Steroids. The fetus and placenta function together in steroid biosynthesis. The placenta synthesizes progesterone from cholesterol. Progesterone then enters the maternal circulation, where it functions to maintain the endometrium, inhibit uterine contractions, and stimulate the lobular unit of the breast. It also enters the fetal circulation, where it is used by the fetal adrenal for synthesis of cortisol and sex steroid precursors, especially DHEA and 16-OH DHEA-S. Conversion of these compounds to estrogen (see Fig. 19-1) cannot proceed in fetal adrenal because the necessary enzymatic machinery is missing. These enzymes are, however, present in the placenta, which converts DHEA to estrone and estradiol and 16-OH DHEA to estriol. Estriol then enters the maternal circulation, where it can be measured in plasma and urine (Fig. 19-6). Normally, maternal plasma and urine estriol levels increase throughout pregnancy, with a surge in the last 4 to 6 weeks.[35,70,71]

Since measurement in plasma is not dependent on maternal renal function or proper collection of a 24-hour urine specimen, plasma levels are generally preferred over urine for measurement of estriol. Serial measurements are necessary in the last 4 to 8 weeks of pregnancy, since the pattern of estriol levels is as important as the absolute level in the diagnosis of

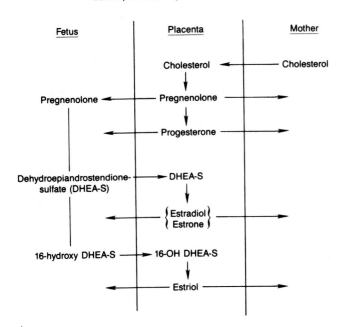

Estriol production by the fetoplacental unit

Fetus	Placenta	Mother
	Cholesterol ←	← Cholesterol
Pregnenolone ←	Pregnenolone	
	↓	
	Progesterone	
Dehydroepiandrostendione-sulfate (DHEA-S)	→ DHEA-S	
	↓	
	{ Estradiol } { Estrone }	
16-hydroxy DHEA-S	→ 16-OH DHEA-S	
	↓	
	Estriol	

▲

Figure 19-6. Estriol synthesized only by the fetoplacental unit moves into both amniotic fluid and maternal blood, allowing assessment of fetal well-being.

fetal problems. Estriol measurement may be useful in the following:

Evaluating the fetoplacental unit in diabetic mothers—A decrease of 40% in estriol levels compared with the mean of three preceding determinations has been shown to be an accurate predictor of fetal distress.

Identifying postdate gestations where the infant is at risk for development of the postmaturity syndrome—In this case, estriol levels rise appropriately but drop off after the 40th week of gestation.

Diagnosis of intrauterine growth retardation—Estriol levels are low in this condition and often fail to show the normal surge.

Estriol levels have not been useful in the management of the patient with Rh isoimmunization.[35,70,71]

Estriol is not produced in significant quantities in the mother and is solely a reflection of fetoplacental function. Thus, estriol measurement provides valuable information about fetal well-being.[35,70,71]

Testes

The testes, like the ovaries, are part of a hypothalamic-pituitary-gonadal axis. FSH stimulates spermatogenesis, whereas LH stimulates the production of testosterone by interstitial (Leydig's) cells. Both LH and FSH are suppressed by high levels of testosterone (negative feedback).

Testosterone is a steroid synthesized from cholesterol. It causes growth and development of the male reproductive system, prostate, and external genitalia. Increased testosterone levels at puberty cause development of pubic, axil-

lary, and facial hair, growth of the internal and external genitalia, rapid musculoskeletal growth and eventual epiphyseal closure, hypertrophy of the larynx with a deepening voice, initiation of spermatogenesis, and development of sex drive and potency. Reference ranges for testosterone and gonadotropins in adult and prepubertal males are listed in Table 19-1. Abnormalities of testosterone synthesis may be primary in the testes or secondary to pituitary or hypothalamic problems.[39]

Hyperandrogenemia. Excessive androgen production in the adult male produces no demonstrable clinical abnormality. In the prepubertal male child, elevated androgens cause early onset of the pubertal changes listed previously. Onset of precocious puberty in male children is more likely to indicate the presence of serious organic disease (60%) that it is in female children (10%). The most common causes of male precocious puberty are hypothalamic tumors, congenital adrenal hyperplasia, and testicular (Leydig's cell) tumors. Androgen and gonadotropin production does not cycle in the male, and the levels that are found in various conditions causing precocious puberty are usually within the reference range. Consequently, measurement of these compounds is not particularly helpful in the differential diagnosis of precocious puberty in males. Instead, physical examination, radiologic evaluation of the hypothalamus and pituitary, and laboratory measurements of serum cortisol and 17-OH progesterone and urinary pregnanetriol levels are indicated.[82]

Hypoandrogenemia. In male children, hypoandrogenemia results in delayed puberty, whereas in the adult, lack of androgen causes impotence and loss of secondary sex characteristics. Primary testicular abnormalities resulting in decreased androgen production include infections, occasionally vascular or neoplastic disorders, and a variety of congenital abnormalities, the most common of which is Klinefelter's syndrome. In this syndrome, the patient has an extra X chromosome (XXY). Many variant forms occur, but classically these patients are tall with long extremities, small firm testes, gynecomastia, a decreased sperm count, and presence of Barr bodies on buccal smear. Gonadotropin levels are elevated. Many of these patients have a low IQ. The amount of androgen secretion and, consequently, degrees of development of secondary sex characteristics vary. Delayed puberty also may result from impaired production of FSH and LH due to a pituitary or hypothalamic lesion, most commonly a pituitary tumor.[39,82]

In adults, primary testicular failure most commonly results from infectious disease, usually mumps, or irradiation. Serum testosterone levels do not usually decrease with age in males, although men do occasionally experience a decline of androgen production, with symptoms similar to those of the female menopause. Alcoholism is frequently associated with testicular atrophy, loss of secondary sex characteristics, and gynecomastia. It has been thought that these symptoms result from a relative excess of estrogen levels due to failure of the diseased liver to inactivate circulating estrogen. However, it has recently been shown that alcohol does have a primary effect on the testes, resulting in decreased androgen production.[39,82]

Pituitary or hypothalamic lesions, with decreased production of FSH and LH, also may result in delayed puberty. Most commonly, the tumors are pituitary in origin, but they may be hypothalamic. In the adult male, secondary testicular failure is most commonly due to a pituitary neoplasm. The problem most often results from compression of a normal pituitary by an expanding pituitary tumor, with loss of ability to produce FSH, LH, and frequently, other hormones as well (panhypopituitarism). Occasionally, prolactin-secreting tumors also produce secondary testicular failure by inhibiting production of FSH and LH.[82]

Infertility

Laboratory evaluation of infertility must involve both partners, since causes of infertility are relatively evenly split between males and females. Evaluation should include a thorough history and physical examination to rule out chronic disease or structural abnormalities that may result in failure to conceive. Schemes for laboratory evaluation of male and female infertility vary, depending on the history and physical findings. Generally, the first step for evaluation of male

infertility is seminal fluid evaluation. If seminal fluid analyses are normal on several (at least four) occasions and survival of sperm in cervical mucus is normal, no further examination of seminal fluid is indicated. If the findings are abnormal, a series of hormonal measurements as outlined in Figure 19-7 are undertaken, and eventually, testicular biopsy may be necessary. In women, the first step for evaluation of infertility is usually the determination of basal body temperature over several menstrual cycles to establish whether or not ovulation appears to be occurring. If the basal body temperature pattern is normal, luteal-phase defects or structural abnormalities of the uterus or fallopian tubes would be likely causes of infertility. Luteal-phase defects occur in 3% to 10% of infertile women and may be diagnosed by serum progesterone levels and endometrial biopsy, in conjunction with basal body temperatures. Low (<5 ng/mL) serum progesterone levels and an endometrial biopsy that is 2 or more days out of phase according to the basal body temperature chart are strongly suggestive of a luteal-phase defect. Progesterone levels and endometrial biopsy are obtained also if the basal body temperature pattern is abnormal (no thermal shift, shortened luteal phase, etc.). In this case, serum estrogen, testosterone, and gonadotropin levels are also measured[22,50,83,88] (Fig. 19-8). If hypogonadotropic hypogo-

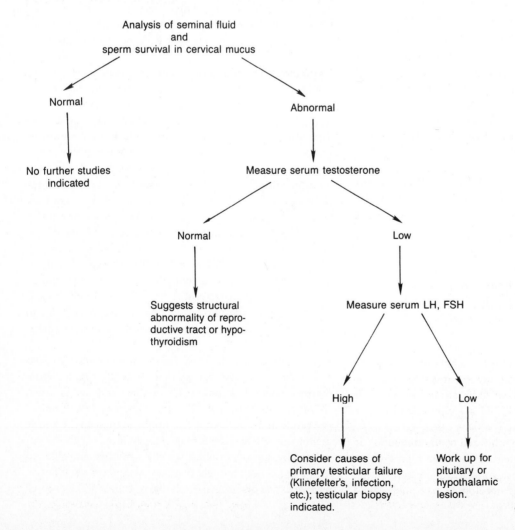

Figure 19-7. Laboratory workup of male infertility.

nadism is suspected, patients may be evaluated using the GnRH stimulation test to evaluate and separate hypothalamic and pituitary abnormalities.[83]

Other Hormones of Clinical Interest

Gastrointestinal Hormones

A variety of hormones, including gastrin, secretin, cholecystokinin, and serotonin, are produced by the cells of the GI tract. Although all these hormones can be measured by immunologic means, only gastrin and serotonin have proved clinically useful.

Gastrin. Gastrin is a peptide secreted by the G cells of the antrum (lower one-third) of the stomach. It is released in response to contact of food, particularly protein, with the antral mucosa. Hypoglycemia and elevation of serum calcium also stimulate gastrin release. Gastrin causes secretion of hydrochloric acid by parietal cells in the body (middle one-third) of the stomach. As the pH of gastric contents in the antrum decreases, gastrin secretion is inhibited, and acid production slows. The reference range for serum gastrin is given in Table 19-1. As with many other immunologic assays, differences in antibody specificity and laboratory techniques may make it difficult to compare values obtained in different laboratories.[23]

Hypergastrinemia. Excessive production of gastrin by the antrum occurs with decreasing acid production by gastric parietal cells (loss of negative feedback). Thus, elevated serum gastrin levels are seen in patients with achlorhydria for any reason (especially chronic atrophic gastritis) or in those who have had a vagotomy. Gastrin levels are also increased in chronic renal failure owing to decreased renal clearance. Measurement of serum gastrin in these situations and in the immediate postprandial period may reveal values up to five times the upper limit of the reference range.[23,37]

Gastrin also may be increased because of hyperplasia of the gastrin-producing cells in the antrum or because of ectopic production of gastrin by an islet-cell carcinoma of the pancreas. The latter was first described by Zollinger and Ellison in 1955. The syndrome produced by these tumors is one of severe peptic ulceration of the stomach, often extending into the duodenum and jejunum. Serum gastrin levels in this condition may be extremely high, but more often they overlap the values seen in the situations described previously. Therefore, an elevated serum gastrin level must be accompanied by gastric hyperacidity to make the diagnosis of Zollinger-Ellison syndrome.[23,37,46]

Gastrin-producing tumors of the islets are malignant, and the majority have metastasized by the time of diagnosis. Because of the presence of metastases and also of the frequency of multiple primary tumors within the pancreas, surgical re-

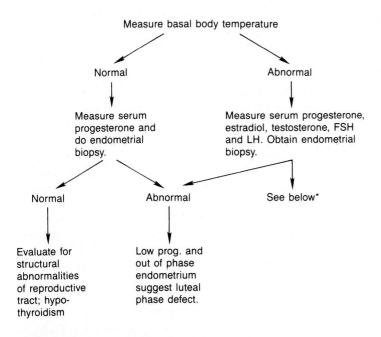

*Estradiol and testosterone low
  1. FSH and LH high suggests primary ovarian failure
  2. FSH and LH low suggests hypothalamic/pituitary abnormality

Estradiol and testosterone normal/low normal
  1. FSH low/normal and LH high suggests polycystic ovary syndrome.
  2. FSH and LH low suggests primary ovarian dysfunction.

Estradiol and testosterone high
  1. FSH and LH high suggests hypothalamic/pituitary abnormality or ectopic production.
  2. FSH low/normal and LH high suggests polycystic ovary syndrome.
  3. FSH and LH normal suggests primary ovarian dysfunction.

Figure 19-8. Laboratory workup of female infertility.

moval of the gastrin-secreting tumor is not usually possible. Instead, surgical therapy of the Zollinger-Ellison syndrome is removal of the target organ (*i.e.,* total gastrectomy). Medical therapy with antiulcer agents is also promising and, in some patients, may make surgery unnecessary. Gastrinomas are relatively slow-growing tumors, and the 5-year survival after surgical treatment has been good (>75%).[23,37]

Hypogastrinemia. No clinical abnormalities associated with lack of gastrin production have been identified.

Serotonin. Serotonin (5-OH tryptamine) is an amine derived from hydroxylation and decarboxylation of tryptophan. It is synthesized by enterochromaffin cells, which are located primarily in the GI tract and, to a lesser degree, in the bronchial mucosa, biliary tract, and gonads. Serotonin is secreted into the blood, where it binds to platelets and is released during coagulation. It is metabolized by the liver to 5-hydroxyindole acetic acid (5-HIAA), which is excreted in the urine.[68,69]

Tumors arising from enterochromaffin cells (carcinoid tumors) occur mainly in the appendix and the ileum, although occasionally such tumors may develop in the lung or ovary. These tumors produce large amounts of serotonin as well as other amines (especially histamine), kallikrein, and prostaglandins. If these substances are released into the circulation, they produce a constellation of symptoms known as the *carcinoid syndrome* (attacks of diarrhea, flushing, tachycardia, and hypotension). Damage of the heart valves also may occur. Detection of elevated plasma serotonin levels and/or urinary metabolites of serotonin (5-HIAA) are diagnostic of the carcinoid syndrome, provided the patient has not taken any drugs or food that produces positive or negative interference with the assay. (Drugs interfering with this test include glyceryl guaiacolate, Mandelamine, and phenothiazines. Foods such as bananas, pineapple, walnuts, and avocados are very high in serotonin content and may produce false-positive results.) The reference range for 5-HIAA in the urine is given in Table 19-1. Elevated 5-HIAA levels occur with some other GI diseases but are not as high as in the carcinoid syndrome.[68,69]

Carcinoid tumors are low-grade malignancies that grow slowly. The 5-year survival may be relatively good even if metastases are present at the time of diagnosis. Treatment is surgical removal of the primary tumor and of isolated liver metastases if present.

Multiple Endocrine Neoplasia

Multiple endocrine neoplasia (*MEN*) is defined as the occurrence of several tumors or hyperplasias involving diverse endocrine organs. Two types are currently recognized. These symptoms are familial, and inheritance is in an autosomal dominant pattern.

MEN I, or Wermer's syndrome, includes: (1) tumors of pancreatic islets (81%), often associated with peptic ulcer disease (Zollinger-Ellison syndrome); (2) pituitary adenomas (65%), either functional or nonfunctional; and (3) parathyroid adenoma or hyperplasia (90%). Other disorders in MEN I include benign and malignant bronchial and intestinal carcinoids and tumors of nerve, adipose tissue, and thymus.[96]

MEN II, Sipple's syndrome, includes: (1) medullary carcinoma of the thyroid; (2) pheochromocytomas, which are usually bilateral, rarely malignant, and often nonfunctional; and (3) parathyroid adenoma or hyperplasia. Mucosal neuromas also may be present (MEN IIb).[16,73,80]

In both MEN I and MEN II, penetrance is high but expression is variable. All patients do not have the complete syndrome, and many patients may present originally with one component, only to develop symptoms of other neoplasms at a later date.

Pathogenesis of MEN has been of much interest and concern since the occurrence of multiple endocrine tumors was first described in 1903. The basis for the present theory of pathogenesis rests on the concept of the APUD (amine precursor uptake and decarboxylation) system proposed by Pearse.[61] Pearse investigated a number of cells that produced low-molecular-weight polypeptide hormones and that were located in various endocrine and nonendocrine tissues. He discovered that these cells had common cytochemical characteristics, in particular, the ability to take up and decarboxylate amine precursors. They have more recently been found to have numerous ultrastructural characteristics in common, the most notable of which is the presence of membrane-bound neurosecretory granules. Originally, all of these cells were described as neuroendocrine and were thought to be derived from the embryonic neural crest. Although it now appears likely that some of these cells derive from other areas, it is apparent that they all share common synthetic abilities and produce similar, and occasionally the same, hormones. The occurrence of neoplasms or hyperplasias involving varying combinations of these cells is, therefore, not unexpected.[2,77,87]

Ectopic Hormone Production

As noted previously for HCG and ACTH, protein hormones may occasionally be produced by cells in sites other than the gland from which they are usually derived. This occurrence is called *ectopic hormone production*. Hormones produced in this manner appear to be immunologically identical to those produced by the usual endocrine gland. Small differences in conformation or the release of larger, inactive forms of the hormone result in varying functional activity. When these hormones are functionally active, they produce syndromes identical to those of excessive hormone production by the endocrine gland where they are normally synthesized. The tumor most commonly associated with ectopic hormone production is oat-cell carcinoma of the lung, which not infrequently produces ACTH or ADH. Other hormones that may be produced ectopically are HCG by carcinoma of lung, GI tract, and pancreas; PTH by squamous cell carcinomas of the lung; and gastrin by islet-cell carcinoma of pancreas.[97]

Cytokines

A new class of hematopoietic proliferation factors has been shown to affect the survival and proliferation of various populations of lymphocytes and their progenitor cells. These factors include a large group of proteins collectively called *interleukins* as well as *colony stimulating factor* (*s*) and *tumor necrosis factor.* Their involvement with hematologic tumors, the inflammatory process, anemia, and other pathological processes in the immune system are just being investigated.[12,33,38,47,53,57,75]

SUMMARY

This chapter has covered basic mechanisms of endocrine diseases as they apply to the anterior and posterior pituitary, parathyroid glands, adrenal glands, islets of Langerhans, reproductive system, and other hormones secreted by organs which are not technically considered part of the endocrine system. Thyroid diseases and diabetes mellitus are covered in Chapter 20, Thyroid Function, and Chapter 14, Carbohydrates.

Evaluation of endocrine function differs considerably from that of other systems. Stimulation and suppression testing provide the cornerstones of endocrine evaluation and require adherence to strict protocols to ensure that results truly reflect the patient's underlying condition. As analytic methods become even more sensitive and molecular diagnostic techniques are applied to diagnosis of endocrine diseases, this field of clinical chemistry will continue to expand and tests currently restricted to reference or research laboratories will become routine.

CASE STUDY 19-1

A 27-year-old woman delivers her third child after 23 hours of labor. Delivery is complicated by severe hemorrhage requiring transfusion of 4 units of packed red cells. Because of the mother's critical condition, the infant is bottle fed for the first few days of life. When the mother's condition improves, she wants to breast-feed her infant as she had done with her other children. She is unable to lactate, however, and the infant continues on formula. Mother and infant are discharged 5 days after delivery.

When the infant is brought to the family practitioner for her 6 week check-up, the physician notices that the mother appears tired and pale. Her blood pressure is slightly low and she complains of loss of appetite, nausea, and vague abdominal pain. She is found to be mildly anemic and iron supplementation is prescribed.

Three months later, the woman returns to her physician complaining of persistent fatigue, weakness, and weight loss (20 pounds below her pre-pregnancy weight). She appears thin and pale and is wearing a sweater despite the outside temperature of 80 degrees. Her menstrual periods have still not resumed and she has noticed thinning of her pubic hair. Physical examination reveals low blood pressure, decreased deep tendon reflexes, and thin dry skin. The patient says that her family thinks she has "post-partum depression."

Questions

1. What is the most likely diagnosis in this patient? Does it relate to her pregnancy and if so, how?
2. Would you expect this patient to have increased skin pigmentation related to her underlying illness? Why or why not?
3. What type of additional testing should be performed on this patient and what results would you expect?

CASE STUDY 19-2

A 64-year-old man presents to his physician for a "check-up." He has been in good health and felt well until the last few weeks when he has complained of being tired and a little short of breath. He has no prior history of heart or lung problems although he does smoke (1 pack of cigarettes per day for 40 years). He has also noticed a 15-pound weight loss in the last 3 months and has developed a nagging cough. Physical examination reveals a thin, pale man who appears somewhat short of breath. Pulse rate is slightly increased and blood pressure is mildly elevated. Laboratory data show hyponatremia (125 mEq/L with reference range 135–145 mEq/L) and mild anemia.

Questions

1. What is the differential diagnosis in this patient?
2. What other diagnostic tests are needed and what do you expect the results to be?
3. What other hormones are often produced under similar conditions and how are these syndromes distinguished from primary abnormalities of endocrine glands?

CASE STUDY
19-3

A 26-year-old nurse presents to her physician complaining of "spells" of weakness, nervousness, and tachycardia. These spells began 2 months ago and have become more and more frequent. Now they occur several times a day and are interfering with her work. The episodes have no relation to meals and have even waked her up at night. She has never experienced any similar problems although her mother has been told that she has "low blood sugar." Blood pressure and heart rate are normal and physical examination shows no abnormalities. Additional questioning reveals that the patient was recently divorced and is involved in a dispute with her ex-husband over child support for their 3-year-old son.

Questions

1. What is the differential diagnosis of this patient's symptoms?
2. What additional testing should be performed?
3. What clues are present in this patient's history that point to the correct diagnosis?

REFERENCES

1. Abboud CF, Laws ER Jr. Diagnosis of pituitary tumors. Endocrinol Metab Clin North Am 1988;17:241.
2. Bolande RP. The neurocristopathies. Hum Pathol 1974;5:409.
3. Bouloux P, Perrett D, Besser GM. Methodological considerations in the determination of plasma catecholamines by high-performance liquid chromatography with electrochemical detection. Ann Clin Biochem 1985;22:194.
4. Buenocore CM, Robinson AG. The diagnosis and management of diabetes insipidus during medical emergencies. Endocrinol Metab Clin North Am 1993;22:411.
5. Burtis WJ. Parathyroid hormone related protein: Structure, function, and measurement. Clin Chem 1992;38:2171.
6. Cara JF. Growth hormone in adolescence. Endocrinol Metab Clin North Am 1993;22:533.
7. Carpenter PC. Diagnostic evaluation of Cushing's syndrome. Endocrinol Metab Clin North Am 1988;17:445.
8. Chang-Demoranville BM, Jackson, IMD. Diagnosis and endocrine testing in acromegaly. Endocrinol Metab Clin North Am 1992;21:649.
9. Chihal J. Office gynecologic endocrinology. Postgrad Med 1982;71:73.
10. Corpas E, Harman SM, Blackman MR. Human growth hormone and human aging. Endocr Rev 1993;14:20.
11. Cryer PE. Glucose homeostasis and hypoglycemia. In: Wilson JD, Foster DW, eds. Williams' textbook of endocrinology. Philadelphia: Saunders, 1992;1223.
12. Culver KW. Clinical applications of gene therapy for cancer. Clin Chem 1994;40:510.
13. Cuneo RC, Salomon F, McGauley GA, Sonksen PH. The growth hormone deficiency syndrome. Clin Endocrinol 1992;37:387.
14. Endres DB, Rude RK. Mineral and bone metabolism. In: Burtis CA, Ashwood ER, eds. Tietz textbook of clinical chemistry. Philadelphia: Saunders, 1994;1887.
15. Faglia G, Arosio M, Bazzoni N. Ectopic acromegaly. Endocrinol Metab Clin North Am 1992;21:575.
16. Farhl F, Dikman SH, Lawson W, Cobin RH, Zak FG. Paragangliomatosis associated with multiple endocrine adenomas. Arch Pathol Lab Med 1976;100:495.
17. Finney DJ. Response curves for radioimmunoassay. Clin Chem 1983;29:1762.
18. Foster DW. Eating disorders: Obesity, anorexia nervosa and bulimia nervosa In: Wilson JD, Foster DW, eds. Williams' textbook of endocrinology. Philadelphia: Saunders, 1992;1335.
19. Griffin JE. Dynamic tests of endocrine function In: Wilson JD, Foster DW, eds. Williams' textbook of endocrinology. Philadelphia: Saunders, 1992;1663.
20. Habener JF. Recent advances in parathyroid hormone research. Clin Biochem 1981;14:223.
21. Hallman H, Farnebo LO, Hamburger B, Jonsson G. A sensitive method for the determination of plasma catecholamines using liquid chromatography with electrochemical detection. Life Sci 1978;23:1049.
22. Hammond MG. Evaluation of the infertile couple. Obstet Gynecol Clin North Am 1987;14:821.
23. Henderson AR, Tietz TW, Rinker AD. Gastric, pancreatic, and intestinal function In: Burtis CA, Ashwood ER, eds. Tietz textbook of clinical chemistry. Philadelphia: Saunders, 1994; 1576.
24. Howanitz JH, Howanitz PJ, Henry JB. Evaluation of endocrine function. In: Henry JB, ed. Clinical diagnosis and management by laboratory methods. Philadelphia: Saunders, 1993;308.
25. Howanitz PJ, Howanitz JH, Henry JB. Carbohydrates. In: Henry JB, ed. Clinical diagnosis and management by laboratory methods. Philadelphia: Saunders, 1993;172.
26. Insogna KL. Humoral hypercalcemia of malignancy—the role of parathyroid hormone-related protein. Endocrinol Metab Clin North Am 1989;18:779.
27. Iramanesh A, Veldhuis JD. Clinical pathophysiology of the somatotropic (GH) axis in adults. Endocrinol Metab Clin North Am 1992;21:783.

28. Jacobs HS. Endocrinology of menstrual disorders. Practitioner 1982;226:213.

29. Jenkins D, Stewart PM. Advances in medical therapy for pituitary disease: Treating patients with growth hormone excess and deficiency. J Clin Pharm Ther 1993;18:155.

30. Kaplan S. The newer uses of growth hormone in adults. Adv Intern Med 1993;38:287.

31. Kaplan N. Endocrine hypertension. In: Wilson JD, Foster DW, eds. Williams' textbook of endocrinology. Philadelphia: Saunders, 1992;707.

32. Karlsson FA, Kampe O, Winqvist O, Burman P. Autoimmune endocrinopathies 5. Autoimmune disease of the adrenal cortex, pituitary, parathyroid glands, and gastric mucosa. J Intern Med 1993;234:379.

33. Kaushansky K, Karplus PA. Hematopoietic growth factors: Understanding functional diversity in structural terms. Blood 1993;82:3229.

34. Klee DG, Kao PC, Heath H III. Hypercalcemia. Endocrinol Metab Clin North Am 1988;17:573.

35. Kochenour NK. Estrogen assay during pregnancy. Clin Obstet Gynecol 1982;25:659.

36. Kovacs L, Robertson GL. Syndrome of inappropriate antidiuresis. Endocrinol Metab Clin North Am 1992;21:859.

37. Krejs GJ. Non-insulin-secreting tumor of the gastroenteropancreatic system. In: Wilson JD, Foster DW, eds. Williams' textbook of endocrinology. Philadelphia: Saunders, 1992;1567.

38. Lange LG, Schreiner GF. Immune mechanisms of cardiac disease. N Engl J Med 1994;330:1129.

39. Lipsett MB. Endocrine mechanisms of reproduction. In: Frohlich ED, ed. Pathophysiology: Altered regulatory mechanisms in disease. Philadelphia: Lippincott, 1984;423.

40. Loriaux DL, Nieman L. Corticotropin-releasing hormone testing in pituitary disease. Endocrinol Metab Clin North Am 1991;20:363.

41. Losa M, Schopohl J, von Werder K. Ectopic secretion of growth hormone-releasing hormone in man. J Endocrinol Invest 1993;16:69.

42. Lufkin EG, Kao PC, O'Fallon WM, Mangan MA. Combined testing of anterior pituitary gland with insulin, thyrotropin-releasing hormone. Am J Med 1983;75:471.

43. Mallette LE. The parathyroid polyhormones: New concepts in the spectrum of peptide hormone action. Endocr Rev 1991;12:110.

44. Manager WM, Gifford RW. Pheochromocytoma: Current diagnosis and management. Cleveland Clin J Med 1993;60:365.

45. Manger WM, Steinland OS, Nakes GG, Wakin KG, Dufton S. Comparison of improved fluorometric methods used to quantitate plasma catecholamines. Clin Chem 1969;15:1101.

46. McCarthy DM. Zollinger-Ellison syndrome. Annu Rev Med 1982;33:197.

47. Means RT, Krantz SB. Progress in understanding the pathogenesis of the anemia of chronic disease. Blood 1992;80:1639.

48. Meikle HW. Secretion and metabolism of the corticosteroids and adrenal function and testing. In: Degroot LI, ed. Endocrinology. Philadelphia: Saunders, 1989;1610.

49. Melby JD. Diagnosis of hyperaldosteronism. Endocrinol Metab Clin North Am 1991;20:247.

50. Moghissi KS, Wallach EE. Unexplained infertility. Fertil Steril 1983;39:5.

51. Molitch ME. Clinical manifestations of acromegaly. Endocrinol Metab Clin North Am 1992;21:597.

52. Molitch ME. Pathologic hyperprolactinemia. Endocrinol Metab Clin North Am 1992;21:877.

53. Moore MA. Clinical implications of positive and negative hematopoietic stem cell regulators. Blood 1991;78:1.

54. New MI, Josso N. Disorders of gonadal differentiation and congenital adrenal hyperplasia. Endocrinol Metab Clin North Am 1988;17:339.

55. Norman RJ, Menabawey M, Lowings C, Buck RH, Chard T. Relationship between blood and urine concentrations of intact human chorionic gonadotropin and its free subunits in early pregnancy. Obstet Gynecol 1987;69:590.

56. Odell WD. Ectopic ACTH secretion. Endocrinol Metab Clin North Am 1991;20:371.

57. Ogawa M. Ectopic ACTH secretion. Endocrinol Metab Clin North Am 1991;20:371.

58. Orth DN, Kovacs WJ, DeBold CR. The adrenal cortex. In: Wilson JD, Foster DW, eds. Williams' textbook of endocrinology. Philadelphia: Saunders, 1992;489.

59. Painter PC. Discordant HCG results in pregnancy. A method in crisis. Diagn Clin Test 1989;27:20.

60. Painter PC, Cope JY, Smith JL. Appendix. In: Burtis CA, Ashwood ER, eds. Tietz textbook of clinical chemistry. Philadelphia: Saunders, 1994;2161.

61. Pearse AGE. The cytochemistry and ultrastructure of polypeptide hormone-producing cells of the APUD series and the embryologic, physiologic and pathologic implications of the concept. J Histochem Cytochem 1969;17:303.

62. Pisano JJ. A simple analysis for normetanephrine and metanephrine in urine. Clin Chim Acta 1960;5:406.

63. Pisano JJ, Crout RJ, Abraham D. Determination of 3-methoxy-4-hydroxymandelic acid in urine. Clin Chim Acta 1962;7:285.

64. Raab GM. Comparison of a logistic and a mass-action curve for radioimmunoassay. Clin Chem 1983;29:1757.

65. Rebar HW. Hypergonadotropic amenorrhea and premature ovarian failure. J Reprod Med 1982;27:179.

66. Reeves WB, Andreoli TE. The posterior pituitary and water metabolism. In: Wilson JD, Foster DW, eds. Williams' textbook of endocrinology. Philadelphia: Saunders, 1992;311.

67. Reichlin S. Neuroendocrinology. In: Wilson JD, Foster DW, eds. Williams' textbook of endocrinology. Philadelphia: Saunders, 1985;492.

68. Roberts LJ. Carcinoid syndrome in disorders of systemic mast-cell activation including systemic mastocytosis. Endocrinol Metab Clin North Am 1988;17:415.

69. Roberts LJ, Oates JA. Disorders of vasodilator hormones: The carcinoid syndrome and mastocytosis. In: Wilson JD, Foster DW, eds. Williams' textbook of endocrinology. Philadelphia: Saunders, 1992;1619.

70. Rock RC, Chan DW, Peristein MT. Endocrine assays in the monitoring of pregnancy. Clin Lab Med 1981;1:157.

71. Rock RC. Endocrinology of the fetoplacental unit: Role of estrogen assays. Lab Med 1984;15:95.

72. Rodd CJ, Sockalosky JJ. Endocrine causes of hypertension in children. Pediatr Clin North Am 1993;40:149.

73. Rose SR, Ross JL, Uriarte M, Barnes KM, Cassoria FG, Cutler GB Jr. The advantage of measuring stimulated as compared with spontaneous growth hormone levels and the diagnosis of growth hormone deficiency. N Engl J Med 1988;319:201.

74. Sacks DB. Carbohydrates. In: Burtis CA, Ashwood ER, eds. Tietz textbook of clinical chemistry. Philadelphia: Saunders, 1994;928.

75. Saper CB, Breder CD. The neurologic basis of fever. N Engl J Med 1994;330:1880.

76. Saudek CD. Endocrine pancreas. In: Noe DA, Rock RC, eds. Laboratory medicine. The selection and interpretation of clinical laboratory studies. Baltimore: Williams & Wilkins, 1994;646.

77. Schimke RN. Multiple endocrine adenomatosis syndromes. Adv Intern Med 1976;21:249.

78. Service FJ. Hypoglycemia. Endocrinol Metab Clin North Am 1988;17:601.

79. Sheps SG, Jiang MD, Klee GG. Diagnostic evaluation of pheochromocytoma. Endocrinol Metab Clin North Am 1988;17:397.

80. Sipple JH. The association of pheochromocytoma with carcinoma of the thyroid gland. Am J Med 1961;31:163.

81. Sisson JC, Gross MD, Frietas JE, et al. Combining provocative agents of calcitonin to detect medullary carcinoma of the thyroid. Henry Ford Hosp Med J 1981;29:75.

82. Smith KD. Endocrine problems and treatment. Clin Obstet Gynecol 1982;25:483.

83. Swerdloff RS, Wang C, Kandeel FR. Evaluation of the infertile couple. Endocrinol Metab Clin North Am 1988;17:301.

84. Szerlip H, Palevsky P, Cox M. Sodium and water. In: Noe DA, Rock RC, eds. Laboratory medicine. The selection and interpretation of clinical laboratory studies. Baltimore: Williams & Wilkins, 1994;692.

85. Taylor AL, Fishman LM. Corticotropin-releasing hormone. N Engl J Med 1988;319:213.

86. Thorner MO, Vance ML, Horvath E, Kovacs K. The anterior pituitary. In: Wilson JD, Foster DW, eds. William's textbook of endocrinology. Philadelphia: Saunders, 1992;221.

87. Tischler AS, Dichter MA, Biales B, Greene LA. Neuroendocrine neoplasms and their cells of origin. N Engl J Med 1977;296:919.

88. Tyson JE. Reproductive endocrinology: New problems call for new solutions. Diagn Med 1984;7:25.

89. Varner MW, Hauser KS. Current status of human placental lactogen. Semin Perinatol 1981;5:123.

90. Varner MW, Hauser KS. Human placental lactogen and other placental proteins as indicators of fetal well-being. Clin Obstet Gynecol 1982;25:673.

91. Vogt W, Sandel P, Langfelden CH, Knedel M. Performance of various mathematical methods for computer-aided processing of radioimmunoassay results. Clin Chim Acta 1978; 87:101.

92. Vokes TJ, Robertson GL. Disorders of antidiuretic hormone. Endocrinol Metab Clin North Am 1988;17:281.

93. Watts NB. Anterior pituitary. In: Noe DA, Rock RC, eds. Laboratory medicine. The selection and interpretation of clinical laboratory studies. Baltimore: Williams & Wilkins, 1994;618.

94. Weisberg L, Cox M. Potassium. In: Noe DA, Rock RC, eds. Laboratory medicine. The selection and interpretation of clinical laboratory studies. Baltimore: Williams & Wilkins, 1994;732.

95. Werbel SS, Ober KP. Acute adrenal insufficiency. Endocrinol Metab Clin North Am 1993;22:303.

96. Wermer P. Genetic aspects of adenomatosis of endocrine glands. Am J Med 1954;16:363.

97. Whitley RJ, Meikle AW, Watts NB. Endocrinology. In: Burtis CA, Ashwood ER, eds. Tietz textbook of clinical chemistry. Philadelphia: Saunders, 1994;1645.

98. Wilson JD, Foster DW. Introduction. In: Wilson JD, Foster DW, eds. Williams' textbook of endocrinology. Philadelphia: Saunders, 1992;1.

99. Woo J, Cannon DC. Metabolic intermediates and inorganic ions. In: Henry JB, ed. Clinical diagnosis and management by laboratory methods. Philadelphia: Saunders, 1984;133.

100. Woodhead JS. The measurement of circulating parathyroid hormone. Clin Biochem 1990;23:11.

101. Yalow RS, Berson SA. Immunoassay of endogenous plasma insulin in man. J Clin Invest 1960;39:1157.

102. Young RL, Goldzieher JW. Clinical manifestations of polycystic ovarian disease. Endocrinol Metab Clin North Am 1988;17:621.

103. Young WF, Klee GG. Primary aldosteronism, diagnostic evaluation. Endocrinol Metab Clin North Am 1988;17:367.

Thyroid Function

H. Jesse Guiles

Objectives

Upon completion of this chapter, the clinical laboratorian should be able to:

- *Discuss the biosynthesis, secretion, transport, and action of the thyroid hormones.*
- *Given a patient's clinical data, correlate the information with regard to the thyroid disorder.*
- *Diagram the anatomy of the thyroid gland.*
- *Describe the regulation of the thyroid hormones.*
- *Relate the various assays used to assess thyroid function and explain the principles behind each of the thyroid function tests.*

KEY WORDS

Follicular Cells
Free Thyroxine Index (FT$_4$I)
Grave's disease
Hashimoto's disease
Hyperthyroidism
Hypothalamic-Pituitary-Thyroid Axis (HPTA)

Hypothyroidism
Myxedema
Parathyroid Glands
Perifollicular Cells
T$_3$ Uptake
T$_3$
T$_4$
Thyroid
Thyroid-Binding Globulin (TBG)

Thyroid-Releasing Hormone (TRH)
Thyroiditis
Thyrotoxicosis
Thyrotropin (TSH)
Thyroxine-Binding Albumin
Thyroxine-Binding Prealbumin (TBPA)

THYROID ANATOMY AND PHYSIOLOGY

The thyroid gland consists of two lobes located in the lower part of the neck. The lobes are connected by a narrow band, called an *isthmus,* and are typically asymmetrical, with the right lobe being larger than the left. The lobes weigh approximately 20 to 30 g each and measure 4.0 cm in length and 2 to 2.5 cm in width and thickness (Fig. 20-1).

Two types of cells form the thyroid: the follicular (or cuboidal) and the perifollicular cells. The follicular cells are secretory and produce thyroxine (T$_4$) and triiodothyronine (T$_3$). Each follicle is in the shape of a sphere and surrounds a material called colloid, which is mainly thyroglobulin. The perifollicular, or C, cells are situated in clusters along the interfollicular or interstitial spaces. The C cells produce the polypeptide calcitonin, which is involved in calcium regulation.

Adjacent to the thyroid gland are four parathyroid glands. Two of these glands are found in the upper portion and two are found near the lower portion of the thyroid gland. The parathyroid glands produce parathyroid hormone, which controls calcium and phosphate metabolism.

BIOSYNTHESIS, SECRETION, TRANSPORT, AND ACTION OF THYROID HORMONES

Iodine is the most important element in the biosynthesis of thyroid hormones. Approximately 150 μg of iodide is absorbed in the intestine each day. The thyroid gland has a

Michael L. Bishop, Janet L. Duben-Engelkirk, and Edward P. Fody.
CLINICAL CHEMISTRY. © 1996 Lippincott-Raven Publishers.

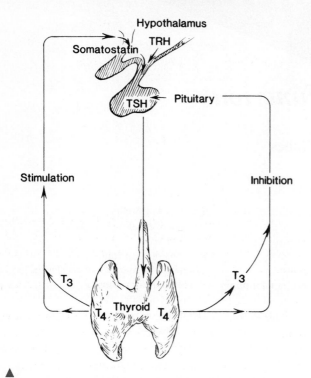

▲
Figure 20-1. The negative feedback system regulating thyroid function.

very high attraction for iodide and traps about 70 μg per day at the base of the cell by active transport. The iodide is then transported to the follicular lumen. The iodide molecule is oxidized by peroxidase, presumably at the interface of the cell and the lumen, to a more reactive form, I^0 or I^+. This form combines with the glycoprotein thyroglobulin. Thyroglobulin acts as a performed matrix containing tyrosyl groups to which the reactive iodine attaches to form the hydroxyl residues of monoiodotyrosine (MIT) and diiodotyrosine (DIT).

The next step is the enzymatic coupling of the iodinated tyrosine molecules, catalyzed by peroxidase, to form T_4 or T_3. T_4 is formed by the coupling of two DIT molecules (Fig. 20-2). The coupling of one DIT molecule and one MIT molecule results in the formation of T_3 (3,5,3'-T_3) (Fig. 20-3) or reverse T_3 (rT_3) (3,3',5'-T_3) (Fig. 20-4). T_3 contains two iodine atoms in the tyrosyl ring and one iodine atom in the phenolic ring. On the other hand, rT_3 contains two iodine atoms in the phenolic ring and one iodine atom

in the tyrosyl ring. These molecules are stored in the thyroglobulin in the thyroid follicle. The stored T_4 and T_3 can be released by enzymatic cleavage of the thyroglobulin molecules. Release of the iodothyronines involves endocytosis of the thyroglobulin from the follicular lumen and hydrolysis of thyroglobulin by a protease and a peptidase in the epithelial cell to liberate free iodoaminoacids (T_3, T_4, rT_3, MIT, DIT). Most iodothyronines (T_3, T_4, rT_3) are secreted into the bloodstream, whereas most iodotyrosines (MIT, DIT) are deiodinated within the thyroid, and the iodide is used in producing iodoaminoacids.

About 80% of circulating T_3 is formed following monodeiodination of T_4 in peripheral tissues, especially the liver and kidney, by 5'-deiodinase. Thus, T_4 can be considered a prohormone for T_3 production. Approximately 33% of the T_4 that is secreted by the thyroid each day undergoes monodeiodination to produce T_3. Another 40% undergoes monodeiodination in the inner ring to produce rT_3. The serum concentration of T_4 is about 50 times greater than that of T_3.

Almost all circulating T_4 (~99.97%) and T_3 (~99.7%) hormones are bound to serum proteins. There are three thyroid-hormone-binding proteins: thyroxine-binding globulin (TBG), thyroxine-binding prealbumin (TBPA), and thyroxine-binding albumin (TBA). T_4 binds predominantly to TBG (70%–75%), to a lesser extent to TBPA (15%–20%), and to a slight extent to albumin (10%). T_3 is bound by TBG and TBA but very little by TBPA. Normally, only one-third of the available binding sites are occupied by the thyroid hormones.

The relationship between the thyroid hormones and the binding proteins can be described by the law of mass action:

$$[FT_4] \times [\text{unbound TBG}] = K \times [T_4 - TBG]$$

(Eq. 20-1)

where $[FT_4]$ = the concentration of free T_4 in serum
$[\text{unbound TBG}]$ = the concentration of unbound sites on TBG
K = reaction rate constant
$[T_4 - TBG]$ = concentration of T_4 occupying TBG binding sites

An increase in the concentration of FT_4 or unbound TBG would drive the reaction to the right. Therefore, in

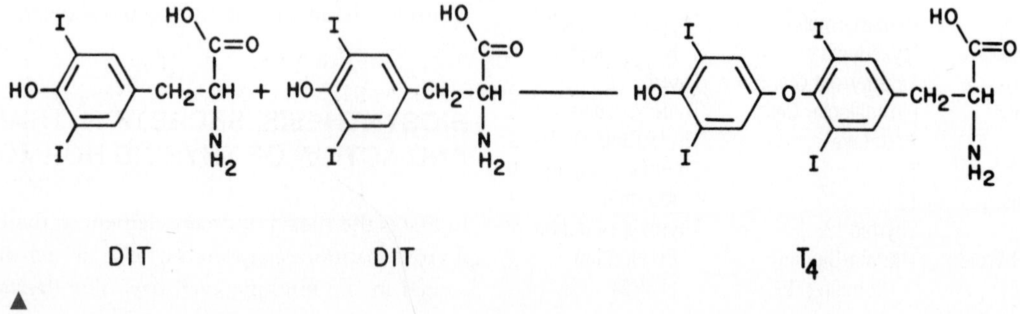

▲
Figure 20-2. The formation of T_4.

DIT   DIT T_4

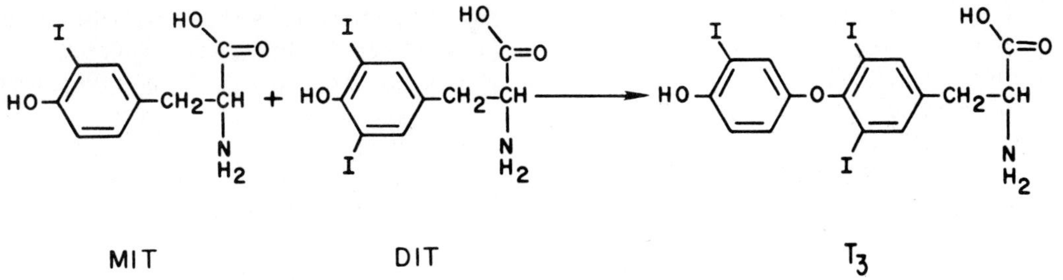

Figure 20-3. The formation of T$_3$.

situations where there is an excess of protein, more T$_4$ will be bound, and in situations where there is little protein available, less T$_4$ will be bound. However, the concentration of free T$_4$ should remain stable. This relationship becomes important when hormones, drugs, or disease states affect protein balance within the body.

Only 0.03% of T$_4$ and 0.3% of T$_3$ are not bound to proteins. These fractions, called free T$_4$ (FT$_4$) and free T$_3$ (FT$_3$), are the physiologically active portions of the thyroid hormones. T$_3$ is the most biologically active thyroid hormone and is three to four times more potent than T$_4$. T$_3$ is more active because it is not as tightly bound to the serum proteins as is T$_4$ and can diffuse into the cell more easily. In addition, the receptor protein within the cell nucleus has a higher affinity for T$_3$ than for T$_4$. rT$_3$ is not biologically active.

Thyroid hormone actions include calorigenesis and oxygen consumption by means of regulation of carbohydrate, lipid, and protein metabolism; central nervous system activity; cardiovascular stimulation; gastrointestinal regulation; growth and development; and sexual maturation.

REGULATION OF THYROID HORMONES

The hypothalamic-pituitary-thyroid axis (HPTA) is the neuroendocrine system that regulates the production and secretion of thyroid hormones (see Fig. 20-1). Regulation of the thyroid begins with the hypothalamus. Thyroid-releasing hormone (TRH) is a tripeptide released by the hypothalamus. It travels along the hypothalamic stalk to the beta cells of the anterior pituitary, where it stimulates the

synthesis and release of thyrotropin or thyroid-stimulating hormone (TSH). At the pituitary level, the secretion of TSH seems to be regulated by an interplay of TRH and somatostatin, an inhibitory factor, as well as by free T$_4$ and T$_3$. T$_4$ and T$_3$ appear to stimulate the secretion of somatostatin by the hypothalamus.[7]

Thyrotropin (TSH) is a glycoprotein consisting of two subunits, alpha and beta, linked non-covalently. The specificity of the hormone action is conveyed by the beta subunit. The alpha subunit is comparable and interchangeable with the alpha subunits of the other glycoprotein hormones, luteinizing hormone (LH), follicle-stimulating hormone (FSH), and human chorionic gonadotropin (HCG). The alpha subunit also influences the overall shape of the glycoprotein and controls the binding to the appropriate receptor on the cell membrane of the thyroid follicle cells. TSH acts like other polypeptide hormones in that when it binds with receptor sites, it activates adenyl cyclase, which then catalyzes the reaction to produce cyclic AMP. TSH is the main stimulus for the uptake of iodide by the thyroid gland and also stimulates the activation of the protease enzymes, which in turn catalyze the hydrolysis of thyroglobulins, the storage forms of T$_4$ and T$_3$. TSH also acts to increase the size and number of follicular cells.

It is the FT$_4$ and FT$_3$ that serve in a negative-feedback system to regulate the amount of TSH that is released from the beta cells in the anterior pituitary and to stimulate the hypothalamus to produce somatostatin. The production of TRH by the hypothalamus does not appear to be involved in the negative-feedback system (see Fig. 20-1).

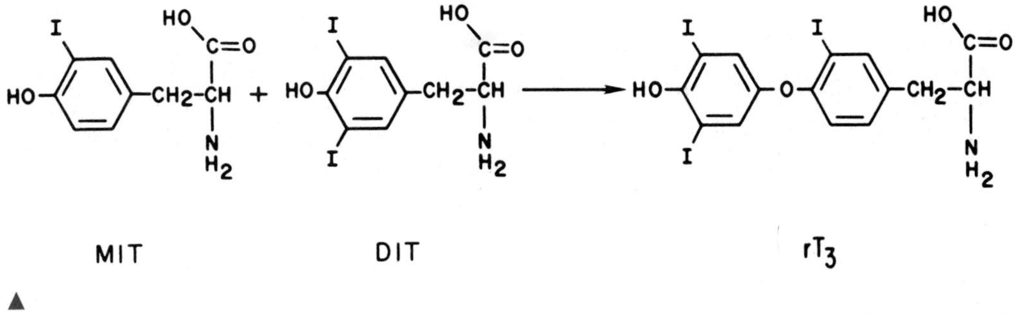

Figure 20-4. The formation of rT$_3$.

Free T_4 and T_3 are thought to enter the target cell, where T_3 binds to receptor sites located in the cell nucleus, probably in the nuclear chromatin. This binding effects changes in the synthesis of particular proteins mediating or reflecting the extranuclear mode of action. There may be thyroid-hormone binding sites in the mitochondria, membrane, and other cytoplasmic cell fractions.[3]

ASSAYS FOR THYROID FUNCTION

Assays for thyroid hormones are used to determine and confirm the nature and extent of hyper- or hypothyroidism. The clinical picture of abnormal thyroid function may be nonspecific and only clarified through laboratory testing. Normal thyroid hormone levels, however, do not rule out thyroid disease, and changes in binding proteins can affect interpretation of results.

Current guidelines of thyroid testing call for an estimation (or direct measurement) of *free thyroxine* (FT$_4$) and a high-sensitivity thyrotropin/thyroid stimulating hormone (s-TSH) assay.[2,23,24,47,48] The estimated free thyroxine index (FT$_4$I) is based upon the total serum thyroxine (TT$_4$) and T_3 uptake (T$_3$U) results. Screening of the general public for thyroid disease is not recommended, but patients in high-risk groups such as those with a family history of thyroid disease, newborns, the elderly, postpartum women, and patients with autoimmune or endocrine disorders are likely candidates for thyroid testing.[14,35,48]

Total Thyroxine (TT$_4$)

T_4 is measured by radioimmunoassay (RIA) or other immunoassay techniques. In the RIA methodology, the sample is incubated with antibody and $^{125}[I]T_4$ in a barbital buffer. After incubation, the bound and the free portions are separated, and the radioactive bound portion is quantified. Standards are run, and the unknowns and controls are extrapolated from a curve of percent binding of standards versus concentration. The detection limit of this assay can be as low as 0.25 μg/dL.

Nonisotopic immunotechniques are also available. The principles of these assays are similar to RIA except that the labeled analyte may be an enzyme as in the enzyme-linked immunosorbent assays (ELISA), a fluorophore, or chemiluminescence. Other immunotechniques may utilize a double-antibody or a solid-phase system. The enzyme-multiplied immunoassay technique (EMIT) is an enzymatic method that does not require separation of the free and bound portions. Reference ranges are given in Table 20-1.

Besides rate of thyroid synthesis and release, serum TT$_4$ levels are dependent on two other major variables, the first of which is the serum concentration of T_4 binding proteins. This concentration varies with pregnancy, acute illnesses, or the effects of some drugs. Generally, when these proteins increase, more T_4 will bind, with a consequential increase of TT$_4$ levels. Conversely, a decrease in these proteins or a

TABLE 20-1. Reference Ranges for Thyroid-Function Tests

Total T$_4$	
Adult	4.5–13 μg/dL
Cord-blood	7.4–13.1 μg/dL
Newborn	7.0–17.0 μg/dL
T$_3$U (Uptake)	25%–30%
T$_3$U Ratio	0.8–1.35
FT$_4$I (Index)	4.5–12
FT$_4$	1.2–2.5 μg/dL
s-TSH	0.5–5.0 μU/mL
TRH stimulation (TSH)	>2 μU/mL after 20 min
Total T$_3$	60–220 ng/dL
rT$_3$	25–62 ng/dL
FT$_3$I (Index)	85–205 ng/dL
FT$_3$	0.2–0.4 ng/dL
TBG	15–24 μg/mL
RAIU	10%–30%
Antithyroglobulin Antibodies	<1:100
Antimicrosomal Antibodies	<1:40
TSH Receptor Antibodies	varies with method

reduction in available binding sites because of competition by other substances such as drugs results in less T_4 binding, realizing a lower TT$_4$ level. Theoretically, the FT$_4$ level should remain within the reference range under these conditions, as explained by the law of mass action.

The other major influence on TT$_4$ levels is the peripheral conversion of T_4 to T_3. Nonthyroidal diseases may suppress the conversion of T_4 to T_3, thus affecting TT$_4$ levels. Diminished peripheral conversion of T_4 to T_3 has been reported in protein-calorie deprivation, in fasting, and after the administration of certain drugs, such as propranolol hydrochloride, glucocorticoids, propylthiouracil, and ipodate (x-ray contrast media).

T$_3$ Uptake (T$_3$U)

The T$_3$U assay is the second major thyroid screening test and must not be confused with the total T_3 assay, which measures endogenous T_3. In fact, to reduce the confusion, the American Thyroid Association has recommended that the name of the test be changed to the Thyroid Hormone Binding Ratio (THBR).[11] The T$_3$U assay is used to measure the number of available binding sites of the thyroxine-binding proteins, most notably TBG. Current T$_3$U assay use iodine-125, enzyme, fluorophore, and chemiluminescent techniques. The T$_3$U assay is illustrated in Figure 20-5.

In this assay, the patient's serum is mixed with labeled T_3 binding material such as a resin, charcoal, silicate, or antibody. Some of the labeled T_3 binds to the available binding sites of the proteins while the remainder binds to the added binding agent. The amount of labeled T_3 taken up by the binding agent is inversely proportional to the number of available binding sites on the TBG. The result is calculated by dividing the matrix-bound counts by the residual serum protein-bound counts.[11]

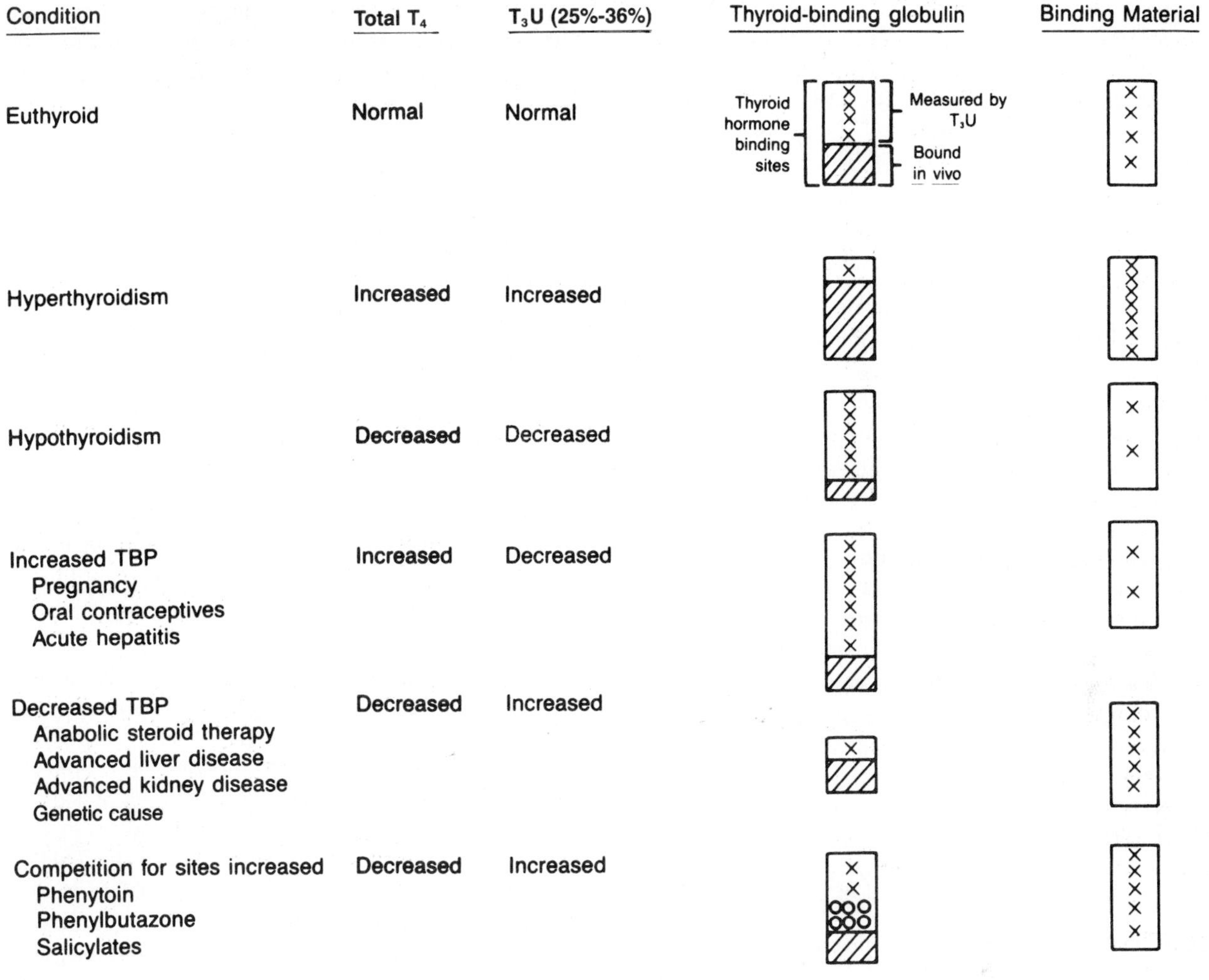

Condition	Total T_4	T_3U (25%-36%)	Thyroid-binding globulin	Binding Material
Euthyroid	Normal	Normal	Thyroid hormone binding sites — Measured by T_3U / Bound in vivo	
Hyperthyroidism	Increased	Increased		
Hypothyroidism	Decreased	Decreased		
Increased TBP Pregnancy Oral contraceptives Acute hepatitis	Increased	Decreased		
Decreased TBP Anabolic steroid therapy Advanced liver disease Advanced kidney disease Genetic cause	Decreased	Increased		
Competition for sites increased Phenytoin Phenylbutazone Salicylates	Decreased	Increased		

Figure 20-5. The T_3U assay and TT_4.

The T_3U is usually reported as a ratio of the patient's T_3U to the average uptake of normal pooled serum. This enables the calculation of a free thyroxine index. Reference ranges are listed in Table 20-1.

The relationship of the T_3U assay and the TT_4 is also illustrated in Figure 20-5. In hyperthyroidism, the TBG binding sites are occupied by the patient's endogenous T_4, resulting in an increased amount of labeled T_3 being bound to the binding agent (or an increased T_3U). The converse is true with hypothyroidism. Protein abnormalities also affect the T_3U. Patients who take oral contraceptives, who are pregnant, or who have acute hepatitis will have a consequent increase of TBG with an increased TT_4 and a decreased T_3U result. Conditions such as an anabolic steroid therapy, advanced liver or kidney disease, or a genetic defect can result in a general decrease of TBG, causing a decreased TT_4 and increased T_3U result. Finally, competition for binding sites by such drugs as phenytoin, phenylbutazone, or salicylate reduce patient TT_4 and reagent labeled T_3 binding to TBG, resulting in an increased T_3U result.

Other assays are available (*e.g.*, "thyroid uptake tests") that use labeled T_4 instead of T_3 to saturate the TBG and generally have a direct relationship between the amount of labeled product detected and the TBG present. These methods may be more specific than the T_3U test in certain abnormal physiological conditions (*e.g.*, dysalbuminemia) but generally give the same type of information as the T_3U.

Free Thyroxine Index (FT_4I)

In theory, the conditions affecting thyroid hormone binding should not affect FT_4 levels. In order to estimate the effects of protein or interfering drugs, a calculation can be performed. The free thyroxine index (FT_4I) is an indirect measure of free hormone concentration and is based on the equilibrium relationship of bound T_4 and FT_4. The FT_4I is calculated by the following formula:

$$FT_4I = T_4 \times T_3U \text{ ratio} \qquad \textit{(Eq. 20-2)}$$

The T_3U ratio is derived by comparing the patient's T_3U with a normal serum pool.

Example:

$$T_4 = 12 \ \mu g/dL$$

$$T_3U = 25\% \ (\text{normal} = 30\%)$$

$$T_3U \ \text{ratio} = \frac{25}{30} = 0.83$$

$$FT_4I = 12 \times 0.83$$

$$= 10.0 \qquad \textit{(Eq. 20-3)}$$

The FT_4I reference range is included in Table 20-1.

The FT_4I is useful in correcting for euthyroid individuals who have altered binding-protein concentrations and who would have been misclassified on the basis of TT_4. For small or moderate variations of binding-protein concentrations, the FT_4I is an adequate indicator of thyroid status and has the advantage of being simple, rapid, and inexpensive. However, there is a curvilinear relationship between the proportion of bound T_4 and percent of free hormone. The direct FT_4 assay is seen as a better indicator of thyroid status than the FT_4I in cases of extreme protein abnormalities, such as congenital TBG excess or deficiency, dysalbuminemia (an inherited disorder where abnormal albumin has a strong affinity for T_4 but not T_3), or severe nonthyroidal illness (NTI), which is sometimes referred to as the euthyroid sick syndrome.[4,22,24,54]

Free Thyroxine (FT$_4$)

The percentage of free T_4 (FT_4) can be directly measured by equilibrium dialysis or ultrafiltration. These are considered the reference methods for FT_4 determination.[2,9,27] The critical step in these procedures is separation of the free hormone from the bound hormone before analysis. Only minimum dilution is permissible, because dilution of sera can distort the equilibrium between free and bound hormone.[24,49] As a result, these methods have been technically demanding, relatively expensive, and time consuming. Other methods of FT_4 determination are considered "indirect" estimates.[24]

The FT_4 analogue assay requires a thyroid hormone tracer that has been chemically modified to prevent it from binding to serum transport protein while, at the same time, enabling it to bind to the reagent antibody.[37] Studies have shown, however, that analogs do bind to patients' albumin thus changing the availability of the analog to bind with the antibody and biasing results.[11,24]

Other indirect methods of FT_4 are the two-step immunoassay techniques where the hormone is immunoextracted from serum onto an antibody. After washing, a second labeled hormone is added, which binds to the unoccupied antibody sites. An inverse relationship between the patient's FT_4 and the amount of bound labeled antibody is measured.

Generally, FT_4 appears to be more sensitive and specific than FT_4I in distinguishing euthyroid from hyper- versus hypothyroid conditions. However, indirect FT_4 assays have been shown to be affected by dysalbuminemia, pregnancy, severe illness, drugs such as dopamine, and glucocorticoids.[4,9,15,24,38] Studies indicate that only equilibrium dialysis

or ultrafiltration techniques, using undiluted sera, can be relied upon for distinguishing between critically ill patients with NTI and those with thyroid disease.[11,49] However, standardization across methods for equilibrium dialysis or ultrafiltration has been poor.[24] The choice of test, therefore, depends on the clinical state of the individual patient.

Thyrotropin or Thyroid-Stimulating Hormone (TSH)

The high-sensitivity thyroid-stimulating hormone (s-TSH) assay has been shown to be extremely useful in the confirmation of suspected primary hypo- and hyperthyroidism (even in borderline cases). Subclinical hyper- or hypothyroidism is defined as an abnormal TSH level and a normal TT_4 or FT_4 in the absence of overt symptoms.[47,51] s-TSH has replaced the traditional monoclonal radioimmunoassay technique for the hormone. The latter test had not been sensitive enough to distinguish between the very low TSH values found in primary hyperthyroidism from those found in some healthy euthyroid patients.

Whereas the traditional radioimmunoassay techniques measured an antigen-antibody complex, the high-sensitivity TSH immunometric assays (IMAs) utilize two or three separate monoclonal antibodies in the methodologies (Fig. 20-6). The requirement of binding to more than one site gives the s-TSH its high specificity. The first antibody is in excess, bound to a solid phase support, and extracts the TSH molecule from the serum. This antibody is usually specific for the beta subunit of the TSH molecule. The second antibody attaches to a different portion of the TSH molecule, usually the alpha subunit, and has a label on it which may be a radioisotope, enzyme, fluorophor, or chemiluminescent compound. The separation of the free and the bound portion is accomplished in one or two steps. The two-step procedure has fewer interferences but is harder to achieve. The double binding results in a labeled insoluble "sandwich" with a concentration directly proportional to the amount of TSH in the specimen.[35,50]

A "generation" of TSH assay is characterized by an approximate tenfold sensitivity increase. Today's second-generation IMAs for TSH have sensitivities from 0.1 to 0.01 $\mu U/mL$. Third-generation TSH tests have detection limits of 0.01 to 0.005 $\mu U/mL$. They permit more reliable distinction between undetectable TSH in patients with hyperthyroidism and subnormal values that are occasionally found in euthyroid patients, and demonstrate superior precision in the critical subnormal range.[27,35,44,50] Currently, work is underway in the development of fourth-generation TSH tests with sensitivity of 0.001 $\mu U/mL$.[50]

It has been shown that a minor rise or fall of thyroid hormones, especially FT_4, elicits a dramatically higher inverse response in TSH by the pituitary.[4,35,44] Thus, the s-TSH is considered to be the most clinically sensitive assay for detection of primary thyroid disease.[2,4,22,35] Furthermore, TSH concentration is not as markedly affected by physiologic alterations resulting from severe nonthyroidal sickness or ad-

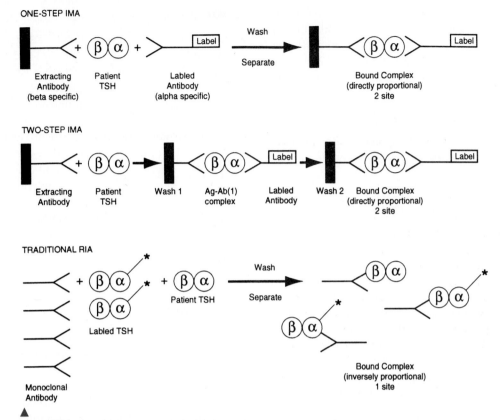

Figure 20-6. Immunometric Assay (IMA) and Radioimmunoassay (RIA) for TSH. *Source:* Taylor CS, Brandt DR. Developments in thyroid-stimulating hormone testing: The pursuit of improved sensitivity. Lab Med 1994;24:337.

ministration of therapeutic drugs as is TT_4 or FT_4.[27,34,35,47,51] Studies have demonstrated that TSH concentration can be effected by severe nonthyroidal illness; pregnancy during the first trimester; treatment with dopamine, glucocorticoids, and certain other drugs; or thyroxine replacement therapy.[2,4,5,9,19,27,35,44,45,47,48]

An additional problem of the increased sensitivity is that it may lead to unnecessary intervention of subclinical thyroid disease, resulting in expensive treatment with uncertain benefits.[14,25,27,51] One study showed that the TSH test was 100 percent sensitive in detecting abnormal thyroid states in hospitalized patients, but had less specificity, due to false-positives, than the FT_4I test.[45] The consensus of opinion is that the test's strengths far outweigh its weaknesses. Because of its exceptional clinical sensitivity (with some sacrifice of clinical specificity) along with its proven chemical sensitivity and specificity, the s-TSH assay, in conjunction with the FT_4 assay or FT_4I, is seen as the primary screening test for thyroid function.[2,5,14,23,24,35,47,48,51]

Thyroid-Releasing Hormone (TRH) Stimulation Test

In the past, due to the insensitivity of the first generation TSH tests, physicians had relied upon the TRH stimulation test to distinguish between euthyroid and hyperthyroid patients who both had undetectable TSH levels. The TRH stimulation test detects residual TSH stores in the pituitary gland. In this test, TRH is injected intravenously under the direction of a physician. In euthyroid patients, the TSH will rise from 10 to 30 µU/mL within 30 minutes. Patients with primary hyperthyroidism will not respond to the TRH and their s-TSH will remain at baseline levels, whereas patients with primary hypothyroidism will have an exaggerated response. However, since baseline TSH levels are typically elevated in primary hypothyroidism, the TRH test adds little information in exchange for the extra cost and pain to the patient.

With the advent of the high-sensitivity TSH assays, the TRH test has become obsolete for the confirmation of primary hyperthyroidism or for detection of subclinical hyperthyroidism.[4] A study of 150 patients found that no patient with baseline TSH <0.1 µU/mL responded to TRH, and all patients with a baseline TSH of >0.4 µU/mL had at least a borderline response. These researchers concluded that, with the availability of the high-sensitivity TSH assay, a large proportion of patients with TSH values <0.1 µU/mL or >0.4 µU/mL do not need the TRH stimulation test because the TRH response can be predicted by s-TSH assay.[29]

The test may still be useful in discriminating between NTI sick and hyperthyroid patients, since euthyroid sick patients will generally respond to TRH and hyperthyroid sick patients will not respond.[35] The TRH test may also be useful in the detection of thyroid hormone resistance syndromes.[32] Furthermore, the TRH test may be utilized in differentiating secondary (pituitary) from tertiary (hypothalamic) hypothyroidism.[4,18,23,26,35] In both conditions, the patients have a low

FT_4 with a borderline or inappropriately normal TSH. When TRH is injected in a patient with secondary hypothyroidism, the TSH level will not increase or will be blunted, whereas in patients having tertiary hypothyroidism the TRH stimulation will have a response. Some studies, however, have shown that even though the pituitary is so damaged that it cannot secrete TSH on a sustained basis, residual thyrotrophs may have a TSH response to TRH that is within normal limits. Conversely, an abnormal response to the TRH test may occur with severe illness and not be diagnostic.

Thyroid-Binding Proteins (TBP)/ Thyroid-Binding Globulin (TBG)

Aside from the assessment of thyroid hormone binding to TBP by means of the T_3U test, thyroid-binding proteins also can be measured directly by means of immunotechniques. Usually, the TBG assay is used to confirm results of FT_3 or FT_4, or abnormalities in the relationship of the TT_4 and T_3U test, or as a postoperative marker of thyroid cancer. The presence of circulating thyroid autoantibodies will interfere with TBG test results.[4] Reference range is included in Table 20-1.

Total T_3 (TT$_3$), Free T_3 (FT$_3$), and Free T_3 Index (FT$_3$I)

Total serum T_3 is measured by immunoassay techniques similar to that of TT_4. TT_3 is not very useful as a primary screening test for hypothyroid disorders because too many nonthyroidal factors affecting binding proteins can influence levels. Rate of formation from T_4 peripheral deiodination declines during almost all significant illnesses and stresses such as cirrhosis of the liver, renal failure, starvation, anorexia nervosa, cystic fibrosis, administration of certain drugs (propranolol, propylthiouracil), and some contrast media. It is estimated that TT_3 levels are decreased in approximately 70% of hospitalized patients who do not have hypothyroidism, whereas as many as 20% to 30% of patients with hypothyroidism have normal TT_3 concentrations.[9,48]

The value of the TT_3 assay is in confirming hyperthyroidism, especially in T_3 thyrotoxicosis, where the TT_4 is within normal limits. Early in hyperthyroidism, TT_3 may be more elevated than TT_4.

A free T_3 index (FT_3I) can be calculated similar to FT_4I by using the formula

$$FT_3I = T_3 \times T_3U \text{ ratio} \qquad \textit{(Eq. 20-4)}$$

A free T_3 (FT_3) level can be performed along the same manner as a FT_4 and also can be helpful in confirming T_3 thyrotoxicosis.

FT_3 and FT_3I are usually not offered as routine tests. The tests offer no advantage over the FT_4I or FT_4 in diagnosing hypothyroidism and give more false-negative and false-positive results.[8] A normal FT_3I does not exclude hypothyroidism, and a low FT_3I does not establish the diagnosis of hypothyroidism. Reference ranges for TT_3, FT_3, and FT_3I are listed in Table 20-1.

Reverse T_3 (rT$_3$)

Metabolically inactive reverse T_3 (rT_3) also can be measured by immunoassay. It is usually only ordered to help in the resolution of borderline or conflicting laboratory results. Almost all rT_3 comes from peripheral deiodination of T_4. It is elevated whenever there is an impairment of conversion of T_4 to T_3.

rT_3 is commonly elevated in patients with NTI where the TT_3 level is often concurrently decreased.[23,30,9] Furthermore, rT_3 is elevated in amniotic fluid at an early stage of pregnancy. The assay for rT_3 in amniotic fluid could be useful in the diagnosis of fetal hypothyroidism and thyroiditis. However, since hypothyroidism can be diagnosed at birth and thyroid replacement therapy *in utero* is unnecessary, this assay has not been extensively used. The rT_3 reference range is listed in Table 20-1.

Thyroid Antibodies

Individuals with certain thyroid diseases may have circulating antibodies against several components of thyroid follicles. These antibodies can act as thyroid stimulators or blockers and are associated with hyper- or hypothyroid states. They include the thyroid-stimulating immunoglobulins (TSI), thyroid antimicrosomal antibodies, and antithyroglobulin antibodies. The American Thyroid Association has proposed the name "TSH receptor antibodies (TRAb)" for TSI because TSI is one of many names linked to a specific assay methodology for the antibodies.[11]

TRAbs are a variety of 7 S gamma-globulins. They can become a functional analogue of TSH and can bind so strongly to the TSH receptor that TSH is blocked from binding on the thyroid membranes. Like TSH, the antibodies can activate the cells to which they bind. However, unlike TSH, they are not subject to feedback control. As a result, there is overproduction of thyroid hormones. Other names for the TRAb include *long-acting thyroid stimulator (LATS), TSH-binding immunoglobulin (TBII), LATS-protector (LATS-P),* and *thyroid-stimulating antibodies (TSAb).*[4,7,11]

The microsomal antibodies bind with the microsomal antigens associated with the follicular cells lining the microsomal membrane. They are mainly IgG and fix complement. It is believed that damage to the thyroid gland is brought about by the interaction of lymphocytic killer cells, which effect tissue destruction. Microsomal antibodies have been associated with patients with hyperthyroid as well as hypothyroid disorders.[52]

Thyroglobulin antibodies are mainly IgG and do not fix complement. These antibodies do not correlate well with thyroid disease and may not damage tissue. Increased levels may indicate only that there is something amiss with the immune system in relation to the gland. Occasionally, another antibody to the colloid is also found, but it seems to have no clinical significance.[52]

Thyroid antibodies are detected by a variety of methods, which are listed in Table 20-2. Correlation of the presence of these antibodies with thyroid disease will be discussed later; however, their detection does not signify the presence of overt disease, and conversely, their absence does not rule out thyroid disease.

Radioactive Iodine Uptake (RAIU)

The radioactive iodine uptake (RAIU) assay is used to measure the ability of the thyroid gland to trap iodine. Radioactive iodine is ingested by mouth, and the radioactivity is counted at various lengths of time, usually 4 to 6 hours and again at 24 hours. Normally, 10% to 30% of the ingested dose can be detected within 24 hours.

The assay is useful in helping to determine the cause of hyperthyroidism. The assay is not useful in the diagnosis of hypothyroidism because there is overlap of euthyroid and hypothyroid responses.

Thyroid Scan

Thyroid scan involves *in vivo* use of either ^{123}I or pertechnetate (^{99m}Tc). The isotope is administered to the patient and is taken up by the thyroid gland. The distribution of the radioactive material is determined by scanning the thyroid. The thyroid scan is useful in determining the cause of hyperthyroidism, as well as detecting extrathyroid iodide-concentrating tissue, such as metastatic carcinoma of the thyroid. The most important use of thyroid scanning is to define areas of increased or decreased uptake within the gland. Nodules as small as 1 cm can be detected.

Nodules are classified according to their ability to take up iodine, *i.e.*, as "cold" or nonfunctioning, "warm" or normal functioning, and "hot" or hyperfunctioning. Theoretically, a malignant lesion should not be able to trap iodine and so should appear as a "cold" area. One study, however, demonstrated surgically confirmed cancer in 17% of cold, 9% of warm, and 4% of hot nodules.[42] Thus, although cold nodules are the ones having the best probability of being

TABLE 20-2. Antigen-Antibody Systems Involved in Humoral Responses of Thyroid Autoimmune Disease

Antigen	Antibody (Function)	Antibody Detection
Thyroglobulin	Thyroglobulin antibody (No clear function)	Precipitin technique; tanned red cell hemagglutination; immunofluorescence on fixed thyroid sections; competitive binding radioimmunoassay; co-precipitation with ^{125}I thyroglobulin; microenzyme-linked immunoassay (ELISA); plaque-forming assay
Microsomal antigen	Microsomal antibody (Cytotoxic in conjunction with lymphocytes)	Complement fixation; immunofluorescence on unfixed thyroid sections; cytotoxicity test on cultured thyroid cells; competitive binding radioimmunoassay; tanned red cell hemagglutination; micro-ELISA
Second colloid component	CA$_2$ antibody (No clear function)	Immunofluorescence on fixed thyroid sections
Cell surface antigen(s)	Membrane antibodies (Cytotoxic with lymphocytes)	Immunofluorescence on viable thyroid cells; mixed hemadsorption; binding assays
Thyroxine and triiodothyronine	Thyroid hormone antibodies (Bind and prevent hormone action)	Antigen-binding capacity
Antigen not defined	Growth-stimulating and -inhibiting antibodies (May induce or inhibit thyroid growth)	Effects on DNA content per thyroid cell nucleus or C6PD activity per cell
TSH receptor-related antigen	TSH receptor antibodies (May stimulate thyroid cells, inhibit TSH effect or exert both or neither)	*Stimulatory assays:* (current terms employed for stimulatory assays include human thyroid stimulator, human thyroid stimulating immunoglobulin [TSI], thyroid stimulating antibody [TSAb]). Long-acting thyroid stimulator (LATS) bioassay; colloid droplet formation in human thyroid slices; stimulation of human thyroid adenylate cyclase in vitro; cytochemical assay *Binding assays:* LATS protector assay; inhibition of ^{125}I-thyrotropin binding to human thyroid membranes (thyrotropin displacement activity [TDA], TSH binding inhibitor immunoglobulin [TBII]); fat cell membrane radioligand assays; fat cell ELISA

Source: Adapted from Volpe R: Autoimmune thyroid disease. Hosp Pract 19:143, 1984.

malignant, most are benign, and even a hot nodule does not exclude malignancy.

Fine-Needle Aspiration

Fine-needle aspiration provides information regarding the malignant potential of a thyroid nodule. The fine-needle aspirate is evaluated by means of histologic and/or cytologic examination.

THYROID DISORDERS AND CORRELATION OF LABORATORY DATA

Hypothyroidism

Hypothyroidism occurs when there are insufficient levels of thyroid hormones to provide metabolic needs at the cellular level. The incidence of hypothyroidism in the United States is about 1% to 2%, and it affects females about four times as often as males. Congenital hypothyroidism exists at a rate of around 1 in 4000 births.[1,6,53,55] However, most patients acquire the disease between 30 and 60 years of age.

Deficiency of thyroid hormones causes many metabolic processes to slow down. Symptoms of hypothyroidism include enlargement of the thyroid gland, or goiter, impairment of cognition (including memory, speech, and attention), fatigue, slowing of mental and physical performance, change in personality, intolerance to cold, exertional dyspnea, hoarseness, constipation, decreased sweating, easy bruising, muscle cramps, paresthesias, and dry skin. Hypothyroidism has been shown to be associated with increased cholesterol, LDL, lipoprotein(a), apolipoprotein B, and increased risk of coronary heart disease.[21,16,36]

Myxedema

As the disease progresses into severe hypothyroidism, these features worsen and the condition is referred to as *myxedema*. *Myxedema* is a term that is used to describe the peculiar nonpitting swelling of the skin. The skin becomes infiltrated by mucopolysaccharides so that the face is puffy, especially around the eyes. The features become coarse and the eyebrows thinned. The tongue may become enlarged and the vocal cords thickened, leading to hoarseness. The speech slows. There is usually a weight gain.

The skin tends to be dry and has a yellowish color. This discoloration may be due to carotene accumulation, since there is a reduced rate of conversion of carotene to vitamin A. Body hair is lost and not replaced. Scalp hair often becomes dry and coarse.

A history of progressive mental status dysfunction including confusion, paranoia, psychotic behavior (myxedema madness), apathy, and antisocial tendencies may occur. Myocardial contractility is reduced, and the pulse slows. Effusions may accumulate in the pericardium. Atherosclerosis is accelerated because of high levels of serum triglyceride and cholesterol.

Anemia is common and is due to a reduced rate of red-cell production resulting from hypometabolism, decreased oxygen requirement, and decreased erythropoietin. The anemia is usually normocytic and normochromic, but it can be macrocytic because of folate deficiency from malabsorption or microcytic from the menorrhagia that results in iron deficiency.

Myxedema coma is a severe presentation of hypothyroidism. The clinical symptoms include coma, cardiac dysfunction, hypothermia, hypoglycemia, hypotension, hyponatremia, and respiratory failure. Although the condition is rare, the mortality rate is high and is associated with irreversible cardiovascular collapse.[20]

Congenital Hypothyroidism/Cretinism

Most cases of *congenital hypothyroidism* result from defects in the development or function of the gland itself. Hormone dysgenesis can also cause congenital hypothyroidism. Such disorders are inherited as autosomal recessive traits that cause impaired thyroid synthesis via production of defective enzymes. Secondary (pituitary) or tertiary (hypothalamic) congenital hypothyroidism is rare, occurring in about 1 in 25,000 to 1 in 60,000 births.[1,6]

Congenital hypothyroidism may result from the ingestion of goitrogens, such as iodine and methimazole, taken for hyperthyroidism by the mother during pregnancy. However, research findings suggest that the benefit of methimazole outweighs any deleterious effect of the drug taken during pregnancy.[6] The maturation of the fetal thyroid can be suppressed by the passage across the placenta of TSH receptor antibodies, thionamides taken for hyperthyroidism, or iodides taken for hyperthyroidism or pulmonary disorders.[6]

Numerous clinical manifestations have been associated with congenital hypothyroidism, also known as *cretinism*. Among these are puffy face; open mouth with enlarged, protruding tongue; hoarse cry; short, thick neck; narrow forehead; pug nose; short legs; distended abdomen; dry, mottled skin; yellowish skin discoloration; hirsutism; and lethargy. Skeletal and mental development is retarded (with IQ typically below 85) if treatment is not begun in early infancy.[1,6,53]

Since most cases of congenital hypothyroidism are not suspected clinically, it is necessary to screen newborns for early diagnosis. This is now a requirement in all 50 states. Either cord blood or capillary blood, usually collected as a filter-paper spot and eluted from the paper prior to the assay, is used. In the United States and Canada, TT_4 is used as the primary screen for neonatal hypothyroidism with TSH the confirmatory test; in other parts of the world TSH is the primary screening test. Utilizing the TSH approach, subclinical cases of hypothyroidism (with less interference from nonthyroidal illness or drugs) are better detected; however, with the TT_4 approach, infants with TBG deficiency or with hypothalamic-pituitary hypothyroidism can be identified. Screening with both tests would probably be the best approach when methods for simultaneous measurement of both hormones on filter paper spot blood become available.[1,53]

Acquired Hypothyroidism

Most cases of *acquired hypothyroidism* result from inadequate secretion of the thyroid hormones from a damaged thyroid gland. There are several causes associated with acquired hypothyroidism: (1) chronic thyroiditis; (2) surgical or radioactive iodine treatment of hyperthyroidism, goiter, or cancer; (3) idiopathic atrophy; and (4) metastatic cancer and other infiltrative disorders.

Chronic Autoimmune Thyroiditis (Hashimoto's Disease). With or without goiter, chronic autoimmune thyroiditis, or Hashimoto's disease, is the most common cause of primary hypothyroidism. The peak incidence of Hashimoto's thyroiditis occurs during the sixth decade of life. It occurs in approximately 2% of the population.[41,43] Patients with Hashimoto's disease are usually euthyroid or hypothyroid. In Hashimoto's disease, there is a massive diffuse infiltration of the thyroid by lymphocytes. In addition, plasma cells are abundant, and there is an increased amount of connective tissue. The condition is often an asymptomatic disease caused by an unidentified abnormality in the immune system. This abnormality appears to be genetically determined, since family members of persons with the condition have a high probability of developing the disease. Furthermore, the disease has been associated with certain histocompatibility antigens.[43] The immune system abnormality leads to the production of several thyroid autoantibodies and lymphocytes sensitized to thyroid antigens and may lead to the destruction of the thyroid tissue. Antimicrosomal or antithyroglobulin antibodies can be detected in about 90% of patients with goitrous Hashimoto's thyroiditis.[41]

Miscellaneous Causes of Hypothyroidism

About one-third of the patients receiving treatment for hypothyroidism have had thyroid surgery or radioactive iodine therapy for hyperthyroidism. Graves' disease, the major cause of thyrotoxicosis, may spontaneously terminate in hypothyroidism, presumably as the result of autoimmune thyroid damage. The incidence of hypothyroidism increases with aging, as do the levels of circulating thyroid autoantibodies.[17] However, symptoms of hypothyroidism present more subtly in the elderly.

Hypothyroidism also may result from a lack of TSH or TRH. In secondary hypothyroidism, the patient usually has a pituitary disorder. A decreased production of TSH may result from pituitary irradiation, intracellular hemorrhage, or Sheehan's syndrome (postpartum pituitary infarction). Tertiary hypothyroidism or hypothalamic failure may result because of tumors, vascular insufficiency, infections, infiltrative processes, or trauma. TRH response should be normal in patients with tertiary hypothyroidism.

Laboratory Evaluation of Hypothyroidism

An algorithm for laboratory testing in the diagnosis of hypothyroidism in ambulatory patients is given in Figure 20-7.[24] The earliest laboratory abnormality noted in primary hypothyroidism is an increased TSH concentration. This may occur even before the patient is symptomatic, as in the case of subclinical hypothyroidism where T_4, T_3, and

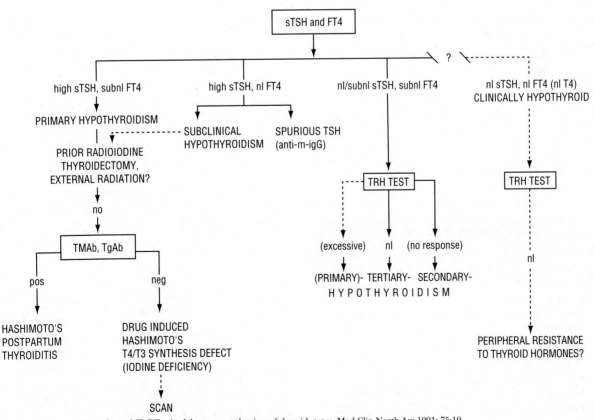

Source: Bayer MF. Effective laboratory evaluation of thyroid status. Med Clin North Am 1991; 75:19.

▲

Figure 20-7. Algorithm for the diagnosis of hypothyroidism in ambulatory patients.

FT_4 levels may be normal. Subclinical hypothyroidism can progress to thyroid gland failure and has been linked to increased levels of LDL cholesterol[30] and coronary heart disease.[14] Furthermore, TSH is probably the best indicator of the possibility of primary hypothyroidism coexisting with nonthyroidal illness.[14,27,34] An elevation of basal TSH, therefore, can be considered the single most sensitive and specific marker for primary hypothyroidism.

As hypothyroidism progresses, the FT_4I and FT_4 levels will decrease. TT_4, TT_3, and rT_3 will also be decreased. Thyroid antimicrosomal and antithyroglobulin antibodies can be measured to confirm or rule out autoimmune diseases. Secondary and tertiary hypothyroidism may present with clinical hypothyroidism and/or decreased s-TSH TT_4 and FT_4 levels. Secondary versus tertiary hypothyroidism may be differentiated by the TRH test.[18,23,26,31,35] A summary of laboratory findings in these conditions can be found in Table 20-3.

If patients are sick, elderly, hospitalized, or on certain medications (corticosteroids, salicylate, phenytoin, dopamine, lithium), laboratory results may be falsely decreased.[2,9,14,48] Although unsuspected thyroid disease occurs in less than 1% of hospitalized patients, thyroid function tests may be abnormal in as many as 70% of these patients.[9,31] The euthyroid sick syndrome is the most common clinical condition responsible for inaccurate test interpretation.[14] Such abnormalities correspond with changes in thyroid hormone secretion, distribution, and metabolism.[14,48] In general, the greater the severity of the non-thyroidal abnormality, the more disparate the laboratory results.

TSH levels may also rise in patients who are recovering from severe illness and in newborns. This is a transient situation, and the TSH generally returns to normal within a few days or weeks.[6,21,31,47] In certain cases of subclinical hypothyroidism, there may be minimal change in FT_4 and TSH, thus necessitating the utilization of the TRH test.[23,55]

Administration of dopamine or high-dose glucocorticoids will suppress the TSH concentrations toward or into the normal range, causing subsequent decreased TT4 and TT3 levels.[9,34,48] Furthermore, if thyroid hormone inhibitors/ antibodies are in circulation, they may also complicate the findings of the profile.[39,55]

Treatment of Hypothyroidism

The treatment of hypothyroidism, except that caused by iodine deficiency, is the administration of L-thyroxine. Proper treatment of subclinical hypothyroidism has been shown to reverse lipid imbalances and improve symptom scoring and psychometric testing.[13,16,30,40] Treatment, however, must be individualized. Cautions regarding treatment of the newborn and children must be taken to avoid overtreatment, which can lead to dangers in heart, bone, and neurological development,[25] and in the elderly, where coronary disease may already exist.[17,23] Because TSH levels may adjust slowly to treatment, FT_4 should be used initially in monitoring recovery from hypothyroidism. Once the patient has reached a near euthyroid status based on clinical assessment and FT_4 levels, continued surveillance of thyroid status should be through the s-TSH assay.[2,4,27,31,35]

Hyperthyroidism

Thyrotoxicosis

Hyperthyroidism is a disorder manifested by excessive circulating levels of thyroid hormones. The term *thyrotoxicosis* is applied to a group of syndromes caused by high levels of free thyroid hormones in the circulation. Thyrotoxicosis means only that the patient is suffering the metabolic consequences of excessive quantities of thyroid hormones. Therapeutic decisions can be made only after the etiology of the thyrotoxicosis has been determined. The symptoms found in thyrotoxicosis are related to the catabolic-hypermetabolic effect or the increased metabolic activity in various tissues and increased sensitivity to catecholamines.[56]

The catabolic-hypermetabolic effect is responsible for the weight loss that occurs even though the patient eats well; loss of muscle mass; loss of fat stores, producing an increase in plasma fatty acids, a decrease in serum cholesterol, LDL, lipoprotein(a) apolipoprotein B,[16,21] and a tendency toward ketosis; a negative nitrogen balance; exercise intolerance; easy fatigability; and dyspnea on exertion related to the intercostal muscle weakness and/or increased oxygen utilization.

The catecholamine sensitivity results in the manifestation of nervousness, irritability, insomnia, fine tremor, excessive sweating, heat intolerance, flushed face, pruritus, tachycardia, palpitations, frequent bowel movements (although diarrhea is unusual), and hyperkinesis (owing to thyroid hormones' effect on the nervous system).

The skin is warm and moist with patchy depigmentation. There may be a recession of the nails from the nail beds, clubbing of the fingers and toes, and swelling of the subcutaneous tissues. Oligomenorrhea, amenorrhea, and decreased libido may occur in patients with thyrotoxicosis. Gynecomastia occurs in about 15% to 20% of male patients. This symptom may be related to the high serum estradiol-to-testosterone ratio.

TABLE 20-3. Summary of Findings in Thyroid Function Tests in Various Conditions

Condition	TT_4	FT_4I	T_3U	T_3	TSH
Euthyroidism	N	N	N	N	N
Hyperthyroidism	↑	↑	↑	↑	↓
T_3 Thyrotoxicosis	N	N	↑	↑	↓
Hypothyroidism					
Primary	↓	↓	↓	↓	↑
Secondary	↓	↓	↓	↓	↓
Tertiary	↓	↓	↓	↓	↓
Increased TBP	↑	N	↓	↑	N
Decreased TBP	↓	N	↑	↓	N
Competition for sites	↓	N	↑	↓	N

Approximately 20% of the patients have hypercalcemia and subsequent polyuria. This probably results from the increase in bone turnover. The hematologic manifestations include a decreased white blood cell count because of a decrease in neutrophils. Lymphocytosis is also present. The patient is anemic in spite of an increased red blood cell mass due to excess demand for oxygen.

Thyroid Storm

Approximately 1% of patients with thyrotoxicosis experience a life-threatening complication called a *thyrotoxic crisis*, or *thyroid storm*. Laboratory testing does not differentiate the thyroid storm from hyperthyroidism. Instead, the diagnosis depends on the clinical manifestations. In addition to the classic symptoms of severe thyrotoxicosis, further signs of unregulated hypermetabolism with fever and tachycardia appear. The patient usually has a temperature greater than 100°F and exhibits prominent cardiovascular symptoms—tachycardia, arrhythmias, and congestive heart failure. Neurologic changes are the most dramatic manifestations of thyroid storm. Symptoms include agitation, delirium, confusion, insensitivity to pain, and coma.[20]

The levels of thyroid hormones are not significantly changed during thyroid storm. Thyroid storm can occur in any patient with uncontrolled thyrotoxicosis who is subjected to stress. Infection is a common precipitating event. General anesthesia, surgery, therapeutic use of iodine-131, or withdrawal of antithyroid drugs also can precipitate the condition. Therapy is directed at reducing thyroid hormone action and treating the precipitating cause.[20]

Causes of Thyrotoxicosis

Causes of thyrotoxicosis can be divided into two subcategories based on results of the RAIU test (Table 20-4). Uptake is increased (or normal) in Graves' disease, toxic multinodular goiter, TSH-secreting tumors, trophoblastic tumors, and toxic adenoma. RAIU is suppressed in subacute thyroiditis, silent thyroiditis, chronic thyroiditis, struma ovarii, metastatic thyroid carcinoma, or as a result of excessive circulating iodine (*e.g.,* from x-ray dyes, medication, or absorption through the skin).[26,33,42,56] Thyrotoxicosis facitia exists when excess exogenous thyroid hormone is ingested.

This can occur when the hormone is used as part of a weight-reducing regimen, for control of the menstrual cycle, or in efforts to increase energy. Community outbreaks of thyrotoxicosis have been reported due to patients eating ground beef that was contaminated with bovine thyroid gland.[28]

Graves' Disease. The most common cause of thyrotoxicosis is Graves' disease (diffuse toxic goiter). Graves' disease occurs six times more commonly in women than in men. It occurs frequently at puberty, during pregnancy, at menopause, or following severe stress. The frequency of Graves' disease in the general population is about 0.4%.

The classic clinical features in Graves' disease consist of a triad: diffuse thyrotoxic goiter, exophthalmos (bulging eyes), and pretibial myxedema.[33] However, the full triad frequently does not develop. Exophthalmos is due to infiltration of the retro-orbital space with lymphocytes, mast cells, and mucopolysaccharides. Other characteristic clinical findings and complications are the same as those listed for thyrotoxicosis.

The pathogenesis of Graves' disease appears to be related to immunologic defects with genetic implications. It is believed to be a autoimmune disorder. TSH receptor autoantibodies and other abnormal immunoglobulins are present in the circulation of patients with this disease. Further evidence for Graves disease being an autoimmune disorder include the fact that it occurs more commonly in certain families, and the increased frequency of HLA-Dr3 haplotypes in patients with Graves' disease than in the general population.[33] Those who are positive (about 57%) never go into remission.[52] However, the relationship between HLA haplotypes, TSH receptor autoantibodies, and the pathogenesis of Graves' disease has recently been challenged.[12]

Graves' disease appears to be related to the production of TRAb that subsequently interacts with the TSH receptor of the thyroid follicular cell membrane. Iodine uptake, synthesis of thyroid hormones, and release of the thyroid hormones into the circulation are consequently stimulated. The production of the thyroid hormones is unrelated to the body's need. TSH production is inhibited.

Neonatal Graves' disease also exists and is almost always associated with maternal disease and TRAbs. The most serious complications of neonatal Graves' disease are premature closing of the fontanel, accelerated bone aging, persis-

TABLE 20-4. Causes of Thyrotoxicosis

Associated with Increased RAIU	Associated with Decreased RAIU
Graves' disease	Subacute thyroiditis
Toxic multinodular goiter	Chronic thyroiditis
Pituitary tumor	Silent thyroiditis
TSH secreting tumors	Struma ovarii
Trophoblastic tumors (hydatidiform mole, choriocarcinoma)	Metastatic thyroid carcinoma
Toxic adenoma	Thyrotoxicosis factitia
	Iodine excess

tent behavior problems, and mental retardation. Mortality in affected neonates can be as high as 10% to 16%.[6,56]

Graves' disease can be diagnosed by clinical manifestations, laboratory assays, and the measurement of antibody titers. The antibody assays may be of value in confirming that hyperthyroidism is caused by Graves' disease, in distinguishing an orbital tumor from Graves' disease with ophthalmopathy, and in following treatment using antithyroid drugs.[33]

Thyroiditis. Thyroiditis is the term used to describe an inflammation of the thyroid gland. The disorder has been classified as acute (suppurative), subacute (nonsuppurative), silent (painless) or nonspecific, lymphocytic or Hashimoto's and fibrous.[41]

Acute thyroiditis is caused by a bacteria, mycobacteria, fungi or parasites, and is characterized by abscess and suppuration. Symptoms include tenderness and swelling of the thyroid, fever, chills, and malaise. This form of thyroiditis is rare apparently because of the gland's inherent resistance to infection manifested by its complete encapsulation, rich blood supply, lymphatic drainage, and high iodide content.[43] TT_4 and T_3 levels are usually normal.

Subacute thyroiditis (SAT) hyperthyroidism presumably results from release of stored hormones from the damaged follicular cells. The etiology of subacute thyroiditis is probably viral in nature, and the gland is very tender. The patient presents with a high fever, inflammation, pain and an elevated erythrocyte sedimentation rate (ESR). Thyroid hormones are usually mild to moderately increased. SAT presents in three phases. Laboratory findings during the 1- to 2-month initial period indicate a mild thyrotoxicosis. During the second phase, lasting 1 to 2 weeks, the patient becomes euthyroid, whereas in the third phase, a transient hypothyroidism may occur that can last for 1 to 2 months until recovery.

In silent or painless thyroiditis, the ESR is normal. The etiology of painless thyroiditis is unclear but is believed to be autoimmune, and has been associated with certain HLA types. The disease is best differentiated from Graves' via RAIU. The clinical course runs through three phases, similar to subacute thyroiditis. Recovery is the rule, but some patients may have persistent goiter, antithyroid antibodies, or chronic thyroiditis.[41,43]

About 10% of the patients with Hashimoto's disease produce excessive secretion of thyroid hormone causing hyperthyroidism, termed *chronic lymphocytic thyroiditis*. These patients have the same clinical manifestations as those with Graves' disease. The hyperthyroid mechanism involved in Hashimoto's disease may be the production of TRAb, which, in turn, stimulates the TSH receptor. The patient may exhibit true exophthalmos that is indistinguishable from that found in Graves' disease.

Riedel's thyroiditis is a condition where sclerosing fibrosis eventually turns the thyroid into a woody or stony-hard mass, which often extends into neighboring structures such as the trachea and esophagus. Thyroid hormones are normal until extensive destruction of the gland has occurred and the patient becomes hypothyroid.[41,43]

Miscellaneous Causes of Hyperthyroidism. Other rare forms of hyperthyroidism exist. In secondary (pituitary) hyperthyroidism, the TSH will be elevated despite a high level of FT_4, due to the presence of a TSH-secreting pituitary adenoma. Excessive TSH secretion also has been found without evidence of a pituitary tumor. This condition may be caused by the hypersecretion of TRH (tertiary hyperthyroidism), loss of the inhibition of TSH secretion normally produced by excessive levels of thyroid hormones, or the presence of thyroid hormone resistance syndrome in the pituitary gland.[32]

Trophoblastic tumors may cause thyrotoxicosis. These tumors, such as hydatidiform mole, produce and secrete human chorionic gonadotropin (HCG). By virtue of the molecular structure (the alpha subunits being identical), HCG may demonstrate thyroid-stimulating ability. The HCG molecule is mistaken for TSH by the TSH receptor on the follicular cell.

In both the toxic multinodular goiter and toxic nodule, the primary defect appears to be the production of thyroid hormone without TSH stimulation. These conditions can be confirmed by thyroid scan. Thyroid nodules are normally found in up to 3% of the general population and are usually not associated with alterations in thyroid function.

Thyroid cancers occur in approximately 0.004% of the population. Patients with solitary, "cold" thyroid nodules on thyroid scan have about a 20% chance of having cancer.[10] Calcitonin levels are elevated in patients with medullary carcinoma of the thyroid. Additionally, although it also can be increased in a number of benign conditions, an elevated serum thyroglobulin level is associated with differentiated thyroid carcinoma. Metastatic thyroid carcinoma and struma ovarii are ectopic tissues that produce thyroid hormone. A struma ovarii is an ovarian teratoma that contains thyroid tissue which, on rare occasions, may cause hyperthyroidism. A radioiodine scan of the pelvis will demonstrate the ovarian tumor.

Laboratory Evaluation of Hyperthyroidism

An expanded algorithm for laboratory testing of ambulatory patients with hyperthyroidism is shown in Figure 20-8.[4] If serum TSH is decreased, the diagnosis is usually primary hyperthyroidism, provided the patient does not have a low TT_4, is not taking thyroid hormone, and is not seriously ill.[11] Other expected laboratory results include elevated TT_4, T_3U_4, FT_4, and FT_4I (see Table 20-3). The TSH level should be clearly subnormal (<0.1mU/L).[46,47,48]

In cases of subclinical hyperthyroidism, look for the s-TSH to be low as high-sensitivity TSH methods can discriminate between euthyroid and hyperthyroid patients. However, the TRH stimulation test of thyrotoxic patients can still be useful as a confirmatory test in these cases.[48] The condition of hyperthyroidism where FT_4I is normal, s-TSH is normal or suppressed, and FT_3I or FT_3 is elevated is termed T_3 *thyrotoxicosis*.[4,26]

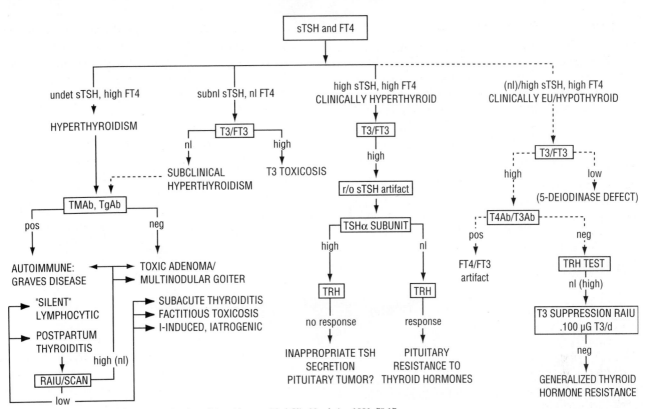

Source: Bayer MF. Effective laboratory evaluation of thyroid status. Med Clin North Am 1991; 75:17.

▲
Figure 20-8. Algorithm for the diagnosis of hyperthyroidism in ambulatory patients.

The diagnosis of hyperthyroidism becomes more complex with sick patients due to changes in thyroid binding proteins and the presence of circulating inhibitors.[47,48] The greater the severity of the disease, the more chance the thyroid hormone levels, including FT_4, will drop into the normal range. The key result in such cases is whether the TSH is below 0.1 μU/mL[47,48] Large doses of dopamine or corticosteroids may also result in decreased TSH levels, but rarely below 0.01 μU/mL.[14,44,47]

Normal TSH levels with elevated FT_4 is uncommon, but may occur in patients with secondary hyperthyroidism, familial dysalbuminemia, or thyroid hormone resistance syndrome. The TRH stimulation test may be useful in the diagnosis of these patients.[4,32]

Treatment of Hyperthyroidism

The therapy for hyperthyroidism depends on the cause, the severity, and the age and general health of the patient. The objective of treatment is to lower the thyroid hormones to normal and to minimize the symptoms. Treatment can be achieved by antithyroid medication, radioiodine ablation, or thyroidectomy.

Current therapy for Graves' disease continues to be directed at the thyroid gland itself. Ablation of the gland is accomplished either by surgery or the use of radioactive iodine. The treatment with radioactive iodine (^{131}I) is cur-

rently the most popular technique. However, radioiodine treatment is absolutely contraindicated during pregnancy.[6] A major long-term side effect of ^{131}I therapy is development of hypothyroidism. As noted earlier, Graves' disease may spontaneously revert to hypothyroidism due to thyroid damage.

Surgery is preferred in some cases of hyperthyroidism, as in adenomatous goiter, whereas in trophoblastic hyperthyroidism the stimulus causing the hypersecretion (*e.g.,* placenta) is removed. Exogenous or factitious hyperthyroidism is successfully treated when the patient eliminates or reduces the intake of thyroid hormones.

Drug therapy also may be employed as a treatment. Antithyroid drugs, such as propylthiouracil, methimazole, and carbimazole, have been used to inhibit: (1) the synthesis, (2) the release, or (3) the peripheral activation of thyroid hormones. However, a lasting remission with drug therapy may only be seen in 20% to 50% of patients with Graves disease.[18] Beta blockers, such as propranolol, may be useful as an adjunct therapy. However, beta blockers alone do not make the patient euthyroid and may precipitate thyroid storm.[33]

Patients that have been treated with antithyroid drugs, thyroidectomy, or radioactive iodine need monitoring tests to ensure euthyroidism. The response to treatment can be best evaluated by measuring FT_4, as the s-TSH levels tend to stay undetectable for extended periods.[2,4]

A 41-year-old woman complained of increased sweating over the previous 3 months. She also stated that she always seemed nervous and had heart palpitations and heat intolerance. She lost 12 pounds over the last 3 weeks and had noncramping diarrhea. On examination, her skin was warm and moist, she had a prominent stare, and she had tachycardia at 140 beats per minute. Her thyroid gland was diffusely enlarged with a suggestion of nodularity.

Laboratory Results

TT_4 = 21.2 μg/dL (273 nmol/L)
RT_3U ratio = 2.0

s–TSH = none detected
FT_4I = 42.4 ng/dL (546 pmol/L)

(Refer to Table 20-1 for reference ranges.)

Questions

1. What is the most probable cause of these symptoms and laboratory results?
2. What additional laboratory testing would be helpful in confirming the diagnosis?
3. What is the significance of the s–TSH test?
4. What causes this disorder?

A 34-year-old woman complained of fatigue, lethargy, constipation, nausea, amenorrhea, and anxiety. Her heart rate was increased. She had brought laboratory results with her from a community screening program that revealed several abnormal results, including increased glucose and alkaline phosphatase levels. The patient's CBC showed a decreased hemoglobin, hematocrit, and erythrocyte count. She stated that she had felt relatively healthy until about 3 months ago. She was not taking oral contraceptives.

Laboratory Results

TT_4 = 18.0 μg/dL (232 nmol/L)

T_3U ratio = 0.45
FT_4I = 8.10
s–TSH = 5.0

(Refer to Table 20-1 for reference ranges.)

Questions

1. From the case history and the laboratory data, what is a likely diagnosis?
2. What factors in the data lead you to the conclusions in question 1?
3. What additional laboratory test(s) would be appropriate?

A 59-year-old man who has been diagnosed with acute respiratory distress syndrome showed symptoms of hypothermia, coolness in the extremities, decreased blood pressure and pulse, fatigue, and delay in reflex time. A thyroid profile was ordered with the following results:

Laboratory Results

TT_4 = 3.6 μg/dL
T_3U ratio = 1.1
FT_4I = 4.0
s–TSH = 1.0 μU/mL

(Refer to Table 20-1 for reference ranges.)

Questions

1. From the case history and the laboratory data, what is a likely diagnosis?
2. What additional laboratory test(s) would be appropriate?
3. What additional information concerning the patient would be helpful in interpreting results? Why?
4. Is treatment for thyroid disorder indicated?

REFERENCES

1. American Academy of Pediatrics. Newborn screening for congenital hypothyroidism: Recommended guidelines. Pediatrics 1993;91:1203.

2. American Thyroid Association. Optimal use of blood tests for assessment of thyroid function. JAMA 1993;269:2736.

3. Baxter JD, Funder JW. Hormone receptors. N Engl J Med 1979;301:1149.

4. Bayer MF. Effective laboratory evaluation of thyroid status. Med Clin North Am 1991;75:1.

5. Beaman JM, Woodhead JS. A simplified strategy for testing thyroid function. Clin Chem 1989;35:828.

6. Becks GP, Burrow GN. Thyroid disease and pregnancy. Med Clin North Am 1991;75:121.

7. Blecher M. Antireceptor autoimmune disease, part 1. Diagn Med 1983;6:58.

8. Blonde L, Riddick FA. Answers to questions on hypothyroidism. Hosp Med 1984;20:13.

9. Cavalieri RR. The effects of nonthyroid disease and drugs on thyroid function tests. Med Clin North Am 1991;75:27.

10. Clark OH, Duh QY. Thyroid cancer. Med Clin North Am 1991;75:211.

11. Committee on Nomenclature of the American Thyroid Association. Revised nomenclature for tests of thyroid hormones and thyroid-related proteins in serum. J Clin Endocrinol Metab 1987;64:1089.

12. Davies TF. New thinking on the immunology of Graves' disease. Thyroid Today 1992;15:1.

13. DeBruin TWA, et al. Lipoprotein(a) and apolipoprotein B plasma concentrations in hypothyroid, euthyroid, and hyperthyroid subjects. J Clin Endocrinol Metab 1993;76:121.

14. Demers, LM. The influence of nonthyroidal factors on thyroid function. Lab Med 1994;24:495.

15. Desai RK, et al. Thyroid function in hospitalized patients: Effects of illness and serum albumin concentrations. Clin Chem 1987;33:1445.

16. Engler H, Riesen WF. Effect of thyroid function on concentrations of lipoprotein(a). Clin Chem 1993;39:2466.

17. Felicetta JV. Thyroid changes with aging: Significance and management. Geriatrics 1987;42:86.

18. Francis T, Wartofsky L. Common thyroid disorders in the elderly. Postgrad Med 1992;92:225.

19. Franklyn EG, et al. Limitations of a sensitive assay for thyrotropin in managing patients with thyroid disease. Clin Chem 1988;34:991.

20. Gavin LA. Thyroid crises. Med Clin North Am 1991;75:179.

21. Hamblin RS, et al. Relationship between thyrotropin and thyroxine changes during recovery from severe hypothyroxinemia of critical illness. J Clin Endocrinol Metab 1986;62:349.

22. Hamburger JI, Kaplan MM. Diagnosis of thyroid dysfunction in ambulatory patients: Primacy of the supersensitive thyroid-stimulating hormone assay. Compr Ther 1990;16:3.

23. Hay ID, Klee GG. Linking medical needs and performance goals: Clinical and laboratory perspectives on thyroid disease. Clin Chem 1993;39:1519.

24. Hay ID, et al. American thyroid association assessment of current free thyroid hormone and thyrotropin measurements and guidelines for future clinical assays. Clin Chem 1991;37:2002.

25. Helfand M, Schmitter J. Screening for thyroid dysfunction: Which test is best? JAMA 1993;270:2297.

26. Isley WL. Thyroid dysfunction in the severely ill and elderly. Postgrad Med 1993;94:111.

27. Kaye TB. Thyroid function tests application of newer methods. Postgrad Med 1993;94:81.

28. Kinney JS. Community outbreak of thyrotoxicosis: Epidemiology, immunogenetic characteristics, and long-term outcome. Am J Med 1988;84:10.

29. Klee GG, Hay ID. Assessment of sensitive thyrotropin assays for an expanded role in thyroid function testing: Proposed criteria for analytic performance and clinical utility. J Clin Endocrinol Metab 1987;64:461.

30. Levy EG. Thyroid disease in the elderly. Med Clin North Am 1991;75:151.

31. Martinez M, et al. Making sense of hypothyroidism. Postgrad Med 1993;93:135.

32. McDermott MT. Thyroid hormone resistance syndromes. Am J Med 1993;94:424.

33. McDougall R. Graves' disease: Current concepts. Med Clin North Am 1991;75:79.

34. Meek JC. Tests of thyroid function: Update in the diagnosis and management of thyroid disease. Comp Ther 1990;16:20.

35. Nicoloff JT, Spencer CA. The use and misuse of the sensitive thyrotropin assays. J Clin Endocrinol Metab 1990;71:553.

36. O'Brien T, et al. Hyperlipidemia in patients with primary and secondary hypothyroidism. Mayo Clin Proc 1993;68:860.

37. Pearce CJ, Byfield PGH. Thyroid hormone assays and thyroid function. Ann Clin Biochem 1986;4:230.

38. Penney MD, O'Sullivan DJ. Total or free thyroxin as a primary test of thyroid function. Clin Chem 1987;33:170.

39. Riter D. Endogenous antibodies that interfere with thyroxine fluorescence polarization assay but not with radioimmunoassay or EMIT. Clin Chem 1993;39:508.

40. Ross DS. Screening thyroid function tests in an acute-care hospital. Am J Med 1994;96:393.

41. Sakiyama R. Thyroiditis: A clinical review. Am Fam Physician 1993;48:615.

42. Sakiyama R. Common thyroid disorders. Am Fam Physician 1988;38(1):227.

43. Singer PA. Thyroiditis acute, subacute, and chronic. Med Clin North Am 1991;75:61.

44. Spencer CA, et al. Applications of a new chemiluminometric thyrotropin assay to subnormal measurement. J Clin Endocrinol Metab 1990;70:453.

45. Spencer CA, et al. Sensitive TSH tests: Specificity limitations for screening for thyroid disease in hospitalized patients. Clin Chem 1987;33:1391.

46. Spencer CA, et al. Thyrotropin secretion in thyrotoxic and thyroxine-treated patients. J Clin Endocrinol Metab 1986;63:349.

47. Surks MI. Guidelines for thyroid testing. Lab Med 1994;24:270.

48. Surks MI, et al. American thyroid association guidelines for use of laboratory tests in thyroid disorders. JAMA 1990;263:1529.

49. Surks MI, et al. Normal free thyroxine in critical nonthyroidal illnesses measured by ultrafiltration of undiluted serum and equilibrium dialysis. J Clin Endocrinol Metab 1988;67:1031.

50. Taylor CS, Brandt DR. Developments in thyroid-stimulating hormone testing: The pursuit of improved sensitivity. Lab Med 1994;24:337.

51. Utiger RD, et al. Thyrotropin measurements: Past, present, and future. Mayo Clin Proc 1988;63:1053.

52. Volpe R. Autoimmune thyroid disease. Hosp Pract 1984;19:141.

53. Willi SM, Moshang T. Diagnostic dilemmas: Results of screening tests for congenital hypothyroidism. Pediatr Clin North Am 1991;38:555.

54. Wong TK. Comparison of methods for measuring free thyroxin in nonthyroidal illness. Clin Chem 1992;38:720.

55. Yeomans AC. Assessment and management of hypothyroidism. Nurse Pract 1990;15:8.

56. Zimmerman D, Gan-Gaisano M. Hyperthyroidism in children and adolescents. Pediatr Clin North Am 1990;37:1273.

Chapter 21

Pancreatic Function

Eileen Carreiro-Lewandowski

Objectives

Upon completion of this chapter, the clinical laboratorian should be able to:

- *Discuss the physiologic role the pancreas plays during digestion.*
- *List the hormones excreted by the pancreas along with their physiologic roles.*
- *Describe the following pancreatic disorders and list the associated laboratory tests which would aid in their diagnosis: acute pancreatitis, chronic pancreatitis, pancreatic carcinoma, cystic fibrosis, and pancreatic malabsorption.*

KEY WORDS

Cholecystokinin
Islets of Langerhans
Pancreatitis
Secretin
Steatorrhea

PHYSIOLOGY OF PANCREATIC FUNCTION

As a digestive gland, the pancreas is only second in size to the liver and weighs about 70 to 105 g. It is located behind the peritoneal cavity across the upper abdomen at about the level of the first and second lumbar vertebrae and, thus, is an inch or two above the umbilicus. It is located in the curve made by the duodenum (Fig. 21-1). The pancreas is composed of two morphologically and functionally different tissues: endocrine tissue and exocrine tissue. The *endocrine* (hormone-releasing) component is by far the smaller of the two and consists of the *islets of Langerhans,* which are well-delineated spherical or ovoid clusters composed of at least four different cell types. The islet cells secrete at least four hormones into the blood: insulin, glucagon, gastrin, and somatostatin. The larger *exocrine* pancreatic component (enzyme secreting) daily secretes about 1.5 to 2.0 L of fluid that is rich in digestive enzymes into ducts that ultimately empty into the duodenum. This digestive fluid is produced by the pancreatic acinar cells (grapelike clusters), which line the pancreas and are connected by small ducts. These ducts empty into progressively larger ones, eventually forming one major pancreatic duct and a smaller accessory duct. The major pancreatic duct and the common bile duct open into the duodenum at the major duodenal papilla (Fig. 21-2). Normal protein-rich pancreatic fluid is clear, colorless, and watery, with an alkaline pH that can reach up to 8.3. This alkalinity is caused by the high concentration of sodium bicarbonate present in pancreatic fluid, which is used to eventually neutralize the hydrochloric acid in gastric fluid from the stomach as it enters the duodenum. The bicarbonate and chloride concentrations vary reciprocally so that they total about 150 mmol/L. Pancreatic fluid has about the same concentrations of potassium and sodium as does serum. The digestive enzymes or their proenzymes secreted by the pancreas are capable of digesting the three major classes of food substances (proteins, carbohydrates, and fats) and include: (1) the proteolytic enzymes trypsin, chymotrypsin, elastase, collagenase, leucine aminopeptidase, and some carboxypeptidases; (2) lipid-digesting enzymes, primarily lipase and lecithininase; (3) carbohydrate-splitting pancreatic amylase; and (4) several nucleases (ribonuclease), which separate the nitrogen-containing bases from their sugar-phosphate strands.

The control of pancreatic activity is under both nervous and endocrine control. Branches of the vagus nerve can cause a small amount of pancreatic fluid secretion when food is smelled or seen, and these secretions may increase as the bolus of food reaches the stomach. However, most of the pancreatic action is under the hormonal control of secretin and cholecystokinin (CCK—formerly called *pancreozymin.*) Secretin is responsible for the production of bicarbonate-rich and therefore alkaline pancreatic fluid, which protects

Michael L. Bishop, Janet L. Duben-Engelkirk, and Edward P. Fody.
CLINICAL CHEMISTRY. © 1996 Lippincott–Raven Publishers.

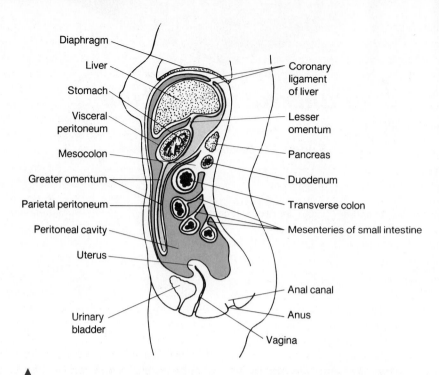

▲

Figure 21-1. Peritoneum and mesenteries. The parietal peritoneum lines the abdominal cavity, and the visceral peritoneum covers abdominal organs. Retroperitoneal organs are covered by the parietal peritoneum. The mesenteries are membranes that connect abdominal organs to each other and to the body wall. (From Akessom. Thompson's Core Textbook of Anatomy, 2nd edition. Philadelphia: JB Lippincott, 1990;115.)

the lining of the intestine from damage. Secretin is synthesized in response to the acidic contents of the stomach reaching the duodenum. It can also effect gastrin activity in the stomach. This pancreatic fluid contains few digestive enzymes. CCK, in the presence of fats and/or amino acids in the duodenum, is produced by the cells of the intestinal mucosa and is responsible for release of enzymes from the acinar cells by the pancreas into the pancreatic fluid.

DISEASES OF THE PANCREAS

The role of the pancreas in diabetes mellitus is discussed in Chapter 14, Carbohydrates, and will not be included here. Other than trauma, only three diseases cause well over 95% of the medical attention devoted to the pancreas. If they impact on the endocrine function of the pancreas, these diseases can result in altered digestion and nutrient metabolism:

1. Cystic fibrosis, also known by various other terms, such as *fibrocystic disease of the pancreas* and *mucoviscidosis,* is inherited as an autosomal recessive disorder and is characterized by dysfunction of mucous and exocrine glands throughout the body. The disease is relatively common and probably affects about 1 in 1600 live births. It has a variety of manifestations and can initially present in such widely varying ways as intestinal obstruction of the newborn, excessive pulmonary infections in childhood, or uncommonly, as pancreatogenous malabsorption in adults. The disease causes the small and large ducts as well as the acini to dilate

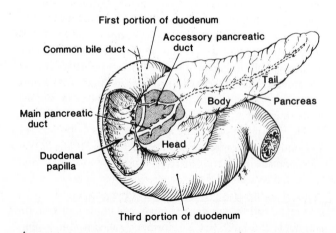

▲

Figure 21-2. Diagram of the pancreas and its relationship to the duodenum.

and convert into small cysts filled with mucous eventually resulting in the prevention of pancreatic secretions reaching the duodenum or, depending on the age of the person, a plug that blocks the lumen of the bowel leading to obstruction. As the disease progresses, there is increased destruction and fibrous scarring of the pancreas and a corresponding decrease in function.

2. Pancreatic carcinoma currently causes about 25,000 deaths each year in the United States, which represents about 5% of all deaths from malignant neoplasms, and is the fifth most frequent form of fatal cancer. The 5-year survival rate after surgery is less than 2%, and more than 90% of patients die within a year of diagnosis. Most pancreatic tumors arise as adenocarcinomas of the ductal epithelial. Since the pancreas has a rich supply of nerves, pain is a prominent feature of the disease. If the tumor arises in the body or tail of the pancreas, tumor detection often occurs during an advanced stage of the disease because of its central location and the associated vague symptoms. Cancer of the head of the pancreas usually is detected earlier because of its proximity to the common bile duct. These tumors make their presence known by jaundice, weight loss, anorexia, and nausea. The jaundice is associated with signs of posthepatic hyperbilirubinemia (intrahepatic cholestasis) and very low levels of fecal bilirubin resulting in clay-colored stools. However, findings are not specific for pancreatic tumors and other causes of obstruction must be ruled out.

Islet cell tumors of the pancreas effect the endocrine capability of the pancreas. If the tumor occurs in the beta cells, hyperinsulinism resulting in very low blood glucose levels followed by insulin shock are the associated findings. Pancreatic alpha cell tumors that overproduce gastrin are called gastrinomas or the Zollinger-Ellison syndrome, named after the individuals who first discovered it, and can also be duodenal in their origin. These tumors are associated with watery diarrhea, recurring peptic ulcer, and marked gastric hypersecretion and hyperacidity. Pancreatic alpha cell glucagon-secreting tumors are rare, and the hypersecretion of glucagon is associated with diabetes mellitus.

3. *Pancreatitis,* or inflammation of the pancreas, ultimately is caused by autodigestion of the pancreas as a result of reflux of bile or duodenal contents into the pancreatic duct. Pathologic changes can include: acute edema with large amounts of fluid accumulating in the retroperitoneal space and the associated decrease in effective circulating blood volume; cellular infiltration leading to necrosis of the acinar cells, with hemorrhage as a possible result of necrotic blood vessels; and intra- and extrahepatic pancreatic fat necrosis. Pancreatitis is classified generally as acute (no permanent damage to the pancreas), chronic (irreversible injury), or relapsing/recurrent pancreatitis, which can also be acute or chronic. It commonly occurs in mid-life. Painful episodes can occur intermittently, usually reaching a maximum within minutes or hours, lasting for several days or weeks, and frequently accompanied by nausea and vomiting. Pancreatitis is often associated with alcohol abuse or biliary tract disease, but patients with hyperlipoproteinemia types I, IV, and V, and those with hyperparathyroidism also have a significantly increased risk of contracting this disease. Other etiologic factors associated with acute pancreatitis include mumps, obstruction due to biliary tract disease, gallstones, pancreatic tumors, tissue injury, artherosclerotic disease, shock, pregnancy, hypercalcemia, hereditary pancreatitis, immunologic factors associated with postrenal transplantation, and hypersensitivity. Symptoms of acute pancreatitis include severe abdominal pain that is generalized or in the upper quadrants and often radiates toward the back or down the right or left flank. The etiology for chronic pancreatitis is similar to that of acute pancreatitis but chronic excessive alcohol consumption appears to be the most common predisposing factor. Laboratory findings include increased amylase, lipase, triglycerides, and hypercalcemia, which is often associated with underlying hyperparathyroidism. Hypocalcemia may be found and has been attributed to the sudden removal of large amounts of calcium from the extracellular fluid due to impaired mobilization or as a result of calcium fixation by fatty acids that are liberated as a result of the increased lipase action on triglycerides. Hypoproteinemia is attributable mainly to the notable loss of plasma into the retroperitoneal space. A shift of arterial blood flow from the inflamed pancreatic cells to less affected or normal cells causes oxygen deprivation and tissue hypoxia in the area of damage, including the surrounding organs and tissues.

All three of these conditions can result in severely diminished pancreatic exocrine function, which can significantly compromise digestion and/or absorption of ingested nutrients. This is the essence of the general malabsorption syndrome, which embodies abdominal bloating and discomfort, the frequent passage of bulky, malodorous feces, and weight loss. Failure to digest and/or absorb fats is known as *steatorrhea* and renders a greasy appearance to feces (more than 5 g of fecal fat per 24 hours). The malabsorption syndrome typically involves abnormal digestion or absorption of proteins, polysaccharides, carbohydrates, and other complex molecules, as well as lipids. Severely deranged absorption and metabolism of electrolytes, water, vitamins (particularly the fat-soluble vitamins A, D, E, and K), and minerals can also occur. Malabsorption can involve a single substance such as vitamin B_{12}, which results in a megaloblastic anemia (pernicious anemia) or lactose caused by a lactase deficiency. In addition to pancreatic exocrine deficiency, the malabsorption syndrome can be caused by biliary obstruction, which deprives the small intestines of the emulsifying effect of bile, and various diseases of the small intestine, which inhibit absorption of digested products.

TESTS OF PANCREATIC FUNCTION[4]

Depending on the etiology and clinical picture, pancreatic function may be suspect when there is evidence of increased amylase and lipase. The reader is referred to Chapter 10,

Enzymes, for a more in-depth discussion of these enzymes. Other laboratory tests of pancreatic function include those used for detection of malabsorption (*e.g.,* microscopic examination of stool for excess fat, starch, and meat fibers, D-xylose test, and fecal fat analysis), tests measuring other exocrine function (*e.g.,* secretin, CCK, fecal fat, trypsin, and chymotrypsin), tests assessing changes associated with extra-hepatic obstruction (*e.g.,* bilirubin), and endocrine-related tests (*e.g.,* gastrin, insulin, glucose, and cortisol) that reflect changes in the endocrine cells of the pancreas.

Direct evaluation of pancreatic fluid may include measurement of the total volume of pancreatic fluid, and the amount or concentration of bicarbonate and/or enzymes, all of which require pancreatic stimulation. Stimulation may be accomplished using a predescribed meal or administration of secretin, which allows for volume and bicarbonate evaluation, or secretin stimulation followed by CCK stimulation which adds enzymes to the pancreatic fluid evaluation. The advantage of these tests, both of which require intubation of the patient, are that the chemical and cytologic examination are performed on actual pancreatic secretions. Cytologic examination of the fluid can often establish the presence or at least the suspicion of malignant neoplasms, although precise localization of the primary organ of involvement (*i.e.,* pancreas, biliary system, ampulla of Vater, or duodenum) is not possible by duodenal aspiration. Because of advances in imaging techniques, these stimulation tests are used less often. None of the tests has proven especially useful in diagnosis of mild or acute pancreatic disease in which the acute phase has subsided. Most of the tests have found their clinical utility in excluding the pancreas from diagnosis. The sweat test used for screening cystic fibrosis is not specific for assessing pancreatic involvement, but when used along with the clinical picture at the time of testing, the sweat test can provide important diagnostic information. The following pancreatic function tests will be reviewed briefly: the Lundh meal, secretin test, fecal fat analysis, sweat chloride determinations, and amylase and lipase interpretation.

Lundh Test

The Lundh test is considered to be an indirect method to stimulate pancreatic secretion because a liquid meal of 5% protein, 6% fat, 15% carbohydrate, and 74% non-nutrient fiber is provided.[7] Lipase or trypsin activity is evaluated on the duodenal aspirate taken in 10 to 20 minute intervals over a 2-hour period. The advantage of this test is its relative simplicity and the fact that a physiologic stimulus is given. The major disadvantage is that abnormal results also occur when disease is present in the small bowel, liver, or biliary tree.

Secretin/CCK Test

The secretin/CCK test is a direct determination of the exocrine secretory capacity of the pancreas. The test in-

volves intubation of the duodenum without contamination by gastric fluid, which would neutralize any bicarbonate. The test is performed after a 6-hour or overnight fast. Pancreatic secretion is stimulated by intravenously administered secretin in a dose varying from 0.25 to 1 U/kg of body weight followed by CCK administration. If a simple secretin test is desired, the higher dose of secretin is given alone.

No one protocol has been uniformly established for the test. Pancreatic secretions are collected variously for 30, 60, or 80 minutes after administration of the stimulant(s), either as 10-minute specimens or as a single, pooled collection. The pH, secretory rate, enzyme activities (*e.g.,* trypsin, amylase, or lipase), and amount of bicarbonate are determined. The average amount of bicarbonate excreted per hour is about 15 mM for males and 12 mM per hour for females, with an average flow of 2.0 mL/kg body weight. Assessment of enzymes must be taken in view of the total volume output. Decreased pancreatic flow is associated with pancreatic obstruction and with an increase in enzyme concentrations. Low concentrations of bicarbonate and enzymes are associated with cystic fibrosis, chronic pancreatitis, pancreatic cysts, calcification, and edema of the pancreas.

Fecal Fat Analysis

Fecal lipids are derived from four sources: unabsorbed ingested lipids, lipids excreted into the intestines (predominantly in the bile), cells shed into the intestines, and metabolism of intestinal bacteria. Individuals on a lipid-free diet will still excrete 1 to 4 g of lipid in the feces in a 24-hour period. Even with a lipid-rich diet, the fecal fat will not normally exceed about 7 g in a 24-hour period. Normal fecal lipid is composed of about 60% fatty acids; 30% sterols, higher alcohols, and carotenoids; 10% triglycerides; and small amounts of cholesterol and phospholipids.

Although significantly increased fecal fat can be caused by biliary obstruction, severe steatorrhea is almost always associated with exocrine pancreatic insufficiency or disease of the small intestines.

Qualitative/Screening Test for Fecal Fat

Various screening tests have been devised for detecting steatorrhea. These tests have in common the use of fat-soluble stains (*e.g.,* Sudan III, Sudan IV, Oil Red 0, or Nile blue sulfate), which will dissolve in and color lipid droplets. Of greater importance than the particular technical procedure is the level of experience and dependability of the clinical laboratorian performing the test.

SUDAN III STAINING FOR FECAL FAT

Principle—Neutral fats (triglycerides) and many other lipids stain yellow-orange to red with Sudan III, because Sudan III is much more soluble in lipid than it is in water or ethanol. Free fatty acids do not stain appreciably unless the specimen is heated in the presence of the stain with 36% acetic acid. The slide may be examined warm or cool and the number of fat droplets assessed. As the slide cools, the fatty acids crystallize out in long colorless needle-like sheaves. Detection of meat fibers is accomplished by a third aliquot of fecal sample mixed on the slide with 10% alcohol and a solution of eosin stained for 3 minutes. The meat fibers should stain as rectangular cross-striated fibers. Splitting of the sample and detecting neutral fats, fatty acids, and undigested meat fibers can provide diagnostic information. Increases in fats and undigested meat fibers are indicative of patients with steatorrhea of pancreatic origin.

Specimen—Representative fecal specimen.

Reagents
1. Ethanol, 95% (v/v)
2. Sudan III stain—saturated solution in 95% ethanol
3. Acetic acid, 36%

Procedure
1. Mix the fecal specimen and place the equivalent of 2 drops on a microscope slide.
2. Add 2 drops of ethanol to the slide and emulsify the specimen.
3. Add 2 drops of Sudan III and mix.
4. Place a coverslip and examine the specimen under the high dry microscope objective for yellow-orange to red lipid droplets.
5. Place another aliquot of sample and Sudan III on a microscope slide and mix with several drops of acetic acid. This slide is them warmed over a flame until a slight boil occurs. The slide is then examined for droplets.
6. Report as normal, borderline, or increased stainable lipid.

Reference Range—Normal feces can have up to 40 or 50 small (1–5 μm) neutral lipid droplets per high-power microscope field. Steatorrhea is characterized by an increase in both the number and size of stainable droplets, often with some fat globules in the 50- to 100-μm range. Fatty acid assessment greater than 100 stained small droplets, along with the presence of meat fibers, is expected in patients with steatorrhea. ■

Quantitative Fecal Fat Analysis

The definitive test for steatorrhea is the quantitative fecal fat determination, usually on a 72-hour stool collection although the collection period may be increased to up to 5 days. There are two basic methods for the quantitation of fecal lipids. In the gravimetric method, fatty acid soaps (predominantly calcium and magnesium salts of fatty acids) are converted to free fatty acids, followed by extraction of most of the lipids into an organic solvent, which is then evaporated so that the lipid residue can be weighed. In titrimetric methods, lipids are saponified with hydroxide, and the fatty acid salts are converted to free fatty acids using acid. The free fatty acids, along with various unsaponified lipids, are then extracted with an organic solvent, and the fatty acids are titrated with hydroxide after evaporation of the solvent and redissolving the residue in ethanol. The titration methods obviously measure only saponifiable fatty acids and, consequently, render results about 20% lower than those from gravimetric methods. A further objection to the titrimetric methods is that they use an assumed average molecular weight for fatty acids to convert moles of fatty acids to grams of lipid.

At one time it was common to measure the amount of free fatty acids as a percentage of total lipids on the presumption that a high percentage of free fatty acids indicates adequate pancreatic lipase activity. This method is no longer considered reliable because of spurious results, particularly caused by lipase produced by intestinal bacteria.

It is essential that individuals be placed on a lipid-rich diet for at least 2 days prior to instituting the fecal collection. The diet must contain at least 50 g, and preferably 100 g, of lipid each day. Fecal collections should extend for 3 or more successive days.

There are various ways to express fecal lipid excretion. Expressing lipid excretion as a percentage of wet or dry fecal weight is open to serious challenge because of the wide variations in both fecal water content and dry residue as a result of dietary intake. The most widely accepted approach is to report the grams of fecal fat excreted in a 24-hour period.

GRAVIMETRIC METHOD OF SOBEL[3] FOR FECAL FAT DETERMINATION (MODIFIED)

Principle—The entire fecal specimen is emulsified with water. An aliquot is acidified to convert all fatty acid soaps to free fatty acids, which are then extracted along with other soluble lipids into petroleum ether and ethanol. After evaporation of the organic solvents, the lipid residue is weighed.

Specimen—All feces for a 3-day period are collected in tared containers. The containers *must not* have wax coating. Keep the specimen refrigerated.

Reagents
1. HCl, concentrated
2. Petroleum ether (*Caution*: Explosion hazard)
3. Ethanol (may be denatured)

Procedure
1. Weigh specimen and containers. Subtract tare weights to obtain total fecal weight in g (W_t). Divide by the number of collection days to determine average fecal weight/day (W_{avg}).
2. Transfer specimen to large blender bowl using a volume of water equal in milliliters to twice the total fecal weight in grams. (The assumption is made that 1 mL of water weighs 1 g.) Emulsify the specimen thoroughly.
3. Transfer approximately 5 mL of emulsified specimen to a tared 50-mL screw-cap glass tube and determine the weight in grams (W_g).
4. Add 4 drops of HCl and mix. Verify that the pH is less than 3.0. If not, add additional HCl.
5. Add 20 mL of petroleum ether and 5 mL of ethanol. (*Caution*: Explosion hazards in steps 5–9.) Cap the tube and shake for 10 minutes.
6. Centrifuge, carefully loosen cap to release pressure, and transfer the ether layer to a tared 100-mL beaker.
7. Extract the aliquot two more times (*i.e.*, repeat steps 5 and 6 twice).
8. Evaporate the ether extract in a hood, using a steam bath or sand bath. (Use no flame. Use only explosion-proof appliances.)
9. Place beaker with lipid residue in a vacuum desiccator containing anhydrous calcium chloride for at least 6 hours.
10. Weigh the beaker with the lipid residue and determine the lipid weight (W_L) to the nearest 0.001 g.

Calculations—Calculate the fecal lipid excretions grams per 24 h as follows:

$$\text{Fecal lipid (g/24h)} = \frac{(W_L)(W_t)(3)}{(W_g)(\text{days of collection})} \qquad \textbf{\textit{(Eq 21-1)}}$$

Report W_{avg} and fecal lipids as g/24 h.

Comments—Total lipid does not change significantly during 5 days' storage of the specimen at refrigerator temperatures. Patients must not ingest castor oil, mineral oil, or other oily laxatives and must not use rectal suppositories containing oil or lipid for 2 days prior to the test and during the test. ■

Reference Range

Fecal lipids (adult)   1–7 g/24 h.

Sweat Electrolyte Determinations

Measurement of the sodium and chloride concentration in sweat is the most useful test for the diagnosis of cystic fibrosis. Significantly elevated concentrations of both these ions occur in more than 99% of affected individuals. The two- to fivefold increases of sweat sodium and chloride are diagnostic of cystic fibrosis in children. Even in adults, no other condition will cause increases in sweat chloride and sodium above 80 mEq/L. Sweat potassium is also increased, but less significantly so, and is not generally relied on for diagnosis. Contrary to some assertions, sweat electrolyte determinations do not distinguish heterozygote carriers of cystic fibrosis from normal homozygotes.

Older methods for acquiring sweat specimens required skilled individuals who performed the test frequently. Induction of sweat included applying plastic bags or wrapping the patient in blankets, which was fraught with serious risks of dehydration, electrolyte disturbances, and hyperpyrexia. In 1959, pilocarpine administration by iontophoresis was reported as an efficient method for sweat collection and stimulation.[2,4] Iontophoresis employed an electric current which caused pilocarpine to migrate into a limited skin area, usually the inside of the forearm, toward the negative electrode from a moistened pad on the positive electrode. A collection vessel was then applied to the skin. The sweat was then analyzed for chloride. For confirmation, the test should be repeated. Commercially available surface electrodes that analyze the sweat chloride are readily available. For details, the reader is referred to Chapter 23, Body Fluids.

It is widely accepted that sweat chloride concentrations in children greater than 60 mmol/L are diagnostic of cystic fibrosis.[5] It is true that sweat sodium and chloride concentrations in females undergo fluctuation with the menstrual cycle and reach a peak 5 to 10 days prior to the onset of menstruation but do not overlap with the ranges associated with cystic fibrosis.

Serum Enzymes

Amylase is the serum enzyme most commonly relied on for detecting pancreatic disease. It is not, however, a function test. Amylase is particularly useful in the diagnosis of acute pancreatitis, in which significant increases in serum concentrations occur in about 75% of patients. Typically, amylase in serum increases within a few hours of the onset of the disease, reaches a peak in about 24 hours and, because of its clearance by the kidneys, returns to normal within 3 to 5 days, often making urine amylase a more sensitive indicator of acute pancreatitis. The magnitude of the enzyme elevation cannot be correlated with the severity of the disease.

Determination of the renal clearance of amylase is useful in detecting minor or intermittent increases in the serum concentration of this enzyme. To correct for diminished glomerular function, the most useful expression is the ratio of amylase clearance to creatinine clearance:

$$\frac{\text{\% Amylase clearance}}{\text{Creatinine}} = 100 \times \frac{\text{UA}}{\text{SA}} \times \frac{\text{SC}}{\text{UC}} \qquad \textbf{\textit{(Eq. 21-2)}}$$

where: UA = urine amylase, SA = serum amylase, SC = serum creatinine, and UC = urine creatinine.

Normal values are less than 3.1%. Significantly increased values, averaging about 8% or 9%, occur in acute pancreatitis.

The use of serum lipase in the clinical detection of pancreatic disease has been compromised in the past by technical problems inherent in the various analytic methods. Improved analytic methods appear to indicate that lipase increases in serum about as soon as amylase in acute pancreatitis and that increased levels persist somewhat longer than do those of amylase. Consequently, some physicians consider lipase more sensitive than amylase as an indicator of acute pancreatitis or other causes of pancreatic necrosis.

Both amylase and lipase may be significantly increased in serum in many other conditions (*e.g.,* opiate administration, pancreatic carcinoma, intestinal infarction, obstruction or perforation, and pancreatic trauma). Amylase levels are also frequently increased in mumps, cholecystitis, hepatitis, cirrhosis, ruptured ectopic pregnancy, and macroamylasemia. Electrophoresis of total amylase, if available, would reveal an increased p-type isoenzyme that, along with the lipase, remains elevated longer than the total amylase in acute pancreatitis. Lipase levels are often significantly increased in fractures of bones and in association with fat embolism.

Other Tests of Pancreatic Function

Table 21-1 summarizes several laboratory tests that might be helpful in diagnosis of pancreatic disorders. Other tests, which differentiate pancreatic from enteric malabsorption, must be performed and are discussed in detail in Chapter 22, Gastrointestinal Function. One such test is the D-xylose absorption test. D-xylose is a pentose sugar that does not require pancreatic enzymes for absorption. In a patient with a suspected malabsorption syndrome, a normal D-xylose points toward pancreatic insufficiency.

The starch tolerance test was devised to differentiate pancreatogenous from intestinal malabsorption. In theory, an individual with inadequate delivery of pancreatic amylase into the small intestines would have a much lower increase

TABLE 21-1. Laboratory Tests in Pancreatic Disorders*

Test	Comments
Acute Pancreatitis	
Glucose	Elevated >200 mg/dL seen in severe acute pancreatitis
Total protein and albumin	Low values due to ascites
Calcium	Must be elevated in terms of the albumin and if elevated or normal, PTH assessment should follow.
Electrolytes	Low chloride, high HCO_3 Hypochloremic alkalosis of vomiting
Lipid/triglycerides	Associated with types I, IV, and V
Bilirubin	
Amylase	Both serum and urine values may be helpful.
Lipase	
BUN, creatinine	Amylase: creatinine ratio may be valuable although not particularly sensitive for acute pancreatitis
Pancreatic Insufficiency	
Fecal fat assessment	Quantitative test is essential to documentation of steatorrhea
D-xylose	Normal
Secretin/CCK	Low bicarbonate and lipase/amylase
Sweat chloride	Chloride >60 mEq/L on two separate testing occasions considered indicative of cystic fibrosis especially in children.
Pancreatic Cancer	
Exocrine Tumors	
Alkaline phosphatase, bilirubin, and cholesterol	Nonspecific for pancreatitis but indicative of obstruction
PTT	Correction of an abnormal PTT with administration of vitamin K most likely indicates biliary obstruction
Secretin test	Low volume is consistent with ductal blockage by tumor
Iron/TIBC	Decreased due to bleeding
Tumor markers	Only detects metastatic cancer, CEA, pancreatic oncofetal antigen, and galatosyl-transferase II
Endocrine Tumors	
Gastrin	Detects gastrinoma, Zollinger-Ellison Syndrome
Glucose, Insulin	Insulinoma; Hypoglycemia
Tolbutamide test	Hypoglycemia with hyperinsulinemia is characteristic of an insulinoma
C-peptide glucagonoma	Insulinoma patients fail to suppress glucagon

*Patients who present with generalized abdominal pain must have the pancreas ruled out as a possible source. Alcoholic patients may also present with other associated changes due to the effects of that disease.

in blood glucose levels following ingestion of a purified starch preparation than would a person with normal amounts of amylase. The results are compared with those of a standard glucose tolerance test. Unfortunately, this reference approach is confused by the fact that the glucose tolerance test is frequently abnormally flat in intestinal malabsorption and also in more than half of patients with pancreatic insufficiency. Other problems include the fact that gelation of the starch upon cooling interferes with digestion and absorption. Also, starch digestion is initiated by salivary secretion, which continues in the intestines of individuals with gastric anacidity. Consequently, this test is rarely used.

Determining proteolytic enzyme activity in feces was formerly a common procedure for evaluating pancreatic exocrine function, particularly in the diagnosis of cystic fibrosis. One such procedure determines fecal proteolytic activity by its ability to digest gelatin on an x-ray film to produce clearing of the film. Unfortunately, the results are of limited reliability because many intestinal bacteria produce proteolytic enzymes, and bacteria also destroy pancreatic enzymes. More specific assays have been devised, but these have received limited acceptance.

Radiographic tests including chest and abdominal x-rays, ultrasound, duodenography, computerized tomography, endoscopy, angiography, and pancreatic biopsy are all essential tools to proper diagnosis of pancreatic disorders.[6]

SUMMARY

The pancreas is a digestive gland and weighs approximately 70 to 105 grams. It is composed of two morphologically and functionally different tissues: endocrine tissue and exocrine tissue. The endocrine component consists of the islets of Langerhans, which secrete at least four hormones into the blood: insulin, glucagon, gastrin, and somatostatin. The larger exocrine pancreatic component secretes digestive enzymes into ducts that ultimately empty into the duodenum. Control of pancreatic activity is under both nervous and endocrine control. Other than trauma, only three diseases cause well over 95% of the medical attention devoted to the pancreas: cystic fibrosis, pancreatic carcinoma, and pancreatitis. The role of the pancreas in diabetes mellitus is discussed in Chapter 14, Carbohydrates. Depending on the etiology and clinical picture, pancreatic function may be suspect when there is evidence of increased amylase and lipase. Tests of pancreatic function include those used for detection of malabsorption (*e.g.,* D-xylose, excess fat, meat fibers, and fecal fat), those tests for measuring other exocrine function (*e.g.,* secretin, CCK, trypsin, and chymotrypsin), tests assessing changes associated with extrahepatic obstruction (*e.g.,* bilirubin), and endocrine-related tests that reflect changes in the endocrine cells of the pancreas (*e.g.,* gastrin, insulin, glucose, and cortisol).

CASE STUDY
21-1

Parents brought their 7-year-old son to the pediatrician with the complaint of frequent fevers and failure to grow. The child had three bouts of pneumonia during the past 2 years and was bothered by chronic bronchitis, which caused him to cough up copious amounts of thick, yellow, mucoid sputum. Despite a big appetite, he had gained only 1 to 2 pounds in the past 2 years and was of short, frail stature. He especially liked salty foods. He usually had 3 or 4 bulky, foul-smelling bowel movements daily. A 9-year-old sister was in excellent health.

Questions

1. What is the most likely disease?
2. What clinical laboratory test would be most informative, and what results would be expected?
3. What other clinical laboratory tests would likely be abnormal?

CASE STUDY
21-2

A 38-year-old man entered the emergency room with the complaint of severe, boring, mid-abdominal pain of 6 hours duration. A friend, who had driven him to the hospital, stated that the patient fainted three times as he was being helped into the automobile. The patient had a 15-year history of alcoholism and drank 1 to 2 pints of whiskey every day. He had last been hospitalized for acute alcoholism 3 months ago, at which time he had relatively minor abnormalities of liver function. Upon admission this time, his blood pressure was 80/40, pulse 110 and thready, and respirations 24 and shallow. Clinical laboratory test results are shown in Case Study Table 21-1.

CASE STUDY TABLE 21-1. Applicable Laboratory Data

Serum amylase	640 units (3.5–260)
Serum sodium	133 mEq/L (135–145)
Potassium	3.4 mEq/L (3.8–5.5)
Calcium	4.0 mEq/L (4.5–5.5)
Blood urea nitrogen	32 mg/dL (8–22)
White blood cell count	16,500
Hemoglobin	12 g/dL

Questions

1. What is the probable disease?
2. What is the cause for the low serum calcium?
3. What is the cause for the increased blood urea nitrogen?

REFERENCES

1. Arvanitakis C, Cooke AR. Diagnostic tests of exocrine pancreatic function and Disease. Gastroenterology 1978;74:932.
2. Gibson LE, Cooke RE. A test for concentration of electrolytes in sweat in cystic fibrosis of the pancreas utilizing pilocarpine by iontophoresis. Pediatrics 1959;23:545.
3. Henry RJ. Clinical chemistry principles and technics. New York: Harper & Row, 1964;881.
4. Kao YS, Liu FJ. Laboratory diagnosis of gastrointestinal tract and exocrine pancreatic disorders. In: Henry JB, ed. Clinical diagnosis and management. 18th ed. Philadelphia: WB Saunders, 1991.
5. Shwachman H, Grand RJ. Cystic fibrosis. In: Sleisinger MH, Fordtran JS, eds. Gastrointestinal disease. 2nd ed. Vol. 2. Philadelphia: Saunders, 1978;1468.
6. Wormsley KG. Test of pancreatic function. Proc R Soc Med 1970;63:431.
7. Henderson RH, Tietz N, Rinker J. Gastric, pancreatic and intestinal function. In: Burtis CA, Ashwood ER, eds. Tietz textbook of clinical chemistry. 2nd ed. Philadelphia: WB Saunders, 1994.

SUGGESTED READINGS

Boat TF, Welsh MJ, Beaudet AL. Cystic fibrosis. In: Scriver CR, Beaudet AL, Sly WS, Valle D, eds. The metabolic basis of inherited disease. 6th ed. Vol 2. New York: McGraw-Hill, 1989;2649.

Meyer JH. Pancreatic physiology. In: Sleisinger MH, Fordtran JS, eds. Gastrointestinal disease. 4th ed. Vol 2. Philadelphia: Saunders, 1989;1777.

Soergel KH. Acute pancreatitis. In: Sleisinger MH, Fordtran JS, eds. Gastrointestinal disease. 4th ed. Vol 2. Philadelphia: Saunders, 1989;1814.

Gastrointestinal Function

Edward P. Fody

Objectives

Upon completion of this chapter, the clinical laboratorian should be able to:

- *Describe the physiology and biochemistry of gastric secretion.*
- *List the tests used to assess gastric function and intestinal function.*
- *Explain the clinical aspects of gastric analysis.*
- *Evaluate a patient's condition given clinical data.*

KEY WORDS

D-Xylose Absorption Test
Gastrin
Lactose Tolerance Test
Pepsin
Secretagogues
Zollinger-Ellison Syndrome

PHYSIOLOGY AND BIOCHEMISTRY OF GASTRIC SECRETION

Gastric secretion occurs in response to various stimuli:

1. Neurogenic impulses from the brain transmitted by means of the vagal nerves (*e.g.,* responses to the sight, smell, or anticipation of food).
2. Distension of the stomach with food or fluid.
3. Contact of protein breakdown products, termed *secretagogues,* with the gastric mucosa.
4. The hormone gastrin, which is the most potent stimulus to gastric secretion and is itself secreted by specialized G cells in the gastric mucosa and the duodenum in response to vagal stimulation and contact with secretagogues.

Inhibitory influences include high gastric acidity, which decreases the release of gastrin by the gastric G cells. Gastric inhibitory polypeptide is secreted by K cells in the mid- and distal duodenum and proximal jejunum in response to food products such as fats, glucose, and amino acids. Vasoactive intestinal polypeptide, produced by H cells in the intestinal mucosa, directly inhibits gastric secretion, gastrin release, and gastric motility.

Gastric fluid has a high content of hydrochloric acid, pepsin, and mucus. Hydrochloric acid is secreted against a hydrogen ion gradient as great as 1 million times the concentration in plasma (*i.e.,* gastric fluid can reach a pH of 1.2–1.3 under conditions of augmented or maximal stimulation). *Pepsin* refers to a group of relatively weak proteolytic enzymes with pH optima from about 1.6 to 3.6 that catalyze all native proteins except mucus. The most important component of gastric secretion in terms of body physiology is intrinsic factor, which greatly facilitates the absorption of vitamin B_{12} in the ileum.

CLINICAL ASPECTS OF GASTRIC ANALYSIS

Gastric analysis is used in clinical medicine mainly to

1. Detect anacidity, which will occur in some cases of advanced carcinoma of the stomach and in all cases of pernicious anemia in adults. In pernicious anemia, the pH of gastric fluid will not fall below 6.0, even with maximal stimulation.
2. Detect hypersecretion characteristic of the Zollinger-Ellison syndrome. This syndrome involves a gastrin-secreting neoplasm, usually located in the pancreatic islets, and exceptionally high plasma gastrin concentrations. Basal 1-hour acid secretion usually exceeds 10 mEq, and

the ratio of basal 1-hour to maximal secretion usually exceeds 60% (*i.e.,* the stomach is not really in the basal state but rather is pathologically stimulated by the high plasma gastrin level).

3. Determine how much acid is secreted as an aid in determining the type of surgical procedure required for ulcer treatment.

4. Verify the completeness of surgical vagotomy for ulcer treatment by determining whether insulin-induced hypoglycemia, which is ordinarily a powerful stimulant to gastric acid secretion by means of impulses transmitted from the brain through the vagal nerves, is in fact capable of causing a significant increase in gastric secretion.

Various substances have been used to stimulate gastric secretion (*e.g.,* caffeine, alcohol, and test meals), but these are submaximal stimuli and thus obsolete. From 1953 until the late 1970s, histamine acid phosphate was used as a maximal stimulus to gastric secretion. Because of side effects, some of them severe, histamine has now been replaced by pentagastrin, which is a synthetic pentapeptide composed of the four C-terminal amino acids of gastrin linked to a substituted alanine derivative. Hypoglycemia is used as a gastric stimulant only to test for completeness of vagotomy. A dose of insulin sufficient to cause a decrease in the level blood glucose to below 50 mg/dL is required.

Normal gastric fluid is translucent, pale gray, and slightly viscous and often has a faintly acrid odor. Residual volume should not exceed 75 mL. Residual specimens occasionally contain flecks of blood or are green, brown, or yellow from reflux of bile during the intubation procedure. The presence of food particles is abnormal and indicates obstruction.

TESTS OF GASTRIC FUNCTION

Measurement of Gastric Acid in Basal and Maximal Secretory Tests

Gastric analysis is usually performed, after an overnight fast, as a 1-hour basal test followed by a 1-hour stimulated test subsequent to pentagastrin administration (6 mg/kg subcutaneously). Test results reveal wide overlap among normal individuals and diseased patients except for anacidity (*e.g.,* in pernicious anemia) and in the extreme hypersecretion found in the Zollinger-Ellison syndrome. Gastric peptic ulcer is usually associated with normal secretory volume and acid output. Duodenal peptic ulcer is usually associated with increased secretory volume in both the basal and maximal secretory tests, but considerable overlap occurs, nevertheless, with the normal range.

BARON METHOD[1] OF MEASURING GASTRIC ACID

Principle—In stimulated-secretion specimens, the ability of the stomach to secrete against a hydrogen ion gradient is determined by measuring the pH. The total acid output in a timed interval is

determined from the titratable acidities and volumes of the component specimens.

Specimen—Following intubation, the residual secretion is aspirated and retained. Secretion for the subsequent 10 to 30 minutes is discarded to allow for adjustment of the patient to the intubation procedure. Specimens are ordinarily obtained as 15-minute collections for a period of 1 h.

The gastrin response to intravenous secretrin stimulation may be used to investigate further patients with mildly elevated serum gastrin levels. In this test, pure porcine secretin is injected intravenously and gastrin levels are collected at 5 minute intervals for the next 30 minutes. In patients with the Zollinger-Ellison syndrome, gastrin level will increase at least 100 pg/mL over the basal level. Patients with ordinary peptic ulceration, achlorhydria, or other conditions show a slight decrease in gastrin concentration.[3]

Reagents
1. Buffer standards for pH range 1–8
2. NaOH, 0.1 N

Procedure
1. For each 15-minute specimen, measure and record the volume (mL). For all specimens, including the initial collection of residual secretion, note any unusual appearance (color or consistency) or odor.
2. Determine the pH of each specimen, using an approximately calibrated glass electrode.
3. Titrate a 5- or 10-mL aliquot of each specimen with 0.1 N NaOH to pH 7.0 measured with a glass electrode.

Calculations—Calculate the titratable acidity and acid output of each specimen:

$$\text{Titratable acidity (mEq/L)} = \frac{(100)\ (\text{mL of NaOH})}{\text{mL of aliquot}}$$

$$\text{Acid output (mEq)} = \frac{(\text{mL volume})(\text{titratable acidity})}{1000}$$

(Eq. 22-1)

The volume, pH, and titratable acidity and the calculated acid output of each specimen are reported, as well as the total volume and acid output for each test period (sum of the component specimens). ■

Reference Ranges

There is considerable variation in gastric acid output among normal individuals in both the basal and maximal secretory tests. Nevertheless, in the basal test, most normal individuals will usually secrete 0 to 6 mEq of acid in a total volume of 10 to 100 mL. In the maximal 1-hour test, using either histamine or pentagastrin as the stimulus, most males will secrete from 1 to 40 mEq of acid in a total volume of 40 to 350 mL. Females and older persons will usually secrete somewhat less acid than will young males.

Plasma Gastrin

Measurement of plasma gastrin levels is invaluable in diagnosing the Zollinger-Ellison syndrome, in which fasting levels typically exceed 1000 pg/mL and can reach 400,000 pg/mL, compared with the normal range of 50 to 150 pg/mL. Gastrin is usually not increased in simple peptic ulcer disease. Increased plasma gastrin levels do occur in most pernicious anemia patients but will decrease toward normal when hydrochloric acid is artificially instilled into the stomach.

INTESTINAL PHYSIOLOGY

Digestion, predominantly a function of the small intestines, is the process whereby starches, proteins, lipids, nucleic acids, and other complex molecules are degraded to monosaccharides, amino acids and oligopeptides, fatty acids, purines, pyrimidines, and other simple constituents. For most large molecules, digestion is necessary for absorption to occur. Each day, the duodenum receives about 7 to 10 L of ingested water and food and secretion from the salivary glands, stomach, pancreas, and biliary tract. The materials then enter the jejunum and ileum, where another 1 to 1.5 L of secretion is added. Ultimately, however, only about 1.5 L of fluid material reaches the *cecum,* which is the first portion of the colon or large intestines. This considerable absorptive capability is possible because the 20 feet or so of small intestines have numerous mucosal folds, minute projections from the luminal

surface called *villi,* and microscopic projections on the mucosal cells called *microvilli,* all of which greatly increase the secretory and absorptive surface to an estimated 200 m². Absorption takes place by passive diffusion for some substances and by active transport for others. In addition, the small intestines actively secrete electrolytes and other metabolic products. The 5 feet or so of large intestines have two major functions: water resorption, whereby the 1.5 L of fluid received by the cecum is reduced to about 100 to 300 mL of feces, and storage of feces prior to defecation. The abdominal structures that constitute the alimentary tract are diagrammatically shown in Figure 22-1.

CLINICOPATHOLOGIC ASPECTS OF INTESTINAL FUNCTION

Clinical chemistry testing of intestinal function focuses almost entirely on the evaluation of absorption and its derangements in various disease states. As discussed in Chapter 21, Pancreatic Function, diseases of the exocrine pancreas and biliary tract also may cause malabsorption. Intestinal diseases that may cause the malabsorption syndrome

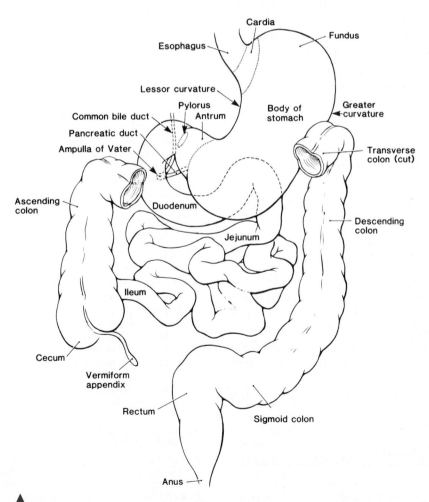

Figure 22-1. The abdominal structures of the alimentary tract.

are highly varied as to etiology, pathogenesis, and severity. These intestinal diseases/disorders include tropical and nontropical or celiac sprue, Whipple's disease, Crohn's disease, primary intestinal lymphoma, small intestinal resection, intestinal lymphangiectasia, ischemia, amyloidosis, giardiasis, and many others. In addition to the malabsorption syndrome, which ordinarily causes impaired absorption of fats, proteins, carbohydrates, and other substances, specific malabsorption states also occur (*e.g.,* acquired deficiency of lactase, which prevents normal absorption of lactose, and the genetic disorder Hartnup disease, which involves deficient intestinal transport of phenylalanine and leucine).

TESTS OF INTESTINAL FUNCTION

Lactose Tolerance Test

The disaccharidases, lactase (which cleaves lactose into glucose and galactose), and sucrase (which cleaves sucrose into glucose and fructose) are produced by the mucosal cells of the small intestine. Congenital deficiencies of these enzymes are rare, but acquired deficiencies of lactase are commonly found in adults. Affected individuals experience abdominal discomfort, cramps, and diarrhea after ingesting milk or milk products. About 10% to 20% of American Caucasians and 75% of African-Americans are affected. Although definitive diagnosis is established by an assay for lactase content in the intestinal mucosa, the lactose tolerance test is of some value in establishing the diagnosis. In one protocol, individuals drink 50 g of lactose in 200 mL of water. Blood specimens are obtained for glucose analysis prior to ingestion and 30, 60, and 120 minutes after ingestion. Increase in the level of blood glucose of 30 mg/dL is considered normal, a 20- to 30-mg/dL increase is borderline, and an increase of less than 20 mg/dL indicates lactase deficiency. However, individuals with normal lactase activity may still test as abnormals.

D-Xylose Absorption Test

D-Xylose is a pentose sugar that is ordinarily not present in the blood in any significant amount. As with other monosaccharides, pentose sugars are absorbed unaltered in the proximal small intestine and thus do not require the intervention of pancreatic lytic enzymes. Therefore, the ability to absorb D-xylose is of value in differentiating malabsorption of intestinal etiology from that of exocrine pancreatic insufficiency. Because only about half the orally administered D-xylose is metabolized or lost by action of intestinal bacteria, significant amounts are excreted unchanged in the urine. Some protocols have used the measurement of only the D-xylose excreted in the urine during the 5 hours following ingestion of a 25-g dose by a fasting adult (0.5 g/kg in a child). Even with normal renal function, false-positive and false-negative results occur frequently. Blood levels

measured at one or more times after ingestion of the D-xylose (*e.g.,* ½, 1, or 2 h) significantly improve the diagnostic reliability of the test. Some protocols use smaller doses of D-xylose in order to avoid the abdominal cramps, intestinal hypermotility, and osmotic diarrhea that frequently accompany the 25-g dose.

D-XYLOSE ANALYTIC METHOD OF ROE AND RICE[2,5]

Principle—Following ingestion of a specified solution of D-xylose, blood specimens are obtained and urine is collected for a 5-h period to determine the extent of absorption of D-xylose. The concentration of D-xylose is determined by heating protein-free supernates of urine and plasma to convert xylose to furfural, which is then reacted with *p*-bromoaniline to form a pink product, the absorbance of which is measured at 520 nm. Thiourea is added as an antioxidant to prevent the formation of interfering chromogens.

Specimens—Following an overnight fast, the patient voids and drinks a D-xylose solution: 25 g of D-xylose in 250 mL of water for adults or 0.5 g/kg for children or other dose as established. The patient drinks an equivalent amount of water during the next hour. No additional food or fluids are to be taken until the test is completed. Collect all urine for 5 h after the D-xylose ingestion. Collect the blood specimen in potassium oxalate at 2 h (commonly, 1 h is chosen for children).

Reagents
1. Stock D-xylose standard, 1 mg/mL. Dissolve 50 mg of D-xylose in, and dilute to 50 mL with, benzoic acid solution, 0.3 g/dL. Store in an amber bottle. Stable 6 months at room temperature.
2. Working D-xylose standard, 0.1 mg/mL. Dilute the stock standard tenfold with benzoic acid solution, 0.3 g/dL.
3. *p*-Bromoaniline solution, 2 g/dL in glacial acetic acid saturated with thiourea. To 4 g of thiourea, add 100 mL of glacial acetic acid. Shake and decant the supernate. Dissolve 1 g of *p*-bromoaniline in, and dilute to 50 mL with, the supernate. Prepare fresh for each use.
4. Zinc sulfate ($ZnSO_4 \cdot 7H_2O$) solution, 5 g/dL. Stable at room temperature in a plastic container.
5. Barium hydroxide solution, 0.15 *M*. Dissolve 23.7 g of $Ba(OH)_2 \cdot 8H_2O$ in approximately 450 mL of water. Boil for 5 minutes, cool, and dilute to 500 mL. Filter. Stable at room temperature.

Procedure
1. Measure and record the 5-h urine volume and dilute with water as follows: If the urine volume is less than 350 mL, dilute 25-fold; if the volume is greater than 350 mL, dilute 12.5-fold.
2. Deproteinize 1 mL of diluted urine or oxalated blood in a test tube as follows (mix 30 seconds on a vortex mixer after addition):
   a. Add 5 mL of water.
   b. Add 2 mL of barium hydroxide solution.
   c. Add 2 mL of zinc sulfate solution.
   d. Centrifuge and use the supernate for step 3.
3. Set up test tubes as shown in Table 22-1. Mix all tubes for 30 sec.
4. Incubate "test" samples at 70°C for 10 minutes. Leave "blank" samples at room temperature.
5. Cool test tubes to room temperature in a water bath and then place all tubes in the dark for 70 minutes.
6. Measure the absorbances of the solution at 520 nm against a water blank. *Calculations*—Calculate the amount of D-xylose excreted in the 5-h period:

TABLE 22-1. Test-Tube Setup for D-Xylose Absorption Test

| | | Deproteinized | | |
Sample	Standard	Blood	Urine	*P*-Bromoaniline
Standard blank (sb)	1.0 mL			5 mL
Standard test (st)	1.0 mL			5 mL
Blood blank (bb)		1.0 mL		5 mL
Blood test (bt)		1.0 mL		5 mL
Urine blank (ub)			1.0 mL	5 mL
Urine test (ut)			1.0 mL	5 mL

$$\text{D-xylose (g/5 h)} = \frac{(A_{ut} - A_{ub})(\text{dil})(0.1)(\text{urine vol})}{(A_{st} - A_{sb})(1000)} \qquad \textbf{(Eq. 22-2)}$$

where A_{ut}, A_{ub}, A_{st}, A_{sb} = the absorbances (see Table 21-1)
  dil = the dilution in step 1 $\times$ the dilution in step 2 ($\times$10) (thus a factor of either 250 or 125)
  0.1 = the standard concentration
  1000 = conversion of mg to g

Calculate the blood concentration:

$$\text{D-Xylose (mg/dL)} = \frac{(A_{bt} - A_{bb})(10)(0.1)(100)}{(A_{st} - A_{sb})} \qquad \textbf{(Eq. 22-3)}$$

where A_{bt}, A_{bb}, A_{st}, A_{sb} = the absorbances. 10, 0.1, and 100 are derived from the solution in step 2, the concentration of the standard, and the conversion of the standard to mg/dL

Comments—Normal blood concentrations of D-xylose in association with decreased urine excretion suggests impairment of renal function or incomplete urine collection. Aspirin therapy diminishes renal excretion of D-xylose, whereas indomethacin decreases intestinal absorption. ■

Reference Range

After ingestion of a 25-g dose of D-xylose, normal adults should excrete at least 4 g in the 5-hour period. For infants and children, the excretion following a dose of 0.5 g/kg for various ages expressed as percentages of ingested dose are shown in Table 22-2. Blood levels for normal adults vary over a wide range, but a blood concentration of less than 25 mg/dL at 2 hours should be considered abnormal following the 25-g dose. With the 0.5 g/kg dose, infants under 6 months should have a blood concentration of at least 15 mg/dL at 1 hour, infants over 6 months and children should achieve levels of at least 30 mg/dL.[2,4]

Serum Carotenoids

Carotenoids are various yellow to orange or purple pigments that are widely distributed in animal tissues, are synthesized by many plants, and impart a yellow color to some vegetables and fruits. The major carotenoids in human serum are lycopene, xanthophyll, and beta-carotene, the chief precursor of vitamin A in humans. Being fat-soluble, carotenoids are absorbed in the small intestine in association with lipids. Malabsorption of lipids typically results in a serum concentration of carotenoids lower than the reference range of 50 to 250 μg/dL. Starvation, dietary idiosyncrasies, and fever also cause diminished serum concentrations. The test does not distinguish among the various etiologies of malabsorption.

Fecal Fat Analysis

As discussed in Chapter 21, Pancreatic Function, increased fecal fat loss or steatorrhea is an integral manifestation of the general malabsorption syndrome. However, neither the presence of steatorrhea nor the documentation of its severity is of benefit in distinguishing among the various etiologies of malabsorption, with the exception that severe steatorrhea is rarely caused by biliary obstruction.

Other Tests of Intestinal Malabsorption

Deficiencies of numerous analytes can occur in association with intestinal malabsorption. Measurement of these analytes is usually of value not so much in confirming the diagnosis of malabsorption as in determining the extent of nutritional deficiency and thus the need for replacement therapy. Diminished appetite and dietary intake are usually more severe in patients with an intestinal etiology for their malabsorption. Body wasting or cachexia may be severe. Frequently, loss of albumin into the intestinal lumen and diminished dietary intake of protein accompany the diminished absorption of oligopeptides and amino acids so that a negative nitrogen balance occurs along with decreased serum total proteins and albumin. A serum albumin of less than 2.5 g/dL is much more characteristic of intestinal disease than of pancreatic disease. In association with severe disease of the small intestines,

TABLE 22-2. D-Xylose Results for Pediatric Patients[2,4]

Under 6 months	11 to 33%
6 to 12 months	20 to 32%
1 to 3 years	20 to 42%
3 to 10 years	25 to 45%
Over 10 years	25 to 50%

deficiencies of the fat-soluble vitamins A,D,E, and K occur. Vitamin K deficiency, in turn, causes deficiencies of the vitamin K–dependent coagulation factors II (prothrombin), VII (proconvertin), IX (plasma thromboplastin component), and X (Stuart-Prower factor), which are reflected in abnormal prothrombin time and partial thromboplastin time tests. In severe small intestinal disease, such as tropical or celiac sprue, malabsorption of folate and vitamin B_{12} can occur so that megaloblastic anemia is rather common and is of some benefit in distinguishing intestinal from pancreatic disease. Absorption of iron is usually diminished, and the tendency toward low serum iron levels may be aggravated by intestinal blood loss. Intestinal absorption of calcium is often diminished as a consequence of calcium binding by unabsorbed fatty acids and accompanying vitamin D deficiency and decreased serum magnesium. Because sodium, potassium, and water absorption and metabolism also may be seriously deranged, serum sodium and potassium levels are decreased, and dehydration occurs. Impaired absorption of carbohydrates in intestinal diseases such as sprue results in decreased to flat blood concentration curves in glucose, lactose, and sucrose tolerance tests.

CASE STUDY
22-1

A 34-year-old man was admitted for diagnostic evaluation with the complaint of epigastric pain, which was variously described as gnawing or burning, of 2 years' duration. He had been diagnosed as having a duodenal peptic ulcer 18 months ago, and at that time, therapy of antacids and dietary revision provided considerable alleviation of symptoms. More recently, however, the pain had become more persistent and awakened the patient four to six times each night. Radiologic studies have revealed a 2.5-cm ulcer crater in the first portion of the duodenum and a 0.5-cm ulcer in the antrum of the stomach. Serum electrolytes were normal. Hemoglobin was 8.3 g/dL with normal red blood cell indices. White blood count was 13,100. Gastric analysis revealed 640 mL of secretion in the basal hour, with an acid output of 38 mEq, and 780 mL of secretion in the 1-hour pentagastrin stimulation test, with an acid output of 48 mEq.

Questions

1. What is the probable disease?
2. In view of the existing data, what other test would be virtually diagnostic for this disease?
3. What is the explanation for the decreased hemoglobin and the increased white blood count?

CASE STUDY
22-2

A 26-year-old woman appeared in the outpatient clinic with the complaint of abdominal discomfort, diarrhea, and an 18-pound unintentional weight loss during the past 2 to 3 years. She related a similar period of 5 or 6 years of abdominal distress and diarrhea in childhood, but this essentially disappeared when she was about 12 to 13 years old. She was now having three to five bowel movements daily, which were described as bulky, malodorous, and floating. She weighed 106 pounds and was 67 inches tall. She had never had any surgical procedures. Physical examination revealed poor skin turgor, general pallor, and a protuberant abdomen. Abnormal clinical laboratory values included those in Case Study Table 22-1.

Fecal examination revealed no ova or parasites, and bacteriologic culture revealed no pathogens.

CASE STUDY TABLE 22-1. Laboratory Results

Hemoglobin	8.1 g/dL
Hematocrit	30
RBC count	$4.1 \times 10^6/\mu L$
Serum sodium	134 mEq/L
Potassium	3.4 mEq/L
Serum carotenoids	14 µg/dL
Fecal fat	22 g/24 h
D-xylose absorption test (25-g dose)	5-h excretion of 1.3 g and blood level at 2 h of 8 mg/dL
Prothrombin time	15.8 s (12–14 s)
Activated partial thromboplastin time	56 s (30–45 s)

Questions

1. What is the disease process?
2. What is the probable etiology in this case?
3. What is the cause of the abnormal coagulation tests?
4. What is the probable major cause for the anemia, and what are other possible contributing causes?

REFERENCES

1. Baron JH. Measurement and nomenclature of gastric acid. Gastroenterology 1961;45:118.
2. Bennett B, Cox R, Duby M, Hairston K, Pappas A, Schoen I, Fody EP. Clinical laboratory handbook for patient preparation and specimen handling. Fascicle VI-chemistry/clinical microscopy. Northfield: College of American Pathologists, 1991.
3. Fody EP, Bennett BD, Richardson LD, Duby MM, Schoen I. Clinical laboratory handbook for patient preparation and specimen handling: Fascicle V-endocrinology/metabolism. Northfield: College of American Pathologists, 1989.
4. Meites S. Pediatric clinical chemistry. 2nd ed. Washington: American Association for Clinical Chemistry, 1981;441.
5. Roe JH, Rice EW. A photometric method for the determination of free pentoses in animal tissue. J Biol Chem 1948;173:507.

SUGGESTED READINGS

Baron JH. Clinical tests of gastric secretion—History, methodology and interpretation. New York: Oxford University Press, 1979.
Cannon DC, Freeman JA. Gastric analysis. In: Freeman JA, Beeler MF, eds. Laboratory medicine/urinalysis and medical microscopy. 2nd ed. Philadelphia: Lea & Febiger, 1983.
Kalser MH. Clinical manifestations and evaluation of malabsorption. In: Berk JF, ed. Bockus gastroenterology. 4th ed. Vol 3. Philadelphia: Saunders, 1985;1667.

Part 4

Specialty Areas of Clinical Chemistry

Body Fluid Analysis

Frank A. Sedor

Objectives

Upon completion of this chapter, the clinical laboratorian should be able to:

- *Identify the source of the following: amniotic fluid, cerebrospinal fluid, sweat, synovial fluid, pleural fluid, pericardial fluid, and peritoneal fluid.*

- *Describe the physiologic purpose of the following: amniotic fluid, cerebrospinal fluid, sweat, synovial fluid, pleural fluid, pericardial fluid, and peritoneal fluid.*

- *Discuss the clinical utility of testing each of the following: amniotic fluid, cerebrospinal fluid, sweat, synovial fluid, pleural fluid, pericardial fluid, and peritoneal fluid.*

- *Given results for a foam stability index, L/S ratio, and sweat test, interpret the patient's status.*

- *Differentiate between a transudate and an exudate.*

KEY WORDS

Amniocentesis	Peritoneal Fluid
Amniotic Fluid	Pleural Fluid
Ascites	Respiratory
Effusion	Distress
Exudate	Syndrome
Hypoglycorrhacia	Rhinorrhea
L/S Ratio	Synovial Fluid
Otorrhea	Thoracentesis
Pericardial Fluid	Transudate

This chapter will attempt to acquaint the reader with several fluids that are often analyzed in the clinical chemistry laboratory. In general, the source, physiological purpose, and clinical utility of laboratory measurements for each of these body fluids will be emphasized.

AMNIOTIC FLUID

The amniotic sac provides an enclosed environment for fetal development. This sac is bilayered as the result of a fusion of the amnionic (inner) and chorionic (outer) membranes at an early stage of development. The fetus is suspended in amniotic fluid (AF) within the sac. This AF provides a cushioning medium for the fetus and serves as a matrix for influx and efflux of constituents.

Obviously, the mother must be the ultimate physiologic source for AF. Depending on the interval of the gestational period, the fluid may be derived from different sources. At initiation of pregnancy, some maternal secretion across the amnion contributes to the volume. Shortly after formation of the placenta, embryo, and fusion of membranes, AF is derived in large part by transudation across the fetal skin. In the last half of pregnancy, the skin becomes substantially less permeable, and fetal micturition, or urination, becomes the major volume source. The fate of the fluid also varies with period of gestation. A bi-directional exchange is presumed to occur across the membranes and at the placenta. Similarly, during early pregnancy, the fetal skin is involved. In the last half of pregnancy, the mechanism of fetal swallowing is the major fate of AF. There is a dynamic balance established between production and clearance; fetal urination and swallowing maintain this balance. The continual swallowing maintains intimate contact of the AF with the fetal gastrointestinal tract, buccal cavity, and bronchotracheal tree. This contact is evidenced by the sloughed material from the fetus that provides us with the "window" to fetal developmental and functional stages.

Cells found in the fluid originate with the fetus, and the chemical content reflects the continual swallowing and clearance of fluid. A sample of fluid is obtained by transabdominal amniocentesis (amniotic sac puncture), which is

Michael L. Bishop, Janet L. Duben-Engelkirk, and Edward P. Fody.
CLINICAL CHEMISTRY. © 1996 Lippincott–Raven Publishers.

performed under aseptic conditions. Before an attempt is made to obtain fluid, the positions of the placenta, fetus, and fluid pockets are visualized using ultrasonography. Aspiration of anything except fluid could lead to erroneous conclusions as well as possible harm to the fetus.

Amniocentesis and subsequent AF analysis is performed to test for: (1) congenital diseases, (2) neural tube defects, (3) hemolytic disease, (4) gestational age, and (5) fetal pulmonary development. The first, diagnosis of genetic abnormality, is accomplished by cell culture. Fluid obtained between 14 and 20 weeks of pregnancy is harvested for cells of fetal origin. The cells are then cultured and collected for chromosomal analysis are lysed, then enzyme contents are determined to evaluate for metabolic defects. This has been largely supplanted in recent years by the use of chorionic villus sampling and cytogenetic analysis.

Screening for neural tube defects (NTDs) is initially performed using maternal serum. The presence of elevated α-fetoprotein (AFP) was originally thought to indicate NTDs such as spina bifida and anencephaly. It is also found that elevated maternal serum AFP could be closely correlated with abdominal hernias into the umbilical cord, cystic hygroma, and poor pregnancy outcome. Low maternal serum AFP was associated with an increased incidence of Down's syndrome and other aneuploidies. The protocol for usage of AFP testing is generally considered to include: (1) maternal serum AFP; (2) repeat, if positive; (3) diagnostic ultrasound; and (4) amniocentesis for confirmation. Interpretation of maternal serum AFP testing is complex, being a function of age, race, weight, gestational age, and level of nutrition.

Testing of amniotic fluid AFP (AFAFP) is the confirmatory procedure. AFP is a product of first the fetal yolk sac and then the fetal liver. It is released into the fetal circulation and presumably enters the amniotic fluid by transudation. Entry into the maternal circulation could be by placenta crossover or from the amniotic fluid. If there were an open defect, *e.g.,* spina bifida, that caused an increase in AFAFP, there would be a concomitant increase in maternal serum AFP. Under normal conditions, AFAFP would be cleared by fetal swallowing and metabolism. An increased presence overloads this mechanism, causing the AFAFP elevation. Assay of the protein is normally done by immunologic means. The lack of treatment for a positive presence and the lack of 100% specificity increase the need for extreme care in analysis. Societal and clinical concerns mandate that the highest levels of quality control be practiced.

This very concern generated the need for a second test to affirm NTDs and abdominal wall defects. The method used is the assay for a CNS-specific acetylcholinesterase (AChE). The NTD allows direct or, at least less difficult, passage of AChE into the AF. Analysis for CNS-specific AChE in the AF then offers a degree of confirmation for AFAFP. The methods used for CNS AChE include enzymatic, immunologic, and electrophoretic with inhibition. The latter includes the use of acetylthiocholine as substrate and BW284C51, a specific CNS inhibitor, to differentiate the serum pseudocholinesterase from the CNS-specific AChE.

Analysis of AF to screen for hemolytic disease of the newborn (erythroblastosis fetalis) was the first recognized laboratory procedure performed on AF. Hemolytic disease of the newborn (HD) is a syndrome of the fetus resulting from ABO incompatibility of the maternal and fetal blood. Maternal antibodies to fetal erythrocytes cause a hemolytic reaction that can vary in severity. The resultant hemoglobin breakdown products, predominantly bilirubin, appear in the AF and provide a measure of the severity of the incompatibility reaction.

The most commonly employed method is a direct spectrophotometric scan of undiluted AF and subsequent calculation of the relative bilirubin amount. Classically, the absorbance due to bilirubin is reported instead of a concentration of bilirubin. The method consisted of scanning AF from 700 to 350 nm against a water blank. The resultant absorbances can be used differently to derive the necessary information. A common method, the method of Liley,[7] requires the plotting of the observations at 5-nm intervals against wavelength, using semilogarithmic paper. A baseline is constructed from 550 to 350 nm; the change at 450 nm is due to bilirubin.

Care must be used in the interpretation of the spectra. A decision for treatment can be made based on the degree of hemolysis and gestational age. The rather limited treatment options are immediate delivery, intrauterine transfusion, or observation. The transfusion can be accomplished by means of the umbilical artery and titrated to desired hematocrit. Several algorithms have been proposed to aid in decision making (Fig. 23-1). An example of an uncomplicated bilirubin scan is shown in Figure 23-2.

The most commonly encountered interferences are maternal urine (from bladder interdiction), fetal or maternal blood, and meconium (fetal fecal material). Although light is not an interferent, *per se,* safeguards against exposure to light, especially sunlight, must be maintained prior to analysis. Light may degrade the bilirubin present, thereby causing an underestimation of HD severity.

Examples of some interferences compared to a normal specimen are given in Figure 23-3. Each laboratory should compile its own catalog of real examples for spectrophotometric analysis. Presence of blood is identified by Soret absorbances of hemoglobin at 410 to 415 nm; urine by the broad curve and confirmed by creatinine, urea, and protein analyses; and meconium by the distinctly greenish color and flat absorbance curve.

In management of pregnancy, it is important to know the gestational age. Laboratory results are often used in age evaluation. The elucidation of an analyte to reflect gestational age has resulted in the cataloging of as many constituents as are known in serum. Four parameters have been used with some degree of frequency: creatinine, urea, uric acid, and osmolality. Creatinine is thought to reflect fetal muscle mass; urea, protein; uric acid, nucleoproteins; and osmolality, a combination of all these. The basic problem in the use of these parameters is the very wide range of concentration for given

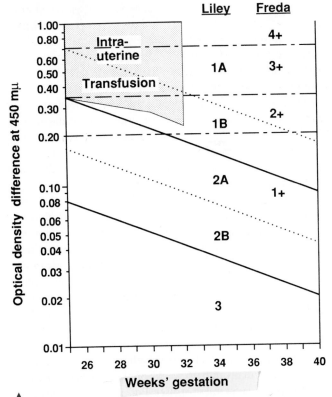

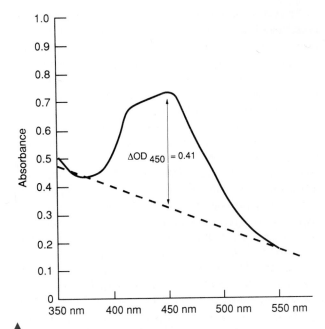

Figure 23-1. Assessment of fetal prognosis by the methods of Liley and Freda. Liley: *1A,* above broken line, condition desperate, immediate delivery or transfusion; *1B,* between broken and continuous lines, hemoglobin less than 8 g/100 mL, delivery or transfusion (stippled area) urgent; *2A,* between continuous and broken lines, hemoglobin 8–10 g/100 mL, delivery 36–37 weeks; *2B,* between broken and continuous lines, hemoglobin 11–13.9 g/100 mL, delivery 37–39 weeks; *3,* below continuous line, not anemic, delivery at term. Freda: *4+,* above upper horizontal line, fetal death imminent, immediate delivery or transfusion; *3+,* between upper and middle horizontal lines, fetus in jeopardy, death within 3 weeks, delivery or transfusion as soon as possible; *2+,* between middle and lower horizontal lines, fetal survival for at least 7–10 days, repeat test, possible indication for transfusion; *1+,* below lower horizontal line, fetus in no immediate danger of death. (Modified from Robertson JG. Am J Obstet Gynecol 1966;95:120.)

Figure 23-2. ΔOD_{450nm} from AF bilirubin scan.

is contemplated due to other risk factors in pregnancy, such as pre-eclampsia, premature rupture of membranes, etc. Risk factors to fetus or mother can be weighed against interventions as delay of delivery or at-risk post delivery therapies as exogenous surfactant therapy, high frequency ventilation, or extracorporal membrane oxygenation (ECMO).

gestational ages. The ranges are so wide that significant overlap occurs leading to inability to make accurate predictions. In practice, creatinine is the parameter used. A fluid with a level of 2.0 g/dL is assumed mature. Note that this is not effective in AF volume aberrations (*e.g.,* oligohydramnios or small AF volume). Ultrasonography has become the better tool to estimate fetal size and gestational age.

The prime reason for testing of amniotic fluid is the need to assess fetal pulmonary maturity. All the organ systems are at jeopardy from prematurity, but the state of the fetal lungs is a priority from the clinical perspective. The availability of laboratory tests that give an indication of maturity also has fostered this emphasis. Consequently, the laboratory is asked if sufficient specific phospholipids are reflected in the AF to prevent atelectasis (alveolar collapse) if the fetus were to be delivered. This question is important when preterm delivery

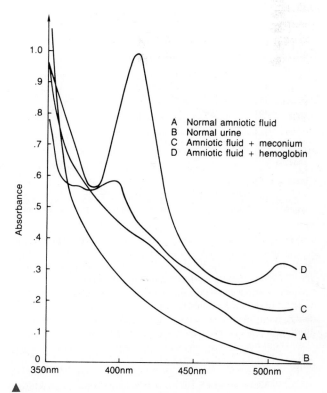

Figure 23-3. Amniotic fluid absorbance scans.

Alveolar collapse in the neonatal lung may occur upon the changeover to air as an oxygen source at birth if the proper quantity and type of phospholipid (surfactant) is not present. The ensuing condition, which may vary in degree of severity, is called *respiratory distress syndrome (RDS)*. It also has been referred to as *hyaline membrane disease* because of the hyaline membrane found in affected lungs. Lung maturation is a function of the differentiation, beginning near the 24th week of pregnancy, of alveolar epithelial cells into type I and type II cells. The type I cells become equipped for gas exchange, and the type II cells become the producers of surfactant. As the lungs mature, increases occur in the concentration of the phospholipids, particularly the compounds phosphatidyl glycerol and lecithin, especially dipalmitoylphosphatidyl choline (DPPC) (Fig. 23-4). It is these two compounds, present in 10% and 70%, respectively, of total phospholipid concentration, that appear most important as surfactants. Their presence in high enough levels acts in concert to allow contraction and re-expansion of the neonatal alveoli. To conceptualize their importance, remember the difficulty in blowing up a new toy balloon relative to a balloon that has been partially inflated. For the newborn, the normal amount of proper surfactant allows contraction of the alveoli without collapse. The next inspiration is the

difference between a partially inflated versus flattened new balloon. Insufficient surfactant allows alveoli to collapse, thus requiring a great deal of energy to re-expand the alveoli upon inspiration. Not only does this create an extreme energy demand on a newborn, but it probably also causes physical damage to the alveoli with each collapse. The damage may lead to "hyaline" deposition, or the newborn may not have the strength to continue inspiration at the energy cost. The end result of either can be fatal.

The approaches to assessing fetal pulmonary status may be divided into functional assays and biochemical assays. Functional assays provide a direct physical measure of AF in an attempt to assess surfactant ability to decrease surface tension. Examples of this group are surface tension, fluorescence polarization, and foam stability index. These tests reflect the gross concentration of surfactants rather than specific phospholipid levels. Belonging to the latter group are assays that quantitate dipalmitoyl phosphatidyl choline (the major lecithin), phosphatidyl glycerols, all or most of the phospholipids, and the classical ratio of lecithins to sphingomyelins (L/S ratio). Each group of tests has its advocacy and extensive citations in the literature. All attempt to indicate the major changes in phospholipid concentrations that occur around 36 weeks of pregnancy and indicate fetal pulmonary maturation (see Fig. 23-4). With all tests, centrifugation of the AF to remove debris is necessary.

Excessive force (anything greater than that needed to remove debris) can change the lipid profile by causing the lipids present to fractionate due to centrifugal force. The difference in ratios observed at 1000 versus 3000 g can radically alter clinical interpretation. Prior to adoption of any method for AF analysis, a protocol for centrifugal separation to include relative force (*not* revolutions per minute) and duration of centrifugation must be adopted and rigorously followed.

The foam stability index,[10] a variant of Clements' original "bubble test,"[2] appears acceptable as a rapid, inexpensive, informative assay. This qualitative, technique-dependent test requires only common equipment. The assay is based on the ability of surfactant to generate a surface tension lower than that of a 0.47 mole fraction ethanol-water solution. If sufficient surfactant is present, a stable ring of foam bubbles remains at the air-liquid interface. As surfactant increases (fetal lung maturity probability increases), a larger mole fraction of ethanol is required to overcome the surfactant-controlled surface tension. The highest mole fraction used while still maintaining a stable ring of bubbles at the air-liquid interface is reported as the FSI. The test is dependent on technique and can also be skewed by contamination of any kind in the AF, *e.g.,* blood, meconium, etc.

FOAM STABILITY INDEX[10]

Specimen—Amniotic fluid free of contamination by blood, vernix, or meconium.

Preparation—Centrifuge at 250 × g for 5 minutes and decant fluid to a second tube.

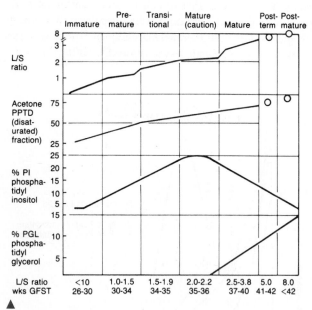

Figure 23-4. The form used to report the lung profile. The four determinations are plotted on the ordinate and the weeks of gestation on the abscissa (as well as the L/S ratio as an "internal standard"). When these are plotted, they fall with a very high frequency into a given grid that then identifies the stage of development of the lung as shown in the upper part of the form. The designation "mature (caution)" refers to the patients other than those with diabetes who can be delivered *if necessary* at this time; if the patient has diabetes, she can be delivered with safety when the values fall in the "mature" grid. (Kulovich MV, Hallman MB, Gluck L. The lung profile I. Normal pregnancy. Am J Obstet Gynecol 1979;135:57. Copyright © 1977 by the Regents of the University of California.)

Materials—95% ethanol, disposable 12 × 75 mM tubes with snap caps, and 1.00-mL graduated pipets.

Procedure—1. Into each 12 × 75 mM tube, pipet 0.50 mL of AF.

2. Titrate to the largest volume fraction of 95% ethanol required to sustain a ring of bubbles at the liquid-tube-air interface after 30 seconds of shaking, 15 seconds of standing (Table 23-1).

Interpretation—Most laboratories have found FSI of 0.47 or 0.48 to represent a borderline maturity status. It is imperative that laboratory-specific reference intervals be determined. Values greater than borderline indicate increasing probability of maturity. Examples of bubble patterns are shown in Figure 23-5. ■

The quantitative tests were given emphasis primarily by the work of Gluck.[4,2] The phospholipids of importance are phosphatidyl glycerol (PG), phosphatidyl choline (lecithin, PC), and sphingomyelin (SP). The relative amounts of PG and PC increase dramatically with pulmonary maturity, whereas the SP concentration is relatively constant. The increases of PG and PC correspond to larger amounts of surfactant being produced by the alveolar type II cells as the fetal lungs mature.

The classical technique for separation and evaluation of the lipids involves thin-layer chromatography (TLC) of an extract of the AF. The extraction procedure removes most interfering substances and results in a concentrated lipid solution. Current practices use either one- or two-dimensional TLC for identification. Described next is a one-dimensional method used without cold acetone precipitation. Laboratories with differing needs should consider a two-dimensional method as an alternative.

DETERMINATION OF L/S RATIO BY ONE-DIMENSIONAL CHROMATOGRAPHY[11]

Specimen—Amniotic fluid, free of contamination.

Preparation—Centrifuge at 250 × g for 5 minutes; decant the supernate.

Materials—Synthetic dipalmitoyl phosphatidyl choline (lecithin, L), glycerol (PG), serine (PS), inositol (PI), ethanolamine (PE), and bovine brain sphingomyelin are used as standards. Solvents must be reagent-grade. Silica gel H plates (20 × 20 cm) with 5% ammonium sulfate are used in glass TLC chambers. Evaluation is accomplished using a densitometer in the transmittance mode at 525 nm.

Procedure—3 mL of AF is extracted with chloroform-methanol (6 mL:3 mL). The chloroform layer is evaporated to just dryness at

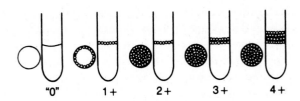

Figure 23-5. Grading system used in evaluating the FS-50 test. (Statland et al. Evaluation of a modified foam stability [FS-50] test. An assay performed on amniotic fluid to predict fetal pulmonary maturity. Reproduced with permission from the Am J Clin Pathol 1978;69:51.)

50°C under a gentle stream of nitrogen. Redissolve the evaporate in 50 μL of chloroform. Spot in duplicate on the activated plates. Flanking spots of combined standards in chloroform should be similarly applied. Allow to dry and then chromatograph in a solvent mixture of chloroform/methanol/water (67/25/3). When the solvent front is within 1 inch of the leading edge of the plate, remove, let dry, spray with 50% sulfuric acid–water and char. After cooling, scan the plate with the densitometer. The phospholipids are separated as shown in Figure 23-6. ■

The classical break point for judgment of maturity has been an L/S ratio of 2.0. The presence of PG is an additional marker. Initially, the presence of PG was presumed to indicate maturity, but later reports have modified this. An initial PG level of at least 3% is usually needed. This also may be method-dependent. There are always exceptions and outliers, however. It is important to remember that the clas-

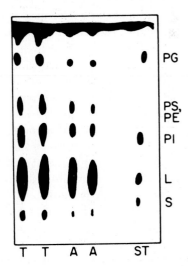

Figure 23-6. Thin-layer chromatogram of amniotic fluid phospholipids. Standard phospholipids (*ST*), total extract (*T*), and acetone-precipitable compounds (*A*) in amniotic fluid are shown. The phospholipid standards contained, per liter, 2 g each of lecithin and P1, 1 g each of PG and sphingomyelin, and 0.3 g each of PS and PE. 10 μL of the standard was spotted. (Tsai MY, Marshall JG. Phosphatidylglycerol in 261 samples of amniotic fluid. Clin Chem 25[5]:683. Copyright 1979 by the American Association for Clinical Chemistry.)

TABLE 23-1. FSI Determination			
	TUBE 1	**TUBE 2**	**TUBE 3**
Vol AF	0.50	0.50	0.50
Vol 95% EtOH	0.51	0.53	0.55
FSI	0.47	0.48	0.49

sical L/S ratio measures total lecithins versus sphingomyelin. Some methods include selective oxidation and removal of the nonsaturated phospholipids so that a direct measurement of dipalmitoylphosphatidyl choline can be measured. It is very difficult to compare these assays empirically.

There are still conflicting reports about decision levels for the phospholipid measurements when applied to certain pathologies, particularly diabetes. Earlier reports suggested that diabetes caused certain stresses that required an L/S ratio of greater than 2.0 and PG greater than 6% to be comparable with a nondiabetic situation. Whether the reports were reflecting the difficulty in management of diabetes or a real effect on fetal maturity is unknown. The most recent reports seem to agree that in diabetes the L/S ratio interpretation is unchanged but that the lecithin fraction may have a lesser proportion of the dipalmitoyl species than is usual.

Because the analysis of AF for an L/S ratio is so terribly technique-dependent, it is absolutely mandatory for the laboratory to thoroughly evaluate the methodology to be used. It is more important to relate that methodology to the particular clinical setting in developing interpretative intervals. It is not unusual to see a nominal L/S ratio maturity level of 2.0 ± 0.3 at one hospital with 2.2 ± 0.2 at another center using the same method. Close cooperation with the medical staff is necessary in defining the ranges to be employed.

Application of fluorescence polarization technique to AF provides another type of measurement. This procedure, as used on a commercially available polarimeter with complete reagent system (Abbott Laboratories), offers a rapidly performed test (<1h) on a versatile analyzer that is relatively technique free. Its use has become so widespread that it has been suggested as the initial test to be used in a testing cascade.[5] The method used is based on the partition of a synthetic lecithin-like fluorescent dye between phospholipid aggregates and albumin. Dye associated with the phospholipid aggregates decreases polarization; with albumin, polarization increases. Total polarization is compared to a set of lipid:albumin standards to yield a unitless value. As the fetal lung matures, the amount of surfactant (phospholipid) increases, and therefore the polarization is affected. The use of this system is widespread, in part, based on the accessibility to all laboratories, straight-forward analytical procedure, and consistency for result interpretation.

The predictive value is compromised by presence of blood or meconium (lipids) or fetal pathologies causing altered albumin levels such as urinary tract anomalies. Additionally, if the AF is centrifuged, not filtered, lipid level is altered and therefore polarization is decreased.

CEREBROSPINAL FLUID

Cerebrospinal fluid (CSF) is the liquid that occupies the spaces of the central nervous system (CNS). As such, it surrounds all facets of the brain and spinal cord. These spaces

are continuous and hence reflect all aspects of the CNS. CSF performs at least four major functions: (1) physical support and protection, (2) method for excretion, (3) provision of a controlled chemical environment, and (4) intra- and extracerebral transport (Fig. 23-7).

The major and most obvious function of CSF is as a buoyant cushion of the brain. The more dense brain floats in the less dense fluid, allowing movement within the skull.

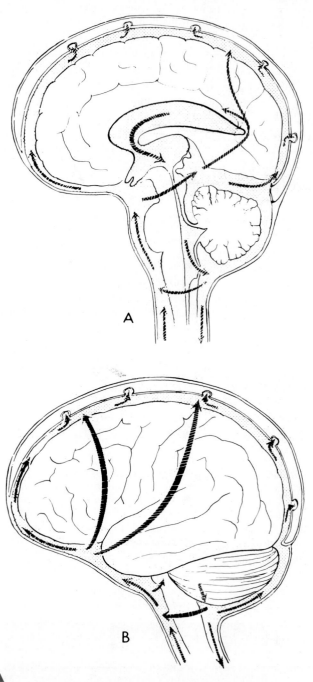

Figure 23-7. Major pathways of cerebrospinal flow. (*A*) Sagittal view; (*B*) lateral view. (Reprinted with permission from Milhorat TH. Hydrocephalus and the Cerebrospinal Fluid. Baltimore: Williams & Wilkins, 1972;25.)

The significance of this is demonstrated by the result of a blow to the head. The initial shock is transferred to the entire brain, instead of inflicting damage to one area. It may be bruised at the side opposite the blow, depending on the force imparted.

The second major function of CSF is the maintenance of a constant gross chemical matrix for the CNS. Serum components may vary greatly, but constituent levels of CSF are maintained within narrow limits.

The excretory function is not well defined, but it is presumed to be effective especially in pathologic states. Because there is no lymphatic system in the brain, only two paths are available for the elimination of wastes: capillary exchange and excretion by means of the CSF. The transport function is described as a neuroendocrine role. The CSF is involved in the distribution of hypophyseal hormones within the brain and the clearance of hormones from the brain to the blood.

The total CSF volume is approximately 150 mL, or about 8% of the total CNS cavity volume. The fluid is formed predominantly at the choroid plexus deep within the brain and by the ependymal cells lining the ventricles.

CSF is formed at an average rate of approximately 0.4 mL per minute, or 500 mL/per day. Formation is a result of selective ultrafiltration of plasma and active secretion by the epithelial membranes. Absorption of CSF occurs at outpouchings in the dura called *arachnoid villi,* and CSF drains into the venous sinuses of the dura. The villi also function to clear particulate matter, *e.g.,* cellular debris. Reabsorption, like formation, is selective and specific. Obviously, if a constant volume is maintained at a formation rate of 500 mL/per day, resorption is constant.

Specimens of CSF are obtained by lumbar puncture, usually at the interspace of vertebra L3 to L4 or lower, using aseptic technique. The fluid obtained is usually aliquoted into three portions: (1) for chemistry and serology, (2) for bacteriology, and (3) for microscopy. It is paramount to remember that this matrix is of limited volume and should be analyzed immediately. Any remaining sample should be preserved because of its limited availability. The order of the tubes reflects the presumed order for minimalization of interference from less than optimal collection technique, with tube 3 presumably least contaminated by cells of intervening tissues.

Laboratory investigation of CSF is indicated for cases of suspected CNS infection, demyelinating disease, malignancy, and hemorrhage in the CNS. Historically, it is usually performed prior to radiologic procedures such as myelography, but the diagnostic utility in these cases is extremely small. As with all patient samples entering the laboratory, visual examination of the specimen is the first and often the most important observation made. The CSF, if normal, is clear, colorless, free of clots, and free of blood. Differences from these standards indicate a probable pathology and merit further examination. Cloudy fluids usually require microscopic examination, while the presence of a yellow to brown or red color may indicate blood.

The two most common reasons for blood and/or hemoglobin pigments to be found in CSF are: (1) traumatic tap and (2) subarachnoid hemorrhage. Traumatic tap is the artifactual presence of blood or derivatives due to interdiction of blood vessels during the lumbar puncture. Hemorrhage results from a breakdown of the barrier of the CNS and circulatory system from, for example, trauma. Obviously, the latter is serious. The two can be differentiated by observation and possibly testing. Bright red color and the presence of erythrocytes in decreasing number as the fluid is sampled indicate a traumatic tap. Xanthochromia or the presence of hemoglobin breakdown pigments indicates erythrocyte lysis and that metabolism has occurred previously, at least 2 hours earlier. Excluding a prior traumatic tap or hyperbilirubinemia (>20 mg/dL), xanthochromia would indicate hemorrhage. Biochemical (chemical) analysis of CSF has led to compilations of the scope of possible constituents. In clinical practice, however, the number of useful indicators becomes very small. The tests of interest are glucose, protein (total and specific), lactate, lactate dehydrogenase, glutamine, and the acid–base parameters. Most often used are glucose, lactate, and proteins. Prior to any analysis, the fluid should be centrifuged to avoid contamination by cellular elements. The enzyme lactate dehydrogenase has been suggested as a tumor marker, but it is relatively nonspecific. It is also elevated in bacterial infection, although other parameters are more specific. The level of glutamine should reflect the level of CNS ammonia removed by glutamine formation from glutamate. This would be elevated in the hepatic encephalopathy of Reye's syndrome. The test has largely been supplanted by the relative ease and simplicity of reliable plasma ammonia determinations. Determination of the specific acid–base parameters obtained using a blood-gas instrument is relevant because of the vital environment of the CSF and the effect on control of respiration. The lack of buffering capacity of the CSF and the stringent requirements for sample integrity in the procedures for collection and analysis render routine usage of this series of tests impractical.

The tests that have been most reliable diagnostically and accessible analytically are those for CSF glucose, total protein, and specific proteins. Glucose enters the spinal fluid predominantly by a facilitative transport as compared with a passive (diffusional) or active (energy-dependent) transport. It is carried across the epithelial membrane by a stereospecific carrier species. The carrier mechanism is responsible for transport of lipid-insoluble materials across the membrane into the CSF. Generally, this is a "downhill" process consistent with a concentration gradient. The CSF glucose concentration is roughly two-thirds that of plasma.

Since an isolated CSF glucose concentration may be misleading, it is recommended that a plasma sample be obtained at the same time so that the plasma/CSF level can be confirmed. Normal CSF glucose is considered to be greater than 45 mg/dL (2.5 mmol/L). Increased glucose levels are not clinically informative, usually providing only confirmation of hyperglycemia. This generality must be tempered by two factors: (1) equilibration after glucose loading usually takes 3 to 4 hours, and (2) with increasing blood glucose levels, the CSF glucose increases, but not proportionally. The first is important in cases of meals taken close to the time of sampling. The second is significant because it implies the plasma/CSF glucose ratio decreases as gross hyperglycemia occurs.

The decreasing CSF/plasma glucose ratio as plasma glucose increases is consistent with a saturable carrier process. It would not be unusual for the ratio to be 0.4 to 0.6 with massive plasma glucose levels (>600 mg/dL), but a CSF glucose of 80 mg/dL, with plasma level of 300 mg/dL, is clinically significant and would merit concern.

Decreased CSF glucose levels (hypoglycorrhacia) can be the result of: (1) disorder in carrier-mediated transport of glucose into CSF, (2) active metabolism of glucose by cells or organisms, or (3) increased metabolism by the CNS. The mechanism of transport decrease is still under intense discussion, but it is speculated to be the cause in tuberculous meningitis and sarcoid. Acute purulent, amoebic, fungal, or trichinotic meningitis are examples of consumption by organism, whereas diffuse meningeal neoplasia and brain tumor are examples of consumption by CNS tissue. Consumption of glucose is usually accompanied by an increased lactate level because of anaerobic glycolysis by organisms or cerebral tissue. An increased lactate with normal to decreased glucose has been suggested as a readily accessible indicator for bacterial versus viral meningitis.[1] Analysis of glucose and/or lactate in CSF is easily accomplished by techniques used for plasma/serum. It is important that provision for analysis of glucose or lactate in CSF be immediate or that the specimen be preserved with an antiglycolytic such as fluoride ion.

Protein levels in CSF reflect both the selective ultrafiltration of the CSF epithelial barrier and their secretory ability. All the protein usually found in plasma is found in CSF, except at much decreased levels. Total protein is about 0.5%, or 1/200 that of plasma. The specific protein concentrations in CSF are not proportional to the plasma levels because of the specificity of the ultrafiltration process. Correlation is best accomplished using hydrodynamic ratios of the protein species rather than molecular weight. Because of the relationship of CSF proteins to serum, serum analysis should accompany specific CSF protein analysis. A decreased level of CSF total protein can arise from: (1) decreased dialysis from plasma, (2) increased protein loss, *e.g.,* removal of excessive volumes of CSF, or (3) leakage of CSF from a tear in the dura, otorrhea, or rhinorrhea. The last reason is most common. A dural tear can occur as a result of a previous lumbar puncture or from severe trauma. *Otorrhea* and *rhinorrhea* refer to leakage of CSF from the ear or into the nose, respectively. Identification of the source of the leak is best done by an analysis for τ-transferrin, a protein unique to the CSF.

An increased level of CSF total protein is a useful nonspecific indicator of pathologic states. Increases may be caused by: (1) lysis of contaminant blood from traumatic tap, (2) increased permeability of the epithelial membrane, (3) increased

production by CNS tissue, (4) obstruction, or (5) decrease in rate of removal. Contamination from blood is significant because of the 200:1 concentration ratio. The presence of any amount of blood can elevate CSF protein levels. The epithelial membrane becomes more permeable from bacterial or fungal infection or cerebral hemorrhage, whereas an increase in CNS production occurs in subacute sclerosing panencephalitis (SSPE) or multiple sclerosis. There also may be combinations of permeability and production, such as the collagen-vascular diseases. An obstructive process, such as tumor or abscess, also would cause increased protein. The last mentioned cause of increased levels of CSF total protein, decreased absorption, is theoretically possible but has not been demonstrated.

Diagnostically more sensitive information can be obtained by analysis of the protein fractions present. A comparison to the serum pattern is necessary for accurate conclusions. Under normal conditions, prealbumin and a form of transferrin (tau protein) are present in CSF in higher concentration than expected. Although the respective proteins can be determined in both serum and CSF, the proteins of greatest interest are albumin and IgG. Because albumin is produced solely in the liver, its presence in CSF must occur by means of membrane transport. IgG, however, can arise by local synthesis from plasma cells within the CSF. The measurement of albumin in both serum and CSF then is used to normalize the IgG values from each matrix to determine the source of the IgG. This IgG/albumin index is primarily used in the diagnosis of demyelinating diseases such as multiple sclerosis and SSPE.

$$\frac{CSF\ IgG/serum\ IgG}{CSF\ albumin/serum\ albumin} = CSF\ index$$

$$Normal = 0.5 \qquad \textit{(Eq. 23-1)}$$

Increases in serum proteins cause increases in the CSF levels because of permeability. However, increased CSF IgG without concomitant CSF albumin increase suggests local production (multiple sclerosis or SSPE). Increases in permeability and production are found with bacterial meningitis. Methods to analyze for IgG and albumin levels are the same as for serum, but optimized for the lower levels found.

Increased CSF protein levels or clinical suspicion usually indicates the need for electrophoretic separation of the respective proteins. At times, this separation will demonstrate multiple banding of the IgG band. This observation is referred to as *oligoclonal proteins* (a small number of clones of IgG from the same cell type with nearly identical electrophoretic properties). This occurrence is usually associated with inflammatory diseases and multiple sclerosis or SSPE. These types of disorders would stimulate the immunocompetent cells. The recognition of an oligoclonal pattern supersedes the report of normal protein levels and is cause for concern if the corresponding serum separation does not demonstrate identical banding.

Another protein thought to be specific for multiple sclerosis was myelin basic protein (MBP). Initial reports suggested high specificity, but it also has been found in non-demyelinating disorders and does not always occur in the presence of demyelinating disorders. MBP levels are used by some to monitor therapy of MS.

SWEAT

The common eccrine sweat glands function in the regulation of body temperature. They are innervated by cholinergic nerve fibers and are a type of exocrine gland. Sweat has been analyzed for its multiple inorganic and organic contents but, with one notable exception, has not proved a clinically useful model. That exception is the analysis of sweat for chloride and sodium levels in the diagnosis of cystic fibrosis (CF). The sweat test is the single most accepted diagnostic tool for the clinical identification of this disease. Normally, the coiled lower part of the sweat gland secretes a "pre-sweat" upon cholinergic stimulation. As the pre-sweat traverses the ductal part of the gland going through the dermis, various constituents are resorbed. In CF, the electrolytes, most notably chloride and sodium ions, are improperly resorbed owing to a mutation in the cystic fibrosis transmembrane conductance regulator (CFTR) gene which controls a cyclic AMP-regulated chloride channel.

CF (mucoviscidosis) is an autosomal recessive inherited disease that affects the exocrine glands and causes electrolyte and mucous secretion abnormalities. This exocrinopathy is present only in the homozygous state. The frequency of the carrier (heterozygous) state is estimated at 1 in 20 in this country. It is predominantly a Caucasian disease. The observed rate of expression ranks CF as the most common lethal hereditary disease in this country, with death usually occurring by the third decade. The primary cause of death is pneumonia, secondary to the heavy, abnormally viscous secretion in the lungs. These heavy secretions cause obstruction of the microairways, predisposing the CF patient to repeated episodes of pneumonia. The third part of the diagnostic triad is pancreatic insufficiency. Again, abnormally viscous secretions obstruct pancreatic ducts. This obstruction ostensibly causes pooling and autoactivation of the pancreatic enzymes. The enzymes would then cause destruction of the exocrine pancreatic tissue.

Diagnostic algorithms for CF continue to rely on abnormal sweat electrolytes, the presence of pancreatic or bronchial abnormalities, and family history. Recently, the use of blood immunoreactive trypsin, a pancreatic product, has been proposed as both a method for newborn screening and a diagnostic adjunct. The rapidly developing area of molecular genetics may eventually provide the definitive methodology. The gene defect causing CF has been localized on chromosome 7, and the most common mutation causing CF has been DNA "fingerprinted." It is anticipated that within the next decade, direct DNA analysis will catalogue all mutations and provide definitive screening and diagnosis. Until that time, however, the sweat chloride test remains the signal laboratory tool for CF diagnosis even though it has been reported that a form of CF with normal sweat chloride exists.[6]

The sweat glands, although affected in their secretion, remain structurally unaffected by CF. Analysis of sweat for both sodium and chloride is valid, but historically, chloride was and is the major element, leading to use of the sweat chloride test. Because of its importance, a standard method has been suggested by the Cystic Fibrosis Foundation. It is based on the pilocarpine nitrate iontophoresis method of Gibson and Cooke.[3] Pilocarpine is a cholinergic-like drug used to stimulate the sweat glands. Other tests (*e.g.,* osmolality or conductivity) that reflect the sodium and chloride concentrations have been proposed, but the sweat chloride test remains the reference method.

SWEAT TEST (GIBSON-COOKE QUANTITATIVE PILOCARPINE IONTOPHORETIC TEST)

Stimulation of Sweating

1. Using forceps, place a 2 × 2 inch ash-free gauze sponge or similarly sized piece of Whatman no. 42 filter paper into a weigh bottle. Close and weigh the bottle, being careful not to handle the bottle directly.
2. Wash an appropriate area of skin on the inside surface of the forearm (adults or older children), or inside the thigh for infants, with distilled water. Place another gauze sponge saturated with a 200 mg/dL solution of pilocarpine nitrate at this site. Place an ECG electrode over the sponge and gently secure it. Place a gauze sponge saturated with isotonic saline on the outside of the forearm. Position and secure a second ECG electrode over it. Be sure that no metal touches bare skin and that the saline sponge is connected to the negative pole of the power supply, pilocarpine to positive.
3. Apply a current of 1 to 2 mA for at least 5 minutes. Remove the electrodes, clean the areas well with distilled water, and dry well with gauze.

Collection of Sweat

1. Rapidly transfer the pre-weighed sponge (paper) with forceps and place over the pilocarpine stimulated site. Cover immediately with a 4 × 4 inch plastic sheet and secure the edges with surgical tape or bandages.
2. After 30 minutes droplets of sweat should be apparent on the plastic sheet. Gently press the sheet so all sweat is absorbed by the gauze. Quickly remove the plastic and transfer the gauze back to the weighing bottle. Cover tightly and re-weigh. The difference in weight is the amount of sweat collected. Usually at least 0.1 g (0.1 mL) is obtained. Larger yields may be obtained by using 2 sponges.

Analysis of Sweat for Chloride and Sodium—Many methods have been suggested, and all are dependent on the requirements of the laboratory. Generally, the sweat is leached into a known volume of distilled water and analyzed for chloride (chloridometer) and sodium (flame photometry). In general, values greater than 60 mmol/L are considered positive for both ions.

Although a value of 60 mmol/L is generally recognized for the quantitative pilocarpine iontophoretic test, it is important to consider several factors in interpretation. Not only will there be analytical variation about the cutoff, there also will occur an epidemiologic borderline area. Considering this, the range of 45 to

65 mmol/L for chloride would be more appropriate in determining the need for repetition. Other variables necessarily must be considered. Age generally increases the limit—so much so that it is increasingly difficult to classify adults. Obviously, the patient's state of hydration also will affect the sweat levels. Because the complete procedure is technically very demanding, expertise should be developed before the test is clinically available. A complete description of sweat collection and analysis including procedural justifications can be found in the recently released document: NCCLS C34-A. Sweat testing: Sample collection and quantitative analysis: Approved guidelines.[9] ∎

SYNOVIAL FLUID

Joints are classified as immovable or movable. Within the movable joint is a cavity enclosed by a capsule whose inner lining is synovial membrane. The cavity contains synovial fluid formed by ultrafiltration of plasma across the synovial membrane. The membrane also secretes into the dialysate a mucoprotein rich in hyaluronic acid, which causes the synovial fluid to be viscous. The membrane is composed of three different cell types. Type A cells are rich in vacuoles and lysosomes and resemble phagocytes in function, type B cells are rich in rough endoplasmic reticulum and are presumed to be secretory in function, and type C cells appear to be hybrids of types A and B in appearance and function. Synovial fluid is assumed to function as a lubricant for the joints and as a transport medium for delivery of nutrients and removal of cell wastes. The volume of fluid found in a large joint, such as the knee, rarely exceeds 3 mL. Normal fluid is clear, colorless to pale yellow, and viscous, and it does not clot. Variations are indicative of pathologic conditions. Collection of a sample is accomplished by arthrocentesis of the joint under aseptic conditions. The sample should be preserved immediately with heparin for culture, with EDTA for microscopic analysis, or with fluoride for glucose analysis. Microscopic examination is the most rewarding in terms of diagnostic significance. Although many analytes have been chemically determined, the ratio of synovial fluid to plasma glucose (normally 0.9–1.0) remains the most useful. Decreased ratios are found in inflammatory (e.g., gout, rheumatoid arthritis, systemic lupus erythematosus) and purulent (bacterial, viral arthritis) conditions. Standard methods for glucose analysis are applicable.

SEROUS FLUIDS

The lungs, heart, and abdominal cavity are surrounded by single-celled membranous bilayered sacs permeable to serum constituents. The fluid formed when *serum* dialyzes across these membranes is called *serous fluid*. Specifically, they are pleural (lung), pericardial (heart), and peritoneal (abdominal) fluids.

The sacs can be pictured as a balloon containing a small amount of water. Push a football into the balloon, and the balloon expels the residual air until only water remains

(Fig. 23-8). The water spreads *to fill* the resultant "balloon bilayer potential space." The serous membranes are analogous to the balloon; the internal organs, to the football. The pleural (lung) sac is continuous at the hilus of the bronchial tree; the pericardium, at the major vessels; and the peritoneum, about each organ. All are expandable and thus present "potential spaces" for fluid or gases to collect. The formation of serous fluid is a continuous process. Formation is driven by the hydrostatic pressure of the systemic circulation and maintenance of oncotic pressure due to protein. The potential space is usually filled, *i.e.,* no gases are present. The fluid reduces or eliminates friction due to expansion and contraction of the encased organs. A disturbance of the dynamic equilibrium that causes an increase in fluid is an abnormal state. An increase in fluid volume is called an *effusion*.

Pleural Fluid

The outer layer of the pleural sac, the parietal layer, is served by the systemic circulation; the inner, visceral layer, by the bronchial circulation. The pleural fluid is essentially interstitial fluid of the systemic circulation. Under normal conditions, there are 3 to 20 mL of pleural fluid in the pleural space. The fluid exits by drainage into the lymphatics of the visceral pleura and the visceral circulation. Any alteration in the rate of formation or removal of the pleural fluid affects the volume, causing an effusion. It is then necessary to classify the nature of the effusion by analysis of the pleural fluid. The fluid is obtained by removal from the pleural space by needle and syringe after visualization by radiology. This procedure is called *thoracentesis*; the fluid is then called *thoracentesis fluid* or *pleural fluid*. Specifically preserved aliquots of the fluid are used for future testing as follows: (1) heparinized for culture, (2) EDTA for microscopy, (3) NaF for glucose and lactate, and (4) untreated for further biochemical testing.

The classification of the fluid as transudate or exudate is critical. *Transudates* are secondary to remote (nonpleural) pathology and indicate that treatment should begin elsewhere. An *exudate* indicates primary involvement of the pleura and lung, such as infection. Exudates demand immediate attention. For example, any mechanical disturbance in

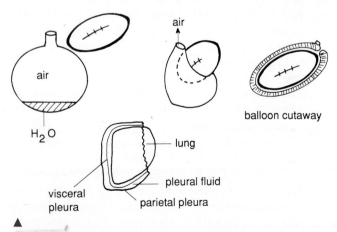

Figure 23-8. Example of formation of serous fluid.

the formation of fluid, *e.g.,* hypoproteinemia causing decreased oncotic pressure, would increase pleural fluid volume. This would be a transudative process. An example of an exudative process would be obstruction of the lymphatic drainage due to malignancy, such as lymphoma (Table 23-2). Further testing, including chemical, microscopic, and culture, is then required to identify the etiology.

The assignment of fluid to either the transudate or exudate category had previously been based on the protein concentration of the fluid. This criterion has been replaced by the use of a series of fluid-plasma (F:P) ratios. Specifically, if the F:P for total protein is greater than 0.5, the ratio for lactate dehydrogenase (LD) is greater than 0.6, or the ratio of pleural fluid LD/upper limit serum LD reference interval is greater than 0.67, the fluid is an exudate. Further characterization of the exudate by the chemistry laboratory may involve analysis for glucose, lactate, amylase, triglyceride, or pH. A decrease in glucose (increase in lactate) would suggest infection or inflammation. An increase in amylase compared with that of serum suggests pancreatitis. Grossly elevated triglyceride levels (2–10 × serum) could indicate chylothorax. The use of pH measurements, performed as one would a blood gas determination, has recently gained favor. Succinctly, pH less than 7.20 suggests infection, and pH greater than 7.40 suggests malignancy. Methodology for these analyses are the same as those employed for the serum/blood constituents and therefore are feasible in the clinical laboratory.

Pericardial Fluid

The relationship of the pericardium, pericardial fluid, and the heart is quite similar to the lungs. Mechanisms of formation and drainage are the same, but the frequency of pericardial sampling and laboratory analysis is rare.

Peritoneal Fluid

The presence of excess (>50 mL) fluid in the peritoneal cavity indicates disease. The presence of excess fluid is called

ascites, and the fluid is called *ascitic fluid*. The process of obtaining samples of this fluid by needle aspiration is called *paracentesis*. Usually, the fluid is visualized by ultrasound to confirm its presence and volume before attempts at paracentesis.

Theoretically, the same mechanisms that cause serous effusions in other potential spaces are operative for the peritoneal cavity. Specifically, a disturbance in the rate of dialysis secondary to a primary, remote pathology is a transudate, as compared with a primary pathology of the peritoneal membrane (exudate). The multiple factors that apply to this large space, including renal function, tend to cloud the distinction. The most common cause of ascites with a normal peritoneum is portal hypertension. Obstructions to hepatic flow, that is, (1) cirrhosis, (2) congestive heart failure, and (3) hypoalbuminemia for any reason, demonstrate the highest incidence.

The exudative causes for ascites are predominantly metastatic ovarian cancer and infective peritonitis. The differentiation of the two is analogous to the measures of protein and LD described for pleural fluid. However, the ability to differentiate has been challenged. Currently, the gradient of albumin concentration (serum albumin–fluid albumin) of 1.1 g/dL or more used to indicate portal hypertension is gaining adherents.

SUMMARY

In addition to serum and plasma, the clinical chemistry laboratory often analyzes other body fluids such as amniotic fluid, cerebrospinal fluid, sweat, synovial fluid, and serous fluids. Amniotic fluid provides a cushioning medium for fetal development.

Amniocentesis and subsequent amniotic fluid analysis are performed to test for congenital diseases, neural tube defects, hemolytic disease, gestational age, and fetal pulmonary development. Cerebrospinal fluid (CSF) is a liquid that occupies the spaces of the CNS, which includes the brain and spinal cord. Functions of the CSF include physical support and protection, method of excretion, provision of a controlled chemical environment, and intra- and extracerebral transport. The total volume of CSF is about 150 mL. Specimens of CSF are obtained by lumbar puncture. Laboratory investigation of CSF is indicated for cases of suspected CNS infection, demyelinating disease, malignancy, and hemorrhage in the CNS. The tests that have been most reliable diagnostically and accessible analytically are those for CSF glucose, total protein, and specific proteins. Sweat is a product of the common eccrine sweat glands, which function in the regulation of body temperature. The sweat test is the single most accepted diagnostic tool for cystic fibrosis (CF). Within the movable joint is a cavity filled with synovial fluid. This fluid is formed by ultrafiltration of plasma across the synovial membrane. Normal fluid is clear, colorless to pale yellow, and viscous; it does not clot. Any variations are indicative of pathologic conditions. Collection of this fluid is accomplished by

TABLE 23-2. Causes of Pleural Effusions

TRANSUDATIVE	EXUDATIVE
Congestive heart failure*	Pneumonia,* bacterial
Nephrotic syndrome	Tuberculosis
Hypoproteinemia	Pulmonary abscess
Hepatic cirrhosis	Malignancy (lymphatic obstruction)
Chronic renal failure	Viral/fungal infection
	Pulmonary infarction
	Pleurisy
	Pulmonary malignancy
	Lymphoma
	Pleural mesothelioma

Most common cause.

arthrocentesis. The lungs, heart, and abdominal cavity are surrounded by single-celled membranous bilayered sac permeable to serum constituents. The fluid formed when serum dialyzes across these membranes is called serous fluid. More specifically, they are pleural (lung), pericardial (heart), and peritoneal (abdominal) fluids. Formation of serous fluid is a continuous process and is driven by the hydrostatic pressure of the systemic circulation and maintenance of oncotic pressure due to protein. Under normal conditions there is from 3 to 20 mL of pleural fluid in the pleural space. Removal of the fluid is called thoracentesis. The relationship of the pericardium, pericardial fluid, and the heart is quite similar to the lungs. The frequency of pericardial sampling in the laboratory is rare. An excess of fluid in the peritoneal cavity indicates disease. This is called ascites. The fluid is ascitic fluid and is obtained by a procedure called paracentesis.

CASE STUDY
23-1

A 26-year-old woman, pregnant for the first time, presented at the emergency room in possible labor. Her blood pressure was 180/110 mmHg, and her temperature was 99.6°F. Her history revealed an anxious woman, with pregnancy of unknown duration, who had not seen a physician before. She admitted to a rapid weight gain the last 2 weeks but said that she had been feeling well until then. Recently, she had felt faint and experienced episodes of vertigo. Urinalysis was significant for 3+ glucose and protein. Hematology revealed a low hemoglobin level and a moderately low platelet count, but leukocyte levels were normal. A chemistry panel revealed Na of 132 mmol/L, K of 3.0 mmol/L, Cl of 100 mmol/L, CO_2 of 29 mmol/L, BUN of 7 mg/dL, creatinine of 0.5 mg/dL, and glucose of 351 mg/dL. A decision was made to admit her and perform several more tests. The following chemistry tests were ordered, with the listed results (upper limit of reference interval given):

L/S ratio = 1.8 with PG = 2%
AF creatinine = 1.5
FSI = 0.47
AST = 40(35), ALT = 40(35), ALP = 250(100)
Mg^{2+} = 1.6(2.2), Ca^{2+} = 9.0(10.5), PO_4 = 4.0(4.3), Alb = 3.2(5.0)

After 24 hours of bed rest, the patient's BP fell to 140/90, and the contractions subsided.

Questions

1. Do the laboratory results indicate a normal pregnancy?
2. Comment on the maturity of the fetus.

CASE STUDY
23-2

A 32-year-old man was in good health until about 1 year ago, when he entered an accelerated training program in computer programming. Within the last year, he began to notice episodic blurring of vision, mild vertigo, and headache. He attributed his complaints of sensory loss in his hands and a feeling of weakness after physical exertion to being "out of shape." He decided to see his doctor upon an attack of blurred vision accompanied by a feeling of paralysis, which was followed by "pins and needles" in his left leg. An optic examination was negative. Neurologic examination led to a spinal tap being performed for laboratory findings and myelography. The latter was negative. The laboratory findings were as follows:

CSF = clear, colorless fluid, apparently free of debris; culture yields no growth
WBC = normal
Glucose = 60 mg/dL (plasma = 80 mg/dL)
IgG/Alb ratio = 1.7
IgG oligoclonal banding present

Questions

1. What is the significance of the normal CSF protein and variant IgG/Alb ratio?
2. What pathology is consistent with these results?

REFERENCES

1. Bailey EM, Domenico P, Cunha BA. Bacterial or viral meningitis. Postgrad Med 1990;88:217.

2. Clements JA, Plataker ACG, Tierney DF, et al. Assessment of the risk of the respiratory distress syndrome by a rapid test for surfactant in amniotic fluid. N Engl J Med 1972;286:1077.

3. Gibson LE, Cooke RE. A test for concentration of electrolytes in cystic fibrosis of the pancreas utilizing pilocarpine by iontophoresis. Pediatrics 1959;23:545.

4. Gluck L, Kulovich MV, Borer RC Jr, et al. Diagnosis of the respiratory distress syndrome by amniocentesis. Am J Obstet Gynecol 1971;109:440.

5. Herbert WNP, Chapman JF, Schnoor MM. Role of the TDX FLM assay in fetal lung maturity. Am J Obstet Gynecol 1993;168:808.

6. Highsmith WE, Burch L, Zhou Z, et al. A novel mutation in the cystic fibrosis gene in patients with pulmonary disease but normal sweat chloride concentrations. N Engl J Med 1994; 331:974.

7. Liley AW. Liquor amni analysis in the management of the pregnancy complicated by rhesus sensitization. Am J Obstet Gynecol 1961;82:1359.

8. Kulovich MV, Hallman MB, Gluck L. The lung profile: I. Normal pregnancy. Am J Obstet Gynecol 1979;135:57.

9. NCCLS. Sweat testing: Sample collection and quantitative analysis approved guidelines. NCCLS Document C34-A. Villanova, PA: National Committee for Clinical Laboratory Standards, 1994.

10. Statland BE, Freer DE. Evaluation of two assays of functional surfactant in amniotic fluid: Surface-tension lowering ability and the foam stability index test. Clin Chem 1979;25:1770.

11. Tsai MY, Marshall JG. Phosphatidylglycerol in 261 samples of amniotic fluid from normal and diabetic pregnancies, as measured by one-dimensional thin layer chromatography. Clin Chem 1979;25:682.

SUGGESTED READINGS

American Society of Human Genetics. Policy statement for maternal serum alpha-fetoprotein screening programs and quality control for laboratories performing maternal serum and amniotic fluid alpha-fetoprotein assays. Am J Hum Genet 1987;40:75.

Brown LM, Duck-Chong CG. Methods of evaluating fetal lung maturity. Crit Rev Clin Lab Sci 1982;17(1):85.

Committee for a Study of Evaluation of Testing for Cystic Fibrosis: Report. J Pediatr 1976;88(4):711.

Daveson H. Physiology of the cerebrospinal fluid. London: Churchill, 1967.

Fairweather DVI, Eskes TKAB, eds. Amniotic fluid research and clinical application. 2nd ed. New York: Excerpta Medica, 1978.

Freeman JA, Beeler MF, eds. Laboratory medicine/urinalysis and medical microscopy. 2nd ed. Philadelphia: Lea & Febiger, 1983.

Freer DE, Statland BE. Measurement of amniotic fluid surfactant. Clin Chem 1981;27:1629.

Guyton AC. Textbook of medical physiology. 5th ed. Philadelphia: Saunders, 1976.

Halsted JA, Halsted CH, eds. The laboratory in clinical medicine. 2nd ed. Philadelphia: Saunders, 1981.

Health and Public Policy Committee, American College of Physicians. Diagnostic thoracentesis and pleural biopsy in pleural effusions. Ann Intern Med 1985;103:799.

Platt PN. Examination of synovial fluid in the role of the laboratory in rheumatology. Clin Rheum Dis 1983;9(1):51.

Queenan JT, ed. Modern management of the Rh problem. 2nd ed. Hagerstown, MD: Harper & Row, 1977.

Rocco VK, Ware AJ. Cirrhotic ascites. Ann Intern Med 1986;105:573.

Rodriguez EM, van Wimersma Greidanus TB. Frontiers of hormone research. Vol 9.

Cerebrospinal fluid and peptide hormones. New York: S Karger, 1982.

Russell JC, Cooper CC, Ketchum CH, et al. Multicenter evaluation of Tdx test for assessing fetal lung maturity. Clin Chem 1989;35:1005.

Teloh HA. Clinical pathology of synovial fluid. Ann Clin Lab Sci 1975;5(4):282.

Webster HL. Laboratory diagnosis of cystic fibrosis. Crit Rev Clin Lab Sci 1983;18(4):313.

Wiswell TE, Mediola J Jr. Respiratory distress syndrome in the newborn: Innovative therapies. Am Fam Physician 1993; 47:407.

Wood JH, ed. Neurobiology of the cerebrospinal fluid. Vol 2. New York: Plenum Press, 1983.

Chapter 24

Exam #3 whole Ch.

Read for take Home test #3

Tumor Markers

Anthony W. Butch, Alex A. Pappas

4/14

Objectives

Upon completion of this chapter, the clinical laboratorian should be able to:

- *Discuss the incidence of cancer in the United States.*
- *Explain the role of tumor markers in cancer management.*
- *Identify the characteristics or properties of an ideal tumor marker.*
- *State the major clinical value of tumor markers.*
- *Name the major tumors and their associated markers.*
- *Describe the major properties, methods of analysis, and clinical use of CEA, AFP, CA125, CA19–9, PSA, β-hCG, and PALP.*
- *Explain the use of enzymes and hormones as tumor markers.*

KEY WORDS

Cancer	Oncogenes	Tumor-Associated
DNA Index (DI)	Proto-Oncogenes	Antigen
Neoplasm	Staging	Tumor-Specific
Oncofetal Antigen	Tumor Marker	Antigen

Cancer is second only to heart disease as a major health issue in the United States, accounting for approximately 24% of all deaths annually. Although the cancer mortality rate has remained relatively constant for several decades, it has been projected that cancer will be the leading cause of death in the United States by the year 2000, supplanting cardiovascular disease, which has had a steadily declining mortality.[61] The American Cancer Society has estimated that in 1995 there will be a total of 1,252,000 new cancer cases and that 547,000 people will die from cancer.[121]

There are at least 50 different types of human malignancies, with more than 60% of new cancer cases occurring in the lung, colon/rectum, breast, prostate, and uterus. Cancers of the lung, prostate, colon/rectum, and urinary tract are most commonly seen in men, whereas breast, lung, colon/rectum, and uterine cancers are the most frequently reported cancers in women.[121] Lung cancer is the most common cause of death from cancer among both sexes, followed by prostate cancer in men, and breast cancer in women.[121]

There are three major approaches taken in the United States to deal with cancer: (1) prevention by identification and elimination of the cancer-causing agent(s) or factor(s), (2) early disease detection when it is localized in primary sites and is more responsive to therapy, and (3) development of more effective therapeutic modalities. Of course, the best

Michael L. Bishop, Janet L. Duben-Engelkirk, and Edward P. Fody.
CLINICAL CHEMISTRY. © 1996 Lippincott–Raven Publishers.

and most ideal way is to prevent the occurrence of cancer, which could be carried out by identifying and eliminating exposure to all carcinogens and/or by enhancing host resistance to cancer. Despite the large amount of research and health care dollars spent to implement this strategy, it is unrealistic and difficult (if not impossible) to avoid exposure to all environmental and biological cancer-associated risk factors. For instance, deaths due to lung cancer could be reduced at least 20-fold if people would abstain from smoking.[3,103] Therefore, early detection is the most promising approach to controlling cancer through early intervention.

There are a number of methods currently available to detect cancer in its early stages. Biophysical methods, such as conventional x-rays, nuclear medicine, rectilinear scanners, ultrasonography, computerized axial tomography (CAT), and magnetic resonance imaging (MRI), all play an important role in early detection and treatment of cancer. Clinical laboratory testing for tumor markers can also be used as an aid in early cancer detection. *Tumor marker tests* as a group measure either tumor-associated antigens or other substances present in cancer patients that aid in diagnosis, staging, disease progression, monitoring response to therapy, and detection of recurrent disease.[47,89]

CLASSIFICATION OF TUMOR MARKERS

Tumor markers can be defined as biological substances synthesized and released by cancer cells or substances produced by the host in response to cancerous tissue. Tumor markers can be present in the circulation, body cavity fluids, cell membranes, or the cytoplasm/nucleus of the cell. Tumor markers differ from substances produced by normal cells in quantity and quality. Tumor markers that circulate in the blood can be measured biochemically by employing chemical reactions, radioimmunoassays, and/or enzyme immunoassays. The markers present on cell membranes or in the cytoplasm are measured routinely by immunohistologic or immunocytologic techniques (*e.g.,* immunofluorescence, immunoperoxidase, and flow cytometry). Nuclear DNA content can be measured routinely by propidium iodine staining and flow cytometry for numerous solid and hematologic malignancies. The detection of chromosomal alterations can now be achieved using molecular diagnostic techniques.[12,21,39,46,49] Clinically useful tumor markers that aid in the diagnosis and management of cancer patients are classified into groups, as shown in Table 24-1.

TUMOR MARKERS IN CANCER MANAGEMENT

Theoretically, tumor markers should be extremely useful in the detection, management, and monitoring of cancer patients, as illustrated in Table 24-2. Unfortunately, almost all available tumor marker tests do not possess sufficient

TABLE 24-1. Classification of Several Clinically Useful Tumor Markers

Oncofetal Antigens
  Carcinoembryonic antigen (CEA)
  Alphafetoprotein (AFP)
  CA 125
  CA 19–9
  CA 15–3
  CA 549
  Tissue polypeptide antigen (TPA)
  Prostate specific antigen (PSA)

Placental Proteins
  Human chorionic gonadotropin (hCG, intact and β-hCG)
  Human placental lactogen (HPL)
  Placental alkaline phosphatase (Regan isoenzyme)

Enzymes & Isoenzymes as Tumor Markers
  Prostatic acid phosphatase (PAP)
  Creatine kinase (CK)-BB isoenzyme
  Alkaline phosphatase
  Neuron-specific enolase (NSE)
  Lactate dehydrogenase-LD1 isoenzyme
  Lysozyme (Muramidase)

Hormones
  Eutopic hormones (normally secreted by the tissue)
    Human chorionic gonadotropin (hCG)-trophoblastic tumors, nonseminomatous testicular tumors
    Epinephrine/norepinephrine-pheochromocytoma and related malignancies
    Gastrin-Zollinger-Ellison syndrome (gastrinoma)
    Calcitonin-medullary carcinoma of the thyroid
  Ectopic hormones (not normally synthesized or secreted by the tissue).
    ACTH-small cell carcinoma of the lung
    Antidiuretic hormone (ADH)
    Parathyroid-related peptide
    Erythropoietin

Monoclonal Immunoglobulins
  IgG, IgA, IgM, IgD, IgE
  Kappa & lambda light chains

Steroid Receptors
  Estrogen and progesterone receptors
  Androgen receptors
  Corticosteroid receptors

Immunophenotyping
  Lymphoid cells
  Myeloid cells

DNA Analysis

Molecular Diagnosis

specificity to be useful screening tools in a cost–effective manner. Even highly specific tumor marker tests suffer from poor predictive value, because the prevalence of a particular cancer is relatively low in the general population. The majority of available tumor marker tests are not useful in diagnosing cancer in symptomatic patients since elevated levels are also seen in a variety of benign diseases. The major clinical value of tumor markers is in tumor staging, moni-

TABLE 24-2. Potential Applications of Tumor Markers in Cancer

Screen for disease in the general population
Diagnosis in symptomatic patients
Adjunct in clinical staging
Indicator of tumor volume
Aid for selecting appropriate therapy
Monitor response to therapy
Prognostic indicator
Early detection of disease recurrence

TABLE 24-3. Characteristics of an Ideal Tumor Marker

Analytical Requirements
High analytical sensitivity
High analytical specificity
Accuracy
Precision
Rapid turnaround time
Easy to measure at a low cost

Clinical Requirements
High sensitivity for disease (no false-negatives, ability to detect micrometastasis)
High specificity for disease (no false-positives, negative in disease-free individuals)
Levels should reflect tumor burden
Levels should remain relatively constant and not fluctuate in patients with stable disease
Should be undetectable or low in patients in complete remission
Should predict outcome in patients with stable disease

toring therapeutic responses, predicting patient outcomes, and detecting cancer recurrence. The usefulness of tumor markers in these clinical settings will vary greatly depending on disease-related variables, such as therapeutic efficacy. For instance, if treatment for a particular cancer is effective, then a tumor marker may be useful for monitoring tumor burden and prognosis, but would not be helpful for monitoring response to therapy. On the other hand, if therapy is highly effective, then a sensitive tumor marker would be of value in monitoring treatment response and disease recurrence, but of little value as a prognostic indicator. In the case of gestational trophoblastic tumors, human chorionic gonadotropin (hCG) is a sensitive and specific marker useful in staging disease and selecting appropriate therapy, but is not helpful in patients undergoing intensive therapy, since hCG levels do not correlate with prognosis. Thus, the application of tumor markers requires a thorough understanding of their capabilities and limitations if they are to be used wisely in the management of cancer patients.

4/15

PROPERTIES OF AN IDEAL TUMOR MARKER

An ideal tumor marker test should possess characteristics that meet both analytical and clinical requirements, as outlined in Table 24-3. Analytically, the method used to measure the tumor marker should be capable of detecting very small quantities (*i.e.,* analytically sensitive) and should not measure unrelated or closely related substances (*i.e.,* analytically specific). The true quantity of the substance should be measured by the test method, and the correct value should be obtained every time the method is performed (the assay should be precise). Operationally, the test should be rapid, easy to perform, and of relatively low cost to the laboratory.

Clinically, an ideal tumor marker should be highly sensitive for disease. In other words, the tumor marker should be positive in all patients with that specific cancer. Elevations in tumor marker levels above normal should also be evident in the presence of micrometastasis, and should predict and precede disease recurrence before it is clinically evident by other parameters used to detect disease. The tumor marker should also have high disease specificity and should be undetectable or at low levels in healthy people, patients with benign dis-

eases, and patients with tumors other than the one in question. If the marker is normally present in the circulation, then it should exist at significantly lower concentrations than those detected in patients with all stages of cancer. The levels should not fluctuate in patients with stable disease and should parallel clinical and pathophysiologic changes in disease activity. For instance, tumor marker levels should correlate with tumor burden, should increase over time as the tumor load increases, and should diminish in patients undergoing successful chemotherapy. Patients in complete remission should have low to undetectable tumor marker levels as long as they are disease-free, whereas increasing levels should reflect relapse. Finally, initial and serial tumor marker levels should have prognostic value when medical treatment for the malignancy is ineffective (Table 24-3).

ONCOFETAL ANTIGENS

During the last two decades, tests based on tumor antigens have greatly contributed to the clinical management of cancer patients. There are basically two types of tumor antigens: tumor-specific antigens and tumor-associated antigens. Tumor-specific antigens are thought to be a direct product of oncogenesis induced by viral oncogenes, radiation, chemical carcinogens, or unknown risk factors. The oncogenic process causes disorganization of the genetic information, leading to synthesis of antigens that are unique for the specific tumor cell. Mechanisms of tumor-specific antigen production have been demonstrated in animal models. Based on these studies, tumor-specific antigens play an important role in clinical oncology.[89,114]

The majority of tumor antigens are not specific for individual tumors, but are usually associated with different tumors derived from the same or closely related tissue. On-cofetal proteins are an example of tumor-associated antigens. They are present in both embryonic/fetal tissue and cancer cells. Oncofetal antigens are produced in large amounts during fetal life and are released into the fetal circulation. After birth, the production of oncofetal antigens is repressed, and only minute quantities are present in the circulation of adults. It has been postulated that oncofetal protein production is reactivated with the onset of malignant cell transformation. This would explain the reappearance of oncofetal proteins in the circulation of cancer patients. Examples of embryonic/fetal antigens that are expressed by tumor cells and can be found in the circulation of patients with cancer are listed in Table 24-1. These tumor markers are routinely used in clinical oncology.[102]

Carcinoembryonic Antigen (CEA)

CEA was identified in 1965 by Gold and Freedman as a tumor-specific antigen present in extracts of tumor tissue from patients with adenocarcinoma of the digestive tract. It is also found in normal fetal gastrointestinal tract epithelial cells. CEA's name is derived from the fact that it is present in both carcinoma and embryonic tissue. CEA was initially considered to be highly specific and sensitive for diagnosing colorectal carcinoma. However, as clinical data and experience accumulated, this marker was also found to be elevated in patients with other malignancies, as well as several non-cancerous disorders. Cigarette smokers also have elevations in serum CEA compared to non-smokers. CEA is the most widely investigated and most frequently used tumor marker in cancer management.[5,36,42,48,68]

Properties

CEA is a large family of cell-surface glycoproteins with common antigenic determinants residing in the protein portion of the molecule. CEA is composed of a single polypeptide chain consisting of 30 amino acids. More than 35 different glycoproteins belong to the CEA family. CEA has a molecular mass between 160 and 300 kD with 45% to 57% of the molecule consisting of carbohydrate. The ratio of protein to carbohydrate in CEA can vary as much as five-fold among different tumors.

CEA is present in both endodermally derived tissues (intestinal mucosa, lung, and pancreas) and nonendodermally derived tissues. It is found in the digestive tract of fetuses after 12 weeks' gestation, and is present in embryonic liver and pancreas. CEA is a component of mucin and other mucopolysaccharides found on the brush border of normal mucosal cells and in the cytoplasm of colonic carcinoma cells. The concentration of CEA in the circulation will depend on the number of cells producing CEA, the rate of synthesis, and hepatic clearance from the circulation. CEA normally disappears from the circulation about 3 weeks after successful removal of a CEA-producing tumor.

Methods of Analysis

The two most commonly used methods for CEA analysis are enzyme immunoassay (EIA) and radioimmunoassay (RIA), both of which are commercially available. Because of the antigenic heterogeneity in CEA molecules from various tumors, different methods do not always produce equivalent results. Therefore, when switching analytical methods, it is important to establish a new baseline value when monitoring cancer patients by serial CEA measurements. In addition, reference ranges provided by the manufacturer for CEA should only be used as a guideline. Each laboratory should carefully evaluate the performance of the method employed and establish their own reference ranges.

Clinical Uses

Although CEA has achieved widespread clinical application, accumulated clinical data have shown that CEA levels are of limited use in detecting colorectal, pancreas, and liver cancer. For instance, approximately 35% of patients with colorectal cancer will not have an elevated CEA level.[52] This lack of sensitivity (high false-negative rate) makes CEA testing impractical for ruling out the presence of cancer. Furthermore, abnormally elevated CEA values have been observed in patients with pulmonary emphysema, acute ulcerative colitis, alcoholic liver cirrhosis, hepatitis, cholecystitis, benign breast disease, rectal polyps, and in heavy smokers.[48,58,66,67,76] Elevated levels of CEA do not normally exceed four to five times the upper limit of the reference range in benign disorders and will return to normal after the acute reaction has subsided. In contrast, continuously elevated CEA levels in serial samples is highly suggestive of a malignant disease process. Elevated CEA values are also seen more often in patients with advanced and disseminated diseases than in patients with localized diseases. This severely limits the use of CEA for establishing an early diagnosis of cancer. Because CEA lacks disease sensitivity and specificity, it cannot be used for screening the general asymptomatic population, a subpopulation with a high risk for malignancies, or for independently diagnosing cancer. However, CEA can be used as an aid in diagnosis, as an adjunct in clinical staging, to detect recurrence in patients who have undergone surgery, as a prognostic indicator, and to monitor the therapeutic response in patients undergoing chemotherapy or radiotherapy. In symptomatic cancer patients, an elevated CEA value, along with other clinical data and diagnostic methods, is strongly suggestive for the presence of cancer.

The CEA level and the degree of elevation has been shown to correlate with the pathologic stage of the cancer lesion. For instance, Wanebo et al. found that only 27.5% of patients with Dukes' stage A colon cancer had elevated CEA levels, whereas 83.8% of stage D patients had an elevation.[116] The degree of CEA elevation also correlated with staging, since only 9% of patients with stage B lesions had CEA elevations greater than four times the upper limit of the reference range and 69% of patients with stage D had similar CEA elevations. Colorectal cancer patients with initially high

CEA values are generally monitored by serial CEA measurements since values normalize after complete and successful tumor resection. CEA values generally remain normal in the absence of recurrent disease; however, a small number of patients (6%) will develop recurrent disease without a rise in CEA.[117] In postoperative patients, a rise in CEA is indicative of recurrent disease in approximately 66% of the cases. An elevation of CEA often precedes other clinical signs in 75% of patients developing recurrent disease by several months.[75,108] It has been suggested that postoperative follow-up should include CEA measurements every 6 weeks for a period of 2 years to increase resectability of recurring tumors, which will extend patient survival.[78]

CEA is also useful as a prognostic indicator. In colorectal carcinoma, patients with normal pretreatment CEA levels had a lower incidence of developing metastasis, whereas the majority of patients with initially high CEA levels developed metastasis. Also, CEA-producing tumors have the tendency to metastasize earlier, exhibit aggressive biologic disease activity, and show shorter disease-free intervals following therapy. Serial CEA measurements in patients undergoing chemotherapy or radiotherapy also show a good correlation between cancer activity and CEA values.[75] The National Cancer Institute Panel on CEA has recommended that CEA cannot be used independently to establish a diagnosis of cancer, and concluded that serial CEA measurements are the best available noninvasive tool for the postoperative monitoring of patients undergoing surgery for colorectal cancer. In addition to colon cancer, CEA has also been reported to be a useful tumor marker for breast cancer, bronchogenic carcinoma of the lung, pancreatic carcinoma, gastric carcinoma, urinary bladder cancer, prostatic cancer, ovarian cancer, neuroblastoma, and carcinoma of the thyroid.[98]

Alpha-Fetoprotein (AFP)

4/16

Alpha-fetoprotein (AFP) was the first tumor-associated antigen to be used as a tumor marker in the clinical management of human malignancies. AFP is an oncofetal antigen that was detected in human fetal sera in 1956.[13] It was later shown to be present in mice with spontaneous or transplanted hepatomas.[1] Interest in AFP became widespread (in 1964) when AFP was found in the sera of human patients with hepatocellular carcinoma. Further investigations revealed that serum AFP levels are also elevated in patients with ovarian, testicular, and presacral teratocarcinomas.[79]

Properties

AFP is a single chain glycoprotein with a molecular mass of 67 to 74 kD comprising 95% protein and 5% carbohydrate.[22] Variations in molecular mass may reflect variations in glycosylation that have been observed in AFP derived from different tumors. AFP is closely related to albumin, both chemically and physically. AFP preparations have considerable charge heterogeneity and numerous forms have been detected on isoelectric focusing in the pH 4.5 to 5.2 range.[22,33]

AFP is synthesized by the liver, yolk sac, and gastrointestinal tract of the developing fetus. AFP is released into the circulation and is normally found in sera from human fetuses greater than 6 weeks of age. The highest concentration of AFP occurs during embryonic fetal life. The fetal serum AFP concentration may reach a peak of 2 to 3 mg/mL by the 12th to 15th week of gestation, constituting about one third of the total fetal serum protein. At birth, the serum AFP declines to approximately 50 μg/mL, and by 1 year of age, the AFP level is at very low concentrations. AFP is found in trace amounts by 2 years of age and is similar to concentrations observed during adult life (usually less than 20 ng/mL).

Methods of Analysis

Initially, large-scale studies were performed to screen for liver cancer in high-risk areas, such as Asia and Africa, using the insensitive technique of immunodiffusion to measure AFP. In these studies, elevated serum AFP values were observed in 50% to 90% of patients with hepatocellular carcinoma, whereas apparently healthy individuals and patients with nonmalignant liver disease or other cancers did not have elevated AFP levels.[65,82] Similar studies in the United States revealed that this method is highly specific but lacks adequate sensitivity. More sensitive methods, such as EIA and RIA, are currently used in the United States and are commercially available. Depending on the assay method selected, each laboratory should establish its own reference ranges for serum AFP.[115]

Clinical Uses   – KNOW

Elevations in AFP (using sensitive and specific RIA or EIA assays) are observed in a wide variety of malignant and benign disorders. For instance, elevated AFP levels are seen in patients with hepatocellular carcinoma, testicular and ovarian germ-cell tumors, pancreatic carcinoma, gastric carcinoma, and colonic carcinoma. AFP is the best tumor marker for managing patients with hepatocellular carcinoma. Unfortunately, elevations of serum AFP, similar to levels observed in malignant disease, are seen in non-malignant hepatic disorders such as viral hepatitis and chronic active hepatitis. Serum AFP values can also be elevated in ataxia telangiectasia and tyrosinemia.[18,74,106,115]

Serum AFP levels are the most specific biochemical test for primary carcinoma of the liver. A National Institute of Health consensus panel has recommended that serum AFP should be measured as a screening test (along with ultrasound) in all patients suspected of liver cancer, cirrhosis, or chronic active hepatitis.[35] AFP levels can also be useful in patients who have a change in clinical status. An elevation or decline in AFP levels are helpful in monitoring the therapeutic response in cancer patients undergoing treatment. Postsurgical AFP levels that fail to return to normal may indicate incomplete surgical resection or metastatic disease. AFP levels also serve as a prognostic indicator, with higher values being associated with a larger tumor mass and a shorter survival time.[80]

AFP is also normally elevated in patients with testicular tumors of germ-cell origin (the most common is testicular cancer).[2] Human chorionic gonadotropin (hCG) is also elevated in testicular cancer; therefore, AFP and hCG are commonly used in combination for managing patients with this type of cancer. AFP (as well as hCG) is frequently not elevated in small testicular tumors, making this test unreliable as a screening tool for cancer.[11]

Germ-cell testicular tumors are classified as either seminomatous or nonseminomatous. AFP is not elevated in seminomatous testicular tumors, whereas the majority of patients with nonseminomatous tumors have elevated levels of AFP. The nonseminomatous tumors may be further divided into four major types: choriocarcinoma, yolk-sac tumor (endodermal sinus tumor), embryonal carcinoma, and teratoma. Testicular germ-cell tumors often contain two or more different tumor types. AFP is specifically elevated in embryonal carcinoma and yolk-sac tumor, whereas hCG is elevated in choriocarcinoma and in a few seminoma cases that contain syncytiotrophoblastic giant cells. The combined use of AFP and hCG as tumor markers for testicular cancer has proved essential in monitoring the response to therapy and in detecting disease recurrence. In one study, AFP was positive in 58% of nonseminomatous testicular cancer cases, and hCG was elevated in 77%. However, in 91% of the cases, either AFP or hCG was elevated.[96,101]

CA 125

Properties

CA 125 is an oncofetal antigen with a high molecular mass (>200 kD) that is produced by ovarian epithelial cells.[8] CA 125 is a glycoprotein, defined in 1981 by reactivity with a monoclonal antibody named OC125. This antibody was developed by immunizing mice with ascitic fluid from a patient with papillary serous cystadenocarcinoma of the ovary.[9] Early studies found that less than 1% of apparently healthy individuals and about 6% of patients with benign disease had elevations of CA 125. More importantly, greater than 80% of patients with ovarian carcinoma had elevated CA 125 values that appeared to correlate with tumor burden and extent of disease.

Methods of Analysis

Numerous commercial kits are available to measure CA 125 by EIA and RIA.[10] An automated EIA from Abbott Laboratories has recently been evaluated and should be available very soon in the United States. However, results obtained by the automated CA 125 method were 20% lower in apparently healthy women and at least 50% higher in patients with benign or malignant ovarian disorders when compared to RIA.[109] This illustrates the importance of establishing a new baseline value when changing the method used to measure a specific tumor marker level.

Clinical Uses

CA 125 is elevated in over 80% of patients with nonmucinous epithelial ovarian cancer. More than 50% of the newly diagnosed cases of ovarian carcinoma are epithelial in origin. Elevated CA 125 levels are also observed in patients with advanced nonovarian gynecologic malignancies, including endometrial cancer and metastatic malignancies of the breast, colon, pancreas, and lung. Mild CA 125 elevations are observed in several benign conditions, including menstruation, pregnancy, benign cysts, endometriosis, and cirrhosis.[62] In patients with a palpable pelvic mass, CA 125 levels higher than 35 U/mL (10 times normal) could be detected in 78% of patients with malignant pelvic tumors and 22% of patients with benign tumors.[111] Other studies have found that CA 125 is not highly sensitive for cancer in the early stages of disease, while others report that CA 125 cannot differentiate between benign and malignant disease.[23]

The major clinical use for CA 125 is in monitoring the treatment response of ovarian carcinoma. CA 125 levels that do not normalize following initial therapy are indicative of residual disease. Rising CA 125 levels after surgery or chemotherapy almost always represent cancer recurrence. In most cases, CA 125 levels above 35 U/mL are indicative of tumor recurrence.[77,84] However, a normal CA 125 does not always indicate the absence of tumor recurrence, since CA 125 elevations are not reliability observed with tumors less than 2 cm in size.[4] CA 125 is also useful as a prognostic indicator, since the degree of elevation correlates with tumor stage and size.[31] Also, increased levels after surgery are associated with shortened survival.

CA 19–9

Properties

CA 19–9 is an asialylated Lewis blood group antigen that was initially identified in a human colorectal carcinoma cell line in 1981.[69] This tumor marker is elevated in numerous malignancies such as colon, gastric, hepatobiliary, and pancreatic cancer. CA 19–9 can also be elevated in benign disorders such as pancreatitis, extrahepatic cholestasis, and cirrhotic liver disease. CA 19–9 is measured by standard RIA and EIA methods.[32]

Clinical Uses

Much like the other oncofetal tumor markers, CA 19–9 is not useful as a screening test for cancer. CA 19–9 has been found to be more sensitive and specific than CEA for pancreatic cancer, but it cannot identify patients with early resectable pancreatic cancer.[7] Using a cutoff of 40 U/mL, CA 19–9 was elevated in 81% of patients with pancreatic cancer, whereas a cutoff of 100 U/mL improved the sensitivity to 96%.[90,119] CA 19–9 is also elevated in 50% to 60% of patients with gastric carcinoma, 60% of patients with hepatobiliary cancer, and 30% of patients with colorectal carcinoma. The specificity of CA 19–9 for pancreatic cancer is further lowered by increased CA 19–9 levels typically found in patients with pancreatitis and liver disease. Since patients with benign pancreatitis usually have serum CA 19–9 elevations that do not normally exceed 120 U/mL, the specificity of CA 19–9 can be improved by increasing the cutoff value for pancreatic cancer.[56] Unfortunately, a cutoff of 120 U/mL would severely limit the use of CA 19–9 for pancreatic cancer, since patients in the early stages of disease (when surgical resection would be most beneficial) have CA 19–9 levels less than 120 U/mL. Nonetheless, CA 19–9 remains the most reliable serum test for differentiating between benign and malignant pancreatic cancer.

CA 19–9 is often used in conjunction with CEA to monitor recurrence of disease and there is good correlation between increasing CA 19–9 levels and recurrence of disease.[94,95] CA 19–9 levels return to normal after successful surgical resection, and increasing levels can detect tumor recurrence 1 to 7 months before a clinical diagnosis of relapse is made.[14] However, because treatment is ineffective after early detection of relapse, the clinical usefulness of CA 19–9 in these settings is somewhat limited.

CA 15–3

CA 15–3 is a high-molecular-weight glycoprotein mucin present on mammary epithelium known as episialin. An RIA method has been developed to detect CA 15–3 using antibodies against a glycoprotein found in membrane-enriched extracts of human breast cancer and human milk fat globule membrane.[70] CA 15–3 is elevated in patients with metastatic breast, pancreatic, lung, colorectal, and liver cancer. It is also elevated in benign diseases of the breast and liver. Several studies have found that CA 15–3 is elevated in less than 33% of patients with stage I or II breast cancer.[120] CA 15–3 is more useful in identifying patients with advanced stages of breast cancer since 79% to 92% of patients with progressive disease have elevations in CA 15–3. CA 15–3 is not recommended for monitoring therapeutic responses, since declining levels do not always occur in patients undergoing successful therapy.[110] Serum CA 15–3 levels correlate with tumor burden and is more sensitive than CEA for detecting disease recurrence.[93]

CA 549

CA 549 is an acidic glycoprotein located in the cell membranes and luminal surfaces of normal breast and other epithelial tissues. This tumor marker is found in serum as part of a cancer-associated mucin. EIA and RIA methods to measure CA 549 have been developed using antibodies against a human breast tumor cell line T 417 and human milk-fat globule membrane.[20] CA 549 is detectable in sera of healthy males and females. While levels can be high in benign disease, increased values are most often seen in a variety of malignant diseases such as breast, lung, prostate,

and colon cancer.[38] Elevations in CA 549 are not normally observed in the early stages of breast cancer but occur during disease progression to a more advanced stage.[27] CA 549 levels also correlate with tumor burden and may prove useful for monitoring treatment and disease progression.

Tissue Polypeptide Antigen (TPA)

TPA is a single-chain polypeptide related to the cytokeratin family. It consists of four protein subunits (A1, B1, B2, and C2) with individual molecular weights between 20 to 45 kD. TPA is synthesized during the S to M phase of the cell cycle and is an indicator of cell generation time and tumor aggressiveness.[16] An RIA method has been developed against the M3-specific epitope of TPA.[15] Serum levels of TPA may be increased in pregnancy and in several nonmalignant diseases such as bacterial/viral infections, acute hepatitis, and autoimmune disorders. TPA levels are also elevated in sera of patients with several malignancies including lung, breast, pancreas, colon/rectum, stomach, prostate, bladder, and uterus. For lung and breast cancer, serum levels of TPA appear to correlate with tumor burden and may be clinically useful in staging and prognosis.[24,41]

Prostate-Specific Antigen (PSA)

Properties

Between 1966 and 1978, three independent laboratories identified proteins in seminal fluid that were named gamma-seminoprotein, E1 antigen, and p30.[59] In 1979, Wang et al. purified a protein from benign and malignant prostatic epithelial cells that was named *prostate-specific antigen*.[118] Later studies revealed that all four substances were either the same or closely related molecules. PSA is a 30 to 34 kD serine protease that is secreted by prostatic epithelial cells. Although PSA was originally thought to be produced only by the prostrate, it is also produced by the periurethral and perirectal glands.[64] PSA is present in the circulation predominately complexed with alpha-2-macroglobulin and alpha-1-antichymotrypsin (ACT).[28] The complexing of PSA with serum proteins probably contributes to its relatively long plasma half-life (2–3 days). PSA is a more sensitive and accurate indicator of disease than prostatic acid phosphatase (PAP) and has largely replaced PAP in the management of known or suspected prostatic cancer patients.[118]

Methods of Analysis

The development of antibodies against PSA has led to several serum assays that are commercially available as EIA, RIA, and chemiluminescent immunoassay. The four major PSA manufacturers are Hybritech, Yang, Abbott, and Ciba-Corning. Although all four assays apparently measure the same analyte, there are considerable differences in recognition of free versus PSA bound to ACT, resulting in discordant results among assays. Differences in basic assay design (monoclonal versus polyclonal antibodies), antibody affinity, and PSA epitopes recognized by the tracer antibody, all contribute to assay discrepancies. For instance, the Hybritech Tandem-R assay recognizes free and complexed PSA in equimolar quantities whereas the other three methods overestimate free PSA by two to three times the actual amount. A lack of standardization in calibrator composition also contributes to differences in results among assays. Therefore, it is imperative that clinical laboratories establish their own reference ranges based on which method is used. In view of the observation that serum PSA levels increase with advancing age, Osterling and associates advocate using age-specific reference ranges.[92] Recently, ultrasensitive assays have been developed that are capable of measuring PSA levels below 0.1 ng/mL. These newer PSA assays have potential advantage over standard, less sensitive assays, for monitoring cancer recurrence, since post-therapeutic PSA levels are usually <0.1 ng/mL.[113]

Clinical Uses

PSA lacks sufficient sensitivity and specificity to be used alone as a screening test for prostate cancer.[30] Besides prostate cancer, elevations in PSA occur in benign prostatic hypertrophy, prostatitis, and intraepithelial neoplasia (a premalignant condition). Significant elevations in PSA do not normally occur following routine digital examination.[19,92] It is generally accepted that PSA can significantly improve prostate cancer detection when used in conjunction with digital rectal exam and transrectal ultrasound.[26] The usefulness of PSA may be further enhanced by the observation that the percentage of PSA bound to ACT (compared to free PSA) may be higher in patients with prostate cancer than in patients with benign prostatic hypertrophy or normal controls.[72,104] Although additional studies are needed, the proportion of bound PSA may facilitate distinction between benign and malignant prostatic diseases.

PSA is also extremely useful for monitoring patients with known prostatic carcinoma. There is a direct relationship between PSA serum concentrations and prostate gland volume. When expressed as PSA divided by prostate volume, patients with cancer were shown to have a ratio of 1.7 compared to 0.1 in normal controls and patients with benign prostatic hypertrophy.[112] The sensitivity of PSA can also be correlated with disease staging: 10%, stage A; 24%, stage B; 53%, stage C; 92%, stage D.[51,54] Following radical prostatectomy, radiation therapy, or androgen therapy, PSA normally declines to undetectable levels. Failure of PSA levels to become undetectable following therapy has been correlated with advanced disease stage and a poor prognosis. A rising PSA level following therapy precedes other indicators of tumor recurrence by 2 to 6 months. The 50% survival rate in postoperative patients (stages A2–D2) was 40 weeks for PSA concentrations one to two times normal and 18 weeks when the PSA was greater than two times normal.[29]

PLACENTAL PROTEINS

Human Chorionic Gonadotropin (hCG)

Properties

Human chorionic gonadotropin (hCG) is a heterodimer with a molecular mass of 40 kD, consisting of a noncovalently linked alpha and beta subunit. The alpha subunit is common to other glycoprotein hormones such as luteinizing hormone, follicle-stimulating hormone, and thyroid-stimulating hormone. The beta subunit of hCG is homologous to the beta subunit of luteinizing hormone except for an additional 24 amino acids at the carboxyl-terminal end that confers immunogenic specificity.[17,105] HCG is synthesized by the syncytiotrophoblastic cells of the placenta during pregnancy. HCG is found in trace amounts in healthy men and nonpregnant women, and becomes markedly elevated in the circulation during pregnancy, reaching a peak 8 to 10 weeks following conception. Serum and/or urine hCG is routinely measured to confirm pregnancy.

Methods of Analysis

Early RIA techniques using antisera against intact hCG and the radioreceptor assay lack specificity and are no longer in use. There are several qualitative and quantitative commercially available EIA and RIA for measuring intact hCG and the beta subunit of hCG. The specificity of the two antibodies (polyclonal or monoclonal) utilized in the EIA generally determine which forms of hCG are measured. For instance, if one antibody is against the alpha subunit and the other is against the beta subunit, then only intact hCG is measured. When both antibodies recognize the beta subunit, then both intact hCG and β-hCG are detected; however, binding to both forms of hCG may not be equal. When intact hCG and β-hCG are recognized with equal affinity, then the assay is usually designated as a total β-hCG assay. These distinctions are extremely important since hCG values for a specific patient can vary tremendously depending on assay design and relative proportions of intact hCG versus free β-hCG present. This issue is further complicated because reference standards used to calibrate hCG assays differ in relative proportions of intact and β-hCG, which can lead to different results depending on which forms are measured and how the assay is calibrated. Nevertheless, even with all of these variables, both intact hCG and β-hCG are used equally as a tumor marker and in other clinical situations.[88]

Clinical Uses

β-hCG is used as a tumor marker for diagnosis and management of gestational trophoblastic disease, which consists of complete and partial hydatidiform mole, gestational choriocarcinoma, and placental-site trophoblastic tumor. β-hCG is also useful for accurate staging and for monitoring the response to treatment in germ-cell tumors of the testis and ovary. The frequency of patients with elevated β-hCG has been estimated to be 15% for seminoma, 50% for embryonal carcinoma, 42% for teratocarcinoma, and approaches 100% for choriocarcinoma. Elevated β-hCG levels have also been observed in ovarian cancer, breast cancer, melanoma, gastrointestinal tract neoplasms, sarcoma, lung cancer, renal cancer, and hematopoietic malignancies. Serial measurements of β-hCG are best for monitoring the clinical course and response to therapy. Persistent elevations of β-hCG after therapy indicates residual disease. HCG (intact and β-hCG) is normally undetectable 4 to 6 weeks postpartum or after abortion. Persistent postpartum elevations may indicate trophoblastic disease.[124]

Human Placental Lactogen (HPL)

Human placental lactogen (HPL) is a trophoblastic hormone that is normally produced by syncytiotrophoblastic cells of the placenta. HPL has lactogenic properties, is present in the serum of pregnant women, and has been used as an index of placental function and fetal well-being. Levels of HPL correlate with fetal and placental weight, reaching a plateau in the last month of pregnancy, and becoming extremely low or undetectable 24 to 48 hours postpartum.

Elevated levels of HPL are associated with trophoblastic neoplasms (gestational, gonadal, or extragonadal), although levels are not as high as those observed in pregnancy. In trophoblastic malignancies, a rapid decline is normally observed after chemotherapy. HPL can be useful for differentiating trophoblastic disease from pregnancy, with low values being associated with disease whereas high values indicate pregnancy.[91]

Placental Alkaline Phosphatase (Regan Isoenzyme)

Properties

Placental alkaline phosphatase (PALP) is an isoenzyme of alkaline phosphatase synthesized by placental trophoblasts. PALP is normally found in the serum of pregnant women. In 1968, the Regan isoenzyme was identified in serum and tumor tissue from a patient with lung cancer.[43] In adults, the PALP gene is normally repressed, but Regan isoenzymes can be produced in various diseases as a slightly modified form of PALP. PALP (Regan isoenzyme) can be differentiated from other isoenzymes of alkaline phosphatase (liver, bone, intestine) by its remarkable heat stability or by using chemical inhibitors such as urea and l-phenylalanine. PALP (Regan isoenzyme) is resistant to urea treatment but sensitive to l-phenylalanine (80% inhibition in activity).[44]

Clinical Uses

PALP is elevated in about 10% to 20% of patients with a wide variety of malignancies such as lung, ovarian, gastrointestinal, and trophoblastic cancers. PALP is not specific for malignancy and is also observed in non-malignant disorders such as cirrhosis, ulcerative colitis, and diverticulitis. Its low incidence in cancer and poor disease specificity limits its usefulness as a tumor marker, although levels may correlate with therapeutic response in individual patients.

ENZYMES AND ISOENZYMES AS TUMOR MARKERS

Various enzymes can be increased in malignancy; however, enzyme elevations are not specific for the type of cancer, its location, or the presence or absence of metastasis. Mechanisms accounting for elevated serum enzyme levels are as follows: altered membrane permeability or necrosis of malignant cells resulting in enzyme leakage into the bloodstream, blockage of the biliary or pancreatic tract resulting in choleostatis, and altered renal clearance. Enzymes that can be used as tumor markers are shown in Table 24-1.

Prostatic Acid Phosphatase

Properties

Since its discovery over 50 years ago, acid phosphatase was one of the first laboratory tests used for the diagnosis of cancer.[55] Although acid phosphatases are produced primarily by the prostate gland, it is also found in bone, liver, spleen, kidney, bone marrow, and several cell types. Prostatic acid phosphatase (PAP) can be distinguished from other sources in enzymatic assays by using substrates, such as thymolphthalein monophosphate, that are more specific for PAP. EIA and RIA are also commercially available to measure PAP. This marker is one of the few enzymes showing organ and cell-type specificity.[87]

Clinical Uses

PAP currently has limited clinical utility since prostate specific antigen (PSA) has emerged as a tumor marker for detecting localized prostate disease and monitoring therapeutic response. PAP can be elevated in prostate cancer, benign prostatic hyperplasia, acute prostatitis, and after prostatic massage. PAP is not sensitive enough to be used as a screening test and is not suitable for detecting prostate cancer in the early stages of disease. For instance, only 12% (stage A disease) and 20% (stage B disease) of patients with localized and potentially curable prostate cancer had elevated PAP levels.[60] The only area where PAP may be superior to PSA is in selecting those patients with metastatic disease who are not candidates for surgery. Elevated PAP values were found in 84% to 100% of patients with prostate cancer exhibiting extracapsular extension and regional lymph node metastases.[6,85] In view of the limited role of PAP in the management of prostate cancer and the proven advantages of PSA, it is likely that PAP will be completely replaced by PSA in the not-too-distant future.

Creatine Kinase (CK)-BB Isoenzyme

Creatine kinase (CK) levels are helpful in establishing a diagnosis of myocardial infarction. CK is a dimer composed of two subunits designated M and B. CK-MM (CK3) is present in skeletal muscle; CK-MB (CK2) is found primarily in cardiac muscle; and CK-BB (CK1) is found in the brain, gastrointestinal tract, uterus, and prostate. Elevations in CK-BB (and total CK) can be found in prostatic carcinoma and metastatic cancer of the stomach.[100]

Alkaline Phosphatase

Alkaline phosphatase (ALP) has been used to detect primary malignancies in bone and liver and to detect metastases to these organs. Osteoblastic lesions in the bone produced by prostate cancer metastases and osteogenic sarcoma give rise to enormous elevations in ALP, whereas osteolytic lesions produced by metastatic breast cancer cause only mild or no elevations in ALP.[107] During successful treatment, there may be a paradoxical rise in ALP due to repair of damaged bone.

Other causes of elevated ALP include extrahepatic obstruction of the biliary tract and usually results in a twofold increase in levels of ALP. Diseases such as leukemia that infiltrate the liver ("space occupying lesions") can cause marked elevations in ALP levels that correlate with the extent of liver involvement. The source of the elevated ALP (bone or liver) can be identified by measuring other liver enzymes such as γ-glutamyltransferase and 5′ nucleotidase, or by measuring ALP activity after heat inactivation of the bone isoenzyme.[83] A much more specific test to measure bone-specific alkaline phosphatase has recently been developed that should be extremely helpful in managing patients with various bone disorders.[86]

Neuron-Specific Enolase

Neuron-specific enolase (NSE) is a glycolytic enzyme that is found in neurons, neuroendocrine cells, and in neurogenic tumors.[97] NSE is an enzyme marker for tumors associated with the neuroendocrine system such as small-cell carcinoma of the lung, neuroblastoma, carcinoid, and pancreatic islet tumors. Elevated levels of NSE are found in 70% of patients with small cell lung cancer. NSE levels decline after successful therapy and rise during recurrence of disease.[25] Elevations of NSE are also found in 90% of children with neuroblastoma.[125] An NSE RIA is commercially available to monitor patients with these types of cancers.

Lactate Dehydrogenase (LD)–1 Isoenzyme

LD is composed of four peptide chains of either the M or H type that combine to produce five isoenzymes. Elevations in LD occur in 70% of patients with liver metastasis and 20% to 60% of cancer patients without liver involvement. The LD5 isoenzyme is normally elevated in breast, stomach, lung, and colon cancers. In leukemias, an increase in total LD occurs in about 50% of patients and is usually associated with LD3 isoenzyme elevations. LD can also be used to monitor response to therapy.[126]

The LD–1 isoenzyme has been shown to be markedly elevated in germ cell tumors such as teratoma and seminoma of the testis. However, LD–1 isoenzyme elevations are also observed in benign conditions, such as acute myocardial infarction, hemolytic disorders, and renal infarction, and should not be used independently to diagnose cancer.

HORMONES

(can be used as tumor markers)

Eutopic hormones are secreted by endocrine tissue under normal physiologic conditions and in excess in certain neoplastic disorders (Table 24-4). Ectopic hormones are produced de novo by endocrine or non-endocrine malignant tissue(s) that normally do not produce the hormone. Inappropriate synthesis and release of hormones by neoplasms produce severe systemic effects that usually occur distant from the primary cancer or metastasis. These specific effects are described as paraneoplastic endocrine syndromes. Clinical manifestations of excess hormonal secretion and the associated malignancies of endocrine paraneoplastic syndromes are detailed in Table 24-5. The symptoms resulting from paraneoplastic syndromes may be a patient's only presenting indication of malignancy and can be life-threatening. Approximately 20% of cancer patients will have a paraneoplastic syndrome. Paraneoplastic syndromes are most commonly associated with ectopic hormone production but may manifest as connective-tissue/dermatologic disorders, neurologic syndromes, or as a consequence of vascular, hematologic, or immunologic complications.[34] Hormones, such as estrogen, progesterone, and testosterone, are now being measured routinely in the clinical laboratory by EIA and RIA techniques that have excellent analytical sensitivity and specificity.

TABLE 24-4. Eutopic Hormone Production

Hormone	Clinical Symptom	Tumor Site
β-hCG	Menstrual irregularities	Trophoblastic tumors
Calcitonin	Hypocalcemia	Thyroid-medullary carcinoma
Gastrin	Zollinger-Ellison syndrome	Non-β-cell pancreatic tumor
Catecholamines	Hypertension	Pheochromocytoma

TABLE 24-5. Ectopic Hormone Production

Hormone	Clinical Symptom	Tumor Site
ACTH	Cushing's syndrome	Lung, thyroid, pancreas
ADH	Hyponatremia	Oat cell carcinoma
Serotonin	Carcinoid syndrome	Enterochromaffin cell
Erythropoietin	Erythrocytosis	Kidney
TSH	Hyperthyroidism	Trophoblastic tumors

STEROID RECEPTORS

4/19

Hormonally dependent organs, such as breast, prostate, and uterus, give rise to malignancies that may in themselves be hormonally dependent. Eliminating hormonal stimulation, by the use of drugs or removal of the hormone-producing organ/tissue, often leads to a regression of the respective malignancy. However, not all cancers arising from hormone-dependent tissues respond to such therapy, and removing the hormonal stimulation carries considerable risk to the patient. Predicting which tumors will most likely respond to hormonal manipulation can be achieved by measuring steroid hormone receptors on tumors.[122]

Properties

Steroid hormones (estrogen, progesterone, testosterone) are transported in plasma by a number of proteins, including testosterone-estradiol binding globulin (TEBG) and corticosteroid-binding globulin (CBG). Unbound steroid enters the cell, apparently by passive diffusion, and complexes with its respective receptor protein. Receptor proteins are found in a variety of concentrations in target organs, but are virtually absent in non-target tissues. Receptor proteins have a high affinity and specificity for their respective steroid hormones. Complexing of the steroid-receptor in the cytoplasm induces translocation into the nucleus of the cell where it associates with chromatin acceptor sites. This initiates RNA synthesis for proteins important in cellular growth and differentiation.

Methods of Analysis

Tissue for cytosolic steroid receptor analysis must be representative of the malignancy. The tissue is first homogenized to release the hormone receptors that can then be quantitated by analyzing the supernatant. Sensitive EIA are presently available to measure estrogen and progesterone receptors and can be performed with considerably less tissue than the older radioreceptor assay.[37]

Clinical Uses

Estrogen (E^R) and progesterone receptor (Pg^R) assays are used to predict the therapeutic response of breast cancer patients to hormonal treatment. Levels greater than 10 fmol/mg for both

E^R and Pg^R are considered positive. Approximately 80% of breast cancer patients that are positive for both E^R and Pg^R respond to hormonal therapy. The higher the concentration of E^R and Pg^R, the more predictable the therapeutic response. About 30% of patients that are positive for E^R and negative for Pg^R will respond to hormonal therapy, while 40% to 50% of patients with E^R negative and Pg^R positive cancers will respond. As expected, only a small fraction of patients (2%) with breast cancer lacking both receptors respond to hormonal therapy.[63,71]

Receptors for other types of cancers are currently being studied and appear to correlate with clinical response to hormonal manipulation. These include androgen and progesterone receptors for prostate cancer, estrogen and progesterone receptors for endometrial cancer, and binding of acute lymphoblastic leukemia cells to dexamethasone.[50,73,123]

IMMUNOPHENOTYPING

With the development of highly specific fluorescent monoclonal antibodies (mAb) and a new generation of flow cytometers, it is now possible to routinely detect a variety of antigens and receptors on the cell surface. This can aid in classifying hematopoietic cells along their respective lineages and in identifying their degree of maturation.[21,40]

Immunophenotyping, using comprehensive antibody panels, can serve as confirmatory evidence of the morphologic diagnosis in leukemias and lymphomas. In general, human leukemic (lymphoid) cells are phenotypically similar to their normal hematopoietic counterpart. In lymphoid leukemias and lymphomas, identification of tumor cells as either T or B lymphocytes can be an important predictor of clinical outcome. The use of monoclonal antibodies in acute nonlymphoid (myelogenous) leukemia is particularly helpful in differentiating acute myelogenous leukemia from acute lymphoblastic leukemia.

Methods of Analysis

Various panels are used to differentiate among the various acute leukemias. Multiple mAb, recognizing different antigens, should be used in order to identify a particular cell lineage and the maturational stage. A listing of commonly used commercially available mAb is shown in Table 24-6. Single-cell suspensions are required for analysis, with blood and

TABLE 24-6. Commonly Used Monoclonal Antibodies

Cluster	Antibody/Common Name	Specificity
Lymphoid Antibodies		
CD1	Leu-6, OKT-6, T6	Thymocytes
CD2	Leu-5, OKT-11, T11	Pan T cell (Sheep RBC receptor)
CD3	Leu-4, OKT-3, T3	Pan T cell (T cell receptor complex)
CD4	Leu-3, OKT-4, T4	T helper/inducer
CD5	Leu-1, OKT-1, T1	Thymocytes, B cell subset
CD8	Leu-2, OKT-8, T8	T suppressor cells
CD10	CALLA	Common acute lymphoblastic leukemia antigen
CD19	Leu-12, B-4	Immature B cells, all malignant B cell neoplasms except multiple myeloma
CD20	Leu-16, B-1	B lymphocytes, all malignant B cells except multiple myeloma
CD24	BA-1	Peripheral B cells
CD25	Tac	Interleukin-2 receptor
CD45	HLe-1	Pan leukocyte antigen
—	PCA1	Plasma cells
—	HLA-DR, DQ	MHC class II antigen
Myeloid Antibodies		
CD11	Leu-15, My-7	Granulocytes, monocytes
CD14	Leu-M3, My-4	Monocytes
CD15	Leu-1, My-1	Granulocytes, monocytes
CD33	Leu-6, My-9	Early myeloid
—	Glycophorin	Erythroblasts
CD34	HPCA, My-10	Early myeloid
CD45	HLe-1	Pan leukocyte

tissue being the most common source of cells. Mononuclear cells can be isolated from whole blood by density gradient separation (centrifugation through Ficoll-Hypaque) or by lysis of red blood cells. Solid or semisolid tissues, such as lymph nodes and bone marrow aspirates, require gentle deaggregation.

Single cell suspensions are stained either directly or indirectly with fluorescent mAb. Direct staining uses an antibody conjugated to a fluorescent label, whereas indirect staining uses a cell surface specific antibody as the first step followed by a fluorescently labeled antibody against IgG or IgM. Commonly used fluorochromes include fluorescein isothiocyanate, phycoerythrin, and rhodamine.

Fluorescently labeled cells are analyzed by a flow cytometer. The flow cytometer quantitates the number of cells that are positive for the antigen of interest (based on the amount of light emitted following excitation of the fluorochrome) and also provides information about cellular size and nuclear complexity. Two or more mAb that recognize different antigens (labeled with different fluorochromes) can be used simultaneously to determine the total number of T lymphocytes (CD3) and the proportion of T lymphocytes expressing the CD4 (T helper cell phenotype) and CD8 (T suppressor cell phenotype) surface antigens.[81]

Clinical Uses

Surface antigen phenotyping is used to classify acute leukemias, to provide prognostic data, and to aid in selecting the appropriate treatment regimen. Other techniques involving mAb and flow cytometry that are currently being used in the clinical laboratory include measuring CD4/CD8 T lymphocyte subset ratios, reticulocyte counting, detection of platelet-associated IgG in immune thrombocytopenia, and anti–neutrophil antibodies.[57]

DNA ANALYSES

Measuring nuclear DNA content (ploidy) and proliferative capacity (S-phase fraction) of malignant cells is useful for predicting the overall survival and disease-free interval in cancer patients. DNA analysis can help discriminate between benign and malignant disease, monitor disease progression, and predict response to treatment. In humans, nonreplicating normal cells in G_0 or G_1 phase of the cell cycle are referred to as diploid. During the synthetic phase (S phase), DNA content increases (aneuploid), which is followed by mitosis (M phase) and cell division. Malignant cells can be nonreplicating (diploid) or contain increased amounts of DNA (aneuploid). Ploidy status or DNA index (DI) of malignancies is defined by the amount of measured DNA in cancer cells relative to that in normal cells and is calculated by the following equation:

$$DI = \frac{\text{peak channel number of aneuploid } G_0/G_1 \text{ peak}}{\text{peak channel number of diploid } G_0/G_1 \text{ peak}}$$

(Eq. 24-1)

A diploid population has a DI equal to 1 by definition. Any DI not equal to 1 is termed aneuploid. A DI of <1 is hypodiploid whereas a DI>1 is considered hyperdiploid. The percentage of cells in S phase of the cell cycle can also be determined by flow cytometry. A %S-phase value >5 is considered elevated.[39,99]

Methods of Analysis

The determination of DNA content is one of the most common clinical applications of flow cytometry. The tissue may be fresh, fixed with alcohol, or removed from paraffin tissue blocks and rehydrated prior to analysis. Cells are permeabilized and DNA within nuclei are stained with DNA intercalating agents, such as propidium iodide and ethidium bromide. The DNA content is determined by fluorescence intensity (channel number).

Clinical Uses

DNA flow cytometry measurements of ploidy (DI) and %S-phase fractions are prognostic indicators in cancer patients. In breast cancer, both aneuploidy and high S-phase fraction correlate with the lack of estrogen and progesterone receptors. These cases are associated with a worse prognosis when compared to diploid, low S-phase, receptor-positive malignancies. In general, aneuploidy is seen in advanced disease and shortened disease-free intervals. Aneuploidy in ovarian cancer is associated with a shorter survival and in prostate cancer it predicts the likelihood of regional spread or metastasis. DNA analysis has also been used to aid in the diagnosis and monitoring of disease and the response to therapy. Flow cytometric results of bladder washings appear to correlate well with bladder biopsies. In colorectal cancer, the majority of malignancies are aneuploid, and those with a high S-phase have a poorer prognosis.

Molecular Diagnosis

Molecular diagnosis employs techniques that can detect alterations at the level of the gene (DNA). It is firmly established that when mutations occur within DNA from chemicals, radiation, or viruses, cancer may arise. DNA analysis in the past decade has made a quantum leap from the research laboratory to the clinical setting and has been used in microbiology, genetics, and forensic pathology. It is rapidly emerging in the area of cancer diagnosis.[53]

Viruses have been associated with a number of cancers in animals and humans. Viral DNA segments that can transform normal cells into malignant cells are known as oncogenes. Normal cellular genes that play essential roles in cell differentiation and proliferation and potentially can become oncogenic are known as proto-oncogenes. The transformation of proto-oncogenes into oncogenes can occur by single-point mutations, translocation, and amplification. Only 20% to 25% of human malignancies are associated with activated oncogenes. In addition, genes that normally repress cellular proliferation can become inactivated resulting in uncontrolled cellular growth. Cancer suppressor genes have been identified in retinoblastoma, Wilm's tumor, and hepatic carcinoma.

Methods of Analysis

Early molecular diagnostic techniques were time consuming, difficult to perform, and involved the use of ^{32}P labeling. With the advent of nonisotopic tags, automation, the polymerase chain reaction, and the growing market of vendors, molecular diagnostic techniques will rapidly become a routine test in the clinical laboratory.

SUMMARY AND FUTURE DEVELOPMENTS

During the last two decades, numerous clinically useful tumor markers have been identified and used in the management of cancer patients. Currently used tumor markers and their associated tumors are shown in Table 24-7. Unfortunately, the majority of available tumor markers are also elevated in nonneoplastic disease processes. In addition, a normal tumor marker value does not necessarily exclude a diagnosis of cancer, since most tumor markers have poor disease sensitivity in the early stages of cancer. Consequently, with few exceptions, none of the tumor markers can be used as a screening tool or can make an independent diagnosis of cancer. Introduction of molecular techniques into the clinical arena, and the rapidly growing number of methods currently available, may make it possible to identify patients at risk for the development of cancer. Perhaps molecular probes will fulfill the criteria of the ideal tumor marker in the not-too-distant future.

TABLE 24-7. Major Tumors and Associated Tumor Markers

Tumor Type	Associated Tumor Marker
Hepatocellular carcinoma	AFP, ferritin, ALP liver isoenzyme, LD5
Germ cell tumors of the testis and ovary	AFP, β-hCG, PTH, calcitonin, CK-BB, PALP (Regan isoenzyme), prolactin
Gastrointestinal cancer	CEA, AFP, fetal sulfoglycoprotein
Breast cancer	CEA, β-hCG, ferritin, casein, mammary tumor–associated glycoprotein (MTGP), calcitonin, TPA
Pancreatic cancer	CEA, pancreatic oncofetal antigen (POA), ACTH
Prostatic cancer	Prostatic acid phosphatase, CEA, LD5 isoenzyme
Trophoblastic diseases (choriocarcinoma or hydatidiform mole)	β-hCG, LD, LD1, SP-1
Renal and bladder cancer	CEA, LD
Multiple myeloma, Waldenstrom's macroglobulinemia	IgG, IgA, IgM, IgD, and IgE immunoglobulins, free lambda and kappa light chains (Bence-Jones protein)

A 72-year-old female complained of intermittent diarrhea and constipation for the last 2 weeks and a weight loss of 22 lb over the last 6 months. Physical examination was unremarkable except for a guaiac positive stool. A colonoscopy was performed, which identified a circumferential mass in the sigmoid colon. A biopsy was performed, which identified the mass as an adenocarcinoma. A carcinoembryonic antigen (CEA) level was obtained as part of her initial surgical workup.

Questions

1. Is the CEA test useful as a screening test for colon carcinoma?
2. What are some other conditions that can result in an elevated CEA level?
3. How is CEA used to monitor patients following surgery for colon cancer?
4. What other laboratory tests should be considered in the initial surgical workup of this patient?

A 25-year-old man was found to have an elevated lactate dehydrogenase (LD) of 350 IU/L (reference range 93–139 IU/L) that was confirmed with a second sample. LD electrophoresis was performed and revealed an elevation in the LD–1 isoenzyme above the normal range. Electrocardiogram (ECG) studies were negative for an acute myocardial infarct. Creatine kinase (CK) electrophoresis showed only a CK-MM band. On physical examination, the man was found to have a tumor mass in the left scrotum. Serum alpha-fetoprotein (AFP) and beta-human chorionic gonadotropin (β-hCG) tests were ordered.

Questions

1. What is the most likely primary diagnosis for this patient?
2. What kind of benign diseases should be considered in the differential diagnosis? Why?
3. Can the type of testicular tumor be determined based on the serum AFP and β-hCG results?
4. Can a final diagnosis be made based only on the tumor marker findings? If not, why not?

A 69-year-old Caucasian male presents to the medicine clinic with complaints of increased frequency of urination and difficulty urinating. Further questioning revealed the need to urinate several times a night. Although the problem has been present for some time, it has recently become worse. A serum prostate-specific antigen (PSA) was obtained and a digital rectal examination was performed. The PSA level was elevated at 15 ng/mL (reference range <4 ng/mL) and the digital exam revealed an enlarged prostate without palpable nodules.

Questions

1. Can serum PSA be used alone as a screening test for prostate cancer?
2. What are some causes for an elevated serum PSA level?
3. If this patient has cancer limited to the prostate, should PSA levels be monitored after radical prostatectomy and/or radiation therapy?
4. What other laboratory tests should be performed in the initial workup of this patient?

REFERENCES

1. Abelev GI, Perova SD, Kharmkova NI, Postinikova ZA, Irlin IS. Production of embryonic α-globulin by transplantable mouse hepatomas. Transplantation 1963;1:174.

2. Altug MU, Akdas A, Ruacan S, et al. Tissue alpha-fetoprotein and human chorionic gonadotrophin in non-seminomatous testicular tumors. Br J Urol 1987;59:458.

3. American Cancer Society. 1989 survey of physicians attitudes and practices in early cancer detection. CA Cancer J Clin 1990;2:77.

4. Atack DB, Nisker JA, Allen HH, et al. CA 125 surveillance and second-look laparotomy in ovarian carcinoma. Am J Obstet Gynecol 1986;154:287.

5. Bacchus H. Serum glycoproteins and malignant neoplastic disorder. Crit Rev Clin Lab Sci 1977;8:333.

6. Bahnson RR, Catalona WJ. Adverse implications of acid phosphatase levels in the upper range of normal. J Urol 1987;137:427.

7. Bakemeier RF, Krachov SF, Adams JT, Poulter CA, Rubin P. Cancer of the major digestive glands: Pancreas, liver, bile ducts, and gallbladder. In: Rubin P. Clinical oncology. American Cancer Society Chicago, 1983;178.

8. Bast RC Jr, Feeney M, Lazarus H, et al. Reactivity of a monoclonal antibody with human ovarian carcinoma. J Clin Invest 1981;68:133.

9. Bast RC, Feeney M, Lazarus H, et al. Reactivity of a monoclonal antibody with human ovarian cancer. J Clin Invest 1981;68:1331.

10. Bast RC, Klug TL, St John E, et al. Radioimmunoassay using a monoclonal antibody to monitor the course of epithelial ovarian cancer. N Engl J Med 1983;309:883.

11. Bates SE, Longo DL. Clinical applications of serum tumor markers in cancer diagnosis and management. Semin Oncol 1987;14:102.

12. Bates SE, Longo DL. Tumor markers: Value and limitations in the management of cancer patients. Cancer Treat Rev 1985;12:163.

13. Bergstrand CG, Czar B. Paper electrophoretic study of human fetal serum protein with demonstration of a new protein fraction. Scand J Clin Lab Invest 1957;9:277.

14. Berretta E, Malesci A, Zerbi A, et al. Serum CA 19-9 in the postsurgical follow-up of patients with pancreatic cancer. Cancer 1987;60:2428.

15. Björklund B, Björklund V, Brunkener M, Gronlund H, Back M. The enigma of human tumor marker: TPA revisited. In: Cimino F, Birkmayer GD, Klavins JV, Pimentel E, Salvatore F, eds. Human tumor markers. Berlin: Walter de Gruyter, 1987;169.

16. Björklund B, Björklund V. Specificity and basis of the tissue of polypeptide antigen. Cancer Detect Prev 1983;6:41.

17. Bohn H, Dati F. Placental-related proteins. In: Ritzman SE, Killingsworth LM, eds. Proteins in body fluids, amino acids, and tumor markers: Diagnostic and clinical aspects. New York: Alan R Liss, 1983;333.

18. Bosl GJ, Lange PH, Fraley EE. Human chorionic gonadotropin and alpha-fetoprotein in the staging of nonseminomatous, testicular cancer. Cancer 1981;47:328.

19. Brawer MK, Schifman RB, Ahmann FR, Ahmann ME, Coulis KH. The effect of digital rectal examination on serum levels of prostate-specific antigen. Arch Pathol Lab Med 1988;112:1110.

20. Bray KR, Koda JE, Gaur C, et al. Serum levels and biochemical characteristics of cancer-associated antigen CA 549, a circulating breast cancer marker. Cancer Res 1987;47:5853.

21. Bray RA, Landay AL. Identification and functional characterization of mononuclear cells by flow cytometry. Arch Pathol Lab Med 1989;113:579.

22. Brummund W, Arvan DA, Mennuti MT, Starkousky NA. Alpha-fetoprotein in the routine clinical laboratory: Evaluation of a simple radioimmunoassay and review of current concepts in its clinical application. Clin Chem Acta 1980;105:25.

23. Buahmah PK, Skillen AW. Serum CA 125 concentrations in patients with benign ovarian tumors. J Surg Oncol 1994;56:71.

24. Buccheri G, Ferrigno D. Prognostic value of the tissue polypeptide antigen in lung cancer. Chest 1992;101:1287.

25. Carney DN, Ihde DC, Cohen MM, et al. Serum neuron specific enolase. Lancet 1982;1:583.

26. Catalona WJ, Smith DS, Ratliff TL, et al. Measurement of prostate-specific antigen in serum as a screening test for prostate cancer. New Engl J Med 1991;321:1156.

27. Chan DW, Beveridge RA, Bhargava A, Wilcox PM, Kennedy MJ, Schwartz MK. Breast cancer marker CA 549: A multicenter study. Am J Clin Pathol 1994;101:465.

28. Christensson A, Laurell C-B, Lilja H. Enzymatic activity of prostrate-specific antigen and its reactions with extracellular serine proteinase inhibitors. Eur J Biochem 1990;194:755.

29. Chu TM, Murphy GP. What's new in tumor markers for prostate cancer. Urology 1986;27:487.

30. Crawford ED, DeAntoni EP. PSA as a screening test for prostate cancer. Urol Clin North Am 1993;20:637.

31. Cruickshank DJ, Fullerton WT, Klopper A. The clinical significance of pre-operative serum CA 125 in ovarian cancer. Br J Obstet Gynecol 1987;94:196.

32. Del Villano BC, Brennan S, Brook P, et al. Radioimmunometri assay for a monoclonal antibody defined tumor marker. Clin Chem 1983;29:549.

33. Deutsch HF. Chemistry and biology of α-fetoprotein. Adv Canc Res 1991;56:253.

34. DeWys WD, Killen JY. The paraneoplastic syndromes. In: Rubin P, Bakemeier RF, Krachov SK, eds. Clinical oncology, A multidisciplinary approach. 6th ed. American Cancer Society Chicago 1983;112.

35. Di Bisceglie AM, Rustgi VK, Hoofnagle JH, et al. Hepatocellular carcinoma. Ann Intern Med 1988;108:390.

36. Dieksheide WC. CEA assay precautions. [Letter]. Diagn Med 1981;4:114.

37. DiFronzo G, Miodini P, Brivio M, et al. Comparison of immunochemical and radioligand binding assays for estrogen receptors in human breast tumors. Cancer Res 1986;46:4278.

38. Dnistrian AM, Schwartz MK, Greenberg EJ, Smith CA, Dorsa R, Schwartz DC. CA 549 as a marker of breast cancer. Int J Biol Markers 1991;6:139.

39. Dressler LG, Bartow SA. DNA flow cytometry in solid tumors: Practical aspects and clinical applications. Semin Diagn Pathol 1989;6:55.

40. Drexler HG, Gignac SM, Minowada J. Routine immunophenotyping of acute leukemias Blut: J Clin Exp Hematology 1988;57:327.

41. Eskelinen M, Hippeläinen, M, Kettunen J, Salmela E, Pentillä I, Alhava E. Clinical value of serum tumor markers TPA, TPS, TAG 12, CA 15-3 and MCA in breast cancer diagnosis; results from a prospective study. Anticancer Res 1994; 14:699.

42. FDA drug bulletin. 1981;11:617.

43. Fishman WH, Inglis NR, Stolbach LL, Krant MJ. A serum alkaline phosphatase isoenzyme of human neoplastic cell origin. Cancer Res 1968;28:150.

44. Fishman WH. Perspectives on alkaline phosphatase isoenzymes. Am J Med 1974;56:617.

45. Fletcher RH. Carcinoembryonic antigen. Ann Intern Med 1986;104:66.

46. Foon KA, Todd RF. Immunologic classification of leukemia and lymphoma. Blood 1986;68:1.

47. Fritsche HA. Tumor marker test in patient monitoring. Lab Med 1982;13:528.

48. Gardner RC, Feinerman AE, Kantrowitz PA, et al. Serial carcinoembryonic antigen (CEA) blood levels in patients with ulcerative colitis. Am J Dig Dis 1978;23:129.

49. Garret PE, Kurtz SR. Clinical utility of oncofetal proteins and hormones as tumor markers. Med Clin North Am 1986; 70:1295.

50. Geller J. Hormone dependency of prostate cancer. In: Thompson BE, Lippman ME, eds. Steroid receptors and the management of cancer. Boca Raton, FL: CRC Press, 1979;113.

51. Gillatt DA, Ferro MA, Gingell JC, Barnes I. Prostate specific antigen in adenocarcinoma of the prostate. 1988;318:992.

52. Go VL. Carcinoembryonic antigen: Clinical application. Cancer 1976;37:562.

53. Grody WA, Gatti WW, Faramarz N. Diagnostic molecular pathology. Mod Pathol 1989;6:553.

54. Guinan P, Bhatti R, Ray P. An evaluation of prostate specific antigen in prostatic cancer. J Urol 1987;137:686.

55. Gutman AB, Gutman EB. "Acid" phosphatase occurring in serum of patients with metastasizing carcinoma of prostate gland. J Clin Invest 1938;17:473.

56. Haglund C, Roberts PJ, Kuusela P, Jalanko H. Tumor markers in pancreatic cancer. Scand J Gastroenterol Suppl 1986; 126:75.

57. Hansen CA. Nontraditional applications of flow cytometry. In: Keren DF, ed. Flow cytometry in clinical diagnosis. Chicago: ASCP Press, 1989;280.

58. Hansen HJ, Snyder LJ, Miller E, et al. Carcinoembryonic antigen (CEA) assay. A laboratory adjunct in the diagnosis and management of cancer. J Hum Pathol 1974;5:139.

59. Hara M, Koyanagi Y, Inoue T, Fukuyana T. Some physicochemical characteristics of gamma-seminoprotein: An antigenic component specific for human seminal plasma. Jpn J Leg Med 1971;25:322.

60. Heller JE. Prostatic acid phosphatase: Its current clinical status. J Urol 1987;137:1091.

61. Henderson BE, Ross RK, Pike MC. Toward the primary prevention of cancer. Science 1991;254:1131.

62. Jacobs I, Bast RC Jr. The CA 125 tumor-associated antigen: A review of the literature. Hum Reprod 1989;4:1.

63. Jaing NS. Breast cancer: Estrogen and progesterone receptor assays as a guide to therapy. Mayo Clin Proc 1983;58:64.

64. Kamashida S, Tsutsumi Y. Extraprostatic localization of prostatic acid phosphatase and prostate-specific antigen: Distrib

ution in cloacogenic glandular epithelium and sex-dependent expression in human anal gland. Hum Pathol 1990;21:1108.

65. Kew MC. Clinical, pathologic, and etiologic heterogenity in hepatocellular carcinoma: Evidence from South Africa. Hepatology 1981;1:366.

66. Khoo SK, MacKay IR. Carcinoembryonic antigen in serum in diseases of the liver and pancreas. J Clin Pathol 1973; 26:470.

67. Khoo SK, Warner NL, Lie JT, MacKay IR. Carcinoembryonic antigen activity of tissue extracts: A qualitative study of malignant and benign neoplasms, cirrhotic liver, normal adult and fetal organs. Int J Cancer 1973;11:681.

68. Klee GG, Go VLW. Serum tumor markers. Mayo Clin Proc 1982;57:129.

69. Koprowski H, Herlyn M, Steplewise Z, Sears HF. Specific antigen in serum of patients with colon carcinoma. Science. 1981;212:53.

70. Kufe DW, Inghirami G, Abe M, Hayes DF, Justi-Wheeler H, Schlomm J. Differential reactivity of a novel monoclonal antibody (DF3) with human malignant versus benign breast tumors. Hybridoma 1984;3:223.

71. Lesser ML, Rosen PP, Senie RT, et al. Estrogen and progesterone receptors in breast carcinoma: Correlation with epidemiology and pathology. Cancer 1981;48:299.

72. Lilja H, Christensson A, Dahlen U, Matikainen M, Nilsson O, Peterson K. Prostate-specific antigen occurs predominantly in a complex with α_1-antichymotrypsin. Clin Chem 1991;37:1618.

73. Lippman ME, Yarbo GK, Leventhal BG, Thompson EB. Clinical correlations of glucocorticoid receptors in human acute lymphoblastic leukemia. In: Thompson BE, Lippman ME, eds. Steroid receptors and the management of cancer. Boca Raton, FL: CRC Press, 1979;67.

74. Liu FJ, Fritsche HA, Trujillo JM, Samuels ML, Logothetis CJ, Trindade A. Serum alpha-fetoprotein (AFP); beta subunits of human chorionic gonadotropin (HCG) and lactate dehydrogenase 1 isoenzyme (LD-1) in patients with germ cell tumors of the testis. Am J Clin Pathol 1983;80:120.

75. Lunde OCH, Havig O. Clinical significance of carcinoembryonic antigen (CEA) in patients with adenocarcinoma in colon and rectum. Acta Chir Scand 1982;148:189.

76. Lurie BB, Lowenstein MS, Zamcheck N. Elevated carcinoembryonic antigen levels and biliary tract obstruction. J Am Med Assoc 1975;233:326.

77. Malkasian GD, Knapp RC, Lavin PT, et al. Preoperative evaluation of serum CA-125 levels in premenopausal and postmenopausal patients with pelvic masses: Discrimination of benign from malignant disease. Am J Obstet Gynecol 1988; 159:341.

78. Martin EW Jr, Cooperman M, Carey LC, Minton JP. Sixty second-look procedures indicated primarily by rise in serial carcinoembryonic antigen. J Surg Res 1980;28:389.

79. Masopust J, Keithier K, Ridl I, Koutecky I, Kotal L. Occurrence of fetoprotein in patients with neoplasms and non neoplastic diseases. Int J Cancer 1968;3:364.

80. Masseyeff R. Human alpha-fetoprotein. Pathol Biol 1972; 20:703.

81. McCoy JP, Lovett EJ. Basic principles in clinical flow cytometry. In: Keren DF, ed. Flow cytometry in clinical diagnosis. Chicago: ASCP Press, 1989;12.

82. McMahon BJ, London T. Workshop on screening for hepatocellular carcinoma. J Natl Cancer Inst 1991;83:916.

83. Narayanan S. Alkaline phosphatase as a tumor marker. Ann Clin Lab Sci 1983;12:133.

84. Niloff JM, Knapp RC, Schaetzl E, et al. CA-125 antigen levels in obstetric and gynecologic patients. Obstet Gynecol 1984;64:703.

85. Oesterling JE. Clinically useful serum markers for adenocarcinoma of the prostate II: Prostate-specific antigen. Am Urol Assoc Update Series 1991;10:137.

86. Panigraphi K, Delmas PD, Singer F, et al. Characteristics of a two-site immunoradimetric assay for human skeletal alkaline phosphatase in serum. Clin Chem 1994;40:822.

87. Pappas AA, Gadsden RH. Acid phosphatase, clinical utility in detection, assessment and monitoring of prostatic carcinoma. Ann Clin Lab Sci 1984;14:285.

88. Pardue MA, Taylor EH, London S, Walls RC, Pappas AA. Clinical comparison of immunoradiometric assays for intact versus beta-specific human chorionic gonadotropin. Am J Clin Pathol 1990;93:347.

89. Reynoso G. CEA, basic concepts, clinical applications. Diagn Med 1981;4:41.

90. Ritts RE, DelVillano BC, Go VLW, Herberman RB, Klug TL, Zurawski VR. Initial clinical evaluation of an immunoradiometric assay for CA 19-9 using the NCI serum bank. Int J Cancer 1984;33:339.

91. Rosen SW, Weintraub BD, Vaitukatis JL, et al. Placental proteins and their subunits as tumor markers. Ann Intern Med 1975;82:71.

92. Ruckle HC, Klee GG, Oesterling JE. Prostate specific antigen: Critical issues for the practicing physician. Mayo Clin Proc 1994;69:59.

93. Safi F, Kohler I, Rottinger E, et al. The value of the tumor marker CA 15-3 in diagnosis and monitoring breast cancer. A comparative study with carcinoembryonic antigen. Cancer 1991;68:574.

94. Safi F, Roscher R, Beger HG. The clinical relevance of the tumor marker CA 19-9 in the diagnosis and monitoring of pancreatic carcinoma. Bull Cancer 1990;77:83.

95. Satake K, Kanazawa G, Kho I, et al. Evaluation of serum pancreatic enzymes, carbohydrate antigen 19-9, and carcinoembryonic antigen in various pancreatic diseases. Am J Gastroenterol 1985;80:630.

96. Scardino PT, Cox PD, Waldman TA, et al. The value of serum tumor markers in the staging and prognosis of germ cell tumors of the testis. J Urol 1977;118:994.

97. Schmechel D, Marangos PJ, Brightman M. Neuron-specific enolase is a molecular marker for peripheral and central neurendocrine cells. Nature 1978;276:834.

98. Seleznick MJ. Tumor markers. Cancer Diagn Treat 1992; 19:715.

99. Shapiro HM. Flow cytometry of DNA content and other indicators of proliferative activity. Arch Pathol Lab Med 1989; 113:591.

100. Silverman LM, Dermer GB, Zweig MH, et al. Creatinine kinase BB: A new tumor associated marker. Clin Chem 1979;25:1432.

101. Skinner DG, Scardino PT, Daniels JR. Testicular cancer. Ann Rev Med 1981;32:543.

102. Statland BE. The challenge of cancer testing. Diagn Med 1981;4:13.

103. Statland BE. Tumor markers. Diagn Med 1981;4:21.

104. Stenman UH, Leinonen J, Alfthan H, Rannikko S, Tuhkanen K, Alfthan O. A complex between prostate-specific antigen and α_1-antichymotrypsin is the major form of prostate-specific antigen in serum of patients with prostatic cancer: Assay of the complex improves clinical sensitivity for cancer. Cancer Res 1991;51:222.

105. Stigbrand T, Engall E. Placental proteins as tumor markers. In: Sell S, Wahren B, eds. Human cancer markers. Clifton NJ: Humana Press, 1982;275.

106. Talerman A, Haije WG, Baggerman L. Alpha-1 antitrypsin (AAT) and alpha-fetoprotein (AFP) in sera of patients with germ cell neoplasms: Value as tumor markers in patients with endodermal sinus tumor (yolk sac tumor). Int J Cancer 1977; 19:741.

107. Tatarinov Y. New data on the embryo-specific antigen components of human blood serum. Vopr Med Khim 1964; 10:584.

108. Teramoto YA, Mariani R, Wunderlich D, Schlom J. The immunochemical reactivity of a human monoclonal antibody with tissue sections of human mammary tumors. Cancer 1982;50:241.

109. Thomas CMG, Massuger LFAG, Segers MFG, Schijf CPT, Doesburg WH, Wobbes T. Analytical and clinical performance of improved Abbott Imx CA 125 Assay: Comparison with Abbott CA 125 RIA. Clin Chem 1995;41:211.

110. Tondini C, Hayes DF, Gelman R, et al. Use of various epithelial tumor markers and a stromal marker in assessment of cervical carcinoma. Obstet Gynecol 1991;77:566.

111. Vasilev SA, Schlaerth JB, Campeau J, Morrow CP. Serum CA 125 levels in preoperative evaluation of pelvic masses. Obstet Gynecol 1988;71:751.

112. Vesey SG, Goble M, Ferro MA, Stower MJ, Hammonds JC, Smith PJB. Quantification of prostatic cancer metastatic disease using prostate-specific antigen. Urology 1990;35:483.

113. Vessella RL. Trends in immunoassays of prostate-specific antigen: Serum complexes and ultrasensitivity. Clin Chem 1993;39:2035.

114. Virji MA, Mercer DW, Herberman RB. Tumor markers in cancer diagnosis and prognosis. CA Cancer J Clin 1988; 38:104.

115. Waldman TA, McIntire KR. The use of a radioimmunoassay for alpha-fetoprotein in the diagnosis of malignancy. Cancer 1974;34:1510.

116. Wanebo HJ, Rao B, Pinsky CM, et al. Preoperative carcinoembryonic antigen level as a prognostic indicator in colorectal cancer. N Engl J Med 1978;299:448.

117. Wang JY, Tang R, Chiang JM. Value of carcinoembryonic antigen in the management of colorectal cancer. Dis Colon Rectum 1994;37:272.

118. Wang MC, Valenzuala LA, Murphy GP, et al. Purification of a human prostate-specific antigen. Invest Urol 1979;17:159.

119. Welfare M, Tanner A. CA 19-9 estimation in the assessment of pancreatic or biliary tract malignancy—What is the most efficient threshold value. Eur J Gastroenterol Hepatol 1991; 3:539.

120. Werner M, Faser C, Silverberg M. Clinical utility and validation of emerging biochemical markers for mammary adenocarcinoma. Clin Chem 1993;39:2386.

121. Wingo PA, Tong T, Bolden S. Cancer statistics, 1995. CA Cancer J Clin 1995;45:8.

122. Witliff JL. Steroid-hormone receptors in breast cancer. Cancer 1984;53:630.

123. Young PCM, Ehrlich CE. Progesterone receptors in human endometrial cancer. In: Thompson BE, Lippman ME, eds. Steroid receptors and the management of cancer. Boca Raton, FL: CRC Press, 1979;135.

124. Zarate A, MacGregor C. Beta subunit HCG and the control of trophoblastic disease. Semin Oncol 1982;9:187.

125. Zeltzer PM, Marangos PJ, Evans AE, Schneider SL. Serum neuron-specific enolase in children with neuroblastoma. Cancer 1986;57:1230.

126. Zondag HA, Klein F. Clinical applications of lactate dehydrogenase isoenzymes: Alterations in malignancy. Ann NY Acad Sci 1968;151:578.

Chapter 25

Porphyrins, Hemoglobin, and Myoglobin

Dennis W. Jay, Louann W. Lawrence

Objectives

Upon completion of this chapter, the clinical laboratorian should be able to:

- *Describe the chemical nature and structure of porphyrins and hemoglobin.*

- *Relate the role of porphyrins in the body.*

- *Outline the biochemical pathway of porphyrin and heme synthesis.*

- *Discuss the clinical significance of the porphyrias.*

- *Correlate the porphyrin disease states with clinical laboratory data.*

- *Compare and contrast the porphyrias with regard to enzyme deficiency and clinical symptoms.*

- *Explain the principles of the basic qualitative and quantitative porphyrin tests, to include PBG, ALA, uroporphyrin, coproporphyrin, and protoporphyrin.*

- *Relate the degradation of hemoglobin.*

- *Discuss the clinical significance and laboratory data associated with the hemoglobinopathies and thalassemias.*

- *Identify the tests used in the diagnosis of the hemoglobinopathies and thalassemias.*

- *Discuss the structure and clinical significance of myglobin in the body.*

KEY WORDS

Cytochrome	Porphyria	Porphyrinuria
Hemoglobinopathy	Porphyrin	Pyrrole
Myoglobin	Porphyrinogens	Thalassemia

Because of chemical similarities, porphyrins, hemoglobin, and myoglobin are grouped together for discussion in this chapter. All of these compounds contain the porphyrin ring, which is composed of four pyrrole groups bonded by methene bridges (Fig. 25-1). Porphyrins are able to chelate metals to form the functional groups that participate in oxidative metabolism.

PORPHYRINS

Role in the Body

Porphyrins are chemical intermediates in the synthesis of hemoglobin, myoglobin, and other respiratory pigments called *cytochromes.* They also form part of the peroxidase and catalase enzymes, which contribute to the efficiency of internal respiration. Iron is chelated within porphyrins to form heme. Heme is then incorporated into proteins to become biologically functional hemoproteins. Porphyrins are analyzed in clinical chemistry to aid in the diagnosis of a group of disorders called the *porphyrias,* which result from disturbances in heme synthesis. The presence of excess amounts of these intermediate compounds in urine, feces, or blood indicates a metabolic block in heme synthesis.

Chemistry of Porphyrins

The porphyrins found in nature are all compounds in which side chains are substituted for the eight hydrogen atoms found in the four *pyrrole* rings that make up porphyrin (see Fig. 25-1). Because of the wide variety of substitutions, many porphyrins have been described in nature. The pigment chlorophyll is a magnesium porphyrin and is essential

Michael L. Bishop, Janet L. Duben-Engelkirk, and Edward P. Fody.
CLINICAL CHEMISTRY. © 1996 Lippincott–Raven Publishers.

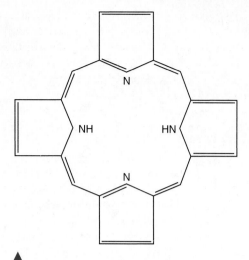

Figure 25-1. Basic structure of porphyrins.

for plants to use light energy to synthesize carbohydrates. Four basic isomers may exist for every porphyrin compound, but only type I and type III occur in nature. The difference between type I and type III isomers is in the arrangement of side chains. Only type III isomers form heme, but in some disorders the functionless type I isomers may be present in excess in the tissues. Porphyrins are stable compounds, red-violet to red-brown in color, that fluoresce red when excited by light near 400 nm. Only three porphyrin compounds are clinically significant in humans: protoporphyrin (PROTO), uroporphyrin (URO), and coproporphyrin (COPRO). Their presence in excess in biologic fluids is a clinical sign of abnormal heme synthesis. The three compounds have different solubility properties and different degrees of ionization determined by the addition of various carboxyl groups to the basic porphyrin structure. This allows for separate assays of each. URO is excreted in urine, PROTO in the feces, and COPRO in either, depending on the rate of formation of the urine and its pH.

The reduced forms of porphyrins are termed *porphyrinogens*. This is the functional form of the compound that must be used in heme synthesis. Porphyrinogens are highly unstable, colorless, and nonfluorescent, which makes them more difficult to analyze. In the presence of light, oxygen, or oxidizing agents, porphyrinogens are readily oxidized to their corresponding porphyrin form. Therefore, the porphyrin form is routinely analyzed in clinical laboratories.

Porphyrin Synthesis

All types of cells contain hemoproteins and can synthesize heme, but the main sites are the cells of the bone marrow and liver. The series of irreversible reactions is outlined in Figure 25-2. Some steps occur in the mitochondria of the cell and some steps in the cytoplasm. The transport of substrates across the mitochondrion membrane is a complex process and a potential point for interruptions in heme synthesis.

Control of the rate of heme synthesis in the cells in the liver is achieved largely through regulation of the enzyme delta-aminolevulinic acid (ALA) synthase. The main mechanism is thought to be repression of synthesis of new enzyme. A negative feedback mechanism exists whereby increases in the pool of hepatic heme diminish the production of ALA synthase. Conversely, ALA synthase production is increased with a depletion of heme. The size of the regulatory heme pool may be affected by the requirement for hemoproteins in the liver. Drugs and other compounds appear to induce ALA synthase production by several different mechanisms, but all result in a depletion of the regulatory heme pool. Therefore, the rate of heme synthesis is flexible and can change rapidly in response to a wide variety of external stimuli. In bone marrow erythrocytes, another mechanism seems to control the rate of heme synthesis. Heme regulates the uptake of iron by reticulocytes, thereby influencing the supply of iron for the enzyme ferrochelatase.[5]

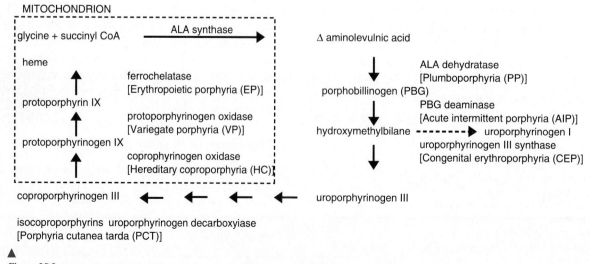

Figure 25-2. Synthesis of heme. Diseases associated with enzyme deficiencies are given in brackets.

Clinical Significance and Disease Correlations

The porphyrias are inherited or acquired enzyme deficiencies, which result in overproduction of heme precursors in the bone marrow (erythropoietic porphyrias) or the liver (hepatic porphyrias). Disease states corresponding to enzyme deficiencies have been identified in every step of heme synthesis except for ALA synthase. Some patients demonstrate an enzyme deficiency but do not show clinical or biochemical manifestations of porphyria, indicating that other factors, such as demand for increased heme biosynthesis, are also important in causing disease expression. An excess of the early precursors in the pathway of heme synthesis (ALA, porphobilinogen, or both) causes neuropsychiatric symptoms including abdominal pain, vomiting, constipation, tachycardia, hypertension, psychiatric symptoms, fever, leukocytosis, and parasthesias. Porphyrias in this category include plumboporphyria (PP) and acute intermittent porphyria (AIP). Excesses of the later intermediates (uroporphyrins, coproporphyrins, and protoporphyrins) may cause cutaneous symptoms including photosensitivity, blisters, excess facial hair, and hyperpigmentation. Porphyria cutanea tarda (PCT), hepatoerythropoietic porphyria (HEP), erythropoietic porphyria (EP), and congenital erythropoietic porphyria (CEP) are associated with cutaneous symptoms. Porphyrin-induced photosensitivity manifests itself either by increased fragility of light-exposed skin, as in PCT, or by burning of light-exposed skin, as in EP. The photosensitizing effects of the porphyrins are attributable to their absorption of light. It is currently believed that light energy excites electrons of porphyrins to an elevated energy level, forming triplet-state porphyrins, which, in turn, react with molecular oxygen forming compounds that damage tissues through several mechanisms.[41] There may also be excesses of both early and late intermediates causing neurocutaneous symptoms. Hereditary coproporphyria (HCP) and variegate porphyria (VP) fall into this category. All of the porphyrias are inherited as autosomal dominant traits except for PP and CEP, which are autosomal recessive.

The diagnosis of porphyrias is made by a combination of history and physical and laboratory findings. The cutaneous porphyrias are easier to diagnose since photosensitivity is usually the presenting symptom. Laboratory diagnosis, if necessary, is made by analysis of the appropriate sample for intermediates in heme synthesis (Table 25-1). The differentiation of neurologic porphyrias from other disorders is more difficult based on history and physical examination and must be verified by laboratory findings.

Inherited plumboporphyria (ALA dehydratase deficiency) is extremely rare, with only three cases having been reported.[28] The pattern of inheritance is autosomal recessive and affected patients are homozygous for the deficiency. Urinary ALA is markedly elevated with normal PBG excretion. The much more common cause of low ALA dehydratase activity is lead poisoning and must be ruled out prior to diagnosis of an inherited disorder.

Acute intermittent porphyria (AIP) usually appears in the third decade of life. Although the true prevalence is unknown, a reasonable estimate of gene frequency is 5 to 10 per 100,000 in the United States.[5] The enzyme deficiency is inherited as autosomal dominant, but only about 10% of the patients who inherit the enzyme deficiency suffer attacks of the disease, so other etiologic factors are involved. The most common precipitating causes of disease are drugs, especially barbiturates and sulfonamides, but a wide variety of drugs are potentially hazardous. This disease is characterized by multiple neurologic symptoms with colicky stomach pain and sometimes fever and vomiting. The characteristic laboratory findings are a marked elevation of ALA and PBG in the urine, but these test results may be normal between attacks. The urines of patients with clinically manifest AIP may turn red or black after exposure to light and air due to nonenzymatic polymerization of PBG to porphyrin or porphobilin.

Congenital erythropoietic porphyria (CEP) is one of the rarest porphyrias and appears shortly after birth. Fewer than 100 cases have been reported. It is also known as *Gunther's disease*. The teeth fluoresce red under ultraviolet light and exhibit a red or brownish discoloration under normal light due to porphyrin deposits in the dentin. The urine is often red because of the presence of URO and COPRO. The major fraction is the type I isomer. Photosensitivity is a major clinical problem, resulting in lesions that may become infected, leaving the patient scarred, or with multiple occurrences, that may lead to mutilation of the ears, nose, or

TABLE 25-1. Metabolites Found in Excess in the Porphyrias

Porphyria	Urine	Feces	Erythrocytes
Plumboporphyria (PP)	ALA	Normal	Normal
Acute intermittent porphyria (AIP)	ALA, PBG	Normal	Normal
Congenital erythropoietic porphyria (CEP)	URO, COPRO	URO, COPRO	PROTO, URO, COPRO
Porphyria cutanea tarda (PCT),	URO	ISOCOPRO	Normal
Hepatoerythopoietic porphyria (HEP)	URO	ISOCOPRO	Normal
Hereditary coproporphyria (HC)	ALA, PBG, COPRO	COPRO	Normal
Variegate porphyria (VP)	ALA, PBG, COPRO	PROTO > COPRO	Normal
Erythropoietic porphyria (EP)	Normal	PROTO	PROTO

digits. Abnormal growth of hair is often seen in exposed areas. It is thought that the disfigurements of this disorder and the tendency to avoid daylight (hence only coming outside at night) led to the legend of the werewolf.[24]

Hepatoerythropoietic porphyria (HEP) is also very rare. It has been described in only 20 patients. Photosensitivity begins in childhood. Patients are severely affected and develop excess facial hair and scarring of the hands and face. The severity of photosensitivity improves somewhat with age, but hepatic disease follows. The liver shows a red fluorescence and a nonspecific hepatitis or portal inflammation. Adults with HEP usually have a mild anemia, and erythrocyte precursors in the bone marrow are fluorescent. Erythrocyte PROTO levels are increased. Urine and fecal porphyrin levels are similar to those found in PCT. In fact, this disease is considered to be a homozygous form of PCT.[5]

Porphyria cutanea tarda (PCT) is the most common of the porphyrias. It presents with cutaneous blistering and fragility in light-exposed areas, typically the hands, along with some abnormal growth of hair. Liver biopsy specimens from these patients show fluorescence, hemosiderosis, fatty infiltration, and variable degrees of necrosis and fibrosis. PCT is often acquired secondary to liver damage due to alcoholism, estrogen therapy, or hexachlorobenzene (insecticide) poisoning. In addition, the disease is now recognized with increasing frequency in young women who use oral contraceptives and in patients on chronic hemodialysis. In the latter, the cause is probably related to iron overload due to multiple blood transfusions. It is well established that excess iron in the liver plays a role in the pathogenesis of PCT. The disease goes into remission if iron stores are decreased. However, the exact role of iron is still unknown. Both alcohol and estrogen are capable of causing modest increases in the activity of ALA synthase, but it is not known if this is the primary cause or a secondary cause due to inhibition of heme synthesis.[5] In the laboratory, PCT is characterized by increased levels of urinary URO and fecal isocoproporphyrins (ISOCOPRO).[14,48]

Hereditary coproporphyria (HCP) is a fairly mild condition, although attacks have been precipitated by exposure to certain drugs. Some carriers of the abnormal gene have normal excretion of porphyrins and porphyrin precursors, similar to silent carriers of AIP. HCP is less common than AIP, but the true prevalence is unknown. The hallmark of HCP is markedly increased excretion of COPRO III in the feces and, to a lesser extent, in the urine. During acute attacks, urinary excretion of ALA and PBG is also increased. Clinical manifestations of HCP may be neurologic, as for AIP, or cutaneous, with lesions resembling those in PCT.

Variegate porphyria (VP) is prevalent in South Africa and can be traced to a single couple that immigrated from the Netherlands in 1688. Clinical manifestations include acute attacks of neurologic dysfunction, such as those in AIP, photodermatitis, as in HCP and PCT, or a combination of both. Hallmark laboratory findings are increased levels of COPRO and PROTO in the feces. During acute attacks, urinary ALA and PBG excretion is increased, but in asymptomatic subjects these are often normal.

Erythropoietic porphyria (EP) is the second most common porphyria. The major clinical symptom is photosensitivity, which is usually present from infancy. Patients complain of burning, itching, or pain in the skin on exposure to sunlight. Some patients also have severe liver disease. The diagnosis of EP is made by demonstrating increased levels of PROTO in erythrocytes, plasma, and stool. Some individuals have no clinical manifestations of the disease but have increased levels of erythrocyte PROTO. They are considered to be clinically unaffected carriers of the gene defect.

Treatment in the inherited porphyrias is aimed at modifying the biochemical abnormalities causing clinical symptoms. The cutaneous symptoms are treated by avoiding sunlight, using sun blocking agents and oral β-carotene, which acts as a singlet oxygen trap preventing skin damage. Reduction of the heme load can be accomplished by phlebotomy or by giving desferroxamine to chelate iron. Intravenous hematin may be used to counteract acute attacks of neurologic dysfunction. Hematin, an enzyme inhibitor, limits synthesis of porphyrins in cells in the bone marrow. Cessation of precipitating factors, such as ingestion of alcohol or estrogens, may be useful in PCT.[15,26]

Secondary porphyrias or *porphyrinurias* is the term given to acquired conditions in which a mild to moderate increase in excretion of urinary porphyrins is seen. In this case, the disorders are not the result of an inherited biochemical defect in heme synthesis but are due to another disorder, toxin, or drug interfering with heme synthesis. Symptoms may be very similar to the inherited porphyrias in some cases. Various anemias, liver diseases, and toxins, such as lead and alcohol, fit into this category. Lead is known to inhibit both the activity of PBG synthase and the incorporation of iron into heme.[40] Secondary porphyrias can be distinguished from true porphyrias by measuring levels of urinary ALA and porphobilinogen. In secondary porphyria, ALA levels are increased in the urine, while PBG excretion usually remains normal. Lead poisoning classically exhibits increased COPRO in the urine and erythrocyte zinc-PROTO, as well as increased ALA. Determination of blood lead is the most sensitive method to detect lead poisoning. If this test is not available, determination of ALA in urine is a sensitive test for subclinical lead poisoning and superior to urinary COPRO.[5,20,52]

Methods of Analyzing Porphyrins

There are individual enzyme assays available for each of the defective enzymes that cause porphyria. However, most are still limited to use in specialized laboratories and will not be discussed here. Screening tests can be performed easily, even in small laboratories, but care should be taken in interpretation since false-negatives and false-positives occur.[8,40] Quantitative assays should follow all screening tests. Quantitative assays of the three porphyrins (URO, PROTO, and COPRO) and two porphyrin precursors (ALA and PBG) will serve to classify most porphyrias.

The two most commonly used screening tests for urinary PBG are the Watson-Schwartz test and the Hoesch test. The Hoesch test is preferred by some because false-positive reactions due to urobilinogen may occur in the Watson-Schwartz test. Owing to the possibility of ambiguity in interpretation of either test, some laboratories prefer to use the Hoesch test to confirm the results of the Watson-Schwartz test.[40]

WATSON-SCHWARTZ TEST FOR URINARY PORPHOBILINOGEN (PBG)[49]

Principle—PBG upon condensation with Ehrlich's reagent (acidic *p*-dimethylaminobenzaldehyde) will form a red-orange colored product.

Special Considerations—A fresh morning urine specimen is preferred. If a quantitative analysis is being done, a 24-hr specimen is better. Specimens may be kept for several weeks under refrigeration, provided they are adjusted to a neutral pH.

Reagents—Ehrlich's reagent—0.7 g reagent-grade *p*-dimethylaminobenzaldehyde plus 150 mL of concentrated hydrochloric acid to 100 mL of distilled water and stored in an amber bottle.

Saturated sodium acetate
Chloroform
n-Butanol
pH test paper

Procedure—Mix 2.5 mL of urine and 2.5 mL of Ehrlich's reagent. Add 5 mL of saturated sodium acetate and check the pH. Adjust pH to between 4 and 5. If no color develops, the test is negative. If a red color develops, add 5 mL of chloroform or butanol. This helps differentiate the color formed with PBG from urobilinogen or indole. PBG-positive color cannot be extracted with chloroform or butanol. A cherry-red color remains in the aqueous phase, indicating a positive test for PBG.

Comments—Sometimes substances will impart a color to the urine prior to performing this test. If this happens, perform the quantitative test instead. This test will remove most interfering chromogens. ■

HOESCH TEST FOR URINARY PORPHOBILINOGEN (PBG)[31,52]

Special Considerations—A freshly voided urine specimen is preferred. If this is not possible, urine may be stored at a pH of 7. At this pH, PBG is stable at 4°C for at least 1 week and at room temperature for at least 2 days.

Reagent—Ehrlich I reagent—Dissolve 2 g of *p*-dimethylaminobenzaldehyde (analytical grade) in 50 mL of concentrated HCl (analytical). Add 50 mL of distilled water and mix. The reagent is stable at room temperature in a tightly stoppered dark container.

Procedure—To approximately 2 mL of Ehrlich I reagent in a 10 × 75 mM test tube, add 1 to 2 drops of urine. If a pink-to-red color develops immediately (within 2 seconds) at the top of the fluid, the test is positive for PBG. The yellow color often seen is due to urea. Urobilinogen concentrations up to 200 mg/dL yield no red color.

Comments—The test becomes positive at PBG concentrations between 20 and 100 mg/dL. A negative test on freshly voided urine excludes the possibility that a patient's symptoms are due to an acute porphyric attack.

Porphobilinogen and ALA are determined quantitatively by isolating the compounds on ion-exchange columns. A fresh sample of urine or a 24-hour sample of urine is obtained. PBG and ALA are separated by absorbing the PBG onto alumina and eluting with dilute acetic acid. ALA that is not absorbed is then isolated on a Dowex-50 column and eluted with sodium acetate. PBG is then determined with Ehrlich's reagent. ALA is also determined with Ehrlich's reagent after first being condensed with acetylacetone to form a pyrrole. The resulting color from both compounds is then measured spectrophotometrically at 552 nm. The readings must be taken at 15 minutes when the color has reached its maximum.[34] ■

QUALITATIVE TESTS FOR PORPHYRINS[12,18]

Principle—Porphyrins have a characteristic red fluorescence when viewed under ultraviolet light because of their chemical properties. The porphyrins are extracted into an organic solvent system such as acetic acid–ethyl acetate and then reextracted into HCl.

Special Considerations—All screening tests should be performed using glass apparatus because synthetic materials may cause quenching of fluorescence. A first morning sample of urine or a 24-hr sample may be used; whole blood is collected in any anticoagulant, and about 1 g is needed for fecal analysis. If the screening test is positive and a quantitative test is necessary, it should be performed on a 24-hr urine specimen collected in a container with 4 to 5 g sodium bicarbonate to maintain the pH at 6 to 8. When porphyrin analyses cannot be performed soon after collection, specimens should be stored in the dark at 4°C or frozen if saved for more than 1 to 2 days.[40]

Reagent—3.0 mL HCl and 1 part of glacial acetic acid to 4 parts of ethyl acetate.

Procedure

Urine—Mix 5 mL of urine with 3 mL of acetic acid–ethyl acetate mixture. Shake and allow to separate. Illuminate the top layer with an ultraviolet (Wood's) lamp for the characteristic orange-red fluorescence. To check for any interfering substances that may affect the color of the reaction, carefully remove this upper layer with a pipet and mix with 0.5 mL HCl. With this mixture, the porphyrins are extracted into the lower layer and most interfering substances remain in the upper layer. Fluorescence in this lower (acidic) layer confirms the presence of porphyrins.

Blood—Mix 1 mL of whole blood with 3 mL of acetic acid–ethyl acetate mixture. Stir, centrifuge, and decant the top layer into a tube containing 0.5 mL of HCl. Mix and allow the two phases to separate, and observe the lower layer with an ultraviolet lamp for characteristic fluorescence.

Feces—Take a small portion (1 g) of the material onto the end of a glass rod and mix in a test tube containing 3 mL of acetic acid–ethyl acetate mixture. Shake and centrifuge. Decant the supernatant into a second tube with 0.5 mL of HCl. Shake and allow to settle or centrifuge. Observe the lower layer for fluorescence.

Most quantitative procedures for porphyrins are based on solvent extraction and measurement by fluorometric or spectrophotometric methods.[44] Electrophoresis and thin-layer chromatography have been used in the past.[23,45] Newer methods include high-pressure liquid chromatography (HPLC)[10,32,33] and high-performance thin-layer chromatography,[30] which are able to identify almost all of the porphyrin intermediates. ■

HEMOGLOBIN

Role in the Body

Hemoglobin has many important functions in the body. Its major role is oxygen transport to the tissues and CO_2 transport back to the lungs. The hemoglobin molecule is designed to take up oxygen in areas of high oxygen tension and release oxygen when in areas of low oxygen tension. Hemoglobin is carried to all tissues of the body by erythrocytes. Hemoglobin is also one of the major buffering systems of the body.

Structure of Hemoglobin

Hemoglobin is a large, complex protein molecule with a molecular weight of approximately 64,000. It is roughly spherical in shape and is composed of two major parts: heme, which makes up 3% of the molecule, and globin proteins, which make up the remaining 97%. The heme portion is composed of a porphyrin ring with iron chelated in the middle. The iron atom is the site of reversible oxygen attachment. The protein portion is composed of two pairs of globins that are twisted together so that heme groups are exposed on the exterior of the molecule (Fig. 25-3). The complete hemoglobin molecule contains four heme groups attached to each of four globin chains and may carry up to four molecules of oxygen. Each globin chain contains 141 or more amino acids.

The structure of each chain is fourfold. The primary structure is the individual amino acids and their sequences. Their sequences vary and are the basis of chain nomenclature: alpha, beta, delta, and gamma. The secondary structure is the three-dimensional arrangement of the amino acids making up the polypeptide chain. Regions of amino acids may form helixes or a pleated structure. The tertiary structure is a larger fold superimposed on the helical or pleated forms. It represents the position taken by each chain or subunit in three-dimensional space. Finally, the quaternary structure represents the relationship of the four subunits to one another, particularly at the points of contact. Mutations at particular points of contact result in altered specific functional properties of the molecule, such as its oxygen affinity.

The majority of hemoglobin in normal adults is designated as hemoglobin A, or A_1, which contains two alpha and two beta chains (Fig. 25-3). Hemoglobin A_2, composed of two alpha and two delta chains, makes up less than 3% of normal adult hemoglobin. The remainder is composed of hemoglobin F, which contains two alpha and two gamma chains. Hemoglobin F is the main hemoglobin during fetal life and is about 60% of normal hemoglobin at birth. There is a gradual switch from production of delta chains to beta chains, and by about 9 months of age, hemoglobin F usually constitutes less than 1% of total hemoglobin. Hemoglobin F has a greater affinity for oxygen than hemoglobin A; therefore, it is a more efficient oxygen carrier for the fetus. Hemoglobin F is more resistant to alkali than hemoglobin A, and this is the basis of one laboratory test to differentiate these two types of hemoglobin.

Two other hemoglobin chains, designated zeta and epsilon, are present only in embryonic life. Production of these chains stops by the tenth week of gestation, and gamma–chain production takes over. The three embryonic hemoglobins are identified as Gower I, two zeta chains, and two epsilon chains; Gower II, two alpha chains and two epsilon chains; and Portland I, two zeta chains and two gamma chains.[24]

Genetic control of hemoglobin synthesis occurs in two areas: control of structure and control of rate and quantity of

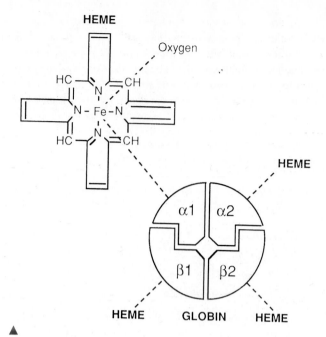

Figure 25-3. Hemoglobin A: structure of the hemoglobin molecule. (From Linman, JW. Hematology. New York: Macmillian, 1975. Reprinted with permission.)

production. Defects in structure produce a group of diseases called the hemoglobinopathies. Defects in rate and quantity of production lead to disorders called the thalassemias. Structurally, each globin chain has its own genetic locus; therefore, it is the individual chains, not the whole hemoglobin molecule, that is under genetic control. The genes for the globin chains can be divided into two major groups: the α-genes, which are on chromosome 16, and the non-α-genes, which are on chromosome 11. In most persons, the α-gene locus is duplicated—there are two alpha-chain genes per haploid set of chromosomes. The α-gene and hence its chain are identical in hemoglobins A, A_2, and F. The non-α-genes for the beta, delta, and gamma chains are sufficiently close in genetic terms to be subjected to nonhomologous crossover, with the resulting production of fused or hybrid globin chains such as hemoglobin Lepore (delta-beta-globin chain) and Kenya (gamma-beta-globin chain).

Based on the genetics of the globin-chain production, the structural abnormalities can be divided into four groups:

1. Amino acid substitutions (*e.g.,* hemoglobins S, C, D, E, O, G, etc.)
2. Amino acid deletion—deletions of three or multiples of three nucleotides in DNA (*e.g.,* hemoglobin Gun Hill)
3. Elongated globin chains resulting from chain termination, frame shift, or other mutations (*e.g.,* hemoglobin Constant Spring)
4. Fused or hybrid chains resulting from nonhomologous crossover (*e.g.,* hemoglobins Lepore and Kenya).

The amino acid substitutions are the most common abnormalities, with several hundred described so far. Approximately two thirds of the hemoglobinopathies have an affected beta chain. They may be clinically silent, or they may cause severe damage, as with hemoglobin S. Amino acid substitutions, for the majority of defects, result from a single-base nucleotide substitution in DNA.

Absent or diminished synthesis of one of the polypeptide chains of human hemoglobin characterizes the thalassemias, a heterogeneous group of inherited disorders. In α-thalassemia, alpha-globin chain synthesis is absent or reduced, and in β-thalassemia, β-globin chain synthesis is absent (β⁰-thal) or markedly reduced (β⁻-thal).

Synthesis and Degradation of Hemoglobin

Hemoglobin synthesis takes place in the immature red blood cells in the bone marrow: 65% in the nucleated cells and 35% in reticulocytes. Normal synthesis is dependent on adequate iron supply as well as normal synthesis of heme and protein synthesis to form the globin portion. Heme is synthesized in the mitochondria of the cells. Iron is transported to the developing red blood cells by transferrin, a plasma protein. Iron traverses the cell membrane and the mitochondria, where it is inserted into the protoporphyrin ring to form heme. Protein synthesis of the globin chains takes place in the cytoplasmic polyribosomes. Heme leaves the mitochondria and is joined to the globin chains in the cytoplasm in the final step.

Hemoglobin is degraded by two possible pathways. The normal pathway is called extravascular because it occurs outside of the circulatory system in the reticuloendothelial, or mononuclear phagocyte, system. Within the splenic phagocytic cells, or macrophages, hemoglobin loses its iron to transferrin, its alpha carbon is expired as CO, the globin chains return to the amino acid pool, and the rest of the molecule is converted to bilirubin, which undergoes further metabolism. Normally, 90% of all hemoglobin is degraded in this manner (Fig. 25-4).

Normally, less than 10% hemoglobin is released directly into the bloodstream and dissociated into alpha and beta dimers. Greater amounts are released during hemolytic episodes. The dimers are bound to haptoglobin, which prevents renal excretion of plasma hemoglobin and stablizes the heme-globin bond. This complex is then removed from the circulation by the liver and processed in a fashion similar to extravascular degradation. If the amount of circulating haptoglobin is decreased, as during a hemolytic episode, the unbound dimers go through the kidneys, reabsorb, and are converted to hemosiderin. If greater than 5 g per day is broken down, free hemoglobin and methemoglobin will appear in the urine.

Hemoglobin that is not entirely bound by haptoglobin, or processed by the kidneys, is oxidized to methemoglobin. Heme groups are released and taken up by the protein hemopexin. The heme-hemopexin complex is cleared by the liver and catabolized. Finally, heme groups present in excess of the binding capacity of hemopexin complex combine with albumin to form methemalbumin and are held by this protein until additional hemopexin becomes available for shuttle to the liver (Fig. 25-5). Laboratory measurement of any of these hemoglobin degradation products can help to determine increased red blood cell destruction, such as in a hemolytic anemia.

Clinical Significance and Disease Correlations

Hemoglobin Qualitative Defects, the Hemoglobinopathies

Hemoglobin S. The amino acid defect in hemoglobin S is at the sixth position on the beta chain, where glutamic acid is substituted by valine giving the hemoglobin a less negative charge than hemoglobin A.

Individuals have either sickle cell trait (HbAS, the heterozygous state) or sickle cell disease (HbSS, the homozy-

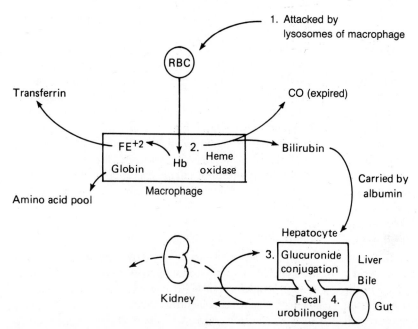

Figure 25-4. Extravascular degradation of hemoglobin.

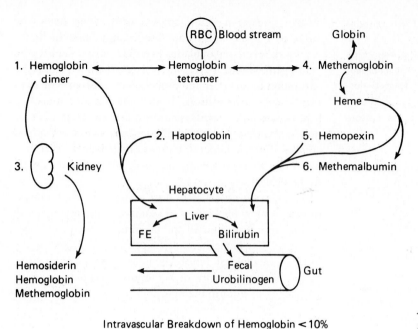

Intravascular Breakdown of Hemoglobin < 10%

Figure 25-5. Intravascular breakdown of hemoglobin.

gous state). Black Africans and African-Americans have the highest incidence: 1 in 500 infants have sickle cell anemia and 8% to 10% carry the HbAS trait. It is also found in Mediterranean countries such as Greece, Italy, and Israel, as well as in Saudi Arabia and India.

Because of the high mortality and morbidity associated with homozygous expression of the gene, the frequency of the mutant gene would be expected to decline in the gene pool. However, a phenomenon known as balanced polymorphism exists, which indicates that the heterozygous state (HbAS) has a selective advantage over either of the homozygous states (HbAA or HbSS). It appears that the heterozygous condition offers protection from parasites, particularly *Plasmodium falciparum,* especially in children. When infected with *Plasmodium falciparum,* children with sickle cell trait have a lower parasite count, the infection is shorter in duration, and the incidence of death is low. It is thought that the infected red blood cells are preferentially sickled and thereby efficiently destroyed by phagocytic cells.[51]

When hemoglobin S is deoxygenated *in vitro* under near-physiologic conditions, it becomes relatively insoluble as compared with hemoglobin A and aggregates into long, rigid polymers called *tactoids.* These cells appear as sickle- or crescent-shaped forms on stained blood films. Sickled cells may return to their original shape when oxygenated; however, after several sickling episodes, irreversible membrane damage occurs and cells are phagocytized by macrophages in the spleen, liver, or bone marrow causing anemia. The severity of the hemolytic process is directly related to the number of damaged cells in circulation. The rigid sickled cells are unable to deform and circulate through small capillaries resulting in blockage. Tissue hypoxia results, causing extreme pain and leading to tissue death. Infarctions in the spleen are common, causing excessive necrosis and scarring

leading to a nonfunctional spleen in most adults with sickle cell anemia. This is referred to as autosplenectomy. The amount of sickling is related to the amount of hemoglobin S in the cells. The reported inhibitory effect of hemoglobins A and F is due to a dilutional effect. There is also less tendency for hemoglobin F to copolymerize with hemoglobin S than with hemoglobin A. This is considered responsible for the observed protective effect of elevated hemoglobin F levels in individuals with sickle cell anemia.

Laboratory findings in the homozygous disease include a normocytic, normochromic anemia, increased reticulocyte count, and variation in size and shape of red blood cells with target cells and sickle cells present. Polychromatophilia and nucleated red blood cells are common. The heterozygous disease is clinically asymptomatic and usually has a normal blood film. The dithionite solubility test will be positive in both homozygous and heterozygous forms, but should always be confirmed with hemoglobin electrophoresis. On cellulose acetate electrophoresis at an alkaline pH, hemoglobin S moves in a position between hemoglobin A and A_2. Eighty-five to 100% of total hemoglobin will be HbS in the homozygous state and usually less than 50% in the heterozygous state. Other hemoglobins that migrate in the same position as hemoglobin S are hemoglobins D and G, but both would be negative with the dithionite solubility test. Electrophoresis on citrate agar at an acid pH is necessary to separate these hemoglobins from hemoglobin S (see Fig. 25-7).

Hemoglobin C. The glutamic acid in the sixth position of the beta chain is replaced by lysine resulting in a net positive charge. Hemoglobin C is found in West Africa in the vicinity of North Ghana in 17% to 28% of the population and in 2% to 3% of African-Americans.

The heterozygous form, hemoglobin AC, is asymptomatic. The homozygous form usually causes a mild, well-compensated anemia characterized by abdominal pain and splenomegaly. The most prominent laboratory feature is the presence of codocytes (target cells). There is a tendency to form large, oblong, hexagonal crystalloid structures within the red cell. These structures are most prominent in patients who have undergone splenectomy.

A differential diagnosis is obtained by cellulose acetate electrophoresis. Hemoglobin C moves with hemoglobin A_2 and is negative with the dithionite solubility test. In the heterozygous form, hemoglobin C falls in the range of 35% to 48%. Hemoglobins E, O, and C_{Harlem}, also known as $C_{Georgetown}$, migrate with hemoglobin C. These hemoglobin variants can be readily distinguished from HbC by citrate agar electrophoresis at an acid pH.

Hemoglobin SC. Hemoglobin SC disease is the most common mixed hemoglobinopathy. One beta gene codes for beta-S chains and the other beta gene codes for beta-C chains, thus leaving no normal beta chains to produce hemoglobin A. Clinically, this disease is less severe than homozygous sickle cell anemia but has similar clinical symptoms. The blood film characteristically shows many target cells and occasional abnormal shapes resembling both the sickle cell, the hexagonal HbC crystal, and a combination of the two. The dithionite solubility test is positive, and electrophoresis on cellulose acetate shows about equal amounts of hemoglobin S and hemoglobin C.

Hemoglobin E. Hemoglobin E is an amino acid substitution of lysine for glutamic acid in the 26th position of the beta chain resulting in a net positive charge. Hemoglobin E is somewhat unstable when subjected to oxidizing agents.

It is found in Asia and is estimated to occur in about 20 million individuals, 80% of whom live in Southeast Asia. In the homozygous form, there is a mild anemia with microcytosis and target cells. In the heterozygous form, the patient is asymptomatic. The differential diagnosis is obtained by electrophoresis. On cellulose acetate, hemoglobin E moves with A_2, C, and O. It is present in the heterozygous form in amounts varying from 30% to 45%, which is somewhat lower than the percentage for hemoglobin C. This is probably due to the somewhat unstable nature of hemoglobin E. On citrate agar, hemoglobin E migrates with A. It is more common to find this defect in association with both α- and β-thalassemia. E-β-thalassemia is a more severe disorder, with moderate anemia and splenomegaly.

Hemoglobin D. The letter D is given to any hemoglobin variant with an electrophoretic mobility on cellulose acetate similar to that of hemoglobin S but which has a negative dithionite solubility test. Hemoglobin D Punjab is the most common, with glutamic acid substituted by glutamine at the 121st position of the beta chain. Hemoglobin D Punjab is found in Northwest India but occasionally can be seen in

English, Portuguese, and French individuals because of the close historical connection of these countries with East India.

The homozygous state is rare. There is no anemia or splenomegaly and only a slight anisocytosis. The oxygen affinity is higher than in normal blood. Heterozygous individuals are asymptomatic. Differential diagnosis is accomplished with electrophoresis. On cellulose acetate, hemoglobin D migrates with hemoglobin S in proportions of 35% to 50%. On citrate agar, hemoglobin D migrates with A.

Hemoglobin Quantitative Defects, the Thalassemias

The thalassemias are a group of diseases in which a defect in the rate of synthesis of one or more of the hemoglobin chains occurs, but the chains are structurally normal. Gene deletion is the most common cause. The two most common types are α-thalassemia, resulting from defective production of alpha chains, and β-thalassemia, resulting from a defect in production in beta chains. Defects in production of the delta and gamma chains have been described, but these are not involved in production of hemoglobin A and therefore are not clinically significant. Rarely, combinations of gene deletions, such as delta and beta, may lead to clinical disease. Any form of unbalanced production of globin chains causes the erythrocytes to be small, hypochromic, and sometimes deformed. Intracellular accumulation of unmatched chains in the developing erythrocytes causes precipitation of the proteins, which leads to cell destruction in the bone marrow. Although erythropoiesis is taking place, it is ineffective because mature cells do not reach the peripheral blood to carry oxygen.

Thalassemia is inherited as an autosomal dominant disorder with heterogeneous expression of the disease. It is one of the most common hereditary disorders and is distributed worldwide. The prevalence of the thalassemia gene has been attributed to the protection it offers against Falciparum malaria. The heterozygous state produces a disorder called *thalassemia minor,* which is clinically asymptomatic and resembles iron deficiency. The homozygous state, *thalassemia major,* is either lethal before birth or in childhood.

α-Thalassemias. α-Thalassemia occurs with high frequency in Asian populations but is also seen in the Black African, African-American, Indian, and Middle Eastern populations. There are four principal clinical types of different severity known to occur in the population, and these four types can be explained, respectively, by deletions of 4, 3, 2, or 1 of the alpha-globin gene loci (Fig. 25-6). The type of α-thalassemia found in Black Africans and African Americans is also associated with deletion of the alpha-globin genes but in a different pattern than is found in the Asian population (see Fig. 25-6). The four clinical types of α-thalassemia in order of most deletions to least deletions are the following:

1. Hydrops fetalis is the most clinically severe form of α-thalassemia because of the total absence of alpha-chain synthesis. Bart's hemoglobin, which is a tetramer of

Alpha thalassemias

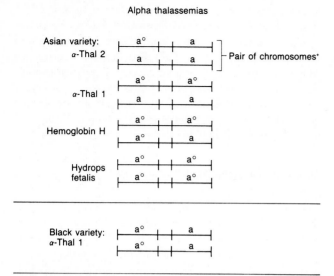

a   is normal alpha chain (normal gene loci)
a° is deleted alpha chain (deleted alpha gene loci)
+   note that a loci is duplicated on each chromosome

▲

Figure 25-6. Deletions of α-globin gene loci in α-thalassemia.

gamma chains, is the main hemoglobin found in the red cells of affected infants. Hb Bart's has an extremely high O_2 affinity and allows almost no oxygen transport to the tissues. These infants are either stillborn or die of hypoxia shortly after birth.

2. Hemoglobin H disease has alpha chain synthesis at about one-third the amount of beta-chain synthesis. As a result, beta chains accumulate and form tetramers, which are called hemoglobin H. Beta-chain precipitates (HbH inclusions) alter the shape and ability of the cell to deform, thus significantly shortening the life span of the cells. These individuals have a moderate hemolytic anemia, with 5% to 30% hemoglobin H, 1% hemoglobin A_2, and the remainder hemoglobin A. Hemoglobin H inclusions can be seen in the red cells with a supravital stain. Cord blood contains 10% to 20% Bart's hemoglobin.

3. α-Thalassemia trait (or α-thalassemia minor) results from two gene deletions, either on the same or on different chromosomes. The deletion on two separate chromosomes is more common in Black Africans and African-Americans (Fig. 25-6). These individuals have a mild microcytic, hypochromic anemia. Occasionally, excess beta chains may form HbH inclusions. Cord blood contains 2% to 10% hemoglobin Bart's, but after 3 months of age, electrophoresis is normal.

4. Silent carriers are missing only one alpha gene, and the remaining genes direct production of sufficient alpha chains for normal hemoglobin production. This state can only be detected by expression of 1% to 2% of hemoglobin Bart's in a neonate. After 3 months of age, it can only be detected by globin-gene analysis. It has been estimated that the frequency of this genetic disorder may be as high as 30% in the African-American population.[27]

β-Thalassemias. In contrast to α-thalassemia, gene deletions usually do not cause β-thalassemia. One of several types of lesions in the beta gene result in failure to produce normal amounts of beta-globin chains. One common type involves faulty processing of the messenger RNA from the gene directing beta-chain production.[2] The β-thalassemias are classically divided into homozygous disease, called thalassemia major or Cooley's anemia, and heterozygous disease, called thalassemia minor. However, the clinical expression of the disease is very heterogeneous, depending on the type of genetic defect and involvement with other gene loci. The disease may be broadly divided into two major subtypes according to genetic expression: $β^+$, in which beta chains are produced in reduced amounts, and $β^0$ which is complete absence of beta chains.

1. $β^+$-Thalassemia is the most common type. There is some synthesis of beta-globin chains but in significantly reduced amounts (5%–30%) of normal. The biochemical defect shows a quantitative deficiency of beta-globin mRNA. The hemoglobin electrophoresis pattern and hemoglobin F and A_2 quantitation show about 2% to 8% hemoglobin A_2 and an elevated but varying amount of hemoglobin F. The MCV is very low, with a severe anemia, reticulocytes, nucleated red blood cells, basophilic stippling, codocytes, extreme poikilocytosis, and anisocytosis.

2. $β^0$-Thalassemia accounts for 10% of homozygous β-thalassemia, with a total absence of beta-chain synthesis but intact synthesis of gamma chains. In the homozygous form, there is 1% to 6% hemoglobin A_2 and 95% hemoglobin F. The hemoglobin concentration is very low, with a severe anemia. The heterozygous form appears clinically the same as homozygous $β^+$-thalassemia.

Homozygous β-thalassemia, β-thalassemia major, either $β^+$ or $β^0$, is a crippling disease of childhood. This is unlike α-thalassemia, in which the child either dies shortly after birth or leads a normal life. The hypochromic, microcytic anemia is due to both the defect of functional hemoglobin tetramer synthesis and the premature destruction of red blood cells, both intra- and extramedullary, due to the increased presence of alpha chains. The bone marrow compensates by expanding enormously in size, sometimes causing structural bone abnormalities. A side effect of iron therapy is hemochromatosis, which is only partially alleviated by drugs designed to bind the excess iron (chelation therapy).

Heterozygous β-thalassemia, thalassemia minor, may be caused by inheritance of one thalassemia gene, either $β^+$ or $β^0$. The other gene directing beta-chain production is normal, and red blood cell survival is not shortened. About 1% of African-Americans are affected. Clinically, the condition is usually asymptomatic but may sometimes cause a mild microcytic anemia. The hematologic laboratory values resemble those of iron-deficiency anemia, and it is important to distinguish the two because quite different treatments are required. Hemoglobin electrophoresis of thalassemia minor characteristically shows an increase in hemoglobin A_2, usually between 3.5% and 7%.

δβ-Thalassemia is a rare type characterized by total absence of both beta-chain synthesis of hemoglobin A and delta-chain synthesis of hemoglobin A₂. Homozygous patients have 100% hemoglobin F. The heterozygous individuals have 93% hemoglobin A, 2% to 3% hemoglobin A₂, and 3% to 10% hemoglobin F.

Patients are anemic and show a thalassemic phenotype because the gamma-chain synthesis of hemoglobin F is not equal to alpha-chain synthesis. There is approximately one-third as much gamma chain produced as alpha chain. Hemoglobin F is heterogeneously distributed among the erythrocytes, as revealed by a Kleihauer-Betke stain.

Hereditary persistence of fetal hemoglobin (HPFH) is genetically and hematologically heterogeneous. In African Blacks and African-Americans there is a total absence of beta- as well as delta-chain synthesis because of deletions in chromosome 11. Gamma-chain synthesis is present in the adult at a high level, and in contrast to synthesis of hemoglobin F in β-thalassemia or δβ-thalassemia, it is uniformly distributed. In heterozygotes, there is no imbalance of globin-chain synthesis. There is 17% to 33% hemoglobin F. The patients are clinically normal. In homozygotes, there is 100% hemoglobin F, with no synthesis of hemoglobin A or A₂. In the Greek type, hemoglobin F production is slightly lower, 10% to 20%, and hemoglobin A₂ production is slightly higher at 1% to 3%. There is hypochromia, microcytosis, anisocytosis, and poikilocytosis. In the Swiss type, hemoglobin F ranges between 1% and 7% and is unevenly distributed (heterogeneously) among the erythrocytes.

Methodology

DITHIONITE SOLUBILITY TEST[38,39] (SCREENING TEST FOR SICKLING HEMOGLOBINS)

Principle—Sickling hemoglobin, in the deoxygenated state, is relatively insoluble, and it forms a precipitate when placed in a high-molarity phosphate buffer solution. The precipitate appears because the deoxygenated hemoglobin molecules form tactoids that refract and deflect light rays, thereby producing a turbid solution.

Special Considerations—This test has commonly employed whole blood. However, to avoid several sources of false-positive or false-negative results, it is recommended that packed red blood cells be used. Run a positive (blood containing HbS) and a negative (blood containing only HbA) control with each run of specimens.

Reagents

Stock Buffer—In a 500-mL volumetric flask, place about 350 mL of water and add 140.94 g of anhydrous K_2HPO_4. Mix until completely in solution. Add 80.24 g KH_2PO_4 crystals slowly over 10 to 15 minutes. Let the crystals dissolve completely, then dilute to 500 mL.

1% Saponin—Put 0.5 g of saponin in a 50-mL volumetric flask and dilute to 50 mL with distilled water.

Working Solution—Put 400 mL of stock buffer in a 500-mL volumetric flask. Add 10 g of sodium dithionite ($Na_2S_2O_4$) and mix

until dissolved. Add 30 mL of 1% saponin. Mix and dilute to 500 mL. This solution is stable for about 1 month under refrigeration.

Procedure

1. Add 2.0 mL of dithionite working solution to a 12×75 mM tube and allow to come to room temperature.
2. Centrifuge blood and remove buffy coat and plasma. Add 10 mL of packed erythrocytes to solution, mix, and allow tube to stand at room temperature for 6 minutes.
3. Place the tube approximately 1 inch in front of a heavy, black-lined index card. If there is no hemoglobin S present, the lines on the card will be easily read. If hemoglobin S is present, the lines will be very indistinct or impossible to read. Report as positive or negative.

Comments—Outdated reagents and reagents not at room temperature interfere with the test. False-negative tests may be due to recent transfusions or may occur at birth because of high concentrations of HbF. If whole blood is used instead of packed erythrocytes, false-positive results may occur due to erythrocytosis, hyperglobulinemia, extreme leukocytosis, or hyperlipidemia, and false-negative results may occur in anemia. Positive results with this test should be confirmed with the hemoglobin electrophoresis test. ■

CELLULOSE ACETATE HEMOGLOBIN ELECTROPHORESIS[7,16]

Principle—A fresh hemolysate is made from a fresh-packed red blood cell sample. The hemolysate is applied to a cellulose acetate plate and electrophoresis is performed. After electrophoresis, the membrane is stained and cleared. Then interpretation is made, comparing the hemoglobin's migration with that of a control.

Special Considerations—The specimen is a 7-mL EDTA anticoagulated blood sample. A hemoglobin AFSC control is made from SS, SC, AA, and FA hemolysates. Perform electrophoresis on the control and adjust the proportions of hemoglobin as necessary. This control should be included with each plate.

Reagents

Buffer—TRIS-EDTA borate buffer (TEB), pH 8.6. Dissolve 12.0 g of TRIS, 1.56 g of EDTA, and 0.92 g of boric acid in deionized water and dilute to 1 L. Store at 4°C. Good indefinitely. This buffer may be purchased prepackaged from several suppliers.

Stain—Ponceau S stain. Dilute 0.5 g of Ponceau S to 100 mL with 5% (w/v) trichloracetic acid. The stain is stable for 2 months at room temperature.

Procedure

1. Make a hemolysate from packed cells by pipetting 0.1 mL of packed red cells into 0.5 mL of water and vortexing vigorously; add 1 drop of KCN and vortex again. Allow to stand 10 minutes before using.
2. Pour buffer into the outer compartments of an electrophoresis chamber. Wet wicks in the buffer and place in the chamber, making sure that the wicks are in contact with the buffer compartments. Slowly place the cellulose acetate membrane into the buffer and allow to soak for at least 5 minutes. Immersion of the plates too rapidly will result in bubbles on the plates, faulty separation patterns, and/or streaking.
3. Apply the hemolysates and control to the plates with an appropriate dispenser washed with water. Other solutions may cause streaking in the patterns.
4. Perform the electrophoresis for the time suggested by the manufacture at a voltage of 450 to 500 V. Place the chamber on cooling blocks to keep heat buildup to a minimum. Excess heat will cause distortions in the plates.

5. After electrophoresis is completed, place the membranes in Ponceau S stain for 3 to 5 minutes. Destain in three successive washes of 5% acetic acid for 2 to 3 minutes each. Dehydrate for 2 minutes in absolute methanol. Clear in a clearing solution of 25% glacial acetic acid and 75% methanol, for 3 minutes.

6. Remove from clearing solution and place plates, Mylar, or plastic side down, on glass plates. Let air dry for about 3 minutes then place in 100°C oven for 2 to 3 minutes to completely dry.

7. Label membrane and store in a plastic envelope.

8. A rough estimate of proportions of different hemoglobins on a plate may be made using a densitometer.

Comments—The order of electrophoretic mobility, from slowest to fastest, is hemoglobins C, S, F, and A (Fig. 25-7). (There are several mnemonic devices for remembering the migration pattern. The most sensible is *A*ccelerated, *F*ast, *S*low, and *C*rawl. The most fun is *A Fat Santa Claus!*) Any hemoglobin that migrates beyond hemoglobin A is termed a fast hemoglobin. Bart's hemoglobin and hemoglobin H both migrate here. Any abnormal hemoglobin or any hemoglobin present in increased amounts, such as hemoglobin A_2 and F, should be confirmed. If there is an abnormal hemoglobin, confirm with a sickling test and citrate agar electrophoresis. If an increased amount of A_2 or F occurs, then quantify the amount (Fig. 25-8). ■

Citrate Agar Electrophoresis

Citrate agar electrophoresis is performed after an abnormal hemoglobin is found on cellulose-acetate electrophoresis.[26] In this method, an important factor in determining the mobility of hemoglobin is solubility. Hemoglobin F, with the fastest cathodal mobility, is also the most soluble, probably because it is most resistant to denaturation at pH 6.0. Adult hemoglobins whose solubility is similar to that of hemoglobin A, such as G, D, E, O, I, etc., move with hemoglobin A. The relatively insoluble hemoglobin S moves behind hemoglobin A, and the even more insoluble hemoglobin C moves behind hemoglobin S (see Fig. 25-7).

Hemoglobin A_2 Quantitation

Hemoglobin A_2 can be estimated by hemoglobin electrophoresis; however, this yields only a rough estimate. Quantitation must be made by microcolumn chromatography[13,22,27] or HPLC,[37] both of which use an ion-exchange resin.

Hemoglobin F Quantitation

Erythrocytes containing an increased amount of hemoglobin F can be distinguished from normal adult cells by the acid-elution technique. Adult hemoglobin is eluted from the erythrocytes by incubation in an acid buffer. Hemoglobin F remains behind and is stained with eosin.[11] Adult cells are negative and have no staining, since they contain no hemoglobin F. Cord blood cells appear as smooth, homogeneous cells with a scarlet-red color and a distinct halo. Very old fetal cells and incompletely laked adult cells stain with a bluish cast. Lymphocyte nuclei stain pale red or pink and should not be mistaken for the fetal red cells.

Fetal hemoglobin is resistant to alkali denaturation in 1.2 M NaOH for 1 minute. Denatured hemoglobin A is precipitated out with ammonium sulfate and removed by filtration. The optical density of the clear supernatant solution is read at 540 nm, and the percentage of fetal hemoglobin is calculated against the optical density of the total hemoglobin solutions.[2,21]

HPLC on an ion exchange resin also appears promising for analysis of fetal hemoglobin.[47]

The average adult has less than 1.5% fetal hemoglobin. However, elevated levels may be found in a number of inherited and acquired diseases. The hereditary persistence of fetal hemoglobin should be suspected in individuals who possess 10% or more of fetal hemoglobin with no other apparent clinical abnormalities.

DNA Probe Technology

The definitive diagnosis of some cases of thalassemia that involve combinations of genetic defects may require more sophisticated testing. Actual measurement of the rate of globin-chain synthesis may be determined by labeling the chains in reticulocytes that are still synthesizing hemoglobin and separating them by electrophoretic techniques. Since alpha-gene deletion is the most common cause of α-thalassemia, DNA probe technology is becoming useful in this group of diseases. Leukocyte DNA from the patient can be

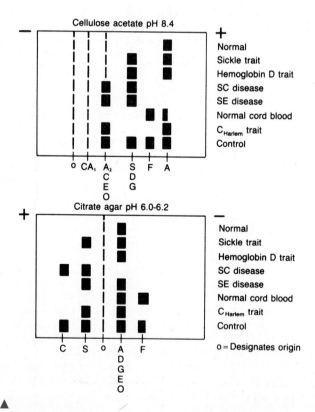

▲
Figure 25-7. Comparison of various hemoglobin samples on cellulose acetate and citrate agar.

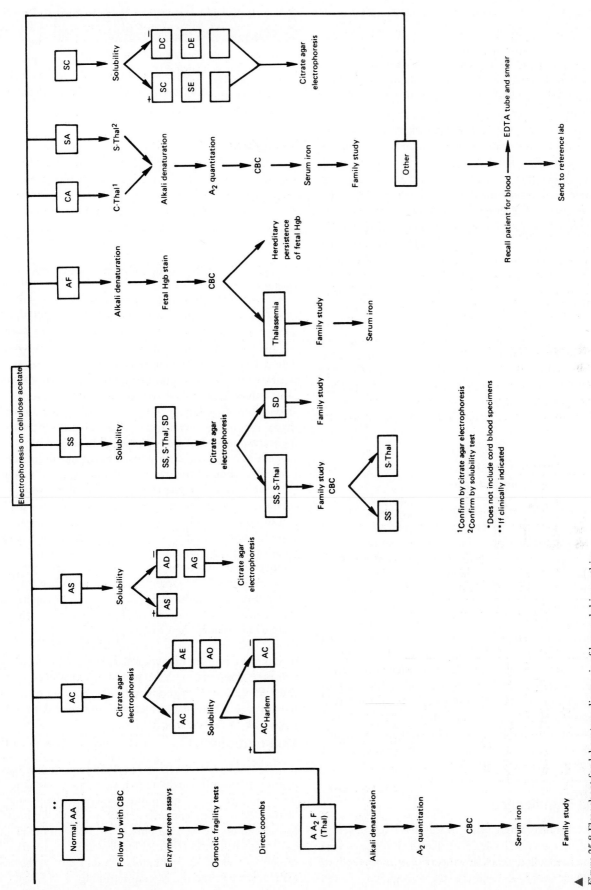

Figure 25-8. Flow sheet for laboratory diagnosis of hemoglobinopathies.

513

cut into well-defined segments with an appropriate restriction endonuclease and subjected to electrophoretic separation based on fragment size. The pieces of DNA that contain the characteristic coding sequence of the alpha-globin genes can be detected using a labeled fragment of alpha DNA as a probe. If the patient has one of the alpha genes missing, the fragment will be shorter than normal. This technology also may be used in the β-thalassemias. In some cases, direct detection of the molecular lesions is possible. A special strength of the DNA technology is in the prenatal diagnosis of thalassemia major. Because the globin genes are represented in all tissues, including those in which they are not active, prenatal diagnosis of thalassemic states may be made by sampling tissues that are relatively easy to obtain, such as chorionic villi or amniotic fluid cells, rather than fetal blood, which is obtained with much greater difficulty and at a much greater risk to the fetus.[3]

DNA probe technology also has been used in the prenatal diagnosis of sickle cell anemia. Fetal cells obtained by amniocentesis or chorionic villus sampling may be analyzed using similar techniques as in the thalassemias. Hemoglobin electrophoresis on hemolysate of fetal blood cells also can be used in cases where DNA probe technology is not available or when rapid results are needed owing to advanced gestational age of the patient.[43]

MYOGLOBIN

Structure and Role in the Body

Myoglobin is a heme protein found only in skeletal and cardiac muscle in humans. It can reversibly bind oxygen in a manner similar to the hemoglobin molecule, but myoglobin is unable to release oxygen, except under very low oxygen tension. Myoglobin is a simple heme protein containing one polypeptide chain and one heme group per molecule. The polypeptide chain contains 153 amino acids, making it slightly larger than one chain in the hemoglobin molecule. Therefore, its size is slightly larger than one-fourth that of a hemoglobin molecule, with a molecular weight of approximately 17,000. The iron atom in the center of the heme group is the site of reversible oxygen binding, identical to the hemoglobin molecule. In the body, myoglobin acts as an oxygen carrier in the cytoplasm of the muscle cell. Transport of oxygen from the muscle cell membrane to the mitochondria is its main role. It also may play a minor role in oxygen storage for the muscle.[47]

Clinical Significance

Damage to muscles often results in elevated levels of serum and urine myoglobin (Table 25-2). Renal clearance is very rapid, and myoglobinemia following a single injury tends to be transient. In crush injuries, the myoglobin filtered

TABLE 25-2. Causes of Elevations of Myoglobin
Acute myocardial infarction
Angina without infarction
Rhabdomyolysis
Multiple fractures; muscle trauma
Renal failure
Myopathies
Vigorous exercise
IM injections
Open heart surgery
Grand mal seizures
Electric shock
Arterial thrombosis
Certain toxins

through the kidneys may be sufficient to cause renal occlusion and death.

Measurement of myoglobin has been advocated to assist in the diagnosis of myocardial infarction. With the recent advent of thrombolytic therapy, the need for rapid and precise diagnosis of myocardial infarction has increased greatly. The damaged heart muscle cells release myoglobin within the first few hours of onset of myocardial infarction, and peak values are reached sooner than peak values of the enzyme creatine kinase (CK). Therefore, an increase of myoglobin in the circulation provides an early indicator of myocardial infarction. Serum myoglobin level also has been suggested to be a good early predictor of size of infarction. Early prediction of the size of tissue damage is also helpful with the newer therapies. Highest levels of myoglobin are found in extensive myocardial infarction, and less marked elevations are found in less severe cases.[17] However, since myoglobin from cardiac muscle may be immunochemically identical to myoglobin liberated from skeletal muscle, false-positive results may occur from accompanying minor injury to skeletal muscle.[50,9] Despite the poor specificity of myoglobin for myocardium, skeletal muscle damage can be ruled out in many cases. Negative myoglobin results within the first few hours after chest pain can be used to rule out myocardial infarction. In cases where thrombolytic therapy is used in the treatment of myocardial infarction, myoglobin is a marker of reperfusion.[53] Myoglobin also has been investigated to aid in the diagnosis and differentiation of the different types of hereditary progressive muscular dystrophy.[42]

Methodology

It is very difficult on visual examination to distinguish myoglobinuria from hemoglobinuria, since the urine is red to brown in both and gives a positive reaction when tested for "blood" with a reagent strip (oxidation of *o*-toluidine). In myoglobinuria, there is no pink or red color in the plasma, since the small myoglobin molecule readily passes into the urine. However, myoglobinuria should always be confirmed by more specific laboratory testing.

There are several methods for measurement and identification of myoglobin. Myoglobin may be separated from hemoglobin as a result of the physicochemical differences in the two molecules by precipitation, electrophoresis, and ultracentrifugation. All these methods have limited sensitivity and may not be able to detect small increases in myoglobin. In addition, denaturation of myoglobin, which readily occurs in urine samples, may cause inaccuracies in these methods. Immunoassay is the most specific method at present and can detect as little as 0.5 ng/mL of myoglobin.[27,46] Latex agglutination, ELISA, immunonephelometry, and fluoroimmunoassays for myoglobin have been developed. These have the advantage of being less time-consuming and easier to perform than radioimmunoassay procedures, making them especially useful as emergency screening tests in the diagnosis of myocardial infarction.[9,19,25,29]

QUALITATIVE MYOGLOBIN, AMMONIUM SULFATE PRECIPITATION[4]

Principle—Because of the molecular weight of hemoglobin and myoglobin, hemoglobin is less soluble in 80% ammonium sulfate solution.

Special Considerations—At least 15 mL of fresh urine is needed.

Reagent—80% ammonium sulfate; reagent strip, which measures blood.

Procedure—With the reagent strip, test for "blood" in the urine. If the reagent strip is negative for blood, then myoglobin is not present, and there is no need to proceed further. If the reagent strip is positive, filter urine with no. 4 filter paper and repeat reagent-strip test. If negative, do not proceed further, since the first positive result was due to erythrocytes. If positive, mix 5 mL of urine with 2.8 g of ammonium sulfate. Mix to dissolve. Filter and repeat reagent strip. A positive reaction for blood on the filtrate indicates myoglobin.

Comments—The test should be carried out on a fresh urine specimen. If this is not possible, adjust the pH to 7.0 to 7.5 and refrigerate at 4°C or freeze. Urines preserved this way may be kept for months.

A urine specimen containing only denatured myoglobin will give a negative test. Both hemoglobin and myoglobin may be present in the urine at the same time, especially if there are erythrocytes in the urine. ■

Myoglobin electrophoresis is done on cellulose acetate at pH 8.6, with myoglobin migrating in a position similar to that of hemoglobin C. If the urine contains hemoglobin only, a band will be seen only at the hemoglobin A position. If myoglobin is present, then a band at the position of hemoglobin C will be seen.[6]

SUMMARY

Porphyrins are the intermediates in the multi-step synthesis of heme, an iron chelating group that binds oxygen. The synthesis of heme is started in the mitochondrion by combination of glycine and succinyl CoA to form δ-aminolevulinic acid (ALA). Porphobilinogen (PBG) is formed from ALA in the cytosol. ALA and PBG are precursors to the formation of porphyrins. Uroporphyrinogen and coproporphyrinogen are formed in the cytosol and protoporphyrinogen is synthesized in the mitochondrion. Protoporphyrin is then formed, which incorporates iron to form heme. ALA synthase is the rate controlling enzyme of heme synthesis and is regulated by the amount of heme. Enzyme deficiencies can occur at almost every step of heme synthesis resulting in a group of inherited disorders called the porphyrias. A buildup of intermediates occurs, which can cause cutaneous symptoms, neuropsychiatric symptoms, or both. Blocks in the early steps of heme synthesis tend to produce cutaneous symptoms, while buildups of later intermediates cause neuropsychiatric symptoms. Most porphyrias can be differentiated by laboratory analysis of ALA, PBG, uroporphyrin, coproporphyrin, and protoporphyrin.

Hemoglobin is synthesized in immature erythroid cells in the bone marrow and functions in carrying oxygen to the tissues. Hemoglobin is composed of two pairs of globin proteins each containing a heme molecule capable of carrying oxygen. There are six types of globin chains; alpha, beta, gamma, delta, epsilon, and zeta, with the latter two present only in the embryo. In the normal adult, three types of hemoglobin are present: A ($\alpha_2\beta_2$), A$_2$ ($\alpha_2\delta_2$), and F ($\alpha_2\lambda_2$). Hemoglobinopathies result when defects are present in the structure of hemoglobin. These are caused by amino acid substitutions or deletions, globin chain elongation, and globin chain fusion or hybridization. The most common hemoglobinopathy is hemoglobin S, which can be inherited as heterozygous (AS, sickle cell trait) or homozygous (SS, sickle cell disease). The deoxygenated form of hemoglobin S is relatively insoluble *in vitro* and can be detected by the dithionite test. Heterozygotes are asymptomatic while homozygotes may have anemia and compromised microcirculation. This and other hemoglobinopathies can be detected by a combination of alkaline and acid electrophoresis. Thalassemias are inherited disorders in the rate of synthesis of one or more globin chains, usually due to gene deletions. Defects in the rate of alpha globin are termed α-thalassemias and β-thalassemia denotes defective beta globin production. The heterozygous state is termed thalassemia minor and the homozygous state is called thalassemia major. These disorders are common in Asian, Black African, African-American, Indian, and Middle Eastern populations and cause anemias of varying degrees.

Myoglobin is a heme-containing protein present in skeletal and cardiac muscle. Its presence in serum or urine is indicative of skeletal or cardiac muscle damage. Because of its relatively small molecular weight, it is exuded into the plasma within a few hours after muscle damage. In this respect, it is useful as a marker of myocardial infarction or as a marker of reperfusion after thrombolytic therapy.

CASE STUDY 25-1

A 32-year-old African-American woman came to the obstetrics and gynecology clinic of her local hospital because she was feeling a little weak. A complete blood count showed a hemoglobin of 9.9 g/dL with an MCV of 87 fl. The physician ordered a hemoglobin electrophoresis as a follow-up. The cellulose-acetate pattern showed a peak of 58% at the hemoglobin A position, a peak of 35% at the hemoglobin S position, and a peak of 5% at the A_2 position. Further studies indicated a positive dithionite solubility test result and a hemoglobin F value of 1%.

Questions

1. What is the best possible diagnosis for this woman?
2. Does this condition require further follow-up and treatment?
3. What implications does this disease have for her unborn child?
4. Are the values for hemoglobins. A_2 and F normal for this condition?

CASE STUDY 25-2

A young nurse from South Africa became emotionally disturbed and appeared to be hysterical a few days after a laparotomy for "intestinal obstruction." Prior to her operation, she had, over a period of a week, taken barbiturate capsules to help her sleep. When first seen, she complained of severe abdominal and muscle pain and general weakness, her tendon reflexes were absent, and she was vomiting and constipated. Her urine was dark in color on standing and gave a brilliant pink-fluorescence when viewed in ultraviolet light. Within 24 hours, she was totally paralyzed, and within 2 days she died.

Questions

1. What possible condition could this young woman have had, and why did it manifest itself at this time?
2. Would any members of her family have a similar disease?
3. What enzyme defect did she have?
4. What other confirmatory tests, if any, could be done?

CASE STUDY 25-3*

A 54-year-old African-American woman was admitted to the hospital with the chief complaint of left hip pain and lethargy. She had a long history of multiple emergency room visits for hip pain requiring medication. She had previously been found to have a positive solubility test for hemoglobin S, but denied a history of sickle cell disease. There was family history of sickle cell trait. She had had a mastectomy for breast cancer 10 years previously. Admission laboratory values are listed below.

Hemoglobin	5.3 g/dL
Hematocrit	17%
MCV	82 fL
MCHC	31 g/dL
WBC	12,000/μL
Platelet count	53,000/μL
Differential	Normal
Reticulocyte count	6.4% (corrected 2.4%)
RBC morphology	Target cells, spherocytes, schistocytes, basophilic stippling, bizarre forms, including elongated, block-shaped, and more densely stained cells

*Case study data provided by Margaret Uthman, MD, Assistant Professor of Pathology, University of Texas Medical School at Houston.

CASE STUDY
25-3*

A hemoglobin electrophoresis was ordered. Patterns from the cellulose-acetate and citrate-agar electrophoresis are shown in Figure 25-9. Chest x-ray showed a right lower lobe infiltrate with pulmonary vascular congestion and an enlarged spleen. Fluid aspirated from the nasogastric tube was positive for blood.

The patient was given medication for an aspiration pneumonia, gastrointestinal bleeding, and congestive heart failure. She was given packed RBCs, fresh-frozen plasma, and platelets, but her condition continued to worsen. Six hours later, laboratory tests confirmed disseminated intravascular coagulation (DIC). A bone marrow biopsy was performed and revealed extensive necrosis of marrow elements. Three hours later the patient died of cardiac arrest.

Questions

1. What hemoglobinopathy is indicated by the hemoglobin electrophoresis patterns?
2. What clinical feature of this disease differs from the typical picture in sickle cell anemia?
3. What other hemoglobins interact with hemoglobin S, and how can these be differentiated from hemoglobin C?
4. Was this patient's death due to the hemoglobinopathy? Is it unusual for HbSC to be life-shortening?

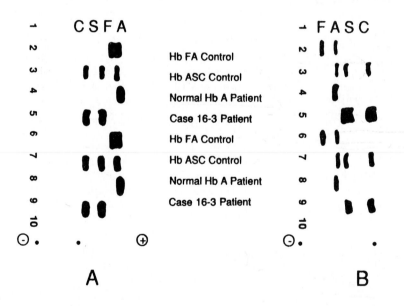

A

B

Figure 25-9. Hemoglobin electrophoretic patterns for Case Study 25–3. (A) cellulose acetate at pH 8.4. (B) citrate agar at pH 6.2. (Courtesy of Margaret Uthman, M.D., University of Texas Medical School of Houston.)

REFERENCES

1. Adams RF, et al. Porphyrin experiments in urine using high pressure liquid chromatography and fluorescence detection. Chromatography Newsletter 1976;4(2):24.
2. Betke K, Marti HW, Shlicht I. Estimation of small percentage of fetal haemoglobin. Nature 1959;184:1877.
3. Beutler E. The common anemias. JAMA 1988;259:2433.
4. Blondheim SH, et al. A simple test for myohemoglobinuria (myoglobinuria). JAMA 1958;167:453.
5. Bloomer JR, Bonkovsky HL. The porphyrias. Dis Mon 1989;35:1.
6. Boulton FE, Huntsman RG. The detection of myoglobin in urine and its distinction from normal and variant hemoglobins. J Clin Pathol 1971;24:816.
7. Briere RO, Golias T, Gatsakis JG. Rapid qualitative and quantitative hemoglobin fractionation. Am J Clin Pathol 1965;44:695.
8. Buttery JE, Chamberlain BR, Beng CG. A sensitive method of screening for urinary porphobilinogen. Clin Chem 1989;35:2311.
9. Castaldo AM, et al. Plasma myoglobin in the early diagnosis of acute myocardial infarction. Eur J Clin Chem Clin Biochem 1994;32:349.
10. Christensen NG, Romsolo I. Stool porphyrins determined by high pressure liquid chromatography and by fractional hydrochloric acid–ether extraction. Scand J Clin Lab Invest 1979;39:223.
11. Clayton E, Foster BE, Clayton EP. New stain for fetal erythrocytes in peripheral blood smears. Obstet Gynecol 1970;35:642.

12. Cripps DJ, Peters HA. Fluorescing erythrocytes and porphyrin screening test in urine, stool and blood. Arch Dermatol 1967;96:712.

13. Efremov GD, Huisman THJ, Bowman K, Wrightstone RN, Schroeder WA. Microchromatography of hemoglobin: II. A rapid method for the determination of hemoglobin A2. J Lab Clin Med 1974;83:657.

14. Elder GH. Differentiation of porphyria cutanea tarda symptomatica from other types of porphyria by measurement of isocoproporphyrin in feces. J Clin Pathol 1975;28:601.

15. Elder GH. The cutaneous porphyrias. Semin Dermatol 1990; 9:63.

16. Graham JL, Grunbaum BW. A rapid method for microelectrophoresis and quantitation of hemoglobins on cellulose acetate. Am J Clin Pathol 1963;39:567.

17. Groth T, Hakman M, Sylven C. Prediction of myocardial infarction size from early serum myoglobin observations. Scand J Clin Lab Invest 1987;47:599.

18. Haining RG, Hulse T, Labbe RF. Rapid porphyrin screening of urine, stool and blood. Clin Chem 1969;15:460.

19. Hangaard J, et al. Early diagnosis of acute myocardial infarction with a rapid latex agglutination test for semiquantitative estimation of serum myoglobin. Acta Med Scand 1987; 221:343.

20. Hindmarsh JT. The porphyrias: Recent advances. Clin Chem 1986;32:1255.

21. Huisman THJ. Normal and abnormal hemoglobins. Adv Clin Chem 1963;6:231.

22. Huisman THJ, Schroeder WA, Brodie AN, Mayson SM, Jakway J. Microchromatography of hemoglobins: II. A simplified procedure for the determination of hemoglobin A2. J Lab Clin Med 1975;86:700.

23. Jackson AH. Modern spectroscopic and chromatographic techniques for the analysis of porphyrin on a microscale. Semin Hematol 1977;14:193.

24. Jandl JH. Blood—Textbook of hematology. Boston: Little, Brown, 1987;85.

25. Juronen EI, Viikmaa MH, Mikelsaar AV. Rapid, simple and sensitive antigen capture ELISA for the quantitation of myoglobin using monoclonal antibodies. J Immunol Methods 1988;111:109.

26. Kiechle FL. The porphyrias. In: Bick RL, ed. Hematology—Clinical and laboratory practice. St. Louis: Mosby, 1993;553.

27. Kagen LJ. Myoglobin: Methods and diagnostic uses. In: Batsakis J, Savory J, eds. Crit Rev Clin Lab Sci Cleveland, OH: CRC Press, 1978;273.

28. Kappas A, Sassa S, Galbraith RA, Nordmann Y. The porphyrias. In: Scriver CR, Beaudet AL, Sly WS, Valle D, eds. The metabolic basis of inherited disease. 6th ed. New York: McGraw-Hill, 1989;1303.

29. Konings CH, Funke Kupper AJ, Verheugt FWA. Comparison of two latex agglutination test kits for serum myoglobin in the exclusion of acute myocardial infarction. Ann Clin Biochem 1989;26:254.

30. Lai C-K, Lam C-W, Chan Y-W. High-performance thin-layer chromatography of free porphyrins for diagnosis of porphyria. Clin Chem 1994;40:2026.

31. Lamon JM, With TK, Redeker AG. The Hoesch test: Bedside screening for urinary porphobilinogen in patients with suspected porphyria. Clin Chem 1974;20:1438.

32. Lim CK, Li F, Peters TJ. High-performance liquid chromatography of porphyrins. J Chromatogr 1988;429:123.

33. Longas MO, Poh-Fitzpatrick MB. High pressure liquid chromatography of plasma free acid porphyrins. Anal Biochem 1980;104:268.

34. Mauzerall D, Granick S. The occurrence and determination of D-aminolevulinic acid and porphobilinogen in urine. J Biol Chem 1956;219:435.

35. McKenzie SM. Textbook of hematology. Philadelphia: Lea & Febiger, 1988;160.

36. Milner PF, Gooden H. Rapid citrate-agar electrophoresis in routine screening for hemoglobinopathies using a simple hemolysate. Am J Clin Pathol 1975;64:58.

37. Muller CJ, Pik C. A simple and rapid method for the quantitative determination of hemoglobin-A2. Clin Chim Acta 1962;7:92.

38. Nalbandian RM, et al. Dithionite tube test—A rapid, inexpensive technique for the detection of hemoglobin. Clin Chem 1971;17:1028.

39. National Committee for Clinical Laboratory Standards. Solubility test for confirming the presence of sickling hemoglobins; approved standard. NCCLS publication H10-A. Villanova, PA: NCCLS, 1986.

40. Nuttall KL. Porphyrins and disorders of porphyrin metabolism. In: Burtis CA, Ashwood ER, eds. Tietz textbook of clinical chemistry. 2nd ed. Philadelphia: WB Saunders, 1994;2073.

41. Paslin DA. The porphyrias. Int J Dermatol 1992;31:527.

42. Poche H, et al. Hereditary progressive muscular dystrophies: Serum myoglobin pattern in patients with different types of muscular dystrophies. Clin Physiol Biochem 1989;7:40.

43. Posey YF, et al. Prenatal diagnosis of sickle cell anemia—Hemoglobin electrophoresis versus DNA analysis. Am J Clin Pathol 1989;92:347.

44. Schwartz SM, et al. Determination of porphyrins in biological materials. In: Glick D, ed. Methods of biochemical analysis. Vol 8. New York: Interscience, 1960;221.

45. Smith SG. The use of thin layer chromatography in the separation of free porphyrins and porphyrin methyl esters. Br J Dermatol 1975;93:291.

46. Stone MJ, Willerson JT, Gomez-Sanchez CE, Waterman MR. Radioimmunoassay of myoglobin in human serum. Results of patients with acute myocardial infarction. J Clin Invest 1975;56:1334.

47. Tan GB, et al. Evaluation of high performance liquid chromatography for routine estimation of haemoglobins A2 and F. J Clin Pathol 1993;46:852.

48. Tefferi A, Solberg LA, Ellefson RD. Porphyrias: Clinical evaluation and interpretation of laboratory tests. Mayo Clin Proc 1994;69:289.

49. Watson CJ, Schwartz S. A simple test for urinary porphobilinogen. Proc Soc Exp Biol Med 1941;47:393.

50. Willerson JT. Clinical diagnosis of acute myocardial infarction. Hosp Pract (Off Ed) 1989;24:65.

51. Williams WJ, et al. Hematology. 3rd ed. New York: McGraw-Hill, 1983;587.

52. With TK. Diagnostic tests for porphyria. Lab Med 1980; 11:446.

53. Zabel M, et al. Analysis of creatine kinase, CK-MB, myoglobin, and troponin T time-activity curves for early assessment of coronary artery reperfusion after intravenous thrombolysis. Circulation 1993;87:1542.

Assessment of Hemostasis

Patricia Hudson

Objectives

Upon completion of the chapter, the clinical laboratorian should be able to:

- *Explain the balance between coagulation and fibrinolysis.*
- *List the four primary elements that contribute to hemostasis.*
- *Identify the hemostasis proteins and inhibitors.*
- *Outline the intrinsic and extrinsic pathways of coagulation.*
- *Describe the clinical laboratory tests used to assess the vessels, platelet function, the coagulation proteins, the fibrinolysis proteins, and the inhibitors.*
- *Describe the therapeutic management of thrombosis.*

KEY WORDS

Anticoagulant Therapy	Fibrinolysis	Serine Proteases
Coagulation Cascade	Hemostasis	Thrombosis
	Hypercoagulability	Vitamin K
Contact Activation	Inhibitors	von Willebrand
Extrinsic Pathway	Intrinsic Pathway	Factor
Fibrin	Platelets	Zymogens

INTRODUCTION

Hemostasis has evolved quickly to its present state of knowledge and understanding. Likewise, the hemostasis laboratory has evolved from a corner in the hematology or chemistry lab into its own independent facility. Coagulation testing has evolved from its early days with a test tube and stopwatch to sophisticated automated analyzers and high technologic procedures. Today, hemostasis has discarded the early image of witchcraft and emerged as a complex independent biological science.

Hemostasis is the property of the blood circulation system that maintains the blood in a fluid state within the vessel walls in combination with an ability to prevent excessive blood loss when injured. This occurs as a result of a delicately balanced interaction between blood vessel walls, the circulating platelets, and the plasma coagulation factors. Too much coagulation at the site of injury leads to excessive thrombosis and too little coagulation at the site of injury leads to persistent bleeding. Normal hemostasis maintains this critical balance. There are four primary elements that contribute to hemostasis: the vessels, the platelets, the coagulation factors, and the fibrinolytic factors.

The vessels are important because the vessel walls play a role in the initiation of coagulation and the prevention of major blood loss from injury. The platelets are important because they provide the initial platelet plug at the site of injury and contribute to the initiation of coagulation with the platelet membrane and other platelet components. The coagulation cascade is important because it generates a fibrin meshwork that provides structural support to the platelet plug. Fibrinolysis is important because it clears the clot from the system to reestablish normal blood flow following an injury and healing (Fig. 26-1).

The Blood Vessels

Normal vessels maintain a barrier to coagulation. The lining prevents coagulation from taking place. It is upon injury, when the lining is damaged, that other tissue is exposed to stimulate the platelets. In severe injury, the vessel itself constricts to reduce the flow of blood.

Michael L. Bishop, Janet L. Duben-Engelkirk, and Edward P. Fody.
CLINICAL CHEMISTRY. © 1996 Lippincott–Raven Publishers.

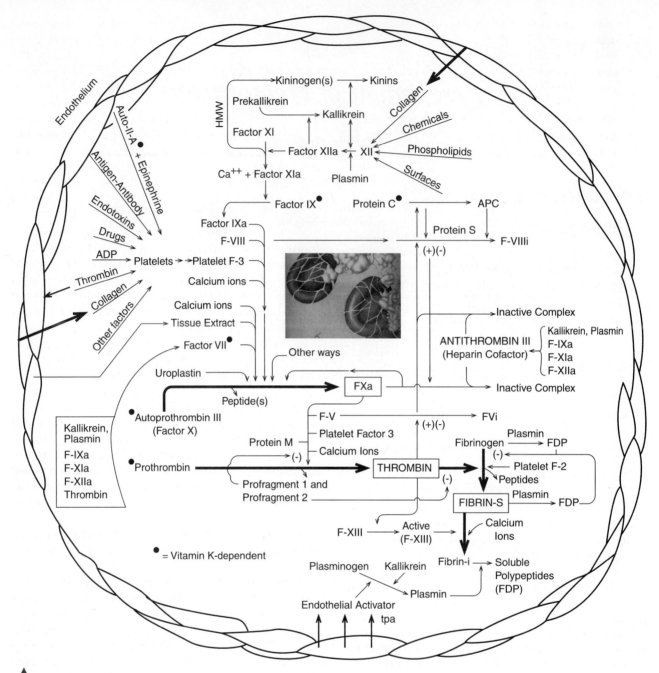

Figure 26-1. Hemostasis Systems Interaction. (Adapted from Seeger Murano, 1979.)

The Platelets

The platelets play several critical roles in hemostasis, so normal hemostasis is dependent upon both an adequate number and properly functioning platelets. Platelets influence coagulation through adherence, aggregation, initiation of coagulation, and clot retraction.

The Coagulation Proteins

The coagulation factors are proteins that interact via a complex series of reactions to form the fibrin meshwork that reinforces the platelet plug. Many of them circulate as unactivated forms of enzymes, or zymogens, which must be converted into active forms for biological function. The

reactions can be initiated via either of two pathways, the intrinsic pathway or the extrinsic pathway, depending on the conditions and elements that create the first reaction. Both pathways progress to the activation of factor X to factor Xa, which is the first reaction of the common pathway, where the pathways converge for the final reactions that end with fibrin formation.

The Fibrinolysis System

The balance to the clot-forming reactions of the coagulation cascade is the clot-dissolving reactions of the fibrinolytic mechanism. Fibrin, formed as a matrix for tissue repair or as a hemostatic plug, is no longer needed and must be removed.

Fibrinolysis removes fibrin by breaking it down into small fragments, which can then be cleared from the circulation by the reticuloendothelial system.

In summary, hemostasis is a delicately balanced equilibrium between two activated systems. Disturbances of the balance lead to hemorrhage or thrombosis. Hemorrhage is the result of impaired clotting and increased fibrinolysis. Thrombosis is the result of increased clotting and impaired fibrinolysis.

THE HEMOSTATIC MECHANISM

The Blood Vessels

Coagulation is initiated when an injury occurs to the vessel wall and the endothelium lining the vessel. Normal vessels maintain an intact lining, the endothelium, which is essential as an inert barrier. This barrier separates the procoagulant proteins and platelets from the procoagulant subendothelial layers. The endothelial lining of intact vessels is antithrombotic. It does not activate platelets or promote coagulation. It provides a smooth surface that facilitates blood flow.

Minor trauma to a vessel, which causes damage to the endothelium, exposes basement membranes, including collagen, to the platelets. With minor vessel trauma, hemostasis is dependent on the number, quality, and function of the platelets.

In the case of severe trauma to a major vessel, the vessel itself constricts to reduce the flow of the blood and thus prevent major blood loss from the injury. Almost simultaneously, with vasoconstriction, damage to the endothelial lining exposes the underlying connective tissue, including collagen, and platelet adherence is initiated.

Upon injury, the smooth endothelium is broken and the subendothelial collagen is exposed. Contact with collagen stimulates the platelets to become attached to it. This action of platelet adherence promotes clot formation by causing the initiation of coagulation. The platelet reaction initiates the contact phase of coagulation leading to the intrinsic pathway of coagulation. The tissue damage releases tissue thromboplastin, which initiates the extrinsic pathway of coagulation. Vessel injury also initiates fibrinolysis by allowing the endothelial cells to release tissue plasminogen activator (tPA). There is a lack of tests for vascular integrity, but the tourniquet test persists as a crude measure of capillary fragility.

The Platelets

Platelets play several critical roles in hemostasis: (1) adhere to injured vessels, (2) aggregate at the injury site, (3) promote coagulation on the phospholipid surface, (4) release biochemicals important to hemostasis, and (5) induce clot retraction. Hemostasis is dependent upon platelets that are normal both in number and in function.

Platelets are non-nucleated disc-shaped elements that circulate in the blood. Circulating platelets are formed from megakaryocytes in the bone marrow. They contain storage granules that hold substances essential to clot formation.

At the site of vessel injury, the platelets adhere to the subendothelial collagen to form a primary platelet plug. This adherence occurs within seconds following stimulation. The platelets adhere to the exposed tissue with the support of von Willebrand factor. With adhesion, the platelets are stimulated to release the contents of the storage granules, alpha granules, and dense bodies. During this secretory process, the alpha granules and dense bodies extrude stored substances to the platelet exterior via the open canalicula system. Substances released include adenosine diphosphate (ADP), adenosine triphosphate (ATP), calcium, serotonin, epinephrine, platelet Factor-4 (PF4), fibrinogen, and platelet Factor-3 (PF3) (Table 26-1).

Following adhesion, platelets undergo an internal transformation and a shape change leading to irreversible aggregation. Pseudopods extend, which present receptor sites for other platelets. Aggregation is induced by the ADP released by the dense bodies, thrombin, collagen, or other platelet substances. Platelet aggregates are the first defense against blood loss in injury.

The platelets initiate the coagulation reactions that lead to fibrin formation for reinforcement of the platelet plug. The platelet products and the platelet surface provide procoagulant activity by accelerating several reactions in the coagulation scheme. Specifically, platelets can initiate the contact phase of intrinsic coagulation in the presence of ADP by activating factor XII on the platelet surface.[33] Platelet Factor 3 facilitates the surface activity that enables assembly of the coagulation factors. ADP is also a potent platelet aggregator. Serotonin and epinephrine act as vasoconstrictors and platelet aggregators. Platelets also release factor V and fibrinogen.

Finally, platelet retraction occurs. This retraction causes the clot reduction in size that contributes to the return to normal blood flow.

Platelet function is severely impaired by aspirin. It interferes with platelet activation and aggregation, and aspirin's effect lasts for the life of the platelet. Aspirin exerts its effect on platelets through blocking prostaglandin and thromboxane production. Patients taking aspirin or aspirin-containing medications may show prolonged bleeding time test results.

Tests for platelets are based on assessing number and function. Platelet counts identify the number of platelets. The bleeding time, the glass bead retention test, and von Willebrand's Factor tests measure platelet adhesion. Platelet aggregation studies measure aggregation function. The clot retraction test demonstrates retraction.

Biochemistry of Hemostatic Proteins

General Structure and Activation Mechanism

All the factors involved in hemostasis, except thromboplastin, calcium, and platelet factor 3, are proteins. All, except factor VIII, are synthesized in the liver. Factors II, VII, IX, and X depend on vitamin K for their production. The factors are either cofactors, which are not enzymes, or precursors of the enzyme group serine proteases.

TABLE 26-1. Summary of Most Important Substances Secreted by Platelets and Their Role in Hemostasis

Role in Hemostasis	Substance	Source	Comments on Principal Function*
Promote coagulation pathway	HMWK	Alpha granules	Contact activation of intrinsic coagulation
	Fibrinogen	Alpha granules	Converted to fibrin for clot formation
	Factor V	Alpha granules	Cofactor in fibrin clot formation
	Factor VIII:vWF	Alpha granules	Assists platelet adhesion to subendothelium to provide coagulation surface
Promote aggregation	ADP	Dense bodies	Promotes platelet aggregation
	Calcium	Dense bodies	Same
	Platelet factor 4	Alpha granules	Same
	Thrombospondin	Alpha granules	Same
Promote vasoconstriction	Serotonin	Dense bodies	Promotes vasoconstriction at injury site
	Thromboxane A_2 precursors	Membrane phospholipids	Same
Promote vascular repair	Platelet-derived growth factor	Alpha granules	Promotes smooth muscle growth for vessel repair
	Beta thromboglobulin	Alpha granules	Chemotactic for fibroblasts to help in vessel repair[33]
Other systems affected	Plasminogen	Alpha granules	Precursor to plasmin, which induces clot lysis
	α_2-Antiplasmin	Alpha granules	Plasmin inhibitor; inhibits clot lysis
	Cl esterase inhibitor	Alpha granules	Complement system inhibitor

*Functions other than those indicated also exist for some substances. (From Lotspeich-Steininger CA, Stiene-Martin EA, Koepke, JA. Clinical hematology. Philadelphia: JB Lippincott, 1992;593.)

These precursors are zymogens that can be activated by a conformational change or a highly specific cleavage of the protein. The conformational activation is accomplished by a physical change resulting in the exposure of a serine residue, thus converting the protein to a serine protease enzyme. The activation requires a phospholipid surface, the zymogen, possible cofactor(s), and a serine protease. The enzyme hydrolyzes the zymogen at a specific peptide bond, cleaving the molecule to expose the active serine residue, thus creating another serine protease. Early reactions in the coagulation sequence are conformational changes, whereas later ones are of the cleavage type.

Classification

Classification of the factors is indicated by Roman numerals, as seen in Table 26-2. The more recently discovered factors of prekallikrein, kallikrein, and high-molecular-weight kininogen have not been assigned Roman numerals.[34]

Factor Proteins

Fibrinogen. Fibrinogen is a glycoprotein.[9] It is composed of three pairs of peptide chains, designated alpha, beta, and gamma. These peptide chains are linked by disulfide bonds. Both immunologic and electron microscopic data indicate its structure to be a complex, tertiary one.[10,16]

Thrombin acts on fibrinogen by catalyzing the removal of polar peptides, which results in the conversion to a fibrin gel. These peptides (fibrinopeptide A and B) make up approximately 3% of the mass of the fibrin clot. Following thrombin's conversion of fibrinogen to fibrin, Factor XIIIa stabilizes the fibrin clot by catalyzing the formation of peptide bonds and cross-linking the gamma and alpha chains.[12]

Plasmin degrades fibrin. A bleeding tendency exists in patients with abnormal fibrinogen or those who have been bitten by certain poisonous snakes. Also, several snake venoms (*Bothrops jararaca, Bothrops atrox,* and ancrod) contain enzymes that hydrolyze fibrinogen, splitting off only fibrinopeptide A, which induces clotting. This unusual fibrinogen structure polymerizes abnormally forming a friable clot. This friable fibrin clot is not able to stabilize and form a solid structure.

Factor II (Prothrombin). Prothrombin is a glycoprotein, which is a single polypeptide chain with a free amino group on the end. The formation of thrombin from prothrombin is mediated by the proteolytic enzyme, Factor Xa, a phospholipid surface, calcium ions, and Cofactor Va. The thrombin produced is a strong proteolytic enzyme. It initiates irreversible platelet plug formation, enhances the efficiency of Factors V and VIII when thrombin is in trace amounts, inhibits the previously mentioned Factors if thrombin is in large amounts, converts fibrinogen to fibrin, activates Factor XIII, and activates the fibrinolytic system.[18,30]

Tissue Factor Tissue Factor, *called tissue thromboplastin,* is found in all tissues, although brain, liver, and placenta are richest in its concentration. It enters the circulation through injury. Tissue Factor is a complex of a phospholipid and protein. Both are required for tissue factor to activate Factor VII in the extrinsic path of coagulation.[40]

Factor V. Factor V is a glycoprotein which is a Cofactor to facilitate activation of prothrombin. Factor V is also found in platelets; however, the structure is slightly different,

TABLE 26-2. International Nomenclature for Coagulation Proteins

Factor	Synonym(s)
I	Fibrinogen
II	Prothrombin, prethrombin
III	Tissue thromboplastin, tissue factor
IV	Calcium
V	Proaccerlerin, labile factor, accelerator globulin
VI	Not assigned
VII	Proconvertin, stable factor, serum prothrombin conversion accelerator (SPCA), autoprothrombin I
VIII	Antihemophilic factor (AHF), antihemophilic globulin (AHG), platelet cofactor I
IX	Plasma thromboplastin component (PTC), Christmas factor, antihemophilic factor B, autoprothrombin II, platelet cofactor II
X	Stuart-Prower factor, Stuart factor, autoprothrombin III
XI	Plasma thromboplastin antecedent (PTA), antihemophilic factor C
XII	Hageman factor
XIII	Fibrin-stabilizing factor, fibrinase, Laki-Lorand factor
Prekallikrein	Fletcher factor
HMWK	High-molecular-weight kininogen, Fitzgerald factor, Flaujeac factor

Source: Corriveau D, Fritsma G. Hemostasis and thrombosis in clinical laboratory. Philadelphia: Lippincott, 1988.

although the procoagulant activity is the same. Factor V is activated by thrombin.[8,18]

Factor VII. Factor VII, a glycoprotein, requires tissue Factor for activation. In a reverse reaction, Xa can attack Factor VII in the presence of phospholipid to enhance Factor VII activity. Glass, Factor XIIa, Factor IXa, and prekallikrein can indirectly activate Factor VII.[2,31]

Factor VIII. Factor VIII is the term used to designate the entire multimeric molecule that can be concentrated as a cryoprecipitate. Factor VIIIa is a cofactor to X activation, reacts to thrombin, and is sensitive to degradation by plasmin. It is also inactivated by protein C.[17]

Upon treatment with a high salt concentration of NaCl or CaCl₂, the complex molecule dissociates into a low-molecular-weight (LMW) portion and a high-molecular-weight (HMW) portion.[39] The LMW component is controlled by sex-linked inheritance. A deficiency of this component results in decreased Factor VIII:C, which results in classic hemophilia A. The site of production for this component is still unclear. The HMW component, the larger portion, is controlled by autosomal inheritance. Its parts consist of VIII:vWF, VIII:Ag, and VIIIRCo, all synthesized by the endothelium and megakaryocytes. A deficiency in these Factors is seen in von Willebrand's disease, in which there is a defect in platelet adhesion and aggregation with the endothelium.[38] Below is the nomenclature for Factor VIII proposed by the International Committee on Thrombosis and Haemostasis. The terminology reflects different methods of testing applied to physical and functional abilities of the factor molecule parts.

VIII/vWF—Factor VIII, a molecule composed of von Willebrand Factor and VIII:C non-covalently bound

VIII/R:RCo—Ristocetin-Cofactor activity, the Factor VIII–related activity required for the aggregation of human platelets

VIII:vWF—von Willebrand Factor activity, bleeding-time Factor

VIIIR:Ag—Factor VIII–related antigen as measured by immunologic techniques

VIII:C—Factor VIII procoagulant activity as measured by a modified APTT Factor assay

VIIIC:Ag—Antigen with procoagulant activity

Factor IX. Factor IX is also controlled by a sex-linked gene, with the deficiency causing hemophilia B. Factor IX is activated by XIa, which cleaves the molecule at two sites. This provides a highly negatively charged enzyme that can aid in the activation of Factor X.[5]

Factor X. Factor X activation begins the common path of coagulation through action of Factor VIIa (extrinsic) or Factors IXa–VIII (intrinsic). It may also be activated by Russell's viper venom and dilute trypsin.[14]

Factor XI. Factor XI circulates in the plasma complexed with high-molecular-weight kininogen (HMWK). Factor XI is activated by both trypsin and Factor XIIa through limited proteolysis. It, in turn, activates Factor IX.

Factor XII. Factor XII is an enzyme central to the intrinsic path of coagulation. It can be activated by exposure to a negatively charged surface such as glass, ellagic acid, collagen, homocysteine, fatty acids, endotoxin, and others. In the contact activation reactions, Factor XII activates prekallikrein to convert it to kallikrein. Once Factor XIIa is present, three systems are activated: the kinin system, the fibrinolytic system, and the coagulation system.[20]

Factor XIII. Factor XIII is the fibrin stabilizing factor. During clot formation it is trapped in the fibrin clot and activated by thrombin. Factor XIIIa produces the cross-linking of fibrin as the final enzymatic step of coagulation. This is accomplished by the forming of intermolecular bridges between fibrin monomers. Some evidence also points to factor XIIIa's involvement in strengthening the attachment of fibrin to platelets and fibroblasts with fibronectin.[7] The deficiency of factor XIII tends to cause keloid formation. These patients should not be subjected to bleeding-time tests unless absolutely necessary.

Prekallikrein (Fletcher Factor). Prekallikrein, has been isolated in two forms of equal weight and function. Factor XIIa hydrolyzes a peptide bond of prekallikrein, converting it to kallikrein. Kallikrein then attacks prekallikrein and also activates plasminogen (fibrinolysis).

High-Molecular-Weight Kininogen (Fitzgerald Factor). High-molecular-weight kininogen (HMWK) is a single chain glycoprotein and is thought to be produced by the liver. HMWK circulates complexed with both prekallikrein and factor XI and acts as a cofactor primarily in the contact activation reactions.

Kallikrein. Kallikrein acts on HMWK to release bradykinin and several other components. Only the light chain of one of the small subunits has procoagulant activity. Characteristics of the clotting factors are summarized in Table 26-3.

Other Factors. Other biological substances, such as phospholipid and calcium, are essential accelerators of blood coagulation by precipitating reactions involving factors Va and VIIIa. Platelets provide a source of both, although calcium circulates freely in large concentration in the plasma. Phospholipid is also contained within tissue.

In addition, inhibitors of both the coagulation and fibrinolytic systems keep the hemostatic mechanism in check. Anti-thrombin III (ATIII) is a naturally occurring inhibitor of thrombin and factor Xa. Along with its cofactor, heparin, which also occurs naturally, ATIII complexes 1:1 with thrombin to inactivate it. Another vitamin K-dependent protein, protein C, functions as an anticoagulant by inhibiting factors VIII, Va, and the bridging of platelets to factor Xa. Protein S functions as a cofactor to these reactions.

The inhibitors to fibrinolysis are also enzymes that block the action of plasmin on fibrinogen. α_2-Antiplasmin, α_2-macroglobulin, and α_1-antitrypsin are the most important ones.

COAGULATION PATHWAYS

Extrinsic Coagulation Pathway

The extrinsic coagulation pathway is so named because its activation requires the involvement of a substance that resides outside the circulation. The extrinsic pathway involves the activity of tissue thromboplastin, which reacts with factor VII and calcium for the rapid proteolytic activation of factor X to factor Xa. Tissue thromboplastin is a glycoprotein that is widely distributed in the cells. When injury occurs to the vessel wall, the damage to the cells results in tissue thromboplastin availability, which, in the presence of calcium, can form a complex with factor VII. This complex acts to convert the factor X to factor Xa, the proteolytically active enzyme. The activity of the tissue factor–factor VII complex seems to be largely dependent on the concentration of tissue thromboplastin.

The next step is the activation of factor X. At this point, the extrinsic system converges with the intrinsic system for the common pathway reactions. Once factor X is converted to its active form, Xa, it can feed back to convert factor VII to VIIa. The common pathway constitutes the final steps to fibrin formation.

Intrinsic Coagulation Pathway

The intrinsic coagulation pathway is so named because its activation can occur with elements that reside entirely within the circulation. Activation of the intrinsic pathway is initiated with the disruption of the intact endothelium and exposure of the negatively charged subendothelial surface, such as collagen. The surface serves as the stage upon which a number of components can interact. Circulating factor XII becomes attached to the surface; a conformational change occurs in the factor XII molecule increasing its susceptibility to activation by specific enzymes, such as kallikrein. Platelets are also thought to participate in this reaction. Factor XIIa, in a positive feedback pathway, converts prekallikrein to additional kallikrein; this reaction is facilitated by the action of high-molecular-weight kininogen (HMWK). Factor XIIa is also thought to be an activator of factor VII. HMWK and the exposed negative surface serve as Cofactors for the proteolytic activation of factor XI to XIa by factor XIIa. After activation, factor XIa remains localized at the activation site while kallikrein can circulate freely.

Factor XIa, with calcium, converts factor IX to its active form, factor IXa, factor IX can also be activated by the tissue factor-factor VII complex, which is the complex that initiates the extrinsic pathway.

The next series of reactions in the intrinsic system involves the formation of a complex from factors IXa, VIII, calcium, and PF3, This complex activates factor X. Factor IXa is the enzyme that actually cleaves factor X, but the reaction is accelerated significantly in the presence of the other components. At this point, the extrinsic and intrinsic systems converge for the common pathway to fibrin formation.

The Common Pathway

The common pathway consists of the final reactions for activation of prothrombin to thrombin and conversion of fibrinogen to the insoluble fibrin gel. It begins with the activation of factor X through either the extrinsic or intrinsic pathways. The extrinsic and intrinsic pathways differ in the way and rate that factor X is activated and also in the amount

TABLE 26-3. Characteristics of Clotting Factors

Factor	Active Form	Molecular Weight ($\times 10^3$)	Pathway Participation	Site of Production	Vitamin K–Dependent?	In vivo Half-Life (Hours)	Plasma Concentration (mg/dL)	Minimum Hemostatic Level (%)*	Present in BaSO₄–Adsorbed Plasma?
I	Fibrin clot	340	Common	Liver	No	90	300–400	100 mg/dL	Yes
II	Serine protease	72	Common	Liver	Yes	60	10–15	20–40	No
V	Cofactor	330	Common	Liver	No	12–36	0.5–1.0	10–25	Yes
VII	Serine protease	63	Extrinsic	Liver	Yes	4–6	0.2	5–10	No
VIII:C	Cofactor	70–240 (VIII/vWF > 1000)	Intrinsic	Uncertain for (VIII:C)**	No	12 22–40†	1–2 (VIII/vWF)	25–30	Yes
IX	Serine protease	62	Intrinsic	Liver	Yes	24	0.3–0.4	15–25	No
X	Serine protease	58.9	Common	Liver	Yes	48–72	0.6–0.8	10–20	No
XI	Serine protease	160	Intrinsic	Liver	No	48–84	0.4	10–20	Yes
XII	Serine protease	80	Intrinsic	Liver	No	48–52	2.9	0–5	Yes
XIII	Transglutaminase	320	Common	Liver	No	3–5 days	2.5	2–3	Yes
Prekallikrein	Serine protease	85	Intrinsic	Liver	No	35	5.0	?‡	Yes
HMW Kininogen	Serine protease	120	Intrinsic	Liver	No	6.5 days	4.7–12.2	?	Yes

*Approximate minimum plasma concentration required for normal coagulation. Note that minimum hemostatic level for factor XII is 0–5% (i.e., factor XII may not be required for normal coagulation).
**vWF portion synthesized by endothelial cells and megakaryocytes.
† 22–40 hours for high-molecular-weight subunit of factor VIII.
‡ None reported.

(From Lotspeich-Steininger CA, Stiene-Martin EA, Koepke, JA. Clinical hematology. Philadelphia: JB Lippincott, 1992;593.)

of prothrombin that can be converted to thrombin. In contrast to the fast clotting of the extrinsic system, fibrin formation takes longer with the intrinsic system.

Factor Xa requires the presence of factor V, a cofactor, calcium ions, and platelet factor 3, to rapidly convert small amounts of prothrombin to its active form, the proteolytic enzyme thrombin. Platelet factor 3 is a phospholipid and serves as a surface for assembly of the coagulation proteins and enzymes that react to enhance the rate of coagulation. Thrombin acts on fibrinogen to convert it to fibrin.

Fibrin Formation and Stabilization

The final step for the coagulation pathways is the conversion of fibrinogen to fibrin. Fibrin formation occurs in three phases: proteolytic, polymerization, and stabilization. Fibrinogen is composed of three polypeptide chains, α alpha, β Beta, and gamma. During proteolysis, thrombin splits bonds from the Aa and BB chains releasing a pair of peptides A and B. The remaining portion of the molecule is the fibrin monomer. The monomers line up end-to-end and side-by-side in a spontaneous self-assembly process to form fibrin polymers. This is the polymerization stage. Stabilization of the fibrin polymers occurs through the action of factor XIIIa. In the presence of calcium ions, thrombin activates factor XIII to XIIIa. Factor XIIIa introduces peptide bonds within the polymerized fibrin meshwork. This cross-linking renders the fibrin more elastic and less susceptible to lysis by fibrinolytic agents.

Interrelation of the Systems

The extrinsic and intrinsic systems of coagulation activation are interrelated. Factor VII can activate factors IX and X in the intrinsic system, and factor XIIa can activate factors VII to VIIa in the extrinsic system. Extrinsic coagulation provides a mechanism for rapid production of small amounts of thrombin, which convert factors V and VIII to more reactive forms, thereby accelerating coagulation. Conversely, further proteolysis of these factors by thrombin destroys their cofactor activity.

Coagulation is regulated by the action of inhibitors, which are primarily enzymes that inactivate activated factors. The principle inhibitor of procoagulant proteins is antithrombin III. Secondary inhibitors include alpha-2-macroglobulin and alpha-1-antitrypsin.

FIBRINOLYSIS

Fibrinolysis has a principle role in the balance that regulates clotting. Fibrinolysis removes unwanted fibrin deposits and facilitates the healing process following inflammation and injury (Fig. 26-2).

The primary reaction is the conversion of plasminogen to the active enzyme plasmin. Other significant elements of fibrinolysis are the plasminogen activators and inhibitors.

Plasminogen, a glycoprotein, is activated to its enzyme form, plasmin, by different activators. The activators of plas-

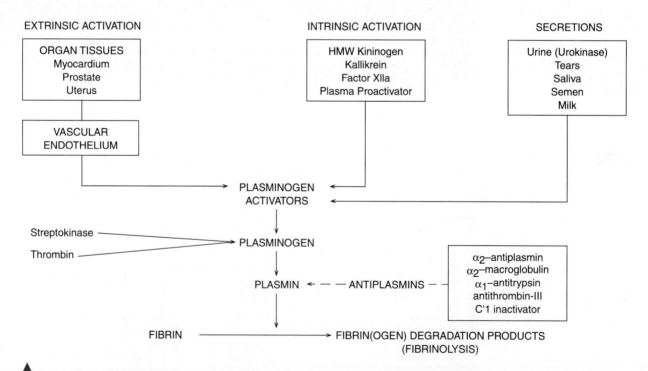

Figure 26-2. Fibrinolytic system may be activated by plasminogen activators from extrinsic sources such as vascular endothelium or by intrinsic sources such as factor XII$_a$ and others as shown. These plasminogen activators convert plasminogen to plasmin. Thrombin also activates plasminogen; streptokinase, administered therapeutically in thrombotic disorders, acts in same way. Plasmin promotes fibrinolysis. Antiplasmins control (inhibit and neutralize) excess plasmin, thus preventing excessive and premature fibrinolysis. (From Lotspeich-Steininger CA, Stiene-Martin EA, Koepke, JA. Clinical hematology. Philadelphia: JB Lippincott, 1992;593.)

minogen can be classified into three groups: (1) intrinsic, (2) extrinsic, and (3) exogenous. The initiation of intrinsic coagulation through factor XIIa and kallikrein activates the fibrinolytic system. The contact system of coagulation therefore serves as an intrinsic activator of fibrinolysis.

Certain tissues, including vascular endothelial cells, contribute to extrinsic activation of fibrinolysis. Tissue plasminogen activator (t-PA) is the most important activator and is produced in the endothelial cells. Stimuli, such as injury, ischemia, exercise, and pyrogens, provide localized extrinsic activation. Exogenous activators come from outside the body and are usually therapeutic agents including streptokinase, urokinase, and tissue plasminogen activator (t-PA).

Fibrinolysis appears to be localized within the formed clot since plasminogen binds strongly to fibrin as the thrombus forms. The tissue plasminogen activators are specifically coupled to fibrin to activate plasminogen to plasmin, which, when released, destroys the fibrin clot.

The action of plasmin on fibrinogen or fibrin breaks the molecules into small fragments called *degradation products*, which are designated by fragment size as X, Y, D, and E. These products block the clotting mechanism by serving as competitive inhibitors of thrombin. They also block platelet aggregation by adsorption onto the platelet surface.

The specificity of plasmin ensures that clot dissolution will occur without widespread proteolysis of proteins. Plasmin also activates the complement system, liberates kinins from kininogen, and can hydrolyze coagulation factors XII, VIII, and V.

The fibrinolytic system, like the coagulation system, is regulated by inhibitors and antiplasmins. Physiologically, the most important plasmin inhibitor is alpha-2-antiplasmin. It has a strong affinity for plasmin and forms an irreversible covalent complex to rapidly inhibit plasmin. If there is saturation of the inhibitor, secondary inhibitors such as alpha-2-macroglobulin and antithrombin III can also neutralize excess plasmin. The most important fibrinolysis inhibitor may well be plasminogen activator inhibitor type 1 (PAI1), which prevents the initial activation from occurring.

Laboratory evaluation of fibrinolysis measures the end result of the system. Fibrinogen degradation products, FDP, and D-Dimer are the most common measures of fibrinolytic activity. The tests are valuable in the diagnosis of disseminated intravascular coagulation (DIC) and in monitoring fibrinolytic therapy. More specific tests include assays for the components of the fibrinolytic system: plasminogen and alpha-2-antiplasmin.

REGULATION

The fibrin-forming and fibrinolytic systems exist in activated states in equilibrium with each other. Maintenance of this delicate balance between fibrin forming and fibrinolytic factors depends on regulation of the systems.

Regulation of both the clotting system and the fibrinolytic system is achieved through the action of naturally occurring inhibitors (Table 26-4)[11,15].

TABLE 26-4. Naturally Occurring Inhibitors of Coagulation and Fibrinolysis

Inhibitor	Molecular Weight ($\times 10^3$)	Plasma Concentration (mg/dL)	Rate of Inhibition	Rate Accelerated By Heparin?	System Inhibited	Function Inhibited
Antithrombin III	67	23–40	Slow	Yes	Coagulation	Thrombin, XII_a, XI_a, X_a, IX_a, and kallikrein
					Fibrinolytic	Plasmin and kallikrein
α_a-Macroglobulin	725	190–330	Variable	No	Coagulation	Thrombin and kallikrein
					Fibrinolytic	Plasmin and kallikrein
α_1-Antitrypsin	40–50	245–335	Slow	No	Coagulation	Potent inhibitor of XI_a; weak inhibitor of thrombin
					Fibrinolytic	Plasmin
C'1 Inactivator	135	14–30	Slow	No	Coagulation	XII_a, XII_f, XI_a, and kallikrein
					Fibrinolytic	Plasmin
					Complement	C'1 Esterase
Protein C	62	0.5	Slow	No	Coagulation	Complexed with protein S, inhibits V_a and VIII_a
					Fibrinolytic (enhanced)	Complexed with protein S, *enhances* fibrinolysis by inactivating inhibitors of plasminogen activators
Protein S	69	1.5 (free form)	Slow	No	Coagulation	Complexed with protein C, inhibits V_a and VIII_a
					Fibrinolytic (enhanced)	Complexed with protein C, *enhances* fibrinolysis by inactivating inhibitors of plasminogen activators
α_a-Antiplasmin	65–70	9.6–13.5	Rapid	No	Fibrinolytic ONLY	Principal inhibitor of plasmin

(From Lotspeich-Steininger CA, Stiene-Martin EA, Koepke, JA. Clinical hematology. Philadelphia: JB Lippincott, 1992;593.)

The clotting system continually generates small amounts of activated factors, which are serine proteases. Unless controlled, the factors could initiate unwanted clot formation. Inhibitors of the activated factors prevent this imbalance. Antithrombin III is the primary inhibitor of the clotting system. It is a slow progressive inhibitor and it keeps the activated enzymes in check by its action against serine proteases. This activity is directed first against thrombin, but also other factors: Xa, XIIa, XIa, and IXa. The activity of antithrombin III is accelerated nearly 1000-fold by the administration of heparin. Decreased levels of ATIII can lead to increases in activated factors and possible thrombosis.

Protein C is another anticoagulant protein. It is activated by thrombin formation and in its active form, in the presence of its cofactor Protein S,[24] it inhibits clotting by the enzymatic degradation of factors V and VIII. The activation is a feedback mechanism where the generation of thrombin results in the formation of activated protein C, which, in turn, prevents further thrombin formation by the inhibition of factors V and VIII.

These regulators are backed up by other minor inhibitors of clotting, which include alpha-2-macroglobulin, heparin cofactor II, and alpha-1-antitrypsin.

The fibrinolytic system in circulation is also maintained in balance.[3] Plasminogen activator converts plasminogen to the active enzyme plasmin. The system is initially controlled by tissue plasminogen activator inhibitor type 1 (PAI-1), which interferes with the activation of plasminogen to plasmin.[23] Nonetheless, if free plasmin is generated, alpha-2-antiplasmin acts quickly to inhibit it. Alpha-2-antiplasmin is the primary inhibitor of plasmin and has a strong affinity for it. Its action is rapid and irreversible. Decreased levels of inhibitors could lead to increased fibrinolytic activity and possible hemorrhage. The laboratory tests available for inhibitors are the specific assays.

PRETHROMBOTIC STATE

While the origins of the study of hemostasis lie in examination of bleeding conditions, a shift to concern for the thrombotic condition has taken place in more recent times. Thrombosis and thrombotic disorders are among the leading causes of disease and death in this country.[19]

Recognition of the incidence and significance of thrombotic disease has lead us to look for means to identify patients at risk of thrombosis before it occurs. Certain alterations in the hemostatic balance may predispose for thromboembolism, which places one in a hypercoagulable or prethrombotic state. The investigation of early changes in the hemostatic mechanism as indicators of potential thrombosis holds promise for prevention of serious thrombotic consequences.

The hemostasis system is in a steady state of activation of both the clotting system and the fibrinolysis system, which is maintained as an equilibrium. An imbalance will result in hemorrhage or thrombosis. In the clotting system, activated factors are controlled by inhibitors, primarily ATIII and the protein C and S system.[24] A decrease in this inhibitor activity will decrease control of the activated factors and may lead to thrombosis.

In the fibrinolysis system, plasminogen and plasmin are also controlled by inhibitors, primarily plasminogen activator inhibitor (PAI-1) and by antiplasmins, primarily alpha-2-antiplasmin. An increase in this inhibitor activity will reduce the fibrinolytic activity and may allow thrombosis to occur.

There is no global assay for the fibrinolytic system that reflects clot breakdown or the components of the system. Therefore, tests that help identify the imbalance of the prethrombotic state measure the specific proteins. In the clotting system, tests include assays for ATIII, Protein C, and Protein S. In the fibrinolysis system, tests include assays for plasminogen, tPA, pAI-1, and alpha-2-antiplasmin.

Another significant indicator of hypercoagulability is the presence of the lupus-like anticoagulant.[31] This is an antibody that functions as an inhibitor through its direction against phospholipids. It was first thought to be a laboratory phenomenon but has been associated with increased thrombotic complications and/or tendencies. The tests that are phospholipid-dependent, *i.e.,* the APTT, PT, or dRVVT, can be prolonged. A common screening test is a mixing study with normal plasma. Confirmatory tests are the platelet neutralization test and, less specifically, the dilute tissue thromboplastin inhibition test.

The study of hypercoagulability has also focused on fibrinogen as a significant factor.[37] Plasma fibrinogen concentrations have been shown to play an important role in atherosclerosis and may be a useful marker for identifying high risk individuals. An increase in fibrinogen level appears to be an early and relatively easy to perform test for prethrombosis.

THERAPY

Thrombosis occurs when the delicate hemostatic balance shifts. Although it is a normal physiologic response designed to maintain hemostasis, thrombosis can create pathological disorders. In fact, it is a leading cause of death in the United States and has become an area of major emphasis in hemostasis.

Therapeutic management of thrombosis has developed and is approached with a variety of drugs. Widely used are coumarins, heparin, and thrombolytic agents such as streptokinase or tissue plasminogen activator (tPA).

Many PT and APTT assays are performed with the purpose of monitoring anticoagulant therapy.

Coumadin

Coumadin is frequently the choice for effective long-term anticoagulation. It functions as an anticoagulant by interfering in the synthesis of a group of coagulation factors, called the prothrombin complex. This group includes factors II, VII, IX, and X, which are all manufactured in the liver. The effect of coumadin is linked to the role of vitamin K during manufacture of the factors, and the result is production of factors that lack the biological coagulant activity.

The prothrombin time test (PT) is used to monitor coumadin therapy. It is sensitive to three of the four factors affected by coumadin, *i.e.*, factors II, VII, and X, and so is a sensitive indicator of the drug's efficacy. The therapeutic range for coumadin therapy has been standardized by the implementation of the International Normalized Ratio (INR) for reporting results.

Heparin

Heparin is used in the treatment of acute thrombosis as a means of preventing further clot formation. Heparin functions as an anticoagulant by catalyzing and accelerating the action of the naturally occurring inhibitor antithrombin III. ATIII is a serine protease and acts by inactivating certain activated factors, primarily thrombin but also XII, XIa, IXa, and Xa. The activity of ATIII is increased 1000-fold in the presence of heparin.

The APTT is often used to monitor heparin therapy. It is sensitive to the levels of factors XII, XI, IX, X, and II (thrombin). The therapeutic range for heparin should be determined by each laboratory. Other laboratory tests used to monitor heparin therapy are the thrombin time, Xa assays, and antithrombin III assays.

Thrombolytic Agents

Newest in use for treatment of thrombosis are the thrombolytic drugs.[28] These are used to dissolve an existing thrombus and restore normal blood flow. The mode of action is the activation of the fibrinolytic mechanism by the activation of plasmin. Streptokinase, urokinase, or tissue plasminogen activator are infused to the region of the clot where they activate the plasminogen to plasmin, which breaks down the fibrin.

Tests available for monitoring thrombolytic therapy have been indirect and are not ideal. Indirect tests, such as the thrombin time or fibrinogen concentration, have been used, as well as tests of the fibrinolytic system, *i.e.,* the plasminogen assay, or the alpha-2-antiplasmin assay.

LABORATORY EVALUATION OF HEMOSTASIS

Laboratory evaluation of hemostasis has expanded as rapidly as the knowledge base in hemostasis. A wide variety of procedures and methods are in use to define the multifaceted hemostatic condition of the patient.

The methods vary from the global fibrin clot endpoint of the PT and APTT, to specific immunological procedures, to complex synthetic substrate studies (Table 26-5).

Review of Classic Methods

Testing for coagulation defects has rapidly evolved from simple, manual procedures using whole blood to the sophisticated methods performed on the fully automated instrumentation used today. The early days of test tube and stopwatch were crude but led the way to learning and discovery of the hemostasis principles we use today. This knowledge facilitated the progressive development of much more sophisticated testing techniques for the routine lab as well as for the special coagulation labs.

TABLE 26-5. Laboratory Evaluation of Hemostasis

The Vessels
Tourniquet test
Bleeding time

The Platelets
Platelet count
Bleeding time
Glass bead retention test
Clot retraction
Platelet factor 3 assay
Platelet factor 4 assay
Beta-thromboglobulin assay
Platelet aggregation studies
von Willebrand factor assay

The Coagulation Proteins
Lee-White whole blood clotting time
Activated clotting time
Partial thromboplastin time
Activated partial thromboplastin time (APTT)
Thromboplastin generation test
Prekallikrein (Fletcher Factor) screening test
Factor VIII related antigen test
Prothrombin time test (PT)
Prothrombin and proconvertin test (P&P)
Stypven time
Factor assays
Thrombin clotting time
Reptilase time
Fibrinopeptide A
Fibrinogen assay
Factor XIII assay
Heparin anti-Xa assay
Protamine sulphate test
Prothrombin fragment F1.2

The Fibrinolysis Proteins
Whole blood clot lysis time
Euglobulin lysis time
Plasminogen assay
Ethanol gelation test
Tanned red cell hemaglutination inhibition immunoassay
Fibrin(ogen) degradation products
D-Dimer

The Inhibitors
Mixing studies screening test
Tissue thromboplastin inhibition test
Platelet neutralization procedure
Dilute Russell's viper venom test (dRVVT)
Bethesda inhibitor assay
Antithrombin III
Protein C
Protein S
Alpha 2 antiplasmin

The first automation of the test tube and stopwatch was with mechanical endpoint analyzers. These instruments employed the use of the conductance of the fibrin strands to close an open electrical circuit as the endpoint. The Fibrometer is the best known and most widely used instrument of this type.

Photo-optical endpoint analyzers applied another principle to endpoint detection and measured the change in light transmittance at fibrin clot formation. Many manufacturers produced instruments with this principle and developed newer systems with additional automation. Over 70% of current coagulation testing is performed on instrumentation based on principles of clot detection by optical or electromechanical changes. All the measurements are related to the activity of factors by measuring the time required for a specimen to clot.

These instruments provide the means to automate the prothrombin time (PT), activated partial thromboplastin time (APTT), thrombin time (TT), and specific factor assays, but they are limited to only these tests. Special testing, using special instruments and materials, is necessary to perform other assays not based on the clot formation, e.g., platelet function studies, fibrin-degradation products, and assay of components of Factor VIII. The last two tests are immunologic, whereas platelet function testing measures the ability of platelets to adhere and aggregate in vitro. Except for the latex-agglutination testing of FDPs, the tests are time-consuming and require a great deal of technical expertise. They can, if performed correctly, offer excellent screening methods for hemorrhagic diatheses.

The two most widely used tests for hemostatic assessment are the PT and APTT. These qualitative assays are measures of the extrinsic and intrinsic systems of coagulation, respectively. They are readily applied to automated instruments that are dedicated to large-volume testing in coagulation, screening of routine and presurgical cases, and monitoring of anticoagulant therapy. The PT and APTT are modified for factor assays which determine specific factor activity through use of deficient factor plasmas. If the suspected factor deficiency is in the extrinsic system (i.e., factor VII), then a PT assay is performed using dilutions of plasma deficient in the factor. For factor assays of the intrinsic system, the APTT based assay is performed. Comparison is made to a factor activity curve established with a normal plasma.

Excellent reviews of accepted, current methodologies are available.[22,27] These classic methods are effective global tools that should not necessarily be replaced in the basic assessment of coagulation problems. In spite of the many new and exciting methods becoming available, they still offer inexpensive screening protocols. Until the clinical utility of new methods has been more fully evaluated, the old tried-and-true methods should not be discarded.

Coagulation Testing in the Chemistry Laboratory

The newly developed molecular markers and synthetic substrates offer a promising future in the testing for specific coagulation proteins. They are readily adaptable to available instrumentation in the chemistry laboratory and can provide a more quantitative, cost-effective, and sensitive procedure for specific factor assay. Currently available are immunologic methods, such as radioimmunodiffusion (RID), laser nephelometry, enzyme-linked immunoassay (EIA), and electroimmunodiffusion (EID); chemistry methods; biochemical methods using synthetic substrates (both chromogenic and fluorogenic); physical methods using light-scatter or fluorescence; physiologic methods; and pharmacologic methods. These methods are amenable to the ACA™, Encore™, ACL™, and kinetic analyzers, discrete analyzers, centrifugal analyzers, and immunoanalytic instruments. The literature contains comprehensive review.[5,6,24,32]

The adaptation of coagulation assays to clinical chemistry analyzers may provide more efficient coagulation profiling for patient benefit. This will become possible as more methods are adapted to current chemistry instrumentation. The benefit to the patient will be in faster and more reliable results that can aid in monitoring of anticoagulant therapy, thrombotic therapy, identification of the specific cause of bleeding, and blood component therapy. Bleeding disorders due to vascular dysfunction will be more readily identified with testing for specific prostaglandins. The ability to quantitate the specific factors involved in hemostasis or factors identified as specific markers for problems offers a clearer picture for interpretation of the hemostatically compromised patient. Standardization of methods and reagents that are not currently uniform will be possible once these new methodologies gain acceptance.

The discussions of methods and clinical significance that follow cover the state of testing as it exists. The methods given are the only commercial ones currently available for chemistry analyzers. Table 26-3 lists the factor concentrations; however, each laboratory should establish its own reference ranges.

Monitoring the Coagulation System
Prothrombin Time. When performed as an amidolytic assay (activity or function measured), the PT is called the extrinsic pathway–generated thrombin test (EPGT). This method may be applied to automated kinetic analyzers (Vivatron AKES, Gilford 3500 or Impact, LKB 8600, and Mark II) and is based on the use of synthetic substrates. Substrates are available from Kabi-Vitrum (S-2160, S-2238), Boehringer-Mannheim (Chromozym TH), and Diagnostica Stago (CBS 34.37). Coagulation enzymology is another term that has been coined to describe the use of synthetic substrates to assay coagulation factors. The principle of these assays involves an active enzyme that cleaves a tag attached to a substrate that has been prepared commercially to mimic the original substrate. The rate of formation of the optically active tag is then measured over a period of time. The amount of activity of the protein (factor) is directly proportional to the amount of released tag. With the EPGT, the measurement is of the activity of the extrinsic pathway and all the proteins involved in its execution.[32]

Plasma + thromboplastin → thrombin
substrate RpNa + extrinsic thrombin →
RCOOH + pNa (optically active tag)

$$(Eq. 26\text{-}1)$$

Another approach to automating the PT on automated instrumentation involves the principle of forward light-scatter. An antibody complexes with an antigen, which produces an amount of scattered light when a laser beam is directed through the solution. This is called nephelometry, and it is a direct measure of the amount of activity of antigen (the activity of extrinsic pathway factors in this case). The ACL (Instrumentation Laboratories) offers this method to perform PTs.

According to Messmore et al.,[26] the PT amidolytic assay is equivalent to clotting methods if the proper activator is used and the prothrombin molecule is normal.[29] Hypofibrinogenemia and inhibitors of fibrin polymerization would not be detected with synthetic substrate analysis of the PT.[29] An abnormal PT could indicate a factor deficiency in the extrinsic system (factor VII) or in the common path of coagulation (factors X, V, II). It would also be abnormal in liver disease and vitamin K deficiency. The PT alone cannot be specific. A normal result can rule out only a problem in the extrinsic system. The APTT must be performed concurrently to help determine the specific factors involved.

Activated Partial Thromboplastin Time. When performed as an amidolytic assay, the APTT is called the intrinsic pathway–generated thrombin test (IPGT). It measures the functional components of the intrinsic pathway of coagulation. Chromogenic substrate methodologies for the APTT have been applied to the ACL (Instrumentation Laboratories), the Encore (J. T. Baker), and the Gilford. Instead of the activator, tissue thromboplastin, a kaolin-type activator is used. The principle and substrates are the same as in the PT amidolytic assay.

$$\text{Plasma} \xrightarrow{\text{Activators like kaolin}} \text{thrombin}$$
$$\text{Substrate RpNA} \xrightarrow{\text{Intrinsic thrombin}} \text{RCOOH} + \text{pNA}$$

$$(Eq. 26\text{-}2)$$

An abnormal APTT indicates a deficiency in one of the factors of the intrinsic or common pathway of coagulation. If the PT is normal, the common pathway factors and extrinsic factors. (factors VII, X, V, II) can be excluded. If both the PT and APTT are abnormal, the defect most likely is one of the factors in the common path, a vitamin K deficiency, liver disease, or an inhibitor. The APTT is useful in monitoring heparin therapy.

Prothrombin-Thrombin Assays. Chromogenic substrates are now under development for prothrombin assays that will be amenable to centrifugal analyzers and discrete analyzers.

Fibrinogen. Laser nephelometry with forward light-scatter can provide measurement of fibrinogen on the ACL cen-trifugal analyzer.[39] The ACA method is based on the tubido-metric assay of fibrin polymer formation.[38]

$$\text{Fibrinogen} + \xrightarrow[\text{(CaCl}_2 + \text{Dextran}]{\text{Thrombin}}$$

fibrin clot (absorbs at 340 nm)

$$(Eq. 26\text{-}3)$$

The rate of increase of absorbance at 340 nm is directly proportional to the fibrinogen concentration.

A decrease or absence of fibrinogen would indicate hypofibrinogenemia or afibrinogenemia, both severe bleeding disorders. Fibrinogen is also consumed with disseminated intravascular coagulation (DIC), in which consumption of all the coagulation factors, including platelets, takes place. Inhibitors such as heparin, ATIII, α_2-macroglobulin, FDPs, and α_1-antitrypsin do not interfere with this method as they do in the conventional thrombin time to measure fibrinogen.

Factor VIII. A variety of synthetic substrate methods are being developed for the individual components of factor VIII. One example is factor VIII:C.

$$\text{Factor X} + \text{IXa-PL-CA} \rightarrow \text{Xa}$$
$$\text{RpNa} + \text{Xa} \rightarrow \text{RCOOH} + \text{pNA} \qquad (Eq. 26\text{-}4)$$

Factor VIII:RAg is now performed by RIA, Laurell "rocket" techniques of IED, and laser nephelometry.[21] For the protein with normal antigenic structure, Martinoli et al. report good correlation ($r = 0.98$) between nephelometric and EID procedures.[25] EIA offers the ability to detect low levels of less than 1 µg/mL of protein. The principle is one of competition or noncompetition between protein and labeled antibody. In the noncompetitive binding EIA method, the process resembles a sandwich in which antigen and antibody bind, and the concentration of protein is directly proportional to the enzyme conjugate. The competitive method (CELIA) hydrolyzes the substrate following competition for the substrate. In this method, the concentration of protein is inversely proportional to measured substrate hydrolysis. Using Asserachrom VIII, factor VIII:RAg can be quantitated. Correlation of EIA with EID or nephelometry is good. The advantages of EIA are: (1) ability to automate; (2) faster, more reliable results; (3) smaller sample size; and (4) practicality of reagent/equipment costs.

Factor VIII:RAg is decreased in von Willebrand's disease. Decreased levels of factor VIII:vWF and usually factor VIII:C are found in this disease. Von Willebrand's disease is inherited most frequently in an autosomal dominant manner and is characterized by a prolonged bleeding time, lack of ristocetin cofactor activity, and abnormal platelet aggregation with ristocetin. Clinically, the patients present with gingival, postoperative, or nasal bleeding, which is mild to moderate. The disease has variable expression, dependent on the level of the various factor VIII components.

The inheritance of factor VIII:C is sex-linked. A deficiency in this factor activity results in hemophilia A in males. Females are carriers of the trait. Since the gene is located on the X chromosome, females carry the defect on only one of their X chromosomes and present in the asymptomatic carrier state. It should be noted that it is possible, but rare, for a woman to have classical hemophilia if her father is a hemophiliac and her mother is a carrier.

Factor X and Other Factors. Using both fluorogenic and chromogenic synthetic substrates, several other Factors (prekallikrein, kallikrein, XI, XII, VII, XIII, and X) can be measured amidolytically on the Gilford, Cobas Bio, Isamat, ACL, and Encore instruments. An example of one of the more established assays is for Factor X.

$$\text{Factor X} + \text{Russell's viper venom} \xrightarrow{\text{Ca}} \text{Factor Xa}$$
$$\text{Factor Xa} + \text{substrate (tagged)} \rightarrow \text{chromophore released}$$

(Eq. 26-5)

Martinoli et al. found excellent correlation between synthetic substrate methods and the manual method or use of deficient plasma with classic methods ($r = 0.98$). Between the Gilford and Cobas Bio instruments, the correlation was $r = 0.83$. The authors caution the user of synthetic substrates about the possible sources of error:

1. Chromogenic markers can be released under nonspecific conditions.
2. Standardization is still needed with automation.
3. Because clotting enzymes are not specific in their actions, synthetic substrates are only as specific as the originals.
4. The amidolytic reaction requires careful control of the kinetic variables. To ensure linearity, incubation times should be well defined and as short as possible. K_{cat}, V_{max}, and K_m, the kinetic constants, should be at their maximum.

Advantages of synthetic substrates include accuracy, reproducibility, possibility for standardization, increased use of automation for testing, better knowledge of the reactions, and results that provide a measure of functional activity of a specific coagulation factor protein.[24] Factor Xa assay specifically provides an excellent tool for the monitoring of coumadin therapy. Factor X is activated by RVV when in the presence of calcium. The second step of this method is the activated factor X acting on a substrate to release the measured chromophore. Therefore, measurement of the amount of chromophore release is an indirect measurement of factor X (PIVKA X), found in patients on Coumadin, and is not activated by RVV.

Coumadin, an oral coagulant, is an antagonist to vitamin K (competing with vitamin K). Factor X, dependent on vitamin K for its synthesis, is decreased when patients are placed on oral anticoagulant therapy. The amidolytic assay, therefore, evaluates the true concentration of factor X and the effect of the anticoagulant. Aiach et al. have used the factor X assay for more than a year in the monitoring of oral anticoagulant therapy.[1] They report earlier prediction of dose levels using this protocol. Heparin and ATIII do not influence plasma factor X levels. The two main reasons for decreased synthesis are vitamin K deficiency or inhibition and decreased synthesis of factor X by the liver.

Monitoring of Platelet Function

Platelets act as the central element in coagulation. Classic methods of evaluating their presence and function include platelet counts, platelet sizing, adhesion tests, platelet survival, aggregation studies, clot retraction, bleeding time, PF3, and PF4 availability.

Platelet activation and release can now be measured more specifically through quantitative assays of PF4, β-thromboglobulin, PGE2, TXB2, and other prostaglandin derivatives by RIA techniques. The procedures are based on the common technique and principle of RIA for chemistry.

$$\beta\text{-Thromboglobulin} + \text{labeled protein-}^{125}\text{I} + \text{Ab}$$
$$= (\text{protein}) (\text{competition for site}) \text{Ab-}^{125}\text{I} \quad \textit{(Eq. 26-6)}$$

The amount of bound antibody is inversely proportional to the concentration of unlabeled protein. Measurement is through use of standards. Through calculation of a curve, the unknown concentration of protein can be extrapolated. Gamma and scintillation counters are readily adaptable to this principle. Semiautomated and fully automated instrumentation is available.

β-Thromboglobulin and PF4 are considered pathologic markers for abnormal platelet release or function. The introduction of these new assays to automated instrumentation is an exciting development in platelet function analysis.

ADP is also released during platelet aggregation. The measurement of its release can now be accomplished through use of more sophisticated testing with the Lumiaggregometer (Chronolog Corp). The ADP released when platelets are activated couples with luciferin-uciferase (firefly extract) and can be semiquantitatively measured by luminescence fluorometry.

Identification of platelet antibodies (DUZO, PL1, PL2, etc.) can be a tedious task. Current methods include agglutination, complement fixation, and indirect immunofluorescence. None, as yet, is automated.

Fibronectin, a noncollagen, glycoprotein of the subendothelium, is located on the surface of fibroblasts and endothelial cells. Recently, it was discovered that this protein plays a vital role in platelet adhesion. Its measurement can be made on the Multistat III by lightscatter principles of nephelometry.

Monitoring Fibrinolysis

Plasminogen. Measurement of the activity of plasminogen has classically been done by casein digestion, fibrin plate lysis, or RIA technique. Recently, synthetic substrates have been introduced, as well as a method for plasminogen assay adapted to a variety of instruments [Gilford 3500, 3800, or Impact, Cobas Bio, Isamat (ISA Biologie), ACA, ABA 100, Abbott VP, Vitatron AKES, Coulter KemoMat, Dade Protopath]. Chromogenic substrates are available from Kabi-Vitrum (Helena Labs, Stockholm, Sweden), Boehringer-Mannheim (Indianapolis, IN), Diagnostica Stago (Asnieres, France), American Diagnostica (Greenwich, CT), or Dade

(Miami, FL). The Multistat III also can measure plasminogen by fluorescence methods.

$$Plasminogen + streptokinase \text{ (excess)}$$
$$\rightarrow plasminogen\text{-}SK \text{ (active complex)}$$
$$Z\text{-}Lys\text{-}SBzl \xrightarrow{\text{Plasminogen-SK}} Z\text{-}Lys + SBzl$$

$$SBzl + DTNB = (405 \text{ nm}) \text{ chromophore}$$

(Eq. 26-7)

All the principles of chromogenic substrates are similar. In this method, the synthetic substrate is α-*N*-carbobenzoxy-l-lysine thiobenzlester (Z-Lys-S-Bzl). The product resulting from the hydrolysis, α-toluene-thiol (SBzl), reacts with DTNB (5,5′ dithiobis-[2-nitrobenzoic acid]) to form an optically active chromophore. The chromophore formed is measured at 405 nm, and the rate of absorbance is directly proportional to plasminogen concentration. This method is similar to the Gilford adaptation of Martinoli et al.[28,35]

The method is not subject to interference by α_1-antitrypsin, fibrinogen, FDPs, heparin, or protamine sulfate. Hemolysis and lipemia have no effect on the determination. Plasminogen assays are useful in the monitoring of thromboembolytic therapy with streptokinase and in the detecting of DIC in the newborn. Plasminogen is increased in pregnancy, use of oral contraceptives, chronic inflammation, malignancy, thrombophlebitis, trauma, and myocardial infarction. Decreased levels are associated with DIC, severe liver disease, coronary bypass operations, and thrombotic episodes.

Fibrin Degradation Products. When breakdown products or fragments of fibrinogen degradation by plasmin are not sufficiently removed from the circulation, they interfere with platelets and coagulation. An easy-to-perform, semiquantitative immunologic method of latex agglutination is widely used to measure fragments D and E. Recently, laser nephelometry procedures for determining amounts of FDPs have been applied to the Hyland and Beckman immunochemistry systems, the Multistat III, and Behring's laser nephelometer. These offer automated methods that are specific, reliable, and sensitive. An increase of FDPs is seen as an indication of disseminated intravascular coagulation occurring.

Monitoring Activators and Inhibitors

The major plasma inhibitors of fibrinolysis and coagulation are α_1-antitrypsin, α_2-macroglobulin, C′1-inactivator, α_2-antiplasmin, and antithrombin III. These proteolytic enzymes can be quantitatively measured by immunochemical, amidolytic, or light-scatter methods. Again, a variety of instruments have been able to adopt these methodologies: the ACA, the Multistat III, Encore, Gilford Instruments, Abbott, and Isamat. Both ATIII and α_2-antiplasmin determinations are being included in the coagulation profiles offered on these instruments. Synthetic substrates, both chromogenic and fluorogenic, are available as kits from Stago, Kabi, and Boehring-Mannheim. Monitoring of plasminogen activators is difficult; therefore, indirect measuring of plasmin activity is the only way to obtain any information about them.

A dynamic equilibrium exists between the activators and inhibitors of fibrinolysis. Decreases of ATIII result in a thrombotic tendency, whereas decreases in α_2-antiplasmin are found in DIC and hemorrhagic diatheses. Both α_2-antiplasmin and plasminogen concentrations are useful tools in the monitoring of streptokinase therapy. Thrombolytic therapy produces decreased levels of ATIII, α_2-antiplasmin, and kininogen and increased levels of plasmin, plasminogen activator, and FDPs.

ATIII. When ATIII is present, it combines with heparin to form a complex that rapidly inactivates an amount of thrombin added in excess. The residual thrombin catalyzes the hydrolysis of a synthetic substrate, 1-*N*-carbobenzoxy-l-lysine thiobenzylester (Z-Lys-SBzl). The product of this reaction, a toluenethiol (SBzl), then reacts with DTNB to form the optically active chromophore.

$$ATIII + thrombin + heparin \rightarrow$$
$$\text{(excess)} \quad \text{(excess)}$$
$$thrombin\text{-}ATIII\text{-}heparin \text{ complex}$$
$$Z\text{-}Lys\text{-}SBzl \xrightarrow{\text{Residual thrombin}} Z\text{-}Lys + SBzl$$
$$SBzl + DTNB \rightarrow \text{chromophore (452 nm)}$$

(Eq. 26-8)

The rate of increase of absorbance at 452 nm is directly proportional to the functional ATIII level.[20]

As with the plasminogen ACA method, the following substances have no effect on ATIII values: hemolysis, lipemia, α_1-antitrypsin, α_2-macroglobulin, fibrinogen, FDPs, heparin, and protamine sulfate.

ATIII is known as the heparin cofactor and inhibitor of factor Xa, thrombin, and factors XIIa, IXa, and XIa. Its main action, however, is against thrombin and factor Xa. The therapeutic efficacy of heparin depends totally on ATIII being present. Without the total complex, each is ineffective separately. ATIII levels are decreased in liver disease, DIC, deep venous thrombosis, hereditary deficiency of ATIII, thrombophlebitis, advanced stages of sickle cell anemia, acute leukemia, gram-negative sepsis, endotoxemia, and oral contraceptive therapy.

NEW DEVELOPMENTS

Expanding understanding of hemostasis has led to ongoing identification of new coagulation proteins. The early theory of the coagulation cascade has exploded into a complex interaction of multiple biological systems (Fig. 26-3).

Clinically, this knowledge has enabled the early identification of a hypercoagulable condition that can prevent thrombosis. It has led to more effective means of treatment for hemostatic disorders and its complications. There is little doubt that hemostasis will continue to expand to even further discovery and understanding of the mechanisms involved.

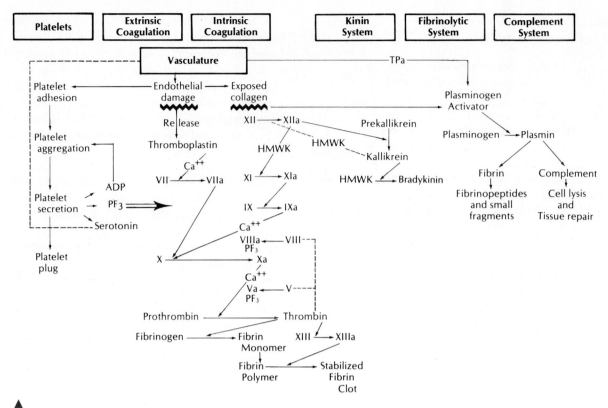

▲

Figure 26-3. Interrelationships of the coagulation, kinin, complement, fibrinolytic, and vascular systems.

CASE STUDY
26-1

A 49-year-old male entered the hospital for surgery. All preoperative test results were normal and the surgery was uncomplicated. Following surgery, he was transferred to the surgical intensive care unit. The patient's APTT was 48 seconds at the first preoperative sample. Additional hemostatic testing was performed (Case Study Table 26-1).

CASE STUDY TABLE 26-1

Test	Patient	Reference Range
Prothrombin time	11.8 s	11.5–13.0 seconds
APTT	48.0 s	24–36 seconds
Fibrinogen	480 mg/dL	200–400 mg/dL
Thrombin time	18 s	11–14 seconds
50:50 Mixing study	41.3 s	
Factor VIII	170%	50–150%
Factor IX	92%	50–150%
Factor XI	75%	60–140%
Factor XII	105%	50–150%
dRVVT	normal	

Questions

1. What could be the causes of the increase in the APTT?

2. What tests could the laboratory have done to identify the problem more quickly? How can the different causes be readily distinguished?

3. Additional investigation into the collection of the sample revealed that a nurse had collected it through an intravascular (IV) line kept open with heparin. What effect would this have on the patient's test results?

CASE STUDY
26-2

A 16-year-old female was evaluated for a possible bleeding disorder. She had had spontaneous nose-bleeds since the age of 3. Her clinical history indicated easy bruising, iron therapy to treat anemia, mucosal bleeding, hematuria, and heavy menstrual periods. She sometimes has painful, swollen, and tender joints. She reported that her uncle had frequent, severe nosebleeds that required transfusion (Case Study Table 26-2).

CASE STUDY TABLE 26-2. Laboratory Data

Test	Patient	Reference Range
Template bleeding time	13.5 min (excessive blood)	2–8 min
PT	11.4 s	11.5–14.0 s
APTT	33.0 s	22–36 s
Platelets	339.000/μL	150–400.000/μL

Platelet function studies:

Retention	40%	31%–83%
Aggregations	(See Fig. 26-4)	
Factor VIII:C	80%	
Factor VIII related ag	40%	45%–185%
Plasma ristocetin vWF	23%	45%–140%

Questions

1. What does this patient have and what tests aided in your interpretation?
2. Is the factor VIII:C level normal? Can it be decreased in von Willebrand's disease?
3. Describe the factor VIII molecule and tell where and how it is produced.
4. What are the methodologies used to test for the bleeding problem that this patient has? State the principles.

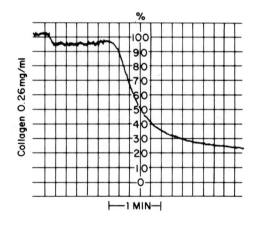
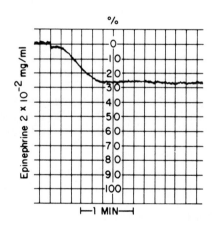
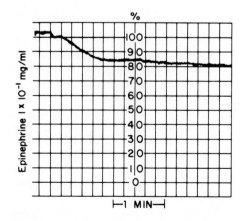
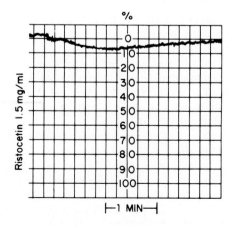

Figure 26-4. Platelet function studies in Case Study 26-2.

CASE STUDY
26-3

A 22-year-old male was admitted through the Emergency Room with injuries sustained in a motorcycle accident. Following surgery, he developed a pulmonary embolism. Hematology was consulted. See Case Study Table 26-3. Family history revealed a sister and uncle with chronic thrombophlebitis. A diagnosis was made and treatment with coumadin was initiated.

CASE STUDY TABLE 26-3. Presurgical Lab Evaluation

Test	Patient	Reference Range
PT	11.8	11.5–13.0 s
APTT	41.3	24–36 s
50:50 mix	38.5	
Fibrinogen	490	200–400 mg/dL
Thrombin Time	13.0	11–14 s
Factor VIII	140%	50–150%
Factor IX	95%	50–150%
Factor XI	97%	60–140%
Factor XII	89%	50–150%

Consultative Lab Evaluation

Test	Patient	Reference Range
Plasminogen	111%	75–130%
alpha 2 antiplasmin	96	88–114%
ATIII (function)	35%	85–111%
ATIII (conc)	12	20–50 mg/dL
Protein C	103%	65–150%
Protein S	91%	65–150%
tPA	5	2–7 ng/mL
PAI-1		

Questions

1. What do the initial laboratory tests suggest?
2. What is revealed through additional testing performed as part of the hematological consultation?
3. What is the normal antithrombin III function and what is the effect of the deficiency?

REFERENCES

1. Aiach M, Leon M, Michand A, Capron L. Adaption of synthetic peptide substrate-based assays on a discrete analyzer. Semin Thromb Hemost 1973;9(3):206.
2. Altman R, Hemker HC. Contact activation in the extrinsic blood clotting system. Thromb Diasth Haemm 1967;18:525.
3. Aoki, N. Natural inhibitors of fibrinolysis. Prog Cardiovasc Dis 1979;21:249.
4. Aurell, et al. A new sensitive and highly specific chromogenic peptide substrate for factor Xa. Thromb Res 1977;11:595.
5. Bertino RM, Veltkamp JJ. Physiology and biochemistry of factor IX. In: Bloom A, Thomas D, eds. Haemostasis and thrombosis. New York: Churchill Livingstone, 1981;98.
6. Bick RL. Clinical hemostasis practice: The major impact of laboratory automation. Semin Thromb Hemost 1983;9(3):139.
7. Curtis CG. Plasma factor XIII. In: Bloom A, Thomas D, eds. Haemostasis and thrombosis. New York: Churchill Livingstone, 1981;192.
8. Dahlback B. Human coagulation factor V purification and thrombin catalyzed activation. J Clin Invest 1980;66:538.
9. Doolittle RF. The structures of fibrinogen and fibrin. In: Mahn KG, Taylor FB, eds. The regulation of coagulation. New York: Elsevier, 1980;501.
10. Edington TS, Plow EE. Conformational and structural modulation of the NH_2 terminal region of fibrinogen and fibrin associated with plasmin cleavage. J Biol Chem 1975;250:3393.
11. Esmon CT. The regulation of natural anticoagulant pathways. Science 1987;235:134.
12. Francis CW, Marder VJ, Barlow GH. Plasmic degradation of cross fibrin: Characterization of new macromolecular soluble complexes model of their structures. J Clin Invest 1980; 66:1033.
13. Fujikawa K, Legaz M, Davies EW. Bovine factor X (Stuart factor): Mechanism of activation by a protein from Russell's viper venom. Biochemistry 1982;11:4892.
14. Griffin JH, Cochran CG. Recent advances in the understanding of contact activation reactions. Semin Thromb Hemost 1979;V:1254.
15. Harpel P. Blood proteolytic enzyme inhibitors: Their role in modulating blood coagulation and fibrinolytic pathways. In: Coleman RW, et al., eds. Hemostasis and thrombosis: Basic principles and clinical practice. Philadelphia: Lippincott, 1982;738.
16. Hawn CV, Porter KR. The fine structure of clots formed from purified bovine fibrinogen and thrombin: A study with the electron microscope. J Exp Med 1947;86:285.
17. Hoyer LW. The factor VIII complex: Structure and function. Blood 1981;58:1.
18. Jackson CM. Biochemistry of prothrombin activation. In: Bloom A, Thomas D, eds. Haemostasis and thrombosis. New York: Churchill Livingstone, 1981;140.
19. Joist HJ. Hypercoagulability: Introduction and perspective. Semin Thromb Hemost 1990;16:2.
20. Kaplan AP. Initiation of the intrinsic coagulation and fibrinolytic pathways of man: The role of surfaces, Hageman factor, prekallikrein, high-molecular-weight kininogen, and factor XI. Prog Hemost Thromb 1977;4:127.
21. Laurell CB. Quantitative estimation of proteins by electrophoresis in agarose containing antibodies. Anal Biochem 1966;15:45.
22. Longberry J. Hemostasis: Part 1, screening tests to evaluate abnormal hemostasis. Am J Med Technol 1982;48(2):99.

23. Loskutoff DJ, Sawdey M, Minura J. Type 1 plasminogen activator inhibitor. Prog Hemost Thromb 1989;9:67.

24. Mannucci PM, et al. Familial dysfunction of Protein S. Thromb Haemost 1989;62:763.

25. Martinoli JL, Amiral J. The impact of automation in the development of reagents and kits for automated methods in coagulation testing. Semin Thromb Hemost 1983;9(3):194.

26. Messmore H, et al. Synthetic substrate assays of the coagulation enzymes and their inhibitors: Comparison with clotting and immunologic methods for clinical and experimental usage. Ann NY Acad Sci 1981;370:785.

27. Palkuti HS. Hemostasis, part II: Investigation of the coagulation factors by use of qualitative and quantitative assay techniques. Am J Med Technol 1982;48(2):109.

28. Robbins KC. Fibrinolytic therapy: Biochemical mechanisms. Semin Thromb Hemost 1991;17:1.

29. Seligsohn U, Osterud B, Brown SF, Griffin JH, Rapaport SI. Activation of human VII in plasma and in purified systems: Roles of activated factor IX, Kallikrein, activated factor XII. J Clin Invest 1979;64:1056.

30. Shapiro SS, McCord S. Prothrombin. Prog Hemost Thromb 1978;4:177.

31. Triplett DA. Laboratory diagnosis of lupus anticoagulants. Semin Thromb Hemost 1990;16:182.

32. Walenga J, et al. Automated instrumentation and the laboratory diagnosis of bleeding and thrombotic disorders. Semin Thromb Hemost 1983;9(3):172.

33. Walsh PN. The role of platelets in the contact phase of blood coagulation. Br J Haematol 1972;22:237.

34. Williams W. Biochemistry of plasma coagulation factors. In: Williams W, Beutler E, Erslev A, Lichtman, eds. Hematology. New York: McGraw-Hill, 1983;1202.

35. Wohl RC, Summaria L, Robbins KC. Kinetics of activation of human plasminogen by different activator species at pH 7.4 and 37 degrees. J Biol Chem 1980;255:2005.

36. Wright TN. Vessel proteoglycans and thrombogenesis. Prog Hemost Thromb 1980;5:1.

37. Wu KK, et al. Fibrinogen and carotid atherosclerosis. AEP 1992;2:4.

38. Zimmerman T, Meyer D. Structure and function of factor VIII/von Willebrand factor. In: Bloom A, Thomas D, eds. Haemostasis and thrombosis. New York: Churchill Livingstone, 1981.

39. Zimmerman TS, Edgington TS. Factor VIII related antigen: Multiple molecular forms in human plasma. Proc Nat Acad Sci 1975;72:5121.

40. Zur M, Nemerson. Tissue factor pathways of blood coagulation. In: Bloom A, Thomas D, eds. Haemostasis and thrombosis. New York: Churchill Livingstone, 1981.

SUGGESTED READINGS

Barnhart MI, Baechler CA. Endothelial physiology, perturbations, and responses. Semin Thromb Hemost 1978;5:50.

Collen D. On regulation and control of fibrinolysis. Thromb Hemost 1980;43:77.

Comp PC. Hereditary disorders predisposing to thrombosis. Prog Hemost Thromb 1986;8:71.

Corriveau D, Fritzma G. Hemostasis and thrombosis in clinical laboratory. Philadelphia: Lippincott, 1988.

Egeberg O. Inherited antithrombin deficiency causing thrombophilia. Thromb Diath Hemorrh 1965;13:516.

Hougie C. The biochemistry of blood coagulation. Triplett DA Laboratory evaluation of coagulation. Chicago: ASCP Press, 1982;6.

Magno C, Joris I. Endothelium: A review. Adv Exp Med Biol 1978;104:196.

Marlar RA, Montgomery RR, Brockmans AW. Diagnosis and treatment of homozygous protein C deficiency. J Pediatr 1989;114:528.

Murano G, Bick RL, eds. Plasma protein functions in hemostasis and thrombosis. FL: CRC Press, Inc 1982;43.

Penner JA. Hemostasis coagulation proteins. Blood coagulation: Clinical and laboratory agents. Orland East Lansing, MI: MSU Press, 1983.

Thompson AR, Harker LA. Manual of hemostasis and thrombosis. Philadelphia: FA Davis Co, 1983.

Triplett DA. Platelet function: Laboratory evaluation and clinical application. Chicago: ASCP, 1978.

Triplett DA. Laboratory evaluation of coagulation. Chicago: ASCP, 1983.

Weiss HJ. Platelet physiology and abnormalities of platelet function—Parts I and II. N Engl J Med 1975;293:11.

Chapter 27

Therapeutic Drug Monitoring

John A. Bittikofer

Drug Disposition
Absorption
Distribution
Metabolism
Excretion
Pharmacokinetics
Atypical Drug Disposition
Age
Pregnancy
Disease
Active Drug Metabolites
Specimen Collection and Handling
Sample Storage
Monitoring Specific Drugs
Quality Control
Cardioactive Drugs

Digoxin
Disopyramide
Procainamide
Propranolol
Lidocaine
Quinidine
Antiepileptic Drugs
Phenobarbital
Phenytoin
Valproic Acid
Carbamazepine
Ethosuximide
Bronchodilators
Theophylline
Antibiotics
Aminoglycosides

Chloramphenicol
Vancomycin
Psychoactive Drugs
Lithium
Tricyclic Antidepressants
Immunosuppressants
Cyclosporine
FK-506
Antineoplastic Drugs
Methotrexate
Summary
Case Studies
References
Suggested Readings

Objectives

Upon completion of this chapter, the clinical laboratorian should be able to:

- *Define the following terms: pharmacokinetics, TDM, and therapeutic range.*
- *Given patient data, evaluate the patient's condition.*
- *State some examples of the different types of drugs, e.g., antiepileptic drugs.*
- *Explain disposition of a drug in the body.*
- *Describe first-pass elimination.*
- *Name three factors that affect drug disposition.*
- *Discuss the importance of specimen collection, transport, and storage in TDM.*
- *Relate when peak and trough drug levels should be drawn.*
- *Compare and contrast the methodologies, e.g., gas chromatograph, RIA, etc., used in TDM.*
- *Identify the action of the drugs listed in the chapter.*

KEY WORDS

Drug Disposition	Therapeutic Drug
Metabolite	Monitoring
Peak Drug Level	Therapeutic Range
Pharmacokinetics	Trough Drug Level

Many drugs may be prescribed with little fear of toxic side effects. Over the years, the most appropriate dosage regimens have been developed by simply increasing or decreasing the amount of dose and interval between doses until the appropriate effect was observed in the patient. Although this trial-and-error method is time-consuming, it is effective and allows for each individual to receive optimal treatment.

A number of drugs, however, cause rather severe toxic side effects at easily attained concentrations. A trial-and-error method of dosing these drugs is an inappropriate way to establish effective drug levels in the majority of patients. Monitoring the serum concentrations of these potentially toxic drugs when trying to establish dosage regimens became common practice and is now a well-established component of modern drug therapy. Unfortunately, not every patient disposes of a drug in a predictable manner, and the typical dose amount and interval may be either ineffective or cause toxic side effects for any one individual. Certain drugs, even though maintained at therapeutic levels, cause side effects ranging from tinnitus (ringing in the ears) and gastrointestinal (GI) disturbances to skin lesions. Consequently, patients may be reluctant to take their prescribed medication in order to avoid the uncomfortable side effects. It has been suggested that 60% of all individuals on drug therapy are noncompliant at one time or another. The

Michael L. Bishop, Janet L. Duben-Engelkirk, and Edward P. Fody.
CLINICAL CHEMISTRY. © 1996 Lippincott–Raven Publishers.

physician encountering a patient who is not responding to drug therapy may effectively evaluate possible noncompliance by requesting serum drug level determinations from the therapeutic drug monitoring (TDM) laboratory.

Of more concern is the fact that many drugs cause severe toxic effects at serum levels that are only slightly different from therapeutic levels or may even be in the generally accepted therapeutic range. Nephrotoxicity is a common toxic effect of some drugs if their concentration exceeds therapeutic levels. Furthermore, drug disposition varies from individual to individual, and this may make drug therapy ineffective or toxic based only on dosage. However, with the availability of appropriate laboratory facilities, the clinician can monitor drug concentrations and adjust dosage and interval to maintain therapeutic levels or to avoid toxic levels.

DRUG DISPOSITION

In order to appreciate the significance of therapeutic drug monitoring, it is appropriate to develop an understanding of drug disposition; that is, the way in which the body handles a foreign compound, the drug. The mechanisms used by the body to handle a drug can be explained in terms of four general processes: absorption, distribution, metabolism, and excretion. Figure 27-1 illustrates these processes and how they interrelate.

Absorption

Typically, there are three routes of administering a drug: orally, rectally, and parenterally [intravenous (IV) and intramuscular (IM)]. Drugs administered orally are absorbed by the GI tract depending, more or less, on many factors, including the chemical characteristics of the drug and the characteristics of the GI tract. A drug must be formulated to withstand the acidity of stomach fluids and to resist decomposition by digestive enzymes. For example, aminophylline is formulated to degrade to two molecules of the active theophylline when it encounters stomach fluids. Penicillin is not very stable in the GI tract, whereas the derivative phenoxymethyl penicillin is much more acid stable.

A pharmaceutic preparation should be soluble in GI fluid to facilitate good absorption. Polar (ionic) compounds typically are more soluble in GI fluid than are nonpolar compounds. However, most drugs are absorbed by passive diffusion of the nonionized or less polar form of the molecule. That is, lipid-soluble compounds are absorbed faster than

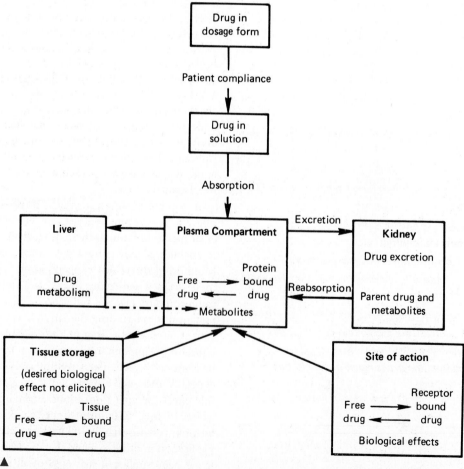

Figure 27-1. The general processes by which the body deals with a drug. (Moyer TP, Boec Kx, eds. Applied therapeutic drug monitoring. Washington, DC: American Association for Clinical Chemistry, 1982;10. Reprinted with permission.)

ionic compounds. Because of the acidic nature of stomach contents, weak acids are absorbed more quickly than weak bases. The requirements that a drug preparation be ionic in nature for best solubility in GI fluid and nonionic for best transport across lipid membranes create a paradox for the pharmacist, but as a rule, the weak acids and weak bases represent optimal characteristics.

Following absorption through the intestinal mucosa, the drug transverses the hepatic portal system and liver before reaching the systemic circulation. Certain drugs, *e.g.,* propranolol, undergo substantial metabolism on this first pass through the liver. This phenomenon, often referred to as *first-pass elimination*, must be taken into consideration when prescribing dosage.

For those drugs that are not easily formulated to overcome poor absorption or first-pass elimination or to withstand the harsh conditions of the GI tract, intramuscular or intravenous administration may be appropriate. Digoxin tablets, for example, are 70% absorbed when taken orally, but the same dose administered IV results in 50% greater blood levels of the drug. Intramuscular and subcutaneous drug administration, however, have their own problems. Some drugs are incompletely absorbed, whereas others, such as digoxin and phenytoin, show erratic and markedly prolonged absorption phases. Water solubility at physiologic pH is a primary determinant of IM absorption.

Of course, there may be significant interactions between a drug taken orally and any food or drink ingested at or near the same time. These may be difficult to predict, leading to the need for TDM.

Distribution

Once the drug, or metabolite, reaches the systemic circulation, it is quickly distributed throughout the vasculature. Water-soluble drugs are easily transported in their ionic forms, whereas insoluble drugs must bind to a carrier, usually a protein. Albumin and α_1-acid glycoprotein (AAG) are the two most significant binders. Albumin binds many of the acidic drugs, whereas cationic drugs are preferentially bound to AAG. Any drug bound to protein must compete for the binding sites with many other compounds, including other drugs, lipids, and other water-insoluble molecules. Protein binding may be quite substantial, as in the case of phenytoin and lidocaine, or practically nonexistent, as is true of lithium. The equilibrium between protein-bound and free drug is quite dynamic but can be influenced by several factors, which are discussed later. Protein-binding information is readily available in a variety of handbooks and is one of the parameters often used to characterize drug disposition.[8,13,18] (See Appendix T for a summary table of pharmacokinetic parameters.)

Once the drug has reached the tissue site where it will have its effect, it must traverse the cell membranes, since most drug receptors are internal cell components. There is overwhelming evidence that the free drug traverses the cell membrane, whereas the protein-bound drug remains in the circulatory system. Since the cell membrane is organic in nature, lipid-soluble compounds will cross the membrane more easily and more quickly than will large-molecular-weight ionic compounds. However, small-molecular-weight ionic drugs, such as lithium, can easily migrate through the membranes.

It is well accepted that the free drug is the therapeutically active drug species in blood. There are several factors that can change the free fraction and free drug concentration. There are several drugs for which there are limited protein binding sites. When the concentration of drug exceeds the binding capacity of the respective protein or proteins, the free drug fraction will increase. Examples are valproate, disopyramide, and salicylate. A decrease in albumin or albumin's affinity for a given drug can occur in a variety of disease states, such as cirrhosis, nephrotic syndrome, malnutrition, etc., which decreases drug binding capacity. This, in turn, results in an increase in the drug's free fraction. Examples are phenytoin, valproic acid, carbemazepine, and disopyramide. An increase in AAG occurs in several disorders since this protein is an acute phase reactant. Consequently, the free fraction of a drug highly bound to AAG will decrease. An example is a possible 200% to 400% increase in AAG and 50% decrease in lidocaine free fraction in major traumatic events.[2] Both endogenous (*e.g.,* hormones, urea, uric acid, bilirubin) and exogenous (*e.g.,* other drugs) can compete with a given drug for binding sites. An increase in endogenous compounds or the presence of exogenous compounds will cause an increase in the free fraction of a competing drug. Examples are valproic acid and salicylate displacing phenytoin from albumin sites[2] and increased free fatty acids in fasting, stress, pregnancy, and infections increasing valproic acid free fraction.[10]

The volume of distribution is a hypothetical volume of body fluid needed to dissolve the total amount of drug required to achieve the concentration found in the blood. This parameter is one of several used to characterize drug disposition and is expressed mathematically for a single dose situation as follows:

$$V_d = D/C_p \qquad \text{(Eq. 27-1)}$$

where V_d = volume of distribution (in L)
    D = dose (in mg or g)
    C_p = concentration in plasma (in mg/L or g/L)

Metabolism

A drug is perceived by the body as foreign material, and mechanisms are quickly initiated to purge this "toxin." Most substances that have been absorbed and distributed by the vascular system are ultimately excreted by the kidneys. Since ionic, water-soluble moieties are more readily excreted, the liver attempts to convert any insoluble material, such as lipid-soluble drugs, into a more water-soluble species. The microsomal portion of hepatocytes contains enzymes capable of converting a wide array of drugs to a more soluble glucuronide, sulfate, or phosphate. These enzymes are nonspecific and catalyze drug conjugation, oxidation, reduction, or hydrolysis, as is appropriate, to render the drug more excretable. Several of these processes are facilitated by microsomal cytochrome P-450. Since there are multiple forms of cytochrome P-450, major variations exist, from individual to individual, in the rate at which a drug is metabolized.

In addition, cytochrome P-450 and other hepatocyte microsomal components are induced by other drugs, with cytochrome P-450 levels increasing by as much as 20 times.[26] Insecticides, barbiturates, other drugs, alcohol, smoking, and diet are all factors that may induce drug metabolism. On the other hand, an enzyme system can be induced to the saturation point, beyond which additional activators have no further effect on metabolism.

Documented genetic variations in drug metabolism have resulted in some individuals demonstrating a significant enhancement in drug-metabolizing capacity ("fast metabolizer"), whereas in other subjects the metabolic rate is reduced ("slow metabolizers").

Of utmost importance for several specific drugs is the fact that the metabolite or metabolites of the parent drug have significant or comparable therapeutic activity. Table 27-1 lists many of the drugs whose metabolites have measurable physiologic activity. N-acetylprocainamide has activity comparable with that of the parent compound, procainamide, and should be measured whenever a procainamide level is determined. Both amitriptyline and nortriptyline should be measured when amitriptyline is the prescribed drug. Likewise, imipramine and desipramine should be measured when imipramine is prescribed.

Excretion

The primary route of drug and metabolite excretion is by means of the kidneys. Excretion by the biliary tract, lungs, and sweat glands usually becomes significant only in the event of severe renal failure. Urinary excretion occurs faster for the more water-soluble materials, and thus metabolites are typically eliminated more easily. The pH of urine can significantly influence the excretion of acidic and basic drugs and their metabolites. An acidic urine will enhance excretion of bases, whereas alkaline urine extracts acidic drugs from plasma.

PHARMACOKINETICS

Pharmacokinetics is an effort to characterize mathematically the disposition of a drug over time in order to better understand and interpret blood levels and to effectively adjust dosage amount and interval for best therapeutic results with minimal toxic effects. Figure 27-2 illustrates the relationship between time and plasma concentration for a drug taken as a single oral dose. The first phase of the curve represents absorption and distribution of the drug, whereas the second phase illustrates the decay of serum drug concentration because of metabolism and excretion. Several assumptions are made to minimize the number of potential variables:

1. The factors affecting drug elimination remain constant.
2. Drug effect is closely linked to serum drug concentration.
3. The body is considered a single compartment.
4. Distribution occurs uniformly and is rapid relative to absorption and elimination.

All or any one of these assumptions could be false for a given individual, but for many drugs they hold true and allow for a much simplified kinetic description of drug disposition. There are several parameters that can be derived

TABLE 27-1. Parent Drugs and Their Metabolites with Physiologic Activity

Drug	Metabolite
Procainamide	N-Acetylprocainamide*
Propranolol	4-Hydroxypropranolol
Lidocaine	Monoethylglycinexylidide
	Glycinexylidide
Quinidine	(3S)-3-Hydroxyquinidine
	2′-Oxoquinidine
	O-Desmethylquinidine
	Dihydroquinidine†
Disopyramide	Nordisopyramide
Digitoxin	Digoxin
Primidone	Phenobarbitone
	Phenylethylmalonamide
Carbamazepine	10 ± 11 epoxide derivative
Amitriptyline	Nortriptyline*
Imipramine	Desipramine*

*Metabolites of sufficient activity and concentration to necessitate monitoring along with the parent drug

†A pharmacologically active contaminant in some pharmaceutic quinidine preparations

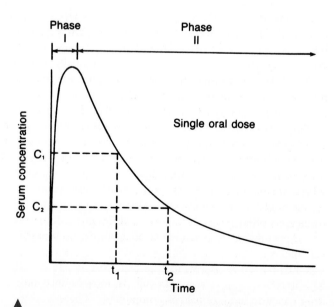

Figure 27-2. The dose-response curve for a drug obeying first-order elimination kinetics taken as a single, oral dose.

from this relationship that facilitate optimal clinical use of drug therapy. The second phase of the dose-response curve for many drugs is characterized by the first-order equation

$$C_t = C_0 e^{-k_d t} \qquad \textbf{(Eq. 27-2)}$$

The serum concentration at time t is represented by C_t, C_0 is the initial dose divided by the volume of distribution, t is time, and k_d represents a disposition or elimination constant. The half-life ($t\frac{1}{2}$) of the drug is the period of time during which the concentration decays in half. This parameter may be determined from phase II of the dose response. In Figure 27-2, an arbitrary serum concentration, such as C_1, is selected, and the corresponding t_1 is determined from the graph. Since $t\frac{1}{2}$ is defined as the time for the concentration to decrease in half, C_2 must be equal to one-half of C_1. From C_2 on the graph, t_2 is determined, and the difference between t_1 and t_2 is the half-life. Any portion of the phase II graph may be used to calculate $t\frac{1}{2}$. Alternately $t\frac{1}{2}$ may be calculated from Equation 27-2, provided k_d is known. (The disposition constant k_d is determined experimentally by administering a single oral dose of drug to a subject and collecting blood samples at measured time intervals. A plot of $\ln C_t$ vs. t will produce a straight line, if the drug follows first-order kinetics, with a slope equal to k_d and a y-intercept equal to $\ln C_0$.) Equation 27-2 is rearranged as follows:

$$C_t / C_0 = e^{-k_d t} \qquad \textbf{(Eq. 27-3)}$$

Then the logarithm is taken on both sides of the equation:

$$\ln C_0 - \ln C_t = k_d t \qquad \textbf{(Eq. 27-4)}$$

Since, at one half-life, $C_t = \frac{1}{2} C_0$ by definition, $\ln C_0 - \ln C_t = 0.693$ and

$$t\frac{1}{2} = \frac{0.693}{k_d} \qquad \textbf{(Eq. 27-5)}$$

The fraction of drug after n half-lives equals $(\frac{1}{2})^n$ (Table 27-2). Thus, the fraction of drug remaining in the bloodstream can be determined at any time if one knows the $t\frac{1}{2}$ of the drug.

Of note is the fact that for first-order kinetics, the $t\frac{1}{2}$ does not vary with concentration. Figure 27-3 is a series of dose curves differing only in C_0, which relates directly to dose. As initial dose increases, $t\frac{1}{2}$ remains the same, and at any given point in time, the fraction of drug left is the same, regardless of initial concentration. The absolute amount of drug present will increase with C_0.

As the dose, or C_0, increases, it may reach a level at which the liver can no longer increase its rate of metabolism. At this point the enzyme metabolism is saturated. Once this has occurred, the disappearance of the drug follows zero-order kinetics, and the *rate* of change of drug concentration is independent of concentration. Figure 27-4 illustrates the relationship between time and serum concentration. In this situation, $t\frac{1}{2}$ changes with concentration.

Ethanol, phenytoin, and salicylate are drugs that typically follow zero-order or saturation kinetics at therapeutic con-

TABLE 27-2. Fraction of Drug Remaining after n Half-Lives

n	% Remaining
1	50
2	25
3	12
4	6
5	3

centrations. Every drug conceivably would reach saturation kinetics if concentrations (doses) were increased sufficiently. For many drugs, the toxic effects would be manifest much sooner than would saturation kinetics.

Total body clearance (TBC) is another pharmacokinetic parameter that can be very useful in projecting effective dosages. TBC is the rate of disappearance of drug from the body by all pathways of elimination at any time, divided by C_t (the concentration of drug at time t):

$$\text{TBC} = \frac{\Delta A}{\Delta t} \div C_t \qquad \textbf{(Eq. 27-6)}$$

This expression, stated slightly differently, is useful in adjusting drug rates when administering a drug IV (*e.g.*, lidocaine, theophyline, and procainamide).

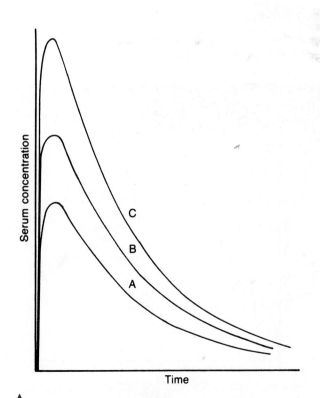

Figure 27-3. The dose-response curves for a drug obeying first-order kinetics taken as a single, oral dose. (*Curve B*) C_0 is 1.5 times greater than curve A. (*Curve C*) C_0 is 2.0 times greater than curve A.

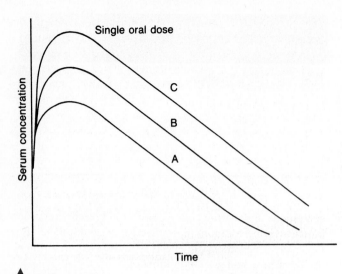

Figure 27-4. The dose-response curves for a drug following zero-order kinetics. (*Curve B*) C_0 is 18% greater than curve A. (*Curve A*) C_0 is 15% greater than curve B.

$$\text{Clearance (L/min)} = \frac{\text{infusion rate (mg/min)}}{C_t \text{ (mg/L)}} \qquad \textbf{(Eq. 27-7)}$$

A multiple oral dose-response curve is represented in Figure 27-5. The dose interval is one half-life. As the second, third, and subsequent doses are taken, the total drug load accumulates until a steady state is reached. For a drug obeying first-order kinetic elimination and demonstrating an absorption-distribution phase much faster than its elimination phase, the drug concentration increases in a pseudo–first-order fashion. It is appropriate to apply the concept of half-lives to this example and state a rule of thumb that a drug will reach steady state after five half-lives. Likewise, if the drug is discontinued, less than 3% of the drug will persist after five half-lives (see Table 27-2).

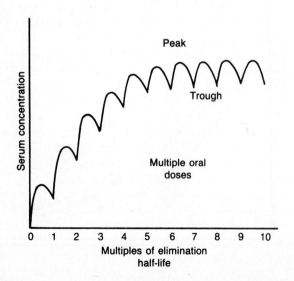

Figure 27-5. Dose-response curve after multiple oral doses given each half-life.

ATYPICAL DRUG DISPOSITION

Although the majority of people will dispose of a drug under normal circumstances according to well-defined pharmacokinetics, there are individuals with varying characteristics that result in unpredictable drug disposition. Furthermore, healthy individuals develop disease, women become pregnant, and infants grow older. All these factors can affect drug disposition. Table 27-3 illustrates many of the factors that have been shown or are suspected to affect drug disposition. Although space does not permit a thorough explanation of each factor, several are important and common enough to discuss here.

Age

Age is one of the more profound factors affecting drug disposition. Figure 27-6 illustrates the relative changes in disposition with age. The most dramatic changes occur in infants, for several well-documented reasons. Not as dramatic, but just as significant, are the changes that occur in the elderly.

The neonate shows little capacity to eliminate a drug owing primarily to immature liver and renal function. Once these functions mature, drug disposition increases and drug half-life decreases. Of utmost importance is the dramatic change in disposition during the first month of life, which makes dosing very difficult and demands close therapeutic drug monitoring, not only to avoid toxicity but to ensure therapeutic levels.

For children between the ages of 6 and 10, it is generally considered appropriate to double the adult dose (per kilogram of body weight) because drug disposition is at its zenith in these prepubescent children and is considerably greater than that seen in adults. However, disposition takes a dip toward adult levels with the initial onset of puberty. This

TABLE 27-3. Factors Affecting Drug Disposition in Humans

Age
Gender
Pregnancy
Exercise
Circadian and seasonal variations
Disease
   altered liver function
   altered renal function
   altered cardiac function
   altered gastrointestinal function
   fever
Dietary factors
   high or low protein diet
   starvation
Alcohol intake
Drugs
Smoking
Occupational exposures

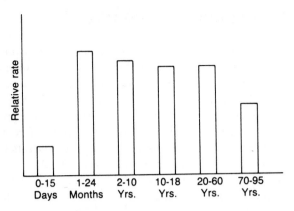

Figure 27-6. The relative rates of drug disposition with changing age. (Adapted from Morselli PL, Pippenger CE. Drug disposition during development: An overview. In: Moyer TP, Boeckx RI, eds. Applied therapeutic drug monitoring. Vol 1: Fundamentals. Washington, DC: American Association for Clinical Chemistry, 1982. Reprinted with permission.)

change occurs during a few months and requires close monitoring of serum drug levels. A possible explanation for this behavior is the increased production of sex hormones, which may compete with drugs for hepatic microsomal receptor sites, thus slowing drug metabolism and increasing half-life.

Protein binding in neonates is compromised because of inherently reduced amounts of protein in the circulation, most importantly albumin. Increased amounts of bilirubin and free fatty acids compete for the available binding sites. This, coupled with reduced blood flow, reduces drug distribution and lengthens drug half-life. Reduced renal blood flow, reduced glomerular filtration rate, reduced tubular secretion, and an acid urine all affect the excretion of drugs in the neonate.

The elderly present a multifaceted picture of drug disposition. Serum albumin concentrations decline with age, providing fewer binding sites for drugs. The protein binding of warfarin, tolbutamide, and phenytoin, among other drugs, has been shown to decrease with age, whereas benzylpenicillin, salicylate, phenobarbital, and sulfadiazine show no change in protein binding with age.[21] Metabolic rate decreases in the elderly, as does renal excretion. Examples of this are a 40% increase in the $t\frac{1}{2}$ of digoxin and a nearly 150% increase in kanamycin's half-life in the elderly.[4] Now, add to the effects of age the effects of diseases prevalent in the aged. Liver disease will further reduce metabolism, whereas renal disease inhibits excretion. Along with disease come the various drugs used to treat disease and the potential drug-drug interferences. The end result is a gradual or perhaps acute decline in drug disposition, which must be accounted for when dosing a drug for the elderly.

Pregnancy

During pregnancy and the immediate postpartum period, several physiologic changes occur, including significant hormonal changes. Many times these changes affect drug disposition in the pregnant patient, necessitating extensive

therapeutic drug monitoring.[22] Factors such as gastric emptying, GI pH, and blood flow can affect drug absorption. An approximately 50% increase in plasma volume occurs in pregnancy and can significantly alter the volume of distribution. Total body water increases and cardiac output increases during pregnancy, both of which affect the volume of distribution. The changes in plasma protein concentration and composition during pregnancy may affect drug binding. Acidic drugs typically bind to albumin, whereas basic drugs bind preferentially to AAG and lipoprotein. Since pregnancy induces a decrease in the concentrations of both albumin and AAG, drug disposition will increase to a greater or lesser extent, depending on the drug's acidic nature. Non-esterified fatty acid concentrations increase in pregnancy. Since they compete for binding sites, free fractions of phenytoin, valproic acid, and disopyramide are altered during pregnancy.[10]

The increases in cardiac output and blood flow that occur in pregnancy increase the drug's residence time in the liver and provide for an increase in drug metabolism. Since hepatic metabolism of drugs is also related to the amount of free drug in plasma, less protein binding results in increased free drug levels and increased drug metabolism. Renal blood flow and glomerular filtration rate may increase as much as 60% during pregnancy, thus increasing drug elimination.

Specific examples of varied drug disposition in pregnancy are the decreased plasma levels of phenytoin and other anticonvulsants, digoxin, and lithium. Of note in the case of lithium is the fact that pregnant women are more sensitive to lithium and thus do not require increased doses to compensate for decreased plasma levels.[8]

Disease

Not all patients on drug therapy are healthy in all other aspects relative to the disease being treated pharmaceutically. Since most classical pharmacokinetic evaluations are described for otherwise healthy individuals, pharmacologists must be concerned with any potential variations in drug disposition caused by varying disease states.

The liver's ability to metabolize a drug is affected by hepatic blood flow, the metabolizing ability of the hepatocyte, and the degree of drug binding to protein. Consequently, various hepatic disorders affecting these factors will influence drug disposition. Not only may the liver's detoxifying capacity be compromised, its ability to manufacture protein will affect drug disposition due to decreased protein binding. Altered nutritional status, which frequently accompanies liver disease, also may affect drug disposition.[17]

Cardiac disease, although not affecting absorption, distribution, metabolism, or elimination directly, may have a profound effect due to variations in blood flow to the GI tract (drug absorption), receptor tissues (distribution), hepatocytes (metabolism), and kidney (elimination).

Renal disease may have a profound direct effect on the elimination of drugs and metabolites in the form of renal failure or changes in urine pH, which affects the extractability of acidic and basic drugs and metabolites into the urine. Uremia creates additional competition for protein-binding sites, which will affect the volume of distribution.

ACTIVE DRUG METABOLITES

As a general rule, the derivatives of drugs, or metabolites formed by the liver to render them more excretable, are not active, or their activity is insignificant relative to the parent drug. However, there are several metabolites of drugs commonly monitored that have comparable therapeutic activity and should be monitored or compensated for when evaluating a drug's therapeutic effects (see Table 27-1).

Procainamide is an effective antiarrhythmic agent, which is about 50% excreted in the urine unchanged. The major metabolite of procainamide is *N*-acetyl procainamide (acecainide, NAPA), which has activity comparable to the parent drug.[11] However, there is significant variability among individuals' hepatic *N*-acetyl transferase activity, which is the enzyme responsible for converting procainamide to NAPA. Thus, there are fast acetylators and slow acetylators. Furthermore, it has been demonstrated that certain arrhythmias respond more favorably to NAPA than to procainamide and vice versa. It has been suggested that NAPA contributes to the toxic side effects of procainamide, and studies have demonstrated that NAPA accumulates in the body in renal failure. Couple this information with the fact that the NAPA/procainamide plasma concentration ratio varies by a factor of 6 among individuals, and it is clear that both procainamide and NAPA must be monitored in patients under procainamide therapy.[9]

Quinidine, an active antiarrhythmic agent, has several metabolites, (3*S*)-3-hydroxyquinidine, 2′-oxoquinidine, and *o*-desmethylquinidine, among others. All three have antiarrhythmic activity to a greater or lesser extent than does quinidine. Some of these metabolites may exist in blood at higher concentrations than does quinidine itself. In addition, some quinidine preparations contain significant amounts of dihydroxyquinidine, which is also pharmacologically active.

The various quinidine metabolites are not used therapeutically, and little or no information is available to characterize the therapeutic or toxic contributions they make to the use of quinidine therapy. What this does point out is that the analyst, when performing quinidine therapeutic monitoring, must select an analytic method that differentiates quinidine from its metabolites.

Several metabolites of propranolol (an effective beta-blocking drug) are known and have been demonstrated to have beta-blocking activity. (Beta-blocking drugs are antagonists to the effects of catecholamines and are used to treat angina pectoris, hypertension, and cardiac arrhythmias.) At this time, convenient methods of measuring the metabolites and well-characterized pharmacokinetic data are not available for a clearer understanding of their contributions to therapy and the avoidance of toxicity. In order to maintain sufficient specificity in propranolol monitoring, the analyst must be aware of these metabolites when selecting analytic methods.

On the first pass through the liver, lidocaine is metabolized to either monoethylglycinexylidide (MEGX) or glycinexylidide (GX). Both are pharmacologically active, although this activity is not well characterized. Specific methods should be used when monitoring lidocaine therapy. Monitoring of lidocaine metabolites is possible but not generally necessary.

Nitroprusside, a potent diuretic, is immediately converted to thiocyanate when it is administered IV. This metabolite is formed so quickly that monitoring thiocyanate is the mechanism of choice in evaluating nitroprusside therapy.

About 8% of digitoxin is metabolized to digoxin, which has similar activity, although its concentrations are not significant in digitoxin therapy. Disopyramide, carbamazepin, and primidone are all commonly monitored drugs with active metabolites; however, the therapeutic or toxic concentrations of the metabolites are considered negligible.

SPECIMEN COLLECTION AND HANDLING

The proper specimen container must be used to collect or transport the specimen. Vacuum collection tubes are suitable for blood specimens, but one must select the proper tube type or additive. Tris(2-butoxyethyl)phosphate (TBEP), a plasticizer used in the manufacture of some types of vacuum-tube stoppers, has been shown to interfere in the analysis of highly protein-bound basic drugs, including tricyclics, lidocaine, and propranolol. Therefore, it is necessary to collect these drugs in containers with special stoppers, which are now available from the major vendors of vacuum tubes (refer to the manufacturer's literature).

It has been demonstrated that TBEP reduces the protein binding of propranolol to α_1-glycoprotein in plasma.[7] With the increase in the amount of free drug, more drug diffuses into the red cells, and plasma drug levels are falsely reduced. Similar behavior is true for several other basic lipophilic drugs: lidocaine, alprenolol, chlorimipramine, imipramine, nortriptyline, meperidine, and quinidine.

A weak binding of several aminoglycoside antibiotics, including gentamicin and tobramycin, to the glass walls of vacuum collection tubes has been demonstrated.[12] Vacuum collection tubes containing a silicone barrier have been evaluated for collecting TDM specimens.[19] Of 12 common drugs tested, the highly lipophilic lidocaine, pentobarbital, and phenytoin showed decreases in serum levels of 32%, 15%, and 9%, respectively, when collected in these tubes. The cause is thought to be absorption of the drug on the separator material, but the problem can be avoided by spinning the specimen and separating the serum within 2 hours after the specimen is collected.

It is inappropriate for every laboratory to evaluate each type of vacuum collection tube for possible interference with any one of the many drugs that are measured routinely. Thus, it is necessary to follow the literature as these studies are reported or refer to the literature provided by vendors of commercial reagent kits to determine the appropriate container for collecting blood specimens for drug analysis.

Anticoagulants are necessary if an analysis must be done on plasma. Most drugs can be monitored adequately in plasma or serum. Serum, when available, is the specimen of choice. If plasma is required, heparin is usually the preferred anticoagulant. Heparin is coated on the tube walls and causes no change in plasma water, whereas other anticoagulants added as solutions to the vacuum collection tube cause a slight dilution effect. The calcium-binding anticoagulants add varying amounts of cations and anions, which may affect drug distribution between plasma and cells.

The timing of specimen collection relative to the dose administration is very critical. A rule of thumb suggests that, for peak drug levels, the specimen should be collected 1 hour after the dose is administered. For an oral dose, this rule assumes that absorption and distribution of the drug is complete in 1 hour. Since the half-life of the drug is assumed to be 2, 3, or more hours, very little drug has been eliminated in 1 hour. Of course, factors affecting absorption and distribution may invalidate this assumption.

All rules have exceptions, and several are important to note. Phenobarbital, phenytoin, digoxin, and lithium are slowly absorbed and distributed when administered subcutaneously. Phenobarbital has a $t\frac{1}{2}$ of 50 to 120 hours, and phenytoin's half-life is 18 to 30 hours. Consequently, it is appropriate to wait 8 or more hours before collecting a peak-level specimen. Trough levels should be drawn immediately before the next dose.

The aminoglycoside antibiotic drugs are usually administered parenterally because they are very poorly absorbed by the GI tract when administered orally. Since distribution is very rapid, it is appropriate to collect specimens for peak drug levels about 30 minutes after the IV infusion is completed and between 30 to 60 minutes following an IM injection.

Correct specimen collection times may be determined for specific drugs from data found in several handbooks and references.[1,3,6,13,15,25] As a general rule, trough levels are the easiest to collect at the most appropriate time. Peak levels require careful coordination of sample collection with dosing.

SAMPLE STORAGE

Traditional blood specimen–handling procedures should be followed (see Chapter 3, Specimen Collection and Processing). Blood should be centrifuged, and the plasma or serum should be removed from the cells. Once serum or plasma has been removed from the cells and refrigerated, most drugs are remarkably stable and may be stored for several days with no noticeable change. For example, lidocaine is stable for at least 3 weeks in serum at room temperature, and disopyramide is stable in serum for at least 100 hours at room temperature and 1 month when frozen.[13]

MONITORING SPECIFIC DRUGS*

Colorimetric methods of analysis represent the backbone of analytic techniques used in the clinical laboratory. Not surprisingly, they were the primary methods used in the early stages of TDM. It soon became apparent that colorimetric methods measured not only the parent drug but often metabolites, both active and inactive, as well as other interferences, making interpretation of results difficult. To improve specificity, extraction procedures preceded the actual color-developing reaction, but analysis time and difficulty increased, and precision often suffered.

Gas chromatography (GC) has been applied to the analysis of several drugs, but the requirement that the analyte be volatile at operating temperatures has limited its applicability or has required derivatization that adds to the difficulties of analysis. When applicable, GC provides tremendous specificity and sensitivity but is not easily automated and is often not applicable to large batches of samples.

The addition of high-performance liquid chromatography (HPLC) and radioimmunoassay (RIA) to the analyst's techniques revolutionized TDM. HPLC provides the necessary specificity and is capable of measuring several drugs simultaneously. The major disadvantage of HPLC is the lack of speed in analyzing large batches of samples. The immunologic characteristics of RIA give this method specificity that is sufficient for many drugs but is capable of measuring only one compound at a time. The technique is easily automated and suitable for handling large batches of samples. The characteristics of radioisotopes represents the primary shortcoming of RIA for TDM. Reagent shelf lives are short, and radioisotope waste disposal is a problem.

Enzyme immunoassay (EIA) and fluorescence polarization immunoassay (FPIA) techniques provide the specificity of RIA, are capable of being automated, and have eliminated the drawbacks of working with radioisotopes. Nephelometric techniques of measuring drug levels are also contributing to the overall array of tools available for TDM.

The immunoassays, although convenient, capable of being automated, accurate, and precise, may be too specific in some cases and insufficiently specific in others. When possible, maximal specificity has been developed for a particular drug. For example, procainamide immunoassays are specific for procainamide and require yet another antibody (kit, procedure) to measure the active metabolite, NAPA. On the other hand, antibodies to quinidine are not yet highly developed or specific for the parent drug, and these EIA procedures measure several metabolites in addition to quinidine.

For maximal specificity and the capacity to separate and measure in one process both parent drug and metabolites, HPLC excels. The principal drawback of HPLC is lack of flexibility in the instrumentation. It is difficult to quickly change columns and column conditions when analyzing a variety of drugs on one unit.

Table 27-4 shows the usage of various analytic techniques in TDM. The predominant techniques are immunoassays including EIA, which has been incorporated into the EMIT system manufactured by Syva Corporation, and FPIA, the technique used by Abbott in the TD_x system.

---

*The reader is referred to Chapter 5, Analytical Techniques and Instrumentation, for in-depth descriptions of the analytic techniques.

TABLE 27-4. Representative Techniques Available for Therapeutic Drug Monitoring

	EIA	FPIA	RIA	GC	HPLC	Atomic Absorption	Ion Specific Electrode	Nephelometry	UV/VIS Spectroscopy	Flame Emission
Acetaminophen	24	61			3				4	
Amikacin	10	87							1	
Carbamazepine	28	56		1	7					
Desiprimine	19			16	66					
Digoxin	18	60	11							
Disopyramide	22	67		4	7					
Ethosuximide	27	61		11	1					
Gentamicin	19	77						2		
Imipramine	19			15	66					
Lithium						20	45			33
Phenobarbital	30	58		3	5					
Phenytoin	28	58		3	6		3			
Primidone	31	61		4	3					
Procainamide	26	66			6					
NAPA	25	65			6					
Quinidine	26	67			4					
Salicylate	2	49							40	
Theophylline	25	49			2			9		
Tobramycin	14	83								
Valproic Acid	27	65		6						
Vancomycin	11	89								

Adapted from various proficiency testing programs. (Numbers expressed as percentage of total participants. "Other" category not included.)

RIA and UV/Vis spectrophotometry have all but disappeared, except for digoxin and salicylate, respectively.

Lithium is a unique therapeutic drug because it is not the typical organic molecule, but rather an elemental cation. The immunoassays are not applicable because of the lack of antibodies to lithium. Furthermore, alternative techniques, which are applicable to lithium, are more convenient. Lithium can be measured accurately, precisely, and conveniently by flame photometry, atomic absorption spectrometry, or ion-selective electrode.

The tricyclic antidepressants, because of their very low therapeutic concentrations (ng/mL), require extra sensitivity of the analytic technique. Initially, the method of choice for TCA determinations was gas chromatography using a nitrogen–phosphorus detector. This detector incorporates a rubidium bromide bead in the exit port, which provides additional sensitivity for compounds containing nitrogen or phosphorus. With improved ultraviolet (UV) absorption detectors, HPLC is capable of accurately measuring tricyclic antidepressants. More recently, GC/MS and FPIA have been applied to TCA analyses. Peak levels are typically measured, and it is necessary to measure active metabolite activities as well as the parent drug when amitriptyline and imipramine are the dosed drugs.

Some clinical laboratory scientists do not consider coagulation testing part of TDM. However, anticoagulant therapy should be monitored to allow these drugs to be used effectively and safely.

Anticoagulant therapy is monitored not so much to avoid the potential toxic effects of the drug itself, but rather to evaluate the overall effect of the anticoagulant and to avoid hemorrhage subsequent to an excess of drug. For clinical purposes, instead of the drug itself being measured, the global effect of the drug is evaluated using primarily the prothrombin time and activated partial thromboplastin time.

Citrated plasma is combined with thromboplastin and calcium ions, and the clotting time is measured. This parameter measures the prothrombin activity that has been altered by the prescribed anticoagulant, warfarin. Heparin therapy is most often measured using the activated partial thromboplastin time. A citrated aliquot of plasma is combined with excess calcium and appropriate clotting factors, and the clotting time is measured. Methods are available for measuring heparin and warfarin concentrations in blood, but the interpretation of these results relative to appropriate drug therapy is not well characterized.

Coagulation testing may be the purest form of drug monitoring because the drug effect is measured. Measuring the drug concentration itself only indirectly measures the effect of the drug, and the relationship between the two may not be linearly proportional.

Occasionally, the TDM laboratory may be asked to measure a bromide level. Bromide-containing compounds were formerly used as sedatives in the concentration range of 80 to 120 mg/dL, and levels above 130 mg/dL may be toxic. Bromide in a protein-free filtrate of serum will react with gold chloride to form a brownish yellow–gold tribromide, which can be measured colorimetrically at 440 nm.[5]

Sodium nitroprusside is an extremely potent and fast-acting antihypertensive drug. It is not recommended for long-term use, and it is necessary to monitor the blood levels. The parent drug is quickly metabolized to thiocyanate, which can be toxic at concentrations greater than 10 mg/dL. Thiocyanate can be measured in a protein-free filtrate of serum by adding ferric nitrate. A red ferric thiocyanate complex forms and can be measured colorimetrically.[14]

Methods for measuring free drug levels first incorporate a means of separating the free fraction from the protein-bound fraction. This can be done by equilibrium dialysis, ultrafiltration, and ultracentrifugation. Ultrafiltration through a membrane of optimum pore size is the only method practical for the routine drug monitoring laboratory. Membranes, such as YMT membranes, have been developed that minimize drug absorption on the membrane (<5%) and protein leakage.[20] Factors such as temperature and sample pH must be closely controlled to maximize accuracy and precision. A fixed angle head centrifuge is most often used to maximize the volume of ultrafiltrate harvested in the shortest time. Once an ultrafiltrate has been obtained, the free drug level can be measured using the same techniques as those used for total drug determinations. Drugs often evaluated by determining free fraction include valproic acid, phenytoin, and carbamazepine.

QUALITY CONTROL

Maintaining a good quality-control (QC) program in the TDM laboratory is as important as in any other area of laboratory medicine. The basic principles are the same (see Chapter 4, Quality Assurance and Quality Control), but the plasma or serum material maybe unique, since drugs are not endogenous to most donors or sources of blood. In the early days of therapeutic drug monitoring, commercial QC material was not available and each laboratory had to prepare its own serum pools and add to them the appropriate drugs. Commercial vendors of QC material now provide a wide variety of drugs in their routine chemistry QC material as well as special material with the newer drugs and drugs that require special preparation procedures. For those drugs still not available in commercial QC material, the TDM laboratory must continue to concoct its own "brew." As with other areas of the clinical chemistry laboratory, there are also proficiency-testing programs available for TDM.

CARDIOACTIVE DRUGS

The prevalence of cardiovascular disease has spawned the development of several drugs capable of correcting arrhythmias. Many of these drugs have significant toxic side effects, their therapeutic windows are narrow, and toxic levels closely approach and overlap therapeutic levels. Frequently, administered dose does not correlate well with effectiveness. Therapeutic drug monitoring is required in order to use these drugs safely and effectively. Cardioactive drugs can be separated into two categories: the cardiac glycosides (digoxin and digitoxin) and the antiarrhythmics.

Digoxin

Digoxin and Digitoxin are two of a group of glycosides harvested from the digitalis plant. Digitoxin is used very infre-

quently, while digoxin is widely used to improve cardiac contraction (inotropic response) in congestive heart failure. Digoxin is also used to correct supraventricular tachycardia. The therapeutic range for the inotropic effect is 0.8 to 2.0 ng/mL, while levels at or near 3 ng/mL are required for decreased ventricular response.

Adverse side effects are GI and CNS toxicity and cardiac dysfunction. Adverse cardiac effects are the most frequent and may occur at serum levels within the generally accepted therapeutic range, whereas GI toxicity usually occurs at levels of 3 mg/mL and above.

The bioavailability of orally administered digoxin varies from 70% for tablets to greater than 90% for capsules. IM administration is unacceptable because of the discomfort from the injection and the erratic and incomplete absorption behavior. Digoxin is highly bound to skeletal and myocardial muscle and has a V_d (volume of distribution) of 5 to 10 L/kg depending on renal function. Protein binding to albumin is 30%. About 25% of a digoxin dose is metabolized, while 75% is excreted by the kidneys unchanged. Half-life is 38 hours normally and longer in nephrotic patients.

The timing of sample collection is critical for digoxin monitoring (as is the case in almost any drug monitoring). Digoxin demonstrates a multiphasic disposition with an initial phase of about 8 hours, when serum levels may reach 8 to 10 mg/mL. Following this initial phase, a second and longer elimination phase occurs. Therefore, it is appropriate to wait at least 8 hours after dosing to evaluate serum digoxin levels. Individuals with hypothyroidism are typically more responsive to digoxin than euthyroid patients. Conversely, hyperthyroid patients are resistant. Drug disposition is compromised in hypothyroid individuals, while lower digoxin levels are seen in hyperthyroid individuals.

Disopyramide

Disopyramide is an effective antiarrhythmic used to treat several ventricular and supraventricular arrhythmias. A major difficulty with disopyramide is the extreme variability within and between patients of the binding of disopyramide to plasma proteins, specifically AAG. The binding behavior is concentration-dependent. Consequently, pharmacokinetic parameters are difficult to determine based on total drug concentrations, and therapeutic drug monitoring is equally difficult to interpret. The generally accepted therapeutic range for total drug measurement is 2 to 5 µg/mL. However, some investigators have reported that there is no relationship between total serum levels and antiarrhythmic utility. Toxic side effects include anticholinergic reactions (dry mouth, urine retention, visual impairment) and cardiovascular alterations (congestive heart failure, myocardial depression, hypertension, hypotension).

The most predictable parameters are those for the unbound fraction of disopyramide, but measuring the unbound drug is technically difficult and certainly not feasible for therapeutic drug monitoring.

Procainamide

Procainamide is used to correct atrial and ventricular arrhythmias. Serum concentrations of 4 to 10 μg/mL are effective therapeutically in most patients. Some patients, particularly those with ventricular tachycardia associated with chronic heart disease, may require and tolerate levels as high as 20 μg/mL. The appropriate therapeutic range for procainamide is capricious owing to the fact that the major metabolite, *N*-acetyl procainamide (NAPA), has antiarrhythmic activity but is less potent than the parent drug. Consequently, NAPA is often measured as well as procainamide when monitoring procainamide therapy. The goal when measuring both procainamide and NAPA is to achieve procainamide concentrations between 8 and 12 μg/mL and procainamide plus NAPA levels below 30 μg/mL. Adverse reactions to procainamide include hypotension, bradycardia, prolongation of ECG intervals, and systemic lupus erythematosus.

Propranolol

Propranolol is one of several beta-blockers, but it is the most popular. It is used to treat hypertension, cardiac arrhythmias, and angina pectoris. Propranolol has multiple mechanisms of effect, but most important, it reduces heart rate and contractility.

The most common side effect of propranolol is congestive heart failure and bronchospasm in certain individuals. Barring these complications, individuals may tolerate levels as high as 100 ng/mL. Levels this high, however, approach total beta blockade. The minimum effective level for angina pectoris therapy is 30 to 50 ng/mL. Consequently, a target range is 50 to 100 ng/mL.

Therapeutic drug monitoring of propranolol is best done by measuring heart rate, blood pressure, and electrocardiographic parameters. Measuring plasma levels may be useful when high doses of propranolol are not effective because of noncompliance or unusual pharmacokinetics.

Lidocaine

Lidocaine corrects ventricular arrhythmias associated with myocardial infarctions, cardiac surgery, and digitalis intoxication. It is used prophylactically to prevent ventricular fibrillation, especially in patients with acute myocardial infarctions. Although the generally accepted therapeutic range is 1.5 to 5.0 μg/mL, some individuals may require levels as high as 8 μg/mL to prevent arrhythmias. Unfortunately, significant CNS toxicity occurs in individuals with serum levels above 5 μg/mL, so there is considerable overlap of therapeutic and toxic ranges. Adverse effects of lidocaine include drowsiness, dizziness, euphoria, tinnitus, and nausea. Since 50% to 75% of an oral lidocaine dose is metabolized on the first pass in the liver, alternate routes of administration are preferred. Bioavailability is nearly complete with IM injection, but the site of injection may influence the rate of absorption. Consequently, intravenous administration is preferred when possible.

Lidocaine is distributed throughout the body and eliminated primarily by the liver. The two primary metabolites, monoethylglycinexylidide (MEGX) and glycinexylidide (GX), show some antiarrythmic activity.

The time to collect a specimen for drug monitoring depends on the conditions of dose administration and the reason for monitoring. Lidocaine is often administered as a loading dose followed by a constant infusion. To evaluate for effective therapeutic levels, specimens should be collected 30 minutes after the loading dose and 5 to 7 hours following initiation of therapy. Levels should be monitored every 12 hours in patients with congestive heart failure and every 24 hours in patients on prophylaxis following myocardial infarction. If toxicity is suspect, collect a specimen as close to the event as possible. Likewise, if ventricular arrhythmias occur during lidocaine therapy, a specimen should be collected as close to the event as possible. Since cimetidine and propranolol inhibit hepatic clearance of lidocaine, it is appropriate to monitor lidocaine levels when these drugs are coadministered.

Quinidine

Quinidine is a broad-spectrum antiarrhythmic used to correct irregular and rapid heart rates. It is a natural alkaloid harvested from cinchona bark. Typical pharmaceutical preparations are quinidine sulfate and the slow-release quinidine gluconate. Most preparations include the naturally occurring analogue dihydroquinone. Quinidine metabolites include 3-hydroxyquinidine and 2′-oxoquinidinone, among others, which show varying degrees of antiarrhythmic activity. The therapeutic range depends on the method of analysis and its specificity for quinidine. Less specific single-extraction fluorescence methods require a range of 2 to 6 μg/mL, while the more specific immunoassays dictate a narrower range of 2.3 to 5 μg/mL for optimal therapeutic effect. Quinidine experiences a significant first-pass effect, and its rate of elimination is affected by both renal and hepatic function. Impairment of either or both will increase elimination half-life and require reduced doses.

By far the most common toxic side effects of quinidine are GI tract irritation–anorexia, diarrhea, nausea, vomiting, and general discomfort. Cardiovascular disorders such as bradycardia, arrhythmias, and hypotension may occur. Toxicity may occur with levels in the upper end of the therapeutic range and higher. Quinidine, when administered alone to correct atrial fibrillation or flutter, may convert the rapid atrial arrhythmia to a rapid ventricular arrhythmia. Consequently, digoxin is often administered before or in conjunction with quinidine therapy.

Most quinidine monitoring is done to ensure therapeutic levels; consequently, trough levels are appropriate and the specimen is drawn shortly before dosing. If toxicity is suspected, peak levels should be evaluated.

ANTIEPILEPTIC DRUGS

Antiepileptic drugs have been classified by many as wonder drugs because of their effectiveness in treating seizures. Part of this enthusiasm over a group of drugs is due to the exposure the general public had to the disease they have prevented—epilepsy.

Part of the difficulty in evaluating the effectiveness and, thereby, the therapeutic range of anticonvulsant drugs is the infrequency and unpredictability of seizures. Once the effectiveness of these drugs was discovered, it became difficult to acquire a control group for further experimentation. Nevertheless, therapeutic ranges have been determined, and guidelines exist as to which levels may result in toxicity.

Phenobarbital

Phenobarbital is effective therapy for febrile seizures and neonatal seizures. The optimal therapeutic range is 15 to 40 µg/mL, with levels below 10 to 15 µg/mL showing little or no effect. Adverse side effects of phenobarbital include fatigue, drowsiness, sedation, depression, and reduced memory concentration. Children, on the other hand, frequently experience hyperactivity and excitation while on phenobarbital therapy. Bioavailability of phenobarbital oral administration is over 90% for adults and children but less (80%) for neonates. The V_d for adults and children is about 0.6 L/kg, and for neonates, it is 0.9 L/kg. Elimination is primarily hepatic, with about 30% excreted as the parent drug.

Phenobarbital is an effective liver enzyme activator and enhances the metabolism of other drugs. Several drugs, however, inhibit phenobarbital metabolism, presumably by liver enzyme inhibition. They include valproic acid, chloramphenicol, cimetidine, and phenytoin. Ethanol is a potent liver enzyme activator and enhances the elimination of phenobarbital.

Trough levels, drawn shortly before dosing, are appropriate because they are the most consistent. If toxicity is suspected, then peak levels are appropriate. With a half-life of 3 to 5 days, timing of sample collection is not critical. Peak concentration occurs 6 to 18 hours after a dose.

Primidone is also an effective antiseizure medication partially because it is metabolized to phenobarbital. It can improve seizure control in some individuals who were previously on phenobarbital therapy, whereas in others it may reduce seizure control.

Phenytoin

Phenytoin is an effective antiepileptic agent often preferred over phenobarbital because of its lack of hypnotic properties. Phenytoin is especially effective at correcting tonic-clonic seizures, and effect correlates well with blood levels, especially free drug levels. Wide interindividual variation exists for the protein binding of phenytoin (87%–93% bound), which makes drug-level analyses very effective. Most patients on phenytoin show good therapeutic effect with plasma levels of 10 to 20 µg/mL. Volume of distribution is 0.5 to 0.8 L/kg, which is very close to total body water.

Phenytoin levels are reduced by several drugs, including phenobarbital, carbamazepine, and alcohol, owing to their inducing effect on metabolic enzymes. Valproic acid and phenylbutazone are known to inhibit phenytoin metabolism and to displace phenytoin from albumin-binding sites. Aspirin also competes with phenytoin for protein-binding sites.

Phenytoin itself is a potent liver enzyme inducer and is known to increase the clearance of multiple drugs, including carbamazepine, phenobarbital, acetaminophen, steroids, vitamins, and so on.

Phenytoin monitoring, like so many other drugs, is appropriate when a change in dosage or preparation formulation is made, in various disease states, when other drugs are coadministered, in suspected toxicity, when noncompliance is suspected, in pregnancy, or when seizures are not well controlled. Although free drug levels are best correlated with therapeutic effect, total plasma levels appear to be satisfactory in most cases and are more conveniently determined.

Valproic Acid

Valproic acid is a recent entrant in the antiseizure arsenal of drugs. Structurally, it is dissimilar to the other anticonvulsants. It is a primary drug in the treatment of absence (petit mal), generalized tonic-clonic, and myoclonic seizures. Therapeutic monitoring is important because of valproic acid's variable and often unpredictable pharmacokinetics. Absorption is nearly 100%, but it varies with time depending on the formulation—capsules, coated tablets, or syrup. Protein binding is concentration-dependent (90%), and free levels are significantly influenced by renal disease, impaired hepatic function, and variable albumin levels.

Hepatic elimination accounts for 97% of valproic acid ingested, and the rate of elimination is increased by other antiepileptic enzyme activators—phenytoin, phenobarbital, and carbamazepine.

Total concentrations of valproic acid in plasma of 50 to 100 µg/mL show good therapeutic effect, while free levels of 4 to 10 µg/mL are optimal. Adverse effects occasionally appear at total concentrations above 100 µg/mL and may include weight gain, anorexia, vomiting, and nausea. Hyperammonemia is common, especially in children, and may cause confusion, stupor, or coma in severe cases.

Carbamazepine

Carbamazepine has been found to be effective in the treatment of psychomotor seizures, generalized tonic-clonic seizures, and mixed seizures. It is often administered in combination with phenytoin for seizure control. Additionally, carbamazepine is the drug of choice for treatment of trigeminal and glossopharyngeal neuralgia pain. The antidiuretic effect of carbamazepine has been used to treat diabetes insipidus.

The therapeutic range for carbamazepine is 6 to 12 µg/mL, but frequently, patients with levels below this range may show therapeutic effect, while others require levels above 8 µg/mL and higher. The unpredictability of therapeutic levels is due to a wide interindividual variation in protein binding. Carbamazepine binds to albumin and AAG and may vary from 10% to 50% free. This variability may be due to fluctuations in AAG concentrations.

Of the several metabolites of carbamazepine, the 10,11-epoxide metabolite is the most prevalent, and it demonstrates both therapeutic and toxic activity. Enzyme-inducing drugs such as phenytoin and phenobarbital, when administered concurrently, cause a higher metabolite-to-parent drug ratio.

Adverse neurosensory effects such as diplopia, drowsiness, blurred vision, nystagmus, unsteady gait, and ataxia are common, especially in the elderly.

Ethosuximide

Ethosuximide is a highly effective treatment for absence (petit mal) seizures. It has no other antiepileptic efficacy and must be complemented when treating mixed seizures. Although valproic acid has similar antiepileptic features, ethosuximide is the primary agent for absence seizures. When ethosuximide is ineffective, valproic acid is the obvious choice, often with good success.

Therapeutic levels fall between 40 and 100 µg/mL, with toxicity frequently occurring above 150 µg/mL. Nausea, vomiting, drowsiness, anorexia, and dizziness are common side effects during initial drug therapy. These effects usually diminish with time.

BRONCHODILATORS

Theophylline

Theophylline is an effective relaxant of smooth muscle in bronchial airways and pulmonary blood vessels. Thus, it acts as a bronchodilator and pulmonary vasodilator and is extremely effective therapy for acute asthmatic attacks. The drug is a diuretic, coronary vasodilator, and cardiac stimulant, and it improves diaphragmatic contractility and reduces diaphragmatic fatigue. This makes it an effective treatment for chronic breathing problems, and it is used prophylactically.

The therapeutic range is 5 to 20 µg/mL. Levels above 20 µg/mL are associated with adverse effects. Interestingly, the relationship between serum concentration and therapeutic effect and toxic side effects is logarithmic. Adverse reactions most often include vomiting, and hematemesis is prevalent in children. Theophylline stimulates the CNS, causing increased respiratory rate, restlessness, agitation, tremor, seizures, hyperthermia, and hyperreflexia. Theophylline has an inotropic and chronotropic effect on the heart and at therapeutic levels causes an increase in heart rate. In severe toxicity, theophylline causes abnormal cardiac rhythms. The half-life of theophylline may be as short as 3 to 4 hours in children. Consequently, slow-release formulations are available. Unfortunately, many of these demonstrate erratic and unpredictable bioavailability. Several preparations use aminophylline (theophylline ethylenediamine) as the parent drug, which quickly converts to theophylline in the stomach. Theophylline is absorbed in the small intestine, and consequently, the rate at which the stomach empties influences the rate at which theophylline is absorbed.

Theophylline is 40% free, and consequently, small changes in albumin, various diseases, competing drugs, and fluctuations in other factors that influence protein binding have minimal effect on free theophylline concentrations.

Metabolism and elimination of theophylline change significantly with age. Adults metabolize about 90% of the drug before elimination, whereas newborns may excrete 40% to 50% of the drug unchanged. The rate of elimination is much accelerated in children ($t\frac{1}{2}$ 4 hours) as compared with adults, in whom the $t\frac{1}{2}$ may range from 6 to 12 hours. Newborns, on the other hand, have a much reduced elimination rate, and half-lives are typically 30 to 40 hours.

Theophylline therapy monitoring is a classic example of effective drug-level monitoring. The efficacy of theophylline therapy is closely correlated with serum drug levels. The therapeutic "window" is narrow, and toxic levels lie close to therapeutic levels. There is a large interindividual variation in elimination rate, which is influenced by several factors, especially age of the patient. Various drug formulations exist with varying release rates and rather erratic absorption characteristics. For those individuals on chronic theophylline therapy who experience an acute asthmatic attack, serum theophylline level determinations are extremely helpful in deciding if and how much theophylline should be administered and whether or not it can be tolerated. Often, a loading dose is required, with follow-up maintenance doses. Periodic serum level determinations are necessary to arrive at the proper maintenance dose while avoiding toxicity. Of course, signs of theophylline toxicity dictate that a serum level be measured to aid in future dosing and treatment.

ANTIBIOTICS

Aminoglycosides

The aminoglycoside group of antibiotics includes gentamicin, tobramycin, netilmicin, amikacin, streptomycin, neomycin, and kanamycin. They consist of two or more amino sugars connected to hexose by means of glycosidic bonds. These drugs are the major components in the arsenal of agents against serious systemic infections of gram-negative aerobic bacteria. The transport of the aminoglycoside antibiotics into cells is oxygen-dependent. Because anaerobic bacteria do not have an oxygen-utilizing transport mechanism, they are resistant. Aminoglycoside antibiotics are used in combination with beta-lactam antibiotics, cephalosporins, and penicillins for their synergistic effects against bacteria such as *Streptococcus faecalis, Staphylococcus aureus, Listeria monocytogenes,* and gram-positive cocci.

Absorption of aminoglycosides following oral administration is less than 2%. Consequently, IV, IM, or intrathecal administration is required, which results in complete absorption. Aminoglycosides are not highly protein bound (<40%) and quickly equilibrate into extracellular compartments. Topical application to wounds and burns is quite effective.

The volume of distribution for aminoglycosides varies dramatically from individual to individual and is influenced by several factors, including hydration status, renal function, congestive heart failure, peritonitis, and edema. Typical values are 0.2 to 0.3 L/kg. Changes in V_d can dramatically influence elimination rate and thus serum drug levels. Elimination is predominantly renal, with 85% to 95% of the administered drug being excreted unchanged.

The aminoglycosides tend to concentrate in specific tissues, most noticeably the renal cortex. This helps explain their major adverse effect of nephrotoxicity, which occurs 2% to 10% of the time. Of course, the nephrotoxicity is exacerbated by the existence of sepsis. Ototoxicity occurs less frequently (0.5%–3.0%), with loss of hearing in the higher frequency sounds. Conversational hearing loss also may occur. Hearing loss is usually bilateral and permanent. Neuromuscular blockade is rare but has serious consequences if present. Respiratory failure is rare but may require therapeutic intervention, especially in seriously ill patients.

Aminoglycoside therapy requires immediate and close surveillance of serum concentrations. Upon initiation of antibiotic therapy, serum determinations are appropriate to ensure therapeutic levels. Subsequent monitoring of drug levels and serum creatinine levels is necessary to prevent renal failure. Evaluation of electrolytes, hemoglobin, and hematocrit may be necessary to determine degree of hydration and impending changes in volume of distribution.

A significant correlation between toxicity and trough levels of aminoglycoside antibiotics has been identified. To prevent nephrotoxicity, it is appropriate to allow the drug level to fall below a given level before the next dose is administered. Vestibular toxicity also may be avoided by monitoring trough levels. Effective therapeutic monitoring requires close observance to optimal sample collection times. Samples for trough levels should be collected immediately before an IV infusion is started. Peak levels occur within a narrow time "window" ½ hour following the completion of an IV infusion. These times are so important for proper interpretation of results that their documentation on the requisition should be a required entry.

Aminoglycosides can be inactivated by other antibiotics, specifically beta-lactams, semisynthetic penicillins, and cephalosporins. A complex forms between the aminoglycoside and the semisynthetic penicillin that inactivates both antibiotics. This interaction does not appear to be significant *in vivo* in patients with normal renal function. However, IV solutions containing combinations should not be prepared, patients with poor renal function should be closely monitored, and samples that cannot be measured soon after collection should be frozen to prevent measurable inactivation.

Chloramphenicol

Although a very popular antibiotic at first, chloramphenicol usage quickly diminished because of the realization of rather serious idiosyncratic and dose-related toxicities, primarily aplastic anemia. Therapeutic levels are 10 to 20 µg/mL.

Chloramphenicol toxicity occurs as hematologic disorders at serum levels above 25 µg/mL. Reversible bone marrow depression is accompanied by increased serum iron levels, reticulocytopenia, decreased hemoglobin levels, thrombocytopenia, and leukopenia. Plasma levels above 50 µg/mL in infants have caused the fatal gray baby syndrome (cardiovascular-respiratory collapse). Aplastic anemia, although rare, is usually fatal.

The bioavailability of an oral dose of chloramphenicol succinate ranges from 93% in premature infants to 64% in children. Its elimination is dependent on the renal excretion of the succinate and the rate of metabolism of the succinate to chloramphenicol. Chloramphenicol is metabolized to a monoglucuronide in the liver or to an amine in the GI tract.

Peak serum levels are most often monitored, and samples should be collected 1.5 to 3.0 hours after an oral dose and 0.5 to 1.5 hours after completion of an IV infusion of chloramphenicol succinate.

Vancomycin

Vancomycin is a glycopeptide that binds irreversibly to the cell wall of many gram-positive cocci and bacilli and prevents cell-wall synthesis. The drug is effective against *Staphylococcus epidermidis* and *Staphylococcus aureus,* including the methicillin-resistant *S. aureus.* Vancomycin has negligible absorption following oral doses and is therefore administered intravenously. It is widely distributed (V_d = 0.5–0.8 L/kg) except in CSF. Protein binding is approximately 50%, and 80% to 90% of the drug is excreted unchanged.

Therapeutic peak concentrations are 30 to 40 µg/mL, and trough levels are 5 to 10 µg/mL. Phlebitis, chills, fever, and nausea are not uncommon. Neutropenia, nephrotoxicity, ototoxicity, and nephritis along with a variety of infusion reactions are also adverse reactions to vancomycin. Toxic levels are greater than 20 µg/mL for trough levels and greater than 80 µg/mL for peak levels. Early preparations of vancomycin contained substantial amounts of impurities, and these impurities are implicated as perhaps causing some of the adverse effects of vancomycin therapy.

PSYCHOACTIVE DRUGS

Lithium

Lithium as lithium carbonate or lithium citrate is used in the treatment of manic depression. For acute mania, effective peak levels are 0.8 to 1.4 mEq/L, while maintenance therapy levels are 0.6 to 1.2 mEq/L. Initial signs of toxicity include nausea, vomiting, diarrhea, muscle weakness, apathy, drowsiness, polyuria, and ataxia, which often occur at therapeutic levels. Concentrations above 1.5 mEq/L are associated with more severe toxicity, including tremors, muscle twitching, slurred speech, and confusion. Levels above 2.0 mEq/L are life-threatening and cause stupor, seizure, and coma.

Lithium is thoroughly water-soluble and totally absorbed within 8 hours, with peak levels occurring 2 to 4 hours after an oral dose. Distribution is throughout the total body water compartment with no protein binding in plasma. Lithium is completely filtered by the glomeruli and 80% reabsorbed by the tubules.

Tricyclic Antidepressants

The tricyclic antidepressants (TCAs) include imipramine and its active metabolite desipramine, amitriptyline and its active metabolite nortriptyline, and doxepin, among others. Their mode of action is to block a wide variety of receptors and to inhibit the uptake of serotonin and norepinephrine. TCAs have several effects on cardiovascular performance, the most significant being postural hypotension. The TCAs are used for depressions with no known societal or organic causes. Symptoms include loss of appetite, anhedonia, guilt, insomnia, weight loss or gain, decreased sex drive, and suicidal tendencies and paranoia in severe cases.

Typical therapeutic levels fall between 100 and 300 ng/mL depending on the individual drug. Levels above 1000 ng/mL are associated with rather severe effects, including seizures, unconsciousness, cardiac arrhythmias, and respiratory depression. Side effects of TCA therapy include dry mouth, constipation, blurred vision, urine retention, memory loss, decreased sweating, and drowsiness.

Ingestion of 1 to 2 g of TCAs may create a life-threatening situation. Consequently, patients with suicidal tendencies must have the amount of drug available to them restricted.

Bioavailability of an oral dose varies from 13% to 90% depending on the TCA. This is partly due to a significant first-pass effect. TCAs are lipid-soluble and thus have a large V_d. Protein binding is high—63% to 96%.

Monitoring of TCA serum levels is appropriate when a patient does not show signs of improvement after 4 weeks of therapy on standard doses. Owing to interindividual variations in drug disposition, the patient may require an increased dose. Knowing serum levels provides for confidence in increasing dose and allows for an optimal adjustment of dose. Because of uncomfortable side effects, patient noncompliance in taking TCA medication may exist. Nortriptyline has a very narrow therapeutic "window," and monitoring is extremely helpful in attaining therapeutic levels and avoiding toxicity in a short period of time after initiation of medication.

Initially, methods for TCA determinations used gas chromatography with a nitrogen-phosphorus detector. The N-P detector was required to provide the sensitivity necessary to measure the nanogram per milliliter therapeutic levels for TCAs. HPLC and FPIA methods are now available for routine monitoring of TCA levels.

IMMUNOSUPPRESSANTS

Cyclosporine

Cyclosporine is a natural 11-amino acid cyclic polypeptide harvested from fungi. It exhibits potent immunosuppressive activity, and cyclosporine has been used as antirejection therapy in all forms of organ transplantation. Cyclosporine inhibits the activation of inactive T-lymphocytes and inhibits the activity of active T-lymphocytes. However, it has no activity toward B cells.

Cyclosporine is water-insoluble but is formulated in an olive oil vehicle for oral administration. This suspension must be further diluted with milk, chocolate milk, or orange juice and taken immediately to facilitate optimal bioavailability.

Clinical response does not correlate well with administered dose. However, there is good correlation between whole-blood levels and immunosuppression and toxicity. Most routine monitoring is done using trough levels by collecting a specimen just before the morning dose. Diurnal variations suggest that collection be done at the same time each day.

Cyclosporine is distributed throughout red blood cells and plasma. Temperature, and to a lesser extent other parameters, influences the equilibrium between cells and plasma. Early attempts to normalize this equilibrium and use plasma or serum samples were minimally successful. Consequently, whole blood is the sample of choice.

Oral bioavailability of cyclosporine is highly variable and erratic depending on several factors, including type of transplant, time after transplant, liver function, GI function, and food in the stomach. It can vary from less than 5% in pediatric liver transplants to nearly 90% in adult kidney transplants. Peak blood concentrations occur 1 to 8 hours after a dose. The volume of distribution averages 4 L/kg. The distribution in whole blood is 55% in RBCs, 10% in WBCs, and 30% in plasma, where it is 98% protein bound.

Cyclosporine is more than 90% metabolized to 12 or more derivatives that are eliminated in bile. Less than 6% of cyclosporine is excreted in urine. The half-life of the drug is 4 to 6 hours.

Toxic side effects of cyclosporine include renal dysfunction, hypertension, hirsutism, and tremors. In renal transplant patients, nephrotoxicity is often difficult to differentiate from allograft rejection.

Several methods are available for measuring cyclosporine in serum, plasma, or whole blood.[23] Original RIA methods used for serum were polyclonal and measured several metabolites as well as the parent drug. Newer RIA methods use a monoclonal antibody specific for the parent drug. Fluorescence polarization immunoassays using polyclonal (pFPIA) and monoclonal (mFPIA) antibodies are now in general use. An EMIT method is also available. HPLC methods separate the parent drug from metabolites and thus are specific for the dosed drug, cyclosporine. There is often substantial variation between methods. All these methods are able to measure levels in whole blood. Therapeutic levels are variable depending on analytic method, type of specimen, transplanted organ, transplantation center preferences, and whether cyclosporine is used in combination with other immunosuppressants. Trough levels in whole blood appear to be universally preferred, with typical values in adult kidney transplants of 100 to 375 ng/mL for HPLC to as high as 1000 ng/mL for mFPIA methods.[24] Levels are typically lowered 60 to 90 days after transplantation to reduce the risk of nephrotoxicity. HPLC levels greater than 400 ng/mL result in increased incidence of overimmunosuppression and renal, hepatic, and neurologic toxicity.

FK-506

New immunosuppressants are being evaluated that promise to have less severe toxicity and greater efficacy than cyclosporine and the other first-generation immunosuppressants. Most of these new agents, except FK-506, suppress the immune response via different mechanisms than cyclosporine; however, by using them in combination with cyclosporine, the hope is that lower levels of cyclosporine can be used with equal therapeutic effect but with lowered incidence of toxicity. FK-506 is an immunosuppressant drug that functions similar to cyclosporine but is more potent and has demonstrated less toxicity. Monitoring FK-506 levels in whole blood will likely become more common as a guide for dosage adjustments.

ANTINEOPLASTIC DRUGS

Methotrexate

Methotrexate is a chemotherapeutic agent used to treat a wide variety of neoplasms. It is a folic acid antagonist that poisons dihydrofolate reductase. This, in turn, blocks the formation of DNA nucleotide thymidine. Methotrexate also inhibits purine synthesis, which adds to its cytotoxicity.

Initially, low doses were used for the treatment of acute leukemia. Later, it was realized that higher doses were effective treatments for a variety of neoplasms. With higher doses, more severe toxic consequences developed until leucovorin was discovered as a rescue mechanism for normal cells following very high dose methotrexate therapy. Present-day methotrexate therapy involves administering extremely high doses of methotrexate, and after a prescribed period of time, leucovorin is administered as an antagonist to methotrexate to minimize normal cell destruction.

Therapeutic monitoring of methotrexate is unique in that serum levels are monitored serially after a bolus dose of drug to determine:

1. Pharmacokinetics to characterize an individual's susceptibility to severe toxicity.
2. Optimal leucovorin therapy to avoid toxicity.
3. Optimal dosage for future doses.

Serum levels of methotrexate in high-risk patients must be monitored early in high-dose therapy, since a delay in leucovorin rescue therapy beyond 48 hours can result in severe and irreversible toxicity. To prevent toxicity, serum levels should decay below 10^{-5} M at 24 hours, 5×10^{-7} M at 48 hours, and 5×10^{-8} M at 72 hours.

The major toxicity of methotrexate deals with the pharmacologic behavior of the drug and deciding on the appropriate leucovorin rescue therapy. The most prevalent adverse reactions to methotrexate are GI irritation with nausea and vomiting and hematologic suppression with leukopenia, anemia, and thrombocytopenia. Nephrotoxicity and ocular irritation also may occur.

Methotrexate may be administered IM, IV, orally, and interthecally depending on the site of the neoplasm. Oral doses are absorbed rapidly with a bioavailability of 30%, and serum levels peak at 1 to 2 hours. It is 50% to 70% protein bound, with a steady-state V_d of 0.75 to 0.80 L/kg.

SUMMARY

Over the years, the most appropriate drug dosage regimens have been developed by simply increasing or decreasing the amount of dose and the interval between doses until the appropriate effect was observed in the patient. However, because a number of drugs can cause rather severe toxic side effects at easily attained concentrations, this trial-and-error method of dosing drugs is an inappropriate way to establish effective drug levels in the majority of patients. Monitoring the serum concentrations of potentially toxic drugs when trying to establish dosage regimens is now a well-established component of modern drug therapy.

It is important that the clinical laboratorian appreciate the significance of therapeutic drug monitoring and the mechanisms used by the body to handle a foreign compound, the drug. The mechanisms used by the body to handle a drug are explained in terms of four general processes: absorption, distribution, metabolism, and excretion.

In order to better understand and interpret blood levels and effectively adjust dosage amount and interval, the principles of pharmacokinetics are used to mathematically characterize the disposition of a drug over time. Figure 27-2 illustrates the relationship between time and plasma concentration for a drug taken as a single oral dose.

Although the majority of people dispose of a drug according to well-defined pharmacokinetics, there are individuals with varying characteristics that result in unpredictable drug disposition. Table 27-3 illustrates many of the factors that have been shown to affect drug disposition.

The derivatives of drugs or metabolites formed by the liver are, as a general rule, not active, or their activity is insignificant relative to the parent drug. There are, however, several metabolites of drugs commonly monitored that have comparable therapeutic activity and should be monitored or compensated for when evaluating a drug's therapeutic effects (see Table 27-1).

Specimen collection and transport are important considerations in therapeutic drug monitoring. The proper specimen container must be used to collect or transport the specimen. Most drugs can be monitored adequately in plasma or serum. Timing of the specimen collection relative to the dose administration is also very critical.

Several methods of analysis are used in therapeutic drug monitoring. These include colorimetric methods (used in the early stages of TDM), GC, HPLC, EIA, FPIA, RIA, and nephelometry.

A good quality-control (QC) program is just as important in the TDM laboratory as it is in other areas of laboratory medicine. Today, commercial vendors of QC materials provide a wide variety of drugs in their routine chemistry QC materials. Proficiency-testing programs are also available for TDM as they are in other areas of the laboratory.

Drugs typically analyzed for therapeutic levels may be classified into the following categories: cardioactive drugs, antiepileptic drugs, bronchodilators, antibiotics, psychoactive drugs, immunosuppressants, and antineoplastic drugs. The clinical laboratorian needs to become familiar with the characteristics and methods of analysis for the major drugs in each of these categories.

CASE STUDY
27-1

A markedly obese, female homemaker sustained a myocardial infarction at the age of 47, followed by congestive heart failure. Four years later, she had an episode of ventricular tachycardia resulting in angina. One year after that, she suffered ventricular fibrillation and was successfully resuscitated without event. The records showed she had a normal potassium level at that time, and she was put on quinidine therapy. About 1 year later, at the age of 53, she lost consciousness while riding in a car and was hospitalized and resuscitated from ventricular fibrillation. Her potassium level was 2.2 mEq/L, and her quinidine level was 7.1 mg/L (therapeutic range 2–6 mg/L).

She was transferred to a tertiary-care hospital 7 days later, where her quinidine level was determined as 1.5 mg/L and potassium level as 4.4 mEq/L. She was maintained on her previous dose of quinidine sulfate (300 mg every 8 hours), but repeat measurements of quinidine level were still around 1.5 mg/L. She underwent cardiac catheterization, and quinidine administration was discontinued in preparation for electrophysiologic studies (EPS).* During these studies she developed ventricular tachycardia and seizures. She was cardioverted to normal rhythm and given a loading dose of 600 mg of quinidine IV and a maintenance infusion of 1.5 mg per minute. The study was continued, but again she developed ventricular tachycardia requiring cardioversion. The quinidine level at this point was 2.3 mg/L.

*Electrophysiologic study (EPS) is a procedure used to stimulate the heart into fibrillation under controlled conditions.

Following the study, she was given 300 mg of quinidine sulfate every 6 hours, which maintained blood levels at about 2.5 mg/L.

Two days later, the patient was given a 300-mg loading dose and a 2.4 mg/min infusion to raise quinidine levels to about 5 mg/L in preparation for a second EPS. During the study, quinidine levels ranged from 6.6 to 5.2 mg/L, and these levels were shown to be effective in controlling heartbeat.

The rest of her hospital stay was spent adjusting her oral dose regimen to maintain quinidine levels close to 5 mg/L (see Case Study Table 27-1).

Questions

1. Why were the quinidine levels in this patient fluctuating so dramatically?
2. Is there a direct relationship between potassium and quinidine levels, and, if so, what is it?
3. What is the usefulness of EPS, and how does TDM relate to these studies?

CASE STUDY TABLE 27-1. Laboratory Data

Oral Dose	Blood Level
400 mg quinidine sulfate q6h	2.8 mg/L
600 mg quinidine sulfate q6h	7.0 mg/L
600 mg quinidine sulfate q8h	5.0 mg/L

**CASE STUDY
27-2**

A 37-year-old female with stage IV breast cancer was admitted and treated with high-dose chemotherapy and autologous bone marrow transplantation. Chemotherapeutics were Cytoxan, Cisplatinum (CisPt), and BCNU. CisPt is nephrotoxic and BCNU is hepatotoxic. High doses of chemotherapy typically cause neutropenia. Consequently, aminoglycoside antibiotics were administered empirically. Case Study Table 27-2 summarizes the laboratory data and the aminoglycoside dosages. This patient had occasional febrile episodes and developed renal insufficiency and hepatic insufficiency, probably due to veno-occlusive disease from the alkalating chemotherapeutics. During a previous chemotherapy event that failed, she had developed liver toxicity.

Questions

1. Why were the aminoglycoside doses changed periodically?
2. Why was this patient given an oral dose (PO) of vancomycin on day 12?
3. Why were spot vancomycin levels measured on day 15 and forward?

CASE STUDY TABLE 27-2

Days After Chemotherapy	Amikacin, >5/20–30*	Amikacin Dosing	Vancomycin, 5–10/30–40*	Vancomycin Dosing	Creatinine 0.7–1.4 mg/dL	BUN, 7–21 mg/dL	LDH, 90-190 IU/L
4		240 mg IV q8h			0.8	11	137
5	3.8/20.3				0.8	10	
6	9.7/27.8	270 mg IV q8h		720 mg IV q24h	1.2	19	
7					1.5	30	
8	2.6/17	270 mg IV q12h	4.4/24.7		1.4	30	
9		300 mg IV q12h			1.3	23	
10				1 g IV q24h	1.3	30	
11			7.0/34.7		1.5	33	
12				250 mg PO q6h	1.8	48	181
13	2.6/3.0				1.9	60	346
14	4.7/23.7		5.5/41.8	discontinued	1.7	61	
15		discontinued	19.9		2.6	97	
16			14.2		2.6	143	883
17			7.7	1 g IV	2.9	159	1111
18			21.3		2.7	200	1091
19			14.6		3.2	192	1260
20			9.3	1 g IV			
21			23.5				

*Pre- and postdose therapeutic ranges in µg/mL.

CASE STUDY
27-3

A 60-year-old female suffering from hepatocellular carcinoma received a liver transplant. Her medications included, among other things, the immunosuppressant cyclosporine. After a 2-month postoperative stay in the hospital, she was discharged with trough cyclosporine levels between 100 and 150 ng/mL. She received doxorubicin (an antineoplastic drug) on an outpatient basis, and approximately 1 month later she returned to the hospital with rapidly developing renal failure, hypotension, and tachycardia. Her cardiac output was very weak. She was given dopamine for hemodynamic support, digoxin to improve cardiac output, and was dialyzed. Her cardiac output improved and her renal function slowly returned; however, she developed thrombocytopenia and later hypertension. She slowly improved and was discharged 3 weeks after

admission and after her cyclosporine regimen had been stabilized. Case Study Table 27-3 is a summary of some of her laboratory results and her cyclosporine dosages.

Questions

1. Did the cyclosporine levels cause the renal failure, or did the renal failure cause the elevated cyclosporine levels?
2. Once cyclosporine was resumed on day 13, attempts were made to titer her dosage to stabilize blood levels in the 100 to 150 range again. It appears, however, that changes in the dosages lag behind the blood levels by a day. Why?
3. Explain cyclosporine dosages and timing.

CASE STUDY TABLE 27-3

Day	K, mEq/L	BUN, mg/dL	Creatinine, mg/dL	Cyclosporine (HPLC), ng/mL	Cyclosporine Doses
1	6.9	100	5.3		
2	5.6	81	5.3	436	400 mg
3	4.7	91	5.2	252	
4	3.5	75	4.1	183	200 mg
6	4.7	72	3.3	194	
8		78	2.7	242	
10		58	1.7	176	
12		35	1.3	148	
13		34	1.2	152	200 mg
14		29	1.5	133	180 mg
15	3.2	30	1.3	105	180 mg
16	4.3	28	1.2	95	200 mg
17	3.9	32	1.3	116	250 mg
18	3.5	29	1.4	132	250 mg
20	4.1	31	1.4	104	250 mg
22				121	250 mg

REFERENCES

1. Angel JE. Physician's desk reference. Oradell, NJ: Medical Economics Company (updated yearly).
2. Barre J, Didey F, Delion F, et al. Problems in therapeutic drug monitoring: free drug level monitoring. Ther Drug Monit 1988;10:133.
3. Baselt RC. Disposition of toxic drugs and chemicals in man. 2nd ed. Davis, CA: Biomedical Publications, 1982.
4. Bender AD, Goldsmith BE. Effect of age on drug disposition. In: Moyer TP, Boeckx RL, eds. Applied therapeutic drug monitoring. Vol 1. Fundamentals. Washington: American Association for Clinical Chemistry, 1982.
5. Blanke RV, Decker WJ. Analysis of toxic substances. In: Tietz NW, ed. Textbook of clinical chemistry. Philadelphia, PA: Saunders, 1986;1716.
6. Boyd JR. Drug facts and comparisons. St Louis: Facts and Comparisons (updated monthly).
7. Cotham RH, Shand DG. Spuriously low plasma propranolol concentration resulting from blood collection methods. Clin Pharmacol Ther 1975;18:535.
8. Creasey WA. Drug disposition in humans—The basis of clinical pharmacology. New York: Oxford University Press, 1979.
9. Drayer D. Methods of analysis for antiarrhythmic drugs in serum. In: Moyer TP, Boeckx RL, eds. Applied therapeutic drug monitoring. Vol 1. Fundamentals. Washington: American Association for Clinical Chemistry, 1982.
10. Echizen H, Nakura M, Saotome T, et al. Plasma protein binding of disopyramide in pregnant and postpartum women, and in neonates and their mothers. Br J Clin Pharmacol 1990;29:423.

11. Elson J, Strong JM, Lee WK, Atkinson AJ. Antiarrhythmic potency of *N*-acetylprocainamide. Clin Pharmacol Ther 1975;17:134.
12. Hosephson L, Houle P, Haggerty M. Stability of dilute solutions of gentamicin and tobramycin. Clin Chem 1979;25:298.
13. Ireland G. Drug monitoring data—Pocket guide. Washington: American Association for Clinical Chemistry, 1980.
14. Kaye S. Handbook of emergency toxicology. 3rd ed. Springfield, IL: Charles C Thomas, 1970;405.
15. Knoben JE, Anderson PO. Handbook of clinical drug data. 6th ed. Hamilton, IL: Drug Intelligence Publication, 1988.
16. Meyer BD, et al. Cyclosporine-associated chronic nephrotoxicity. N Engl J Med 1984;311:699.
17. Piafsky KM. Disease-induced changes in the plasma binding of basic drugs. Clin Pharmacokinet 1980;5:246.
18. Pribor HC, Morrell G, Scherr GH. Drug monitoring and pharmacokinetic data. Park Forest South, IL: Pathotox Publishers, 1980.
19. Quattrocchi F, Karnes HT, Robinson JD, Hendeles L. Effect of serum separator blood collection tubes on drug concentrations. Ther Drug Monit 1983;5:359.
20. Ratnaraj N, Goldberg VD, Hjelm M. A micromethod for the estimation of free levels of anticonvulsant drugs in serum. Clin Biochem 1989;22:443.
21. Richey DP, Bender AD. Pharmacokinetic consequences of aging. Annu Rev Pharmacol Toxicol 1977;17:49.
22. Rodvold KA, Zaske DE. Effects of pregnancy on pharmacokinetics: Clinical considerations for patient management. In: Moyer TP, Boeckx RI, eds. Applied therapeutic drug monitoring. Vol 1. Fundamentals. Washington: American Association for Clinical Chemistry, 1982.
23. Shaw LM Chairman. Critical issues in cyclosporine monitoring: Report of the task force on cyclosporine monitoring. Clin Chem 1987;33:1269.
24. Shaw LM. Cyclosporine-based immunosuppression: A look at the past and present and a glimpse into the future. Ther Drug Monit Toxicol In-Service Train Cont Ed 1994; 15(7):157.
25. Wade A. Martindale—The extra pharmacopoeia. 27th ed. London: The Pharmaceutical Press, 1977.
26. White A, Handler P, Smith EL. Principles of biochemistry. 5th ed. New York: McGraw-Hill, 1973;404.

SUGGESTED READINGS*

Evans WE, Schentag TT, Tusko WJ. Applied pharmacokinetics— Principles of therapeutic drug monitoring. 3rd ed. Vancouver, WA: Applied Therapeutics, 1992.

Gilman AG, Rall TW, Nies AS, Taylor P. Goodman and Gilman's the pharmacological basis of therapeutics. 8th ed. New York: Pergamon Press, 1990.

Mungall DR. Applied clinical pharmacokinetics. New York: Raven Press, 1983.

Richens A, Marks V. Therapeutic drug monitoring. Edinburgh: Churchill Livingstone, 1981.

Sadee W, Beelen GCM. Drug level monitoring. New York: Wiley, 1980.

Lin ET, Sadee W. Drug level monitoring. Vol 2. New York: Wiley, 1986.

Taylor WJ, Diers Caviness MH. A textbook for the clinical application of therapeutic drug monitoring. Irving, TX: Abbott Laboratories, 1986.

*The reader is referred to Chapter 5, Analytical Techniques and Instrumentation, for in-depth descriptions of the analytic techniques.

Chapter 28

Toxicology

John A. Bittikofer

Objectives

Upon completion of this chapter, the clinical laboratorian should be able to:

- *Define the term toxicology.*
- *Identify those drugs most often involved in overdose or poisoning situations.*
- *Describe "chain of custody" in relationship to toxicology.*
- *Outline the management of a drug-overdose patient.*
- *Name some of the antidotes used in drug-overdose management.*
- *Explain the role of the clinical laboratory in the evaluation of drug overdose.*
- *Given drug-overdose information for a patient, evaluate the patient's condition.*
- *Describe the analytical techniques used in toxicology, e.g., GLS, TLC immunoassays, etc.*
- *List several examples of the major classes of drugs discussed in the chapter.*
- *Discuss the difference between quantitative and qualitative tests in toxicology.*

KEY WORDS

Analgesic	Forensic
Antidote	Narcotic
Chain-of-Custody	Toxicology
Done Nomogram	

*D*orland's Illustrated Medical Dictionary defines *toxicology* as "the sum of what is known regarding poisons; the scientific study of poisons, their actions, their detection, and the treatment of the conditions produced by them."* This chapter will discuss only one or two aspects of toxicology, specifically, the detection of poisons and, briefly, the actions of poisons, with an occasional mention of the treatment of poisoned patients.

Next, we must establish what a poison is. Referring again to *Dorland's Illustrated Medical Dictionary*, we find that a poison is "any substance which, when ingested, inhaled, or absorbed . . . in relatively small amounts, by its chemical action may cause damage to structure or disturbance of function."* The key phrase is "in relatively small amounts." Almost anything taken in sufficient amounts can be toxic, but for practical purposes, one must limit the discussion to those chemicals that are most potent and highly prevalent or available for ingestion by people.

The patterns of poisonings seen in emergency rooms demonstrate the overwhelming preponderance of drugs as toxins.[1,40] Drug abuse is a common occurrence, and drug overdose is one of the most common modes of suicide attempt.[14,29] Victims of drug overdose range in age from infants to senior citizens, with an increased prevalence in the 20- to 30-year-old age group.[1,40] The drugs most often involved in overdose or poisoning situations are identified in Table 28-1.

Reasons for drug overdosing fall into three categories: (1) suicide gestures, (2) suicide attempts, and (3) recreational

*Dorland's illustrated medical dictionary. 26th ed. Philadelphia: Saunders, 1981. Used with permission.

Michael L. Bishop, Janet L. Duben-Engelkirk, and Edward P. Fody.
CLINICAL CHEMISTRY. © 1996 Lippincott–Raven Publishers.

TABLE 28-1. Substances Most Frequently Involved in Human Exposure and Deaths—1993

	Exposures	% of Total	Deaths	% of Deaths
Cleaning substances	180161	10.3	19	3
Analgesics	167762	9.6	129	20.6
acetaminophen	(70104)	(4.0)	(74)	(11.8)
aspirin	(15557)	(0.9)	(27)	(4.3)
Cosmetics	143861	8.2	1	0.2
Cough and cold preparations	105588	6	5	0.8
Plants	94725	5.4		
Bites/envenomations	72637	4.1	1	0.2
Topicals	64697	3.7		
Pesticides	64298	3.7	7	1.1
Foreign bodies	61640	3.5		
Antimicrobials	60435	3.5	2	0.3
Food products, food poisoning	59997	3.4		
Hydrocarbons	58636	3.3	20	3.2
Sedatives/hypnotics/antipsychotics	54521	3.1	36	5.8
Alcohols	46594	2.7	20	3.2
ethanol	(27203)	(1.6)	(12)	(1.9)
isopropanol	(7774)	(0.4)		
methanol	(813)	(.05)	(8)	(1.3)
Chemicals	44240	2.5	29	4.6
ethylene glycol	(501)	(.02)	(10)	(1.6)
Vitamins	41547	2.4	2	0.3
Antidepressants	40549	2.3	121	19.3
Gases and fumes	32078	1.8	33	5.3
Cardiovascular drugs	25999	1.4	57	9.1
Stimulants and street drugs	23305	1.3	65	10.4
Asthma therapies	16699	0.9	25	4
Total	1749136		626	

Categories of poisons are left out between Antidepressants and Asthma therapies and after Asthma therapies, since they are not drugs or poisons easily monitored with laboratory techniques. All poisons are included in this table between Cleaning substances and Antidepressants to illustrate the prevalence of drugs in overall poisonings. (Adapted from Litovitz TL, Clark LR, Soloway RA. 1993 Annual report of the American association of poison control centers toxic exposure surveillance system. Am J Emerg Med 1993;12:546.)

drug abuse. People who make a suicide gesture typically provide an accurate history if they are not comatose. On the other hand, the other two groups would incriminate themselves by providing accurate information, and therefore, the information they may provide must be regarded as highly suspect.[25] In all three situations, more information is helpful, and occasionally, the laboratory can help provide this corroborative data. Diagnosing a drug overdose must be done using all information, including the history, physical examination, and laboratory data. Diagnosing drug overdose based only on laboratory findings is extremely risky. With so many drugs on the market, it is mandatory that the analytical toxicologist use screening methods and concentrate on the most prevalent drugs. Consequently, the laboratory may not include in its armamentarium the drug(s) in a particular overdose case or may be capable of providing only a qualitative answer when a quantitative result is necessary for proper patient management. Thus, the clinician must understand what the laboratory can do and what it cannot do.

Occasionally, the toxin in a poisoning case is not a drug but perhaps a metal such as lead or mercury. Various alcohols in addition to ethanol, such as methanol, isopropanol, and ethylene glycol, may be the poison, or perhaps a gas such as carbon monoxide, cyanide, or halogens may be implicated. The mode of exposure may not always be intentional ingestion but rather industrial exposure or passive exposure in the case of homicides.

The analytical toxicologist must design the toxicology laboratory according to the type of facility to be serviced and the nature of the poisoning cases. The characteristics of the laboratory, such as hours of operation, turnaround time, quantitative versus qualitative measurements, and legal responsibilities, must be well defined.

A clinical toxicology laboratory that services an emergency room will be asked to aid in overdose situations. A 1- to 2-hour turnaround time is required for results to be effective as diagnostic aids. The service may need to be available 24 hours a day. Screening tests are appropriate as a first phase of testing, since a wide variety of poisons may be involved. To maintain short turnaround times, qualitative analyses may have to suffice unless sufficient clinical information is available to narrow the poison to a single compound, in which case a quantitative level may be requested and provided quickly.

The laboratory involved in drug-abuse screening must rely on screening tests as a first phase, but all positive results must be confirmed by an alternative method. When a laboratory's findings may prevent an employee from retaining his or her position, positive drug findings must be accurate. This type of service need not be available around the clock, and result turnaround time is not critical.

A forensic laboratory, perhaps one associated with a coroner's office, also must maintain maximal accuracy and specificity and must pay special attention to documentation of all procedures. The laboratory must document that the specimen analyzed was the specimen collected. This is done by properly completing a chain-of-custody document, which records each step and each person who handled the sample from time of collection to time of analysis. Such a laboratory need not be open 24 hours a day, and turnaround time of data is not critical. Quantitative measurements often are as important as qualitative determinations, and accuracy is of utmost importance.

Occasionally, the clinical laboratory is called on to aid in identifying or correcting a hazardous situation in the workplace that is due to toxic chemicals. Although this is an area of industrial hygiene, the clinical laboratory can assist with its expertise in testing biologic samples. Quantitative analyses typically are called for, but quick turnaround time and availability around the clock are not necessary.

Although the clinical toxicology laboratory may cross all these areas, the discussion in this chapter will concentrate primarily on the laboratory's interface with the emergency room and the testing of biologic specimens for the most prevalent toxins. Many of the following analytic techniques and procedures are applicable to all phases of laboratory toxicology, including drug overdose, industrial hygiene, and forensic toxicology.

CLINICAL MANAGEMENT OF THE DRUG-OVERDOSE PATIENT

Treatment of the drug-overdose patient first focuses on stabilizing the patient by establishing or maintaining an open airway, proper breathing, adequate circulation, and correcting any life-threatening compromise of the central nervous system (CNS). Further treatment involves preventing the absorption of the drug from the gastrointestinal (GI) tract, hastening its elimination from the body, applying appropriate antidotes, if available, and continuing care and disposition. Few specific antidotes are available but those that do exist are very helpful.

Naloxone (Narcan) is a safe, specific opiate antagonist that acts in 1 to 3 minutes to reverse toxicity. Since the clinical response to naloxone is a good estimate of opiate toxicity, the need for laboratory tests for opiates is confirmatory.

N-acetylcysteine is an antidote to acetaminophen poisoning. However, to be effective, it must be administered as soon after acetaminophen ingestion as possible. Therefore, the laboratory must be prepared to quantitatively measure serum acetaminophen levels at all times of the day or night on a rapid turnaround basis if the clinician is waiting to administer N-acetylcysteine.

Naloxone and N-acetylcysteine often are administered empirically, because they have few side effects and because rapid treatment is essential. However, in acetaminophen poisoning, unlike opiate poisoning, the results of treatment are not immediately apparent. Therefore, even if a physician has treated a patient with N-acetylcysteine for acetaminophen poisoning, the physician may still request a *STAT* acetaminophen determination because hospital admission and observation for hepatic toxicity will be mandatory for the patient who has ingested a toxic amount of the drug.

Organophosphate and carbamate insecticides are other toxins for which specific antidotes (atropine and paralidoxime) are available. Confirmation of poisoning by demonstration of decreased serum pseudocholinesterase levels is desirable before these agents are administered.

Heavy metal poisoning may be specifically treated with chelating agents. This treatment, however, carries some risk of nephrotoxicity, so confirmation of metal poisoning by laboratory means is highly desirable prior to therapy.

With these exceptions, specific antidotal therapy for poisoning is not available. For most poisons, ipecac or gastric lavage is employed to empty the stomach. Activated charcoal is then administered to absorb drugs from the GI tract. In some cases, such as mercury, iodine, iron, strychnine, nicotine, or quinine poisoning, a neutralizing agent is preferable to activated charcoal. In some cases, a cathartic is also employed.

Toxic affects of methanol, isopropanol, and ethylene glycol (antifreeze) poisoning may be minimized by administering ethanol—usually intravenously (IV). This therapy is appropriate if administered early enough after ingestion because alcohol dehydrogenase (ADH) is responsible for metabolizing, *in vivo*, small-molecular-weight alcohols but has an overwhelming preference for ethanol. The toxic effects of methanol, isopropanol, and ethylene glycol are caused primarily by their metabolites. By saturating ADH with ethanol, before a toxic amount of the ingested alcohol is metabolized, the ingested alcohol can be excreted with little or no toxic effect.

The role of the laboratory in the evaluation of drug overdose is still important even if the drug detected has no specific antidote. For example, coma in an alcoholic may be due to drugs (*e.g.,* alcohol, barbiturates), metabolic conditions (*e.g.,* hypoglycemia), trauma (*e.g.,* subdural hematoma), or some condition totally unrelated to the patient's alcoholism (*e.g.,* myocardial infarction). The laboratory may assist with other tests besides a drug screen for this type of patient and should be prepared to supply the clinician with the necessary information.

ETHANOL

Alcoholism is a serious illness, and estimates are that 9.3 to 10 million adult Americans (7% of the adult population) suffer from alcoholism.[26] The disease crosses all age, sex, racial, and economic categories. Ethanol is one of the most easily obtained toxins, and yet is one of the most lethal in an indirect way. More than 50% of highway fatalities in the United States are alcohol-related. Most state laws have well-defined criteria for determining when a person is too intoxicated to operate a motor vehicle, and much of the evaluation centers on the driver's blood alcohol level. In many states, a person with a blood alcohol level of 0.080 g/dL or greater is defined by law as too intoxicated to drive. The most recent rules from the Department of Transportation, which went into effect January 1, 1995, prohibit performance of safety-sensitive functions while having a blood alcohol concentration of 0.04 g/dL or greater. Of course, many drunk-driving cases are brought to court, and the integrity of the blood analysis may be challenged. Therefore, the laboratory making these measurements must maintain strict proficiency. Furthermore, the laboratory must provide chain-of-custody documentation for the specimens. In some states, any laboratory doing this type of testing must be licensed.

A commonly used chemical method of quantitating ethanol uses the enzyme alcohol dehydrogenase (ADH). This enzyme catalyzes the reaction of ethanol with NAD^+ to form acetaldehyde and NADH.

$$CH_3CH_2OH + NAD^+ \xrightarrow{ADH} CH_3CHO + NADH + H^+$$

(Eq. 28-1)

This reaction can be used to quantitatively measure ethanol by spectrophotometrically measuring at 340 nm the amount of NADH produced. This reaction is not entirely specific for ethanol. Table 28-2 illustrates the relative rates of dehydration for various alcohols catalyzed by ADH.

Gas chromatography (GC) has been appropriately applied to the measurement of ethanol and several other small-molecular-weight alcohols.[23] Figure 28-1 is a representative chromatogram of several volatile compounds, and it illustrates the specificity and sensitivity of this method. By measuring the retention times and comparing them with stan-

TABLE 28-2. Relative Rates of Dehydration of Various Alcohols Using Alcohol Dehydrogenase	
Alcohol	**Relative Rate of Dehydration**
Ethanol	100
1-Propanol	36
Butanol	17
Ethylene glycol	9
2-Propanol	6
Methanol	<1

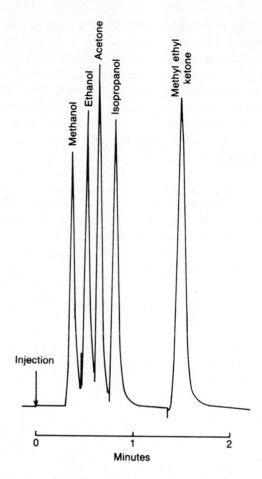

▲

Figure 28-1. A relative chromatogram of several volatile compounds. The concentration of each component is 100 mg/dL. The six-foot column is packed with 60/80 Carbopack™ C/0.2% Carbowax ®1500. Methylethyl ketone is the internal standard.

dards, the analyst can identify each volatile compound. Peak height or peak area measurements of unknowns compared with standards provide for an accurate quantitative measurement of each compound. Although, theoretically, interferences exist for any or all of these compounds, GC is considered the most specific, appropriate method for the analysis of any or all of these compounds in blood. GC is used primarily in forenic analyses.

Ethylene glycol is the predominant ingredient in antifreeze and is occasionally drunk by alcoholics as a substitute for ethanol. Being an alcohol, it satisfies the craving for alcohol, but its aldehyde metabolite is highly toxic. Ethylene glycol has a sweet taste, and given the opportunity, young children are easily tempted to drink it. A convenient method of detecting and quantitating ethylene glycol is the use of GC, although there are no exceptionally good columns yet developed.[27,28,41]

DRUGS

Those compounds included in the very broad class of poisons called *drugs* include the entire gamut of pharmaceutic preparations and certain other materials that may be made at home, such as phencyclidine (PCP) and lysergic acid (LSD). Although there are hundreds of drugs and drug combinations, certain groups of drugs and specific drugs are found more often than others in overdose situations. Table 28-1 illustrates the drugs most often implicated in suspected ingestions. Space does not allow discussion of each drug; however, certain compounds deserve individualized attention.

Analgesics

Two of the most commonly used drugs, salicylate and acetaminophen, are also most often taken in toxic amounts (Table 28-1). The laboratory is an integral part of the diagnostic and therapeutic system because quantitative measurements of serum levels of both drugs are immediately useful in treating the patient.

Salicylate

Acute salicylate intoxication results in stimulation of the respiratory system, which causes an initial respiratory alkalosis. As time goes on, the acidic nature of salicylate causes gastric irritation and contributes to impending metabolic acidosis. Salicylate inhibits enzymes in the Krebs cycle, resulting in the conversion of pyruvate to lactic acid. In addition to inhibiting oxidative phosphorylation, salicylate stimulates the breakdown of fatty acids, which results in the production of ketoacids. All these factors contribute to metabolic acidosis.

The severity of salicylate toxicity can be accessed by measuring the blood level of salicylate and referring to a Done nomogram (Fig. 28-2). The time since ingestion must be known to interpret the nomogram; however, this time period is not always known. Furthermore, it is mandatory that the specimen be drawn at least 6 hours after ingestion to ensure that absorption and distribution are complete.

Since salicylate is an acid, it is possible to increase the renal clearance of salicylate by rendering the urine alkaline. This treatment, which tends to extract acid, is referred to as *forced alkaline diuresis*. Whereas it is effective in reducing salicylate burden, it has not been shown to decrease morbidity and mortality.

SALICYLATE DETERMINATION BY THE TRINDER METHOD[38]

Principle—Salicylate has traditionally been measured by the Trinder method.[38] An aliquot of serum is combined with an acidic ferric chloride reagent. Salicylate chelates with ferric iron to form a blue complex, which can be measured colorimetrically.

Reagents
1. Color reagent—With the aid of heat, dissolve 40 g of mercuric chloride in 850 mL of deionized water. Cool the solution and add 130 mL of concentrated HCl and 40 g of ferric nitrate [$Fe(NO)_3 \cdot 9 H_2O$]. When everything has dissolved, adjust the volume to 1.0 L. This solution is stable indefinitely.
2. Stock salicylate standard—Dissolve 580 mg of sodium salicylate in deionized water in a 250-mL volumetric flask. Add a few drops of $CHCL_3$ as preservative and add deionized water to the mark. The salicylate concentration is 200 mg/L.
3. Salicylate standard—Dilute 20 mL of stock solution to 100 mL in a volumetric flask with deionized water. The salicylate concentration is 40 mg/100 mL. This solution is stable for 6 months if refrigerated.

Procedure
1. Add 1.0 mL of serum and 5.0 mL of color reagent to a test tube and mix well by shaking. Set up a tube using deionized water and color reagent for a "blank." Set up a tube using standard and color reagent for a calibration curve.
2. Centrifuge the serum tube at $2000\times$ for 2 minutes and transfer the supernatant to a cuvet.
3. Transfer the "blank" solution and the standard solution to cuvets and read the absorbance of the solutions at 540 nm. The purple color is stable for at least 60 minutes.
4. Subtract the "blank" from the serum absorbance and the standard absorbance.
5. Compare the serum absorbance to the standard absorbance and calculate salicylate concentration in the unknown:

$$C_{unk} = \frac{A_{unk} \times C_{std}}{A_{std}} \qquad \textit{(Eq. 28-2)}$$

Comments—A calibration curve may be plotted by making a series of salicylate standards. The relationship between absorbance and salicylate is linear up to 50 mg/mL on most spectrophotometers.

Urine and spinal fluid may be analyzed similarly. It may be necessary to dilute the urine to obtain concentrations within the linear region of the calibration curve. This procedure provides a quick answer and is sufficiently accurate for emergency purposes. The analyst must be aware of the many interferences that can cause background readings equivalent to 0 to 5 mg of salicylate/dL. A more specific method incorporates the extraction of salicylic acid from acidified serum into ethylene chloride and then back into acidic ferric chloride to form the blue ferric–phenolic complex.[34] High-performance liquid chromatograph (HPLC) has been used

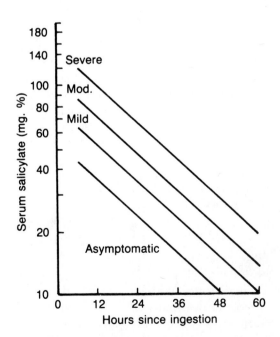

Figure 28-2. Nomogram relating serum salicylate. Concentration and expected severity of intoxication at varying intervals following the ingestion of a single dose of salicylate. (Done AK. Salicylate intoxication. Pediatrics 1960;26:800.)

to measure salicylate in serum, and immunoassays for salicylate are available on several automated analyzers. ∎

Acetaminophen

When taken as directed, acetaminophen presents no risk of toxicity, because the toxic levels are far removed from therapeutic concentrations. Since acetaminophen is incorporated in more than 250 pharmaceutic preparations, most of which can be obtained without prescription, acetaminophen overdose is a common occurrence. The toxic effect of acetaminophen is delayed hepatocystic necrosis, which develops some 3 or 4 days after an overdose. The toxic agent in acetaminophen overdose is a metabolite and not the parent compound. The free radical metabolite acetamidoquinone is conjugated with glutathione and further converted to cysteine or mercapturic acid conjugate before finally being excreted. With an overwhelming amount of acetaminophen present, the liver's ability to conjugate acetamidoquinone becomes saturated. Acetamidoquinone accumulates, covalently binds to tissue cells, and causes cell death.

There is an effective antidote in *N*-acetylcysteine, which, if administered shortly after ingestion, will prevent hepatotoxicity. Speculation is that the sulfhydryl groups of *N*-acetylcysteine (Mucomyst) act as glutathione substitutes and bind to the toxic free radical metabolite, rendering it nontoxic.

The decision to administer the antidote is usually based on the determination of plasma acetaminophen levels. Nomograms are available (Fig. 28-3) to aid in interpreting acetaminophen plasma levels and predicting possible hepatotoxicity. One must know the time since ingestion in order to read the nomogram. Often this datum is not known accurately, and an alternative method of predicting toxicity is applied. Although the elimination kinetics of acetaminophen are not well characterized, it is possible to use a measured half-life to predict impending toxicity. The half-life of acetaminophen changes with the amount present. But by measuring $t^{1/2}$ experimentally, it has been determined that if $t^{1/2}$ is greater than 4 hours, there is 99% chance that significant hepatotoxicity will occur.[32] A suggested mechanism for measuring $t^{1/2}$ is to draw two specimens, 4 hours apart. If the second specimen concentration is less than half that of the first specimen level, $t^{1/2}$ is less than 4 hours, and vice versa. However, the first specimen must not be collected until absorption and distribution of the orally ingested medication is complete (about 4 h). Those patients who are toxic enough to require the antidote should be followed closely by monitoring hepatic transaminases, alkaline phosphatase, serum bilirubin, and prothrombin time.

Acetaminophen Assays

Several analytic techniques have been applied to the assay of acetaminophen in serum. A colorimetric method uses a trichloroacetic acid precipitation of serum proteins followed by reacting the acetaminophen with sulfamic acid to form a yellow-colored derivative, which when measured at 430 nm is proportional to the acetaminophen concentration.[19]

HPLC has been applied to the measurement of acetaminophen. A typical procedure includes the addition of an internal standard, acetoacetanilide, to an aliquot of serum. Ethyl acetate is used to extract the acetaminophen and internal standard from the protein matrix, and the extract is separated on a reverse-phase column. Acetaminophen is quantitated using a UV absorbance detector.[20]

Several immunoassay methods using enzyme immunoassay and fluorescence polarization immunoassay are available on automated analyzers. These assays are specific, take little time, and require no sample-preparation procedures. Consequently, they have become very popular.

Reference Ranges

Acetaminophen

Therapeutic	10–25 mg/mL
Toxic	100–250 mg/mL

The Drug Screen

Controversy continues over whether the drug screen is necessary or not in managing the drug overdose patient. Many argue that management of the patient often occurs quickly and before the drug screen results are available and that management of the patient is seldom altered by the results of the drug screen. In one study, 209 cases of intentional drug overdose were evaluated.[5] In 47% of the cases, the results of the drug screen agreed exactly with the clinical impression. Unexpected drug screen results altered patient management in only three cases but these changes had no major impact on outcome. Yet, many emergency department physicians

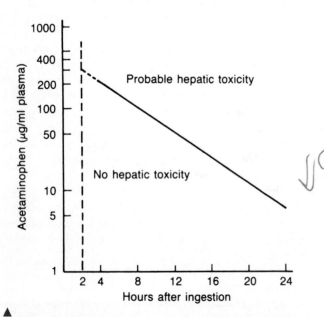

Figure 28-3. Semilogarithmic plot of plasma acetaminophen levels versus time. (Rumack BH, Matthew H. Acetaminophen poisoning and toxicity. Pediatrics 1975;55:871.)

demand the availability of drug screening around the clock and when available, use it extensively.

The analyst has a tremendous challenge in trying to identify the drug or drugs present in blood, urine, or gastric contents of a toxic patient when there are so many possibilities. A screening mechanism would be helpful, and several have been devised, although none will detect everything. Some procedures are more specific than others, and sensitivity will vary from procedure to procedure. Reliability of the results is important and usually improves with the cost of the analytic technique applied.

Specimen selection is important. When possible, blood, urine, and gastric contents should be collected. In addition, if any empty medication containers are found near the patient, they should be brought to the emergency room and sent on to the laboratory. Acidic drugs are best found in detectable concentrations in the blood. Neutral and basic drugs are best detected in urine, as are metabolites. In either situation, it is helpful to know when the specimens were obtained relative to when the suspected ingestion occurred. This information is necessary to evaluate whether the drug(s) has had time to be absorbed and distributed throughout the vasculature and excreted. If the patient reaches the ER shortly after ingesting a drug, the gastric lavage is an appropriate sample. When analyzing lavage fluid, the analyst must look chemically for the parent drug but also may identify the drug by the color and form of the yet undissolved pills.

Chromatography

Thin-layer chromatography (TLC) will detect a very wide variety of drugs but suffers from an inability to separate closely related compounds. TLC plates may be very difficult to interpret and require exceptional expertise on the part of the analyst. Several procedures have been developed with specific application to the clinical toxicology laboratory.[3,9,10,18] There is at least one commercial TLC kit available that is making a major contribution to drug screening (Analytical Systems, Kansas City, MO 64114). Several broad-spectrum stains have been developed that provide the capacity to detect a tremendous number of drugs. Using two or three stains in combination with, perhaps, long-wavelength fluorescence, the experienced analyst can often accurately identify the drug(s) present. Sensitivity is adequate for most drugs, especially when they are taken in toxic amounts, and in some cases, therapeutic levels can be detected. Because of the many drugs reacting with the limited number of stains and because of the limitations of separability on the plate, drugs may overlap, making identification impossible. The analyst may then use variations of the TLC procedure to better separate the drugs. This might be done by changing the pH of the developing solvent, which would further separate a portion of the original chromatogram, or by using alternate stains, which could differentiate two or more overlapping drugs. TLC is able to detect many acidic and basic drugs. However, it does not reliably detect the water-soluble glucuronide metabolites of many drugs (*e.g.,* opiates). For these drugs, sample pretreat-

ment with glucuronidase or hydrolyzing methods must be employed.

Because of TLC's sensitivity to so many drugs, a negative result is a very meaningful piece of information for the physician treating the patient. If TLC is being used to screen for drugs of abuse, it is mandatory to use alternate techniques to confirm those drugs detected by TLC. This is necessary because of the appreciable number of uninterpretable and false-positive chromatograms.

GC has been applied to drug screening, and several protocols have been devised.[2,13,15,30] GC provides for better identification of drugs than does TLC and, for some drugs, better sensitivity. However, GC is more involved procedurally. A particular GC column will separate only a limited number of compounds, and it is necessary to operate two or more columns to provide a broad-spectrum screen. This usually requires more than one instrument, since switching columns and changing temperatures are disruptive to a column. GC detectors are quantitative, allowing for both qualitative and quantitative information in one run.

GC provides an excellent confirmatory technique for drugs detected by TLC. The analyst should have a good idea what drug is present on a TLC separation and can optimize the GC parameters for maximal efficiency.

Gas chromatography/mass spectroscopy (GC/MS) is the method of choice for confirming drugs detected by screening techniques. GC provides the separation of drugs, while MS provides further specificity and extreme sensitivity, thus making GC/MS the most powerful technique available for measuring drugs. The equipment is expensive and requires a knowledgeable operator. Consequently, it is typically not used as a screening tool.

Colorimetric Screening Tests

Colorimetric procedures, or *spot tests*, as they are often called, have been available for quite some time, and not long ago, they represented the largest part of the analyst's armamentarium. Although the chromatographic techniques are more commonly used today, the toxicology laboratory may still find these colorimetric tests helpful. For a few drugs, a colorimetric test may be an excellent confirmatory test or may be a satisfactory quantitative procedure.[7,16,17] Colorimetric screening tests are also available.[33] As a rule, colorimetric tests are quickly and easily performed, and instrumentation is available in most analytical laboratories. Colorimetric tests are not highly specific when compared with chromatographic and immunoassay techniques. Sensitivity is adequate for most applications. Drugs for which colorimetry is commonly used include salicylate, acetaminophen, and phenothiazines, among others.

Immunoassays

By design, immunoassays were developed to provide maximal specificity for a given analyte. However, with the production of antibodies with less specificity, designed to respond to a class of drugs, the immunoassays are the method of choice for much of the drug-overdose testing. For example, EMIT kits (Syva Company), which use an enzyme immunoassay technique, are available for acetaminophen, amphetamines, barbiturates, benzodiazepines, cannabinoids, cocaine metabolite, methadone, methaqualone, opiates, phencyclidine, propoxyphene, and tricyclic antidepressants. Abuscreen RIA kits (Roche Diagnostics) are available for amphetamine, barbiturates, cocaine metabolite (benzoylecgonine), methaqualone, morphine, and phencyclidine (PCP).

Immunoassays have more than adequate sensitivity for drug-screening applications, are easily automated, and provide results in a matter of minutes. Disadvantages include the expense of reagents, the need for several different antibodies to detect an array of potential drugs, and the inability to detect drugs for which no antibodies are available. Nevertheless, more and more laboratories are using a battery of immunoassays for initial screening.

Drug testing capability has moved out of the laboratory and to the patient. Point of care drug screen testing is now possible by using handheld devices that incorporate immunoassay principles and reagents, require no instrumentation, and include a wide range of drugs. Examples are the OnTrak® (Roche Diagnostic Systems) and Triage® (Biosite Diagnostics) systems which can detect PCP, benzodiazepines, cocaine, amphetamines, cannabinoids, opiates, and barbiturates.

Barbiturates

Barbiturates are only occasionally found in overdose cases, although, in years past, they were one of the most frequently encountered toxic drugs. Although specific antidotes are not available, laboratory identification of the barbiturates present can be useful in managing the patient.

Barbiturates are typically divided into three groups: short-acting (pentobarbital, secobarbital), intermediate-acting (amobarbital, botabarbital, butalbital), and long-acting (phenobarbital, barbital). The toxic effects are almost all due to CNS depression, with respiratory depression and cardiac insufficiency being the usual cause of death in severe overdose.

Treatment of barbiturate overdose is primarily to attend to the "ABCs"—airway, breathing, and circulation—by establishing an open airway, supporting ventilation, and assessing cardiac output.

A wide variety of analytic methods have been applied to measuring barbiturates in serum and urine: thin-layer, gas, and liquid chromatography; ultraviolet (UV) absorption spectroscopy; and various immunoassays. One of the early methods using UV absorption spectroscopy took advantage of the change in spectrum of the ionized compound compared with that of the hydrogenated compound to minimize the effects of interference that also absorbs in the UV region of the spectrum.[37] More recently, the various chromatographic techniques have been used to further increase specificity. With the advent of the immunoassays and their ap-

plication to barbiturates, many laboratories are changing techniques because of ease of operation, quick turnaround time, specificity, and low cost.

Narcotics

Narcotics, or the opioid group, encompass not only heroin, morphine, and codeine, but also several synthetic compounds such as meperidine, methadone, propoxyphene, pentazocine, and others. All have some potential for addiction. Symptoms of overdose include the classic triad of coma, respiratory depression, and miosis (pinpoint pupils). The CNS depression causes coma, whereas the respiratory suppression results from decreased sensitivity to carbon dioxide, resulting in slow, shallow breathing.

Narcotics are one of the few groups of compounds for which there is an effective antidote. Naloxone (Narcan) is a narcotic antagonist that can quickly reverse CNS depression. This drug has no significant side effects and thus is used both diagnostically and therapeutically. Following intravenous injection, naloxone will produce a marked improvement in level of consciousness, respiratory rate, and blood pressure within 1 or 2 minutes. The effect lasts for 2 or 3 hours, but naloxone may have to be readministered to counteract long-lasting narcotics, such as methadone.

Because of the existence of an effective antidote, laboratory support for narcotic overdoses is not essential. However, the laboratory is often called upon to assess possible abuse, and specific procedures are required for this. EMIT is quite applicable for screening purposes.

Other Drugs

There is a plethora of other drugs available for which there are no specific treatments other than supportive therapy. Of course, new medications are continually becoming available and eventually show up in drug screening.

Several compounds of note that are not manufactured commercially but are common "street" drugs are the hallucinogens phencyclidine (PCP, "angel dust," etc.) and lysergic acid (LSD). Their prevalence is due to the ease with which they are manufactured in the home. Cannabinoids (marijuana) are very common but of little concern to the ER clinician because of their mild effects.

PESTICIDE POISONING

Pesticides are somewhat unique poisons because they are deliberately added to the environment. They are compounds designed to kill specific organisms and cells, some of which are similar to human tissue. Although the long-term effects to humans of low-level exposure are not well characterized, it is known what happens physiologically in acute exposures. The most common victims of acute poisonings are people who are applying pesticides and do not take proper precautions. The mortality rate attributed to pesticide poisonings has been estimated at 1 in 1.54 million in the United States. There are 100 nonfatal poisonings for every fatal one.[22]

Pesticides come in many forms, ranging from heavy metal salts and organic complexes to large organic molecules, including organophosphates, carbamates, and halogenated hydrocarbons. Organophosphates represent the largest single group of pesticides used and cause approximately one-third of pesticide poisonings.

Organophosphates and carbamates inhibit the action of acetylcholinesterases, which are responsible for cleansing nerve endings of the impulse transmitter acetylcholine. Once acetylcholine has transmitted a nerve impulse, it must be cleared from the nerve ending before another impulse can be received. The accumulation of acetylcholine results in stimulation of smooth-muscle receptors in the heart and exocrine glands, which, in turn, results in tightness in the chest, wheezing, increased bronchial secretions, increased salivation and lacrimation, sweating, and GI disruptions manifested by vomiting, cramps, and diarrhea. Bradycardia, involuntary urination and defecation, and miosis also result. Excess acetylcholine at motor nerve ends leading to skeletal muscles causes weakness, involuntary twitching, cramps, fasciculation, and weakness of respiratory muscles. CNS effects include tension, anxiety, restlessness, insomnia, headache, neurosis, confusion, nightmares, apathy, slurred speech, tremors, and so on. Death results from respiratory failure.

The laboratory may help identify individuals who have become toxic with organophosphates. The method of screening is to measure serum pseudocholinesterase activity, which will be depressed in the presence of organophosphates. This test is truly a screen and cannot identify the poison, but if available can be helpful since specific antidotes (atropine and pralidoxime) for these poisons are available. (Individual pesticide testing is well developed but not warranted in a clinical toxicology laboratory because of the infrequency of pesticide poisonings seen in the average emergency room and the expense of such testing.)

There are two major categories of cholinesterases. Acetylcholinesterases are commonly referred to as *true* or *specific cholinesterases*. There are two isoenzymes, and they are found predominately in red blood cells and tissues. Eleven isoenzymes found in serum are collectively called *pseudocholinesterases* because of their relative nonspecificity and widely differing characteristics.

PSEUDOCHOLINESTERASE DETERMINATION METHOD[11]

Principle—Several analytical techniques have been applied to measuring pseudocholinesterase, including manometry, electrometric titration, and colorimetry. The following procedure is photometric and uses acetylthiocholine as a substrate. Acetylthiocholine is converted to thiocholine and acetate. Thiocholine then reacts with dithiobisnitrobenzoic acid to form the yellow-colored 2-nitro-5-mercaptobenzoate.

$$\text{Acetylthiocholine} \rightarrow \text{thiocholine} + \text{acetate}$$

$$\text{Thiocholine} + \text{dithiobisnitrobenzoic acid} \rightarrow \text{thionitrobenzoic acid}$$

Reagents—Cholinesterase reagent set (Boehringer-Mannheim Diagnostics, 7800 Westpark, Houston, TX 77063).

1. Buffer/chromogen reagent—Dissolve the contents of 1 vial of buffer/chromogen in 100 mL of distilled or deionized water. This solution is stable for 6 weeks at 2 to 8°C.
2. Substrate reagent—Dissolve the contents of 1 vial of substrate with 3.0 mL of distilled or deionized water. This solution is stable for 6 weeks at 2 to 8°C.

Procedure

1. Add to a test tube 3.0 mL of buffer/chromogen reagent and 0.10 mL of substrate reagent and mix gently.
2. Bring the solution to the reaction temperature of 25°C.
3. Add 0.02 mL of serum or plasma, mix gently by inversion, and immediately transfer the reaction mixture to a thermostat cuvet.
4. Read the absorbance at 405 nm and record the change per minute (ΔA/min).

Calculations—The enzyme activity is calculated using the following equation: mU/mL specimen

$$= \frac{\Delta A/\text{min} \times \text{total assay volume (mL)} \times 1000}{\text{abs coeff} \times \text{light path (cm)} \times \text{spec vol (mL)}}$$

$$= \frac{\Delta A/\text{min} \times 3.12 \times 1000}{13.3 \times 1\ \text{cm} \times 0.02\ \text{mL}}$$

$$= \Delta A/\text{min} \times 11,700 \qquad \textit{(Eq. 28-3)}$$

where 1000 = factor for conversion of units/mL to mU/mL
abs coeff = absorbancy coefficient = 13.3 cm²/μmol thionitrobenzoic acid at 405 nm

Comments—The reaction may be used at temperatures between 25°C and 37°C. If temperatures >25°C are used, a table of conversion factors must be used to convert enzyme activities to 25°C, or new reference ranges must be determined.

This procedure and substrate may be used to measure "true" acetylcholinesterase in red cells. ■

Reference Ranges
Pseudocholinesterase

Serum   1900 to 3800 mU/mL
Plasma   1000 to 3500 mU/mL at 25°C

This reference range, based on a population of individuals, is quite wide. A given individual may have a normal pseudocholinesterase level at the high end of this range. After exposure to an organophosphate or carbamate, the enzyme activity may fall to the low end of the reference range. To the clinician, having only the population reference range as a guide, such a level would appear normal, and a toxic condition might be overlooked.

CARBON MONOXIDE

Carbon monoxide is an odorless, colorless, tasteless gas with a 210 times greater affinity for hemoglobin than that of oxygen. In carbon monoxide poisoning, as greater amounts of hemoglobin are converted to carboxyhemoglobin, the ability of the hemoglobin pool to exchange oxygen is diminished. With each exposure to CO, suffocation and, ultimately, death occurs. Table 28-3 illustrates the symptoms of carbon monoxide poisoning at varying concentrations of carboxyhemoglobin and gives examples of the types of exposures necessary to create a given carboxyhemoglobin concentration.

The treatment for carbon monoxide poisoning is to remove the source of carbon monoxide or remove the victim from the source. Three hours of breathing normal air or 30 minutes of breathing oxygen reduces carboxyhemoglobin concentrations by half. There are no lasting effects unless brain damage has occurred during exposure.

Blood containing sufficient amounts of carboxyhemoglobin will take on a cherry-red appearance. This is the basis for a spot test to detect carbon monoxide poisoning. By adding 5 mL of 40% NaOH to 5 mL of a 1:20 dilution of blood, one can observe a persistent pink color if carboxyhemoglobin is present at a level of 20% or higher.

Quantitative determinations of carbon monoxide concentrations are better made using spectrophotometry or gas chromatography.[8] Spectrophotometric measurements are based on the observation that the absorbance spectrum of oxyhemoglobin and carboxyhemoglobin are different. An aliquot of blood is added to a hemolyzing agent to release all the heme pigments. Sodium hydrosulfite is added, which reduces oxyhemoglobin and methemoglobin but not carboxyhemoglobin. The resulting mixture contains only two heme pigments—carboxyhemoglobin and deoxygenated hemoglobin. The absorbances at the carboxyhemoglobin maximum, 541 nm, and the deoxygenated hemoglobin maximum, 555 nm, are measured. The ratio of the two absorbance readings is proportional to the %CO saturation. To establish a calibra-

tion curve, a blood sample must be saturated with CO. This is a hazardous procedure and must be done under a hood. A second sample is saturated with oxygen, and a series of solutions are prepared by mixing proportional amounts of the CO- and oxygen-saturated blood samples.

A more practical approach is to use a cooximeter, which is used in combination with many blood gas analyzers for measuring oxygen saturation.[36] The cooximeter is capable of measuring carboxyhemoglobin, oxyhemoglobin, deoxyhemoglobin, and methemoglobin simultaneously. Each hemoglobin species has a unique spectrum. After a hemolyzing agent is added to an aliquot of whole blood to release all the hemoglobin, the cooximeter automatically measures the absorbance of the solution at four wavelengths: 535.0, 585.2, 594.5, and 626.6 nm. The amount of carboxyhemoglobin, oxyhemoglobin, methemoglobin, and deoxyhemoglobin is calculated electronically by applying Beer's Law at each wavelength and solving four simultaneous equations using absorptivity coefficients stored in memory.

CARBON MONOXIDE DETERMINATION[4]

Venous blood should be collected in EDTA or heparin.

Reagents

1. Buffer—KH_2PO_4/K_2HPO_4, 0.1 mol/L, pH 6.85.
2. Hemolyzing solution—Prepare a 1:10 dilution of the buffer weekly.
3. CO-Hb diluting solution—Add 25 mg of sodium hydrosulfite to 20 mL of buffer. Prepare just before use.

Procedure

1. Add 25 μL of blood to 3 mL of hemolyzing solution and mix.
2. After 5 minutes, pipet 0.1 mL of this mixture into 1.15 mL of CO-Hb diluting solution.
3. Mix by gently inverting, and allow to stand 10 minutes.
4. Place in a cuvet and read the absorbance at 420 and 432 nm against a matched cuvet containing CO-Hb diluting solution.

Calculation—The percent carboxyhemoglobin is calculated using the formula

$$S_{CO} = \frac{1 - (A_R \times F_1)}{A_R(F_2 - F_1) - F_3 + 1}$$ *(Eq. 28-4)*

where $A_R = A420/A432$
$F_1 = 1.3330$
$F_2 = 0.4787$
$F_3 = 1.9939$

The constants F_1, F_2, and F_3 are computed from the molar absorptivities of reduced hemoglobin and carboxyhemoglobin at 420 and 432 nm. ∎

METALS

Occasionally, the toxicology laboratory is asked to aid in diagnosing possible heavy metal (mercury, arsenic, lead) toxicity, and if the diagnosis proves positive, quantitative determinations of blood or urine levels are very helpful in following the course of therapy. Sources of exposure to

TABLE 28-3. Exposure Levels, Carboxyhemoglobin Concentrations, and Symptoms of CO Poisoning in Humans		
% CO (V/V) In Air	**% Carboxyhemoglobin**	**Symptoms**
	0.5	Typical in nonsmokers
	5	Typical in smokers
0.04	10	Shortness of breath with exercise
	10–20	Shortness of breath, slight headache
	20–40	Severe headache, irritation, impairment of judgment
0.1	50	
	60–70	Unconsciousness, respiratory failure, death from long exposure
0.4	80	Immediately fatal

these metals range from house paints for lead to homicidal attempts using arsenic.

Lead

Lead oxides have been used extensively as paint pigment, with concentrations occasionally exceeding 40%. With the discovery that lead poisoning was afflicting children exposed to and eating paint, the Consumer Product Safety Commission, in 1978, banned the manufacture of paint containing more that 0.06% lead for residential use, toys, and furniture. Nevertheless, in older buildings, residue remains from earlier coats of paint, and unsupervised children who eat the paint show signs of peripheral neuropathies that relate to lead poisoning.

Lead additives in gasoline provide chronic exposure to everyone in the form of air pollutants. However, with the elimination of leaded gasoline beginning in the late 1970s, the lead levels in air have dropped precipitously, as have the average blood lead levels in the general population.

Earthenware made of clay rich in lead salts can be a potent source of lead exposure. Regulations now require that pottery made in the United States be kilned at temperatures high enough to seal in the lead in clay. Some insecticides contain lead and represent potential sources of intoxication. Homemade spirits (moonshine) distilled through car radiators provide a potent source for lead ingestion.

Symptoms of acute lead poisoning include GI tract irritation, avitaminosis, loss of weight, hematologic changes (including red blood cell stippling), kidney damage, and CNS damage. Lead accumulates in the body because elimination is very slow. These deposits are primarily phosphates laid down in bone tissue. Low blood calcium levels and mild acidosis mobilize lead from the bones. Increased exposure to sunlight increases vitamin D production, which also mobilizes lead and explains the increased symptoms of lead poisoning in the summer. Since lead is found primarily in red blood cells, whole blood is the specimen of choice for detecting lead poisoning.

The treatment for lead poisoning includes the IV administration of EDTA and BAL (British anti-Lewisite). These chelaters bind lead, releasing it from bone and other tissue stores, and allow it to be excreted by the kidneys. Both of these chelating agents are nephrotoxic, and kidney function tests must be monitored during therapy. Urine lead levels are very helpful in monitoring the effectiveness of chelation therapy. The methods of choice for measuring lead levels in blood and urine are atomic absorption spectroscopy using a heated graphite furnace and electrochemical methods, specifically anodic stripping voltammetry and induction coupled plasma (ICP).[35]

DETERMINATION OF BLOOD LEAD BY ATOMIC ABSORPTION SPECTROSCOPY[12,39]

Principle—A small aliquot of whole blood is diluted with a nitric acid detergent solution and injected into the graphite furnace for measurement.

Specimen—Draw blood specimens in brown-stoppered vacuum tubes, which are specially manufactured to minimize lead contamination.

Reagents
1. Diluent—Mix 20 mL of Triton X-100 and 10 mL of concentrated nitric acid in a 1-L volumetric flask and fill to the mark with deionized water. Mix thoroughly. This solution is stable for 1 year at room temperature.
2. Lead standards—Dilute a 1000 ppm Pb standard solution (available commercially) with deionized water in a 100-mL volumetric flask according to Table 28-4.

Procedure
1. Dilute 200 µL of standard or whole-blood sample with 1.0 mL of diluent in a polystyrene tube. Cap the tube and mix by vortexing for 30 seconds.
2. Remove the foam with a disposable plastic pipet.
3. Introduce 10 µL of diluted standard or sample into the graphite furnace and initiate measurement using standard procedures for the particular atomic absorption spectrometer.

Calculations

µg Pb/100 mL

$$= \frac{\text{peak height of the unknown} \times 50 \ \mu\text{g Pb/100 mL}}{\text{pk ht of 50 std}}$$

(Eq. 28-5)

Comments—This procedure should be linear to at least 200 µg Pb/100 mL whole blood. The sensitivity is 1 µg Pb/100 mL. Ultrapure nitric acid must be used to avoid contamination. The operator should not look at the graphite furnace when it is in the atomization phase, since it emits intense UV light, which is harmful to the retina of the eye. ■

The hematologic effects of lead poisoning provide the toxicologist with another mechanism for diagnosing lead poisoning. Figure 28-4 diagrammatically illustrates the biosynthesis of heme. Of note is the poisoning effect of lead on several of the enzymes in this progression of reactions, especially δ-aminolevulinic acid dehydrogenase (ALAD) and heme synthetase. The result of the poisoning is the accumulation of δ-aminolevulinic acid (ALA) in serum and urine and protophyrins in red blood cells, respectively. This poisoning effect may last as long as a year following an acute poisoning. Thus, measuring urinary ALA levels and free

TABLE 28-4. Lead Standards

Stock Pb Std, µL	HNO$_3$ (mL)*	Total Vol (mL)	µG Pb/100 mL
0	1	100	0
20	1	100	20
50	1	100	50
100	1	100	100

Ultrapure nitric acid must be used to avoid contamination.

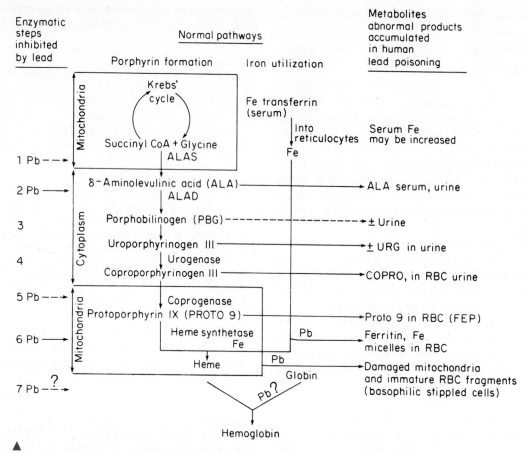

Figure 28-4. Lead interferes with the biosynthesis of heme at several enzymatic steps, with the utilization of iron, and, in erythrocytes, with globin synthesis. Inhibition of ALAD and heme synthetase (steps 2 and 6), which are SH-dependent enzymes, is well documented, and accumulation of the substrates of these enzymes (ALA and PROTO 9) is characteristic of human lead poisoning. Inhibition of ALAS (step 1) is based on experimental evidence only. Whether there is enzymatic inhibition or if other factors operate at step 5 is not clear; nevertheless, increased urinary excretion of COPRO is prominent in human lead poisoning. Minor increases in PBG and URO in urine are occasionally reported in severe lead poisoning. The utilization of iron is impaired; although the *in vivo* mechanisms are unclear, nonheme iron (ferrin and iron micelles) accumulates in red blood cells, with damaged mitochondria and other fragments not found in normal mature erythrocytes. Serum iron may be increased in humans with lead poisoning, but without iron-deficiency states. Heme synthesis is reduced and, in red blood cells, globin synthesis is apparently impaired, although the mechanisms responsible for reduced globin synthesis remain unknown. (Lead: Airborne Lead in Perspective. Publication number ISBN 0-309-01941-9, Committee on Biologic Effects of Atmospheric Pollutants, National Research Council—National Academy of Sciences, Washington, DC, 1972.)

erythrocyte protoporphyrin (FEP) levels can identify lead poisoning that occurred in the past after blood and urine lead levels have returned to normal.

There have been many studies and considerable controversy over what is a normal blood lead level. For a long time, normal adult levels were considered to be 0 to 80 µg/dL, but more recent studies showed that the upper limit was too high. Lead levels in blood taken from a representative group of blood donors and measured in this author's laboratory all fell below 20 µg/dL.

Lead poisoning in young children is a chronic problem that has existed for years but was not thoroughly appreciated.[6] Blood lead levels in children down to 20 µg/dL can cause alterations in heme synthesis,[31] while levels as low as 10 ug/dL can cause cognitive deficits, learning difficulties, delays in achieving development goals, and a lowered IQ.[42] Currently, the Centers for Disease Control recommend a threshold of 10 µg/dL. Children with levels above this threshold should be evaluated for possible corrective action. Once the CNS has developed, higher lead levels can be tolerated with no adverse effects.

Specimens must be collected in lead-free containers, and airborne contaminates must be avoided. Clean plastic urine containers are appropriate but may require an acid wash before use. The typical clot tubes used for blood collection are contaminated, and it is necessary to use specially manufactured tubes when collecting blood lead specimens. For example, Becton-Dickinson (Rutherford, NJ 07070) distributes a brown-stoppered Vacutainer containing sodium heparin made specifically for lead testing.

Mercury

Many antibacterial preparations contain mercury salts in the form of disinfectants, antiseptics, preservatives, and so on. Pesticides, paints, photographic reagents, laboratory equipment, batteries, and solder all contain mercury compounds. Since elemental mercury is quite volatile at room temperature, air pollution may be a source of exposure. Even with the surprising prevalence of mercury, there are few mercury poisonings. However, they do occur, and occasionally, it is appropriate to evaluate blood and urine mercury levels.

Mercury poisoning is cumulative and results in GI tract irritation, kidney damage, and disorders such as peripheral neuropathies, among others. An effective therapy for mercury poisoning is the use of EDTA, N-acetyl-d,l-penicillamine, or BAL, which chelates mercury, releases it from its tissue stores, and allows it to be excreted by the kidneys.

Typical mercury levels in blood are 0 to 5 µg/100 mL (0.25 µg/dL), and urine levels of 5 to 25 µg/L are considered normal.

The method most commonly used for both blood and urine mercury determinations is cold vaporization atomic absorption spectroscopy.[21,24] A very simple but crude test to detect large amounts of mercury in urine is the Reinsch test. This test will also detect antimony, selenium, bismuth, and arsenic but is not very sensitive to any of these metals. The method of choice is gold-film potentiometry.

SIMPLE SCREENING METHOD FOR MERCURY

Principle—A clean copper wire is submerged in boiling urine to which has been added a small amount of HCl. Antimony, arsenic, bismuth, mercury, and selenium, if present, will be deposited on the copper wire.

Procedure
1. Add to a 250-mL Erlenmeyer flask 100 mL of urine and 10 mL of concentrated HCl.
2. Submerge into the solution a copper wire coil that has been thoroughly cleaned with HCl.
3. Bring the solution to a boil and continue boiling it until 20 mL are left.
4. Remove the copper wire, rinse in tap water, and observe any coating on the wire (Table 28-5).
5. To confirm arsenic, put the copper wire in 10% KCN. Arsenic will dissolve, whereas bismuth and antimony will persist. ■

Arsenic

A wide array of pesticides, including rodenticides, insecticides, and weed and tree killers, contain arsenic compounds. Paints are another source of arsenic exposure. Because arsenic compounds are odorless, tasteless, and colorless, they

TABLE 28-5. Reinsch Test

Metal	Appearance of Wire	Sensitivity
Antimony	Dark purple sheen	1 mg/L
Selenium	Gray to silver	500 µg/L
Mercury	Silvery upon rubbing	500 µg/L
Bismuth	Shiny black	250 µg/L
Arsenic	Gray to black	25 µg/L

are difficult to detect and have been (and are still) used, on occasion, as a homicidal poison.

Symptoms of acute arsenic poisoning include explosive gastroenteritis, blistering of the epithelium, sloughing of intestinal lining, shredded stomach lining, and Mees lines in the fingernails. Arsenic combines with sulfhydryl groups on proteins and is quickly cleared from the blood. It has a high selectivity for keratin and thus appears in hair and nails. Mees lines are 1-mm-wide white bands across each finger and toenail. These bands contain very high concentrations of arsenic.

Chronic poisoning results in malaise, fatigue, conjunctivitis, tracheitis, peripheral neuropathy, and hyperkeratosis of hands and feet, often resulting in skin pigmentation.

Because arsenic is quickly cleared from the blood, urine is the specimen of choice for diagnosing arsenic poisoning. Arsenic will persist in the urine for about a week after an acute poisoning and for as long as a month following chronic exposure. Occasionally, hair and nails are analyzed to detect the long-term effect of arsenic poisoning. Pubic hair is often most appropriate because it grows so slowly compared with scalp hair and provides a measurement of arsenic poisoning over a period of months. Scalp hair and nails are so easily contaminated, despite copious washing, that they should be avoided as specimens for arsenic evaluations.

Reference Range

Various reports in the literature indicate that typical As levels for people with no unusual exposures to As fall between 40 and 80 µg/24 h (534–1068 nmol/day). Normal hair/fingernail levels are less than 1.0 µg/g of tissue.

DRUGS OF ABUSE TESTING

Substance abuse in this country is not a new or shocking topic anymore. What is shocking is the fact that the problem appears to be growing despite increasing public awareness and multifaceted attempts to reduce the incidence of drug abuse. One aspect of the war on drugs is the growing utilization of drug testing in the workplace. The defense department has conducted random drug testing for several years, with a drastic reduction in the number of drug-positive individuals. More and more companies are using preemployment drug screening to avoid hiring individuals who abuse drugs or have a drug-addiction problem. The Department of Transportation and the Department of Justice have initiated drug-testing programs to discourage drug abuse and protect the public from drug-related mistakes by federal employees.

Whether doing preemployment testing, random testing of existing employees, or testing of employees who display signs or symptoms of drug abuse (reasonable cause), the outcome may be the same—"You test positive, you don't work here."

The forensic ramifications of drug-abuse testing are obvious and require an entirely different approach to drug testing by the laboratory. The number of drug-testing laboratories has steadily increased to meet the escalating demand. Some are strictly commercial laboratories, while others exist as a spinoff and subset of clinical laboratories. In any case, strict procedures and protocols with documentation must be followed not only in the analysis of drugs, but also in the collection of specimens, handling of samples, and reporting and archiving of results. An in-depth treatment is inappropriate here, but the following should give the reader a flavor for the additional requirements of forensic urine drug testing as compared with clinical drug-overdose testing.

Sample Collection

Urine is the specimen of choice for several reasons. Urine collection is nontraumatic compared with venipuncture for blood collection. There is tremendous resistance to drug testing on the basis that it is an invasion of privacy. Requiring a venipuncture of an already resistant person would only make matters worse. For many drugs, detection is best accomplished by measuring metabolites in urine because the parent compound is short-lived in blood or the concentration is extremely low and hard to measure accurately in blood. Furthermore, urine provides a mechanism to look at drug presence over a longer period of time following drug ingestion (*e.g.,* cannabinoids may be identified in urine as long as 3 to 4 weeks after cessation of marijuana smoking).

Because a drug test may incriminate the individual being tested, the subject may make attempts to adulterate the specimen to invalidate the test. Therefore, it is mandatory to ensure a satisfactory specimen. This may require that the subject be observed urinating into the container. Short of direct observation, the subject should be asked to remove all outer garments such as a coat, jacket, or hat to minimize the possibility of the subject concealing a container of drug-free urine or water that could be substituted for the person's urine. In preferably a small restroom, there should be no running water at the sink, and water in the toilet should be stained with a dye to prevent the subject from diluting his or her urine. Once the urine is collected, it should be placed in a container, sealed with a tamper-proof closure, and labeled with appropriate identification, such as name or ID number and date. A chain-of-custody form must be completed showing sample identification, date, location, who processed the sample, and the signature of the subject acknowledging that the sample contained is the specimen submitted. Any further processing by anyone else should be documented with that person's signature, date, and reason for handling, on the chain-of-custody form.

Laboratory Processing

Once the sample reaches the laboratory, it is opened and processed in a secured area that, ideally, is isolated from the rest of the laboratory and restricted by a locked door to only those individuals who process samples. The ID on the container must match the information on the chain-of-custody form, and the seal must not be broken. From here on, all individuals processing the sample and performing the various analyses must be identified on an internal chain-of-custody form that documents each person who interacted with the sample, when, and what they did with the sample. Once the sample is opened, it is evaluated for possible tampering by observing any unusual color, appearance, or odor. The pH, specific gravity, or creatinine level may be measured and compared with expected values. At the time of collection, sample temperature may have been measured for the same reason. Once aliquots are withdrawn, the sample must be stored (optimally refrigerated) in a secured location. All this security and documentation are to avoid tampering with the sample, ensure that each result relates to the correct individual, and guarantee that all documents and results will stand up in court if a lawsuit should occur in the future.

The laboratory is obligated to supply each client with a detailed set of instructions on how to obtain specimens properly. Optimally, the laboratory should provide the client with a collection kit containing the collection instructions, container with tamper-proof sealing mechanism, chain-of-custody and order form, and instructions on how to prepare the package for mailing or delivery to the laboratory. Mailing biologic samples has become extremely difficult, so many laboratories rely on commercial carriers or their own couriers.

Analysis

All samples are screened first using immunoassays, such as EIA, FPIA, or chromatography, *e.g.,* TLC, GC, or HPLC. Immunoassays are by far the most popular owing primarily to their speed of analysis, high throughput, and small demand on operator time. They provide adequate sensitivity and specificity in most instances. Once a drug(s) is identified as present in a sample, based on its concentration being above the appropriate threshold or cutoff concentration, it is labeled "Presumptive positive."

Confirmation of a presumptive positive is mandatory in forensic urine drug testing. The analytical technique of choice is gas chromatography with mass spectroscopy detection (GC/MS). GC provides excellent specificity by separating a mixture into its individual components, while MS provides a very sensitive detector, with an added degree of specificity with its ability to measure the mass fragment spectrum of each component passing from the GC into the MS. This "fingerprint" of each component is a powerful and highly specific means of identifying a drug. Many mass spectrometers come with a computer and a library of mass spectra. As each spectrum is measured, it is compared with the library for identification.

Most commonly abused drugs or their metabolites are not sufficiently volatile to be separated by gas chromatography. To make them more volatile, they must be derivatized subsequent to extraction from the biologic matrix, urine, and prior to injection into the GC.

A presumptive positive that is confirmed positive by GC/MS is reported positive based on a threshold or cutoff concentration level, which should be reported along with the result. A presumptive positive not confirmed positive by GC/MS is reported negative. All reports should be returned to the medical review officer (MRO) for the client.

Reporting Results

The laboratory and the client should develop a reporting system that provides for the timely reporting of results and the confidentiality of the results. The laboratory report should include the date of specimen receipt in the laboratory, drugs analyzed, threshold concentrations, positive or negative results, and date of report. Reports should be returned to the client by mail or electronically. Reporting results by telephone should only be done under strict, written protocol.

The laboratory should store frozen all positive urines for 6 months. Records should be archived for 2 years and include chain-of-custody documents, sample identification logs, standardization data for each assay run, instrument maintenance records, instrument outputs (*e.g.,* chromatograms), identification of analyst, standard operating procedures, quality-control and proficiency-testing records, documentation of supervisory review, and final reports. In addition, personnel records must be kept that include documentation of each individual's training and experience, as well as continuing-education records.

Two accreditation programs are available for laboratories performing drugs-of-abuse testing. The Substance Abuse and Mental Health Services Administration (SAMHSA, formerly the National Institute on Drug Abuse) conducts a program that requires a laboratory to pass a biannual, on-site inspection and satisfactory performance in a quarterly proficiency-testing challenge. The joint program operated by the College of American Pathologists (CAP) and the American Association for Clinical Chemistry (AACC) requires that a laboratory pass an annual, on-site inspection and perform satisfactorily in a quarterly proficiency-testing program and includes a quarterly educational newsletter. To attract federal contracts, a laboratory must be approved by the SAMHSA, whereas the AACC/CAP program is voluntary. Both programs demand stringent adherence to the standard operating procedures described above. The SAMHSA program evaluates each laboratory on its ability to test for five drug classes with specified cutoff concentrations (Table 28-6).

The AACC/CAP program requires proficiency in testing these drugs as well as any other drugs that a laboratory may test for through contract with various clients. Satisfactory performance on proficiency testing requires that each drug be identified correctly with no false-positive results, and each drug present must be measured quantitatively. Furthermore, the level determined must agree with the target value within certain limits.

SUMMARY

Toxicology is the study of poisons, their actions, their detection, and the treatment of the conditions produced by them. This chapter primarily discusses the detection of poisons and, briefly, the actions of poisons.

TABLE 28-6. SAMHSA Cutoff Concentrations

	Initial Test Level (ng/mL)		Confirmatory Test Level (ng/mL)
Marijuana metabolites	50	Marijuana metabolite	15
Cocaine metabolites	300	Cocaine metabolite	150
Opiate metabolites	300	Opiates:	
		Morphine	300
		Codeine	300
Phencyclidine	25	Phencyclidine	25
Amphetamines	1000	Amphetamines:	
		Amphetamine	500
		Methamphetamine	500

From: Federal Register, Vol. 59, No. 110, Thursday, June 9, 1994. Notices.

A poison is any substance which, when ingested, inhaled, or absorbed in relatively small amounts, may cause damage to structure or disturbance of function as a result of its chemical action. The laboratorian should note, however, that almost anything taken in sufficient amounts can be toxic.

Drug abuse is a common occurrence, and drug overdose is one of the most common modes of suicide attempt. The drugs most often involved in overdose or poisoning situations are identified in Table 28-1. Reasons for drug overdosing fall into three categories: (1) suicide gestures, (2) suicide attempts, and (3) recreational drug abuse. Diagnosing a drug overdose must be done using as much information as possible, including the history, physical examination, and laboratory data. Diagnosing drug overdose based only on laboratory findings is extremely risky.

Although the toxin in a poisoning is usually a drug, it may also be a metal, such as lead or mercury, an alcohol, such as ethanol or methanol, or a gas, such as carbon monoxide. Toxicology laboratories are generally designed according to the type of facility to be serviced and the nature of the poisoning cases.

The laboratory involved in drug-abuse screening must rely on screening tests as a first phase, but all positive results must be confirmed by an alternative method. Positive drug findings must be accurate.

Treatment of the drug-overdose patient first focuses on stabilizing the patient by establishing or maintaining an open airway, proper breathing, adequate circulation, and correcting any life-threatening compromise of the central nervous system. Further treatment involves preventing the absorption of the drug from the GI tract, hastening its elimination from the body, utilizing appropriate antidotes (if available), and continuing care and disposition. Although few specific antidotes are available, those that do exist are helpful.

Substance abuse appears to be growing in this country despite increasing public awareness and mutifaceted attempts to reduce the incidence of drug abuse. One aspect of the war on drugs is the growing utilization of drug testing in the workplace.

Urine is the specimen of choice for drug testing because of the nontraumatic means of collection in comparison to venipuncture. For many drugs, detection is best accomplished by measuring metabolites in urine because the parent compound is either short-lived in the blood or the concentration is extremely low and difficult to measure. Urine also provides a mechanism to assess drug presence over a longer period of time following drug ingestion.

Once submitted to the laboratory, a specimen must be opened and processed in a secure area isolated from the rest of the laboratory and restricted by a locked door. A "chain of custody" must be established.

All samples are screened first using immunoassays, such as EIA, FPIA, or chromatography. If a drug or drugs are identified as present in a sample, confirmation procedures may be performed. Confirmation of a presumptive positive is mandatory in forensic urine drug testing. The analytical technique of choice is gas chromatography with mass spectroscopy detection.

A reporting system that provides for timely reports and the confidentiality of the results is used by the toxicology laboratory. Positive urines should be stored frozen for 6 months and records archived for 2 years.

CASE STUDY 28-1

A 53-year-old male was seen in the emergency room with metabolic acidosis, chronic alcoholism, and a history of recent ingestion of excess amounts of ethanol and other substances. His family explained that he had not eaten for the last month. He was unconscious and unresponsive, and his temperature was 32°C. (See Case Study Table 28-1.)

Drug screen—Glutethimide in urine, possibly methaqualone or PCP, isopropanol = 11 mg/100 mL

Question

In most situations of organic drug overdoses a qualitative result is adequate. Why is a quantitative measurement not necessary?

CASE STUDY TABLE 28-1. Laboratory Data

	Patient's Results	Reference Range
Na	134 mmol/L	135–145
K	5.2 mmol/L	3.2–4.8
Cl	81 mmol/L	98–108
Bicarb	4 mmol/L	21–30
Ketones	neg	neg
Ethylene glycol	neg	neg
Ethanol	279 mg/100 mL	neg
Glucose	214 mg/100 mL	75–100
Calcium	5.3 mg/100 mL	8.7–10.2
Hct	18%	39–49
WBC	5,100	$3.2–9.8 \times 10^9$
pH	6.64	7.31–7.42

CASE STUDY
28-2

A 37-year-old male was brought to the emergency room with a 3- to 4-day history of weakness and dizziness and a 1-week history of side and abdominal pain. The man was an alcoholic with a history of multiple medical problems, including ethanol abuse, alcoholic liver disease with coagulopathy, chronic GI bleeding, multiple gunshot wounds and stab wounds, and macular degeneration. He had a cardiac arrest in the emergency room and was defibrillated. He was severely acidotic. (See Case Study Table 28-2.)

He was admitted to the medical intensive care unit. A repeat EG was 118 mg/100 mL, and lactate was 280 mg/100 mL.

He was given sodium bicarbonate, which raised the pH to 7.2. He also received blood products. He was hemodialyzed and given an ethanol drip to reverse EG toxicity.

The patient died 24 hours after admission as a result of irreversible shock with pulmonary edema and severe hypoxemia, followed by cardiopulmonary arrest.

CASE STUDY TABLE 28-2. Laboratory Data

	Patient's Results	Reference Range
Hct	8%	39–49
WBC	15,600/L	$3.2–9.8 \times 10^9$
Glucose	20 mg/100 mL	75–110
BUN	27 mg/100 mL	7–21
Sodium	145 mmol/L	135–145
Potassium	6.2 mmol/L	3.2–4.8
Chloride	96 mmol/L	98–108
CO_2	6 mmol/L	21–30
Anion gap	50	10–20
Ethylene glycol (EG)	49 mg/100 mL	neg
Bilirubin	3.9 mg/100 mL	0.2–1.2
Calcium	10.7 mg/100 mL	8.7–10.2
Amylase	166 IU/L	35–130
Ethanol	0	neg
pH	6.66	7.31–7.42
PO_2	19 mmHg	30–50
PCO_2	41 mmHg	39–55
Methanol	8 mg/100 mL	neg

Question

Why is ethanol administered as therapy in ethylene glycol and methanol poisonings?

CASE STUDY
28-3

A 26-year-old male was brought to the emergency room at 3:22 P.M. by a friend. He was ambulatory, alert, oriented, and explained that he had taken 26 Motrin, 56 Extra-Strength Tylenol, and 12 analgesics tablets along with a bottle of wine about noon because he had a "falling out" with his male lover. His pulse was 100, respiration 20, and BP 130/100. The CBC and electrolytes were normal. His ethanol level 109 mg/100 mL, and acetaminophen level was 458 µg/mL. He was given an oral dose of 30 mL of ipecac, followed by 2 to 8 oz of water to induce vomiting. At 4:00 P.M., he received 2 g of charcoal PO and one bottle of magnesium citrate and was admitted to the hospital. His acetaminophen level at 9:15 P.M. was 294 µg/mL. At 9:30 P.M., he received a loading dose of 11.2 g of Mucomyst (Mead Johnson) and 5.6 g q4h for 17 doses.

His LD and AST values during his 5-day hospital stay are shown in Case Study Table 28-3.

CASE STUDY TABLE 28-3. Laboratory Data

Day	LD (U/L)	AST (U/L)
1/18	198	21
1/19	200	24
1/20	197	21
1/21	259*	125*
1/22	205	74*
Reference range	90–250	5–35

*Abnormal value.

Question

Why is it appropriate to quantitate acetaminophen in serum in overdose situations rather than just identifying whether it is present or not?

REFERENCES

1. Bailey DN, Mansquerra AS. Survey of drug-abuse patterns and toxicology analysis in an emergency room population. J Anal Toxicol 1980;4:199.

2. Barrett MJ. An integrated gas chromatographic program for drug screening in serum and urine. Clin Chem News 1985.

3. Berry DJ, Grove J. Emergency toxicology screening for drugs commonly taken in overdose. J Chromatogr 1973;80:205.

4. Beutler E, West C. Simplified determination of carboxyhemoglobin. Clin Chem 1984;30:871.

5. Brett AS. Implications of discordance between clinical impression and toxicology analysis in drug overdose. Arch Intern Med 1988;148:437.

6. Cherry FF. Childhood lead poisoning: Prevention and control. New Orleans: Department of Health and Human Resources, 1981.

7. Clark EGC. Isolation and identification of drugs. London: The Pharmaceutical Press, 1974;(1) and 1975;(2).

8. Collison HA, Rodkey FL, O'Neal JD. Determination of carbon monoxide in blood by gas chromatography. Clin Chem 1968;14:162.

9. Davidow B, LiPetri N, Quame B. A thin-layer chromatographic screening procedure for detecting drug abuse. Am J Clin Pathol 1968;50:714.

10. Dole VP, Kim WK, Eglitis I. Detection of narcotic drugs, tranquilizers, amphetamines, and barbiturates in urine. JAMA 1966;198:115.

11. Ellman GL, Courtney KD, Andres V Jr, Featherstone RM. A new and rapid colorimetric determination of acetylcholinesterase activity. Biochem Pharmcol 1961;7:88.

12. Evenson MA, Pendergast DD. Rapid ultramicro direct determination of erythrocyte lead concentration by atomic absorption spectrophotometry, with use of a graphite-tube furnace. Clin Chem 1974;20:163.

13. Finkle BS, Cherry EJ, Taylor DM. A GLC bases system for the detection of poisons, drugs, and human metabolites encountered in forensic toxicology. J Chromatogr 1971;9:393.

14. Finkle BS, McCloskey KL, Goodman LS. Diazepam and drug-associated deaths: A survey in the United States and Canada. JAMA 1979;242:429.

15. Forester EH, Hatchett D, Garriott JC. A rapid, comprehensive screening procedure for basic drugs in blood or tissues by gas chromatography. J Anal Toxicol 1978;2:50.

16. Forrest IS, Forrest FM. Urine color test for the detection of phenothiazine compounds. Clin Chem 1960;6:11.

17. Frings CS, Cohen PS. Rapid colorimetric method for the quantitative determination of ethchlorvynol (Placidyl) in serum and urine. Am J Clin Pathol 1970;54:833.

18. Frings CS. Drug screening. Crit Rev Clin Lab Sci Dec 1973; 297.

19. Glynn JP, Kendal SE. Paracetamol measurement. Lancet 1975;1:1147.

20. Gotelli GR, Kabra PM, Marton LJ. Determination of acetaminophen and phenacetin in plasma by high-pressure liquid chromatography. Clin Chem 1977;23:957.

21. Hatch WR, Ott WL. Determination of sub-microgram quantities of mercury by atomic absorption spectrophotometry. Anal Chem 1968;40:2085.

22. Hayes WJ Jr. Pesticides and human toxicity. Ann NY Acad Sci 1969;160:40.

23. Jain NC. Direct blood injection method for gas chromatographic determination of alcohols and other volatile compounds. Clin Chem 1971;17:82.

24. Kubasik NP, Sine HE, Volosin MT. Rapid analysis for total mercury in urine and plasma by flameless atomic absorption analysis. Clin Chem 1972;18:1326.

25. McCarron MM. The use of toxicology tests in emergency room diagnosis. J Anal Toxicol 1983;7:131.

26. Michigan Substance Abuse Information Clearinghouse, Lansing, MI 48901:1994.

27. Nyanda AM, Cross RE, Phillips JC. A simple and rapid method for determination of ethylene glycol and propylene glycol using capillary column gas chromatography and ultrafiltration for sample preparation. Clin Chem 1992;38:1017.

28. Miceli JN, Vroon DH, Harris C, Poor J. Ethylene glycol: A continuing problem. Clin Chem 1992;38:995.

29. O'Brien JP. Increase in suicide attempts by drug ingestion. Arch Gen Psychiatry 1977;34:1165.

30. Pierce WO, Lamoreaux TC, Urry FM. A new, rapid gas chromatography method for the detection of basic drugs in postmortem blood using a nitrogen phosphorus detector: I. Qualitative analysis. J Anal Toxicol 1978;2:26.

31. Piomelli S, Seaman C, Zullow D, Curran A, Davidow B. Threshold for lead damage to hema synthesis in urban children. Proc Natl Acad Sci USA 1982;79:3335.

32. Prescott LF, Wright N, Roscoe P, Brown SS. Plasma-paracetamol half-life and hepatic necrosis in patients with paracetamol overdosage. Lancet 1971;1:519.

33. Rio JG, Hodnett CN. Evaluation of a colorimetric screening test for basic drugs in urine. J Anal Toxicol 1981;5:267.

34. Routh JI, Dryer RL. Salicylate. Standard methods. Clin Chem 1961;3:194.

35. Searle B, Chan W, Davidow B. Determination of lead in blood and urine by anodic stripping voltammetry. Clin Chem 1973;19:76.

36. Severinghous JW, Astrup PB. History of blood gas analysis: VI. Oximetry. J Clin Monit 1986;2:270.

37. Tietz NW. Textbook of clinical chemistry. Philadelphia: Saunders, 1986;1719.

38. Trinder P. Rapid determination of salicylate in biological fluids. Biochem J 1954;57:301.

39. Volosin MT, Kubasik NP, Sine HE. Use of the carbon rod atomizer for analysis of lead in blood: Three methods compared. Clin Chem 1975;21:1986.

40. Wiltbank TB, Sine HE, Brody BB. Are emergency toxicology measurements really used? Clin Chem 1974;20:116.

41. Improved detection and separation of glycols and ethylene oxide residues using GC. Bulletin 789. Bellfonte, PA: Supelco, Inc, 1980.

42. Air quality criteria for lead. US Environmental Protection Agency. EPA-600/8-83/028aF, 1986.

SUGGESTED READINGS

Amdur MO, Doull J, Klassen CD, eds. Casarett and Doull's toxicology: The basic science of poisons. 4th ed. New York: Pergamon Press, 1991.

Baselt RC. Analytical procedures for therapeutic drug monitoring and emergency toxicology. 2nd ed. Littleton MA: PSG Publishing, 1987.

Baselt RC. Disposition of toxic drugs and chemicals in man. 3rd ed. Davis, CA: Year Book Medical Publishers, 1989.

Clark EGC. Isolations and identification of drugs. Vol 2. London: The Pharmaceutical Press, 1986.

Ellenhorn MJ, Barceloux DG. Medical toxicology—Diagnosis and treatment of human poisoning. New York: Elsevier, 1988.

Haddad LM, Winchester JF, eds. Clinical management of poisoning and drug overdose. 2nd ed. Philadelphia: W. B. Saunders, 1990.

Wilford BB. Drug abuse: A guide for the primary care physician. Chicago: American Medical Association, 1981.

Vitamins

Sharon Monahan Miller

Objectives

Upon completion of this chapter, the clinical laboratorian should be able to:

- *Describe the biochemical roles played by vitamins.*

- *For each vitamin, list the signs and symptoms that characterize suboptimal vitamin balance as well as conditions of deficiency and toxicity.*

- *Correlate alterations in vitamin status with circumstances of increased metabolic requirements, age-related physiologic changes, and/or pathologic conditions.*

- *Describe drug-nutrient interactions that influence vitamin status.*

- *State the clinical evidence and biochemical bases for therapeutic and prophylactic use of vitamin supplements.*

- *Recognize how vitamins, in excess, may affect biochemical tests for various analytes.*

- *Delineate and appraise the clinical laboratory procedure(s) used in the assessment of vitamin status.*

- *Recognize and be able to correct for variables that interfere with vitamin measurements by the clinical laboratory.*

- *Discuss the present and projected role of the laboratory in nutritional assessment and monitoring as a function of population demographics and health care economics.*

KEY WORDS

Activity Coefficient	Megavitamin	RDA
Antioxidant	Nyctalopia	Rickets
Beriberi	Osteomalacia	Vitamer
Hypervitaminosis	Pellagra	Vitamin
Hypovitaminosis		

Clinical laboratory tests of vitamin status help physicians confirm existing deficiency, identify developing vitamin imbalance, and monitor nutritional therapy. Overt vitamin deficiencies are not routinely encountered among the general population in developed countries. Marginal deficiencies, as identified by laboratory tests, are more common and, if prolonged, may pose a major health threat. Groups at greatest risk of deficiency include pregnant women, infants and children, and the elderly. Patients who cannot or will not ingest, digest, or absorb sufficient nutrients to prevent malnutrition are likely to have inadequate levels of vitamins, and the laboratory plays a key role in assessing the effectiveness of nutrition support therapy.[181,236] Drug-nutrient interactions can adversely affect vitamin status, as can smoking and the excessive consumption of alcohol. Prompt identification of toxic levels of vitamins before irreversible pathologic changes occur is also a critical issue. Both medical and popular literature report the therapeutic nonnutritional value of vitamins.[203] In 1993, approximately 43% of Americans purchased $4.2 billion worth of vitamin and mineral supplements.[54] Laymen eager to promote optimal health and fitness may ingest large doses of vitamins in the hope of forestalling aging, increasing fertility, preventing illness, or treating disease. Only in a limited number of clinical conditions, such as inherited metabolic disorders, has megavitamin therapy been shown unequivocally to be beneficial (Table 29-1).[59] Evidence from numerous recent, well-publicized reports suggests that supplementation with certain vitamins may help prevent specific birth defects or reduce the risk of certain types of cancer and of cardiovascular diseases. These studies, while not conclusive, have identified important areas for additional research. Major clinical trials may confirm preliminary findings on the

Michael L. Bishop, Janet L. Duben-Engelkirk, and Edward P. Fody.
CLINICAL CHEMISTRY. © 1996 Lippincott–Raven Publishers.

TABLE 29-1. Inborn Errors of Metabolism Responding to Pharmacologic Doses of Vitamins and Cofactors*[59]

Vitamin Dependency	Metabolic Disorder	Disease Type	Enzyme Defect/Protein Deficiency
Biotin	Biotinidase deficiency	Propionic acidemia	Biotinidase
	Multiplecarboxylase deficiency	Lactic acidosis, β-methylcrotonic acidemia	Holocarboxylase synthetase
Calciferol	Rickets	Familial hypophosphatemia	Renal phosphate transport
		Renal resistant rickets	1,25-Dihydroxy vitamin D-3-hydroxylase
Cobalamin	Homocystinuria with hypomethioninemia		N-5-Methyltetra-hydrofolate:homocystine methyl transferase
	Methylmalonic aciduria		Cob(1)alamin adenosyl transferase
Pyridoxine	Homocystinuria	Pyridoxine dependent	Cystathionine β-synthase
	Cystathionuria		γ-Cystathionase
Riboflavin	Ethylmalonic-adipic aciduria	Short chain fatty aciduria	Multiple acyl-CoA dehydrogenase
Tetrahydrobiopterin	Hyperphenylalaninemia	Biopterin deficiency type I	Guanosine triphosphate cyclohydrolase
		Biopterin deficiency type II	6-Pyruvoyltetrahydropterin synthase
		Biopterin deficiency type III	Dihydropteridine reductase
Thiamin	Maple syrup urine disease	Thiamine responsive	Branched-chain α-keto acid dehydrogenase
Vitamin D	Pseudohypoparathyroidism		Parathyroid cAMP metabolism
Vitamin E	Abetalipoproteinemia	Bassen-Kornzweig syndrome	Lipoprotein and lipid transport

The various therapeutic regimens do not include the use of drugs. Only foods and nutrients are listed.

benefits of antioxidant supplementation. As vitamin usage becomes more commonplace, there is growing concern regarding the development of harmful side-effects from excessive vitamin intake, whether self-prescribed or therapeutic overdose. Currently, the U.S. Food and Drug Administration (FDA) is considering the regulation of over-the-counter sales of high-dose vitamin supplements.

NOMENCLATURE AND CHEMICAL CHARACTERISTICS

Vitamin is a contraction of the term first applied by Casimir Funk in 1911 to a crystalline substance he had isolated from the outer husks of rice.[57] This substance (thiamin) was effective in curing and preventing the deficiency disease *beriberi*. Because the compound was essential for life and contained nitrogen in an amino group, Funk coined the noun *vita/amine*. However, not all vitamins are amines, and the word was later shortened and applied generically.

Vitamins are organic molecules required by the body in amounts ranging from micrograms to milligrams per day for maintenance of structural integrity and normal metabolism. They perform a variety of functions in the body. B complex vitamins enhance the activity of enzymes by serving as components or precursors of coenzymes or prosthetic groups. Vitamins E, C, and beta-carotene (a precursor of vitamin A) are antioxidants that scavenge cell-damaging free radicals. A metabolite of vitamin D is a calcium and phosphate regulating hormone.

Thirteen organic compounds are known to serve as vitamins in humans. Mainly supplied by the diet, some vitamins are also synthesized by the body and its intestinal microflora (Table 29-2). Although all vitamins share a designation as es-

sential micronutrients, these organic molecules do not possess a common chemical structure. Vitamins were identified as essential nutrients before their chemical structures were determined. They were therefore referred to by letters and numbers supplied at the time of discovery. Official (trivial) names were assigned following chemical identification. Current accepted practice is to use the trivial name of a vitamin rather than the cumbersome systematic chemical name as established by the Commission on Biological Nomenclature (CBN) of the International Union of Pure and Applied Chemistry (IUPAC) (Table 29-3).

Related compounds that interconvert to or substitute for the functional form of the vitamin are known as *vitamers*. A generic descriptor may be applied to all compounds qualitatively exhibiting the biological activity of the vitamin. For example, the term *folate* would be used to include folic acid and its relatives.[3] Pyridoxine and the related compounds, pyridoxal and pyridoxamine, are all vitamers of B_6.

For convenience and simplicity, vitamins of diverse chemical structure are unified on the basis of solubility characteristics. Those soluble in water are the eight members of

TABLE 29-2. Sources of Vitamins

DIETARY	PARTIALLY ENDOGENOUS
Vitamin A	Vitamin D
Thiamin	Niacin
Riboflavin	
Vitamin B_6	
Vitamin B_{12}	INTESTINAL MICROFLORA
Folates	Vitamin K
Pantothenic acid	Biotin
Vitamin C	
Vitamin E	

TABLE 29-3. Fat-Soluble Vitamins Required in Humans

Generic Descriptor	Plasma Adult Reference Ranges*	Trivial Name* of Major Vitamer	Structural Formula
A	250–750 μg/L 50–300 mg/dL (varies with diet)	All-*trans* retinol β Carotene (provitamin A)	Vitamin A 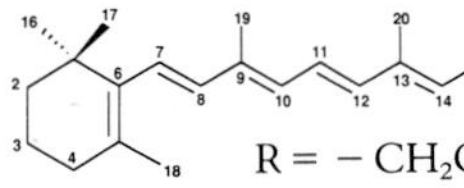 R = − CH$_2$OH
D	1–5 ng/mL 10–50 ng/mL 15–45 pg/mL	Cholecalciferol (D$_3$) 25-hydroxycholecalciferol 1,25-dihydroxycholecalciferol	Vitamin D 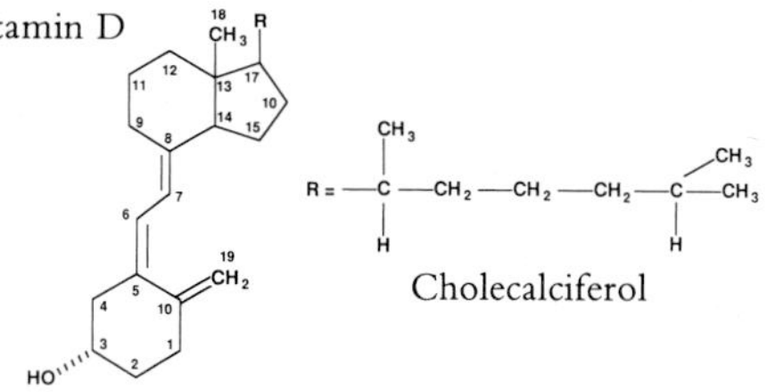 Cholecalciferol
E	5–12 mg/L	Tocopherol (alpha)	Vitamin E 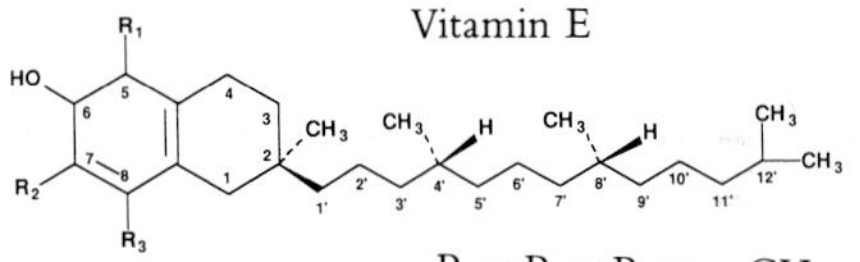 R$_1$ = R$_2$ = R$_3$ = − CH$_3$
K	250–2500 ng/L	Phylloquinone (K$_1$)	Vitamin K$_1$ 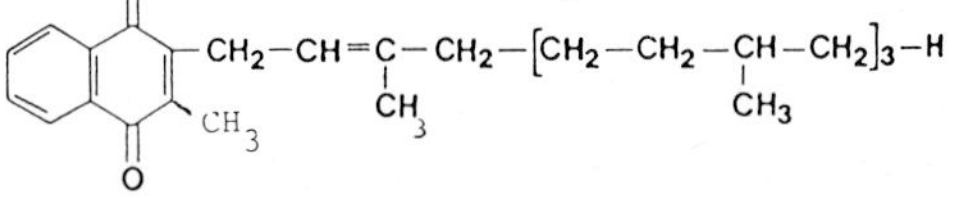

Varies with assay technique.

the B complex of vitamins (thiamin, riboflavin, niacin, B$_6$, pantothenic acid, B$_{12}$, biotin, and folate) and vitamin C. Lipid-soluble vitamins, all isoprene derivatives, include A, D, E, and K. The 5-carbon isoprenoid unit is readily identifiable in the structural formulas of vitamins A, E, and K (Figs. 29-1 and 29-2).

Water-soluble vitamins are readily excreted in the urine and are less likely to accumulate to toxic levels in the body than fat-soluble vitamins, which are stored in the liver or adipose tissue. However, cases of toxicity from ingestion of large amounts of water-soluble vitamins (*e.g.,* niacin, B$_6$) are reported in the medical literature. As the percentage of the population over 65 years of age continues to increase, awareness of the special risks to vitamin status associated with aging is attracting increasing attention among clinicians and laboratorians. In the elderly, age-associated deterioration of renal function increases the retention time of water-soluble vitamins and increases the potential for harm if large quantities of these vitamins are ingested. With aging, the percentage of body fat increases while lean body mass declines. Lipid-soluble vitamins, deposited in adipose tissue, may therefore accumulate to toxic levels more readily in seniors than in young adults.[167]

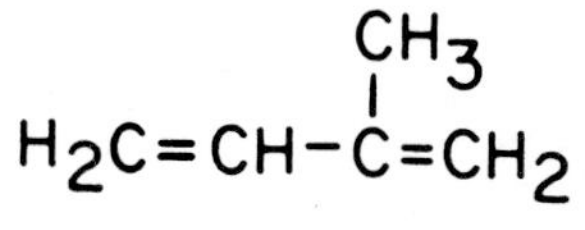

DIETARY RECOMMENDATIONS

The clinical signs and symptoms of vitamin depletion in humans have been observed by physicians for centuries. Correlation of diet and disease was made by James Lind in 1753, when he described the use of citrus fruits to prevent development of scurvy among British sailors during long sea voyages.[9] Systematic research at the beginning of the 20th century employed dietary manipulation to induce deficiency diseases in animals. In the late 1920s and 1930s, organic chemists isolated, characterized, and synthesized the majority of vitamins. The most recently discovered vitamin, B$_{12}$, was isolated in 1948.[206] Today, inexpensive, synthetic vitamins

$$H_2C = CH - \underset{\underset{CH_3}{|}}{C} = CH_2$$

▲

Figure 29-1. Isoprene (2-methylbutadiene).

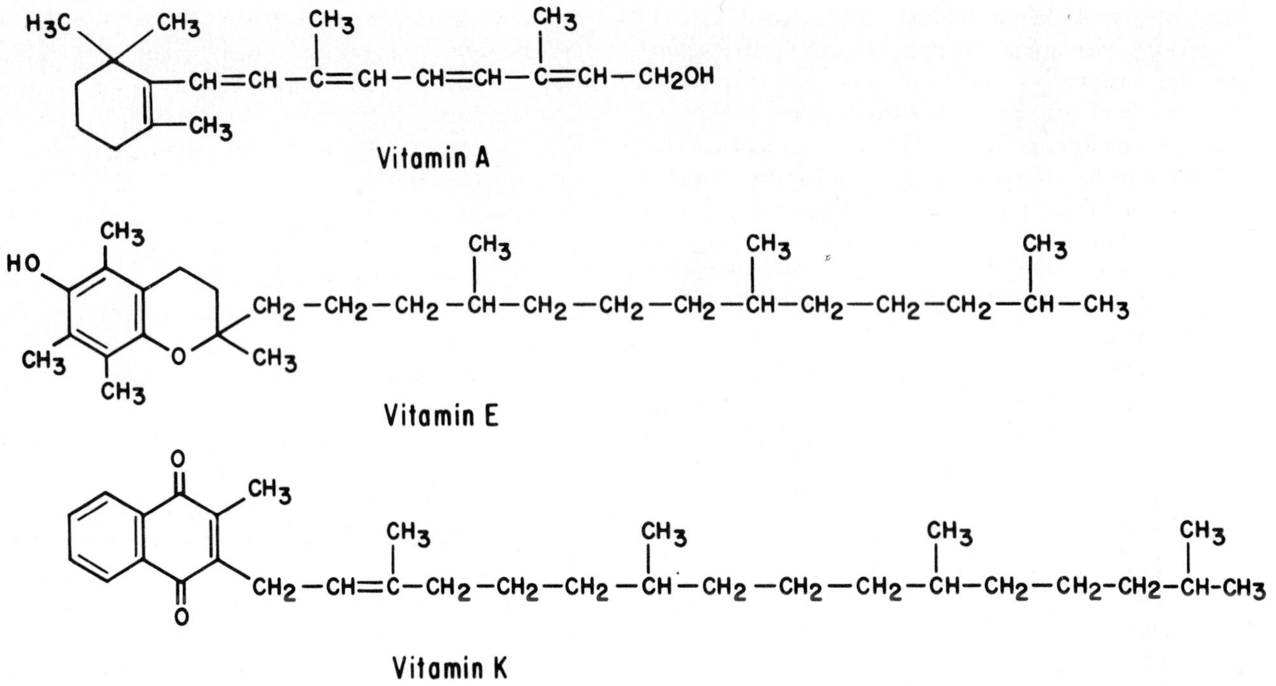

Figure 29-2. Fat-soluble vitamins composed of isoprene subunits.

TABLE 29-4. Food and Nutrition Board, National Academy of Sciences–National Research Council
Designed for the maintenance of good nutrition of practically all healthy people in the United States

Category	Age (years) or Condition	Weight[b] (kg)	Weight[b] (lb)	Height[b] (cm)	Height[b] (in)	Protein (g)	Vitamin A (μg RE)[c]	Vitamin D (μg)[d]	Vitamin E (mg a-TE)[e]	Vitamin K (μg)	Vitamin C (mg)	Thiamin (mg)
Infants	0.0–0.5	6	13	60	24	13	375	7.5	3	5	30	0.3
	0.5–1.0	9	20	71	28	14	375	10	4	10	35	0.4
Children	1–3	13	29	90	35	16	400	10	6	15	40	0.7
	4–6	20	44	112	44	24	500	10	7	20	45	0.9
	7–10	28	62	132	52	28	700	10	7	30	45	1.0
Males	11–14	45	99	157	62	45	1,000	10	10	45	50	1.3
	15–18	66	145	176	69	59	1,000	10	10	65	60	1.5
	19–24	72	160	177	70	58	1,000	10	10	70	60	1.5
	25–50	79	174	176	70	63	1,000	5	10	80	60	1.5
	51+	77	170	173	68	63	1,000	5	10	80	60	1.2
Females	11–14	46	101	157	62	46	800	10	8	45	50	1.1
	15–18	55	120	163	64	44	800	10	8	55	60	1.1
	19–24	58	128	164	65	46	800	10	8	60	60	1.1
	25–50	63	138	163	64	50	800	5	8	65	60	1.1
	51+	65	143	160	63	50	800	5	8	65	60	1.0
Pregnant						60	800	10	10	65	70	1.5
Lactating	1st 6 months					65	1,300	10	12	65	95	1.6
	2nd 6 months					62	1,200	10	11	65	90	1.6

[a]The allowances, expressed as average daily intakes over time, are intended to provide for individual variations among most normal persons as they live in the United States under usual environmental stresses. Diets should be based on a variety of common foods in order to provide other nutrients for which human requirements have been less well defined. See Reference 181 for detailed discussion of allowances and of nutrients not tabulated.

[b]Weights and heights of Reference Adults are actual medians for the U.S. population of the designated age, as reported by NHANES II. The median weights and heights of those under 19 years of age were taken from Hamill et al. (1979). [181] The use of these figures does not imply that the height-to-weight ratios are ideal.

are readily available. Chemically, there is no difference in reactivity of identical preparations of naturally occurring and synthetic vitamins. However, a vitamin present in food will be a mixture of vitamers, while a supplement is likely to contain only one form of the vitamin.[97] Investigations over the past 50 years have focused on determining the physiologic functions and therapeutic uses of vitamins. The popularization of vitamins and interest by laymen in megavitamin therapy began in 1970 with the publication of *Vitamin C and the Common Cold*, by Linus Pauling.[197]

The amount of a vitamin that should be ingested by a healthy individual to meet routine metabolic needs and allow for biologic variation, maintain normal serum concentrations, prevent depletion of body stores, and thus preserve normal function and health is reflected in the RDAs (Recommended Dietary Allowances) published by the Food and Nutrition Board of the National Academy of Sciences—National Research Council (NAS/NRC) (Table 29-4).[181] Knowledge of daily vitamin requirements is not as precise as RDA figures may suggest. The required levels of vitamins have not been established conclusively. Other countries employ different criteria to establish recommended nutrient intakes (Table 29-5).[177] For example, various countries have

recommended adult intake of vitamin C ranging from 30 to about 100 mg/day for nonsmoking adult males. In recent years, there has been growing interest in the consumption of vitamins for "special purposes." Vitamin supplements are defined by the American Medical Association (AMA) as amounts 0.5 to 1.5 times greater than the U.S. RDA. The effectiveness of vitamin supplements in reducing the risk of developing diseases such as cancer, cardiovascular disease, and neural tube defects is currently an area of considerable controversy.[203] When used therapeutically, *e.g.,* to treat deficiency states or existing pathologic conditions, vitamins are generally taken in amounts ranging from 2 to 10 times the RDA.

Although based on current scientific knowledge, RDAs serve only as guidelines for assessing the adequacy of vitamin intake. The Food and Nutrition Board acknowledges that RDAs are neither minimal requirements nor necessarily optimal levels of intake. Requirements for many vitamins vary with protein-calorie intake and are altered by illness, injury, pregnancy, and other conditions of physiologic stress. Drug/nutrient interactions impact vitamin status through a variety of biochemical mechanisms (Table 29-6).[131] The physiologic changes of normal aging and the occurrence of

Recommended Daily Dietary Allowances, Revised 1989

Water-Soluble Vitamins					Minerals						
Riboflavin (mg)	Niacin (mg ne)f	Vitamin B6 (mg)	Folate (µg)	Vitamin B12 (µg)	Calcium (mg)	Phosphorus (mg)	Magnesium (mg)	Iron (mg)	Zinc (mg)	Iodine (µg)	Selenium (µg)
0.4	5	0.3	25	0.3	400	300	40	6	5	40	10
0.5	6	0.6	35	0.5	600	500	60	10	5	50	15
0.8	9	1.0	50	0.7	800	800	80	10	10	70	20
1.1	12	1.1	75	1.0	800	800	120	10	10	90	20
1.2	13	1.4	100	1.4	800	800	170	10	10	120	30
1.5	17	1.7	150	2.0	1,200	1,200	270	12	15	150	40
1.8	20	2.0	200	2.0	1,200	1,200	400	12	15	150	50
1.7	19	2.0	200	2.0	1,200	1,200	350	10	15	150	70
1.7	19	2.0	200	2.0	800	800	350	10	15	150	70
1.4	15	2.0	200	2.0	800	800	350	10	15	150	70
1.3	15	1.4	150	2.0	1,200	1,200	280	15	12	150	45
1.3	15	1.5	180	2.0	1,200	1,200	300	15	12	150	50
1.3	15	1.5	180	2.0	1,200	1,200	280	115	12	150	55
1.3	15	1.6	180	2.0	800	800	280	15	12	150	55
1.2	13	1.6	180	2.0	800	800	280	10	12	150	55
1.6	17	2.2	400	2.2	1,200	1,200	320	30	15	175	65
1.8	17	2.2	280	2.6	1,200	1,200	355	15	19	200	75
1.7	20	2.1	260	2.6	1,200	1,200	340	15	16	200	75

c Retinol equivalents. 1 retinol equivalent = 1 µg retinol or 6 µg β-carotene. See text for calculation of vitamin A activity of diets as retinol equivalents.

d As cholecalciferol. 10 µg cholecalciferol = 400 IU of vitamin D.

e α-Tocopherol equivalents. 1 mg d-α tocopherol = 1 α-TE. See Reference 181 for variation in allowances and calculation of vitamin E activity of the diet as α-tocopherol equivalents.

f 1 NE (niacin equivalent) is equal to 1 mg of niacin or 60 mg of dietary tryptophan.

TABLE 29-5. Recommended Dietary Allowances for Vitamin C for Various Countries

Country	Allowance (mg/day)[a]	Country	Allowance (mg/day)
Australia	30	Japan	50
Bolivia	60	Malaysia	30
Brazil	75	Mexico	50
Bulgaria	95	New Zealand	60[b]
Canada	60[b]	Norway	60
Colombia	40	Philippines	75
Czechoslovakia	60	South Africa	60
Denmark	60	South Korea	55
England	30	Spain	60
FAO/WHO	30[c]	Sweden	60
Finland	60	Switzerland	75
France	80[b]	Taiwan	60
Holland	50	Thailand	30
Hungary	30	Uruguay	30
India	50	U.S.A.	60
Israel	70	*U.S.S.R.	78[a]
Italy	60	Venezuela	60

[a]*Recommendations are for moderately active, adult males.*
[b]*Higher intake recommended for smokers: Canada 80–85, France 120, New Zealand 75.*
[c]*Recommended intake by World Health Organization (WHO): 30.*
**Prior to breakup of Soviet Union.*

age-dependent diseases may significantly affect the body's need for specific micronutrients.[74] RDAs for adults 50 years of age and older have been established for some vitamins. However, many investigators advocate establishment of RDAs for seniors based upon narrower chronological intervals, for example, 50 to 64, 65 to 74, 75 to 84, and 85 years of age or older. Along with refinement of RDA requirements for the elderly, there is a need to define age-related vitamin reference ranges.[215]

Amounts of vitamins found in tissue samples and body fluids are expressed in terms of weight or international units (IU), which reflect biologic activity of an extract under a specified set of conditions. IUs are an older system of measurement originating at a time when vitamin assessment was possible only with bioassays. Quantitation by international units has been officially discontinued, but their use persists, especially in the popular press, for vitamins A, D, and E.

There are tremendous differences in normal circulating levels of the vitamins. Assay methods must be capable of measuring very small amounts of substances in body fluids or tissues. Concentrations vary from picogram (10^{-12}g) quantities of the active metabolite of vitamin D, B_{12}, and biotin, to milligram amounts (10^{-3}g) of vitamins C and E per milliliter of serum.[130,250]

ANALYTICAL METHODS

In judging the vitamin status of a patient, the physician may choose to rely on observation of overt clinical signs of deficiency or intoxication. Marginal or subclinical problems will not be discovered if this is the sole means of assessment. Dietary studies may also be conducted in which records are kept of a patient's food consumption, usually over a period of several days. The vitamin content of the foods ingested is determined by reference to a nutrient data base and the total amount of vitamin ingested is calculated. Vitamin status is then inferred from estimated dietary intake compared to a standard reference such as the RDAs. Because of their objectivity, specificity, and sensitivity, laboratory procedures play a key role in assessing vitamin status.[116] Vitamin deficiency may be viewed as developing through several stages (Table 29-7). Laboratory findings can permit the detection of vitamin imbalance before cellular damage occurs and clinical symptoms appear. Evaluation of vitamin status by biochemical tests requires interpretation of laboratory data. Calibration standards and quality control materials are not readily available for some vitamins. Reference ranges are needed for all vitamins, but collection of appropriate reference value data has been challenging. Assessment of vitamin nutriture should take into account the physiologic continuum from obvious deficiency to toxicity. The optimal vitamin concentration to support normal metabolism should lie somewhere in between (Fig. 29-3).[131] Reference values are affected by age and gender and are method-specific. Particularly in the elderly, there may be a lack of correlation among clinical manifestations, dietary intake, and biochemical measures of nutrients.[127] It would be premature to conclude that biochemical measures exclusively should be used to assess vitamin status. They are powerful tools for acquiring critical

TABLE 29-6. Drug/Nutrient Interaction Mechanisms

Biochemical Mechanism	A Drug/Nutrient Example	Effect on Nutrient
Solubilization	Mineral oil/tocopherol	Decreased absorption
Derivitization	Isoniazid/pyridoxal phosphate	Deactivation and excretion
Adsorption	Cholestyramine/retinol	Decreased absorption
pH Change	Sodium bicarbonate/folate	Decreased absorption
Enzyme inhibition	Methotrexate/folate	Decreased utilization
Enhanced utilization	Alcohol/thiamin	Increased requirement

Adapted from Labbe RF, Veldee M. Nutrition in the clinical laboratory. Clin Lab Med 1993;13:313.

TABLE 29-7. Changes in the Development of Vitamin Imbalance	
Stage	**Response**
1	Nutrient availability inadequate
2	Nutrient stores depleted
3	Biochemical (or physiologic) processes affected
4	Cellular or tissue changes (symptoms) appear

Source: Labbé RF, Veldee M. Nutrition in the clinical laboratory. Clin Lab Med 1993;13(2):313.

information on a patient's vitamin nutriture. Continuing research is essential to maximize their clinical usefulness.

Biochemical methods of vitamin assessment may be categorized as direct or indirect. Direct methods measure the level of the vitamin, its precursor, or metabolities in body fluids and tissues. Such static tests provide information on the level of vitamin in the particular body "pool" from which the sample was obtained. Indirect procedures quantitate the vitamin by evaluating a particular physiologic response or biochemical action that depends on availability of the vitamin. Such functional tests are usually enzymatic or metabolic.[242]

Vitamin levels are measured by a variety of procedures, used singly or in combination, including biological and microbiological assays, "loading tests," enzyme activation tests, colorimetric, spectrophotometric, and fluorometric methods, high-performance liquid chromatography, electrochemical methods, radioimmunoassays, and competitive protein-binding techniques. Specimens in which vitamin levels are routinely determined include whole blood, plasma, serum, or special preparations of blood cells. A fasting specimen is preferable, since recent intake of food or supplements can affect blood levels of vitamins with short plasma elimination half-lives (e.g., pyridoxine and folic acid). If plasma is evaluated, recommended anticoagulants have been lithium heparinate for thiamin, riboflavin, vitamins A, E, D, and beta-carotene. Heparin has been reported to modestly reduce B_{12} levels. Ethylenediaminetetraacetic acid (EDTA) has been suggested for use with specimens for vitamin B_6 and folate determinations.[101] For vitamins that are subject to oxidation (e.g., ascorbic acid, alpha-tocopherol, and beta-carotene), or photodecomposition (e.g., alpha-tocopherol and beta-carotene), special collection, handling, and storage procedures are generally necessary. However, a recent study of vitamin stability in sera subjected to repeated freezing and thawing suggests no adverse effects on levels of retinol, tocopherol, and carotenoids.[113] Urine is also an acceptable specimen for assessment of vitamin nutriture, especially when vitamin metabolites or abnormal metabolic products arising from a specific vitamin deficiency are to be measured. Ideally, a 24-hour specimen should be collected.

Historically, biological and microbiological assays were the principal means of determining vitamin content of specimens. Although bioassays require a large specimen volume and are cumbersome, time-consuming, expensive, hard to standardize, and often difficult to evaluate, they remain the standard of reference against which all other test methods are judged. Rats or chicks are the usual experimental animals in bioassays. The animal is placed on a carefully defined diet that will produce deficiency of one specific vitamin. When the animal has reached a depleted condition, it is fed the test material, and any subsequent improvement in its condition is noted. Bioassays measure all chemical substances in the sample that possess the biologic activity of that vitamin. The amount of activity present in the test material is quantitated by comparing its biopotency with that of a graded series of preparations whose vitamin content is known. Reversal of skeletal deformities in animals with rickets is a classic method for determining vitamin D levels in test doses.[5]

Microbiological methods of assessment are based on stimulation of growth in such vitamin-dependent microorganisms as *Lactobacillus casei* or *Lactobacillus leichmanii*.[5] The organism is cultured in the presence of the test material, and density of the cell population is determined turbidimetrically after a specified interval, usually several days. The amount of

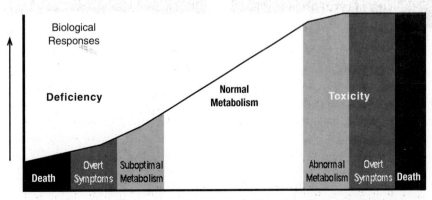

Figure 29-3. Physiologic continuum of an essential nutrient. Source: Labbe RF, Veldee M. Nutrition in the clinical laboratory. Clin Lab Med 1993;13(2):313.

lactic acid liberated into the culture medium also can be taken as a measure of cell population. Although such procedures are generally both sensitive and specific, the time requirement reduces their clinical usefulness. No microorganism is known to require ascorbic acid. Therefore, the only water-soluble vitamin for which a microbiological assay has not been used is vitamin C.[227] Blood levels of folate, biotin, and pantothenic acid are still considered by many to be most reliably measured by microbiological assay.

Loading tests are based on the accumulation of metabolic intermediates or abnormal metabolites in the body fluids of patients following oral administration of a large or "loading" dose of a specific substrate. A deficiency of vitamin cofactor impairs enzyme-catalyzed substrate utilization by normal metabolic pathways. Products of this biochemical lesion are measured in body fluids, and levels of these metabolites are correlated with the vitamin status of the patient. Loading tests with the amino acids tryptophan and histidine have been used routinely to assess patient vitamin B_6 and folate, respectively.[6]

Chemical methods of vitamin assessment require only small samples and are technically simple and rapidly performed. Thus, they lend themselves to automation. Older conventional colorimetric and spectrophotometric procedures depend on the formation of a colored complex in solution. Quantitation is possible by reference of solution absorbance to a standard curve. A colorimetric method for determination of ascorbic acid based on its reducing properties has been successfully automated for centrifugal analysis.[32] The primary disadvantage of many chemical procedures is lack of specificity and/or sensitivity. Fluorometric methods have proven to be extremely useful because of their sensitivity. Vitamin A is a highly fluorescent compound that may be measured by a rapid, simple, and inexpensive assay for serum retinol that is especially well suited for use in physician's office laboratories (POLs) or clinics where sophisticated laboratory facilities may not be available.[243,261] To minimize interference from fluorescent naturally occurring or pharmacologic substances in body fluids, specimen pretreatment may be required in some fluorometric analyses.

High-performance liquid chromatography (HPLC) is already employed in the clinical laboratory for the rapid analysis of drugs. For many vitamin assays, it is now the technique of choice because of its fast analysis times, good limits of detection, and ease of automation.[116,223] HPLC makes possible the separation and quantitative recovery of closely related chemical forms of the vitamins.[158] Simultaneous resolution of multiple vitamins and vitamers in the same specimen is possible.[39,281] Commercial systems are available for use in the clinical laboratory that are fully automated, from sample injection to data reduction. HPLC is capable of quickly and accurately resolving and reproducibly quantitating nanogram quantities of vitamins. Preliminary purification and concentration of the specimen are usually necessary. Either reverse-phase or normal-phase procedures are coupled with fluorometric, ultraviolet, or electrochemical detection.

HPLC has been successfully applied to the quantitation of ascorbic acid; vitamins A, beta-carotene, D, E, and K; thiamin; and riboflavin.[116,149,271]

Electrochemical analytical techniques generally lack specificity. However, in combination with liquid chromatography (LC), electrochemical detection (ED) has proved useful for quantitation of the small amounts of vitamins present in plasma or serum, with a minimum of sample preparation. Determinations are based on analysis of electrical potentials arising from electrochemical oxidation of purified vitamin fractions at various types of indicator electrodes. The detector's output signal may be displayed on a strip chart, an integrator, or a computer-based data station. A flow-through coulometric fritt type electrode is not simply a higher-efficiency variant of the traditional amperometric approach. Cells designed with multiple electrodes in series allow for increased resolution when the reaction potentials of the individual electrodes are set at different values. The sensor surface in the coulometric cell is more than 50 times the area of conventional amperometric sensors. Coulometric sensors are reported to provide greater resolution and consistency (W. Matson, ESA Inc., Bedford, MA. Personal Communication, 1994) than is possible with amperometric sensors.[38] LC/ED assays have been used to rapidly and efficiently measure levels of ascorbic acid, beta-carotene, vitamins E, D, and K, and several B complex vitamins, including B_6 and folic acid.[47,92,223,227]

Radioligand-binding assays are saturation-competition procedures employing proteins with high affinity and high specificity for particular vitamins. The protein utilized may be an antiserum prepared by animal immunization, as in radioimmunoassay. Naturally occurring transport or receptor proteins are employed in competitive protein-binding assays. The vitamin molecule (ligand) associates reversibly with the binding protein. A reaction mixture containing a fixed quantity of binding protein, trace amounts of radioactive vitamin, and the sample, containing unlabeled vitamin, is incubated. Since the concentration of binding protein is held constant, competition for binding sites will occur between the labeled and nonlabeled vitamin. The greater the concentration of nonradioactive vitamin in the test material, the less radioactive ligand will be bound. Bound and unbound vitamins are separated, and the radioactivity of either phase may be measured. Quantitation of the test material is possible by reference to a calibration curve prepared using ligand standards. Radioligand-binding assays are extremely sensitive procedures. They are suitable for detecting picogram amounts of vitamins in biologic materials. Folates and vitamin B_{12} are routinely measured by means of commercially available radioassay kits.[232] Competitive protein-binding methods have been particularly useful in assays for metabolites of vitamin D.[122] A fully automated, chemiluminescent, competitive receptor assay using an acridinium ester label and paramagnetic particle separation has been developed for quantitating vitamin B_{12} that correlates well with RIA assays.[20]

Erythrocyte enzyme activation tests are commonly used to assess levels of thiamin, riboflavin, and vitamin B_6 in patient specimens.[6,116,223] Hemolysates of whole blood or erythrocytes are prepared. Enzyme activity of sample aliquots are compared *in vitro* before and after addition of saturating amounts of vitamin cofactor. Enhancement of enzyme activity upon saturation with the vitamin is expressed as an *activity coefficient (AC)*. The greater the AC, the more likely the patient is to be deficient in the vitamin. Guidelines are available for correlating the stimulatory effect of the added vitamin with the adequacy of *in vivo* vitamin levels. Enzymes whose activities are routinely assessed include transketolase for thiamin, glutathione reductase for riboflavin, and aspartate aminotransferase (AST) or alanine aminotransferase (ALT) for vitamin B_6. A variant of this type of assay is enzyme reactivation *in vitro* using highly purified apoenzymes from bacteria. Activity of the vitamin-requiring apoenzyme will be restored in proportion to the amount of specific vitamin present in the test material. Enzyme activation analyses are highly specific. Only the metabolically active form of vitamin is capable of stimulating enzyme activity *in vitro*. All other vitamers will be undetected, since they require prior *in vivo* conversion to the active form.

Until recently, limitations in our knowledge and technology have prohibited direct assessment of vitamin K. Vitamin E assessment was, at one time, similarly restricted. If more sophisticated testing is unavailable, the laboratory may choose to determine levels of these vitamins by indirect biologic methods. Clinical examination and a detailed dietary history are essential to support these functional tests. Vitamin K is required for synthesis of clotting Factors II, VII, IX, and X. A prothrombin time prolonged more than 2 seconds beyond the control may indicate a deficiency in vitamin K.[116,223] Vitamin E status in a patient may be estimated by aggregation of platelets in response to ADP addition or by the erythrocyte hemolysis test. Sensitivity of washed erythrocytes to a 2% solution of the oxidant, hydrogen peroxide, is determined by the amount of hemoglobin liberated from cells incubated for 3 hours. The amount of hemoglobin liberated is compared with that produced by hemolysis in distilled water. The result is expressed as a percentage, with increased RBC fragility indicated by values greater than 20%. This suggests vitamin E deficiency.[116,223]

Measurement of breath pentane output also has been suggested as a sensitive, noninvasive functional test for assessing vitamin E status. Excretion of pentane, a product of lipid peroxidation, is increased in subjects with plasma vitamin E deficiency. Pentane exhalation decreases following supplementation with the vitamin.[138]

To understand the clinical significance of vitamin-assessment procedures, it is necessary to be familiar with the basic chemistry and physiologic functions of these chemically diverse, essential micronutrients. It is also helpful to know the clinical manifestations of deficiency and toxicity. The following sections provide an overview of vitamins for the clinical laboratorian.

FAT-SOLUBLE VITAMINS

Vitamin A

Vitamin A occurs as vitamin A_1 and vitamin A_2. A_1 is found predominantly in mammals and marine fish; A_2 occurs in freshwater fish.[151] The molecules possess an unsaturated substituted cyclohexane ring, 11-membered hydrocarbon side chain, and polar terminal group (see Table 29-3). The carbocyclic ring of A_2 contains one more double bond than does the A_1 ring.

Biologically active forms of vitamin A include the alcohol, aldehyde, acid, and esters referred to, respectively, as *retinol, retinal, retinoic acid,* and *retinyl esters* (Fig. 29-4). Vitamin A and its derivatives along with structurally related naturally occurring and synthetic compounds possessing vitamin A activity are collectively termed retinoids.[181,190] Dietary sources of vitamin A include fish, liver, butter, whole milk, cheese, and eggs.[147] Animal products contain the vitamin predominantly as retinyl esters. All trans–beta carotene is the major source of vitamin A for most of the world's population.[188,262] Of the over 400 carotenoids widely distributed among plants, approximately 50 serve as provitamins or precursors of vitamin A (Fig. 29-5). These yellow-orange pigments are particularly abundant in yellow and dark green leafy vegetables such as carrots, squash, broccoli, and spinach.[147]

For nutritional purposes, there are several different ways of reporting vitamin A activity. Data may be expressed in terms of IUs, RE (retinol equivalents), or the actual amount of substance (*e.g.,* retinol, beta carotene) present. Vitamin A activity is now conventionally stated in terms of microgram retinol equivalents (RE). One RE equals 1 μg of all-*trans* retinol, 6 μg of all-*trans* beta carotene, or 12 μg of other provitamin A carotenoids.[110] The RDA of vitamin A for adult males is 1000 retinol equivalents (RE). The allowance for adult females is 800 RE and approximately 400 RE for infants and young children.[181] However, it is not uncommon to find vitamin A activity in vitamin supplements and foods expressed in the older system of international units (IU). One RE equals 3.33 IU vitamin A activity for retinol. One IU equals the biologic activity of 0.3 μg of all-*trans* retinol or 0.6 μg of all-*trans* beta carotene.

Any condition, disease, or drug that impairs fat absorption decreases intestinal uptake of all lipid-soluble vitamins (Fig. 29-6).[189a,212] A common indicator of such malabsorption is steatorrhea. Long-term fat malabsorption due to intestinal, hepatobiliary, and pancreatic disorders, including cystic fibrosis, or loss of bowel surface can produce a deficiency of vitamin A and carotenes.[29] Chronic use of laxatives, as can be found among elderly concerned with "regularity of bowel movements" and adolescent females with eating disorders, may reduce micronutrient uptake by damaging the intestinal mucosa and shortening the time available for absorption in the GI.

Dietary retinal must be reduced to retinol or converted to retinoic acid before it can be further metabolized. Retinol is

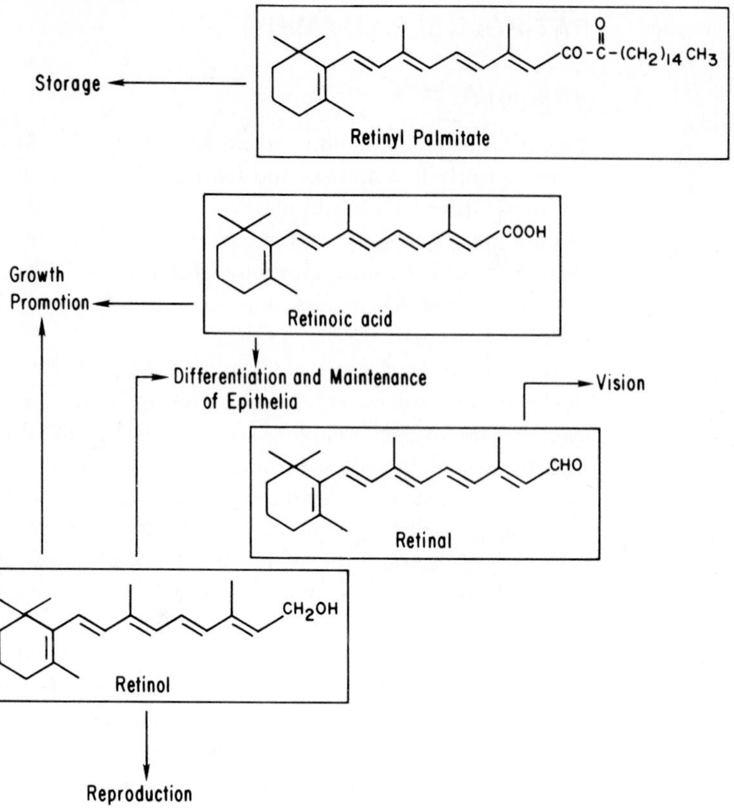

Figure 29-4. Physiologic function of vitamin A compounds.

enzymatically esterified during intestinal absorption. Retinyl esters (primarily retinyl palmitate) are transported in chylomicra via the lymphatics to the general circulation.[19,212] Plasma contains essentially no retinyl esters, except in cases of excessive vitamin A intake.[189] Using HPLC techniques, serum retinyl ester values exceeding 11 μg/dL or greater than 10% of total circulating vitamin A indicates excessive liver stores of the vitamin.[116] When chylomicron remnants are cleared from the blood, most of the esterified retinol is taken up by and stored in the liver. Typically, 80% to 90% of the body's total retinol (retinol plus retinyl esters) is normally present in this organ.[19] Retinoic acid (RA) is absorbed without modification into the portal circulation and is bound to albumin for transport. RA exhibits biologic activity in differentiation and maintenance of epithelia and in growth promotion (Fig. 29-6). However, it is not stored and is rapidly metabolized and excreted in the bile and urine.[151] Beta-carotene (Fig. 29-5), the primary precursor of vitamin A, is enzymatically cleaved to retinal and reduced by the mucosal cells of the small intestine to retinol.[19,262] Carotenes are also absorbed in the intestine without conversion to retinol and incorporated into chylomicrons. Because of inefficiencies in absorption and utilization, biologic activity of beta-carotene is only slightly greater than one-half that of the preformed vitamin A.[181]

The plasma carotene reference range by HPLC is 50 to 300 μg/dL (0.93–5.58 μmol/L). Less than 50 μg/dL would be consistent with steatorrhea.[29] Production of the enzyme beta carotene-15,15′-dioxygenase that catalyses the cleavage of the provitamin to yield two molecules of retinal is inhibited by a rising concentration of vitamin A. Because of this feedback inhibition, excess beta carotene (>300 μg/dL) does not produce toxic levels of vitamin A.[201,250] Copious consumption of carotene-rich fruits and vegetables may impart a yellow color to skin and body fluids, but the sclera is not discolored as in jaundice.[151] Hypercarotenemia is also associated with hypothyroidism and hyperlipidemia, as occurs in diabetes mellitus.[29]

Before being released from storage in the liver, retinyl esters are enzymatically hydrolyzed. The liberated retinol combines with retinol-binding protein (RBP), which stabilizes the vitamin against oxidation. Retinol deficiency inhibits the secretion of RBP. In the circulation, most retinol-RBP re-

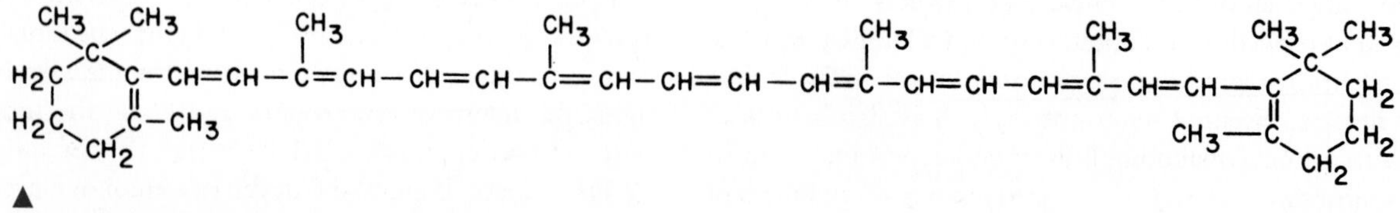

Figure 29-5. All-*trans* beta carotene (provitamin A).

VITAMIN A

UPTAKE ⁓ STORAGE ⁓ TRANSPORT ⁓ ACTION

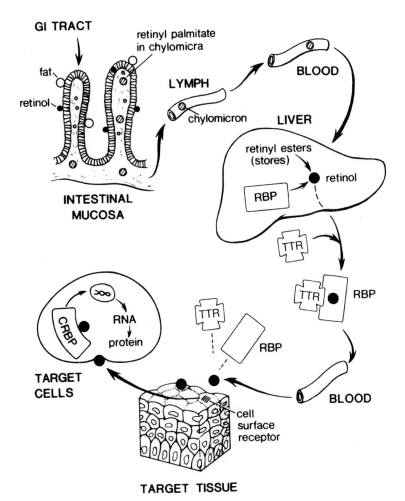

◄
Figure 29-6. Vitamin A—uptake, storage, transport, and action. TTR = transthyretin; RBP = retinol binding protein; CRBP = cellular retinol binding protein.

versibly complexes with an additional protein, transthyretin (TTR). In addition to participating in the transport of vitamin A, TTR binds and transports thyroxine in the plasma (see Fig. 29-6). Formation of this trimolecular 1:1:1 aggregate protects the vitamin from glomerular filtration and excretion in the urine.[81] Having reached its target tissues, retinol-RBP binds to a specific membrane receptor, dissociates from the complex, and enters the cell. Intracellularly, it is conveyed by cell RBP (CRBP) to the nucleus, where it alters gene activity and, thus, cell differentiation. In hypovitaminosis A, synthesis of mRNA is diminished. The roles of other cellular retinoid-binding proteins in intracellular retinol metabolism are being aggressively investigated.[19,191] Upon internalization of the retinol, RBP and TTR separate.[189a]

At birth, plasma concentrations of retinol and RBP, as well as liver stores of vitamin A, are lower in premature than in full-term infants.[101,145] Preterm neonates are, consequently, at risk for vitamin A deficiency. Diseases or nutritional disorders that reduce plasma proteins lower circulating levels of retinol despite adequate stores of vitamin A. If

patients with severe protein malnutrition are given massive protein supplements, vitamin A will be rapidly released from the liver, and hypovitaminosis A is likely to follow depletion of liver stores. Infections and parasitic infestations lower plasma retinol levels. Contrary to their depression of circulating levels of vitamins such as B_6 and folate, oral contraceptive steroids increase retinol levels by 30% to 50% by promoting RBP synthesis.[264] Plasma retinol and carotenoid concentrations increase moderately with aging. Values for males are slightly higher than for females up to age 60.[190]

The best understood physiologic function of vitamin A vitamers is in the visual cycle.[273] Rod cells of the human retina contain stacks of membrane-bound disks. These membranes contain the photosensitive pigment, rhodopsin. Rhodopsin is a complex of the prosthetic group, retinal, and the protein opsin. During synthesis of this pigment, retinol is oxidized to retinal. Absorption of light initiates molecular and electrochemical changes. "Bleached" rhodopsin dissociates, an electrical potential develops across the disk membranes, and a nerve impulse is generated. In a complex sequence of chemical reactions involving isomerization and oxidation–reduction, retinal is re-formed and again combines with opsin to create rhodopsin. Small amounts of retinol are continually lost in the visual process and must be replaced from the body's supply of vitamin A. It is the rod cells of the retina that are responsible for vision in conditions of reduced light. An early sign of vitamin A deficiency is poor vision in dim light. This condition is known as *nyctalopia,* or *night blindness.* Deterioration of vision, even in the presence of adequate supplies of the vitamin, may be linked to production of granules of the lipopigment lipofuscin in the retina. The process of lipopigment formation may be related to an increase in intracellular lipid peroxidation. Such retinal changes are usually observed among the elderly. Macular degeneration, in which the patient is unable to focus and to discern colors, may subsequently develop. Retardation of macular degeneration, a common cause of blindness among the elderly, may be linked to vitamin A status.[234a]

Less is known of the detailed role of vitamin A in such complex processes as cell differentiation and proliferation, bone development, cellular adhesiveness of epithelial tissue, reproduction, and resistance to infection.[121,151] Retinol's function in the differentiation of epithelial cells is presumably related to its role in the synthesis of specific glycoproteins. Biosynthesis of glycoproteins, including the mucoproteins of mucous secretions and cell-surface glycoproteins, requires vitamin A. In its absence, loss of goblet cells occurs in the intestinal mucosa, ocular conjunctiva, and tracheal respiratory epithelium. Transformation from mucus secreting to stratified keratinized epithelia occurs in vitamin A deficiency.

The involvement of vitamin A in cell differentiation is the basis for its relationship to cancer.[180,251] Experimental and epidemiologic evidence suggests that increased intake of carotenoids and vitamin A reduces the risk of neoplastic changes in epithelial tissues such as those of the lung, oral cavity, and urinary bladder, perhaps by prevention of chemical carcinogenesis.[180] Retinoids promote redifferentiation of metaplastic tissue and prevent the malignant transformation of epithelial tissues. Beta-carotene's function as an antioxidant may be the basis for its reported anticarcinogenic properties.[1,7] It is a strong antioxidant, capable of scavenging oxygen-free radicals and thus preventing damage to DNA.[161,278] There are conflicting reports on the relationship between serum beta-carotene and the risk of lung and colon cancer.[7,84,260,278]

Vitamin A is also necessary for synthesis of the acid mucopolysaccharide chondroitin sulfate found in bone and cartilage. In its absence, disruption of the intercellular matrix leads to skeletal deformities. In a reproductive role, vitamin A may be linked to modulation of gonadotropin receptors on cell membranes in the testes.[155] Clinical evidence supports a role for the vitamin in maintenance and proper function of the immune system.[182] Vitamin A has been called the "anti-infective" vitamin. Treatment with vitamin A reduces morbidity and mortality in measles, and supplementation for all children seriously ill with measles has been advocated. In many developing countries, efforts to increase vitamin A intake among preschool children is a major public health initiative.[259] Vitamin A deficiency is estimated to affect >124 million children under 5 years of age worldwide and is implicated in 1 to 2.5 million deaths from associated diarrheal and respiratory illnesses.[72]

Signs of vitamin deficiency include drying (xerosis), degeneration, and accompanying increased risk of infection in the conjunctiva, cornea, skin, and mucous membranes (Table 29-8).[121] Worldwide, severe vitamin A deficiency, with associated xerophthalmia, corneal ulcerations, and keratomalacia, is an important cause of blindness in young children.[145,190] Effective therapy with retinoids in severe dermatologic conditions such as acne, psoriasis, and skin cancer depends on their promotion of normal growth and development of epithelia.[21] The synthetic analogs of vitamin A offer a therapeutic advantage, since they are less toxic than the naturally occurring vitamin.

Therapeutic use of retinoids in relatively high doses carries the risk of side-effects. For example, isotretinoin (13-*cis*-retinoic acid), used in the treatment of acne, has proven to be teratogenic. Ingestion of therapeutic doses of 0.5 to 1.5 mg/kg of body weight during the first trimester of pregnancy has been linked with an increased incidence of spontaneous abortion and congenital malformations.[181]

TABLE 29-8. Deficiencies of Fat-Soluble Vitamins

Vitamin	Clinical Condition	Major Signs and Symptoms of Deficiency
A	Xerophthalmia	Xerosis (drying of conjunctiva); keratomalacia (corneal softening)
	Nyctalopia	Night blindness
	Hyperkeratosis	Accumulation of keratin in hair follicles
D	Rickets	Failure of growing bones to calcify; skeletal deformities; hypocalcemic tetany; dental abnormalities
	Osteomalacia	Softening of mature bones; fractures; muscle spasms
	Osteoporosis	Brittle bones; fractures; decreased bone density
E	Hemolytic disease of premature neonates	Increased hemolysis under conditions of oxidative stress
K	Hypoprothrombinemia	Increased prothrombin time; hemorrhage

Hypercalcemia associated with oral isotretinoin use in the treatment of severe cystic acne has been observed.[268]

Topical application of tretinoin (all-*trans* retinoic acid) has proven to be highly successful in reversing photoaging of skin. Studies continue on the antineoplastic action of topical tretinoin in treatment of actinic keratoses and basal cell carcinoma of the skin.[190,277]

Vitamin A intoxication may arise from therapeutic overdose or excessive self-medication. Toxicity occurs in adults who ingest more than 15,000 RE (50,000 IU) daily for prolonged periods.[181] Hypervitaminosis A of dietary origin is uncommon. A wide variety of toxicity symptoms occur including lethargy, drowsiness, irritability, vomiting, severe headaches, itching peeling skin, brittle nails, hair loss, bone and joint pain, hepatosplenomegaly, and benign intracranial hypertension.[238] Hypercalcemia caused by chronic hypervitaminosis A has been reported.[268] Psychiatric symptoms may be manifested.[82,121]

Older colorimetric methods for determining serum retinol, based on the measurement of transient blue complex formed at 620 nm when vitamin A is dehydrated in the presence of Lewis acids, have largely been replaced. These procedures are useful primarily in field studies where more sophisticated methods are not available. Some fluorometric assays continue to be used.[243] High carotene levels in plasma interfere with colorimetric and spectrophotometric techniques. Use of the carotenoid, canthaxanthine, as a "quick tanning agent" colors serum orange and interferes with determinations of provitamin and vitamin A. Phytofluene, a serum-contaminating carotene, interferes with fluorometric assessment of vitamin A and must be removed prior to assay.

Currently the most flexible analytical technique available for assessing vitamin A status is high-performance liquid chromatography (HPLC).[116,244] It permits the separation and individual quantitation of retinol and its esters, as well as therapeutic retinoids.[159] Simultaneous quantitation of retinoids and tocopherols is also possible. HPLC is also suitable for measuring beta carotene in serum.[125,244]

Vitamin A status assessment is usually performed on a non-hemolyzed, fasting specimen that has been protected from light. Either serum or plasma specimens can be utilized. Retinol must be released from the protein-protein complex on which it is transported and extracted into a lipid solvent for assay. Adult serum concentrations of vitamin A above 30 to 35 µg/dL suggest adequate nutriture; less than 25 µg/dL is considered to indicate deficiency with probable functional impairment; a level of 25 to 30 µg/dL is considered "dangerously low."[243] Plasma levels of vitamin A in excess of 200 µg/dL are diagnostic of intoxication. In the absence of HPLC capability, plasma retinol-binding protein levels may be used to assess serum retinol since the two values show strong correlation. Plasma RBP concentration can be determined with commercially available radial immunodiffusion plates.[116]

Patients receiving long-term or high-dose vitamin A therapy may be receiving vitamin E supplements concur-

rently. If alpha-tocopherol is also included as an internal standard, vitamin A and E levels in the patient specimen may be determined simultaneously by HPLC.[39] This capability is particularly useful, given the clinical observation that a large intake of vitamin E appears to protect against some of the effects of vitamin A intoxication.[63]

Other tests that may be used to assess vitamin A status include conjunctival impression cytology, testing for night blindness or poor dark adaptation, or the relative dose response (RDR) test.[72] In the RDR test, serum or plasma retinol is measured before and 5 hours after administration of an oral or IV retinol preparation. A percentage increase (RDR) of <10% is normal; >20% indicates low body stores of vitamin A.[116]

Vitamin D

Vitamin D is derived from a steroid whose B ring has been opened, thus making it a secosteroid. Naturally occurring vitamin D family members differ from one another in their side chain structure.[52] The vitamer synthesized in animals is D_3 (cholecalciferol); that found in plants is D_2 (ergocalciferol) (see Table 29-3). Nutritional supplements may contain either vitamer but pharmacological preparations usually contain vitamin D_2.[192] Both forms are effective in preventing or curing rickets in humans and appear to be metabolized identically by the body. Some evidence suggests that in the same doses, cholecalciferol supplementation raises serum 25-OH vitamin D levels more than ergocalciferol supplementation.[43] The RDA of vitamin D for children and pregnant or lactating women is 10 µg as cholecalciferol (400 IU). For adults of both sexes after the age of twenty-four, the RDA is 5 µg.[181] Past 50 years of age, the vitamin D requirement may again rise due to life-style and physiologic changes among older adults.[80,213]

Vitamin D_3 is provided in the diet by fish, fish oils, eggs, and liver. Milk is fortified to supply 10 µg (400 IU) of vitamin D per quart.[86] Infant formula, breakfast cereals, pasta, margarine, and butter are also routinely fortified. The fat-soluble vitamin is absorbed from the small intestine and transported in the chylomicra of the lymph. Endogenous vitamin D is derived from cholesterol. Cholecalciferol is produced in skin by photolytic action of ultraviolet light (wavelength, 285 to 310 nm) on the provitamin, 7-dehydrocholesterol, forming previtamin D_3. The previtamin slowly equilibrates with vitamin D_3.[107] Estimates are that less than ½ hour of exposure to sunlight daily permits adequate synthesis of vitamin D. Especially in older adults, increased exposure to sunshine improves vitamin D status, since cutaneous levels of the provitamin decline with aging.[149a] As the vitamin is formed, it is bound by D-binding protein (DBP) and albumin and transported in the blood. Up to 80% of both exogenous and endogenous vitamin D_3 is cleared by the liver, with storage primarily in adipose tissue. In the liver, microsomal enzymes catalyze 25-hydroxylation of the molecule to produce

25-hydroxyvitamin D_3 (25-OH-D_3), known as *calcidiol*. Calcidiol is the major circulating form of vitamin D_3. Plasma reference values for this metabolite are 10 to 50 ng/mL, whereas those for concentrations of vitamin D_3 are 1 to 5 ng/mL.[70] Because of the role of sunlight in vitamin D_3 biosynthesis, levels of these substances vary significantly with season of the year, geographic location, atmospheric conditions, and length of time spent outdoors.[107] Long-term use of sunscreens with a protection factor of 8 or higher can prevent the synthesis of previtamin D_3 in the skin. In the elderly, chronic sunscreen use can result in decreased levels of 25-OH-D_3.

Vitamin D-binding protein (DBP) transports calcidiol to the kidney. DBP binds all metabolites of vitamin D but has the greatest affinity for 25-OH-D_3. In renal mitochondria, 25-OH-D_3 undergoes a final enzymatic hydroxylation, either at the alpha position of carbon 1 or on carbon 24 (Fig. 29-7A and B). The most potent physiologically active form of vitamin D_3 is 1,25-dihydroxyvitamin D_3 [1,25-(OH)$_2D_3$], known as *calcitriol*. The activity of calcidiol and calcitriol is 1.5 and 5 to 10 times greater, respectively, than that of vitamin D.[181,202]

The adult plasma reference range for 1,25-(OH)$_2D_3$ is 15 to 45 pg/mL; for children 18 to 60 pg/mL.[181] While circu-

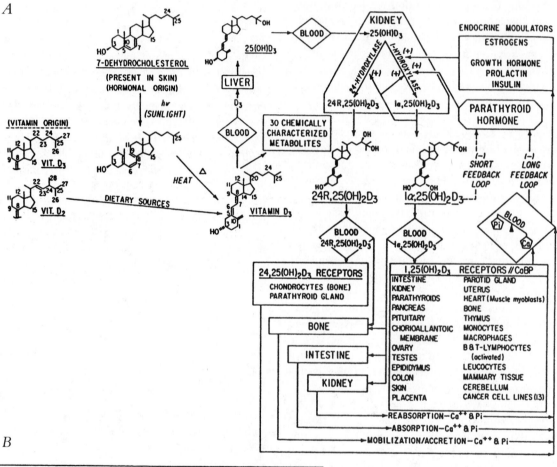

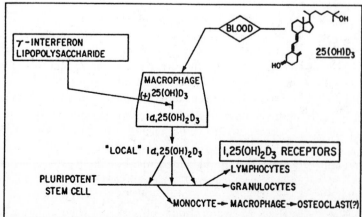

Figure 29-7. A & B. Overview of the vitamin D endocrine system. (*Adapted from* Collins ED, Norman AW. Vitamin D. In: Machlin LJ, ed. Handbook of vitamins. 2nd ed. New York: Marcel Dekker, 1991.)

lating levels of 25-OH-D have been known to show seasonal variations in all adults, only recently has seasonal variation in 1,25-$(OH)_2D_3$ levels been shown that is unique to the elderly. Calcitriol acts as a hormone, being synthesized in one location and traveling in the blood to act elsewhere. Intracellular receptor proteins present in the target tissues bind calcitriol and enable it to exert its metabolic effect by mediating gene expression (Fig. 29-8).[52,86] Intestinal uptake of calcium and phosphorus, renal reabsorption of calcium, and bone calcium and phosphorus mobilization are affected. Adequacy of 1,25-$(OH)_2D_3$ is essential for normal mineralization of bone and neuromuscular activity. When calcium and phosphorus levels in the body are within normal limits, the predominant dihydroxylated form of vitamin D that is synthesized is the less active metabolite, 24,25-$(OH)_2D_3$. In calcium-deficient individuals, the kidney produces mainly calcitriol, and plasma levels of 1,25-$(OH)_2D_3$ increase. Renal hydroxylases are indirectly regulated by blood calcium levels. Parathyroid hormone (PTH) mediates calcium's effect on the hydroxylases. The parathyroid gland responds to changing levels of calcium and phosphorus by varying PTH secretion. PTH, in turn, stimulates 1α-hydroxylase activity. A decrease in extracellular ionized calcium or a deficiency in vitamin D triggers an increase in parathyroid hormone and, subsequently, increased synthesis of 1,25-$(OH)_2D_3$. As the calcium level is restored, PTH secretion diminishes, and calcitriol synthesis declines. This feedback regulation exerted at the level of the renal hydroxylases permits fine-tuning of calcium homeostasis.[61] Low levels of inorganic phosphate directly stimulate 1α-hydroxylase activity and enhance production of calcitriol. Cells of the intestinal mucosa respond to the hormone by increased production of phosphate transport protein. Calcitriol is both a calcium and phosphate-mobilizing hormone (see Fig. 29-7A).

Assay of 1,25-$(OH)_2D_3$ is often pivotal to the differential diagnosis of disorders of calcium and phosphorus metabolism (Table 29-9).[4,106] Recent evidence suggests that calcitriol is also involved with differentiation and proliferation in the hematopoietic immune system.[52,202] Macrophages from certain tissues are able to synthesize calcitrol when activated by γ-interferon. Calcitrol appears to suppress immunoglobulin production by the activated B lymphocytes and to inhibit DNA synthesis and proliferation of both activated B and T lymphocytes (Fig. 29-7B). It also has been postulated that calcitriol may have an inhibitory effect on growth of cancer cells and plays a role in the differentiation of epidermal cells.[52,202]

A variety of diseases have been shown to have some association with vitamin D status. Rickets is the classic vitamin D deficiency disease of children. It may be nutritional or metabolic in origin (Fig. 29-9). In adults, vitamin D deficiency produces osteomalacia. Inadequate dietary vitamin D intake and insufficient exposure to sunlight are major factors contributing to poor vitamin D nutriture among the elderly.[196] Children whose diet is low in vitamin D or who have little exposure to sunlight are at high risk for rickets. This disease was particularly widespread throughout the industrialized nations at the turn of the century. Child labor and environmental pollution deprived urban youngsters of the opportunity for adequate exposure to sunlight. Nutritional requirements could, therefore, not be satisfied by endogenous vitamin D synthesis. Rickets is still encountered

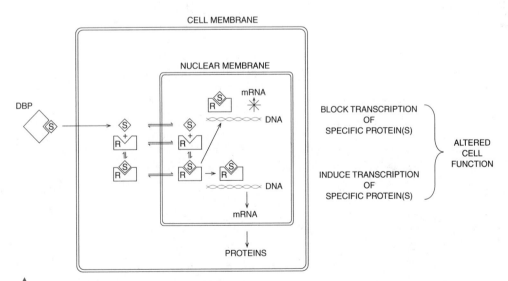

Figure 29-8. General model for the mechanism of action of steroid hormones illustrated using 1α,25$(OH)_2D_3$ as the steroid hormone. Target tissues contain receptors for the steroid which confer upon them the ability to modulate genomic events. S = steroid; R = receptor protein, which may be present inside the cell in either the cytosol or nuclear compartment; SR = steroid-receptor complex; DBP = serum vitamin D-binding protein, which functions to transport the steroid hormone from the endocrine gland to its various target tissues. (*Source:* Collins ED, Norman AW. Vitamin D. In: Machlin LJ, ed. Handbook of vitamins. 2nd ed. New York: Marcel Dekker, 1991.)

TABLE 29-9. Assay of Choice for Assessing Vitamin D Status	
1,25-Dihydroxyvitamin D	**25-Hydroxyvitamin D**
Vitamin D deficiency due to impaired renal function	Vitamin D deficiency due to nutritional causes, impaired gastro-intestinal absorption, nephrotic syndrome, or inadequate exposure to sunlight
Monitoring compliance of patients, especially renal dialysis patients, treated with 1,25-dihydroxyvitamin D (calcitriol)	Vitamin D intoxication
Differential diagnosis of primary hyperparathyroidism from the hypercalcemia of malignancy	Impaired 25-hydroxylation due to liver disease or to long-term anticonvulsant therapy
Differentiation of vitamin D–dependent rickets from vitamin D–resistant rickets	
May be of use in patients with sarcoidosis, absorptive hypercalciuria associated with kidney stones, or with parathyroid disease	

Source: From Ashby CD. Laboratory consultant-metabolites of vitamin D. Clin Chem News 1986;12(6):14.

today, particularly among children living in poor socioeconomic conditions. Skeletal abnormalities due to failure of the bones to calcify are the predominant feature of rickets. Legs are bowed and ribs are deformed; teeth are slow to erupt and have defective enamel; and abdominal muscles are weak, and in severe cases, hypocalcemic tetany occurs.[220] Osteomalacia is a condition of defective new bone mineralization in the adult skeleton due to vitamin D deficiency. Failure of newly synthesized organic matrix to mineralize leads to bones becoming soft and pliable. Seasonal osteomalacia has been reported to be common among the elderly.[86,107] Clinical features of osteomalacia include skeletal pain and tenderness accompanied by muscular weakness (see Table 29-8). Simple vitamin D deficiency rickets and osteomalacia are treated by short-term oral administration of high doses of vitamin D and calcium supplements. Rickets also may occur from failure of the body to produce calcitriol or insensitivity of target tissues to the hormone. Rickets is frequently seen in patients with chronic renal and hepatic disorders.[60] Dysfunctions of the parathyroid also lead to altered vitamin D metabolism. Drugs whose long-term use is known to disrupt vitamin D metabolism and induce metabolic bone disease include the anticonvulsants phenobarbital and phenytoin.[212,213]

Bone remodeling occurs throughout life. When resorption exceeds formation, bone density is reduced. The outcome is osteoporosis, or porous, easily fractured bones. Osteoporosis is an age-related disease that occurs in both men and women.[153,186a] In women, the process starts at or around menopause and affects nearly one-fourth of all women past the age of 50 in the United States.[152] Hip fracture is the most serious consequence of osteoporosis. Among older women, approximately 20% do not survive the first year after fracture, and an additional 20% are never able to walk again without assistance. A number of studies suggest that vitamin D has a protective effect for osteoporotic fractures by retarding bone loss.[43,107,126] Intestinal calcium absorption may decrease with age, insensitivity of target tissues to parathyroid hormone may develop, and in

postmenopausal women, lowered estrogen production is often accompanied by declining plasma levels of vitamin D metabolites. An essential role for calcitriol in regulation of osteoblast function may account for its therapeutic effectiveness at low doses in improving the body's calcium balance and reversing bone loss. The synthesis by osteoblasts of osteocalcin (bone Gla-protein), a vitamin K–dependent noncollagenous bone-specific protein, is stimulated by calcitriol. Although the precise function of osteocalcin is not known, it is incorporated into the bone matrix and binds to the bone mineral, hydroxyapatite.[176,202]

Vitamin D is stored in the body and may accumulate to toxic levels. Very rarely, improper fortification of food or beverages may cause intoxication.[118] Toxicity usually arises from long-term therapeutic use or excessive self-medication. Sustained daily intake of as little as 100 µg (4000 IU) in adults and 45 µg (1800 IU) in children have been reported as toxic. Serum values >150 ng/mL (375 nmol/L) usually indicate vitamin D intoxication.[106] Early signs and symptoms of vitamin D intoxication are associated with hypercalcemia and hyperphosphatemia, and include nausea, vomiting, diarrhea, headache, polydipsia, and polyuria. Prolonged hypervitaminosis D leads to hypertension, nephrolithiasis, and metastatic calcifications of heart, lung, kidney, and other soft tissues. Despite prompt withdrawal of the vitamin, tissue damage may be irreversible.[71]

Laboratory assessment of vitamin D metabolites is difficult because of their presence at very low levels in biologic fluids. Methods currently in use are HPLC and radioligand-binding assays. Typically, assays of vitamin D metabolites require three steps: extracting the serum with organic solvent, chromatographic purification of the specimen extract, and quantitation by radioligand (competitive binding assay), radioimmunoassay, or ultraviolet detection.

The most useful measurement for assessing vitamin D status is the determination of serum 25-hydroxyvitamin D (Table 29-9).[106] Many assays have been developed for measuring circulating levels of calcidiol. These methods are predominantly competitive protein-binding assays that make

TABLE 29-10. Autoxidation and Antioxidant Reactions Involving Vitamin E

1. Initiation (formation of a free radical)

$$LH \xrightarrow{\text{Initiators}} L\cdot$$

2. Reaction of radical with oxygen

$$L\cdot + O_2 \longrightarrow LO_2\cdot$$

3. Propagation

$$LO_2\cdot + LH \longrightarrow L\cdot + ROOH$$

4. Antioxidant reaction

$$LO_2\cdot + E \longrightarrow E + LOOH$$

5. Regeneration

$$E\cdot + C \longrightarrow E + C\cdot$$

$$C\cdot + NADPH \xrightarrow{\text{Semidehydro ascorbate reductase}} C + NADP$$

$$E\cdot + 2GSH \xrightarrow{\text{Enzyme?}} E\cdot + GSSG$$

$$GSSG + NADPH \xrightarrow{\text{Glutathione reductase}} 2GSH + NADP$$

6. Termination

$$E\cdot + E\cdot \longrightarrow E\text{-}E \text{ (dimer)}$$

$$E\cdot + LO_2\cdot \longrightarrow EOOL \text{ (?)}$$

Abbreviations: LH, fatty acid; L·, fatty acid radical; LO$_2$·, peroxy radical; E, tocopherol; E·, tocopheroxy radical; LOOH, hydroperoxide; C, ascorbic acid; C·, ascorbyl radical; GSH, reduced glutathione; GSSG, oxidized glutathione.
Source: Machlin LJ. Vitamin E. In: Machlin LJ, ed. Handbook of vitamins. 2nd ed. New York: Marcel Dekker 1991.

in the lipid phase (see Fig.29-10). Vitamin E, like ascorbic acid, inhibits the formation of nitrosamines in the stomach.[251] Intensive investigation of the anticarcinogenic role of tocopherol continues. Low serum levels of vitamin E have been related to an increased incidence of lung cancer in some, but not all, studies.[161,260] There is disagreement as to the protective value of alpha-tocopherol against prostate, breast, and colorectal cancers.[53,84,114]

It has been proposed that in circumstances where the cell's ability to cope with oxidative stress has been compromised, therapeutic use of vitamin E would be beneficial.[7] Vitamin E is found in red cell membranes. There is clinical evidence to support the therapeutic use of vitamin E as an antioxidant in cases of genetic hemolytic anemias due to G-6-PD deficiency or glutathione synthetase deficiency.[17] Epidemiologic studies suggest a role for vitamin E in reducing the risk of cataract development among older adults.[150,208,210,234] In neonates with respiratory distress syndrome, administration of vitamin E may reduce the incidence of bronchopulmonary dysplasia. Reduced risk of retinopathy of prematurity (ROP), which can result in blindness, is reported in infants receiving high concentrations of oxygen when vitamin E is administered. However, except in rare cases of vitamin E deficiency, the value of high-dose therapy in preventing tissue damage from oxygen toxicity in premature infants has not been confirmed.[12,199] Pharmacologic doses of vitamin E may decrease the inci-

dence of intraventricular hemorrhage, a common cause of death in premature infants. However, the potential for complications, such as hemorrhage due to hypervitaminosis E, leaves the value of vitamin E in neonatology still under discussion.

Megavitamin E supplementation has reportedly been useful in protecting against or in treatment of a variety of conditions and disorders. The therapeutic benefits of megavitamin E supplementation are speculative at this time. Scientific research and clinical trials are in progress to assess the long-term effects and benefits of megavitamin therapy. Research is in progress to assess the merits of megadose therapy in Parkinsonism to delay the appearance and progression of tremors, rigidity, and loss of balance, thereby postponing the need for dopamine therapy. The action of the vitamin may be related to its ability to reduce free radicals in the brain.[68,272] Patients with intermittent claudication improved after 12 to 18 months of vitamin E therapy. This may be related to an anticoagulant role played by the vitamin. Platelet aggregation and release of platelet factor 3 can be inhibited by tocopherol. Numerous studies have demonstrated that vitamin E inhibits platelet aggregation. It is suggested that the vitamin may inhibit reactions leading to the formation of prostaglandins or thromboxane, a potent platelet-aggregating factor. A preventive and therapeutic role for vitamin E in stroke seems plausible. Vitamin E, as an antioxidant, interferes with peroxidation of arachidonic acid during the synthesis of prostaglandins. Significant immune-enhancing effects of vitamin E have been reported, which could have major public health significance. Supplementation of healthy elderly people with 800 mg/d of vitamin E for 30 days has been shown to significantly improve immune responsiveness. Changes noted included increased delayed-type hypersensitivity (DTH) skin test response, increased interleukin-2 (IL-2) production, increased T cell proliferation response to certain mitogens, decreased synthesis of prostaglandin E2 synthesis, and reduced overall lipid peroxidation. The improvements noted in the DTH response following short-term supplementation were maintained when 400 mg/d of vitamin E were provided over a period of 6 months.[163,164]

There is cautious optimism over the beneficial effects of long-term use of vitamin E in significantly decreasing the risk of coronary artery disease, although supplementation has produced inconsistent effects on serum lipid and lipoprotein levels.[15,208,247,252] One explanation of the possible preventive action of the vitamin is based on a substantially greater atherogenic effect of oxidized as compared to native LDL. As an antioxidant, vitamin E could prevent the oxidation of lipoproteins that play a key role in atherogenesis[7] (Fig. 29-11). Administration of pharmacologic doses of vitamin E (900 mg/d for 4 months) seems to improve insulin action in healthy subjects and non–insulin dependent diabetic patients.[194,195] It has also been proposed that vitamin E may reduce protein glycation in diabetes and thereby prevent diabetic complications.[40] Because chronic degenerative diseases of muscle in humans have many characteristics in common with hypovitaminosis E myopathy in animals, it has been suggested that vitamin E therapy might be beneficial in the long-term treatment of patients with such severe muscle diseases as Duchenne's muscular dystrophy.[115]

Toxic effects associated with the ingestion of large doses of vitamin E by adults appear limited to gastrointestinal disturbances such as nausea or diarrhea. Oral vitamin E supplementation has resulted in few side effects, even in doses as high as 3200 mg per day (3200 IU).[15] Intestinal absorption of vitamins A and K is reduced by a large intake of vitamin E.[17] In vitamin K-deficient individuals, coagulation defects are exacerbated when high levels of vitamin E are ingested.[15] In some neonates, use of intravenous vitamin E solutions has been associated with adverse reactions and, rarely, death.

Determination of vitamin E status is routinely accomplished by measuring tocopherol levels in serum or plasma, or by use of the erythrocyte hemolysis test. A more exotic method of assessing vitamin E status is by breath pentane analysis—an index of lipid peroxidation and reflective of vitamin E nutriture.[138,223] Blood should be collected after an overnight fast; serum or plasma whose level of alpha tocopherol is to be determined should be protected from light prior to analysis. Membrane fragility, as evidenced by increased erythrocyte susceptibility to hydrogen peroxide, provides an estimate of plasma tocopherol levels. Hemolysis of greater than 20% of RBCs after washing in a solution of 2% hydrogen peroxide for 3 hours is considered to represent vitamin E deficiency.[116,226] Older spectrophotometric and fluorometric methods have largely been replaced. The

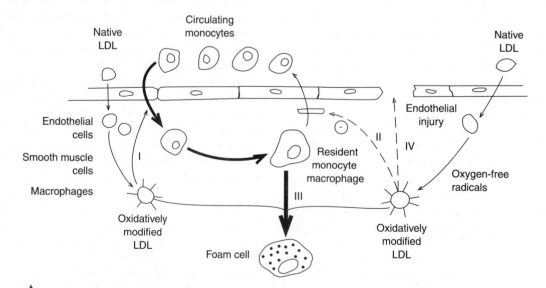

Figure 29-11. Proposed mechanisms by which oxidized LDL contributes to atherogenesis. (I) Oxidized LDL, through its chemotactic actions, recruits circulating monocytes to the intima; (II) oxidized LDL inhibits the motility of monocytes/macrophages and therefore prevents their return to circulation; (III) macrophages are converted to foam cells as a result of oxidized LDL uptake; (IV) endothelial injury caused by the cytotoxic effect of oxidized LDL. (*Source:* Quinn MT, Parthasarathy S, Fong LG, et al. Oxidatively modified low density lipoproteins. A potential role in recruitment and retention of monocyte/macrophages during atherogenesis. Proc Natl Acad Sci USA 1987;84:2995. With permission.)

method of choice for assessing vitamin E nutriture is reversed-phase HPLC for separation and ultraviolet or electrochemical detection. This makes possible the separation and quantitation of tocopherols in plasma, red blood cells, platelets, and leukocytes.[93,149,244] However, vitamin E levels are rarely used in clinical practice; these methods are primarily for research applications.

Vitamin K

The name *vitamin K* derives from the Danish designation of this substance as an antihemorrhagic or "Koagulation vitamin." Several naturally occurring compounds show vitamin K activity.[187] All are naphthoquinone derivatives and possess side chains of differing length. Vitamin K_1 has a phytyl side chain and is known as *phylloquinone* according to IUPAC nomenclature (see Table 29-3). The International Union of Pure and Nutritional Sciences (IUNS) applies the name *phytlmenaquinone* to this compound. Vitamin K_1, produced in plants, represents the bulk of dietary vitamin K. Major food sources are green leafy vegetables, dairy products, and meat. Cow's milk contains more vitamin K than does human milk. Cruciferae (mustard family) such as kale (750 µg/100 g), Chinese cabbage (175 µg/100 g), broccoli (200 µg/100 g), turnip greens (300 µg/100 g), and brussels sprouts (220 µg/100 g) are particularly rich in vitamin K_1.[256] The vitamer found in animal tissue and bacteria was originally called vitamin K_2. It is now characterized as a series of menaquinones with seven isoprenoid units in the side chain. In mammals, vitamin K_2 is synthesized by the intestinal microflora. It is estimated that bowel microflora contribute from 10% to 30% of the daily requirement of vitamin K.[181] Once absorbed in the intestine, vitamin K is transported in the chylomicra and stored briefly in the liver. The body pool of vitamin K is small and turns over rapidly. Other organs in which vitamin K is concentrated include the adrenal glands, lungs, bone marrow, kidneys, and lymph nodes.

An RDA has recently been established based on estimates that an average adult requires approximately 1 µg/kg of body weight per day.[181] Suggested phylloquinone plasma reference range values have varied substantially. Overall, K_1 circulating levels are substantially lower than those of the other fat-soluble vitamins—25-hydroxyvitamin D and vitamins A and E. The mean serum concentrations in healthy adults have been reported to range from 0.26 ng/mL to 2.6 ng/mL! Some evidence has suggested a normal plasma range of 0.29 to 2.64 nmol/L, with a geometric mean of 0.87 nmol/L (1 nmol/L = 0.45 ng/mL).[217] Other investigators suggest that the plasma vitamin K_1 concentration in adults is approximately 1 ng/mL (2.2 nmol/L) and that a vitamin K intake >100 µg/d but <550 µg/d is likely to maintain this concentration. Overnight fasting levels of most individuals are closer to 0.5 ng/mL.[256,257] In children age 1 to 6 years, serum concentrations have ranged from 0.4 to 0.8 ng/mL.[178]

Although vitamin K is not appreciably stored in the body, deficiency of the vitamin is uncommon. It may be seen in patients on long-term total parenteral nutrition (TPN), in those with chronic fat malabsorption, or in patients receiving prolonged antibiotic therapy. Newborns are at high risk of hypovitaminosis, owing to absence of an intestinal flora at birth, low hepatic stores, poor placental transfer, and low vitamin K levels in human milk. Poor maternal nutrition or use of coumarins, anticonvulsants, or antitubercular drugs during pregnancy can also jeopardize the newborn's vitamin K status. It is now common to administer vitamin K_1 prophylactically at birth to prevent development of hemorrhagic disease.[145] The only major sign of vitamin K deficiency is an increased tendency to bleed. Nosebleeds, easy bruising, hematuria, and gastrointestinal bleeding are all common clinical manifestations of hypovitaminosis (see Table 29-8).[187]

Vitamin K functions as a co-enzyme for γ-glutamyl carboxylase, which posttranslationally modifies protein-bound glutamic acid (Glu) residues. This vitamin K–dependent carboxylation reaction converts specific precursor proteins into their functionally active form. For example, the Glu residues in preprothrombin are converted to gamma-carboxyglutamic acid (Gla) residues, thus forming the coagulation protein, prothrombin or Factor II (Fig. 29-12).[253] Gla residues allow the protein to bind calcium. In the absence of vitamin K, a nonfunctional clotting factor is synthesized and hemorrhage can result.[256]

Synthesis of active coagulation Factors VII, IX, and X and plasma proteins C and S also requires participation of vitamin K–dependent carboxylase (Fig. 29-13). Other vitamin K–dependent calcium-binding proteins have been identified in the blood, bone, lung, spleen, liver, and testes. For example, osteoblasts synthesize the Gla-containing protein osteocalcin. Formation of osteocalcin, the most abundant noncollagenous protein in mature bone matrix, is stimulated by 1,25-dihydroxyvitamin D and may influence normal bone mineralization. Preliminary findings suggest a relationship between vitamin K intake and bone mass. If so, this could explain the

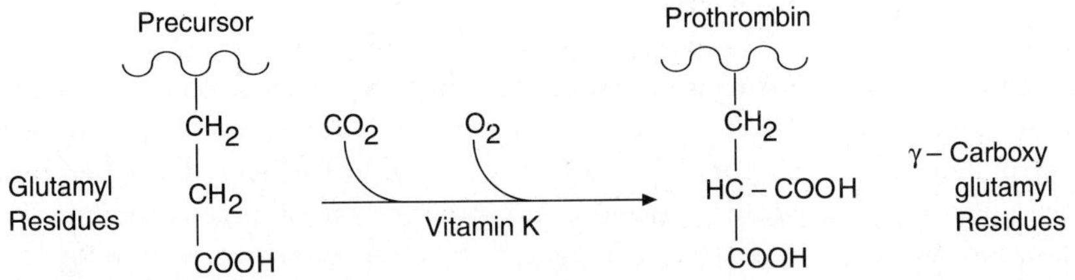

Figure 29-12. The vitamin K–dependent carboxylation reaction. (*Source*: Suttie JW. Vitamin K. In: Machlin LJ, ed. Handbook of vitamins. 2nd ed. New York:Marcel Dekker, 1991.)

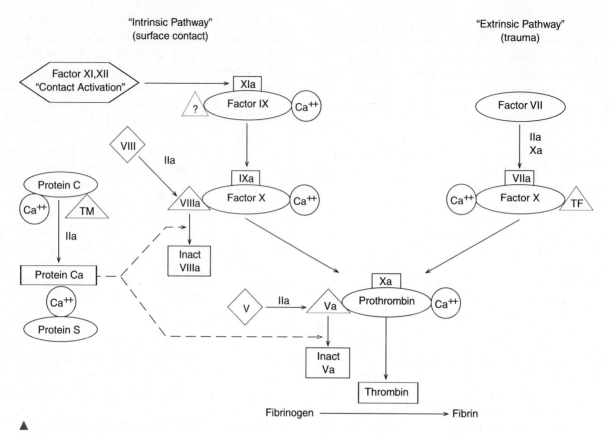

"Intrinsic Pathway"
(surface contact)

"Extrinsic Pathway"
(trauma)

Figure 29-13. Involvement of vitamin K–dependent clotting factors in coagulation. Vitamin K–dependent proteins (shown as ellipses) circulate as inactive forms of serine proteases until coverted to their active (subscript a) forms. TM refers to an endothelial cell protein called thrombomodulin; TF is a tissue factor. (*Modifed from:* Suttie JW. Vitamin K. In: Machlin LJ, ed. Handbook of vitamins. 2nd ed. New York:Marcel Dekker, 1991.)

observed relationship of vitamin K status and occurrence of bone abnormalities such as osteoporosis.[92]

The action of vitamin K is antagonized by selected drugs, including 4-hydroxycoumarins, salicylates, certain broad-spectrum antibiotics, and pharmalogic doses of vitamins A and E. Regeneration of vitamin K from its metabolite, 2-3-epoxide, occurs enzymatically. The hepatic enzyme responsible, epoxide reductase, is inhibited by the anticoagulant drug warfarin.[156] The name *warfarin* is derived from initials of the Wisconsin Alumni Research Foundation (WARF). The metabolic consequence of administration of Coumadin, a proprietary name of a warfarin preparation, is a drug-induced vitamin K deficiency. Interference with vitamin K–dependent hepatic synthesis of coagulation factors leads to hypo-prothrombinemia and hemorrhage. Severe hypoprothrom-binemia secondary to vitamin K deficiency may occur from excessive intake of the fat-soluble vitamins, particularly vita-min A.[130] Vitamin A inhibits bacterial synthesis of vitamin K_2 and functions as a vitamin K antagonist in the liver. Imma-ture or damaged liver cells cannot effectively utilize vitamin K. Vitamin K may be administered therapeutically to low-birth-weight infants to raise the concentration of blood-clotting factors to a level equal to that of a full-term neonate. Patients receiving TPN or antibiotic therapy are at increased risk of vitamin K deficiency. Phylloquinone and men-

aquinone are nontoxic. The parent compound, menadione, once known as vitamin K_3, is seldom used in clinical practice because of its potential to cause hemolytic anemia, as well as liver and kidney damage. The resulting hyperbilirubinemia and kernicterus can be fatal to newborns. Patients receiving warfarin anticoagulants may develop a resistance to the drugs if they ingest excessive amounts of vitamin K. Megadoses of vitamin E can exacerbate coagulation defects arising from vi-tamin K deficiency—whether due to malabsorption or anti-coagulant therapy.[17] Warfarin activity may thus be amplified by administration of vitamin E. Large doses of salicylates may depress vitamin K–dependent factors by inhibiting enzymes in the vitamin K cycle.

The standard clinical laboratory method for evaluating vitamin K adequacy has been by clotting function tests, such as prothrombin time (PT). Prolongation of the PT by 2 sec-onds, or more, beyond the normal range is compatible with a diagnosis of hypovitaminosis K.[223] This is an indirect, non-specific means of estimating vitamin levels.

Direct, rapid quantitative assessment of phylloquinone in plasma is now possible with HPLC separation methods and a variety of detector systems.[116] A reversed-phase HPLC assay with fluorometric detection has been developed that permits isolation and quantitation of plasma vitamin K_1 down to 0.05 μg/L.[91] Electrochemical detection of the vitamin has also been reported. Newer procedures require a single liquid-chromatographic step and smaller volumes of plasma than previous assays. Immunochemical methods and modified chromogenic assays for the determination of plasma levels of partially carboxylated prothrombin molecules have been proposed as sensitive assays for vitamin K.[257] Unfortunately, the relationship of plasma phylloquinone to vitamin K intake and status is not clear.

With further procedural simplification, reliable assays for all fat-soluble vitamins should find routine application in the clinical laboratory.

WATER-SOLUBLE VITAMINS

Vitamin C

Ascorbic acid (AA), commonly known as *vitamin C,* is an internal ester of a hexonic acid (Table 29-11).[192] The double bond between carbons 2 and 3 in the 5-membered lactone ring is readily oxidized. Oxidation at this enediol linkage produces dehydroascorbic acid (DHAA). Ascorbic acid and dehydroascorbic acid are the only biologically active forms of vitamin C (Fig. 29-14). Both are effective in treatment of the deficiency disease scurvy. Further oxidation of dehydroascorbic acid results in a loss of antiscorbutic activity. Approximately 90% of vitamin C present in animal tissues is in the form of ascorbic acid. Most animals, with the notable exceptions of guinea pigs, flying mammals, and primates, synthesize L-ascorbic acid from D-glucose or D-galactose. The animals that do not synthesize it lack the final enzyme in the biosynthetic pathway.[9]

Vitamin C is abundant in fruits and most vegetables, particularly green leafy vegetables. Unfortunately, vitamin C is the most labile vitamin in foods, and much of its activity is lost through oxidation during preparation, cooking, and storage.[181] Based upon local, regional, and national surveys, it is conservatively estimated that 25% of the U.S. adult population use vitamin supplements on a daily basis. As many as 30% to 40% of U.S. children may take vitamin supplements.[239] Vitamin C is the most common single vitamin among these supplements. It is also the vitamin most likely to be taken in amounts substantially in excess of the RDA.[209,254] Absorption of vitamin C occurs predominantly in the distal portion of the small intestine. Pectin and zinc may impair its absorption. It is widely distributed among body tissues rather than being stored in one specific organ. Leukocytes, in particular, contain high levels of the vitamin. When tissue stores are saturated, plasma levels rise, and excess vitamin C is excreted by the kidney. The RDA for ascorbic acid is 60 mg for both sexes.[181] It has been suggested that a different RDA should be considered for men and women.[76] Clinical trials show that men and women consuming the same amounts of vitamin C have different plasma levels of the vitamin. Elderly men consuming the RDA for vitamin C (60 mg/d) had a mean plasma level of

0.5 mg/dL; elderly women had a level of 0.9 mg/dL.[270] Reference values range from 0.3 to 2.0 mg/dL of serum or plasma and from 15 to 30 mg/dL in leukocytes.[64,116,181,223] Plasma vitamin C levels >1.0 mg/dL are believed to reflect a maximum body pool. Plasma vitamin levels <0.2 mg/dL (11 µmol/L) or leukocyte concentration <7 mg/dL indicate a high risk of vitamin deficiency. Leukocyte levels, while technically more difficult to determine, more closely reflect tissue levels and the body pool than do plasma values. In both sexes, plasma and leukocyte levels of vitamin C decrease with increasing age. Vitamin C requirements are increased in disease and conditions of prolonged physiologic stress. Depletion of body reserves and hypovitaminosis may occur in chronic diseases such as alcoholism and cancer. Drug-induced deficiencies are also reported, particularly among cigarette smokers and users of oral contraceptive steroids.[211] Urinary excretion of ascorbic acid is promoted by the use of such medications as aspirin and barbiturates. Excessive intake of ascorbic acid interferes with the action of amphetamines, anticoagulants, and tricyclic antidepressants. When long-term megadoses of vitamin C are abruptly discontinued, a rebound phenomenon is reported. Metabolic demands for the vitamin have been artificially increased by its oversupply. A reduction in the amount available may precipitate hypovitaminosis C.

The classic deficiency disease, scurvy, occurs after chronic deprivation of vitamin C. Early symptoms of the disease can appear in apparently healthy women after less than 4 weeks of depletion. The first signs of scurvy generally appear when plasma levels of vitamin C fall below 0.2 mg/dL.[64] Clinical manifestations of extreme deficiency include profound behavioral changes, severe emotional disturbances, skeletal lesions, necrosis of gums with loss of teeth, ecchymoses, and bleeding into muscles, joints, the kidneys, gastrointestinal tract, and pericardium. Anemia develops because of blood loss through hemorrhage. More commonly, hypovitaminosis C is characterized by moderately lowered levels of the vitamin. Consequently, the deficiency may be subclinical or associated with less dramatic signs and symptoms, including anorexia, fatigue, bone and joint pain, bleeding and tender gums, nosebleeds, and irritability.[104]

A wide variety of physiologic roles are suggested for vitamin C.[9,177] Depending upon its chemical role, ascorbate can donate either one or two of its electrons to an electron acceptor. Ascorbic acid is a potent reducing agent, capable of donating electrons to cupric ions, ferric ions, metal ions bound to various cytochromes, and oxygen.[192] In the oxidation-reduction reactions of basic metabolism, two electrons are donated by the vitamin to an acceptor. Ascorbic acid is required for the optimal activity of a number of different enzymes (Table 29-12). It functions to keep prosthetic metal ions, such as iron or copper, in the required reduced form.[193] Ascorbate is also a good scavenger of oxidants including H_2O_2, O_2^-, OH·, and ROO· (Table 29-13). The vitamin functions as an antioxidant in aqueous environments, such as the cytosol and extracellular fluids, by donating one

TABLE 29-11. Water-Soluble Vitamins Required in Humans

Generic Description	Trivial Name	Adult Reference Ranges*	Structural Formula	Cofactor Form
C	Ascorbic acid	Plasma—3–20 mg/L	Ascorbic acid	None
Thiamin	Thiamin (B₁)	Plasma—4–20 µg/L Whole blood—32.7–68.7 µg/L RBC—75–153 µg/L RBC—TPP stimulation effect <15%	Thiamin	Thiamin pyrophosphate (TPP)
Riboflavin	Riboflavin (B₂)	RBC—EGR activity coefficient <1.2 Serum—5.5–14.4 µg/L Urine—36–349 µg/g of creatinine Whole blood—40–150 ng/mL	Riboflavin	Flavin mononucleotide (FMN) Flavin adenine dinucleotide (FAD)
Niacin	Nicotinic acid	Whole blood—4–9 µg/mL Urine—2-PY/NIMN >2.0	Nicotinic acid	Nicotinamide adenine dinucleotide (NAD) Nicotinamide adenine dinucleotide phosphate (NADP)
Pantothenic acid	Pantothenic acid	Serum—15–40 µg/dL	Pantothenic acid	Coenzyme A (CoA-SH)
B₆	Pyridoxine	Plasma—4–20 ng/mL RBC—EAST index <1.5 Whole blood—30–80 ng/mL	Pyridoxine	Pyridoxal-5′-phosphate (PLP)
Biotin	Biotin	Plasma—300–800 pg/mL	Biotin	N-carboxybiotinyl lysine
Folate	Folic acid	Serum—3–25 ng/mL RBC—150–600 ng/mL CSF—15–135 ng/mL	Folic acid	Tetrahydrofolate (THF) coenzymes *e.g.,* 5-methyl THF
B₁₂	Cobalamin	Serum—200–900 pg/mL	R = —CN cobalamin (cyanocobalamin) = —CH₃ (methylcobalamin) = —5′-deoxyadenosyl (adenosylcobalamin)	Adenosylcobalamin Methylcobalamin

Varies with assay technique.

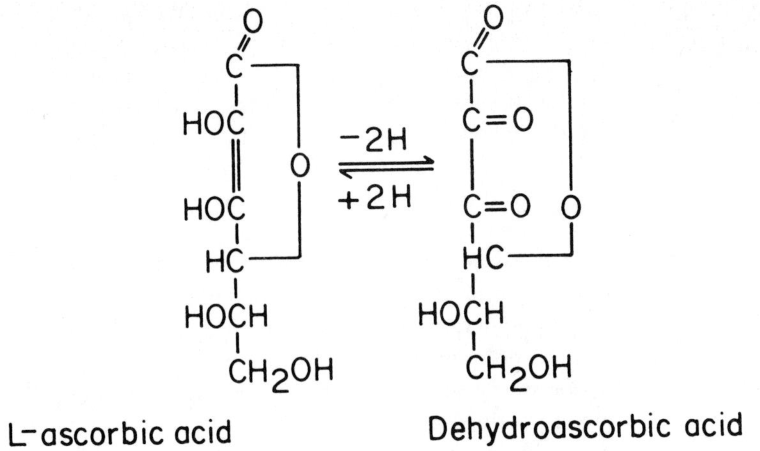

L-ascorbic acid   Dehydroascorbic acid   *Figure 29-14.* Biologically active forms of vitamin C.

of its electrons to these free radicals. This action "quenches" or eliminates the condition of an odd or unpaired electron in the acceptor ions. It is the unpaired electron characteristic of free radicals that accounts for their high chemical reactivity and ability to damage cellular constituents.[148,177] Approximately one quarter of circulating peroxyl radicals are scavenged by vitamin C.[7] Ascorbic acid can also affect the balance between oxidative products and antioxidant defense mechanisms by donating an electron to the tocopherol free radical, thus regenerating the major fat-soluble antioxidant, vitamin E.

Ascorbic acid plays a major role in biologic hydroxylation reactions by serving as a reducing agent for a number of hydroxylases. Vitamin C is essential for the formation of the protein collagen. The amino acids proline and lysine must be hydroxylated after they have been incorporated into the polypeptide procollagen. Unless this occurs, the final triple helix structure of collagen cannot be formed. This unique structure of collagen provides the structural integrity characteristic of connective tissue. In hypovitaminosis C, collagen synthesis is reduced, existing collagen fibers are destabilized, and there is general dissolution of intercellular ground substance. Reduction in the effective amount of collagen and

matrix breakdown are the basis for the connective-tissue lesions and impaired wound healing seen in hypovitaminosis C. At least two other hydroxylation reactions require the participation of ascorbic acid as a reducing agent. Evidence suggests that vitamin C is essential for carnitine synthesis in muscle. Carnitine stimulates fatty acid oxidation in mitochondria. Decreased levels of vitamin C could lead to lowered energy levels and muscle weakness. Ascorbic acid also serves as a reducing agent in the biosynthesis of the neurotransmitters noradrenaline and adrenaline. Behavioral changes associated with hypovitaminosis C may be related to this function. Vitamin C is identified as promoting or facilitating, but not essential for, cholesterol, histamine, and tyrosine metabolism, lymphocytic proliferation and transformation of B and T lymphocytes, functioning of macrophages, neutrophil chemotaxis, folate reduction, and dietary iron absorption, distribution, and storage.[9] Megadoses of vitamin C have been advocated by various proponents for both therapeutic and prophylactic purposes.[33] Ascorbic acid has been shown to lower serum total and LDL cholesterol and triglyceride concentrations in some individuals.[64,120] A large body of evidence suggests a role for vitamin C in cholesterol homeostasis.[239] Hypovitaminosis C could increase the risk of

TABLE 29-12. Enzymes Dependent on Ascorbic Acid for Maximum Activity

Enzyme	Function
Proline hydroxylase (EC 1.14.11.2)	*trans*-4-Hydroxylation of proline in procollagen biosynthesis
Procollagen-proline 2-oxoglutarate 3-dioxygenase (EC 1.14.11.7)	*trans*-3-Hydroxylation of proline in procollagen biosynthesis
Lysine hydroxylase (EC 1.14.11.4)	5-Hydroxylation of lysine in procollagen biosynthesis
gamma-Butyrobetaine 2-oxoglutarate 4-dioxygenase (EC 1.14.11.1)	Hydroxylation of a carnitine precursor
Trimethyllysine-2-oxoglutarate dioxygenase (EC 1.14.11.8)	Hydroxylation of a carnitine precursor
Dopamine β-monooxygenase (EC 1.14.17.1)	Dopamine β-hydroxylation in norepinephrine biosynthesis
Peptidyl glycine α-amidating monooxygenase activity	Carboxyterminal α-amidation of glycine-extended peptides in peptide hormone processing
4-Hydroxyphenylpyruvate dioxygenase (EC 1.13.11.27)	Hydroxylation and decarboxylation of a tyrosine metabolite

Source: Moser U, Bendich A. Vitamin C. In: Machlin LJ, ed. Handbook of vitamins. 2nd ed. New York: Marcel Dekker, 1991.

TABLE 29-13. Common Oxygenated Free-Radicals

Species	Common Name	Half-life (37°C)
O_2·	Superoxide anion radical	Unstable
HO_2·	Hydroperoxyl radical	—
HO·	Hydroxyl radical	1 nanosecond
RO·	Alkoxyl radical	1 microsecond
ROO·	Peroxyl radical	7 seconds
RNO·	Nitroxyl radical	5 seconds

R = lipid, for example, linoleate.

Source: Bankson DD, Kestin M, Rifai N. Role of free radicals in cancer and atherosclerosis. Clin Lab Med 1993;13:463.

hypercholesterolemia. Approximately 20% of Americans over 65 years of age have cataracts. Evidence has been presented of a decreased risk of cataract formation with high levels of vitamin C and carotenoids.[30,119,234a]

Excessive quantities of the vitamin are readily excreted in the urine, and it is generally assumed that adverse effects from megadoses are minimal. However, reports have been published of gastrointestinal disturbances, increased hemolysis in subjects with G-6-PD deficiency, interaction with warfarin anticoagulants, increased destruction of vitamin B_{12}, excessive absorption of dietary iron, and uricosuria or oxaluria with increased potential for kidney stone formation. Some investigators cite evidence of the harmful, even lethal effect of vitamin C supplementation in the over 10% of Americans having high iron stores.[97] It appears that at pharmacologic or megadoses, in the presence of iron, vitamin C is one of the most potent oxidants known and drives iron-catalyzed free radical generation. Gastric intrinsic factor and vitamin B_{12} can be damaged or destroyed if subjected to excessive concentrations of vitamin C.[96] Large doses of ascorbic acid are reported to affect a number of biochemical parameters. The advantages and disadvantages of megavitamin therapy is currently under debate. Scientific research and clinical trials are in progress, but the definitive answer is not available at this time. The effect may be physiologic or methodologic.[285] When methodologic, the nature and extent of interference varies with the analytical procedure used (Table 29-14). The prophylactic value of large intakes of vitamin C is questionable.[114] Anticancer capabilities of vitamin C are most clearly linked to its ability to inhibit formation of carcinogenic *N*-nitroso compounds.[180] High-dose vitamin C therapy has not been shown to be beneficial in the treatment of advanced malignancies.[175] A number of epidemiologic studies have addressed the relationship of vitamin C status and the likelihood of cancer.[251] Limited evidence associates a lowered risk of cancers of the stomach and esophagus with consumption of vitamin C–containing foods. As a free radical scavenger, vitamin C may inhibit the oxidative modification of low-density lipoprotein, thereby lowering atherogenic risk.[7] Some, but not all, correlation studies report a strong inverse

association between vitamin C intake and the risk of ischemic heart disease and strokes, mortality from coronary heart disease, and overall mortality.[67,216,252] A positive relationship has also been reported between HDL cholesterol and plasma ascorbic acid.[88,120] Vitamin C has been used therapeutically in treatment of caustic corneal burns, Chediak-Higashi syndrome, and in hypercholesterolemia.[27,181]

Vitamin C levels in plasma, leukocytes, and urine may be assessed by the clinical laboratory.[116,222,223] The majority of analytical procedures are based on either the oxidation-reduction potential of the vitamin or the formation of chromogenic derivatives, such as hydrazones or fluorophors, following oxidation of the reduced form of the vitamin. Reduction of the dye, 2,6–dichlorophenolindophenol (DCIP), or formation of a colored dinitrophenylhydrazine derivative are the two most commonly used colorimetric methods.[228] The DCIP method, which assays only the reduced form of the vitamin, has been automated. Plasma ascorbic acid values are slightly higher by this method than by HPLC. At low plasma ascorbate levels (<4 mg/L) used to identify individuals suspected to be at moderate risk of vitamin C deficiency, this difference can become clinically significant.[75,271] The high incidence of significant concentrations of ascorbic acid in urine as a consequence of vitamin supplementation has serious implications for routine urinalysis. A recent study found that urinary concentrations >2280 micromoles/L (41 mg/dL) were not uncommon.[26] Dipstick tests for urinary glucose and hemoglobin that involve peroxidase-redox indicator systems are subject to interference by ascorbic acid. Strip manufacturers have introduced a variety of modifications to minimize this interference. To prevent false-negative results for hemoglobin and glucose in screening urinalysis protocols employing reagent stick technology, it is important to screen with a strip that has a vitamin C indicator or to use strips that show the greatest resistance to ascorbic acid interference. One reagent strip method available for use in assessing urine levels of vitamin C is based upon reduction of phosphomolybdate to "molybdenum blue." Another popular chemical test depends on the reduction of ferric to ferrous iron and its subsequent determination as a red-orange α, α′-dipyridyl complex.[223] An automated pro-

TABLE 29-14. Biochemical Parameters Affected by Excessive Levels of Vitamin C

Aminotransferases (AST and ALT)
Total bilirubin
Urobilinogen
Glucose
Serum bicarbonate
Vitamin B_{12}
Cholesterol
Uric acid
Lactate dehydrogenase (LD)
Alkaline phosphatase (ALP)
Creatinine

cedure based on reduction of ferric ions and formation of a complex with the chromogenic reagent ferrozine has been adapted to the centrifugal analyzer.[32] HPLC techniques with electrochemical detection are now widely applied to provide rapid, quantitative measurements of L-ascorbic acid, dehydroascorbic acid, and other vitamin C metabolites simultaneously.[271] Measurement of vitamin C in a buffy coat preparation may reflect the distribution of cells as well as the vitamin C status.[116,134,135] Different types of leukocytes and platelets vary greatly in their ascorbic acid content. Mononuclear cells have substantially higher levels of ascorbic acid than do neutrophils. This is a technically difficult procedure and is unsuitable for use with small volumes of blood (<2 mL). It is suggested that leukocyte levels of ascorbic acid reflect body stores of the vitamin, whereas plasma and whole blood levels are related to dietary intake.

The B Complex of Vitamins

The B complex of vitamins is a collection of chemically diverse molecules originally grouped together on the basis of their shared solubility in water and isolation from the same dietary sources. B vitamins are particularly abundant in whole grain cereals, liver, and yeast. Because of their presence in the same foods, deficiencies of B vitamins are often multiple rather than singular. Many, but not all, studies indicate that the typical diet of the elderly in the United States is low in the B vitamins thiamin, riboflavin, B_{12}, folate, and B_6.[75,123,127,198,216] Some geriatric affective disorders and cognitive dysfunctions may be due to inadequate supplies of these vitamins.[13] Thiamin, riboflavin, B_6, B_{12}, and folate are critical for the synthesis of neurotransmitters, including norepinephrine, dopamine, serotonin, gamma aminobutyric acid (GABA), and acetylcholine. Since carbohydrate metabolism is the primary source of energy for nerve cells, deficiencies of the B vitamins lead to a variety of neurologic diseases. There is also similarity of deficiency symptoms, with lesions of the skin, lips, and tongue being common clinical manifestations. Laboratory identification or confirmation of specific hypovitaminosis is essential for appropriate therapy.

The B vitamins are absorbed in the intestine and transported in the portal circulation. Tissue stores of most B vitamins are minimal, and depletion occurs over several weeks in response to dietary restriction, drug-induced deficiencies, or increased requirements due to physiologic stress. Body stores of folate and vitamin B_{12} are more extensive than those of other B vitamins. Although water soluble, some of the B vitamins can cause toxic side effects when taken in excess.

Unlike fat-soluble vitamins, whose biochemical functions are not always clearly defined, B vitamins are known to function as enzyme cofactors.[192] Those vitamin derivatives readily dissociable from the protein apoenzyme are called *coenzymes*. Their function is to transport individual atoms or groups between molecules in coupled reactions. They may be considered co-substrates to the many different enzymes with which they are associated. On the other hand, prosthetic groups are vitamin derivatives that remain firmly attached to enzyme proteins. They also function to facilitate the transfer of atoms and groups between molecules. Table 29-11 identifies vitamin derivatives comprising coenzyme and prosthetic groups and the biochemical function of each.

Deficiency of any B vitamin may have wide-ranging adverse effects on the body's metabolism because of requirements for these vitamins in every major biosynthetic and energy-liberating pathway (Fig. 29-15).[58] Requirements for the B vitamins vary with dietary nutrient and total caloric intake. For example, a high-carbohydrate diet greatly increases the metabolic demand for thiamin as the coenzyme thiamin pyrophosphate (TPP); daily utilization of niacin and riboflavin vary with total energy requirements because of their participation in oxidative reactions of the electron transport chain; and pyridoxine needs are related to protein intake, since B_6 is an essential coenzyme in amino acid metabolism.

Thiamin

The "vita/amine" identified by Casimir Funk was vitamin B_1, thiamin. Thiamin is identified as serving two separate biochemical functions. In the tricarboxylic acid (Krebs) cycle and the pentose phosphate pathway (hexose monophosphate shunt) it functions in its coenzyme phosphorylated form [(TPP) thiamin pyrophosphate or thiamin diphosphate] in the metabolism of alpha-keto acids and 2-keto sugars, respectively.[192] Because Mg^{++} is an essential cofactor for TPP function, thiamin deficiency is aggravated by magnesium deficiency. TPP can be further phosphorylated to thiamin triphosphate (TTP), which is believed to play a key role in nerve conduction. Functional defects in thiamin-dependent enzymes have been reported in patients with Alzheimer's disease.[78] The liver, erythrocytes, and cells of the cerebral cortex are the primary sites of thiamin phosphorylation. The RDA for adult males is 1.2 to 1.5 mg; for adult females, the RDA is 1.0 to 1.1 mg.[181] Some authorities believe that thiamin deficiency is the most common vitamin deficiency among adults in Western societies.[35] Subclinical or borderline thiamin deficiency, though believed to be common, is difficult to recognize clinically because of its non-specific presentation. It has been suggested that laboratory evaluation of thiamin status is indicated when sleep disorders, depression, forgetfulness, and poor coordination are present in the elderly. Significant correlation between thiamin status and neuropsychologic function, indicated by cognitive performance and brain electrophysiology, has been reported.[266] It has also been recommended that the thiamin status of an individual be evaluated before the physician prescribes sedative hypnotics for depression and sleep disorders in older adults.[240] Chronic deficiency of thiamin produces the disease beriberi.[183] Two forms of beriberi are recognized. *Wet beriberi* affects predominantly the cardiovascular system and is characterized by extensive edema and high-output cardiac failure. *Dry beriberi* affects the nervous system and is associated

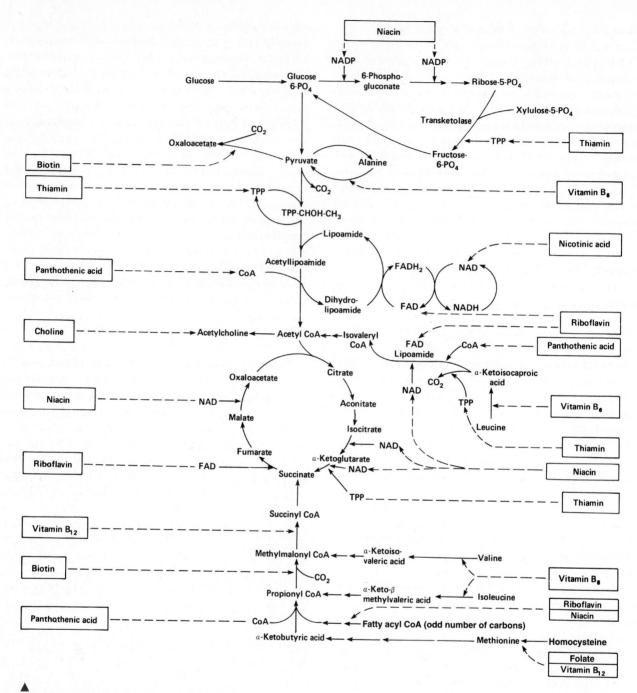

Figure 29-15. Water-soluble vitamins required as cofactors in selected metabolic pathways. (*Adapted from* Gilman AG, Goodman LS, Gilman A. Goodman and Gilman's The pharmacological basis of therapeutics. 6th ed. copyright © New York: Macmillan Publishing Co, Inc, 1980;1562.)

with neurologic symptoms such as polyneuritis, hyperesthesia, and muscle wasting. Extreme thiamin deficiency may lead to neuronal degeneration and development of Wernicke's encephalopathy, Korsakoff's psychosis, or combined Wernicke-Korsikoff syndrome.[82] Wernicke's disease is characterized by ocular disturbances and ataxia. Korsakoff's syndrome includes defective memory and impaired learning ability. Patients on prolonged TPN, those receiving long-term hemodialysis or diuretic therapy, and individuals with chronic febrile infections may develop severe thiamin defi-

ciency.[207] In the U.S., however, deficiency is most often associated with alcoholism and the accompanying malnutrition. Thiamin deficiency has been detected by biochemical tests in 25% of alcoholics admitted to U.S. hospitals. Deficiency of thiamin prevents the oxidative decarboxylation of pyruvate to acetyl CoA by a multienzyme complex that includes a TPP-dependent pyruvate decarboxylase. Pyruvate, unable to be metabolized via the Krebs cycle, is converted to lactate. Severe metabolic acidosis due to lactic acid accumulation may develop. An acute fulminant form of beriberi (shoshin beriberi) may be the most common cause of curable lactic acidosis (LA) that presents with a blood lactate >10 mM/L (reference range 0.6–2.2 mM/L). If untreated, fatality approaches 100%. Beriberi is probably under-diagnosed and should be considered in the differential diagnosis if an alcoholic patient develops lactic acidosis of unknown origin.[174] Conditions of excessive alkalinity destroy thiamin, and long-term use of antacids may lead to B_1 deficiency. Diets high in raw fish (containing a thiaminase) or consumption of large amounts of tea (containing tannic acid, which is a thiamin inactivator) can place an individual at increased risk.[181] High dose therapy with thiamin has been effective in the treatment of certain inborn errors of metabolism, including one variant of Maple Syrup Urine Disease (branched-chain keto-aciduria) (Table 29-1).[59,85] Even with high oral doses of thiamin, no toxic effects have been reported.

Thiamin levels may be assessed by direct measurement of thiamin in blood or urine or by evaluation of the activity of a thiamin-dependent enzyme erythrocyte transketolase (ETK). Blood contains only about 0.8% of the body's total thiamin and the concentration is very low (6–12 μg/dL) Most blood thiamin is found in the blood cells, with the concentration in leukocytes about 10 times greater than the concentration in RBCs.[85] Direct determination of thiamin may combine: (1) alkaline oxidation of the vitamin by ferricyanide into the fluorescent compound thiochrome, (2) HPLC for isolation, and (3) fluorometric detection and quantitation.[283] HPLC methods permit the separate determination of thiamin and its phosphate esters.[219] A low-cost assay sufficiently sensitive for pediatric application has been described in which cyanogen bromide, rather than potassium ferricyanide, is used to oxidize thiamin to thiochrome.[283] Thiamin reference ranges for plasma of 4 to 20 μg/L, 32.7 to 68.7 μg/L for whole blood, and 75 to 153 μg/L for red blood cells are widely accepted (see Table 29-11).[46]

More commonly in the clinical laboratory, ETK activity is evaluated to determine thiamin nutriture.[275] Because the transketolase is especially labile, blood specimens should be kept cold and analyzed promptly.[116] The thiamin-dependent enzyme catalyzes two reactions in the utilization of pentose sugars.[192] One of these reactions, the conversion of xylulose-5-phosphate and ribose-5-phosphate (R5P) to sedoheptulose-7-phosphate and glyceraldehyde-3-phosphate (GAP), forms the basis for the assay procedure. Erythrocyte hemolysates are incubated with ribose-5-phosphate in the presence and absence of TPP for a timed interval, after which proteins are precipitated by addition of trichloroacetic acid. Only the substrate ribose-5-phosphate is provided, since enzymes present in the erythrocytes generate the second substrate, xylulose-5-phosphate, from R5P. Either or both substrate consumption and product formation may be followed. Assessment of the remaining pentose and sedoheptulose may be by conversion to furfural and furfural derivatives, with subsequent formation of colored complexes. Spectrophotometric determination of the concentrations of these derivatives provides an indirect measure of ETK activity. An alternative assessment methodology, which has been adapted for a semiautomated continuous-flow procedure, separates the hemolysate and indicator reactions by dialysis and removes hemoglobin interference.[274] This makes possible the use of an NADH-coupled indicator reaction with glyceraldehyde-3-phosphate dehydrogenase. Decreasing absorbance at 340 nm is followed as NADH is oxidized to NAD^+ in proportion to the concentration of GAP in the hemolysate reaction mixture. In thiamin deficiency, ETK activity decreases, but its *in vitro* stimulation by TPP increases. This percentage of stimulation is referred to as the *TPP effect*. A TPP effect <15% denotes adequate thiamin status. Values between 15% and 25% indicate marginal thiamin deficiency, and values greater than 25% indicate severe deficiency.[6,85,112,223]

Riboflavin

Vitamin B_2 is also known as *riboflavin*. The primary naturally occurring forms of riboflavin are the prosthetic groups riboflavin-5′-phosphate (flavin mononucleotide, or FMN) and flavin adenine dinucleotide (FAD). FAD and FMN function in the biologic oxidation-reduction reactions of the electron-transport chain, lipid metabolism, sterol biosynthesis, and intermediary metabolism of amino acids, purines, and pyrimidines.[192] Riboflavin is also essential for the biochemical conversion of vitamin B_6 into its active form, pyridoxal-5′-phosphate.[50] During digestion, the cofactors are largely converted to free riboflavin. Both FMN and free riboflavin are readily absorbed in the small intestine and bind to albumin for transport in the blood. Very limited storage occurs in the liver with enzymatic formation of FAD and FMN.

Riboflavin, along with a variety of oxidation products, is excreted in the urine.[44] When ingested in large doses, toxicity is minimal because of rapid urinary excretion of excess vitamin. The RDA for adults is 1.3 to 1.7 mg.[181] Dietary surveys indicate that inadequate intake of riboflavin is not uncommon, particularly among the elderly and adolescents, although overt clinical manifestations of hypovitaminosis are rare. Pregnancy, lactation, and the use of oral contraceptive steroids increase the body's requirement for riboflavin.[181] Phenothiazine or tricyclic antidepressant use has been shown to interfere with activation of riboflavin and to accelerate its excretion by the kidneys.[13]

Clinical conditions associated with negative nitrogen balance are characterized by increased urinary excretion of riboflavin.[181] Ariboflavinosis is suggested by cheilosis, angular stomatitis, glossitis, seborrheic dermatitis, and ocular disturbances, including photophobia.[64,82] Laboratory assessment of riboflavin status may be by direct measurement of riboflavin levels in body fluids, commonly urine, or by a functional test of erythrocyte glutathione reductase (EGR) activity *in vitro* with and without added FAD.[185] The concentration of riboflavin in the blood and urine is dependent on recent dietary intake. Evaluation of EGR activity is believed to better reflect tissue reserves and the metabolic availability of riboflavin. HPLC has been used to isolate serum and urinary riboflavin with quantitation by fluorometry.[77] Enzymatic evaluation of riboflavin nutriture is based on the following reaction:

$$NADPH + H^+ + GSSG \xrightarrow{\text{erythrocyte glutathione reductase-FAD}}$$

$$NADP^+ + 2\ GSH \qquad \textit{(Eq. 29-1)}$$

The enzyme erythrocyte glutathione (GSSG) reductase has an absolute requirement for FAD as a prosthetic group. The stimulatory effect of exogenous FAD on aliquots of hemolyzed erythrocytes depends on the degree of saturation of the erythrocyte enzyme *in vivo*. Decrease in absorbance at 340 nm is a function of the oxidation of NADPH, and an activity coefficient (AC) is calculated by dividing the decrease in absorbance with added FAD by the decrease in absorbance without added FAD at the end of 10 minutes. An AC of less than 1.20 is interpreted as a low risk of hypovitaminosis; medium risk is associated with an AC of 1.20 to 1.40, and high risk of deficiency with an AC greater than 1.40.[186]

The reference intervals for riboflavin established by liquid chromatographic measurements are: (1) in urine, 36 to 349 μg/g of creatinine excreted, and (2) in serum, 5.5 to 14.4 μg/L.[132]

Niacin

Niacin is the generic name for nicotinic acid and its amide nicotinamide. Metabolic activity is dependent on the formation of the coenzymes nicotinamide adenine dinucleotide (NAD) and nicotinamide adenine dinucleotide phosphate (NADP). By virtue of its role in the formation of these coenzymes, niacin is essential for a wide variety of biochemical oxidation-reduction reactions catalyzed by dehydrogenases. Deficiency of niacin interferes with biologic oxidations in carbohydrate, lipid, and protein metabolism. Niacin is derived from the diet but is also made endogenously by the liver from the amino acid tryptophan.[192]

The extent of conversion of the amino acid to the vitamin is under hormonal influence and appears to increase during pregnancy or when oral contraceptive steroids are used.[181] Protein is digested and absorbed in the small intestine along with niacin that is liberated from ingested pyridine nucleotides. Following transport in the portal circulation to the liver, synthesis of NAD and NADP occur in the liver from both niacin and its precursor, tryptophan (Fig. 29-16). Because of its dual origin, the RDA for niacin is expressed in niacin equivalents (NE), one NE being equal to 1 mg of niacin or 60 mg of dietary tryptophan.[181] The

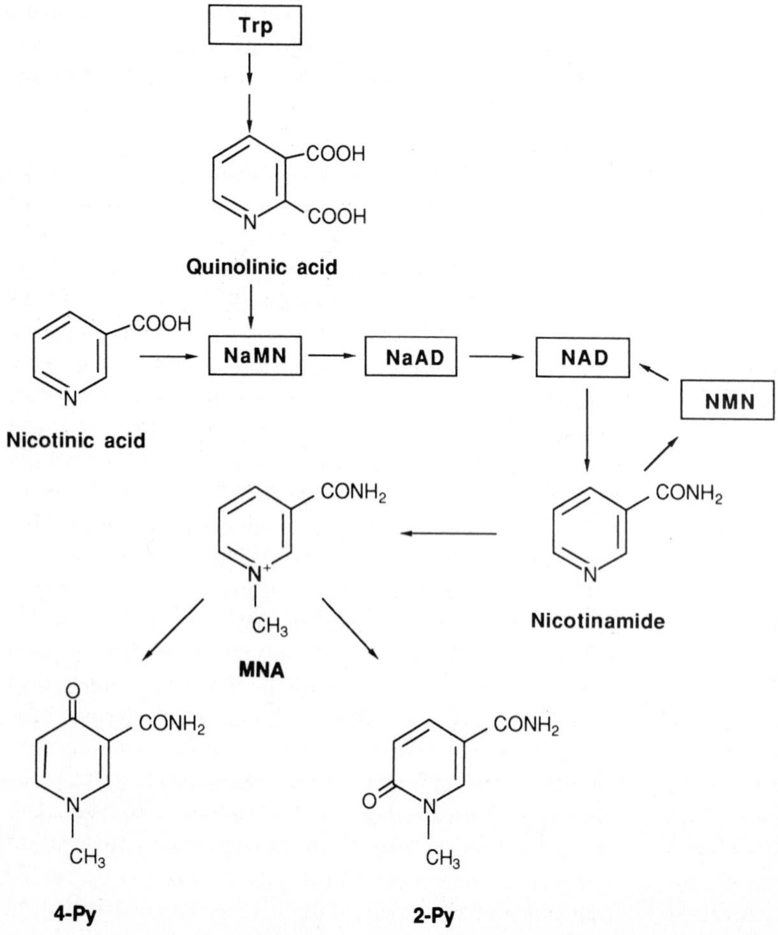

Figure 29-16. The metabolism of niacin. Trp, tryptophan; NaMN, nicotinic acid mononucleotide; NaAD, nicotinic acid adenine dinucleotide; MNA, N'-methylnicotinamide; 2-py, N'-methyl-2-pyridone-5-carboxamide; 4-py; N'-methyl-4-pyridone-3-carboxamide; NMN, nicotinamide mononucleotide. *From:* Shibata K, Matsuo H. Correlation between niacin equivalent intake and urinary excretion of its metabolites, N'-methylnicotinamide, N'-methyl-2-pyridone-5-carboxamide, and N'-methyl-3-4-pyridone-carboxamide, in humans consuming self-selected food. Am J Clin Nutr 1989;50(114):115.

RDA ranges from 13 to 20 NE, depending on age and sex (see Table 29-4).[181] Niacin deficiency leads to the development of pellagra. Average diets in the United States typically supply adequate amounts of both niacin and tryptophan. However, protein found in corn contains only 0.6% tryptophan as compared with approximately 1.4% in animal protein and 1.0% in most other proteins of vegetable origin. In addition, corn and wheat contain niacin in the form of nicotinyl esters, which the body cannot hydrolyze during digestion. A diet predominantly composed of corn meal places the individual at risk of hypovitaminosis unless the corn meal is pretreated before cooking to liberate nicotinic acid from the esters.

Body stores of niacin are small. Despite normal dietary intake of tryptophan and niacin, subclinical niacin deficiency and even the frank deficiency disease pellagra have been reported to occur in association with alcoholism, upon treatment of Parkinson's disease and TB with hydrazine-related drugs, in Hartnup's disease (an inherited disorder of intestinal and renal transport of certain amino acids), and in carcinoid syndrome.[117] Initial signs of pellagra include anorexia, headaches, weakness, irritability, indigestion, and sleeplessness. This progresses to the classic 4Ds of advanced pellagra: dermatitis, diarrhea, dementia, and death.[14,64]

Blood cholesterol and triglyceride levels are lowered in response to nicotinic acid therapy.[203,269] Hypolipidemic actions of nicotinic acid (but not nicotinamide) are due to inhibition of secretion of VLDL from the liver and reduction in the liberation of fatty acids from adipose tissue. An increase in HDL cholesterol and a reduction of Lp(a), a variant of LDL, have also been reported. A strong correlation has been observed between a positive Lp(a) level and a family history of ischemic heart disease.[105] Ingestion of large amounts of nicotinic acid produces the annoying side effects of peripheral vasodilatation or flushing and itching. The effectiveness of pharmacologic doses of the vitamin (3–9 g/day) in lowering serum lipid levels has made it a popular choice as an over-the-counter drug among individuals concerned with reducing their risk of coronary heart disease. Adverse effects observed with such large dosages of nicotinic acid (but *not* nicotinamide) include "niacin hepatitis," gout, and impaired glucose tolerance. Even low-dose niacin therapy (<0.5 g/d) for a period of a few months can pose a serious threat for hepatic damage, especially when sustained-release (SR) forms of nicotinic acid (usually labeled niacin) are consumed.[169] It is essential that hepatic enzyme assays (*e.g.*, AST and ALT) be monitored in conjunction with niacin therapy.

Laboratory assessment of niacin status involves determination of the major urinary metabolites of the vitamin, N'-methylnicotinamide (NMN), N'-methyl-2-pyridone-5-carboxamide (2-py), and N'-methyl-4-pyridone-3-carboxamide (4-py), in a 24-hour urine sample.[117,223] Niacin deficiency reduces the total amounts of metabolites excreted, but measurements of the nicotinamides, 2-py and 4-py, are reported to be better indices of niacin status than is NMN excretion.[237] Following preliminary purification, the metabolites are separated and quantified by HPLC with UV detection. Until the recent development of a method for simultaneous determination of 2-py and 4-py, little information on 4-py was available. Data from a study of young Japanese women showed urinary excretion of 4-py ranged from 2.4 to 15.8 µmol/day.[237] Most chemical assessments still focus on determinations of the more abundant urinary metabolites, NMN and 2-py. Urinary excretion of less than 1.2 mg a day for either NMN or 2-py represent low dietary intake and deficient niacin status in adults.[116] Older literature suggests that a ratio of 2-py to NMN >2.0 indicates adequate niacin nutriture, whereas a ratio <2.0 indicates deficiency.[6] Blood tests for evaluating niacin nutriture have not proven to be reliable, although plasma levels of 2-pyridone have been suggested as useful.[116,117] Research on better biochemical markers of niacin status continue.[117] Microbiologic reference procedures to assess niacin levels in whole blood are based on a growth requirement for either nicotinic acid or nicotinamide by *Lactobacillus plantoides*, strain 8014, from the American Type Culture Collection.[14]

Pantothenic Acid

Pantothenic acid is a dihydroxydimethylbutyric (pantoic) acid derivative of the amino acid β-alanine (see Table 29-11). Its biochemical function is as a constituent of coenzyme A (CoASH) and within the 4′-phosphopantetheine moiety of the acyl carrier protein of fatty acid synthetase.[69] Coenzyme A is required for biologic acetylation reactions or other acyl group transfers. In this capacity, pantothenic acid is required in the synthesis and β-oxidation of fatty acids; the acetylation of choline, amines, and amino acids; the synthesis of cholesterol, porphyrin, and sterols; gluconeogenesis; and the oxidative metabolism of carbohydrates.[181,192] Lack of pantothenic acid presumably has far-reaching effects on the body's metabolism, but no clinical disease is specifically associated with inadequate intake in humans. Deficiency has been experimentally induced in humans. Volunteers on a pantothenic acid–deficient diet developed nausea, vomiting, a burning sensation in the feet, muscular weakness, cramps, malaise, headache, and insomnia. Vitamin deficiency has been found to lead to lowered blood sugar levels and increased sensitivity to insulin. Impaired antibody production has also been reported.[224] Administration of pantothenic acid relieved the signs and symptoms. In endemic pantothenic acid deficiency, as reported in a Japanese study, the occurrence of hypertension increased.[234] Generally, it seems unlikely that a specific deficiency of the vitamin would occur, given its widespread distribution in foods. It is more likely that any inadequacy would be part of a generally malnourished condition or deficiency of the B complex as a whole. No RDA has been established, but a daily ingestion of 4 to 7 mg is suggested as adequate for adults (see Table 29-4B). Pantothenic acid appears to be relatively nontoxic, with only occasional diarrhea and other minor GI disturbances reported when doses up to 20 g are administered.[181]

Pantothenic acid is found in whole blood, plasma, serum, and erythrocytes. In the red blood cells the vitamin exists predominantly as coenzyme A; serum and urine contain free pantothenate. Pantothenic acid levels in blood and other tissues may be assessed by radioimmunoassay or microbiologically by growth of vitamin-dependent *Lactobacillus plantarum* or *Pediococcus acidilactici,* NCIB 6990.[241,284] Cell population is estimated turbidimetrically or, in the diffusion method, by measuring average growth-zone diameters around cups cut in the inoculated agar into which test solutions have been placed. Standards are run with each assay. Whole blood levels <100 μg/dL and urinary excretion <1 mg/d for total pantothenic acid may be interpreted as evidence of dietary deficiency.[69] Pantothenic acid intake and serum levels of the vitamin do not correlate in all studies. A serum reference range of 15 to 40 μg/dL has been proposed.[112]

Biotin

Biotin is a bicyclic derivative of urea containing imidazole and thiophene rings. A valeric acid side chain is attached to the sulfur-containing thiophene ring (see Table 29-11).

Humans cannot synthesize biotin and thus must rely on three major sources of supply: (1) diet, (2) intestinal microflora, and (3) recycled vitamin liberated from biotin-containing enzymes.

Biotin is found in the same wide variety of foods that are rich in other B complex vitamins. Even in the richest dietary sources, the absolute amount of biotin is low. Dietary biotin may be free or protein-bound. If protein-bound, the vitamin must be released by enzymatic action in the GI prior to absorption.[173]

A conventional diet is generally sufficient to meet the body's daily needs. Peculiar dietary practices may induce biotin deficiency. The egg white glycoprotein avidin has long been known to bind biotin in a complex that resists digestion and prevents vitamin absorption in the small intestine. Rare instances of biotin deficiency due to avidin toxicity from ingestion of large numbers of raw eggs by food faddists over long periods of time have been reported primarily as curiosities in the medical literature. Heating denatures avidin and destroys its ability to complex with biotin.[23] Human milk contains only about one-tenth as much biotin as cow's milk. Infants exclusively breast-fed may have inadequate dietary intake. Vitamin deficiency can also arise as a complication of long-term TPN. Although prolonged use of anticonvulsants other than valproate may reduce circulating biotin levels and lead to excretion of abnormal organic acids, no clinical symptoms of deficiency have been reported in epileptic patients.[22]

Biotin-producing microorganisms exist in the large intestine. Bacteria known to synthesize biotin include *Escherichia coli, Proteus vulgaris,* and *Streptococcus faecalis.* Large doses of antibiotics, particularly sulfonamides, or massive bowel resection may lead to biotin deficiency by radically altering intestinal microflora and absorptive surface availability. Low-birth-weight infants are at considerable risk of a suboptimal supply of the vitamin due to the minimal development of enteral microflora at the time of birth.[18] In all cases, intestinally synthesized biotin is not sufficient to maintain normal biotin levels.[22]

No RDA has been established for biotin. Safe and adequate intake ranges from 30 to 100 μg per day for adolescents and adults depending on body weight and energy consumption.[181] Biotin appears to be nontoxic even in pharmacologic doses.[214]

Unlike most B vitamins that function as coenzymes for a wide variety of enzymes, biotin serves as a prosthetic group for a small number of enzymes. When the vitamin is covalently linked to the epsilon amino acid group of a lysine residue in the carboxylase's active site, the apoenzyme is converted to a holoenzyme. Enzymatic action is required to attach as well as to detach the biotin. The enzyme that catalyzes the attachment of biotin to the apocarboxylase is holocarboxylase synthetase (HCS). Once bound to the holoenzyme, further enzymatic action is required to release the vitamin. Proteolytic degradation of the carboxylases does not liberate free biotin. Breakdown products are small biotin-containing peptides and biocytin—the lysly derivative of biotin. Liberation of the free vitamin requires the hydrolytic action of the enzyme biotinidase. In the absence of biotinidase to recycle the vitamin, biotin deficiency is likely to develop unless therapeutic doses of the vitamin are provided. Hereditary defects in the metabolism of biotin may be attributed to quantitative, structural, or functional deficiencies in the enzymes holocarboxylase synthetase or biotinidase. In these instances, biotin treatment is effective therapy. When the inherited alteration is within the apocarboxylase itself, enzyme activity is unresponsive to increased levels of biotin and the clinical picture is not improved by vitamin therapy.[55]

By catalyzing the transfer of CO_2 as a carboxyl group to various acceptor molecules, the four biotin-requiring carboxylases play key roles in fat and carbohydrate metabolism (Fig. 29-17).[4,59,214] One biotin-dependent enzyme, acetyl-CoA carboxylase (ACC), is mainly a cytoplasmic enzyme. The remaining three enzymes, pyruvate carboxylase (PC), propionyl-CoA carboxylase (PCC), and beta-methylcrotonyl-CoA carboxylase (MCC), are found in mitochondria. ACC catalyzes carboxylation of acetyl-CoA to form malonyl-CoA and is thus essential for the *de novo* synthesis of saturated straight-chain fatty acids. PC catalyzes the formation of oxaloacetate from pyruvate—a vital step in gluconeogenesis, the sequence of reactions which produce "new" glucose from noncarbohydrate sources. Defective functioning of this enzyme suggests an increased risk of hypoglycemia. MCC catalyzes the degradation of the ketogenic amino acid leucine. PCC plays a key role in the degradation of isoleucine, methionine, threonine, cholesterol, and odd-chain fatty acids. Biochemical indications of biotin deficiency/dependence may be traced to disruptions in these metabolic pathways.[24,83]

Classic manifestations of biotin deficiency encompass a wide range of cutaneous, ophthalmic, and neurologic signs and symptoms. Biochemical abnormalities may include metabolic acidosis, organic aciduria, and mild hyperammonemia.[214] The diversity of manifestations of hypovitaminosis and the clinical similarities to zinc and essential fatty

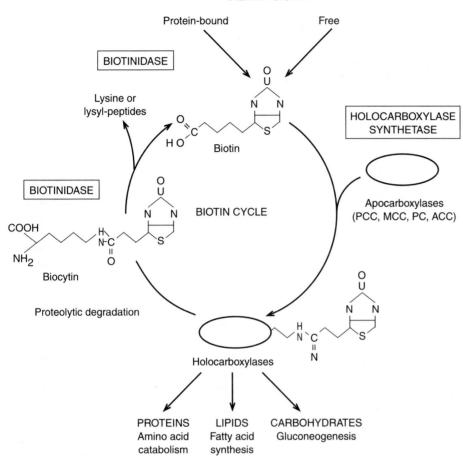

DIETARY BIOTIN

Figure 29-17. The biotin cycle demonstrates the metabolism of biotin. The two major enzymes involved in this cycle are holocarboxylase synthetase, which covalently attaches biotin to the various apocarboxylases to form holocarboxylases, and biotinidase, the hydrolase that cleaves biotin from biocytin or short biotinyl peptides that are formed from the proteolytic degradation of holocarboxylases and possibly from dietary protein bound sources. (*From* Wolf B, Mitchell PL, Lambert FW, et al. Newborn screening for biotinidase deficiency. *In:* Carter T. Wiley AM, eds. Birth defect symposium XVI genetic diseases: Screening and management. New York: Alan R Liss, 1986;176; with permission.)

acid deficiencies may delay recognition and initiation of vitamin therapy.[166,173,215]

Biotin-responsive inherited disorders of metabolism may be due to either isolated or combined defects in vitamin-dependent enzymes (see Table 29-1).[59] Isolated deficiencies are rare, though they have been reported for each of the three mitochondrial carboxylases. Multiple carboxylase defects (MCD) are far more common. MCD are characterized by excretion of abnormal organic acid metabolites, and there is dramatic resolution of clinical manifestations of dysfunction following administration of pharmacologic doses of biotin. The organic aciduria will clear within 1 to 2 days; neurologic and cutaneous manifestations will clear within 1 to 2 weeks. It appears that combined carboxylase deficiency can be attributed to either a defect of holocarboxylase synthetase, *e.g.*, a low V_{max} and/or abnormally high K_m for biotin, or a deficiency of biotinidase. HCS abnormalities result in an inability to bind biotin to the apocarboxylase despite normal serum biotin concentrations. Biotinidase deficiency creates a hypovitaminosis due to recycling failure and inability to liberate the vitamin from dietary proteins. Plasma and urinary biotin concentrations, in this condition, are below normal limits.[282]

The two forms of MCD that have been characterized are initially distinguished by time of symptom onset. In neonatal or early-onset MCD, manifestations appear shortly after birth and include vomiting, lethargy, hypotonia, later development of hair loss, and erythematous skin rash. Biochemical tests show severe metabolic ketoacidosis, hyperammonemia, and organic aciduria. However, serum biotin levels are within reference range. The primary enzymatic abnormality associated with this clinical disorder is decreased HCS activity. Consequently, carboxylase activities are also reduced. Continuing administration of therapeutic doses of biotin (10 mg/day) enhance HCS ability to biotinylate apocarboxylase, and normal carboxylase functions are restored.

Juvenile or late-onset MCD usually results from a deficiency in biotinidase.[94,214] In these individuals, HCS activity is unaffected as long as biotin levels are adequate. Because of variations in dietary intake, composition of intestinal flora, and level of vitamin stores, the age of onset of biotin deficiency varies widely—from weeks up to 2 years. Clinical signs and symptoms of hypovitaminosis are frequently present by 6 months. Low-birth-weight infants are at greater risk for early onset of symptoms. Neurologic symptoms are always present and are often the first signal of a problem. Symptoms include seizures, hypotonia, alopecia, seborrheic skin rash (often with candidiasis), ataxia, often developmental delay, conjunctivitis, and, in approximately 50% of patients, hearing loss. Biochemical findings include excretion of large quantities of abnormal organic acids and development of metabolic ketolactic acidemia. However, prolonged biotin deficiency may be required before ketoacidosis and organic aciduria develop. Plasma biotin concentration is low; 10 to 20 mg of biotin administered daily has become standard treatment. However, a recent report suggests that symptoms of biotin deficiency can be controlled by a daily dose of less than 100 μg. As the body's biotin needs are met, patient improvement is dramatic.[94] Neonatal screening for biotinidase has been proposed. The incidence of biotinidase deficiency in the general population is estimated to be 1 per 50,000 to 75,000. Incidence of phenylketonuria (PKU) is approximately 1 in 11,000; for neonatal hypothyroidism, about 1 in 4000 births; and galactosemia, between 1 per 18,000 and 1 per 180,000 births. Given the high cost of long-term health care or institutionalization, it is advocated by some that screening for abnormalities in biotin nutriture can be economically, as well as medically, justified.

The need for assessment of biotin status by the laboratory has increased as a variety of biotin-responsive inborn errors of metabolism have been identified and knowledge of clinical conditions contributing to biotin deficiency has expanded.[90] A relatively simple competitive protein-binding assay for determination of plasma and urinary biotin has been proposed.[221]

Evaluation of biotin status can involve direct measurement of serum biotin levels. Serum values range from 300 to 800 pg/mL.[221] Concentrations <200 pg/mL are considered indicative of deficiency. Investigation of biotin-responsive inborn errors of metabolism lead to the development of accurate biotin assays for use with plasma and urine. Radioligand assays employing a variety of radiotracers and separation agents have been used. Liquid chromatography with fluorimetric detection has been described.[22] An RIA procedure has been reported that is simple, sensitive, and allows the handling of 100 samples in less than 4 hours.[144]

Biotin quantification based upon the specific biotin-avidin interaction has been utilized in several assays.[62] A simple, fast, accurate, and reproducible method—an enzyme-linked ligand sorbant assay (ELLSA)—has been described that permits the quantification of biotin without use of isotopes or immunologic reagents.[90] The procedure is based on the competition between biotin in the serum and immobilized biotin for binding to horseradish peroxidase (HRP)-avidin conjugate. HRP-avidin bound to immobilized avidin is measured. The reagents employed have a long shelf-life, and the procedure lends itself to automation.

An alternative is to evaluate the activity of the biotin-dependent enzymes. HCS and carboxylase activities can be measured in cultured skin fibroblasts or in lymphocyte preparations. Biotinidase activity in serum can be measured by determining the amount of biotin liberated from biocytin or from an artificial substrate such as *N*-1-biotinyl-*p*-aminobenzoic acid.[94] Subsequent color development by diazotization is quantitated colorimetrically. Using a modification of the quantitative colorimetric assay, a semiquantitative colorimetric screening test of newborns for biotinidase activity can be performed on a spot of blood dried on the filter paper discs (Guthrie papers) that are employed in most neonatal screening programs.[282] A semiautomated colorimetric method on the Roche Cobas Bio Centrifugal Analyzer for quantitating biotinidase activity using twenty μL of serum has been described.[200]

Vitamin B₆

Like niacin, vitamin B_6, is a pyridine derivative. Six major forms of the vitamin are identified (Fig. 29-18).[192] Pyridoxine (PN) is a primary alcohol, pyridoxal (PL), the corresponding aldehyde, and pyridoxamine (PM) contains an aminomethyl group. These compounds differ in substituents on carbon number four of the pyridine nucleus. The 5'-phosphorylated derivatives of each, PLP, PNP, and PMP, constitute the majority of vitamin B_6 in tissues. Vitamin B_6 is now the recommended generic descriptor for this family of vitamers. Pyridoxal kinase, the enzyme that catalyzes the phosphorylation reaction, requires zinc for activation. Zinc status therefore plays a role in vitamin B_6 metabolism. PLP, synthesized primarily in the liver, accounts for more than 50% of B_6 vitamers in plasma. In the circulation, almost all PLP is bound to albumin. The RBCs also appear to play an important role in the transport and metabolism of vitamin B_6.[136] Limited capacity for B_6 storage exists in the liver and skeletal muscles. Over 90% of PLP is present in skeletal muscle in association with the enzyme glycogen phosphorylase.[140] PLP-binding proteins are also present in the liver. Clinical symptoms of deficiency may appear in as few as 3 weeks if a patient is placed on a B_6-deficient diet. The principal excretory product of vitamin B_6 is 4-pyridoxic acid (4PA).[8]

As with all B vitamins, B_6 is particularly abundant in liver, yeast, and whole-grain cereals, as well as in fruits and veg-

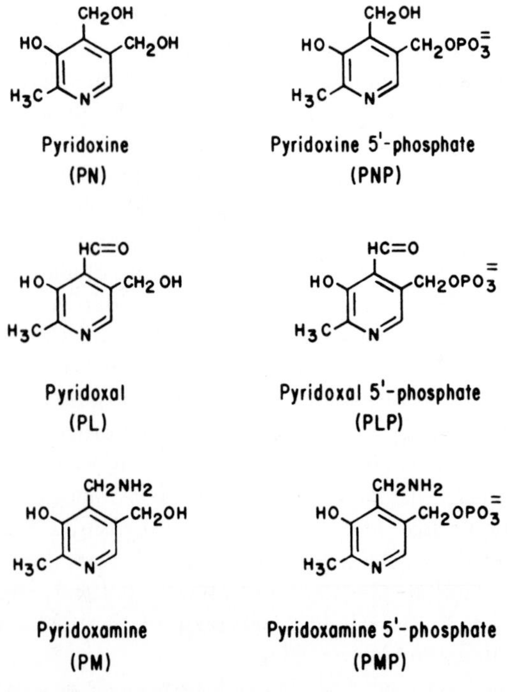

Figure 29-18. Vitamers of vitamin B_6.

etables.[82] As a group, the B_6 vitamers are labile and thermal processing considerably reduces the vitamin content of foods. Dietary studies indicate that changes in food-consumption habits and restricted diets frequently encountered in the elderly are likely to produce deficiencies in vitamin B_6.[167,223] Drug-induced deficiencies are also common, with as high as 30% of alcoholic populations exhibiting hypovitaminosis.[139] Use of estrogen-containing oral contraceptives produces a temporary decrease in plasma levels of pyridoxal phosphate. It has been suggested that this hypovitaminosis accounts, at least in part, for the mental depression, impaired glucose tolerance, and general malaise commonly reported among users of oral contraceptive agents.[276]

The RDA for adult males is 2.0 mg, and it is 1.6 mg for adult females. During pregnancy and lactation, the daily allowance is 2.2 and 2.1 mg, respectively.[181] Establishment of an RDA is complicated by the fact that dietary factors influence the requirement for vitamin B_6. As protein intake increases, there is an increased need for PLP. Plasma PLP values in excess of 7.5 ng/mL (30 nmol/L) are considered to reflect adequate vitamin status.[136]

Manifestations of vitamin B_6 deficiency involve many physiologic functions and organ systems. Eczema, seborrheic dermatitis, cheilosis, glossitis, and angular stomatitis are often noted. Disruptions of the nervous system include some forms of mental depression, hyperirritability, excessive sensitivity of sounds, abnormal ECG patterns, and convulsions. Hematologic and immunologic abnormalities include microcytic, hypochromic anemia, and impairment of both humoral and cell-mediated immunity.[8,82,162,165,258] It has been suggested that low levels of B_6 increase the risk of coronary heart disease.[279] The enzyme cystathionine beta-synthase that catalyzes conversion of methionine to cystathionine is B_6-dependent. Reduced enzyme activity, due to unavailability of its coenzyme, leads to accumulation of the atherogenic amino acid homocysteine.[235,248,267] Studies have shown a high prevalence of marginal vitamin B_6 nutriture among older adults.[146] A large proportion of U.S. elderly do not consume

adequate levels of vitamin B_6 as judged by the RDA standards or by the elevated plasma levels of homocysteine.[123,205,249] In addition, a positive association between PLP levels and HDL cholesterol has been reported.[128] Plasma reference values for pyridoxine range from approximately 4 to 20 ng/mL, while for whole blood the range is 30–80 ng/mL.[56,250]

The metabolically active vitamer of B_6 is pyridoxal phosphate. All vitamin B_6-dependent enzymes utilize PLP as the coenzyme.[8] PLP is covalently bound to enzymes via a Schiff base with an ϵ-amino group of lysine in the enzyme. It has been reported to be a cofactor in over 100 enzymatic reactions, and plays a key in the decarboxylation and transamination of amino acids.[136] The coenzyme readily condenses with the alpha amino group of a donor amino acid to form a Schiff base, or aldimine (Fig. 29-19). This may lead to decarboxylation of the donor amino acid, forming an amine, or transfer of the amino group to the alpha carbon of an alpha ketoacid with the formation of a new amino acid. Among the most important B_6-dependent amino acid decarboxylases are histidine decarboxylase, required for the synthesis of the vasoactive compound histamine; aromatic amino acid decarboxylase, which catalyzes formation of dopamine from dopa (dihydroxyphenylalanine) and serotonin (5-hydroxytryptamine) from 5-hydroxytryptophan; and glutamic acid decarboxylase, which acts on glutamic acid to form the neurotransmitter gamma-amino butyric acid (GABA).[8] The antagonist action of B_6 on the drug L-DOPA, administered to patients with Parkinson's disease, is due to enhancement of decarboxylation of the drug outside its target tissue, the central nervous system. Patients receiving L-DOPA should avoid ingestion of megadoses of vitamin B_6.[8] Pyridoxal phosphate is also required by enzymes catalyzing side-chain modifications of amino acids. Thus, PLP functions as a coenzyme in transamination, deamination, desulfuration, decarboxylation, racemization, cleavage, and synthesis reactions of amino acid metabolism. In addition, B_6 is required for conversion of tryptophan to nicotinamide (Fig. 29-20), synthesis of heme and other porphyrins, gluconeogenesis, carnitine synthesis, and the conversion of linoleic to arachidonic acid, a precur-

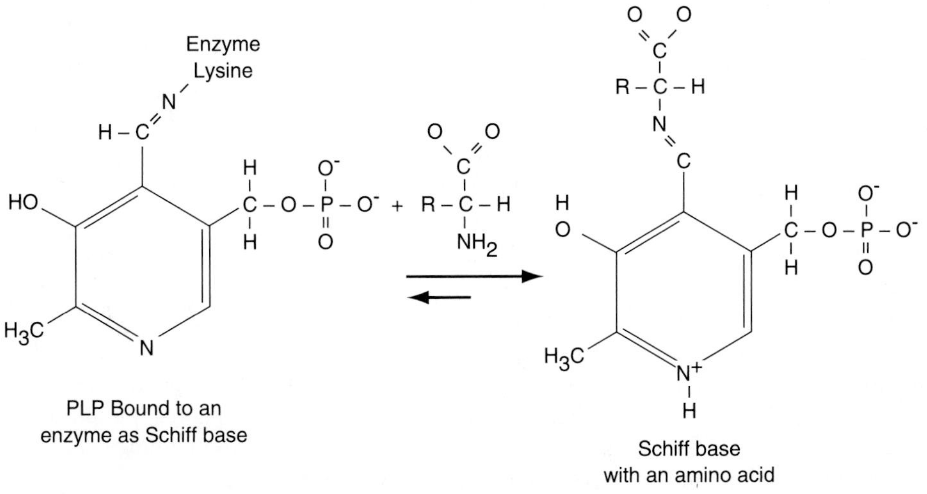

Figure 29-19. Schiff base formation between pyridoxal-5'-phosphate and an amino acid. *Source:* Leklem JE. Vitamin B_6. In: Machlin LJ, ed. Handbook of vitamins. 2nd ed. New York: Marcel Dekker, 1991.

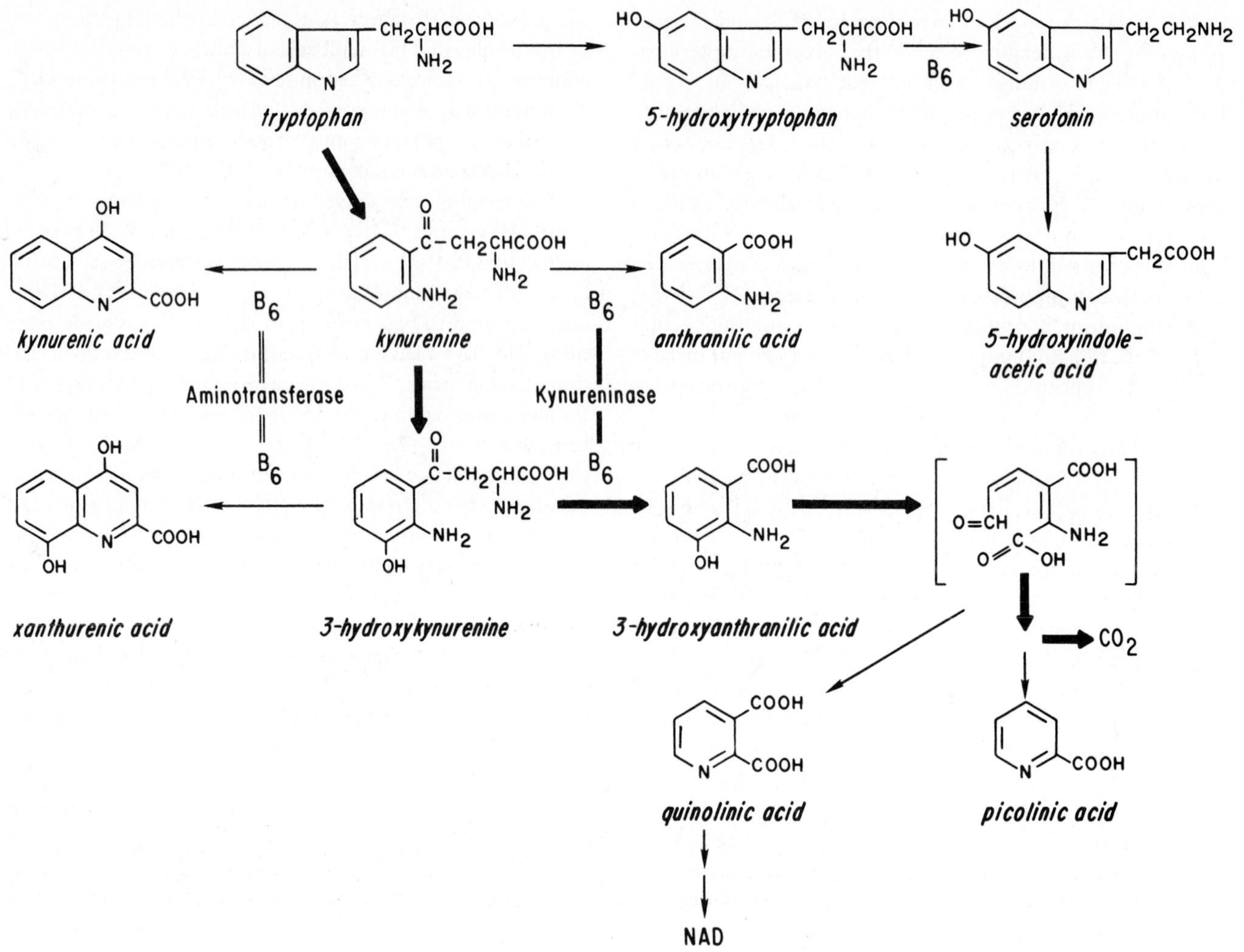

Figure 29-20. Major metabolites of the amino acid tryptophan. B_6 indicates the sites of action of enzymes requiring PLP as a cofactor. (Adapted from Brown RR. The tryptophan load test as an index of vitamin B-6 nutrition. In: J Leklem, Reynolds R, eds. Methods in vitamin B-6 nutrition. Analysis and status assessment. New York: Plenum Press, 1981.)

sor of the prostaglandins.[192] It has been proposed that PLP at physiologic concentrations serves as a modulator of steroid hormone action by its inhibition of the binding of intracellular steroid-receptor complexes to DNA (Fig. 29-21).[136]

Deficiency of B_6 is most common among alcoholics because of impaired intestinal absorption, pregnant females because of the effect of elevated estrogen levels on tryptophan metabolism, infants on commercial formulas subjected to high-temperature autoclaving, and patients receiving a wide variety of drugs (Table 29-15).[212] Plasma PLP levels have been shown to decrease with advancing age, often due to chronic illnesses and medical intervention.[146] An unusual population at risk of hypovitaminosis B_6 are personnel working on submarines. Decreased vitamin B_6 status has been reported in crews during prolonged patrols, despite supplementation.[204]

Megavitamin therapy with B_6 has been successful in the treatment of a number of genetically determined vitamin-

dependent disorders.[98,272] Pyridoxine dependency is presumed to be the result of either an enzyme deficiency or decreased enzyme activity attributed to reduced affinity of the apoenzyme for PLP. Vitamin B_6–responsive disorders include homocystinuria, cystathioninuria, xanthurenic acidurias, hyperoxalurias, and infantile convulsions.[8,59] At least some studies have reported beneficial effects on symptoms of premenstrual syndrome (PMS) following high dose supplementation with vitamin B_6.[136] Oral doses of up to 1000 mg per day reportedly produce no adverse side effects.[58] Recent reports, however, indicate sensory and peripheral neuropathy from megadoses (2–6 g) of B_6. A variety of toxicity symptoms have been reported to be associated with long-term high doses of vitamin B_6 (Table 29-16).[137]

Indices used to assess vitamin B_6 status may be direct or indirect. Direct indices include plasma pyridoxal-5'-phosphate concentration, plasma total vitamin B_6 concentration, urinary 4-pyridoxic acid excretion, and urinary total

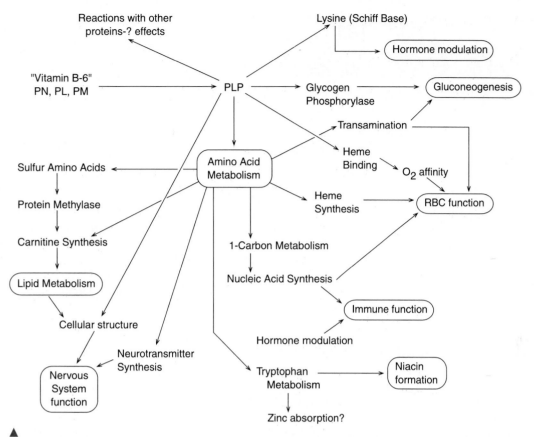

Figure 29-21. Cellular processes in which pyridoxal 5′-phosphate (PLP) acts as a coenzyme or binds with proteins and modifies the action of the protein. The primary biological systems subsequently influenced are circled. Heme binding refers to PLP binding to hemoglobin. *Source:* Leklem, LE. Vitamin B-6 metabolism and function in humans. In: Leklem JE, Reynolds RD, eds. Clinical and Physiological Applications of Vitamin B-6. New York: Alan R. Liss, 1988.

vitamin B_6 excretion. Indirect indices include trytophan loading measuring urinary xanthurenic acid, methionine loading measuring urinary cystathionine, RBC aminotransferase activity coefficient, and urinary oxalate excretion. The most common method for clinical laboratory assessment of vitamin B_6 status is measurement of PLP–dependent erythrocyte transaminase activity *in vitro* in the presence and absence of exogenous PLP.[6,116,136,225] *In vivo* unsaturation of as-

partate transaminase (AST) or alanine aminotransferase (ALT) with coenzyme PLP is reflected in the percentage increase in enzyme activity following PLP addition to incubating hemolysate aliquots. An activity coefficient is obtained by dividing enzyme activity with added PLP by enzyme activity in the absence of added cofactor. Ratios for AST of less than 1.5 and for ALT of less than 1.25 reflect normal saturation of the enzyme and adequate B_6 levels. In

TABLE 29-15. Drug-Vitamin B_6 Interactions

Drug or Drug Type	Examples	Mechanism of Interaction
Hydrazines	Iproniazed, isonicotinyhydrazine hydralazine	React with pyridoxal and PLP to form a hydrozone
Antibiotic	Cycloserine	Reacts with PLP to form an oxime
L-DOPA	L-3,4-dihydroxyphenylalanine	Reacts with PLP to form tetrahydroquinoline derivatives
Chelator	Penicillamine	Reacts with PLP to form thiazolidine
Oral contraceptives	Ethinyl estradiol, mestranol	Increased enzyme levels in liver other tissues; retention of PLP
Alcohol	Ethanol	Increased catabolism of PLP, low plasma levels

Source: Leklem LE. Vitamin B6. In: Machlin LJ, ed. Handbook of vitamins. 2nd ed. New York: Marcel Dekker, 1991.

TABLE 29-16. Toxicity Symptoms Reported to Be Associated with Chronic High Doses of Pyridoxine

Symptoms

Motor and sensory neuropathy; vesicular dermatosis on regions of the skin exposed to sunshine.

Peripheral neuropathy; loss of limb reflexes; impaired touch sensation in limbs; unsteady gait, impaired or absent tendon reflexes; sensation of tingling that proceeds down neck and legs

Dizziness; nausea; breast discomfort or tenderness

Photosensitivity on exposure to sun

Source: Leklem LE. Vitamin B6. In: Machlin LJ, ed. Handbook of vitamins. 2nd ed. New York: Marcel Dekker, 1991.

conditions of vitamin B_6 deficiency, the enzyme is not saturated by coenzyme *in vivo*, and the activity ratio will exceed 1.5 and 1.25, respectively.[223] An elevated erythrocyte AST (EAST) index or ratio is a commonly accepted indicator of inadequate B_6 nutriture.

An older procedure for determination of B_6 nutritional status is the tryptophan loading test. Urine is collected for 24 hours after ingestion of 2 to 5 g of l-tryptophan, and output of xanthurenic acid is measured. In vitamin B_6 deficiency, kynureninase activity is decreased, and kynurenine and 3-hydroxykynurenine accumulate. There is a resultant increase in excretion of tryptophan metabolites, including xanthurenic acid (see Fig. 29-20). A similar protocol is employed in the methionine loading test, with assessment of cystathionine excreted being used to evaluate B_6 status. Other widely used methods for vitamin assessment have included microbiologic and fluorometric assays.[225]

The concentration of plasma PLP is considered to be the best indicator of vitamin B_6 status, including tissue stores. The 4-pyridoxic acid content of a 24-hour urine reflects the production and excretion of the major metabolite of B_6. Reduced excretion of this urinary metabolite is one of the earliest indicators of a B_6 deficiency.

Direct assessment of B_6 levels is complicated by photosensitivity of the vitamers. HPLC methods for measurement of 4-pyridoxic acid levels in the urine or B_6 vitamers in the plasma are rapid, specific, and sufficiently sensitive to be clinically useful.[8,65,218] A sensitive and reliable procedure for determination of PLP by HPLC with electrochemical detection has been described.[47]

Also of interest is a radioenzymatic assay for direct measurement of PLP, based on activity of the PLP-dependent enzyme tyrosine decarboxylase from *Streptococcus fecalis*.[34] The commercially available apoenzyme is incubated with tritiated tyrosine and patient plasma. PLP in the specimen provides the required coenzyme, and the decarboxylated metabolite formed ([^{3}H]tyramine) is extracted and quantified by liquid scintillation counting.

Folates

Folates comprise a family of compounds derived from folic or pteroylglutamic acid.[192] All members of the family possess the double-ring structure pteridine (2-amino-4-hydroxy-6-methylpterin) joined by a methylene bridge to *para*-aminobenzoic acid (PABA). This parent compound is called *pteroic acid (Pte)*. PABA, in turn, is linked through a peptide bond to one molecule of glutamic acid, forming folic acid (FA) or pteroylglutamic acid (PteGlu; PGA) (Fig. 29-22). Conjugation with additional glutamic acid residues produces a series of polyglutamates. The bulk of the vitamin is present in the diet as folate polyglutamates. Enzymes requiring folic acid as a coenzyme catalyze chemical reactions involving the transfer and utilization of single carbon units. Nitrogen atoms at the 5 and 10 positions in the pteridine ring portion of the molecule are active in these single carbon unit transfers. The polyglutamate chain attaches the coenzyme to the apoenzyme. Double bond reduction and presence of various substituents serve to differentiate the various analogs of folic acid. Reduction of double bonds between ring positions 5–6 and 7–8 converts folic acid into tetrahydrofolic acid (THFA, or FH_4). The term *folate* is applied generically to the entire group of compounds. Use of the older generic descriptor, folacin, is no longer acceptable.

The most recently published folate RDAs are 180 μg for adult females and 200 μg for adult males.[181] For adolescents, 150 μg is recommended. The minimal daily requirement for folate is approximately 50 μg for adults.[100] Use of oral contraceptive steroids can increase urinary excretion of folate. Increased vitamin intake may be required to offset the loss.[212,229] In pregnancy, the RDA is raised to 400 μg to maintain maternal folate reserves and adequately support normal fetal growth (see Table 29-4).[181] Megaloblastic anemia of pregnancy is commonly due to folate deficiency. Folic acid, even as much as 15 mg daily over several years, is reportedly not toxic in humans.[103] However, some data suggest that excessive intake of supplemental folate may interfere with intestinal absorption of zinc.[31,170]

The name *folate*, like the word foliage, is derived from the Latin word for leaf. Cruciferous vegetables, such as spinach, turnip greens, asparagus, broccoli, and brussels sprouts, are

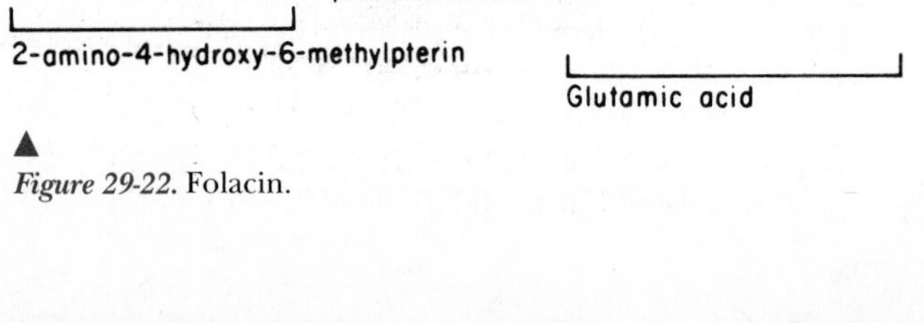

Figure 29-22. Folacin.

rich in folate. Folate is abundant in liver, kidney, whole-grain cereals, yeast, and mushrooms.[82] The vitamin is also synthesized by intestinal microflora. Prolonged cooking, particularly steaming and boiling, destroys most folate in foods. Infants receiving boiled formulas prepared with pasteurized, sterilized, or powdered cow's milk require folate supplementation.

Following ingestion, polyglutamates are enzymatically hydrolyzed to monoglutamates by action of conjugates in the mucosa of the small intestine (Fig. 29-23). Folate monoglutamates are rapidly absorbed and transported in the circulation mainly as the tetrahydrofolate (FH_4) derivative. The major form of folate in serum and red cells is 5-methyl-tetrahydrofolate (N^5-methyl-FH_4). Dihydrofolate reductase

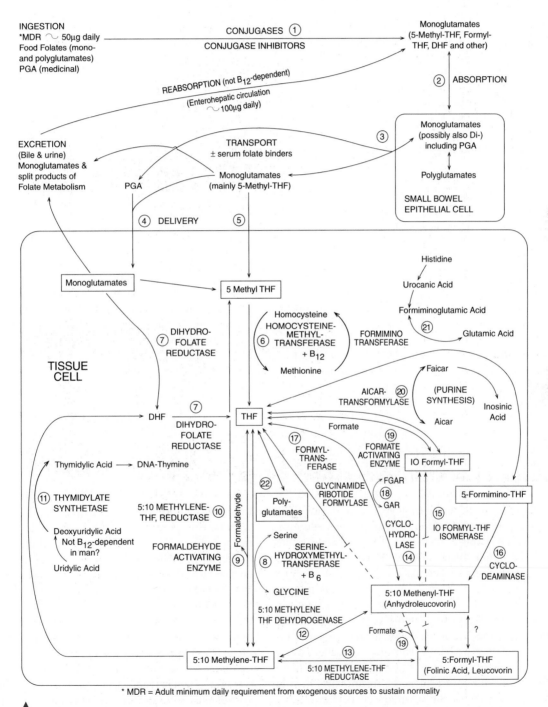

Figure 29-23. Flow chart of folate metabolism in humans. Circled numbers indicate individual steps in folate metabolism. *Source:* Herbert V, Das KC. Folic acid and vitamin B12. In: Shils ME, Olson JA, Shike M, eds. Nutrition in health and disease. 8th ed. Vol.1. Philadelphia, PA: Lea & Febiger, 1994.

* MDR = Adult minimum daily requirement from exogenous sources to sustain normality

catalyzes the enzymatic reduction reaction.[42] Folate may circulate in the free form or attached to low-affinity or high-affinity binders. Approximately two-thirds of folate is loosely bound to plasma proteins, including albumin, α_2-macroglobulin, and perhaps, transferrin. High-affinity folate-binding proteins have been purified from serum, milk, and cerebrospinal fluid. The role these specific proteins play in overall folate nurture is not clear. The milk protein could facilitate intestinal uptake of folate. Presence of a folate-binding protein in the choroid plexus may account for the high CSF/serum ratio of the vitamin. Serum folate levels range from 3 to 25 ng/mL.[42] Marginal deficiency is suggested by concentrations from 3 to 5 ng/mL; levels above 5 ng/mL are interpreted as indicating adequate folate.[28,100] Recent efforts to establish a pediatric reference range show folic acid concentrations to be higher in children, especially in those less than 1 year of age, than in adults. During adolescence, a significant decrease in serum folic acid concentration has been noted.[102] Folate concentration in CSF ranges from 15 to 35 ng/mL.[42] While folate monoglutamates are the circulating and transport forms, polyglutamates are the primary intracellular storage forms of the vitamin.[100] Hepatic stores are believed to account for approximately 50% of the body's reserve of folate, predominantly as pentaglutamates. Other tissues with high concentrations of folate are the kidney and blood cells. RBC folate is almost entirely in the form of methylfolate pentaglutamates. Negative folate balance is indicated by erythrocyte vitamin levels less than 200 ng/mL; tissue depletion occurs when folate levels fall below 160 ng/mL.[100] Tissue folate-binding proteins are reported in granulocytes as well as the brush border of intestinal mucosa. Leukocyte folate ranges from 60 to 123 μg/L of WBCs.[42] Folate-requiring enzymes serve as intracellular folate binders. Excretion occurs in the bile and urine (Fig. 29-23).

Vitamin deficiency may be dietary in origin, associated with malabsorption, or drug-induced (Table 29-17). Nutritional folate deficiency is seen in infants raised on goat's milk, which has only about 10% of the concentration of the vitamin found in human or cow's milk.[181] Inborn errors of folate metabolism (*e.g.*, dihydrofolate reductase deficiency and congenital folate malabsorption) give rise to folate deficiency. Total parenteral nutrition using amino acid solutions, unsupplemented by folate, has been reported to induce acute depression of serum folate, marked by pancytopenia and megaloblastic anemia.[73] Folate malabsorption may occur in conditions such as Crohn's disease or ulcerative colitis. Ironically, sulfasalazine, which is used in the treatment of inflammatory bowel disease, impairs folate absorption. Vitamin deficiency may arise during anticonvulsant therapy with phenytoin or phenobarbital.[31] Other drugs that affect folate status include cycloserine, metformin, and cholestyramine. Antifolate medications are used in the treatment of a wide range of malignant and nonmalignant disorders.[100] Folate antagonists appear to bind irreversibly to the enzyme dihydrofolate reductase. Examples of such drugs are triamterene, a diuretic; pyrimethamine, an antimalarial; trimethoprim, an antimalarial as well as a potentiator of sulfonamides in the

TABLE 29-17. Diseases Treated with Drugs Known to Interfere with Folate Metabolism	
Disease	**Drug**
Cancer, leukemia	Methotrexate
Psoriasis	Methotrexate
Rheumatoid arthritis	Methotrexate
Bronchial asthma	Methotrexate
Bacterial infection	Trimethoprim
Malaria	Pyrimethamine
Hypertension	Triamterene
Crohn's disease	Sulfasalazine
Gout	Colchicine
Epilepsy	Phenytoin
AIDS	Trimetrexate

Source: From Butterworth CE, Tamura T. Folic acid safety and toxicity: A brief review. Am J Clin Nutr 1989;50:353.

treatment of bacterial infections; and pentamidine, used in treatment of trypanosomiasis and leishmaniasis.[212] Pentamidine is also employed in the treatment of pneumonia, presumably due to protozoal infection. Pulmonary disease caused by *Pneumocystis carinii* occurs in 65% to 85% of all AIDS patients. The most common manifestation of this infection is pneumonia. Among the adverse reactions arising from standard pentamidine therapy in the treatment of this pneumonia is the development of folate deficiency. The cancer chemotherapeutic agent methotrexate (MTX) is an especially potent folate antagonist.[212] MTX may also be utilized in the treatment of psoriasis and rheumatoid arthritis. The acute toxicity of folate antagonists is due to their impairment of DNA synthesis. A pharmacologic amount (>0.4 mg/d) of folic acid may be administered as a "rescue dose" to patients receiving cancer chemotherapy.

In the U.S., inadequate folate nurture is particularly common among those in lower socioeconomic groups.[255] Folic acid deficiency has been reported as the most common nutritional deficiency among low-income and institutionalized elderly.[28,167,216] Exposure to ethanol may alter the activity of intestinal brush border folate hydrolase (conjugase), causing malabsorption of the vitamin. Alcohol also interferes with hepatic processing of folate. Chronic alcoholism is a major cause of folate deficiency in the United States.

Tetrahydrofolate (THF) derivatives serve as cofactors for enzymes catalyzing the transfer of 1-carbon groups in methylation reactions necessary for a variety of biochemical reactions. The coenzyme forms of the vitamin include the following tetrahydrofolates: N^5-formyl-FH$_4$; N^{10}-formyl-FH$_4$; N^5-formimino-FH$_4$; N^5, N^{10}-methenyl-FH$_4$; N^5, N^{10}-methylene-FH$_4$; and N^5-methyl-FH$_4$.[192] The carbon units transferred by the coenzymes are present in varying states of reduction. Coenzyme activity appears to be greater with polyglutamate, rather than monoglutamate, forms of folate. Metabolic reactions requiring THF coenzymes include interconversion of serine and glycine; methionine synthesis from homocysteine (also a B$_{12}$-dependent pathway); histidine degradation to glutamic acid by means of formiminoglutamic acid (FIGLU); purine biosynthesis; synthesis of the pyrimidine thymidylate, required in DNA synthesis; and the methylation of biogenic amines, including dopamine, tryptamine, serotonin, adrenaline, noradrenaline, and the generation/activation of formate.[42,192] A number of studies have suggested a role for folate in the reversal of preneoplastic conditions of cervical and lung cancers.[255]

Both biochemical and hematologic changes (Table 29-18) are characteristic of poor folate nutriture. The principal clinical feature of folate deficiency is megaloblastic anemia, but folate depletion may precede anemia by months. Other signs and symptoms of deficiency include anorexia, glossitis, nausea, diarrhea, hepatosplenomegaly, and hyperpigmentation of the skin.[42,64] Neurologic disorders also have been attributed to folate deficiency, although this is not routinely part of the clinical picture. Serum folate levels fall below normal after as few as three weeks of folate deprivation.[28] Deficiency of folate leads to inadequate synthesis of DNA and abnormal cell division. Morphologic evidence of the biochemical inadequacy includes bone marrow megaloblastosis, appearance of hypersegmented neutrophils in the peripheral blood, and macrocytosis of reticulocytes and platelets.[95] When red cell folate levels are less than 100 ng/mL (226.6 nmol/L), morphologic abnormalities in mature circulating red blood cells are detected with development of a macrocytic, normoblastic, or megaloblastic anemia. An elevated mean red cell volume and low hemoglobin are consequences of long-standing folate deficiency.[100] Elevation of certain metabolites in the serum serves as an early indicator of suboptimal levels of folate. For example, poor folate status can result in higher plasma levels of the atherogenic amino acid homocysteine.[49,123,246,267] Because of a lack of 5-methyltetrahydrofolate in amounts sufficient for the remethylation of homocysteine to methionine, homocysteine accumulates in the plasma. Toxic effects arising from excess homocysteine may be due to its interference with normal cross linking of collagen molecules, thereby disrupting or damaging the intimal surface of arteries. Low normal serum folate concentrations could therefore place an individual at increased risk of cardiovascular disease. Supplementation with modest doses of folate (1 to 5 mg/d) can often normalize elevated homocysteine concentrations.[25,124,216]

Approximately 6000 infants are born each year in the U.S. with neural tube defects. Maternal folic acid supplementation in early pregnancy reduces the risk of giving birth to an infant with a neural tube defect (e.g., spina bifida or anencephaly) by as much as 75%. Because closure of the embryonic neural tube normally occurs by the sixth week of pregnancy, there is no deterrent advantage reported for women who begin supplementation after that point in time.[160,203,280] Folic acid fortification of basic foods, such as wheat flour, has been advocated. While this action would address the issue of women of child-bearing age receiving the vitamin in amounts sufficient to reduce the risk of fetal neural tube defects, it could create a medical dilemma for the elderly. It is estimated that pernicious anemia (PA) caused by malabsorption of vitamin B_{12} effects approxi-

TABLE 29-18. Sequence of Events in Developing Folate Deficiency. Earliest Abnormalities in Each Stage are Boxed

	STAGE II Excess*	STAGE I Early Positive Folate Balance	Normal	STAGE I Early Negative Folate Balance	STAGE II Folate Depletion	STAGE III Damaged Metabolism: Folate Deficiency Erythropoiesis	STAGE IV Clinical Damage: Folate Deficiency Anemia
Serum Folate (ng/ml)	>10	>10	>5	<3	<3	<3	<3
RBC Folate (ng/ml)	>400	>300	>200	>200	<160	<120	<100
Diagnostic dU Suppression	Normal	Normal	Normal	Normal	Normal	Abnormal*	Abnormal*
Lobe Average	<3.5	<3.5	<3.5	<3.5	<3.5	>3.5	>3.5
Liver Folate (µg/g)	>5	>400	>3	>3	<1.6	<1.2	<1
Erythrocytes	Normal	Normal	Normal	Normal	Normal	Normal	Macroovalocytic
MCV	Normal	Normal	Normal	Normal	Normal	Normal	Elevated
Hemoglobin (g/dL)	>12	>12	>12	>12	>12	>12	>12
Plasma Clearance of Intravenous Folate	Normal	Normal	Normal	Normal	Normal	Increased	Increased

*Dietary excess of folate reduces zinc absorption.

Due to hormonal effects (on receptors?), there may be folate deficiency (i.e. Stage III-IV negative balance) in cervical epithelial cells (a reversible lesion) (possibly precancerous?) when there is only early negative balance (i.e. Stage I-II negative balance) in the erythron (Ran et al. Blood, November 1990).

Source: Herbert B, Das K. Folic acid and vitamin B_{12}. In: Shils ME, Olson JA, Shike M, eds. Modern nutrition in health and disease. 8th ed. Philadelphia: Lea & Febiger, 1994.

mately 1 million Americans. Most of these individuals are older adults. A deficiency of either vitamin B_{12} or folic acid will create the same hematologic picture (*i.e.,* macrocytic, megaloblastic anemia). However, only a B_{12} deficiency will produce irreversible neurologic lesions. Folic acid supplementation can mask or delay diagnosis of B_{12} deficiency by restoring a normal hematologic picture without preventing the B_{12}-induced neurological disease. There is serious concern among health practitioners that widespread fortification of foods with folic acid would significantly increase the incidence of delayed diagnosis of vitamin B_{12} deficiency.[286]

Microbiologic assays of folates in serum, erythrocytes, and urine have been conducted in the clinical laboratory for many years.[110,171] Although not as rapid or convenient as newer radioassay procedures, microbiologic assay remains the reference method. The organism of choice is *Lactobacillus casei* (ATCC 7469), which utilizes all monoglutamate forms of folate, including the reduced form, 5-methyltetrahydrofolate, for growth. RBC folate is present as polyglutamates and must be converted to monoglutamates for analysis. Interference by antibiotics in the patient specimen presents a serious problem.

Indirect measurement of folate status has been attempted by employing a histidine loading test. Histidine is metabolized to glutamic acid by way of the intermediate formiminoglutamic acid (FIGLU). The final enzyme of this pathway, formiminotransferase, is folate-dependent. If folate is deficient, FIGLU accumulates. When an oral 2- to 15-g dose of histidine is administered to a folate-depleted patient, the amount of FIGLU excreted in the urine in the 8-hour period following the load is at least 5 to 10 times greater than the amount excreted by a folate-replete individual under the same conditions.[229]

Another approach in evaluating the adequacy of tissue folate to support normal biochemical function is by means of the deoxyuridine (dU) suppression test, which reflects slowed *de novo* DNA synthesis. The final step in the conversion of deoxyuridylate to thymidylate for DNA synthesis is folate-dependent. This test is generally abnormal in megaloblastic anemia due to both folate and B_{12} deficiency.[36]

There is concern over falsely low serum values for folate arising from oxidative destruction of the vitamin prior to analysis. To avoid vitamin loss, serum may be stored frozen or a reducing substance such as ascorbic acid may be added to the specimen. A recent study on the effect of light on serum folate concluded that specimens to be tested can be stored at room temperature for up to 8 hours in either a gel separator collection tube or in a polypropylene storage tube without substantial loss (<7%) of the vitamin. Folate specimens exposed to light for more than 8 hours should be redrawn.[154] Since folate levels of erythrocytes exceed serum levels by approximately 40-fold, it is essential that hemolyzed samples not be accepted for assay of serum folate. To measure erythrocyte folate levels, a hemolysate, prepared with an aqueous 1% ascorbic acid solution, is tested.[28]

Folate assessment by competitive protein-binding radioassay techniques is common. Tracers used are [^{125}I]folate

or ^{3}H-PGA. The weak binding of folate to plasma proteins necessitates pretreatment or a denaturation step to liberate the vitamin before application of CPB techniques. Denaturation may be by heat (boiling) or by pH inactivation (no-boil). Incomplete denaturation of interfering proteins is sometimes experienced with a no-boil protocol. Radioligand assay procedures have been adapted for automated systems to permit simultaneous assays of serum folate and B_{12} after manual heat denaturation of endogenous protein binders.[45] Because these two vitamins are so closely linked in terms of biochemistry and metabolic function, it is important that they be evaluated together.[116]

HPLC is particularly useful in separating the various folate compounds. A competitive enzyme-linked ligand sorbent assay (ELLSA) for quantitation of folates has been described that offers promise for application in the clinical laboratory.[89]

Individuals with a folate deficiency will have a reduced capacity to convert homocysteine to methionine. Measurement of serum levels of homocysteine by modified techniques using capillary-gas chromatography and mass spectrometry have proven useful as a means of identifying suboptimal folate nutriture.[230] Totally automated methods, including a C_{18}-based HPLC assay and an FPIA requiring no pretreatment or chromatographic step, have been reported in the literature recently.[237a]

Vitamin B_{12}

In 1948, vitamin B_{12} was isolated and crystallized for the first time by both American and British researchers.[206] IUPAC recommendations call for generic use of the name *cobalamin* for those vitamins that possess a cobalt-containing corrin ring attached to the nucleotide 5,6-dimethylbenzimidazole (see Table 29-11). Dimethylbenzimidazole is similar in structure to riboflavin. The corrin nucleus contains four substituted pyrrole rings and resembles the porphyrin nucleus of heme. Various ligands may be covalently linked to the cobalt atom, including cyanide anion (cyanocobalamin), hydroxyl group (hydroxocobalamin), methyl group (methylcobalamin), or 5′-deoxyadenosyl group (adenosylcobalamin).[192] The coenzyme forms of B_{12}, adenosylcobalamin and methylcobalamin, function as transmethylating agents.[42] Methylcobalamin accounts for approximately 75% of plasma vitamin B_{12}, whereas a similar percentage of liver B_{12} is in the form of adenosylcobalamin. B_{12} in erythrocytes and the kidney is also largely present as adenosylcobalamin. Smaller amounts of hydroxocobalamin and cyanocobalamin exist in body fluids and tissues.

The RDA for vitamin B_{12} is 2 μg for adults and adolescents of both sexes. In pregnancy and lactation, the requirement is increased, respectively, to 2.2 μg and 2.6 μg daily.[181] Although vegetable matter is devoid of vitamin B_{12}, it is present in animal products such as meat and dairy foods, including liver, eggs, milk, and cheese.[82] Microorganisms alone synthesize the vitamin, and animals, including humans, ultimately depend on this activity to furnish preformed B_{12}. Enteric microorganisms, mainly actinomycetes, synthesize B_{12} in the human colon, but it is not absorbed through the mucosa in this region of the gastrointestinal tract.[29] B_{12} deficiency is rarely caused by poor nutrition. However, strict vegetarians, unless they receive B_{12} as a contaminant in food or supplement the diet, will develop a clinical deficiency.[96] The liver stores 50% to 90% of the body's B_{12}.[99] Reserves are relatively large, and it may require literally years for the classic features of deficiency to appear, even in complete absence of vitamin intake.

Vitamin B_{12} is absorbed in the intestine, depending primarily on the availability of intrinsic factor (IF), a glycoprotein secreted by gastric parietal cells (Fig. 29-24). These same cells secrete hydrochloric acid. Impaired absorption due to lack of intrinsic factor in gastric secretions gives rise to the clinical condition known as *pernicious anemia (PA)*.

Achlorhydria, which diminishes B_{12} absorption, and PA, associated with atrophy of the gastric mucosa, are most common among individuals over 60 years of age. An extremely small percentage of vitamin B_{12}, probably less than 1%, is absorbed passively throughout the intestine, independent of IF complex formation. A diffusion-type mechanism for vitamin uptake, not mediated by IF, also seems to operate when large amounts (100–300 µg) of B_{12} are supplied.

The four common forms of cobalamin bind equally well to IF.[41] In the ileum, IF–B_{12} complex binds to specific membrane receptors of the mucosal brush border. A pH above 6 and the presence of calcium ions are required to promote vitamin absorption. Upon transiting the mucosal cell, vitamin B_{12} is released into the portal circulation. Plasma B_{12} is

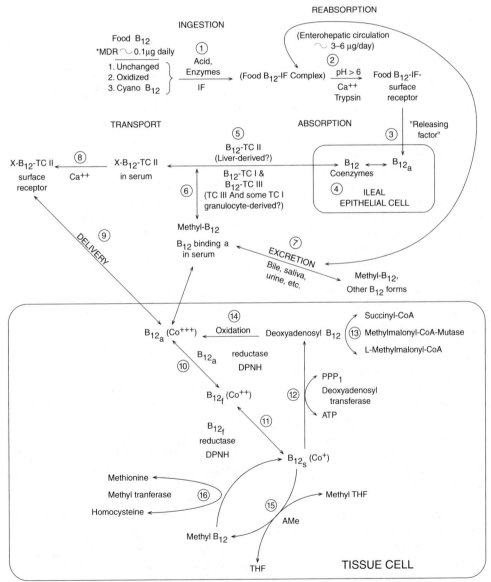

* MDR = Adult minimum daily requirement from exogenous sources to sustain normality

Figure 29-24. Flow chart of cobalamin (B_{12}) metabolism. Circled numbers identify individual metabolic steps. *Source*: Herbert B, Das K. Folic acid and vitamin B_{12}. In: Shils ME, Olson JA, Shike M, eds. Modern nutrition in health and disease. 8th ed. Philadelphia: Lea & Febiger, 1994.

bound by members of a group of carrier globulins, the transcobalamins (TC). Transcobalamin II (TCII) serves as primary transport protein for distribution of newly absorbed vitamin B_{12} to the tissues.[99] All cells that synthesize DNA possess surface receptors for TCII. One of the earliest detectable signs of a negative B_{12} balance is reportedly a decrease in serum holotranscobalamin (TCII + cobalamin).[96] Vitamin B_{12} also binds to haptocorrin, a circulating storage protein. The only receptors for haptocorrin are on B_{12} storage cells (*e.g.*, liver and reticuloendothelial cells). Other protein binders of B_{12} have been identified in body fluids, including serum, saliva, tears, milk, colostrum, cerebrospinal fluid, and gastric juice, as well as in blood cells. These endogenous proteins (TCI and TCIII) have been collectively designated as *R proteins* because of their rapid migration during electrophoresis. R proteins bind both biologically active cobalamin and inactive analogs. The physiologic function of these binding proteins is not clear, but they do not facilitate ileal absorption of the vitamin.[231] The therapeutic form of vitamin B_{12} is cyanocobalamin. If it is administered subcutaneously or intramuscularly, the need for IF-mediated intestinal absorption is bypassed. While the treatment of PA with oral B_{12} megadose therapy is more common in Europe than in the U.S., it has proven to be successful.[133] When given orally, in excess, enough B_{12} is absorbed even in the absence of IF to meet the requirements of most patients. Even in megadoses, cobalamin is reported to be nontoxic.

In humans, two enzymes are known to be vitamin B_{12}–dependent: 5-methyltetrahydrofolate (5-methyl-THF) homocysteine methyltransferase and methylmalonylcoenzyme A mutase.[42] Methylcobalamin functions as coenzyme for a methyltransferase reaction in methionine synthesis. The coenzyme form of folate, 5-methyl-THF, donates a methyl group to cobalamin, which transfers it to homocysteine, forming a new amino acid, methionine. Methionine is subsequently metabolized to succinyl-CoA. Thus, both folate and B_{12} participate in methionine synthesis. In the process, tetrahydrofolate (THF), required for synthesis of thymidylate in DNA, is regenerated from 5-methyl-THF. Vitamin B_{12} acts as a methyl receiver to prevent "trapping" of folate as the methylated tetrahydrofolate.[192] Interference with nucleotide synthesis impairs erythropoiesis and leads to development of megaloblastic anemia due either to deficiency of B_{12} or folate. The interrelationship of folate and B_{12} is also seen in a cobalamin requirement for folate uptake by cells.[82] In this instance, a folate deficiency may occur indirectly because of inadequate levels of B_{12}.

Adenosylcobalamin is required by the enzyme methylmalonyl-coenzyme A mutase for rearrangement of 1-methylmalonyl-CoA to succinyl-CoA. Succinyl-CoA is further metabolized through the tricarboxylic acid cycle. In states of B_{12}, but not folate, deficiency, methylmalonyl-CoA is not converted to succinyl-CoA, and methylmalonic acid (MMA) excretion in the urine is increased. In summary, B_{12} functions in oxidative degradation of amino acids and, since methionine is a glycogenic amino acid, in carbohydrate metabolism.[192] Fatty acids with odd numbers of carbon atoms are oxidized by a pathway requiring methylmalonyl-CoA mutase activity.

Thus, B_{12} is also essential for normal lipid metabolism.[192] Inadequate supplies of cobalamin will disrupt lipid synthesis. This, along with decreased availability of adenosyl methionine needed for myelin protein formation, could explain the neurologic complications, including demyelination and degeneration of the central nervous system and the optic and peripheral nerves, seen in B_{12} deficiency.

Deficiency of B_{12} may be due to dietary absence, as among strict vegetarians, increased requirements, as in pregnancy, malabsorption due to disease, drug-induced interference (Table 29-19), or intrinsic factor and transport protein inadequacies.[99] Low cobalamin plasma levels are reported in patients with sprue, Crohn's disease, regional enteritis, pernicious anemia, gastric or intestinal resection, multiple myeloma, IF-blocking antibodies, or serum gastric parietal cell autoantibodies. Gastric and intestinal bacterial overgrowth may contribute to cobalamin malabsorption. Up to 25% of the geriatric population may be afflicted with chronic atrophic gastritis. Occurrence of this condition increases with age and may account for the widely reported low serum cobalamin concentrations among the elderly.[2,143,198,263] Studies have shown low serum cobalamin in 10% to 50% of elderly, depending upon variables such as the specific population of older adults examined, assay techniques employed, and cut-off values used to define risk of deficiency. The prevalence of cobalamin deficiency was found to be at least 12% in a large sample of ambulatory older adults when deficiency was defined by a serum cobalamin concentration <258 pmol/L and elevation of one or both of the metabolites MMA and homocysteine. Many elderly with "normal" serum vitamin levels were metabolically deficient in B_{12} or folate.[142] Impaired intestinal absorption of B_{12} has been reported in patients taking anticonvulsants, neomycin, *para*-aminosalicylic acid, phenformin, and cholestyramine, and also has been reported in alcoholics.[212] Controversy continues over reports that megadoses of ascorbic acid may lead to inactivation of vitamin B_{12} and destruction of IF.[27,96]

Clinical features of B_{12} deficiency generally include both hematologic (*e.g.*, macrocytic anemia, megaloblastosis, hypersegmentation of neutrophils) and neurologic (*e.g.*,

TABLE 29-19. Cobalamin-Drug Interactions

Drug	Result of Interaction
Aminosalicylic acid (PAS)	Decreased absorption
Colchicine	Malabsorption
Neomycin	Malabsorption
Guanidines	Decreased absorption
Metformin	Decreased absorption
Phenformin	Decreased absorption
Potassium chloride	Decreased absorption
Nitrous oxide	Interferes with B_{12} metabolism
Fiber	Enhances excretion

Source: Ellenbogen L, Cooper BA. Vitamin B_{12}. In: Machilin LJ, ed. Handbook of Vitamins. 2nd ed. New York: Marcel Dekker, 1991.

peripheral nerve degeneration) manifestations. The hematologic picture is identical in both B_{12} and folate deficiency due to abnormal replication of DNA in hematopoietic tissue. Especially among the elderly, neuropsychiatric disorders may be the primary or only indication of cobalamin deficiency.[141] Numbness, tingling, and weakness of extremities are frequent early neurologic symptoms of vitamin B_{12} deficiency. Vision may be impaired. Spinal cord degeneration leads to changes in tendon reflexes and difficulty in walking. Cognitive dysfunctions include poor memory, loss of mental alertness and confusion, marked personality and mood changes, and, in rare instances, delusions and hallucinations may develop. Research is in progress to determine what, if any, relationship exists among serum cobalamin levels, normal aging, and the occurrence of dementia or Alzheimer's disease.[10,51] Some cognitive and hematopoietic dysfunctions found in AIDS patients have been reversed by vitamin B_{12} therapy. Elevated serum homocysteine concentrations due to vitamin deficiency may play a part since, in excess, the amino acid is both neurotoxic and vasculotoxic.[96]

Limited observations suggest that osteoblast activity depends on cobalamin and that bone metabolism is affected by cobalamin deficiency. Cobalamin-deficient patients were reported to have lower alkaline phosphatase and osteocalcin levels than controls. Osteocalcin, a vitamin K−dependent bone-specific protein, is synthesized by osteoblasts. Its concentration in plasma reflects the rate of bone formation. If so, not only bone marrow cells but also adjoining skeletal cells could be affected in B_{12} deficiency.[37] The osteopenia of aging may be related to an inadequate supply of vitamin B_{12}.[37]

Pernicious anemia (PA), a common cause of vitamin B_{12} deficiency, primarily affects the elderly. Diagnosis of PA by assessment of B_{12} intestinal absorption may be accomplished by measuring urinary excretion of [57]Co-labeled vitamin in the Schilling test.[42] An oral dose of [57]Co-B_{12} is administered along with a parenteral injection of nonlabeled B_{12}. Labeled B_{12} absorbed in the intestine enters the pool of unlabeled vitamin in the plasma, and both forms are excreted in the urine. The percentage of the oral dose appearing in the urine in 24 hours is calculated. Normal B_{12} absorption is indicated when more than 10% of the oral dose is excreted by the patient. Reduced excretion of radioactive B_{12} is seen in pernicious anemia. If repetition of the test with addition of IF results in increased radioactivity in the urine, lack of functional IF is confirmed. Decreased glomerular filtration, due to either renal disease or aging, and improper urine collection invalidate the test results. With elderly patients, collection and evaluation of a 48-hour urine specimen will improve the accuracy of the test.

A recent study evaluated the effect of light on serum B_{12} concentrations (111–812 ng/L). Under typical storage conditions encountered in a clinical laboratory, B_{12} was not affected by light for up to 24 hours after collection when stored at room temperature (20–25°C).[154] Depending on the assessment method employed, serum levels of B_{12} range from approximately 200 to 900 pg/mL.[112] B_{12}-deficient erythropoiesis is associated with levels less than 100 pg/mL (74 pmol/L).[66,95] Serum folate and vitamin B_{12} levels must be determined in patients with megaloblastic anemia to pinpoint its etiology.

Large-dose folate therapy may bring about transient improvement of megaloblastic anemia associated with B_{12} deficiency, but neurologic damage will develop or progress, often irreversibly. It is essential to distinguish the true nature of the underlying disorder (e.g., folate or B_{12} deficiency) so that appropriate therapy may be provided as quickly as possible.

Some patients with serum B_{12} in the lower portion of the reference range may still develop PA. B_{12} deficiency may be by assessment of serum methylmalonate and homocysteine concentrations (Table 29-20).[95,96,168,245,246] Elevated levels of methylmalonic acid (MMA) and total homocysteine are detected in over 90% of cases of cobalamin deficiency. Measurement of urinary MMA excretion is also diagnostically useful. Increase in these metabolites often occurs before any other clinical evidence of deficiency is manifested. Serum MMA levels >950 nmol/L (110–950 nmol/L) and total homocysteine concentrations >29 micromoles/L (6–29 micromoles/L) indicate B_{12} deficiency even in the presence of normal hematologic parameters. An automated assay of MMA in serum and urine by derivatization with 1-pyrenyldiazomethane, liquid chromatography, and fluorescence detection has recently been described.[233] The risk factor for occlusive atherosclerosis is increased by hyperhomocysteinemia. Improved vitamin B_{12} status normalizes homocysteine levels within weeks, thereby reducing the patient's risk of coronary artery disease.

Cobalamin determinations may be by microbiologic or radioligand assays. Although a variety of vitamin B_{12}-dependent test organisms have been used, including *Euglena gracilis*, *Lactobacillus leichmannii* (ATCC 7830) remains the microorganism of choice.[79] Microbiological assay is used as the reference method or in a research setting. In the clinical laboratory, radioassays are routinely used for determination of serum B_{12} levels. Differential radioassays measure cobalamin content more accurately than do microbiologic assays, since noncobalamin corrinoids not utilized by humans will support microbial growth.[95] Plasma transcobalamins must be heat denatured (boiling) or subjected to alkaline pH inactivation (noboil) prior to either microbiologic or radioassay of the specimen to release the cobalamin for measurement.

Radioisotope dilution methods are the most widely used assays for cobalamin. These competitive inhibition radioassays measure the extent to which cobalamin, after being freed from bound materials, competes with radioactive cyanocobalamin for binding sites on a protein.[66] Radioligand assays may be either RIA or CBP procedures. In the case of CBP assays, purified IF has been strongly recommended as the cobalamin-binding protein. A semiautomated radioassay system makes possible simultaneous assessments of serum B_{12} and folate, following off-line denaturation of endogenous binding proteins.[45] Purified IF is used as the competitive binding protein, with solid-phase adsorbent separating free and bound [57]Co. Recently, it has been reported that no boiling or other pretreatment of patient specimen is required when a non-intrinsic factor blocking agent is used along with a magnetizable solid-phase separation system.[111] This assay is highly specific for cobalamin. With elimination of a pretreatment requirement, and ease of separation

TABLE 29-20. Sequential Stages of Vitamin B-12 Status. Biochemical and Hematological Sequence of Events as Negative Vitamin B-12 Balance Progresses. [© 1990, 1993 Victor Herbert (Modified 1993 to Include Homocysteine).]

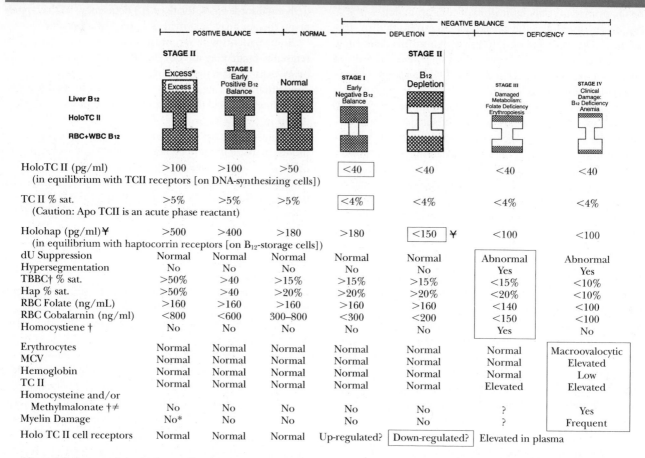

	POSITIVE BALANCE		NORMAL	NEGATIVE BALANCE — DEPLETION		DEFICIENCY	
	STAGE II Excess*	STAGE I Early Positive B12 Balance	Normal	STAGE I Early Negative B12 Balance	STAGE II B12 Depletion	STAGE III Damaged Metabolism: Folate Deficiency Erythropoiesis	STAGE IV Clinical Damage: B12 Deficiency Anemia
HoloTC II (pg/ml) (in equilibrium with TCII receptors [on DNA-synthesizing cells])	>100	>100	>50	<40	<40	<40	<40
TC II % sat. (Caution: Apo TCII is an acute phase reactant)	>5%	>5%	>5%	<4%	<4%	<4%	<4%
Holohap (pg/ml)¥ (in equilibrium with haptocorrin receptors [on B$_{12}$-storage cells])	>500	>400	>180	>180	<150 ¥	<100	<100
dU Suppression	Normal	Normal	Normal	Normal	Normal	Abnormal	Abnormal
Hypersegmentation	No	No	No	No	No	Yes	Yes
TBBC† % sat.	>50%	>40	>15%	>15%	>15%	<15%	<10%
Hap % sat.	>50%	>40	>20%	>20%	>20%	<20%	<10%
RBC Folate (ng/mL)	>160	>160	>160	>160	>160	<140	<100
RBC Cobalarnin (ng/ml)	<800	<600	300–800	<300	<200	<150	<100
Homocystiene †	No	No	No	No	No	Yes	No
Erythrocytes	Normal	Normal	Normal	Normal	Normal	Normal	Macroovalocytic
MCV	Normal	Normal	Normal	Normal	Normal	Normal	Elevated
Hemoglobin	Normal	Normal	Normal	Normal	Normal	Normal	Low
TC II	Normal	Normal	Normal	Normal	Normal	Elevated	Elevated
Homocysteine and/or Methylmalonate †≠	No	No	No	No	No	?	Yes
Myelin Damage	No*	No	No	No	No	?	Frequent
Holo TC II cell receptors	Normal	Normal	Normal	Up-regulated?	Down-regulated?	Elevated in plasma	

*Cyanocobalamin excesses (injected or intranasal) produce transient rise in B$_{12}$ analogues on B$_{12}$ delivery protein (TC II); the significance of such rises is unknown (Herbert et al., 1987). Cyanocobalamin acts as an anti-B$_{12}$ in a rare congenital defect in B$_{12}$ metabolism.

≠ In serum and urine.

† TBBC = Total B$_{12}$ binding capacity.

¥ Low holohaptocorrin correlates with liver cell B$_{12}$ depletion. There may be hematopoietic cell and glial cell B$_{12}$ depletion prior to liver cell depletion, and those cells may be in STAGE III or IV negative B$_{12}$ balance while liver cells are still in STAGE II.

---

achieved in a magnetic radioassay, a fully automated continuous-flow procedure can be realized. Assay automation of B$_{12}$ on the Abbott IM$_x$ provides rapid results in a nonradioisotopic format.[129] B$_{12}$ deficiency can be detected and quantitated by measuring methylmalonic acid in urine or assessing its serum level using capillary gas chromatography/mass spectrometry.[245,246]

SUMMARY

The Joint Commission on Accreditation of Healthcare Organizations (JCAHO) is mandating more stringent nutritional review of all patients. There can be no doubt that this will impact the clinical laboratory. The clinical laboratorian will be required to know more about vitamins, their biochemical functions and physiologic roles, and the best assay methodologies to use to provide the clinician with timely information on the patient's nutritional status. There are financial implications to optimizing a patient's nutritional status, thereby hastening the desired medical outcomes and reducing the patient's length of stay in the hospital.

The general public is also increasingly concerned with health promotion and disease prevention. Supplemental use of vitamins to increase longevity and improve the quality of life is regularly advocated in the media. Vitamin sales is a multi-billion-dollar commercial enterprise in this country. While there is strong support for the beneficial effects of vitamins in the prevention of certain cancers and cardiovascular disease, there is also concern over the possibility of toxicity from overly aggressive vitamin supplementation.[203]

Historically, medicine has focused more attention on conditions of vitamin deficiency than excess. Despite the high standard of living in this country, significant numbers of individuals are characterized by an overall vitamin status that is suboptimal or overtly deficient. Nutritional requirements in special physiological states such as growth, pregnancy and lactation, and aging may not be met by dietary consumption. For example, age-related changes in vitamin status due to altered dietary practices, physiologic changes, and drug-nutrient interaction contribute to the risk for deficiency of one or more vitamins among the 32 million Americans who are over 65 years of age.

Biochemical determinations of vitamin status and the monitoring of nutritional support will increase in the years ahead. In the future, vitamin assays will not be viewed as esoteric reference laboratory procedures; rather they will be acknowledged as essential for the promotion of wellness and for the cost-effective provision of quality health care.

CASE STUDY 29-1

During her most recent physical examination, a blood pressure of 175/96 had been recorded for a widowed, 65-year-old female. Over the past 3 years, her blood pressure as recorded on annual physical examinations had gradually risen, but this report was the first clear indication of hypertension. Her physician prescribed 150 mg of hydralazine per day, administered orally. In follow-up office visits, her physician noted that the dosage prescribed was not producing a satisfactory lowering of the patient's blood pressure. Adjustment of dosage was attempted, and satisfactory results were finally achieved with 400 mg of hydralazine daily. Several months after initiation of therapy, the patient's daughter called the physician to report pronounced changes in her mother's personality. Her usual optimism had been replaced by depression and irritability. In addition, her daughter indicated that the woman no longer appeared interested in her house or her family. She was reluctant to cook for herself but had purchased a supply of high-protein supplement, which she consumed for nourishment. Such a lack of responsibility was not in keeping with her mother's traditional behavior. These changes,

coupled with the appearance of a rash on her mother's forehead, prompted the daughter to bring her mother to the clinic. Upon review of the medication record and noting signs of peripheral nerve inflammation in the patient, the physician requested the laboratory to evaluate the patient's vitamin B_6 status.

Questions

1. What type of assessment procedure will the laboratory be most likely to employ in evaluating the patient's vitamin B_6 status?
2. Identify the patient specimen required for testing and any special precautions to be taken in its handling or processing.
3. What clinical manifestations suggested a vitamin B_6 deficiency to the physician?
4. In what way is it likely that the patient's medication and dietary practices contributed to development of a B_6 deficiency?
5. A marginal or deficient vitamin B_6 status is indicated by laboratory values of what magnitude?

CASE STUDY 29-2

A 62-year-old male had been admitted to the hospital with a diagnosis of acute myocardial infarction. Anticoagulant therapy was initiated in an attempt to reduce the incidence of secondary thromboembolism. While he was hospitalized, heparin therapy had been initiated, and, upon discharge, the patient was switched to Coumadin. For 3 months after leaving the hospital, the patient had been completely stable on a Coumadin regimen of 30 mg per week. During a follow-up visit to his physician, the man's prothrombin time was reported as 12 seconds, as compared with previously obtained PT times of 22 to 24 seconds. Effective oral anticoagulant therapy calls for maintenance of a prothrombin time that exceeds "normal" by 1.5 to 1.7

times. Review of the patient's medication record did not suggest drug interference as the basis for the decreased anticoagulant effect. A careful dietary history provided an explanation for the newly acquired warfarin resistance.

Questions

1. Excessive intake of what vitamin is likely to account for the observed shortening of prothrombin time?
2. Describe the physiologic function of this vitamin.
3. Suggest possible dietary practices that could induce warfarin (Coumadin) resistance.

CASE STUDY 29-3

An 18-year-old male with a 3-year history of severe cystic acne was seen by his family physician for a routine physical examination prior to the start of school. Because the patient's acne had been unresponsive to a variety of oral antibiotics and brief periods of corticosteroid therapy, the physician had referred the young man to a dermatologist for additional therapy 4 months earlier. At that time, results of a complete blood cell count and a chemistry profile on the patient were all within normal limits. In the course of conducting the physical examination, the physician commented on the substantial improvement in the patient's acne. While pleased with his skin's appearance, the young man complained of feeling "tired, achy, and queasy" despite the fact that he believed he was getting sufficient rest and not engaging in particularly strenuous activities. He mentioned that he had been taking a new prescription drug ordered by the dermatologist for about 2 months.

It seemed to the young man that since taking the medication he had begun to feel "out of sorts." The patient was sent to the laboratory for a CBC and a chemistry workup.

Questions

1. Bone pain, nausea, and lethargy are consistent with a diagnosis of disruption of metabolism of what mineral?
2. What medication has proven to be a potent inhibitor of sebum secretion and therefore is particularly effective in the treatment of severe nodulocystic acne?
3. Side effects associated with use of this drug are similar to signs of intoxication from excessive consumption of what vitamin?

CASE STUDY 29-4

A health-conscious, physically active 59-year-old male with a family history of heart disease was seen by his family physician for an annual physical examination. The patient reported no health problems and none was noted by the physician. However, because of the history of heart disease in the patient's family, he was advised by his physician to schedule an appointment with the laboratory, and an order for a lipid profile was called in. The patient carefully followed directions regarding fasting prior to specimen collection.

The following laboratory results were sent to the physician:

CHOLESTEROL: 288 mg/dL (NCEP Suggested Levels: <200 Desirable, >240 Increased risk).
TRIGLYCERIDE: 212 mg/dL (20–200).
HDL cholesterol: 32 mg/dL (Normal Mean: 62 mg/dL, Range: 50–92).
LDL cholesterol: 201 NCEP Suggested Levels: <130 Desirable, >160 Increased risk.

The patient was informed of the laboratory results and their significance by the physician. The patient was told to report to the laboratory for retesting in 2 weeks and the instructions for fasting before specimen collection

were repeated. Results of the second test series did not differ significantly from the earlier findings.

The physician's personal knowledge of the patient's life-style and dietary practices prompted a clinical judgment that drug treatment should be initiated immediately rather than first attempting dietary therapy to reduce the patient's risk for coronary heart disease (CHD). One month after beginning drug therapy, the patient was retested and a substantial improvement in the hyperlipidemia was noted.

Questions

1. Pharmacologic doses of what vitamin have proven to be effective in reducing a patient's risk for CHD?
2. Describe the anti-atherogenic effects of nicotinic acid therapy.
3. What minor side effects are often associated with nicotinic acid therapy?
4. Describe the possible serious side effects associated with high-dose consumption of nicotinic acid.
5. What laboratory tests would be ordered to monitor a patient's response to nicotinic acid therapy?

REFERENCES

1. Allard JP, Royall D, Kurian R, Muggli R, et al. Effects of beta-carotene supplementation on lipid peroxidation in humans. Am J Clin Nutr 1994;59:884.

2. Allen LH, Casterline J. Vitamin B-12 deficiency in elderly individuals: Diagnosis and requirements. Am J Clin Nutr 1994;60:12.

3. Anonymous. Nomenclature policy: Generic descriptors and trivial names for vitamins and related compounds. J Nutr 1987;117:7.

4. Ashby CD. Laboratory consultant: Metabolites of vitamin D. Clin Chem News 1986;12(6):14.

5. Association of Official Analytical Chemists. Official methods of analysis of the Association of Official Analytical Chemists. 13th ed. Horwitz W, ed. Washington, DC: Association of Official Analytical Chemists 1980.

6. Bamji MS. Laboratory tests for the assessment of vitamin nutritional status. In: Briggs MH, ed. Vitamins in human biology and medicine. Boca Raton, FL: CRC Press, 1981.

7. Bankson DD, Kestin M, Rifai N. Role of free radicals in cancer and atherosclerosis. Clin Lab Med 1993;13:463.

8. Barker BM, Bender DA. Vitamin B_6. In: Barker BM, Bender DA, eds. Vitamins in Medicine. 4th ed. Vol 1. London: William Heinemann, 1980.

9. Basu TK, Schorah CJ. Vitamin C in health and disease. Westport, CT: AVI, 1982.

10. Basun H, Fratiglioni L, Winblad B. Cobalamin levels are not reduced in Alzheimer's disease: Results from a population-based study. J Am Geriatr Soc 1994;42:132.

11. Bauernfeind J. Tocopherols in foods. In: Machlin LJ, ed. Vitamin E: A comprehensive treatise. New York: Marcel Dekker, 1980.

12. Bell EF. History of vitamin E in infant nutrition. Am J Clin Nutr 1987;46:183.

13. Bell IR, Edman JS, Morrow FD, et al. B complex vitamin patterns in geriatric and young adult inpatients with major depression. J Am Geriatr Soc 1991;39:252.

14. Bender DA. Niacin. In: Barker BM, Bender DA, eds. Vitamins in medicine. 4th ed. Vol 1. London: William Heinemann, 1980.

15. Bendich A, Machlin LJ. Safety of oral intake of vitamin E. Am J Clin Nutr 1988;48:612.

16. Bertelloni S, Baroncelli GI, Benedetti U, et al. Commercial kits for 1,25-dihydroxyvitamin D compared with liquid-chromatographic assay. Clin Chem 1993;39:1086.

17. Bieri JG, Corash L, Hubbard VS. Medical uses of vitamin E. N Engl J Med 1983;308(18):1063.

18. Bitsch R, Toth-Dersi A, Hoetzel D. Biotin deficiency and biotin supply. Ann NY Acad Sci 1985;447:133.

19. Blomhoff R. Transport and metabolism of vitamin A. Nutr Rev 1994;52(2):Suppl 13.

20. Boland J, Carey G, Krodel E, Kwiatkowski M. The Ciba Corning ACS:180 ™ benchtop immunoassay analyzer. Clin Chem 1990;36:1598.

21. Bollag W. Vitamin A and retinoids: From nutrition to pharmacotherapy in dermatology and oncology. Lancet 1983;1(8329):860.

22. Bonjour JP. Biotin. In: Machlin LJ, ed. Handbook of vitamins. 2nd ed. New York:Marcel Dekker, 1991.

23. Bonjour JP. Biotin in man's nutrition and therapy—A review. Int J Vit Nutr Res 1977;47:107.

24. Bonjour JP. Biotin in human nutriture. Ann NY Acad Sci 1985;447:97.

25. Brattstrom LE, Isrealsson B, Jeppsson JO, Hultberg BL. Folic acid—An innocuous means to reduce plasma homocysteine. Scand J Clin Lab Invest. 1988;48:215.

26. Brigden ML, Edgell D, McPherson M, Leadbetter A, et al. High incidence of significant urinary ascorbic acid concentrations in a West Coast population—implications for routine urinalysis. Clin Chem 1992;38:426.

27. Brin M. Nutritional and health aspects of ascorbic acid. In: Seib PA, Tolbert BM, eds. Ascorbic acid: Chemistry, metabolism, and uses. Washington: American Chemical Society, 1982.

28. Brody T. Folic acid. In: Machlin LJ, ed. Handbook of vitamins. 2nd ed. New York: Marcel Dekker, 1991.

29. Brondos GA, Lyford CL. Clinical laboratory tests in the diagnosis of malabsorption. Diagn Dialog 1982;4(3):10.

30. Bunce GE. Antioxidant nutrition and cataracts in women: A prospective study. Nutr Rev 1993;51:84.

31. Butterworth CE Jr, Tamura T. Folic acid safety and toxicity: A brief review. Am J Clin Nutr 1989;50:353.

32. Butts WC, Mulvihill HJ. Centrifugal analyzer determination of ascorbate in serum or urine with Fe^{3+} Ferrozine. Clin Chem 1975;21:1493.

33. Cameron E, Pauling L, Leibovitz B. Ascorbic acid and cancer: A review. Cancer Res 1979;39:663.

34. Camp VM, Chipponi J, Faraj BA. Radioenzymatic assay for direct measurement of plasma pyridoxal 5′-phosphate. Clin Chem 1983;29(4):642.

35. Campbell CH. The severe lactic acidosis of thiamine deficiency: Acute pernicious or fulminating beriberi. Lancet 1984;2:446.

36. Carmel R, Karnaze DS. The deoxyuridine suppression test identifies subtle cobalamin deficiency in patients without typical megaloblastic anemia. JAMA 1985;253(9):1284.

37. Carmel R, Lau K-HW, Baylick DJ, et al. Cobalamin and osteoblast-specific proteins. N Engl J Med 1988;319(2):70.

38. Castle MC, Cooke WJ. Measurement of vitamin E in serum and plasma by high-performance liquid chromatography with electrochemical detection. Ther Drug Monit 1985;7:364.

39. Catignani GL, Bieri JG. Simultaneous determination of retinol and alpha-tocopherol in serum or plasma by liquid chromatography. Clin Chem 1983;29(4):708.

40. Ceriello A, Giugliano D, Quatraro A, Donzella C, et al. Vitamin E reduction of protein glycosylation in diabetes: New prospect for prevention of diabetic complications? Diabetes Care 1993;14:68.

41. Chanarin I. The cobalamins (vitamin B_{12}). In: Barker BM, Bender DA, eds. Vitamins in Medicine. 4th ed. Vol 1. London: William Heinemann, 1980.

42. Chanarin I. The folates. In: Barker BM, Bender DA, eds. Vitamins in medicine. 4th ed. Vol 1. London: William Heinemann, 1980.

43. Chapuy MC, Arlot ME, Duboeuf F, et al. Vitamin D3 and calcium to prevent hip fractures in elderly women. New Engl J Med 1992;327:1637.

44. Chastain JL, McCormick DB. Flavin catabolites: Identification and quantitation in human urine. Am J Clin Nutr 1987;46:832.

45. Chen I-W, Silberstein EB, Maxon HR, Volle CP, et al. Semiautomated system for simultaneous assays of serum vitamin B_{12} and folic acid in serum evaluated. Clin Chem 1982; 28(10):2161.

46. Cheng KS, Chou PP. Measurement of total thiamine in whole blood, RBC and plasma by HPLC with fluorometric detection. Clin Chem 1988;34:1308.

47. Chou PP, Bailey JL. Quantitative analysis of vitamin B6 in plasma by HPLC/EC. Clin Chem 1988;34(6):1308.

48. Chow CK. Vitamin E. In: Chen LH, ed. Nutritional aspects of aging. Vol I. Boca Raton, FL: CRC Press, 1986.

49. Chu RC, Hall CA. The total serum homocysteine as an indicator of vitamin B12 and folate status. Am J Clin Pathol 1988;90:446.

50. Clements JE, Anderson BB. Pyridoxine pyridoxamine phosphate oxidase activity in the red cell. Biochim Biophys Acta 1980;613:401.

51. Cole MG, Prochal JF. Low serum vitamin B12 in Alzheimer-type dementia. Aging 1984;13:101.

52. Collins ED, Norman AW. Vitamin D. In: Machlin LJ, ed. Handbook of vitamins. 2nd ed. New York: Marcel Dekker, 1991.

53. Comstock GW, Bush TL, Helzlsouer K. Serum retinol, beta-carotene, vitamin E and selenium as related to subsequent cancer of specific sites. Am J Epidemiol 1992;135:115.

54. Condon R. To pop or not. Chicago Tribune, Business Section, November 28, 1994;1.

55. Council on Scientific Affairs, American Medical Association. Vitamin preparations as dietary supplements and as therapeutic agents. JAMA 1987;257(14):1929.

56. Dakshinamurti K, Chauhan MS. Chemical analysis of pyridoxine vitamers. In: Leklem JE, Reynolds RD, eds. Methods in Vitamin B-6 nutrition: Analysis and status assessment. New York: Plenum Press, 1980.

57. Danford DE, Munro HN. The vitamins: Introduction. In: Gilman AG, Goodman LS, Gilman A, eds: The pharmacological basis of therapeutics. 6th ed. New York: Macmillan, 1980.

58. Danford DE, Munro HN. Water-soluble vitamins: The vitamin B complex and ascorbic acid. In: Gilman AG, Goodman LS, Gilman A, eds. The pharmacological basis of therapeutics. 6th ed. New York: Macmillan, 1980.

59. Dashman T, Sansaricq C. Nutrition in the management of inborn errors of metabolism. Clin Lab Med 1993;13(2):407.

60. DeLuca HF. The vitamin D system: A view from basic science to the clinic. Clin Biochem 1981;14(5):213.

61. DeLuca HF, Ghazarian JG. The role of vitamin D and its metabolites in calcium and phosphate metabolism. In: Kuhlencordt F, Bartelheimer H, eds. Handbuch der inneren medizin VI/1A. Knochen. Gelenke. Muskeln. Berlin: Springer-Verlag, 1980.

62. Diamandis EP, Christopoulos TK. The biotin-(strep) avidin system: Principles and applications in biotechnology. Clin Chem 1991;37:625.

63. Draper HH. Nutrient interrelationships. In: Machlin LJ, ed. Vitamin E: A comprehensive treatise. New York: Marcel Dekker, 1980.

64. Driskell JA. Water-soluble vitamins. In: Chen LH, ed. Nutritional aspects of aging. Vol I. Boca Raton, FL: CRC Press, 1986.

65. Edwards P, Liu PKS, Rose GA. A simple liquid-chromatographic method for measuring vitamin B6 compounds in plasma. Clin Chem 1989;35:241.

66. Ellenbogen L, Cooper BA. Vitamin B12. In: Machlin LJ, ed. Handbook of vitamins. 2nd ed. New York: Marcel Dekker 1991.

67. Enstrom JE, Kanim LE, Klein MA. Vitamin C intake and mortality among a sample of the United States population. Epidemiology 1992;3:194.

68. Farrell PM. Deficiency states, pharmacological effects, and nutrient requirements. In: Machlin LJ, ed. Vitamin E: A comprehensive treatise. New York: Marcel Dekker, 1980.

69. Fox HM. Pantothenic acid. In: Machlin LJ, ed. Handbook of vitamins. 2nd ed. New York: Marcel Dekker 1991.

70. Fraser DR. Biochemical and clinical aspects of vitamin D function. Br Med Bull 1981;37(1):37.

71. Fraser DR. Vitamin D. In: Barker BM, Bender DA, eds. Vitamins in medicine. 4th ed. Vol 1. London: William Heinemann, 1980.

72. Fuchs GJ, Ausayakhun S, Ruckphaopunt S, Tansuhaj A, et al. Relationship between vitamin A deficiency, malnutrition, and conjunctival impression cytology. Am J Clin Nutr 1994; 60:293.

73. Further studies of acute folate deficiency developing during total parenteral nutrition. Nutr Rev 1983;41(2):51.

74. Garry P. Nutrition and aging. In: Faulkner W, Meites S, eds. Geriatric clinical chemistry reference values. Washington DC: AACC Press, 1994.

75. Garry PJ, Goodwin JS, Hunt WC, et al. Nutritional status in a healthy elderly population: Dietary and supplemental intakes. Am J Clin Nutr 1982;36:319.

76. Garry PJ, Goodwin JS, Hunt WC, et al. Nutritional status in a healthy elderly population: Vitamin C. Am J Clin Nutr 1982;36:332.

77. Gatautis VJ, Naito HK. Liquid-chromatographic determination of urinary riboflavin. Clin Chem 1981;27(10):1672.

78. Gibson GE, Kwan-Fu RS, Blass JP, et al. Reduced activities of thiamine-dependent enzymes in the brains and peripheral tissues of patients with Alzheimer's disease. Arch Neurol 1988;45:836.

79. Gijzen AHJ, deKock HW, Meulendijk PN, Schmidt NA, et al. The need for a sufficient number of low level sera in comparisons of different serum vitamin B_{12} assays. Clin Chim Acta 1983;127:185.

80. Gloth FM III, Tobin JD, Sherman SS, et al. Is the recommended daily allowance for vitamin D too low for the homebound elderly? J Am Geriatr Soc 1991;39:137.

81. Goodman D. Retinoid-binding proteins in plasma and in cells. In: DeLuca LM, Shapiro SS, eds. Modulation of cellular interactions by vitamin A and derivatives (retinoids). Ann NY Acad Sci 1981;359:69.

82. Grant JP. Handbook of total parenteral nutrition. Philadelphia: Saunders, 1980.

83. Gravel RA, Robinson BH. Biotin-dependent carboxylase deficiencies (propionyl-CoA and pyruvate carboxylases). Ann NY Acad Sci 1985;447:225.

84. Greenberg ER, Baron JA, Tosteson TD, Freeman D, et al. A clinical trial of antioxidant vitamins to prevent colorectal adenoma. N Engl J Med 1994;331(3):141.

85. Gubler CJ. Thiamin. In: Machlin LJ, ed. Handbook of vitamins. 2nd ed. New York: Marcel Dekker, 1991.

86. Haddad JG. Vitamin D—solar rays, the milky way, or both? N Engl J Med 1992;326:1213.

87. Hallfrisch J, Muller DC, Singh VN. Vitamin A and E intakes and plasma concentrations of retinol, beta-carotene and alpha-tocopherol in men and women of the Baltimore Longitudinal Study of Aging. Am J Clin Nutr 1994;60:176.

88. Hallfrisch J, Singh VN, Muller DC, Baldwin H, et al. High plasma vitamin C associated with high plasma HDL and HDL2 cholesterol. Am J Clin Nutr 1994;60:100.

89. Hansen SI, Holm J. A competitive enzyme-linked ligand sorbent assay (ELLSA) for quantitation of folates. Anal Biochem 1988;172:160.

90. Hansen SI, Holm J. Quantification of biotin in serum by competition with solid-phase biotin for binding to peroxidase-avidin conjugate. Clin Chem 1989;35(8):1721.

91. Haroon Y, Bacon DS, Sadowski JA. Liquid-chromatographic determination of vitamin K_1 in plasma, with fluorometric detection. Clin Chem 1986;32(10):1925.

92. Hart JP, Shearer MJ, Klenerman L, et al. Electrochemical detection of depressed circulating levels of vitamin K_1 in osteoporosis. J Clin Endocrinol Metab 1985;60:1268.

93. Hatam LJ, Kayden HJ. A high-performance liquid chromatographic method for the determination of tocopherol in plasma and cellular elements of the blood. J Lipid Res 1979;20:639.

94. Heard GS, McVoy JRS, Wolf B. A screening method for biotinidase deficiency in newborns. Clin Chem 1984;30(1):125.

95. Herbert V. The 1986 Herman Award Lecture. Nutrition science as a continually unfolding story: The folate and vitamin B_{12} paradigm. Am J Clin Nutr 1987;46:387.

96. Herbert V. Staging vitamin B12 (cobalamin) status in vegetarians. Am J Clin Nutr 1994;59(Suppl):1213S.

97. Herbert V. The antioxidant supplement myth. Am J Clin Chem 1994;60:157.

98. Herbert V. The vitamin craze. Arch Intern Med 1980;140:173.

99. Herbert V, Colman N, Jacob E. Folic acid and vitamin B_{12}. In: Goodhart RS, Shils ME, eds. Modern nutrition in health and disease. 6th ed. Philadelphia: Lea & Febiger, 1980.

100. Herbert V, Das KC. Folic acid and vitamin B 12. In: Shils ME, Olson JA, Shike M, eds. Nutrition in health and disease. 8th ed. Vol. 1. Philadelphia, PA: Lea & Febiger, 1994.

101. Herbeth B, Zittoun J, Miravet L, et al. Reference intervals for vitamins B_1, B_2, E, D, retinol, beta-carotene, and folate in blood: Usefulness of dietary selection criteria. Clin Chem 1986;32(9):1756.

102. Hicks JM, Cook J, Godwin ID, Soldin SJ. Vitamin B12 and folate. Pediatric reference ranges. Arch Pathol Lab Med 1993;117:704.

103. Hillman RS. Vitamin B_{12}, folic acid, and the treatment of megaloblastic anemias. In: Gilman AG, Goodman LS, Gilman A, eds. The pharmacological basis of therapeutics. 6th ed. New York: Macmillan, 1980.

104. Hodges RE. The vitamins. Ascorbic acid. In: Goodhart RS, Shils ME, eds. Modern nutrition in health and disease. 6th ed. Philadelphia: Lea & Febiger, 1980.

105. Hoefler G, et al. Lipoprotein Lp(a): A risk factor for myocardial infarction. Atherosclerosis 1988;18:398.

106. Holick MF. The use and interpretation of assays for vitamin D and its metabolites. J Nutr 1990;120(11S):1464.

107. Holick MF. McCollum Award Lecture, 1994. Vitamin D— New horizons for the 21st century. Am J Clin Nutr 1994;60:619.

108. Hollis BW. Assay of circulating 1,25-dihydroxyvitamin D involving a novel single-cartridge extraction and purification process. Clin Chem 1986;32:2060.

109. Hollis BW, Kamerud JQ, Selvaag SR, et al. Determination of vitamin D status by radioimmunoassay with an 125-I-labeled tracer. Clin Chem 1993;39:529.

110. Horne DW, Patterson D. *Lactobacillus casei* microbiological assay of folic acid derivatives in 96-well microtiter plates. Clin Chem 1988;34(11):2357.

111. Houts TM, Carney JA. Radioassay for cobalamin (vitamin B_{12}) requiring no pretreatment of serum. Clin Chem 1981;27(2):263.

112. Howard L, Meguid MM. Nutritional assessment in total parenteral nutrition. Clin Lab Med 1981;1(4):611.

113. Hsing AW, Comstock GW, Polk BF. Effect of repeated freezing and thawing on vitamins and hormones in serum. Clin Chem 1989;35(10):2145.

114. Hunter DJ, Manson JE, Colditz GA, Stampfer MJ, et al. A prospective study of the intake of vitamins C, E and A and the risk of breast cancer. N Engl J Med 1993;329:234.

115. Jackson MJ, Jones DA, Edwards RHT. Vitamin E and muscle disease. J Inherit Metab Dis 1985;8(Suppl 1):84.

116. Jacob RA, Milne DB. Biochemical assessment of vitamins and trace metals. Clin Lab Med 1993;13:371.

117. Jacob RA, Swendseid ME, McKee RW, et al. Biochemical markers for assessment of niacin status in young men: Urinary and blood levels of niacin metabolites. J Nutr 1989;119:591.

118. Jacobus CH, Holick MF, Shao Q, Chen TC, et al. Hypervitaminosis D associated with drinking milk. N Engl J Med 1992;326:1173.

119. Jacques PF, Hartz SC, Chylack LT, et al. Nutritional status in persons with and without senile cataract: Blood vitamin and mineral levels. Am J Clin Nutr 1988;48:152.

120. Jacques PF, Sulsky SI, Perrone GA, Schaefer EJ. Ascorbic acid and plasma lipids. Epidemiology 1994;5:19.

121. Jones G. Vitamins A and D. In: Chen LH, ed. Nutritional aspects of aging. Vol I. Boca Raton, FL: CRC Press, 1986.

122. Jongen MJM, Van Ginkel FC, van der Vijgh WJF, et al. An international comparison of vitamin D metabolite measurements. Clin Chem 1984;30(3):399.

123. Joosten E, van der Berg A, Riezler R, Naurath HJ, et al. Metabolic evidence that deficiencies of vitamin B-12 (cobalamin), folate, and vitamin B-6 occur commonly in elderly people. Am J Clin Nutr 1993;58:468.

124. Kang SS, Wong PWK, Norusis M. Homocysteinemia due to folate deficiency. Metabolism 1987;36:458.

125. Khachik F, Beecher GR, Goli MD, et al. Separation and quantitation of carotenoids in human plasma. Methods Enzymol 1992;213:205.

126. Khaw K-T, Scragg R, Murphy S. Single-dose cholecalciferol suppresses the winter increase in parathyroid hormone concentrations in healthy older men and women: A randomized trial. Am J Clin Nutr 1994;59:1040.

127. Koehler KM, Garry PJ. Nutrition and aging. Clin Lab Med 1993;13:433.

128. Kok FJ, Schrijver J, Hofman A, et al. Low vitamin B_6 status in patients with acute myocardial infarction. Am J Cardiol 1989;63:513.

129. Kuemmerle SC, Boltinghouse GL, Delby SM, Lane TL, et al. Automated assay of vitamin B-12 by the Abbott IMx analyzer. Clin Chem 1992;38:2073.

130. Kutsky RJ. Handbook of vitamins, minerals and hormones. 2nd ed. New York: Van Nostrand Reinhold, 1981.

131. Labbé RF, Veldee M. Nutrition in the clinical laboratory. Clin Lab Med 1993;13:313.

132. Lambert WE, Cammaert PM, DeLeenheer AP. Liquid-chromatographic measurement of riboflavin in serum and urine with isoriboflavin as internal standard. Clin Chem 1985;31(8):1371.

133. Lederle FA. Oral cobalamin for pernicious anemia. Medicine's best kept secret? JAMA 1991;265(1):94.

134. Lee W, Davis KA, Rettmer RL, et al. Ascorbic acid status: Biochemical and clinical considerations. Am J Clin Nutr 1988;48:286.

135. Lee W, Hamernyik P, Hutchinson M, et al. Ascorbic acid in lymphocytes: Cell preparation and liquid-chromatographic assay. Clin Chem 1982;28(10):2165.

136. Leklem JE. Vitamin B6. In: Machlin LJ, ed. Handbook of vitamins. 2nd ed. New York: Marcel Dekker, 1991.

137. Leklem JE, Reynolds RD, eds. Introduction. In: Clinical and physiological applications of vitamin B6. New York: Alan R Liss, 1988.

138. Lemoyne M, Van Gossum A, Kurian R, et al. Breath pentane analysis as an index of lipid peroxidation: A functional test of vitamin E status. Am J Clin Nutr 1987;46:267.

139. Li T-K. Factors influencing vitamin B_6 requirement in alcoholism. In: Human vitamin B_6 requirements. Washington: National Academy of Sciences, 1978.

140. Li T-K, Lumeng L. Plasma PLP as indicator of nutrition status: Relationship to tissue vitamin B-6 content and hepatic metabolism. In: Leklem JE, Reynolds RB, eds. Methods in vitamin B-6 nutrition. Analysis and status assessment. New York: Plenum Press, 1980.

141. Lindenbaum J, Healton EB, Savage DG, et al. Neuropsychiatric disorders caused by cobalamin deficiency in the absence of anemia or macrocytosis. N Engl J Med 1988;318:1720.

142. Lindenbaum J, Rosenberg IH, Wilson PWF, et al. Prevalence of cobalamin deficiency in the Framingham elderly population. Am J Clin Nutr 1994;60:2.

143. Linstedt G, Lundberg P-A, Johansson P-M. High prevalence of atrophic gastritis in the elderly: Implications for health-associated reference limits for cobalamin in serum. Clin Chem 1989;35(7):1557.

144. Livaniou E, Evangelatos GP, Ithakissios DS. Biotin radioligand assay with an ^{125}I-labeled biotin derivative, avidin, and avidin double-antibody reagents. Clin Chem 1987;33:1983.

145. Lockitch G. Perinatal and pediatric nutrition. Clin Lab Med 1993;13:387.

146. Lowik MRH, van den Berg H, Westenbrink S, et al. Dose-response relationships regarding vitamin B_6 in elderly people: A nationwide nutritional survey (Dutch Nutritional Surveillance System). Am J Clin Nutr 1989;50:391.

147. Lui NST, Roels OA. The vitamins. Vitamin A and carotene. In: Goodhart RS, Shils ME, eds. Modern nutrition in health and disease. 6th ed. Philadelphia: Lea & Febiger, 1980.

148. Lunec J. Oxygen radicals: Their measurement and role in major disease. Journal of International Federation of Clinical Chemistry 1992;4:58.

149. MacCrehan WA, Schoenberger E. Determination of retinol, alpha-tocopherol, and beta-carotene in serum by liquid chromatography with absorbance and electrochemical detection. Clin Chem 1987;33(9):1585.

149a. MacLaughlin J, Holick M. Aging decreases the capacity of human skin to produce vitamin D_3. J Clin Invest 1985; 76:1536.

150. Machlin LJ. Vitamin E. In: Machlin LJ, ed. Handbook of vitamins. 2nd ed. New York: Marcel Dekker, 1991.

151. Mandel HG, Cohn VH. Fat-soluble vitamins. Vitamins A, K, and E. In: Goodman AG, Gilman LS, Gilman A, eds. The pharmacological basis of therapeutics. 6th ed. New York: Macmillan, 1980.

152. Marcus R. The relationship of dietary calcium to the maintenance of skeletal integrity in man: An interface of endocrinology and nutrition. Metabolism 1982;31(1):93.

153. Marcus R. Understanding osteoporosis. West J Med 1990; 155:53.

154. Mastropaolo W, Wilson MA. Effect of light on serum B12 and folate stability. Clin Chem 1993;39:913.

155. Mather JP. Vitamin A response of testicular cells in culture. In: DeLuca LM, Shapiro SS, eds. Modulation of cellular interactions by vitamin A and derivatives (retinoids). Ann NY Acad Sci 1981;359:412.

156. Matschiner JT, Zimmerman A, Bell RG. The influence of warfarin on vitamin K epoxide reductase. Thromb Diat Haemorrh 1974;57(Suppl):45.

157. Matson W. ESA Inc, Bedford, MA. Personal communication, October 24, 1994.

158. McCormick AM, Napoli JL, DeLuca HF. High-pressure liquid chromatography of vitamin A metabolites and analogs. In: Methods in Enzymology 67, Part F:220. New York: Academic Press, 1980.

159. McLean SW, Ruddel ME, Gross EG, et al. Liquid-chromatographic assay for retinol (vitamin A) and retinol analogs in therapeutic trials. Clin Chem 1982;28(4):693.

160. Medical Research Council Vitamin Study Research Group. Prevention of neural tube defects: Results of the Medical Research Council Vitamin Study. Lancet 1991;338:131.

161. Menkes MS, Comstock GW, Vuilleumier JP, et al. Serum beta-carotene, vitamins A and E, and selenium, and the risk of lung cancer. N Engl J Med 1986;315:1250.

162. Merrill AH, Henderson JM. Diseases associated with defects in vitamin B_6 metabolism or utilization. Annu Rev Nutr 1987;7:137.

163. Meydani M, Meydani SN, Leka L, et al. Effect of long-term vitamin E supplementation on lipid peroxidation and immune responses of young and old subjects. FASEB J 1993;7:A415.

164. Meydani S, Barklund P, Liu S, et al. Vitamin E supplementation enhances cell-mediated immunity in healthy elderly subjects. Am J Clin Nutr 1990;52:557.

165. Meydani SN, Ribaya-Mercado JD, Russell RM, et al. Vitamin B-6 deficiency impairs interleukin 2 production and lymphocyte proliferation in elderly adults. Am J Clin Nutr 1991;53:1275.

166. Miller SM. Abnormalities in biotin nutriture. AACC Nutr Division Newsletter 1988;6(2):1.

167. Miller SM. Aging and changes in vitamin status. Clin Lab Sci 1988;1(6):342.

168. Miller SM. Old and new rationales for serum B12 and folate determinations. Clin Lab Sci 1993;6:272.

169. Miller SM. Potential perils of niacin therapy. Clin Lab Sci 1991;4:156.

170. Milne DB, Canfield WK, Mahalko JR, Sanstead HH. Effect of oral folic acid supplements on zinc, copper and iron absorption and excretion. Am J Clin Nutr 1984;39:535.

171. Milne DB, Gallagher SK. Microbiological and radioimmunological assays for folic acid in whole blood compared: Effect of zinc nutriture. Clin Chem 1983;29:2117.

172. Milunsky A, Jick H, Jick SS, et al. Multivitamin/folic acid supplementation in early pregnancy reduces the prevalence of neural tube defects. JAMA 1989;262(20):2847.

173. Mistry SP. Biotin. In: Barker BM, Bender DA, eds. Vitamins in medicine. 4th ed. Vol 1. London: William Heinemann, 1980.

174. Mizock BA. Lactic acidosis. Dis Mon 1989;35:1.

175. Moertel CG, Fleming TR, Creagan ET. High-dose vitamin C versus placebo in the treatment of patients with advanced cancer who have had no prior chemotherapy. N Engl J Med 1985;312(3):137.

176. Monaghan DA, Power MJ, Fottrell PF. Sandwich enzyme immunoassay of osteocalcin in serum with use of an antibody against human osteocalcin. Clin Chem 1993;39:942.

177. Moser U, Bendich A. Vitamin C. In: Machlin LJ, ed. Handbook of vitamins. 2nd ed. New York: Marcel Dekker, 1991.

178. Moussa F, Dufour L, Didry JR, et al. Determination of transphylloquinone in children's sera. Clin Chem 1989;35:874.

179. Muller DPR. Vitamin E: Its role in neurological function. Postgrad Med J 1986;62:107.

180. National Research Council. Diet, nutrition and cancer. Washington: Committee on Diet, Nutrition and Cancer, Assembly of Life Sciences, 1982.

181. National Research Council (U.S.), Subcommittee on the Tenth Edition of the RDAs, Food and Nutrition Board, Commission on Life Sciences. Recommended dietary allowances. 10th ed. Washington: National Academy Press, 1989.

182. Nauss KM. Influences of vitamin A status on the immune system. In: Bauernfeind JC, ed. Vitamin A deficiency and its control. Orlando, FL: Academic Press, 1986.

183. Neal RA, Sauberlich HE. The vitamins. Thiamine. In: Goodhart RS, Shils ME, eds. Modern nutrition in health and disease. 6th ed. Philadelphia: Lea & Febiger, 1980.

184. Nelson JS, Fischer VW. Vitamin E. In: Barker BM, Bender DA, eds. Vitamins in Medicine. 4th ed. Vol 1. London: William Heinemann, 1980.

185. Nicholalds GE. Assessment of status of riboflavin nutriture by assay of erythrocyte glutathione reductase activity. In: Selected methods of clinical chemistry. Washington: American Association for Clinical Chemistry, 1977;8:199.

186. Nicholalds GE. Riboflavin. Clin Lab Med 1981;1(4):685.

186a. Odell WD, Heath H. Osteoporosis: Pathology, prevention, diagnosis, and treatment. Disease-a-Month 1993;39:789.

187. Olson RE. The vitamins. Vitamin K. In: Shils ME, Olson JA, Shike M. eds. Modern nutrition in health and disease. 8th ed. Philadelphia: Lea & Febiger, 1994.

188. Olson JA. 1992 Atwater lecture: The irresistible fascination of carotenoids and vitamin A. Am J Clin Nutr 1993;57:833.

189. Olson JA. Physiologic and metabolic basis of major signs of vitamin A deficiency. In: Bauernfeind JC, ed. Vitamin A deficiency and its control. Orlando, FL: Academic Press, 1986.

189a. Olson JA. The vitamins. Vitamin A, retinoids, and carotenoids. In: Shils ME, Olson JA, Shike M, eds., Modern nutrition in health and disease, 8th ed. Philadelphia: Lea & Febiger, 1994.

190. Olson JA. Vitamin A. In: Machlin LJ, ed. Handbook of vitamins. 2nd ed. New York: Marcel Dekker, 1991.

191. Ong DE. Cellular transport and metabolism of vitamin A: Roles of the cellular retinoid-binding proteins. Nutr Rev 1994;52(2)[Suppl II]:S24.

192. Orten JM, Neuhaus OW. Human biochemistry. St Louis: Mosby, 1982.

193. Padh H. Vitamin C: Newer insights into its biochemical functions. Nutr Rev 1991;49(2):65.

194. Paolisso G, Di Maro G, Galzerano D, Cacciapuoti F, et al. Pharmacologic doses of vitamin E and insulin action in elderly subjects. Am J Clin Nutr 1994;59:1291.

195. Paolisso G, DiAmore A, Giugliano D, Ceriello A, et al. Pharmacologic doses of vitamin E improve insulin action in healthy subjects and non-insulin dependent diabetic patients. Am J Clin Nutr 1993;57:650.

196. Parfitt AM, Gallagher JC, Heaney RP, et al. Vitamin D and bone health in the elderly. Am J Clin Nutr 1982;36:1014.

197. Pauling L. Vitamin C and the common cold. San Francisco: Freeman, 1970.

198. Pennypacker LC, Allen RH, Kelly JP, et al. High prevalence of cobalamin deficiency in elderly outpatients. J Am Geriatr Soc 1992;40:1197.

199. Phelps DL. Current perspectives on vitamin E in infant nutrition. Am J Clin Nutr 1987;46:187.

200. Post GR, Boon SE. Semi-automated method for quantitating biotinidase activity in newborns. Clin Chem 1988;34(6):1309.

201. Prince MR, Frisoli JK. Beta-carotene accumulation in serum and skin. Am J Clin Nutr 1993;57:175.

202. Reichel H, Koeffler HP, Norman AW. The role of the vitamin D endocrine system in health and disease. N Engl J Med 1989;320(15):980.

203. Reynolds RD. Vitamin supplements: Current controversies. J Am Coll Nutr 1994;13(2):118.

204. Reynolds RD, Styer DJ, Schlichting CL. Decreased vitamin B_6 status of submariners. Am J Clin Nutr 1988;47:463.

205. Ribaya-Mercado JD, Russell RM, Sahyoun N, et al. Vitamin B-6 requirements of elderly men and women. J Nutr 1991;121:1062.

206. Rickes EL, Brink NG, Koniuszy FR, et al. Crystalline vitamin B_{12}. Science 1948;107:396.

207. Riley L, Apple FS. Beriberi. Clin Chem News 1990;16:12.

208. Rimm EB, Stampfer M, Ascherio A, Giovannucci E. Vitamin E consumption and the risk of coronary heart disease in men. N Engl J Med 1993;328:1450.

209. Rivers J. Safety of high-level vitamin C ingestion. Ann Acad Sci 1987;498:445.

210. Robertson JM, Donner AP, Trevithick JR. Vitamin E intake and risk of cataracts in humans. Ann NY Acad Sci 1990; 503:372.

211. Roe DA. Drug interference with the assessment of nutritional status. Clin Lab Med 1981;1(4):647.

212. Roe DA. Drug-induced nutritional deficiencies. Westport, CT:AVI, 1976.

213. Roe DA, Campbell TC. Drugs and nutrients: The interactive effects. New York: Marcel Dekker, 1984.

214. Roth KS. Biotin in clinical medicine: A review. Am J Clin Nutr 1981;34:1967.

215. Russell RM. Micronutrient requirements of the elderly. Nutr Rev 1992;50(38):463.

216. Russell RM, Suter PM. Vitamin requirements of elderly people: An update. Am J Clin Nutr 1993;58:4.

217. Sadowski JA, Hood SJ, Dallal GE, Garry PJ. Phylloquinone in plasma from elderly and young adults: Factors influencing its concentration. Am J Clin Nutr 1989;50:100.

218. Sampson DA, O'Connor DK. Analysis of B6 vitamers and pyridoxic acid in plasma, tissues and urine using high performance liquid chromatography. Nutr Res 1989;9:259.

219. Sander S, Hahn A, Stein J, et al. Comparative studies on the high-performance liquid chromatographic determination of thiamin and its phosphate esters with chloroethylthiamine as an internal standard using pre-column and post-column derivatization procedures. J Chromatogr 1991;558:115.

220. Sandstead HH. Clinical manifestations of certain classical deficiency disease. In: Goodhart RS, Shils ME, eds. Modern nutrition in health and disease. 6th ed. Philadelphia: Lea & Febiger, 1980.

221. Sanghvi RS, Lemons RM, Baker H, Thoene JG. A simple method for determination of plasma and urinary biotin. Clin Chim Acta 1982;124:85.

222. Sauberlich HE. Ascorbic acid (vitamin C). Clin Lab Med 1981;1(4):673.

223. Sauberlich HE. Methods for the assessment of nutritional status. In: Chen LH, ed. Nutritional aspects of aging. Vol I. Boca Raton, FL:CRC Press, 1986.

224. Sauberlich HE. The vitamins. Pantothenic acid. In: Goodhart RS, Shils ME, eds. Modern nutrition in health and disease. 6th ed. Philadelphia:Lea & Febiger, 1980.

225. Sauberlich HE. Vitamin B-6 status assessment: Past and present. In: Leklem JE, Reynolds RB, eds. Methods in vitamin B-6 nutrition. Analysis and status assessment. New York: Plenum Press, 1980.

226. Sauberlich HE, Dowdy RP, Skala JH. Laboratory tests for the assessment of nutritional status. Crit Rev Clin Lab Sci 1973; 4:288.

227. Sauberlich HE, Green MD, Omaye ST. Determination of ascorbic acid and dehydroascorbic acid. In: Seib PA, Tolbert BM, eds. Ascorbic acid: Chemistry, metabolism and uses. Washington: American Chemical Society, 1982.

228. Sauberlich HE, Goad WC, Skala JH, Waring PP. Procedure for mechanized (continuous flow) measurement of serum ascorbic acid (vitamin C). In: Selected methods of clinical chemistry. Washington:American Association for Clinical Chemistry, 1977;8:191.

229. Sauberlich HE, Kretsch MJ, Skala JH, et al. Folate requirement and metabolism in nonpregnant women. Am J Clin Nutr 1987;46:1016.

230. Savage DG, Lindenbaum J, Stabler SP, Allen RH. Sensitivity of serum methylmalonic acid and total homocysteine determinations for diagnosing cobalamin and folate deficiencies. Am J Med 1994;96:239.

231. Schilling RF. Vitamin B$_{12}$: Assay and absorption testing. Lab Manage 1982;20:31.

232. Schilling RF, Fairbanks VF, Miller R, Schmitt K, et al. Improved vitamin B$_{12}$ assays: A report on two commercial kits. Clin Chem 1983;29(3):582.

233. Schneede J, Ueland PM. Automated assay of methylmalonic acid in serum and urine by derivatization with 1-pyrenyldiazomethane, liquid chromatography, and fluorescence detection. Clin Chem 1993;39:392.

234. Schwabedal PE, Pietrzik K, Wittkowski W. Pantothenic acid deficiency as a factor contributing to the development of hypertension. Cardiology 1985;72(Suppl 1):187.

234a. Seddon J, Ajani U, Sperduto R, et al. for the Eye disease case-control study group. Dietary carotenoids, vitamins A, C, and E, and advanced age-related macular degeneration. JAMA 1994;272:1413.

235. Selhub J, Jacques PF, Wilson PWF, Rush D, Rosenberg IH. Vitamin status and intake as primary determinants of homocysteinuria in an elderly population. JAMA 1993;270:2693.

236. Shamberger RJ. The subtle signs of chronic vitamin undernutrition. Diag Med 1984;7(3):75.

237. Shibata K, Matsuo H. Correlation between niacin equivalent intake and urinary excretion of its metabolites, N'-methylnicotinamide, N'-methyl-2-pyridone-5-carboxamide, and N'-methyl-4-pyridone-3-carboxamide, in humans consuming a self-selected food. Am J Clin Nutr 1989;50:114.

237a. Shipchandler MT, Moore EG. Rapid, fully automated measurement of plasma homocyst(e)ine with the Abbott IMX® analyzer. Clin Chem 1995;41:991.

238. Silverman AK, Ellis CN, Voorhees JJ. Hypervitaminosis A syndrome: A paradigm of retinoid side effects. J Am Acad Dermatol 1987;16:1027.

239. Simon JA, Schreiber GB, Crawford PB, Frederick MM, et al. Dietary vitamin C and serum lipids in black and white girls. Epidemiology 1993;4:537.

240. Smidt LJ, Cremin FM, Grivetti LE, et al. Influence of thiamin supplementation on the health and general well-being of an elderly Irish population with marginal thiamin deficiency. J Gerontol 1991;46:M16.

241. Solberg O, Hegna IK. Microbiological assay of pantothenic acid. In: McCormick DB, Wright LD, eds. Methods in enzymology. Part D. Vitamins and coenzymes. New York: Academic Press, 1979;62:201.

242. Solomons NW, Allen LH. The functional assessment of nutritional status: Principles, practice and potential. Nutr Rev 1983;41(2):33.

243. Sowell A. Simple, rapid method for assessment of serum vitamin A concentration. Clin Lab Sci 1994;7(3):160.

244. Sowell AL, Huff DL, Yaeger PR, Caudill SP, et al. Retinol, alpha-tocopherol, lutein/zeaxanthin, lycopene, alpha-carotene, trans-beta-carotene, and four retinyl esters in serum determined simultaneously by reversed-phase HPLC with multiwave detection. Clin Chem 1994;40:411.

245. Stabler SP, Marcell PD, Podell ER, et al. Assay of methylmalonic acid in the serum of patients with cobalamin deficiency using capillary gas chromatography–mass spectrometry. J Clin Invest 1986;77:1606.

246. Stabler SP, Marcell PD, Podell ER, et al. Elevation of total homocysteine in the serum of patients with cobalamin or folate deficiency detected by capillary gas chromatography–mass spectrometry. J Clin Invest 1988;81:466.

247. Stampfer MJ, Hennekens CH, Manson JE, Colditz GA, et al. Vitamin E consumption and the risk of coronary disease in women. N Engl J Med 1993;328:1444.

248. Stampfer MJ, Malinow R, Willett WC, Newcomer, LM, et al. A prospective study of plasma homocyst(e)ine and risk of myocardial infarction in US physicians. JAMA 1992;268:877.

249. Stampfer MJ, Willett WC. Homocysteine and marginal vitamin deficiency: The importance of adequate vitamin intake. JAMA 1993;270:2726.

250. Statland B. Vitamins and minerals: Passing fads or keys to health? Med Lab Observ 1993;25(12):21.

251. Statland BE. Nutrition and cancer. Clin Chem 1992; 38(8):1587.

252. Steinberg D. Antioxidant vitamins and coronary heart disease. N Engl J Med 1993;328:1487.

253. Stenflo J, Fernlund P, Egan W, Roepstorff P. Vitamin K dependent modifications of glutamic acid residues in prothrombin. Proc Natl Acad Sci USA 1974;71:2730.

254. Subar AF, Block G. Use of vitamin and mineral supplements: Demographics and amounts of nutrients consumed. The 1987 Health Interview Survey. Am J Epidemiol 1990;132:1091.

255. Subar AF, Block G, James LD. Folate intake and food sources in the U.S. population. Am J Clin Nutr 1989;50:503.

256. Suttie JW. Vitamin K and human nutrition. J Am Diet Assoc 1992;92:585.

257. Suttie JW, Mummah-Schendel LL, Shah DV, et al. Vitamin K deficiency from dietary vitamin K restrictions in humans. Am J Clin Nutr 1988;47:475.

258. Talbott MC, Miller LT, Kerkvliet NI. Pyridoxine supplementation: Effect on lymphocyte response in elderly persons. Am J Clin Nutr 1987;46:659.

259. Tanumihardjo SA, Permaesih D, Dahro AM, Rustan E, et al. Comparison of vitamin A status assessment techniques in children from two Indonesian villages. Am J Clin Nutr 1994; 60:136.

260. The Alpha-Tocopherol, Beta-Carotene Cancer Prevention Study Group. The effects of vitamin E and beta-carotene on the incidence of lung cancer and other cancers in male smokers. N Engl J Med 1994;330:1029.

261. Thompson JN, Erdody P, Brien R, et al. Fluorometric determination of vitamin A in human blood and liver. Biochem Med 1971;5:67.

262. Thurnham DI. Carotenoids: Functions and fallacies. Proc Nutr Soc 1994;53:77.

263. Tietz NW, Shuey DF, Wekstein DR. Laboratory values in fit aging individuals—sexagenarians through centenarians. Clin Chem 1992;38(6):1167.

264. Tonkin SY. Vitamins and oral contraceptives. In: Briggs MH, ed. Vitamins in human biology and medicine. Boca Raton, FL: CRC Press, 1981.

265. Traber MG, Ingold KU, Burton GW, Kayden HJ. Absorption and transport of deuterium-substituted 2R, 4′, 8′R-alpha-tocopherol in human lipoproteins. Lipids 1988;23:791.

266. Tucker DM, Penland JG, Sandstead HH, et al. Nutrition status and brain function. Am J Clin Nutr 1990;52:93.

267. Ubbink JB, Vermaak WJH, van der Merwe A, Becker PJ. Vitamin B-12, vitamin B-6, and folate nutritional status in men with hyperhomocysteinemia. Am J Clin Nutr 1993;57:47.

268. Valentic JP, Elias AN, Weinstein GD. Hypercalcemia associated with oral isotretinoin in the treatment of severe acne. JAMA 1983;250(14):1899.

269. Van Eys J. Nicotinic acid. In: Machlin LJ, ed. Handbook of vitamins. 2nd ed. New York: Marcel Dekker, 1991.

270. Vanderjagt DJ, Garry PJ, Bhagavan HN. Ascorbic acid intake and plasma levels in healthy elderly people. Am J Clin Nutr 1987;46:290.

271. VanderJagt DJ, Garry PJ, Hunt WC. Ascorbate in plasma as measured by liquid chromatography and by dichlorophenolindophenol colorimetry. Clin Chem 1986;32(6):1004.

272. Wahlqvist ML. Vitamin use in clinical medicine. Med J Aust 1987;146:30.

273. Wald G. The molecular basis of visual excitation. Nature 1968;219:800.

274. Waring PP, Fisher D, McDonnell J, et al. A continuous-flow (AutoAnalyzer II) procedure for measuring erythrocyte transketolase activity. Clin Chem 1982;28(11):2206.

275. Warnock LG. Transketolase activity of blood hemolysate, a useful index for diagnosing thiamine deficiency. In: King DS, ed. Selected methods of clinical chemistry. Vol 8. Washington: American Association for Clinical Chemistry, 1977.

276. Webb JL. Nutritional effects of oral contraceptive use. A Review. J Reprod Med 1980;25(4):150.

277. Weiss JS, Ellis CN, Headington JT. Topical tretinoin improves photoaged skin. JAMA 1988;259:527.

278. Whitlock EP. Selenium, vitamin E and vitamin A: Nutritional and physiologic findings. Lab Manage 1987;25(5):49.

279. Willett WC. Does low vitamin B6 intake increase the risk of coronary heart disease? In: Reynolds RD, Leklem JE, eds. Vitamin B6: Its role in health and disease. New York: Alan R. Liss, 1985.

280. Willett WC. Folic acid and neural tube defect: Can't we come to closure? Am J Public Health 1992;82:666.

281. Williams ATR. Simultaneous determination of serum vitamin A and E by liquid chromatography with fluorescence detection. J Chromatogr 1985;341:198.

282. Wolf B. Biotinidase deficiency. Lab Manage 1987;25(10):31.

283. Wyatt DT, Lee M, Hillman RE. Factors affecting a cyanogen bromide-based assay of thiamin. Clin Chem 1989;35(10):2173.

284. Wyse BW, Wittwer CW, Hansen RG. Radioimmunoassay for pantothenic acid in blood and other tissues. Clin Chem 1979;25:108.

285. Young DS. Effect of vitamin C on laboratory tests. Lab Med 1983;14(5):278.

286. Zimmerman MB, Shane B. Supplemental folic acid. Am J Clin Nutr 1993;58:127.

SUGGESTED READINGS

Bronner F, ed. Nutrition and health: topics and controversies. Boca Raton, FL: CRC Press, 1995.

Chen LH, ed. Nutritional aspects of aging. Vols 1 and 2. Boca Raton FL: CRC Press, 1986.

Faulkner WR, Meites S. Geriatric clinical chemistry reference values. Washington: AACC Press, 1994.

Groff JL, Gropper SS, Hunt SM. Advanced nutrition and human metabolism. 2nd ed. Minneapolis: West Publishing, 1995.

Herbert V, Subak-Sharpe GJ, eds. Total nutrition: the only guide you'll ever need: from the Mount Sinai School of Medicine. New York: St. Martin's Press, 1995.

Jacob MM, ed. Vitamins and minerals in the prevention and treatment of cancer. Boca Raton FL: CRC Press, 1991.

Labbe RF, Veldee M. Nutrition in the clinical laboratory. Clin Lab Med 1993;(13).

Machlin LJ, ed. Handbook of vitamins. 2nd ed. New York: Marcel Dekker, 1991.

Machlin LJ, Sauberlich HE, eds. Beyond deficiency: New views on the function and health effects of vitamins. Ann NY Acad Sci 1992;669.

National Research Council (U.S.), Subcommittee on the Tenth Edition of the RDAs, Food and Nutrition Board, Commission on Life Sciences. Recommended dietary allowances. 10th ed. Washington: National Academy Press, 1989.

Shils ME, Olson JA, Shike M, eds. Modern nutrition in health and disease. Vols. 1 and 2. 8th ed. Philadelphia: Lea & Febiger, 1994.

Nutritional Assessment

A. Michael Spiekerman

Objectives

Upon completion of this chapter, the clinical laboratorian should be able to:

- *Define the following terms: anthropometric, kwashiorkor, marasmus, nutritional assessment, and total parenteral nutrition (TPN).*
- *Name the groups or populations at risk for malnutrition.*
- *Identify the anthropomorphic measures of nutritional status.*
- *Relate the plasma proteins of interest in nutritional assessment.*
- *Discuss the guidelines used to monitor patients on TPN.*
- *Describe some of the electrolyte and mineral abnormalities associated with TPN.*
- *Given clinical data, evaluate a patient's nutritional status.*

KEY WORDS

Anthropometric
  Methods
Kwashiorkor
Malnutrition
Marasmus

Nutritional
  Assessment
Total Parenteral
  Nutrition (TPN)

Nutritional assessment has become increasingly important in medical care, especially in the treatment of critically ill patients in hospitals and in the treatment of chronically emaciated patients. Despite this fact, medical school curricula are slow to embrace nutritional assessment or nutrition related courses.

Systematically identifying patients who are at risk of developing nutritionally related complications in the course of an illness is uncommon, and there is no standard for monitoring the effectiveness of feeding plans to replace nutritional losses. This situation exists because of a current lack of understanding about the metabolic management of stressed hospitalized patients.

Because this is a chapter on nutritional assessment, a general definition of malnutrition should be provided. Malnutrition is a state of decreased intake of calories or micronutrients (vitamins and trace elements) resulting in a risk of impaired physiologic function associated with increased morbidity and mortality.[99,107,108] Patients who are chronically calorie malnourished suffer from a loss of both adipose and muscle tissue as a result of increased lipolytic and gluconeogenic activity, respectively. These patients, however, do not demonstrate protein deficiency as reflected by normal levels of serum transport proteins. This type of malnourished state is referred to as *marasmus*. Acute protein calories malnutrition, known as *kwashiorkor*, is associated with physiologic stress and results in impaired protein synthesis and lower serum transport proteins.[13]

Decreases in vitamin and trace metal intake are usually associated with calorie and protein deficiencies. They may occur independently, however, as a result of dietary inadequacy, as in alcoholism, or as a result of malabsorption syndromes.

The malnourished states can be assessed by the measurements of anthropomorphic methods. The anthropomorphic methods are the common measurements that dietitians use in evaluating patients. These methods are shown in Table 30-1.[144]

Nutritional assessment, which is the evaluation of a patient's metabolic and nutritional needs, is performed by

Michael L. Bishop, Janet L. Duben-Engelkirk, and Edward P. Fody.
CLINICAL CHEMISTRY. © 1996 Lippincott–Raven Publishers.

TABLE 30-1. Anthropomorphic Measures of Nutritional Status

Height
Weight
Skinfold thickness
Mid-arm muscular circumference
Wrist circumference

diurnal, empirical, laboratory, and other objective measurements of nutritional status.

THOSE AT RISK FOR MALNUTRITION

Malnutrition has been estimated to affect over 1.3 billion people worldwide.[160] It is a major co-morbid cause of death. The estimates of malnutrition in the United States range from 10 to 20 million individuals.

Poverty is certainly a major predisposing factor for the risk of malnutrition in the non-hospitalized individual, especially worldwide. However, malnutrition is erroneously believed to be associated only with low socioeconomic status. On the contrary, the prevalence of significant malnutrition in the United States among nursing home patients, the elderly, and medical and surgical hospitalized patients is no longer challenged. Those at risk for malnutrition are listed in Table 30-2.

Recently, it has been documented that 30% to 50% of hospitalized patients may be malnourished.[119,144] Malnutrition in the hospital setting may occur when oral nutritional intake is inadequate or impossible because of treatment procedures or complications that follow surgery. Chronic malnutrition may occur prior to hospitalization as a result of a chronic disease process or a medical condition that leads to poor food choices, protein depletion, or low protein intake. Chronic diseases leading to chronic malnutrition include malignancy, gastroin-

TABLE 30-2. Groups at Risk for Malnutrition

Depressed or mentally ill populace that fail to eat.
Elderly, with emphasis on those in nursing homes.
Low socio-economic status.
Population with loss of fluids and nutrients from long-term diarrhea, draining fistula, or wounds.
Stroke victims.
Post-operative state and ileus with long periods without oral intake.
Hypermetabolic conditions: head trauma, multiple trauma, major burns, organic failure, and sepsis.
Cancer of the gastrointestinal tract or cachexia.
Unintentional weight loss exceeding 10% in 6 months.
Chronic disease, long-term dialysis.
Pancreatitis.

testinal disease including malabsorption syndrome, infectious disease such as acquired immune deficiency syndrome (AIDS) and tuberculosis, alcoholism, and liver disease.[13,99,144]

The hospitalized patient may undergo acute malnutrition as a result of decreased food intake secondary to the disease, chemotherapy, or depression, or reduced intake due to nausea.[156] The hypermetabolic state may cause acute malnutrition secondary to an increased energy expenditure. The hypermetabolic condition is associated with major trauma, sepsis, surgery, burns, multiple organ failure, and acute renal failure.[12]

A severely ill patient with hypermetabolism (kwashiorkor-like) and multi-organ failure is metabolically different from the starved hypometabolic patient (marasmus-like), although a clear distinction in clinical presentation is not seen frequently. The long-term malnourished hypometabolic patient with primary loss of body fat utilizes ketogenic fatty acid, not carbohydrates, as a primary fuel. Such individuals tend to be thin, with a wasted appearance, and have little adipose tissue reserves.

The hypermetabolic stressed patient may become malnourished, regardless of adipose tissue reserves, because of the reliance of these patients on gluconeogenic fuel precursors obtained initially from muscle proteolysis and later from new protein, resulting in a decreased hepatic production of serum proteins. These patients may be overlooked as malnourished because they may have adequate stores of body fat and may have normal weight and appearance. The difference in the metabolism of the hypo- and hypermetabolic patients should be recognized and treated with different modalities.

In the hypometabolic starved patient, aggressive feeding, especially with carbohydrates, may not be desirable. In these patients, with the body's ability to adapt to chronic starvation by reducing the basal metabolic rate, cardiac output, temperature, and physical activity, over-zealous nutritional support may induce a re-feeding syndrome.[120] In the re-feeding syndrome, life-threatening fluid and electrolyte shifts occur after the initiation of aggressive nutritional support therapies. Re-feeding syndrome can reduce cardiac mass leading to cardiac insufficiency conditions such as hypophosphatemia and hypokalemia. It can also reduce respiratory muscle mass and decrease ATP content leading to respiratory failure.[23]

The hypermetabolic patient, on the other hand, has an increased energy need and may require aggressive nutritional support to minimize catabolism and protein losses and to support immune response. When malnutrition is not detected and treated in patients during their hospital stay, there is increased morbidity and mortality.[88]

Patients with protein-energy malnutrition have impaired wound healing and increased rates of infection[119] with extended lengths of hospital stay and higher mortality rates. Of the patients who are identified as malnourished, too few receive nutritional support early enough.

The early identification of malnourished individuals and the establishment of nutritional intervention have become a quality of care issue.[11,18]

NUTRITIONAL ASSESSMENT

Many of the methods used for nutritional assessment are designed to provide information regarding body composition abnormalities. In addition to clinical evaluation, they include straightforward physical measurement (body weight and anthropometric measurements), biochemical or immunologic determinations, and sophisticated laboratory measurements.

Some authors suggest that clinical judgment is a desirable method to evaluate nutritionally associated complications. This type of nutritional assessment is based on past nutritional intake, disease process, the extent of catabolic disease, functional status, edema, skin rash, and neuropathology.[7,40] Other groups have shown that clinical judgment alone could not reliably identify the malnourished patient.[112,116]

One method used for nutritional assessment is anthropometric indices. In this procedure, the thickness of the biceps skinfold of the nondominant arm is taken to represent body fat mass. Despite its limitations, some usefulness has been established for anthropometrics and skinfold thickness.[43] They demonstrate that the density of the fat-free (lean) body mass is a reasonably constant value and that the equations to calculate density and total body fat can be derived for various age and sex groups solely from the sum of four (sometimes even one) skinfold thickness measurements. Validated by hydrodensitometric measurements, their equations permit an estimate of density and total body fat from skinfold thickness measurements with an accuracy of $\pm$ 5% for men and $\pm$ 3.5% for women.

Measurements from arm muscle circumference are thought to provide an indication of the body's muscle mass and hence its main protein reserve. Arm muscle circumference can be derived from total arm muscle circumference and triceps skinfold thickness, but a 15% to 25% overestimation of lean body mass by anthropometry in arm muscle can be incorrectly obtained.[70] Other studies found no correlation between the circumference of arm muscle and the initial total body nitrogen content in patients receiving parenteral nutrition. These studies further showed no changes in total body nitrogen were found after completion of total parenteral nutrition.[80,32]

Differences in arm muscle circumference did occur between well-nourished groups of differing ethnic origins and changes occurred throughout adult life in both sexes. A secular trend has been observed in the values obtained in large population studies. No sound basis has been established for classifying as malnourished those men with arm circumference of <23 cm or those women with circumference of <22 cm, as is the common practice in many articles that were reviewed.

Since it was shown that a preoperative weight loss of more than 20% was associated with 33% mortality, whereas a 4% mortality rate occurred in those with <20% weight loss, the estimation of weight loss is a key indicator in the nutritional assessment of surgical patients.[146]

The weight loss is usually calculated by subtracting the present observed weight from either the recall "usual" weight or from the predicted normal weight taken from standard tables or equations. However, a cross-sectional study of 105 patients found that patient recall of weight was associated with an error as great as 3.6 kg.[104]

Recent weight change is a commonly used index of malnutrition. More than 10% loss over any time period has been taken as evidence of malnutrition and a useful correlation has been proposed between the extent of weight loss and the time over which it develops,[104] as illustrated in Table 30-3. A weight loss of > 4.5 kg is highly predictive of surgical mortality.[14,132]

Creatinine/Height Index

Since creatine is present almost entirely within muscle (as creatine phosphate) and is converted to creatinine at a relatively constant rate, the excretion rate of urinary creatinine may be indicative of total muscle mass.[63] This is reasonably accurate in patients without renal failure, if it is based on multiple measurements.[51] Urinary creatinine excretion correlates with the total body nitrogen content except in cancer patients. In these patients, creatinine excretion remains stable when total body nitrogen decreases, indicating increased creatinine loss as the disease progresses. Other studies, however, were unable to demonstrate that the creatinine/height index was of value in predicting complications.[108]

Immunologic Testing

The peripheral blood absolute lymphocyte count and the skin response to injected antigen have been used as indicators of status of nutrition. A lymphocyte count below 300/mm^3 reflects immune deficiency,[14] and such a count is associated with a negative response to an injected antigen.[83]

A strong association between morbidity and decreased skin test reactivity has been shown.[37,64] The proportion of patients with negative skin test reactivity (no response) who died (50%) was significantly higher than the proportion of those with normal skin test reactivity who died (23%).[66] In surgical patients studied, those who were anergic and showed no skin response to injected antigens preoperatively

TABLE 30-3. Weight Change Evaluation		
Time	**Significant Weight Loss**	**Severe Weight Loss**
1 week	1–2%	>2%
1 month	5%	>5%
3 months	7.5%	>7.5%
6 months	10%	>10%

Values used are percent weight change (%W): %W = (Usual weight − Actual Weight)/(Usual Weight) × 100

had a 29% incidence of sepsis and a 29.9% mortality rate, compared with 7.5% and 4.6%, respectively, for those classified as immunologically normal.[29] Correcting anergy and providing a positive immunologic response with parenteral nutrition have been shown to improve the outcome of nutritionally depleted patients.[34] Immunologic testing has been useful in predicting morbidity and identifying the effects of malnutrition in disease states.

Multivariable Index

Multiple regression equations from a number of anthropomorphic and laboratory tests have been used to predict surgical outcomes as affected by nutritional variables. Prognostic nutritional index (PNI) is based on serum albumin, triceps skinfold, serum transferrin, and cutaneous reactivity to three recall antigens.[107] This index has been applied to 100 patients undergoing major elective gastrointestinal surgery. The incidence of complications was 8% for the low-risk population (PNI <40%), 30% for the intermediate group (PNI = 40–49%), and 46% for the high-risk group (PNI>50%). Mortality rate in the low-risk group is 3%, for the intermediate group 4.3%, and for the high-risk group 33%. The best use of PNI is in relating complications and assessing hazards of individual patients.[19]

A discriminant function test for the treatment of cancer patients has been developed based on serum albumin, skin test response, and the presence of sepsis at the time of assessment. This index had a predictive accuracy of 72% for subsequent hospital mortality with a sensitivity of 74% and a specificity of 66%.[66]

Body Composition

Measuring body composition has proven useful in monitoring nutritional states because of marked changes in body composition that occur with malnutrition. Body composition can be described in either biochemical or cellular terms. The composition of an adequately nourished 75-kg adult man is illustrated in Figure 30-1 in both biochemical and cellular terms.[103]

The values may vary from one individual to another. Body mass described in cellular terms is divided into two mass components. These are lean body mass (LBM, tissue devoid of all fat) and body fat (sum of all ether-extractable substances). Lean body mass comprises the metabolically active tissues that are known as *body cell mass* and metabolically inactive components, which are known as the *extracellular mass*. The body cell mass is the portion of the body responsible for all of the oxygen consumption and carbon dioxide production. This is directly linked to energy requirements.[54,103] The primary function of the extracellular mass is intracellular transport and structural support.

Malnourished hospitalized patients have a body cell mass that is 40.5% less than healthy non-hospital patients of the same age group. This large percentage of difference in the two groups is disproportionate to a mean difference in body weight of 16% (due to the expansion of extracellular mass in the malnourished state). This demonstrates the insensitivity of body weight as a measure of body composition. Obesity is associated with very little increase in body cell mass and extracellular mass and greatly expanded body fat. Body cell mass declines with age and is greater in men than women. Body cell mass is affected by race, heredity, and stature.[135]

Body Component Status

The body is composed of three major compartments: fat, protein, and water. Although much body fat is subcutaneous, the abdomen contains a considerable quantity that is not easily measured and this is not comparably utilized during starvation. Endocrine studies show that abdominal fat is utilized more slowly during long-time starvation. The relationship between components within each compartment

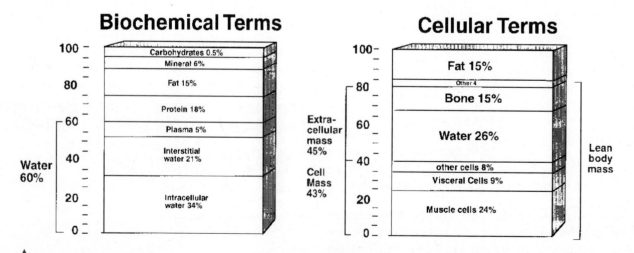

▲

Figure 30-1. Biochemical and cellular terms used to describe body mass and the percentage contribution to total mass in a well-nourished 75-kg male.[112]

may be defined during health, but these relationships change during disease and lean muscle depletion.

Body composition determinations utilize a variety of techniques including multiple isotope detection methods, proton-gamma analysis, and neuron activation to measure total body nitrogen.[103] These methods are generally complicated and time consuming and thus are not suitable for routine use in patient care or for measurement of relatively short-term changes. However, they are the standard against which simpler measurements must be validated.[50]

Total body potassium in nourished and malnourished patients was determined as an index of body cell mass. At the same time, the usually used nutrition indices were measured. The results indicate that the correlation coefficients were poor, the 95% confidence limits were wide, and only gross changes from values within the normal range could be determined.[3] Other studies have demonstrated that biochemical parameters do not correlate significantly with body composition data derived from measurements of total body potassium.

Bioelectrical Impedance Analysis (BIA)

Standard body composition methods are complex and are not widely available, and for this reason, bedside techniques to assess body composition are constantly sought. The body impedance analysis procedure is based on a measurement of changes in the conduction of an applied electrical current through the body. Intracellular and extracellular fluids are thought to act as electroconductors and all membranes are considered to act as electrocapacitors. However, both are regarded as imperfect reactive elements. Since lean body tissue has far greater electrolyte content than fat, this marked difference in ionic content permits the estimation of lean body mass by the assessment of body electrical impedance.[95,97] Studies have shown a correlation between *in vivo* bioelectrical impedance analysis measurement of body composition and measurement by densitometry and deuteron dilutor techniques.[99] Body composition estimates from anthropometric-derived formulas and from BIA measurements were compared in a normal adult population to test the potential usefulness of the bioelectrical impedance analysis method as an adjustment to nutritional assessment in surgical patients.[130] In this population of healthy individuals under steady-state conditions, bioelectric impedance measurements were found to be very reliable in the estimation of body density, total body water, and total body fat. Bioelectric impedance is a new technique that has been found to be reliable in determining nutritional state and hydration status in the sick surgical patient.[20,91]

Functional Tests

Muscle function has been evaluated as an index of nutritional status, because it is susceptible to the effects of withdrawing nutrients and refeeding. One method of evaluating muscle function is a hand grip dynamometry.[89] In this new approach, grip strength had a sensitivity of 90% in predicting postoperative complications. The hand grip dynamometry involved some motivation, which may or may not be constant from patient to patient or with the same patient during the course of an illness. To obviate this problem in very ill patients, an electrical stimulation technique of the ulnar nerve of the wrist has been adapted and is shown in Figure 30-2. The result is the contraction of the abductor pollicis longus muscle, which negates the cooperation of the patient and does not appear to be affected by sepsis, drugs, trauma, surgical intervention, or anesthesia.[124,125] Figure 30-3 shows the anatomical location of the abductor pollicis longus muscle in the wrist.

Starvation has long been known to cause forced muscle contraction to decrease and to cause the development of fatigue. Recently, surgeons have shown that postoperative anorexia and reduced food intake have an effect on muscle function and serum creatinine phosphokinase (CK) in patients undergoing cardiac and general abdominal operations.[5,52] A close correlation existed between rises in CK and prolongation of the relaxation rate of the abductor pollicis longus muscle in both groups of patients. However, when the cardiac and abdominal surgical patients started to eat and meet daily energy requirements, muscle function and CK levels both normalized.[38]

The rate of relaxation of the electrically stimulated abductor pollicis longus muscle is a sensitive and specific measurement of nutrient status in both chronic states and anorexia, and during nutrient withdrawal and refeeding in the postoperative period. Muscle power is not dependent on body nitrogen, body potassium, or muscle bulk, as assessed by the anthropometric measure of arm muscle circumference. Muscle power in the abductor pollicis longus can be doubled by providing the patient with nutritional support, without any detectable change in the arm muscle circum-

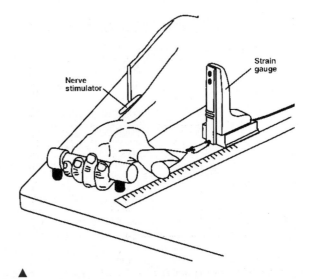

▲
Figure 30-2. Muscle function testing device.

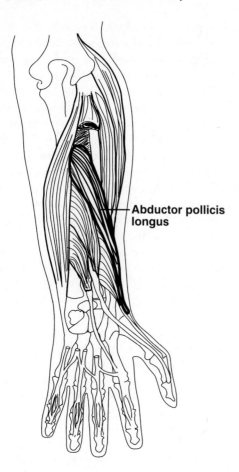

Abductor pollicis longus

▲
Figure 30-3. Anatomical site of abductor pollicis longus muscle.

ference. Relevant and specific information in assessing nutritional status may be generated through future studies of muscle function.[98]

Protein Markers in Nutritional Assessment

The primary objective of nutritional assessment is to identify the patient who is malnourished and then, through nutritional therapy, to preserve or replenish the protein component of the body. Laboratory nutritional assessment is best accomplished by monitoring selected serum proteins.

These proteins should have a short biological half-life and should reflect protein status by measurable concentration changes in the serum. The proteins should have a concentration that is relatively small, a rapid rate of synthesis, a constant catabolic rate, and should be responsive only to protein and energy restrictions.

Protein deficiency states in humans will be prolonged and severe if insensitive markers are used, because it takes a long period before significant change occurs in concentrations of these proteins. The concentration of insensitive proteins may be affected, not only by protein deficiency but by other factors such as hepatic density, renal disease, and severe infection.

Measurements of the concentration of selected individual proteins may not only provide a sensitive index on protein states but may also offer a valuable indication of morbidity. Visceral protein measurements also remain a reliable and relatively easy method of assessing the energy state of the patient.[45] Table 30-4 offers detailed information about these protein markers.

Albumin

Albumin has long been used in the assessment of hospitalized patients. Low levels of serum albumin may reflect low hepatic production or protein loss from the vascular component. Low albumin levels have been identified as a common abnormality in patients in long-term care facilities.

Hospitalized patients with low serum albumin levels experienced a fourfold increase in morbidity and a sixfold increase in mortality.[131] If a low serum albumin is obtained on a patient on the nephrology service, it can be correlated with a longer stay in the hospital and with an increased number of infections.[49]

The albumin concentration in the body is influenced by albumin synthesis, degradation, and distribution. While albumin has the highest concentration of any protein in the vascular system, over 60% of the protein is present in the extravascular spaces. Albumin from the extravascular pool can be mobilized during periods of protein depletion resulting from the stress of major surgery or infection so that serum albumin concentrations may not decline for a long period.[36,68,74] The degeneration rate of the albumin is propor-

Protein	Molecular Weight (KD)	Half-Life	Reference Range
Albumin	65,000	20 days	33–48 g/L
Fibronectin	250,000	15 hours	220–400 mg/L
Prealbumin (transthyretin)	54,980	48 hours	160–350 mg/L
Retinol binding protein	21,000	12 hours	30–60 mg/L
Somatomedin C (Insulin growth factor-1)	7,650	2 hours	0.10–0.40 mg/L
Transferrin	76,000	9 days	1.6–3.6 g/L

TABLE 30-4. Characteristics of Plasma Proteins of Nutritional Interest

tional to the size of the extravascular pool, which allows the concentration in the serum to be relatively constant. The long biological half-life of albumin (20 days) allows changes in the serum concentration only after long periods of malnutrition. Since there are many mechanisms that can potentially produce a depressed albumin concentration, an isolated serum albumin level may be of limited value in evaluating liver synthesis rates in patients who are critically ill.[118]

Serum albumin is not a good indicator of short-term protein and energy deprivation. However, albumin levels are good indicators of chronic deficiency. Traditionally, albumin has been used to help in determining two important nutritional states. First, it helps identify chronic protein deficiency under conditions of adequate non-protein calorie intake, which leads to marked hypo-albuminemia. This may result from the net loss of albumin from both the intravascular and extravascular pools, causing kwashiorkor. Secondly, albumin concentrations may help define the condition called marasmus. This is caused by caloric insufficiency without protein insufficiency, so that the serum albumin level remains normal but there is considerable loss of body weight.

Studies have classified various levels of malnutrition by using albumin levels. Serum albumin levels of 35 g/L or greater are considered normal.[4] Albumin levels of 28 to 35 g/L indicate mild malnutrition, 21 to 27 g/L indicate moderate malnutrition, and <21 g/L indicate severely depleted levels of albumin.[145] Serum albumin is an accurate marker of the catabolic stress of infection.[62,121] This finding in the pediatric population is in agreement with studies in the adult population. Levels of albumin have been interpreted in various ways, including a level of ≤32 g/L as an indicator that a patient has a 75% chance of developing decubitus ulcers if he or she is in the hospital up to 10 days. Serum albumin levels <25 g/L can be used as an accurate measure of predicting survival prognosis in 90% of critically ill patients.[6]

Transferrin

Transferrin is a glycoprotein with a molecular weight of approximately 76,000. Its biological half-life is 9 days, which is shorter than that of albumin. It is synthesized in the liver and binds and transports ferric iron. Ninety-nine percent of iron in the serum is bound to about one-third of the transferrin pool. Transferrin synthesis is regulated by iron stores. When hepatocyte iron is absent or low, transferrin levels rise in proportion to the deficiency. It is an early indicator of iron deficiency, and the elevated transferrin is the last analyte to return to normal when iron deficiency is corrected.[77] Because the body needs iron for cell maintenance, growth, and multiplication, membranes of cells contain transferrin receptors. The number of receptors is related to the iron needs of the cells. Transferrin molecules are captured by the plasma membrane, drawn inside the cell by an active transport mechanism, acidified, and stripped of iron. The transferrin molecule then passes to the outside of the cell to con-

tinue its iron transport function. The iron in the serum represents the iron received by transferrin from the reticuloendothelial cells, intestinal mucosa, and hepatocytes, with the exception of the iron delivered to the body cells.[77,87]

The half-life of transferrin is one-half that of albumin, and the body pool is smaller than that of albumin, so transferrin is more likely to indicate protein depletion before serum albumin concentration changes. However, the usefulness of transferrin in diagnosis of subclinical, marginal, or moderate malnutrition is questionable, because a wide range of values have been reported in various studies. Conditional status can alter transferrin levels.[76]

Birth control pills, pregnancy, iron deficiency, and acute hepatitis are associated with increased levels of transferrin. Protein losing states and end-stage liver disease are to known to decrease transferrin. Large fluid shifts and postoperative severe stress in patients with complications have effects on the level of transferrin that have not been clearly defined. Transferrin levels can be lowered by factors other than protein or energy deficiency, such as nephrotic syndrome, liver disorders, anemia, and neoplastic disease.[122]

In hospital and nursing home settings, transferrin levels have been used as indices of morbidity and mortality.[107,117] Important levels of transferrin have been established for assessing nutritional status, as shown in Table 30-5.

However, information shows that serum transferrin concentrations are not sufficiently sensitive to detect a change in nutritional status that occurs after 2 weeks of total parenteral nutrition. In malnourished patients whose nutritional status improved significantly during 2 weeks of total parenteral nutrition, serum transferrin concentration did not reflect this nutritional improvement.[55]

In addition to being responsive to serum iron concentrations, transferrin is uniquely sensitive to some antibiotics and fungicides. Transferrin levels are altered when patients are taking high doses of aminoglycosides, tetracycline, and some cephlosporins.[123] Transferrin is unreliable as a nutritional marker when the patient has these medications prescribed. When the levels of albumin and transthyretin (prealbumin) remain unchanged, the transferrin levels fall rapidly in patients on these medications. Albumin and transthyretin remain normal, indicating an unchanged nutritional status, and the patient's weight remains constant during this time.

TABLE 30-5. Transferrin Levels Used in Assessing Nutritional Status

Transferrin Concentrations	Indication
<100 mg/dL	Severe visceral protein deficiency
100–150 mg/dL	Moderate protein depletion
150–200 mg/dL	Mild nutritional deficiency
>200 mg/dL	Adequate nutritional state

Transthyretin

Transthyretin is sometimes called prealbumin. Prealbumin is so named because it migrates ahead of albumin in the customary electrophoresis of serum or plasma proteins. Transthyretin is a stable and symmetrical tetramer composed of four identical subunits; it has a molecular weight of 54,000 daltons. In normal situations, each transthyretin subunit contains one binding site for retinol binding protein. Transthyretin and retinal binding protein are considered the major transport proteins for thyroxine and vitamin A, respectively.

The transthyretin and retinol-binding protein (RBP) complex circulates in a 1:1 molar ratio. This complex is very sensitive to ionic strength with major binding occurring at physiological pH.[137] Transthyretin has a high concentration of the amino acid tryptophane, and tryptophane plays a major role in the initiation of protein synthesis. Transthyretin has one of the highest proportions of essential to nonessential amino acids of any protein in the body.[60] Because of its high tryptophane content, short half-life, high proportion of essential to nonessential amino acids, and small body pool, transthyretin is a better indicator of visceral protein status and positive nitrogen balance than are albumin and transferrin.[140] Transthyretin and other carrier proteins are better indicators for monitoring short-term effects of nutritional therapy. The concentration of transthyretin and RBP complex is greatly decreased in protein energy malnutrition and returns toward normal values after nutritional replenishment. Transthyretin has a low pool concentration in the serum, a half-life of 2 days, and a rapid response to low energy intake, even when protein intake is inadequate for as few as 4 days.[10] Serum transthyretin concentrations are decreased postoperatively by 50 to 90 mg/L in the first week and have the ability to double in a week or at least increase 40 to 50 mg/L in response to adequate nutritional support. If the transthyretin response increases less than 20 mg/L in one week as an outcome measure, this indicates either an inadequate nutritional support or an inadequate response.[74,92]

When transthyretin decreases to levels <80 mg/L, there is development of severe protein calorie malnutrition; however, nutritional support can cause a daily increase of transthyretin of up to 10 mg/L.[144] These concentrations do not appear to be significantly influenced by fluctuations in the hydration state. Although end-stage liver disease appears to affect all protein levels in the body, liver disease does not affect transthyretin as early or to the same extent as it affects other serum protein markers, particularly retinal binding protein. Although transthyretin levels may be elevated in patients with renal disease, if a trend in the direction of change is noted, the changes are likely to reflect alteration in nutritional status and nitrogen balance. Steroids can cause a slight elevation in transthyretin, but the nutritional trend can still be followed because transthyretin will respond to both overfeeding and underfeeding.[143]

Serum transthyretin concentration above 110 mg/L is an entropy value that has to be obtained for any patient being transitioned from parenteral to enteral or oral feeding. A new usage of serum transthyretin is its ability to be a predictor or indicator of the adequacy of a nutritional feeding plan.[144] Transthyretin concentration in serum increases or decreases in relationship to the severity or the absence of energy defects. Serum concentrations of this protein increase when more than 60% of the assessed protein and energy needs are met. When the concentrations do not increase or remain below 110 mg/L and nutritional support is being provided, the amount and composition of nutrients, the mode of feeding, and the presence of disease in the attainment of nutritional goals should be examined. Changes in plasma protein have also been correlated with nitrogen balance. Transthyretin concentrations have been shown to increase significantly during a week of intensive tube feeding when the patient was in positive nitrogen balance.[152] Serum albumin concentrations, because of the long half-life of albumin and its regulation by osmotic conditions in the body, do not change during this same period. Transthyretin concentrations increase in patients with positive nitrogen balance and decrease in patients with negative nitrogen balance. When the transthyretin level is at 180 mg/L or more, this correlates with a positive nitrogen balance and indicates a return to adequate nutritional status. Transthyretin has been shown in both the pediatric and neonate population to be a highly accurate and relatively inexpensive marker for nutritional status (Fig. 30-4).[56,58] It is also accurate in high-risk, sick infants. Transthyretin has been found to be the most sensitive and helpful indicator when looking at the nutritional status of very ill patients.[106] Studies have shown that transthyretin concentrations reflect improved changes in body composition by increasing twofold when patients are converted from negative nitrogen balance to positive nitrogen balance.[17]

Transthyretin and retinol binding protein have been used to monitor nutritional therapy during the transition from total parenteral nutrition to oral or enteral feeding.[84] After malnourished patients have 1 week of adequate oral or enteral feeding, only 36% of the patients have normal values of serum albumin, 78% have normal values of retinol binding protein, 80% have normal values of transferrin, and 98% have normal values of transthyretin.[159] This shows that transthyretin is the quickest marker to indicate adequate nutrition delivered by parenteral nutrition, and this improvement occurred during the transition from a week of total parenteral nutritional to oral or enteral feeding.[93] A rise in transthyretin over a 1-week period has been documented to have a positive predictive value of 93% and this has been correlated to the development of a positive nitrogen balance.[30] In summary, transthyretin effectively demonstrates an anabolic response to feeding and is a good marker for this visceral protein synthesis in patients receiving metabolic or nutritional support.[150]

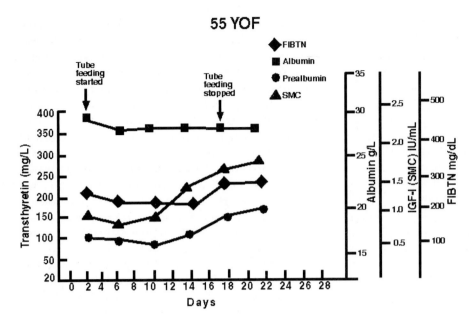

Figure 30-4. Sequential measurements of transthyretin, albumin, insulin-like growth factor-I (somatomedin C, SMC), and fibronectin (FIBTN) in a 55-year-old female with poor nutritional status. Short half-life proteins: transthyretin and SMC.

Retinol Binding Protein

Retinol binding protein is composed of 181 amino acid residues and has a molecular mass of 21,000 daltons. It has been used in monitoring short-term changes in nutritional status.[25] Its usefulness as a metabolic marker is based on its biological half-life of 12 hours and its small body pool size. Retinol binding protein reportedly responds quickly to both energy and protein deprivation and has been correlated with nitrogen balance in severe burn patients.[24,75]

As a single polypeptide chain, retinol binding protein interacts strongly with plasma transthyretin and circulates in the plasma as a 1:1 mol/L transthyretin-retinol binding protein complex (TRBPC). The interaction between retinol binding protein and transthyretin is sensitive to ionic strength, with the maximum binding occurring near a physiological pH. It has a binding site for one molecule of retinol, which is the alcohol form of vitamin A.[139] Vitamin A is metabolized from liver stores and transported in the plasma in the alcohol form bound to retinol binding protein. The binding of retinol binding protein appears to be stabilized by the formation of the TRBPC. No interdependence exists for the binding of thyroxin as a stabilizer for the TRBPC. Only the free form of the retinol binding protein, which has no affinity for transthyretin, undergoes glomerular filtration unhindered as a result of its low molecular weight.[140] It is then reabsorbed by the tubule cells and catabolized there. This may explain the extremely elevated serum level of retinol binding protein in advanced chronic renal insufficiency. A potential problems exists in using retinol binding protein as a nutritional marker. Although retinol binding protein has a shorter half-life than transthyretin (12 hours compared with 2 days), retinol binding protein is excreted in urine and its concentration increases more significantly than transthyretin in patients with renal failure. In chronic renal failure, the plasma concentrations of retinol binding protein are elevated because of its decreased catabolism by the kidney. In contrast to retinol binding protein, the catabolism of transthyretin occurs only to a limited extent in the kidney, which accounts for only moderately elevated concentrations being measured in advanced chronic renal insufficiency. In a patient with chronic renal failure, the marked increase in concentration of retinol binding protein may make it unreliable in monitoring nutritional status.[142] The initial high levels of retinol binding protein and the moderately elevated transthyretin levels in patients with acute renal failure may be difficult to interpret with respect to nutritional status.[57] In patients with relatively stable renal failure, the rate of change of the concentration of retinol binding protein and transthyretin, rather than an isolated value, can be better used to monitor the adequacy of nutritional therapy. The kidneys may play a major role in homeostasis, influencing the release of retinol binding protein from the liver into circulation. In end-stage liver disease, the levels of retinol binding protein and vitamin A are decreased in relationship to the severity of the disease.[138] In early and mid-stage liver disease, patients who had a drug-induced hepatitis had retinol binding protein levels that did not parallel prealbumin levels.[132] The retinol binding protein levels tend to increase for a period of time and then decrease in an unpredictable manner. Several physiologic changes may explain this initial increase, then decrease. The liver has a very high store of retinol binding protein, and in early liver damage and disease these stores may be released. Retinol binding protein levels in serum are principally controlled by the need to maintain the visual cycle, and in acute liver disease these levels may be increased initially.[137,139]

Somatomedin C

Somatomedin C (SMC) is also known as insulin growth factor I. It is a 7500 dalton single-chain peptide that plays an important role in the stimulation of biological growth. The molecular size and structure of SMC is similar to pro-insulin.[9] SMC serum concentrations are regulated by both growth hormone and nutritional intake. Growth hormone stimulates liver and other tissues to produce SMC. SMC is bound in the serum to specific proteins. These binding proteins are responsible for maintaining a relatively constant concentration of SMC in the serum. These proteins also help to minimize any rapid decreases in this short half-life protein. The binding proteins modulate the biological effect of SMC under certain circumstances causing both decrease and increase in its biological activity.[65]

Most of SMC in the serum is bound to a protein of 150,000 daltons and it also binds very rapidly to a protein with a molecular weight of 50,000 daltons. More than 99% of SMC circulates in the bound form. This complex dissociates completely at a low pH, yielding free SMC. The carrier protein has acid-labile and acid-stable components. The free SMC concentration is comparable to that of other peptide hormones, but the concentration of the bound SMC is several magnitudes greater than other hormones produced in the liver. SMC has been shown to be eliminated from the serum more slowly than other hormones, and this probably reflects the nature of the carrier protein binding. The similarity between SMC and pro-insulin has interesting evolutionary implications. These peptides may have risen from a single ancestral gene whose product may have had combined metabolic and growth regulatory activities. The two main biological effects may have diverged following gene duplication.[8] Insulin and SMC have specific cell receptors but do cross react at the receptor sites to varying degrees. The appropriate ligand, however, is more effective in binding to its own receptor.

SMC is thought to mediate the many effects of growth hormone and is not subject to diurnal variation or acute influences of stress, sleep, exercise, or changes in circulating levels of nutrients. For these reasons, SMC may be a more sensitive index of rapid changes in nitrogen balance than albumin, transferrin, and possibly transthyretin, all of which have a longer half-life. The nutritional intake and plasma concentration of SMC appear tightly controlled by dietary uptake. An important property of SMC that makes it a good nutritional marker is the 2 to 4 hour half-life of the bound form.[114]

SMC is produced by many types of cells, but most circulating SMC is of hepatic origin. Although the concentration of SMC in circulation can be estimated by RIA on nonextracted plasma or serum samples using high-affinity antibodies, the binding proteins interfere with measurements in these direct assays. In healthy patients, the binding proteins have less effect on the calculated values if these values are calculated against a serum pool standard that has been treated in a manner similar to that of the unknown sample.

Assays to measure insulin-like growth factor I must have the capability of performing some method to remove binding proteins that might interfere with the assay. These extraction procedures have consisted of either acid gel chromatography, acid ethanol extraction, or chromatography of samples on octadecasalicylic acid columns. The acid-gel chromatography involves multiple scientific apparatus to separate binding protein from insulin-like growth factor I, and it is difficult to standardize. Acid ethanol extraction of binding protein is simple to perform and usually is reproducible from laboratory to laboratory. In acid column chromatography, the conditions of extraction must be closely controlled and the correction for losses must be accurate or this extraction method can give highly variable results.[115] In some patients, the binding proteins can interfere to a variable degree with the quantitation of SMC in serum. For this reason, the measurement of SMC after extraction is recommended.[28,78]

SMC has strong metabolic properties. Interestingly, SMC correlates better with activity in acromegaly than does the level of growth hormone itself. Circulating SMC levels are regulated by nutritional status. These levels are particularly altered by the protein content of the diet. SMC levels fall during protein starvation and rise rapidly upon re-feeding.[31] This relationship of SMC level to malnutrition has been thoroughly investigated in undernourished children who fail to grow.[151] Low circulating levels of SMC in children rise rapidly within 1 day of re-feeding. SMC levels in healthy humans respond rapidly to dietary changes in protein and calorie intake, and SMC levels are much more reliable than albumin, transferrin, and lymphocyte count in documenting a malnourished patient's replenishment of nutritional status.[67,151] SMC levels can be variable in patients with liver disease, kidney failure, and other autoimmune diseases unless the carrier proteins are completely removed during the chromatography or extraction procedure in the assay. SMC remains the most reliable nutritional marker in children. Evidence of this is that children who undergo severe protein catabolism due to trauma, infection, and sepsis have SMC levels that can decrease dramatically, indicating a poor prognosis (Fig. 30-5).[86]

Fibronectin

Fibronectin has a molecular weight of approximately 250,000 daltons and is composed of subunits. It is an opsonic glycoprotein that has a half-life of approximately 15 hours in humans. This protein has been classified as a paradigm of a modular protein. It has been shown to have repeating blocks of a homogeneous sequence of amino acids. It is an alpha-2 glycoprotein that serves important roles in cell-to-cell adherence and tissue differentiation, wound healing, microvascular integrity, and opsonization of particulate matter. It is considered the major protein regulating phagocytosis. Sites of synthesis include endothelial cells, peritoneal macrophages, fibroblasts, and the liver. Fibronectin concentrations may decrease after physiological damage caused by severe shock, burns, or infection. The fibronectin level returns to normal on recovery. Fibronectin is useful in nutritional panels because it is one of the few nutritional markers that is not exclusively synthesized in the liver.[105] Fibronectin showed response to both caloric deprivation

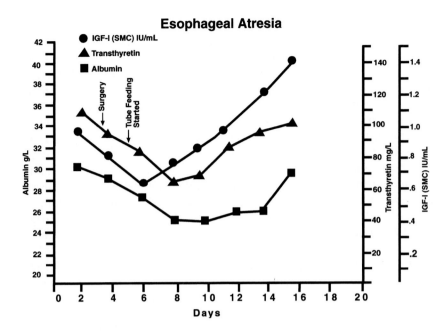

Esophageal Atresia

● IGF-I (SMC) IU/mL
▲ Transthyretin
■ Albumin

◄

Figure 30-5. Sequential protein markers on a 4-week-old that had a surgical correction for a congenital birth defect, partial esophageal atresia. The protein markers utilized were transthyretin or prealbumin, albumin, and SMC. Note that upon feeding on day 8, SMC increased its concentration and showed nutritional repletion much sooner than did albumin or transthyretin. SMC is a very sensitive protein nutritional marker in the pediatric population.

and repletion in healthy obese subjects who fasted 3 weeks and then were re-fed.[96] Concentrations decreased to a low consistent value after 1 week of starvation and stayed low for the remaining 2 weeks. It returned to normal 5 days after re-feeding.[27] Albumin remained essentially the same throughout the 3 weeks, and transferrin took considerably longer to return to prestarvation level than did fibronectin. In another study, fibronectin significantly declined in normal men after 2 to 3 days of starvation followed by an increase in concentration within 5 days of refeeding.[129]

Fibronectin levels have been shown to decrease during infection or severe stress due, in part, to its opsonic property. The infection or trauma, however, does not significantly decrease fibronectin concentration.[15] Fibronectin does not decrease during mild fever and only increases in severe infection where it remains flat and does not increase as soon as SMC does. There are two factors that potentially can diminish fibronectin's usefulness as a nutritional marker. First, heavy glucose composition in a feeding formula has been shown to decrease fibronectin concentration.[126] Also, bacterial infections, such as *Staphylococcus aureus,* which may cause tissue degradation at the site of the injury, can inhibit the concentration of fibronectin by reducing its ability to bind to actin, fibrin, collagen, and gelatin.[72] Fibronectin is a reliable index in monitoring short-term nutritional support in infants with protein malnutrition.[161] In recent nutritional therapy, severely malnourished children have been treated with intravenous fibronectin as an adjunct nutritional treatment. The fibronectin apparently helped in the immune competence of these patients, because they had increased rates of survival over children who were not treated with the fibronectin.[127]

Nitrogen Balance

Another nutritional evaluation tool, *nitrogen balance*, is the difference between nitrogen intake and nitrogen excretion. It is one of the most widely used indicators of protein

change. In the healthy population, anabolic and catabolic rates are in equilibrium and the nitrogen balance approaches zero. During stress, trauma, or burns, the nutritional intake decreases and the nitrogen loss may exceed intake, leading to a negative nitrogen balance. During recovery from illness, the nitrogen balance should become positive with nutritional support. In humans, 90% to 95% of the daily nitrogen loss is accounted for by elimination through the kidneys. About 90% of this loss is in the form of urea. Therefore, the determination of 24-hour urinary urea nitrogen (UUN) is a method for estimating the amount of nitrogen excretion. The nitrogen balance is calculated as follows.[141]

$$\text{Nitrogen balance} = \frac{24\text{-h protein intake (g)}}{6.25}$$

$$- \frac{24\text{-h UUN} + 4}{\text{Total volume (L)}}$$

(Eq. 30-1)

Protein intake includes grams of protein that are provided either by intravenous (IV) amino acids or by enteral feeding. Protein intake is converted into grams of nitrogen by dividing by 6.25. The factor of 4 in the equation represents an estimation of non-urinary nitrogen loss (*e.g.,* from skin, feces, hair, nails, and other non-urea nitrogen losses).[141] Nitrogen balance, as calculated by this equation, is not valid in patients with severe stress or sepsis, as can be seen in critical care areas or in patients with renal disease. Determining the validity of this equation in other clinical conditions involving normally high nitrogen losses may be difficult or even incorrect.

C-Reactive Protein

C-reactive protein (CRP) is an acute phase protein that increases dramatically under conditions of sepsis, inflammation, and infection. CRP can increase dramatically up to

1000 times after tissue injury, which is more than two or three orders of magnitude greater than any other acute phase reactant. C-reactive protein rises in concentration 4 to 6 hours before other acute phase reactants begin to rise.[39,71]

A generally accepted concept now is that major trauma or critical illness can be a direct cause of problematic malnutrition. From this knowledge has come the current medical practice of selective nutritional support. Immediately after the onset of trauma, sepsis, or critical illness, there is a breakdown of cells with the loss of intracellular ions and nitrogen.[158] Critical care physicians now characterize the injured patient with early shock as in the *ebb phase* followed by the *catabolic flow phase*. The flow phase is a period of marked catabolism that presents itself clinically with tachycardia, fever, increased respiratory rate, and increased cardiac output.[26] During this time, synthesis rates of C-reactive protein and other acute phase proteins increase and albumin and prealbumin decrease. Even with this increase in acute phase proteins, a significant negative nitrogen balance usually occurs secondary to the greater protein catabolism. Whether this catabolic state produces a clinically defined malnutrition or a separate entity is unknown, but it certainly produces weight loss and decreases albumin and prealbumin levels. In events of limited time sequence, such as major non-fatal trauma or controlled sepsis, the flow phase generally lasts 3 to 7 days and gradually develops into an anabolic phase. The anabolic phase is characterized by a decrease in the clinical signs of hypermetabolism. Body temperature and respiratory and heart rates decrease toward normal levels.[155] The synthesis of acute phase reactants, such as C-reactive protein, shows a decreased rate. At this point, the short half-life proteins such as somatomedin C, prealbumin, and fibronectin begin to be synthesized from available substrate and become useful as nutritional markers. With this understanding, a need arises for first monitoring the C-reactive protein in these 3 to 7 days of sepsis and major trauma until it begins to decrease. At this time, monitoring the shorter half-life proteins will help define the nutritional status of the patient.[46] An example of a condition known as acute phase reaction is shown in Figure 30-6.

Interleukins

Recently, nutrition research has focused on the interleukins, which are a complex group of proteins and glycoproteins that can exert pleiotropic effects on a number of different target cells. Most of the interleukins are produced by macrophages and T lymphocytes, in response to antigenic or mitogenic stimulation, and affect primary T lymphocyte function. Most of the nutritional investigations have been performed on interleukin-1 (IL-1), which is initially formed as a 31,000-d molecule, then processed to a smaller 17,500-d molecule. IL-1 is the focal point in T lymphocyte mitogenesis and provides a specific signal to a T lymphocyte cell subpopulation that produces IL-2.[157] IL-1 is therefore directly related to the availability of IL-2. IL-1 is present on activated macrophages and is able to activate thymocytes by cell-to-cell contact. IL-1 is chemotactic to neutrophils, monocytes, and T lymphocytes.[41] IL-1 is involved in the induction of fever, slow wave sleep, bone resorption, and muscle proteolysis. Recent research has shown that interleukins need adequate amounts of fat-soluble vitamins and pyridoxin to exert adequate biologic effects.[100,109]

It has been postulated that if IL-1 and IL-2 are present in adequate amounts, then the fat-soluble and B complex vitamins have an adequate concentration in the body.[22] This is an interesting hypothesis that has yet to be proven. IL-1 has a half-life of 2 to 4 hours and has a small body pool that makes it a good nutritional marker. Currently, however, the cost of performing the assay for IL-1 is extremely high and is not practical as a clinical test for nutritional assessment.

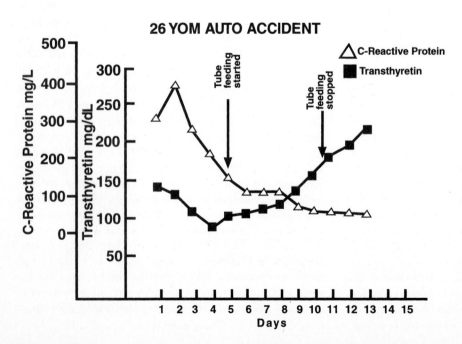

Figure 30-6. Sequential levels of C-reactive protein and transthyretin were obtained on a 26-year-old man who was involved in an automobile accident. This pattern shows an initial elevation of C-reactive protein, which is increasing because of the inflammatory response due to the automobile accident. As it begins to decrease, transthyretin, an inverse acute phase reactant, decreases. At day 5, acute phase response is over and transthyretin becomes a nutritional marker. At this time, tube feeding is started because the patient shows both biochemical and physical signs of malnutrition. Tube feeding is stopped once the transthyretin reaches a level of 180 mg/L. This illustrates the use of transthyretin in an acute phase response situation.

TOTAL PARENTERAL NUTRITION (TPN)

Total parenteral nutrition (TPN) is a widely used means of intense nutritional support for patients who are malnourished, or in danger of becoming malnourished, because they are unable to consume required nutrients. Those requiring TPN include: (1) presurgical patients in a poor nutritional state who require repletion of nutrients to prevent or limit postsurgical complications (*e.g.,* impaired wound healing and infection); (2) patients with injuries that result in impaired mental and physical capacities (trauma or sepsis) who may require TPN to maintain adequate nutrition when basal metabolic rates are greatly increased due to a hypercatabolic state[1]; and (3) patients with chronic gastrointestinal (GI) disease (*e.g.,* cancer, short bowel syndrome, or surgical procedures, including bypass or resection) who are often candidates for long-term therapy.[21]

Parenteral nutrition therapy involves not only administering appropriate amounts of carbohydrate, amino acid, and lipid solutions, but also of electrolytes, vitamins, minerals, and trace elements to meet the caloric, protein, and nutrient requirements while maintaining water and electrolyte balance.[42]

Parenteral nutritional preparations usually are administered by enteral or parenteral routes. Enterally, a nasogastric tube is placed to deliver nutrients directly into the stomach or duodenum, or patients having GI surgery may have a feeding gastrostomy or jejunostomy catheter put in place during the surgical procedure. Feeding solutions are then administered through the catheter. The parenteral route uses an IV catheter (most often a Hickman catheter) that is threaded through the right subclavian vein to the superior vena cava and positioned above the right atrium.[113,102] This is shown in Figure 30-7.

Since TPN administration bypasses the normal absorption and circulation routes, careful laboratory monitoring of these patients is critical. By whatever routes nutrients are administered, the goal is to provide optimal nutritional status. Often this requires administering high levels of nutrients during short periods of time.

When a patient is considered for TPN, accurate dietary and medical histories are needed to ensure adequate nutritional support. Anthropometric information is used to assess the general nutritional status of the patient. These determinations include arm circumference and triceps skinfold measurement to evaluate muscle mass and fat stores. The patient's weight is used as a general indicator of possible disease or malnutrition. An unintended weight loss of >10% to 12% leads one to suspect either disease or malnutrition. The patient's height, age, and activity level also are considered.[73]

An important part of the initial TPN work-up involves acquiring laboratory measurements that provide information relating to the patient's initial protein and lipid status, electrolyte balance, and renal and liver function. From this information, a TPN solution is prepared to meet specific caloric (lipid or glucose) and nutritional (protein) needs, which may change weekly or even daily. It is important to monitor the TPN patient to avoid possible complications.

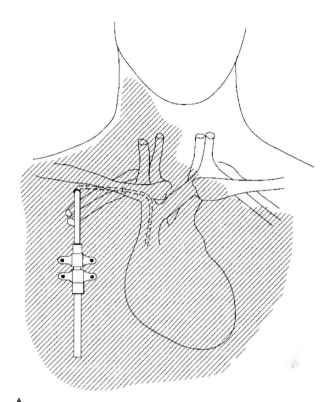

▲
Figure 30-7. Insertion path of catheter for total parenteral nutrition.

Such laboratory monitoring provides necessary information to properly administer TPN therapy. Suggested guidelines for frequency of monitoring of metabolic variables are shown in Table 30-6.

Urine Testing

Many patients receiving TPN have normal kidney function, so laboratory monitoring of urine composition provides valuable information. Routine orders normally include volume, specific gravity or osmolarity, and an estimate of glycosuria every 8 hours. Glycosuria should be minimal. In very small premature infants, glycosuria during the early phase of parenteral nutrition is a signal that glucose infusion is too rapid. If glucose appears in the urine of small infants after glucose tolerance has been established, however, one should question the presence of respiratory disease, sepsis, or cardiovascular changes.[162]

Suggestions have been made to monitor sodium concentrations in spot urine specimens in conjunction with changes in body weight. Normal urine sodium content varies from 20 to 50 mmol/L in adults and from 5 to 10 mmol/L in newborns and infants. Unfortunately, this wide variation requires that each patient serve as his own control. Changes in urine sodium may indicate excessive or deficient infusion. In sodium deficiency, extracellular volume decreases and the renin-angiotensin-aldosterone mechanism is stimulated for renal conservation of sodium.[16] Urinary sodium concentration thus falls below 10 mmol/L and the patient loses weight. However, if a decrease in urinary sodium is accom-

TABLE 30-6. Guidelines for Test Monitoring Frequency in Patients on TPN

Baseline Studies

CBC, glucose
Sodium, potassium, chloride, CO_2
Bilirubin, aspartate aminotransferase, alkaline phosphatase
BUN, creatinine, ammonia
Calcium (or ionized calcium), phosphorus, magnesium
Thyroid-binding prealbumin (transthyretin)
Cholesterol, triglycerides
Urinalysis, serum acid-base status
Urinary urea nitrogen (UUN)
Zinc, copper, iron (optional)

Measurements Obtained Twice Weekly

Glucose, BUN, creatinine
Sodium, potassium, chloride, CO_2
Thyroid-binding prealbumin (transthyretin)
Calcium (or ionized calcium), phosphorus, magnesium
Bilirubin, aspartate aminotransferase, alkaline phosphatase

Results Obtained Every Week

CBC, urinalysis, serum acid-base status
Zinc, copper, iron, magnesium
Cholesterol, triglycerides (if infusing lipids)
UUN
Ammonia, creatinine

Measurements Obtained Every Two Weeks

Zinc, copper, iron, folate, vitamin B_{12}
In unstable clinical conditions, laboratory studies may be
    required several times a day

CBC = complete blood cell count; BUN = blood urea nitrogen; CO_2 = carbon dioxide; UUN = urinary urea nitrogen

panied by weight gain and polyuria, the reason may be excessive water intake. On the other hand, if urinary output is normal, fluid may be accumulating in other body spaces.

Tests to Monitor Electrolyte Disturbances

Sodium regulation during TPN is a problem in pediatric patients.[111] Daily sodium requirements may vary depending on renal maturity and the ability of the child's body to regulate sodium.[33] Factors that increase the amount of sodium necessary to maintain normal serum sodium concentrations in both pediatric and adult patients are: glycosuria, diuretic use, diarrhea or other excessive GI losses, and increased post–operative fluid losses.[85]

Hyperkalemia is a common problem in pediatric patients when blood is obtained by heel stick. The squeezing of the heel may cause red cell hemolysis resulting in a falsely high serum potassium. Although adequate nutrition may be supplied to promote anabolism, hypokalemia may develop as the extracellular supply is used for cell synthesis.

The primary function of chloride is osmotic regulation. Hyperchloremia metabolic acidosis is a problem when crystal amino acid solutions are used, but such acidosis can be prevented or treated by altering the amount of chloride salt

in the parenteral nutrition solution. Supplying some of the sodium and potassium requirements as acetate or phosphate salts can reduce the required amount of chloride.[116] The reformation of synthetic amino acid solutions by commercial manufacturers has helped to avoid this serious complication. If hyperchloremia metabolic acidosis does occur, treatment with sodium and potassium acetate solutions is used, since acetate is metabolized rapidly to bicarbonate. Acetate salts are compatible with all other common parenteral nutrition components and thus are ideal to use when acidosis is present. Sodium bicarbonate cannot be used in parenteral nutrition solutions containing calcium because calcium carbonate readily precipitates. The use of acetate not only increases serum bicarbonate but also decreases the amount of chloride delivered to the patient.[48]

Mineral Tests to Monitor

One of the most important aspects of TPN monitoring is determining deficiencies and excesses of calcium, phosphorus, and magnesium. When regulated inadequately, these minerals not only affect bone mass but can precipitate life-threatening situations. Calcium and phosphorus are related closely in the important role of bone mineralization. Calcium is present in serum in two forms—protein-bound, or nondiffusible calcium and ionized diffusible calcium. Ionized calcium is the physiologically active form and constitutes only 25% of total serum calcium. Regardless of the total serum calcium, a decrease in ionized calcium may result in tetany.[44] A decrease in ionized calcium often is caused by an increase in blood pH (alkalosis). It is important to monitor ionized serum calcium and blood pH, especially in a patient with TPN who is receiving calcium supplementation along with ingredients in the TPN solution that may alter blood pH.[35]

While calcium imbalance is frequent in newborns undergoing TPN, it is much less common in the adolescent and adult.[85] Hypercalciuria with nephrolithiasis has been reported, however, as a complication in patients on long-term TPN.[2,101] Gluconate is the recommended calcium salt for use in parenteral nutrition solutions, because this salt dissociates less than the chloride salt. This fact allows more calcium and phosphorus to remain soluble in the solution, thus making it possible to give more of these two minerals to newborns and children. Daily maintenance calcium should be given even when serum calcium is normal. The serum calcium may be normal or even high when hypophosphatemia is present. It also may be normal at the expense of bone demineralization.[149]

A reciprocal relationship exists between calcium and phosphorus. In cells, most phosphorus exists in the organic form, with a smaller amount found in serum as inorganic phosphate. Intracellular phosphate is necessary to promote protein synthesis and other cellular functions. Calcium and phosphorus must be monitored carefully to maintain the correct balance between these two minerals.

Severe hypophosphatemia has been reported in patients undergoing prolonged TPN. Hypophosphatemia also can cause respiratory failure. While hypophosphatemia can affect skeletal muscle and cardiac function, and may even be responsible for hemolysis, the most important function of phosphate is in the pulmonary generation of 2,3 diphosphoglycerate (2,3 DPG), which facilitates the unloading of oxygen from hemoglobin. When levels of 2,3 DPG are inadequate, oxygen is bound firmly to hemoglobin with decreased availability to peripheral tissues.[147] The combined effect of such decreased function may result in respiratory insufficiency, which is corrected readily by administering phosphate.

Patients with pulmonary problems related to hypophosphatemia are particularly susceptible to infusions with high glucose concentrations. Carbon dioxide production is linked directly to the amount of oxygen consumed, by the respiratory quotient. At a given level of oxygen consumption, the amount of CO_2 produced depends on the mixture of dietary carbohydrate and fat. In some situations, infusions of high concentrations of carbohydrate can increase the respiratory quotient to values > 1, resulting in an increased burden on the ventilatory system to excrete the increased CO_2. Oxygen consumption itself increases in patients receiving parenteral nutrition and also directly results in increased CO_2 production. Consequently, the standard glucose concentrations of TPN may contribute to respiratory insufficiency. Other consequences of acute hypophosphatemia include: myocardial dysfunction; depressed chemotactic, phagocytic, and bactericidal activity of granulocytes; central nervous system symptoms compatible with metabolic encephalopathy; metabolic acidosis; and osteomalacia.[90,148]

Phosphorus usually is given as a potassium or sodium phosphate salt. Serum phosphorus should be monitored frequently during phosphorus repletion. Calcium supplements may be required during phosphate administration, but administering concentrated calcium solutions with phosphate-containing solutions may result in precipitation.

Some electrolyte and mineral abnormalities associated with parenteral nutrition are shown in Table 30-7.

Magnesium, as a TPN solution additive, is related closely to calcium and phosphorus. A reciprocal relationship exists between magnesium and calcium and, in certain situations, between magnesium and phosphorus. Low levels of magnesium can cause tetany, while high levels can increase cardiac atrioventricular conduction time.[116]

Serum alkaline phosphatase has been used to detect bone demineralization, since the enzyme is elevated in many osteopathic conditions. Alkaline phosphatase also is increased in obstructive jaundice, parenchymal liver disease, hyperparathyroidism, and pregnancy, so it cannot be used reliably as an osteopathic indicator in TPN.

Trace Elements to Monitor

The diets of most patients on TPN must be supplemented to maintain optimal levels of several trace elements; these elements also must be monitored to prevent deficiency or toxicity. Copper and zinc are the most common trace elements added to TPN solutions.[134]

Pallor, decreased pigmentation, vein enlargement, and rashes resembling seborrheic dermatitis are the major clinical signs of copper deficiency, which at times go unnoticed.[133] Some other abnormalities include recurrent leukopenia (WBC $<5.0 \times 10^9/L$) and neutropenia (neutrophils $<1.5 \times 10^9/L$). Often, an iron-resistant anemia precedes the fall in the leukocyte count.[82,163] Radiologic findings of copper deficiency include: osteoporosis with flaring and thickening of the ribs, periosteal reactions, spur formation, and spontaneous fractures.[69]

The diagnosis of copper deficiency is confirmed when both serum copper and ceruloplasmin (the copper-binding glycoprotein) are low. It is difficult to make this diagnosis in premature infants, however, because their serum copper levels remain depressed until about 9 weeks of age. Low copper levels also have been reported in malabsorption syndrome, protein-wasting intestinal diseases, nephrotic syndrome, severe trauma, and burns.[133]

Patients on TPN may develop acute zinc deficiency.[61,79] They initially suffer from a massive urinary loss of zinc during phases of catabolism. When weight gain begins, the zinc-deficient patient may experience diarrhea, perioral dermatitis, and alopecia. A study[154] of adult patients receiving

TABLE 30-7. Electrolyte and Mineral Abnormalities Associated with TPN

Abnormality	Manifestations	Usual Causes
Hypernatremia	Edema, hypertension, thirst, intracranial hemorrhage	Inappropriate sodium intake in relation to water intake, especially with abnormal losses (diarrhea, diuretic use)
Hyponatremia	Weakness, hypotension, oliguria, tachycardia	Inadequate sodium intake relative to water intake
Hyperkalemia	Weakness, paresthesia, cardiac arrhythmias	Acidosis, renal failure, excessive potassium intake
Hypokalemia	Weakness, alkalosis, cardiac abnormalities	Insufficient potassium intake associated with protein anabolism
Hyperchloremia	Metabolic acidosis	Excessive chloride intake, amino acid solutions with high chloride content
Hypercalcemia	Renal failure, aberrant ossification	Inadequate phosphorus intake, excessive vitamin D intake
Hypocalcemia	Tetany, seizures, rickets, bone demineralization	Inadequate calcium, phosphorus, and/or vitamin D intake
Hypophosphatemia	Weakness, bone pain, bone demineralization	Insufficient phosphorus intake in relation to calcium intake
Hypomagnesemia	Seizure, neuritis	Inadequate intake of magnesium

TPN for GI disorders produced evidence that the appearance of acne and seborrheic lesions coincided first with decreases in both serum alkaline phosphatase and zinc levels. This was followed by perioral erythematous weeping rashes when the deficiency developed fully.

Premature newborns are particularly predisposed to zinc deficiency, since zinc normally is acquired at the rate of ~500 µg/day during the final month of gestation. To compensate for this deficiency, zinc supplements for the premature newborn should be 50% higher than for the full-term newborn. This concentration is then decreased gradually until it is the same as that for a full-term infant.[128]

Serum zinc and copper levels should be monitored weekly, or at least bimonthly. Zinc should be monitored even more frequently in patients with ongoing GI loss, even if they are receiving zinc supplements in their parenteral solutions. Monitoring zinc and copper levels in the premature newborn may be confusing because their levels can vary greatly. Information has been reported[94] on zinc and copper ranges in newborns of varying gestational ages that are being infused with different levels of these trace elements.

Chromium deficiency has been described in patients on long-term parenteral nutrition.[53] Initial signs and symptoms include weight loss, increased carbohydrate intolerance, and neuropathy.[81] The diagnosis is supported by low levels of serum chromium and by the clinical response to chromium administration. Evidence indicates that plasma chromium levels are reduced not only in deficiency, but also in acute illnesses.[110] They are elevated by increased insulin and glucose loads.[59] Molybdenum deficiency has been reported and symptoms include headache, night blindness, and lethargy leading to coma. Hypouricemia, hypouricosuria, increased serum methionine, and excessive urinary excretion of thiosulfite and sulfite indicate deficient activities of the molybdenum-containing enzymes, xanthine oxidase and sulfite oxidase, respectively. These clinical and laboratory abnormalities disappear after molybdenum supplementation.[1]

Selenium deficiency has been described in long-term TPN.[153] It has been associated with both cardiomyopathy and malabsorption. The cardiomyopathy can be severe and death has been reported.[47,136] More work is needed to establish its significance in TPN, however.

Use of Panel Testing

To regulate the metabolic state of critically ill patients on TPN, it is necessary to consistently monitor their nutritional therapy. The constant demand for monitoring these patients produces an ideal situation for using laboratory panels. By requesting these panels, a group of tests can be performed in the most cost-efficient manner. This allows the laboratory the option of test batching and automated analysis, which results in less specimen processing, more efficient chemical analysis, and therefore aids in cost containment. These panels should consist only of routine tests that are ordered repeatedly.

Two such panels designed for intensive care in newborns and adults on TPN are shown in Table 30-8. These panels should be ordered on a weekly basis, or more frequently if the need arises.

SUMMARY

Nutritional assessment has become increasingly important in medical care, especially in the treatment of critically ill patients and in the treatment of chronically emaciated patients. Unfortunately, identifying patients who are at risk of developing nutritionally related complications in the course of an illness is uncommon, and there are no standards for monitoring the effectiveness of feeding plans to replace nutritional losses.

Malnutrition is a state of decreased intake of calories or micronutrients (vitamins and trace elements) resulting in a risk of impaired physiologic function associated with increased morbidity and mortality.[99,107,108] It has been estimated that from 10 to 20 million individuals in the United States are malnourished.

There are several methods used for nutritional assessment. These include clinical evaluation, physical measurement (body weight and anthropometric measurements), biochemical or immunologic determinations, and sophisticated laboratory measurements.

Total parenteral nutrition (TPN) is a widely used means of intense nutritional support for patients who are malnourished, or in danger of becoming malnourished, because they are unable to consume required nutrients. Parenteral nutrition therapy involves the administration of the appropriate amounts of carbohydrate, amino acid, and lipid solutions, as well as electrolytes, vitamins, minerals, and trace elements. Parenteral nutrition therapy is usually administered by enteral or parenteral routes. However, because TPN administration bypasses the normal absorption and circulation routes, careful laboratory monitoring of these patients is critical.

TABLE 30-8. Panel Testing for Patients Receiving TPN

Adults	Newborns
CBC	CBC
Prothrombin time	Prothrombin time
Glucose	Glucose
Electrolytes	BUN
Calcium	Electrolytes
Magnesium	Calcium
Phosphate	Magnesium
Copper	Phosphate
Zinc	Aspirate aminotransferase
Transthyretin (prealbumin)	Bilirubin (total and direct)
	Alkaline phosphatase
	Triglycerides
	Cholesterol
	Transthyretin (prealbumin)

CASE STUDY
30-1

A 62-year-old farmer received a severe abdominal puncture wound in a farm-related accident. His medical history revealed long-term respiratory difficulties. Surgical repair of the wound required excision of a short segment of small intestine. Because of postsurgical complications he was placed on postoperative TPN. He received, via central vein, a TPN solution comprising 8.5% crystalline amino acid solution, 42% dextrose, plus various electrolytes, minerals, and vitamins. In addition, 500 mL of 10% lipid solution was administered every other day. Initial glucose, electrolyte, mineral, and liver function tests were normal. After 1 week the patient's profile showed an elevated glucose level (172 mg/dL [9.5 mmol/L]), normal electrolyte and mineral values, and a decreased phosphate level (2.1 mg/dL [0.70 mmol/L]; N = 2.5–4.0 mg/dL [0.80–1.30 mmol/L]). The patient developed respiratory problems and was placed on a respirator. After three weeks of continuous TPN, his serum enzyme and liver function tests are shown. On cessation of TPN, his enzyme and liver function returned to normal (Case Study Table 30-1).

CASE STUDY TABLE 30-1. Laboratory Test Data For Case Study 30-1

Tests	Results* (SI)	Reference Ranges* (SI)
ALP	260	30–115 U/L
SGOT (AST)	178	0–40 U/L
SGPT (ALT)	192	0–40 U/L
GGTP (GT)	132	0–65 U/L
LDH (LD)	175	60–200 U/L
CPK (CK)	172	0–225 U/L
Total bilirubin	1.7	0.2–1.0 mg/dL
	(30)	(4–18 µmol/L)

Results and reference ranges for enzymes are the same in conventional and SI units.

Questions

1. What is the relationship among the elevated serum glucose, low phosphate, and respiratory problems experienced after 1 week on TPN?
2. What are some of the common liver profile abnormalities that occur in patients on TPN?

CASE STUDY
30-2

At birth, a full-term infant (weight 3.1 kg) had excessive salivation and respiratory difficulty. Inability to pass a nasogastric tube led to a diagnosis of esophageal atresia with tracheoesophageal fistula. This diagnosis was confirmed by x-ray examinations that showed a significant amount of gas in the stomach and intestine and an upper blind esophageal pouch. The infant underwent surgical repair of these congenital abnormalities. On the sixth postoperative day, an x-ray study demonstrated a leak at the esophageal anastomosis with recurrence of the tracheoesophageal fistula. The baby then was started on central vein TPN. To monitor visceral protein synthesis, serum albumin and transthyretin measurements were performed. The serum values obtained for these two analytes are shown in Case Study Table 30-2.

Based on the transthyretin levels and an average weight gain of 25 g/day, the patient was weaned from IV nutritional therapy after 16 days to a regimen of completely oral feeding.

CASE STUDY TABLE 30-2. Serum Albumin and Transthyretin Measurements for Case Study 30-2

Month/Day	Albumin g/dL (SI)	Transthyretin mg/L (SI)
10/8	2.7 (27 g/L)	90 (0.09 g/L)
10/10	TPN started	
10/12	2.5 (25 g/L)	80 (0.08 g/L)
10/16	2.6 (26 g/L)	100 (0.10 g/L)
10/20	2.6 (26 g/L)	130 (0.13 g/L)
20/24	2.8 (28 g/L)	180 (0.18 g/L)
Reference Intervals:	2.8–5.2 g/dL (28–52 g/L)	120–250 mg/L (0.12–0.25 g/L)

Questions

1. What is the half-life of transthyretin and of albumin?
2. Why is transthyretin a more sensitive indicator of visceral protein synthesis than albumin?

CASE STUDY
30-3

The following is a simple cost-analysis procedure involving malnutrition and complications that result from not identifying malnutrition. The total hospital cost for uncomplicated elective GI surgery is $4,500 including hospital and laboratory tests. If complications related to the poor nutritional status of the patient occur, the typical cost goes up dramatically to $15,000. The additional cost of $10,500 comprises antibiotics ($1,500), total parenteral nutrition ($1,500), reoperation ($4,000), laboratory charges ($500), and the remainder is non-ICU care.

The laboratory panel of tests to identify malnutrition consists of albumin and transthyretin at a total cost of $50. This hospital also offers a panel of questions and anthropometric measurement for $100. This laboratory panel clearly identifies 50% of those at risk for the malnutrition. This hospital panel of questions and anthropometric measurements is one-half as effective as the laboratory panel of tests for identifying malnutrition.

The hospital has 30 uncomplicated elective GI surgeries a year. Epidemiological studies show that 20% of this hospital population is at risk for malnutrition.

Questions

1. What is the cost of the non-ICU care?
2. What percentage of the additional cost of uncomplicated elective surgery of $10,500 is the laboratory charges?
3. How much would be saved in this hospital by identifying malnutrition using the laboratory test panel before the uncomplicated elective surgery was performed. In these patients, would nutritional therapy be undertaken before surgery was performed?
4. How many more malnourished patients would be identified using laboratory tests over questions and anthropometrics?

CASE STUDY
30-4

A 72-year-old man underwent a surgical resection of his lower intestine. He was maintained on peripheral parenteral nutrition postoperatively. At day 12, he developed nausea and vomiting associated with a partial iliojejunal obstruction, which was surgically corrected and total parenteral nutrition was started. He resumed oral feeding at 30 days and was discharged soon after this. The albumin level was essentially unchanged over the entire stay. In contrast, the prealbumin reflects the course of treatment.

The initial value on day 1 reflects the acute phase reaction to surgery. The albumin and transthyrectin on days 5 and 7 demonstrate a recovery from the acute phase and an adequate nutritional status. The values on days 13, 14, and 15 reflect the compromised nutrition associated with the partial blockage and related stress. His recovery is reflected in the increased values of days 16 and 26 (see Case Study Table 30-3).

CASE STUDY TABLE 30-3. Serum Albumin and Transthyretin Measurements for Case Study 30-4

Month/Day	Albumin g/dL(SI)	Transthyretin mg/L
5/1	2.7 (27 g/L)	82 (0.082)
5/5	2.5 (25 g/L)	150 (0.15)
5/7	2.7 (27 g/L)	182 (0.182)
5/12	Surgery	Surgery
5/13	2.5 (25 g/L)	122 (0.122)
5/14	2.7 (27 g/L)	112 (0.112)
5/15	2.5 (25 g/L)	136 (0.136)
5/16	2.5 (25 g/L)	161 (0.161)
5/26	2.7 (27 g/L)	205 (0.205)

Questions

1. If transthyretin is correctly reflecting the course of the treatment, explain why days 13 and 14 levels are lower than days 5 and 7.
2. Why did the transthyretin continually increase on days 5 and 7?

REFERENCES

1. Abumrad NN, Schneider AJ, Steel DR, et al. Acquired molybdenum deficiency. Clin Res 1979;27:774A.

2. Adelman RD, Abern SB, Merten D. Hypercalciuria with nephrolithiasis: A complication of total parenteral nutrition. Pediatrics 1977;59:473.

3. Almond DJ, Burkinshaw L, Laughland A, McMahon M. Potassium depletion in surgical patients: Intracellular cation deficiency is independent of loss of body protein. Clin Nutr 1987;6:45.

4. Anderson CF, Wochos DN. The utility of serum albumin values in the nutritional assessment of hospitalized patients. Mayo Clin Proc 1982;57:181.

5. Antonas KN, Curtas S, Meguid MM. Use of serum CPK-MM to monitor response to nutritional intervention in catabolic surgical patients. J Surg Res 1987;42:219.

6. Apelgren KN, Rombeau JL, Twomey PL, et al. Comparison of nutritional indices and outcome in critically ill patients. Crit Care Med 1982;10:305.

7. Baker JP, Detsky AS, Wesson DE, at al. Nutritional assessment: A comparison of clinical judgment and objective measurements. N Engl J Med 1982;306:969.

8. Ballard J, Baxter R, Binoux M, et al. On the nomenclature of IgF binding proteins. Acta Endocrinologica 1989;121:751.

9. Baxter RC. The somatomedins: Insulin-like growth factors. Adv Clin Chem 1986;25:49.

10. Bernstein LH, Leukhardt-Fairfield CJ, Pleban W, et al. Usefulness of data on albumin and prealbumin concentrations in determining effectiveness of nutritional support. Clin Chem 1989;35:271.

11. Bernstein LH, Shaw-Stiffel TA, Schorow M, Brouillette R. Financial implications of malnutrition. In: Labbe R, ed. Clinics in Laboratory Medicine. Philadelphia, PA: WB Saunders, 1993:1.

12. Bistrian BR, Blackburn GL, Hollwell H, Heddle R. Protein status of general surgical patients. JAMA 1974;230:858.

13. Bistrian BR, Blackburn GL, Vitale J, Cochrane M. Protein status in general medical patients. JAMA 1976;235:1567.

14. Blackburn GL, Bistrian BR, Maini BS, et al. Nutritional and metabolic assessment of the hospitalized patient. J Parenter Enteral Nutr 1977;1:11.

15. Blackburn GL, Bistrian BR, Harvey K. Indices of protein-calorie malnutrition as predictors of survival. In: Levenson SM, ed. Nutritional assessment—Present status, future directions and prospects. Columbus, OH: Ross Laboratories, 1981:131.

16. Borresen HC. Urine electrolytes and body weight changes in the routine monitoring of total intravenous feeding. Scand J Clin Lab Invest 1979;39:591.

17. Bourry J, Milano G, Caldani C, et al. Assessment of nutritional proteins during the parenteral nutrition of cancer patients. Ann Clin Lab Sci 1982;12:158.

18. Buzby GP, et al. Perioperative total parenteral nutrition in surgical patients. VA TPN Cooperative study group. N Engl J Med 1991;325:525.

19. Buzby GP, Mullen JP, Matthews DC, Hobbs CL, Rosato EF. Prognostic nutritional index in gastrointestinal surgery. Am J Surg 1980;139:160.

20. Campos ACL, Chen M, Meguid MM. Comparisons of body composition derived from anthropomorphic and bioelectrical impedance methods. J Am Coll Nutr 1989;8:484.

21. Canizaro PC. Methods of nutritional support in surgical patients. In: Yarborough MF, ed. Surgical nutrition. New York: Churchill Livingstone, Inc, 1981;13.

22. Cannon JG, Meydani SN, Fielding RA, et al. Acute phase response in exercise. II. Associations between vitamin E, cytokines, and muscle proteolysis. Am Physiol Soc 1991;260:R1235.

23. Cannon PR, Wissler RW, Woolridge RL, et al. The relationship of protein deficiency to surgical infection. Ann Surg 1944;120:514.

24. Carlson DE, Cioffi WG Jr, Mason AD Jr, et al. Evaluation of serum visceral protein levels as indicators of nitrogen balance in thermally injured patients. J Parenteral Enteral Nutr 1991;15:440.

25. Cavarocchi NC, Au FC, Dalal FR, et al. Rapid turnover proteins as nutritional indicators. World J Surg 1986;10:468.

26. Cerra FB. Hypermetabolism, organ failure, and metabolic support. Surgery 1987;101:1.

27. Chadwick SJD, Sim AJW, Dudley HAF. Changes in plasma fibronectin during acute nutritional deprivation in healthy human subjects. Br J Nutr 1986;55:7.

28. Chatelain PG, Van Wyk JJ, Copeland KC, et al. Effect of in vitro action of serum proteases or exposure to acid on measurable immunoreactive somatomedin-C in serum. J Clin Endocrinol Metab 1982;56:376.

29. Christou NV, Meakins JL. Neutrophil function in anergic surgical patients: Neutrophil adherence and chemotaxis. Ann Surg 1979;190:557.

30. Church JM, Hill GL. Assessing the efficacy of intravenous nutrition in general surgical patients: Dynamic nutritional assessment with plasma proteins. J Parenteral Enteral Nutr 1987;11:135.

31. Clemmons DR, Underwood LE, Dickerson RN, et al. Use of plasma somatomedin-C/insulin-like growth factor I measurements to monitor the response to nutritional repletion in malnourished patients. Am J Clin Nutr 1985;41:191.

32. Collins JP, McCarthy ID, Hill GL. Assessment of protein nutrition in surgical patients—The value of anthropometrics. Am J Clin Nutr 1979;32:1527.

33. Committee on Nutrition of the American Academy of Pediatrician. Sodium intake of infants in the United States. Pediatrics 1981;68:444.

34. Copeland EM, MacFadyen BV, Dudrick SJ. Effect of intravenous hyperalimentation on established delayed hypersensitivity in the cancer patient. Ann Surg 1976;184:60.

35. Crottogini AJ, Siggaard-Anderson O. Plasma ionized calcium in the critically ill on total parenteral nutrition. Scand J Clin Lab Invest 1981;41:49.

36. Dahn MS, Jacobs LA, Smith S, et al. The significance of hypoalbuminemia following injury and infection. Am Surg 1985;51:340.

37. Daly JM, Dudrick SJ, Copeland EM. Effects of protein depletion and repletion on cell mediated immunity in experimental animals. Ann Surg 1978;188:791.

38. DeLone JB, Curtas S, Jeejeebhoy JN, Meguid MM. Effect of operation and nutrient intake on muscle function and enzymes. Surg Forum 1987;37:36.

39. Deodhar SD. C-Reactive protein: The best laboratory indicator available for monitoring disease activity. Cleveland Clin J Med 1989;56:126.

40. Detsky AS, Baker JP, Mendelson RA, Wolman SL, Wesson DE, Jeejeebhoy KN. Evaluation of accuracy of nutritional assessment techniques applied to hospitalized patients: Methodology and comparisons. J Parenter Enteral Nutr 1984;8:153.

41. Dinarello CA. Biology of interleukin 1. FASEB J 1988;2:108.

42. Dudrick SJ. A clinical review of nutritional support of the patient. J Parenter Enteral Nutr 1979;3:444.

43. Durnin JUGA, Womersley J. Body fat assessed from total body density and its estimation from skinfold thickness: Measurements on 481 men and women aged from 16 to 72 years. Br J Nutr 1974;32:77.

44. Eggert LD, Rusho WJ, MacKay MW, et al. Calcium and phosphorus compatibility in parenteral nutrition solutions for neonates. Am J Hosp Pharm 1982;39:49.

45. Fischer JE. Plasma proteins as indicators of nutritional status. In: Levenson SM, ed. Nutritional assessment, present status, future direction and prospects. Report of the second cross conference on medical research. Columbus, OH: Ross Laboratories, 1982;25.

46. Fleck A. Acute phase response: Implications for nutrition and recovery. Nutrition 1988;4:109.

47. Fleming CR, Lie JT, McCall JT, et al. Selenium deficiency and fatal cardiomyopathy in a patient on home parenteral nutrition. Gastroenterology 1982;83:689.

48. Forlaw L. Parenteral nutrition in the critically ill child. Crit Care Quart 1981;3:1.

49. Forse RA, Shizgal HM. Serum albumin and nutritional status. J Parenteral Enteral Nutr 1980;4:450.

50. Forse RA, Shizgal HM. The assessment of malnutrition. Surgery 1980;88:17.

51. Forbes GB, Bruining GJ. Urinary creatinine excretion and lean body mass. Am J Clin Nutr 1976;20:1359.

52. Fraser IM, Russel DMcR, Whittaker JS, et al. Skeletal and diaphragmatic muscle function in malnourished patients with chronic obstructive lung disease. [Abstract] Am Rev Respir Dis 1984;129:A269.

53. Freund H, Atamian S, Fischer JE. Chromium deficiency during total parenteral nutrition. JAMA 1979;241:496.

54. Friis-Hansen B. Body composition in growth. Pediatrics 1971;47:264.

55. Georgieff MK, Amarnath UM, Murphy EL, et al. Serum transferrin levels in the longitudinal assessment of protein-energy status in preterm infants. J Pediatr Gastroenterol Nutr 1989;8:234.

56. Georgieff MK, Sasanow SR, Pereira GR. Serum transthyretin levels and protein intake as predictors of weight gain velocity in premature infants. J Pediatr Gastroenterol Nutr 1987;6:775.

57. Gerlach TH, Zile MH. Metabolism and secretion of retinol transport complex in acute renal failure. J Lipid Res 1991;32:515.

58. Giacoia GP, Watson S, West K. Rapid turnover transport proteins, plasma albumin, and growth in low birth weight infants. J Parenter Enteral Nutr 1984;8:367.

59. Glinsmann WH, Feldman FJ, Metz W. Plasma chromium after glucose administration. Science 1966;152:1243.

60. Gofferje H. Prealbumin and retinol-binding protein[hm3u]=m[hm3u] Highly sensitive parameters for the nutritional state in respect to protein. Med Lab 1978;5:38.

61. Gordon EF, Gordon RC, Passal DB. Zinc metabolism: Basic clinical and behavioral aspects. J Pediatr 1981;99:341.

62. Grant JP, Custer PB, Thurlow J. Current techniques of nutritional assessment. Surg Clin North Am 1981;61:437.

63. Greenblatt DJ, Ransil BJ, Harmatz JS, Smith W, Duhme DW, Koch-Weser J. Variability of 24-hour urinary creatinine excretion by normal subjects. J Clin Pharmacol 1976;16:321.

64. Haffejee AA, Angorn IB. Nutritional status and the nonspecific cellular and humoral immune response in esophageal carcinoma. Ann Surg 1979;189:475.

65. Hall K, Tally M. The somatomedin-insulin-like growth factors. J Intern Med 1989;225:47.

66. Harvey KB, Moldawer LL, Bistrian BR, Blackburn GL. Biological measures for the formulation of a hospital prognostic index. Am J Clin Nutr 1981;34:2013.

67. Hawker FH, Stewart PM, Baxter RC, et al. Relationship of somatomedin-C/insulin-like growth factor I levels to conventional nutritional indices in critically ill patients. Crit Care Med 1987;15:732.

68. Hay RW, Whitehead RG, Spicer CC. Serum-albumin as a prognostic indicator in dematous malnutrition. Lancet 1975;ii:427.

69. Heller RM, Kirchner SG, O'Neill JA, et al. Skeletal changes of copper deficiency in infants receiving prolonged total parenteral nutrition. J Pediatr 1978;92:947.

70. Heymsfield SB, Olafson RP, Utner MH, Nixon DW. A radiographic method of quantifying protein-calorie malnutrition. Am J Clin Nutr 1979;32:693.

71. Hokama Y, Nakamura RM. C-Reactive protein: Current status and future perspectives. J Clin Lab Anal 1987;1:15.

72. Horowitz GD, Groeger JS, Legaspi A, et al. The response of fibronectin to differing parenteral caloric sources in normal man. J Parenter Enteral Nutr 1985;9:435.

73. Howard L, Meguid MM. Nutritional assessment in total parenteral nutrition. Clin Lab Med 1981;1:611.

74. Ingenbleek Y, De Visscher M, De Nayer P. Measurement of prealbumin as an index of protein-calorie malnutrition. Lancet 1972;ii:106.

75. Ingenbleek Y, Van Den Schrieck H-G, De Nayer P, et al. The role of retinol-binding protein in protein-calorie malnutrition. Metabolism 1975;24:633.

76. Ingenbleek Y, Van Den Schrieck H-G, De Nayer P, et al. Albumin, transferrin and thyroxine-binding prealbumin/retinol-binding protein (TBPA-RBP) complex in assessment of malnutrition. Clin Chim Acta 1975;63:61.

77. Irie S, Tavassoli M. Transferrin-mediated cellular iron uptake [review]. Am J Med Sci 1987;293:103.

78. Isley WL, Lyman B, Pemberton B. Somatomedin-C as a nutritional marker in traumatized patients. Crit Care Med 1990;18:795.

79. Jeejeebhoy KN: Zinc and chromium in parenteral nutrition. Bull NY Acad Med 1984;60:118.

80. Jeejeebhoy KN, Baker JP, Wolman SL, et al. Critical evaluation of the role of clinical assessment and body composition studies in patients with malnutrition and after total parenteral nutrition. Am J Clin Nutr 1982;35 (Suppl):1117.

81. Jeejeebhoy KN, Chu RC, Marliss EB, et al. Chromium deficiency, glucose intolerance, and neuropathy reversed by chromium supplementation in a patient receiving long-term total parenteral nutrition. Am J Clin Nutr 1977;30:531.

82. Joffe G, Etzioni A, Levy J, et al. A patient with copper deficiency anemia while on prolonged intravenous feeding. Clin Pediatr 1981;20:226.

83. Johnson WC, Ulrich F, Meguid MM, et al. Role of delayed hypersensitivity in predicting postoperative morbidity and mortality. Am J Surg 1979;137:536.

84. Katz MD, Lor E, Norris K, et al. Comparison of serum prealbumin and transferrin for nutritional assessment of TPN patients: A preliminary study. Nutr Support Serv 1986;6:22.

85. Kerner JA, Sunshine P. Parenteral alimentation. Semin Perinatol 1979;3:417.

86. Kirschner BS, Sutton MM. Somatomedin-C levels in growth-impaired children and adolescents with chronic inflammatory bowel disease. Gastroenterology 1986;91:830.

87. Klausner RD. From receptor to genes—Insights from molecular iron metabolism. Clin Res 1988;36:494.

88. Klidjian AM, Archer TJ, Foster KG, et al. Detection of dangerous malnutrition. J Parenter Enteral Nutr 1982;6:119.

89. Klidjian AM, Foster KJ, Kammerling RM, Cooper A, Karran SJ. Anthropometric and dynamometric variables to serious postoperative complications. Br Med J 1980;281:899.

90. Knochel JP. Hypophosphatemia. West J Med 1981;134:15.

91. Kushner RF, Scholler DA, Estimation of total body water by bioelectrical impedance analysis. Am J Clin Nutr 1986;44:417.

92. Large S, Neal G, Glover J, et al. The early changes in retinol-binding protein and prealbumin concentrations in plasma of protein-energy malnourished children after treatment with retinol and an improved diet. Br J Nutr 1980;43:393.

93. Leider Z. Nutritional assessment present and future. Open Forum Nutritional Support Services 1988;8:8.

94. Lockitch G, Godophin W, Pendray MR, et al. Serum zinc, copper, retinol-binding protein, prealbumin and ceruloplasmin concentrations in infants receiving intravenous zinc and copper supplement. J Pediatr 1983;102:304.

95. Lukaski HC, Bolonchuk W, Hall CB, Siders WA, Estimation of fat free mass in humans using the bioelectrical impedance method: A validation study. J Appl Physiol 1986;60:1327.

96. McKone TK, Davis AT, Dean RE. Fibronectin: A new nutritional parameter. Am Surg 1985;51:336.

97. Meguid MM, Campos AC, Lukaski HC, Kjell C. A new single cell *in vitro* model to determine volume and sodium concentration changes by bioelectrical impedance analysis. Nutrition 1988;4:363.

98. Meguid MM, Curtas S, Chen M, Nole E. Adductor pollicis muscle tests to detect and correct subclinical malnutrition in preoperative cancer patients. [Abstract]. Am J Clin Nutr 1987;45:843.

99. Meguid MM, Mughal MM, Meguid V, Terz JJ. Risk-benefit analysis of malnutrition and preoperative nutrition support: A review. Nutr Int 1987;3:25.

100. Meydani SN, Ribaya-Mercado JD, Russell RM, et al. Vitamin B-6 deficiency impairs interleukin 2 production and lymphocyte proliferation in elderly adults. Am J Clin Nutr 1991;53:1275.

101. Miller RR, Menke JA, Mentser MI. Hypercalcemia associated with phosphate depletion in the neonate. J Pediatr 1984;105:814.

102. Mitchel L, Serrano A, Mart RA. Nutritional support of hospitalized patients. N Engl J Med 1981;304:1147.

103. Moore FD, Oleson KH, McMurphy JD, Parker HV, Ball MR, Boyden CM. The body cell mass and its supporting environment. Body composition in health and disease. Philadelphia:W B Saunders, 1963.

104. Morgan DB, Hill GL, Burkenshaw L. The assessment of weight loss from a single measurement of body weight: The problems and limitations. Am J Clin Nutr 1980;33:210.

105. Mosher DF. Physiology of fibronectin. Ann Rev Med 1984;35:561.

106. Moskowitz SR, Pereira G, Spitzer A, et al. Prealbumin as a biochemical marker of nutritional adequacy in premature infants. J Pediatr 1983;102:749.

107. Mullen JL, Buzby GP, Waldman MT, Gertner MH, Hobbs CL, Rosato EF. Prediction of operative morbidity and mortality by preoperative nutritional assessment. Surg Forum 1979;30:80.

108. Mullen JL, Gertner MH, Buzby GP, Goodhart GL, Rosato EF. Implications of malnutrition in the surgical patient. Arch Surg 1979;114:121.

109. Payette H, Rola-Pleszezynski M, Ghadirian P. Nutrition factors in relation to cellular and regulatory immune variables in a free-living elderly population. Am J Clin Nutr 1990;52:927.

110. Pekarek RS, Haver EC, Rayfield EJ, et al. Relationship between serum chromium concentrations and glucose utilization in normal and infected subjects. Diabetes 1975;24:350.

111. Perkins RM, Levine DL. Common fluid and electrolyte problems in pediatric intensive care unit. Pediatr Clin North Am 1980;27:567.

112. Pettigrew RA, Charlesworth PM, Farmilo RW, Hill GL. Assessment of nutritional depletion and immune competence: A comparison of clinical examination and objective measurements. J Parenter Enteral Nutr 1984;8:21.

113. Phillips GD, Odgers CL. Parenteral nutrition: Current status and concepts. Drugs 1982;23:276.

114. Phillips LS. Nutrition, metabolism and growth. In: Daughaday WH, ed. Endocrine control of growth. New York: Elsevier, 1981;121.

115. Powell DR, Rosenfeld RG, Baker BK, et al. Serum somatomedin levels in adults with chronic renal failure: The importance of measuring insulin-like growth factor I (IGF-I) and IGF-II in acid-chromatographed uremic serum. J Clin Endocrinol Metab 1986;63:1186.

116. Quinby GE, Nowak MM, Andrew BF. Parenteral nutrition in the neonate. Clin Perinatol 1975;2:59.

117. Rainey-Macdonald CG, Holliday RL, Wells GA, et al. Validity of a two-variable nutritional index for use in selecting candidates for nutritional support. J Parenter Enteral Nutr 1983;7:15.

118. Reeds PJ, Laditan AAO. Serum albumin and transferrin in protein-energy malnutrition. Br J Nutr 1976;36:255.

119. Reilly JJ Jr, Hull SF, et al. Economic impact of malnutrition: A model system for hospitalized patients. J Parenter Enteral Nutr 1988;12:371.

120. Reinhardt GF, Myskofski JW, Wilkins DB, et al. Incidence and mortality of hypoalbuminemic patients in hospitalized veterans. J Parenter Enteral Nutr 1980;4:357.

121. Royle GT, Kettlewell MGW. Liver function tests in surgical infection and malnutrition. Ann Surg 1980;192:192.

122. Roza AM, Tuitt D, Shizgal HM. Transferrin—A poor measure of nutritional status. J Parenter Enteral Nutr 1984;8:523.

123. Rubin J, Deraps GD, Walsh D, et al. Protein losses and tobramycin absorption in peritonitis: Treated by hourly peritoneadialysis. Am J Kidney Dis 1986;8:124.

124. Russell DMcR, Leiter LA, Whitwell J, Marliss EB, Jeejeebhoy KN. Skeletal muscle function during hypocaloric diets and fasting: A comparison with standard nutritional assessment parameters. Am J Clin Nutr 1983;37:133.

125. Russell DMcR, Walker PM, Leiter LA, et al. Metabolic and structural changes in skeletal muscle during hypocaloric dieting. Am J Clin Nutr 1984;39:503.

126. Saba TM, Blumenstock FA, Shah DM, et al. Reversal of opsonic deficiency in surgical, trauma, and burn patients by infusion of purified human plasma fibronectin. Am J Med 1986;80:229.

127. Sandberg LB, Owens AJ, VanReken DE, et al. Improvement in plasma protein concentrations with fibronectin treatment in severe malnutrition. Am J Clin Nutr 1990;52:651.

128. Sann L, Rigal D, Galy G, et al. Serum copper and zinc concentration in premature and small-for-date infants. Pediatr Res 1980;14:1040.

129. Scott RL, Sohmer PR, MacDonald MG. The effect of starvation and repletion on plasma fibronectin in man. JAMA 1982;248:2025.

130. Segal KR, Butin B, Presta E, Wang J, VanItallie TB. Estimations of human body composition by bioelectrical impedance methods: A comparative study. J Appl Physiol 1985;58:1565.

131. Seltzer MH, Bastidas JA, Cooper DM, et al. Instant nutritional assessment. J Parenter Enteral Nutr 1979;3:157.

132. Shamberger RJ. Vitamin A and retinol binding protein alterations in disease. In: Brewster MA, ed. Nutritional elements and clinical biochemistry. New York: Plenum Publishing Corp, 1980;117.

133. Shike M. Copper in parenteral nutrition. Bull NY Acad Med 1984;60:132.

134. Shils ME, Burke AW, Greene HL, et al. Guidelines for trace element preparation for parenteral use. JAMA 1979; 241:2051.

135. Shizgal HM. Nutritional assessment with body composition measurements. J Parenter Enteral Nutr 1987;11:42S.

136. Sjils ME, Levander OA, Alcock NW. Selenium levels in long term TPN patients. Am J Clin Nutr 1982;35:838.

137. Smith FR, Goodman DS, Zaklama MS, et al. Serum vitamin A, retinol-binding protein, and prealbumin concentrations in protein-calorie malnutrition. I. A functional defect in hepatic retinol release. Am J Clin Nutr 1973;26:973.

138. Smith FR, Goodman DS. The effects of disease of the liver, thyroid, and kidney on the transport of vitamin A in human plasma. J Clin Invest 1971;50:2426.

139. Smith FR, Suskind R, Thanangkul O, et al. Plasma vitamin A, retinol-binding protein and prealbumin concentrations in protein-calorie malnutrition. III. Response to varying dietary treatments. Am J Clin Nutr 1975;28:732.

140. Smith JE, Goodman DS. Retinol-binding protein and the regulation of vitamin A transport. Federation Proc 1979; 38:2504.

141. Spiekerman AM. Laboratory tests for monitoring total parenteral nutrition (TPN). Clin Chem 1987;27:1.

142. Spiekerman AM. Retinol binding protein and prealbumin levels in renal failure, abstracted. American Society of Parenteral and Enteral Nutrition, Sixth Clinical Congress, San Francisco, 1982;6:341.

143. Spiekerman AM. Some recognized nutritional markers in liver and kidney disease. Poster session at the 1989 American Association of Clinical Chemistry Meeting in San Francisco, CA, August 1989 (Beckman Brochure).

144. Spiekerman AM, Rudolph RA, Bernstein LH. Determination of malnutrition in hospitalized patients with the use of a group-based reference. Arch Pathol Lab Med 1993;117:184.

145. Starker PM, Gump FE, Askanazi J, Elwyn DH, Kinney JM. Serum albumin levels as an index of nutritional support. Surgery 1982;91:194.

146. Studley HO. Percentage of weight loss: A basic indicator of surgical risk in patients with chronic peptic ulcer. JAMA 1936;106:458.

147. Takala J, Neuvonen P, Klossner J. Hypophosphatemia in hypercatabolic patients. Acta Anaesthesiol Scand 1985;29:65.

148. Tovey SJ, Benton KGF, Lee HA. Hypophosphatemia and phosphorus requirements during intravenous nutrition. Postgrad Med 1977;53:289.

149. Tsang RC, Steichen JJ, Brown DR. Perinatal calcium homeostasis: Neonatal hypocalcemia and bone mineralization. Clin Perinatol 1977;4:385.

150. Tuten MB, Wogt S, Dasse F, et al. Utilization of prealbumin as a nutritional parameter. J Parenter Enteral Nutr 1985;9:709.

151. Unterman TG, Vazquez RM, Slas AJ, et al. Nutrition and somatomedin. XIII. Usefulness of somatomedin-C in nutritional assessment. Am J Med 1985;78:228.

152. Vanlandingham S, Spiekerman AM, Newmark SR. Prealbumin: A parameter of visceral protein levels during albumin infusion. J Parenter Enteral Nutr 1982;6:230.

153. Van Rij AM, Thompson CD, McKenzie JM, et al. Selenium deficiency in total parenteral nutrition. Am J Clin Nutr 1979;32:2076.

154. Van Vloten WA, Bos LP. Skin lesions in acquired zinc deficiency due to parenteral nutrition. Dermatologica 1978; 156:175.

155. Watters JM, Bessey PQ, Dinarello CA, et al. Both inflammatory and endocrine mediators stimulate host responses to sepsis. Arch Surg 1986;121:179.

156. Weinsier RL, Hunker EM, Krumdieck GL, et al. A prospective evaluation of general medical patients during the course of hospitalization. Am J Clin Nutr 1979;32:419.

157. Whicher JT, Evans SW. Cytokines in disease. Clin Chem 1990;36:1269.

158. Wilmore DW, Black PR, Muhlbacher F. Injured man: Trauma and sepsis. In: Winters RW, Greene M, eds. Nutritional support of the seriously ill patient. New York: Academic Press 1983;33.

159. Winkler MF, Pomp A, Caldwell MD, et al. Transitional feeding: The relationship between nutritional intake and plasma protein concentrations. J Am Dietetic Assoc 1989; 89:969.

160. 1991 World Health Statistics Annual. Geneva, Switzerland. WHO, 1992.

161. Yoder MC, Anderson DC, Gopalakrishna GS, et al. Comparison of serum fibronectin, prealbumin, and albumin concentrations during nutritional repletion in protein-calorie malnourished infant. J Pediatr Gastroenterol Nutr 1987;6:84.

162. Zlotkin SH, Stallings VA, Pencharz PB. Total parenteral nutrition in children. Pediatr Clin North Am 1985;32:381.

163. Zidar BL, Shadduck RK, Zeigler Z, et al. Observations on the anemia and neutropenia of human copper deficiency. Am J Hematol 1977;3:177.

Chapter 31

Pediatric Clinical Chemistry

John E. Sherwin, Juan R. Sobenes

Objectives

Upon completion of this chapter, the clinical laboratorian should be able to:

- *Define the following terms: extrauterine, infancy, neonate, and pediatric.*

- *Describe some of the adaptive changes that occur upon the birth of an infant.*

- *Discuss some of the problems with blood collection from pediatric patients.*

- *Summarize some of the changes that occur in pediatric patients with regard to electrolyte and water balance, energy metabolism, hormone balance, and humoral and cellular immunity.*

- *Explain how drug treatment and pharmacokinetics are different in pediatric patients and adult patients.*

- *Relate the procedures used to identify inherited metabolic disorders in pediatric patients.*

KEY WORDS

Adaptive Immune System
Common Variable Agammaglobulinemia
Congenital
Agammaglobulinemia
Extrauterine
Hyperoxemia
Hypoxemia
Infancy
Innate Immune System
Microchemistry
Neonate
Pediatric

AGE-RELATED DEVELOPMENT CHANGES

Much of laboratory medicine's value in diagnosis and therapy is based on the fact that specific organ system function can be assessed by serum concentrations of substances produced by the organ system. This premise relies on the fact that normal individuals have consistent reference values for body-fluid analyte concentrations ("normal values"). The fact that human beings grow and develop throughout life suggests that reference values may vary with age. This is particularly true during infancy and adolescence because significant growth and developmental changes occur during this period.

The birth of an infant is attended by rapid adaptation to extrauterine life. The infant adapts by initiating active respiration, increasing kidney function, closure of the patent ductus arteriosus of the heart, and eating. Generally, infants lose weight following delivery because of insensible water loss. This loss is quickly offset by weight gain due to caloric intake. Weight gain is about 15 g per day for the first year, and the infant's birth weight of about 3200 g doubles in 4 to 6 months. Kidney function increases during the first year until it approximates adult values. Liver function matures and portal circulation becomes patent within 30 to 60 days. Motor development increases linearly during at least the first 12 months. This development is accompanied by changes in the electroencephalogram, growth and differentiation of the brain, and increasing visual acuity. In addition, several specific changes are of diagnostic importance.

Michael L. Bishop, Janet L. Duben-Engelkirk, and Edward P. Fody.
CLINICAL CHEMISTRY. © 1996 Lippincott–Raven Publishers.

Thyroid function changes abruptly following birth. Serum thyrotropin-stimulating hormone concentration decreases rapidly following birth. Hemoglobin shifts over a period of 90 days from fetal to adult hemoglobin. This, in turn, frequently causes increased bilirubin concentrations.

During the adolescent growth spurt, bone growth results in a large increase in serum alkaline phosphatase. Sexual maturation in females is associated with chronic blood loss due to menstruation. The cyclic production of female sexual hormones also requires special attention. Sexual maturation in males is associated with fewer large changes in serum analytes, but nonetheless, the changes do result in differentiation of male and female values for such analytes as creatine kinase and uric acid, as well as the steroid hormones. Therefore, it is important that evaluation of illness in infants include the use of sex- and age-related reference values.

PHLEBOTOMY AND MICROCHEMISTRY FOR THE PEDIATRIC PATIENT

Blood collection from the pediatric patient is complicated not only by the patient's size but very often by his or her ability to communicate. This can complicate identification of the patient. The frequent presence of parents during blood collection from infants also can make blood collection more difficult because of behavior problems, however. Venipuncture in pediatric patients varies little, if any, from venipuncture in adults.

It is important to remember that the availability of appropriate veins for therapy is often minimal in infants because of their size. Therefore, they must be reserved for therapy rather than phlebotomy. This is a particular problem in the newborn. For this reason, blood collection by skin puncture is often preferable to venipuncture.

Skin puncture for blood collection is done on a finger in older children or on the heel in infants. The National Committee for Clinical Laboratory Standards (NCCLS) has developed a standard protocol for skin puncture. Prior to performing skin puncture, the skin should be "arterialized" using either a warm towel or a histamine cream. Care should be taken not to use dry heat or a water temperature greater than 42°C. Some reports have questioned the efficacy of histamine creams in inducing local hyperemia, but they are still commonly used.

Fingers are commonly used in children for skin puncture. The palmar surface of the distal phalanx of the middle finger should be used. In infants, the heel is preferable to the fingers because of size. The specimen should be collected on the outer portions of the heel to avoid the calcaneus. In larger infants, skin puncture can be done using red lancets, whereas the blue neolet should be used for small or premature infants because of the shorter puncture point.

Punctures with these devices generally yield 300- to 400-μL specimens after the first drop of blood has been discarded due to tissue contamination. Several studies have demonstrated that with proper technique, hemolysis and excessive tissue fluid contamination can be avoided.

Upon completion of the skin puncture, it is important that the phlebotomist not leave the patient until the bleeding has stopped. In small children and infants, the usual adhesive bandages should be avoided because of the possibility of ingestion or aspiration by the patient.

Collection of microsamples (less than 500 μL) presents unique problems for the laboratory with regard to labeling, centrifugation, and separation. The production of microcollection systems, such as those of Sarstedt or Becton Dickinson, has simplified the task of specimen collection. However, no completely satisfactory system for labeling of these small containers has been developed. If adhesive labels are used, care must be taken that the size permits the container to fit into the centrifuge. Because of the smaller tube size, mixing of anticoagulated specimens is particularly important to prevent formation of small clots. This can be a particularly vexing problem with blood-gas specimens collected in capillary tubes.

Once the specimens have been separated, they can be analyzed using any of a variety of automated instruments. Historically, microchemistry was a separate discipline because the majority of analyzers required several milliliters of sample for a complete analysis. Today, however, virtually all analyzers require only a few microliters of sample for each test. Therefore, the analysis of microsamples is no longer a large problem. Nonetheless, it should be kept in mind that some instruments do require additional specimen to account for "dead volume."

ACID-BASE BALANCE IN THE INFANT

Acid-Base Balance Disorders

The trauma of labor during delivery often induces an acidosis in the newborn as a result of significant asphyxia. In such cases, blood pH may be an indication for surgical intervention if the clinical presentation warrants it. For this reason, some practitioners have advocated the collection and analysis of fetal scalp blood specimens as a means of monitoring fetal distress. In the hour following birth, the pH will rapidly return to the normal value (Table 31-1). Newborn infants with acidosis require additional evaluation. Many will be normal, since the acidosis was the result of labor. However, a small number of infants will be acidotic and clinically depressed. These infants may be suffering from inborn errors of metabolism. On the other hand, clinically depressed infants with a normal pH also present a diagnostic challenge.

Infants who are born premature and have respiratory distress syndrome will rapidly become acidotic and exhibit an elevated PCO_2. In most cases they will respond to oxygenation. However, therapy generally cannot be effectively designed to return the blood-gas parameters to normal, so the patients remain slightly acidotic with a pH of 7.20 to

TABLE 31-1. Reference Ranges for pH and Blood-Gas Parameters in the Newborn

	Normal Values	Panic Values
pH		
Arterial	7.27–7.47	Less than 7.10, greater than 7.55
Venous	7.33–7.43	—
PCO_2 (mmHg)		
Arterial	27–40	Greater than 80
Venous	35–40	—
PO_2 (mmHg)		
Arterial	65–80	Less than 40, greater than 250
Venous	30–50	—
Bicarbonate (mmol/L)		
Arterial	16–23	—
Serum, venous	18–25	—

7.25 and an elevated PCO_2. The PO_2 is kept above 40 mmHg. In most nurseries, the oxygenation status is measured with the transcutaneous O_2 (tcO_2) monitor. During the years since the introduction of a successful transcutaneous oxygen electrode, the very close correlation between the PaO_2 and the tcPO_2 has been repeatedly verified. The PO_2 measured by skin puncture is approximately 5 mmHg lower than the tcPO_2.

Continuous transcutaneous oxygen monitoring also has provided surprises. Transient hyperoxemia ($PO_2 > 100$ mmHg) and hypoxemia ($PO_2 < 50$ mmHg) are induced very easily by infant handling, which is routine in many neonatal nurseries. Hypoxemia can be induced by crying, bowel movements, feeding, diaper changes, blood sampling, and physical examination. The relatively banal treatment of rubbing of the heel with a gel is sufficient in some infants to cause a prolonged decrease in the tcPO_2. However, in one infant, this same treatment actually improved the oxygen status. Hyperoxemia also can result from abrupt changes in shunt blood-flow patterns and improvement in lung function owing to resolving respiratory distress and from briefly breathing concentrated O_2 to alleviate an apneic spell.

There are some instances when the transcutaneous monitor does not accurately reflect arterial PO_2. Fortunately, these instances are relatively rare, being restricted to infants in whom the blood pressure falls 2 standard deviations below normal and to infants given tolazoline, a pulmonary vasodilator, or anesthetics such as nitrous oxide and halothane. These changes are transient, of only 5 to 10 minutes duration, and sometimes less. These changes, which can be as large as 20 mmHg, would be missed by traditional phlebotomy and laboratory analysis practices. For this reason, the transcutaneous monitors are in widespread use.

The transcutaneous CO_2 (tcCO_2) monitors also have been developed and are in routine use. The tcCO_2 exhibits smaller changes than the tcO_2 but can provide valuable trending information. The use of these monitors has reduced the need for frequent blood-gas analysis. In many intensive care units it has resulted in a 50% decrease in the number of requests for blood-gas analysis.

The calculation of bicarbonate from the pH and PCO_2 using the Henderson-Hasselbalch equation is generally adequate for therapeutic intervention. Most blood-gas instruments do this automatically using a pK of 6.1 and a solubility coefficient of 0.0302. While these parameters have been reported to vary with acute illness, there is no compelling evidence that for medical practice the constants are inadequate.

If the acidosis is not respiratory in nature, as described above, the cause(s) needs to be defined and treated rather than merely treating the acidosis. Metabolic acidosis in infants can have any of several etiologies (Table 31-2). Some of these have ready cures, such as antibiotic therapy. Others represent inborn errors of metabolism, for which no cure is known.

Analysis of Acid-Base Deficits

The analyses for pH, PCO_2, and PO_2 are really independent of the patient's age and size. However, few of the blood-gas analyzers easily accommodate capillary blood samples. This can be a particular problem in a busy neonatal intensive care unit, which may generate as many as 200 such samples daily. Capillary samples can easily clot and coat electrode membranes with protein, thereby distorting values and increasing the required maintenance. Our experience suggests that manufacturers' recommended maintenance periods must be reduced by at least half if performance is to be optimized.

ELECTROLYTE AND WATER BALANCE

The immediate postnatal period is attended by growth and developmental changes that adapt the infant to extrauterine life. The ability of the infant to adapt is a function of the intrauterine environment and the infant's developmental maturity. The normal infant adapts with little difficulty. On the other hand, a 26-week-gestation infant weighing 600 g will require intensive care. Fluid balance in these infants is precarious and requires careful monitoring.

The neonate has a relatively high surface-to-volume ratio. This leads to increased insensible water loss at birth. Renal function is immature, and glomerular filtration rates are low. Additionally, water balance is complicated by the

TABLE 31-2. The Etiologies of Metabolic Acidosis in the Infant

Birth trauma	Excess intralipid
Liver dysfunction	Fanconi's syndrome
Lactic acidosis	Renal tubular acidosis
Shock	Sepsis
Diarrhea	Glycogen storage diseases
Diabetes	Hyperthyroidism

fact that the body water content is normally being reduced by contracting the extracellular fluid volume at this time. In fact, this accounts in large part for the observed normal weight loss following birth. Electrolyte balance is also precarious in these infants not only because of fluid volume changes, but also because of immaturity in renal filtration and reabsorption of electrolytes.

Developmental Changes in Body Fluid and Electrolyte Composition During the Neonatal Period

The total body water content of the normal newborn is about 80%: 55% intracellular fluid and 45% extracellular fluid. The extracellular fluid is about 20% plasma water and 80% interstitial fluid. During the first weeks of postnatal life, the total body water content decreases to about 60%, primarily at the expense of interstitial fluid.

Renal function is immature even in the normal neonate. The glomerular filtration rate is only about 25% of the adult and matures slowly during the first 6 months of life (Table 31-3). The maximal concentrating ability of the kidney is only about 78% of the adult, and this is due to the fact that the loop of Henle is shorter and does not penetrate the medulla as deeply as in the adult kidney. Response to antidiuretic hormone in the neonatal period appears to be normal but is limited by the renal concentrating ability.

Insensible water loss is inversely proportional to birth weight or gestational age. This is due to increased permeability of the skin epidermis to water, larger body surface per unit of weight, and greater skin blood flow relative to metabolic rate. The higher respiratory rate of infants also may be a contributing factor. Elevation of ambient temperature by 1°C results in a three- to fourfold increase in insensible water loss. Use of a radiant warmer or phototherapy also may increase insensible water loss by as much as 50%. In sick newborn infants, these considerations are very important, since the infant's fluid intake is derived solely from parenteral sources. Fluid overload may lead to congestive heart failure secondary to patent ductus arteriosus with a left-to-right shunt. Factors that may reduce insensible water loss include the use of head shields, high relative humidity in the ambient environment, maintaining an infant in a thermonentral environment, and keeping an infant in an isolette.

In those infants receiving enteral fluids, water and electrolytes are absorbed in the proximal jejunum. Intestinal disorders resulting in diarrhea can rapidly cause dehydration in the neonate. Administration of hyperosmotic enteral fluids also may result in an increased rate of water secretion by the mucosa of the small intestine. This may in turn result in necrotizing enterocolitis. Administration of parenteral fluids does not ordinarily alter the secretion of water by the gastrointestinal tract in the infant.

The electrolytes of concern here are the four major ones: sodium, potassium, chloride, and bicarbonate. The reference ranges for these electrolytes are nearly the same as for the adult, although the concentrations are less rigidly controlled in the neonate (Table 31-4). Serum sodium concentrations change most commonly in response to alterations in water homeostasis. Hypernatremia is usually associated with water loss, but it may be accompanied by an increased extracellular fluid. It is particularly common in the preterm infant because of high insensible water loss, a reduced ability to respond to shifts in sodium balance, and a physical inability to replace water loss *ad libitum*. Iatrogenic hypernatremia can result from osmotic diuresis due to hyperosmolar fluid or glucose solution administration. Hypernatremia also can result from excessive sodium intake from formula, parenteral fluids, or sodium bicarbonate administration. These sodium-overload states are often accompanied by increased extracellular fluid volume, and the infant becomes edematous.

Hyponatremia with hypovolemia results in large part from the fact that the kidneys of premature infants have a low capacity to retain sodium when they are in negative

TABLE 31-3. Glomerular Filtration Rates of Infants and Adults

Age	Glomerular Filtration Rate (ml/min/1.73 m²)	
	Mean	Range
1 day	24	3–38
2–8 days	38	17–60
10–22 days	50	32–69
37–95 days	58	30–86
1.5–2 years	115	95–135
Adult	100	85–115

Note: Adult values are achieved by 12 to 18 months of age.

TABLE 31-4. Age-Related Reference Ranges for Serum Electrolytes

Analyte	Age	Concentration Range (mmol/L)
Sodium	2–5 days	135–148
	1–12 months	130–145
	Greater than 12 months	135–147
Potassium	Newborn to 3 years	3.5–5.1
	3–8 years	3.6–5.0
	8–15 years	3.5–4.9
	Greater than 12 years	3.5–4.7
Chloride	Newborn to 3 years	100–110
	3–10 years	99–109
	10–19 years	99–107
	Greater than 19 years	100–106
Bicarbonate	Infant	19–24
	Child	21–27
	Adult	23–32

sodium balance. In this instance, the urinary sodium will be high. It is important to distinguish this renal immaturity from congenital adrenal hyperplasia, which may present with similar electrolyte values. Enteric fluid loss by vomiting or diarrhea due to systemic infection or other cause also can result in hyponatremia with a decreased extracellular volume. Hyponatremia accompanied by edema due to increased extracellular volume is generally a result of inappropriate antidiuretic hormone secretion or administration of hypo-osmolar or dextrose solutions. Hyperkalemia can result from cell destruction, impaired renal function, or excessive intake. Symptoms of hyperkalemia include muscle weakness and cardiac conduction changes manifested in the electrocardiogram. Clinical manifestations of hyperkalemia are apparent at 6.5 mEq/L. Hypokalemia may be due to decreased intake, increased urinary losses due to renal tubular acidosis or hyperaldosteronism, or increased gastrointestinal losses due to vomiting, diarrhea, or diuretics. Hypokalemia results in muscle weakness, loss of abdominal sounds, and abdominal distension and can be accompanied by polyuria.

Under normal circumstances, sodium and chloride are excreted in equimolar amounts. In most instances, sodium is absorbed as neutral sodium chloride across the intestinal brush border. Hypochloremia can accompany states of hyponatremia, e.g., inappropriate ADH syndrome, but it also may selectively occur in infants who vomit excessively. This hypochloremia is due to loss of chloride in the stomach acid. Hyperchloremia occurs primarily as a result of increased intake and is associated with hypernatremia.

Newborn infants are very susceptible to acidosis because of a limited capacity to excrete hydrogen ion. Renal reabsorption of bicarbonate appears to be only slightly reduced in the neonate, which is reflected in the somewhat lower plasma bicarbonate concentration of 22 mmol/L. This approaches the adult normal of 24 to 27 mmol/L by 1 to 2 months of age. The proximal tubule reabsorbs the majority of bicarbonate, and any remaining bicarbonate is filtered by the distal tubule. Therefore, two types of renal tubular acidosis can occur, one involving the proximal tubule and the other, more common, involving the distal tubule. Neonatal acidosis is usually treated with sodium bicarbonate.

After the first 12 to 18 months of life, the electrolyte balance approximates that of the adult. Nonetheless, in preadolescent children it is important to consider fluid and electrolyte balance in any serious illness. This stems from the fact that their condition tends to be more fragile and communication of fluid loss may be incomplete. The approach to managing fluids and electrolytes is similar in all age groups.

Fluid and Electrolyte Management

Maintenance of fluid and electrolyte balance requires assessment of three parameters: (1) estimation of the quantity of the fluid and electrolyte deficit, (2) calculation of the amounts required to replace losses and correct the abnormal deficit, and (3) implementation of therapy and monitoring and modifying, if necessary, this therapy. If an accurate measure of weight loss is available, this can be used as a direct reflection of fluid loss. If not, clinical assessment of dehydration can be used. The physical signs of skin turgor, blood pressure, activity, and mucous membrane hydration are used to differentiate hypotonic (serum sodium less than 130 mmol/L), isotonic, and hypertonic (serum sodium greater than 150 mmol/L) dehydration. Laboratory measurements of hematocrit, blood urea nitrogen, creatinine, and total protein are useful in conjunction with serum electrolytes for assessment of dehydration. Replacement of fluids is ordinarily done in the same manner as the fluid was lost; e.g., if dehydration is of rapid onset, fluid replacement also should be rapid. The one exception is in hypertonic dehydration, where rapid rehydration may cause convulsions. The assessment of electrolyte deficit is based on physical examination and laboratory measurements of serum electrolytes.

Replacement of fluids is ordinarily done over at least 48 hours, with 50% of the fluids administered in the first 8 hours, except in the case of hypertonic dehydration, in which replacement should be done evenly over 3 to 5 days. Calculations are done to return the plasma osmolality to normal, 280 mOsmol/kg. It must be remembered that 50% of the osmolality deficit is due to cations and 50% is due to anions.

Once the fluid and electrolyte deficit has been estimated and maintenance requirements have been considered, therapy can be initiated using appropriate combinations of the readily available fluids. Maintenance therapy in the neonate requires an additional 75 to 80 mL of water per kilogram per 24 hours containing 2 to 3 mmol of potassium and sodium. Obviously, this fluid therapy also will require the administration of calories as glucose. Abnormal losses are added to the maintenance requirements for the total amount required.

Monitoring therapy in the neonate is crucial to an appropriate outcome, since the neonate has higher insensible water loss due to increased permeability of the epidermis, greater skin blood flow, and a higher respiratory rate, as well as a larger surface-to-volume ratio, than the adult. Fluid overload in these patients may lead to congestive heart failure secondary to a patent ductus arteriosus with a left-to-right shunt.

Therapy should be monitored regularly. Laboratory tests include serum sodium, potassium, chloride, osmolality, and urine specific gravity. Other laboratory tests that may be required in specific disorders include blood urea nitrogen, serum creatinine, and urinary electrolytes. (These parameters should be monitored periodically until they return to normal.)

ENERGY METABOLISM

Carbohydrates

The carbohydrates hold an important place in the production of metabolic calories. They are usually ingested as complex polymers (starch, saccharose, etc.) and are broken down into monosaccharides, which are absorbed by the intestine and enter the portal vein to be transformed into glucose-6-phosphate in the hepatocyte. This compound can follow any of five different metabolic pathways depending on the physiologic needs of the moment.

1. It can be hydrolyzed to glucose and phosphate by the enzyme glucose-6-phosphatase, which is present exclusively in the hepatocyte. The glucose leaves the liver cells and circulates in blood, nourishing the cells of all the organs.
2. A small portion is metabolized in the hepatic cell to produce energy.
3. It may follow the pentose-phosphate pathway to supply the nucleotides of pyrimidine-reduced pyridine (NADPH), which will be utilized in the synthesis of certain compounds such as cholesterol and fatty acids.
4. It may be stored in the hepatocytes as glycogen, which is composed of thousands of glucose molecules, but exerts the same osmotic pressure as a single mole of glucose. Thus the storage of glucose in the liver is done without altering the body's intracellular osmotic pressure.
5. When all the preceding avenues are saturated by an excessive availability of glucose, it is transformed into lipids after going through the two-carbon acetyl-CoA stage.

The glucose that gains intracellular access to any organ is immediately transformed into glucose-6-phosphate, which is unable to traverse the cellular membrane. Only hepatocytes contain glucose-6-phosphatase and are capable of freeing glucose.

The blood glucose is captured by all cells, which utilize it either to derive energy by metabolizing it to CO_2 and water or transform it into polymers and store it (glycogen of muscle cells) or incorporate it into macromolecules such as glycoproteins. Thus, the importance of maintaining a normal glucose level and its narrow hormonal regulation is clear.

During periods of muscular contraction, the myocytes utilize chemical energy derived from ATP and convert glycogen into glucose, which is catabolized to lactic acid. Lactic acid leaves the muscle and travels by means of the blood to the liver, where it is resynthesized to glucose (Cori's cycle).

Between meals, when the glycemia tends to decrease, the hepatic glycogen frees units of glucose (glycogenolysis). If the degradation of glycogen is insufficient to maintain a normal glycemia to satisfy the organs that are glucose-dependent (brain, red cells), the process of gluconeogenesis goes into effect. Gluconeogenesis is the synthesis of glucose in the liver utilizing noncarbohydrate substances such as alanine and other glycogenic amino acids or lactic acid that comes from muscles or from glycerol (lipids).

Hormonal Control of Carbohydrate Metabolism

Many hormones are involved in the control of carbohydrate metabolism. Their effects are aimed at ensuring that glucose leaves the liver in exactly the same amount as glucose being utilized by other organs. This balance is based on maintaining the level of blood glucose within a relatively narrow range. The main two hormones that control glycemia and which have apparent antagonistic effects are glucagon and insulin. Glucagon produces hyperglycemia and insulin produces hypoglycemia.

Glucagon produces glycogenolysis, activating the hepatic phosphorylase under the influence of cyclic AMP. A second mechanism is the hyperglycemic effect of the hypophysis through growth hormone and ACTH, which activates the gluconeogenesis mechanism and the production of glucose from amino acids, glycerol, and triglycerides. At the same time, the fatty acids provide the necessary energy. Note that these are the "stress" hormones and as such can influence test results done under nonbasal conditions.

Insulin plays an essential and continuous role in the utilization of glucose derived from the reactions just described. It is secreted by the beta cells of the islets of Langerhans of the pancreas, where it is primarily synthesized as a larger inactive precursor composed of a single polypetide chain called *proinsulin*. This polypeptide is cleaved at several points by peptidases, which break up the polypeptide in a N-terminal fraction and a central protein called *peptide C*.

The amount of peptide C freed into blood is equimolecular with the amount of insulin produced. The C-peptide is not metabolized by the liver. Its assessment allows us to evaluate the rate of endogenous insulin secretion.

The molecule of active insulin is composed of two polypeptide chains A and B joined by two disulfide bonds. It attaches to specific receptors in the membrane of peripheral cells and activates the process of membrane penetration of glucose. It also inhibits the migration of hepatic glucose by the action of glucose-6-phosphatase and activates the synthesis of glycogen.

Diabetes
Classification

TYPE I: Insulin-dependent diabetes mellitus
TYPE II: Non-insulin-dependent diabetes mellitus
TYPE III: Gestational diabetes mellitus

These disorders affect more than 6 million Americans and generate direct and indirect costs of more than $12 billion a year.

Screening for disease is only justified when early treatment in asymptomatic patients is more effective than treatment begun after symptoms have appeared. Currently, early treatment is available only for gestational diabetes.

Gestational diabetes is glucose intolerance that appears for the first time during pregnancy. It occurs in approximately 3% of all pregnancies. Untreated gestational diabetes mellitus (GDM) can cause macrosomia of the fetus, increased risk of birth trauma, hyperbilirubinemia, hypercalcemia, hypoglycemia, or respiratory distress syndrome (RDS). Control of plasma glucose in the mother usually can be accomplished by diet alone, although insulin treatment is required in 10% to

15% of patients with GDM (oral hypoglycemic agents are contraindicated in GDM because of possible teratogenesis). Desirable fasting blood glucose levels are less than 115 mg/dL.

The etiologic considerations of diabetes suggest an underlying hereditary role. Hyperglycemic syndromes can be the result of pancreatic insults that reduce its capacity to secrete insulin (*e.g.*, pancreatitis, carcinoma of the pancreas, hemochromatosis, pancreatectomy) or can be the result of hyperfunction of antagonistic hormones (*e.g.*, acromegaly, hypercortisolism, pheochromocytoma, glucagonoma).

Hyperglycemias also can be iatrogenic as a result of corticotherapy or the use of progestational estrogens or certain drugs that sometimes can elicit a latent anomaly to glucose tolerance.

There are two main forms of diabetes: (1) insulin-dependent diabetes mellitus (IDDM) and (2) non-insulin-dependent diabetes mellitus (NIDDM). Usually, IDDM is seen in young individuals (infantile or juvenile diabetes); occasionally it is seen in older people. It appears to be related to diverse insults, particularly viral infections of the beta cells of the islets of Langerhans that exhibit an exaggerated immunologic reaction and a genetic background characterized by the predominance of certain HLA groups (Dw3 Dw4 B8 Bw15). NIDDM, type II, or maturity-onset diabetes mellitus, is by far the more common (80% of the cases) and is accompanied in the majority of cases by obesity. It is manifested as a peripheral resistance to insulin associated with an anomaly of insulin secretion that is responsible for glucose intolerance.

The end result of diabetes, regardless of its type, is related to diffuse tissue damage involving chiefly the vascular basal membranes and is responsible for the microangiopathy in the retina and glomeruli and involvement of the nervous system. Additionally, diabetes and glucose intolerance are factors in the risk for atheroma formation. It has been postulated that glycosylated hemoglobin binds oxygen with great affinity and probably deprives the endothelium of adequate oxygenation.

The concept of reversible functional angiopathy with normalization of glycemia reinforces the need to control diabetes mellitus as close as possible.

Diabetic Ketoacidosis. One of the consequences of the altered glucose metabolism is ketosis, with the possibility of ketoacidosis. In coma states it is related to the acute deficiency of insulin and to the absence of glycolysis and the resulting absence of oxaloacetic acid.

Nitrogen Metabolism

The liver plays a central role in nitrogen metabolism. It is involved with the metabolic interconversions of amino acids and the synthesis of nonessential amino acids. The liver synthesizes most body proteins, such as albumin, beta-macroglobulin, and orosomucoid, but does not synthesize the immunoglobulins. Hemoglobin is synthesized in the liver of the neonate before birth but not in the liver of the

adult. The liver is also largely responsible for the production of metabolic end-products, such as creatinine, urea, ammonia, and uric acid, which can be more easily excreted. The liver is also involved with the detoxification of bilirubin produced from hemoglobin. The metabolic reactions primarily are concerned with adding glucuronic acid to bilirubin to increase its aqueous solubility. These metabolites are usually excreted more easily in the urine or intestinal lumen.

Table 31-5 lists the reference ranges for creatinine, ammonia, urea, and uric acid and the effect of age on their concentration. Blood ammonia decreases after the neonatal period because of the completion of the development of the hepatic portal circulation. Urea and uric acid increase because of dietary changes during development. Creatinine increases slightly owing to growth, which increases muscle mass. This analyte is relatively independent of diet and thus can be used as a measure of renal function.

Bilirubin

The breakdown of heme-containing proteins, primarily hemoglobin, results in the production of about 250 mg bilirubin per day. This occurs in the spleen as a result of hydrolysis of heme to release the iron and form the intermediate biliverdin, which is reduced to bilirubin. The iron is bound to transferrin and is reutilized for heme synthesis. The bilirubin is bound to albumin and is transported to the liver. This serves to detoxify bilirubin during transport. In some instances, bilirubin can become covalently bound to albumin, which is referred to as *delta-bilirubin*. This complex is apparently nontoxic and has been suggested as a measure of the recovery phase in obstructive liver disease.

Bilirubin binding occurs at two to three sites on albumins, one with a high affinity for bilirubin and one or two weak sites. This bilirubin can be displaced by drugs such as salicylate, which can cause nerve damage.

Once the bilirubin is transported to the liver, it is released from albumin and reacts with one or two molecules of glucuronic acid. This increases the water solubility of bilirubin so that it can be excreted in the bile. The majority of bilirubin (about 85%) is excreted as diglucuronide and the remainder as monoglucuronide. Following excretion in the bile, the conjugated bilirubin is cleaved to bilirubin, which is converted by intestinal bacteria to a series of urobilino-

TABLE 31-5. Reference Ranges for Selected Nitrogenous Metabolites in Plasma Analyte

Age (Years)	Ammonia (µmol/L)	Urea (mg/L)	Creatinine (mg/L)	Uric Acid (mg/L)
0–1	—	60–450	2–10	10–76
1–5	10–40	50–170	2–10	18–50
5–19	11–35	80–220	4–13	30–60
Adult male	11–35	100–210	5–12	40–90
Adult female	11–35	100–210	4–10	30–60

gens. These metabolic steps result in steady-state concentrations in the blood and urine of bilirubin and its metabolites, as indicated in Table 31-6.

Increases in bilirubin result from either overproduction or impaired excretion. Overproduction is the result of hemolysis, whereas impaired excretion signals liver dysfunction. Increased production is the result of accelerated red-cell hemolysis. Total serum bilirubin generally does not exceed 50 mg/L in adults but may be higher in infants. It is virtually all unconjugated bilirubin. The laboratory analysis should include a complete blood count as a means of identifying the source of the hemolysis. The distinction between these alternatives is accomplished by clinical assessment and laboratory evaluation.

If the clinical features of the disease suggest liver dysfunction, bilirubin analysis can be helpful in differentiating the cause of jaundice. Prehepatic jaundice results in a large increase in unconjugated bilirubin because of the increased release and metabolism of hemoglobin after hemolysis. No increase or only a slight increase in conjugated bilirubin is observed because the transport of bilirubin into the liver and the formation of the glucuronide conjugated become rate limiting. Additionally, because of the increased levels of unconjugated bilirubin excreted by the liver, urinary urobilinogen and fecal urobilin concentrations are elevated, but urinary bilirubin (which is only the freely soluble, conjugated form) is absent. In contrast, posthepatic obstructive jaundice is characterized by large increases in serum-conjugated bilirubin. The accumulation of bilirubin in the serum is the result of decreased biliary excretion after the conjugation of bilirubin in the liver rather than the result of an increased bilirubin load caused by hemolysis. Excretion of bilirubin metabolites is low, and urinary bilirubin can usually be demonstrated. Hepatic jaundice presents an intermediate pattern wherein both conjugated and unconjugated serum bilirubin levels are increased to the same degree and conjugated bilirubin is present in the urine. However, the fecal concentration of urobilin is generally decreased.

Urea and Ammonia Metabolism in the Liver

Metabolic pools of amino acids are present in the hepatocytes of the liver. From these pools, amino acids are drawn for the synthesis of proteins. When a protein is degraded, the bulk of the constituent amino acids is returned to these intracellular pools. The released amino acids also can be used in gluconeogenesis, transamination, or deamination reac-

tions or be reincorporated into new proteins. Important transamination reactions are catalyzed by the enzymes alanine aminotransferase (ALT or SGPT) and aspartate aminotransferase (AST or SGOT). In the normal person, the amino groups of excess amino acids in serum are converted to ammonia or urea for excretion, and the carbon skeletons are used for glycogen formation or gluconeogenesis. Some amino acids are also excreted unchanged in the urine.

Urea, creatinine, and ammonia account for 70% to 75% of the serum nonprotein nitrogen; urea accounts for 60% of the total. Most of the metabolism of nonprotein nitrogen occurs in the liver. Urea is produced in the liver from ammonia by means of the Krebs-Henseleit urea cycle. Production of urea is restricted to the liver because arginase, the enzyme that converts arginine to urea and ornithine, is present only in the liver. The majority of excess nitrogen is converted to urea. Since urea is the principal nitrogenous compound excreted by the body, the majority of this formed urea is eliminated in the urine.

Urea Metabolism in Liver Disease

Since urea is synthesized in the liver, liver disease without renal impairment results in a low serum urea nitrogen, although the urea-to-creatinine ratio may remain normal. Other causes of a decreased urea include malnutrition, normal pregnancy, inappropriate ADH secretion, and overhydration. An elevated serum urea nitrogen level does not necessarily imply renal damage, since dehydration may result in a urea nitrogen level as high as 600 mg/L. Infants receiving high-protein formulas may exhibit a urea nitrogen level of 250 to 300 mg/L. Other causes of an elevated urea level include congestive heart failure, hypotension, and renal diseases such as acute glomerulonephritis, chronic nephritis, polycystic kidney, and renal necrosis.

Urea-creatinine ratio is normally between 15:1 and 24:1. Measurement of this ratio can be helpful in assessing the source of azotemia. In cases of retention of urea due to prerenal causes, such as intestinal bleeding, the ratio will be increased sometimes as high as 40:1 because creatinine clearance is normal but the urea load is elevated. In cases of severe renal tubular damage, the ratio will fall to as low as 10:1.

Ammonia

Although blood ammonia concentration is normally quite low, less than 35 μmol/L, ammonia is an important intermediate in amino acid synthesis. Sources of ammonia include hepatic oxidation of glutamate to ketoglutarate, the transamination and oxidative deamination of amino acids and catecholamines, and bacterial breakdown of urea in the gut. The primary mechanism for the metabolic disposal of ammonia is the synthesis of glutamate, glutamine, and carbamyl phosphate. Glutamate and glutamine are both excreted into the urine when in excess. However, if required, they can be reabsorbed and returned to the amino acid pool. Carbamyl phosphate may be used to synthesize orotic acid and ultimately pyrimidines for nucleic acids or to synthesize urea.

TABLE 31-6. Reference Values for Bilirubin

Age	Total Bilirubin (mg/L)	Conjugated Bilirubin (mg/L)	Delta-Bilirubin (mg/L)
Up to 1 week	1–126	0–12	Not detected
1 week to 1 year	2–12	0–5	3–6
Greater than 1 year	2–14	0–2	3–6

In addition to being converted to urea, ammonia is also excreted in the urine as the ammonium ion (NH_4^+). This ammonium salt formation from ammonia serves as a significant mechanism for excretion of excess protons produced during metabolism. This acid–base balance function of ammonia becomes important particularly in diseases associated with acidosis.

Blood ammonia concentration is higher in infants than adults because the development of the hepatic circulation is not completed until after birth. Hyperammonemia results infrequently from congenital defects of the urea cycle. The most common of these inborn errors of metabolism is a deficiency of the enzyme ornithine transcarbamylase. A much more frequent cause of hyperammonemia in infants is hyperalimentation. Most hyperalimentation fluids contain amino acid concentrations above the infant's nutritional nitrogen requirements. The amino acids are used for energy, and the excess ammonia is excreted in the urine. Ammonia concentrations can be as high as 60 μmol/L. Reye's syndrome is frequently diagnosed by an elevated blood ammonia level in the absence of any other demonstrable cause. Reye's syndrome is usually seen in children rather than infants. This may be the result of maternally acquired immunity in the infant. Elevation of plasma ammonia above five times normal is associated with significant mortality in Reye's syndrome.

Some patients exhibit elevated blood ammonia levels in the terminal stages of liver cirrhosis, hepatic failure, and acute and subacute liver necrosis. This is due to the failure of the liver to produce urea. Ammonia is a toxin that acts as a CNS depressant and can cause coma. Urinary ammonia excretion is increased in acidosis and decreased in alkalosis, since ammonia salt formation is a significant mechanism for excretion of excess protons. Damage to the renal distal tubules, as occurs in renal failure, glomerulonephritis, hypercorticoidism, and Addison's disease, results in decreased ammonia excretion in the urine.

Creatinine is formed by the dehydration of creatine. Creatine is produced by the reversible transamidination of the guanidino group of arginine to glycine. Creatine is produced in the liver and in muscle tissue. It is phosphorylated with ATP to produce creatine phosphate, a reserve energy source. The irreversible dehydration of creatine to produce creatinine results in a metabolic end-product that is filtered by the kidney and excreted in the urine.

Creatinine concentration is relatively constant in the serum at 5 to 12 mg/L. The amount excreted daily is a function of individual body muscle mass. Therefore, women and children tend to have lower total urinary creatinine values than men. Serum and urine creatinine measurements are most useful in evaluating renal function. Calculation of a creatinine clearance by determining the ratio of serum-to-urine creatinine and correcting for body surface area is a good measure of kidney function, since, as noted earlier, creatinine does not vary appreciably with diet. The urea clearance is no longer used for this purpose because of dietary induced fluctuations in clearance results.

Uric Acid and Cell Turnover

Uric acid is the end-product of purine metabolism. As cells die and are replaced, a portion of the DNA and RNA is degraded into the individual purines and pyrimidines. The purines are converted to uric acid as shown in Figure 31-1.

The production of uric acid in humans is relatively constant. However, dietary intake of foods high in purines, such as liver and shell fish, can increase serum uric acid concentrations. Men, because of their increased muscle mass, exhibit a higher concentration of uric acid than do women and children. Continuous urinary excretion of uric acid is important, since uric acid is not highly water-soluble and the normal concentration of 55 to 65 mg/L is near the solubility limit for uric acid. Therefore, increased dietary load or renal dysfunction can quickly result in crystallization of uric acid in the bladder and kidney resulting in stones or in crystallization in the joints of the extremities resulting in gout.

Uric acid is useful in assessing cellular turnover and destruction. Therefore, uric acid is increased in idiopathic hyperuricemia, gout, chronic renal disease, tissue neurosis, metabolic acidosis, and hematologic conditions such as leukemia, lymphoma, and anemia.

Lipid Metabolism

The common American diet contains about 40% fat. The fat is predominately animal fat (*i.e.*,) saturated fat while the remainder is vegetable or polyunsaturated fat. These fats are

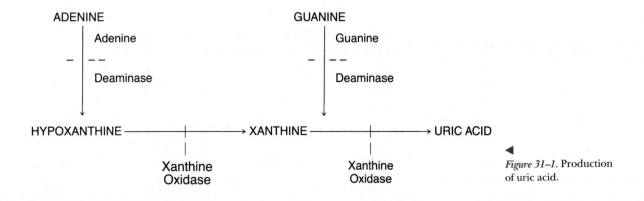

Figure 31–1. Production of uric acid.

TABLE 31-7. The Plasma Apolipoproteins

Apolipoprotein	Plasma Concentration (mg/mL)	Lipoprotein Class
AI	1.2–1.4	Chylomicrons, HDL
AII	0.25–0.40	HDL
AIV	0.14	Chylomicrons, HDL
BI	0.9	VLDL, LDL
BII	?	Chylomicrons, VLDL
CI	0.07	Chylomicrons, VLDL, HDL
CII	0.04	Chylomicrons, VLDL, HDL
CIII	0.100	Chylomicrons, VLDL, HDL
D	0.100	HDL
E	0.045	VLDL, HDL

composed primarily of triglycerides. The remainder is cholesterol, steroids, phospholipids, and mono- and diglycerides. The triglycerides contain primarily the fatty acids myristic, palmitic, stearic, oleic, and linoleic acid. Three of the polyunsaturated fatty acids are considered to be essential because they cannot be synthesized by the human body.

Ingested fat absorption occurs in three stages: the intraluminal phase, in which fats are hydrolyzed; the cellular or absorptive phase, in which the hydrolyzed materials are absorbed by mucosal cells and then the triglycerides are re-formed; and the transport phase, during which the lipids are transported throughout the body.

Bile acids produced from cholesterol in the liver and stored in the gallbladder play an important role in the intraluminal phase by emulsifying the lipids and allowing attack by lipases. The lipids are converted to glycerides and fatty acids. During the absorptive phase, these compounds are taken up by mucosal cells of the small intestine and are reformed into triglycerides. Once the triglycerides are re-formed, they are packaged with apolipoproteins (Table 31-7). These lipoproteins, primarily chylomicrons, are transported in the lymphatic vessels and then in the systemic circulation.

Metabolism of chylomicrons is complex and involves sequential delipidation of the chylomicrons. This, in turn, results in the formation of the various classes of lipoproteins. Chylomicrons and very-low-density lipoproteins (VLDLs) are depleted of triglycerides by a lipoprotein lipase. VLDLs are sequentially depleted of apoprotein C and triglycerides and ultimately form low-density lipoproteins (LDLs). LDLs and other lipoprotein remnants are then converted to high-density lipoproteins (HDLs) by action of lipoprotein lipase (CII). This conversion occurs by two pathways, the receptor pathway and the Scavenger cell pathway. The receptor pathway ultimately transfers LDLs to liposomes of adipose tissue. Both apoproteins B and E are important in this pathway. LDLs are probably the major vehicle for the catabolism of cholesterol esters resulting from lecithin–cholesterol acyl transferase (AI) action. Esters are transferred from either VLDLs or HDLs and this is mediated by apoprotein D.

Fatty acids are metabolized by the liver in the beta-oxidation cascade. This, in turn, results in the production of acetyl-CoA and, subsequently, ATP and NADH. In cases of starvation or diabetes, the formation of acetyl-CoA exceeds the ability to form citrate in the tricarboxylic acid cycle. This excess acetyl-CoA is shunted into forming the ketone bodies acetone acetoacetate and beta-hydroxybutyrate, and metabolic acidosis develops. Subsequent replenishment of oxaloacetate by carbohydrate metabolism allows normal metabolism of acetyl-CoA and relief of the acidosis.

The metabolism of lipids in infants is essentially the same as in adults. However, the expected values are different (Table 31-8). In the case of atherosclerosis, it is now established that lower values of cholesterol (<170 mg/dL) are required in childhood if atherosclerosis is to be minimized in adulthood. Additionally, in early infancy, hydroxybutyrate can be used by the brain as an energy source. This is not so in adults.

Disorders of lipid metabolism in childhood are either genetic or acquired. Acquired disorders include hepatitis, which causes a rise in cholesterol in the early stages of the disease. Other causes of obstructive liver disease cause the same increase. Malabsorption results in decreased triglycerides and cholesterol. The syndrome is characterized by abnormal fecal fat loss (>6 g/day). The clinical diagnosis is often confirmed by a 72-hour fecal fat measurement. Malabsorption, in turn, can result in impaired absorption of the fat-soluble vitamins A, D, E, and K. Three general defects cause malabsorption in infants: (1) a defect in intraluminal

TABLE 31-8. Age-Related Reference Ranges for Serum Lipids

Age (Years)	Cholesterol (mg/dL)	HDL Cholesterol (mg/dL)	LDL Cholesterol (mg/dL)	Triglycerides (mg/dL)
<1	90–260	—	—	20–120
1–5	95–215	35–82	55–160	20–120
5–10	95–240	35–65	75–165	20–190
10–20	110–230	35–65	50–160	25–270

absorption, which can result from pancreatic insufficiency, bacterial overgrowth, or hepatobiliary disease; (2) a defect in mucosal cell transport resulting from disaccharidase deficiency, beta-lipoproteinemia, vitamin B_{12} malabsorption, or endocrine disorders; and (3) intestinal lymphatic obstruction due to congestive heart failure, constrictive pancarditis, or intestinal lymphangiectasia.

Genetic disorders of lipid metabolism include the sphingolipid storage diseases (Table 31-9), mitochondrial and electron-transport defects, defects of beta-oxidation, and the dyslipoproteinemias. The storage diseases are all autosomal recessive mutations, except Fabry's disease, which is x-linked. They are almost all degenerative diseases with mental retardation and early demise. Table 31-9 summarizes the storage diseases. Defects of mitochondrial transport of lipids and electron transport, although rare, have been reported. In some instances, treatment with carnitine may have a palliative effect. Defects of beta-oxidation are known to occur and are the result of mutation in the enzymes of the beta-oxidation cascade.

The dyslipoproteinemias include both the hypolipidemias and the hyperlipidemias. Hypolipidemias include abetalipoproteinemia, which results in degeneration of the cerebellar tract. Tangiers disease is due to the absence of HDL and results in the accumulation of cholesterol in the tissues. Malabsorption often results in hypolipidosis.

Hypercholesterolemia has a strong genetic component, and evidence is accumulating that early intervention and therapy can lower the incidence of coronary heart disease. Mass screening of children for hypercholesterolemia remains controversial. The American Academy of Pediatrics recommends that only children older than 2 years and with a family history of myocardial infarction at an early age be screened for hypercholesterolemia. Children of this group with serum cholesterol levels greater than 170 mg/dL require follow-up care. Prior to initiating either dietary or drug therapy, secondary causes of hypercholesterolemia should be ruled out. These include liver disease, hypothyroidism, nephritic syndrome, anorexia nervosa, and systemic lupus erythematosus. If dietary therapy is to be used, other coronary heart disease risk factors such as obesity, smoking, high blood pressure, and sedentary lifestyle should be assessed and modified if possible. Drug therapy should be reserved only for the most severe cases where dietary intervention has not been effective.

DEVELOPMENTAL CHANGES IN HORMONE BALANCE

Hormones are substances that are formed by one group of cells, secreted into blood, and carried to other cells, where they manifest their effects. Some of them are proteins and others are steroids or amino acid derivatives. The endocrine system is involved in the control of many aspects of growth and development.

TABLE 31-9. The Sphingolipidoses

Disease	Signs and Symptoms	Enzyme Defect	Mode of Inheritance	Variants Identified
Farber's disease	Hoarseness, dermatitis, skeletal deformation, mental retardation	Ceramidase	Autosomal recessive	—
Gaucher's disease	Spleen and liver enlargement, erosion of long bones and pelvis, mental retardation only in infantile form	Glucocerebroside, β-glucosidase	Autosomal recessive	3
Niemann-Pick disease	Liver and spleen enlargement, mental retardation, about 30% with red spot in retina	Sphingomyelinase	Autosomal recessive	5
Krabbe's disease (globoid)	Mental retardation, almost total absence of myelin	Galactocerebroside, β-galactosidase	Autosomal recessive	—
Metachromatic leukodystrophy	Mental retardation, psychological disturbances in adult form, nerves stain yellow-brown with cresyl violet dye	Sulfatidase	Autosomal recessive	4
Fabry's disease	Reddish purple skin rash, kidney failure, pain in lower extremities	Ceramidetrihexoside, α-galactosidase	X-linked	1
Tay-Sachs disease	Mental retardation, red spot in retina, blindness, muscular weakness	Hexosaminidase A	Autosomal recessive	1
Sandhoff's disease	Same as Tay-Sachs disease, but progressing more rapidly	Hexosaminidase A and B	Autosomal recessive	1
Generalized gangliosidosis	Mental retardation, liver enlargement, skeletal deformities, about 50% with red spot in retina	β-Galactosidase	Autosomal recessive	2
Fucosidosis	Cerebral degeneration, muscle spasticity, thick skin	α-Fucosidase	Autosomal recessive	—

Regulation of Secretion

The endocrine output can be stimulated or suppressed by other hormones or regulatory factors. The production of cortisol and thyroid and sex hormones is under the influence of trophic hormones from the anterior pituitary. These, in turn, are regulated by hypothalamic factors—for example, thyrotropin-releasing hormone (TRH) from the hypothalamus stimulates the pituitary to produce and secrete thyroid-stimulating hormone (TSH) which, in turn, stimulates the production of thyroid hormones that serve as a signal to the pituitary to increase or decrease the production of TSH. Thyroid hormone evidently also mediates the inhibition of TRH. This kind of control is called *negative feedback*, and many endocrine glands are regulated in this fashion. For example, parathyroid hormone secretion is directly stimulated by the level of Ca^{2+}.

Hormones act on target cells by interacting with specific receptors located in the cell membrane in the protein hormone systems, whereas receptors for most of the nonprotein hormones are intracellular and interact directly with chromosomal acceptors. Depending on the system, specific DNA is transcribed and mRNA for that message is increased in the cell, resulting in the synthesis of a specific protein.

Peptide and Polypeptide Hormones

These hormones include TRH, GnRH (gonadotropin-releasing hormone), CRH (corticotropin-releasing hormone), and somatostatin, as well as the protein hormones from the anterior pituitary: ACTH, GH, TSH, FSH, LH, and MSH (melanocyte-stimulating hormone). Also included are the pancreatic hormones insulin and glucagon; parathyroid hormone (PTH); the medullary cells of the thyroid, calcitonin; and the gonads, inhibin. Many of these hormones exert their effects by increasing intracellular cyclic adenosine monophosphate (cAMP) in their respective target cells. The binding of the hormone to the receptor site leads to activation of membrane-bound adenylate cyclase, which catalyzes the conversion of ATP to cAMP.

Derivatives of Amino Acids

Both thyroid hormone and catecholamines are derivatives of the amino acid tyrosine. Epinephrine interacts with adrenergic receptors on the cell membrane and alters adenylate cyclase in a manner similar to peptide hormones. Thyroid hormones bind to nuclear receptors, increasing the synthesis of specific mRNAs and the production of specific proteins. T_3 and T_4 are carried in the blood bound reversibly to thyroid-binding globulin (TBG) and thyroxine-binding prealbumin (TBPA). Only the small free hormone fractions interact with their specific receptors.

Steroid Hormones

Steroids are lipid-soluble hormones that are carried in the bloodstream bound to protein carriers and readily cross the outer cell membrane to enter the target cell, where specific receptors are intracellular.

There are six general categories of steroid hormones: vitamin D and its derivatives, glucocorticoids, mineralocorticoids, estrogens, progestins, and androgens. They are all derived from cholesterol and all, except for vitamin D, are synthesized by a common metabolic pathway.

Transcortin (cortisol-binding globulin, CBG) has high affinity for both progesterone and cortisol. Sex hormone binding globulin (SHBG) binds dihydroxytestosterone, testosterone, and estradiol with high affinity.

Role of Hormones in Normal Growth

Different tissues mature at different rates; thus the growth of a child is a complex series of changes. One of the problems pediatricians face is the child who is very short and who may have a treatable disorder of growth.

Standardized growth curves are available for heights and weights. The 50th percentile at a given age is that height at which 50% of the children fall above the curve and 50% fall below the curve.

The major factor contributing to growth is lengthening of the skeleton. Several tables exist to predict adult height from a comparison of height and bone age, but such predictions are only rough estimates.

Most of the hormones involved in growth (GH, TSH, glucocorticoids, sex hormones, and somatomedins) are partly or completely regulated by the hypothalamic neurons and are secreted and carried to the anterior pituitary. The factors involved in controlling pituitary synthesis and secretion of growth hormone have been isolated, purified, and their structures elucidated. Somatostatin, TRH, and GnRH are available for clinical studies. CRH and GHRH are currently available only for research.

The inhibiting factors somatostatin and dopamine inhibit release of the appropriate pituitary hormones (GH, prolactin). Those pituitary trophic hormones acting on their respective target tissues activate the synthesis and release of the final hormone: somatomedin C (SmC), GH, TSH, glucocorticoids (ACTH), and sex hormones (LH, FSH). The final hormone products produce a negative-feedback effect on the hypothalamus and/or pituitary gland, thus controlling formation of the releasing factors and the pituitary trophic hormones.

Growth Hormone

Growth hormone (GH) promotes normal growth from birth to adulthood. GH is present and produced in the fetus but is not indispensable, as seen in normal weight anencephalics born without pituitary glands.

Children who are deficient in GH grow at a much lower rate and reach an adult height of about 130 cm with normal body proportions. Growth hormone appears to act indirectly on target tissues, by means of somatomedins. The terms *somatomedin* and *insulin-like growth factor (IGF)* are now used as synonyms. Human serum contains IGFI, which is immunologically identical to SmC and IGFII. Their biologic responses are mediated by interaction with the insulin receptor or with specific IGF receptors, depending on the tissue. The IGF is tightly bound to carrier proteins. IGFI-related peptides reach a peak during puberty and fall during old age. It is known that the production of IGFI-related peptides is regulated by GH; the trophic hormone for IGFII peptides is unknown.

In addition to IGF peptides, many other growth factors are in the tissues and blood. Factors such as epidermal growth factor, erythropoietin, nerve growth factor, platelet-derived growth factor, and fibroblast pneumocyte factor are all involved with growth.

Thyroid Hormone

Although it does not appear to play a direct role in growth of the fetus, thyroid-stimulating hormone (TSH) is essential for protein synthesis in the brain and for normal development of the CNS and skeleton and dental maturation. Severe hypothyroidism results in nearly complete lack of growth. T_3 increases the concentration of mRNA for GH in GH-secreting cells. In general, thyroid hormone stimulates metabolism, which results in growth and development of the organism.

Sex Hormones

Testosterone and its metabolite, dihydrotestosterone (DHT), are powerful anabolic hormones that promote growth, weight gain, and muscle mass. The presence of growth hormone is necessary for these effects. Estrogens in physiologic amounts stimulate growth and epiphyseal maturation; in excessive amounts, their effect may be inhibitory. Administration of pharmacologic doses of estrogen to excessively tall girls leads to alteration of their growth and acceleration of epiphyseal maturation and fusion.

Glucocorticoids

Glucocorticoids, as opposed to the growth-promoting effects of other hormones, have an inhibitory effect of growth. It takes only a two- or threefold amount above the average of cortisol to arrest growth. GH administered under this set of circumstances will not reverse the growth retardation. Glucocorticoids impede growth by inhibiting thymidine incorporation into DNA of a variety of tissues.

Insulin

There is evidence that suggests that insulin may function as a stimulator of growth in addition to its role in carbohydrate metabolism. Insulin deficiency or insulin resistance is associated with intrauterine growth retardation. In tissue cultures, high concentrations of insulin are required to stimulate DNA synthesis and mitosis. Insulin promotes the uptake of glucose and amino acids, and its role in growth may be to provide the energy and building blocks for growth rather than directly influencing it.

Primary Disorders of Growth—Short Stature

Skeletal Disorders

Achondroplastic dwarfism is due to mutation of a single gene and is inherited as an autosomal dominant condition. Such individuals have normal intelligence and health.

Chromosomal Abnormalities

Growth is retarded, but bodily proportions are normal or close to it.

Abnormalities of Autosomes. The most common chromosomal disorder is trisomy of chromosome 21 (Down's syndrome). Shorter stature, mental retardation, and characteristic physical features are seen.

Abnormalities of X Chromosomes. The lack of the second X chromosome produces Turner syndrome. Any girl with growth retardation should be tested.

Intrauterine Growth Retardation

Chromosomal and Genetic Abnormalities, Congenital Infections, and Congenital Anomalies. Abnormal implantation of the placenta, vascular malformation or disease or malnutrition, toxemia, severe diabetes, or drug abuse can be the cause for small children.

Genetic Short Stature

This should be considered only when all other causes of growth retardation have been excluded. When the diagnosis is made, the question of GH therapy is usually raised. Currently, the policy is to reserve human GH for those children demonstrated to lack biologically active GH.

Secondary Disorders of Growth—Endocrine

Growth Hormone Deficiency

A deficiency of GH secretion or diminished action on the target organ may be the result of impaired pituitary function, hypothalamic dysfunction, or target-cell unresponsiveness. Secretion of GH occurs in a series of irregular bursts throughout the day and night. Sleep, exercise, a rapid fall in glucose concentration, administration of certain amino acids, and stress can bring about bursts of GH release. Sleep-associated GH release is closely associated with the development of slow-wave rhythm in sleep. Grandma was probably right when she said, "If you don't take your nap, you will not grow."

Neurotransmitters play an important role in the regulation of GH secretion. The administration of 500 mg of L-DOPA by mouth leads to the secretion of GH between 45 minutes and 2 hours later. Administration of 5-hydroxy-tryptophan, a serotonin precursor, increases GH release.

Pituitary hypofunction secondary to hypothalamic damage may be caused by infections, granulomata, hydrocephalus, and hypothalamic tumors. The prototype of the GH-resistant syndrome is Loran dwarfism, believed to have an autosomal recessive mode of transmission. These children have elevated basal GH and exaggerated GH response to provocative stimuli. The somatomedin concentrations are low and do not increase in response to GH.

Diagnosis. Resting levels of GH are low. Provocative tests are laborious and expensive.

Exercise Stimulation Test. The child is asked to exercise vigorously (running up and down stairs) for 20 minutes after an overnight or 4-hour fast. A raise of 7 mg/mL or more is considered normal.

Probably one of the most reliable provocative tests is the induction of hypoglycemia with intravenous insulin. The risk of severe hypoglycemia necessitates the presence of a physician while performing the test. An IV line with concentrated glucose should be available during this test.

The nadir of blood glucose occurs 20 to 30 minutes after IV insulin, but GH response is observed later. Blood is collected every 15 minutes for 2½ hours, and blood glucose is monitored by the use of chemstrips during the procedure. The samples obtained are analyzed for glucose, GH, and cortisol. LH, FSH, and thyroid hormone should be checked in the initial blood sample.

Arginine is also an excellent stimulus for GH secretions. Recently, GHRH has become available, but it is not as readily available as TRH. There is a poor correlation between levels of serum SmC and growth rate. Very low values (<0.25 U/mL) should be of some concern. Malnutrition also lowers SmC.

Hypercortisolism (excess amounts of glucocorticoids, either endogenous or exogenous) will suppress linear growth. Sex hormone deficiency and pseudohypoparathyroidism type I are two other conditions where short stature can be manifested.

Excessive Growth

Genetic tall stature, Klinefelter syndrome, Marfan syndrome, and homocystinuria patients are usually well above average in height.

Growth Hormone Excess

Elevation of the mean GH level usually above 10 ng/mL is accompanied by signs and symptoms. Elevations are usually caused by somatotropic adenomas of the pituitary gland (eosinophilic or chromophobic). When manifest in childhood, this produces gigantism. Excess GH after fusion of the epiphyses results in acromegaly.

Impaired glucose tolerance is present in about 50% of the cases of acromegaly. Insulin resistance is present in varying degrees in all patients.

Radiographs of the skulls, hands, and feet are very helpful in making the diagnosis of acromegaly. CAT scans of the area of the sella turcica may show the tumor.

Fetal Gigantism

Fetal gigantism is characterized by excessive growth *in utero* with no evidence of endocrine abnormalities, although some patients have hypoglycemia, probably due to hyperinsulinemia. The infant of a diabetic mother is another example of fetal overgrowth.

Hyperthyroidism

Linear growth may be accelerated somewhat in hyperthyroidism, and the bone age may be advanced.

Precocious Secretion of Sex Hormones

The rapid growth is accompanied by an inappropriate bone maturation for age. The final adult stature is diminished. Regardless of the origin of excess sex hormones when secondary to tumor (gonad, adrenal, hypothalamus), a surgical approach should be employed for removal of the tumor.

Posterior Pituitary Disorders

These conditions are characterized by either excessive or deficient secretion of vasopressin or a deficient response of target cells. However, clinical disorders of oxytocin secretion have not been identified.

The hormones of the posterior pituitary [oxytocin and vasopressin (ADH)] are synthesized in the hypothalamic nuclei and are then transported along the neurohypophyseal tract to the posterior pituitary, where they are stored and released.

Vasopressin

Osmoreceptors located near the supraoptic nuclei and the thirst center influence both drinking behavior and vasopressin (VP) secretion; increased osmolality results in ADH (VP) release and increased osmolality of the urine. VP release is also influenced by intravascular volume, which inhibits release of VP during hypervolemia, while baroreceptors stimulate release during hypotension.

VP half-life is about 15 to 20 minutes. VP receptors are present in cells of the collecting duct of the kidney. Deficiency of VP causes diabetes insipidus, which is diagnosed when urine osmolality is inappropriately low, when serum osmolality is raised.

Oxytocin

Oxytocin is primarily involved in milk ejection, since it binds to myoepithelial cells of the mammary gland. Naturally, its secretion is stimulated by suckling. Oxytocin has been used in obstetrics to induce labor because of its uterus-contracting properties, but oxytocin-deficient women present normal labor.

Thyroid Gland

The synthesis and release of T_4 and T_3 are controlled by TSH from the anterior pituitary, which, like other glycoprotein hormones, is composed of an alpha and a beta subunit. The beta subunit confers specificity on the hormone. The normal range is 0.5 to 4.5 μU/mL.

The synthesis and release of TSH is governed by TRH (thyrotropin-releasing hormone), which is stimulatory, and thyroxine, which is inhibitory. TRH in the bloodstream has a half-life of 5½ minutes, and appropriate doses (200–500 mg) will cause a rise in serum TSH, with a peak in 30 minutes; when T_4 concentration in the blood is low, the response of thyrotrophs to TRH is exaggerated.

T_4 and T_3 circulate bound to TBG (70%), TBPA (20%), thyroxine-binding prealbumin, and albumin. Only about 0.03% of T_4 and 0.3% of T_3 are not bound and are the most "bioavailable" hormones for entry into target cells. Hormone carried by albumin is weakly bound and readily provides additional bioavailable hormone.

Testing Thyroid Function

Serum total T_4 includes both free and protein-bound protein. An increase of TBG will increase total T_4, and in these cases estimation of the free thyroid hormone is necessary.

$$FTI = T_4 \times \% \text{ uptake}$$

or

$$\text{"Normalized" FTI} = T_4 \times \frac{\text{patient uptake}}{\text{pooled normal serum uptake}}$$

(Eq. 31-1)

TBG can be measured directly by RIA if hereditary defects are suspected. Free T_3 and free T_4 determinations also can be performed by equilibrium dialysis or immunoassay. High-sensitivity T_5 assays are increasingly used to evaluate thyroid function.

Thyroid Function Disorders

Hypothyroidism

Hypothyroidism is a clinical state of thyroid hormone insufficiency. Complete or partial lack of thyroid hormone may occur, and it can be congenital or acquired.

Congenital hypothyroidism can be due to dysgenesis, ectopic positioning, inborn errors of thyroid hormone synthesis (dyshormonogenesis), and defects of the hypothalamic-pituitary system. Also, iodine deficiency, as seen in high-altitude areas such as the Andes and Himalayas, goitrogens, or peripheral resistance to thyroid hormone can cause hypothyroidism. Congenital hypothyroidism occurs in about 1 in 3500 to 4000 infants. Twice as many females as males are affected.

Screening programs for hypothyroidism were begun in 1973. Since then, many countries have made this screening mandatory. Usually, it is done using dry blood specimens in filter paper obtained within the first days of life. Infants with low concentrations of T_4 undergo further measurements to assess TSH levels. Once the testing of T_4 and TSH both suggest hypothyroidism, the physician is informed of this result. The infant should be checked for signs of hypothyroidism, and blood should be drawn for additional thyroid hormone analysis, generally T_4, T_3, FTI, and TSH. In congenital TBG deficiency, the T_4 is low, but the FTI is usually normal. The deficiency can be confirmed by TBG analysis.

Infants with hypothyroidism should receive L-thyroxine therapy. Therapy should not be delayed awaiting further testing (scan). If the scan is normal, the thyroid levels are borderline low normal, and the child is asymptomatic, a period of close observation can be instituted, with frequent periodic assessment of thyroid function and reinstitution of therapy as soon as indicated.

Recent review of results of early therapy in those hypothyroid babies picked up in screening programs are encouraging. Early therapy appears to be important in preventing mental retardation, and the expense-benefit ratio of screening programs is heavily in favor of patient benefit and state and federal savings by avoiding potential underemployment and/or need for institutional care.

DRUG THERAPY AND PHARMACOKINETICS

The goal of drug treatment for pediatric patients is the same as in adults—a therapeutic event that improves patient outcome. However, drug therapy in pediatric patients is markedly different than in adults. Adjustment of adult drug dosages solely on the basis of the pediatric patient's weight is inadequate to account for the observed differences in clinical response between pediatric and adult patients. Children and infants exhibit different pharmacodynamics for most drugs than do adults.

These therapeutic changes stem from developmental differences between adults and children. Drug absorption is often different because gastric pH is higher in infants than in adults and gastric emptying time is prolonged in infants relative to adults. At birth, the gastric pH is near neutral, but it rapidly falls to below 3. Gastric acidity is not stable in infants until nearly 3 years of age owing to development of the gastric mucosa. This difference in pH affects absorption of acid-labile drugs such as the penicillins. Delayed gastric emptying can increase the concentration of circulating drug because of an increase in drug absorption. This effect is generally observed only in the first 3 to 6 months of life, at which time gastric emptying approaches that of adults.

Distribution of drugs also can be different in infants and children when compared with adults. This is due primarily to differences in protein binding of drugs and in body water content and compartmentalization. Protein binding is generally reduced in infants and neonates compared with adults. In the neonate, binding of drugs to albumin may be decreased because of bilirubin binding to albumin. An alternative concern relates to drug displacement of bilirubin and the associated increased risk for neurologic damage due to hyperbilirubinemia. Other proteins that bind drugs may actually be present in smaller concentrations in the neonate and infant than in the adult. This decreased binding implies that smaller drug dosages will result in a therapeutic outcome and that the standard dosage may have toxic results.

Drug distribution is also altered by body water and fat content. Following birth, the total body water decreases and fat tissue is only about 15% of body mass, whereas in the adult it is nearer 20%. Skeletal muscle mass is about 25% of body weight. These differences generally result in larger apparent volumes of distribution in infants and children.

Elimination of drugs is also altered in children, as one might expect. For those drugs excreted in the urine without transformation, the elimination rate is determined primarily by renal function. Renal function develops during the first 12 to 18 months of life, when it approaches the adult value. This altered renal function can result in a significantly prolonged half-life of some drugs. At the same time, drugs such as the barbiturates that require hepatic biotransformation by the cytochrome P-450 enzyme system also exhibit a prolonged half-life. This is due to the immaturity of the liver in infants. In addition to decreased drug hydroxylation, infants exhibited a decreased ability to inactivate drugs by reaction with carbohydrates. However, sulfation reactions seem to be similar to those of adults.

It is obvious from these observed differences that a comprehensive program of therapeutic drug monitoring is an essential aspect of pediatric biochemistry. Table 31-10 indicates the suggested specimen collection time following drug administration for several commonly administered drugs.

TABLE 31-10. Suggested Drug Specimen Collection Times for Commonly Prescribed Drugs

Drug Formation	Administration Route	Collection Time
Amikacin	If administered IV, infuse over 30 min	Peak—60 minutes after beginning of dose (60 minutes after IM dose)
		Trough—just prior to next dose
Carbamazepine	Oral dosage	Trough—just prior to next dose
Digoxin	N/A	6–24 hours after dose
Disopyramide	PO for 2 days	Trough—just prior to next dose
Ethosuximide	PO for 1 week	Peak—2–4 hours after dose
		Trough—just prior to next dose
Gentamicin/tobramycin	If administered IV, infuse over 30 min	Peak—60 minutes after beginning of dose (60 minutes after IM dose)
		Trough—just prior to next dose
Lidocaine	IV	12 hours after initiating therapy and every 12 hours thereafter
Phenobarbital	N/A	Patient should be at steady state (10–25 days)
Phenytoin	N/A	Patients should be at steady state (2 days)
Primidone	N/A	Trough—just prior to next dose
Procainamide	PO	Peak—60–75 minutes after dose
		Trough—just prior to next dose
Propranolol	N/A	Patient should be at steady state (minimum 24 hours)
		Trough—just prior to next dose
Quinidine	N/A	6–24 hours after dose
Theophylline	IV therapy	Prior to intravenous infusion if patient has received theophylline 30 minutes after completion of loading dose
		12–18 hours after beginning infusion
		Every 24 hours thereafter
	Oral therapy	Peak—2 hours after liquid or rapid dissolution dose regardless of name of drug
		4 hours after Slo-Phylline dose
		4–6 hours after Theodur dose
		Trough—just prior to next dose
Valproic acid	PO (since syrup is absorbed more rapidly than tablets, use the shorter peak draw time if syrup is used)	Peak—60–180 minutes after dose
		Trough—just prior to next dose
Chloramphenicol	IV	Peak—IV push: 60 minutes after dose IV infusion: 60 minutes after infusion was begun (30-minute drug infusion)
	PO	2 hours after dose
	All	Trough—just prior to next dose (toxic trough concentrations have not been established)
Vancomycin	IV	Peak—2 hours after start of dose
		Trough—just prior to next dose

Table 31-11 summarizes the therapeutic ranges from several drug classes.

INBORN ERRORS OF METABOLISM

Metabolic Screening Procedures

Procedures for identifying inherited metabolic disorders span the spectrum from broad screening procedures to specific tests for individual defects. Screening tests are generally oriented toward biochemical systems, and the first-line specimen of choice is urine. Timed urine specimens are the most useful because they permit estimation of the amount of metabolite being produced per unit time. Blood specimens are generally used for follow-up testing. Many metabolic disorders only become apparent during periods of significant stress. Urine specimens collected during these periods can be particularly enlightening.

The clinician is presented with a difficult problem in differentiating metabolic causes of disease from other organic causes because of the nonspecificity of the clinical signs and symptoms. Symptoms may include lethargy, convulsions, fever, and hepatosplenomegaly, among others. Clearly, these symptoms also can be present in serious infections.

Once a clinical suspicion has developed that an inherited metabolic disorder is the source of the illness, a routine urinalysis can be very helpful. The odor of the urine can be distinctive (Table 31-12), such as the sweet odor associated with maple syrup urine disease. Urine color may change upon standing, such as occurs in alcaptonuria. Crystals present in urine also may be instructive. Cystinuria can easily be recognized by the presence of characteristic crystals of cystine in the urine. Other useful information can be the

TABLE 31-12. Odors Associated with Inherited Metabolic Disorders

Disorder	Description
Diabetic ketoacidosis	Decomposing fruit, acetone-like
Hawkinsinuria	Swimming pool (chlorine-like)
Isovaleric acidema	Sweaty feet
Maple syrup urine disease	Pancake syrup or burnt sugar
Methionine malabsorption (Oasthouse syndrome)	Malt or hops
β-Methylcrotonic aciduria	Tomcat urine
Methylmalonic acidemia	Ammoniacal
Phenylketonuria	Musty; mouse urine
Propionic acidemia	Ammoniacal
Trimethylaminuria	Fishy
Tyrosinemia	Methionine (cabbage-like)
Urea-cycle defects	Ammoniacal

anion gap, since the organic acids will increase the anion gap in the presence of normal electrolyte balance. Blood ammonia that is persistently elevated in the absence of identifiable causes should lead to consideration of inherited urea-cycle defects, such as argininosuccinic acidemia. There are often physical signs of inherited metabolic diseases. In the case of argininosuccinic acidemia, patients have peculiarly kinky hair that is quite brittle.

Hypoglycemia, particularly with vomiting, should lead the clinician to consider galactosemina, fructose intolerance, or branched-chain amino acid disorders. The presence of hypercalcemia and/or hypophosphatemia may herald the presence of renal tubular disorders. Abnormal liver function also may indicate disorders of the urea cycle, galactosemia, one of the glycogenoses, or aminoacidopathies.

Most urinalysis includes a test for ketones using nitroprusside. Positive results commonly indicate diabetic ketoacidosis but also may indicate maple syrup urine disease or glucose-6-phosphate dehydrogenase deficiency.

Screening tests for inborn errors of metabolism are now mandated by many states. Screening for phenylketonuria (PKU) has been common for many years. Many mandated programs include testing for hypothyroidism and galactosemia, as well as PKU. These tests are generally performed on all newborns about 3 days after birth and following the beginning of protein feedings. These tests have identified and successfully treated these three inborn errors of metabolism. There are, however, a multitude of other inherited metabolic diseases that have been recognized and require diagnosis.

These diseases require broad-spectrum screening programs. The ferric chloride test is usually included because in addition to identifying phenylketonuria, it can identify tyrosinemia, maple syrup urine disease, ketosis, and alcaptonuria (homogentisic acid). The interpretation of results of this test is complicated by the fact that both salicylates and phenothiazines react with ferric chloride. Certainly, a test for reducing substances is required. The Clinitest tablet is the

TABLE 31-11. Therapeutic Ranges (in μg/mL) for Commonly Used Drugs in Pediatric Biochemistry

Anticonvulsants
Phenobarbital	15–40
Valproic acid	50–100
Carbamazepine	8–12
Phenytoin	10–20 (neonates 5–15)

Antimicrobials
Tobramycin	Peak—6–10; trough—up to 2
Gentamicin	Peak—6–10; trough—up to 2
Chloramphenicol	Peak—10–20; trough—up to 10

Antidepressants
Diazepam	0.10–1.50
Propranolol	0.05–0.10
Procainamide	5–10

Antiasthmatics
Theophylline	Neonatal apnea	5–15
Caffeine	Asthma	10–20
	Neonatal apnea	5–20

most commonly used and can be helpful in identifying those diseases listed in Table 31-13. Nitroprusside is included to identify cystinemia or homocystinemia. A test for glycosaminoglycans to identify the mucopolysaccharidoses is an essential aspect of a good screening program. These tests involve color changes due to dye binding with cetylpyridinium chloride or azure A. Dinitrophenylhydrazine is used to identify the organic acidemias. This test has poor sensitivity, and contrast dyes cause significant interference.

Positive results from the screen require confirmation and follow-up testing to identify the specific defect. Organic acids are most commonly identified by gas chromatography and increasingly includes mass spectrometry for specific identification. On the other hand, amino acid disorders are confirmed by any of several techniques, including high-voltage electrophoresis, thin-layer chromatography, either one- or two-dimensional, and high-performance liquid chromatography using either pre- or postcolumn derivatization with a variety of compounds such a *ortho*-phthaldehyde or ninhydrin. Normal ranges for amino acids are shown in Table 31-14.

Many inherited metabolic disorders can only be identified by measuring the specific activity of the enzyme. This often requires cultured cells, either white blood cells or skin fibroblasts. The complexity of many of these tests has restricted their performance to a few centralized laboratories.

TABLE 31-13. Metabolic Diseases with Positive Reducing Substances

Glucosuria
Diabetes mellitus
Renal glycosuria
Cystinosis
Fanconi syndrome
Thyrotoxicosis
Hyperadrenal states
Phenochromocytoma

Galactosuria
Galactosemia
Severe liver damage

Fructosuria
Hereditary fructose intolerance
Essential fructosuria
Severe liver damage

Lactosuria
Lactose intolerance
Severe liver damage
Severe intestinal disease

Positive Benedict's Test Only
Alcaptonuria (oxidation product of homogentisic acid)
Essential pentosuria
Pentosuria ingestion of fruit
Sialic acid
Cephalothin
Ampicillin

These laboratories are not only the repository of the majority of experience with these assays, but very often include the most extensive clinical experience with these rare disorders. Therefore, most laboratories refer these specimens and often patients for work-up by these specialized centers.

Cystic fibrosis is one of most common inborn errors of metabolism, with an incidence of 1 in 2400 live births. Therapy for this disease is palliative and includes both pulmonary lavage and antibiotic therapy, as well as physical therapy. Diagnosis of cystic fibrosis is made by the presence of clinical symptoms of repeated respiratory infections, meconium ileus at birth, and pancreatic insufficiency. If some of these are present, a sweat chloride test should be performed according to the Cystic Fibrosis Foundation guidelines. The results of this test should be positive on two separate occasions.

The gene deletion responsible for cystic fibrosis has been identified. This has made the diagnoses of cystic fibrosis more comprehensive, and polymerase chain-reaction assays are already being developed. Certainly, this identification of the cystic fibrosis gene location holds the promise of permitting identification of carriers for cystic fibrosis and perhaps prenatal diagnosis of cystic fibrosis. This will permit effective genetic counseling. This assay may be as significant a tool as the sweat test in the diagnosis of cystic fibrosis.

HUMORAL AND CELLULAR IMMUNITY

Basic Concepts

Human beings are continually exposed to infectious agents. It is evident that the great majority of infections in normal individuals are of limited duration and leave very little permanent damage. This is due to the individual's immune system. The immune system is divided into two functional divisions, namely, the innate immune system and the adaptive immune system. The *innate immune system* is the first line of defense. The *adaptive immune system* produces a specific reaction to each infectious agent that normally eradicates that agent and remembers that particular infectious agent, preventing it from causing disease later. For example, measles and diphtheria produce a lifelong immunity following an infection. The functions of innate and adaptive immunity are depicted in Table 31-15.

The Innate Immune System

Skin. Skin is an effective barrier to most microorganisms. Most infections enter the body by means of the nasopharynx, gut, lungs, and genitourinary tract. A variety of physical and biochemical defenses protect these areas; for example, lysozyme is an enzyme distributed widely in different secretions that is capable of splitting a bond found in the cell walls of many bacteria.

Phagocytes. When an organism penetrates an epithelial surface, it encounters phagocytic cells or the reticuloendothe-

TABLE 31-14. Reference Values (µmol/L; Mean ± SD) for Amino Acids in Serum

Year Amino Acid, P.S	Age			
	1–12 Months (N = 31)	1–5 Years (N = 30)	5–12 Years (N = 32)	12–18 (N = 22)
Phosphoserine	34 ± 17	32 ± 16	36 ± 14	26 ± 10
Taurine	100 ± 64	104 ± 61	104 ± 56	114 ± 70
Phosphoethanolamine	11 ± 21	8 ± 14	7 ± 9	20 ± 29
Aspartic acid	31 ± 25	23 ± 14	25 ± 11	30 ± 16
Threonine	155 ± 82	127 ± 43	122 ± 34	146 ± 83
Serine	270 ± 208	188 ± 52	181 ± 54	164 ± 67
Asparagine	40 ± 17	42 ± 22	20 ± 14	48 ± 33
Glutamic acid	206 ± 148	184 ± 106	227 ± 61	177 ± 114
Glutamine	379 ± 162	331 ± 116	255 ± 76	311 ± 127
Proline	206 ± 129	197 ± 74	223 ± 105	209 ± 82
Glycine	292 ± 153	260 ± 46	243 ± 56	249 ± 103
Alanine	514 ± 249	492 ± 128	485 ± 134	468 ± 184
Citruline	24 ± 15	30 ± 15	27 ± 16	29 ± 18
α-Aminobutyric acid	10 ± 5	20 ± 24	16 ± 21	9 ± 5
Valine	186 ± 77	210 ± 82	191 ± 68	186 ± 51
Cystine	10 ± 10	9 ± 10	10 ± 12	11 ± 12
Methionine	31 ± 16	21 ± 13	19 ± 10	20 ± 11
Isoleucine	65 ± 27	65 ± 29	66 ± 27	66 ± 26
Leucine	135 ± 53	142 ± 55	140 ± 47	137 ± 48
Tyrosine	94 ± 35	78 ± 27	75 ± 29	74 ± 25
Phenylalanine	73 ± 37	63 ± 22	63 ± 22	70 ± 26
β-Alanine	11 ± 10	7 ± 10	4 ± 5	6 ± 10
Tryptophan	95 ± 29	85 ± 40	75 ± 33	65 ± 54
Ornithine	187 ± 128	154 ± 53	179 ± 59	157 ± 60
Lysine	167 ± 76	174 ± 54	184 ± 55	178 ± 51
Histidine	107 ± 52	103 ± 29	98 ± 24	99 ± 26
Arginine	54 ± 36	46 ± 31	29 ± 20	39 ± 25

Source: Meites S, ed. Pediatric clinical chemistry. 3rd ed. Washington, DC: American Association for Clinical Chemistry, 1988.

lial system. Phagocytes are of different types, but they are all derived from bone marrow. They engulf particles, including infectious agents.

The blood phagocytes include the polymorphonuclear cells (PMNs) and monocytes. Both can migrate out of the blood into tissues in response to suitable stimuli. The PMNs are short-lived, while the monocytes develop into tissue macrophages.

Natural Killer (NK) Cells and Soluble Factors. NK cells are leukocytes capable of recognizing cell-surface changes on virally infected cells. The NK cells bind to these target cells and can kill them. The natural killer cells are activated by interferons, which are themselves components of the innate immune system. Interferons are produced by virally infected cells and sometimes also by lymphocytes. Apart from their action on NK cells, interferons induce a state of viral resistance in uninfected tissue cells. Interferons are produced very early in infection and are the first line of resistance against many viruses.

In bacterial infections, some of the serum proteins increase manyfold (2- to 100-fold). These are called *acute-phase*

TABLE 31-15. Functions of Innate and Adaptive Immunity

	Innate Immune System	Adaptive Immune System
	Resistance not improved by repeated infection	Resistance improved by repeated infection
Soluble factors	Lysozyme, complement, acute-phase proteins, *e.g.*, CRP interferon	Antibody
Cells	Phagocytes, natural killer (NK) cells	T-lymphocytes

proteins. An example of these is C-reactive protein, so called because of its ability to bind the C protein of pneumococci. C-reactive protein bound to bacteria promotes the binding of complement, which facilitates their uptake by phagocytes. This process is called *opsonization*.

Another group of proteins (about 20) that interact with each other and with other components of the innate and adaptive immune systems is called *complement*. The complement system, which is composed of a cascade similar to the blood-clotting system, is activated spontaneously by the surface of a number of microorganisms by the alternative complement pathway. Following activation, some complement components can cause opsonization, while others attract phagocytes to the site of infection.

An additional group of complement components causes direct lysis of the cell membranes of bacteria by the *lytic pathway*. All these mechanisms, although described separately, *in vivo* act in concert.

For the purposes of simplification, the immune system can be divided into four groups or compartments.

The antibody or B lymphocyte system
The T lymphocyte system
The complement system
The phagocytic system

Predisposition to recurrent infections may occur when any one of the immune systems is deficient. Table 31-16 is a list of immunodeficiencies.

The Antibody or B Lymphocyte System

B Cell Compartment. B cells are lymphocytes characterized by the presence of surface immunoglobulins and by the ability to differentiate into plasma cells that produce immunoglobulins. The activation, proliferation, and differentiation of B cells is aided by signals provided by T cells. Upon binding antigen, antibodies can activate the comple-

---

TABLE 31-16. Immunodeficiency Diseases

Predominant Antibody Defect
- X-linked agammaglobulinemia
- X-linked hypogammaglobulinemia with growth hormone deficiency
- Autosomal recessive agammaglobulinemia
- Immunoglobulin deficiency with increased IgM (and IgG)
- IgA deficiency
- Selective deficiency of other immunoglobulin isotypes
- Kappa-chain deficiency
- Antibody deficiency with normal gammaglobulin levels or hypergammaglobulinemia
- Immunodeficiency with thymoma
- Transient hypogammaglobulinemia of infancy
- Common variable immunodeficiency with predominant B cell defect:
    - Nearly normal B cell number with $\mu^+\delta^+$, without $\mu^+\delta^-$, $\mu^+\gamma^+$, γ, or α^+ cells
    - Very low B cell numbers
    - $\mu^+\gamma^+$ or γ^+ "nonsecretory" B cells with plasma cells
    - Normal or increased B cell number with $\mu^+\delta^+\gamma^+$, $\mu^+\delta^+\alpha^+$, γ^+, and α^+ B cells
- Common variable immunodeficiency with predominant immunoregulatory T-cell disorder:
    - Deficiency of helper T cells
    - Presence of activated suppressor T cells
- Common variable immunodeficiency with autoantibodies to B or T cells

Predominant Defects of Cell-Mediated Immunity
- Combined immunodeficiency with predominant T cell defect
- Purine-nucleoside phosphorylase deficiency
- Severe combined immunodeficiency with adenosine deaminase deficiency
- Severe combined immunodeficiency:
    - Reticular dysgenesis
    - Low T and B cell numbers
    - Low T cell and normal B cell numbers (Swiss)
    - "Bare lymphocyte" syndrome
- Immunodeficiency with unusual response to Epstein-Barr virus

Immunodeficiency Associated with Other Defects
- Transcobalamin 2 deficiency
- Wiskott-Aldrich syndrome
- Ataxia telangiectasia
- Third- and fourth-pouch/arch (DiGeorge's) syndrome

Source: Rosen FS, Cooper WD, Wedgewood RJP. WHO classification of primary immunodeficiencies. N Engl J Med 1984;311:235.

ment cascade, producing lysis of the organisms (by fixing C1–C9) and/or facilitating their phagocytosis by neutrophils and monocytes.

Immunoglobulins are of five major isotypes based on their structure and function: IgG, IgM, IgA, IgD, and IgE. Their main properties are summarized in Table 31-17. The IgG group is the major antibody in blood and tissues and represents the main antibody response to antigenic stimulation. IgG is the only immunoglobulin isotype that can cross the placenta and represents an important temporary maternal contribution to the immunologic defense of the newborn, as discussed later.

IgM antibody consists of five immunoglobulin units linked by disulfide bonds. These antibodies are usually secreted as a primary response to an immunologic challenge. The IgM-producing B cells later switch to the production of IgG, IgA, and IgE isotypes.

IgA is the predominant antibody in secretions. Two such classes of IgA, IgA$_1$ and IgA$_2$, are recognized. In serum, the predominant IgA is IgA$_1$ (90%), and IgA$_2$ represents only 10%. In secretions, both IgA$_1$ and IgA$_2$ are almost equal.

IgD is present in very small quantities in serum. Its role is not well understood.

IgE antibodies have the capacity to bind to high-affinity receptors on mast cells and basophils and promote degranulation of these cells. Degranulation releases preformed mediators such as histamine, eosinophil chemotactic factors of anaphylaxis, and platelet-activating factor, and also triggers the synthesis and release of leukotrienes. These substances cause allergic reactions by increasing vascular permeability and smooth-muscle contraction.

Neonatal and Infant Antibody Production. Antibodies synthesized by the human fetus are mainly IgM and, to a much lesser degree, IgA. In the normal fetus, the bulk of circulating antibodies represent transfer of maternal IgG by means of the placenta. Virtually no IgA and IgM cross the placenta.

Maternal IgG is slowly catabolized, with a half-life of approximately 30 days, and by the third month, the infant has lost approximately 87% of maternal IgG. Because of a delayed onset of IgG synthesis in the human infant, a physiologic hypogammaglobulinemia stage occurs between 4

and 9 months. Adult levels are attained by the end of the first year of life. Severe hypogammaglobulinemia may be seen in premature infants owing to birth before completion of transplacental transfer of gammaglobulin.

Humoral Immunity Disorders

Transient hypogammaglobulinemia of infancy is characterized by an abnormal prolongation and degree of the physiologic hypogammaglobulinemia. This condition affects both male and female infants, and sometimes there is a familial occurrence.

Affected infants present with recurrent infections of the upper and lower respiratory tracts and rarely of skin or meninges. In some cases, this disorder is associated with food allergies.

Lymph nodes from these patients display very small or no germinal centers with marked reduction in the number of plasma cells. Since these patients have a normal number of circulating B cells, the defect presumably involves terminal differentiation of B cells into antibody-producing plasma cells.

In patients with recurrent infection, gammaglobulin replacement therapy is indicated, usually for 12 to 36 months. It is discontinued when IgG synthesis is adequate, which occurs spontaneously around 4 years of age.

X-linked agammaglobulinemia is also called *Bruton disease* or *congenital agammaglobulinemia*. This disorder presents with early onset of recurrent pyogenic infections. These patients have no circulating B cells, low concentrations of all circulating immunoglobulin classes, and absence of plasma cells in all lymphoid tissue. T cells are not involved. This disease is transmitted in an X-linked recessive pattern. Patients are usually asymptomatic during the first 6 months of life, probably because of protection conferred by maternal gammaglobulin.

The most common infections are of the lower and upper respiratory tracts, causing pneumonia, otitis, purulent sinusitis and bronchiectasis, meningitis, sepsis, pyoderma, and osteomyelitis. Without early replacement of gammaglobulin, many of these children develop bronchiectasis and die of pulmonary complications.

Cellular immunity in these patients is intact with normal delayed hypersensitivity, normal T cell number, and normal *in vitro* response to mitogens, antigens, and allogenic cells.

TABLE 31-17. Properties of Immunoglobulins

| Property | IgG | | | | IgA | IgM | IgD | IgE |
	IgG$_1$	IgG$_2$	IgG$_3$	IgG$_4$				
Molecular weight ($\times10^{-3}$)	150	150	150	150	150	150–600	900	190
Present in secretions					++	−		
Crosses placenta	+	±	+	+	−	−	−	−
Fixes complement	++	+	++	−	−	++		−
Binds to mast cells	−	−	−	+	−		−	++
Binds to macrophages	++	+	++	±	−			

Acquired agammaglobulinemia is also known as *common variable agammaglobulinemia*. This disorder represents a group of disorders characterized by hypogammaglobulinemia. Patients in this category have immunodeficiency that varies in time of onset as well as clinical and immunologic pattern.

The etiology of common variable agammaglobulinemia is unknown. There is no clear genetic influence. Both sexes are affected equally, and symptoms rarely occur before the age of 6 years. Any patient with chronic progressive bronchiectasis should be considered suspect. Noncaseating granulomas of the lung are also a frequent occurrence.

IgG levels are usually less than 500 mg/100 mL, and IgA and IgM levels are less than 50 mg/100 mL. Although most patients have normal numbers of circulating B cells, plasma cells are rarely found. In addition to B cell deficiency, many of these patients have T cell abnormalities. Over 50% eventually exhibit cutaneous anergy.

Besides the preceding immunodeficiencies, there are also selective IgE subclass deficiencies, selective IgA deficiency, and immunodeficiency with increased IgM. In the latter, IgA and IgG are deficient but IgM is normal or elevated.

The T Lymphocyte System

T Cell Compartment. T cells (thymus-derived) are involved in several immune mechanisms. One subset of T cells, called *helper/inducer cells*, helps B cells in the process of differentiation to antibody-secreting cells. In addition, helper/inducer T cells are involved in expanding another subset of cytotoxic and suppressor T cells. The function of cytotoxic cells is to bind and destroy target cells, such as tumor cells or virus-infected cells.

T cells are responsible for contact sensitivity (*e.g.,* poison ivy dermatitis), delayed hypersensitivity reactions (*e.g.,* tuberculosis test), immunity to intracellular organisms (viruses, tuberculosis, *Brucella*), immunity to fungal organisms, and graft–versus–host disease.

The T cell population can be subdivided into two subsets, one of which has helper T function and another that has suppressor T activity. Release of soluble factors by activated T cells is important in the immune response. These lymphokines include macrophage chemotactic and activation factors, B cell growth and differentiation factors, and gamma (8) interferons. Activated suppressor T cells, in turn, release soluble factors that suppress helper T cells, dampening the immune response.

T Cell Immunodeficiency Diseases

T cell immunodeficiency can result in severe fungal and viral infections. Since T cells pass through multiple stages in thymic development, and since T cell activation is a multistage process, we can find defects in both maturation and/or activation that ultimately would result in functional T cell deficiencies.

The investigation of T cell function includes a CBC, since lymphopenia is often seen. Chest x-rays to assess thymic shadow are also useful. Delayed hypersensitivity skin tests are helpful in ensuring T cell function. (Tetanus toxoid, *Monilia,* and mumps are the usual skin tests performed.)

Total T cell and T cell subset distribution can be readily seen by use of monoclonal antibodies: T3 for total T cells, T4 for helper cells, and T8 for suppressor T cells.

Primary T Cell Deficiency. When T cell deficiency is severe, both T and B cell compartments are deficient, resulting in severe combined T and B cell deficiency. This immunodeficiency can be inherited as an autosomal recessive or an X-linked recessive trait; 75% of patients with these disorders are male.

In some cases, the T cells do not express HLA-A and HLA-B antigens. These "bare" lymphocytes are functionally incompetent. These patients usually reveal absence of tonsils and lymph nodes, evidence of wasting and oral thrush, absence of thymic shadow on x-rays, lymphopenia, and the immunoglobulins are markedly decreased. Delayed skin tests are always negative. Proliferative responses to mitogens and antigens are absent. AIDS has to be ruled out in the differential diagnosis. Unless bone marrow transplantation is performed, death usually occurs within the first few weeks of life.

DiGeorge Syndrome (Thymic Hypoplasia, Third and Fourth Pouch Syndrome). This syndrome is probably a developmental rather than an inherited disease. It is manifested by dysmorphogenesis of the third and fourth pharyngeal pouches, which results in aplasia or severe hypoplasia of the thymus and parathyroid glands. It is associated with abnormalities of the great vessels, atrial and ventricular septal defects, esophageal atresia, bifid vulva, short philtrum, mandibular hypoplasia, hypertelorism, and low-set notched ears. These patients usually develop hypocalcemic seizures. Cardiac defects for which these children are usually evaluated initially are the most common cause of death.

Graft-versus-Host Disease. When a T cell–deficient host receives a transfusion of HLA-incompatible immunocompetent T cells, necrosis of skin, liver, and gastrointestinal tract can result. This can be precipitated by intrauterine transfusions, blood transfusions, or more commonly, after bone marrow transplantation.

The acute disease begins 7 to 14 days after grafting and usually starts with a maculopapular skin eruption or with "scalded skin" syndrome and alopecia. The epidermis shows coagulation necrosis. Liver changes can result in liver failure. Chronic diarrhea and malabsorption are poor prognosticators. Latent viral infections and opportunistic infections can manifest and may be lethal.

Recent approaches to prevent acute graft-versus-host disease associated with bone marrow transplant consist of depleting the donor bone marrow of T cells by use of monoclonal antibodies and complement by absorption on lectin columns. Patients who recover from this acute episode may present dermatofibrosis, progressive hepatitis, or Sjögren's syndrome weeks or years later.

Chronic Mucocutaneous Candidiasis. Chronic mucocutaneous candidiasis is not a single clinical entity but a collection of related syndromes. Cell-mediated immunity is most frequently impaired. Other deficiencies include low secretory IgA, abnormal complement system, and impaired macrophage/monocyte function. Some children show concomitant polyendocrine failure.

Wiskott-Aldrich Syndrome. This syndrome is characterized by recurrent infections, thrombocytopenia, and eczema. It is X-linked and is present in infancy. The basic defect has not been identified.

Initially, patients present with otitis, upper respiratory tract infections, sepsis, and meningitis. When T cell function diminishes, viral, fungal, and *Pneumocystis carinii* infections may occur. *In vitro* mitogen responses are often normal. Responses to antigens are usually absent. The diagnosis is confirmed by the presence of platelets that are half the normal size. IgG levels are normal or slightly low, IgM is slightly low, and IgA and IgE are elevated.

Ataxia Telangiectasia. This is an autosomal recessive disorder characterized by progressive cerebellar ataxia associated with increasing telangiectasia beginning in the bulbar conjunctiva and later spreading to the skin; infections are frequent. Tumors, commonly of the lymphoreticular system, are the most common cause of death. Numerous chromosomal breaks and translocations involving chromosomes 7 and 14 have been reported.

Hyper-IgE Syndrome. Hyper-IgE syndrome is a primary immunologic deficiency characterized by recurrent staphylococcal abscesses and markedly elevated IgE. Patients have a lifelong history of severe staphyloccal infections involving the skin, lungs, joints, and other sites. There is an abnormally low anamnestic antibody response to booster immunizations and poor antibody and cell-mediated responses to neoantigens. Serum IgE is usually greater than 10,000 IU/mL. Lymphocyte-proliferative responses to the antigens *Candida albicans* and tetanus toxoid are absent or very low in those tested.

AIDS in Children. Transplacental acquisition of AIDS-causing virus results in a syndrome that usually involves B cell dysfunction. These children grow poorly and present with lymphadenopathy. Often, parotid hypertrophy and otitis, sepsis pneumonitis, as well as pneumonia due to Epstein-Barr virus and cytomegalovirus, hypergammaglobulinemia, and HIV antibody are present.

The Complement System
The primary deficiencies of the complement system include

C1q deficiency
C1r deficiency
C4 and C2 deficiency
C3 deficiency
C5 deficiency
C6, C7, and C8 deficiency
C9 deficiency
C1 inhibitor deficiency (hereditary)
Angioneurotic edema.

The main clinical features of these inherited diseases are summarized in Table 31-18.

The Phagocytic System
Phagocytes include neutrophils and monocyte macrophages. Their primary role is to engulf and kill microbial organisms. Phagocytosis is enhanced by opsonization. Phagocytic disorders can be classified into quantitative disorders and functional disorders.

Quantitative Disorders

When the neutrophil count is below $500/mm^3$, there is increased risk of infection. This decrease can be secondary to antimetabolite therapy or to autoimmune disease. Neutropenia may be congenital.

Qualitative Disorders

Defects in Chemotaxis
Defects in chemotaxis result in superficial infections of skin and mucous membranes. Disorders of chemotaxis can be secondary to circulating inhibitors, as seen in some patients with Hodgkin's disease and hepatic cirrhosis. It is also seen in immotile cilia syndrome and Schichman syndrome.

Defects in Phagocytosis
Three cell-surface molecules have been described: MO1 (on monocytes), LFA1 (lymphocyte functional on T and B cells), and P150/90 (on monocytes). They are all involved in cell-cell adhesion. MO1 has been shown to be the receptor for the fragment of the third component of complement (C3bi), and its expression is essential for bacterial phagocytosis, chemotaxis, and adherence. LFA1 is important for T cell-monocyte interaction.

Patients with defective phagocytosis have severe recurrent pyogenic infections. The diagnosis is made by demonstrating absence of MO1 antigen by fluorescence technique and by the failure of neutrophils to adequately ingest opsonized bacteria. The severity of the defect is variable. Patients with the lowest expression of LEV-CAM suffer the most severe, potentially fatal, recurrent infections.

Defects in Intracellular Killing
The prototype is chronic granulomatous disease. Opsonized bacteria are phagocytized, but the mechanism of producing superoxide (which is needed for intracellular killing) is defective. This condition is usually inherited in an X-linked recessive manner and is extremely rare.

TABLE 31-18. Inherited Deficiencies in Complement and Complement-Related Protein

Deficient Protein	Observed Pattern of Inheritance at Clinical Level	Reported Major Clinical Correlates*
C1q	Autosomal recessive	Glomerulonephritis, systemic lupus erythematosus
C1r	Probably autosomal recessive	Syndrome resembling systemic lupus erythematosus
C1s	Found in combination with C1r deficiency	Systemic lupus erythematosus
C4	Probably autosomal recessive (two separate loci, C4A and C4B)†	Syndromes resembling systemic lupus erythematosus
C2	Autosomal recessive, HLA-linked	Systemic lupus erythematosus, discoid lupus erythematosus, juvenile rheumatoid arthritis, glomerulonephritis
C3	Autosomal recessive	Recurrent pyogenic infections, glomerulonephritis
C5	Autosomal recessive	Recurrent disseminated neisserial infections, systemic lupus erythematosus
C6	Autosomal recessive	Recurrent disseminated neisserial infections
C7	Autosomal recessive	Recurrent disseminated neisserial infections, Raynaud's phenomenon
C8 beta-chain or C8 alpha-γ chains	Autosomal recessive	Recurrent disseminated neisserial infections
C9	Autosomal recessive	None identified
Properdin	X-linked recessive	Recurrent pyogenic infections, fulminant meningococcemia
Factor D̄	?	Recurrent pyogenic infections
C1 inhibitor	Autosomal dominant	Hereditary angioedema, increased incidence of several autoimmune diseases‡
Factor H	Autosomal recessive	Glomerulonephritis
Factor I	Autosomal recessive	Recurrent pyogenic infections
CR1	Autosomal recessive§	Association between low numbers of erythrocyte CR1 and systemic lupus erythematosus
CR3	Autosomal recessive¶	Leukocytosis, recurrent pyogenic infections, delayed umbilical-cord separation

*Note that many people with complement deficiencies, especially of C2 and the terminal components, are clinically well. A substantial number of patients with defects in C5 through C9 have had autoimmune disease.

†The deficiency in persons lacking C4A or C4rB is referred to as q = 0, for quantity zero. Thus, such a deficiency can be designated C4Aq0 or C4bq0. Persons with such deficiencies are reported to have a higher than normal incidence of autoimmune diseases. Similarly, heterozygous C2-deficient persons are reported to have an increased incidence of autoimmune disease.

‡Approximately 85% of cases involve silent alleles, and 15% involve alleles encoding for dysfunctional variant C1-inhibitor protein.

§Homozygosity for a low (not absent) numerical expression of CP1 on erythrocytes is detectable in vitro and appears to be associated with systemic lupus erythematosus. An acquired defect in the number of CP1 receptors also may be operative.

¶Low but not absent levels of leukocyte CR3 are detectable in both parents of most CRF3-deficient children.

Source: From Henry JB: Clinical Diagnosis and Management by Laboratory Methods, 18th ed. Philadelphia, Saunders, 1991.

The organisms most prone to be involved in this type of defect are the ones that do not produce hydrogen peroxide and/or produce catalase, such as *Klebsiella, Serratia, Salmonella,* and certain fungi such as *Aspergillus* and *Candida.* When the organism involved is not killed by the neutrophils, macrophages are summoned to the scene and form a granuloma. When these reach large sizes, they can obstruct lumens. Pulmonary disorders occur in nearly all children with chronic granulomatous disease and include hilar lymphadenopathy, bronchopneumonia empyema, and lung abscess; gastrointestinal disorders are frequent. Perianal fistula, steatorrhea, and B_{12} malabsorption are seen. Osteomyelitis, particularly of small bones of the hands and feet, is also seen.

Chronic granulomatous disease is distinguished by an inability of leukocytes to change nitroblue tetrazolium (NBT) from colorless to deep blue during phagocytosis (NBT test).

Bone marrow transplantation has been successful in correcting the defect.

Chediak-Higashi Syndrome. The clinical features of this syndrome consist of partial albinism, hypopigmentation, recurrent fever and infection, mostly with *S. aureus*, and abnormal leukocyte morphology. There is also hepatosplenomegaly, lymphadenopathy, neutropenia, anemia, and perivascular infiltration. Death usually occurs as a result of infection (65%), bleeding (15%), or respiratory failure due to lymphohistiocytic and reticular cell infiltrate. Ascorbic acid and agents that increase cyclic AMP have both been found to benefit some patients.

SUMMARY

Pediatrics is a branch of medicine that deals with the treatment of children. The birth of an infant is attended by rapid adaptation to extrauterine life. The infant adapts by initiating active respiration, increasing kidney function, closure of the patent ductus arteriosus of the heart, and eating. Generally there is a loss of weight after delivery followed by a weight gain. Kidney function increases during the first year and liver function matures within 30 to 60 days. Thyroid function changes abruptly following birth and hemoglobin shifts over a period of 90 days from fetal to adult hemoglobin. Many other hormonal related changes occur during adolescence. Because of all of these changes, it is important that evaluation of illness in infants include the use of sex- and age-related reference values. Treatment of the pediatric patient often requires the collection of blood samples, which may be complicated by the patient's size and inability to communicate. Clinical laboratorians need to make themselves aware of the special requirements and precautions regarding blood collection in pediatric patients. There are several aspects of infant development or physiology that also need to be considered in the treatment of pediatric patients. These include: acid-base balance, electrolyte and water balance, energy metabolism, developmental changes in hormone balance, drug therapy and pharmacokinetics, inborn errors of metabolism, and humoral and cellular immunity. Details of each of these physiologic aspects are discussed in the chapter.

CASE STUDY 31-1

At 36 weeks of uneventful pregnancy, a 22-year-old primagravida woman goes into spontaneous labor. Delivery is unremarkable, and the apparently normal male infant is taken to the special care nursery because of mild respiratory insufficiency. Oxygen, fluids, and electrolytes are administered.

At 3 days of age, the infant is doing well except for mild volume depletion. However, mild hyperkalemia and hyponatremia are noted on a routine chemistry profile.

Questions

1. What are the possible causes of these findings?
2. What items in the infant's history should be investigated?
3. What additional testing should be performed?

CASE STUDY 31-2

A normal-appearing term infant is found to have a low thyroxine level on routine neonatal screening.

Questions

1. What additional testing should be performed?
2. What therapy is indicated?

3. What are the consequences of untreated hypothyroidism?

**CASE STUDY
31-3**

A 9-year-old boy is brought to his pediatrician by his parents because of growth failure. The past history is unremarkable except for an episode of head trauma suffered 2 years previously, when he was struck by an automobile. Neurosurgical intervention was required because of a basalar skull fracture and intracranial bleeding. However, the child seemed to recover fully. He continued to do well at school, but his parents state that he has grown very little since the accident. Measurement of his height and weight reveal that he is indeed approximately 2 years below the median for his age.

Questions

1. What conditions may be associated with growth retardation?
2. What condition should be primarily considered because of this child's history?
3. What laboratory tests should be performed to confirm this?
4. What additional testing should be done?

SUGGESTED READINGS

Aperia A, Broberger O, Herin P, Thodenius K, Zetterstrom R. Postnatal control of water and electrolyte homeostasis in preterm and full-term infants. Acta Paediatr Scand 1983;305(Suppl):61.

Bickel H, Gutherie R, Hammerson G, eds. Neonatal screening for inborn errors of metabolism. New York: Springer-Verlag, 1980.

Cohn R, Roth K. Metabolic disease. Philadelphia: Saunders, 1983.

Evans SE, Durbin GM. Aspects of the physiological and pathological background to neonatal clinical chemistry. Ann Clin Biochem 1983;20:193.

Evans WE, Schentag JJ, Jusko WJ. Applied pharmacokinetics. San Francisco: Applied Therapeutics, Inc, 1980.

Faulkner WR, Meites S. Selected methods of clinical chemistry. Vol 9. Washington, DC: American Association for Clinical Chemistry, 1982;3.

Hicks JM, Boeckx RL, eds. Pediatric clinical chemistry. Philadelphia: Saunders, 1984.

Hirschhorn N. Oral rehydration therapy for diarrhea in children: A basic primer. Nutr Rev 1982;40:97.

Johnson TR, Moore WM, Jeffries JE, eds. Children are different: Development physiology. 2nd ed. Columbus, OH: Ross Laboratories, 1978.

Kaplan L, Pesce A, eds. Clinical chemistry. St. Louis: Mosby, 1988.

Koren G, Prober CG, Golb R, eds. Antimicrobial therapy in infants and children. New York: Marcel Dekker, 1988.

Lebenthal E, Hietlinger LA. Impact of development of the gastrointestinal tract on infant feeding. J Pediatr 1983;102:1.

Levin DL, Perkin RM. Abnormalities in fluids, minerals and glucose. In: Levin DL, Morriss FC, Moore GC, eds. A practical guide to pediatric intensive care. 2nd ed. St. Louis: Mosby, 1984;93.

Meites S, ed. Pediatric clinical chemistry. 3nd ed. Washington, DC: American Association for Clinical Chemistry, 1988.

Milla PJ. Disorders of electrolyte absorption. Clin Gastroenterol 1982;11:31.

Natelson S, Natelson EA. Principles of applied clinical chemistry: Maintenance of fluid and electrolyte balance. Vol 1. New York: Plenum Press, 1975;109.

NCCLS.ASH-3, Standard procedure for the collection of diagnostic blood specimens by venipuncture. Villanova, PA: NCCLS, 1978.

NCCLS. TSH-4, Standard procedures for the collection of diagnostic blood specimens by skin-punctures. 2nd ed. Villanova, PA: NCCLS, 1986.

Oli W. Fluid and electrolyte management. In: Avery GB, ed. Neonatology. Philadelphia: Lippincott, 1975;471.

Olsson K. Central control of vasopressin release and thirst. Acta Paediatr Scand 1983;305(Suppl):36.

Pagliaro LA, Pagliaro AM, eds. Problems in pediatric drug therapy. 2nd ed. Hamilton, IL: Drug Intelligence Publications, 1987.

Pendergraph GE. Handbook of phlebotomy. 2nd ed. Philadelphia: Lea & Febiger, 1988.

Philip AGS. Neonatology: A practical guide. New York: Medical Examination Publishing, 1977;129.

Rowland M, Tozer TM. Clinical pharamacokinetics. 2nd ed. Philadelphia: Lea and Febiger, 1986.

Seashore JH. Metabolic complications of parenteral nutrition in infants and children. Surg Clin North Am 1980;60:1239.

Soldin SJ, Rifai N, Hicks SMB, eds. The biochemical bases of pediatric disease. Washington DC: American Association of Clinical Chemistry, 1992.

Stanbury J, Wynguarden, J, Fredrickson D, eds. The metabolic basis of inherited disease. New York: McGraw-Hill, 1983.

Tilton RC, Balows, Hohnadel DC, Reiss RF, eds. Clinical laboratory medicine. St. Louis: Mosby-Yearbook, 1992.

Vaughn VC, McKay RJ, Behrman RE, eds. Nelson's textbook of pediatrics, 11th ed. Philadelphia: Saunders, 1979.

Chapter 32

Geriatric Clinical Chemistry

Janet L. Duben-Engelkirk

Objectives

*Upon completion of this chapter, the clinical laboratorian
should be able to:*

- *Define the following terms: aging, gerontology, geriatrics,
  and osteoporosis.*

- *Delineate and appraise the physiologic changes that occur
  with aging.*

- *Discuss the problems associated with establishing reference ranges for the elderly.*

- *Describe the effects of medications on clinical chemistry
  results in the elderly.*

- *Correlate patient data with age-related physiologic
  changes and/or pathologic conditions.*

KEY WORDS

Aging
Atherosclerosis
Geriatrics
Gerontology
Homeostasis
Osteoporosis

AGING

Aging has been defined as "a progressive unfavorable loss of
adaptation, leading to increased vulnerability, decreased
viability, and decreased life expectancy."[11] The study of the
aging process in the human body is known as *gerontology.*
Geriatrics, however, is a branch of general medicine dealing
with remediable and preventable clinical problems in the
elderly, as well as the social consequences of such illness.[11]

In the United States, the proportion of elderly individuals in the population has been increasing at a more rapid rate
than the general population since the turn of the century.
Since 1960, the older population has grown at a rate twice
that of the total U.S. population.[12] In 1950, approximately
12 million people, or 7.7% of the total population, were
considered to be elderly (age 65 and older).[5] Projections for
the years 2000, 2020, and 2040 are 13.0%, 17.3%, and
21.7%, respectively.[12] The increase in the number of elderly
individuals will undoubtedly present a major challenge to
our nation's health care and social systems. The focus of this
chapter will be on the changes in clinical chemistry test
results associated with the normal aging process, not with
any associated disease or illness.

Impact on the Clinical Laboratory

Clinical laboratorians must familiarize themselves with
problems unique to or especially common in the geriatric
population. They must become aware of special considerations regarding the diagnosis of diseases in the elderly. Most
important, however, they must thoroughly understand the
effects of aging on laboratory values. Although this chapter
is directed toward the latter category, all clinical laboratorians are urged to seek out and obtain education in the field
of gerontology, with special emphasis on the laboratory
aspects of providing health care services to elderly patients.
For example, as part of their education, clinical laboratory
science students should receive specialized instruction and
gain experience in the collection of blood from elderly
patients. Geriatric patients are notoriously difficult to draw
blood from as a result of calcified veins, paralyzed limbs, or
permanent fetal positions.[3] All clinical laboratorians should
help in the development of clearly legible and totally un-

Michael L. Bishop, Janet L. Duben-Engelkirk, and Edward P. Fody.
CLINICAL CHEMISTRY. © 1996 Lippincott–Raven Publishers.

derstandable instructions for kits and instruments to be used in "home testing" or other alternative sites for laboratory testing. An interesting model has been proposed for a laboratory-sponsored geriatric education program.[4] In this program, clinical laboratorians would educate the elderly on such topics as medical/laboratory terminology, abbreviations, anatomy and physiology, the rationale for performing certain laboratory procedures, interpretation of certain laboratory results, medications and therapeutic regimens, and medical services/agencies available to the aged.[4]

ESTABLISHING REFERENCE RANGES FOR THE ELDERLY

Several problems exist with establishment of reference ranges for geriatric patients. Many biochemical test values show significant changes with age and often also between the sexes—independent of disease. It is difficult, therefore, to separate truly abnormal results from results that may be "normal" for a particular age group.[7] The lack of well-defined comparison studies and a lack of "healthy" elderly subjects for studies contribute to the difficulty in establishing reference ranges for geriatric patients. In general, however, it is suggested that reference ranges be adjusted for the elderly.[1,14] Factors known or thought to induce age-related changes in reference ranges are shown in Table 32-1.

Interpretation of laboratory data for elderly patients is complex and requires careful evaluation. Comparing laboratory data from elderly individuals with reference values for younger adults can be dangerous, since "normal" conditions may appear abnormal or pathologic, thereby leading to unnecessary or even harmful treatment.[17] Reference ranges used in test interpretation should always be derived from subjects of the appropriate age and sex.[8] It is recommended that baseline values be obtained when an elderly patient first enters an extended-care facility. Physicians frequently rely on such values when they attempt to evaluate subsequent test results that fall outside established reference ranges.[3]

TABLE 32-1. Factors Known or Thought to Induce Age-Related Changes in Reference Ranges

Alcohol and tobacco use
Changes in diet and nutrient intake (impaired ingestion, digestion, absorption; achlorhydria)
Decreased renal function
Hormonal changes
Osteoclastic bone destruction
Prescription medications and other drugs
Sedentary lifestyles
Underlying (perhaps asymptomatic) pathology/disease, often with multiple organ involvement

PHYSIOLOGIC CHANGES IN THE ELDERLY

In general, aging is associated with decreasing efficiency in maintenance of homeostasis, a decrease in cell water, a reduction in muscle mass, and a gradual decline in respiratory, cardiovascular, kidney, liver, immune system, neurologic, and endocrine system functions.[7] Carbohydrate, protein, lipid, and calcium metabolism all decline with age. Aging is also associated with an accelerated incidence of many diseases, including diabetes mellitus, artherosclerosis, hypertension, and osteoporosis.[11]

Tables 32-2 through 32-4 have been compiled from several sources and summarize the more common clinical chemistry values that increase, decrease, or remain essentially unchanged in the elderly. In reviewing these tables, the reader should note that there may be variability in the results of studies on aging and laboratory values. For example, some investigators may report a decrease in an analyte, whereas another may report no significant change for the same analyte in the aging process. Undoubtedly, there needs to be more research into this important area of clinical laboratory science.

Endocrine

Several complex endocrine-related changes occur in the aging process. The most notable of these occur in the thyroid and gonadal hormones.

Although there is little evidence of thyroid deterioration in the elderly, the incidences of hypothyroidism and hyperthyroidism both increase.[5,11,14] It has been reported that approximately 15% to 20% of all cases of hyperthyroidism occur in individuals above age 60.[5,14] Hypothyroidism may be easily missed in elderly patients because of its subtle clinical presentation and similarity to normal aging.[1,11] Thyroxine (T_4), thyroid-binding globulin (TBG), and reverse T_3 (rT_3) levels essentially do not change. However, there is a significant decrease in triiodothyronine (T_3) after age 50, with a slight increase in thyroid-stimulating hormone (TSH) levels (perhaps as a normal response to the low T_3).[1,17] Elderly patients with signs of hypothyroidism should be screened and periodically tested with T_4, T_3, rT_3, and TSH tests.[11] However, it is suggested that the upper limits of the reference range be adjusted for age.[2,14]

A variety of other significant and complex hormonal changes relating to gonadal function occur in both men and women. In women, these changes are primarily related to menopause, when there is a cessation of ovarian estrogen production. In men, the rate of testosterone production declines gradually with age. Between the ages of 60 and 80, the ratio of testosterone to estradiol falls from 12:1 to 2:1.[11] The decrease in testosterone is linked primarily to diminished testicular function.[17]

Other aspects of endocrine function, such as the hypothalamus–anterior pituitary system and adrenal glands, exhibit less dramatic change. There is essentially no change in cortisol production. However, efficiency of homeostatic

TABLE 32-2. Clinical Chemistry Values Reported to Increase in the Elderly*

Analyte	Comments/Reference

Alkaline phosphatase (ALP)
   Bone isoenzyme of ALP (Aldrich, 1989)
   Liver isoenzyme of ALP (Aldrich, 1989)
	Higher values in females (Jeppesen, 1986; Tietz, 1992); a significant increase in females during the years spanning menopause (Aldrich, 1989); shows a small but distinct decrease in men after age 70 (Aldrich, 1989); increases 8–10 U/L (Garner, 1989)

Ages 60–69:	Males	10.21 (+/−3.25) KA units	
		72.49 (+/−23.08) U/L	
	Females	10.66 (+/−3.47) KA units	
		75.69 (+/−24.64) U/L	
Ages 70–79:	Males	8.64 (+/−3.31) KA units	
		61.34 (+/−23.50) U/L	
	Females	10.51 (+/−3.42) KA units	
		74.62 (+/−24.28) U/L	

Analyte	Comments/Reference
Amylase	(Tietz, 1992)
Aspartate aminotransferase (AST)	Higher values in males (Jeppesen, 1986; small but steady increase with age (Aldrich, 1989); no change (Garner, 1989); slight increase (Tietz, 1992)
Beta-globulin	Slight increase (Garner, 1989)
Bicarbonate	Slight increase (Aldrich, 1989)
Blood urea nitrogen (BUN)	Higher values in males

Ages 60–69:	Males	16.80 (+/−3.08) mg/dL	
		5.99 (+/−1.10) mmol/L	
	Females	16.39 (+/−3.01) mg/dL	
		5.85 (+/−1.07) mmol/L	
Ages 70–79:	Males	16.87 (+/−3.51) mg/dL	
		6.02 (+/−1.25) mmol/L	
	Females	15.33 (+/−2.64) mg/dL	
		5.47 (+/−0.94) mmol/L	

(Jeppesen, 1986; Aldrich, 1989; Garner, 1989); Tietz, 1992)

Analyte	Comments/Reference
Cholesterol	Levels vary with diet; maximum values at about age 65; values decrease after age 65, but generally remain higher than in young adults (Aldrich, 1989; Tietz, 1992); the magnitude of the average increase with age is significant (Aldrich, 1989)
Creatine kinase (CK)	Linear, clinically insignificant increase with age (Aldrich, 1989; Garner, 1989; Tietz, 1992)
Creatinine	Higher values in males

Ages 60–69:	Males	1.05 (+/−0.18) mg/dL	
		92.82 (+/−15.9) µmol/L	
	Females	0.89 (+/−0.17) mg/dL	
		78.68 (+/−15.03) µmol/L	
Ages 70–79:	Males	1.10 (+/−0.22) mg/dL	
		97.24 (+/−19.45) µmol/L	
	Females	0.84 (+/−0.16) mg/dL	
		74.26 (+/−14.14) µmol/L	

(Jeppesen, 1986; Rock, 1984; Aldrich, 1989; Garner, 1989)

Analyte	Comments/Reference
Ferritin	(Aldrich, 1989; Jeppesen, 1986)
Fibrinogen	(Aldrich, 1989)
Free fatty acids	(Jeppesen, 1986)
Gamma glutamyl transferase (GGT)	(Jeppesen, 1986)
Globulins	Slightly higher in males after age 55; values for females tend to increase until menopause and then level off (Aldrich, 1989)
Glucose (fasting)	(Tietz, 199; Aldrich, 1989; Jeppesen, 1986)
Glucose tolerance	Rises more quickly in the first 2 hours, then drops to baseline more slowly (Garner, 1989)
Glycosylated hemoglobin (HgbA$_{1c}$)	From about 7% in healthy 25-year-olds to >9% in healthy people above age 70 (Aldrich, 1989)
High-density lipoprotein (HDL) cholesterol	Increases significantly for males aged 60–79 years, compared to the levels in males aged 25–44 years and 45–59 years, but the value for females is essentially constant in all three age groups (Aldrich, 1989)
Immunoglobulins IgA	(Tietz, 1992)
IgG	(Tietz, 1992)
Inorganic phosphate	(Aldrich, 1989)
Insulin	(Tietz, 1992)
Ionized calcium	(Tietz, 1992)
Lactate dehydrogenase (LD)	Small but steady increase with age (Aldrich, 1989; Garner, 1989 Tietz, 1992)
Oral glucose tolerance test (OGTT)	Produces a higher peak at 1–2 hours, followed by a slower decline (Jeppesen, 1986)
Potassium	Slight increase (Garner, 1989; Jeppesen, 1986)

(continued)

TABLE 32-2. (*Continued*)

Analyte	Comments/Reference
Proteinuria	More prevalent in the elderly (Jeppesen, 1986)
Serum osmolality	Slight increase (Tietz, 1992)
Thyroid-stimulating hormone (TSH)	Increases slightly (Aldrich, 1989; Miller, 1989)
Triglycerides	(Aldrich, 1989; Tietz, 1992)
	Ages 60–79: Males 150 (62–338) mg/dL
	1.70 (0.70–3.82) mmol/L
	Females 137 (56–572) mg/dL
	1.55 (0.63–6.46) mmol/L
Urea	Increases markedly (Rock, 1984)
Uric acid	Increases in females; may be unchanged for males (Aldrich, 1989; Jeppesen, 1986)
Urine albumin	Slight increase (Aldrich, 1989)
Urine glucose	Slight increase (Aldrich, 1989)
Urine protein	Slight increase (Garner, 1989)

*Serum values unless otherwise stated; applies to both males and females unless otherwise stated.

regulation may decline, as well as aldosterone and 17-ketosteroids levels. Although the prevalence of hypertension increases with age, this is more likely the result of renal factors.[11]

Enzymes

Enzyme changes during the aging process are also varied and complex. For example, enzyme production may be affected by changes in hormones that occur with aging, or enzymes may lose their recognition mechanisms. Also, inhibitors of enzyme activity may develop or organ failure may affect enzyme production.[9] The most significant enzyme change in the elderly is the increase in alkaline phosphatase. This change occurs in both men and women (but at a greater rate in women) and may be attributed to liver disease or renal changes affecting vitamin D and the demineralization of bone.[9]

For creatine kinase (CK), many investigators report a slight increase in activity in the 60- to 70-year-old population, but a decrease for those older than 70 years. The CK-MB isoenzyme decreases drastically in the population over 90 years.[17] Other changes in enzyme concentrations are listed in Tables 32-2 through 32-4.

TABLE 32-3. Clinical Chemistry Values That Are Essentially Unchanged in the Elderly*

Analyte	Comments/Reference
Acid phosphatase	(Garner, 1989; Jeppesen, 1986)
Arterial PCO_2	(Jeppesen, 1986); increases approximately 2% per decade after age 50 (Aldrich, 1989)
Arterial pH	(Aldrich, 1989)
Bilirubin	(Aldrich, 1989; Jeppesen, 1986)
Carbon dioxide	(Garner, 1989; Jeppesen, 1986)
Chloride	No change or slight increase (Tietz, 1992; Garner, 1989; Jeppesen, 1986)
Cortisol	(Goldman, 1986; Liew, 1986; Tietz, 1992)
CSF glucose	(Jeppesen, 1986)
CSF protein	(Jeppesen, 1986)
Folate	(Aldrich, 1989); decreases age 60–90, same >90; (Tietz, 1992)
Growth hormone	No change or slightly decreased (Goldman, 1986)
Reverse T_3	(Aldrich, 1989)
Sodium	(Tietz, 1992)
Thyroid-binding globulin (TBG)	(Aldrich, 1989; Tietz, 1992)
Thyroxine (T_4)	(Aldrich, 1989; Goldman, 1986; Rock, 1984)
Total protein	Concentration virtually constant after age 55 (Aldrich, 1989; Garner, 1989); slight decrease (Tietz, 1992)
Vitamin E	(Miller, 1988)

*Serum values unless otherwise stated; applies to both males and females unless otherwise stated.

TABLE 32-4. Clinical Chemistry Values Reported to Decrease in the Elderly*

Analyte	Comments/Reference
Adrenal androgens	(Goldman, 1986)
Alanine aminotransferase (ALT)	(Jeppesen, 1986)
Albumin	(Jeppesen, 1986; Tietz, 1992; Aldrich, 1989); before age 65, levels in males higher than females, but after 65, values equalize and decline at the same rate (Garner, 1989)
Aldosterone	(Goldman, 1986)
Apolipoproteins	(Aldrich, 1989)
Arterial PO_2	Decreases approximately 5% every 15 years starting in the thirties (Aldrich, 1989; Jeppesen, 1986)
Calcium	Decreases in males, but shows no change or increases in females (Jeppesen, 1986); males have only a slight decreasing level during their entire adult life, but after reaching a peak level in their sixties, female total calcium drops to pre-menopausal levels (Aldrich, 1989; Tietz, 1992)

Ages 60–69:	Males	9.89 (+/−0.41) mg/dL 2.47	
		2.47 (+/−0.10) mmol/L	
	Females	10.05 (+/−0.42) mg/dL	
		2.51 (+/−0.11) mmol/L	
Ages 70–79:	Males	9.90 (+/−0.49) mg/dL	
		2.48 (+/−0.12) mmol/L	
	Females	9.88 (+/−0.40) mg/dL	
		2.47 (+/−0.10) mmol/L	

Analyte	Comments/Reference
Creatinine clearance	Decreases steadily until age 65; decreases markedly thereafter (Aldrich, 1989)
Estrogens	Decreased in females, no change in males (Goldman, 1986)
Folic acid	(Jeppesen, 1986); has been reported to be the most common nutritional deficiency in the elderly (Miller, 1988)
Glucose-6-phosphate dehydrogenase (G6PD)	(Jeppesen, 1986)
Immunoglobulins IgM	(Aldrich, 1989)
Iron	(Aldrich, 1989; Jeppesen, 1986; Tietz, 1992)
Parathyroid hormone (PTH)	(Goldman, 1986)
Phosphorus	Decreases in males, no change in females (Tietz, 1992)
Testosterone	(Hodkinson, 1984; Goldman, 1986; Tietz, 1992)
Transferrin	(Aldrich, 1989; Jeppesen, 1986; Tietz, 1992)
Triiodothyronine (T_3)	(Goldman, 1986); shows a significant decrease between ages 50–65 that accelerates in the following 15 years, amounting to a total drop of 20% (Aldrich, 1989); decreases 25% (Garner, 1989)
Urinary calcium	Decreases in males, increases in females (Jeppesen, 1986)
Urinary potassium	Decreases in males, increases in females (Jeppesen, 1986)
Urinary sodium	Decreases in males, increases in females (Jeppesen, 1986)
Urine specific gravity	(Jeppesen, 1986); declines to 1.024 by age 80 (Garner, 1989)
Vitamin A	Reduced vitamin A intake presents a serious health hazard among the elderly in rural areas (Miller, 1988)
Vitamin B_1 (thiamin)	Most often linked to alcohol abuse (Miller, 1988)
Vitamin B_{12}	(Aldrich, 1989; Jeppesen, 1986; Miller, 1988, Tietz, 1992)
Vitamin C	(Jeppesen, 1986; Miller, 1988)
Vitamin D	(Jeppesen, 1986; Miller, 1988)
Vitamin K	(Miller, 1988)

Serum values unless otherwise stated; applies to both males and females unless otherwise stated.

Electrolytes

The electrolytes, sodium, and potassium, all show little change from values seen in younger adults. Bicarbonate is reported by some investigators to increase slightly, while pH may decrease slightly.[10,17]

Metabolic Processes

Several metabolic functions tend to decline with age. The most significant of these include metabolism of calcium, car-

bohydrates, lipids, and proteins. Important aspects of each of these will be discussed in the following sections.

Calcium Metabolism

Calcium is particularly important in a discussion of geriatrics because of its role in bone formation and strength. Total serum calcium shows some change with age and gender. Calcium levels tend to fall slightly in men, whereas they remain normal or slightly increased in women age 60 to 90 years.[1,17] After about 90 years, levels are reported to fall slightly.[17] The decrease in calcium in men may be related to the fall in the level of albumin to which calcium is bound.[8,11]

Osteoporosis is a problem in both older men and women, but particularly in women over 50 years of age. Osteoporosis is actually a group of diseases that may have several causes. It involves a gradual loss of bone mass in which the skeleton becomes weak and less dense. The actual mechanism behind the age-related loss of bone mass is not totally understood.[5] At menopause it seems that bone resorption exceeds bone formation. Calcium is therefore needed to maintain bone formation and skeletal strength. Postmenopausal Caucasian women with a life history of a calcium deficit are most likely to be affected.

Ionized calcium values were found to increase slightly with age.[17] This might be explained by an overall decrease in pH and an increase in parathyroid hormone in the aging population.[17]

Carbohydrate Metabolism

There is a significant change in carbohydrate metabolism in the elderly. Glucose tolerance declines with age, with elderly individuals having a slightly higher fasting serum glucose level than younger adults.[1] The renal threshold (point at which glucose spills in the urine) likewise increases with age.[2] There is also an altered insulin response to glucose.[5,11] Fasting glucose increases approximately 2 mg/dL (0.11 mmol/L) per decade throughout life.[1] Glycosylated hemoglobin (HgbA$_{lc}$) increases only slightly with age.[2] Diabetes mellitus is a common problem in the elderly, with a prevalence of about 17% at age 65 and 26% by age 85.[11]

Lipid Metabolism

Lipid metabolism also changes in the elderly, with cholesterol levels increasing dramatically in both sexes. In one study of healthy adults, the median cholesterol level for individuals between the ages of 60 and 79 showed an increase of about 40 mg/dL (0.45 mmol/L).[1] Triglyceride levels also rise with age (but to a lesser extent than cholesterol), with males showing a more rapid rise between the ages 30 to 50 and then paralleling those of women 50 years old or over.[11] High-density lipoprotein (HDL) also rises in men from ages 25 to 79, but stays relatively constant in women during the same age range.[1] Atherosclerosis, a disease in which there is an accumulation of lipid material in the veins and arteries, affects 1 in 4 men at age 60 versus 1 in 8 men at age 40.[11]

Protein Metabolism

There is a general decline in muscle mass and bone protein in the elderly. This is generally attributed to the reduction in anabolic hormones.[11] Total serum protein remains essentially unchanged, but this may be due to an increase in globulins that is offset by a decrease in albumin.[1,6,17] Approximately 3% of people over 70 years of age are reported to have a benign increase in a monoclonal protein. This is referred to as benign monoclonal gammopathy (BMG).

Generally, a monoclonal protein is found in the serum in concentrations of less than 3 gm/dL (30 g/L), with no physical or laboratory signs of multiple myeloma or macroglobulinemia.[2] However, these patients should be carefully monitored, since about 20% of individuals with a BMG who survive 10 years develop a malignant disease.[2] Other changes in various serum proteins and immunoglobulins are listed in Tables 32-2 through 32-4.

Hepatic Function

Some functions of the liver are affected by the aging process. Although the liver performs many functions, three of importance (the synthetic, excretory/secretory, and detoxification/drug metabolism) will be addressed in this discussion.

The synthetic function of the liver can be monitored by the concentrations of plasma proteins. The total serum protein level remains virtually constant after age 55. Tietz reported a slight decrease with age.[17] This is due to the offset in albumin and globulin concentrations explained previously (*i.e.,* an increase in globulins with a corresponding minor decrease in albumin). The decrease in albumin should be considered in therapeutic drug monitoring. There is also a decrease in other carrier proteins with aging, such as transferrin and the apolipoproteins.[1]

Certain enzymes that are part of the liver's excretory and detoxification functions also change. Alkaline phosphatase, for example, shows a significant increase during menopause in women (1.5 times the upper limit of the reference range for younger adult women), but not in men.[17] Also, although aspartate aminotransferase (AST) and lactate dehydrogenase (LD) do not change significantly in men, they do change slightly in elderly women.[1]

Some authors report a slight decrease in bilirubin, another indicator of the liver's excretory function, while others report that it remains essentially the same.[8] Bilirubin also does not seem to vary by gender.

Pulmonary Function

The better-known changes related to pulmonary function in the elderly include PO_2, and PCO_2 values. These changes reflect the decreased vital capacity found in elderly individuals. Arterial PO_2 decreases about 5% every 15 years beginning in the thirties. The PCO_2, however, increases about 2% every decade after age 50. There is also a very slight increase in bicarbonate, with a slight decrease or no change in pH.[1,16,17]

Renal Function

Renal function also changes with age. Creatinine clearance, glomerular filtration rate, and renal plasma flow all decrease. Analytes [*e.g.,* creatinine, blood urea nitrogen (BUN), uric acid, and inorganic phosphate] that reflect the glomerular filtration rate are found to increase.[17] The albumin/creatinine ratio markedly increases in very old individuals.[17] This decrease in renal function should be taken into consideration in therapeutic drug monitoring.

Nutritional Aspects

Although relatively few studies regarding the nutritional needs of the elderly have been conducted in the past, there is now an increasing body of information on their needs. Nutritional problems of the elderly are related to, but not solely dependent on, proper nutrient intake (nutritional balance). Nutritional balance is affected by several factors, including disorders or diseases that modify or impair intestinal absorption, increased urinary and fecal excretion, hindrance of food intake, and interference with storage and utilization.[18]

Although the nutritional requirements of the elderly are complex and controversial and beyond the scope of this chapter, it is important to mention the role of vitamins and aging. Vitamins are required by all individuals to maintain proper body functions. However, vitamin requirements change with age owing to altered dietary practices, physiology, disease, and/or medications. The elderly, then, are particularly prone to nutritional imbalances or malnutrition.[13] Because laboratorians can play a major role in nutritional assessment, they should be familiar with the alterations that can occur in the elderly. Some of the more significant vitamin changes are listed in Tables 32-2 through 32-4. The reader is referred to Chapter 29, Vitamins, and Chapter 30, Nutritional Assessment, for more detailed information on these topics and the aging process.

MEDICATIONS AND THE ELDERLY

Elderly individuals are far more likely to be taking medications (and often, several medications together) than any other age group.[9] With the normal process of aging, there are many changes in how the body handles drugs. The absorption, distribution, metabolism, and excretion of a drug are all affected by the aging process.[15] In general, the ability to metabolize drugs decreases with age. Decreased metabolic capacity in the elderly has been found for such drugs as ac-

etaminophen, phenytoin, quinidine, and theophylline.[5] In light of this information, therapeutic drug monitoring should be done more frequently than in younger adults.

As with other age groups, clinical laboratorians also need to be aware of the effect of drugs on laboratory results in the elderly. Some drugs may affect laboratory values by altering the regulation, production, or excretion of the analyte being tested. Others may affect laboratory values by interfering with the test methodology itself.[9] Because of the numerous medications and laboratory tests available, it would be impractical to discuss the effects of drugs on laboratory tests in this chapter. See Appendix H for the effects of common drugs on selected clinical chemistry tests.

SUMMARY

The aging process has been described as a loss of adaptation with a decrease in viability and life expectancy. In the United States, the proportion of elderly individuals in the population continues to rise. Projections for the year 2000 indicate that approximately 13% of the population will be considered elderly. This increase will undoubtedly present a major challenge for our health care system. It is important, therefore, that clinical laboratorians understand the effects of aging on clinical chemistry values. One of the major problems in geriatric clinical chemistry is establishing reference values for geriatric patients. It is generally difficult to separate truly abnormal results from those that may simply be a result of the aging process. Several factors are thought to induce age-related changes in reference values including: changes in diet and nutrition, hormonal changes, alcohol and tobacco use, and medications. Aging is associated with several physiologic changes. There is a decreasing efficiency in maintenance of homeostasis; a reduction in muscle mass; and a decline in respiratory, cardiovascular, kidney, liver, immune system, neurologic, and endocrine system functions. Carbohydrate, protein, lipid, and calcium metabolism all decline with age. Specific analytes that increase, decrease, or remain the same during the aging process are summarized in the chapter tables.

**CASE STUDY
32-1**

A 70-year-old woman with mild diabetes has a hemoglobin A_{1c} value of 8% (normal is less than 5%). She manages her diabetes by diet alone and reports persistently negative urine glucose using an in-home urine-testing method. Her fasting blood glucose is normal, and a random blood glucose is 170 mg/dL (9.4 mmol/L).

Questions

1. What is the significance of the elevated hemoglobin A_{1c}?
2. Why were the patient's urine glucose results consistently negative?

CASE STUDY 32-2

A 75-year-old man reported to his physician for a routine examination and was found to have a total serum protein of 8.8 g/dL (88 g/L) (normal: 6.3–8.2 g/dL; 63–82 g/L) and an albumin of 3.8 g/dL (38 g/L) (normal: 3.7–5.1 g/dL; 37–51 g/L). His CBC, other blood chemistries, and physical examination were all found to be normal. The man had been relatively healthy all his life.

Question

Is there any significance to his elevated total serum protein?

CASE STUDY 32-3

An otherwise healthy 75-year-old woman was found to have a thyroid-stimulating hormone (TSH) value of 10.0 μU/mL (normal: 0.8–4.8 μU/mL).

Question

What is the significance of this result, if any?

REFERENCES

1. Aldrich JE. Geriatric changes in laboratory results. In: Davis B, Bishop M, Mass D, eds. Clinical laboratory science: Strategies for practice. Philadelphia: Lippincott, 1989.
2. Anhronheim JC. Case studies in geriatrics for the house officer. Baltimore: Williams & Wilkins, 1990.
3. Broka J, Standifer R. Filling the special needs of extended-care facilities. Med Lab Observer 1990;22:29–32.
4. Ciesla B. Proposed laboratory-sponsored educational program for the senior health-care consumer. Clin Lab Sci 1988;1(6):353.
5. Faulkner WR, Meites S. Geriatric clinical chemistry: Reference values. Washington, DC: AACC Press, 1994.
6. Garner BC. Guide to changing lab values in elders. Geriatr Nurs 1989;10(3):144.
7. Goldman R. Aging changes in structure and function. In: Carnevali DL, Patrick M, eds. Nursing management for the elderly. 2nd ed. Philadelphia: Lippincott, 1986;73.
8. Hodkinson M, ed. Clinical biochemistry of the elderly. New York: Churchill-Livingstone, 1984.
9. Jeppesen ME. Laboratory values for the elderly. In: Carnevali DL, Patrick M, eds. Nursing management for the elderly. 2nd ed. Philadelphia: Lippincott, 1986;102.
10. Kane RL, Ouslander JG, Abrass IB. Essentials of clinical geriatrics. 3rd ed. New York: McGraw-Hill, Inc., 1994.
11. Liew CC. Biochemical aspects of aging. In: Gornall AG, ed. Applied biochemistry of clinical disorders. 2nd ed. Philadelphia: Lippincott, 1986;558.
12. Longino CF, Soldo BJ, Manton KG. Demography of aging in the United States. In: Ferraro KF, ed. Gerontology: Perspectives and issues. New York: Springer, 1990;19.
13. Miller SM. Aging and changes in vitamin status. Clin Lab Sci 1988;6(1):342.
14. Miller SM. Changes in endocrine function with aging. Clin Lab Sci 1989;2(1):19.
15. Ostrom J. Medications and the elderly. In: Carnevali DL, Patrick M, eds. Nursing management for the elderly. 2nd ed. Philadelphia: Lippincott, 1986;143.
16. Rock RC. Effects of age on common laboratory tests. Geriatrics 1984;39(6):57.
17. Tietz NW, Shuey DF, Wekstein DR. Laboratory values in fit aging individuals—Sexagenarians through centenarians. Clin Chem 1992;38:1167.
18. Worthington-Roberts B, Karkeck JM. Nutrition. In: Carnevali DL, Patrick M, ed. Nursing management for the elderly. 2nd ed. Philadelphia: Lippincott, 1986;189.

SUGGESTED READINGS

Nahemow L, Pousada L. Geriatric diagnostics: A case study approach. New York: Springer, 1983.

Rai GS, Murphy PJ, Wright G. Case presentations in clinical geriatric medicine. Boston: Butterworths, 1987.

Rossman I. Clinical geriatrics. 3rd ed. Philadelphia: JB Lippincott Co., 1986.

Young DS, Pestaner LC, Gibberman V. Effects of drugs on laboratory tests. Clin Chem 1975;21(5):240D.

Appendices

For more detailed or additional information, refer to the latest edition of one of the following excellent references:

Bold AM, Wilding P. Clinical chemistry companion. Oxford: Blackwell Scientific Publications.

Weast RC. CRC handbook of chemistry and physics, Boca Raton, FL: CRC Press.

Werner M. CRC handbook of clinical chemistry. Vol. 1. Boca Raton, FL: CRC Press.

Michael L. Bishop, Janet L. Duben-Engelkirk, and Edward P. Fody.
CLINICAL CHEMISTRY. © 1996 Lippincott–Raven Publishers.

APPENDIX A. Selected Laboratory Abbreviations

A	Ampere, current	LFT	Liver function test
ABG	Arterial blood gases	M.	*misce, mistura* (Latin), mix, mixture
AC	Alternating current	mo	months
a.c.	*ante cibum* (Latin), before meals	NA	Not applicable
ad lib	*ad libitum* (Latin), at will	n/m	Not measured
A/G	Albumin-globulin ratio	NPO	*non per orum* (Latin), nothing by mouth
AMAP	As much as possible	o.d.	*omni die* (Latin), everyday
AOAP	As often as possible	OR	Operating room
BBT	Basal body temperature	p.c.	*post cibum* (Latin), after meals
BFP	Biologic false positive	PO	*per os* (Latin), by mouth
b.i.d.	*bis in die* (Latin), twice daily	ppm	Parts per million
BMR	Basal metabolic rate	Ppt.	Precipitate
BP	Blood pressure	p.r.n.	*pro re nata* (Latin), as required
bp	Boiling point	psi	Pounds per square inch
BS	Blood sugar	Px	Physical examination
BSA	Body surface area	QA	Quality Assurance
C	Coulomb, quantity of electricity	q.a.m.	*quaque ante meridiem* (Latin), every morning
CCU	Coronary care unit	QC	Quality Control
CLIA	Clinical Laboratory Improvement Act	q.d.	*quaque die* (Latin), every day
DC	Direct current	q.h.	*quaque hora* (Latin), every hour
DOA	Dead on arrival	q.i.d.	*quater in die* (Latin), 4 times a day
DRG	Diagnostic-related group	q.n.s.	*quantum non satis* (Latin), quantity not sufficient
Dx	Diagnosis	q.s.	*quantum satis* (Latin), quantity sufficient
ECG,EKG	Electrocardiogram	rad	Dose of radiation absorbed by tissue
EEG	Electroencephalogram	RDA	Recommended dietary allowance
EM	Electron microscopy	rem	Roentgen equivalent man
ER	Emergency Room	R/O	Rule out
FBS	Fasting blood sugar	RR	Recovery room
Fp	Freezing point	Rx	*recipe* (Latin), prescription
FVU	First voided urine	SBP	Systolic blood pressure
GTT	Glucose tolerance test	SE	Standard error
gtt.	*guttae* (Latin), drops	SI	Système International d'Unités
h	hour	SQ	Subcutaneous
h.s.	hour of sleep, at bedtime	stat.	*statim* (Latin), immediately
Hx	history	TBW	Total body weight
IB	Inclusion body	T/D	Treatment discontinued
ICU	Intensive care unit	t.i.d.	*ter in die* (Latin), 3 times a day
ICCU	Intensive coronary care unit	TLV	Threshold limit value
ID	Initial dose	TO	Telephone order
I & O	Intake and output	Tx	Treatment
IR	Infrared	UV	Ultraviolet
IV	Intravenous	V	Volt or volume
J	Joule, energy	W	Watt, power
KUB	Kidney, ureter, bladder	WNL	Within normal limits
LD50	Lethal dose (median)	W/V	Weight by volume
LED	Light-emitting diode		

APPENDIX B. Basic SI Units

Measure	Name	Symbol
Length	Meter	m
Mass	Kilogram	kg
Quantity of substance	Mole	mol
Time	Second	s
Electric current	Ampere	A
Thermodynamic temperature	Kelvin	K
Luminous intensity	Candela	cd

Note: SI (Système International d'Unités) are units having a definition recognized by international agreement. Note that some SI units have capitalized symbols. This is to avoid confusion with SI prefixes using the same letter symbol.

APPENDIX C. Prefixes to be Used with SI Units

Factor	Prefix	Symbol
10^{-18}	atto	a
10^{-15}	femto	f
10^{-12}	pico	p
10^{-9}	nano	n
10^{-6}	micro	μ
10^{-3}	milli	m
10^{-2}	centi	c
10^{-1}	deci	d
10^{1}	deka	da
10^{2}	hecto	h
10^{3}	kilo	k
10^{6}	mega	M
10^{9}	giga	G
10^{12}	tera	T
10^{15}	peta	P
10^{18}	exa	E

Note: Prefixes are used to indicate a subunit or multiple of a basic SI unit.

APPENDIX D. Basic Clinical Laboratory Conversions

Length, Volume, Weight Conversions

To Convert	Into	Multiply By
Inches	Centimeters	2.54
Centimeters	Inches	.39
Yards	Meters	.91
Meters	Yards	1.09
Gallons (US)	Liters	3.78
Liters	Gallons (US)	.26
Fluid ounces (US)	Milliliters	29.6
Milliliters	Fluid ounces (US)	.034
Ounces	Grams	28.4
Grams	Ounces	.035
Pounds	Kilograms	.45
Kilograms	Pounds	2.2

Temperature Conversions

To Convert	Into	Use
Centigrade (°C)	Kelvin (°K)	$°K = °C + 273$
Centigrade (°C)	Fahrenheit (°F)	$°F = (°C \times 1.8) + 32$
Fahrenheit (°F)	Centigrade (°C)	$°C = (°F - 32) \times 0.556$

Concentration Conversions

To Convert	Into	Use
% w/v	Molarity (M)	$M = \dfrac{\% \text{ w/v} \times 10}{\text{GMW}}$
% w/v	Normality (N)	$N = \dfrac{\% \text{ w/v} \times 10}{\text{eq wt}}$
mg/dL	mEq/L	$\text{mEq/L} = \dfrac{\text{mg/dL} \times 10}{\text{eq wt}}$
Molarity	Normality	$N = M \times \text{valence}$

APPENDIX E. Conversion of Traditional Units to SI Units for Common Chemistry Analytes*†

	Conventional/ Current Unit	SI Unit	Conversion Factor
Albumin	g/100 mL	g/L	10
Aspartate aminotransferase (AST)	U/L (mU/mL)	μkat/L	0.0167
Ammonia	μg/dL	μmol/L	0.587
Bicarbonate (HCO_3)	mEq/L	mmol/L	1.0
Bilirubin	mg/dL	μmol/L	17.1
BUN	mg/dL	mmol/L	0.357
Calcium	mg/dL	mmol/L	0.25
Chloride	mEq/L	mmol/L	1.0
Cholesterol	mg/dL	mmol/L	0.026
Cortisol	μg/dL	μmol/L	0.0276
Creatinine	mg/dL	μmol/L	88.4
Creatinine clearance	mL/min	mL/s	0.0167
Folic acid	ng/mL	nmol/L	2.27
Glucose	mg/dL	mmol/L	0.0555
Hemoglobin	g/dL	g/L	10
Iron	mg/dL	μmol/L	0.179
Lithium	mEq/L	μmol/L	1.0
Magnesium	mEq/L	mmol/L	0.5
Osmolality	mOsm/kg	mmol/kg	1.0
Phosphorus	mg/dL	mmol/L	0.323
Potassium	mEq/L	mmol/L	1.0
Sodium	mEq/L	mmol/L	1.0
Thyroxine (T_4)	μg/dL	nmol/L	12.9
Total protein	g/dL	g/L	10
Triglyceride	mg/dL	mmol/L	0.0113
Uric acid	mg/dL	mmol/L	0.0595
Vitamin B_{12}	ng/mL	pmol/L	0.0738
PCO_2	mm/Hg	kPa	0.133
PO_2	mm/Hg	kPa	0.133

* To obtain SI unit, multiply current unit by conversion factor. To obtain the conventional or current unit, divide the SI unit by the conversion factor.

† See SI Unit Pocket Guide, by McQueen MJ. Chicago: ASCP Press, 1990. A computerized version and slide chart are also available from ASCP Press.

APPENDIX F. Greek Alphabet Table

Greek Letter Upper, Lower Case	Greek Name	English Equivalent
A α	Alpha	a
B β	Beta	b
Γ γ	Gamma	g
Δ δ	Delta	d
E ε	Epsilon	e
Z ζ	Zeta	z
H η	Eta	ē
Θ θ	Theta	th
I ι	Iota	i
K κ	Kappa	k
Λ λ	Lambda	l
M μ	Mu	m
N ν	Nu	n
Ξ ξ	Xi	x
O o	Omicron	o
Π π	Pi	p
P ρ	Rho	r
Σ σ	Sigma	s
T τ	Tau	t
Υ υ	Upsilon	ü
Φ φ	Phi	ph
X χ	Chi	ch
Ψ ψ	Psi	ps
Ω ω	Omega	ō

Representative examples:

α_1 *antitrypsin* ϵ *(molar absorptivity)* Σ *(sum of)*

β *globulin* λ *(wavelength)* π *(= 3.14. . .)*

ΔA *(change in)* X μL *(volume)* Ω *(Ohm, resistance)*

APPENDIX G. Concentrations of Commonly Used Acids and Bases with Related Formulas

Chemical Substance	Chemical Formula	Formula Weight/GMW	Equivalent Weight	Average Specific Gravity Conc Reagent*	Average % Purity CONC Reagent*	Normality CONC Reagent (Approximate)
Ammonium hydroxide	NH_4OH	35.05	35.05	0.90	28.0	15
Acetic acid (glacial)	CH_3COOH	60.05	60.05	1.06	99.5	18
Formic acid	$HCOOH$	46.03	46.03	1.20	88.0	23
Hydrochloric acid	HCL	36.46	36.46	1.19	37.0	12
Nitric acid	HNO_3	63.02	63.02	1.42	70.0	16
Perchloric acid	$HCLO_4$	100.46	100.46	1.67	71.0	12
Phosphoric acid	H_3PO_4	98.00	32.67	1.69	85.0	44
Sulfuric acid	H_2SO_4	98.08	49.04	1.84	96.0	36

* *Varies according to lot and/or manufacturer.*

Related formulas:

$$Molarity\ (M) = \frac{g/L}{GMW}$$

$$Equivalent\ weight\ (eq\ wt) = \frac{GMW}{valence}$$

$$Normality\ (N) = \frac{g/L}{eq\ wt}$$

Specific gravity (sp gr):
$$sp\ gr \times \%\ purity\ (in\ decimal\ form) = g\ of\ solute/mL$$

APPENDIX H. Effects of Some Common Drugs on Selected Clinical Chemistry Tests*†

Test	Drug	Effect	Test	Drug	Effect
Albumin, S	Acetaminophen	—		Cefoxitin	—
	Caffeine	—		Codeine	—
	Aspirin	D		Diazepam	—
	Penicillin	D		Digoxin	—
	Sulfonamides	D		Phenobarbital	—
Alkaline Phosphatase, S	Acetaminophen	—		Salicylate	—
	Caffeine	—		Amphotericin B	I
	Digoxin	—		Bromides	I
	Ibuprofen	—		Cefotaxime	I
	Theophylline	D		Iodides	I
ALT (alanine amino transferase), S	Acetaminophen	—		Tetracycline	I
	Caffeine	—		Ampicillin	D
	Ibuprofen	—		Ascorbic acid	D
	Erythromycin	I		Bromides	D
AST (aspartate amino transferase), S	Acetaminophen	—		Methyldopa	D
	Ampicillin	—		Rifampin	D
	Caffeine	—	Creatine kinase (CK), S	Acetaminophen	—
	Digoxin	—		Caffeine	—
	Ibuprofen	—		Cefoxitin	—
	Erythromycin	I		Ibuprofen	—
	Metronidazole	D		Methicillin	—
	Rifampin	D		Phenobarbital	—
Amylase, S	Acetaminophen	—		Tetracycline	—
	Ascorbic acid	—		Theophylline	—
	Ibuprofen	—	Creatinine, S	Acetaminophen	—
	Methicillin	—		Ampicillin	—
	Tetracycline	—		Ascorbic acid	—
	Theophylline	—		Aspirin	—
Bilirubin, S	Acetaminophen	—		Caffeine	—
	Ampicillin	—		Chloramphenicol	—
	Ascorbic acid	—		Codeine	—
	Caffeine	—		Digoxin	—
	Digoxin	—		Ibuprofen	—
	Ibuprofen	—		Methicillin	—
	Methicillin	—		Tetracycline	—
	Phenobarbital	—		Theophylline	—
	Theophylline	—		Cefoxitin	I
	Rifampin	I		Levodopa	I
Calcium, S	Acetaminophen	—	Glucose, S	Acetaminophen	—
	Ampicillin	—		Ampicillin	—
	Barbiturates	—		Aspirin	—
	Caffeine	—		Caffeine	—
	Codeine	—		Cefoxitin	—
	Digoxin	—		Chloramphenicol	—
	Ibuprofen	—		Codeine	—
	Theophylline	—		Digoxin	—
	Cefotaxime	I		Ibuprofen	—
	Aspirin	D		Methicillin	—
	Sulfonamides	D		Phenobarbital	—
Chloride, S	Acetaminophen	—		Tetracycline	—
	Ampicillin	—		Theophylline	—
	Caffeine	—		Ascorbic acid	I
	Diazepam	—		Cefotaxime	I
	Digoxin	—		Metronidazole	I
	Isoniazid	—		Isoniazid	D
	Methicillin	—		Levodopa	D
	Salicylates	—		Tolbutamide	D
	Theophylline	—	Iron, S	Acetaminophen	—
	Bromides	I		Caffeine	—
	Cefotaxime	I		Codeine	—
	Iodides	I		Ibuprofen	—
	Sodium bromide	I		Methicillin	—
	Fluorides	D		Phenobarbital	—
	Liposol	D		Tetracycline	—
	Potassium oxalate	D		Theophylline	—
Cholesterol, S	Acetaminophen	—		Cefotaxime	I
	Ampicillin	—		Deferoxamine	D
	Caffeine	—			

Test	Drug	Effect	Test	Drug	Effect
Lactate dehydrogenase (LD), S	Acetaminophen	—		Theophylline	—
	Caffeine	—		Carbenicillin	I
	Cefoxitin	—		Acetylsalicylic acid	D
	Codeine	—		Levodopa	D
	Digoxin	—		Methotrexate	D
	Ibuprofen	—		Methyldopa	D
	Phenobarbital	—		Rifampin	D
	Theophylline	—	Urea nitrogen (BUN), S	Acetaminophen	—
	Fluosol-DA	I		Acetylsalicylic acid	—
	Triamterene	I		Ampicillin	—
	Acetylsalicylic acid	D		Aspirin	—
	Ascorbic acid	D		Caffeine	—
	Cefotaxime	D		Cefoxitin	—
Potassium, S	Acetaminophen	—		Codeine	—
	Ampicillin	—		Digoxin	—
	Ascorbic acid	—		Erythromycin	—
	Chloramphenicol	—		Ibuprofen	—
	Digoxin	—		Methicillin	—
	Erythromycin	—		Phenobarbital	—
	Gentamicin	—		Theophylline	—
	Penicillin	—		Chloramphenicol	I
	Salicylate	—		Sulfonamides	I
	Theophylline	—		Tetracycline	I
	Cefotaxime	I		Ascorbic acid	D
	Fluosol-DA	I		Cefotaxime	D
	Procainamide	I		Levodopa	D
Protein, S	Acetaminophen	—		Streptomycin	D
	Ascorbic acid	—	Uric acid, S	Acetaminophen	—
	Caffeine	—		Ampicillin	—
	Cefoxitin	—		Ascorbic acid	—
	Digoxin	—		Aspirin	—
	Ibuprofen	—		Caffeine	—
	Phenobarbital	—		Carbenicillin	—
	Salicylate	—		Cefoxitin	—
	Theophylline	—		Codeine	—
	Ampicillin	I		Digoxin	—
	Carbenicillin	I		Erythromycin	—
	Chloramphenicol	I		Ibuprofen	—
	Oxacillin	I		Methicillin	—
	Penicillin G	I		Oxacillin	—
	Caproxamine	D		Penicillin	—
	Cefotaxime	D		Phenobarbital	—
	Fluosol-DA	D		Rifampin	—
Sodium, S	Acetaminophen	—		Salicylate	—
	Ampicillin	—		Theophylline	—
	Ascorbic acid	—		Acetylsalicylic acid	I
	Codeine	—		Isoniazid	I
	Digoxin	—		Levodopa	I
	Methicillin	—		Methyldopa	I
	Penicillin	—		Rifampin	I
	Phenobarbital	—		Tetracycline	I
	Theophylline	—		Cefotaxime	D
	Cefotaxime	I		Chloramphenicol	D
	Fluosol-DA	I		Methyldopa	D
	Liposol	D		Xanthine	D
Thyroxine (T4), S	Carbamazepine	—			
	Phenytoin	—			
	Danazol	D			
	Diflunisal	D			
	Iothiouracil	D			
	Oxyphenbutazone	D			
Triglycerides, S	Acetaminophen	—			
	Ampicillin	—			
	Ascorbic acid	—			
	Caffeine	—			
	Codeine	—			
	Digoxin	—			
	Ibuprofen	—			
	Methicillin	—			
	Phenobarbital	—			

*Effect of drug is on the analytical test or procedure. Effect of drug may vary according to methodology, concentration, and other factors. Reader should refer to the reference for more complete information.

† Key:

S = serum

— = no effect

I = increased analytical value

D = decreased analytical value

(Source for this partial listing is Young, DS. Effects of drugs on clinical laboratory tests. 3rd ed. Washington, DC: AACC Press, 1990. For more detailed information see this reference.)

APPENDIX I. Examples of Incompatible Chemicals

Chemical	Is Incompatible With
Acetic acid	Chromic acid, nitric acid, hydroxyl compounds, ethylene glycol, perchloric acid, peroxides, permanganates
Acetylene	Chlorine, bromine, copper, fluorine, silver, mercury
Acetone	Concentrated nitric and sulfuric acid mixtures
Alkali and alkaline earth metals (such as powdered aluminum or magnesium, calcium, lithium, sodium, potassium)	Water, carbon tetrachloride or other chlorinated hydrocarbons, carbon dioxide, halogens
Ammonia (anhydrous)	Mercury (in manometers, for example), chlorine, calcium hypochlorite, iodine, bromine, hydrofluoric acid (anhydrous)
Ammonium nitrate	Acids, powdered metals, flammable liquids, chlorates, nitrites, sulfur, finely divided organic or combustible materials
Aniline	Nitric acid, hydrogen peroxide
Arsenical materials	Any reducing agent
Azides	Acids
Bromine	See Chlorine
Calcium oxide	Water
Carbon (activated)	Calcium hypochlorite, all oxidizing agents
Carbon tetrachloride	Sodium
Chlorates	Ammonium salts, acids, powdered metals, sulfur, finely divided organic or combustible materials
Chromic acid and chromium trioxide	Acetic acid, naphthalene, camphor, glycerol, alcohol, flammable liquids in general
Chlorine	Ammonia, acetylene, butadiene, butane, methane, propane (or other petroleum gases), hydrogen, sodium carbide, benzene, finely divided metals, turpentine
Chlorine dioxide	Ammonia, methane, phosphine, hydrogen sulfide
Copper	Acetylene, hydrogen peroxide
Cumene hydroperoxide	Acids (organic or inorganic)
Cyanides	Acids
Flammable liquids	Ammonium nitrate, chromic acid, hydrogen peroxide, nitric acid, sodium peroxide, halogens
Fluorine	Everything
Hydrocarbons (such as butane, propane, benzene)	Fluorine, chlorine, bromine, chromic acid, sodium peroxide
Hydrocyanic acid	Nitric acid, alkali
Hydrofluoric acid (anhydrous)	Ammonia (aqueous or anhydrous)
Hydrogen peroxide	Copper, chromium, iron, most metals or their salts, alcohols, acetone, organic materials, aniline, nitromethane, combustible materials
Hydrogen sulfide	Fuming nitric acid, oxidizing gases
Hypochlorites	Acids, activated carbon
Iodine	Acetylene, ammonia (aqueous or anhydrous), hydrogen
Mercury	Acetylene, fulminic acid, ammonia
Nitrates	Sulfuric acid
Nitric acid (concentrated)	Acetic acid, aniline, chromic acid, hydrocyanic acid, hydrogen sulfide, flammable liquids, flammable gases, copper, brass, any heavy metals
Nitrites	Acids
Nitroparaffins	Inorganic bases, amines
Oxalic acid	Silver, mercury
Oxygen	Oils, grease, hydrogen, flammable liquids, solids, or gases
Perchloric acid	Acetic anhydride, bismuth and its alloys, alcohol, paper, wood, grease, oils
Peroxides, organic	Acids (organic or mineral), avoid friction, store cold
Phosphorus (white)	Air, oxygen, alkalis, reducing agents
Potassium	Carbon tetrachloride, carbon dioxide, water
Potassium chlorate	Sulfuric and other acids
Potassium perchlorate (see also chlorates)	Sulfuric and other acids
Potassium permanganate	Glycerol, ethylene glycol, benzaldehyde, sulfuric acid
Selenides	Reducing agents
Silver	Acetylene, oxalic acid, tartaric acid, ammonium compounds, fulminic acid
Sodium	Carbon tetrachloride, carbon dioxide, water
Sodium nitrite	Ammonium nitrate and other ammonium salts
Sodium peroxide	Ethyl or methyl alcohol, glacial acetic acid, acetic anhydride, benzaldehyde, carbon disulfide, glycerin, ethylene glycol, ethyl acetate, methyl acetate, furfural
Sulfides	Acids
Sulfuric acid	Potassium chlorate, potassium perchlorate, potassium permanganate (similar compounds of light metals, such as sodium, lithium)
Tellurides	Reducing agents

(Reprinted with permission from National Research Council, Committee on Hazardous Substances in the Laboratory. Prudent practices for handling hazardous chemicals in laboratories. Washington: National Academy Press, 1981. For additional information see Pipitone, DA. Safe storage of laboratory chemicals. 2nd ed. New York: John Wiley & Sons, 1991.

APPENDIX J. Nomogram for the Determination of Body Surface Area

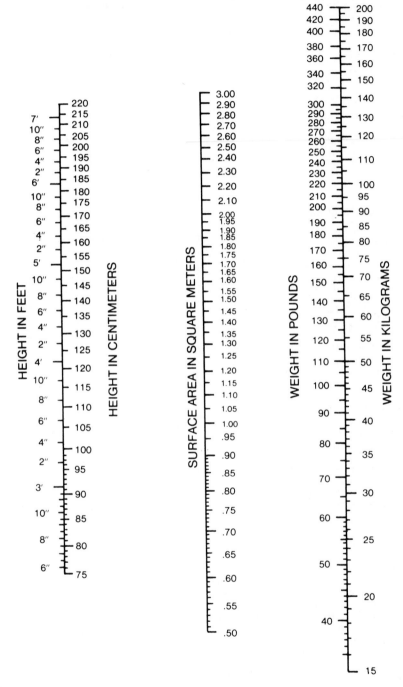

HEIGHT IN FEET   HEIGHT IN CENTIMETERS   SURFACE AREA IN SQUARE METERS   WEIGHT IN POUNDS   WEIGHT IN KILOGRAMS

(Reprinted by permission from The New England Journal of Medicine, 1921; 185:337.)

APPENDIX K. Relative Centrifugal Force Nomograph

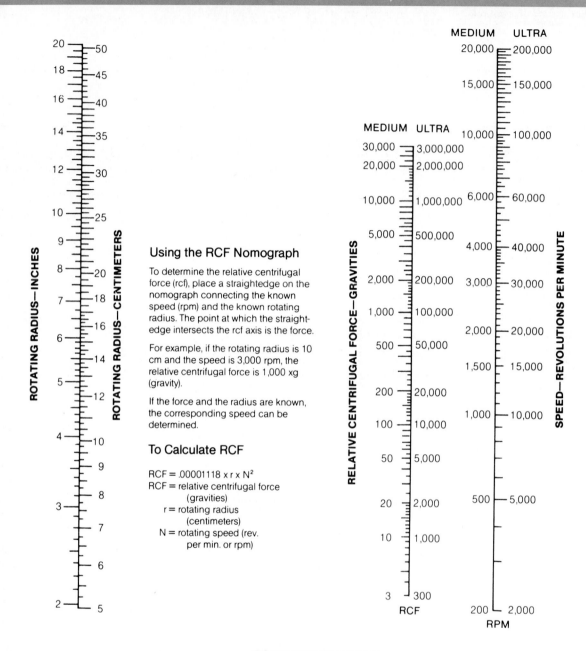

Using the RCF Nomograph

To determine the relative centrifugal force (rcf), place a straightedge on the nomograph connecting the known speed (rpm) and the known rotating radius. The point at which the straight-edge intersects the rcf axis is the force.

For example, if the rotating radius is 10 cm and the speed is 3,000 rpm, the relative centrifugal force is 1,000 xg (gravity).

If the force and the radius are known, the corresponding speed can be determined.

To Calculate RCF

$RCF = .00001118 \times r \times N^2$
RCF = relative centrifugal force
　　　　(gravities)
　　r = rotating radius
　　　　(centimeters)
　　N = rotating speed (rev.
　　　　per min. or rpm)

ROTATING RADIUS—INCHES

ROTATING RADIUS—CENTIMETERS

RELATIVE CENTRIFUGAL FORCE—GRAVITIES

SPEED—REVOLUTIONS PER MINUTE

RCF

RPM

ROTATING TIP RADIUS

The distance measured from the rotor axis to the tip of the liquid inside the tubes at the greatest horizontal distance from the rotor axis is the rotating tip radius.

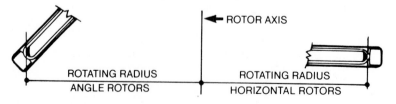

ROTOR AXIS

ROTATING RADIUS
ANGLE ROTORS

ROTATING RADIUS
HORIZONTAL ROTORS

(Reprinted by permission from International Equipment Co., Damon Corporation)

APPENDIX L. Centrifugation—Speed and Time Adjustment

Use the following formula to calculate the new speed necessary to achieve the relative centrifugal force specified in the original procedure, given a different rotor with a different radius.

$$RPM = 1000 \sqrt{\frac{RCF}{1.12\ r}}$$

where RPM = revolutions per minute (speed)
RCF = relative centrifugal force
r = rotating tip radius in millimeters (check manufacturer's rotor specifications)

Use this formula to calculate the new centrifugation time necessary when a different rotor is used.

$$t_N = \frac{t_0 \times RCF_0}{RCF_N}$$

where t_N = centrifugation time necessary with different rotor
t_0 = time (minutes) specified in original procedure
RCF_N = relative centrifugal force of different rotor
RCF_0 = relative centrifugal force specified in original procedure

APPENDIX M. Percent Transmittance–Absorbance Conversion Table

%T	Absorbance* .00	.25	.50	.75	%T	Absorbance* .00	.25	.50	.75
1	2.000	1.903	1.824	1.757	52	.284	.282	.280	.278
2	1.690	1.648	1.602	1.561	53	.276	.274	.272	.270
3	1.523	1.488	1.456	1.426	54	.268	.266	.264	.262
4	1.398	1.372	1.347	1.323	55	.260	.258	.256	.254
5	1.301	1.280	1.260	1.240	56	.252	.250	.248	.246
6	1.222	1.204	1.187	1.171	57	.244	.242	.240	.238
7	1.155	1.140	1.126	1.112	58	.237	.235	.233	.231
8	1.097	1.083	1.071	1.059	59	.229	.227	.226	.224
9	1.046	1.034	1.022	1.011	60	.222	.220	.218	.216
10	1.000	.989	.979	.969	61	.215	.213	.211	.209
11	.959	.949	.939	.930	62	.208	.206	.204	.202
12	.921	.912	.903	.894	63	.201	.199	.197	.196
13	.886	.878	.870	.862	64	.194	.192	.191	.189
14	.854	.846	.838	.831	65	.187	.186	.184	.182
15	.824	.817	.810	.803	66	.181	.179	.177	.176
16	.796	.789	.782	.776	67	.174	.172	.171	.169
17	.770	.763	.757	.751	68	.168	.166	.164	.163
18	.745	.739	.733	.727	69	.161	.160	.158	.157
19	.721	.716	.710	.704	70	.155	.153	.152	.150
20	.699	.694	.688	.683	71	.149	.147	.146	.144
21	.678	.673	.668	.663	72	.143	.141	.140	.138
22	.658	.653	.648	.643	73	.137	.135	.134	.132
23	.638	.634	.629	.624	74	.131	.129	.128	.126
24	.620	.615	.611	.606	75	.125	.124	.122	.121
25	.602	.598	.594	.589	76	.119	.118	.116	.115
26	.585	.581	.577	.573	77	.114	.112	.111	.100
27	.569	.565	.561	.557	78	.108	.107	.105	.104
28	.553	.549	.545	.542	79	.102	.101	.100	.098
29	.538	.534	.530	.527	80	.097	.096	.094	.093
30	.523	.520	.516	.512	81	.092	.090	.089	.088
31	.509	.505	.502	.498	82	.086	.085	.084	.082
32	.495	.491	.488	.485	83	.081	.080	.078	.077
33	.482	.478	.475	.472	84	.076	.074	.073	.072
34	.469	.465	.462	.459	85	.071	.069	.068	.067
35	.456	.453	.450	.447	86	.066	.064	.063	.062
36	.444	.441	.438	.435	87	.061	.059	.058	.057
37	.432	.429	.426	.423	88	.056	.054	.053	.052
38	.420	.417	.414	.412	89	.051	.049	.048	.047
39	.409	.406	.403	.401	90	.046	.045	.043	.042
40	.398	.395	.392	.390	91	.041	.040	.039	.037
41	.387	.385	.382	.380	92	.036	.035	.034	.033
42	.377	.374	.372	.369	93	.032	.030	.029	.028
43	.367	.364	.362	.359	94	.027	.026	.025	.024
44	.357	.354	.352	.349	95	.022	.021	.020	.019
45	.347	.344	.342	.340	96	.018	.017	.016	.014
46	.337	.335	.332	.330	97	.013	.012	.011	.010
47	.328	.325	.323	.321	98	.009	.008	.007	.006
48	.319	.317	.314	.312	99	.004	.003	.002	.001
49	.310	.308	.305	.303	100	.0000	.0000	.0000	.0000
50	.301	.299	.297	.295					
51	.292	.290	.288	.286					

*Absorbance = 2 − log %T.

APPENDIX N. Selected Atomic Weights

Element	Symbol	Atomic Number	Atomic Weight*	Valence
Aluminum	Al	13	26.98	3
Antimony	Sb	51	121.75	3,5
Argon	Ar	18	39.95	0
Arsenic	As	33	74.92	3,5
Barium	Ba	56	137.34	2
Beryllium	Be	4	9.01	2
Bismuth	Bi	83	208.98	3,5
Boron	B	5	10.81	3
Bromine	Br	35	79.90	1,3,5,7
Cadmium	Cd	48	112.40	2
Calcium	Ca	20	40.08	2
Carbon	C	6	12.01	2,4
Cerium	Ce	58	140.12	3,4
Cesium	Cs	55	132.91	1
Chlorine	Cl	17	35.45	1,3,5,7
Chromium	Cr	24	51.99	2,3,6
Cobalt	Co	27	58.93	2,3
Copper	Cu	29	63.55	1,2
Fluorine	F	9	18.99	1
Gold	Au	79	196.97	1,3
Helium	He	2	4.00	0
Hydrogen	H	1	1.01	1
Iodine	I	53	126.90	1,3,5,7
Iron	Fe	26	55.85	2,3
Lead	Pb	82	207.19	2,4
Lithium	Li	3	6.94	1
Magnesium	Mg	12	24.31	2
Manganese	Mn	25	54.94	2,3,4,6,7
Mercury	Hg	80	200.59	1,2
Molybdenum	Mo	42	95.94	3,4,6
Nickel	Ni	28	58.71	2,3
Nitrogen	N	7	14.01	3,5
Oxygen	O	8	16.00	2
Phosphorus	P	15	30.97	3,5
Platinum	Pt	78	195.09	2,4
Potassium	K	19	39.10	1
Rubidium	Rb	37	85.47	1
Selenium	Se	34	78.96	2,4,6
Silicon	Si	14	28.09	4
Silver	Ag	47	107.87	1
Sodium	Na	11	22.99	1
Strontium	Sr	38	87.62	2
Sulfur	S	16	32.06	2,4,6
Tellurium	Te	52	127.60	2,4,6
Tin	Sn	50	118.69	2,4
Tungsten	W	74	183.85	6
Uranium	U	92	238.03	4,6
Xenon	Xe	54	131.30	0
Zinc	Zn	30	65.37	2
Zirconium	Zr	40	91.22	4

Atomic weights are based on C^{12} and have been rounded to two decimal places.

APPENDIX O. Selected Radioactive Elements

The following list of radioactive elements represents those of diagnostic or clinical interest.

Element	Symbol	Atomic No.	Atomic Weight	Half-Life, $t_{1/2}$	
Bismuth	Bi	83	210	5	days
Calcium	Ca	20	45	165	days
Carbon	C	6	14	5730	years
Chromium	Cr	24	51	27.8	days
Cobalt	Co	27	57	270	days
			58	72	days
			60	5.3	years
Hydrogen	H	1	3	12.3	years
Iodine	I	53	125	60	days
			131	8.0	days
Iron	Fe	26	59	45.6	days
Mercury	Hg	80	203	46.9	days
Nickel	Ni	28	63	125	years
Phosphorus	P	15	32	14.3	days
Potassium	K	19	42	12.4	hours
Rubidium	Rb	37	86	18.7	days
Selenium	Se	34	75	127	days
Sodium	Na	11	22	2.6	years
			24	15	hours
Strontium	Sr	38	85	64	days
			90	27.7	years
Sulfur	S	16	35	87.9	days
Technetium	Te	43	99	6.0	hours
Tin	Sn	50	113	115	days
Xenon	Xe	54	133	5.27	days
Zinc	Zn	30	65	245	days

APPENDIX P. Characteristics of Types of Glass

Glass Category	Type of Material	Common or Brand Names	Routine Uses	Limitations
High thermal resistance	Borosilicate with low alkaline content	Pyrex Kimax	All purpose, all types of beakers, flasks, etc. Can tolerate heating and sterilization for lengthy periods of time to 510°C	Should not be cooled too quickly after heating May cloud after use with a strong alkali Subject to scratching
	Aluminosilicate	Corex	Centrifuge tubes and thermometers Extremely strong and hard Temperature stability to 672°C; short-term use to 850°C	Resists scratching Subject to some acid or alkali attack at temperature of 100°C
		Vycor	Ashing and ignition techniques. Can withstand very high temperature (900–1200°C), as well as drastic changes in temperature. Most are alkali resistant in this category.	
High silica	96% silica		Cuvets and thermometers Can be used at high temperatures (900–1200°C) and withstand a sharp change in temperature Can be considered optically pure (cuvets, thermometers)	
High resistance to alkali	Aluminosilicate		Can be used with strong alkali and suffer minimal attack (0.09 mg/cm² vs. 1.4 mg/cm² for borosilicate or 0.35 mg/cm² for regular aluminosilicate	Must be heated and cooled with care Highest temperature for safe use is 578°C

APPENDIX Q. Characteristics of Types of Plastic

Plastic	Temperature Limit (°C)	Transparency	Autoclavable	Flexibility	Usage Examples
Polystyrene (PS)	70	Clear	No	Rigid	Disposables
Polyethylene Conventional (CPE)	80	Translucent	No	Excellent	All-purpose Reagent bottles Test-tube rack Carboys Droppers
Linear (LPE)	120	Opaque	With caution	Rigid	Specimen transport containers Reagent bottles
Polypropylene (PP)	135	Translucent	Yes	Rigid	Screw-cap closures Bottles
Tygon	95	Translucent	Yes	Excellent	Tubing
Teflon FEP	205	Clear Translucent	Yes	Excellent	Stopcocks Wash bottles Beakers
Polycarbonate (PC)	135	Very clear	Yes	Rigid	All-purpose Large reagent containers Carboys Test-tube rack
Polyvinylchloride* (PVC)	70	Clear	No	Rigid	Bottles/tubing

PVC tubing can be heated to 120°C, can be autoclaved, and is very flexible.

APPENDIX R. Chemical Resistance of Types of Plastic

Plastic	Chemical Resistance*
Polystyrene	Useful with water and aqueous salt solutions. It is not recommended for use with acids, aldehydes, ketones, ethers, hydrocarbons, or essential oils. Alcohols and bases can be used, but storage beyond 24 h is discouraged.
Polyethylene	Both classifications of polyethylene (*i.e.,* conventional and linear) have similar chemical resistances. They have excellent chemical resistance to most substances, with the exception of aldehydes, amines, ethers, hydrocarbons, and essential oils. For conventional polyethylene, the exceptions should also include lubricating oil and silicones. The usage of any of the above-named chemical groups should be limited to 24 h at room temperature.
Polypropylene	Has the same chemical resistance as linear polyethylene.
Teflon	This resin possesses excellent chemical resistance to almost all chemicals used in the clinical laboratory.
Polycarbonate	Very susceptible to damage by most chemicals. It is resistant to water, aqueous salts, food, and inorganic acids for a long period of time.

It should be noted that this information is based on room temperature (22°C) and normal atmospheric pressure. Resistance to chemicals decreases as the temperature of the resin nears its maximum. Chemical resistance will also vary as the concentration of the chemical increases.

APPENDIX S. Cleaning Labware

Glassware "Problem"	Cleaning Technique
General usage (procedure 1 is recommended for routine washing needs)	1. *Dirty* glassware should be immediately placed in a soapy or dilute bleach solution and allowed to soak. Wash using any detergent designed for labware. Rinse with tap water 3 times, followed by 1 rinse with distilled water. Dry in an oven at temperature less than 140°C. 2. *Acid dichromate.* Dissolve 50 g technical-grade sodium dichromate in 50 mL of distilled water. Add this mixture to 500 mL of technical-grade concentrated sulfuric acid. This solution is useful until a green color develops. Store in a covered glass jar. Soak glassware overnight and then rinse with dilute ammonia. Rewash glassware according to procedure 1. 3. *Nitric acid* (20%). Soak for 12–24 h. Wash according to procedure 1.
Blood clots	4. *Sodium hydroxide* (10%). Soak for 12–24 h; then follow routine procedure. Dry micropipets using an acetone rinse.
New pipets (S1 alkaline)	5. *Rinse* with 5% hydrochloric acid or 5% nitric acid. Wash following routine procedure.
Metal ion determinations	6. *Acid soak* (20% nitric acid), for 12–24 h. Rinse with distilled water 3 to 4 times. Water should be fresh for each rinsing step. Dry.
Grease	7. *Soak* in any organic solvent. 8. *Dissolve* 100 g potassium hydroxide in 100 mL of distilled water. Allow to cool. Add 900 mL commercial-grade 10% ethanol. Not to be used for delicate glassware. 9. *Contrad 70* (manufactured by Harleco).
Permanganate stains	10. *50% Hydrochloric acid.* Rinse with tap water. Wash. 11. *Dissolve* 1% ferrous sulfate in 25% sulfuric acid.

APPENDIX T. Summary Table of Pharmacokinetic Parameters

	Therapeutic Range, per mL Plasma	Toxic Concentrations, per mL Plasma	Time to Peak Conc., Hours	Half-Life, Hours	% Protein Bound	Volume of Distribution, L/kg	Oral Bioavailability, %	% Excreted in Urine Unchanged
Cardioactive drugs								
Amiodarone	1.0–3 µg		2–6	15–100 d	95–98	70–150	22–88	0
Digitoxin	15–30 ng	>35		2.4–16.4 d	90	0.6	95	30–50
Digoxin	0.8–2 ng	>2.4	1–5	36–51	20–40	5–10	Tablets, 60–75 Elixir, 80 Capsules, 95 IM, 80	60–80
Disopyramide	2–5 µg	>7	0.5–3.0	5–6	10–80	0.8–2	80	
Lidocaine	1.5–5 µg 0.5–1.5 free	>5	15–30[a]	1–2	70	1.3	25–50[b]	5–10
Procainamide NAPA Total	4–10 µg 15–25 µg 5–30 µg	>12	1–2	2.5–4.7 4.3–15	15	1.7–2.2	70–95	50
Propranolol	50–100 ng	Variable	1–2	2–6	90–95	4–6	20–40[b]	1–4
Quinidine	2–6 µg	>6	1–2 4–8[c]	6–8	70–90	2–3	70–80[b]	10–30
Antiepileptic drugs								
Carbamazepine	6–12 µg	>15	6–12	18–54[d] 10–25[e]	72–75	0.8–1.4	75–85	2
Ethosuximide	40–100 µg	>150	1–4	40–60	<10	0.6–0.9	100	10–20
Phenobarbital	15–40 µg	>40	6–18	50–120	49–58	0.6	80–100	10–30
Phenytoin	10–20 µg	>20	4–8	7–42	87–93	0.5–0.8	85–95	5

(*continued on next page*)

APPENDIX T. Summary Table of Pharmacokinetic Parameters (*continued*)

	Therapeutic Range, per mL Plasma	Toxic Concentrations, per mL Plasma	Time to Peak Conc., Hours	Half-Life, Hours	% Protein Bound	Volume of Distribution, L/kg	Oral Bioavailability, %	% Excreted in Urine Unchanged
Primidone	5–12 µg	>15	2–4	3.3–19	0–20	0.6–1	80–90	45–50
Valproic Acid	50–100 µg	>100	1–2	8–20	85–95	0.1–0.5	85–100	3
Bronchodilator								
Theophylline	10–20 µg	>20	2–3	6–12	55–65	0.3–0.7	95–100	9–11
Antibiotics								
Aminoglycosides			0.5–IM[a]					
Amikackin								
Peak	20–25 µg	>32						
Trough	1–4 µg	>5						
Gentamicin								
Peak	5–10 µg	>12						
Trough	0.5–1.5 µg	>2						
Kanamycin								
Peak	20–25 /µg	>30						
Trough	1–4 µg	>10						
Netilmicin								
Peak	5–12 µg	>12						
Trough	0.5–1.5 µg	>2						
Streptomycin								
Peak	20–25 µg	>30						
Trough	1–3 µg	>10						
Tobramycin								
Peak	5–12 µg	>12						
Trough	0.5–1.5 µg	>2						
Vancomycin								
Peak	30–40 µg	>80		3–9	50	0.5–0.8	<2	80–90
Trough	5–10 µg	>20						
Chloramphenicol	10–20 µg	>25	2	1.5–3	50	0.5–1	90	10
Psychoactive drugs								
Amitriptylline	125–250 ng	>500	1–5	17–40	82–96	6.4–36	56–70	
Nortriptylline	50–150 ng	>500	3–12	16–88	87–95	14–38	46–70	2–5
Imipramine	150–250 ng	>500[f]	1.5–3	6–34	63–96	9–23	29–77	1–4
Desipramine	150–300 ng	>500	3–6	11–46	73–92	15–60	31–51	1–4
Doxepin	110–250 ng		1–4	8–36	68–82	9–52	13–45	
Protriptylline	70–260 ng		6–12	54–198	90–94	15–31	75–90	
Lithium	0.8–1.4 µEq	>2	1–3	8–35[g]	0	0.5–1.0	85–95	100
Immunosuppressants								
Cyclosporine (HPLC)	100–300 ng	>400	1–8	4–60	98	3.5–4.5	4–90	<6
Antineoplastics								
Methotrexate	After 24 h > 10^{-5} M After 48 h > 10^{-6} M After 72 h > 10^{-7} M		1–2	Variable	50–70	0.75–0.8	30	90

[a] *Varies dependent on dosage regimen, immediately after IV infusion.*

[b] *Much of the drug metabolized on first pass through the liver.*

[c] *Slow-release preparation.*

[d] *After single dose.*

[e] *After multiple doses.*

[f] *Imipramine + desipramine.*

[g] *Variable with renal function.*

APPENDIX U. Selected Information on Commonly Abused Drugs

Drug	Street Name (Trade Name)	Route of Ingestion	Duration of Effect (h)	Half Life (h)	Excreted Unchanged in Urine	Principal Urinary Metabolites	Symptomatology
Stimulants							
Cocaine	Coke, crack, snow, flake	Nasal, oral, IV, smoked	1–2	2–5	<10%	Benzoyl ecogonine; ecogonine; ecogonine methyl ester	Anesthesia, euphoria, confusion, depression, convulsions, cardiotoxicity
Amphetamine	Bennies, dexies, uppers	Oral, IV	2–4	4–24	~30%	Benzoic acid; *p*-hydroxyamphetamine; *p*-hydroxynorephedrine; phenylacetone	Insomnia, anorexia, euphoria, tolerance and dependence, paranoid psychosis
Methamphetamine	Meth, speed, crystal	Oral, IV	2–4	9–24	10–20%	4-Hydroxymethamphetamine; amphetamine; 4-hyroxyamphetamine; norephedrine	Euphoria, agitation, psychosis, depression, exhaustion
Narcotics							
Heroin	Horse, smack, white lady, scag	IV, nasal, smoked	3–6	1–1.5	<1%	6-Acetylmorphine; morphine; morphine glucuronide	Euphoria, drowsiness, respiratory depression, convulsions, coma
Codeine	Junk, white stuff, morpho, M	Oral, IV, IM	3–6	2–4	5–20%	Morphine; norcodeine; conjugates	Sedation, convulsions, respiratory failure
Morphine		IV, IM, oral, smoked	3–6	2–4	<10%	Morphine-3-glucuronide; morphine-6-glucuronide; morphine sulfate; normorphine; codeine	Analgesia, euphoria, nausea, respiratory coma
Methadone	Methadose	Oral, IV, IM	12–24	15–60	5–50%	2-Ethylidene-1, 5-dimethyl-3,3-diphenyl-pyrroline; 2-ethyl-5methyl-3,3-diphenyl-pyrroline methadol; normethadol; conjugates	Analgesia, sedation, respiratory depression, coma
Meperidine	(Demerol)	IV, oral	3–6	2–5	5%	Normeperidine; meperidinic acid; normeperidinic acid	Analgesia, stupor, respiratory depression, hypotension, coma
Propoxyphene	Yellow footballs (Darvon)	Oral	1–6	8–24	<1%	Norpropoxyphene; dinorpropoxyphene	Analgesia, stupor, respiratory depression, coma
Hallucinogens							
Phencyclidine (PCP)	PCP, angel dust, hog; killer weed	IV, oral, nasal, smoked	2–4; psychoses may last weeks	7–16	30–50%	4-Phenyl-4-piperidinocyclohexanol; 1-(1-phenylcyclohexyl)-4-hydroxy-piperidine; glucuronide conjugates	Dissociative anesthesia, depression, psychosis, stupor, coma, seizures
LSD	Acid, LSD-25, white lightning, microdots	Oral	8–12	3–4	1%	*N*-Desmethyllysergide; 13-hydroxylysergide	Hallucinations, flashbacks, psychosis, vomiting, paralysis, respiratory depression
Marijuana, hashish	Pot, THC, mary jane, grass, hash	Oral, smoked, IV	2–4	14–38	<1%	11-Nor-9-carboxy-Δ^9-THC; 11-hydroxy-tetrahydrocannabinol	Altered perception, memory loss, disorientation, psychosis

(continued on next page)

Drug	Street Name (Trade Name)	Route of Ingestion	Duration of Effect (h)	Half Life (h)	Excreted Unchanged in Urine	Principal Urinary Metabolites	Symptomatology
Benzodiazepines							
Chlordiazepoxide	(Librium)	Oral, IM	4–8	6–27	<1%	Norchlordiazepoxide; demoxepam; nordiazepam; oxazepam; glucuronide conjugates	Drowsiness, muscle relaxation, coma
Diazepam	(Valium)	Oral, IV, IM	4–8	20–50	<1%	Nordiazepam; oxazepam; 3-hydroxy-diazepam; glucuronide conjugates	Drowsiness, dizziness, muscle relaxation
Sedatives/Depressants							
Pentobarbital	Yellow, nembies, yellow jackets	Oral, IV, IM	3–6	15–48	1%	3-Hydroxypentobarbital; N-hydroxpento-barbital; 3-carboxypentobarbital	Sedation, respiratory collapse
Amobarbital	Rainbows, blues, bluebirds	Oral, IV, IM	3–24	12–60	<1%	3-Hydroxyamobarbital; N-glucosyl amobarbital; 3-carboxyamobarbital	Exhilaration, sedation, disorientation, respiratory depression, coma
Secobarbital	Reds, seccies, red devils, M & M's	Oral, IV, IM	3–6	15–40	5%	3-Hydroxysecobarbital secodiol; 5-(1-methylbutyl) barbituric acid	Sedation, lethargy, coma, respiratory collapse
Ethanol		Oral	2–6	2–14	2–10%	Acetaldehyde; acetic acid	Slurred speech, loss of equilibrium, drowsiness, coma, respiratory collapse
Methaqualone	Ludes, soapers	Oral	4–8	20–60	<1%	3',4'-, and 6-hydroxymethaqualone; respective glucuronide conjugates	Sedation, dizziness, paresthesias, convulsions, respiratory and circulatory depression
Chloral hydrate	Joy juice	Oral, rectal	5–8	<1	<1%	Trichloroethanol; trichloroacetic acid; conjugates	Sedation, GI distress, hypotension, respiratory depression

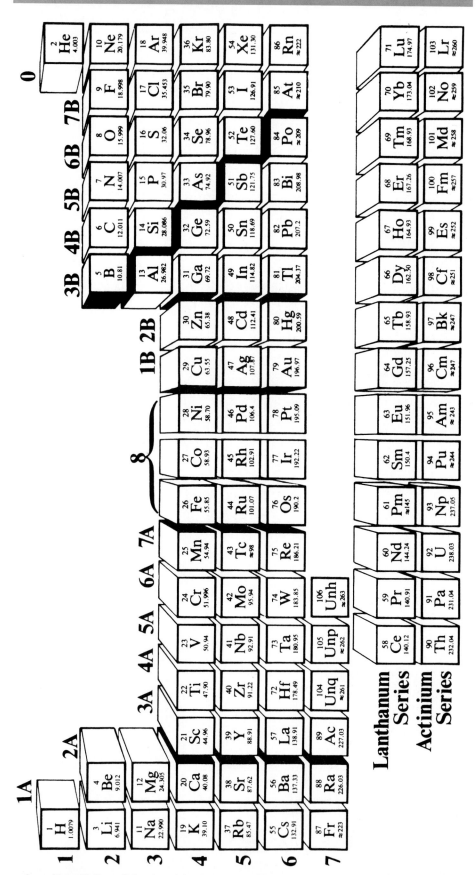

Source: Monroe M, Abrams K. Experimental chemistry: A laboratory course. Belmont, CA: Star Publishing Company, 1991. Reprinted with permission.

Glossary

ACCURACY   without error; closeness to the true value.

ACID   a substance that can yield a hydrogen ion or hydronium ion when dissolved in water.

ACIDEMIA   a blood pH less than the reference range.

ACIDOSIS   a pH below the reference range.

ACTH   adrenocorticotropic hormone; a peptide hormone secreted by the anterior pituitary. It stimulates the cortex of the adrenal glands to produce adrenal cortical hormones.

ACTIVATION ENERGY   energy required to raise all molecules in one mole of a compound at a certain temperature to the transition state at the peak of the energy barrier. At the transition state, each molecule is equally likely either to participate in product formation or to remain an unreacted molecule. Reactants possessing enough energy to overcome the energy barrier participate in production formation.

ACTIVE TRANSPORT   a mechanism that requires energy in order to move ions across cellular membranes.

ACTIVITY COEFFICIENT (AC)   relating to the study of vitamins, an expression indicating the enhancement of enzyme activity upon saturation with a vitamin. The greater the AC, the more likely the patient is to be deficient in the vitamin.

ACUTE RENAL FAILURE   a sudden, sharp decline in renal operation due to an acute toxic or hypoxic insult to the kidneys. This has been defined as occurring when the GFR is reduced to <10 mL/min.

ADAPTIVE IMMUNE SYSTEM   one of two functional parts of the immune system; it produces a specific reaction to each infectious agent, then normally eradicates that agent and remembers that particular infectious agent, preventing it from causing disease later. For example, measles and diphtheria produce a lifelong immunity following an infection.

ADH   antidiuretic hormone (vasopressin). The supraoptic and paraventricular nuclei of the hypothalamus produce ADH and oxytocin.

ADRENAL GLANDS   paired organs located one at the upper pole of each kidney. Each gland consists of an outer cortex and an inner medulla, which have different embryological origins, different mechanisms of control, and different products.

AGING   maturing; a progressive loss of adaptation leading to decreased viability and life expectancy and increased vulnerability.

AIRBORNE PATHOGEN   any infectious agent transmissible by air, *e.g.,* tuberculosis.

ALBUMIN   the main protein in plasma.

ALDOSTERONE   the principal mineralocorticoid (electrolyte-regulating hormone) produced by the zona glomerulosa of the adrenal cortex.

ALKALEMIA   a blood pH greater than the reference range ($\sim$7.35–7.45).

ALKALOSIS   a pH above the reference range.

ALLANTOIN   produced by the oxidation of uric acid by the enzyme uricase. It is the end-product of purine metabolism.

ALLOGRAFT   transplant tissue from the same species (*e.g.,* kidney).

AMENORRHEA   cessation of menstruation.

AMINE   any one of a group of nitrogen-containing organic compounds. The amine hormones include epinephrine, norepinephrine, thyroxine, and triiodothyronine.

AMINO ACID   small biomolecules with a tetrahedral carbon covalently bound to an amino group, a carboxyl group, a variable (R) group, and a hydrogen atom.

AMINOACIDOPATHIES   inherited disorders of amino acid metabolism.

AMMONIA NH_3; formed *in vivo* from breakdown of amino acids.

AMNIOCENTESIS   puncture of the amniotic sac to obtain fluid for analysis.

AMNIOTIC FLUID   a fluid in which the fetus is suspended; it provides a cushioning medium for the fetus and serves as a matrix for influx and efflux of constituents.

Michael L. Bishop, Janet L. Duben-Engelkirk, and Edward P. Fody.
CLINICAL CHEMISTRY. © 1996 Lippincott–Raven Publishers.

AMPHOTERIC   having two or more ionizable sites that can result in either a net positive or a net negative charge depending on the pH of the environment.

ANALGESIC   a drug that relieves pain.

ANALYTE   a biologic solute or constituent.

ANALYTICAL ERROR   errors due to the instrument's (or laboratorian's) handling of a specimen.

ANALYTICAL VARIATIONS   non-identical measurements due to diverse causes, including instrument, reagent, and operator variations.

ANDROGEN   any substance (hormone) stimulating the development of male characteristics (*e.g.,* testosterone).

ANGIOTENSIN   (vasopressin) produced when renin is released from the kidney; a vasopressor substance.

ANHYDROUS   without water.

ANION   a negatively charged ion; anions move toward the anode (positive pole) because of its positive charge.

ANION GAP   the difference between unmeasured anions and unmeasured cations.

ANTECUBITAL FOSSA   area of the forearm at the bend of the elbow; most commonly used for venipuncture.

ANTERIOR PITUITARY   adenohypophysis. The pituitary is located in a small cavity in the sphenoid bone of the skull called the sella turcica. The tropic hormones of the anterior pituitary are mediated by negative feedback, which involves interaction of the effector hormones with the hypothalamus, as well as with cells of the anterior pituitary.

ANTHROPOMETRIC METHODS   used to assess the general nutritional status of a patient. Includes skin fold test, arm circumference, and height and weight measurements.

ANTIBODY   glycoproteins (immunoglobulins) secreted by plasma cells, which, in turn, are under the control of many lymphocytes and their cytokines. Antibodies are produced in response to antigens.

ANTICOAGULANT   inhibits the blood's clotting action; yields specimens containing intact clotting factors.

ANTICOAGULANT THERAPY   treatment with drugs that interfere with blood coagulation to prevent diseases caused by venous thrombosis. Widely used drugs include: coumarins, heparin, and thrombolytic agents such as streptokinase or tissue plasminogen activator (tPA).

ANTIGEN   agents that are recognized as foreign by the immune system. The immune system produces antibodies in response.

ANTIGENIC DETERMINANT   a part of an antigen's structure that is recognized as foreign by the immune system. This structural domain is also referred to as an epitope.

ANTIOXIDANT   substance (*e.g.,* vitamin) that inhibits or prevents oxidation.

ARTERIAL BLOOD   blood from arteries.

ARTERIOSCLEROSIS   includes a number of pathological conditions in which there is a thickening or hardening of the walls of the arteries.

ASCITES   excess fluid in the peritoneal cavity; the fluid is called ascitic fluid.

ASTM E 1238 STANDARD   addresses standard specifications for transferring clinical observations between independent computer systems; defines the logical format and encoding rules for messages intended for the interchange of any clinical information that can be reported in a textual form.

ATHEROSCLEROSIS   a disease in which there is an accumulation of lipid material in the veins and arteries.

ATOMIC ABSORPTION   analytical technique that measures concentration of analyte by detecting absorption of electromagnetic radiation by atoms rather than by molecules. Instrument is atomic absorption spectrophotometer.

AUTOIMMUNE DISORDER   disease or disorder in which the body produces antibodies (immunological response) against itself.

AUTOMATION   mechanization of the steps in a procedure. Manufacturers of clinical chemistry analyzers design their instruments to mimic the manual techniques in a procedure.

AZOTEMIA   elevated level of urea in blood.

BAR CODE   a set of vertical bars of varying width used to encode information. Used most frequently in the clinical laboratory for patient and specimen information.

BASAL STATE   early morning before the patient has eaten or become physically active. This is a good time to draw blood specimens because the body is at rest and food has not been ingested during the night.

BASE   a substance that can yield hydroxyl ions (OH^-).

BASE EXCESS (BE)   the theoretical amount of titrable acid or base required to return the plasma pH to 7.40 at a PCO_2 of 40 mmHg at 37°C. A positive value (base excess) indicates an excess of bicarbonate or relative deficit of noncarbonic acid and suggests metabolic alkalosis. A negative value (base deficit) indicates a deficit of bicarbonate or relative excess of noncarbonic acids and suggests metabolic acidosis.

BEER'S LAW   mathematically establishes the relationship between concentration and absorbance in photometric determinations; expressed as: $A = abc$.

BERIBERI   chronic deficiency of the vitamin thiamin produces the disease beriberi. Two forms of beriberi are recognized. Wet beriberi affects predominantly the cardiovascular system and is characterized by extensive edema and high-output cardiac failure. Dry beriberi affects the nervous system and is associated with neurologic symptoms such as polyneuritis, and hyperesthesia, and muscle wasting.

BICARBONATE   the HCO_3^- anion.

BILE   a fluid produced by the liver and composed of bile acids or salts, bile pigments (primarily bilirubin esters), cholesterol, and other substances extracted from the blood. Total bile production averages about 3 L per day, although only 1 L is excreted.

BILIRUBIN the principal pigment in bile; derived from the breakdown of hemoglobin when aged red blood cells are phagocytized by the reticuloendothelial system, primarily in the spleen, liver, and bone marrow. About 80% of the bilirubin formed daily comes from the degradation of hemoglobin. The remainder comes from the destruction of heme-containing proteins (myoglobin, cytochromes, catalase) and from the catabolism of heme.

BIOHAZARD anything harmful or potentially harmful to man, other organisms, or the environment. Examples include blood or blood products and contaminated laboratory waste.

BLOODBORNE PATHOGEN any infectious agent or pathogen transmissible by means of blood or blood products.

BUFFER a substance that minimizes any change in hydrogen ion concentration; a weak acid or base and its conjugate salt.

BUN/CREATININE RATIO mg/dL plasma or serum urea nitrogen: mg/dL plasma or serum creatinine.

BURET a wide, long, graduated pipet with a stopcock at one end.

CAH congenital adrenal hyperplasia; results from the lack of an enzyme necessary for the production of cortisol.

CALIBRATION standardization or the determination of the accuracy of an instrument.

CANCER the uncontrolled growth of cells from normal tissue. Cancer cells can grow and spread thereby killing the host.

CAPILLARY BLOOD blood from minute blood vessels.

CARBOHYDRATE polyhydroxy aldehydes or polyhydroxy ketones, or multimeric units of such compounds. The general formula of a carbohydrate is $(CH_2O)_n$.

CARBONIC ACID H_2CO_3.

CARCINOGEN cancer causing agent.

CATION a positively charged ion; cations migrate in the direction of the cathode because of their positive charge.

CENTRIFUGAL ANALYSIS analytical technique that uses the force generated by centrifugation to transfer and then contain liquids in separate cuvets for measurement at the perimeter of a spinning rotor.

CENTRIFUGATION a process whereby centrifugal force is used to separate solid matter from a liquid suspension.

CEREBRAL SPINAL FLUID (CSF) a selective ultrafiltrate of the plasma that surrounds the brain and spinal cord.

CERULOPLASMIN an alpha-2-glycoprotein to which copper is attached; more than 90% to 95% of the circulating copper in the plasma is bound to ceruloplasmin.

CHAIN-OF-CUSTODY records each step and each person who handled a sample for analysis from time of collection to time of analysis.

CHANNEL on an automated analyzer, a path or passage for reagents, specimens, or electrical impulses. Automated analyzers may be single or multiple channel analyzers.

CHARACTER the number to the left of the decimal point in a logarithmic expression.

CHEMICAL HYGIENE procedures and work practices for regulating exposure of laboratory personnel to hazardous chemicals.

CHEMILUMINESCENCE light produced as a result of a chemical reaction. Most important chemiluminescence reactions are oxidation reactions of luminol, acridinium esters, and dioxetanes and are characterized by a rapid increase in intensity of emitted light followed by a gradual decay.

CHOLECYSTOKININ CCK, formerly called pancreozymin; a hormone produced by the pancreas. CCK, in the presence of fats and/or amino acids in the duodenum, is produced by the cells of the intestinal mucosa and is responsible for release of enzymes from the acinar cells by the pancreas into the pancreatic juice.

CHOLESTEROL an unsaturated steroid alcohol of high molecular weight, consisting of a perhydrocyclopentanthroline ring and a side chain of eight carbon atoms. In its esterified form, it contains one fatty acid molecule.

CHRONIC RENAL FAILURE a clinical syndrome that occurs when there is a gradual decline in renal operation over time.

CHYLOMICRONS large triglyceride rich particles.

CIRRHOSIS derived from the Greek word that means "yellow." However, in current usage, cirrhosis refers to the irreversible scarring process by which normal liver architecture is transformed into abnormal nodular architecture.

CLEARANCE volume of plasma filtered by glomeruli per unit time.

CLOSED COLLECTION SYSTEM a type of collection system in which blood is taken directly from a patient's vein into a stoppered tube; the sample is completely contained, thereby reducing the risk of outside contaminants to the sample and reducing the hazard of the collector's exposure to the blood. The system consists of a needle, plastic sleeve or tube holder, and the evacuated tube; also known as the evacuated-tube system.

COAGULATION CASCADE complex series of reactions in the blood coagulation process. Includes the intrinsic and extrinsic coagulation systems. The early theory of the coagulation cascade has exploded into a complex interaction of multiple biological systems.

COFACTOR a nonprotein molecule that may be necessary for enzyme activity.

COLLIGATIVE PROPERTY the properties of osmotic pressure, freezing point, boiling point, and vapor pressure.

COMMON VARIABLE AGAMMAGLOBULINEMIA also known as acquired agammaglobulinemia. This disorder represents a group of disorders characterized by hypogammaglobulinemia.

COMPENSATION   the body's attempt to return the pH toward normal whenever an imbalance occurs.

COMPETITIVE PROTEIN BINDING   also competitive assay. Labeled and unlabeled antigen compete for limited antibody sites. Labeled antigen bound to antibody will be inversely proportional to the concentration of antigen.

CONDUCTIVITY   relates to the ease in which electricity passes through a solution.

CONGENITAL AGAMMAGLOBULINEMIA   x-linked agammaglobulinemia, also called Bruton disease. This disorder presents with early onset of recurrent pyogenic infections. These patients have no circulating B cells, low concentrations of all circulating immunoglobulin classes, and absence of plasma cells in all lymphoid tissue. T cells are not involved.

CONJUGATED BILIRUBIN   bilirubin diglucuronide; it is water-soluble, and is secreted from the hepatic cell into the bile canaliculi and then passes along with the rest of the bile into larger bile ducts and eventually into the intestines.

CONJUGATED PROTEIN   composed of a protein (amino acids) and a non-protein moiety.

CONN'S SYNDROME   aldosterone-secreting adrenal adenoma.

CONTACT ACTIVATION   initiation of the coagulation process. The contact group of coagulation factors include XI, XII, HMWK, and prekallikrein.

CONTINUOUS FLOW   an approach to automated analysis in which liquids (reagents, diluents, and samples) are pumped through a system of continuous tubing. Samples are introduced in a sequential manner, following each other through the same network. A series of air bubbles at regular intervals serve as separating and cleaning media.

CONTROL   a substance or material of determined value, used to monitor the accuracy and precision of a test. Controls are run with the patients' specimens.

CORROSIVE CHEMICAL   chemicals injurious to the skin or eyes by direct contact or to the tissues of the respiratory and gastrointestinal tracts if inhaled or ingested. Examples include acids (acetic, sulfuric, nitric, and hydrochloric) and bases (ammonium hydroxide, potassium hydroxide, and sodium hydroxide).

CORTISOL   a steroid hormone produced by the adrenal glands.

COUPLED ENZYMATIC METHOD   use of several sequential enzymatic reactions that produce a product which can be detected spectrophotometrically and whose concentration will be related to the analyte in question.

CREATINE   compound found in muscle synthesized from several amino acids. It combines with high energy phosphate to form creatine phosphate, which functions as an energy compound in muscle.

CREATININE   compound formed when creatine or creatine phosphate spontaneously loses water or phos-

phoric acid. It is excreted into the plasma at a relatively constant rate in a given individual and excreted in the urine.

CREATININE CLEARANCE   rate of removal of creatinine from plasma. Calculated as urine creatinine concentration times 24-h urine volume divided by serum creatinine concentration or UV/P, expressed in mL/min. Usually corrected to normal body surface area.

CRH   corticotropin-releasing hormone. A hormone released from the hypothalamus that acts on the anterior pituitary to increase ACTH secretion.

CRYOGENIC MATERIAL   material brought to low temperatures, such as liquefied gases.

CYCLOSPORIN   a cyclic polypeptide of fungal origin. It consists of 11 amino acids and has a molecular weight of 1203. It inhibits the immune response selectively by inhibiting the interleukin-2 dependent proliferation of activated T cells, which destroy the allograft. The immune response is frozen and unresponsive.

CYTOCHROME   a pigment that plays a role in respiration, such as hemoglobin or myoglobin.

CYTOKINES   extracellular factors produced by a variety of cells including monocytes, lymphocytes, and other non-lymphoid cells. They are important in controlling local and systemic inflammatory responses.

D-XYLOSE ABSORPTION TEST   an analytical technique that assesses the ability to absorb d-xylose; it is of value in differentiating malabsorption of intestinal etiology from that of exocrine pancreatic insufficiency.

DATA TYPES   numerical or textual information contained in a database field.

DEIONIZED WATER   water purified by ion exchange.

DELIQUESCENT SUBSTANCES   compounds that absorb enough water from the atmosphere to cause dissolution.

DELTA CHECK   an algorithm in which the most recent result of a patient is compared with the previously determined value.

DELTA ABSORBANCE   difference in absorbance, known as delta absorbance or ΔA.

DENSITY   weight of a substance compared with a standard; expressed in terms of mass per unit volume.

DESCRIPTIVE STATISTICS   statistics or values (*e.g.,* mean, median, mode) used to summarize the important features of a group of data.

DESICCANT   causing dryness; materials that can remove moisture from the air as well as from other materials.

DESICCATOR   a closed chamber for drying substances.

DHEA   dehydroepiandrosterone. An androgen primarily derived from the adrenal gland.

DIABETES MELLITUS   a diverse group of hyperglycemic disorders with different etiologies and clinical pictures.

DIALYSIS   a method for separating macromolecules from a solvent.

DIFFUSION the movement of the molecules of a substance from a location of higher concentration to one of lesser concentration; as in gel diffusion precipitation.

DILUTION a dilution represents the ratio of concentrated or stock material to the total final volume of a solution and consists of the volume or weight of the concentrate plus the volume of the diluent (the concentration units remaining the same).

DISACCHARIDE separate carbohydrates or monosaccharides joined together.

DISCRETE ANALYSIS an approach to automated analysis in which each sample and accompanying reagents are in a separate container. Discrete analyzers have the capability of running multiple tests one sample at a time, or multiple samples one test at a time.

DISPERSON the spread of data; most simply estimated by the range, the difference between the largest and smallest observations. The most commonly used statistic for describing the dispersion of groups of single observations is the standard deviation, which is usually represented by the symbol *s*.

DISTAL TUBULE portion of renal tubule extending from ascending limb of loop of Henle to the collecting duct.

DISTILLED WATER water solely purified by distillation results.

DNA PROBE a known fragment of DNA molecule used to join with or locate an unknown or comparable DNA strand. Human papilloma virus DNA has been detected using DNA probes.

DNA INDEX (DI) ploidy status of malignancies; the amount of measured DNA in cancer cells relative to that in normal cells.

DONE NOMOGRAM a chart that approximates drug toxicity given the time of ingestion and blood drug level.

DRUG DISPOSITION the way in which the body handles a foreign compound, the drug. The mechanisms used by the body to handle a drug can be explained in terms of four general processes: absorption, distribution, metabolism, and excretion.

DRY CHEMISTRY SLIDE a multi-layered film technology (dry chemicals) used by the Kodak series of automated analyzers. All reagents necessary for a particular test are contained on the "slide."

FLAGS on automated analyzers, a computerized printed warning of an error or instrument problem.

DYSLIPIDEMIAS diseases associated with abnormal lipid concentrations.

ECTOPIC HORMONE PRODUCTION hormones produced by cells in sites other than the gland from which they are usually derived.

ECTOPIC in an abnormal position or location.

EFFUSION abnormal accumulations of pleural or pericardial fluid.

ELECTROCHEMISTRY the use galvanic and electrolytic cells (electrochemical cells) for chemical analysis.

Examples include potentiometry, amperometry, coulometry, and polarography.

ELECTRODES electronic sensing devices to measure PO_2, PCO_2, and pH.

ELECTROLYTE ions capable of carrying an electric charge.

ELECTRONIC MEDICAL RECORD a centralized (computer) database of all patient information.

ELECTROPHORESIS migration of charged solutes or particles in an electrical field.

EMBDEN-MEYERHOF PATHWAY series of steps involving the anaerobic metabolism of glucose, glycogen, or starch to lactic acid; principal means of producing energy in humans.

ENCEPHALOPATHY disorder or dysfunction of brain.

ENDOCRINE GLAND a ductless gland that produces a secretion (hormone) into the blood or lymph to be carried by the circulation to other parts of the body.

ENDOGENOUS triglycerides synthesized in liver and other tissues.

ENZYME specific biologically synthesized proteins that catalyze biochemical reactions without altering the equilibrium point of the reaction or being consumed or undergoing changes in composition.

ENZYME-SUBSTRATE COMPLEX a physical binding of a substrate to the active site of an enzyme.

EPINEPHRINE an amine hormone. The adrenal medulla primarily produces epinephrine.

EQUIVALENT WEIGHT equal to the molecular weight of a substance divided by its valence.

ESSENTIAL ELEMENT an element that is absolutely necessary for life. Deficiency or absence of the element will cause a severe alteration of function and/or eventually lead to death.

ESTRIOL an estrogenic hormone; estriol is not produced in significant quantities in the mother and is solely a reflection of fetoplacental function. Thus, estriol measurement provides valuable information about fetal well-being.

ESTROGEN any of the group of substances (hormones) that induces estrogenic activity; specifically the estrogenic hormones, estradiol and estrone, produced by the ovary.

EVACUATED TUBE sample collection tube with a vacuum.

EXOCRINE GLAND any gland that excretes externally through a duct. Secretions of exocrine glands do not directly enter the circulation.

EXOGENOUS triglycerides obtained from dietary sources.

EXTRACELLULAR FLUID water (fluid) outside the cell; can be subdivided in the intravascular extracellular fluid (plasma) and the interstitial cell fluid (ISCF) that surrounds the cells in the tissues.

EXTRAUTERINE outside the uterus or womb.

EXTRINSIC PATHWAY involves the activity of tissue thromboplastin, which reacts with factor VII and calcium for the rapid proteolytic activation of factor X to factor Xa. Tissue damage releases tissue thromboplastin, which initiates the extrinsic pathway of coagulation. The extrinsic coagulation pathway is so named because its activation requires the involvement of a substance that resides outside the circulation.

EXUDATE   accumulation of fluid in a cavity; also the production of pus or serum. In comparison with a transudate, an exudate contains more cells and protein. Exudates demand immediate attention.

F-TEST   statistical test used to compare the features of two or more groups of data.

FASTING SPECIMENS   a blood specimen taken after the patient has not eaten for at least 12 hours.

FATTY ACIDS   major constituents of triglycerides and phospholipids. There are short-chain (4–6 carbon atoms), medium-chain (8–12 carbon atoms), and long-chain fatty acids carbon atoms ($>$12).

FERRITIN   a spherical protein shell composed of 24 subunits with a molecular mass of 500 kDa. The protein can bind up to 4,000 iron molecules, making it a large potential source of iron.

FIBRIN   a filamentous protein formed by the action of thrombin on fibrinogen. It is the basis for clotting of the blood. Fibrinolysis removes fibrin by breaking it down into small fragments, which can then be cleared from the circulation by the reticuloendothelial system.

FIBRINOLYSIS   a process that removes fibrin by breaking it down into small fragments that can be cleared from the circulation by the reticuloendothelial system.

FIELD   a part of a database record specified for a particular type of data (*e.g.,* name, telephone, zip code).

FILTRATE   liquid that passes through filter paper is called the filtrate.

FILTRATION   separation of solids from liquids.

FiO$_2$ the fraction of inspired oxygen; can be as much as 100% when oxygen is being supplied.

FIRE TETRAHEDRON   a three-dimensional pyramid representing the element of fire; previously the fire triangle.

FISHER PROJECTION   model that can be used to represent carbohydrates. The Fisher projection of a carbohydrate has the aldehyde or ketone at the top of the drawing. The carbons are numbered starting at the aldehyde or ketone end and the compound can be represented either as a straight chain or cyclic, hemiacetal form.

FLAME PHOTOMETRY   analytical technique that measures the wavelength and intensity of light emitted from a burning solution (patient specimen).

FLOW CELL CYTOMETRY   the use of immunofluorescent labels to identify specific antigens on live cells in suspension. Stained cell suspensions are transported under pressure past a laser beam, and emitted fluorescence (at 90° relative to the beam) is measured and computer analyzed with this technique. Using multiple labels, cells

can be identified and either electronically or physically sorted. The technique has been used to analyze subpopulations of lymphocyte cells in various clinical diagnoses.

FLUORESCENCE POLARIZATION (FPIA)   a technique in which a fluorescein labeled antigen rotates very fast in solution and, when excited, does not emit polarized light. After binding to antibody the label rotates much slower and will emit polarized light. When polarized light is used as a source of excitation, emitted fluorescence is measured in two planes. The difference in fluorescence polarization (calculated) before and after the addition of labeled analyte is inversely proportional to the concentration of unknown analyte. This methodology is most useful for small antigens (drugs, hormones, etc.)

FLUOROMETRY   analytical technique used to measure fluorescence (light emitted as a result of energy absorbed).

FOLLICULAR CELLS   (or cuboidal) one of two types of cells of the thyroid; they are secretory cells and produce thyroxine (T$_4$) and triiodothyronine (T$_3$).

FOLLICULAR PHASE   the first half of the menstrual cycle, when estrogen effect is unopposed by progesterone, also called the proliferative phase.

FORENSIC MEDICINE   medical knowledge or results that apply to questions of law affecting life or property. A specimen result might be used as evidence in a court of law to prove cause of death, innocence, or guilt of an accused individual, or possibly to prove alcohol or drug abuse.

FORENSIC   pertaining to the law or legal matters (*e.g.,* toxicology and the law).

FRACTIONAL OXYHEMOGLOBIN   (FO$_2$Hb) is the ratio of the concentration of oxyhemoglobin to the concentration of total hemoglobin (ctHb).

FREE THYROXINE INDEX (FT$_4$I)   an indirect measure of free hormone concentration and is based on the equilibrium relationship of bound T$_4$ and FT$_4$. The FT$_4$I is calculated by the following formula: $FT_4I = T_4 \times T_3U$ ratio.

FRUCTOSAMINE   any glycated serum protein.

FSH   follicle stimulating hormone; a protein hormone secreted by the anterior pituitary.

GAS CHROMATOGRAPHY   analytical technique used to separate mixtures of compounds that are volatile or can be made volatile. Gas chromatography may be gas–solid chromatography (GSC), with a solid stationary phase, or gas–liquid chromatography (GLC), with a nonvolatile liquid stationary phase. Gas–liquid chromatography is commonly used in clinical laboratories.

GASTRIN   a gastrointestinal hormone. It is a peptide secreted by the G cells of the antrum (lower one-third) of the stomach. It is released in response to contact with food.

GERIATRICS   branch of general medicine dealing with remedial and preventable clinical problems in the elderly, as well as the social consequences of such illness.

GERONTOLOGY study of the aging process in the human body.

GH growth hormone; a peptide composed of 191 amino acids. In contrast to most of the other protein hormones, GH does not act through cAMP. Binding of GH to its membrane receptor leads to glucose uptake, amino acid transport, and lipolysis. Secreted by the anterior pituitary.

GLOBULINS heterogeneous group of proteins that can be separated by electrophoresis into alpha-1, alpha-2, and gamma fractions.

GLOMERULAR FILTRATE filtrate of plasma containing water and small molecules, but lacking cells and large molecules, such as most proteins.

GLOMERULAR FILTRATION RATE (GFR) rate at which plasma is filtered by glomerulus expressed in mL/minute.

GLOMERULONEPHRITIS inflammation of the glomeruli of the kidney. May be acute, subacute, or chronic.

GLOMERULUS small blood vessels in the nephron, which project into the capsular end of the proximal tubule and serve as a filtering mechanism.

GLUCAGON a protein hormone that is involved in the regulation of carbohydrate, fat, and protein metabolism. It is synthesized by the α cells of the pancreatic islet and is composed of 29 amino acids.

GLUCOCORTICOID a general classification of hormones synthesized in the zona fasciculata of the adrenal cortex. Cortisol is the principal glucocorticoid hormone.

GLUCONEOGENESIS the conversion of amino acids by the liver and other specialized tissues such as the kidney to substrates that can be converted to glucose. Gluconeogenesis also encompasses the conversion of glycerol, lactate, and pyruvate to glucose.

GLYCOGEN a polysaccharide, similar to starch; form in which carbohydrates are stored.

GLYCOGENOLYSIS process by which glycogen is converted back to glucose-6-phosphate for entry into the glycolytic pathway.

GLYCOLYSIS hydrolysis of glucose by an enzyme into pyruvate or lactate; the process is anaerobic.

GLYCOLYTIC INHIBITOR a substance that prevents the hydrolysis of sugar. Sodium fluoride is a glycolytic inhibitor.

GOUT arthritis associated with increased levels of uric acid in the blood, which then become deposited in the joints or tissues causing painful swelling. It is more common in men than in women.

GRAFT tissue that is transplanted.

GRAPHICAL USER INTERFACE also GUI; computer interface software (object-oriented) written to enhance the operator's usability of a computer. Pictures or icons are used to represent files or processes.

GRAVES' DISEASE diffuse toxic goiter. Graves' disease occurs six times more commonly in women than in men. It occurs frequently at puberty, during pregnancy, at menopause, or following severe stress. The frequency of Graves' disease in the general population is about 0.4%.

HASHIMOTO'S DISEASE chronic autoimmune thyroiditis; it is the most common cause of primary hypothyroidism.

HAWORTH PROJECTION represents glucose in a cyclic form, which is more representative of the actual structure. When glucose is drawn in a Haworth projection, the α form of D-glucopyranose is represented by the hydroxy group of carbon one oriented downward or below the plane of the paper.

HAZARD COMMUNICATION based on the fact that all employees must be informed of any health risks involving the use of chemicals; from the Hazardous Communication Standard of 1987 (Right to Know Law).

HAZARDOUS WASTE any potentially dangerous waste material.

HbA$_{1c}$ largest sub-fraction of normal HbA in both diabetic and non-diabetic subjects. It is formed by the reaction of the β chain of HbA with glucose. It reflects the concentration of glucose present in the body over a prolonged time period related to the 60-day half-life of erythrocytes.

HCG human chorionic gonadotropin. A protein hormone produced by the placenta and consisting of α and β subunits.

HDL high-density lipoproteins; a "clean-up crew" that gathers up extra cholesterol for transport back to the liver.

HEMODIALYSIS a technique or procedure that provides the function of the kidneys when one or both are damaged. The patient's blood is circulated through membranes to remove wastes.

HEMOFILTRATION an ultrafiltration procedure or technique (similar to hemodialysis) used to remove an excess accumulation of normal metabolic products from the blood.

HEMOGLOBIN OXYGEN (BINDING) CAPACITY the maximum amount of oxygen that can be carried by hemoglobin in a given quantity of blood.

HEMOGLOBIN-OXYGEN DISSOCIATION CURVE a graphic representation ("S"-shaped) of oxygen content as percent oxygen saturation against PO_2; based on the principle that oxygen dissociates from adult hemoglobin in a characteristic fashion.

HEMOGLOBINOPATHY a disorder associated with the presence of an abnormal hemoglobin.

HEMOLYSIS damage to erythrocyte membranes, causing release of cellular constituents (*e.g.,* hemoglobin) into the blood plasma or serum.

HEMOSTASIS the property of the blood circulation system, which maintains the blood in a fluid state within the vessel walls in combination with an ability to prevent excessive blood loss when injured.

HENDERSON-HASSELBALCH EQUATION equation that mathematically describes the dissociation characteristics of weak acids and bases and the effect on pH.

HEPA FILTER   High Efficiency Particulate Air filter; a respirator.

HEPARIN   an anticoagulant used in blood collection.

HEPATITIS   "inflammation of the liver"; may be caused by a virus, bacteria, parasites, radiation, drugs, chemicals, or toxins. Among the viruses causing hepatitis are hepatitis types A, B, C, D (or delta), and E, cytomegalovirus, Epstein-Barr virus, and probably several others.

HEPATOMA   primary malignant tumors of the liver; also known as heptocellular carcinoma or hepatocarcinoma.

HETEROGENEOUS ASSAY   a technique (*e.g.,* radioimmunoassay) where it is necessary to physically separate labeled antigen or hapten bound to antibody from a labeled antigen or hapten that remains free in solution.

HISTOGRAM   graphical representation of data where the number or frequency of each result is placed on the *y*-axis and the value of the result is plotted on the *x*-axis.

HL7 STANDARD   Health Level 7 Standard; system for electronic data exchange in health care environments, especially inpatient acute care facilities/hospitals. The term "Level 7" refers to the top level (7) of the OSI Reference Model.

HOMEOSTASIS   state of equilibrium in the body maintained by dynamic processes.

HOMOGENEOUS ASSAY   a technique (*e.g.,* EMIT) that does not require the physical separation of the bound and free labeled antigen.

HORMONE   a chemical substance that is produced and secreted into the blood by an organ or tissue and has a specific effect on a target tissue.

HPL   human placental lactogen is a protein hormone that is structurally, immunologically, and functionally very similar to growth hormone and prolactin. Like HCG, it is produced by the placenta and can be measured in maternal urine and serum as well as amniotic fluid.

HYDRATE   compound and its associated water.

HYDROLASE   an enzyme that catalyzes the hydrolysis of ether, ester, acid–anhydride, glycosyl, C–C, C–halide, or P–N bonds.

HYGROSCOPIC   substances that take up water on exposure to atmospheric conditions.

HYPERALDOSTERONISM   excessive production of aldosterone by the adrenal gland. It may be primary (due to an adrenal lesion) or secondary to abnormalities in the renin-angiotensin system.

HYPERCALCEMIA   elevated levels of calcium in the blood.

HYPERCHLOREMIA   elevated levels of chloride in the blood.

HYPERCOAGULABILITY   increased ability (*i.e.,* of blood) to coagulate.

HYPERGLUCAGONEMIA   excessive production of glucagon. Hyperglucagonemia due to tumors must be differentiated from other causes of increased glucagon, which include diabetes, pancreatitis, and trauma.

HYPERGLYCEMIC   increased level of blood sugar.

HYPERINSULINEMIA   excessive and/or inappropriate release of insulin.

HYPERKALEMIA   elevated levels of potassium in the blood.

HYPERMAGNESEMIA   elevated levels of magnesium in the blood.

HYPERNATREMIA   elevated levels of sodium in the blood.

HYPEROXEMIA   increased oxygen in the blood.

HYPERPARATHYROIDISM   condition resulting from the increased activity of the parathyroid glands.

HYPERPHOSPHATEMIA   elevated levels of phosphorus in the blood.

HYPERPROLACTINEMIA   excess secretion of prolactin due to hypothalamic-pituitary dysfunction. Generally due to a pituitary neoplasm.

HYPERPROTEINEMIA   total protein level in the blood that is higher than the reference interval.

HYPERTHYROIDISM   excessive secretion of the thyroid glands.

HYPERURICEMIA   plasma levels of uric acid greater than 7.0 mg/dL in males and 6.0 mg/dL in females.

HYPERVITAMINOSIS   a condition that results from excessive intake of a vitamin(s).

HYPOALDOSTERONISM   decreased level of aldosterone.

HYPOCHLOREMIA   decreased levels of chloride in the blood.

HYPOGLYCEMIC   decreased level of blood sugar.

HYPOGLYCORRHACIA   decreased CSF glucose levels.

HYPOINSULINEMIA   decreased level of insulin.

HYPOKALEMIA   decreased levels of potassium.

HYPOMAGNESEMIA   decreased levels of magnesium in the blood.

HYPONATREMIA   decreased levels of sodium in the blood.

HYPOPARATHYROIDISM   most often due to destruction of the adrenal glands and, as such, is associated with glucocorticoid deficiency.

HYPOPHOSPHATEMIA   decreased levels of phosphorus in the blood.

HYPOPROTEINEMIA   total protein level in the blood that is below the reference interval.

HYPOTHALAMIC-PITUITARY-THYROID AXIS (HPTA)   the neuroendocrine system that regulates the production and secretion of thyroid hormones.

HYPOTHALAMUS   the portion of the brain located in the walls and floor of the third ventricle. It is directly above the pituitary gland and is connected to the posterior pituitary by the pituitary stalk.

HYPOTHYROIDISM   decreased thyroid secretion.

HYPOURICEMIA   plasma levels of uric acid less than 2.0 mg/dL.

HYPOVITAMINOSIS   a condition that results from a lack or deficiency of vitamins in the diet.

HYPOVOLEMIA decreased blood volume.

HYPOXEMIA insufficient or decreased oxygenation of the blood.

ICTERUS jaundice; coloration of blood (also tissues, membranes, and secretions) with bile pigments; yellowish pigmentation in the skin or sclera. A result of excess bilirubin concentration in the blood.

ICTERUS INDEX a test that involves diluting serum with saline until it visually matches the color of a 0.01% potassium dichromate solution. The number of times the serum must be diluted is called the icterus index.

IMMUNE SYSTEM a complex series of events in the body that protect the individual from external harmful agents. Individual survival depends on a properly functioning system.

IMMUNOASSAY a technique that measure the rate of immune complex formation. Immunoassays can be labeled or non-labeled.

IMMUNOFIXATION or IMMUNOFIXATION ELECTROPHORESIS (IFE) a technique in which an immune precipitate is trapped (fixed) on the electrophoretic support medium. This method has superseded immunoelectrophoresis because of ease and speed.

INDIRECT IMMUNOFLUORESCENCE a technique in which serum antibody reacts with antigen fixed on a slide and the antibody, in turn, reacts with conjugated antihuman globulins. If the patient serum contains antibody of interest, this will be observed under a fluorescent microscope. Antinuclear antibodies (ANA) and antibodies to cytomegalovirus, rubella virus, and *Toxoplasma gondii* are but a few examples of the use of this technique.

INFANCY infant; period in life where child is unable to walk or feed itself.

INFERENTIAL STATISTICS values or statistics used to compare the features of two or more groups of data.

INHIBITORS that which inhibits. With regard to coagulation, primarily enzymes that inactivate activated factors. Coagulation is regulated by the action of inhibitors. The principle inhibitor of procoagulant proteins is Antithrombin III. Secondary inhibitors include alpha-2-macroglobulin and alpha-1-antitrypsin.

INNATE IMMUNE SYSTEM one of two functional divisions of the immune system; it is the first line of defense.

INSULIN a peptide hormone that is synthesized in the β cells of the islets of Langerhans in the pancreas.

INTERNATIONAL UNIT the amount of enzyme that will catalyze the reaction of one micromole of substrate per minute under specified conditions of temperature, pH, substrates, and activators.

INTRACELLULAR FLUID (ICF) fluid inside the cells.

INTRINSIC PATHWAY one of two coagulation pathways. Platelet reaction initiates the contact phase of coagulation leading to the intrinsic pathway of coagulation. The intrinsic coagulation pathway is so named because its activation can occur with elements that reside entirely within the circulation.

INULIN an exogenous plant polysaccharide derived from artichokes and dahlias. It is completely filtered by the glomeruli and neither secreted nor reabsorbed by the tubules. It is the most accurate of all GFR assays.

ION SELECTIVE ELECTRODES the half-cell or electrode (indicator) that responds to a specific ion in a solution.

IONIC STRENGTH concentration or activity of ions in a solution or buffer.

ISLETS OF LANGERHANS clusters of cells (alpha, beta, and delta) in the pancreas; insulin is a peptide hormone that is synthesized by the beta cells of the pancreatic islets.

ISOELECTRIC POINT (pI) the pH at which the molecule has no net change.

ISOENZYME different forms of an enzyme that may originate from genetic or nongenetic causes and may be differentiated from each other based on certain physical properties such as electrophoretic mobility, solubility, or resistance to inactivation.

KETONE a compound containing a carbonyl group ($C=O$) attached to two carbon atoms.

KINETIC ASSAY a type of test or procedure in which there is a reaction proceeding at a particular rate.

KINETIC METHOD OF MEASUREMENT quantitation by determining the rate at which a reaction occurs or a product is formed.

KUPFFER'S CELLS phagocytic macrophages capable of ingesting bacteria or other foreign material from the blood that flows through the sinusoids.

KWASHIORKOR acute protein calories malnutrition.

L/S RATIO a classic test that assesses the ratio of lecithins to sphingomyelins to determine fetal lung maturity.

LABORATORY INFORMATION SYSTEM (LIS) a network of computers and laboratory instrumentation for the purpose of storing and processing data.

LACTESENCE resembling milk; milky appearance.

LACTOSE TOLERANCE TEST an assay to determine the lactase content in the intestinal mucosa. The enzyme lactase is essential to the absorption of lactose from the intestinal tract.

LANCET a surgical device used to puncture the skin for blood collection.

LATERAL to the side.

LD FLIPPED PATTERN situation in which serum levels of LD-1 increase to a point where they are present in greater concentration than LD-2.

LDL low-density lipoproteins; rich in cholesterol, are the "empty tankers" that remain after the triglycerides have been deposited.

LH leutinizing hormone. A glycoprotein hormone secreted by the anterior pituitary.

LIPOPROTEIN protein bound with lipid components (*i.e.,* cholesterol, phospholipid, and triglyceride). May be classified as very-low-density (VLDL), low-density (LDL), and high-density (HDL).

LIQUID CHROMATOGRAPHY separation technique in which the mobile phase is a liquid.

LOBULE forms the structural unit of the liver, which measures 1 to 2 mm in diameter. It is composed of cords of liver cells (hepatocytes) radiating from a central vein. The boundary of each lobule is formed by a portal tract made up of connective tissue containing a branch of the hepatic artery, portal vein, and bile duct.

MALNUTRITION a state of decreased intake of calories or micronutrients (vitamins and trace elements) resulting in a risk of impaired physiologic function associated with increased morbidity and mortality.

MANTISSA that portion of the logarithm to the right of the decimal point, derived from the number itself.

MARASMUS a condition caused by caloric insufficiency without protein insufficiency so that the serum albumin level remains normal; there is considerable loss of body weight.

MECHANICAL HAZARD any potential danger from equipment such as centrifuges, autoclaves, and homogenizers.

MEDIAL in the middle.

MEGAVITAMIN intake of a vitamin or vitamins in extreme excess of daily requirements.

MEN multiple endocrine neoplasia is the occurrence of several tumors or hyperplasias involving diverse endocrine organs.

METABOLIC (NONRESPIRATORY) ACIDOSIS AND ALKALOSIS a disorder due to a change in the bicarbonate level (a renal or metabolic function).

METABOLITE any product of metabolism, as in the derivative of a drug.

METANEPHRINE a metabolic product of epinephrine and norepinephrine.

MICROALBUMINURIA small quantities of albumin in the urine; microalbumin concentrations are between 20–300 mg/d.

MICROCHEMISTRY clinical chemistry analyses involving only several microliters of sample.

MINERALOCORTICOID a group of substances produced by the adrenal cortex. Aldosterone is the principal mineralocorticoid (electrolyte-regulating hormone) produced by the zona glomerulosa.

MOLALITY represents the amount of solute per one kilogram of solvent.

MOLARITY number of moles per liter of solution.

MONOSACCHARIDE a simple carbohydrate; it cannot be decomposed by hydrolysis. Examples include glucose, galactose, and fructose.

MSDS Material Safety Data Sheets; a major source of safety information for employees who may use hazardous materials in their occupations.

MYOGLOBIN myoglobin is a heme protein found only in skeletal and cardiac muscle in humans. It can reversibly bind oxygen in a manner similar to the hemoglobin molecule, but myoglobin is unable to release oxygen, except under very low oxygen tension.

MYXEDEMA condition resulting from hypofunction of the thyroid. The term is used to describe the peculiar nonpitting swelling of the skin.

NARCOTIC any group of substances (drug) that produces the opioid group of substances; encompass not only heroin, morphine, and codeine, but also several synthetic compounds such as meperidine, methadone, propoxyphene, pentazocine, and others. All have some potential for addiction.

NCCLS National Committee for Clinical Laboratory Standards; an agency that establishes laboratory standards.

NEONATE a newborn up to 6 weeks of age.

NEOPLASM a new and abnormal growth of cells or tissue; also tumor.

NEPHELOMETRY analytical technique that measures the amount of light scattered by particles (immune complexes) in a solution. Measurements are made at 5–90° incident to the beam.

NEPHROTIC SYNDROME glomerular injury; an abnormally increased permeability of the glomerular basement membrane. Can be a result of many different etiologies.

NEUROBLASTOMA malignant tumors of the adrenal medulla that occur in children. They produce catecholamines and may occasionally be associated with hypertension.

NFPA National Fire Protection Association.

NITROGEN BALANCE equilibrium between protein anabolism and catabolism.

NON PROTEIN NITROGEN (NPN) nitrogen containing compounds remaining in a blood sample after the removal of protein constituents.

NOREPINEPHRINE a hormone (and catecholamine) produced by the adrenal medulla. As hormones, both norepinephrine and epinephrine serve to mobilize energy stores and prepare the body for muscular activity (increase heart rate and blood pressure, increase blood sugar, etc.). They are secreted in increased amounts with stress (pain, fear, etc.).

NORMALITY number of gram equivalent weights per liter of solution.

NUTRITIONAL ASSESSMENT the evaluation of a patient's metabolic and dietary (nutritional) needs.

NYCTALOPIA poor vision in dim light due to vitamin A deficiency. This condition is also known as night blindness.

ONCOFETAL ANTIGEN a protein produced in large amounts during fetal life and released into the fetal circulation. After birth, the production of oncofetal antigens is repressed, and only minute quantities are present in the circulation of adults.

ONCOGENES viral DNA segments that can transform normal cells into malignant cells.

ONE POINT CALIBRATION (or calculation) a term that refers to the calculation of the comparison of a known standard/calibrator concentration and its corresponding absorbance to the absorbance of an unknown value.

OPEN ARCHITECTURE design (software and hardware) to allow multiple disparate computer platforms a common means of communicating with each other in order to transfer data.

OPEN SYSTEM INTERCONNECTION (OSI) an open architecture approach to allow multiple disparate computer platforms used in the health care environment a common means of communicating with each other to transfer data.

OSHA Occupational Safety and Health Act; enacted by Congress in 1970. The goal of this federal regulation was to provide all employees (clinical laboratory personnel included) with a safe work environment.

OSMOLAL GAP difference between the measured osmolality and the calculated osmolality. The osmolal gap indirectly indicates the presence of osmotically active substances other than sodium, urea, or glucose, such as ethanol, methanol, ethylene glycol, lactate, or β-hydroxybutyrate.

OSMOLALITY physical property of a solution, based on the concentration of solutes (expressed as millimoles) per kilogram of solvent.

OSMOLARITY concentration of osmotically active particles in solution reported in milliosmoles per liter; not routinely used.

OSMOMETER laboratory instrument used to measure osmolality, or concentration of solute per kilogram of solvent.

OSMOTIC PRESSURE pressure that allows solvent flow between a semipermeable membrane to establish an equilibrium between compartments of different osmolality.

OSTEOMALACIA condition resulting from a deficiency of vitamin D. Deficiency of vitamin D causes bones to be soft and brittle. It is the adult form of rickets.

OSTEOPOROSIS a disease involving the gradual loss of bone mass resulting in a less dense and weak skeleton.

OTORRHEA discharge from the ear; also leakage of CSF from the ear.

OXIDIZING AGENT substance that accepts electrons.

OXIDOREDUCTASE an enzyme that catalyzes an oxidation-reduction reaction between two substrates.

OXYGEN CONTENT sum of the oxygen bound to hemoglobin as O_2Hb and the amount dissolved in the blood.

OXYGEN SATURATION (SO_2) represents the ratio of oxygen that is bound to the carrier protein-hemoglobin compared with the total amount that the hemoglobin could bind.

OXYTOCIN a nonapeptide produced in the paraventricular nucleus of the hypothalamus. It stimulates contraction of the gravid uterus at term and also results in contraction of myoepithelial cells in the breast, causing ejection of milk.

P_{50} represents the partial pressure of oxygen at which the hemoglobin oxygen saturation (SO_2) is 50%. The P_{50} is a measure of the O_2Hb binding characteristics and identifies the position of the oxygen-hemoglobin dissociation curve at half saturation.

PANCREATITIS inflammation of the pancreas ultimately caused by autodigestion of the pancreas as a result of reflux of bile or duodenal contents into the pancreatic duct.

PANHYPOPITUITARISM a condition resulting from pituitary hormone deficiencies; it usually involves more than one and eventually all the anterior pituitary hormones. These hormones are lost in a characteristic order, with growth hormone and gonadotropins disappearing first, followed by TSH, ACTH, and prolactin.

PARACENTESIS aspiration of fluid through the skin; as in removal or aspiration of pericardial, pleural, and peritoneal fluids.

PARATHYROID GLANDS four glands adjacent to the thyroid gland. Two of these glands are found in the upper portion and two are found near the lower portion of the thyroid gland. The parathyroid glands produce parathyroid hormone, which controls calcium and phosphate metabolism.

PARTIAL PRESSURE the pressure exerted by an individual gas in the atmosphere; equal to the BP at a particular altitude times the appropriate percentage for each gas.

PCO_2 the partial pressure of carbon dioxide.

PEAK DRUG LEVEL the time after administration until a drug reaches peak concentration in the body. A rule of thumb suggests that, for peak drug levels, the specimen should be collected 1 hour after the dose is administered.

PEDIATRIC regarding the treatment of children.

PELLAGRA a condition resulting from a deficiency of niacin. Initial signs of pellagra include anorexia, headaches, weakness, irritability, indigestion, and sleeplessness. This progresses to the classic 4Ds of advanced pellagra: dermatitis, diarrhea, dementia, and death.

PEPSIN a group of relatively weak proteolytic enzymes with pH optima from about 1.6 to 3.6 that catalyze all native proteins except mucus.

PEPTIDE BOND linkage combining the carboxyl group of one amino acid to the amino group of another amino acid.

PEPTIDES compounds formed by the cleavage of peptones and which contain two or more amino acids. The peptide hormones include insulin, glucagon, parathyroid hormone, growth hormone, and prolactin.

PERCENT SOLUTION the amount of solute per 100 total units of solution.

PERICARDIAL FLUID fluid surrounding and protecting the heart. The frequency of pericardial sampling and laboratory analysis is rare.

PERIFOLLICULAR CELLS   one of two types of cells forming the thyroid gland. Also C cells; they are situated in clusters along the interfollicular or interstitial spaces. The C cells produce the polypeptide calcitonin, which is involved in calcium regulation.

PERITONEAL FLUID   a clear to straw-colored fluid secreted by the cells of the peritoneum (abdominal cavity). It serves to moisten the surfaces of the viscera.

pH   represents the negative or inverse log of the hydrogen ion concentration.

PHALANX   any of the bones of the fingers or toes.

PHARMACOKINETICS   mathematical characterization of the disposition of a drug over time in order to better understand and interpret blood levels and to effectively adjust dosage amount and interval for best therapeutic results with minimal toxic effects.

PHEOCHROMOCYTOMA   tumors of the adrenal medulla or sympathetic ganglia that produce and release large quantities of catecholamines.

PHLEBOTOMIST   an individual who obtains or draws blood samples.

PHOSPHOLIPIDS   formed by the conjugation of two fatty acids and a phosphorylated glycerol. Phospholipids are amphipathic, which means they contain polar hydrophilic (water-loving) head groups and non-polar hydrophobic (water-hating) fatty acid side chains.

PIPET   utensils made of glass or plastic that are used to transfer liquids; they may be reusable or disposable.

pK   the negative log of the ionization constant.

PLACENTA   a structure in the uterus through which the fetus derives nourishment. The placenta synthesizes and secretes a variety of protein hormones, as well as the steroids estrogen and progesterone. Evaluation of maternal serum and urine concentration of these hormones may be of value not only in diagnosing pregnancy but also in monitoring placental development and fetal well-being.

PLASMA RENAL FLOW   renal secretory capability.

PLASMA   liquid portion of blood which contains clotting factors.

PLATELETS   round or oval anuclear disks or cells found in the blood. Platelets influence coagulation through adherence, aggregation, initiation of coagulation, and clot retraction.

PLEURAL FLUID   essentially interstitial fluid of the systemic circulation; it is contained in a membrane that surrounds the lungs.

PO₂   the partial pressure of oxygen.

POLYDIPSIA   excess H_2O intake due to chronic thirst.

POLYSACCHARIDE   complex carbohydrates; polysaccharides are the most abundant organic molecules in nature.

PORPHYRIA   disorders that result from disturbances in heme synthesis.

PORPHYRIN   chemical intermediates in the synthesis of hemoglobin, myoglobin, and other respiratory pigments

called cytochromes. They also form part of the peroxidase and catalase enzymes, which contribute to the efficiency of internal respiration.

PORPHYRINOGENS   the reduced forms of porphyrins.

POST-RENAL   obstruction in the flow of urine from kidney to bladder and its excretion.

POSTERIOR PITUITARY   a portion of the pituitary; also the neurohypophysis.

POSTHEPATIC   extrahepatic disturbance in the excretion of bilirubin. Posthepatic jaundice results from the impaired excretion of bilirubin caused by mechanical obstruction of the flow of bile into the intestines. This may be due to gallstones or a tumor.

PRE-RENAL   prior to plasma reaching the kidney.

PRE-ANALYTICAL ERROR   mistakes introduced during the collection and transport of samples prior to analysis.

PRECISION   the closeness of repeated results; quantitatively expressed as standard deviation or coefficient of variation.

PRECOCIOUS PUBERTY   onset of puberty (normal sexual development) earlier than the expected time (generally 10 to 14 years old).

PREDICTIVE VALUE THEORY   referring to diagnostic sensitivity, specificity, and predictive value. The predictive value of a test can be expressed as a function of sensitivity, specificity, and disease prevalence.

PREHEPATIC   before the liver. Prehepatic jaundice results when an excessive amount of bilirubin is presented to the liver for metabolism, such as in hemolytic anemia. This type of jaundice is characterized by unconjugated hyperbilirubinemia.

PRIMARY STANDARD   a highly purified chemical that can be measured directly to produce a substance of exact known concentration.

PROBE   on an automated analyzer, a mechanized device that automatically dips into a sample cup and aspirates a portion of the liquid.

PROGESTERONE   a steroid hormone produced by the corpus luteum and placenta. Progesterone serves to prepare the uterus for pregnancy and the lobules of the breast for lactation.

PROHORMONE   a precursor to the active hormone.

PROINSULIN   precursor to insulin; it is packaged into secretory granules, where it is broken down into equimolar amounts of insulin and an inactive C-peptide.

PROLACTIN   a protein hormone whose amino acid composition is similar to that of GH. It is produced by the pituitary gland. In humans, it appears to function solely in the initiation and maintenance of lactation.

PROTEIN-FREE FILTRATE   clear filtrate of whole blood or plasma prepared by adding acid or other ions to remove proteins.

PROTEINURIA   protein in the urine.

PROTO-ONCOGENES   normal cellular genes that

play essential roles in cell differentiation and proliferation and potentially can become oncogenic. The transformation of proto-oncogenes into oncogenes can occur by single-point mutations, translocation, and amplification.

PROXIMAL TUBULE   portion of renal tubule beginning at Bowman's capsule and extending to loop of Henle.

PTH   parathyroid hormone; synthesized as a prohormone containing 115 amino acids. The active form of the hormone contains 84 amino acids and is secreted by the chief cells of the parathyroid glands.

PYRROLE   a heterocyclic structure or compound that is the basis for substances such as hemoglobin.

QUALITY ASSURANCE   system or process that encompasses (in the laboratory) pre-analytical, analytical, and post-analytical factors. Quality control is part of a quality assurance system.

QUALITY CONTROL   system for recognizing and minimizing (analytical) errors. The purpose of the quality control system is to monitor analytical processes, detect analytical errors during analysis, and prevent the reporting of incorrect patient values. Quality control is one component of the quality assurance system.

RADIOACTIVE MATERIAL   any material capable of emitting radiant energy (rays or particles).

RANDOM ACCESS   the capability of an automated analyzer to process samples independently of other samples on the analyzer. Random access analyzers may be programmed to run individual tests or a panel of tests without operator intervention.

RANDOM ERROR   a type of analytical error; random error affects precision and is the basis for disagreement between repeated measurements. Increases in random error may be caused by factors such as technique and temperature fluctuations.

RANDOM URINE SPECIMEN   a urine specimen in which collection time and volume are not important. The first morning void is often requested because it is the most concentrated; generally used for routine urinalysis (pH, glucose, protein, specific gravity, and osmolality).

RDA   recommended dietary allowance. The amount of a vitamin that should be ingested by a healthy individual to meet routine metabolic needs and allow for biologic variation, maintain normal serum concentrations, prevent depletion of body stores, and thus preserve normal function and health.

REACTIVE CHEMICAL   substances that, under certain conditions, can spontaneously explode or ignite or that evolve heat and/or flammable or explosive gases.

RECEPTOR   site of hormone action on a cell; receptors provide for target-organ specificity, since not all cells have receptors for all hormones.

REDOX POTENTIAL   a measure of a solution's ability to accept or donate electrons.

REFERENCE INTERVAL   the usual values for a healthy population; also normal range.

REFERENCE METHOD   an analytical method used for comparison. It is a method with negligible inaccuracy in comparison with its imprecision.

RENAL TUBULAR SECRETION   process that transports substances from plasma into tubular filtrate for excretion into urine.

RESPIRATORY ACIDOSIS AND ALKALOSIS   a disorder due to ventilatory dysfunction (a change in the PCO_2, the respiratory component).

RESPIRATORY DISTRESS SYNDROME   a condition that may occur upon the changeover to air as an oxygen source at birth if the proper quantity and type of phospholipid (surfactant) is not present. It is also referred to as *hyaline membrane disease* because of the hyaline membrane found in affected lungs.

RHINORRHEA   discharge from the nose; also leakage of CSF into the nose.

RICKETS   the classic vitamin D deficiency disease of children. It may be nutritional or metabolic in origin.

ROBOTICS   front end automation to "handle" a specimen through the processing steps and load the specimen onto the analyzer.

ROTOR   on some automated analyzers, a round device that holds sample cups and is capable of spinning.

RS-232   a type of physical electronic connection to allow data transfer between a computer and another piece of equipment (*i.e.,* printer, modem, mouse); also RS-232 port or serial port.

SECONDARY STANDARD   a substance of lower purity whose concentration is determined by comparison with a primary standard.

SECRETAGOGUES   an agent that stimulates or causes secretion (*e.g.,* protein breakdown products).

SECRETIN   secretin is synthesized by cells in the small intestine in response to the acidic contents of the stomach reaching the duodenum. It can control gastrin activity in the stomach and cause the production of alkaline bicarbonate rich pancreatic juice thereby protecting the lining of the intestine from damage.

SEGMENT   data fields in logical groupings.

SELLA TURCICA   a small cavity in the sphenoid bone of the skull; the pituitary is located in this cavity.

SEPARATOR CHARACTER   divides fields in a data base; commas and quotation marks are commonly used.

SERIAL DILUTION   multiple progressive dilutions ranging from more concentrated solutions to less concentrated solutions.

SERINE PROTEASES   coagulation factors that function as enzymes in coagulation.

SEROTONIN   (5-OH tryptamine) an amine derived from hydroxylation and decarboxylation of tryptophan. It is synthesized by enterochromaffin cells, which are located primarily in the gastrointestinal tract and, to a lesser degree, in the bronchial mucosa, biliary tract, and gonads.

SERUM   liquid portion of the blood without clotting factors.

SHIFT   a sudden change in data and the mean.

SIADH   syndrome of inappropriate ADH; it results when ADH is released despite low serum osmolality in association with a normal or increased blood volume.

SIGNIFICANT FIGURES   the minimum number of digits needed to express a particular value in scientific notation without loss of accuracy.

SIMPLE PROTEIN   composed only of amino acids.

SINUSOIDS   spaces between the cords of liver cells; they are lined by endothelial cells and Kupffer's cells.

SKIN PUNCTURE   an open collection system. Blood is brought to the surface of the skin by applying pressure to the site. The sample is dripped into a capillary blood collector (instead of vacuum pressure pulling the sample into the tube). Specimen contains both venous and arterial blood.

SOLUTE   a substance that is dissolved in a liquid or solvent.

SOLUTION   a liquid containing a dissolved substance; the combination of solute and solvent.

SOLVENT   liquid in which the solute is dissolved.

SPECIFIC GRAVITY   term used to express density.

SPECIFICITY   in regard to quality control, the ability of an analytical method to quantitate one analyte in the presence of others in a mixture such as serum.

SPECTROPHOTOMETRY   analytical technique to measure the light absorbed by a solution. A spectrophotometer is used to measure the light transmitted by a solution in order to determine the concentration of the light-absorbing substance in the solution.

STAGING   determination of the period in the course of a disease. The major clinical value of tumor markers is in tumor staging, monitoring therapeutic responses, predicting patient outcomes, and detecting cancer recurrence.

STANDARD REFERENCE MATERIALS (SRM)   established by authority; used for comparison of measurement.

STANDARD   a substance or solution in which the concentration is determined. Standards are used in the calibration of an instrument or method.

STEATORRHEA   failure to digest and/or absorb fats.

STEROIDS   classification of hormones; they are all synthesized from cholesterol, using the same initial biochemical pathways. The end result depends on the enzymatic machinery that is predominant in a particular organ.

SYNOVIAL FLUID   fluid formed by ultrafiltration of plasma across the synovial membrane of a joint. The membrane also secretes into the dialysate a mucoprotein rich in hyaluronic acid, which causes the synovial fluid to be viscous.

SYSTEMATIC ERROR   a type of analytical error that arises from factors that contribute a constant difference, either positive or negative, and directly affects the estimate of the mean. Increases in systematic error can be caused by poorly made standards or reagents, failing instrumentation, poorly written procedures, and so on.

SYSTEME INTERNATIONAL D'UNITES (SI)   internationally adopted system of measurement. Established in 1960 and is the only system used in many countries. The units of the system are referred to as SI units.

_t_-TEST   used to determine whether there is a statistically significant difference between the means of two groups of data.

T_3   triiodothyronine; a hormone produced by the thyroid gland.

T_3 UPTAKE   assay used to measure the number of available binding sites of the thyroxine-binding proteins, most notably TBG. It should not be confused with the T_3 assay.

T_4   thyroxine; a hormone produced by the thyroid gland.

TERATOGEN   anything that may cause abnormal development of an embryo.

TESTOSTERONE   a steroid hormone synthesized from cholesterol. It causes growth and development of the male reproductive system, prostate, and external genitalia.

TETANY   irregular muscle spasms.

THALASSEMIA   disorder involving a defect in rate and quantity of production of hemoglobin.

THERAPEUTIC DRUG MONITORING   the determination of serum drug levels in order to produce a desirable effect.

THERAPEUTIC RANGE   concentration range of a drug that is beneficial to the patient without being toxic.

THERMISTOR   electronic thermometer.

THORACENTESIS   removal of fluid from the pleural space by needle and syringe after visualization by radiology.

THROMBOSIS   the formation or development of a blood clot or thrombus in the circulatory system.

THYROID   a gland consisting of two lobes located in the lower part of the neck. The lobes are connected by a narrow band called an isthmus and are typically asymmetrical, with the right lobe being larger than the left. The lobes weigh approximately 20 to 30 g each and measure 4.0 cm in length and 2 to 2.5 cm in width and thickness.

THYROID-BINDING GLOBULIN (TBG)   a protein that binds thyroid hormones. The TBG assay is used to confirm results of FT_3 or FT_4 or abnormalities in the relationship of the TT_4 and T_3U test; or as a postoperative marker of thyroid cancer.

THYROID-RELEASING HORMONE (TRH)   a tripeptide released by the hypothalamus. It travels along the hypothalamic stalk to the beta cells of the anterior pituitary, where it stimulates the synthesis and release of thyrotropin or thyroid-stimulating hormone (TSH).

THYROIDITIS   inflammation of the thyroid gland.

THYROTOXICOSIS   a group of syndromes caused by high levels of free thyroid hormones in the circulation. Thyrotoxicosis means only that the patient is suffering the metabolic consequences of excessive quantities of thyroid hormones.

THYROTROPIN (TSH)   thyrotropin, or thyroid stimulating hormone (TSH), is a glycoprotein consisting of two subunits, alpha and beta, linked non-covalently. It is released by the anterior pituitary.

TIMED URINE SPECIMEN specimens collected at specific intervals, such as before and after meals, or specimens to be collected over specific periods of time. For example, discrete samples collected over a period of time are used for tolerance tests (*e.g.,* glucose).

TITER the highest dilution of serum that shows a positive reaction in the presence of antigen (*e.g.,* precipitation of antigen-antibody complex).

TOTAL PARENTERAL NUTRITION (TPN) a widely used means of intense nutritional support for patients who are malnourished, or in danger of becoming malnourished, because they are unable to consume required nutrients.

TOTAL IRON BINDING CAPACITY (TIBC) an estimate of serum transferrin levels; obtained by measuring the total iron binding capability of a patient's serum. Since transferrin represents most of the iron-binding capacity of serum, TIBC is generally a good estimate of serum transferrin levels.

TOXICOLOGY the study of poisons, their actions, their detection, and the treatment of the conditions produced by them.

TRACE ELEMENT an element that occurs in biological systems at concentrations of mg/kg amounts or less (parts per million). Typically the daily requirement of such an element is a few milligrams per day.

TRANSFERASE an enzyme that catalyzes the transfer of a group other than hydrogen from one substrate to another.

TRANSFERRIN SATURATION percent of transferrin molecules that have iron bound. A ratio of serum iron (actual iron in the serum) and serum transferrin or TIBC (potential quantity of iron that can be bound). Also Percent Saturation.

TRANSUDATE fluid that passes through a membrane; in comparison to an exudate, it has fewer cells and is of lower specific gravity. Transudates are secondary to remote (nonpleural) pathology and indicate that treatment should begin elsewhere.

TREND a gradual change in data and the mean.

TRH thyrotropin releasing hormone; a peptide hormone primarily synthesized in the hypothalamus. It stimulates TSH production.

TRIGGER EVENT an event in the health care environment that creates a need for data exchange.

TRIGLYCERIDES consists of one molecule of glycerol with three fatty acid molecules attached (usually three different fatty acids including both saturated and unsaturated molecules). Triglycerides containing saturated fatty acids without bends pack together closely and tend to be solids at room temperature. Unsaturated fatty acids do not pack together closely and tend to be liquids at room temperature.

TRIOSE a monosaccharide having three carbons.

TROUGH DRUG LEVEL the lowest concentration of drug obtained in the blood. Trough levels should be drawn immediately before the next dose.

TSH thyroid stimulating hormone; secreted by the anterior pituitary and stimulates the thyroid.

TUBULES tubes or canals that make up a part of the kidney; as in convoluted tubules.

TUMOR MARKER biological substances synthesized and released by cancer cells or substances produced by the host in response to cancerous tissue. Tumor markers can be present in the circulation, body cavity fluids, cell membranes, or the cytoplasm/nucleus of the cell.

TUMOR-ASSOCIATED ANTIGEN an antigen associated with a tumor; it is derived from the same or closely-related tissue. Oncofetal proteins are an example of tumor-associated antigens. They are present in both embryonic/fetal tissue and cancer cells.

TUMOR-SPECIFIC ANTIGEN a tumor antigen; thought to be a direct product of oncogenesis induced by viral oncogenes, radiation, chemical carcinogens, or unknown risk factors.

TURBIDIMETRY an analytical technique that measures the decreased amount of light transmitted through a solution as a result of light scatter by particles. Measurements are made at 180° to the incident beam (unscattered light).

ULTRATRACE ELEMENT present in tissues at concentrations of mg/kg amounts or less (parts per billion) and has extremely low daily requirements (usually less than 1 mg).

UNSOLICITED UPDATE trigger events for data exchange that are initiated by the application system. These transactions are sent to other systems by the application without any request from other systems. Since not all systems are interested in receiving all unsolicited update events, a *subscription* system may be implemented. An application would subscribe to the trigger events that it is interested in receiving.

UREA compound synthesized in the liver from ammonia and carbon dioxide and excreted in the urine.

UREMIA OR UREMIC SYNDROME very high levels of urea in blood accompanied by renal failure.

URIC ACID end-product of the breakdown of purines from nucleic acids in humans. Crystallization of urate in the joints causes gout.

URINALYSIS consists of a group of screening tests generally performed as part of a patient's admission work-up or physical examination. It includes assessment of physical characteristics, chemical analyses, and a microscopic examination of the sediment from a (random) urine specimen.

UROBILINOGEN colorless product or derivative of bilirubin formed by the action of bacteria.

VALENCE mass of material that can combine with or replace one mole of hydrogen ions.

VENIPUNCTURE   puncture of a vein. For example, to obtain blood for analysis.

VENOUS BLOOD   blood obtained from a vein.

VITAMER   related compounds that interconvert to or substitute for the functional form of the vitamin.

VITAMIN   organic molecules required by the body in amounts ranging from micrograms to milligrams per day for maintenance of structural integrity and normal metabolism. They perform a variety of functions in the body.

VLDL   very-low-density lipoproteins; they carry triglycerides assembled in the liver out to cells for energy needs or storage as fat.

VMA   vanillylmandelic acid; epinephrine and norepinephrine are metabolized by the enzymes monoamine oxidase and catechol-O-methyl transferase to form metanephrines and vanillylmandelic acid.

VON WILLEBRAND FACTOR   (vWF) an adhesive protein that links platelets to platelets and platelets to endothelial cells.

WESTERN BLOT   transfer technique used for analyzing protein antigens. Antigens are separated by electrophoresis and transferred to a new medium by absorption or covalent bonding and then detected by a wide range of antibody probes. The probe could be labeled with a radioactive tag, an enzyme that can produce a visual product, or a fluorescent or chemiluminescent label. Detection would be accomplished with specialized instrumentation, spectrophotometer, fluorometer, and luminometer, respectively. Used to detect the presence of human immunodeficiency virus (HIV).

WHOLE BLOOD   complete blood, containing the liquid portion (plasma) and the cellular elements.

WIDE AREA NETWORK (WAN)   several local area networks (LANs) linked through communication network protocols.

ZERO-ORDER KINETICS   point in enzyme reaction when product is formed and the resultant free enzyme immediately combines with excess free substrate; the reaction rate is dependent on enzyme concentration only.

ZOLLINGER-ELLISON SYNDROME   a gastrin-secreting neoplasm, usually located in the pancreatic islets, associated with exceptionally high plasma gastrin concentrations. The fasting plasma gastrin levels typically exceed 1000 pg/mL and can reach 400,000 pg/mL, compared with the normal range of 50 to 150 pg/mL.

ZYMOGENS   unactivated forms of enzymes. Many of the coagulation factors circulate as unactivated forms of enzymes, or zymogens; they must be converted into active forms for biological function.

Answers to Exercises, Problems, and Case Studies

CHAPTER 1

1. a. 2.19 M (NaCl = 58.5 g)
   b. Normality = molarity
   c. 12.8%
   d. 1/7.8
2. a. 5 mg/dL
   b. 4.50×10^{-4} (CaCl$_2$ = 111 g)
   c. 9.01×10^{-4}
3. 5.83 mL/L
4. 5.75×10^{-5}
5. a. ~15/1
   b. 4.46
6. 4.03
7.

Std (g/dL)	(A) Dilution Factor	(B) Volume of Stock (mL)	(B) Volume of Diluent (mL)
1.0	1/30	0.5	14.5
2.0	2/30(1/15)	1	14
4.0	4/30(2/15)	2	13
6.0	5/30(1/5)	3	12
8.0	8/30(4/15)	4	11
10.0	10/30(1/3)	5	10

Total volume desired is 15 mL.

8. 8×10^6; 4×10^{-2}; 0.13; 0.2; 5×10^{-3}; 5×10^{-2}; 40
9. 0.114 mL
10. a. 160 mg/dL
    b. 1.6 mg/mL
    c. 1.92 g/24 h
11. 312.5 g
12. 363 U

CHAPTER 2

1. Both the employer and the employee share responsibility for safety.
2. Chemicals should be stored based on the quantity needed and the nature or type of chemicals.
3. Gloves are worn to protect hands and lower arms.
   Liquid-resistant lab coats protect clothing and the major part of the body.
   Goggles, face shields, and plexiglass shields protect against exposure of bloodborne pathogens to the face and mucous membranes.
   Respirators are worn for protection against airborne pathogens.
4. Laboratory personnel are unknowingly in frequent contact with potentially biohazardous materials and therefore must consider all specimens as infectious.
5. The Hazardous Communication Policy must provide employees with the following:
   A written hazard communication program
   Material Safety Data Sheet for each hazardous material
   Annual continuing education relating to hazardous material
6. Material Safety Data Sheets are to be located in a central location in the laboratory for use by all personnel.
7. Flammable/combustible chemicals—possibility of fire or explosion
   Corrosive chemicals—irritant to skin, eyes, or tissues
   Reactive chemicals—may spontaneously ignite or explode
   Carcinogenic chemicals—cancer-causing agents
8. Class A—paper, wood, plastic, or fabric
   Class B—flammable liquids/gases and combustible petroleum products

Michael L. Bishop, Janet L. Duben-Engelkirk, and Edward P. Fody.
CLINICAL CHEMISTRY. © 1996 Lippincott–Raven Publishers.

Class C—electrical equipment

Class D—combustible/reactive metals

9. Chemical waste—some chemicals may be flushed down the drain, acids and bases must be neutralized, other chemicals may be filtered or redistilled.

Radioactive waste—must be collected and disposed of according to NRC regulations.

Biohazardous waste—incineration, steam sterilization, burial, thermal inactivation, chemical disinfection, or encapsulation in a solid matrix.

10. Accidents must be reported immediately to a supervisor. An accident report must be filled out followed by an accident investigation report.

CHAPTER 3

1. Antecubital fossa of the arm.

The hand.

Foot and ankle may be used after consulting physician.

2. The heel.

The finger.

See Figures 3-8 and 3-9, page 51.

3. Tourniquet.

Sterile venipuncture needle of appropriate size.

Plastic tube holder or sleeve.

Evacuated tubes.

Sterile alcohol pads.

Gloves.

2×2 inch gauze pads.

Adhesive tape or bandages.

4. Transport in a timely manner.

Primary leak proof container in a secondary leak proof container.

Transported in an upright position.

Agitated as little as possible.

Temperature between 22 and 25° C (unless chilling is required).

Specimen protected from light, if necessary.

5. List steps for venous collection.

1. Identify patient.
2. Select supplies.
3. Choose site.
4. Position patient.
5. Apply tourniquet.
6. Palpage vein.
7. Cleanse site.
8. Inspect needle, allow alcohol to dry.
9. Fix the selected vein.
10. Align needle.
11. Anchor needle.
12. Engage and fill all required tubes.
13. Withdraw needle and apply pressure to site with gauze pad.
14. Mix tubes with additives.
15. Check site for bleeding, bandage if necessary.

16. Dispose of contaminated supplies.
17. Label tubes.
18. Package samples according to handling instructions.
19. Log collection if required.
20. Transport specimen to lab.

6. Steps for Skin Puncture

1. Identify patient.
2. Select supplies.
3. Choose site.
4. Warm site for 3–5 minutes.
5. Position patient.
6. Cleanse site.
7. Puncture heel or finger.
8. Firmly and slowly apply pressure toward the puncture site.
9. Wipe away the first drop of blood.
10. Position site downward, touch collector to next drop of blood, and allow blood to flow into the collector.
11. Seal collection tube when filled.
12. Apply pressure to site once collection is completed.
13. Dispose of contaminated supplies.
14. Label tubes.
15. Package samples according to handling instructions.
16. Log collection if required.
17. Transport specimen to lab.

CHAPTER 4

Problem 4-1

$$\text{Sensitivity} = \frac{100TP}{TP + FN} = \frac{100 \times 5}{5 + 3} = 62.5\%$$

$$\text{Specificity} = \frac{100TN}{TN + FP} = \frac{100(843)}{843 + 4} = 99.5\%$$

$$PV^+ (\%) = \frac{100TP}{TP + FP} = \frac{100(5)}{5 + 4} = 55.5\%$$

$$\text{Efficiency} = \frac{100(TP + TN)}{TP + TN + FP + FN}$$

$$= \frac{100(5 + 843)}{5 + 843 + 4 + 3}$$

$$= 99.2\%$$

Problem 4-2

1. See Figure A4-1.
2. The values of the high concentration control material appear to be shifted. You are not getting an even distribution about the mean.
3. Systematic error is occurring. This is indicated by 2_{2_s} rule violation.
4. Because the low concentration control is within the control limits of the test and all the patient values are in that

HIGH   JANUARY

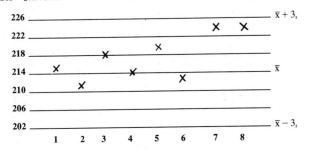

LOW

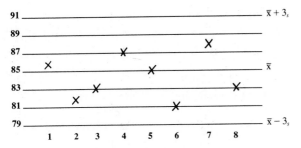

▲
Figure A4-1. Plot of daily control values for January.

range, you could make the decision to report the patient data. Meanwhile, troubleshoot the systematic error.

Problem 4-3

1. The problem appears to be in the preparation of the syringes for drawing a heparinized sample or in the mixing of the samples after collection.
2. If you are the laboratory supervisor or section head, you need to discuss the situation with the supervisor of the MICU. The problem needs to be explained, and the laboratory supervisor and MICU supervisor should solve the problem. If you are the blood gas technologist, you need to discuss the situation with the supervisor in your section and explain the problem. Then the problem should be addressed supervisor-to-supervisor. This is a quality-management problem.
3. This is an extra-analytical problem, and the problem is occurring in the preanalytical phase of sample collection. Quality-control systems that use reference samples detect only analytical errors in the testing procedure.

Problem 4-4

$$\overline{Y}_1 = 120.3 \text{ mg/dL}$$
$$s_{TM1} = 2.43 \text{ mg/dL}$$
$$CV_1 = 2.0\%$$

$$\overline{Y}_2 = 300.5 \text{ mg/dL}$$
$$s_{TM2} = 5.45 \text{ mg/dL}$$
$$CV_2 = 1.8\%$$

Problem 4-5

Concentration added =	50 mg/dL	100 mg/dL
	Recovery	Recovery
	102%	97%
	98%	97%
	100%	99%
	96%	96%
	90%	95%
Percent recovery:	97.2%	96.8%

Average recovery: $R = 97\%$

Problem 4-6

Ascorbic acid concentration tested = 15 mg/dL
Interference for each patient:

A	−9 mg/dL
B	−8 mg/dL
C	−10 mg/dL
D	−10 mg/dL
E	−11 mg/dL
Average:	−9.4 mg/dL

Problem 4-7

1. See Figure A4-2.

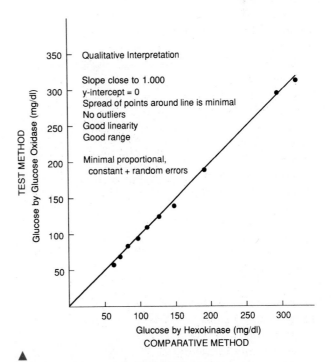

▲
Figure A4-2. Graph of Problem 4-7 data.

2. a.

x_i	y_i	x_i^2	y_i^2	$x_i y_i$
191	192	36,481	36,864	36,672
97	96	9,409	9,216	9,312
83	85	6,889	7,225	7,005
71	72	5,041	5,184	5,112
295	299	87,025	89,401	88,205
63	61	3,969	3,721	3,843
127	131	16,129	17,161	16,637
110	114	12,100	12,996	12,540
320	316	102,400	99,856	101,120
146	141	21,316	19,881	20,586

b.

Σx_i	Σy_i	Σy_i^2	Σy_i^2	$\Sigma x_i y_i$
1,503	1,507	300,759	301,505	301,032

c. $\bar{x} = 150.3$ mg/dL

$\bar{y} = 150.7$ mg/dL

$$m = \frac{10(301,032) - 1,503(1,507)}{10(300,759) - (1,503)^2}$$

$$= \frac{3,010,320 - 2,265,021}{3,007,590 - 2,259,009} = \frac{745,299}{748,581}$$

$$= 0.996$$

$y_0 = 150.7 - (0.996)(150.3)$

$$= 1.001 \text{ mg/dL}$$

d. $Y_i = 1.001 + 0.996 x_i$

Y_i	$y_i - Y_i$	$(y_i - Y_i)^2$
191	1	1
98	−2	4
84	1	1
72	0	0
295	4	16
64	−3	9
127	4	16
111	3	9
320	−4	16
146	−5	25

$$\Sigma(y_i - Y_i)^2 = 97$$

e. $S_{y/x} = \sqrt{\dfrac{\Sigma(y_i - Y_i)^2}{n-2}} = \sqrt{\dfrac{97}{8}} = 3.5$

f. $r = \dfrac{10(301,032) - (1,503)(1,507)}{\sqrt{[10(300,759) - (1,503)^2][10(301,505) - (1,507)^2]}}$

$$= \frac{3,010,320 - 2,265,021}{\sqrt{(3,007,590 - 2,259,009)(3,015,050 - 2,271,049)}}$$

$$= \frac{745,299}{\sqrt{(748,581)(744,001)}}$$

$$= \frac{745,799}{\sqrt{5.5694501 \times 10^{11}} = 746,287}$$

$$= \frac{745,799}{746,287}$$

$$= 0.999$$

3. $y_0 = 1.001$ (mg/dL)

$m = 0.999$

$S_{y/x} = 3.5$ (mg/dL)

$r = 0.996$

Problem 4-8

1. RE $= 1.96 s_{TM}$
   X_{c1}: RE $= 4.76$ mg/dL
   X_{c2}: RE $= 10.68$ mg/dL
2. CE $= y$-intercept $= 1.001$ mg\dL
   PE $= 1.000 - bX_c$
   X_{c1}: PE $= 0.48$ mg/dL
   X_{c2}: PE $= 1.20$ mg/dL
3. SE $= (y_o + mX_c) - X_c = (1.001 + 0.996X_c) - X_c$
   X_{c1}: SE $= 0.52$ mg/dL
   X_{c2}: SE $= 0.20$ mg/dL
   X_{c1}, CE $(1.001 > 0.48)$
   X_{c2}, PE $(1.2 > 1.001)$
4. X_{c1}: TE $= 5.28$ mg/dL
   X_{c2}: TE $= 10.88$ mg/dL
5. a. $s_{y/x}$
   b. $s_{y/x} = 3.5$ mg/dL
6. a. The following are the E_a for glucose:
   $X_{c1} = 120$ mg/dL $E_{a1} = 10$ mg/dL
   $X_{c2} = 300$ mg/dL $E_{a2} = 25$ mg/dL
   At X_{c1} all errors are <10 mg/dL.
   At X_{c2} all errors are <25 mg/dL.
   b. The test method is acceptable.

Problem 4-9

1. Serving on this team might be: laboratory medical director, laboratory manager/director, and quality assurance representatives from the laboratory, medical-surgical unit, and nursing, as well as representation from individuals who will be performing the test. Instrument manufacturers should have representatives who can interface with this group.
2. The following must be in place before implementing any point of care testing: (Modified with permission from Jones BA, Testing at the patient's bedside. Clin Lab Med 1994;14:473.)
   a. Clinical laboratory oversight for the organization, administration and quality assurance of the testing program
   b. Characterization of the analytical limitations of the POC device including the bias
   c. Procedure manual for all aspects of the program
   d. Training program and continuing education for instrument operators
   e. Quality control testing on each instrument
   f. System to monitor and document operator competence
   g. Internal proficiency testing to compare the POC result with main laboratory's
   h. External proficiency testing
   i. Scheduled instrument maintenance and cleaning
3. The following are desirable features of POC analyzers (Reprinted with permission from Cembrowski GS, Kiechle FL. Point of care testing: Critical analysis and practical application. Adv Pathol Lab Med 1994;7:3).

Speed and ease of use

Accuracy and precision approaching central laboratory analyzer

Small, portable, and affordable

Should analyze small volumes of whole blood

Flexible test menu

Broad analytical range to minimize dilutions, repeats, etc.

Lock-out ability to prevent testing that is either unauthorized or improperly identified

Automatic calibration and quality control interpretation

Seamless interface with laboratory or hospital information system

Low maintenance

Bar code reading

Use of either no reagent or ready-for-use reagent

Recyclable disposables

CHAPTER 5

1. E. A photodiode array can be used in place of a PM tube.
2. B. Stray light usually limits the absorbance range to 2.0, and places limits on the upper range of linearity.
3. A. Atomic absorbance spectrometry uses hollow cathode lamps, while flame photometry does not use a light source.
4. D. Fluorescence is about 1000 times more sensitive than absorption.
5. A. Chemiluminescence has sensitivities in the sub-picomolar range.
6. A. A power supply is needed for all lasers.
7. B. In amperometry, an oxidation-reduction reaction at a fixed applied voltage takes place at the electrode surfaces to produce a flow of electrons (current).
8. D. Electroendosmosis occurs when there are charged groups attached to the medium and can interfere with the migration of analyte molecules.
9. A. Reverse-phase LC enables use of aqueous samples.
10. D. Current instrumentation requires one sample analysis at a time.
11. D. Isocratic refers to use of a mobile phase that is unchanging with respect to pH, ionic strength, and concentration of organic modifiers.
12. A. MS/MS instruments enable mass fragmentation of molecular ions to form daughter ions, which can be used for qualitative identification.

CHAPTER 6

1.

Advantages of RIA	Disadvantages of RIA
Sensitivity	Short reagent half-life
Resistance to interference	Necessity of separation step
	Requires radioactivity license
	Time and cost of waste disposal

2. Standard dilutions of known material are tested and results graphed against concentration. Unknowns are evaluated against the standard line. Depending on the characteristics of the assay and analyte, various mathematical curve fits may be applied to the standard curve.
3. Protection of the human body from foreign agents is the major function of the immune system. Protection mechanisms include intact skin, cellular defenses, and adaptive defenses.
4. In immunoelectrophoresis, electrophoresis is used to separate proteins. Then, these proteins are allowed to diffuse toward troughs containing antibodies. At the equivalence point, a precipitin line forms. The plates are then washed and stained.
5. Electrophoretically separated antigens on polyacrylamide are transferred to a nitrocellulose media. Antigens are fixed on the new media by absorption or covalent bonding. Detection is accomplished by a wide range of antibody probes.
6. The major purpose for using a flow cell cytometry is to identify specific antigens on live cells in suspension using immunofluorescent labels.
7. Two diffusion techniques used to detect ag–ab reactions are:
   a. passive gel diffusion or single immunodiffusion.
   b. double immunodiffusion or Ouchterlony.

CHAPTER 7

1. Automation = mechanization of the steps in a manual procedure.
   Continuous flow = liquids (reagents, diluents, and samples) are pumped through a system of continuous tubing. Samples are introduced in a sequential manner, following each other through the same network.
   Discrete analysis = separation of each sample and accompanying reagents in a separate container.
   Random access = the ability to program different tests and/or profiles, on any number of patient samples, in any order.
2. Continuous flow.
   Centrifugal analysis.
   Discrete analysis.
3. Specimen preparation and identification.
   Specimen measurement and delivery.
   Reagent systems and delivery.
   Chemical reaction phase.
   Measurement phase.
   Signal processing and data handling.
4. B
   B
   A
   C
5. True
6. Several answers are possible: cost, mode of acquisition, open vs. closed reagent system, labor, performance characteristics, space, and user friendliness.

CHAPTER 8

1. Automation of total patient care information.
   Sharing of critical data.
   Centralized total electronic medical record.
   Integration of separate automated systems.
2. American National Standard Institute
   Health Information Systems Planning Panel
   International Standards Organization
3. Network = linkage of individual computer systems.
   LIS = laboratory information system; stores, transmits and generates patient data.
   WAN = wide area network.
   GUI = graphical user interface; generally an operating system or program that uses icons or graphics, for operations.
   Field = one part of a record in a database.
4. The HL7 Standard®, or Health Level 7 Standard®, is a set of standards for electronic data exchange in health care environments. Level 7 refers to the top level of the OSI Reference Model.
5. Consolidation or centralization of all patient care information.

CHAPTER 9

Case Study 9-1

1. The mother has an increased α_2-globulin. This may be due to a residual increase of ceruloplasmin and α_2-macroglobulin that occurred in pregnancy. The cord-blood serum, representing the baby, has an abnormally low α_1-globulin fraction. This is associated with α_1-antitrypsin deficiency.

   This case illustrates α_1-antitrypsin deficiency. The mother was known to be heterozygous prior to admission to the hospital. Her serum α_1-antitrypsin concentration is within the expected range. The reason for this concentration even in the MZ phenotype was her pregnancy. Pregnant women are reported to have concentrations that may reach twice the reference values.
2. The baby had α_1-antitrypsin deficiency of a severe phenotype, ZZ. Although the electrophoretic pattern with no discernible α_1-globulin fraction may be found in premature infants, term newborns have concentrations of α_1-antitrypsin in the range of adult levels. This deficiency is associated with juvenile pulmonary emphysema and infantile hepatitis. Most babies with α_1-antitrypsin deficiency, however, do not develop symptoms until later in life, when the predisposition to pulmonary emphysema materializes.
3. Other tests that were performed to confirm the α_1-antitrypsin deficiency were the quantitation by radial immunodiffusion and phenotyping of α_1-antitrypsin by isoelectric focusing. The α_1-antitrypsin concentration obtained by radial immunodiffusion of the mother's blood was 225 mg/dL, while that of the baby's blood was 42 mg/dL.

   Subsequently, the mother was phenotyped by isoelectric focusing and found to be of the MZ phenotype, while the baby girl was found to be ZZ phenotype. The baby girl is now 10-years-old and is thriving, although she has been slightly slower than average in the development of motor skills.

Case Study 9-2

This case serves to illustrate the importance of investigating abnormalities even though they are small, especially if experienced clinical laboratory scientists who are very familiar with protein electrophoretic patterns sense that a peak is outside the usual typical deviation. This is particularly the case when the bone-marrow aspirate is unsatisfactory or impossible to obtain. In this case, several bone-marrow aspirates were taken before one that was satisfactory for the pathologist to examine was obtained.

Another point that can be made here is that the proteins that are normally present in very low concentrations in the serum do not make a remarkable difference in the quantitative values or even in the electrophoretic pattern itself, even though they may be increased tenfold.

One finding that led to the persistence in quantitating IgD was the low concentrations that were found for the immunoglobulins that are normally present in the largest concentrations. This is a fairly common event in the case of a myeloma or the presence of a monoclonal protein.[3]

Case Study 9-3

1. Increases in serum aminotransferases and blood ammonia levels are the most prominent laboratory indicators of Reye's syndrome. Leukocytosis, hypoglycemia, ketonuria, and an elevated BUN may also be present. In this case, the patient's serum aminotransferases were elevated, but the blood ammonia level remained in normal range.
2. With the serum and urine showing the presence of ketoacids, the patient is acidotic. Investigation of the blood gas results shows an acidic pH and a decreased PCO_2. Multiplying the PCO_2 by the solubility coefficient for CO_2 (0.0306), the amount of dissolved CO_2 (dCO_2) is found to be 0.70 mmol/L. The HCO_3^- content can be calculated by subtracting the dCO_2 from the TCO_2. The HCO_3^- content is 9.3 mmol/L (normal values: 23–28 mmol/L). Therefore, the calculated ratio of HCO_3^-/dCO_2 is decreased ($\approx 13/1$) due to a HCO_3^- deficit, indicating that the patient is in a state of metabolic acidosis. Since ketosis is more commonly seen in impaired carbohydrate metabolism, the serum glucose does not agree with these findings. A defect in protein metabolism is then suspected.

CASE STUDY TABLE 9-3B. Amino Acid Values (Reference Range)

Urine metabolic screen: Ketoacids 2+ (negative)
(Presence of valine and leucine or isoleucine)

Plasma quantitation

Isoleucine	70 μmol/dL	(3.7–14)
Leucine	163 μmol/dL	(7.0–17)
Valine	134 μmol/dL	(16–35)
Alloisoleucine	35 μmol/dL	(0–trace)

3. A metabolic screen of the urine followed by a quantitation of plasma amino acids would be needed. High levels of the amino acids valine, leucine, and isoleucine were found (Case Study Table 9-3B). The results are seen in maple syrup urine disease.

Maple syrup urine disease (MSUD) is a rare (1 in 216,000 live births) autosomal recessive trait in which there is a branched-chain amino acid disorder that causes the build-up of ketoacids in plasma and urine. In the classic form of the disease, this buildup results in mental retardation and eventually leads to death in the early stages of infancy. However, several variants of MSUD have been found that do not present this clear picture. These variants have been classi-fied as "intermittent" MSUD, "intermediate" MSUD, "thiamine-responsive" MSUD, and "E₃ deficiency" MSUD.

In intermittent MSUD, the disease is manifested as a result of a variety of stresses on relatively normal children. However, each episode is potentially fatal. In intermediate MSUD, there is a continuous accumulation of ketoacids and branched-chain amino acids, but the neurologic problems are not as severe and there is an increase in longevity among untreated cases. When the level of thiamine within the body drops below a critical point, thiamine-responsive MSUD will appear. The diet must then be supplemented with overdoses of thiamine for treatment and prevention. The multi-enzyme complex responsible for the conversion of branched-chain amino acids contains three known enzymes. It is the lack of the third enzyme (E3) in the complex that results in the accumulations of ketoacids in E3 deficiency MSUD. The prognosis for such patients is poor after seven months of age.

The patient in this case report has intermediate MSUD. The presence of an upper respiratory tract infection coupled with ketonuria and hypoglycemia produced from the unde-tected MSUD gave initial symptoms which resembled Reye's syndrome. There have been other documented cases in which patients had initial symptoms that did not suggest MSUD. Only after further testing was the disease dis-covered.

One important point to keep in mind is that the ur-ine in MSUD will have a distinct maple syrup odor—whether the form of the disease is classic or variant. Noting this odor may provide a clue to early detection of these variants.

CHAPTER 10

Case Study 10-1

1. The working diagnosis is not myocardial infarction.
2. The lack of the classic enzyme activity of AMI rules out a working diagnosis of AMI. Specifically, the CK-MB is normal, and total CK, AST, and LD activities do not re-veal the typical rise and fall over the 48-hour time period during which blood was sampled. The LD flipped isoen-zyme pattern is present, but this can be seen in conditions other than AMI.
3. CK elevation is due to CK-MM, most probably because of intramuscular injection, since CK-BB is not detected, and CK-MB is normal. Increased ALT activity and LD-5 levels reflect liver congestion secondary to inadequate car-diac output. The elevation of LD-1 is presumably due to intravascular hemolysis of red blood cells, and LD-3 may be derived from lung tissue. Overall, the enzyme data are more suggestive of pulmonary embolism.

Case Study 10-2

1. The elevations are caused by muscle injury from the box-ing match.
2. The LD takes longer to rise in myocardial infarction. Also, the ALT would not be elevated.
3. The levels should return to normal within 4 days.

Case Study 10-3

1. With respect to the symptoms and physical findings, the most probable diagnosis is acute pancreatitis. One might also suspect hepatitis, but there was no evidence of jaun-dice. Another possibility is appendicitis, but the pain would likely be lower.
2. The most likely elevations would be in serum amylase and lipase. Amylase and lipase are produced and stored in the pancreas. The enzyme elevation is due to release of the en-zymes into the circulation following cellular damage to the pancreas. When an individual arrives in an emergency room complaining of the same symptoms as this woman, it is common for the physician to order a STAT amylase to differentiate pancreatitis from flu, appendicitis, or hepatitis.

CHAPTER 11

Case Study 11-1

1. Acid–base status (pH) is within the normal range, as are the HCO_3^- and carbonic acid values.
2. Yes, the curve is shifted to the right, since the $P50$ is higher than the reference value.
3. The hemoglobin is elevated. A secondary polycythemia is not unusual in chronic respiratory disease to compensate for the low SO_2.

4. No, due to the high level of COHb. This is not considered in a "calculated" SO_2 from a blood gas analyzer and reconfirms the need for an actual SO_2 measurement. This level of carboxyhemoglobin is abnormal; it could be due to continued heavy smoking or exposure to unventilated combustion or automobile exhaust. Elevated levels of hemoglobin (secondary polycythemia) are common in patients with chronic pulmonary disease.

Case Study 11-2

1. Secobarbital depresses the breathing center; consequently, CO_2 is not effectively eliminated nor is sufficient O_2 taken into the lungs.
2. The patient's blood gas values indicate a respiratory *and* metabolic acidosis. Respiratory acidosis is indicated by the elevated PCO_2; metabolic acidosis is indicated by the decreased bicarbonate level.
3. A negative base excess (actually base deficit) confirms a metabolic acidosis and not a *compensatory* metabolic alkalosis that could be seen with just a primary respiratory acidosis. This is consistent with acute respiratory insufficiency and, in this case, the poor circulation that accompanies the shocklike state. With tissue hypoxia, lactic acid is produced by anaerobic metabolism and causes metabolic acidosis.
4. Metabolic acidosis. Appropriate ventilation would correct the respiratory component, lowering the PCO_2 and restoring the PO_2 to normal. The decreased bicarbonate level due to the metabolic disturbance will take longer to return to normal.

Case Study 11-3

1. Partially compensated metabolic acidosis.
2. The HCO_3^- level is the primary contributor to the metabolic acidosis. The PCO_2 (H_2CO_3) level is low due to the patient hyperventilating and blowing off CO_2. The negative BE/D also indicates a metabolic acidosis.
3. A major and serious complication of long-bone fractures in the elderly is pulmonary fat emboli. Tachycardia, tachypnea, and low arterial oxygen tension along with chest pain are the classic signs. Globules of fatty marrow enter small veins in the area of the fracture and travel to the lung, hindering the pulmonary circulation. Depending on the extent of the damage, the output of the right and left ventricles can drop dramatically, compromising the oxygen supply to the tissues.

CHAPTER 12

Case Study 12-1

1. The diagnosis is diabetic ketoacidosis. This is suggested by the patient's history of diabetes mellitus and failure to take

her insulin and is confirmed by the findings of glycosuria, hyperglycemia, ketonuria, and acidosis.
2. The anion gap may be calculated as follows:

$$ANION\ GAP = [Na^+] + [K^+] - [HCO_3^-] - [CL^-]$$

The anion gap is normally less than 17 mmol/L. The value in this patient is 58.8 mmol/L. The anion gap is not a true gap but represents unmeasured anions (*e.g.*, phosphate, sulfate, organic acids) in the plasma. Any condition causing an increase in these anions will cause an increased anion gap. In this patient, the cause of the elevated anion gap is diabetic ketoacidosis.
3. The elevation of the plasma osmolality is due to water loss from osmotic diuresis and to the increased amount of glucose in the plasma. Hyperosmolality is a common finding in patients with diabetic ketoacidosis.
4. Diabetic ketoacidosis is a form of metabolic acidosis, which results in a decrease in plasma bicarbonate. Patients breathe rapidly in an attempt to compensate for this acidosis, which results in a low PCO_2. PO_2 is usually normal. Chloride is typically increased to maintain neutrality for the lost bicarbonate.

The plasma sodium value is deceptive. Hyperglycemia causes a marked osmotic diuresis, with water being lost in excess of sodium. This would tend to cause hypernatremia. However, the hyperosmolality stimulates the thirst mechanism, leading to replacement of water but not sodium. Thus, plasma sodium may be elevated, normal, or decreased in diabetic ketoacidosis, even though such patients may be severely depleted of water, sodium, and other electrolytes.

The potassium value is also deceptive. Only 2–3% of total body potassium is present in the extracellular fluid. Hyperkalemia in this patient is due to the severe acidosis and, to a lesser extent, to volume depletion. Most patients in diabetic ketoacidosis are potassium-depleted. The potassium level will fall as the acidosis, fluid depletion, and hyperglycemia are corrected, and supplemental potassium may be needed to prevent hypokalemia.

Case Study 12-2

1. The factors that contributed to the hypernatremia are diarrhea, insensible fluid loss, and diminished thirst without fluid replacement.
2. When adequate fluid is available to drink, our thirst mechanism is capable of preventing hyponatremia. While diarrhea certainly contributed to the hyponatremia, the loss of the thirst mechanism is most likely the major factor in development of hypernatremia.
3. Low blood volume (or low blood pressure) will activate the renin-angiotensin-aldosterone mechanism, which promotes Na retention in the blood, leaving less Na in the urine. Because these reabsorbed Na ions carry water, they help maintain blood volume. The relatively hyperosmolar urine results from the ADH response to hypernatremia

(hyperosmolality), which promotes H_2O reabsorption by the kidneys, producing a concentrated urine.

Case Study 12-3

Several factors appear to be contributing to the hypokalemia:
Metabolic alkalosis promotes cellular uptake of K, which lowers plasma K.
Diuretics enhance renal tubular flow, which enhances renal K loss.
Increased aldosterone promotes renal loss of K. The patient may have hyperaldosteronism caused by adrenal disease.

Case Study 12-4

Case A is (1) primary hyperparathyroidism because of the moderately elevated ionized calcium with moderately elevated PTH. PTH should be completely suppressed with hypercalcemia. The slightly decreased phosphate is also consistent with this diagnosis.

Case B is (3) hypomagnesemic hypocalcemia. The Mg deficit inhibits release of PTH, which is much lower than it should be for hypocalcemia.

Case C is (2) malignancy. Malignancy is usually associated with the greatest elevations in calcium. The diminished parathyroid response is appropriate for hypercalcemia. The low hematocrit is also consistent with this diagnosis.

CHAPTER 13

Case Study 13-1

1. Iron deficiency anemia is characterized by low serum iron and ferritin and increased TIBC and transferrin.
2. There are several possible factors that may have led to the development of iron deficiency anemia in this person. Iron deficiency anemia commonly occurs due to loss of iron, particularly chronic blood loss due to GI bleeding. In addition, decreased dietary intake of iron over a period of time can lead to overall iron deficiency. Given the loss of weight and possible poor dietary habits, this could also be a contributing factor. Finally, iron deficiency anemia occurs as a result of decreased absorption of dietary iron. For example, drugs like Zantac® can interfere with iron absorption in the small intestine.
3. Proper medical treatment would include dietary supplementation of iron with ferrous sulfate. In addition, treatment of his recurring gastritis is required to prevent further gastrointestinal loss of iron due to blood loss.
4. Common causes of microcytic anemia include iron deficiency, β-thalassemia trait, and anemia of chronic disease. In β-thalassemia, serum ferritin would be normal or even increased and there would not be any alteration in transferrin or TIBC levels. In the anemia of chronic disease, serum TIBC and transferrin would be decreased due to

decreased protein production by the liver and/or increased protein loss by the kidneys and serum ferritin would be normal.

Case Study 13-2

1. The most likely cause of this woman's microcytic anemia is heterozygous β-thalassemia. The characteristics of this disease include minimal or no anemia with the presence of significant microcytosis. The incidence of this disease is high (1–2%) in person's of Southeast Asian ancestry. Affected individuals commonly show a slightly elevated erythrocyte count. Iron deficiency anemia can be ruled out because of the normal results of the iron studies, in particular ferritin.
2. Further tests that could be performed would include hemoglobin electrophoresis to rule out an abnormal hemoglobin as the cause of the microcytosis. In addition, the presence of elevated hemoglobin A_2 would be supportive of the diagnosis of heterozygous β-thalassemia.
3. Because of the woman's desire to become pregnant, two issues should be discussed with her. Because of her heterozygous β-thalassemia, it would be important that the status of her husband be identified. If he too is a heterozygote, the parents would have a 1:4 chance of having a child with β-thalassemia, which is a severe life-threatening condition. In addition, pregnant women are at risk for the development of iron deficiency anemia. It would be important to educate this patient about proper diet during pregnancy.
4. Iron supplementation in this patient is contraindicated. Since this patient does not have iron deficiency, iron therapy will not have any direct benefit. In fact, supplementation with iron in patients who are misdiagnosed with heterozygous β-thalassemia can, over a period of time, lead to possible iron overdose and toxicity.

Case Study 13-3

1. Low albumin and transferrin values are hallmarks for malnutrition. Transferrin is a very sensitive indicator of inadequate nutritional status. With cases of severe malnutrition, levels are often lower than 50 mg/dL. Low serum and/or copper are commonly observed in patients on long-term parenteral nutrition (due to inadequate supplementation of administered fluids).
2. Zinc and/or copper are commonly observed in patients receiving long-term parental nutrition. It can take several months for these deficiencies to become evident. They can produce impairment of several biological functions. Low zinc is associated with impaired growth, particularly in children, diarrhea, dermatitis, and other abnormalities including hair loss (as observed in this case). Copper deficiency commonly produces defective pigmentation of hair and skin and mental cognitive impairment.

3. Treatment would be aimed at ensuring adequate nutritional intake of all essential components. Patients on total parenteral nutrition are susceptible to deficiency of essential fatty acids, protein, trace metals, and vitamins.

CHAPTER 14

Case Study 14-1

1. Although one would suspect that this patient is diabetic, a confirmatory test is required to ensure that the patient is not exhibiting nondiabetic glucose intolerance. Two fasting glucose results greater than 140 mg/dL (7.8 mmol/L) or an OGTT with 200 mg/dL (11.1 mmol/L) at 2 hours and one other time is diagnostic.
2. The patient probably has IDDM, due to the time course for onset and her age. NIDDM patients often have an insidious onset and sometimes present with another illness that can be a complication of the diabetes itself.
3. Her anion gap is increased due to the increased presence of the unmeasured anions, β hydroxybutyrate, and acetoacetate. Acetone is uncharged and does not contribute to the anion gap. Lactate and free fatty acids can contribute to an anion gap along with phosphate and other inorganic anions. Exogenous substances that contribute to an anion gap include metabolites of alcohols and salicylates.

Case Study 14-2

1. A glucose greater than 600 mg/dL is a definitive marker for the hyperosmolar state. The elevated BUN is also useful and the absence of ketones confirms the diagnosis. A diabetic with normal renal function would be able to match glucose production with glucose losses to urine. In the patient with compromised renal function this safety mechanism is lost. The glucose concentration in SI units is 62.3 mmol/L. Convert from mg/dL to mmol/L by obtaining mg/L (×10), converting to mmols (wt/mw; x/180). Therefore the, conversion factor is 10/180 or 1/18.
2. The BUN and creatinine values are indicative of impaired renal function. The patient has no muscle tone on his right side. As with many diabetic patients, there are vascular complications. In this case, the patient may have had another stroke and this will have to be investigated while the patient is being treated for hyperosmolar coma.
3. The clinician noted that respiration was shallow. Compensated metabolic acidosis is corrected by an increased respiratory drive. Kussmaul respirations assist in the removal of carbon dioxide and this, in turn, elevates the pH of the patient. Provided the patient had intact neurological function in the respiratory drive center of the brain, deep breathing would automatically occur in metabolic acidosis. The test results that can tentatively rule out acidosis in this individual are the normal anion gap and the

normal bicarbonate. If the patient had metabolic acidosis, then the anion gap should be elevated. Respiratory acidosis would be indicated by an elevated carbon dioxide. Blood gases with blood pH would confirm that the patient was not acidotic.

Case Study 14-3

1. The infant's mother may have been a gestational diabetic (GDM). GDM fetuses are exposed to excessive amino acids, glucose, and lipids. This stimulates the infant to overproduce insulin. Insulin acts as a growth hormone.
2. The infant's pancreas has been responding to the substrates delivered by the mother across the placenta. The β cells of the pancreas have been overproducing insulin. After the delivery, the infant is in a hyperinsulinemic state. Infants are prone to hypoglycemia after delivery from diabetic mothers.
3. Whole blood glucose concentrations are usually 15% less than plasma glucose concentrations. Increased hematocrits (polycythemia) will cause the differences to be even greater. It is common for neonates to be polycythemic.
4. Although many gestational diabetics revert back to normoglycemic control after delivery, it would be worthwhile to monitor her glucose, first by measuring the fasting glucose and then by an OGTT if the fasting is abnormal.
5. Pregnant women should be tested for GDM at the initial visit and then around 24 weeks, when GDM is most prevalent.

CHAPTER 15

Case Study 15-1

1. The triglyceride and HDL-cholesterol values.
2. The estimated LDL cholesterol, 167 mg/dL (4.3 mmol/L), is considered high.
3. One could not estimate the LDL cholesterol concentration, since the triglycerides were >400 mg/dL (4.5 mmol/L). Lipoproteins would need to be measured in a specialty laboratory following ultracentrifugation.

Case Study 15-2

1. Cardiac enzyme CK and CK isoenzyme, LD and LD isoenzyme would be useful. Fasting total cholesterol, triglycerides, HDL cholesterol, and LDL cholesterol could also be performed, but the values should be interpreted with caution, since the patient is under stress and may have suffered an MI. The fasting lipids should be repeated after he has been released from the hospital. If he has suffered an MI, it will be approximately 6 weeks before the values represent his true levels.

2. With a total cholesterol level of approximately 280 mg/dL (7.2 mmol/L) and normal triglycerides and HDL cholesterol, his LDL cholesterol is estimated to be >190 mg/dL (4.9 mmol/L), which is extremely elevated. His physician presumably would have started him on a Step II diet prior to leaving the hospital, based on the in-hospital values. If the 6 to 8 week values do not show significant improvement, the physician would probably start him on medication. The patient has a family history of CHD, and presumably the patient has now been diagnosed as having CHD himself. Therefore the goal would be to reduce his LDL cholesterol value to <100 mg/dL (2.6 mmol/L).

Case Study 15-3

1. Every assay except potassium, cholesterol, and glucose.
2. Her kidney function is grossly abnormal, and the associated hypertriglyceridemia is classically seen when the kidney is malfunctioning. Because the patient is a 60-year-old woman, however, it should also be determined if she is taking replacement estrogen, since that can also elevate triglycerides in some women.
3. Nephrotic syndrome.

CHAPTER 16

Case Study 16-1

1. The most likely cause of the patient's elevated BUN is *prerenal* because the BUN-creatinine ratio is greater than 10, and the creatinine is only minimally elevated. This is substantiated by the return to normal with improved cardiac output.

Case Study 16-2

1. The most likely cause is chronic renal disease. Supporting data are the essentially normal BUN-creatinine ratio and the marked elevation of all NPN values. There was no marked improvement when cardiac function improved, thus further eliminating congestive heart failure as a cause of the elevated BUN.
2. If the patient had an elevated level of acetone and other alpha-ketoacids, as might be found in a diabetic, her elevated creatinine levels could have been an erroneous result, since alpha-ketoacids are known to cause a positive bias when creatinine is measured by a kinetic Jaffee reaction, the most commonly used assay method. However, the normal glucose level and abnormal values for other NPN substances make this unlikely.

Case Study 16-3

1. The increased uric acid is due to the marked increase in nuclear breakdown in the presence of a very high WBC. It is not from renal disease because the BUN and creatinine are normal.

2. Chemotherapy has reduced the WBC to below normal levels, and the patient is taking allopurinol.
3. It is probably due to decreased intake (patient is unable to eat); a determination of total serum protein and albumin would be helpful.

CHAPTER 17

Case Study 17-1

1. Urine formation which implied a good blood flow in the kidney.
2. Immediately as reflected by the formation of urine.
3. Yes, peritoneal dialysis is much less efficient than the traditional hemodialysis, which in turn is much less efficient than a normal kidney.
4. Cholesterol and triglyceride levels are often elevated in dialysis. It was also ordered for a baseline to compare post-transplant values.
5. Yes, renal failure is associated with increased PTH due to increased secretion caused by a decreased Ca^{++} and also reduced renal clearance, particularly of the C-terminal fragment, which is normally degraded by the renal tubules after filtration by the glomeruli.
6. The decreasing BUN, CR, K^+, and P, β_2-microglobulin and increasing urine volumes indicated graft success.

Case Study 17-2

1. Because he was immunocompromised from cyclosporine therapy.
2. A mild renal failure is present in this case.
3. Renal tubular cell injury.
4. To determine the functional capability of the kidneys to clear myoglobin.
5. As urine myoglobin greatly exceeds serum values, adequate clearance of myoglobin is observed. In this case, acute renal failure is not due to myoglobin precipitation within the tubules.
6. Liver and pancreas, however moderate increases in amylase are seen in renal insufficiency since this is a major route of amylase excretion.
7. The ratio indicated that the patient did not have pre-renal azotemia.

Case Study 17-3

1. The patient is in diabetic ketoacidosis.
2. Yes. Both the BUN and creatinine levels are elevated, albeit not dramatically.
3. No. Uremia also may be a cause of coma, but respiration is stertorous, without hyperventilation. Also, in uremia, the creatinine is markedly elevated, not just slightly increased, as it is here.
4. Note: The urine osmolality is reported as being within normal range; this does not, however, mean that the urine osmolality is unaffected by the patient's condition.

Rather, this "normal" value, within the reference range, is the outcome of a balance between two opposing forces: the depression of urine concentration resulting from diabetic polyuria and the increase of glucose concentration in urine due to the overflow of the renal threshold.

Exercises

1. The creatinine clearance is calculated by multiplying the ratio of urine to serum creatinine by the urine volume excreted in 24 hours (expressed in mL/min).

$$\frac{(150 \text{ mg/dL})(2000 \text{ mL/day})(1 \text{ day}/1440 \text{ min})}{1.5 \text{ mg/dL}}$$
$$\times (1.73/1.90) = 126 \text{ mL/min}$$

2. B. Interleukin-2 and interferon are lymphokines secreted by antigen-stimulated T lymphocytes. Prednisone is a glucocorticoid used as an adjunctive anti-inflammatory agent. FK506 is a recently approved new generation immunosuppressant currently being studied in transplant patients.

3. A. Hemodialysis is typically performed three times a week for the removal of toxic waste products and requires several hours to complete.

4. E. Although most renal disease can be characterized by variable amounts of proteinuria, nephrotic syndrome is characterized by increased permeability of the glomerular basement membrane and profound proteinuria.

5. D. Renin, aldosterone, ADH and prostaglandins all regulate water and electrolyte balance. Erythropoietin acts on erythroid progenitor cells in the bone marrow to increase red cell mass.

6. C. A substance that is completely removed on first pass through active secretion (e.g., p-aminohippurate) estimates the plasma renal flow.

7. C. Much of the filtered bicarbonate is reabsorbed back into the circulation for the maintenance of blood pH.

8. E. Creatinine clearances decrease with age by about 6.5 mL/min/1.73 m^2/decade.

9. D. Contaminations can occur through the air and skin.

10. E. Abnormal concentrations of β_2-microglobulin in serum are seen in clinical conditions characterized by increased cellular turnover, and the inability of renal tubules to reabsorb it.

CHAPTER 18

Case Study 18-1

1. These laboratory test results suggest jaundice due to extrahepatic obstruction either from a stone in the common bile duct, a carcinoma of the head of the pancreas, or, possibly, a postoperative stricture.

Case Study 18-2

1. This laboratory profile suggests hepatocellular damage resulting from either viral hepatitis, alcoholic hepatitis, or a toxin such as carbon tetrachloride or phosphorus. A specific diagnosis of viral hepatitis can be made with additional serologic tests such as HepBSAg determination, etc. Other causes of suspected hepatocellular jaundice may require a needle biopsy of the liver for confirmation.

Case Study 18-3

1. The presence of hepatitis B surface antigen and IgM hepatitis core antibody and the absence of hepatitis B surface antibody all point to the diagnosis of acute hepatitis B. The hepatitis A studies indicate remote infection and current immunity to that disorder. The test for hepatitis C is negative.

2. The prognosis is good. Approximately 90% of patients with hepatitis B recover uneventfully.

3. Less than 1% of patients with acute hepatitis B develop the fulminant form of the disease. Many of these patients die. Therefore, the patient should be observed carefully until improvement occurs. Of greater concern is the fact that about 10% of patients with acute hepatitis B will develop chronic hepatitis. While many of these patients will eventually recover, those who do not are at risk to develop cirrhosis, liver failure, and hepatocellular carcinoma.

Case Study 18-4

1. The chemical and serologic findings are typical of hepatitis A infection. The enzymes associated with liver function, serum bilirubin, and the urine bilirubin are all elevated. The patient's history is typical for hepatitis A infection.

2. The patient should be questioned for recent contact with other infected individuals, foreign travel, or exposure to unsafe drinking water. In this case, the patient gave a history of a recent camping trip to Mexico.

3. The prognosis is good. The patient can be reassured that she will gradually begin to feel better. No further therapy is indicated.

Case Study 18-5

1. These findings are diagnostic of hepatitis B infection. It is not possible on the serologic basis to distinguish acute from chronic hepatitis B. However, by definition, since the hepatitis B surface antigen has been present for 6 months, this patient has chronic hepatitis B. It should also be noted that the patient has a previous infection with hepatitis A and is now immune. This plays no part in his current illness.

2. The prognosis for hepatitis B is variable. A liver biopsy may be necessary to establish the degree of damage and to determine whether interferon therapy is needed.

3. Patients with chronic hepatitis B infection are at high risk for the development of cirrhosis and also hepatocellular carcinoma.

4. A test for hepatitis Be antigen and antibody should be performed. The presence of hepatitis Be antigen indicates that the patient is highly infectious and has a poor prognosis. More favorable prognosis is indicated by the presence of antibody to hepatitis Be antigen. The patient should also be tested for Delta hepatitis infection. Co-infection with Delta hepatitis worsens the prognosis of hepatitis B.

CHAPTER 19

Case Study 19-1

1. This patient most likely has panhypopituitarism secondary to severe hemorrhage during delivery (Sheehan's syndrome). The pituitary is enlarged during pregnancy and is especially susceptible to ischemia at the time of delivery. Failure to lactate after delivery is very suggestive of this condition when taken in conjunction with the history of hemorrhage and successful breast feeding after previous deliveries.

2. No. Increased skin pigmentation is seen in patients with primary adrenal insufficiency because of the melanocyte stimulating activity of ACTH. In primary adrenal insufficiency, ACTH levels in plasma are increased in an effort to stimulate the unresponsive adrenal cortex. This patient has secondary adrenal insufficiency characterized by low ACTH. Therefore, increased pigmentation would not be present.

3. This patient should receive a complete pituitary endocrine work-up beginning with determination of thyroid and adrenal hormone levels. In a patient with panhypopituitarism, thyroxine, free thyroxine, and cortisol would all be decreased as would TSH and ACTH. Stimulation testing should then be performed to confirm the diagnosis. Although several different agents could be used, this patient could be a good candidate for an insulin tolerance test combined with administration of TRH and GnRH. This combination of agents should normally result in increased levels of growth hormone, cortisol and ACTH, thyroxine and TSH, prolactin, and FSH. Since the suspicion of panhypopituitarism is very strong in this patient, if the insulin tolerance test is used the physician would have to watch this patient very closely to prevent complications of hypoglycemia. Other combinations of tests would take longer to perform but have less risk of complication. These would include arginine infusion (GH), TRH infusion (thyroxine and prolactin), GnRH infusion (FSH), and CRH infusion or an ACTH stimulation test (ACTH or cortisol).

Case Study 19-2

1. This 64-year-old smoker presents with symptoms of fatigue, weight loss, and shortness of breath. The differential diagnosis in this setting would include congestive heart failure (due to coronary artery disease, hypertension, etc.), lung disease (emphysema or bronchitis), or malignancy (most likely lung).

2. Evaluation of the patient's cardiac status would require a stress test and possibly additional radiologic studies, up to and including cardiac catheterization. Other less invasive and costly procedures can be done first to evaluate the possibility of pulmonary disease or a lung tumor. Chest x-ray would be a good place to start and in this patient showed a widening of the mediastinum with a probable mass in the hilum of the left lung.

   Additional chemistry tests that should be performed include evaluation of other electrolytes, serum osmolality, urine osmolality, and urine sodium excretion. In this case, the serum osmolality was decreased while urine osmolality and urine sodium were increased. These data are strongly suggesting of the syndrome of inappropriate ADH (SIADH). Cytological examination of the patient's sputum revealed a diagnosis of small cell undifferentiated carcinoma of the lung.

3. Numerous examples of ectopic hormone production have been described. The most common are associated with lung neoplasms, especially small cell carcinoma. These tumors frequently produce ACTH or ADH resulting in Cushing's syndrome or SIADH, respectively. Other hormones that may be produced ectopically include HCG, PTH, gastrin, growth hormone, etc. Differentiating the ectopic production of a hormone from a primary endocrine abnormality usually involves a combination of suppression or stimulation testing and radiologic procedures. In patients with Cushing's syndrome, for example, due to carcinoma of the lung, ACTH levels would be increased and would not suppress with dexamethasone. With CRH administration, however, pituitary production of ACTH increases whereas ectopic production of ACTH does not. These tests make the diagnosis of ectopic ACTH production and can be combined with radiologic and cytological examination to localize and diagnose the neoplasm.

Case Study 19-3

1. The patient's symptoms are most suggestive of hypoglycemia although the possibility of the hypertensive episodes of pheochromocytoma or the symptoms of the carcinoid syndrome cannot be excluded from the history. Observation of the patient and determination of blood pressure during an episode would be helpful in resolving this differential.

2. To rule out the possibility of pheochromocytoma or carcinoid syndrome, a 24-hour urine should be collected and analyzed for metanephrines and 5-HIAA.

   Given the fact that the patient's symptoms do not occur in relation to meals, the diagnosis of reactive hypoglycemia is unlikely. The possibility of organic hypoglycemia due to an insulin producing neoplasm versus factitious hypoglycemia due to insulin or oral hypoglycemic ingestion must be evaluated. Since fasting may provoke those symptoms, the patient should be hospitalized and placed on a 72-hour fast. Blood should be drawn whenever symptoms develop and the specimens analyzed for glucose and insulin. An insulin to glucose ratio of >0.3 suggests that excess insulin is present. True organic hypoglycemia can be distinguished from insulin injection by analysis of C-peptide. C-peptide is the connecting peptide present in the proinsulin molecule, which is removed when insulin is secreted. Excess endogenous insulin release will be associated with increased levels of C-peptide. This peptide is not present in commercial insulin preparations, however, and will be absent or normal if the patient is injecting insulin to produce the hypoglycemic symptoms.

In this case, the patient did have an elevated insulin/glucose ratio but C-peptide levels were low. When presented with these findings, the patient admitted injecting insulin.

3. Patients who inject insulin to produce the symptoms of hypoglycemia do so to get attention from family and medical providers. They are frequently health care workers who have access to insulin preparations and needles and syringes. Relatives of insulin-dependent diabetics (who also have access to required materials) may also exhibit this type of behavior. Frequently these people have some type of acute stress in their personal or professional lives and the attention that they receive because of their activities answers a psychological need.

CHAPTER 20

Case Study 20-1

1. Graves' disease causing hyperthyroidism.
2. Thyroid antibody tests, FT_4.
3. The high-sensitivity TSH is undetected. This is a good indication of primary hyperthyroidism and not another disease or drug effect.
4. Graves' disease is believed to be an autoimmune disorder whereby antibodies attach to TSH receptor sites causing increased production of thyroid hormones.

Case Study 20-2

1. Pregnancy.
2. Symptoms and increased TT_4, decreased T_3U, and normal FT_4I and TSH. The abnormal tests are due to pregnancy. Also, the abnormal glucose result may be significant.
3. Pregnancy test, beta HCG. If negative, then a more detailed history is in order to exclude other artifactual causes.

Case Study 20-3

1. Euthyroid sick syndrome (or NTI).
2. FT_4 by equilibrium dialysis; perhaps FT_3 and rT_3.
3. Whether the patient is on drug therapy, especially dopamine or corticosteroids. These drugs decrease thyroid hormone levels.
4. Treatment with L-thyroxine is not indicated.

CHAPTER 21

Case Study 21-1

1. Cystic fibrosis.
2. Sweat electrolyte determination—greatly increased chloride and sodium concentrations.
3. Increased levels of fecal lipids, decreased levels of serum total proteins and albumin, diabetic glucose tolerance test, prolonged prothrombin time, and greatly diminished response to secretin.

Case Study 21-2

1. Acute pancreatitis.
2. Enzymatic fat necrosis and digestion to result in free fatty acids in the adipose tissue of the abdomen. The fatty acids then bind calcium as they form fatty acid salts.
3. Shock resulting in prerenal azotemia.

CHAPTER 22

Case Study 22-1

1. Zollinger-Ellison syndrome (high acid output and volume of secretion in both the basal and stimulated test and the high ratio of basal acid output to stimulated acid output).
2. Plasma gastrin, which should be very high.
3. Hemoglobin is decreased as a result of blood loss. White blood count is slightly high as a reaction to stress and blood loss.

Case Study 22-2

1. Malabsorption syndrome.
2. Intestinal malabsorption characteristic of nontropical or celiac sprue.
3. Malabsorption of vitamin K and resulting deficiencies of coagulation factors II, VII, IX, and X.
4. Iron deficiency (hypochromic, microcytic and red blood cell indices) is the probable cause for the anemia. This is caused by malabsorption of iron. Poor dietary intake of iron may be a contributing factor. Intestinal blood loss is unlikely to be an important factor. Other possible contributing causes are deficiencies of folate and vitamin B_{12} as a result of malabsorption.

CHAPTER 23

Case Study 23-1

1. The patient presented with a combination of two common complications of pregnancy, hypertension and diabetes. The two are often linked. The rapid weight gain, proteinuria, and electrolyte values suggest edema as well, consistent with hypertension. A first pregnancy appears especially prone to these two complications.
2. Delivery of the fetus at this time would be with some risk. The L/S ratio is not at a safe decision level for the complications. The PG%, FSI, and creatinine all suggest borderline situations. Since the contractions subsided, no further action was taken. The enzymes are relatively normal for the situation. Magnesium levels will be carefully monitored for hypertension, and dietary intervention will be used to treat the diabetes, which was probably secondary to the pregnancy.

Case Study 23-2

1. The CSF total protein is at the upper limit of the reference interval. It may or may not be elevated for this specific patient. The ratio elevation, however, definitely suggests increased permeability.
2. The results are consistent with demyelinating processes such as MS. The elevated ratio and oligoclonal banding present in the CSF but not serum electrophoresis support this hypothesis.

CHAPTER 24

Case Study 24-1

1. No. Like the majority of tumor markers, CEA is not sensitive or specific enough to be used as a screening test for malignancy. In colorectal cancer, less than 33% of patients present with CEA elevations that are two times the upper limit of normal, and 50% of patients present with normal levels of CEA.
2. CEA is a glycoprotein that is found in normal fetal gastrointestinal tract epithelial cells. Elevations of CEA also occur in colorectal, pancreas, and liver cancer. CEA is also elevated in pulmonary emphysema, acute ulcerative colitis, alcoholic liver cirrhosis, hepatitis, cholecystitis, benign breast disease, rectal polyps, and in heavy smokers.
3. CEA levels should return to normal after complete and successful resection of colorectal tumors in patients with initially elevated CEA levels. CEA values generally remain normal in the absence of recurrent disease, and a rise in serial CEA levels is indicative of recurrent disease in approximately 66% of the cases. An elevated CEA often precedes other clinical signs in 75% of patients developing recurrent disease by at least several months.

4. Any patient who has long-term changes in bowel habits, especially diarrhea, should have their electrolyte levels monitored. Potassium levels can rapidly decrease in patients with diarrhea. Blood urea nitrogen and creatinine can also be used to help determine if dehydration is present. A complete blood count (CBC) should be performed and may show a low hemoglobin and hematocrit from blood loss (if the patient was guaiac positive) due to anemia or chronic disease. Liver function tests (ALT, AST, alkaline phosphatase) may be elevated if the cancer has metastasized to the liver.

Case Study 24-2

1. The most likely primary diagnosis for this patient is a testicular germ-cell tumor. Given the ECG findings and the age of the patient, an acute myocardial infarct is highly unlikely. Before surgery, this patient's serum AFP level was normal, whereas his β-hCG was elevated. After surgical removal of the tumor, the elevated serum β-hCG as well as the elevated LD–1 isoenzyme returned to normal. Histologic examination of the tumor mass showed a seminoma containing syncytiotrophoblastic giant cells.
2. Any benign condition associated with an elevated LD–1 isoenzyme value should be considered in the differential diagnosis. A tumor mass of the testicle could be a benign nodule or cyst (although rare). Benign conditions associated with an abnormal LD–1 isoenzyme may be observed in patients with a myocardial infarct, in vivo or in vitro hemolytic disorders, ineffective erythropoiesis, renal infarct, or muscular dystrophy.
3. Yes. If serum AFP and β-hCG levels are normal and all other possible benign conditions associated with an elevated LD–1 are ruled out, the germ-cell tumor is most likely a pure seminoma. An elevated serum AFP level would suggest an embryonal carcinoma and/or yolk-sac tumor. Elevations of serum β-hCG would suggest a pure seminoma containing syncytiotrophoblastic giant cells, a choriocarcinoma, or a mixture of these two types. Combined elevations of both serum AFP and β-hCG usually indicate a mixed germ-cell tumor.
4. No. A final tumor diagnosis can be made only by histopathologic examination of the tumor tissue removed during surgery. None of the currently available tumor marker tests can be used independently to establish a diagnosis because they lack high specificity for disease.

Case Study 24-3

1. No. PSA lacks sufficient sensitivity and specificity to be used alone as a screening test for prostate cancer. However, when used in conjunction with digital rectal exam and transrectal ultrasound, PSA can significantly improve prostate cancer detection.
2. Besides prostate cancer, elevations in PSA occur in benign prostatic hypertrophy, prostatitis, and intraepithelial neo-

plasia. Significant elevations in PSA do not normally occur following routine digital examination. A diagnosis of cancer can only be made by histological examination of prostate tissue.

3. Yes. PSA is extremely useful for monitoring cancer recurrence in patients following therapeutic intervention for prostate cancer. After therapy, PSA levels should be extremely low or undetectable, with rising levels indicating recurrent disease. There is a direct relationship between the serum PSA concentration and the volume of the prostate gland.

4. Additional laboratory tests, such as creatinine and blood urea nitrogen measurements, should be performed to assess renal function. Prolonged obstruction of the urinary tract can damage the renal parenchyma. Electrolytes should also be measured. Urine analysis may also be performed to rule out a possible urinary tract infection, which is often seen in patients with obstructive symptoms.

CHAPTER 25

Case Study 25-1

1. This woman has sickle cell trait.
2. No; other than a persistent anemia, this patient should live a normal life.
3. If the father of the child is hemoglobin AA, then there is a 1:2 chance that the child will also acquire sickle cell trait. If the father has hemoglobin AS, there is a 1:4 chance that the child will have sickle cell anemia and a 1:2 chance that the child will have the trait.
4. Yes, these values are normal. A slightly elevated hemoglobin A_2 is not uncommon in sickle cell trait.

Case Study 25-2

1. She had variegate porphyria and had an acute attack because of her ingestion of the barbiturates for a week prior to the surgery.
2. Yes; since this disease is inherited as an autosomal dominant trait, a number of her family and relatives could have this disease.
3. The enzyme that causes this disease is protoporphyrinogen oxidase.
4. The feces could be tested for the presence of coproporphyrin and protoporphyrin. Family history also revealed that she had skin that was photosensitive.

Case Study 25-3

1. The cellulose-acetate electrophoresis shows two bands of about equal magnitude migrating in the areas of HbS and HbC, indicating hemoglobin SC disease. The citrate-agar electrophoresis confirmed these findings.

2. In sickle cell anemia (homozygous HbSS), infarction in the spleen is so common that after childhood the spleen usually becomes very small and nonfunctional because of scarring (autosplenectomy). The spleen was enlarged in this patient due to sequestration of sickled cells.
3. Hemoglobins E and A_2 migrate in the same position as hemoglobin C on cellulose acetate at an alkaline pH. These hemoglobins may be differentiated by hemoglobin electrophoresis on citrate agar at an acid pH.
4. Hemoglobin SC disease has a wide range of severity but generally is usually less severe than hemoglobin SS disease, and life expectancy is only slightly shortened. HbSC patients usually are less anemic and have fewer symptoms than HbSS patients, and some cases go undiagnosed throughout life. The mixture of hemoglobin S and hemoglobin C within the erythrocyte does not allow it to sickle as readily. However, of all the sickling syndromes, HbSC disease is the most common cause of sudden death, generally ascribed to infarction of vital tissue. Because HbSC patients tend to be less anemic, the resulting increased blood viscosity can cause more massive infarcts in larger vessels if sickling does occur. Examination of lung postmortem in this patient showed multiple, small emboli of necrotic bone marrow. The presence of large amounts of dead tissue initiated DIC. The hip pain in this patient was due to necrosis of the femoral head, which is common in HbSC disease.

CHAPTER 26

Case Study 26-1

1. In general, increased APTTs are a result of factor deficiency, the presence of an inhibitor (either a factor-specific inhibitor or a lupus-like inhibitor), the presence of an anticoagulant such as heparin, or the effect of a compromised sample.
2. In this case, the history of presurgical screening tests that were normal suggests that the patient is not likely to have a true factor deficiency that appeared after the operation. Transient inhibitors do occur, but the mixing study and the dRVVT does not suggest the presence of an inhibitor. Heparin contamination is a significant possibility, especially for patients in intensive care that are likely to have a heparin line. Heparin contamination is relatively easy to identify by treating the sample with a heparin neutralizer and repeating the abnormal tests. In this case, the APTT and thrombin time should be repeated on the neutralized sample. If the results return to normal, heparin contamination is likely. Samples that have been compromised are always a problem and can be identified by a repeat, properly collected sample.
3. This sample was collected through an arterial line that was kept open by the administration of heparin. Heparin is widely used in keeping intravascular (IV) lines open and

is nearly impossible to eliminate adequately for the purpose of coagulation studies. Enough blood can be drawn and discarded to dilute heparin sufficiently for chemistry analysis, but heparin-sensitive tests, like the APTT and the TT, can be affected by even minute amounts of heparin. This is one of the most common causes of artificially prolonged APTTs in the coagulation lab. A fresh sample was properly collected from a venipuncture in the other arm and the results were normal.

Case Study 26-2

1. This is a classic case of von Willebrand's disease with a defective response of platelet aggregation to ristocetin and decreased levels of factor VIII–related antigen and von Willebrand's factor. The prolonged bleeding time with excessive bleeding is a strong indicator of a vascular or platelet problem. Von Willebrand's disease can be an autosomal dominant disorder or recessive with more severe symptoms. Several variants of the disease also have been found to exhibit mild depression of factor VIII:C. Severe symptoms with chronic joint problems, that are similar to the clinical signs of hemophilia, may be seen.

2. Yes, factor VIII:C is normal in this case, but it can be decreased in some cases, usually ranging between 24% and 40%. The other portions of the factor VIII molecule would not be decreased if the diagnosis were hemophilia. Only the procoagulant, factor VIII:C portion of the molecule is defective in hemophilia A. When a hemophilia A patient is given cryoprecipitate, the factor VIII:C level remains elevated for 8 to 12 hours, whereas the von Willebrand's-infused patient will have a longer elevation of 2 to 4 days.

3. The factor VIII molecule is a large, multimeric structure, composed of three distinct parts. Factor VIII:C is the procoagulant portion, and its synthesis is sex-linked. Where it is produced has yet to be determined. The larger portion, which is autosomal in inheritance, consists of factors VIII:Ag and VIIIR:RCo or von Willebrand's factor. The latter is produced by endothelial cells and platelets.

4. Test methods include immunoprecipitation, radioimmunodiffusion, electrophoresis, and (now) synthetic substrates to determine the specific component deficient or defective. The APTT is prolonged in both hemophilia A and von Willebrand's disease.

Case Study 26-3

1. The abnormal APTT is partially corrected in the mixing study suggesting a factor deficiency could be present. Although inconsistent with the clinical presentation, factor assays were performed. Fibrinogen and factor VIII are elevated as acute phase reactants.

2. The family history suggested hypercoagulability and the patient was thus evaluated. The AT-III test results confirmed a familial AT-III deficiency.

3. AT-III is the primary inhibitor of activated coagulation proteins. It functions as a balance for procoagulant reactions and so helps maintain normal hemostasis. The deficiency leaves the patient at risk for thrombosis.

CHAPTER 27

Case Study 27-1

1. The difficulty of ensuring therapeutic drug levels by applying typical dosing regimens is exemplified in this case, since this patient lost consciousness at one point with a quinidine level above the therapeutic range. Congestive heart failure can increase the volume of distribution and prolong the elimination half-life of quinidine. Drug half-lives in the elderly are often increased because of a significant decrease in total clearance.

2. The effect of quinidine is enhanced by potassium and reduced if hypokalemia is present. With these factors and many others changing so quickly in this case, it is not surprising that quinidine levels fluctuated so dramatically and the patient's condition was not under control.

3. It is very apparent that the availability of quinidine monitoring laboratory facilities was instrumental in characterizing this individual's loss of consciousness. Providing immediate quinidine results during an EPS greatly facilitates the study and is a mandatory component of the study in determining the most effective quinidine levels. Subsequent to the EPS, quinidine monitoring was necessary in establishing the proper oral dosing regimen that would maintain optimal therapeutic blood levels and would correct the ventricular fibrillations.

Case Study 27-2

1. On days 5 and 8, the amikacin levels were borderline to low therapeutic. Consequently, dose amount was increased in an attempt to keep levels in the therapeutic midrange—the mid 20's. Likewise, vancomycin was increased on day 10 to maintain therapeutic levels. The amikacin post dose level on day 13 looked suspicious, but a phone call to the ward revealed that the patient was agitated that day and tried to pull out her IV line.

2. The oral vancomycin dose given on day 12 was a gesture to minimize the febrile episodes thought to be due to GI infection. Oral vancomycin doses are not absorbed and pass completely in the stool.

3. Vancomycin is slowly eliminated in the presence of hepatic insufficiency, which developed in this patient (note LD activity). The protocol is to allow the levels to decay below 10 μg/mL before giving another, but single, dose, as occurred on days 17 and 20.

Case Study 27-3

1. Hypertension is the most prevalent adverse effect of cyclosporine. This individual presented with hypotension. However, she had other more severe complications, such as poor cardiac output, which may have overwhelmed any renal hypertension due to cyclosporine. This patient's cyclosporine levels, as an outpatient, were below 100 ng/mL as late as 12 days before admission. There is no evidence that cyclosporine levels below 100 ng/mL cause acute renal failure. Although levels at 300 to 350 ng/mL (immunoassay) can cause renal toxicity, it is chronic and typically takes several months to a year to develop.* The actual cause of the acute renal failure was never completely identified. The suspicion was that her cardiomyopathy caused the renal failure. The cause of the cardiomyopathy was attributed either to the doxorubicin chemotherapy or perhaps to a herpes or CMV infection that was suspected.
2. Cyclosporine doses are usually given at 9:00 A.M., and trough levels drawn just before each dose are most often monitored, thus causing a 24-hour delay in dosage adjustment based on blood levels.
3. The 400-mg dose on day 2—based on previous doses and without the knowledge of the blood level. The 200 mg dose on day 4—based on falling blood level and improving renal and cardiac function. No doses on days 5 through 12—based on blood levels remaining therapeutic. The 200 mg dose on day 13—based on falling blood levels and renal function returning to normal. Subsequent doses—adjusted to maintain 100 to 150 mg/mL blood levels.

CHAPTER 28

Case Study 28-1

Typically, antidotes do not exist, and supportive therapy is not altered by knowing the amount(s) of drug present. Therapy is warm lavage to increase temperature, administration of dopamine to increase blood pressure, thiamine, Narcan, dextrose 50, bicarbonate to raise pH, and intubation.

The patient died 9 days after admission of irreversible shock due to multiple toxin ingestion.

Final Diagnoses

1. Severe metabolic acidosis, secondary to lactic acidosis and multiple toxin ingestion.
2. Hypothermia.
3. Anemia and GI bleeding.
4. Adult respiratory distress syndrome with severe hypoxia.
5. Acute and chronic ethanol abuse.

* See reference 16 at the end of Chapter 27, Therapeutic Drug Monitoring.

6. Pressor-dependent refractory shock.
7. Abnormal liver function tests, probably secondary to ethanol abuse.
8. Electrolyte abnormalities, including hypomagnesemia, hypoglycemia, and hypokalemia.
9. *E. coli* sepsis, terminally.

Case Study 28-2

This patient demonstrates the classic signs of ethylene glycol intoxication—an alcoholic in severe metabolic acidosis. The treatment with an ethanol drip is effective in less severe cases. Ethanol and EG are both metabolized by the liver enzyme alcohol dehydrogenase, but the enzyme has an overwhelming preference for ethanol. The therapeutic ethanol saturates the enzyme, which prevents the metabolism of EG to the toxic aldehyde products and allows the EG to be excreted unchanged.

Final Diagnoses

1. Severe metabolic acidosis secondary to ethylene glycol intoxication and lactic acidosis.
2. History of ethanol abuse.
3. Alcoholic liver disease.
4. Adult respiratory distress syndrome.
5. Severe anemia.
6. Gastrointestinal bleeding.
7. Status post cardiac arrest.
8. Hypothermia.
9. Hypoglycemia.
10. History of multiple traumatic lesions:
    a. 1960 GSW w/o removal of 22 cal bullet
    b. 1961 GSW to face
    c. Late 1960s 32 cal GSW R thigh
    d. Stab wounds L thorax L forearm
    e. 1973 shotgun to legs

Case Study 28-3

Although the initial acetaminophen level of 498 µg/mL was easily interpretable from a nomogram as being highly toxic, a second level was measured about 4 hours later. Since the acetaminophen concentration was greater than half the original level, the $t1/2$ was greater than 4 hours and also indicated that the antidote was necessary. Although this patient's liver enzymes did exceed reference levels after 3 days, they were not grossly elevated and quickly returned to normal. This case demonstrates the effectiveness of acetylcysteine as an antidote for acetaminophen poisonings.

CHAPTER 29

Case Study 29-1

1. The clinician would most likely request the measurement of erythrocyte transaminase (ALT and AST) activity and the calculation of an activity index.

2. An EDTA whole blood specimen is required. It should be placed on ice immediately after collection, and the erythrocytes should be separated from the plasma as soon as possible.
3. The physician would notice dermatitis and the patient's family reporting personality changes and hyperirritability.
4. A restricted diet and excessive ingestion of protein have increased the body's need for vitamin B_6, which serves as a cofactor for enzymes of amino acid metabolism. The drug, hydralazine, may induce a vitamin B_6 deficiency.
5. The activity index of AST would be greater than 1.5, and the activity index of ALT would be greater than 1.25.

Case Study 29-2

1. The excessive intake of vitamin K will cause a shortening of the prothrombin time.
2. Vitamin K is required for the synthesis of coagulation proteins, particularly factors II, VII, IX, and X.
3. Vegetarians often exhibit warfarin resistance. Excessive ingestion of leafy green vegetables such as Chinese cabbage, broccoli, turnip greens, kale, and other cruciferous vegetables will cause warfarin resistance.

Case Study 29-3

1. Disruption of calcium metabolism, specifically hypercalcemia, commonly presents with these symptoms.
2. Isotretinoin or 13-*cis*-retinoic acid.
3. Isotretinoin is a derivative of vitamin A, and side effects of drug use are also seen in chronic hypervitaminosis A.

Case Study 29-4

1. Nicotinic acid (niacin).
2. Nicotinic acid is valuable in treating high blood cholesterol in patients with low HDL cholesterol levels or when combined hyperlipidemia (elevated cholesterol and triglyceride) is present. In normal subjects, 3 g of nicotinic acid taken daily for 1 month has lowered serum cholesterol by 15% and triglycerides by 27%. The major drop is seen in lipids transported as LDL and VLDL. HDL cholesterol increases by 23%.
3. Transient flushing and itching of the skin are common and are related to the vitamin's effect on vascular tone. This accounts for the use of nicotinic acid in a variety of conditions presumed to be due to impaired circulation.
4. Toxic effects include liver damage, impaired glucose tolerance, and hyperuricemia.
5. The physician will periodically request a hepatic profile as well as a lipid profile. Microscopic examination of the urine would reveal uric acid crystals as evidence of increased uric acid production. Changes in the patient's glucose tolerance could be measured by a glucose tolerance test.

CHAPTER 30

Case Study 30-1

1. At a given level of oxygen consumption, the amount of CO_2 produced is dependent on the mixture of carbohydrate and fat in the diet. In particular situations, infusions of high concentrations of glucose or carbohydrate can actually increase the respiratory quotient to values greater than one and cause marked burden on the respiratory system to excrete the increased CO_2. Oxygen consumption itself is increased in patients receiving parenteral nutrition and also results directly in the increased CO_2 production. The standard glucose concentrations of TPN may thus contribute to respiratory insufficiency.

   Hypophosphatemia can also be responsible for respiratory failure by a number of parameters. While hypophosphatemia can affect muscle function and cardiac function, the most important respiratory function of phosphate is in the generation of 2,3 diphosphoglycerate (2,3 DPG), which facilitates the unloading of oxygen from hemoglobin. When inadequate levels of 2,3 DPG are present, oxygen is bound firmly to hemoglobin with decreased availability to peripheral tissues. The combined effect of all of these abnormalities can occasionally result in respiratory insufficiency.
2. One of the side effects of long-term TPN after 1 week is that cholestasis occurs with excess bile within the hepatocytes and kupffer cells of the liver, and some of the liver enzymes become elevated. Those enzymes showing the earliest indication of liver and bile duct inflammation are alkaline phosphatase, SGOT (or aspartate aminotransferase), and SGPT (or alanine aminotransferase), and gammaglutamyl transpeptidase. Also, in this case, total bilirubin is elevated.

Case Study 30-2

1. The half-life of albumin is 21 days. The half-life of transthyretin is 2 days.
2. Serum proteins that have a short half-life and are present in smaller concentration are better indicators of nutritional assessment than are long half-life proteins such as albumin.

Case Study 30-3

1. 3,000
2. 500
   $10,500 \times 100 = 4.76\%$
3. 300 patients
   .20 @ risk

   60 patients @ risk $\times$ \$10,500 $-$ (300 $\times$ \$50)*

   \$630,000 $-$ \$15,000 = \$615,000 savings potential
   ↑   ↑
   Cost   Reward

*Remember, all 300 patients must be screened to identify 20 @ risk.

Case Study 30-4

1. On day 12, surgery was performed to correct the partial intestinal obstruction. Day 13 and day 14 reflect the acute phase reaction that occurs after surgery, and prealbumin (transthyretin) is an inverse acute phase reactant causing the concentrates of transthyretin to decrease.
2. The patient's nutritional status was increasing over these days and transthyretin, as a short half-life nutritional marker, reflected this improvement in nutritional status.

CHAPTER 31

Case Study 31-1

1. Hyponatremia, often with hypovolemia, is fairly common in premature infants. The immature kidney has a poor capacity to retain sodium. Vomiting, diarrhea, and infection can cause these findings. Hyponatremia with increased volume may result from excessive fluid therapy with hypoosmotic or glucose solutions or from the syndrome of inappropriate antodiuretic hormone secretion.

   Hyperkalemia is less common and may result from renal failure, cellular destruction, or excessive therapy.

   An important consideration when these findings are observed is the adrenogenital syndrome.
2. The fluid and electrolyte therapy were carefully reviewed and found to be in order.

   Renal function was normal for a neonate. There was no clinical or laboratory evidence of infection. The infant had not experienced vomiting or diarrhea.
3. At this point, it was decided to begin a workup for the adrenogenital syndrome. Several forms exist, but the most common is due to deficiency of the enzyme 21-hydroxylase, which is involved in cortisol synthesis. Cortisol deficiency results in lack of negative feedback on pituitary ACTH secretion, resulting in androgen excess. Cortisol and aldosterone deficiency occur, resulting in the electrolyte abnormalities observed in this case. The diagnosis may be confirmed by finding elevated levels of progesterone (a cortisol precursor) and adrenal androgen (DHEA) in the infant's serum.

   The adrenogenital syndrome (also known as congenital adrenal hyperplasia) occurs in about 1 in 10,000 live births. It is due to an abnormality of the p-450 c21 gene. Affected male infants are usually normal in appearance, while females undergo pronounced virilization due to the excessive androgens and may have ambiguous genitalia. Therapy is based on cortisol replacement. Because of the genetic basis of this disorder, parental counseling and investigation of siblings is indicated.

Case Study 31-2

1. Serum thyroxine T3, TSH, and the FTI should all be measured. Some screening T4 results are not confirmed by low serum levels. In hypothyroidism, the T3 level is decreased and the TSH level is elevated. Rarely, hypothyroidism may be caused by pituitary insufficiency, in which case the TSH level will be low.

   TBG deficiency, a rare condition, may be excluded by direct measurement of TBS. A thyroid scan may be useful to detect any anatomic defects.
2. Hypothyroidism is treated with thyroxine.
3. Neonatal screening is necessary because even severe hypothyroidism is not evident at birth, since the infant receives maternal hormone. Untreated, neonatal hypothyroidism leads to severe growth and mental retardation, a syndrome known at cretinism. Once these changes have developed, they are usually irreversible, but they may be entirely prevented by thyroxin therapy.

Case Study 31-3

1. Many different conditions may be associated with growth retardation. Various skeletal dysplasias such as anchondroplasia, chromosomal abnormalities such as gonadal dysgenesis, and dysmorphic syndromes such as the Prader-Willi syndrome may be associated with short stature. Many children with these disorders have a characteristic physical appearance or other findings which may serve as diagnostic clues. Additionally, nutritional deficiencies, malabsorption, renal failure, hemoglobinopathy, diabetes mellitus and a variety of inborn errors of metabolism may cause growth retardation.
2. This child's history points to an intracranial cause, specifically, growth hormone deficiency.
3. A random growth hormone level is of limited value; indeed, it was normal in this case. However, after insulin and arginine stimulation, the serum growth hormone did not increase. This is diagnostic of growth hormone deficiency.
4. A deficiency of growth hormone may be secondary to pituitary or hypothalamic failure, in this case, as a result of previous head trauma. Infusion of GHRH resulted in no increase in serum growth hormone levels, indicating pituitary failure. Synthetic growth hormone was administered to this patient, resulting in a growth spurt. Since the pituitary produces other hormones, they may also be deficient. Thyroid and adrenal function should be assessed due to possible deficiencies of TSH and ACTH. If puberty does not occur in a few years, gonadotropin levels should be determined.

CHAPTER 32

Case Study 32-1

1. Although HgbA$_{1c}$ values do increase slightly with age, this patient's value is definitely in the diabetic range. This indicates a lack of control of the patient's diabetes. The elevated random blood glucose level also indicates a lack of control.

2. The point at which glucose spills into the urine (the renal threshold) varies from patient to patient and has been found to increase with age. In elderly diabetic patients, it is not uncommon for blood glucose levels to be over 200 mg/dL (11.11 mmol/L), with no glucose appearing in the urine. The patient also may be performing the home test incorrectly or may have difficulty reading the results because of vision problems. Medications and other substances, such as aspirin, vitamin C, and levodopa, also may be affecting the patient's urine glucose results (*i.e.,* producing false-negative results). Home blood glucose testing is recommended for diabetic patients of any age.

Case Study 32-2

In general, total serum protein is reported to decrease slightly with age. It has also been found that 3% of people over 70 years of age have been found to have a monoclonal protein in their serum in concentrations less than 3 gm/dL (30 g/L). This is known as benign monoclonal gammopathy (BMG). Typically, these people do not show any physical or laboratory indications of multiple myeloma, macroglobulinemia, or related diseases. A benign monoclonal gammopathy is much more common among elderly individuals than among younger ones. There is some potential, however, for a BMG to become a malignant condition if the patient survives long enough for such a disease to manifest itself. If a confirmatory immunoelectrophoresis and urinalysis for Bence Jones protein confirm the diagnosis of BMG, then no treatment is necessary. There should be a careful follow-up of patients with BMG, however. Nearly 20% of individuals with a BMG who live at least another 10 years will likely develop a malignant condition such as multiple myeloma or Waldenstrom's macroglobulinemia. However, the diagnosis of BMG can only be made after thorough investigation (including bone marrow examination) fails to show evidence of multiple myeloma, leukemia, or other conditions associated with monoclonal proteins. The fact that the protein is less than 3 gm/L does not mean that it is not associated with a neoplasm.

Case Study 32-3

While some investigators report an increase in TSH levels, others report no change. In more than 7% of women over 60 years of age, the TSH level may be greater than 10 μU/mL. This prevalence rises in women over age 75 and in institutionalized elderly patients. The prevalence is somewhat less in men than in women. In general, aging is associated with an increased incidence of primary hypothyroidism characterized by elevated elevated levels of TSH.[*] Increased TSH levels may be accompanied by either normal or low levels of the thyroid hormones T_4 and T_3. Patients with increased TSH and persistently normal thyroid hormone levels are said to have a failing thyroid syndrome. If a patient does not show any symptoms, treatment is generally not indicated. However, the patient should be periodically monitored, since thyroid function may decline further with age. A trial of low-dose thyroxine therapy may normalize the TSH. The signs and symptoms of hypothyroidism in the elderly may be subtle, and one should not rely on laboratory tests to detect them.

[*] See reference 5 at the end of Chapter 32, Geriatric Clinical Chemistry.

Index

Lowercase *f* following a page number indicates a figure; *t* following a page number indicates tabular material; *c* following a page number indicates material in case studies.

For 2/18 - Read electrophoresis : p